THE

BANTU

LANGUAGES

Written by an international team of experts, this comprehensive volume presents grammatical analyses of individual Bantu languages, comparative studies of their main phonetic, phonological and grammatical characteristics and overview chapters on their history and classification.

It is estimated that some 300 to 350 million people, or one in three Africans, are Bantu speakers. Van de Velde and Bostoen bring together their linguistic expertise to produce a volume that builds on Nurse and Philippson's first edition.

The Bantu Languages, 2nd edition is divided into two parts; Part 1 contains 11 comparative chapters, and Part 2 provides grammar sketches of 12 individual Bantu languages, some of which were previously undescribed. The grammar sketches follow a general template that allows for easy comparison.

Thoroughly revised and updated to include more language descriptions and the latest comparative insights.

New to this edition:

- new chapters on syntax, tone, reconstruction and language contact
- 12 new sketch grammars
- thoroughly updated chapters on phonetics, aspect-tense-mood and classification
- exhaustive catalogue of known languages with essential references

This unique resource remains the ideal reference for advanced undergraduate and postgraduate students of Bantu linguistics and languages. It will be of interest to researchers and anyone with an interest in historical linguistics, linguistic typology and grammatical analysis.

Mark Van de Velde is a researcher at LLACAN (CNRS-INaLCO) in Paris, of which he has been the director since 2015. He works on the grammatical analysis and documentation of the north-western Bantu languages and the languages of the Benue valley in Nigeria, especially those currently classified as Adamawa. He is also interested in linguistic typology and in the comparative study and reconstruction of Bantu grammar, particularly in the domain of the noun phrase.

Koen Bostoen is Professor of African Linguistics and Swahili at Ghent University and member of the UGent Centre for Bantu Studies. His research focuses on Bantu languages and interdisciplinary approaches to the African past. He obtained an ERC Starting Grant for the KongoKing project (2012–2016) and an ERC Consolidator's Grant for the BantuFirst project (2018–2022). Apart from several research articles, he is the author of *Des mots et des pots en bantou: une approche linguistique de l'histoire de la céramique en Afrique* (2005) and co-editor of *Studies in African Comparative Linguistics, with Special Focus on Bantu and Mande* (2005), *The Kongo Kingdom: The Origins, Dynamics and Cosmopolitan Culture of an African Polity* (2018) and *Une archéologie des provinces septentrionales du royaume Kongo* (2018).

Derek Nurse edited the first edition of *The Bantu Languages* and has worked on historical linguistics, language contact, phonological change, Bantu and (East) African languages, Swahili, ethnolinguistics, the interface of linguistics, archaeology, and history and tense/ aspect systems in Bantu.

Gérard Philippson edited the first edition of *The Bantu Languages* and is emeritus Professor of Bantu Languages at the Institut National des Langues et Civilisations Orientales (Paris) and member of the Laboratoire Dynamique du Langage (Lyon). He has worked mostly on East African Bantu Languages, Bantu comparative tonology and diachronic phonology, as well as culture history.

ROUTLEDGE LANGUAGE FAMILY SERIES

Each volume in this series contains an in-depth account of the members of some of the world's most important language families. Written by experts in each language, these accessible accounts provide detailed linguistic analysis and description. The contents are carefully structured to cover the natural system of classification: phonology, morphology, syntax, lexis, semantics, dialectology, and sociolinguistics.

Every volume contains extensive bibliographies for each language, a detailed index and tables, and maps and examples from the languages to demonstrate the linguistic features being described. The consistent format allows comparative study, not only between the languages in each volume, but also across all the volumes in the series.

Other titles in the series can be found at www.routledge.com/languages/series/SE0091

THE

BANTU

LANGUAGES

Second edition

Edited by
Mark Van de Velde, Koen Bostoen,
Derek Nurse, and Gérard Philippson

Routledge
Taylor & Francis Group
LONDON AND NEW YORK

Second edition published 2019
by Routledge
2 Park Square, Milton Park, Abingdon, Oxon, OX14 4RN

and by Routledge
605 Third Avenue, New York, NY 10017

First issued in paperback 2022

Routledge is an imprint of the Taylor & Francis Group, an informa business

Publisher's Note
The publisher has gone to great lengths to ensure the quality of this reprint but points out that some imperfections in the original copies may be apparent.

First edition published by Routledge 2004

British Library Cataloguing-in-Publication Data
A catalogue record for this book is available from the British Library

Library of Congress Cataloging-in-Publication Data
Names: Velde, Mark L. O. van de, 1976– editor, author. | Bostoen, Koen A. G., editor,
    author. | Nurse, Derek, editor, author. | Philippson, Gérard, editor, author.
Title: The Bantu languages / edited by Mark Van de Velde, Koen Bostoen, Derek
    Nurse, and Gérard Philippson.
Other titles: Routledge language family series.
Description: Second edition. | New York : Routledge, 2018. | Series: Routledge
    language family series | Includes index.
Identifiers: LCCN 2018029329 | ISBN 9781138799677 (hardback : alk. paper) | ISBN
    9781315755946 (ebook) | ISBN 9781317628675 (mobi)
Subjects: LCSH: Bantu languages.
Classification: LCC PL8025 .B35 2018 | DDC 496.39—dc23
LC record available at https://lccn.loc.gov/2018029329

ISBN: 978-1-03-240166-9 (pbk)
ISBN: 978-1-138-79967-7 (hbk)
ISBN: 978-1-315-75594-6 (ebk)

DOI: 10.4324/9781315755946

Typeset in Times New Roman
by Apex CoVantage, LLC

Visit the eResources: www.routledge.com/9781138799677

MIX
Paper | Supporting
responsible forestry
FSC
www.fsc.org   FSC® C013985

Printed in the United Kingdom
by Henry Ling Limited

CONTENTS

CONTRIBUTORS

Lotta Aunio, PhD, is Senior Lecturer of Bantu Languages at the Department of World Cultures, University of Helsinki. Her research interests include Bantu languages, Nilotic languages, tone, language contact and language description.

Koen Bostoen is Professor of African Linguistics and Swahili at Ghent University and member of the UGent Centre for Bantu Studies. His research focuses on Bantu languages and interdisciplinary approaches to the African past. He obtained an ERC Starting Grant for the KongoKing project (2012–2016) and an ERC Consolidator's Grant for the BantuFirst project (2018–2022). Apart from several research articles, he is the author of *Des mots et des pots en bantou: une approche linguistique de l'histoire de la céramique en Afrique* (2005, Peter Lang) and co-editor of *Studies in African Comparative Linguistics, with Special Focus on Bantu and Mande* (2005, RMCA), *The Kongo Kingdom: The Origins, Dynamics and Cosmopolitan Culture of an African Polity* (2018, Cambridge University Press) and *Une archéologie des provinces septentrionales du royaume Kongo* (2018, Archaeopress).

Robert Botne is Professor of Linguistics at Indiana University. He is a general linguist whose research focuses primarily on the Bantu languages and on the morphology and typology of tense, aspect, mood systems.

Thera Crane is a researcher in the Department of Languages at the University of Helsinki. She is currently investigating morphosemantic variation in South African Ndebele varieties as part of the project "Stability and Change in Language Contact: The Case of Southern Ndebele (South Africa)."

Maud Devos is a researcher at the Royal Museum of Central Africa and a Visiting Professor of Swahili at Ghent University. Her main research interest lies in descriptive linguistics, focusing on coastal Mozambican Bantu languages that were once part of the larger Swahili world. She also investigates grammaticalisation processes in Bantu languages, mainly pertaining to the expression of negation (negative stacking, prohibition, negative existence and not yet), mood and modality, focus and location.

Laura J. Downing has been Professor of African Languages at the University of Gothenburg since 2012. Her research since her dissertation has focused mainly on the prosody of Bantu languages, including topics like tone, reduplication, the syntax-phonology interface and information structure. She has numerous scholarly publications to her credit on these topics, including, most recently, a book on *The Phonology of Chichewa* (co-authored with Al Mtenje) and an edited volume on *Tone and Intonation in African Languages* (co-edited with Annie Rialland).

Helen Eaton received a PhD in Linguistics from the University of Reading in 2002. Her doctoral research was on focus in Sandawe. She is a linguistics consultant for SIL International and is based in Mbeya, southern Tanzania.

Rebecca Grollemund, PhD, is a Professor of Linguistics at the University of Missouri. She is specialised in historical linguistics, African languages (including Bantu languages) and phylogenetic methods used to study language evolution.

Rozenn Guérois is a postdoctoral researcher at the UGent Centre for Bantu Studies of the Department of Languages and Cultures at Ghent University. Her three-year research project consists of comparing passive constructions in Bantu languages, and developing a typology of this voice phenomenon within Bantu. Prior to this, she was a postdoctoral research assistant in the Leverhulme-funded project Morphosyntactic Variation in Bantu: Typology, Contact and Change. As part of her PhD research, she wrote a grammar of Cuwabo (2015, Université Lyon 2). Her research focuses on the morphosyntactic study of Bantu languages from a descriptive and typological point of view.

Harald Hammarström is Senior Lecturer in General Linguistics at Uppsala University. He has educational background in both Computer Science and Linguistics. He has a very broad linguistic interest spanning all areas of the world, including Africa and the Bantu languages. His research activities span documentary fieldwork, classical linguistic analytic work, typological databases and NLP for lesser-known languages. He is currently focussing on large-scale empirical and computational approaches to linguistic diversity, genealogical/areal relationships and language universals.

Larry Hyman has since 1988 been Professor of Linguistics in Berkeley's Department of Linguistics, which he chaired from 1991 to 2002. He has worked extensively on phonological theory and other aspects of language structure, including several books (e.g., *Phonology Theory and Analysis*, *A Theory of Phonological Weight*) and numerous theoretical articles in such journals as *Language*, *Linguistic Inquiry*, *Natural Language and Linguistic Theory*, *Phonology*, *Studies in African Linguistics* and *Journal of African Languages and Linguistics*. His current interests centre around phonological typology, tone systems and the descriptive, comparative and historical study of Niger-Congo languages, especially Bantu. He is also currently Executive Director of the France-Berkeley Fund.

Charles Kisseberth is a retired professor of phonology at the Universities of Illinois and Tel Aviv. He has published widely on a variety of Eastern Bantu languages, most notably Chimwiini, and on theoretical issues in phonology.

Joseph Koni Muluwa obtained his PhD from the Université Libre de Bruxelles (Belgium) in 2011 and subsequently was a postdoctoral researcher at the Royal Museum for Central Africa in Tervuren (Belgium) and at Ghent University (Belgium). In 2016–2017 he was a visiting professor at Ghent University. He currently is lecturer in African Linguistics at the Institut Supérieur Pédagogique of Kikwit (Democratic Republic of the Congo). His research focuses on Guthrie's B80 West-Coastal Bantu languages, many of which are endangered, and encompasses language documentation, historical-comparative linguistics, ethnobotany and ethnozoology.

Constance Kutsch Lojenga is a senior linguistics consultant within SIL International, and a retired associate professor at the Department of Languages and Cultures of Africa at Leiden University. Her research interests are phonetics, phonology, vowel harmony, tone (analysis, orthography, teaching), morphology and language and orthography development, with focus on Bantu and Central Sudanic languages.

Ian Maddieson is currently Adjunct Research Professor at the University of New Mexico. His interest in the phonetics of African languages can be traced back to four years spent teaching at the University of Ibadan in Nigeria. After many years associated with the Phonetics Laboratory at UCLA, he moved to UC Berkeley in 2000. His research on Bantu languages is situated in the context of an interest in global patterns in phonetic and phonological systems, as exemplified in the book *Sounds of the World's Languages* (co-authored with Peter Ladefoged) and the UPSID and LAPSyD databases.

Michael Marlo is Associate Professor of English at the University of Missouri. He has broad interests in African linguistics and Bantu languages, and has carried out extensive research on varieties of the Luyia macrolanguage of western Kenya and eastern Uganda. His research often focuses on tone, and involves in-depth synchronic studies of individual languages as well as comparative/historical work across Bantu. He has also carried out a series of large-scale micro-typological studies of object marking in Bantu. His current project involves a comprehensive, collaborative description of Luyia language varieties.

Lutz Marten is Professor of General and African Linguistics at the School of Oriental and African Studies (SOAS), University of London. His research focuses on the description and analysis of structural, social and functional aspects of language, with a specific focus on African languages. His current work includes theoretically informed linguistic analysis (morphosyntax, semantics, pragmatics) as well as language description, comparative and historical linguistics (especially Bantu languages), language contact and questions of language and identity. He has conducted fieldwork in East, Central and Southern Africa, and has led a number of research and collaboration projects relating to comparative linguistics and language variation. His publications include *At the Syntax-Pragmatics Interface* (OUP 2002), *A Grammatical Sketch of Herero* (with Wilhelm Möhlig and Jekura Kavari, Köppe 2002), *The Dynamics of Language* (with Ronnie Cann and Ruth Kempson, Elsevier 2005), and *Colloquial Swahili* (with Donovan McGrath, Routledge 2003/2012).

Kassim Mohamed-Soyir completed his PhD in Linguistics at Université Paris 7 in 2015. His dissertation discussed the phonological, morphological, syntactic and semantic properties of the noun in the Bantu language Ngazidja, and his work focuses on the morphology of Bantu and Semitic languages. He teaches French literature in high school and Arabic studies at Paris 8.

Maarten Mous is Professor of African Linguistics at Leiden University. He does research on Cushitic languages (notably Iraqw, Alagwa, Konso and Somali), language and identity, diathesis and derivation, and on Bantu languages. His research on language and identity derives from his interest in Creole languages and includes a study of the mixed language Ma'a/Mbugu as well as typological work on African urban youth languages. The Bantu languages that he has published on are Tunen and Nyokon in Cameroon; and Mbugu, Pare and Mbugwe in Tanzania.

Elisabeth Njantcho Kouagang obtained in 2018 a PhD degree at INALCO (Paris) where she was working on a descriptive study of Kwakum (Bantu A90). For her Maitrise and Master's degrees, she carried out descriptive work on Kari and Karang, two Adamawa languages spoken in Cameroon. From 2009 to 2013, she worked as a second language teacher in Cameroon where she taught English to French-speaking students and French to English-speaking students. Her current interests center around the descriptive and comparative study and documentation of endangered languages in Africa.

Derek Nurse has worked on historical linguistics, language contact, phonological change, Bantu and (East) African languages, Swahili, ethnolinguistics, interface of linguistics, archaeology, and history and tense/aspect systems in Bantu.

David Odden is Professor Emeritus of Linguistics at Ohio State University. His areas of research include phonological theory and language description, especially tone and the structure of African languages, with a focus on Bantu. He has published numerous descriptive works on Shona and Matumbi, as well as reporting the essentials of the tone systems of Shambaa, Kuria, Bakweri, Yao, Tachoni, Kotoko and multiple dialects of Makonde and Taita. His current primary research project includes a descriptive phonology of Logoori.

Cédric Patin completed his PhD in Linguistics at Université Paris 3 in 2007. His thesis examined the tonal system of the Bantu language Ngazidja. After a postdoctorate at the Laboratoire de Linguistique Formelle (CNRS/Université Paris 7), he became Assistant Professor in French phonetics and phonology at the University of Lille in 2009. His research focusses on the phonology of Bantu languages, with emphasis on the prosody-syntax interface.

Malin Petzell, PhD, is Researcher in African Linguistics at the Department of Languages and Literatures, University of Gothenburg. Her research interests include language description (documentation and analysis), Bantu languages, nominal and verbal morphology, aspectual classification and field methods.

Gérard Philippson is emeritus Professor of Bantu Languages at the *Institut National des Langues et Civilisations Orientales* (Paris) and member of the *Laboratoire Dynamique du Langage* (Lyon). He has worked mostly on East African Bantu Languages, Bantu comparative tonology and diachronic phonology, as well as culture history.

JeDene Reeder currently works as a senior literacy and education consultant with SIL International in West Africa and is based in Burkina Faso after several years in Togo. She worked with SIL in the Democratic Republic of the Congo from 1993 to 1997 as a literacy specialist, learning the language and culture of the Pagibete beginning in 1994. She obtained an Ed.D. in Educational Leadership in 2011 from Simon Fraser University and an MA in linguistics in 1998 from the University of Texas at Arlington.

Holly Robinson, MA, works with SIL in Mara, Tanzania in linguistic research and orthography development. Her research interests include Bantu vowel systems, orthography development, linguistic descriptions for speakers of minority languages and historical-comparative linguistics.

Tim Roth is also a linguistics consultant for the Uganda-Tanzania Branch of SIL International. His current research interests include historical/comparative linguistics, TAM and verbal semantics. In 2018 he obtained a PhD degree at the University of Helsinki with a doctoral dissertation titled "Aspect in Ikoma and Ngoreme : A comparison and analysis of two Western Serengeti Bantu languages". His master's thesis, *The genetic classification of Wungu: implications for Bantu historical linguistics*, was completed at the Canada Institute of Linguistics in 2011.

Bonny Sands is an Adjunct Professor in the Department of English at Northern Arizona University. She has conducted fieldwork on languages in Eastern and Southern Africa, with a special focus on the phonetics and phonology of click consonants. Additional research interests include African linguistic prehistory, classification and reconstruction and language endangerment.

Thilo C. Schadeberg is Professor Emeritus of African Linguistics at Leiden University, Netherlands. His main interest has been the descriptive and comparative study of Bantu and Kordofanian languages. He has published (sketch) grammars on the Bantu languages Swahili, Nyamwezi, Umbundu and Koti, articles on phonology (including tone) and on comparative-historical and typological issues. He is co-author of the *Bantu Lexical Reconstructions* website (BLR3). His current project is the extraction of grammatical information from the New Testament in the Kordofanian (Niger-Congo) language Ebang or Heiban from the Nuba Mountains (Sudan).

Oliver Stegen, PhD, works with SIL in Eastern Africa, mainly advising minority language projects in orthography development and discourse analysis. His research interests include language description (esp. Bantu), tone, literary studies and their application to translation.

Mark Van de Velde is a researcher at LLACAN (CNRS-INALCO) in Paris, of which he has been the director since 2015. He works on the grammatical analysis and documentation of the north-western Bantu languages and the languages of the Benue valley in Nigeria, especially those currently classified as Adamawa. He is also interested in linguistic typology and in the comparative study and reconstruction of Bantu grammar, particularly in the domain of the noun phrase.

John B. Walker works as Linguistics Consultant for the Uganda-Tanzania Branch of SIL International and is currently based in Musoma, Mara, Tanzania. His main research interests are lexicography, historical and comparative linguistics and TAM. He completed his MA in Linguistics at the Canada Institute of Linguistics in 2013 where he wrote his master's thesis entitled *Comparative tense and aspect in the Mara Bantu languages: Towards a linguistic history*.

Vera Wilhelmsen defended her PhD of Linguistics at Uppsala University, Sweden in June 2018. She is doing research on Mbugwe (F34), and has done fieldwork in Tanzania. She did her undergraduate studies at University of Bergen, Norway and earned her master's degree from Trinity Western University, BC, Canada. Her research is focused on language description with special interest in tone, the verbal system (TAM) and typology.

ABBREVIATIONS

1	noun class 1
2	noun class 2
1PL	first person plural
2PL	second person plural
3PL	third person plural
1SG	first person singular
2SG	second person singular
3SG	third person singular
ACP	adnominal concord prefix
ADD	additive
ADJ	adjective / adjectiviser
ADV	adverb
AG	agentive
ANA	anaphoric
ANT	anterior
ANTIP	antipassive
AP	aspect prefix / adjectival prefix
APPL	applicative
ASSOC	associative
ATM	aspect tense mood
ATR	advanced tongue root
AUG	augment
AUX	auxiliary
B	verbal base
BHH	back height harmony
BLR	Bantu Lexical Reconstructions
C	consonant
CAUS	causative
CE	counterexpectational
CERT	certain
CF	counterfactual
CJ	conjoint
CM	Comparative Method
CMPL	completive
CNT	continuous
COM	comitative
COMP	complementiser
CON	connective
COND	conditional

CONS	consecutive
CONTR	contrastive
COP	copula
CP	connective prefix
DEF	definite
DEM	demonstrative
DIST	distal
DJ	disjoint
DP	demonstrative (concord) prefix
DTP	definite tone pattern
EGIDS	Expanded Graded Intergenerational Disruption Scale
EMPH	emphatic
END	endophoric
EP	numeral prefix
EPG	electropalatography
EXPL.SM	expletive subject marker
F	final (verb) suffix
FDEM	far demonstrative
FHH	front height harmony
FOC	focus
FUT	future
FV	final vowel
FVS	final vowel shortening
H	high tone
HAB	habitual
HOD	hodiernal
HTA	high tone anticipation
HYP	hypothetical
IAV	immediate after the verb (focus position)
IBV	immediate before the verb (focus position)
IDS	identificational suffix
IMM	imminent
IMP	imperative
INAN	inanimate
INF	infinitive
INT	intensive
INTR	intransitive
IPA	International Phonetic Alphabet
IPFV	imperfective
IRR	irrealis
ITR	iterative
L	low tone
LNK	linker
LOC	locative
MED	medial
MOT	motional
μ	mora
N	nasal / noun

NARR	narrative
NDEM	near demonstrative
NEG	negation
NEUT	neuter
NLNK	nominaliser-linker
NP	nominal (concord) prefix / noun phrase
NP$_3$	nominal prefix of noun class 3
NPST	non-past
NT	Namibian Totela
NUM	numeral
OBJ	object
OCP	obligatory contour principle
OC	object enclitic
OP	object prefix
OP$_4$	object prefix of noun class 4
PART	particle
PASS	passive
PB	Proto-Bantu
PER	persistive
PFV	perfective
PI	pre-initial
PL	plural
PLA	plural addressee
POSS	possessive
POT	potential
PP	pronominal (concord) prefix
PPR	personal pronoun
PREHOD	prehodiernal
PRF	perfect
PRIOR	priorative
PRO	pronoun
PROC	process
PROG	progressive
PROH	prohibitive
PROX	proximal
PRS	present
PST	past
Q	question word
QUAL	qualifier
QUANT	quantifier
RECP	reciprocal
RED	reduplication
REF	referential
REFL	reflexive
REL	relative (marker)
REM	remote
REP	repetitive
REPA	repetitive animate

1 THE BANTU LANGUAGES: DELIMITATION, SPEAKERS AND GEOGRAPHICAL DISTRIBUTION

The most recent inventory of the Bantu languages (Hammarström, Chapter 2, this volume) lists 555 distinct Bantu languages. It uses the same language versus dialect divisions as the 18th edition of the Ethnologue (Lewis *et al*. 2015). Hammarström observes that a stricter adherence to the criterion of mutual intelligibility would decrease the number of Bantu languages by about 15%, to more or less 472. In order to keep track of so many languages, Bantuists make use of a referential classification, devised by Malcolm Guthrie, in which every language is identified by means of a so-called Guthrie code, which gives an indication of the language's geographical location (see Chapter 11, Section 3.1.1). The emergence and discovery of new languages and the extinction of others are minor factors to account for, but the difficulty in drawing a discrete line between a language and

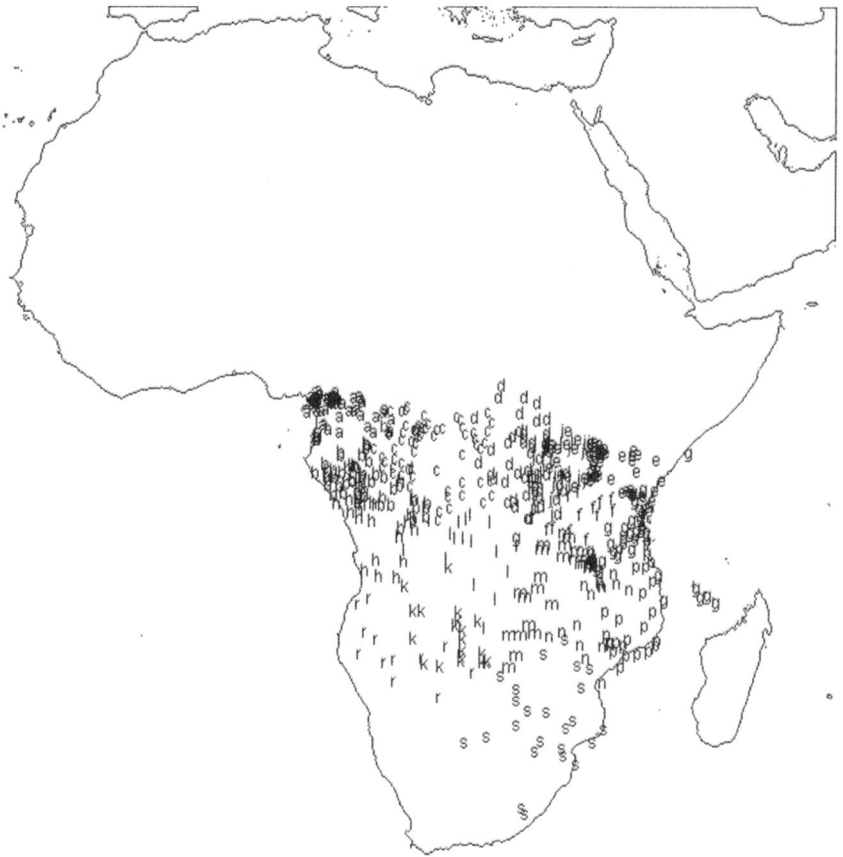

FIGURE 1.1 THE BANTU LANGUAGES REPRESENTED BY THE LETTER CORRESPONDING TO THEIR GUTHRIE ZONE

CHAPTER 1

INTRODUCTION

Koen Bostoen and Mark Van de Velde

This second edition of *The Bantu Languages* consists of two parts. Part 1 contains general chapters that provide an overview of the state of the art in the study of the sound systems and morphosyntactic structures of the Bantu languages and of their classification, reconstruction and different contact situations. Part 2 contains short grammatical analyses of individual Bantu languages for which no book-size grammar is available. Together, these chapters provide a thorough introduction to the grammatical structures of the Bantu languages and to the historical evolutions that have shaped them. The focus on language structure and history means that this volume does not aim at giving an exhaustive overview of contemporary Bantu studies. Such an overview would have to include work on documentation, orthography creation, lexicography, youth languages and so on. This edition is very similar in structure and approach to the highly influential first edition (Nurse & Philippson 2003a), but its contents are almost entirely new. Some chapters from Part 1 have been thoroughly revised. Bonny Sands has renewed the chapter on the Sounds of the Bantu Languages (Chapter 3) in consultation with Ian Maddieson, the author of this chapter in the first edition. Larry Hyman has updated his chapter on Segmental Phonology (Chapter 4). David Odden and Michael Marlo have revised the chapter on Tone (Chapter 5), originally written by Charles Kisseberth and David Odden. The chapter on Derivation was renamed Word Formation (Chapter 6) after an update by Koen Bostoen in close collaboration with its primary author, Thilo C. Schadeberg. The chapter on Aspect and Tense has been revised by its first author Derek Nurse and complemented with a section on mood/modality by Maud Devos to become a new chapter on Aspect, Tense and Mood (Chapter 7). The six other chapters of Part 1 are entirely new. For Part 2, we chose to invite chapters on a new set of languages, because sketch grammars are both extremely useful and relatively hard to get published. For some of the languages included in Part 2, such as Chimpoto N14 (Chapter 23) and Pagibete C401 (Chapter 15), hardly any other published information is available, while others, such as Ngazidja G44a (Chapter 20), have a rich literature but no reference publication that provides a coherent overview of the basic grammatical features of the language. The sketch grammars of the first edition are available on the companian website of this second edition. They have not been revised for this second edition.

The remainder of this chapter is a brief portrait of the Bantu family, starting with its delimitation, number of speakers and geographical distribution (Section 1), and some of its main typological characteristics (Section 2). It finishes with a concise history of its scholarly study, including some early attempts at external classification (Section 3).

REPI	repetitive inanimate
RES	resultative
RNDEM	reference near demonstrative
RTR	retracted tongue root
σ	syllable
S	sonorant
SBJV	subjunctive
SEQ	sequential
SFX	suffix
SG	singular
SIT	situative
SM	stem marker
SP	subject prefix
SP$_5$	subject prefix of noun class 5
STAB	stabiliser
STAT	stative
SUBJ	subject
SUBS	subsecutive
TAM	tense aspect mood
TBU	tone-bearing unit
TP	tense prefix
TPRT	temporal particle
TR	transitive
UN	uncertain
V	vowel
VB	verbal base
VENT	ventive
VHH	vowel height harmony
VOC	vocative
VOT	voice onset time
ZT	Zambian Totela

a dialect is the main reason why numbers diverge considerably in earlier inventories of Bantu languages or language varieties, e.g., 440 in Guthrie (1971), approximately 680 in Mann and Dalby (1987), 542 in Bastin *et al.* (1999) and 660 in Maho (2003). In the latest online version of his New Updated Guthrie List, Maho (2009) inventories 950 different varieties with a unique extended Guthrie code. Many of these are recognised as different varieties of a single language, i.e., those ending in a lower-case or upper-case letter. The number of distinct varieties without such a final letter in Maho (2009) is 631 (Harald Hammarström, pers. comm.). The 555 languages listed in Chapter 2 are represented by the letter corresponding to their Guthrie zone on the map in Figure 1.1.

The boundary between the Bantu languages and their closest relatives of the Bantoid family spoken in Cameroon and Nigeria is to a certain extent established by convention, rather than by a set of shared innovations that are attested in all and only the Bantu languages. According to convention, languages are considered to belong to the Bantu family if they have a Guthrie code. An area where this convention is most likely to be at odds with reality is the little-studied group of Jarawan languages of Nigeria and northern Cameroon, which lack a Guthrie code but which have been argued to be Bantu, possibly Bantu A60 (Gerhardt 1982, Blench 2015). Several studies in genealogical classification based on basic vocabulary (cf. Bastin & Piron 1999, Grollemund 2012: 349, Bostoen *et al.* 2015, Grollemund *et al.* 2015) do indeed recognise Jarawan Bantu languages (as well as certain other Bantoid languages) as being most closely related to the languages of the Mbam-Bubi group, which do have Guthrie codes (in groups A30, A40, A50 and A60), without a discrete cut-off point between Bantu and Bantoid or so-called "Narrow Bantu" and "Wide Bantu."

As for the number of speakers of Bantu languages in Africa, Nurse and Philippson (2003b: 1) estimate that about 240 million Africans speak one or more Bantu languages, multilingualism being the rule rather than the exception in Africa. In 2003, this meant that one African out of three to four spoke a Bantu language (given a total African population of about 875 million at that time). This is more than half of all Niger-Congo speakers, which Nurse and Philippson (2003b: 1) estimate at about 400 million. Patin *et al.* (2017) estimate that there are about 310 million Bantu speakers. This would correspond to about one African in four, the number of Africans in 2018 being around 1.2 billion. Of the 556 Bantu languages in the 20th edition of the *Ethnologue* (Simons & Fennig 2017), 529 have a population estimate, whose sum is 276,513,509 speakers (Harald Hammarström, personal communication). This number could be extrapolated to about 290 million for the 556 Bantu languages. However, one could also argue for a higher contemporary number, taking into account that according to the 20th edition of the *Ethnologue*, the total African population including Madagascar is 929,932,101. This estimate is based on 2039 languages out of the 2178 present in Africa that have a population estimate (from well before 2018). The proportion of this estimate with respect to the actual number of 1.2 billion Africans today is 0.775. Applied inversely to the sum of Bantu speakers estimated in the *Ethnologue* 20, their number would amount to about 350 million today (Harald Hammarström pers. comm.), which is even more than what Patin *et al.* (2017) propose without explaining on what their estimate is based.

The Bantu languages are mainly spoken between Cameroon's South-West region (4°8'N and 9°14'E) in the North-West, southern Somalia's Barawe (Brava) area (1°6'N and 44°1'E) in the North-East and Cape Agulhas (34°48'S and 20°E), the continent's southernmost tip, in the Western Cape province of South Africa. Their distribution area is contiguous – some very rare languages surrounded by non-Bantu languages notwithstanding – and spans 23 countries on the African mainland. In alphabetical order, these are Angola, Botswana, Burundi, Cameroon, Central African Republic, Congo-Brazzaville, Congo-Kinshasa, Equatorial Guinea, Gabon, Kenya, Lesotho, Malawi, Mozambique, Namibia, Rwanda, Somalia, South Africa, Southern Sudan, Swaziland, Tanzania, Uganda, Zambia and Zimbabwe. In some of these, such as Burundi, Malawi and Rwanda, Bantu languages are the only indigenous African languages. Long-standing Bantu speech communities are also found on the islands of Bioko (part of Equatorial Guinea), Mayotte (an overseas department of France) and the Comoros (see also Nurse & Philippson 2003b, Hammarström et al. 2017). Nurse and Hinnebusch (1993: 14) report a variety of the Bantu language Swahili spoken on the small island of Nosse-Be, off the northwest coast of Madagascar, with another pocket further down the west coast of this island. In other African countries, especially those in the northern and southern borderlands, Bantu languages coexist with languages that belong to other families, such as Central Sudanic, Nilotic, Cushitic and Omotic, which are part of the wider Nilo-Saharan or Afro-Asiatic phyla, or that are considered isolates today, such as Hadza and Sandawe (formerly considered to be Khoisan). Yet in others, such as the Central African Republic, Southern Sudan and Somalia, Bantu languages are sporadic, not to say nearly absent. All in all, Bantu is the predominant language group in Central, Eastern and Southern Africa. Other Niger-Congo languages – apart from Adamawa-Ubangi and Kordofanian – predominate in Western Sub-Saharan Africa, but have a distribution area that is about one-third to one-half of the Bantu area. Thanks to the vastness of the Bantu area, Niger-Congo is by far Africa's most widespread language phylum.

The massive spread of the Bantu languages is striking, especially in consideration of the group's estimated age of no more than 4,000–5,000 years (Vansina 1995: 52, Blench 2006: 126). This time depth is quite shallow compared to the 10,000–12,000 years that have been proposed for the Niger-Congo phylum (Blench 2006: 126). The geographic distribution of Niger-Congo minus Bantu is much smaller than the spread zone of its tardive Bantu offshoot. Bantu languages would have gradually split off from their closest South-Bantoid relatives in the borderland of South-Eastern Nigeria and Western Cameroon, an area of high linguistic diversity within the Bantoid subgroup of Benue-Congo, one of the major Niger-Congo branches (cf. Blench 2015). Ever since Greenberg (1972), there is great unanimity to consider this area as the Bantu homeland. It is from this ancestral homeland that the concurrent dispersal of Bantu languages and Bantu-speaking people across Central, Eastern and Southern Africa started. This phenomenon is commonly referred to as the Bantu Expansion (Oliver 1966, Bouquiaux 1980, Vansina 1995, Ehret 2001, Bostoen 2018). During their initial migration between roughly 5,000 and 1,500 years ago, Bantu speech communities not only introduced new languages in the areas where they immigrated, but also new lifestyles, in which technological innovations such as pottery making and the use of large stone tools originally played an important role, as did farming and metallurgy subsequently. Wherever early Bantu speakers settled down, they left an archaeologically visible culture (Phillipson 2005, de Maret 2013, Bostoen et al. 2015). New insights from the field of evolutionary genetics show that the Bantu Expansion was not just a spread of languages and technology through cultural contact, as was once thought (Lwanga-Lunyiigo 1976, Gramly 1978, Schepartz 1988, Vansina 1995, Robertson & Bradley 2000), but involved the actual migration of people (Pakendorf et al. 2011, Li et al. 2014, Patin et al. 2017). Moreover, Bantu-speaking newcomers strongly interacted with resident hunter-gatherers, as can still be observed in the gene pool (Destro-Bisol et al. 2004, Wood et al. 2005, Quintana-Murci et al. 2008,

Verdu *et al.* 2013, Patin *et al.* 2014), and/or the languages of certain present-day Bantu speech communities (Herbert 2002, Bostoen & Sands 2012, Gunnink *et al.* 2015, Pakendorf *et al.* 2017). The driving forces behind what is the principal linguistic, cultural and demographic process in Late Holocene Africa are still a matter of debate, but it is increasingly recognised that a climate-induced crisis of the Central African rainforest around 2,500 years ago boosted the scale and pace of the Bantu Expansion (Schwartz 1992, Brncic *et al.* 2009, Ngomanda *et al.* 2009, Maley *et al.* 2012, Neumann *et al.* 2012, Oslisly *et al.* 2013, Bostoen *et al.* 2015, Grollemund *et al.* 2015, Hubau *et al.* 2015).

2 TYPOLOGICAL CHARACTERISATION

As pointed out in the conclusion of Chapter 3 and in the introduction of Chapters 4 and 9, one of the most attractive features of the Bantu family is that it allows for the comparative study of linguistic variation in a huge set of closely related languages. There is a marked typological divide between the North-Western Bantu languages and the others. The North-Western languages are spoken close to the Proto-Bantu homeland and in a spread zone called the "Macro-Sudan belt" (Güldemann 2008) or "Sudanic belt" (Clements & Rialland 2008). They typically have dense tone systems, with an equipollent opposition between low and high tones, few or no tonally underspecified morphemes and a high number of floating tones. At the other end of the typological spectrum are the few Eastern languages that have lost tone. In languages with intermediate tonal density, many morphemes are tonally underspecified and receive their surface tone through the application of rules. Chapter 5 discusses many more typological differences between the tone systems of the Bantu languages, such as the nature of the tone-bearing unit or the way in which rules like tone spreading and tone shift work. The high amount of floating tones in the Northwest is due to the loss of segmental material, which is itself due to the existence of maximality constraints on the size of stems (see, e.g., Hyman 2004). These same constraints also explain why the verbal derivational suffixes discussed in Chapter 6 can hardly be stacked in many North-Western languages, whereas they typically can in the East of the Bantu domain, sometimes exuberantly.

The Bantu languages are well known for their rich noun class systems. They have on average about 15 noun classes. On top of those, most languages outside of the North-West also have three locative classes. The few Bantu languages that have considerably reduced or lost their noun class system are either contact languages or spoken in the North of the Bantu domain. For some of the latter, loss of noun classes has been argued to be due to contact with languages from other families, notably Central Sudanic (see Chapters 8 and 12). Another well-known characteristics of the Bantu languages is their high number of past and future tense distinctions, discussed in Chapter 7. Probably less well-known is the pragmatically conditioned freedom of constituent order on the clause level in some languages (see Chapter 9), as well as typologically unusual word order patterns in the noun phrase (Chapter 8).

3 A BRIEF HISTORY OF THE DESCRIPTION AND CLASSIFICATION OF THE BANTU LANGUAGES

Bantu languages started to kindle the scholarly curiosity of Europeans as early as the late 15th century, when Portuguese sailors began their voyages along the coasts of Central,

Southern and Eastern Africa. Bantu words and phrases slipped into the writings of European seafarers, merchants, soldiers and missionaries. In their endeavours to spread the holy word among the peoples of Africa, missionaries had the most direct utilitarian interest in acquiring knowledge of Bantu and other African languages. It is therefore not surprising that the oldest extant Bantu language text is a Kongo translation of the catechism by the Portuguese Jesuit Mattheus Cardoso (1584–1625) from 1624 (see also Bontinck & Ndembe Nsasi 1978). The oldest Bantu (Latin-Spanish-Kongo) dictionary was compiled through close collaboration between the Kongo priest Manuel Roboredo (†1665) and several Spanish Capuchins. It was subsequently hand copied by the Flemish Capuchin Joris Van Gheel (1617–1652), and his copy from 1652 is the only one available to us (Van Gheel 1652, Van Wing & Penders 1928, De Kind *et al.* 2012). The oldest Bantu grammar, on Kongo too, is also a product of clerical scholarship. It was written by the Italian Capuchin priest Giacinto Brusciotto (1601–1659) – "Brusciotto" would be a misspelling of "Brugiotti" according to Pacchiarotti (2017: 8) based on Turchetta (2007) – and published in 1659. This grammar was translated into English and made available to a wider public by Guinness (1882). All three documents pertain to one and the same variety of South Kongo, i.e., the court language spoken at Mbanza Kongo, the capital of the Kongo kingdom, the direct ancestor of present-day Sikongo H16a (Bostoen & de Schryver 2018). Detailed historical accounts of early missionary and other research into the Bantu languages are to be found in Johnston (1919: 1–14) and Doke (1935, 1959).

The birth of Bantu linguistics as a scientific discipline is commonly attributed to the Rhenan (German) philologist Wilhelm Heinrich Bleek (1827–1875), who established Bantu as a family and gave it the name it still has today. As pointed out by Fodor (1980: 127–128), the unity of the Bantu languages was actually recognised more than a decade earlier by the American scholar H. E. Hale (1817–1896), who collected African vocabularies during his voyage around the globe as part of the United States Exploring Expedition commanded by Charles Wilkes (1838–1842). Hale divided what would become known as the Bantu languages into two distinct branches, i.e., the Congo-Makua and the Caffrarian languages, but did not propose a common label for the group. It was Bleek himself who introduced the label *Bantu* in a book volume published in 1958. Further elaborating on ideas that he had started to develop in his doctoral thesis (Bleek 1851), he proposed to use that name for designating those African languages which are "prefix-pronominal" (cf. Chrétien 1985: 46), in contrast to the "suffix-pronominal" or "sex-denoting" languages, including what he called the "Hottentot" and "Bushman" languages according to the parlance of that time, i.e., those known as "KhoiSan" today. Koelle (1854) had also noticed that concord prefixes equally occurred in many languages of West Africa. Bleek defined "prefix-pronominal" languages as those "in which the pronouns were originally identical with the derivative prefixes of the nouns" and situated them in the "Tropical Regions of Africa, and probably also of the Islands in the Indian Ocean and in the Pacific" along with two families in "the African or Continental Section of this Class," i.e., "the *Bántu* and the *Gor* Family" (Bleek 1858: 35), already previsioning there what would later become Niger-Congo. Bleek situated his "Gor" family in sub-Saharan Western Africa. Bantu, on the other hand, had both a South-African and West-African division. The South-African division comprised languages still considered to be Bantu today, subdivided in three distinct branches: South-Eastern, including all known languages of Southern Africa (except KhoiSan), North-Eastern, including languages spoken along the Eastern Coast of Africa and South-Western, including those spoken along the Western Coast of Central and Southern Africa, such as Herero R31. The West-African division, however, occupied parts of the territory of the "Gor" family and included languages such as "[t]he Otshi dialect of Ashanti-land, and the Bullom and Timneh of Sierra Leone" (Bleek 1858: 36), which are classified today in other branches of the Niger-Congo phylum. Hence, Bleek's original definition of Bantu was territorially much wider than ours today. Moreover, it was not really genealogically founded. Bleek's answer to the question heading this section was rather typologically oriented and took the feature of pronominal agreement through prefixes as its point of departure. In his *Comparative Grammar of South African Languages*, Bleek (1862, 1869) was also the first to come up with a noun class prefix system, whose structure and numbering system are still used by current-day Bantuists – and scholars of Niger-Congo more generally – some changes notwithstanding (cf. Katamba 2003: 104).

The scholar who truly defined Bantu as a genealogical unity, i.e., a family of languages descending from a common ancestor that can be reconstructed through the establishment of regular sound correspondences among modern-day languages, was the Prussian (German) philologist Carl Meinhof (1857–1944). Meinhof postulated the existence of *Ur-Bantu* or Proto-Bantu and stated that "*Die Gesetze des Ur-Bantu sind nur aus den heute gesprochenen Bantusprachen zu erschliessen. Da sie aber in allen Bantusprachen ihre Spuren hinterlassen haben, ist ihre Kenntnis unerlässlich für die Erforschung der einzelnen Sprachen*" (Meinhof 1899: 7) ["The laws of Proto-Bantu can only be deduced from Bantu languages spoken today. However, as they have left their traces in all Bantu languages, their understanding is imperative for the study of the individual languages," our translation]. Meinhof is not really explicit on the territorial spread of the Bantu languages, but from the map at the end of his 1899 treatise, it is obvious that his idea of the family's distribution was much narrower than Bleek's. Excluding Western Africa, Meinhof's conception comes close to our own current conception of the Bantu area, except for his northern extensions into the Darfur and Kordofan regions. In his comparative work aiming at the reconstruction of Proto-Bantu, Meinhof also incorporated North-Western Bantu languages, such as Duala A24 from Cameroon.

Johnston (1919: 15) conceived the geographical delimitation of the Bantu languages along the same lines as Meinhof, i.e., "the whole of the southern third of Africa, with the exception of very small areas in the south-west (still inhabited sparsely by Hottentot and Bushman tribes) and a few patches of the inner Congo basin." He was only hesitant about "[t]he northern boundary of the Bantu field," which he considered to be "still a little uncertain and not easy to delineate geographically. It may be said to start on the west coast of Africa in the Bight of Biafra (due north of the island of Fernando Po), at the mouth of the Rio del Rey in the southern portion of the Bakasi peninsula, which flanks the estuary of the Old Calabar river." He situated the north-western extremity of the Bantu area in the borderland between present-day Nigeria and Cameroon, but recognised that the languages spoken there miss some of the distinctive characteristics of a typical Bantu language:

> There is no mistaking a Bantu language for a member of any other African speech family. A momentary glance at the numerals, at a dozen word-roots with their prefixes or suffixes, determines the fact whether it is or is not a member of the Bantu family. The phonology also is as a rule distinctive, though appearances may be deceptive in the case of a few languages of the north-western part of the Bantu field. The semi-Bantu languages on this north-west borderland have a vocabulary which contains a greater or smaller amount of Bantu roots, and farther north and west there are other language families which display obvious resemblances and affinities with

what may have been the Bantu mother tongue; but outside the Bantu family there is no known speech group in Africa which displays all the characteristic features of Bantu word-construction and syntax and at the same time shows unmistakable affinity in word-roots.

(Johnston 1919: 17)

For a language to be qualified as Bantu according to Johnston, it thus had to fulfil at least two conditions: (1) possess a sufficient number of distinctive word-roots cognate with word-roots found elsewhere in the family, and (2) manifest certain characteristic phonological, morphological and syntactic features, such as simple vowel systems and open syllables, agglutination, invariable word-roots, noun class system, the absence of sex-based gender distinctions, pronominal agreement, verb extensions, decimal numeration and the use of prepositions rather than postpositions (Johnston 1919: 18–20). The 12 so-called "propositions" laid down by Johnston to "define the special or peculiar features of the Bantu languages" were actually a critical reassessment of the 12 structural parameters originally proposed by Lepsius (1880). These were long considered "an authoritative outline of Bantu criteria" and also quoted and used by Cust (1880) and Werner (1919) in their reference sketches of African and Bantu languages, respectively (Guthrie 1948: 9).

Johnston (1919) qualified a language only fulfilling one of the two conditions mentioned above as "semi-Bantu." Although some of these "semi-Bantu" are what we consider today as "Bantoid" or "Wide Bantu" languages (cf. infra), Johnston (1919: 17–18) also identified "semi-Bantu" languages elsewhere in the current-day Niger-Congo area, very much in line with Bleek's West-African division of Bantu:

Curiously enough, there are languages in southern Kordofan, in Nigeria, at the back of the Gold Coast, or in the Sierra Leone region, the syntax or construction of which frequently recalls the Bantu idiosyncrasy ; but the word-roots of the vocabulary would be found wholly dissimilar. Or there are others, again, in West Central Africa that exhibit a decided likeness to Bantu in their word-roots, yet in syntax and word construction are quite unlike the Bantu.

Johnston (1919: 17–18)

Johnston's distinction between Bantu and semi-Bantu languages persisted in the comparative work of Malcolm Guthrie (1948, 1971). Building on the Bantu scholarship discussed above, Guthrie adhered for his referential classification to two principal (1–2) and two subsidiary (3–4) criteria to define what a proper Bantu language is: (1) A system of grammatical genders (or noun classes), usually at least five, corresponding to four more features which we cannot recall here for reasons of space; (2) a vocabulary, part of which can be related by fixed rules to a set of hypothetical common roots; (3) a set of invariable cores, or radicals, from which almost all words are formed by an agglutinative process, these radicals having five more features which we also cannot recall here for reasons of space; (4) a balanced vowel system in the radicals, consisting of one open vowel 'a' with an equal number of back and front vowels (Guthrie 1948: 11–12). Guthrie distinguished between two categories of "languages which are incompletely Bantu," viz. "Bantoid" and "Sub-Bantu" languages. In Guthrie's view, Bantoid languages are those spoken in Cameroon and south-eastern Nigeria, which "have a system of grammatical genders and agreements operated by means of prefixes," but "show little or no relationship of vocabulary with full Bantu languages" and also "do not display even the rudiments of the structural features laid down in the third criterion; moreover their vowel system is frequently complicated" (Guthrie 1948: 19). Sub-Bantu languages, on the other hand, are those responding to all criteria set out by Guthrie except the first, i.e., a system of grammatical genders. They still manifest traces of the noun classes and grammatical agreement, but these systems have become very fragmentary, if not completely defunct. Most of Guthrie's Sub-Bantu languages, such as the Congolese language Bira D32, occur in the northern Bantu borderland. Others are vehicular languages spoken further south, such as Lingala C36d. While Guthrie excluded Bantoid languages from his referential classification, he did include Sub-Bantu languages. As has been said, this slightly arbitrary delimitation of the Bantu family at its North-Western border is still in place and is used to define the subject of this book.

ACKNOWLEDGEMENTS

We wish to thank Sara Pacchiarotti for her critical reading of an earlier draft of this introduction and for signalling us the mention of H. E. Hale in Fodor (1980). The usual disclaimers apply.

REFERENCES

Bastin, Y., A. Coupez & M. Mann (1999) *Continuity and Divergence in the Bantu Languages: Perspectives from a Lexicostatistic Study*. Tervuren: Royal Museum for Central Africa.

Bastin, Y. & P. Piron (1999) Classifications lexicostatistiques: bantou, bantou et bantoïde. De l'intérêt des "groupes flottants." In Hombert, J. M. & L. M. Hyman (eds.), *Bantu Historical Linguistics: Theoretical and Empirical Perspectives*, 149–164. Stanford: CSLI Publications.

Bleek, W. (1851) *De nominum generibus linguarum Africae Australia*. Bonnae: Formis Caroli Georgii.

Bleek, W. H. I. (1858) *The Library of His Excellency Sir George Grey, K.C.B. Philology. Vol. 1 – Part I. South Africa (within the Limits of British Influence)*. London: Trübner & Co.

Bleek, W. H. I. (1862) *A Comparative Grammar of South African Languages. Part I. Phonology*. London: Trübner & Co.

Bleek, W. H. I. (1869) *A Comparative Grammar of South African Languages. Part II. The Concord*. London: Trübner & Co.

Blench, R. (2006) *Archaeology, Language and the African Past*. Lanham: Altamira Press.

Blench, R. (2015) "The Bantoid Languages." In *Oxford Handbooks Online*. Oxford: Oxford University Press, www.oxfordhandbooks.com/view/10.1093/oxfordhb/97801999353 45.001.0001/oxfordhb-9780199935345-e-17.

Bontinck, F. & D. Ndembe Nsasi (1978) *Le catéchisme kikongo de 1624. Ré-édition critique*. Bruxelles: Académie Royale des Sciences d'Outre-Mer.

Bostoen, K. (2018) The Bantu Expansion. In Spear, T. (ed.), *Oxford Research Encyclopedia of African History*. Oxford: Oxford University Press, Oxford Research Encyclopedias Online.

Bostoen, K., B. Clist, C. Doumenge, R. Grollemund, J.-M. Hombert, J. Koni Muluwa & J. Maley (2015) Middle to Late Holocene Paleoclimatic Change and the Early Bantu Expansion in the Rain Forests of West Central-Africa. *Current Anthropology* 56(3): 354–384.

Bostoen, K. & G.-M. de Schryver (2018) Seventeenth-Century Kikongo Is Not the Ancestor of Present-Day Kikongo. In Bostoen, K. & I. Brinkman (eds.), *The Kongo Kingdom: The Origins, Dynamics and Cosmopolitan Culture of an African Polity*, 60–102. Cambridge: Cambridge University Press.

Bostoen, K. & B. Sands (2012) Clicks in South-Western Bantu Languages: Contact-Induced Vs. Language-Internal Lexical Change. In Brenzinger, M. & A.-M. Fehn (eds.), *Proceedings of the 6th World Congress of African Linguistics Cologne 2009*, 129–140. Cologne: Rüdiger Köppe.

Bouquiaux, L. (ed.) (1980) *L'expansion bantoue: Actes du Colloque International du Centre National de la Recherche Scientifique, Viviers (France), 4–16 avril 1977 – Volume 2: L'expansion bantoue*. Paris: Société des Etudes Linguistiques et Anthropologiques de France.

Brncic, T. M., K. J. Willis, D. J. Harris, M. W. Telfer & R. M. Bailey (2009) Fire and climate change impacts on lowland forest composition in northern Congo during the last 2580 years from palaeoecological analyses of a seasonally flooded swamp. *Holocene* 19(1): 79–89.

Brusciotto, G. (1659) *Regulae quaedam pro difficillimi Congensium idiomatis faciliori captu, ad grammaticae normam redactae. A F. Hyacintho Brusciotto a Vetralla Concionatore Capuccino Regni Congi Apostolicae Missionis Praefecto*. Rome: Sacra Congregatio de Propaganda Fide.

Cardoso, M. (1624) *Doutrina christãa [. . .] De novo traduzida na lingua do Reyno de Congo*. Lisboa: Geraldo da Vinha.

Chrétien, J.-P. (1985) Les Bantous, de la philologie allemande à l'authenticité africaine. Un mythe racial contemporain. *Vingtième siècle* 8: 43–66.

Clements, N. & A. Rialland (2008) Africa as a Phonological Area. In Heine, B. & D. Nurse (eds.), *A Linguistic Geography of Africa*, 36–85. Cambridge; New York: Cambridge University Press.

Cust, R. N. (1880) *A Sketch of the Modern Languages of Africa*. London: Trübner.

De Kind, J., G.-M. de Schryver & K. Bostoen (2012) Pushing Back the Origin of Bantu Lexicography: The Vocabularium Congense of 1652, 1928, 2012. *Lexikos* 22: 159–194.

de Maret, P. (2013) Archaeologies of the Bantu Expansion. In Mitchell, P. & P. Lane (eds.), *Oxford Handbook of African Archaeology*, 319–328. Oxford: Oxford University Press.

Destro-Bisol, G., F. Donati, V. Coia, I. Boschi, F. Verginelli, A. Caglià, S. Tofanelli, G. Spedini & C. Capelli (2004) Variation of Female and Male Lineages in Sub-Saharan Populations: The Importance of Sociocultural Factors. *Molecular Biology and Evolution* 21: 1673–1682.

Doke, C. M. (1935) Early Bantu Literature: The Age of Brusciotto. *Bantu Studies* 9(2): 87–114.

Doke, C. M. (1959) Early Bantu Literature – The Age of Brusciotto. *African Studies* 18(2): 49–67.

Ehret, C. (2001) Bantu Expansions: Re-Envisioning a Central Problem of Early African History. *The International Journal of African Historical Studies* 34(1): 5–27.

Fodor, I. (1980) H. E. Hale and His African Vocabularies (1846). *Sprache und Geschichte in Afrika* 2: 127–171.

Gerhardt, L. (1982) Jarawan Bantu: The Mistaken Identity of the Bantu Who Turned North. *Afrika und Übersee* 65: 75–95.

Gramly, R. M. (1978) Expansion on Bantu-Speakers Versus Development of Bantu Language in Situ. An Archaeologist's Perspective. *South African Archaeological Bulletin* 33: 107–112.

Greenberg, J. H. (1972) Evidence Regarding Bantu Origins. *Journal of African Languages and Linguistics* 13(2): 189–216.

Grollemund, R. (2012) *Nouvelles approches en classification: Application aux langues bantu du Nord-Ouest.* Lyon: Université Lumière Lyon 2, thèse de doctorat.

Grollemund, R., S. Branford, K. Bostoen, A. Meade, C. Venditti & M. Pagel (2015) Bantu Expansion Shows That Habitat Alters the Route and Pace of Human Dispersals. *Proceedings of the National Academy of Sciences of the United States of America* 112(43): 13296–13301.

Guinness, H. G. (1882) *Grammar of the Congo language as spoken two hundred years ago, translated from the Latin of Brusciotto.* London: Hodder & Stoughton.

Güldemann, T. (2008) The Macro-Sudan Belt: Towards Identifying a Linguistic Area in Northern Sub-Saharan Africa. In Heine, B. & D. Nurse (eds.), *A Linguistic Geography of Africa,* 151–185. Cambridge; New York: Cambridge University Press.

Gunnink, H., B. Sands, B. Pakendorf & K. Bostoen (2015) Prehistoric Language Contact in the Kavango-Zambezi Transfrontier Area: Khoisan Influence on Southwestern Bantu Languages. *Journal of African Languages and Linguistics* 36(2): 193–232.

Guthrie, M. (1948) *The Classification of the Bantu Languages.* London; New York: Oxford University Press for the International African Institute

Guthrie, M. (1971) *Comparative Bantu: An Introduction to the Comparative Linguistics and Prehistory of the Bantu Languages. Volume 2: Bantu Prehistory, Inventory and Indexes.* London: Gregg International.

Hammarström, H., R. Forkel & M. Haspelmath (eds.) (2017) *Glottolog 3.0.* Jena: Max Planck Institute for the Science of Human History (Available online at http://glottolog. org, Accessed May 03, 2017).

Herbert, R. K. (2002) The Sociohistory of Clicks in Southern Bantu. In Mesthrie, R. (ed.), *Language in South Africa,* 297–315. Cambridge: Cambridge University Press.

Hubau, W., J. Van den Bulcke, J. Van Acker & H. Beeckman (2015) Charcoal-Inferred Holocene Fire and Vegetation History Linked to Drought Periods in the Democratic Republic of Congo. *Global Change Biology* 21(6): 2296–2308.

Hyman, L. M. (2004) How to Become a Kwa Verb. *Journal of West African Languages* 30(2): 69–88.

Johnston, H. H. (1919) *A Comparative Study of the Bantu and Semi-Bantu Languages.* Oxford: Clarendon Press.

Katamba, F. (2003) Bantu Nominal Morphology. In Nurse, D. & G. Philippson (eds.), *The Bantu Languages,* 103–120. London New York: Routledge.

Koelle, S. W. (1854) *Polyglotta Africana, or a Comparative Vocabulary of Nearly Three Hundred Words and Phrases in More Than One Hundred Distinct African Languages.* London: Church Missionary House.

Lepsius, K. R. (1880) *Nubische Grammatik: mit einer Einleitung über die Völker und Sprachen Afrika's.* Berlin: W. Hertz.

Lewis, M. P., G. F. Simons & C. D. Fennig (eds.) (2015) *Ethnologue: Languages of the World, Eighteenth edition.* Dallas: SIL International. Online version available at www.ethnologue.com.

Li, S., C. Schlebusch & M. Jakobsson (2014) Genetic Variation Reveals Large-Scale Population Expansion and Migration During the Expansion of Bantu-Speaking Peoples. *Proceedings of the Royal Society B (Biological Sciences)* 281: 20141448.

Lwanga-Lunyiigo, S. (1976) The Bantu Problem Reconsidered. *Current Anthropology* 17(2): 282–286.

Maho, J. F. (2003) A Classification of the Bantu Languages: An Update of Guthrie's Referential System. In Nurse, D. & G. Philippson (eds.), *The Bantu Languages*, 639–651. London; New York: Routledge.

Maho, J. F. (2009) NUGL Online: The Online Version of the New Updated Guthrie List, a Referential Classification of the Bantu Languages (4 Juni 2009) (Available online at http://goto.glocalnet.net/mahopapers/nuglonline.pdf, Accessed December 13, 2010).

Maley, J., P. Giresse, C. Doumenge & F. Charly (2012) Comment "on Intensifying Weathering and Land Use in Iron Age Central Africa." *Science* 337: 1040.

Mann, M. & D. Dalby (eds.) (1987) *A Thesaurus of African Languages: A Classified and Annotated Inventory of the Spoken Languages of Africa*. London, München, New York: Paris Hans Zell Publishers and K G Saur Verlag for the International African Institute.

Meinhof, C. (1899) *Grundriß einer Lautlehre der Bantusprachen nebst Anleitung zur Aufnahme von Bantusprachen – Anhang: Verzeichnis von Bantuwortstämmen* Leipzig: F.A. Brockhaus.

Neumann, K., M. K. H. Eggert, R. Oslisly, B. Clist, T. Denham, P. de Maret, S. Ozainne, E. Hildebrand, K. Bostoen, U. Salzmann, D. Schwartz, B. Eichhorn, B. Tchiengué & A. Höhn (2012) Comment on "Intensifying Weathering and Land Use in Iron Age Central Africa." *Science* 337(6098): 1040.

Ngomanda, A., K. Neumann, A. Schweizer & J. Maley (2009) Seasonality Change and the Third Millennium BP Rainforest Crisis in Southern Cameroon (Central Africa). *Quaternary Research* 71: 307–318.

Nurse, D. & T. J. Hinnebusch (1993) *Swahili and Sabaki: A Linguistic History*. Berkeley: University of California Press.

Nurse, D. & G. Philippson (eds.) (2003a) *The Bantu languages*. London New York: Routledge.

Nurse, D. & G. Philippson (2003b) Introduction. In Nurse, D. & G. Philippson (eds.), *The Bantu Languages*, 1–13. London; New York: Routledge.

Oliver, R. (1966) The Problem of the Bantu Expansion. *Journal of African History* 7: 361–376.

Oslisly, R., L. White, I. Bentaleb, C. Favier, M. Fontugne, J.-F. Gillet & D. Sebag (2013) Climatic and cultural changes in the west Congo Basin forests over the past 5000 years. *Philosophical transactions of the Royal Society of London. Series B* 368(1625).

Pacchiarotti, S. (2017) *Bantu Applicative Construction Types Involving *-ɪd: Form, Functions and Diachrony*. Eugene: University of Oregon, PhD dissertation.

Pakendorf, B., K. Bostoen & C. de Filippo (2011) Molecular Perspectives on the Bantu Expansion: A Synthesis. *Language Dynamics and Change* 1: 50–88.

Pakendorf, B., H. Gunnink, B. Sands & K. Bostoen (2017) Prehistoric Bantu-Khoisan Language Contact: A Cross-Disciplinary Approach. *Language Dynamics and Change* 7(1): 1–46.

Patin, E., M. Lopez, R. Grollemund, P. Verdu, C. Harmant, H. Quach, G. Laval, G. H. Perry, L. B. Barreiro, A. Froment, E. Heyer, A. Massougbodji, C. Fortes-Lima, F. Migot-Nabias, G. Bellis, J.-M. Dugoujon, J. B. Pereira, V. Fernandes, L. Pereira, L. Van der Veen, P. Mouguiama-Daouda, C. D. Bustamante, J.-M. Hombert & L.

Quintana-Murci (2017) Dispersals and Genetic Adaptation of Bantu-Speaking Populations in Africa and North America. *Science* 356(6337): 543–546.

Patin, E., K. J. Siddle, G. Laval, H. Quach, C. Harmant, N. Becker, A. Froment, B. Régnault, L. Lemée, S. Gravel, J.-M. Hombert, L. Van der Veen, N. J. Dominy, G. H. Perry, L. B. Barreiro, P. Verdu, E. Heyer & L. Quintana-Murci (2014) The Impact of Agricultural Emergence on the Genetic History of African Rainforest Hunter-Gatherers and Agriculturalists. *Nature Communications* 5: 3163.

Phillipson, D. W. (2005) *African Archaeology*. Cambridge; New York: Cambridge University Press.

Quintana-Murci, L., H. Quach, C. Harmant, F. Luca, B. Massonnet, E. Patin, L. Sica, P. Mouguiama-Daouda, D. Comas, S. Tzur, O. Balanovsky, K. K. Kidd, J. R. Kidd, L. van der Veen, J. M. Hombert, A. Gessain, P. Verdu, A. Froment, S. Bahuchet, E. Heyer, J. Dausset, A. Salas & D. M. Behar (2008) Maternal Traces of Deep Common Ancestry and Asymmetric Gene Flow between Pygmy Hunter-Gatherers and Bantu-Speaking Farmers. *Proceedings of the National Academy of Sciences of the United States of America* 105(5): 1596–1601.

Robertson, J. H. & R. Bradley (2000) A New Paradigm: The African Early Iron Age without Bantu Migrations. *History in Africa* 27: 287–323.

Schepartz, L. A. (1988) Who Were the Later Pleistocene Eastern Africans. *African Archaeological Review* 6(1): 57–72.

Schwartz, D. (1992) Assèchement climatique vers 3000 B.P. et expansion bantu en Afrique centrale atlantique: quelques réflexions. *Bulletin de la Societé Géologique de France* 163(3): 353–361.

Simons, G. F. & C. D. Fennig (eds.) (2017) *Ethnologue: Languages of the World, Twentieth edition*. Dallas: SIL International. (Available online at www.ethnologue.com).

Turchetta, B. (2007) *Missio Antiqua. Padre Giacinto da Vetralla missionario in Angola e in Congo: Un cappuccino italiano del secolo XVI tra linguistica e antropologia*. Viterbo: Sette Citta.

Van Gheel, J. (1652) *Vocabularium Latinum, Hispanicum, e Congense. Ad Usum Missionariorû transmittendorû ad Regni Congo Missiones*. Rome: National Central Library, Fundo Minori 1896, MS Varia 274.

Van Wing, J. & C. Penders (1928) *Le plus ancien dictionnaire bantu. Het oudste Bantu-Woordenboek. Vocabularium P. Georgii Gelensis*. Louvain: J. Kuyl-Otto.

Vansina, J. (1995) New Linguistic Evidence and the Bantu Expansion. *Journal of African History* 36(2): 173–195.

Verdu, P., N. S. A. Becker, A. Froment, M. Georges, V. Grugni, L. Quintana-Murci, J.-M. Hombert, L. J. Van der Veen, S. Le Bomin, S. Bahuchet, E. Heyer & F. Austerlitz (2013) Sociocultural Behavior, Sex-Biased Admixture, and Effective Population Sizes in Central African Pygmies and Non-Pygmies. *Molecular Biology and Evolution* 30(4): 918–937.

Werner, A. (1919) *Introductory Sketch of the Bantu Languages*. London: Trübner.

Wood, E. T., D. A. Stover, C. Ehret, G. Destro-Bisol, G. Spedini, H. McLeod, L. Louie, M. Bamshad, B. I. Strassmann, H. Soodyall & M. F. Hammer (2005) Contrasting Patterns of Y Chromosome and mtDNA Variation in Africa: Evidence for Sex-Biased Demographic Processes. *European Journal of Human Genetics* 13: 867–876.

PART 1

CHAPTER 2

AN INVENTORY OF BANTU LANGUAGES

Harald Hammarström

1 INTRODUCTION

The present chapter aims to provide an updated list of all Bantu languages known at present and to provide individual pointers to further information on the inventory. As a language inventory, it pretends to be complete only on the language level, rather than on the level of dialect, village lect or ethnic group. For the purposes of this inventory we have adopted the language vs dialect divisions used in the ISO 639-3 standard[1] reflected in the Ethnologue (18th edition, Lewis *et al.* 2015). This particular division has some advantages compared to other possibilities. The Ethnologue (henceforth E18) has (unsourced, but) detailed information associated with each speech variety, such as speaker numbers and map location. E18 is widely used for language inventories outside Bantu linguistics and thus allows some comparability. But foremostly, the language/dialect divisions of E18 have a stated aim to follow certain principles and generally exhibit less inconsistencies relative to other alternatives.

The E18 definition of language versus dialect is based on mutual intelligibility, but allows for deviations in either direction following what speakers themselves consider their ethnolinguistic identity. In actual practice, the listing of languages found in E18 appears to follow a more lenient splitting principle than this definition would admit. A language/dialect division based strictly on mutual intelligibility would, with high probability, yield a smaller number of languages, on the order of 85% of the present number (Hammarström 2015: 732–733). Using E18-like language/dialect divisions, the number of Bantu languages listed in this chapter is 555. But, as explained, a more strict adherence to mutual intelligibility would likely yield a number on the order of 85% of this, i.e., approximately 472.

A long tradition among Bantuists is to use what is popularly called Guthrie codes, in which the Bantu area is first divided into sections given letters A-S and further specification within each area is indicated with numbers of two digits. This yields Guthrie codes like A81 or R12 which are used either simplex or, more commonly, alongside an ethnonym or glossonym (e.g., Ngundi C11), meaning the language called Ngundi in the C11 location. The area division has some correlation with what are perceived genealogical relations between Bantu languages, but they are not defined as such and do not change whenever there is an update in our understanding of genealogical relations. Guthrie himself added suffixed lower-case letters to denote what he perceived as dialects of the same language, e.g., Yasa A33a and Kombe A33b. Later researchers have extended the coding system after Guthrie, assigning analogous codes to new varieties, sometimes with suffixed digits or capital letters, e.g., S51A or R214 (cf. Maho 2003: 639–640). The use of Guthrie codes has been successful because it provides a researcher with a rough indication of the location (and sometimes more) of a mentioned language. This is a significant timesaver given the bewildering array of Bantu ethnonyms in circulation and, with experience, adds a level of precision over alternative indications (such as a country-based geographical specification, e.g., South-Eastern Angola).

While the Guthrie codes serve this purpose well, they are less well suited for identification. Ideally, for identification, names of languages should be tied to a well-specified geographical area and/or a specimen of data, which the Guthrie coding system and its extensions do not really amount to, either explicitly or implicitly. Furthermore, there is no stated principle regards what level (language/dialect/ethnolect/idiolect?) of varieties to distinguish, meaning that arbitrariness can easily enter when a researcher has to decide whether a new variety fits into an existing code or merits a new one of its own. On the more practical side, improved knowledge on the ground has led to revision of many old Guthrie codes and the addition of many new ones. Different researchers have adopted different approaches to extending the original system even for large-scale enterprises (e.g., Bastin *et al.* 1999 versus Maho 2003) so that non-trivial code mapping tables would be required to maintain complete referential integrity. Given the popularity of Guthrie codes in Bantu linguistics, our listing also features a complete mapping to Guthrie codes. For this we have chosen the most extensive backwards-compatible Guthrie coding scheme of (Maho 2003) in its latest incarnation, the 2009 online version (Maho 2009). Every variety featured there with a Guthrie code is accounted for in the present listing.

For convenience, the listing of languages is organised by Guthrie groups, which are not argued to be genealogical (for a complete attempt at a genealogical classification of the Bantu languages piecing together arguments from the literature, see Hammarström *et al.* 2017). For every language listed, we given its ISO 639-3 code, its Guthrie code correspondence(s) as per Maho (2009), a list of the most important alternative names, subvarieties and spelling variants as well as a minimal set of literature references. In the cases where no ISO 639-3 code is given, this indicates that there is no ISO 639-3 code for the language in question, but arguably there should be one, according to the ISO 639-3 definition. Similarly, when more than one ISO 639-3 code is given, this indicates that there is some factual confusion manifested in those ISO 639-3 codes, and that according to the definition, there should only be one. The language entries are mapped to one or more Guthrie-coded varieties from Maho (2009). Most of the time the mapping isstraightforward, but since the definition used for language entries is based on intelligibility, varieties that fall halfway between two unintelligible poles may be arbitrarily assigned. The most important alternative names, subvarieties and spelling variants are given for each language, though such lists are necessarily incomplete and reflect some degree of arbitrary selection. Languages can be renamed, so names are of little importance for the language inventory as such. Our choice of names here is little more than convenience and the aim here is not to promote a certain spelling or naming convention. Literature references are provided that account for the status of each language, i.e., that testify to its existence, geographical location and difference to the other languages. This could be the introductory pages of a grammar, a lexicostatistical survey or even a primarily ethnographic publication. For reasons of space, we only cite the bare necessity in terms of justifying references, with primacy given to those which provide original data from the field and/ or contain overarching information on the delimitation of the language in question. Thus, these references are chosen for their role for the language inventory, not for description of the actual language (i.e., its phonology, morphology, syntax or lexicon). For the latter type of references, the reader may consult extensive bibliographies such as Maho (2008) (indexed by Guthrie code), Hammarström *et al.* (2017) (indexed by ISO 639-3 code) and van Bulck (1948) (dated, but invaluable for its detail and lists of unpublished materials in the former Belgian Congo area). Johnston (1922) contains word lists from (almost) the entire Bantu area, and Bastin *et al.* (1999) is the largest published lexicostatistical study of Bantu languages.

The language inventory listed here excludes sign languages used in the Bantu area, speech registers, pidgins, drummed/whistled languages and urban youth languages. Pointers to such languages in the Bantu area are included in the continent-wide overview in Hammarström (2017).

2 THE BANTU LANGUAGE INVENTORY

A10: LUNDU-BALONG GROUP

NUG	ISO	Names	References
A11	bdu	Batanga (Lotanga, Dotanga, A113), Bima (A112), Koko (Lokoko), Londo ba Diko (A115), Londo (Murundo), Balondo Ba Nanga), Longolo (Ngolo, A111), Lundu, Oroko (A101), Oroko-East, Oroko-West	Kuperus 1985: 17–18
A12	bdu	Lolue (Barue, Babue, Lue, Balue), Bakundu (A122, Lokundu, Bekunde, Kundu, Nkundu, Lakundu), Western Kundu, Bakundu-Balue, Ekombe (Bekombo, Ekumbe), Mbonge (A121, Mbɛ)	Atta Ebongkome 1993
A13	bwt	"Bafaw" (A141, Bafo, Lefo,' Afo), Bafaw-Balong, Balúun, Balong, Ngoe	Hedinger 1987
A14	bvg	Bongken (Bonkeng, Bonkeng-Pendia, Bonkenge)	Hedinger 1987: 132–164
A15A	mbo	Mbo of Ekanang-Mbouroukou (A15, Mbo of Ekanang-Mbouroukou), Mbo of Mboebo-Kekem (Mbo of Mboébo-Kekem), Mbo of Ngwatta, Mbo, Mbo'o (Mboo), Nlaa Mboo, Nle Mbuu, Melon (Melong, Eho Mbo), Sambo, Santchou, North Eastern Manenguba	Hedinger 1987
A15B	bsi	Mienge (A15, Lower Mbo), Asobse, Bassossi (Nsose, Nswose, Nswase, Swose, Sosi), North-western Manenguba	Hedinger 1987
A15C	bqz	Babong (Ihobe Mbog, Ihobe Mboong), Bafun (Mbwase Nghuy, Miamilo, Pendia), Bakaka (Ehob Mkaa, Mkaa,' Kaa, Kaka), Balondo (Ehobe Belon), Baneka (Mwaneka), Manehas (Mvae, Mwahed, Mwahet), Ngoten, Eastern Central Manenguba	Hedinger 1987
A15	bss	Akoose (Akosi, Bakossi, Bekoose, Koose, Kosi, Nkoosi, Nkosi), Elung (Along, Elong, Nlong), Mwambong, Mwamenam (Mouamenam), Ninong (Nninong), Northern Bakossi, Southern Bakossi, Western Bakossi, Nhalemoe, Western Central Manenguba	Hedinger 1987
A151	nkc	Kinkwa, Nkongho (Lekongo)	Hedinger 1987

A20: DUALA GROUP

NUG	ISO	Names	References
A21	bqm	Bamboko (Bambuku, Mboko, Womboko, Wumboko)	Ebobissé 2014
A22	bri	Bakpwe (Kpe, Mokpe, Mopkwe), Bakueri (Vakweli, Bakwedi, Bakwele, Bakweri, Bakwiri, Bekwiri, Kwedi, Kweli, Kwili, Kwiri), Ujuwa, Vambeng	Ebobissé 2014
A221	bbx[2]	Bobe (A221), Bubia (Bobea, Wovea, Wuvia)	Ebobissé 2014
A23	szv	Isu (Isu of Fako Division, Isubu, Isuwu, Su, Subu), Bimbia	Ebobissé 2014
A231	kme	Kole (Bakole, Bakolle)	Ebobissé 2014
A24	dua	Duala (Diwala, Douala, Dualla, Dwala, Dwela), Mongo (A261, Mungo, Mungu, Muungo), Pongo (A26)	Ebobissé 2014
A25	-[3]	Ewodi (Oli, Wuri, Ouri, Wuri, Uli, Wouri), Bodiman (Bidiman, Budiman)	Ebobissé 2014
A27	mzd	Malimba (Lemba, Limba, Mudima, Mulimba)	Lamberty 2009

A30: BUBI-BENGA GROUP

NUG	ISO	Names	References
A31	bvb	Bobe (Boobe, Boombe, Bube, Bubi), Adeeyah (Adija, Ediya), Banapa (Banapá), Banni, North Bobe, Southeast Bobe, Southwest Bobe, Ureka	Tessmann 1923
A32	bnm	Batanga (Batanga at Fifinda, Tanga, A32C), Balangi of Great Batanga, Bano'o (Bano'o, Banoho, Banoo, Noho, Nohu, Noko, A32a, Noku), Bapoko (Bapuku, Bapuu, Poko, Puku, A32B)	Ebobissé 2014
A33a	yko	Yasa (Iyasa, Iyassa, Lyaasa, Yassa)	Ebobissé 2014
A33b	nui	Kombe (Combe, Kombe, Ngumbi)	Fernandez Galilea 1951
A34	bng	Benga, Boumba, Ndowe	Idiata 2007 Salvadó y Cos 1891

A40

NUG	ISO	Names	References
A41	bbi	Balombi (Barombi, Barumbi, Lambi, Lombe, Lombi, Rambi, Rombi)	Lamberty 2002
A42	abb	Abaw (Abo, Bo, Bon), Bankon (Bankong, Mankon)	Lamberty 2002
A43a	bas	Basa (Basaa, Bassa, Basso, Bisaa, Mbele (Mvele), Mbene, South Kogo (A43c)	Bitjaa Kody 1988
A43b	bkh	Bakoko (Kogo, A43c), Adie (Basoo Ba Die, Basoo D'edea, Elog Mpoo), Adiangok), Mbang (Dimbambang), Yabyang (Yabyang-Yapeke), Yakalak (Yakalag), Yapoma, Yassuku (Yasoukou, Yasuku, Yasug)	Dodo-Bounguendza 1988
A44	tvu	Nen (Banen, Tunen), Aling'a (A441, Alinga, Tuling, Eling)	Mous and Breedveld 1986

NUG	ISO	Names	References
A45	nvo	Nyo'o (Nyð'ð, Nyo'on, Nyokon, Njokon), Fung (Hung)	Mous and Breedveld 1986
A46	lem	Lemande (Mande, Mandi, Nomaande, Noomaante, Numaand, Numand)	Mous and Breedveld 1986
A461	ttf	Bonek (Ponek), Otomb (Tuotom, Tuotomb, Tuotomp)	Mous and Breedveld 1986
A462	yat	Yambeta (Yambetta)	Mous and Breedveld 1986

A50: BAFIA GROUP

NUG	ISO	Names	References
A501	hij	Hijuk	Bradley 1992a
A51	lfa	Lefa' (Fak, Lefa, Fa'), Balom	Mbongue 1999
A52	dii	Kaalong (Kalong, Lakaalong), Lambong (Mbong, Bumbong, Dimbong)	Dieu and Renaud 1983
A53	ksf	Kpa (Bekpak, Rikpa'), Bafia	Guarisma 2000
A54	ngy	Tibea, (Zangnte, Djanti, Njanti, Minjanti), Ngayaba	Bradley 1992b

A60

NUG	ISO	Names	References
A601	bag	Tuki (Baki, Ki, Oki), Tiki, Bacenga (A601F, A64, Batchenga, Tocenga, Cenga), Mbere (A601G, Mvele, Tumbele, Bamvele, Bambele, Mbele), Ngoro (A601A, A61, Tu Ngoro), Aki, Uki, Cangu (A601B, Tucangu), Kombe (A601C, Wakombe, Tukombe, Bakombe), Tsinga (A601D, Tutsingo, Chinga, Betsinga, Betzinga, Tsingo, Batsingo), Tonjo (A601E, Bondjou, Bounjou, Bunju, Bounjou)	Biloa 2013
A62A	yav	Yangben, Central Yambassa, Kalong (Nukalonge)	Paulian 1986
A62B	mmu	Mmaala (Mmala, Numaala)	Paulian 1986
A62C	ekm	Elip (Belibi, Belip, Libie, Nulibie (A62C, Nulibié), Nulipie)	Paulian 1986
A621	baf	Baca (Nu Baca, Nubaca), Southern Yambassa	Paulian 1986
A622	yas	Gounou (Gunu, Nu Gunu, Nugunu), Gunu Nord, Gunu SudNugunu	Paulian 1986
A623	mlb	Mbule (Dumbule, Mbola, Mbule of Cameroon, Mbure)	Boone 1992
A63	leo	Leti (A63)	Dieu and Renaud 1983 Biloa 2013: 37–38
A65	btc	Bati (Bati Ba Ngong, Bati de Brousse, Pati)	Grant 1992

A70

NUG	ISO	Names	References
A71	eto/ mct[4]	Eton (Iton), Northern Eton (Iton Ekwe, Lower Eton), Southern Eton (Upper Eton, Iton Nke), Mengisa (Mengissa)	van de Velde 2008 Geslin-Houdet 1984

NUG	ISO	Names	References
A72(a)	ewo	Ewondo (Ewundu), Badjia (A72c, Bakjo, Bakja), Bafeuk, Yesoum (Yezum, Yesum), Fok (Fök, Fong), Jaunde (Yaounde, Yaunde), Mvete (A72b, Mvele, Mwele), Mvog-Niengue, Omvang, Yabeka, Yabekanga, Yabekolo (Yebekolo), Yangafek (A72d), Evuzok, Bané	Essono 2000
A73a	beb	Bamvele (Bebele, Bembele), Eki, Manyok	Djomeni 2014
A73b	bxp[5]	Gbigbil, Bobili	Wega Simeu 2004b Dugast 1949: 88–89
A74	bum	Bulu (A74a, Boulou, Bulu), Bene (A74b)	Alexandre and Binet 1958
A75	fan	Fang (Fan), Pangwe, Pahouin, Atsi (Batsi, A75D), Make (Meke, A75C), Mveny (A75F), Ntum (Ntumu, A75A), Nzaman (Zaman, A75E), Okak (A75B), South-West Fang (A751)	Alexandre and Binet 1958 Andeme Allogo 1985 Wilson 1849 Alexandre 1965

A80

NUG	ISO	Names	References
A801	gyi	Bagielli (Bagiele, Bagyele, Bagyeli, Bajele, Bajeli, Bogyel, Bogyeli, Bondjiel, Giele, Gieli, Gyele, Gyeli, Gélé), Bapindi, Bakola	Rénaud 1976
A802	ukh	Ukhwejo (Ukwedjo)	Thornell 2009
A803	-	Shiwa (Swa, Shiwe, Oshieba), "Fang Makina"	Idiata 2007
A81	nmg	Kwasio, Mabea (Mabi, Magbea), Mvumbo, Ngoumba (Ngumba), Bujeba	Echegaray 1960
A82	sox	So (Emvane So, Melan So, Shwo, Sô, Sso)	Bradley and Bradley 1992
A83	mcp	Makaa (Mekaa), South Makaa (South Mekaa), Bebend (A83A, Bebent), Mbwaanz (A83B), (A831), Sekunda (A83C, Shekunda, Shikunda	Cheucle 2014
A831	mkk	Byep, North Makaa, Maka (Makya, Meka, Mekae, Mekay, Mekey, Mekye, Mika, Moka)	Etter 1988
A832	biw	Kol (Bekol, Bikele, Bikele-Bikay, Bikele-Bikeng, Bikeng, Bokol), Kol North, Kol South	Henson 2007
A84	njy	Njyem (Djem, Dzem, Ndjem, Ndjeme, Ndzem, Ngyeme, Njem, Njeme, Nyem)	Cheucle 2014
A842	ozm	Koonzime (Koozhime, Koozime, Nzime, Zimu, Djimu, Dzimou, Kooncimo), Badwe'e (A841, Bajwe'e, Badjoue, Bajue, Badwe'e)	Cheucle 2014
A85	bkw	Bekwel (A85b, Bakouli, Bakwel, Bakwele, Bakwil, Bekwil), Konabem (A85a, Nkumabem, Nkonabeeb, Konabem, Kunabeeb, Konabembe, Kunabembe)	Beavon and Johnson 2011

NUG	ISO	Names	References
A86b	mgg	Mpompon (Mpongmpong, Mpopo, Mpumpoo, Pongpong, Bombo, Mbombo, Mpompo), Bageto (Northern Bangantu, Baagato, Bangantu), Medjime (A86a, Menzime, Mendzime, Medzime, Mezime), Mpomam (Boman, Mboman)	Johnson and Beavon 1989
A86c	mcx	Mpiemo (Bimu, Mbimou, Mbimu, Mbyemo, Mpyemo), Bidjuki (Bidjouki)	Thornell 2004
A87	bmw	Bomali (Bomwali, Boumali, Boumoali, Bumali), Lino, Sangasanga (Sangha-Sangha, Sanghasangha)	Bruel 1911
-	-[6]	Klieman 1997	Köhler 1964

A90: KAKA GROUP

NUG	ISO	Names	References
A91	kwu	Abakoum (Abakum, Akpwakum, Bakum, Kpakum, Kum/Bakum, Kwakum, Pakum), Til, Bheten, Baki	Belliard 2007
A92	pmm	Pol (A92a, Polri, Pori), Kinda Pol, Asum Pol, Pomo (A92b), Boungondjo, Kweso (A92C)	Ballif 1977 Wega Simeu 2012
A93	kkj	Kako (Bokaka, Kaka, Kaka-Kadei, Mkako), Kako du Gabon	Medjo Mvé 2009

B10: MYENE GROUP

NUG	ISO	Names	References
B11	mye	Myene (Omyene), Adjumba (B11d, Ajumba, Adyumba, Dyumba), Enenga (B11F), Galwa (B11c, Galua, Galloa, Galoa), Mpongwe (Npongué, Mpongwé, Npongwe, Pongoué, B11a, Mpongoué, Mpungwe), Nkomi (N'komi, B11e), Orungu (B11b, Rungu, Rongo)	Jacquot 1983

B20: KELE GROUP

NUG	ISO	Names	References
B201	nda	Ndasa (Ndasha, Andasa, Undaza)	Alewijnse et al. 2007 Idiata 2007
B202	sxe	Sighu (Sigu, Lesighu, Lisighu, Lisiwu)	Alewijnse et al. 2007 Idiata 2007
B203	syx	Osamayi (Sama, Samay, Samaye, Shamay, Shamaye, Shamayi)	Matimi 1998 Alewijnse et al. 2007 Idiata 2007
B204	nxo	Ndambomo (Ndambono)	Mvé 2013
B21	syi	Seki (Baseque, Seke, Seki-ani, Baseke, Sekiana, Sekiani, Sekiyani, Sekyani, Seseki, Sheke, Shekiyana), Bulu	Alewijnse et al. 2007 Echegaray 1959a Idiata 2007
B211	bxc	Molengue (Alengue, Alengué, Balengue)	Echegaray 1959b
B211	bxc	Alengue (Alengué), Balengue, Molendji, Molengue	de Granda Gutiérrez 1984
B22a	keb	Kele (Akele, Kélé, Dikele, Kili), Metombolo (B205, Metombola), Western Kele	Alewijnse et al. 2007 Idiata 2007

NUG	ISO	Names	References
B22b	nra	Angom (Ngom, Ongom, Ungom, Ngomo, Bangom, Bangomo), Bakoya variant of Bungom	Alewijnse *et al.* 2007 Idiata 2007 Medjo Mvé no date Medjo Mvé 2011
B22c	B305		
B22D	-	Tombidi (Muntumbudie, Ntumbidi, Tumbidi)	Idiata 2007 Alewijnse *et al.* 2007
B22E	-	Mwesa (Mesa, Yesa)	Alewijnse *et al.* 2007 Idiata 2007
B23	zmn	Mbangwe, Mbaouin (M'Bahouin)	Alewijnse *et al.* 2007 Idiata 2007
B24	wum	Wumbu (Wumbvu, Wumvu)	Alewijnse *et al.* 2007 Idiata 2007
B25	koq	Kota (Ikota, Ikuta, Kotu)	Alewijnse *et al.* 2007 Idiata 2007
B251	sak	Asake (Sake, Shake)	Alewijnse *et al.* 2007 Idiata 2007
B252	mhb	Mahongwe	Alewijnse *et al.* 2007 Idiata 2007

B30: TSOGO GROUP

NUG	ISO	Names	References
B301	gev	Viya (Avija, Evia, Eviya, Gevia, Geviya, Ia (γe-βia), Ivea, Ivéa)	Idiata 2007 Alewijnse *et al.* 2007
B302	sbw	Simba (E-Himba, γe-himba, Gehimba, Ghehimba, Ghehimbaka, Himba, Himbaka)	Alewijnse *et al.* 2007 Idiata 2007 Alewijnse *et al.* 2007
B304	pic	Pinzi (Apindje, Apindji, Apindzi, Apingi, Apinji, E-Pinzi (γe-pinzi), Gapinji, Ghepinzi, Pindji, Pinji)	Alewijnse *et al.* 2007 Idiata 2007
B305	buw	Vove (B22C, Mpovi, Ge-Vove, Gevove, Ghevove, Pove), "Bubi" (Bhubhi, Ibhubhi, Ibubi, Pubi)	Alewijnse *et al.* 2007 Idiata 2007
B31	tsv	Tsogo (Getsogo, Ghetsogo, Itsogho, Mitsogo, γe-tsγ), Babongo-Tsogo (B303, Ebongwe, Bongwe of Raponda-Walker, Ebongo)	Idiata 2007 Raponda Walker 1937
B32	kbs	Kande (Kanda, O-Kande, Okande)	Alewijnse *et al.* 2007 Idiata 2007

B40: SHIRA-PUNU GROUP

NUG	ISO	Names	References
B401	bwz	Bwisi (I-Bwisi, Ibwiisi, Ibwisi, Mbwisi)	Jacquot 1978 Bastin *et al.* 1999 Yenguitta 1991
B402	bbg	Barama (γi-βarama, Bavarama, Ghibarama, Gibarama, Givarama, Varama), Bwali (B411)	Idiata 2007 Alewijnse *et al.* 2007 Bastin *et al.* 1999
B403	vum	Vungu (Givoungou, Vumbu, Vungu, Givungu), Yivoumbou	Idiata 2007
B404	-	Ngubi (Ngove)	Idiata 2007 Puèch 1988 Aleko and Puèch 1988 Agadji Ayele 2002: 79–83

NUG	ISO	Names	References
B41	swj	Shira (Eshira, Ashira, Gisir, Gisira, Ichira, Ishira, Isira, Shire, Sira, Yichira)	Alewijnse *et al.* 2007 Idiata 2007
B42	snq	Sangu (Ashango, Chango, I-Sangu, Isangu, Massangu, Shango, Yisangou, Yisangu), Babongo-Sangu	Alewijnse *et al.* 2007 Idiata 2007 Bonhomme *et al.* 2012
B43	puu	Punu (Ipunu, Pouno, Pounou, Puno, Punu, Yipounou, Yipunu), Babongo-Rimba	Alewijnse *et al.* 2007 Andersson 1983 Idiata 2007
B44	lup	Lumbu (Baloumbou, Ilumbu, Lumbu, Yilumbu), Igama	Alewijnse *et al.* 2007 Idiata 2007 Klieman 2003

B50: NZEBI GROUP

NUG	ISO	Names	References
B501	wdd	Wanzi (Bawandji, Liwanzi, Wandji)	Alewijnse *et al.* 2007 Idiata 2007
B503	-	Vili (Ivili, Bavili de la Ngouié)	Alewijnse *et al.* 2007 Idiata 2007
B51	dma	Duma (Adouma, Aduma, Liduma, Badouma, Douma)	Alewijnse *et al.* 2007 Idiata 2007
B52	nzb	Njebi (Injebi, Inzebi, Bandzabi, Ndjabi, Ndjevi, Njabi, Njavi, Nzebi, Yinjebi, Yinzebi), Ibongo-Nzebi, Yangho (B603)	Alewijnse *et al.* 2007 Mouele 1997 Klieman 2003 Idiata 2007
B53	tsa	Tsaangi (Icaangi, Itsangi, Itsengi, Tcengui, Tchangui, Tsaangi, Tseengi, Tsengi), Mwele (B502)	Idiata 2007 Mouele 1997

B60: MBETE GROUP

NUG	ISO	Names	References
B602	kzo	Bakanike (Bakaningi, Likaningi, Kaningi, Lekaningi)	Alewijnse *et al.* 2007 Idiata 2007
B61	mdt	Mbete (Limbede, Mbede, Mbédé, Mbere, Mbété), Nkomo-Kelle, Nkomo-Ololi, Obaa, Oyuomi Mbama, Oyuomi Tcherre, Yaba-Mbeti	Idiata 2007 Lane 1989
B62	mbm	Mbaama (B601, Obamba, Ombamba, Lembaama, Lembaamba, Lim-Bamba, M-Bamba, Mbama, Mbamba), Liweme, Sibiti, Mpini, Ndouba, Obeli, Oyabi	Idiata 2007 Lane 1989
B63	nmd	Dumu (Ondoumbo, Ondumbo, Lindumu, Mindoumou, Mindumbu, Minduumo, Lendumu, Ndumu, Ndumbo, Ndumbu, Ndumu, Nduumo, Bandoumou, Doumbou, Dumbu), Epigi, Kanandjoho, Kuya, Nyani (Nyangi)	Idiata 2007 Lane 1989

B70: TEKE GROUP

NUG	ISO	Names	References
B701	tck	Tsitsege (Latsitsege, Lintsitsege, Tchitchege)	Alewijnse *et al.* 2007 Idiata 2007
B71	teg	Teghe (Tege, Iteghe, Latege) Keteghe (Kali, B71A, Ketego), Katege (B71B, Kateghe, Njining'i, Nzikini), Teke of Gabon, Teke of the Upper Alima	Adam 1954
B72	ngz	Northeastern Teke, Engungwel (B72a, Ngungwel, Ngangoulou, Ngangulu, Ngungulu), Mpu (Mpumpum), Mpumpu, B72b)	Rurangwa 1982 Bouka 1989b
B73a	tyi	Tsaayi (Caayi, Getsaayi, Tsaya, Tsaye, Tsayi)	Jacquot 1978
B73b	lli	Laali (Gibongo-Ilaali, Ilaali)	Klieman 1997 Bouka 1989b
B73c	iyx	Yaa (Bayaka, Iyaka, Yaka, Ibongo-Iyaa)	Mouandza (2002) Klieman 1997 Bouka 1989b
B73d	tyx	Tyee (Tye, Tee, Teke-Tyee), Kwe	Bouka 1989b Nkara 2007
B74	ebo/ nzu[7]	Central Teke, Teke-Eboo (B74b, Boma, Babuma, Bamboma, Boo, Boð, Eboom, Iboo), Nzikou (B74a, Ndzindziu, Njiunjiu, Njyunjyu, Nzinzu, Enjyunjyu)	Raharimanantsoa 2012 Nsuka Nkutsi 1990
B75– 6	tek	Bali (Ambali, Ibali Teke, Ibali), Kiteke (B76, East Teke, Mosieno, Ng'ee, Bamfunuka), Nunu (B822, B80Nu)	Sims 1886:i-xii Bastin 1978 Boone 1973: 295–306 Bastin *et al.* 1999
B77a	kkw	Kukuya (Chikuya, Cikuya, Kikuwa, Koukouya, Kukua, Kukuya, Kukwa)	Paulian 1975
B77b	ifm	Fumu (Fuumu, Mfumu, Ifuumu, Teke du Pool), Wuumo (B78, Ewuumo, Wũ, Mpuon (Mpuono, B84A), Mpuun (B84B)	Calloc'h 1911 Jacquot 1965 Makouta-Mboukou 1977 Bastin *et al.* 1999: 13, 183, 193, 202

B80: TIENE-YANZI

NUG	ISO	Names	References
B81	tii	Tiene (Kitiene, Kitiini, Tende, Tiene), Nunu (C31C, Banunu, Kenunu)	Motingea Mangulu 2004 Ellington 1977 Hulstaert 1951: 32–33
B82	boh	Boma (Kiboma, Boma Kasai, Buma)	Hochegger 1972
B821	-	Mpe (Kempee)	Detienne 1984 Bastin *et al.* 1999: 202
B83	zmf	Mfinu (Funika, Emfinu, Mfununga), Ntsiam, Ntswar	Bostoen and Koni Muluwa 2014 Boone 1973
B84	see B77b		
B85	yns	Yans (Yanzi), West Yansi (B85a, Mbien), East Yans (B85b), Yeei (B85c)	Mayanga 1985: 1–22 Bostoen and Koni Muluwa 2014
B85d	soo	Nsong (Tsong), Mpiin (B863)	Bostoen and Koni Muluwa 2014 Boone 1973
B864	noq[8]	Ngongo (Ngong, Ngoongo)	Bostoen and Koni Muluwa 2014 Boone 1973 Torday and Joyce 1907

NUG	ISO	Names	References
B85F (= L12a)	smx	Ntsambaan (Sambaan)	Bostoen and Koni Muluwa 2014 Boone 1973 Maes 1934
B86	diz	Dzing (Di, Din, Ding, Dinga), Mpur (B85e, Mput)	Boone 1973 Muluwa and Bostoen 2015 Mertens 1939
B861	nlo	Ngul (Banguli, Ingul, Ngoli, Ngul, Nguli, Ngulu), Ngwi	Maes 1934 Boone 1973 Bostoen and Koni Muluwa 2014
B862	-	Lwel (Balori)	Khang Levy 1979 Maes 1934 Koni Muluwa and Bostoen 2011
B865	-	Nzadi (Bandjari)	Maes 1934 Crane et al. 2011 Boone 1973
B87	zmp[9]	Mbuun (Mbunda, Kimbuun, Gimbunda, Ambuun)	Bostoen and Koni Muluwa 2014 Boone 1973

C10

NUG	ISO	Names	References
C101	bvx	Babole (Dibole), Dzeke, Ebambe, Edzama, Kinami, Mossengue, Mounda	Gardner 1990 Leitch 2004
C102	ngd	Ngando (Bagandou, Bangandou, Dingando, Mouna-Bagandou) Bodzanga, Dikota (C103, Kota), Dikuta, Ngando-Kota, Bodzanga	Périquet 1915 Bouquiaux and Thomas 1994 Guillaume and Delobeau 1978
C104	axk	Aka (Yaka, Aka de la Lobaye), Benzele (Bendjelle, Mbenzele), 'Babinga'	Bruel 1911 Bahuchet 1989
C105	-	Mikaya, Bambengangale, Baluma	Klieman 1997 Thomas and Bahuchet 1991
C11	ndn	Goundi (Gundi, Ingundi, Ngondi, Ngundi)	Ouzilleau 1911 Périquet 1915
C12	bkj	Pande (C12a, Ipande), Bukongo (C12b, Bogongo, Gongo)	Ouzilleau 1911 Bruel 1911 Périquet 1915 Hauser 1954
C13	mdn	Mbati, Isongo (Issongo, LIsongo, Lissongo)	Guillaume and Delobeau 1978 Mbalanga 1996 Bouquiaux and Thomas 1994
C14	zmx	Bomitaba (Mbomotaba), Itanga	Gardner 1990
C143	bok	Bonjo (Mbonzo), Impfondo (C143)	Gardner 1990 Samarin 1984 Hauser 1954
C15	bui	Bongili ([E18], Bongili-Pikounda, Bongiri, Bungili, Bungiri), Inyele (C141), Bondongo (C142)	Motingea Mangulu 2008 Gardner 1990 Hauser 1954
C16	loq	Lobala, Likoka, Tanda	Gardner 1990 Hauser 1954
C16	bkp	Boko (Iboko)	Motingea Mangulu 1996
C161	bml	Bomboli	de Boeck 1948
C162	bzo	Bozaba (Budjaba, Budzaba, Buzaba)	de Boeck 1948

C20: MBOSHI

NUG	ISO	Names	References
C21 (= C23)	mdu	Mboko (Mboxo, Mbuku), Ngare (Ngáre)	Ndinga Oba 2003
C22	akw	Akwa (Akwá)	Ndinga Oba 2003 Aksenova and Toporova 2002
C24	koh	Koyo (Ekoyo, Kouyou, Koyó)	Ndinga Oba 2003 Poupon 1919
C25	mdw	Mbosi (C25A, Embosi, Mbochi, Mboshi, Mboshe, Mbonzi), Olee (C25B), Ondinga (C25C), Ngolo (C25D), Eboi (C25E)	Fontaney 1989: 87–89 Ndinga Oba 2003
C26	kwc	Kouala (Likouala, Ekwala, Kwala, Likwala, Likwála)	Ndinga Oba 2003
C27	kxx	Kuba (Likuba), Bwenyi (C201, Buegni)	Ndinga Oba 2003 Adoua 1981

C30: BANGI-NTOMBA

NUG	ISO	Names	References
C302	bzm	Bolondo	de Boeck 1948 Motingea Mangulu 2002: 149–150
C31a	biz	Loi (Baloi, Boloi), Makutu, Likila, Mampoko, Mpundza (C36c, Mbudza, Mpundza)	de Boeck 1948 Motingea Mangulu 1996
C311	mmz	Mabale (Mabaale, Lomabaale), Mabembe	Motingea Mangulu 1996
C312	ndw	Ndobo (Ndoobo)	Motingea Mangulu 1990a de Boeck 1948
C313	-	Litoka	Motingea Mangulu 1996
C314	lie	Balobo[10]	de Boeck 1948 Motingea Mangulu 1996
C32	bni	Bangi (Bobangi, Bubangi, Lobobangi)	Whitehead 1899
C32 (= C38)	mow	Moi (Lemoi, Moyi, Moye), Liku, Rebu	Motingea Mangulu and Biako Montanga Mayika 2016
C321	liz	Binza (Libindja, Libinja, Libinza])	van Leynseele 1977 Motingea Mangulu 1996
C322	-	Zamba (C322, Dzamba, Jamba)	de Boeck 1948, Kamanda Kola 1991
C323	-	Mpama	Motingea Mangulu 1996 Hulstaert 1984
C33	szg	Sengele (Kesengele, Sengere)	Hulstaert 1951 Motingea Mangulu 2001
C34	skt	Sakata (C34A), Djia (C34B, Dja, Dia, Kidjia, Wadia), Bai (C34C, Kibay, Kibai, Tuku-Ketu-Batow (C34D)	Tylleskär 1987
C35a	nto	Ntomba (Lontomba, Luntumba, Ntumba), Sakani (C35C, Lotsakani, Sakanyi)	Mamet 1955 Hulstaert 1993
C35b	bli	Bolia (Bulia)	Mamet 1960

NUG	ISO	Names	References
C36	lse	Lusengo (Losengo), Poto (C36a, Pfoto, Yakata), Mpesa (C36b, Limpesa), Kangana (C36f), Enga (C315, Baenga, Bolombo), Eleku	Motingea Mangulu 2008 Motingea Mangulu 1993a Motingea Mangulu 1996
C36e	bkt	Boloki (Baloki, Boleki, Buluki), Ngala (C36d, Mangala)	Motingea Mangulu 2002 Motingea Mangulu 1996
C36g	ndl	Ndolo (Ndoolo), Kula (C415, Likula)	de Boeck 1948 Motingea Mangulu 1996
C37	bja	Budja (Buja, Ebudja, Budza, Ebuja, Embudja, Embuja, Limbudza, Mbudja, Mbuza), Bosambi, Mbila, Monzamboli, Yaliambi	Motingea Mangulu 1996
C371	tmv	Tembo (Motembo, Litembo) de la Mongala, Motembo des grandes îles du Fleuve, Motembo septentrional (C374, Buja, Babale de Bosô-Njanoa), Kunda (C372, Motembo de Budzala, Motembo de la rivière Mbanga)	Motingea Mangulu 1996
C373	-	Egbuta	Motingea Mangulu 2003

C40: NGOMBE GROUP

NUG	ISO	Names	References
C401	pae	Pagibete (Apagibete, Apagibeti, Apakabeti, Apakibeti, Pagabete)	Boone and Olson 2004 Reeder 1998
C403	kty	Kango (Bacango, Kango of Bas-Uélé District, Likango)	McMaster 1988 de Calonne-Beaufaict 1912
C41	ngc	Ngombe (Lingombe), Ngombe of the Congo river (C41A), Ngombe at Bosobolo (C41B, North Ngombe), Ngombe at Libenge (C41C, North-West Ngombe), Binza (C41D, East Ngombe)	Motingea Mangulu 1988 Motingea Mangulu 1996 de Rop 1960 Burssens 1958: 32–36
C411	bws	Bomboma, Bangiri, Likaw (Likao), Lingundu (Lingonda), Ebuku	de Boeck 1948 Motingea Mangulu 1996
C412	bmg	Bamwe, Libobi, Lifonga, Likata	de Boeck 1948 Motingea Mangulu 1996
C413	dzn	Dzando (Djando), Ngiri (C31b)	Hackett and van Bulck 1956: 73–74 de Boeck 1948
C414	lgz	Gendja (Gendza-Bali, Ligendza, Ligenza)	Motingea Mangulu 2001 Vanhouteghem 1947
C42	bwl	Bwela (Buela, Ebwela), Doko around Lisala (C301), Lingi	Motingea Mangulu 1996
C44	bww	Bwa (Boa, Boua, Bobwa, Kibua, Kibwa, Leboa-le, Leboale, Libua, Libwali), Yewu (C402), Leangba, Baati (C43A, Bati, Lebaati), Benge (C43B, Libenge, Mobenge), Boganga (C43C, Boyanga), Ligbe (C43D)	Boone and Olson 2004 Motingea Mangulu 2005 Czekanowski 1924 Halkin and Viaene 1911 McMaster 1988 Bostoen and Grégoire 2007

NUG	ISO	Names	References
C441	bbm	Bango (Babango, Mobango, Southwest Bwa	Motingea Mangulu 1995 Motingea Mangulu 1995: 5
C45	agh	Ngelima (Kingbelima, Bangalema, Bangelima), Beo (C45A, Lebeo), Buru (C45B), Tungu (C45C)	Gérard 1924

C50: SOKO-KELE GROUP

NUG	ISO	Names	References
C502	-	Elinga (Loselinga)	Motingea Mangulu 1994a
C51	zms	Mbesa ([E18], Mobesa, Mombesa)	Hulstaert 1951 de Boeck 1951
C52	soc	So (Basoko, Eso, Gesogo, Heso, Soa, Losoko)	Czekanowski 1924 Harries 1955 Bastin *et al.* 1999: 208
C53	pof	Poke (Pfoke, Poke, Puki, Tofoke, Topoke, Tovoke)	Torday and Joyce 1922 Bastin *et al.* 1999: 208
C54	loo	Lombo (Olombo, Tu-Rumbu, Turumbu, Ulumbu)	Carrington 1947
C55	khy/ fom	Kele (Ekele, Lokele, Kili, Likelo, Lokele), Yakusu, Foma (C56, Fuma, Li-Foma, Lofoma)[11] Likile (C501)[12]	Carrington 1943 Bastin *et al.* 1999: 208

C60: MONGO-NKUNDO

NUG	ISO	Names	References
C61	lol/ ymg	Mongo (Lomongo, Bololo, Lolo), Nkundo (Lonkundo, Lokonda, Nkundu, Nkundo-Twa), Bakutu (C61A), Bokote (C61B, Bokote Tswa, Ngata, Wangata), Booli (C61C), Bosaka (C61D, Bolandá), Konda (C61E, Ekonda, Ekonda-Bosanga, Ekonda-Twa), Ekota (C61F), Emoma (C61G), Ikongo (C61H, Lokalo-Lomela), Iyembe (C61I, Iyembe de la Lokolo), Ntomba (C61J, Lionje, Nsongo, Bikoro), Yamongo (C61K), Mbole (C61L, Nkengo, Yenge, Yongo, Bosanga-Mbole, Mangilongo, Lwankamba, Lonkembe, Liinja), Nkole (C61M), South Mongo (C61N, Belo, Bolongo, Acitu, Panga), Yailima (C61O, Yalima, Yajima), Ngombe-Lomela (C61P, Longombe, Ngome à Múná), Kitwa-Inongo, Ndombe Tswa, Nkundo Batswa, Twa of Ekonda, Mpombo (C351, Mpombo), Yamongeri (C36H)[13]	Engels (1911), Hiernaux *et al.* (1976), Hulstaert (1948), Hulstaert (1999), Motingea Mangulu (1994b), Müller (1964), Nsenga Diatwa (2009)
C611	-	Bafotó, Batswa de l'Equateur	Hulstaert 1978
C63	nxd/ lal	Bongando (Longandu, Ngando, Ngandu), Lalia (C62)[14]	Hulstaert 1987

C70

NUG	ISO	Names	References
C71	hba	Hamba	Kasongo 1993, Labaere and Shango 1989
C71	tll	Tetela (Otetela, Batetela), Batetela du Nord, Otetela-Olembe, Sungu Batetela	Labaere and Shango 1989 Jacobs 1964: 9–12
C72	ksv	Kusu (Bakusu, Kikusu), Kusu of Kimbombo	Motingea Mangulu 1989 Jacobs 1964: 9–12
C73	nkw	Nkutu (Ankfucu, Bankutu, Nkuchu, Nkutshu, Lonkutshu, Lonkucu)	Jacobs and Omeonga 2004 Hulstaert 1951: 23 Motingea Mangulu 1989
C74/C75	yel/kel	Yela (C74, Boyela), Kela (C75)[15]	Hulstaert 1942, 1999
C76	oml	Ombo (Hombo, Loombo)	Meeussen 1952

C80

NUG	ISO	Names	References
C81	dez	Dengese (Ndengese, Londengese), Nkutu	Hulstaert 1951 Goemaere 1984
C82	soe	Hendo (Ohendo, Lohendo, Ohendo), "Songomeno," Hendo	Hulstaert 1951 Motingea Mangulu 1990b
C83	buf	Bushoong (Bushong, Bushongo, Busoong), Bakuba (Kuba, Bacuba), Batwa du Kasayi, Cwa du royaume Kuba), Ngeende (C83A, Ngende, Ngendi), Ngongo (C83B), Pyaang (C83C, Pianga, Piong, Panga), Shuwa (C83D, Shobwa, Shoba, Loshoobo)	Hiernaux 1966 Vansina 1959: 5
C84	lel	Lele (Bashilele, Usilele)	Douglas 1963 Maes 1934 Boone 1973
C85	won	Wongo (Bawongo), Gongo (Bakong, Tukkongo, Tukongo, Tukungo)	Maes 1934 Boone 1973

D10

NUG	ISO	Names	References
D11	mdq	Mbole (Lombole, Mbole of the Lomami), Yaikole, Yaisu, Yangonda, Lokaló de la haute Jwafa, Langa (C701)	Motingea Mangulu 1993b de Rop 1971 Hulstaert 1988
D12	lej	Lengola (Kilengola, Lengora)	Stappers 1971
D13	zmq	Mituku (Metoko, Kinya-Mituku)	Stappers 1973 McMaster 1988
D14	gey	Enya (Genya, Tsheenya, Wagenya), Baena, Enya at Kisangani (D14A), Enya at Kongolo (D14B)	Spa 1973
-	-	Mokpá, Bamanga	Motingea Mangulu 1990

D20: LEGA-HOLOHOLO

NUG	ISO	Names	References
D201	lik	Liko (Lika, Balika, Kilika, Liliká, Lilikó)	van Geluwe 1960 Harvey 1997
D21	bcp	Bali (Baali, Kibaali, Kibala, Kibali, Libaali, Mabali), Bafwandaka, Bakundumu, Bekeni, Bemili	Harvey 1997 van Geluwe 1960 McMaster 1988
D211	kzy	Kango (Dikango), 'Dibatchua'	Schebesta 1952 Vorbichler 1964 Harvey 1997
D22	rwm	Amba (Baamba, Kivamba, Ku-Amba, Kuamba, Kwamba, Rwaamba, RwaAmba, Rwamba)	Winter 1953 Joset 1952
D23	kmw	Komo (Kumu, Kikomo, Kikumo, Kikumu, Kikuumu, Kumo, Kuumu)	Harries 1958 McMaster 1988
D24	sod	Songola (Kesongola, Kisongoola, Songoora, Wasongoa, Watchongoa), Binja, Gengele (Kegengele)	Schoenbrun 1994 Delhaise 1909 Hackett and van Bulck 1956: 90
D25	lgm	Lega-Mwenga, Lega-Ntara, Isile (Ileka Ishile, Ishile, Isile, Kisile, Sile), Iwanyabaale, Kileega of Bakabango, Eastern Lega	Biebuyck 1973 Bastin et al. 1999: 209–210 Botne 2003: 422–423
D251	lea	Lega-Shabunda, Lega-Malinga, Kitila, Kiliga, Kiyoma, Kigala, Gonzabale, Beya-Munsange, Banagabo, Pangi Lega	Biebuyck 1973 Bastin et al. 1999: 209–210 Botne 2003: 422–423
D251	khx	Kanu (Kaanu, Kano, Kanu, Kikaanu, Likanu)	Biebuyck 1973
D251	ktf	Kwame (Kikwame, Kikwami, Kwami)	Biebuyck 1973
D26	zmb	Zimba, ḰɛsÍɛ, kisémolo, kyɛnyɛ-mánila, kikwángɛ	Kabungama 1992
D27	bnx	Bangubangu (Bango-Bango, Bangobango, Kibangobango)	Meeussen 1954
D28	hoo	Holoholo (Horohoro), West Holoholo (D28a, Guha, Kalanga), East Holoholo (D28b), Tumbwe (D281), Lumbwe (D282)	Coupez 1955 Maho and Sands 2004 Bastin et al. 1999: 210

D30

NUG	ISO	Names	References
D301	kbj	Kare (Kare, Akare, Likarili)	Dijkmans 1974 Santandrea 1964
D302	bqu	Guru (Boguru, Kogoro, Bukur)	Costermans 1953 van Bulck 1952 McMaster 1988: 242, 263
D303	nbd/ myc[16]	Bangbinda (Bungbinda, Ngbinda, Ngminda, Ngenda), Mayeka (D307)	Hackett and van Bulck 1956 Johnston 1922 Santandrea 1948 McMaster 1988
D304	hom	Homa (Hôma, Huma)	Santandrea 1948 Santandrea 1963 Hackett and van Bulck 1956
D305	nyc/ gti[17]	Nyanga-li, Gbati-ri (D306)	(Hackett and van Bulck 1956: 74)

NUG	ISO	Names	References
D308	boy	Bodo	Santandrea 1963 Santandrea 1948 McMaster 1988: 242, 263
D31	bhy	Bhele (Bili, Ebhele, Ipere, Kipere, Kipili, Pere, Peri, Pili, Piri)	Meyer and Raymond 1981 de Wit 1994
D311	bip	Bila (Babila, Ebila, Forest Bila, Forest Bira, Kibila)	Kutsch Lojenga 2003
D312	kkq	Kaiku (Ikaiku, Kaiko)	de Wit 1994
D32	brf	Bera (Bira, Plains Bira, Babira, Wawira, Grassland Bavira, Lower Bera), Sese	van Geluwe 1956
D33	nlj	Nyali (Banyari, Linyali), "Huku"	Harries 1959 van Geluwe 1960
D331	vau	Vanuma (Bvanuma, Libvanuma, Livanuma), Mbuttu (D313, Pygmy language spoken in Avakubi district on the north side of the Ituri between its confluence with the Epulu and with the Nepoko)	Schebesta 1966, Johnston 1922: 484–495
D332	buu	Budu (Badimbisa Budu, Bodo, Ebudu, Kibudu, Mabudu	Asangama 1983
D333	ndk	Ndaaka (Ndaka, Bombo Ndaka, Indaaka)	Harvey 1997 van Geluwe 1960
D334	zmw	Mbo (Imbo, Kimbo, Mbo-Beke)	van Geluwe 1960
D335	bkf	Beeke (Beke, Ibeeke, Beeke)	Hackett and van Bulck 1956: 84
D336	jgb	Ngbele (Lingbe, Lingbee, Mangbele, Ngbee)	Hackett and van Bulck 1956 Liesenborghs 1932 Verhulpen 1936b
-	-	Belueli, Pygmies of the Apare river	Vorbichler 1964 Schebesta 1953

D40: NYANGA

NUG	ISO	Names	References
D41	see JD41		
D42	see JD42		
D43	nyj	Nyanga (Inyanga, Kinyanga)	Hackett and van Bulck 1956 Mateene 1980

D50: BEMBE-BUYI

NUG	ISO	Names	References
D501-D531	see JD50		
D54	bmb	Bembe (Babembe of Kivu, Beembe, Bembe, Ebembe, Ibembe)	Meeussen 1953 Verhulpen 1936b Biebuyck 1973
D55	byi	Buyu (Boyo, Bujwe, Buyi, Kibuyu)	Meeussen 1953 Biebuyck 1973

E40: TEMI

NUG	ISO	Names	References
E401-E45	see JE40		
E46	soz	Temi (GiTemi), Ketemi), "Sonjo"	Nurse and Rottland 1992

E50: KIKUYU-KAMBA GROUP

NUG	ISO	Names	References
E51	kik	Kikuyu (Gekoyo, Gikuyu)	Mutahi 1981 Sim 1977
E52	ebu	Embo (Embu), Mbeere (Mbere, Kimbeere)	Mwaniki 2014 Mutahi 1981 Möhlig 1974
E53	mer	Meru (Kimeru, Mero), Imenti	Möhlig 1974
E531	mws	Mwimbi-Muthambi	Möhlig 1974
E54	thk	Tharaka (Tharaka, Central Tharaka, Saraka, Sharoka), Thagicu	Möhlig 1974
E541	cuh	Chuka (Chuku, Cuka, Suka)	Möhlig 1974
E55	kam	Kamba (Kenya), Kekamba, Kikamba)	Lindblom 1926: 3–18
E56	dhs	Daisu (Dhaisu, Daiso, Kidhaiso), "Segeju"	Nurse 2000

E60: CHAGA GROUP

NUG	ISO	Names	References
E61	See West Kilimanjaro E621		
E62a	See West Kilimanjaro E621, Central Kilimanjaro E622		
E62b	See Central Kilimanjaro E622		
E621A (=E61)	rwk	Rwa (Rwo), Meru (Mero)	Philippson 1984 Winter 1980
E621B (=E62a)	jmc	Machame (Caga-Machame, Kimachame, Kimashami, Macame, Machambe, Madjame, Mashami), Siha (E621C, Kisiha), Narumu, Ng'uni (E621F), "Hai," Masama (621E, Masdama)	Philippson 1984 Winter 1980
E622A (=E62a)	old	Mochi (Kimochi, Kimoshi, Moshi, Mosi), Mbokomu (E622B), Uru (E622D, Oru), Kuma (Okuma, "Rusha," E63)	Philippson 1984 Winter 1980

NUG	ISO	Names	References
E622C (=E62b)	vun	Vunjo (Wunjo, Kivunjo, Kiwunjo, Wuunjo), Lema (Kilema), Kiruwa, Mamba, Woso (Kiw'oso, Kibosho), Kindi, Kombo, Mweka (Mwika), Marangu (Morang'u), Kiruwa	Philippson 1984 Winter 1980
E623 (=E62c)	rof	Rombo (Kirombo), Keni (E623D), Mashati (E623B), Mkuu (E623C), Usseri (E623A, Kiseri, Useri), Mriti	Philippson 1984 Winter 1980
E64	hka	Kahe, Kikahe	Kahigi 2008 Winter 1980
E65	gwe	Gweno (Ghonu, Gweno, Ki-Gweno, Kighonu, Kigweno)	Winter 1980 Philippson and Nurse 2000

E70: NYIKA-TAITA GROUP

NUG	ISO	Names	References
E701	mlk	Elwana (Ilwana, Kiwilwana), 'Malakote'	van Otterloo and van Otterloo 1980 Nurse 2000
E71	pkb	Pokomo (Pfokomo), Buu, Upper Pokomo (E71A), Lower Pokomo (E71B, Malachini)	Nurse and Hinnebusch 1993
E72a,e,G,H	nyf	Giryama (Agiryama, Giriama, Giryama, Kigiriama), Nyika (Nika, Kinyika), Kambe (E72G), Chwaka, Rabai (E72e), Ribe (E72H)	van Otterloo and van Otterloo 1980 Möhlig 1986
E72b,c,F	coh	Kauma (E72b), Chonyi (E72c, Conyi, Chichonyi), Jibana (E72F, Dzihana, Chidzihana)	van Otterloo and van Otterloo 1980 Möhlig 1986
E72d	dug	Duruma	Möhlig 1986 van Otterloo and van Otterloo 1980
E73	dig	Digo (Kidigo), Degere (E732 Madhaka)[18]	Nicolle 2013: 1–5
E731	seg	Segeju (Kisegeju, Sageju, Segeju, Sengeju)	Nurse 1982
E74a,C	dav	Taita, Dawida (Dabida, Davida, E74a), Bura, Chawia, Kasigau (E74C, Kisighau)	van Otterloo and van Otterloo 1980
E74b[19]	tga	Sagalla (Kisagalla, Sagala, Sagalla Taita, Saghala), Dambi, Kishamba	van Otterloo and van Otterloo 1980

F10: TONGWE-BENDE GROUP

NUG	ISO	Names	References
F11	tny	Tongwe (Kitongwe, Kitoongwe, Sitongwe)	Masele 2001
F12	bdp	Bende, Gongwe	Abe 2011

F20: SUKUMA-NYAMWEZI

NUG	ISO	Names	References
F21	suk	Sukuma, Kemunasukuma (F21A, North Sukuma), Kemunangweeli (F21B), Kiiya (F21C, Kiya, JinaKiiya), Kemunadakama (F21D, South Sukuma), Nasa (F21E, Kinaanasa), Sumaabu (F21F, SumaaBu, KisumaaBu), Nelaa (F21G, Kinaanelaa), Ntuzu (Gina-Ntuzu, Kimunantuzu, F21H)	Masele 2001
F22A, B,D,E, F,G,H, I,J,K	nym	Nyamwezi (Kinyamwezi), Galaganza (F22A, Sigalagaanza, Garaganza), Mweri (F22B), Nyanyembe (F22D, Unyanyemba, Unyanyembe), Takama (F22E, Dakama), Nangwila (F22F), Ilwana (Kilwana, Ilwana-Nyamwezi F22G), Uyui (F22H), Rambo (F22I), Ndaala (F22J, Ndala), Nyambiu (F22K)	Masele 2001
F22C	kcz	Konongo (Kikonoongo), East Nyamwezi	Masele 2001
F23	suw	Sumbwa (Kisumbwa, Shisumbwa, Shumbwa), Sisiloombo, Siyoombe	Masele 2001
F24	kiv	Kimbu (Kikimbu, Kikiimbu), North KiKiimbu, South Kikiimbu	Masele 2001
F25	wun	Bungu (Ici-Wungu, Iciwungu, Kibungu, Ungu, Wungu)	Roth 2011

F30: NILAMBA-RANGI

NUG	ISO	Names	References
F31A, C,D,E	nim	Nilamba, Ilamba, Central Laamba (F31A, Kinilaamba), Ushoola (F31C, Ushoola), Nyambi (F31D, Iambi), Mbuga (F31E)	Masele 2001
F31B	isn	Isanzu (Ihaanzu, Issansu)	Masele 2001
F32	rim	Limi (Rimi, Keremi, Kilimi), Ahi (GiAhi, Chahi), Ginyamunyinganyi), Rwana (Girwana), Nyaturu (Kinyaturu, Turu)	Masele 2001
F33	lag	Rangi (Langi, Irangi, Kelangi, Kilaangi), Kondoa	Dunham 2007
F34	mgz	Mbugwe (Buwe, Kimbugwe, Kiumbugwe, Mbuwe)	Mous 2004

G10: GOGO-KAGULU

NUG	ISO	Names	References
G11	gog	Gogo (Chigogo, Cigogo, Gogo, Ki-Gogo, Kigogo)	Gonzales 2002
G12	kki	Kagulu (Chikagulu, Cikagulu, Kagulu, Kaguru, Kigaguru, Kikagulu)	Petzell 2008

G20: SHAMBALA GROUP

NUG	ISO	Names	References
G21[20]	tvs	Taveta ("Tubeta")	van Otterloo and van Otterloo 1980 Kitetu 2012

NUG	ISO	Names	References
G22	asa	Pare, Asu (Ashu, Athu, Casu, Chasu, Chiasu), South Pare (G22B, Gonja, Mbaga (G22B), North Pare (G22A)	van Otterloo and van Otterloo 1980 Kitetu 2012 Mreta 2000
G23	ksb	Shambala (Šambala, Kisambaa, Kishambaa, Kishambala, Sambaa, Sambala, Sambara, Schambala, Shambaa)	Nurse 1988
G24	bou	Bondei (Bonde, Boondei, Kibondei)	Merlevede 1995 Legère 1992

G30: ZIGULA-ZARAMO

NUG	ISO	Names	References
G301	doe	Doe (Dohe, Ki-Doe)	Gonzales 2002
G31	ziw	Zigula (Chizigula, Kizigua, Kizigula, Msegua, Seguha, Wazegua, Zeguha, Zegura, Zigoua, Zigua, Zigwa)	Petzell and Hammarström 2013
G311	xma	Mushungulu (Kimushungulu, Mushunguli, Mushungulu), Zigua in Somalia	Hout 2012 Williams 2012 Grottanelli 1953
G32	cwe	Ngwele (Kakwere, Kikwere, Kinghwele, Kwele, Kwere, Ng'were, Ngh'wele, Nhwele, Nwele, Tsinghwele)	Gonzales 2002 Legère 2003
G33	zaj	Zaramo (Dzalamo, Saramo, Zalamo [E18)	Gonzales 2002
G34	ngp	Ngulu (Kingulu, Ngulu, Nguru, Nguu)	Petzell and Hammarström 2013
G35	ruf	Ruguru (Luguru, Cilugulu, Guru, Ikiruguru, Kiluguru, Kiruguru, Lughuru, Lugulu)	Gonzales 2002 Mkude 1974
G36	kcu	Kami (Kikami)	Petzell and Hammarström 2013
G37	kdc	Kutu (Khutu, Kikutu, Kixutu, Kutu	Petzell and Hammarström 2013
G38	vid	Vidunda (Chividunda, Kividunda, Ndunda, Vidunda)	Gonzales 2002 Legère 2007
G39	sbm	Sagala (Saghala, Ki-Sagara, Kisagala, Kisagara) Kondoa, Solwe, Kweny, Nkwifiya (Kwiva, Kwifa)	Gonzales 2002

G40: SWAHILI GROUP

NUG	ISO	Names	References
G402	ymk	Makwe (Coastal Makwe, Interior Makwe, Kimakwe, Macue)	Devos 2007
G403	wmw	Mwani (Cap Delgado, Kimwani, Muane, Mwane, Mwani, Quimuane)	Schadeberg 1997
G412	-	Mwiini (Chimwiini, Mwini, Barawa)	Nurse and Hinnebusch 1993

NUG	ISO	Names	References
G41–43	swh/ swc/ ccl	Swahili (Kiswahili), Tikuu (G41, Bajuni, Bagiuni), Socotra Swahili (G411), Amu (G42a, Pate, Siu, Siyu), Mombasa area Swahili (G42b, Mvita, Ngare, Jomvu, Changamwe, Kilindini), Mrima (G42c, Mtang'ata, Lugha ya Zamani), Unguja (G42d, Kiunguja), Malindi (G42E, Mambrui), Fundi (G42F, Chifundi), Chwaka (G42G), Vumba (G42H), Madagascar Swahili (G42I, Nosse Be), Pemba (G43a, Southern Pemba, Northern Pemba), Tumbatu (G43b), Makunduchi (G43c, "Hadimu," Kae, Kikae), Mafia (G43D, Mbwera, Kingome), Kilwa (G43E), Mgao (G43F, G401, Kimgao), Sidi (G404), Cutchi Swahili, Congo Swahili (Lubumbashi Swahili, Shaba Swahili, Kingwana)	Ashton 1944 Nurse and Hinnebusch 1993 Lodhi 2008 Samarin 2014 Neale 1974
G44(D)	swb	Maore Comorian (Comoro, Komoro, Shimaore)	Full 2006
G44a	zdj	Ngazija (Shingazidja)	Full 2006
G44b	wni	Njuani (Ndzwani, Shindjuani)	Full 2006
G44C	wlc	Mwali (Shimwali)	Full 2006

G50: POGOLO-NDAMBA GROUP

NUG	ISO	Names	References
G51	poy	Pogolo (Chipogolo, Chipogoro, Cipogolo, Pogolu, Pogora, Pogoro)	Hendle 1907
G52	ndj	Ndamba (Kindamba)	Novotná 2005

G60: BENA-KINGA GROUP

NUG	ISO	Names	References
G61	sbp	Sango (Mahango, Eshisango, Kisangu), Lori (Rori)	Mumford 1934 Nurse 1988
G62	heh	Hehe (Ehe, Ekiehe, Kihehe)	Mumford 1934 Nurse 1988
G63	bez	Bena (Ekibena, Ikibena	Mitterhofer 2013
G64	pbr	Pangwa (Ekipangwa, Kipangwa)	Stirnimann 1983
G65	zga	Kinga (Ekikinga, Kikinga)	Schadeberg 1971 Nurse 1988
G651	gmx	Magoma (Kimagoma)	Schadeberg 1971: 1, 190
G66	wbi	Wanji (Kivwanji, Kiwanji, Vwanji, Wanji)	Nurse 1988
G67	kiz	Kisi (Kese, Kikisi)	Fülleborn 1906 Nurse 1988

H10: KIKONGO GROUP

NUG	ISO	Names	References
H10A	ktu	Kituba (Kikongo-Kutuba), Ikeleve, Kibulamatadi, Kikongo Commercial, Kikongo Simplifie (Kikongo Simplifié), Kikongo Ya Leta, Kileta	Mufwene 2009
H10B	mkw	Munukutuba (Monokutuba)	Mufwene 2009
H11	beq	Beembe (Bembe), Keenge, Yari (Kiyari)	Jacquot 1981

NUG	ISO	Names	References
H112A	xku	Kaamba (Kamba, Kikaamba)	Bouka 1989a de Schryver *et al.* 2015
H112B	dde	Doondo (Dondo, Kidoondo)	Even 1931 Mfoutou 1985 Lumwamu 1974
H12	vif	Vili (Civili, Civili ci Loango, Vili of Mayumbe, Loango)	Mavoungou *et al.* 2010
H13	njx	Kunyi (Kuni, Kikunyi, Kugni)	Lumwamu 1974 Bastin *et al.* 1999
H131	sdj	Suundi (Kissoundi, Kisuundi, Sunde), Kifouma, Kimongo	Mabiala 1999
H16a-b	kng	South Kikongo, Kisikongo (H16a), Kisolongo, Kimboma, Central Kikongo, Manyanga (H16b, Kimanyanga), Mazinga, Ndibu (Bandibu), Bwende (H16e), Zoombo (H16h, Zombo, Pende)	Boone 1973 de Schryver *et al.* 2015 Laman 1936:lxxxv-lxxxvii
H16c-d	yom/ kwy	Yombe (Yoombi, Kiyombe), Ciwoyo, Cizobe, Kakongo, Ndingi (H14), Mboka (H15), Cimbala, Cizali	Boone 1973 de Schryver *et al.* 2015
H16f	ldi	Laadi (Laari, Kiladi, Kilari), Ghaangala (H111, Kighaangala, Hangala)	
H16g-h	kng	Eastern Kikongo, Ntandu (Kintandu), Mbeko, Nkanu (H16h), Mbata, Mpangu	Boone 1973 de Schryver *et al.* 2015

H20: KIMBUNDU GROUP

NUG	ISO	Names	References
H21	kmb	Mbundu (H21a, Bunda, Bundo, Kimbundo, Quimbundo, Kimbundu), Kimbamba (H21b, Njinga, Ginga, Jinga)	Vieira-Martinez 2006
H22	smd	Kisama (Kissama, Quissama, Sama)	Heintze 1970 Price 1872
H23	blv	Kibala (Bolo, Llbolo, Libollo, Libolu, Lubalo, Lubolo), Haka (Haco), "Ngoya"	Angenot *et al.* 2011
H24	nsx	Songo (Nsongo, Basongo, Massongo, Songu, Sungu)	Jaspert and Jaspert 1930 Vansina 2004

H30: YAKA GROUP

NUG	ISO	Names	References
H31	yaf/ ppp/ lnz[21]	Yaka (Bayaka, Iaka, Iyaka, Kiyaka), Pelende (Phelende), Lonzo	van den Eynde 1968 Ntoya Maselo 2014
H321 (= L101)	shc	Sonde (Soonde, Kisonde), Lua (Luwa)	Boone 1973 Torday and Joyce 1907
H32	sub	Suku (Kisuku)	Piper 1977

NUG	ISO	Names	References
H33	-	Hungu, Tsotso, Pombo[22]	Atkins 1955
H34	mxg	Mbangala (H34, Bangala, Cimbangala), Shinji (H35, Sinji, Yungo)	Chatelain and Summer 1894

H40: MBALA-HUNGANNA

NUG	ISO	Names	References
H41	mdp	Mbala (Bambala of Kwango, Gimbala, Rumbala)	Boone 1973 Torday and Joyce 1922
H42	hum	Hunganna (Huana, Hungaan, Hungana, Kihungana), Saamba (Tsaamba, Tsamba)	Torday and Joyce 1922 Boone 1973 Takizala 1974 Muluwa and Bostoen 2015

JD41: KONZO-NANDE GROUP

NUG	ISO	Names	References
JD41	koo	Konzo (Kikondjo, Konjo, Lhukonzo, Olukonjo, Olukonzo, Rukonjo, Rukonzo, Rukoonzo), Sanza (Ekisanza)	Schoenbrun 1994
JD42	nnb	Nande (Banande, Nandi, Ndandi, Kinande, Kinandi), Shu (Ekishu), Yira (Ekiyira)	Schoenbrun 1994
JD43	see under D40		

JD50: SHI-HUNDE GROUP

NUG	ISO	Names	References
JD501	nyg	Nyindu	Igunzi 2013
JD51	hke	Hunde (Kihunde, Kihuunde), Kobi (Rukobi)	Schoenbrun 1994
JD52	hav	Haavu (Havu, Kihaavu)	Schoenbrun 1994
JD53	shr	Shi (Mashi), Nyabungu	Schoenbrun 1994
JD531	tbt	Tembo (Chitembo, Kitembo)	Schoenbrun 1994
JD54	see under D50		
JD55	see under D50		
JD56	kcw	Bwari (Kabwari)	Schoenbrun 1994

JD60: RWANDA-RUNDI GROUP

NUG	ISO	Names	References
JD61	kin	Rwanda (Kinyarwanda, Hutu), Kisoro Pygmies, Rwanda-Batwa, Fumbira (Rufumbira), Kirashi (JE221), Yaka (JD502)[23]	Schumacher 1950 Munderi 2009 Schoenbrun 1994 Nzita 1992 (Jouannet 1983)
JD62	run	Rundi (Kirundi)	van Bulck 1957
JD63	flr	Fuliiru (Fulero, Fuliiro, Fuliru, Kifuliiru)	Otterloo and Otterloo 2011 Schoenbrun 1994
JD631	job	Vira (Bavira, Kiviira), Joba (Kijoba)	Schoenbrun 1994

NUG	ISO	Names	References
JD64	suj	Subi (Shuubi, Kishubi, Kisubi, Shubi, Shuwi), "Sinja"	Nurse and Philippson 1980
JD65	han	Hangaza, Kihangaza (KiHangaza)	Schoenbrun 1994
JD66	haq	Ha (Kiha)	Schoenbrun 1994
			Harjula 2004
JD67	vin	Kivinza (KiVinza), Vinza)	Schoenbrun 1994
-	-	Rundi Kitwa, Batwa of Burundi	van der Burgt 1902

JE10: NYORO-GANDA GROUP

NUG	ISO	Names	References
JE101	rub	Gungu (Lugungu, Rugungu)	Schoenbrun 1994
JE102	tlj	Talinga-Bwisi, Kitalinga (Talinga), (Bwissi, Lubwisi, Lubwissi, Mawissi, Olubwisi)	Paluku 1998
JE103	ruc	Ruli (Eciruri, Luduuli, Ruli, Ruluuli, Ruluuli-Runyala, Ruuli)	Ladefoged et al. 1972
JE11	nyo	Nyoro (Lunyoro (LuNyoro, Orunyoro, Runyoro)	Schoenbrun 1994
JE12	ttj	Tooro (Orutoro, Rutooro, Rutoro, Tooro, Toro, Urutoro)	Rubongoya 1999
JE121	nix	Hema (Bahima, Hema-Sud, Kihema), Congo Nyoro, Hema	Thiry 2004
JE13	nyn	Nkore (Nkole, Nyankole, Nyankore, Olunyankole, Runyankole, Runyankore)	Schoenbrun 1994
JE14	cgg	Chiga (Ciga, Kiga, Ruciga)	Schoenbrun 1994
JE15	lug	Ganda (Luganda), Sese (Olusese), Vuma (Luvuma)	Schoenbrun 1994
			Cunningham 1905
JE16	xog	Soga (Busoga), Diope (Ludiope), Gabula (Lugabula), Gweri (Lugweri), Kigulu (Lukigulu), Lamogi (Lulamogi, Lulamoogi, Lamoogi), Luuka, Nholo (Lunholo), Siki (Siginyi, Lusiginyi, Lusiki), Tembe (Tembé, Lutembe), Tenga (Lutenga)	Fallers 1968, Schoenbrun 1994
JE16	lke	Kenyi (Kenye, Kene, Lukenhe)	Ladefoged et al. 1972
			Roscoe 1915
JE17	gwr	Gwere (Lugwere)	Ladefoged et al. 1972
JE18	-	West Nyala	Kanyoro 1983 Ochwaya-Oluoch 2003: 7–8

JE20: HAYA-JITA GROUP

NUG	ISO	Names	References
JE21	now	Nyambo (Ekinyambo), "Karagwe"	Schoenbrun 1994
JE22	hay	Haya, Kyamutwara (JE22A, Kjamtwara, Kiamutwara), Bugabo (JE22B, Bugabu), Bukara (JE22C), Ziba (JE22D, Kiziba), Hanja (JE22E, Kianja, Kihanja, Kjanja), Hangiro (JE22F, Ihangiro), Missenyi (JE22G, Misenyi)	Herrmann 1904 Schoenbrun 1994
JE23	zin	Zinza (Dzinda, Dzindza, Echidzindza, Jinja)	Schoenbrun 1994
JE24	ked	Kerebe (Ecikerebe, Ekikerebe, Ikikerebe, Kerewe)	Schoenbrun 1994
JE25	jit	Jita (Ecejiita, Kijita)	Schoenbrun 1994
JE251, 253	kya	Kwaya (JE251, Kikwaya), Ci-Ruri (JE253, Ruri, Rori, Kirori, Eciruri, Luri, Kiruri, Eciruuri, Ciruri)	Schoenbrun 1994
JE252	reg	Kara (Kikara), Regi (Kilegi, Kiregi)	Schoenbrun 1994

JE30: MASABA-LUHYA GROUP

NUG	ISO	Names	References
JE31	myx	Masaba (Masaaba, Lumasaba), Gisu (JE31a, Gesu, Lugisu), Kisu (JE31b), Buya (J41G, Buuya, Lubuya), Dadiri (JE31F, Ludadiri, North Lumasaaba), Lufumbo, Luteza, Luwalasi, Luyobo	Schoenbrun 1994 la Fontaine 1959
JE31c	bxk	Bukusu (Vugusu)	Wagner and Mair 1970 Austen 1975: 1–37 Kanyoro 1983
JE31D	-	Syan (Orusyan), Bumett (Pumit)	Huntingford 1965 Schoenbrun 1994
JE31E	lts	Tachon (Tachoni, Tatsoni)	Kanyoro 1983
JE32a	lwg	Hanga (Wanga, Luwanga, Luhanga)	Kanyoro 1983
JE32b	lto	Tsotso (Olutsotso)	Kanyoro 1983
JE32C	lrm	Marama (Olumarama)	Kanyoro 1983
JE32D	lks	Kisa (Olukisa)	Kanyoro 1983
JE32E	lkb	Kabras (Kabarasi)	Kanyoro 1983
JE32F	nle	East Nyala (Nyala North, Lunyala 'K'), Kakalewa	Kanyoro 1983
JE33	nyd	Nyole (Abanyole, Lunyole, Lunyore, Nyoole, Nyore, Olunyore)	Kanyoro 1983
JE34	lsm	Saamia (Samia, Lusaamia), Bagwe, Gwe (Lugwe), Songa (JE343)	Kanyoro 1983
JE341	lko	Khayo (Xaayo)	Kanyoro 1983 Marlo 2008
JE342	lri	Marachi (Lumarachi)	Kanyoro 1983
JE35	nuj	Nyole (Lunyole, Nyule, Nyuli), Hadyo (Luhadyo), Sabi (Lusabi), Wesa (Luwesa)	Schadeberg 1989 Morris 1963

JE40: LOGOOLI-KURIA GROUP

NUG	ISO	Names	References
JE401	ngq	Ngurimi (Dengurume, Ikingurimi, Kingereme, Kingoreme, Ngoreme)	Schoenbrun 1994
JE402, 404	ikz	Ikizu (JE402, Ikiikiizu, Ikikizo, Ikikizu), Shashi (JE404, Ikishashi, Kishashi)	Schoenbrun 1994
JE403	sxb	Suba (Egisuba, Luo Abasuba)	Rottland 1993
JE405	cwa	Kabwa (Kikabhwa, Kikabwa)	Walker 2013 Chuo Kikuu cha Dar es Salaam 2009
JE406–7	sgm	Singa (E406, Lusinga), Cula (Chula), Ware (E407)[24]	Hobley 1902
JE41	rag	Logooli (Llogoori, Logoli, Lugooli, Luragoli, Maragoli, Maragooli, Ragoli, Uluragooli)	Wagner and Mair 1970
JE411–3	ida	Idakho (JE411, Itakho), Isukha (JE412, Isuxa, Lwisukha), Tiriki (JE413, Tirichi)	Kanyoro 1983
JE42	guz	Gusii (Ekegusii, Guzii, Kisii), Kosova	Cammenga 2002: 17–33
JE43	kuj	Kuria (Egikuria, Ekikuria, Gikuria, Igikuria, Ikikuria, Kikuria Cha Juu, Kikuria Cha Mashariki, Kikuria, Koria, Kulia)	Schoenbrun 1994 Cammenga 2004

NUG	ISO	Names	References
JE431–4	ssc	Hacha (JE432, Haacha, Kihacha), Ikisimbete (IkiSimbete), Iryege (Iregi, Kiiryege), Kine (Kikine, Kiine), Kiroba, Kironi (Kikirone), Kiseru, Kisimbiti, Kisingiri, Rieri (Ryeri), Simbiti (JE431, Suba-Simbiti) Surwa (JE433, Kisurwa), Sweta (JE434, Kisweta)	Kihore 2000 Schoenbrun 1994
JE44	zak	Zanaki (Ekizanaki, Ikizanaki, Ilizanaki, Kizanaki)	Walker 2013 Schoenbrun 1994
JE45	ntk	Ikoma (Koma, Kiikoma), Isenye (Issenye, Isenyi, Kiisenye), Nata (Ekinata, Natta, Kinatta, Kinata, Ikinata)	Walker 2013 Schoenbrun 1994
E46	see under E40		

K10: CIOKWE-LUCHAZI GROUP

NUG	ISO	Names	References
K11	cjk	Chokwe (Batshokwe, Ciokwe, Kioko, Kiokwe, Quioca, Quioco, Shioko, Tschiokloe, Tshokwe)	Vansina 2004
K12a	lum	Lwimbi (Chiluimbi, Luimbe, Luimbi, Lwimbe)	Jaspert and Jaspert 1930
K12b	nba	Gangela (Ganguela, Ganguella, Ngangela) Nyemba (Nhemba)	Fleisch 2009 Maniacky 2003
K13	lch	Lucazi (Chiluchazi, Luchazi, Lujash, Lujazi, Lutchaz, Lutshase, Luxage)	Fleisch 2009
K14	lue	Lwena (Luena), Luvale (Chiluvale, Lovale, Lubale, Luena, Luvale)	Papstein 1978 Horton 1949
K15	mck	Mbuunda (Chimbunda, Mbunda)	McCulloch 1951 Fleisch 2009 Papstein 1994
K16	nye	Nyengo (Nhengo, Nyenko)	Fortune 1963
K17	mfu	Mbwela (Ambuela, Ambuella, Ambwela, Mbuela, Mbwera, Shimbwera)	Fleisch 2009 Bostoen 2007 McCulloch 1951 Maniacky 2003
K18	nkn	Nkangala (Cangala, Ngangala)	Maniacky 2003

K20: LOZI GROUP

NUG	ISO	Names	References
K21	loz	Lozi (Silozi), "Sikololo," "Kolololo"	Fortune 2001 Jalla 1917 Burger 1960

K30: LUYANA GROUP

NUG	ISO	Names	References
K31	lyn	Luyana (Ca-Luiana, Esiluyana, Louyi, Lui, Rouyi), Kwandi (K371), Kwangwa (K37)	Lisimba 1982 Fortune 1963
K32	mxo	Mbowe (Esimbowe), Liyuwa (K322), Liyuwa	Lisimba 1982 Fortune 1963

NUG	ISO	Names	References
K33	kwn	Kwangali (Kwangare, Kwangari, Rukwangali, Rukwangari, Sikwangali)	Möhlig 1997
K331, 332, 334	diu	Dciriku (K332, Gciriku, Diriku), Sambyu (K331), Rumanyo (K334, Manyo)	Möhlig 2007
K333	mhw	Mbukushu (Thimbukushu)	Möhlig 1997 Larson 1981
K34	mho	Mashi (Masi), Kwandu	Laranjo Medeiros 1981 Lisimba 1982
K35	sie	Simaa, Mulonga (K351), Mbume (K321, Mbumi), Imilangu (K354), Mwenyi (K352)	Bostoen 2007 Fortune 1963 Lisimba 1982
K36	-	Shanjo	de Luna 2008 Bostoen 2009

K40: SUBIYA-TOTELA GROUP

NUG	ISO	Names	References
K402	fwe	Fwe	Bostoen 2009 Seidel 2005 de Luna 2008 Pretorius 1975
K41	ttl	Totela (Echitotela), Totela of Namibia (K411)	Seidel 2005 Crane 2011 de Luna 2008
K42, K401	sbs	Subiya (Echisubia, Subia), Ikuhane (Chikuahane, Chikwahane, Ciikuhane, Ikuhane, Mbalangwe (K401, Mbalanwe)	Seidel 2005 Ohly 1994 de Luna 2008

L10: PENDE GROUP

NUG	ISO	Names	References
L11	pem	Pende (Bapende, Gipende, Giphende, Kipende, Pheende, Phende)	Weeckx 1938 Torday and Joyce 1922 Bittremieux 1939
L12b	hol	Holo (Holu, Kiholo, Kiholu), Yeci	Boone 1973 Atkins 1955
L13	kws	Kwese (Gikwezo, Kikwese, Kwezo, Ukwese)	Boone 1973 Forges 1983 Torday and Joyce 1907

L20: SONGE GROUP

NUG	ISO	Names	References
L201	-	Budya (L201), Yazi (L202)	
L21	kcv	Kete (Lu-Kete, Lukete, Ciket), East Kete (L21A), North Kete (L21B, Kete-Kuba), Kete-Lulua (L21C, South-West Kete)	Kamba Muzenga 1980: 3–4 Maes 1934 Boone 1973
L221	lwa	Lwalwa (Lwalu)	Boone 1973 Timmermans 1967 Bastin et al. 1999
L23	sop	Songe (Basonge, Kisonge, Kisongi, Kisongye, Luba-Songi, Lusonge), East Songye, North Songye (Ikaleebwe), South Songye, Beelandé	van Overbergh 1908 Stappers 1964: 3 Torday and Joyce 1922

NUG	ISO	Names	References
L22/ L231	bpj	Mbagani, Binji (Bindji, Babindji)[25]	van Coillie 1949 Bastin *et al.* 1999: 21
L24	luj	Luna-Inkongo, Northern Luba	Westcott no date

L30: LUBA GROUP

NUG	ISO	Names	References
L31	lua	Luba-Lulua, Luba-Kasai (L31a, Tshiluba, Ciluba), West Luba (L31b, Lulua, Luluwa, Bena-Lulua)	de Clercq 1903:i-vi
L32	kny	Kanioka, Kanyok, Kanyoka	de Clercq 1900 Stappers 1986:xiv-xv
L33	lub	Luba-Katanga, Kiluba, East Luba, Luba-Shaba	Verhulpen 1936a
L34	hem	Hemba (Kiemba, Kihemba), Yazi (L202), Zela (K331, Kizela, Kimbote, Mbote), Eastern Luba, Luba-Hemba, Kebwe (L301)	Vandermeiren 1912 Bastin *et al.* 1999 van Bulck 1948 Kabange Mukala 2005
L35	sng	Sanga (Kisanga), Garengaze (Garenganze), South Luba, Luba-Sanga	Clarke 1911

L40: KAONDE GROUP

NUG	ISO	Names	References
L41	kqn	Kaonde (Chikahonde, Chikaonde, Kahonde, Kawonde, Luba Kaonde), Solwezi, Kasempa	Kashoki and Mann 1978

L50: LUNDA GROUP

NUG	ISO	Names	References
L51	slx	Salampasu (Basala Mpasu, Chisalampasu), Luntu (L511, Luuntu, Bakwa Luntu)	Maes 1934 Denolf 1954
L52	lun	Lunda (Balunda, Chilunda, Southern Lunda), Ndembu	McCulloch 1951 Kawasha 2003
L53	rnd	Ruund (Ruwund, Chiluwunda, Lunda Kambove, Lunda-Kamboro, Luunda, Luwunda, Northern Lunda), Kanincin	Stappers 1954 Nash 1992

L60: NKOYA GROUP

NUG	ISO	Names	References
	nka	Nkoya (L62), Mbwela (L61, Mbwera, Mbowera),	Simwinga 2006 Gluckman
L60– 62		Kolwe (L601, Lukolwe), Lushangi (L602, Lushange), Shasha (L603, Mashasha), Mbowela (Mbwera, Shimbwera, Mbwela), Nkoya, Shasha, Shinkoya	1951 McCulloch 1951: 93–101

M10: FIPA-MAMBWE GROUP

NUG	ISO	Names	References
M11	piw	Pimbwe (Cipimbwe, Ichipimbwe, Icipimbwe, Kipimbwe)	Abe 2011 Chomba 1975 Maurice 1938
M12	rnw	Rungwa (Ichirungwa, Icilungwa, Kirungwa, Lungwa, Nyalungwa, Runga)	Walsh and Swilla 2001
M13	fip	Fipa (Fiba, Cifipa), Fipa-Sukuma (M13A), South Fipa (M13B), Kandaasi (M13C), Siiwa (M13D), Nkwaamba (M13E), Kwa (M13F), Kwaafi (M13G), Ntile (N13H, Cile, "Yantili"), Kuulwe (M131), Peemba (M13I)	Walsh and Swilla 2001 Willis 1966
M14/15	mgr	Lungu (M14, Rungu), Mambwe (M14)	Walsh and Swilla 2001

M20: NYIHA-SAFWA GROUP

NUG	ISO	Names	References
M201	lai	Lambya (Rambya), North Lambya (M201A), Central Lambya (M201B), South Lambya (M201C)	Walsh and Swilla 2001
M21	wbh	Wanda (Wandia), Sichela	Walsh and Swilla 2001 Lindfors et al. 2009b
M22	mwn	Inamwanga (Ichinamwanga, Namwanga, Nyamwanga), Iwa (M26), Tambo (M27, Tembo)	Walsh and Swilla 2001
M23A	nkt	West Nyika, Nyika of Sumbawanga	Lindfors et al. 2009a
M23B	nih	Central Nyiha, Nyiha of Mbozi	Lindfors et al. 2009a
M23C	nyr	East Nyika, Nyika of Rungwe Lindfors et al. 2009a Walsh and Swilla 2001	
M23D	nkv	South Nyiha	Lindfors et al. 2009a Walsh and Swilla 2001
M24	mgq	Malila (Ishimalilia, Kimalila, Malela, Malilia)	Walsh and Swilla 2001
M25	sbk	Safwa, Safwa of Mbeya (M25A), Mbwila (M25B, Uleenje), Soongwe (M25C), Polooto (M25D, Poroto), Guruka (M25E)	Walsh and Swilla 2001 Msanjila 2004

M30: NYAKYUSA-NGONDE GROUP

NUG	ISO	Names	References
M301	ndh	Ndali (Chindali), Sukwa (Chisukwa)	Botne 2008 Kershner 2002
M31	nyy	Nyakyusa-Ngonde, Nyakyusa (M31A, Nyekyosa, Iki-Nyikusa), Kukwe (M31B, Ngumba), Mwamba (M31C, Lugulu, "Sokelo"), Ngonde (M31D), Selya (M31E, Kaaselya), Penja (M302)	Walsh and Swilla 2001

M40: BEMBA GROUP

NUG	ISO	Names	References
M401	bwc	Bwile	Brelsford 1956 Bastin et al. 1999
M402	auh	Aushi (Avaushi, Ushi, Usi, Uzhil, Vouaousi)	Doke 1933 Whiteley 1950

NUG	ISO	Names	References
M41	tap	Tabwa (Taabua, Taabwa, Ichitaabwa), Shila	Whiteley 1950
M42	bem	Bemba (Icibemba, Wemba)	Whiteley 1950 Kashoki and Mann 1978

M50

NUG	ISO	Names	References
M51–52	leb	Lala-Bisa, Bisa (M51, Wisa), Lala (M52, Ichilala), Ambo (M521), Unga, Twa of Bangweulu, Luano (M522)	Brelsford 1946 Whiteley 1950 von Rosen 1916
M54	lam	Lamba, Swaka (M53), Lima (M541, Bulima), Temba (M542)	Doke 1922 Whiteley 1950 Brelsford 1956 Bastin *et al.* 1999: 219
M55	kdg	Seba, Shishi	Whiteley 1950

M60: LENJE-TONGA GROUP/BANTU-BOTATWE GROUP

NUG	ISO	Names	References
M61	leh	Lenje (Chilenje, Lengi, Lenji), Batwa of the Lukanga (M611), Ciina (Chinamukuni, Ciina Mukuni, Mukuni)	Shekleton 1908 de Luna 2008
M62	sby	Soli (Chisoli)	de Luna 2008
M63	ilb	Ila (Chiila [E18], Batwa of the Kafue (M633), Lundwe (M632)	Smith 1907 de Luna 2008 Doke 1928 Torrend 1931
M631	shq	Sala (M631)	de Luna 2008
M64	toi	Tonga (Chitonga, Tonga of Zambia and Zimbabwe), Ndawe, Plateau Tonga, Valley Tonga, We-Zambezi [E18]	de Luna 2008 Hachipola 1991
M651–652	dov	Toka (M651), Leya (M652, Reya), "Dombe"[26]	Hachipola 1998 de Luna 2008

N10: MANDA GROUP

NUG	ISO	Names	References
N101	dne	Ndendeule (Kidendauli, Kindendeule, Kindendeuli, Ndendeuli)	Nurse 1988 Booth 1905
N102	nxi	Nindi (Kinindi, Manundi)	Maho and Sands 2004 Booth 1905
N11	mgs	Manda (Kimanda), Matumba	Fülleborn 1906 Nurse 1988
N12	ngo	Ngoni of Tanzania (Angoni, Chingoni, Kingoni), Kisutu	Spiss 1904 Ebner 1955 Miti 1996
N13	mgv	Matengo (Chimatengo, Ki-Matengo, Kimatengo)	Yoneda 2000 Nurse 1988
N14	mpa	Mpoto (Chimpoto, Cimpoto, Kimpoto, Mpoto)	Nurse 1988
N15	tog	Tonga (Chitonga), Tonga of Malawi, Siska, Sisya	Turner 1952 Young 1933 Turner 1952

N20: TUMBUKA GROUP

NUG	ISO	Names	References
N21	tum	Tumbuka (N21a, Chitumbuka, Citumbuka), Poka (N21b, Chipoka, Phoka), Kamanga (N21b, Nkhamanga, Henga, Ci-Henga), Senga (N21d, Senga in the Luangwa Valley in the Lundazi District), Yombe (N21e), Fungwe (N21f), Wenya (N21g)	Young 1923 Brelsford 1956

N30: CHEWA-NYANJA GROUP

NUG	ISO	Names	References
N31	nya/ mjh	Chewa-Nyanja, Nyanja (N31a, Chinyanja), Chewa (N31b, Achewa, Cewa), Manganja (N31c, South Nyanja), Mozambique Nyasa (N31D, Nyasa-Cewa), Maravi (Malawi), Peta, Mwera of Mbamba Bay(N201)[27]	Stigand 1909 of Malawi 2006 of Malawi 2009

N40: SENGA-SENA GROUP

NUG	ISO	Names	References
N41	nse	Nsenga (Cinsenga, Chinsenga, Nsenga on the lower Luangwa in the Petauke-Lusaka-Feira districts), Kunda of Mambwe district	Miti 2004 Brelsford 1956
N42	kdn	Kunda (Achikunda, Chikunda, Cikunda)	Hachipola 1998 Brelsford 1956
N43	nyu	Nyungwe (Chinyungwe, Chinyungwi, Cinyungwe, Nyongwe, Yungwe), Tete (Tetense, Teta), Pimbi (Phimbi, Pimbe)	van der Mohl 1904
N44	seh	Sena (Chisena, Cisena, Mocambique Sena), Podzo (N46, Ci-Podzo, Phodzo, Shiputhsu, Chipodzo, Puthsu)	van der Mohl 1904
N441	swk	Sena (Chisena, Cisena, Malawi Sena)	Funnell 2004
N45	bwg	Rue (Barwe, Chirue), Balke (Cibalke)	Macalane 2000 Mangoya 2012 Hachipola 1998

P10: MATUUMBI GROUP

NUG	ISO	Names	References
P11/12	ndg/ rui	Ndengereko (Kindengereko, Kingengereko, Ndengeleko, Ndengeleko), Rufiji (Kirufiji)[28]	Ström 2013
P13	mgw	Matuumbi (Kimatumbi, Kimatuumbi, Matumbi)	Nurse 1988
P14	nnq	Ngindo (Ci-Ngindo, Cingindo, Gindo, Kingindo, Njindo, Njinjo)	Cross-Upcott 1956 Nurse 1988
P15	mgy	Mbunga (Bunga, Kimbunga)	Nurse 1988

P20: YAO GROUP

NUG	ISO	Names	References
P21	yao	Yao (Chiyao, Ciyao, Djao, Jao)	Whiteley 1966: xiii-xv
P22	mwe	Mwera (Cimwela, Cimwera, Kimwera, Mwela)	Harries 1950
P23/	kde/	Makonde (Simakonde, Chimakonde, Chinimakonde,	Kraal 2005: 1–7
P24	njd	Kimakonde, Konde, Makonde), "Mavia" (P25, Mávia, Maviha, Cimabiha, Mabiha, Kimawiha, Mawia, Chimaviha, Mawiha), Maraba (Chimaraba), Tambwe (Matambe, Matambwe, Chimatambwe), Chinnima, Ndonde (Chindonde)[29]	
P23	mvw	Machinga	Steere 1876

P30: MAKHUWA GROUP

NUG	ISO	Names	References
P31A,E	vmw	Central-Makhuwa, Makhuwa Enahara (P31E), Nampula	Kröger 2005 Prata 1960
P31B,G	mgh	Meetto (Metto), Makhuwa-Meetto, Masasi, Ruvuma Makhuwa, Imithupi, Ikorovere	Kröger 2005 Prata 1960
P31C	vmk	Chirima (Shirima)	Kröger 2005 Prata 1960
P31C	kzn	Kokola	Shrum and Shrum 2001
P31C	llb	Lolo	Reiman 2002 Shrum and Shrum 2001 Vinton and Vinton 2001
P31C	mny	Manyawa	Reiman 2002 Shrum and Shrum 2001 Vinton and Vinton 2001
P31C	vmr	Marenje (Emarendje, Marendje, Marenji)	Shrum and Shrum 2001 Reiman 2002
P31C	tke	Takwane (Thakwani)	Shrum and Shrum 2001 Reiman 2002
P31D	xmc	Marrevone (Marevone), Nampamela	Prata 1960 Kröger 2005
P31F	xsq	Esaka (Esaaka, Saaka, Saanga, Saka, Sanga)	Katupha 1991
P311	eko	Koti (Ekoti, Coti), "Angoje"	Schadeberg and Mucanheia 2000 Lyndon and Lyndon 2007
P312	nte	Sakati (Sangaji, Esakaji), Enatthembo	Prata 1960 Lyndon and Lyndon 2007
P32	ngl	Lomwe (Ilomwe, Lomue), Western Makua	Prata 1960
P33,331	lon	Malawi Lomwe, Nguru (Ngulu, Anguru), Mihavane	Kayambazinthu 2004
P34	chw	Central Chuwabo (Chichwabo, Chuabo, Chuwabo, Chuwabu, Txuwabo, Chwabo, Cuabo, Cuwabo, Echuabo), Quellimane	Vinton and Vinton 2001
P34	cwb	Maindo, Badoni, Mitange	Vinton and Vinton 2001
P341	mhm	Moniga (Emakhuwa-Emoniga, Emoniga)	Kröger 2005

R10: UMBUNDU GROUP

NUG	ISO	Names	References
R11	umb	Umbundu (M'bundo), Bailoundou (Mbalundu), Bihe, Hanya, Nano, Ovimbundu, South Mbundu, Mbali (R103, Kimbari, Olumbali[30])	Hambly 1934 Jaspert and Jaspert 1930 Lopes Cardoso 1966
R12	ndq	Ndombe (Dombe, Bandombe, Mondome)	Ferreira Diniz 1918
R13	nyk	Nyaneka (Nhaneca, Lunyaneka), Mwila (Olumuila, Huila, Muila), Ngambwe (Olungambwe)	Lang and Tastevin 1937
R14	khu	Khumbi (Lun'cumbi, Lunkumbi, Ngumbi, Nkhumbi, Nkumbi), Ndongwena (R215), Kwankwa (R216)	Estermann 1979 Nogueira 1885 Lusakalalu 2001

R20: WAMBO GROUP

NUG	ISO	Names	References
R21	kua	Kwanyama (Kuanyama), Kafima (R211), Evale (R212), Northeastern Wambo	Baucom 1972
R214	lnb	Mbalanhu, Dombondola (R217), Mbandja (R213, Mbadja), Esinga (R218)	Fourie 1997 Baucom 1972
R22	ndo	Ndonga (Oshindonga), South-Eastern Wambo	Baucom 1972
R23	kwm	Kwambi, South-Central Wambo	Baucom 1972
R24	nne	Ngandjera (Ngandyera), Kwaluudhi (R241), Kolonkhadi (R242), Eunda (R242, Oshiunda)	Baucom 1972

R30: HERERO GROUP

NUG	ISO	Names	References
R101	olu	Kuvale (Kuroka), Kwisi (R102)	Jordan 2015
-	-	Kwandu	Jordan 2015, Laranjo Medeiros 1981
-	-	Ngendelengo	Jordan 2015, Laranjo Medeiros 1981
R31A,B, R312	her	Herero (Otjiherero), Central Herero (R31A), Mbandieru (R31B, Mbanderu, East Herero), Botswana Herero (R312, Mahalapye Herero)	Andersson and Janson 1997 Möhlig 2009
R311	dhm	North-West Herero, Kaokoland Herero, Dhimba (Zemba, Zimba), Tjimba Herero, Tjimba-Tjimba, Himba	Kunkel and Cameron 2002 Malan 1974

R40: YEYI GROUP

NUG	ISO	Names	References
R41A,B	yey	Yei (Yeyi, Shiyei), "Kuba," Bayeye, Ciyei, East Caprivi Yeyi (R41A), Ngamiland Yeyi (R41B)	Larson 1992 Seidel 2008

S10: SHONA GROUP

NUG	ISO	Names	References
S11	twl	Shona, Korekore, Tavara (Tawala), Shangwe, Budya, Gova	Doke 1931
S12, S14	sna	Shona, Zezuru (S12, Shawasha, Harava, Hera), Karanga (S14, Duma, Govera, Jena, Nyubi)	Doke 1931
S13	mxc	Manyika (Bamanyeka, Chimanyika), Bocha (Boka), Guta, Jindwi, Hungwe	Doke 1931 Fortune 2004
S13	twx	Tewe (Teve), Chiute (Ciute, Ciutee)	Fortune 2004 Doke 1931
S15	ndc	Ndau (Chindau, "Sofala"), Garwe, Danda, Shanga, Xinyai	Mkanganwi 1972 Doke 1931
S16A,C, D,E,F,G, H,I,J, K,L	kck	Kalanga (Ikalanga, West Shona), Lilima (S16C, S16K, Humba, Humbe, Limima, Peri), Nyai (S16D, Abanyai, Banyai, Rozvi, Wanyai), Lemba (S16E, Remba), Lembethu (S16F, Rembethu), Twamamba (S16G, Xwamamba), Pfumbi (S16H), Jawunda (S16I), Romwe (S16J), Talahundra (S16L)	Wentzel 1983 Doke 1931
S16B	nmq	Nambya (Nambzva)	Hachipola 1998 Borland 1984
-	dmx	Dema	Hachipola 1998

S20: VENDA GROUP

NUG	ISO	Names	References
S21	ven	Venda (Central Venda), Phani (S21A), Ilafuri (S21B, West Venda), Manda (S21C, Central Venda), Mbedzi (S21D, East Venda), Tavhatsindi (S21E), Ronga (S21F, South-East Venda)	Mulaudzi 2010

S30: SOTHO-TSWANA GROUP

NUG	ISO	Names	References
S31	tsn	Tswana (Setswana, Sechwana), Central Tswana (S31a, Rolong, Ngwaketse), East Tswana (S31b, Kgatla, Tlokwa), North Tswana (S31c, Tawana, Ngwato, Kwena), South Tswana (S31E, Thlaping, Thlaro)	Malepe 1966 Schapera 1952
S311 (=S31d)	xkv	Kgalagadi (Kgalagarhi, Kgalagari, Qhalaxarzi)	Kalasi 2003, Krüger and de Plessis 1977
S32	nso	Northern Sotho (Sesotho sa Leboa), Pedi (S32a, Sepedi, Masemola, Tau, Komi, Transvaal Sotho), Lobedu (S32b, Kgaga), Gananwa (S32C, Xananwa, Hananwa), Kopa (S32D, Ndebele-Sotho), Eastern Sotho (Kutswe S302, Pai S303, Pulana S304), Phalaborwa (S301)	Doke 1954, Ziervogel 1954
S32E	brl	Birwa (S32E, Sebirwa, Virwa)	Batibo 1998 Andersson and Janson 1997: 41–42
S32F	two	Tswapong (Setswapong)	van Wyk 1969 Batibo 1998
S33	sot	Southern Sotho (Sesotho, Sisutho)	Jacottet 1927:i-xiii

S40: NGUNI GROUP

NUG	ISO	Names	References
S407	nbl	South Ndebele (Isindebele, Nrebele), Ndzundza (Isindundza), Southern South African Ndebele, Southern Transvaal Ndebele	Skhosana 2009
S408	-	Northern Transvaal Ndebele, Mbo, Moledlhana, Langa, Lidwaba, Sumayela Ndebele, Northern South African Ndebele	Ziervogel 1959
S41	xho	Xhosa (Isixhosa, "Kaffir"), Mpondo (S41A, Pondo), Xesibe (S41B), Bomwana (S41C), Gaika (S41D), Gcaleka (S41E), Thembu (S41F), Mpondomise (S41G), Ndlambe (S41H), Hlubi (S403), Hlubi-Ciskei (S41I)	Mzamane 1962 Cantrell 1946 Ownby 1985
S42	zul	Zulu (Isizulu), KwaZulu-Natal Zulu (S42A), Transvall Zulu (S42B), Qwabe (S42C), Cele (S42D), Ngoni of Malawi (N121), Bhaca (S402, Baca), Lala, Ingwavuma, Lala (S406, North Lala, South Lala), Old Mfengu (S401, Fingo)	Ownby 1985 van Dyk 1960 Miti 1996
S43	ssw	Swati (Swazi, Siswati), Ngwane, Nhlangwini (S405), Phuthi (S404)	Ownby 1985 Mzamane 1948
S44	nde	Ndebele of Zimbabwe (Sindebele, Ndebele, Tebele)	Ownby 1985

S50: TSWA-RHONGA GROUP

NUG	ISO	Names	References
S51	tsc	Tswa (Xitswa), Dzibi (S51A), Dzonga (S51B), Hlengwe (S511, Khambana-Makwakwe, Khambani, Lengwe, Lhengwe, Makwakwe-Khambana, Shilengwe)	Persson 1932
S52		see Changana S53	
S52/53	tso	Tsonga (Songa, Thonga, Tonga, Tsonga), Changana (Xichangana, Shangaan, Shangana), Xiluleke (S53A, Mhinga, Makuleke), N'walungu (S53B, Shingwalungu), Hlave (S53C), Nkuna (S53D), Gwamba (S53E (=S52), Gwapa), Nhlanganu (S53F, Shihlanganu), Djonga (S53G, Jonga), Bila (S53H)	Baumbach 1970
S54	rng	Ronga (Gironga, Shironga), Landim (Landina), Konde, Xonga	Baumbach 1970

S60: COPI GROUP

NUG	ISO	Names	References
S61	cce	Copi (Chopi, Cicopi, Shichopi, Txopi), Lenge (S611, Xilenge, Lengue, Kilenge)	Bailey 1976
S62	toh	Tonga (Gitonga), Shengwe	Lanham 1955

ACKNOWLEDGEMENTS

Dozens of individuals have helped at one time or another by answering, clarifying and confirming information. In particular I wish to thank Bonny Sands and Koen Bostoen for help in accessing hard-to-find references and Jouni Maho for his immense ground work on Bantu language varieties and their literature. Two anonymous reviewers provided helpful comments.

NOTES

1   See http://www-01.sil.org/iso639-3/ (accessed 1 Aug 2015).
2   According to Ardener (1956: 10–12, 30–31, 35) Bubia is in a dialectal relationship to Mboko, but this is difficult to reconcile with the lexicostatistical figures in Ebobissé 2014: 395–404.
3   Lewis *et al.* (2015) count Oli and Budiman as varieties of Duala, but this is difficult to reconcile with the lexicostatistical figures in Ebobissé 2014: 395–404 and probably reflects diglossia rather than inherent intelligibility.
4   Leti [leo] is the ancestral language of the Mengisa ethnic group who are switching to Eton [eto]. The Eton that most Mengisa now speak is not linguistically remarkable and therefore we count it as an Eton variety here (Geslin-Houdet 1984).
5   Many sources treat Bembele [beb] and Bobili [bxp] as the same language (e.g., Dugast 1949: 88–89). However, in view of the lexicostatistical figures in Wega Simeu (2004a: 3–4, 110–113) separating the two appears arguable, although based only on a 20-word list.
6   The 200-word list of Yambe in Klieman (1997) appears to so different from Bomwali [bmw] that we count it as a separate language here.
7   Teke-Eboo and Teke-Nzikou are commonly enumerated separately but the recent comparison by Raharimanantsoa (2012) shows that such a distinction is untenable, wherefore we count them as the same here.
8   The ethnographic and linguistic record (e.g., Bostoen and Koni Muluwa 2014, Boone 1973: 237–243, Torday and Joyce 1907, Bastin *et al.* 1999: 13, van Bulck 1948: 318) attests an ethnolinguistic group called Ngongo (and variants thereof) in dispersed settlements around the Luzubi and Gobari rivers (this is approximately halfway between the cities of Bandundu and Kikwit) in Bandundu province. E18 and Maho (2009: 54) include an ethnolinguistic group named Ngongo which E18 (contradictorily) places genealogically in the Yaka (H.31) group and geographically in the far north of Kasai Occidental province, inside what is Dengese [dez]-speaking territory. There is an ethnic group with the name Ngongo attested at this location but they have assimilated linguistically to their Bushoong overlords (van der Kerken 1944: 1033) –Maho (2009: 32) thus has this Ngongo as C83B with Bushoong– and there is no linguistic data to connect them with the Ngongo language around the Luzubi and Gobari rivers (van Bulck 1948: 506). We have to assume here that the Ngongo [noq] entry refers to the Ngongo language of around the Luzubi and Gobari but that the location given in E18 is simply a mistake, otherwise that language would be missing from E18 and the entry in question would represent an ethnicity whose (original?) language cannot be asserted as separate. The Ngongo language appears to be most closely related to Nsong and Mpiin (Bostoen and Koni Muluwa 2014) but at least one attested variety shows such lexicostatistical figures that it is likely not intelligible with Mpiin-Nsong (Bastin *et al.* 1999: 101, 203).
9   Maho (2009: 24) merges B87 Mbuun with the B84 varieties named Mpuon and Mpuun of Guthrie (1967: 38) and places it in the Mbuun location (east of Kikwit). Similarly, Lewis *et al.* (2015)'s entry Mpuono [zmp] is classified there as B84 and given the name Mpuono, but located in the Mbuun (B87) area on the map. While

Mpuon and Mpuun are not explicitly located by Guthrie, the word lists referenced in Bastin *et al.* (1999: 13) locate B84, named Mpuono, northwest of Kikwit in the south of the Teke plateau and the lexicostatistical figures there allow Mpuono to be filed as a Teke-Fumu [ifm] variety (Bastin *et al.* 1999: 185, 193, 202).

10 E18 and Maho (2009: 27) attach the name Likila to the location specified in the [lie]-iso. Specialist sources do not recognise such a name for this group (de Boeck 1948, Motingea Mangulu 1996) but reserve the name Likila for the Baloi kindred group west of the Ngiri. Perhaps the propagation of the name Likila for the Balobo is because some outsiders call Likila Balobo.

11 Hulstaert 1951: 21, Vinck 1993: 578, 581.

12 Bastin *et al.* 1999: 208, Carrington 1977: 67.

13 Hulstaert (1999) lists several hundred local varieties.

14 E18 and Maho (2009: 30) recognise a separate entry Lalia [lal]. But studies in the field emphasise that Lalia is simply a variety in the Bangando area with no special status vis-a-vis other Bongando varieties (Hulstaert 1951: 22, Lingomo 1995: 340, Hulstaert 1987: 205–207), wherefore we merge the two here.

15 E18 and Maho (2009: 31) differentiate between Yela and Kela corresponding to two extant geographical enclaves. Hulstaert (Hulstaert 1951: 23, Hulstaert 1999: 20–21), who had studied the varieties in the field, argues that this distinction is not particularly salient linguistically, wherefore we merge them here.

16 Mayeka [myc], a Bantu language of the Democratic Republic of the Congo, is very likely to be the same as Ngbinda [nbd] since de Calonne-Beaufaict (1921: 114, 120, 247, 251) calls Mayeka a clan of Abangbinda, though no actual linguistic data is presented. The information on Mayeka [myc] in E18 presumably derives from Hackett and van Bulck (1956: 74) who took down a Mayeka word list but were not able to find out the location of Mayeka, and did not have access to a Ngbinda word list (van Bulck 1954: 74–75) to check if Mayeka and Ngbinda were the same. Though, at the present time, if Hackett and van Bulck's unpublished Mayeka word list still exists in an archive somewhere, it should be checked whether the data is actually Ngbinda or a different, but related, language.

17 E18 lists Gbati-ri [gti] as a separate language, but the only source on this language (Hackett and van Bulck 1956: 74) has it as a dialect of Nyanga-li [nyc].

18 The Degere, a former hunting and gathering group, have assimilated culturally and linguistically to the Southern Mijikenda, i.e., to Digo and/or Duruma. There is no evidence of any other Bantu language spoken by the Degere, though some words possibly attributable to a former language of the Degere can be identified as Oromoid (Walsh 1990), i.e., Eastern Cushitic.

19 Maho (2009: 42) offers a new code E741 to replace E74b, presumably to indicate that E74a Dawida and E74b Sagalla are too divergent to be considered dialects of the same language. But such considerations do not alter the referential integrity of the codes already at hand, so we retain the E74b code here.

20 By mistake, Maho (2009: 42) considers this the same language as E74a Dawida and places the merge of the two in the E70 group, but the language is most closely related to G22 Asu and may even be considered a dialect of it.

21 Lewis *et al.* (2015) list Pelende [ppp] and Lonzo [lnz] as separate languages. But Pelende and Lonzo denote political rather than ethnolinguistic sub-entities of Yaka [yaf] (Denis 1964: 20, Lamal 1965: 15), as explicit linguistic data, whenever available, confirms (Bastin *et al.* 1999: 215).

22 Some authors (e.g., Maho 2009: 54, 67) fuse Hungu and Holo into one language. They are spoken in similar locations but they are not the same language or even each other's closest relatives (Atkins 1954, 1955).

23 Maho (2009: 57) has an entity Yaka (JD502), reflecting a vocabulary represented in Bastin et al. (1999: 19). The vocabulary was collected at 29.3E/1.8S, in the north-eastern corner of Lake Kivu, on the Rwandan side. No other inventory of ethnolinguistic groups in this area have such an entity at this location, and this is relatively well-surveyed area (e.g., Schumacher 1949) with only varieties of Rwanda JD61 attested. The vocabulary itself indeed resembles other varieties of Rwanda the most, being up to 62% lexicostatistically similar (Bastin et al. 1999: 103), but the cognacy pattern with other languages related to Rwanda makes it difficult to interpret as an early branch of a greater clade around Rwanda (Bastin et al. 1999: 223). We therefore take this vocabulary to reflect a difficult elicitation session rather than an ethnolinguistic entity separate from Rwanda.

24 Maho (2009: 62) and various other sources separate Ware from Singa, but our only first-hand source on these two entities say they spoke the same language (Hobley 1902: 50–51, 92–95).

25 E18, and similarly Maho (2009: 68), has an entry Binji [bpj] said to be in two (disjoint) locations: Kazumba territory in Kasai-Occidental province and Sankura (sic!. Sankuru) district in Kasai-Oriental province, but the map shows Bindji only in the Kasai-Oriental location and places Songe [sop] in the Kasai occidental location. The name Mbagani is given as a dialect under Songe [sop]. An ethnolinguistic group Mbagani with the alternative name Binji (and spelling variants) is known from the linguistic literature in the Kasai occidental location (van Coillie 1949, Bastin et al. 1999: 21). We therefore assume that the Binji [bpj] entry intends to refer to Mbagani-Binji in the Kasai-Occidental location but that there has been some confusion.

26 E18 has a separate Dombe [dov] entry, but Dombe is a derogatory nickname for Tonga found in Hwange district of Zimbabwe (Hachipola 1998). Their variety is closest to the Toka-Leya dialects spoken on the northern side of the Victoria Falls in Zambia.

27 E18 and Maho (2009: 80) have a separate entity for Mwera of Mbamba Bay [mjh]/ N201 on the eastern side of Lake Malawi. While these Mwera speak a language different from their neighbours, the examination by Ebner (1955: 41–43) shows the language to be a variant of Chewa/Nyanja on the opposite side of the lake. I wish to thank Rasmus Bernander for alerting me to this paper.

28 Some authors split Ndengereko (a glossonym) from Rufiji (the name of a river and the area around it), but under any division of the area in question onto those two names, the varieties are mutually intelligible and may even include Matumbi (Ström 2013, Chuo Kikuu cha Dar es Salaam 2009).

29 E18 and Maho (2009: 80) differentiate between Ndonde and Makonde, but the field research of Kraal (2005: 1–7) finds this distinction untenable from both a linguistic and ethnographic point of view.

30 Olumbali are an ethnic group consisting of former slaves mainly from the Kimbundu area who ended up in the Umbundu area. They are sometimes listed as speaking an intertwined Kimbundu-Umbundu language (Smith 1995: 371). However, the data in Lopes Cardoso (1966: 48–73) show that some localities in the Olumbali area speak Umbundu while others speak Kimbundu, and there is little evidence for a mixed language. I wish to thank Peter Bakker for discussing this case with me.

REFERENCES

Abe, Y. (2011) The Continuum of Languages in West Tanzania Bantu: A Case Study of Gongwe, Bende, and Pimbwe. In Hieda, Osamu, König, Christa & Nakagawa, Hirosi (eds.), *Geographical Typology and Linguistic Areas: With Special Reference to Africa*, 177–188. Amsterdam; Philadelphia: John Benjamins.

Adam, J. J. (1954) Dialectes du Gabon: la famille des langues téké. *Bulletin de l'Institut d'Etudes Centrafricaines, nouvelle série* 7/8: 33–107.

Adoua, J.-M. (1981) Description phonologique du likuba (langue bantu du Congo). Brazzaville: Université Marien Ngouabi, mémoire de licence.

Agadji Ayele, S. (2002) Éléments de description du ngubi (B40). Libreville: Université Omar Bongo, mémoire de licence.

Aksenova, I. S. & I. N. Toporova (2002) *Grammatika Yazyka Akva*. Moskva: Rossijskaya Akademiya Nauk, Institut Yazykozaniya.

Aleko, H. & G. Puèch (1988) Notes sur la langue ngové et les Ngubi. *Pholia* 3: 257–269.

Alewijnse, B., J. Nerbonne, L. J. van der Veen & F. Manni (2007) A Computational Analysis of Gabon Varieties. In Osenova, Petya & others (eds.), *Proceedings of the Recent Advances in Natural Language Processing Workshop on Computational Phonology*, 3–17. Borovetz: RANLP.

Alexandre, P. & J. Binet (1958) *Le groupe dit pahouin: fang-boulou-beti*. Paris: Presses Universitaires de France; International African Institute.

Alexandre, P. (1965) Proto-histoire du groupe beti-bulu-fang: Essai de synthèse provisoire. *Cahiers d'études africaines* 5(20): 503–560.

Andeme Allogo, M.-F (1985) Esquisse phonologique du ntumu (dialecte fang de Bitam). Université de la Sorbonne Nouvelle (Paris 3), mémoire de licence.

Andersson, E. (1983) *Les Bongo-Rimba*. Uppsala: Almqvist & Wiksell.

Andersson, L.-G. & T. Janson (1997) *Languages in Botswana: Language Ecology in Southern Africa*. Gaborone: Longman Botswana.

Angenot, J.-P., N. Mfuwa & M. A. Ribeiro (2011) As classes nominais do Kibala-Ngoya, um falar Bantu de Angola não documentado, na intersecção dos grupos Kimbundu [H20] e Umbundu [R10]. *Papia* 21: 253–266.

Ardener, E. W. (1956) *Coastal Bantu of the Cameroons: The Kpe-Mboko, Duala-Limba and Tanga-Yasa groups of the British and French trusteeship territories of the Cameroons* London: International African Institute.

Asangama, N. (1983) Le budu: langue bantu du nord-est du Zaïre: Esquisse phonologique et grammaticale. Université de la Sorbonne Nouvelle (Paris 3), thèse de doctorat.

Ashton, E. O. (1944) *Swahili Grammar (including intonation)*. London: Longmans, Green & Co.

Atkins, G. (1954) An Outline of Hungu Grammar. *Garcia de Orta* 2(2): 145–164.

Atkins, G. (1955) A Demographic Survey of the Kimbundu-Kongo Language Border in Angola. *Boletim da Sociedade de Geografia de Lisboa* 73(7/9): 325–347.

Atta Ebongkome, S. (1993) The Phonology of Lukundu (Bakundu). Université de Yaoundé I, MA thesis.

Austen, C. L. (1975) Aspects of Bukusu Syntax and Phonology. Indiana University, PhD dissertation.

Bahuchet, S. (1989) Les pygmées Aka et Baka: contribution de l'ethnolinguistique à l'histoire des populations forestières d'Afrique centrale. Université de Paris V – René Descartes, thèse de doctorat.

Bailey, R. A. (1976) Copi Phonology and Morphtonology. University of Witwatersrand, MA thesis.

Ballif, N. (1977) *Observations effectuées chez les Pomo: Mission Ogooué-Congo, 1946.* Paris: Ecole des Hautes Etudes en Sciences Sociales.

Bastin, Y., A. Coupez & M. Mann (1999) *Continuity and Divergence in the Bantu Languages: Perspectives from a Lexicostatistic Study.* Tervuren: Royal Museum for Central Africa.

Bastin, Y. (1978) Les langues bantoues. In Barreteau, Daniel (ed.), *Inventaire des études linguistiques sur les pays d'Afrique noire d'expression française et sur Madagascar,* 123–186. Paris: Conseil International de la Langue Française.

Batibo, H. B. (1998) A Lexicostatistical Survey of the Bantu Languages of Botswana. *South African Journal of African Languages* 18(1): 22–28.

Baucom, K. L. (1972) The Wambo Languages of South West Africa and Angola. *Journal of African Languages* 11(2): 45–73.

Baumbach, E. J.M. (1970) 'N Klassifikasie van die Tsongadialekte van die Republiek van Suid-Afrika. University of South Africa (UNISA), doktorale *proefskrif.*

Beavon, K. & A. E. Johnson (2011) *Sociolinguistic Survey Among the Bekwel* (SIL Electronic Survey Reports 2011–045). Dallas: SIL International.

Belliard, F. (2007) *Parlons kwàkúm: langue bantu de l'est Cameroun.* Paris: L'Harmattan.

Biebuyck, D. P. (1973) The Lega. In *Lega Culture; Art, Initiation, and Moral Philosophy Among a Central African People,* 1–25. Berkeley: University of California Press.

Biloa, E. (2013) *The Syntax of Tuki: A Cartographic Approach.* Amsterdam; Philadelphia: John Benjamins.

Bitjaa Kody, Z. D. (1988) Le système verbal du bàsàa. Université de Yaoundé I, thèse de doctorat.

Bittremieux, L. (1939) De Baphende's van Luanda (Opper-Kasai). *Congo: Revue générale de la colonie belge* 20: 1.

de Boeck, L. B. (1948) La classification des langues en Afrique. *Bulletin des séances de l'Institut Royal Colonial Belge* 19(4): 846–873. Plus Notes by van Bulck 874–882.

de Boeck, L. B. (1951) Een greep uit de Mombesa-taal. *Aequatoria* 14: 136–143.

Bonhomme, J., M. De Ruyter & G.-M. Moussavou (2012) Blurring the Lines: Ritual and Relationship Between Babongo Pygmies and Their Neighbours (Gabon). *Antropos* 107: 387–405.

Boone, D. W. (1992) *A Sociolinguistic Survey in Mbola: Mbule Survey Report.* SIL Cameroon.

Boone, D. W. & Kenneth S. Olson (2004) *Bwa Bloc Survey Report* (SIL Electronic survey reports (SILESR) 2004–013). Dallas: SIL International.

Boone, O. (1973) *Carte ethnique de la république du Zaïre: quart sud-est.* Tervuren: Musée royal de l'Afrique centrale.

Booth, J. (1905) Die Nachkommen der Sulukaffern (Wangoni) in Deutsch-Ostafrika. *Globus* 88: 197–201, 222–226.

Borland, C. H. (1984) Conflicting Methodologies of Shona Dialect Classification. *South African Journal of African Languages* 4(1): 1–12.

Bostoen, K. A. G (2007) Bantu Plant Names as Indicators of Linguistic Stratigraphy in the Western Province of Zambia. In Payne, Doris L. & Peña, Jaime (eds.), *Selected Proceedings of the 37th Annual Conference on African Linguistics,* 16–29. Somerville: Cascadilla Proceedings Project.

Bostoen, K. & C. Grégoire (2007) La question bantoue: bilan et perspectives. *Mémoires de la Société de linguistique de Paris*, *N.S.* 15: 73–91.

Bostoen, K. & J. K. Muluwa (2014) Umlaut in the Bantu B70/80 Languages of the Kwilu (RDC). *Transactions of the Philological Society* 112(2): 209–230.

Bostoen, K. (2009) Shanjo and Fwe as Part of Bantu Botatwe: A Diachronic Phonological Approach. In Ojo, Akinloye & Moshi, Lioba (eds.), *Selected Proceedings of the 39th Annual Conference on African Linguistics*, 110–130. Somerville: Cascadilla Proceedings Project.

Botne, R. (2003) Lega (Beya Dialect) (D25). In Nurse, Derek & Philippson, Gérard (eds.), *The Bantu Languages*, 422–449. London; New York: Routledge.

Botne, R. (2008) *A Grammatical Sketch of Chindali (Malawian Variety)*. Philadelphia: American Philosophical Society.

Bouka, L. Y. (1989a) Éléments de description du kaamba: Parler bantou de la République du Congo (groupe koongo, H17b). Université Libre de Bruxelles, mémoire de licence MA thesis.

Bouka, L. Y. (1989b) Teke and Its Dialects in Congo: Status of the Research. In T.G. Bergman (ed.), *Proceedings of the Round Table on Assuring the Feasibility of Standardization Within Dialect Chains, Noordwijkerhout, The Netherlands, September 1988*, 63–75. Nairobi: SIL International.

Bouquiaux, L. & J. M. C. Thomas (1994) Quelques problèmes comparatifs de langues bantoues C10 des confins oubanguiens: le cas du mbati, du ngando et de l'aka. In Geider, Thomas & Kastenholz, Raimund (eds.), *Sprachen und Sprachzeugnisse in Afrika: eine Sammlung philologischer Beiträge Wilhelm J.G. Möhlig zum 60. Geburtstag zugeeignet*, 87–106. Köln: Rüdiger Köppe.

Bradley, D. & K. Bradley (1992) Sso Survey Report. Ministry of Scientific and Technical Research and SIL, B.P. 1299, Yaoundé, Republic of Cameroon.

Bradley, D. P. (1992a) *Hijuk Survey Report*. SIL Cameroon; Ministry of Higher Education, Computer Services and Scientific Research, Cameroon.

Bradley, D. P. (1992b) Tibea Survey Report. Ministry of Scientific and Technical Research and SIL, B.P. 1299, Yaoundé, Republic of Cameroon.

Brelsford, W. V. (1946) *Fishermen of the Bangweulu Swamps: A Study of the Fishing Activities of the Unga Tribe*. Livingstone: The Rhodes-Livingstone Institute.

Brelsford, W. V. (1956) *The Tribes of Northern Rhodesia*. Lusaka: Government Printer.

Bruel, G. (1910, 1911) Les Populations de la Moyenne Sanga. *Revue d'ethnographie et de sociologie* 1 (2): 3–32, 111–125.

Burger, J.P. (1960) *An English-Lozi vocabulary*. Morija- Basutoland: Morija Printing Works.

Burssens, H. (1958) *Les Peuplades de l'entre Congo-Ubangi (Ngbandi, Ngbaka, Mbandja, Ngombe, et Gens d'Eau)* Tervuren: Musée royal du Congo belge.

Calloc'h, J. (1911) *Vocabulaire Français-ifumu (Batéké) précédé d'éléments de grammaire*. Paris: Librarie Paul Geuthner.

Cammenga, J. (2002) *Phonology and Morphology of Ekegusii: A Bantu Language of Kenya*. Cologne: Rüdiger Köppe.

Cammenga, J. (2004) *Igikuria Phonology and Morphology, A Bantu Language of south-west Kenya and north-west Tanzania*. Cologne: Rüdiger Köppe.

Cantrell, J. V. (1946) Some Aspects of Mpondo and Its Relation to Xhosa and Zulu. University of South Africa (UNISA), MA thesis.

Carrington, J. F. (1943) The Tonal Structure of Kele (Lokele) *African Studies* 2(4): 193–209.

Carrington, J. F. (1947) Notes sur la langue Olombo (Turumbu). *Aequatoria* 10: 102–113.

Carrington, J. F. (1977) Esquisse morphologique de la langue likile (Haut-Zaïre). *Africana Linguistica* 7: 65–88.

Chatelain, H. & W.R. Summer (1894) Bantu notes and vocabularies, 2: Comparative tables and vocabularies of Lange, Songe, Mbangala, Kioko, Lunda, etc. *Journal of the American Geographical Society of New York* 26: 208–240.

Cheucle, M. (2014) Étude comparative des langues makaa-njem (bantu A80): phonologie, morphologie, lexique Vers une reconstruction du proto-A80. Université Lumière Lyon 2, thèse de doctorat.

Chomba, D.S. (1975) The verbal morphology of Pimbwe. University of Dar es Salaam, MA thesis.

Chuo Kikuu cha Dar es Salaam (2009) *Atlasi ya Lugha za Tanzania*. Dar es Salaam: Mradi wa Lugha za Tanzania.

Clarke, J. A. (1911) *Luba-Sanga grammar*. Koni Hill, Katanga.

Costermans, B. J. (1953) *Mosaïque Bangba: notes pour servir à l'étude des peuplades de l'Uele*. Bruxelles: Academie Royale des Sciences d'Outre-Mer.

Coupez, A. (1955) *Esquisse de la langue holoholo* Tervuren: Musée royal du Congo belge.

Crane, T. M. (2011) Beyond Time: Temporal and Extra-Temporal Functions of Tense and Aspect Marking in Totela, a Bantu Language of Zambia. University of California at Berkeley, PhD dissertation.

Crane, T. M., L. M. Hyman & S. Nsielanga Tukumu (2011) *A Grammar of Nzadi [B865]: A Bantu Language of Democratic Republic of Congo*. Berkeley: University of California Press.

Cross-Upcott, A. R. W. (1956) The Social Structure of the KiNgindo-Speaking Peoples. University of Cape Town, PhD dissertation.

Cunningham, J. F. (1905) *Uganda and Its People*. London: Hutchinson & Co.

Czekanowski, J. (1924) Sprachenaufnahmen. In *Forschungen im Nil-Kongo-Zwischengebiet: Zweiter Band: Ethnographie Uele/Iturl/Nil-Länder*, 575–714. Leipzig: Klinkhardt & Biermann.

de Clercq, A. (1900) *Eléments de la langue kanioka*. Paris: Imprimerie Franciscaine Missionaire.

de Clercq, A. (1903) *Grammaire de la langue luba*. Louvain: J.B. Istas.

de Calonne-Beaufaict, A. (1912) *Études Bakango*. Liège: M. Thone.

de Calonne-Beaufaict, A. (1921) *Azande: introduction à une ethnographie générale des bassins de l'Ubangi-Uele et de l'Aruwimi*. Bruxelles: Maurice Lamertin.

Delhaise, C. (1909) Chez les WaSongola du Sud: Bantu ou Ba=Bili. *Bulletin de la Société royale belge de géographie* 33: 32–58, 109–135, 159–214.

de Granda Gutiérrez, G. (1984) Perfil lingüístico de Guinea Ecuatorial. In *Homenaje a Luis Flórez: estudios de historia cultural, dialectología, geografía lingüística, sociolingüística, fonética, gramática y lexicografía*, 119–195. Colombia: Instituto Caro y Cuervo.

Denis, J. (1964) *Les Yaka du Kwango: contribution à une étude ethno-démographique*. Tervuren: Musée royal de l'Afrique centrale.

de Luna, K. (2008) Collecting Food, Cultivating Persons: Wild Resource Use in Central African Political Culture, c. 1000 B.C.E. to c. 1900 C.E. Northwestern University, PhD dissertation.

Denolf, P. (1954) *Aan de rand van de Dibese*. Bruxelles: Institut Royal du Congo Belge.

de Schryver, G.-M., R. Grollemund, S. Branford & K. Bostoen (2015) Introducing a State-of-the-Art Phylogenetic Classification of the Kikongo Language Cluster. *Africana Linguistica* 21: 87–162.

Detienne, P. (1984) *Dialectes du Maindombe: essai de géographie linguistique*. Kinshasa: Centre de Recherches Pédagogiques.

Devos, M. (2007) *A Grammar of Makwe*. Munich: LINCOM.

de Rop, A. J. (1960) Les langues du Congo. *Aequatoria* 23: 1–24.

de Rop, A. J. (1971) Esquisse de grammaire mbɔ́lɛ́. *Orbis* 20(1): 34–78.

de Wit, G. (1994) *Bira-Huku Group of Bantu*. Bunia: SIL Eastern Zaire Group.

Dieu, M. & P. Renaud (1983) *Situation linguistique en Afrique centrale- inventaire préliminaire: le Cameroun* (Atlas linguistique de l'Afrique centrale (ALAC)). Paris & Yaoundé: Agence de Coopération Culturelle et Technique (ACCT); Centre Régional de Recherche et de Documentation sur les Traditions Orales et pour le Développement des Langues Africaines (CERDOTOLA); Direction Générale de la Recherche Scientifique et Technique (DGRST), Institut des Sciences Humaines. Carries the date 1983 but did not come out of the presses until 1985.

Dijkmans, J. J. M. (1974) *Kare-Taal: Lijst van woorden gangbaar bij het restvolk Kare opgenomen in de jaren 1927–1947*. Sankt Augustin: Anthropos-Institut- Haus Völker und Culturen.

Djomeni, G. D. (2014) *A Descriptive Grammar of Bàmbəlɔ̀* Munich: LINCOM.

Dodo-Bounguendza, E. (1988) Le koko de Sogeland: Langue bantoue du Caméroun (A43); Elements de description phonologique et morphologique. Université Libre de Bruxelles mémoire de licence.

Doke, C. M. (1954) *The Southern Bantu Lanaguages*. Oxford: Oxford University Press.

Doke, C. M. (1922) *The Grammar of the Lamba Language*. London: Kegan Paul, Trench and Trübner. Publication of MA U College, Johannesburg.

Doke, C. M. (1928) Twa of the Kafue River Swamps. *Bantu Studies* 3(2): 148–150.

Doke, C. M. (1931) *A Comparative Study of Shona Phonetics*. Johannesburg: University of the Witwatersrand Press.

Doke, C. M. (1933) A Short Aushi Vocabulary. *Bantu Studies* 7(3): 285–295.

Douglas, M. (1963) *The Lele of the Kasai*. London: Oxford University Press.

Dugast, I. (1949) *Inventaire ethnique du sud-Cameroun*. Dakar: Institut Fondamental d'Afrique Noire.

Dunham, M. (2007) Le Langi et la vallée du Rift Tanzanien: Contacts et convergences. *Bulletin de la Société de Linguistique de Paris* CII(1): 399–427.

Ebner, Fr. E. P. (1955) *The History of the Wangoni and Their Origin in the South African Bantu Tribes*. Ndanda & Peramiho: Benedictine Publishers.

Ebobissé, C. (2014) *Sawabantu: Eine vergleichende Untersuchung der Küstensprachen Kameruns (Bantu A.20 und A.30)*. Frankfurt: Peter Lang.

Echegaray, C. G. (1959a) La lengua baseque (Guinea Española). In *Estudos guineos*, 73–104. Madrid: Consejo Superior de Investigaciones Científicas.

Echegaray, C. G. (1959b) Las lenguas de la provincia española de Guinea. In *Estudos guineos*, 13–72. Madrid: Consejo Superior de Investigaciones Científicas.

Echegaray, C. G. (1960) *Morfologia y Sintaxis de la Lengua Bujeba*. Madrid: Instituto de Estudios Africanos, Consejo superior de investigaciones científicas.

Ellington, J. E. (1977) Aspects of the Tiene Language. University of Wisconsin, PhD dissertation.

Engels, A. (1911) Note sur les Batshua (Batua). *La revue congolaise* 2: 214–217.

Essono, J.-J. [Marie] (2000) *L'ewondo, langue bantu du Cameroun: phonologie, morphologie, syntaxe.* Yaoundé: Presses de l'Université Catholique d'Afrique Centrale; Agence de Coopération Culturelle et Technique (ACCT).

Estermann, C. (1979) *The Ethnography of Southwestern Angola, 2: The Nyaneka-Nkumbi Ethnic Group.* New York: Africana Publishers.

Etter, M. (1988) Language Assessment Survey Among the Byep (Northern Mekaa). Yaoundé: SIL.

Even, A. (1931) Quelques coutumes des tribus Badondos et Bassoundis. *Bulletin de la Societe de Recherche Congolaise* 7: 77–94.

Fallers, M. C. (1968) *The Eastern Lacustrine Bantu (Ganda and Soga).* London: International African Institute.

Fernandez Galilea, L. (1951) *Diccionario español-kômbé.* Madrid: Instituto de Estudios Africanos.

Ferreira Diniz, J. O. (1918) *Populações indígenas de Angola.* Coimbra: Imprenta da Universidade.

Fleisch, A. (2009) Language History in SE Angola: The Ngangela-Nyemba Dialect Cluster. In Möhlig, Wilhelm J.G., Seidel, Frank & Seifert, Marc (eds.), *Language Contact, Language Change and History: Based on Language Sources in Africa* (Sprache und Geschichte in Afrika 20), 97–111. Cologne: Rüdiger Köppe.

Fontaney, Louise (1988, 1989) Mboshi: Steps Towards a Grammar. *Pholia* 3(4): 87–167, 71–132.

Forges, G. (1983) *Phonologie et morphologie du kwezo* Tervuren: Musée royal de l'Afrique centrale.

Fortune, G. (1963) A Note on the Languages of Barotseland. In *Proceedings of the Conference on the History of Central African Peoples*, 1–27. Lusaka: Rhodes-Livingstone Institute.

Fortune, G. (2001) *An Outline of Silozi Grammar.* Lusaka: Bookworld Publishers.

Fortune, G. (2004) *Essays on Shona Dialects.* Ms.

Fourie, D. J. (1997) The Linguistic Position of Oshimbalanhu within the Wambo Group. In Haacke, Wilfrid & Elderkin, Edward Derek (eds.), *Namibian Languages: Reports and Papers*, 277–306. Cologne: Rüdiger Köppe; University of Namibia (UNAM).

Full, W. (2006) *Dialektologie des Komorischen.* Köln: Rüdiger Köppe.

Fülleborn, F. (1906) *Das Deutsche Njassa- und Ruwuma-Gebiet, Land und Leute: nebst Bemerkungen über die Schire-Länder.* Berlin: Dietrich Reimer.

Funnell, B. J. (2004) *A Contrastive Analysis of Two Standardised Varieties of Sena.* Pretoria: University of South Africa.

Gardner, W. L. (1990) Language Use in the Epena district of Northern Congo. University of North Dakota, MA thesis.

Gérard, [R.P.] (1924) *La langue lebéo, grammaire et vocabulaire.* Bruxelles: A. Vromant & Co.

Geslin-Houdet, F. (1984) Esquisse d'une description du njɔwi, parler des mengisa (Cameroun). Université de la Sorbonne Nouvelle (Paris 3), thèse de doctorat.

Gluckman, M. (1951) The Lozi of Barotseland in North-Western Rhodesia. In Elizabeth Colson & Max Gluckman (eds.), *Seven Tribes of British Central Africa*, 3–93. Manchester: Manchester University Press.

Goemaere, A. (1984) *Grammaire du londengese.* Bandundu: Centre d'Etudes Ethnologiques de Bandundu.

Gonzales, R. M. (2002) Continuity and Change: Thought, Belief, and Practice in the History of the Ruvu Peoples of Central East Tanzania, C. 200 B.C. To A.D. 1800. University of California at Los Angeles, PhD dissertation.

Grant, C. A. (1992) *Bati survey report*. Yaoundé: SIL; Mininistry of Higher Education, Computer Services and Scientific Research, Cameroon.

Grottanelli, V. L. (1953) I Bantu del Giuba nelle tradizioni dei Wazegūa. *Annali Lateranensi* 8: 249–260.

Guarisma, G. (2000) *Complexité morphologique simplicité syntaxique: Le cas du bafia, langue bantoue (A 50) du Cameroun*. Paris: Peeters.

Guillaume, H. & J.-M. Delobeau (1978) Une mosaïque ethnique et linguistique en milieu rural: enquête de démographie linguistique dans la sous-préfecture de Mongoumba (ECA). In Caprile, Jean-Pierre (ed.), *Contacts de langues et contacts de culture: 1-Démographie linguistique: approche quantitative*, 11–65. Paris: Société des Etudes Linguistiques et Anthropologiques de France.

Guthrie, M. (1967) *An Outline of Bantu Prehistory*. Letchworth; Brookfield, VT: Gregg International.

Hachipola, S. J. (1991) A Historico-Comparative Study of Zambian Plateau Tonga and Seven Related Lects. University of London, PhD dissertation.

Hachipola, S. J. (1998) *A Survey of the Minority Languages of Zimbabwe*. Harare: University of Zimbabwe Publications.

Hackett, P. & G. [Vaast] van Bulck (1956) Report of the Eastern Team: Oubangi to Great Lakes. In Guthrie, Malcolm & Tucker, Archibald N (eds.), *Linguistic Survey of the Northern Bantu Borderland*, volume 1, 63–122. Oxford: Oxford University Press.

Halkin, J. & E. Viaene (1911) *Les Ababua (Congo Belge)*. Bruxelles: Albert de Wit, Libraire-Éditeur.

Hambly, W. D. (1934) *The Ovimbundu of Angola*. Chicago: Field Museum of Natural History.

Hammarström, H., R. Forkel & M. Haspelmath (2017) Glottolog 3.0. Jena: Max Planck Institute for the Science of Human History (Available online at http://glottolog.org, accessed January 4, 2017).

Hammarström, H. (2015) Ethnologue 16/17/18th Editions: A Comprehensive Review. *Language* 91(3): 723–737. Plus 188pp online appendix.

Hammarström, H. (2018) A Survey of African Languages. In Güldemann, Tom (ed.), *African Languages and Linguistics*, 1–57. Berlin: De Gruyter Mouton.

Harjula, L. (2004) *The Ha Language of Tanzania: Grammar, Texts and Vocabulary*. Cologne: Rüdiger Köppe.

Harries, L. (1950) *A Grammar of Mwera*. Johannesburg: Witwatersrand University Press.

Harries, L. (1955) Grammar of Gesogo. *Kongo-Overzee* 11: 420–440.

Harries, L. (1958) Kumu, a sub-Bantu language. *Kongo-Overzee* 24: 265–296.

Harries, L. (1959) Nyali, a Bantoid Language (Belgian Congo). *Kongo-Overzee* 25: 174–205.

Harvey, T. K. (1997) The Bali of Northeastern Congo-Kinshasa: Uncovering the History of a People Shrouded by the Ituri Rain Forest. University of Texas at Arlington, MA thesis.

Hauser, A. (1954) La frontière linguistique bantoue-oubanguienne entre le Bas-Oubangui et ses affluents de droite. *Zaire* 8(1): 21–26.

Hedinger, R. (1987) *The Manenguba Languages (Bantu A.15, Mbo Cluster) of Cameroon*. London: SOAS. Publication of Hedinger, Robert (1984) A Comparative-Historical

Study of the Manenguba Languages (Bantu A.15 Mbo cluster) of Cameroon., PhD dissertation. University of London.

Heintze, B. (1970) Beiträge zur Geschichte und Kultur der Kisama (Angola). *Paideuma* 16: 159–186.

Hendle, P.J. (1907) *Die Sprache der Wapogoro (Deutsch-Ostafrika) nebst einem Deutsch-Chipogoro und Chipogoro-Deutschen Wörterbuch*. Berlin: Georg Reimer.

Henson, B. (2007) The Phonology and Morphosyntax of Kol. University of California at Berkeley, PhD dissertation.

Herrmann, C. (1904) Lusíba, die Sprache der Länder Kisíba, Bugábu, Kjamtára, Kjánja und Ihángiro, speziell der Dialekt der "Bayōssa" im Lande Kjamtára. *Mittheilungen des Seminars für Orientalische Sprachen* 7(3): 150–200.

Hiernaux, J., E. Vincke & D. Commelin (1976) Les Oto et les Twa des Konda (Zone de l'Équateur, Zaïre). *L'Anthropologie* 80: 449–465.

Hiernaux, J. (1966) Les Bushong et les Cwa du royaume Kuba (Congo-Kinshasa): Pygmées, Pygmoïdes et pygméisation; anthropologie, linguistique et expansion bantoue. *Bulletins et Mémoires de la Société d'Anthropologie de Paris* 9(3): 299–336.

Hobley, C. W. (1902) *Eastern Uganda: An Ethnological Survey*. London: Anthropological Institute of Great Britain and Ireland.

Hochegger, H. (1972) *Dictionnaire buma-français, avec un aperçu grammatical* Bandundu: Centre d'Etudes Ethnologiques de Bandundu.

Horton, A. E. (1949) *A Grammar of Luvale*. Johannesburg: Witwatersrand University Press.

Hout, K. (2012) The Vocalic Phonology of Mushunguli. University of Kansas MA thesis.

Hulstaert, G. (1941, 1942, 1942) Over het dialekt der Boyela. *Aequatoria* 4, 5, 5: 95–98, 15–19, 41–43.

Hulstaert, G. (1948) Le dialecte des Pygmoides Batswa de l'Equateur. *Africa* 18(1): 21–28.

Hulstaert, G. (1951) Les langues de la cuvette centrale congolaise. *Aequatoria* 14: 18–24.

Hulstaert, G. (1978) Notes sur la langue des Bafotó (Zaïre). *Anthropos* 73: 113–132.

Hulstaert, G. (1984) La Langue des Mpama. *Annales Æquatoria* 5: 5–32.

Hulstaert, G. (1987) Les Parlers des Bangandó Méridionaux. *Annales Æquatoria* 8: 205–287.

Hulstaert, G. (1988) Le parler des Lokalo orientaux. *Annales Æquatoria* 9: 133–170.

Hulstaert, G. (1993) Douze dialectes mongo, 1: le dialecte des losakani. *Annales Aequatoria* 14: 15–38.

Hulstaert, G. (1999) Éléments pour la dialectologie Móngɔ. *Annales Æquatoria* 20: 9–322.

Huntingford, G. W. B. (1965) The Orusyan Language of Uganda. *Journal of African Languages* 4: 145–169.

Idiata, D. F. (2007) *Les langues du Gabon: Données en vue de l'élaboration d'un atlas linguistique*. Paris: L'Harmattan.

Igunzi, M. M. (2013) Expériences de revitalisation de la langue kinyindu. In *Proceedings of the 17th FEL Conference*, 195–196. Berkshire: Foundation for Endangered Languages.

Jacobs, J. & B. Omeonga (2004) Lonkucu: Texte et lexique (Iwaji, Kɔlɛ, Kasai Oriental, R. D. du Congo). *Annales Æquatoria* 25: 303–316.

Jacobs, J. (1962, 1964) *Tetela-Grammatica (Kasayi, Kongo)*. Gent: Wetenschappelijke Uitgeverij en Boekhandel. 2 vols.

Jacottet, É. (1927) *A Grammar of the Sesuto language*. Johannesburg: University of the Witwatersrand Press.

Jacquot, A. (1965) Précisions sur l'inventaire des langues teke du Congo. *Cahiers d'études africaines* 5(18): 335–340.

Jacquot, A. (1978) Le Gabon. In *Inventaire des études linguistiques sur les pays d'Afrique Noire d'expression française et sur Madagascar*, 493–503. Paris: Conseil international de la langue française.

Jacquot, A. (1981) *Études Beembe (Congo): Esquisse Linguistique, Devinettes et Proverbes*. Paris: *Office de la recherche scientifique et technique outre-mer*.

Jacquot, A. (1983) *Les classes nominales dans les langues bantoues des groupes B.10, B.20, B.30 (Gabon-Congo)*. Paris: *Office de la recherche scientifique et technique outre-mer*.

Jalla, A. (1917) *Elementary Grammar of the Sikololo Language*. Torre Pellice: Imprimérie Alpine.

Jaspert, F. & W. Jaspert (1930) *Die Völkerstämme Mittel-Angolas*. Frankfurt am Main: Baer.

Johnson, A. E. & K. H. Beavon (1989) *Sociolinguistic Survey Among the Mpompo and Related Peoples*. SIL Cameroon.

Johnston, H. H. (1919, 1922) *A Comparative Study of the Bantu and Semi-Bantu Languages*. Oxford: Oxford University Press.

Jordan, L. (2015) A Comparison of Five Speech Varieties of Southwestern Angola: Comparing OluHumbe, OluCilenge, OluKwandu, OluNgendelengo, and OluKuvale in the Kamucuio Municipality, Namibe Province. *SIL Electronic Survey Reports* 2015–017: 1–29.

Joset, P. E. (1952) Les Baamba et les Babwizi du Congo Belge et de l'Uganda Protectorate. *Anthropos* 47: 369–387, 909–946.

Jouannet, F. (1983) Introduction. In Francis Jouannet (ed.), *Le kinyarwanda, langue bantu du Rwanda: études linguistiques*, 15–20. Paris: Société des Etudes Linguistiques et Anthropologiques de France; Groupe d'Etudes et de Recherches en Linguistique Appliquée, Université Nationale du Rwanda.

Kamba Muzenga, J. G. (1980) *Esquisse de grammaire kete* Tervuren: Musée royal de l'Afrique centrale.

Kabange Mukala, E. (2005) Elements de grammaire du kizééla (L.33a) phonologie et morphologie: approche structuraliste. Lubumbashi: Université de Lubumbashi (UNILU), mémoire de licence.

Kabungama, Y. (1992) Analyse des formes nominales en kisɛmbɔmbɔ. *Annales Aequatoria* 13: 431–452.

Kahigi, K. K. (2008) *Kikahe: Msamiati wa Kikahe-Kiswahili-Kiingereza na Kiingereza-Kikahe-Kiswahili (Kahe-Swahili-English and English-Kahe-Swahili Lexicon)*. Dar es Salaam: Departement. of Foreign Languages and Linguistics, University of Dar es Salaam.

Kalasi, N. G. (2003) A Lexical Study of Šekgalagari Dialects in Southern and Kgalagadi Districts. University of Botswana, MA thesis.

Kamanda Kola, R. (1991) Eléments de description du zamba, langue bantoue (C31e) du Zaïre. Université Libre de Bruxelles, mémoire de licence.

Kanyoro, R. M. A. (1983) *Unity in Diversity: A Linguistic Survey of the Abaluyia of Western Kenya*. Wien: Afro-Pub. Beiträge zur Afrikanistik 20, Institut für Afrikanistik.

Kashoki, M. E. & M. Mann (1978) A General Sketch of the Bantu Languages of Zambia. In Ohannessian, Sirarpi & Kashoki, Mubanga E (eds.), *Language in Zambia*, 47–100. London: International African Institute.

Kasongo, G.B. Ohanu wa (1993) A propos de l'Otetela-Hamba de Lomela. *Annales Aequatoria* 14: 535–538.

Katupha, J. M. M. (1991) The Grammar Emakhuwa Verbal Extensions: An Investigation of the Role of Extension Morphemes in Derivational Verbal Morphology and in Grammatical Relations. University of London, PhD dissertation.

Kawasha, B. K. (2003) Lunda Grammar: A Morphosyntactic and Semantic Analysis. Eugene: University of Oregon, PhD dissertation.

Kayambazinthu, E. (2004) The Language Planning Situation in Malawi. In Baldauf, Richard B., Jr. & Kaplan, Robert B. (eds.), *Language Planning and Policy in Africa, 1: Botswana, Malawi, Mozambique, and South Africa*, 79–149. Clevedon: Multilingual Matters.

Kershner, T. L. (2002) The Verb in Chisukwa: Aspect, Tense, and Time (Malawi). Indiana University, PhD dissertation.

Khang Levy, N. (1979) Elements de grammaire morphologique de la langue Lwel. Lubumbashi: Université Nationale du Zaïre (UNAZA), mémoire de licence.

Kihore, Y. M. (2000) Historical and linguistic aspects of Kihacha. In Kahigi, K. K., Kihore, Y. M. & Mous, M. (eds.), *Lugha za Tanzania/Languages of Tanzania: Studies Dedicated to the Memory of the Late Prof. C. Maganga* Leiden: Research School of Asian, African and Amerindian Studies, University of Leiden.

Kitetu, C. (2012) Mutual Intelligibility Among the Taita, Taveta and Pare Languages Spoken in the Cross-Border Area of South-Western Kenya and North-Eastern Tanzania. In Ogechi, Nathan Oyori, Ngala Oduor, Jane A. & Iribemwangi, Peter (eds.), *The Harmonization and Standardization of Kenyan Languages. Orthography and Other Aspects*, 64–80. Cape Town: The Centre for Advanced Studies of African Society.

Klieman, K. A. (1997) Hunters and Farmers of the Western Equatorial Rainforest: Economy and Society, 3000 B.C. to A.D. 1880. University of California at Los Angeles, PhD dissertation.

Klieman, K. (2003) Toward a History of Pre-colonial Gabon: Farmers and Forest Specialists along the Ogooué river, 500 B.C.- A.D. 1000. In Reed, Michael C. & Barnes, James F. (eds.), *Culture, Ecology, and Politics in Gabon's Rainforest*, 99–135. Lewistown: Edwin Mellen Press.

Köhler, H. (1964) Mode de capture chez les différentes tribus de chasse et de pêche au bord de la Rivière Sangha. In *Varia 1*, 43–93. Uppsala: Almkvist & Wiksell.

Koni Muluwa, J. & K. Bostoen (2011) Umlaut in the Bantu B70/80 Languages of the Kwilu (RDC): Where Did the Final Vowel Go? Paper Presented at the 41st CALL, Leiden, 2011.

Kraal, P. J. (2005) A Grammar of Makonde (Chinnima, Tanzania). Leiden University, PhD dissertation.

Kröger, O. (2005) Report on a Survey of Coastal Makua Dialects. *SIL Electronic Survey Reports* 2005–020: 1–45.

Krüger, C. J. H. & J. A. de Plessis (1977) *Die Kgalagadi dialekte van Botswana.* Potchefstroomse University vir Christelike Hoër Onderwys.

Kunkel, J. & B. Cameron (2002) *A Sociolinguistic Survey of the Dhimba Language.* SIL Electronic Survey Reports, 2002–070. Dallas: SIL International.

Kuperus, J. (1985) *The Londo Word: Its Phonological and Morphological Structure* Tervuren: Royal Museum for Central Africa.

Kutsch L., Connie (2003) Bila (D32). In Derek Nurse & Gérard Philippson (eds.), *The Bantu languages*, 450–474. London; New York: Routledge.

la Fontaine, J. S. (1959) *The Gisu of Uganda*. London: Oxford University Press for the International African Institute.

Labaere, H. & W. W. L. Shango (1989) Les dialectes otetela. *Annales Æquatoria* 10: 253–267.

Ladefoged, P., R. Moser Glick & C. Criper (1972) The Languages of Uganda. In *Language in Uganda* (Ford Foundation's Language Survey), 31–84. London; New York: Oxford University Press.

Lamal, F. (1965) *Basuku et Bayaka des districts Kwango et Kwilu au Congo* Tervuren: Musée royal de l'Afrique centrale.

Laman, K. E. (1936) *Dictionnaire kikongo-français avec une étude phonétique décrivant les dialectes les plus importants de la langue dite kikongo*. Bruxelles: Institut Royal Colonial Belge.

Lamberty, M. (2002) A Rapid Appraisal Survey of the Abo and Barombi Speech Communities. *SIL Electronic Survey Reports* 2002–075: 1–28.

Lamberty, M. (2009) A Rapid Appraisal Survey of Malimba in Cameroon. SIL International. SIL Electronic Survey Report 2009–004.

Lane, B. D. (1989) Toward Understanding Multilingualism Among the Mbete of Northern Congo (Bantu, Africa). University of Texas at Arlington, MA thesis.

Lang, A. & C. Tastevin (1937) *Ethnographie: La Tribu des Va-Nyaneka*. Corbeil: Imprimerie Crété.

Lanham, L. W. (1955) *A Study of Gitonga of Inhambane*. Johannesburg: Witwatersrand University Press.

Laranjo Medeiros, Carlos (1981) *VaKwandu: History, Kinship and Systems of Production of an Herero People of South-West Angola*. Lisboa: Junta de Investigações Científicas do Ultramar.

Larson, T. J. (1981) The Mbukushu. In Gibson, Gordon D., Larson, Thomas John & McGurk, Cecilia R. (eds.), *The Kavango Peoples*, 211–268. Wiesbaden: Franz Steiner Verlag.

Larson, T. J. (1992) *The Bayeyi of Ngamiland*. Gaborone: The Botswana Society.

Legère, K. (1992) Language Shift in Tanzania. In Matthias Brenzinger (ed.), *Language Death: Factual and Theoretical Explorations with Special Reference to East Africa*, 99–116. Berlin De Gruyter Mouton.

Legère, K. (2003) Trilingual Ngh'wele-Swahili-English and Swahili-Ngh'wele-English Wordlist. Department of Oriental and African Languages, Göteborg University.

Legère, K. (2007) Vidunda (G38) as an Endangered Language?. In Payne, Doris L. & Peña, Jaime (eds.), *Selected Proceedings of the 37th Annual Conference on African Linguistics*, 43–54. Somerville: Cascadilla Proceedings Project.

Leitch, M. (2004) Langue et dialecte au sud du district d'Epena. *SIL Electronic Survey Reports* 2004–007: 1–56.

Lewis, P. M., G. F. Simons & C. D. Fennig (2015) *Ethnologue: Languages of the World*. 18th edn. Dallas: SIL International.

Liesenborghs, O. (1932) Twee Gevallen van Wisselwerking Tusschen Soedaneesche en Bantucultuur. *Congo* 2(8): 69–74.

Lindblom, G. (1926) *Notes on Kamba Grammar*. Uppsala: Appelbergs Boktryckeri.

Lindfors, A.-L., L. Nagler & M. Woodward (2009a) *A Sociolinguistic Survey of the Nyiha and Nyika Language Communities in Tanzania, Zambia and Malawi* (SIL Electronic Survey Reports (SILESR) 2009–012). Dallas: SIL International.

Lindfors, A.-L., M. Woodward & L. Nagler (2009b) *A Sociolinguistic Survey of Sichela Speech Community* (SIL Electronic Survey Reports (SILESR) 2009–014). Dallas: SIL International.

Lingomo, B. (1995) Nkolo Loholi, un peuple Bongando. *Annales Aequatoria* 16: 339–354.

Lisimba, M. (1982) A Luyana Dialectology (Africa, Zambia). Madison: University of Wisconsin, PhD dissertation.

Lodhi, A. Y. (2008) Bantu Origins of the Sidis of India. In Prasad, Kiran Kamal & Angenot, Jean-Pierre (eds.), *TADIA: The African Diaspora in Asia – Explorations on a Less Known Fact*, 301–313. Bangalore, India: Jana Jagrati Prakashana (on behalf the Tadia Society).

Lopes Cardoso, C. (1966) Olumbali do distrito de Moçâmedes (Achegas para o seu estudo). *Boletim do Instituto de Investigação Científica de Angola* 3(1): 37–73.

Lumwamu, F. (1974) Eléments pour un lexique kongo-français (dialectes dondo-kamba-kuni). *Dimi: Bulletindu Centre de Linguistique Appliquée et de Littérature Orale* 2: 20–87.

Lusakalalu, P. (2001) Languages and Glossonymic Units: Contribution to the Assessment of the Linguistic Diversity of Angola and Namibia. *Afrikanistische Arbeitspapiere* 66: 47–65.

Lyndon, C. & A. Lyndon (2007) Enatthembo: An Appraisal of Linguistic and Sociolinguistic Factors. SIL International, Dallas. SIL Electronic Survey Reports 2007–003.

Mabiala, J.-N. Nguimbi (1999) Phonologie comparative et historique du kongo (groupe H10). Université Lumière Lyon 2, thèse de doctorat.

Macalane, G. (2000) Proposta da ortografia da língua Cibalke. In Sitoe, Bento & Ngunga, Armindo S.A. (eds.), *Relatório do II seminário sobre a padronização da ortografia de línguas moçambicanas*, 117–128. Núcleo de Estudo de Línguas Moçambicanas; Centro de Estudos das Línguas Moçambicanas, Universidade Eduardo Mondlane.

Maes, J. (1934) Vocabulaire des populations de la région du kasai-lulua-sankuru [par L. Achten]. *Journal de la Société des Africanistes* 4(2): 209–268.

Maho, J. F. (2009) *NUGL Online: The Online Version of the New Updated Guthrie List, a Referential Classification of the Bantu Languages*. Gothenburg: Department of Oriental and African Languages.

Maho, J. F. & B. E. Sands (2002–2004) *The Languages of Tanzania: A Web Links Collection = Web Appendix to The Languages of Tanzania: A Bibliography* (Maho & Sands 2002).

Maho, J. F. (2008) *The Bantu Bibliography*. Cologne: Rüdiger Köppe.

Maho, J. (2003) A Classification of the Bantu Languages: An Update of Guthrie's Referential System. In Nurse, Derek & Philippson, Gérard (eds.), *The Bantu Languages*, 639–651. London; New York: Routledge.

Makouta-Mboukou, J.-P. (1977) Étude descriptive du fumu, dialecte teke de Ngamaba, Brazzaville. Université de la Sorbonne Nouvelle (Paris 3), thèse de doctorat.

Malan, J. S. (1974) The Herero-Speaking Peoples of Kaokoland. *Cimbebasia (B)* 2(4): 113–129.

Malepe, A. T. (1966) A dialect-geographical survey of the phonology of the central, eastern and southern dialects of Tswana. University of South Africa (UNISA), MA thesis.

Mamet, M. (1955) *La Langue ntomba telle qu'elle est parlée au Lac Tumba et dans la région avoisinante (Afrique Centrale)* Tervuren: Musée royal du Congo belge.

Mamet, M. (1960) *Le langage des Bolia (Lac Léopold II)* Tervuren: Musée royal du Congo belge.

Mangoya, E. (2012) Segmental Phonology of Barwe with Some Articulartory Phonetics. Universidade Eduardo Mondlane, PhD dissertation.

Maniacky, J. (2003) *Tonologie du ngangela: variété de Menongue (Angola)*. Munich: LINCOM.

Marlo, M. (2008) Tura Verbal Tonology. *Studies in African Linguistics* 37(2): 152–243.

Masele, B. F. Y. P. (2001) The Linguistic History of SiSuumbwa, KISukuma and KI Nyamweezi in Bantu Zone F. Memorial University of Newfoundland, PhD dissertation.

Mateene, K. C. (1980) *Essai de grammaire générative et transformationalle de la langue nyanga*. Kinshasa: Presses Universitaires du Zaïre.

Matimi, J.-C. (1998) Tradition et innovation dans la construction de l'identité chez les Shamaye (Gabon) entre 1930 et 1990. Québec: Université Laval, thèse de doctorat.

Maurice, M. (1935, 1935, 1935, 1936, 1936, 1936, 1936, 1937, 1937, 1938) Le pays des Bapimbwe. *La Géographie* 64, 64, 64, 66, 67, 67, 67, 68, 68, 69, 69, 70: 22–31, 228–242, 309–317, 171–289, 86–95, 147–165, 209–221, 224–236, 289–296, 18–33, 264–270, 83–102.

Mavoungou, P. A., Ndinga-Koumba-Binz & S. Hughes (2010) *Civili, langue des Baloango: Esquisse historique et linguistique*. Munich: LINCOM.

Mayanga, Tayeye (1985) *Grammaire yansi (République du Zaïre)* Bandundu: Centre d'Etudes Ethnologiques de Bandundu.

Mbalanga, A. (1996) Une minorité bantu en Centrafrique: les Issongo de la Lobaye. *Zo: revue d'anthropologie* 1: 33–40.

Mbongue, J. (1999) Premiere évaluation globale de la situation sociolinguistique de la langue lefa. Yaoundé: SIL.

McCulloch, M. (1951) *The Southern Lunda and Related Peoples (Northern Rhodesia, Angola and Belgian Congo)*. London: Oxford University Press for the International African Institute.

McMaster, M. A. (1988) Patterns of Interaction: A Comparative Ethnolinguistic Perspective on the Uele Region of Zaïre ca. 500 A.D. to 1900 A.D. University of California at Los Angeles, PhD dissertation.

Medjo Mvé, P. (2009) Les Kakɔ et le kakɔ de la région de Bitam, données préliminaires sur une communauté et une langue du Gabon peu connues. In Idiata, Daniel F (ed.), *Éléments de description des langues du Gabon*, volume 2, 69–104. Libreville: CENAREST.

Medjo Mvé, P. (2011) *Introduction à la langue et la culture des chasseurs-cueilleurs Bakoya (Région de Mékambo, Gabon) avec un petit dictionnaire*. Cologne: Rüdiger Köppe.

Medjo Mvé, P. (no date) A la recherche des Pygmées Bakoya de la région de Mekambo. Ms.

Meeussen, A. E. (1952) *Esquisse de la langue ombo (Maniema- Congo Belge)* Tervuren: Musée royal du Congo belge.

Meeussen, A. E. (1953) De talen van Maniema (Belgisch-Kongo). *Kongo-Overzee* 19: 385–391.

Meeussen, A. E. (1954) *Linguistische Schets van het Bangubangu* Tervuren: Koninklijk Museum voor Belgisch Congo.

Merlevede, A. (1995) Een schets van de fonologie en morfologie van het Bondei gevolgd door een Bondei-Engels en Engels-Bondei woordenlijst. Rijksuniversiteit te Leiden, doctoraatsthesis.

Mertens, J. (1935, 1938, 1939) *Les Ba Dzing de la Kamtsha*. Bruxelles: Campenhout. 3 vols.

Meyer, J. & T. Raymond (1981) *La phonologie de la langue ebhele*. Lubumbashi: Université Nationale du Zaïre (UNAZA).

Mfoutou, J.-A. (1985) Esquisse phonologique du kidoondo, un dialecte koongo de la République populaire du Congo. Brazzaville: Université Marien Ngouabi, mémoire de licence.

Miti, L. M. (1996) Subgrouping Ngoni Varieties within Nguni: A Lexicostatistical Approach. *South African Journal of African Languages* 16(3): 83–93.

Miti, L. M. (2004) *A Grammar of Cingoni-Nsenga: A Central Bantu Language Spoken in Zambia*. Frankfurt am Main: Peter Lang.

Mitterhofer, B. (2013) Lessons from a Dialect Survey of Bena: Analysing Wordlists. *SIL Electronic Survey Reports* 2013–020: 1–131.

Mkanganwi, K. G. (1972) The Relationship of Coastal Ndau to the Shona Dialects of the Interior. *African Studies* 31(2): 111–137.

Mkude, D. J. (1974) A Study of Kiluguru Syntax with Special Reference to the Transformational History of Sentences with Permuted Subject and Object. University of London, PhD dissertation.

Möhlig, W. J. G. (1974) *Die Stellung der Bergdialekte im Osten des Mt. Kenya*. Berlin: Dietrich Reimer.

Möhlig, W. J. G. (1997) A Dialectometrical Analysis of the Main Kavango Languages: Kwangali, Gciriku and Mbukushu. In Haacke, W. H. G. & Elderkin, Edward Derek (eds.), *Namibian Languages: Reports and Papers*, 211–234. Cologne: Rüdiger Köppe.

Möhlig, W. J. G. (2007) Linguistic Evidence of Cultural Change: The Case of the Rumanyo Speaking People in Northern Namibia. *Sprache und Geschichte in Afrika* 18: 131 154.

Möhlig, W. J. G. (2009) Historiography on the Basis of Contemporary Linguistic Data: The Herero Case. In Möhlig, Wilhelm J. G., Seidel, Frank & Seifert, Marc (eds.), *Language Contact, Language Change and History: Based on Language Sources in Africa*, 159–186. Cologne: Rüdiger Köppe.

Möhlig, W. J. G. (1986) Les parlers bantous côtiers du nord-est. In Guarisma, G. & W. J. G. Möhlig (eds.), *La méthode dialectométrique appliquée aux langues africaines*, 45–92. Berlin: Dietrich Reimer.

Morris, H. F. R. (1963) A Note on Lunyole. *The Uganda Journal* 27(1): 127–134.

Motingea Mangulu, A. & B. B. M. Mayika (2016) Notes sur les parlers mɔ́yɛ́ du Moyen Congo (Bantou C38). *Journal of Asian and African Studies* 92: 5–122.

Motingea Mangulu, A. (2001–2002) Situation actuelle des parlers minoritaires au Nord-Ouest de la République Démocratique du Congo. *Bulletin of the International Committee on Urgent Anthropological Ethnological Research* 41: 147–154.

Motingea Mangulu, A. (1988) *Eléments de grammaire lingombe, avec une bibliographie exhaustive*. Bamanya: Centre Aequatoria.

Motingea Mangulu, A. (1989) Sur les parlers nkutsu. *Annales Æquatoria* 10: 269–280.

Motingea Mangulu, A. (1990) Esquisse de la langue des Mokpá (Haut-Zaïre). *Afrika und Übersee* 73: 67–100.

Motingea Mangulu, A. (1990a) Esquisse du parler des Ndobo. In *Parlers riverains de l'entre Ubangi-Zaire: éléments de structure grammaticale*, 254–273. Bamanya: Centre Aequatoria.

Motingea Mangulu, A. (1990b) Esquisse du parler des Ohendo. *Annales Aequatoria* 11: 115–152.

Motingea Mangulu, A. (1993a) Esquisse du parler des Yakata (République du Zaïre). *Afrika und Übersee* 76: 209–246.

Motingea Mangulu, A. (1993b) Le lombuli du Kasai Oriental est-il un dialect mongo? *Afrikanistische Arbeitspapiere* 33: 61–82.

Motingea Mangulu, A. (1994a) Esquisse de la langue des Elíngá: Le parler de lɔsélinga. *Annales Æquatoria* 15: 293–340.

Motingea Mangulu, A. (1994b) Notes sur le parler des Pygmées d'Itɛndɔ (Zone de Kiri/ Maindombe). *Annales Æquatoria* 15: 341–382.

Motingea Mangulu, A. (1995) Esquisse de l'ebango: langue bantoue du groupe C.40. *Afrikanistische Arbeitspapiere* 41: 5–50.

Motingea Mangulu, A. (1996) Étude comparative des langues ngiri de l'entre Ubangi-Zaïre. Université de Leyde, thèse de doctorat.

Motingea Mangulu, A. (2001) Notes sur le parler séŋgɛlɛ de Mbélɔ (Maindombe-R. D. Congo). *Annales Æquatoria* 22: 259–326.

Motingea Mangulu, A. (2001) Note sur la langue des Genja (bantou C.40). *Afrika und Übersee* 84: 101–138, 185–121.

Motingea Mangulu, A. (2002) Aspects du boloki de Monsembe: Le ngala de W. H. Stapleton (Moyen-Congo). *Annales Æquatoria* 23: 285–328.

Motingea Mangulu, A. (2003) Esquisse de l'egbuta, une langue en passe d'extinction au nord du Congo-Kinshasa. *Studies in African Linguistics* 32(2): 25–98.

Motingea Mangulu, A. (2004) Divergences en dialectes tiene (bantou B.81): origine disparate ou perte séparée d'identité linguistique? *Annales Aequatoria* 25: 151–202.

Motingea Mangulu, A. (2005) *Leboale et lebaati: langues bantoues du plateau des Uélé, Afrique Centrale*. Tokyo: ILCAA.

Motingea Mangulu, A. (2008) *Aspects du bongili de la Sangha-Likouala, suivis de l'esquisse du parler énga de Mampoko, Lulonga* Tokyo: ILCAA.

Mouandza, J.-D. (2002) Éléments de description du iyaa (parler bantu du Congo-Brazzaville). Université de Nice, thèse de doctorat.

Mouele, M. (1997) Étude synchronique et diachronique des parlers duma (groupe bantu B.50). Université Lumière Lyon 2, thèse de doctorat.

Mous, M. & A. Breedveld (1986) A Dialectometrical Study of Some Bantu Languages (A. 40 – A. 60) Cameroon. In Guarisma, Gladys & W. J. G. Möhlig (eds.), *La méthode dialectometrique appliquée aux langues africaines*, 177–241. Berlin: Dietrich Reimer.

Mous, M. (2004) *A Grammatical Sketch of Mbugwe: Bantu F34, Tanzania*. Cologne: Rüdiger Köppe.

Mreta, A. Y. (2000) The Nature and Effects of Chasu-Kigweno Contact. In Kahigi, Kulikoyela Kanalwanda, Kihore, Yared Magori & Mous, Maarten (eds.), *Lugha za Tanzania/Languages of Tanzania: Studies Dedicated to the Memory of the Late Prof. C. Maganga* 177–189. Leiden: Research School of Asian, African and Amerindian Studies.

Msanjila, Y. P. (2004) The Future of the Kisafwa Language: A Case Study of Ituha Village in Tanzania. *Journal of Asian and African Studies/Ajia Afuriku gengo bunka kenkyu* 68: 161–172.

Mufwene, S. S. (2009) Kituba, Kileta, or Kikongo? What's in a Name? In de Féral, C. (ed.), *Le nom des langues III. Le nom des langues en Afrique sub-saharienne: pratiques dénominations, catégorisations* (BCILL 124), 211–222. Louvain: Peeters.

Mulaudzi, P. A. (2010) *A Linguistic Description of Language Varieties in Venda.* Saarbrücken: Lambert Academic Publishing.

Müller, E. W. (1964) Die Batwá: Eine kleinwüchsige Jägerkaste bei den Móngo-Ekonda. *Zeitschrift für Ethnologie* 89: 206–215.

Muluwa, J. K. & K. Bostoen (2015) *Lexique comparé des langues bantu du Kwilu (République démocratique du Congo): Français – anglais – 21 langues bantu (B, C, H, K, L)* . Cologne: Rüdiger Köppe.

Mumford, W. B. (1934) The Hehe-Bena-Sangu Peoples of East Africa. *American Anthropologist* 36: 203–222.

Munderi, J. (2009) *Ingingo z'Urufumbira: Rufumbira Grammar.* Kampala: Fountain.

Mutahi, E. K. (1981) *Sound Change and the Classification of the Dialects of Southern Mt. Kenya.* Berlin: Dietrich Reimer. Publications of EK Mutahi 1977 Sound Change and the CLF of the Dialects of Southern Mt Kenya, University of Nairobi, PhD dissertation.

Mvé, P. M. (2013) *Langage et identité chez les Ndambomo du Gabon.* Paris: L'Harmattan.

Mwaniki, D. G. (2014) A Synchronic Survey of Kiembu Dialects. University of Nairobi, MA thesis.

Mzamane, G. I. M. (1948) A Concise Treatise on Phuthi with Special Reference to Its Relationship with Nguni and Sotho. University of South Africa (UNISA), MA thesis.

Mzamane, G. I. M. (1962) A Comparative Phonetic and Morphological Study of the Dialects of Southern Nguni Including Lexical Influences of Non-Bantu Languages. University of South Africa (UNISA), PhD dissertation.

Nash, J. A. (1992) Aspects of Ruwund Grammar. University of Illinois at Urbana-Champaign, PhD dissertation.

Ndinga Oba, A. (2003) *Les langues bantoues du Congo-Brazzaville: étude typologique dea langues du groupe C20 (Mbosi ou Mbochi).* Paris: L'Harmattan. 2 volumes.

Neale, B. (1974) Kenya's Asian Languages. In Whiteley, Wilfred Howell (ed.), *Language in Kenya,* 69–85. Nairobi: Oxford University Press East Africa.

Nicolle, S. (2013) *A Grammar of Digo: A Bantu Language of Kenya and Tanzania.* Dallas: SIL International.

Nkara, J.-P. (2007) Teke Ways of Speaking: An Ethnographic and Sociolinguistic Study. Lancaster University, PhD dissertation.

Nogueira, A. F. (1885) O lu'n kunbi: dialecto do grupo o'n bundo que se falla no interior de Mossamedes. *Boletim da Socidade de Geographia de Lisboa* 5(4): 175–262.

Novotná, J. (2005) A Grammar of Ndamba. Memorial University of Newfoundland, PhD dissertation.

Nsenga Diatwa, I. M. (2009) Descriptive Survey of Some Languages of the Pygmies of the Democratic Republic of Congo. Paper presented at the 6th World Congress of African Linguistics, August 17–21, 2009, Cologne.

Nsuka Nkutsi, F. (1990) Note sur les parlers teke du Zaïre. *Pholia* 5: 147–173.

Ntoya Maselo, E. (2014) Degré de distanciation/rapprochement des variantes du kiyaka: Essai d'analyse lexicostatistique. *Scientia (Revue de sciences, lettres et pédagogie appliquée, Institut Supérieur Pédagogique de Mbanza-Ngungu, Bas-Congo)* 19(2): 89–103.

Nurse, D. & F. Rottland (1991/1992) Sonjo: Description, Classification, History. *Sprache und Geschichte in Afrika (SUGIA)* 12/13: 171–289.

Nurse, D. & G. Philippson (1980) The Bantu Languages of East Africa: A Lexicostatistical Survey. In Polomé, Edgar C. & Hill, C. P. (eds.), *Language in Tanzania*, 26–67. Oxford: Oxford University Press.

Nurse, D. & T. Hinnebusch (1993) *Swahili and Sabaki: A Linguistic History*. Berkeley: University of California Press.

Nurse, D. (1982) Segeju and Daisū: A Case Study of Evidence from Oral Tradition and Comparative Linguistics. *History in Africa* 9: 175–208.

Nurse, D. (1988) The Diachronic Background to the Language Communities of Southwestern Tanzania. *Sprache und Geschichte in Afrika* 9: 15–115.

Nurse, D. (2000) *Inheritance, Contact and Change in Two East African Languages*. Cologne: Rüdiger Köppe.

Nzita, R. (1992) A History of the Batwa in Kisoro District: 1900–1990. Kampala: Makerere University, MA thesis.

Ochwaya-Oluoch, Y. (2003) A Study of the Phonology and Morphology of Lunyala. University of Sheffield, PhD dissertation.

Ohly, R. (1994) The Position of the Subiya Language in Caprivi. *Afrika und Übersee* 77: 105–127.

Ouzilleau, F. (1911) Notes sur les langues des Pygmées de la Sanga: Suivies de vocabulaires. *Revue d'ethnographie et de sociologie* 2: 75–92.

Ownby, C. P. (1985) Early Nguni History: The Linguistic Evidence and Its Correlation With Archaeology and Oral Tradition (South Africa). University of California at Los Angeles, PhD dissertation.

Paluku, A. M. (1998) *Description grammatical du kitalinga (Langue bantu du Nord-Est du Zaïre)*. Munich: LINCOM.

Papstein, R. J. (1978) The Upper Zambezi: A History of the Luvale People, 1000–1900. University of California at Los Angeles, PhD dissertation.

Papstein, R. J. (1994) *The History and Cultural Life of the Mbunda Speaking Peoples*. Cheke Cultural Writers Association.

Paulian, C. (1975) *Le kukuya, langue teke du Congo: phonologie- classes nominales*. Paris: Société des Etudes Linguistiques et Anthropologiques de France.

Paulian, C. (1986) Les parlers yambasa du Cameroun (bantou A.62), dialectométrie lexicale. In Guarisma, Gladys & W. J. G. Möhlig (eds.), *La méthode dialectométrique appliquée aux langues africaines*, 243–279. Berlin: Dietrich Reimer.

Périquet, L. (1915) *Linguistique* (Rapport général sur la mission de délimitation Afrique-Équatoriale Française-Cameroun (1912–1913–1914) III). Paris: Imprimérie Nationale.

Persson, J. A. (1932) *Outlines of Tswa Grammar*, 2nd edition. Cleveland: Central Mission Press.

Petzell, M. & H. Hammarström (2013) Grammatical and Lexical Subclassification of the Morogoro Region, Tanzania. *Nordic Journal of African Studies* 22(3): 129–157.

Petzell, M. (2008) *The Kagulu Language of Tanzania: Grammar, Text and Vocabulary* (East African Languages and Dialects 19). Cologne: Rüdiger Köppe.

Philippson, G. & D. Nurse (2000) Gweno, a Little Known Bantu Language of Northern Tanzania. In Kahigi, K.K., Kihore, Yared M. & Mous, Maarten (eds.), *Lugha za Tanzania/Languages of Tanzania: A Study Dedicated to the Memory of the Late Prof. C. Maganga*, 233–284. Leiden: CNWS.

Philippson, G. (1984) *"Gens des bananeraies": contribution linguistique à l'histoire culturelle des chaga du Kilimanjaro*. Paris: Editions Recherches sur les Civilisations.

Piper, K. (1977) Elemente des Suku: zur Phonologie und Morphologie einer Bantusprache. Universität te Leiden, Doktorarbeit.

Poupon, A. (1918–1919). Étude ethnographique de la tribu Kouyou. *L'Anthropologie* 29: 53–88.

Prata, A. P. (1960) *Gramática da língua macua e seus dialectos*. Cucujães: Escola Tipográfica das Missões pelos Sociedade Portuguesa das Missões Católicas.

Pretorius, J. L. (1975) The Fwe of Eastern Caprivi Zipfel. Stellenbosch University, MA thesis.

Price, F.G.H (1872) A Description of the Quissama Tribe. *Journal of the Anthropological Institute of Great Britain and Ireland* 1(2): 185–193.

Puèch, G. (1988) Augment et préfixe nominal en ngubi. *Pholia* 3: 247–256.

Raharimanantsoa, Ruth (2012) Aspects of Phonology in Eboo-Nzikou. University of Gothenburg, MA thesis.

Raponda Walker, A. (1937) Initiation à l'ébongwé (langage des Négrilles). *Bulletin de la Société des Recherches Congolaises* 23: 129–155.

Reeder, J. (1998) Pagibete, A Northern Bantu Borderlands Language: A Grammatical Sketch (Congo). University of Texas at Arlington, MA thesis.

Reiman, D. (2002) Findings from the Sociolinguistic Survey of the Lolo People. *SIL Electronic Survey Reports* 2002–001: 1–44.

Rénaud, P. (1976) Description phonologique et éléments du morphologie nominale d'une langue Pygmée du Sud-Cameroun: les Bajɛle (Bipindi). Université de la Sorbonne Nouvelle (Paris 3), thèse de doctorat.

Roscoe, J. (1915) *The Northern Bantu: An Account of Some Central African Tribes of the Uganda Protectorate*. Cambridge: Cambridge University Press.

Roth, T. (2011) The Genetic Classification of Wungu. Implications for Bantu Historical Linguistics. Trinity Western University, MA thesis.

Rottland, F. (1993) "Suba": Searching for Linguistic Correlates to an Ethnic Notion. *Afrikanistische Arbeitspapiere* 33: 7–36.

Rubongoya, L.T. (1999) *A Modern Runyoro-Rutooro Grammar*. Cologne: Rüdiger Köppe.

Rurangwa, I. M. (1982) Eléments de description du ngungwel, langue bantoue du Congo. Université Libre de Bruxelles, mémoire de licence.

Salvadó y Cos, F. (1891) *Colección de apuntes preliminares sobre la lengua Benga: Ó sea, instrucción a una gramática de este idioma*. Madrid: A. Pérez Dubrull.

Samarin, W. J. (1984) Bondjo Ethnicity and Colonial Imagination. *Canadian Journal of African Studies/Revue Canadienne des Études Africaines* 18(2): 345–365.

Samarin, W. J. (2014) Swahili in Central African Contact and Colonization. In de Feral, C., M. Kossmann & M. Tosco *In and Out of Africa. Languages in Question in Honour of Robert Nicolai: Volume 2. Language Contact and Language Change in Africa*, 209–249. Louvain: Peeters.

Santandrea, S. (1948) Little Known Tribes of the Bahr El Ghazal. *Sudan Notes and Records* 29: 78–106.

Santandrea, S. (1963) Short Notes on the Bɔdɔ, Huma and Kare Languages. *Sudan Notes and Records* 44: 82–99.

Santandrea, S. (1964) A Note on Kare Grammar. *Sudan Notes and Records* 45: 103–112.

Schadeberg, T. C. (1971) Zur Lautstruktur des Kinga (Tanzania). Marburg: Philipps-Universität, Doktorarbeit.

Schadeberg, T. C. (1989) The Velar Nasal in Nyole (E35). *Annales Aequatoria* 10: 169–179.

Schadeberg, T. C. (1997) De Swahili-talen van Mozambique. *Mededelingen van de Afdeling Letterkunde, Koninklijke Nederlandsche Akademie van Wetenschappen* 60(2): 57–84.

Schadeberg, T. C. & Francisco Ussene Mucanheia (2000) *Ekoti: The Maka or Swahili language of Angoche*. Cologne: Rüdiger Köppe.

Schapera, I. (1952) *The Ethnic Composition of Tswana Tribes*. London: London School of Economics and Political Science.

Schebesta, P. J. (1952) *Les Pygmées du Congo Bel*. Bruxelles: Institut Royal Colonial Belge. 2 vols.

Schebesta, P. (1953) Die Belueli vom Apare (Ituri). *Kongo-Overzee* 19: 357–374.

Schebesta, P. (1966) Die Süd-Nyali oder bafuaNuma am Albertsee. *Wiener Völkerkundliche Mitteilungen* 8: 37–54.

Schoenbrun, D. L. (1994) Great Lakes Bantu: Classification and Settlement Chronology. *Sprache und Geschichte in Afrika* 15: 91–152.

Schumacher, Peter (1949–1950) *Expedition zu den zentralafrikanischen Kivu-Pygmäen*. Bruxelles: Librairie Falk fils/Boekhandel Falk zoon. 2 vols.

Seidel, F. (2005) The Bantu Languages of the Eastern Caprivi: A Dialectometric Analysis and Its Historical and Sociolinguistic Implications. *South African Journal of African Languages* 26(4): 207–242.

Seidel, F. (2008) *A Grammar of Yeyi: A Bantu Language of Southern Africa*. Cologne: Rüdiger Köppe.

Shekleton, C. (1908) The Inhabitants of the "Utwa" or great Lukanga Swamp. *Proceedings and Transactions of the Rhodesien Scientific Association* 7(2): 43–54.

Shrum, J. & M. Shrum (2001) Western Zambezia Language Survey in Mozambique: A Sociolinguistic Survey of the Manyawa, Takwane, Marenje, Kokola and Lolo Peoples of Mozambique. *Working Papers* 2: 1–30.

Sim, R. J. (1977) *A Sociolinguistic Profile of the Mt Kenya Bantu*. Dallas: SIL International.

Sims, A. B. (1886) *Vocabulary of the Kiteke, as Spoken by the Bateke (Batio) and Kindred Tribes on the Upper Congo*. London: Hodder & Stoughton. 2 vols.

Simwinga, J. (2006) The Impact of Language Policy on the Use of Minority Languages in Zambia with Special Reference to Tumbuka and Nkoya. University of Zambia, PhD dissertation.

Skhosana, P. B. (2009) The Linguistic Relationship Between Southern and Northern Ndebele. University of South Africa (UNISA), PhD dissertation.

Smith, E. W. (1907) *A Handbook of the Ila Language (Commonly Called the Seshukulumbwe) Spoken in North-Western Rhodesia, South-Central Africa: Comprising Grammar, Exercises, Specimens of Ila Tales, and Vocabularies*. London: Oxford University Press.

Smith, N. (1995) An Annotated List of Creoles, Pidgins, and Mixed Languages. In Arends, Jacques, Muysken, Pieter & Smith, Norval (eds.), *Pidgins and Creoles: An Introduction*, 331–374. Amsterdam; Philadelphia: John Benjamins.

Spa, J. J. (1973) *Traits et tons en enya: phonologie générative d'une language bantoue* Tervuren: Musée royal de l'Afrique centrale.

Spiss, C. (1904) Kingoni und Kisutu. *Mittheilungen des Seminars für Orientalische Sprachen* 7(3): 270–414.

Stappers, L. (1954) Een Ruund Dialekt: De Taal der Beena Tubeya. *Kongo-Overzee* 20: 369–375.

Stappers, L. (1964) Morfologie van het Songye. Leuven: Katholieke Universiteit, doctoraatsthesis.

Stappers, L. (1971) Esquisse de la langue lengola. *Africana Linguistica* 5: 255–307.

Stappers, L. (1973) *Esquisse de la langue mituku*. Tervuren: Musée royal de l'Afrique centrale.

Stappers, L. (1986) *Kanyok, eine Sprachskizze*. Cologne: Institut für Afrikanistik.

Steere, E. (1876) *Collections for a handbook of the Makonde language* (Unpublished Manuscript). Zanzibar.

Stigand, C. H. [Maj.] (1909) Notes on the Tribes in the Neighbourhood of Fort Manning, Nyassaland. *Journal of the Royal Anthropological Inst. of Great Britain and Ireland* 39: 35–43.

Stirnimann, H. (1983) *Praktische Grammatik der Pangwa-Sprache (SW-Tansania)/ Indaki cha luchovo lwa vaPANGWA*. Schweiz: Universitätsverlag Freiburg.

Ström, E.-M. (2013) The Ndengeleko language of Tanzania. Gothenburg University PhD dissertation.

Takizala, A. (1974) Studies in the Grammar of Kihungan. University of California at San Diego, PhD dissertation.

Tessmann, G. (1923) *Die Bubi auf Fernando Poo: völkerkundliche Eiszelbeschreibung eines westafrikanischen Negerstammes*. Hagen: Folkwang.

Thiry, E. (2004) *Une introduction à ethnohistoire des Hema du Nord (Congo du Nord-Est)*. Tervuren· Musée royal de l'Afrique centrale.

Thomas, J. M. C. & S. Bahuchet (1991) *Les Pygmées Aka: La langue*. Paris: Société des Etudes Linguistiques et Anthropologiques de France.

Thornell, C. (2004) Minioritetsspråket Mpiemos Sociolingvistiska Kontext. *Africa & Asia* 5: 167–191.

Thornell, C. (2009) The Central African Language Cluster Ukhwejo: Geographic Localization and Linguistic Identification. Paper presented at WOCAL, August 2009, Cologne.

Timmermans, P. (1967) Les Lwalwa. *Africa-Tervuren* 13(3/4): 73–90.

Torday, E. & T. Athol Joyce (1907) On the Ethnology of the South-Western Congo Free State. *Journal of the Royal Anthropological Inst. of Great Britain and Ireland* 37: 133–156.

Torday, E. & T. Athol Joyce (1922) *Notes ethnographiques sur les populations habitant les bassins du Kasai et du Kwango oriental: 1. peuplades de la forêt; 2. peuplades des prairies*. Tervuren: Musée royal du Congo belge.

Torrend, J. (1967 [1931]) *An English Vernacular Dictionary of the Bantu-Botatwe Dialects of Northern Rhodesia*. London: Kegan Paul, Trench, Trübner & Co.

Turner, W. Y. (1952) *Tumbuka-Tonga-English Dictionary*. Blantyre: Hetherwick Press.

Tylleskär, T. (1987) Phonologie de la langue sakata (BC34): langue bantoue du Zaïre, parler de Lemvien Nord. Université de la Sorbonne Nouvelle (Paris 3), mémoire de maîtrise.

University of Malawi (2006) *Language Mapping Survey for Northern Malawi*. Zomba: University of Malawi.

University of Malawi (2009) *Language Mapping Survey for the Southern and Central Regions of Malawi*. Zomba: University of Malawi.

van Bulck, G. (1948) *Les recherches linguistiques au Congo Belge: résultats acquis, nouvelles enquêtes à entreprendre*. Bruxelles:Institut Royal Colonial Belge.

van Bulck, V. (1952) Taalstudie op de Bantoetaalgrens. *Kongo-Overzee* 18: 35–49.

van Bulck, G. (1954) *Mission Linguistique (1949–1951)*. Bruxelles: Institut Royal Colonial Belge.

van Bulck, G. (1957) La dialectique des Barundi. *Zaïre: Revue congolaise* 11: 1021–1029.

van der Burgt, J. M. M. (1902) Langue des Watwa (kitwa) = Pygmées. *Mittheilungen des Seminars für Orientalische Sprachen* 79–108.

Vandermeiren, J. (1912) *Grammaire de la langue kiluba-hemba, telle qu'elle est parlee par les Baluba de l'est (Katanga)*. Bruxelles: Ministere des Colonies.

van der Kerken, G. (1944) *L'ethnie mongo*. Bruxelles: Georges van Campenhout.

van der Mohl, A. (1904) Praktische Grammatik der Bantu-Sprache von Tete, einem Dialekt des Unter-Sambesi mit Varianten der Sena-Sprache. *Mittheilungen des Seminars für Orientalische Sprachen* 7(3): 32–85.

van Dyk, P. R. (1960) 'n Studie van Lala: sy fonologie, morfologie en sintaksis. Universiteit van Stellenbosch, doktorale proefskrif.

van den Eynde, K. (1968) *Éléments de grammaire yaka: Phonologie et morphologie flexionelle*. Kinshasa: Publications Universitaires, Université Lovanium.

van de Velde, M. L. O. (2008) *A Grammar of Eton* Berlin: De Gruyter Mouton.

Vanhouteghem, C. (1947) Overzicht der Bantu dialekten van het Distrikt-Lisala. *Aequatoria* 10: 41–50.

van Coillie, G. (1948, 1949) Korte Mbagani-Spraakkunst: De Taal van de "Babindji" in Kasayi. *Kongo-Overzee* 14, 15: 257–279, 172–194.

van Geluwe, H. (1956) *Les Bira et les peuplades limitrophes*. London: International African Institute.

van Geluwe, H. (1960) *Les Bali et les peuplades apparentées (Ndaka-Mbo-Beke-Lika-Budu-Nyari)*. London: International African Institute.

Vansina, J. (1959) *Esquisse de grammaire bushong* Tervuren: Musée royal du Congo belge.

Vansina, J. (2004) *How Societies are Born: Governance in West Central Africa Before 1600*. University of Virginina Press.

van Leynseele, H. (1977) *An Outline of Libinza Grammar*. Leiden University, MA thesis.

Van Otterloo, K. & R. Van Otterloo (2011) *The Kifuliiru Language*. Dallas: SIL International. 2 vols.

Van Otterloo, R. & K. Van Otterloo (1980) *A Sociolinguistic Study of the Bantu Groups of the Kenya Coastal Area*. Dallas: SIL.

van Overbergh, Cyrille (1908) *Les Basonge (Etat Indépendant du Congo)*. Bruxelles: Librairie Albert de Wit pour l'Institut International de Bibliographie.\

van Wyk, E. B. (1969) Die indeling van die Sotho-taalgroep. In *Ethnological and Linguistic Studies in Honour of N.J. van Warmelo: Essays Contributed on the Occasion of His Sixty-Fifth Birthday 28 January 1969*, 169–179. Pretoria: Government Printer;

Ethnological Section of the Departement. of Bantu Administration and Development, South Africa.

Verhulpen, E. (1936a) *Baluba et balubaïsés du Katanga*. Anvers: L'Avenir belge.

Verhulpen, E. (1936b) Comparaison des langues Luba avec des langues voisines. In *Baluba et balubaïsés du Katanga*, 423–443. Anvers: L'Avenir belge.

Vieira-Martinez, C. E. (2006) Building Kimbundu: Language Community Reconsidered in West Central Africa, c. 1500–1750. University of California at Los Angeles, PhD dissertation.

Vinck, H. (1993) John Carrington. *Annales Aequatoria* 14: 565–583.

Vinton, J. E. & V. C. Vinton (2001) A Linguistic Survey of the Chuwabu Language Cluster. *Working Papers* 2: 31–49.

von Rosen, E. (1916) *Träskfolket*. Stockholm: Albert Bonnier.

Vorbichler, A. (1964) Die sprachlichen Beziehungen zwischen den Waldnegern und Pygmäen in der Republik Kongo-Léo. In *Ethnologie* (Comptes rendus du 6ème congrès international des sciences anthropologiques et ethnologiques, Paris 1960 2:2), 85–91. Musée de l'Homme, Université de Paris.

Wagner, G. & L. P. Mair (1970) *The Bantu of Western Kenya, with Special Reference to the Vugusu and Logoli*. London: Oxford University Press; International African Institute.

Walker, J. B. (2013) Comparative Tense and Aspect in the Mara Bantu Languages: Towards a Linguistic History. Trinity Western University, MA thesis.

Walsh, M. T. (1990) The Degere: Forgotten Hunter-Gatherers of the East African Coast. *Cambridge Anthropology* 14(3): 68–81.

Walsh, M. T. & I. N. Swilla (2001) Linguistics in the Corridor: A Review of Research on the Bantu Languages of South-West Tanzania, North-East Zambia, and North Malawi. *Journal of Asian and African Studies* 61: 275–302.

Weeckx, G. (1937, 1938, 1938) La peuplade des Ambundu. *Congo: revue générale de la colonie belge* 18, 19, 19(2, 1, 2): 353–373, 13–35, 150–166.

Wega Simeu, A. (2004a) Etude phonologique du pólrí. Université de Yaoundé I, mémoire de maîtrise.

Wega Simeu, A. (2004b) Étude comparative et lexicostatistique du Polri avec les langues voisines. In *Etude phonologique du pólrí*, 110–113. Yaoundé: University of Yaoundé.

Wega Simeu, A. (2012) Grammaire descriptive du Pólrì: Éléments de phonologie, de morphologie et de syntaxe. Université de Yaoundé I, thèse de doctorat.

Wentzel, P. J. (1983) *Nau dzabaKalanga/A history of the Kalanga*. Pretoria: University of South Africa (UNISA). 3 vols.

Westcott, W. H. (no date) *Concise Grammar of Luna Inkongo*. Bristol: Henry Hill.

Whitehead, J. (1964 [1899]) *Grammar and Dictionary of the Bobangi Language: As Spoken Over a Prt of the Upper Congo, West Central Africa*. Farnborough: Gregg Press. Originally published 1899 for the Baptist Missionary Society's Mission in the Congo Independent State.

Whiteley, W. H. (1950) *Bemba and Related Peoples of Northern Rhodesia/Peoples of the Luapula Valley*. London: Oxford University Press; International African Institute.

Whiteley, W. H. (1966) *A Study of Yao Sentences*. London: Clarendon Press.

Williams, S. T. (2012) Gender Conflict Resolution in Mushunguli. Ohio State University, MA thesis.

Willis, R. G. (1966) *The Fipa and Related Peoples of South-West Tanzania and North-East Zambia*. London: Oxford University Press; International African Institute.

Wilson, J. L. (1849) Comparative Vocabularies of Some Principal Negro Dialects of Africa. *Journal of the American Oriental Society* 1(4): 337–381.

Winter, E. H. (1953) Bwamba: A Structural Analysis of a Patrilineal Society. Harvard University, PhD dissertation.

Winter, J. C. (1980) Internal Classifications of Kilimanjaro Bantu Compared: Towards an East African Dialectometry. In Guarisma, Gladys & Platiel, Susanne (eds.), *Dialectologie et comparatisme en Afrique noire: actes des journées d'étude tenue au Centre de Recherche Pluridisciplinaire du CNRS, Ivry (France), 2–5 juin 1980* (Oralité-documents 2), 101–132. Paris: Société des Etudes Linguistiques et Anthropologiques de France.

Yenguitta, C. (1991) Approche phonologique du ibwiisi (parler bantu du Congo). Brazzaville: Université Marien Ngouabi, mémoire de licence.

Yoneda, N. (2000) A Descriptive Study of Matengo, a Bantu Language of Tanzania, with Focus on the Verbal Structure. Tokyo University of Foreign Studies, PhD dissertation.

Young, T. C. (1923) *Notes on the Speech and History of the Tumbuka-Henga Peoples*. Livingstonia Mission Press.

Young, T. C. (1933) Tribal Intermixture in Northern Nyasaland. *Journal of the Royal Anthropological Institute of Great Britain and Ireland* 63: 1–18.

Ziervogel, D. (1954) *The Eastern Sotho: A Tribal, Historical and Linguistic Survey (with Ethnographical Notes) of the Pai, Kutswe and Pulana Bantu Tribes in the Pilgrim's Rest District of the Transvaal Province*. Pretoria: J.L. van Schaik.

Ziervogel, D. (1959) *A Grammar of Northern Transvaal Ndebele*. Pretoria: J.L. van Schaik.

CHAPTER 3

THE SOUNDS OF THE BANTU LANGUAGES

Ian Maddieson and Bonny Sands

1 INTRODUCTION

Among phoneticians, the Bantu languages have a reputation as not having many interesting features, with the exception of the clicks introduced in some languages of the southern area. Although it's true that many languages within the Bantu group are phonetically quite similar to each other, there is considerably more diversity in their phonetic patterns than is often believed. Some of this diversity may be disguised by the widespread use of simplifying transcriptions and orthographies which normalise away variation within and between languages or underrepresent distinctions. Part of the aim of the present chapter is therefore to draw greater attention to this diversity. Since the Bantu languages have received very extensive historical analysis, this group of languages also provides a fertile field for examining inferences about the nature of phonetic sound change. The phonetic differences which exist between closely related languages provide opportunities for testing theories about phonological organisation.

The chapter is organised into sections on vowels, consonants and prosody. There are many important interactions between these three aspects of phonetic structure and some of these will be taken up at the point where it seems appropriate to do so. For example, the Bantu languages provide very striking examples of vowels affecting consonant realisations, particularly considered diachronically, and the nature of particular segments also has significant impacts on prosodic quantity and on tonal patterns. Special attention is paid to consonants with complex articulations, including clicks and the so-called "whistling fricatives." It is hoped that the brief discussions of selected issues here will encourage more attention to be paid to phonetic aspects of these languages.

2 VOWELS

2.1 Vowel spacing

The majority of Bantu languages – with some notable exceptions, particularly in the North-West – have simple-looking systems of five or seven vowels in which the expected relationships between the features of vowel height, backness and rounding hold. That is, the back non-low vowels are rounded, and the low and front vowels are unrounded. The vowels of the five-vowel systems are therefore usually transcribed as /i e a o u/ and the seven-vowel systems are most often transcribed as /i e ɛ a ɔ o u/ (Hyman 1999). However, these standardised transcriptions may disguise significant differences between languages, especially with respect to the nature of the vowels written /e/ and /o/.

In the five-vowel system of Xhosa S41, for example, /e o/ are genuinely mid in character. The positions of vowels in an acoustic space are often shown by plotting values of the
first two formants. Readers unfamiliar with acoustic analysis might see Ladefoged (2000)
for an introduction to the concept of a formant. The mean formant values for Xhosa S41
vowels given by Roux and Holtzhausen (1989) are plotted in this way in Figure 3.1. In
this and following figures of the same type, the origin of the axes is in the upper right,
with first formant (F1) values increasing down from the origin, and second formant (F2)
values increasing to the left. The distances along the axes are scaled to reflect auditory/
perceptual intervals; F2 is plotted using a logarithmic scale. This kind of display closely
parallels the traditional auditorily based vowel space based on perceived "height" and
"backness" values used, for example, in the IPA Handbook (1999), but has the advantage
of being based on verifiable measurement. In Figure 3.1, it can be seen that in Xhosa
S41 /e o/ are located almost equidistant from the high vowels /i u/ and the low vowel /a/.
There is a raising process in Xhosa S41, which results in higher variants of /e o/ when
/i u/ occur in the next syllable. The means for /e o/ plotted here do not include tokens of
these raised variants.

Compare the spacing of Xhosa vowels with those of Kalanga S16, shown in Figure 3.2.
The maxima in Figure 3.2 are higher compared to Figure 3.1 due to male/female differences in formant range. Note that there are different ways to normalise vowels across

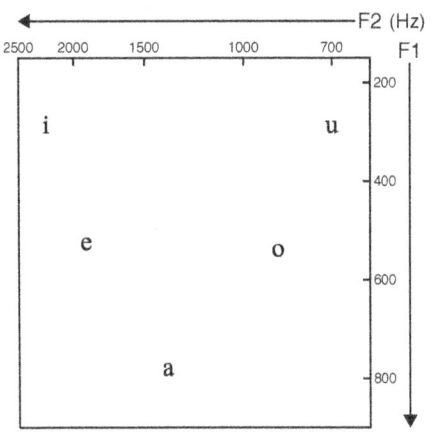

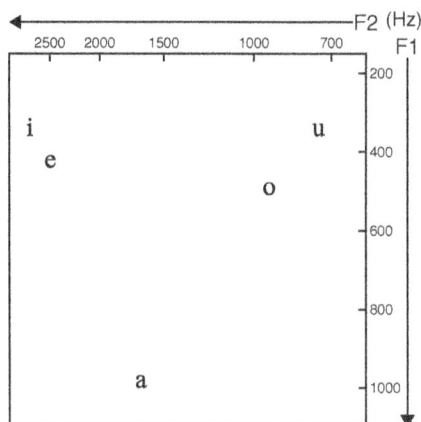

FIGURE 3.1 XHOSA S41 VOWEL FORMANT MEANS (ROUX & HOLTZHAUSEN 1989). EACH POINT REPRESENTS THE AVERAGE OF MEASUREMENTS OF AT LEAST 30 TOKENS OF THE VOWEL FROM ONE MALE SPEAKER READING A TEXT.

FIGURE 3.2 KALANGA S16 VOWEL FORMANT MEANS ACCORD- ING TO MEASUREMENTS BY DONE THE FIRST AUTHOR. EACH POINT REPRESENTS THE AVERAGE OF AT LEAST 28 TOKENS OF THE VOWEL IN PENULTIMATE POSITION IN A WORD LIST SPOKEN BY A FEMALE SPEAKER.

speakers; eight different methods are evaluated by Wissing and Pienaar (2014) for a corpus of Southern Sotho S33 vowels, for instance. In the case of Kalanga S16, the "mid" vowels /e o/ are relatively close to the high vowels /i u/ and far from /a/. As a rough rule of thumb, vowels with a first formant lower than 400 Hz may be considered high vowels in a female voice. On this basis these particular vowels would not quite justify being considered high, but they are clearly markedly higher than those of Xhosa S41.

There have been relatively few acoustic studies of other Bantu five-vowel systems, but Swahili G42 (Nchimbi 1997) has a pattern similar to Xhosa S41, while Bemba M42 (Hamann & Kula 2015), Ndebele S44 and the Zezuru variety of Shona S12 (Manuel 1990) have a pattern similar to Kalanga. The distribution seen in Xhosa S41 or Swahili G42 is similar to that most typically found cross-linguistically in five-vowel systems transcribed /i e a o u/, such as Spanish, Hadza or Hawaiian. This is also the pattern predicted by computational models of vowel system structure from Liljencrants and Lindblom (1972) to Schwartz et al. (1997) based on the principle that vowels should be expected to be roughly equally dispersed in a space defined by the major formant resonances. In Kalanga S16, on the other hand, the vowels are crowded into the upper part of the vowel space, with the front pair in particular being very close together.

A Bantu five-vowel system consisting of /i ɛ a ɔ u/ has been described for Soga JE16 (Nabirye et al. 2016) and Fwe K402 (Gunnink 2016). Means of Fwe vowel formants are shown in in Figure 3.3. The high vowels /i/ and/u/ are lower and more centralised than those in Xhosa S41 and Kalanga S16 and could be transcribed [ɪ] and [ʊ], respectively. Both Soga JE16 and Fwe K402 have a vowel length contrast. Impressionistically, there appear to be no differences in vowel quality between pairs that differ in length in the two languages.

Variations in the structure of seven-vowel systems occur which are similar to those of the five-vowel systems. A plot of vowel distribution in Nyamwezi F22 is shown in Figure 3.4. In this language, the vowels are to a large degree placed where they might be expected, given a respect for dispersion principles. This is particularly apparent for the front vowels, which are equally spaced from each other. The word list available for measurement included a more balanced sample of front than of back vowels, and the back vowels are probably in reality more separated than this plot indicates. These data suggest that transcription of this vowel set as [i e ɛ a ɔ o u], as in Figure 3.4, is appropriate rather than the [i ɪ e a ʊ o u] preferred by Maganga and Schadeberg (1992). For Sukuma F21, Batibo (1985) also provides acoustic evidence for a relatively wide separation of the seven vowels, with /e ɛ ɔ o/ all being clearly mid vowels.

The relationship between the seven vowels of Vove B305 is notably different, as demonstrated in Figure 3.5. Here a pair of vowels in the front and a pair of vowel in the back have such low values of F1 that they are all appropriately considered to be high vowels. The means are 248 Hz for /i/, 313 Hz for /ɪ/, 277 Hz for /u/, and 334 Hz for /ʊ/. The next lower vowels are markedly lower. In addition we may note that the front pair /i/ and /ɪ/ and the back pair /u/ and /ʊ/ have F2 values which are identical or nearly so, whereas Nyamwezi F22 /e o/ have F2 values intermediate between the higher and lower vowels in the system.

The acoustic phonetic characteristics of the eight- and nine-vowel systems of some Mbam languages (A40+A60) are detailed in Boyd (2015). Mande A46, Nen A44 and Gunu A622 all have an eight-vowel system with [-ATR] /ɪ a ɔ ʊ/ and [+ATR] /i ə o u/. Nine-vowel languages in the Mbam group, such as Mmala A62B and Baca A621, have a contrast between /e/ and /ɛ/ not found in the eight-vowel systems. The F1 averages of /ɪ/ and /ʊ/ in Mbam languages is typically higher than that of /e/ and /o/. The [-ATR] high

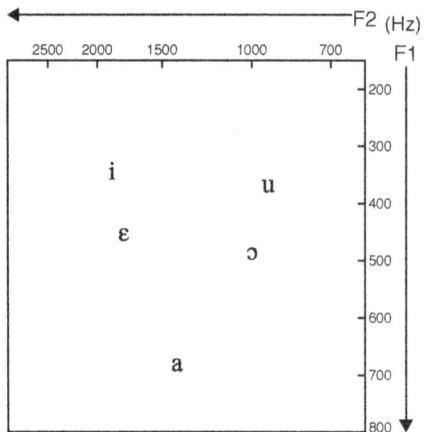

FIGURE 3.3 FWE VOWEL FORMANT MEANS ACCORDING TO MEASUREMENTS BY THE SECOND AUTHOR ON RECORDINGS MADE AVAILABLE BY HILDE GUNNINK. EACH POINT REPRESENTS THE MEAN OF BETWEEN SEVEN AND 27 TOKENS OF UNREDUCED STEM-INITIAL VOWELS SPOKEN BY A MALE SPEAKER.

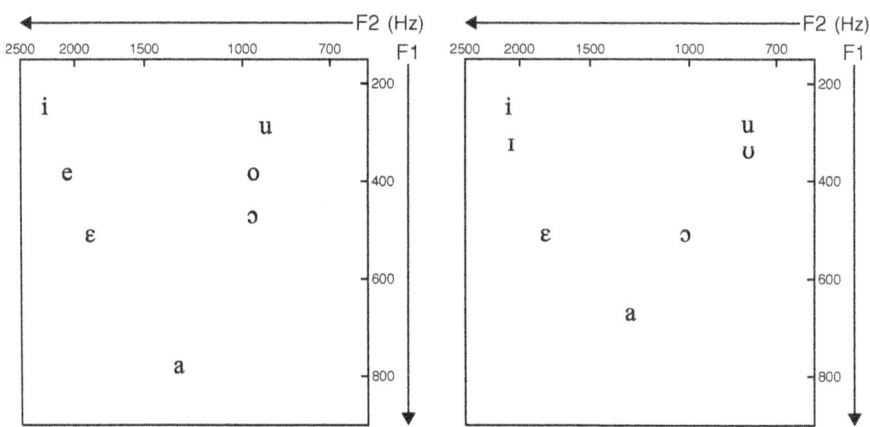

FIGURE 3.4 NYAMWEZI F22 VOWEL FORMANT MEANS ACCORDING TO MEASUREMENTS BY THE FIRST AUTHOR. EACH POINT REPRESENTS THE MEAN OF BETWEEN NINE AND 23 TOKENS OF UNREDUCED FINAL OR PENULTIMATE VOWELS IN A WORD LIST SPOKEN BY A MALE SPEAKER.

FIGURE 3.5 VOVE B305 VOWEL FORMANT MEANS ACCORDING TO MEASUREMENTS BY THE FIRST AUTHOR ON A RECORDING MADE BY JEAN-MARIE HOMBERT, MADE AVAILABLE BY LOLKE VAN DER VEEN. EACH POINT REPRESENTS THE MEAN OF 20 OR 30 MEASUREMENTS ON MINIMAL SETS OF WORDS DIFFERING ONLY IN THE PENULTIMATE VOWEL, SPOKEN BY A MALE SPEAKER.

vowels may thus be misinterpreted as being lower than the [+ATR] mid vowels, but the high F1 values may be instead attributed to a retracted tongue root position. Another nine-vowel Bantu language is Liko D201 (De Wit 2015: 45).

2.2 Tongue root position in harmony systems

Co-occurrence restrictions of a harmonic nature between vowels, very typical of sub-Saharan African languages, are quite commonly found in Bantu languages, though often limited in extent, e.g., only applying in certain morphological contexts, such as between verb roots and extensions. Bantu vowel harmony constraints do not seem to be a survival of an older Benue-Congo or even Niger-Congo harmony (Stewart 2000), but to be mostly more or less local innovations with diverse patterns of implementation (Hyman 1999). Vowel height, backness and rounding can all be factors in control of Bantu harmony. Vowel harmony in Africa often involves the independent use of pharyngeal cavity size, that is, adjustments of pharynx volume which cannot be accounted for as a function of the height and frontness of the tongue body (see Ladefoged & Maddieson 1996 for discussion). This is usually discussed as a contrast between advanced and retracted (or neutral) tongue root position, i.e., ± ATR. An interesting issue is therefore whether the Bantu languages, particularly those with seven or more vowels, make use of the ATR feature in this phonetic sense. Phonologists often use [ATR] as a diacritic feature, even to distinguish pairs of vowels such as i/ɪ in English 'beat'/'bit' where tongue root position is not the phonetic mechanism involved. The question of the role of ATR interacts with the question of the nature of the high vowels, as the *super-high/*high contrast might have been an expression of an ATR contrast or transformed into one in daughter languages.

It is difficult to be certain that ATR contrasts exist in a language unless direct articulatory data on the vocal tract configuration during vowel production is available. There are very few studies of this type available so far for Bantu languages, but one data set is shown in Figure 3.6. These pictures are magnetic resonance images of sustained vowels produced by Pither Medjo Mvé, a speaker of the Bitam variety of Fang A75 (Demolin *et al.* 1992). Figure 3.6 shows very clearly that independent tongue root adjustment does not contribute to the distinctions between any members of the front vowel set /i e ɛ/, nor the back vowel set /u o ɔ/. The pharynx width, measured as the distance from the tongue root surface to the back wall of the pharynx at the height of the top of the epiglottis, in /e/ is intermediate between that in /i/ and /ɛ/, and that in /o/ is intermediate between /u/ and /ɔ/. It can be predicted from tongue body position: front vowels have wider pharynx than back vowels, lower vowels have narrower pharynx than higher vowels. The three front vowels and the three back vowels can therefore be distinguished one from another solely by height.

The Bitam variety of Fang A75 has eight vowels and seven peripheral vowels, plus mid central /ə/ (Medjo Mvé 1997). An acoustic plot of these vowels is given in Figure 3.7. Note particularly the slope of a line connecting the back vowels which points roughly to the position of the central vowel /a/, similar to that seen in Figure 3.1 and Figure 3.2, and attributable to the fact that F1 and F2 frequencies co-vary in these vowels. This pattern is typical of that found in vowel systems where the back series is distinguished by degrees of height with no other factors being significantly involved. In this variety, lexical stems are marked by a strong tendency for V1 and V2 to be identical except if V2 is /a/, when /i ə a o u/ are all relatively common as V1, but /e ɛ ɔ/ are not. Note that as many PB

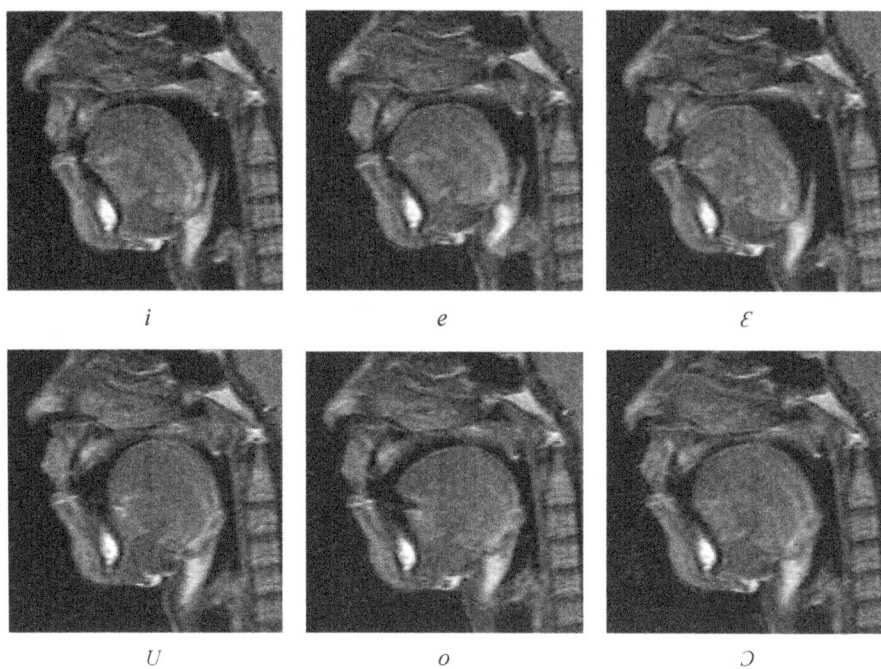

FIGURE 3.6 ARTICULATORY POSITIONS OF SIX OF THE VOWELS OF FANG A75 (VARIETY OF BITAM). TOP ROW, FRONT VOWELS /i/, /e/, /ɛ/; BOTTOM ROW, BACK VOWELS /u/, /o/, /ɔ/.

Source: Mid-sagittal MRI scans of isolated vowels, made available by Didier Demolin.

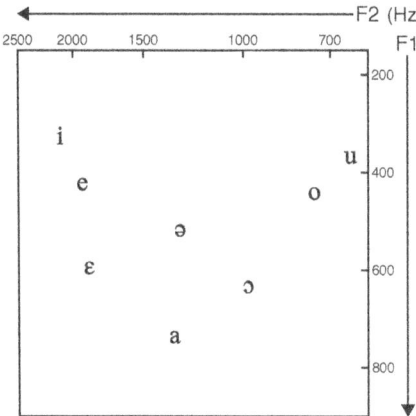

FIGURE 3.7 FANG A75 VOWEL FORMANT MEANS. EACH POINT REPRESENTS THE MEAN OF SIX MEASUREMENTS, THREE OF ISOLATED VOWEL TOKENS, PLUS THREE TOKENS IN FINAL VOWELS IN /alV/ NONSENSE WORDS.

Source: Recording by Pither Medjo Mvé made available by D. Demolin; measurements by the first author.

*CVCV items have become monosyllabic in Fang, the V2 in these cases is often not the *V2 of the reconstructed form. This pattern of co-occurrences is not one which suggests a phonological role for ATR.

Nande JD42 contrasts with Bitam Fang A75 in that it uses ATR for phonetic distinctions. Ultrasound images clearly show differences in tongue root position across vowel pairs (Gick 2002, Gick *et al.* 2006), as seen for the ATR /ẹ/ and RTR /e̱/ vowels in parts a) and b) of Figure 3.8. Note that the back of the mouth is found on the left side of an ultrasound image, but on the right side of an MRI image. The ATR vowel /ẹ/ and the RTR vowel /e̱/ differ both in the shape of the tongue body and in the amount of tongue root retraction, which can be estimated by the volume of tongue mass which occurs to the left of the white dotted line. The RTR vowel shows more tongue root retraction than the ATR vowel while the ATR vowel (on top) shows more of a bunched tongue shape.

The ATR/RTR contrast in Nande JD42 is also suggested by the harmonic behaviour and acoustic characteristics of vowels. Mean formant values of the 10 surface vowels for one speaker are plotted in Figure 3.9. Harmonically related pairs are noted by the use of the same symbol with and without a -ATR diacritic. In each case the putatively [-ATR] vowel has a substantially higher first formant (hence a lower position on the chart) than its harmonic counterpart. Most strikingly, the "high" vowels /i u/ are placed lower than the "mid" vowels /e o/. Narrowing the pharynx raises the first formant, other things being equal. The pair /u u̱/ where F2 is the same are thus quite likely (almost) solely different in pharynx width. The other back vowel pair /o o̱/ shows a smaller than expected F2 difference given the size of the difference between their first formants; a substantial pharynx width difference coupled with a degree of opening of the oral constriction may be inferred. Note that a sloping line can be fit to the vowel set /u o a/ and a second roughly parallel lower one to the set /u̱ o̱ a̱/, but a straight line cannot be fit to the set /u o o̱/ as is possible for Fang A75 /u o ɔ/. The pattern for the front vowels suggests a greater interaction of the major features of vowel height and backness with pharynx width. This is not surprising, as retracting the tongue root is more likely to pull the tongue back and down when the tongue body position is front.

We may now revisit the Kalanga S16 and Vove B305 high vowels in Figure 3.2 and Figure 3.5. The members of the high vowel pairs /i ɪ/ and /u ʊ/ in Vove B305 have virtually the same second formant values as each other and differ only in F1. The Kalanga S16 vowel pairs transcribed /i e/ and /u o/, which are acoustically equally as high as the Vove B305 pairs, differ in both F1 and F2. Plausibly, the Vove B305 vowel pairs differ phonetically in pharynx width, which is consistent with the auditory impression they create, while the Kalanga S16 pairs differ in height and to a lesser degree in backness, which is consistent with the auditory impression they create. Although these acoustic measurements are suggestive, it should be borne in mind that inferences from simple formant measures concerning vowel articulation must be made with caution.

Acoustic evidence for tongue root retraction of vowels in several Bantu languages has been provided by Starwalt (2008). She found that [-ATR] vowels with a constricted voice quality tend to have higher center of gravity values, while [+ATR] vowels with a "hollow" quality have lower center of gravity values (Starwalt 2008: 441). In her study, F1, B1 (F1 bandwidth), center of gravity and A1-A2 (relative amplitudes of F1 and F2) help distinguish vowel pairs that differ in [ATR] value to varying degrees depending on the vowel pair and speaker. This suggests that speakers of the same language may differ in the degree to which they use tongue root position to contrast vowels that are described as differing in the phonological feature [ATR].

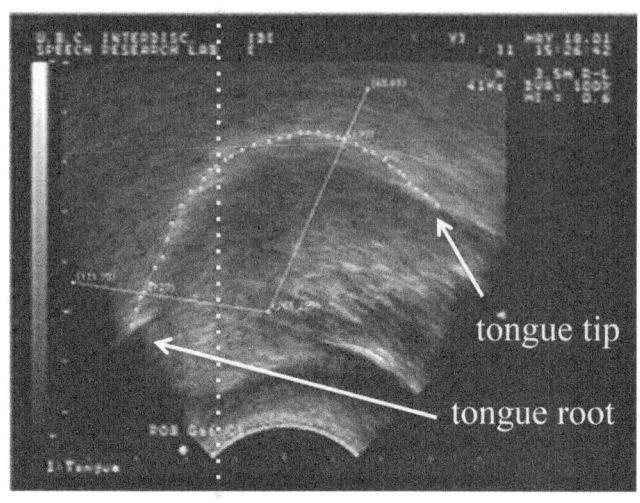

a

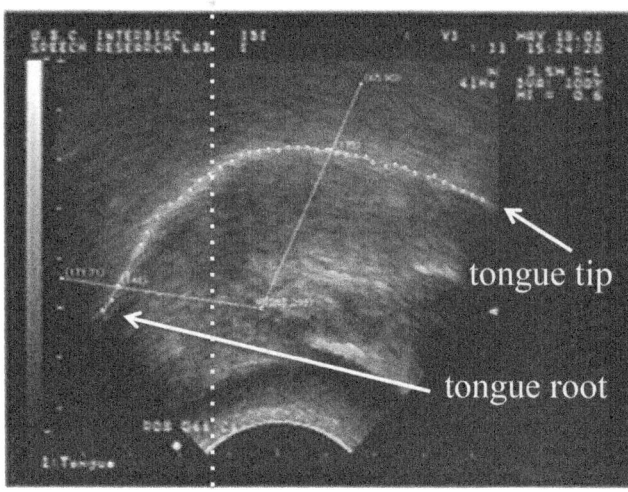

b

FIGURE 3.8 ULTRASOUND IMAGES OF NANDE JD42 VOWELS A) ATR /ẹ/ B) RTR /ɛ̣/, TAKEN ALONG THE MID-SAGITTAL PLANE. NOTE THAT THE TONGUE TIP IS ON THE RIGHT AND THE TONGUE ROOT ON THE LEFT, THE REVERSE OF THE IMAGES IN **FIGURE 3.6**. THE TONGUE SURFACE APPEARS AS A CURVED WHITE LINE. A VERTICAL WHITE DOTTED LINE HAS BEEN ADDED TO FACILITATE COMPARISON BETWEEN THE TWO IMAGES.

Source: Images made available by Bryan Gick (cf. Gick 2002: 118).

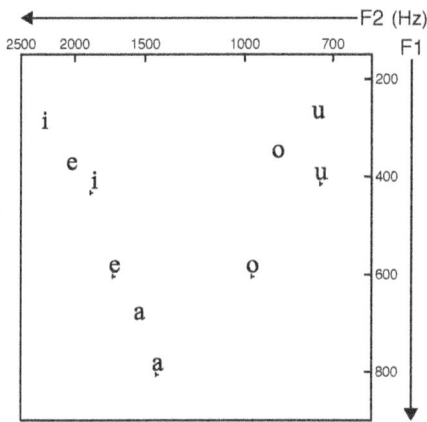

FIGURE 3.9 NANDE JD42 VOWEL FORMANT MEANS. EACH POINT REPRESENTS THE MEAN OF BETWEEN SIX AND 21 TOKENS OF PHONETICALLY LONG VOWELS IN PENULTIMATE POSITION IN WORDS SPOKEN BY A MALE SPEAKER.

Source: Data from Ngessimo Mutaka; measurements by the first author.

2.3 Pharyngealised vowels

Pharyngealised vowels /iˤ uˤ ɛˤ oˤ aˤ/ have been reported for Kwasio A81 (Duke & Martin 2012). These vowels are produced with a retracted tongue root, causing a constriction in the upper pharynx. The upper pharyngealised vowels of Kwasio A81 do not have the harsh voice quality associated with lower pharyngealised vowels, i.e., epiglottalised or aryepiglottalised vowels, as found in Tuu, Kx'a and Khoe languages of southern Africa (cf. Miller 2007, Miller *et al.* 2009a). Pharyngealised vowels occur in a few other Bantu languages including Gyele A801 (Blench 2011) and Jarawan Bantu (Rueck *et al.* 2009, cited in Blench 2015).

Kwasio A81 pharyngealised vowels differ significantly in vowel quality compared to their non-pharyngealised counterparts. For instance, /uˤ/ and /oˤ/ are produced as the lower and more centralised vowels [θˤ] and [ɵˤ], respectively (Duke & Martin 2012: 220). Figure 3.10 shows a typical example of /o/ in the word /ko/ 'to go'; /o/ has a low F2 (below 1000 Hz). The pharyngealised /oˤ/ in Figure 3.11 in the word /koˤ/ 'avarice' has a higher F2 (above 1000 Hz), and the higher formants are much more prominent than those of /o/. The pharyngealised vowel is longer than the plain vowel, which reflects the origin of the pharyngealisation from a reduced velar stops in C_2 position in roots of the shape C_1VC_2V (Duke & Martin 2012: 220). Pharyngeals have developed from velars in other Niger-Congo languages. In the Bantoid language Mundabli (Voll 2012: 535), pharyngealised vowels correspond to final /k/ and /ʔ/ in cognates in its close relative Mufu. In the Gur language Minyanka, the pharyngeal fricative [ʕ] is a variant of /g/ (Dombrowsky-Hahn 1999: 52).

Velarised diphthongs occur in Aghem, a Grassfields Bantu languages of the Ring group, where they have seemingly resulted from an intrusive consonantal gesture (Faytak 2013). This differs from Kwasio A81 pharyngealisation which likely results from the reduction of a consonant.

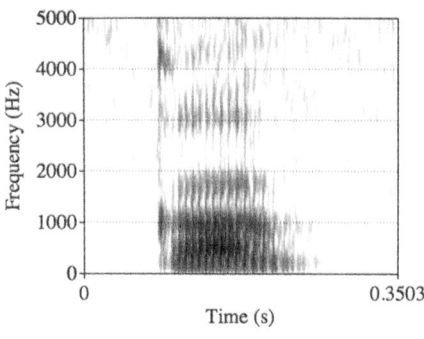
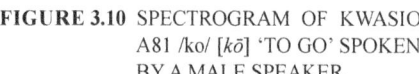

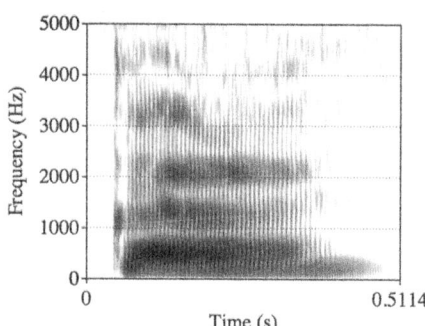

FIGURE 3.10 SPECTROGRAM OF KWASIO A81 /ko/ [kō] 'TO GO' SPOKEN BY A MALE SPEAKER.

FIGURE 3.11 SPECTROGRAM OF KWASIO A81 /ko⁷/ [kə̄ˤ] 'AVARICE' SPOKEN BY A MALE SPEAKER.

Source: Recording made available by Daniel Duke and Marieke Martin.

Source: Recording made available by Daniel Duke and Marieke Martin.

2.4 Nasalised vowels

Nasal vowels are not particularly common in the Bantu languages, but are found in certain mostly western areas, for example in Ngungwel B72a of the Teke group (Paulian 1994), in Umbundu R11 (Schadeberg 1982), in Gyele A801 (Renaud 1976) and in a few words in the Bitam variety of Fang A75 (Medjo Mvé 1997). As is generally the case cross-linguistically, there are fewer nasal vowels than oral ones, at least in lexical stems. In Ngungwel B72a, there are three oral and three nasal vowels in prefixes [e a o ẽ ã õ]. Lexical stems have a system of seven oral vowels but only five nasal vowels. Nasal vowels in the stem are reported to have the qualities [ĩ ẽ ã ɔ̃ ũ] and to be invariably long. The article of Paulian (1994) does include a few words with short nasalised vowels in stems, but these may be misprints. There are thus seven *phonetic* qualities among the nasalised vowels, but no contrast between all seven in any environment. Examples are given in Table 3.1.

TABLE 3.1 NASAL VOWELS IN NGUNGWEL

Vowel	Word	Gloss
ĩ: (stem)	dzĩ̃	tooth
ẽ (prefix)	ẽbèl	kola nut trees
ẽ̃: (stem)	ntsíẽ̃ẽ̃	horn
ã (prefix)	ãpăb	wing
ã: (stem)	bã́ã̂	children
ɔ̃: (stem)	ekúɔ̃ɔ̃	broom
õ (prefix)	ṍkáa	woman
ũ: (stem)	ŋkũ̆ũ̆	name

A role for vowel nasalisation in the transmission of nasal consonant harmony across intervening vowels seems likely in the history of Bantu (Greenberg 1951, Hyman 1995).

2.5 Vowel length

Vowel length contrasts occur in some Bantu languages, which may or may not be accompanied by changes in vowel quality and/or various processes of vowel lengthening (cf. Sections 6-7). Bemba M42 short vowels /i e a o u/ tend to be lax compared to their long vowel counterparts /i: e: a: o: u:/ (Hamann & Kula 2015): short high and mid vowels tend to be lower and more centralised than long ones, while /a/ is higher than /a:/. Rwanda JD61 contrasts long and short vowels yet also has vowel lengthening before NC as well as after a consonant-glide sequence (Myers 2005). Another language with a vowel length contrast, Vili H12, lengthens vowels before liquids (in the environment /C_L/) and before nasals /C_N/ (Roux & Ndinga-Koumba-Binza 2011), but not before NC (Ndinga-Koumba-Binza 2011).

2.6 Super-close vowels

Fricated vowels occur in Kom and Oku, two Grassfields Bantu languages of the central Ring group (Faytak 2014, Faytak & Merrill 2014), as well as in several Bantoid languages of the northern Cameroon Grassfields (Faytak 2015). These vowels bring to mind the super-high or super-close vowels /i̧ u̧/ used by Meeussen (1967, 1969) and Guthrie (1967, 1970a, 1970b, 1971) and notated as /î û/ by Meinhof (1899), in addition to "normal" high /i u/. Super-close vowels were reconstructed in order to account for the set of sound changes known as "Bantu Spirantization," but recent reconstructions have abandoned this explanation (Schadeberg 1995, Bastin et al. 2002, Bostoen 2008). Faytak (2014) reconstructs back vowels *u, *ʊ, *o for the Central Ring group of Grassfields Bantu languages. He argues that, in two of these languages, Kom and Oku, *ʊ raised to /u/ and *u became fricated, sometimes occurring with a schwa [ə] offglide (Faytak & Merrill 2014).

Matumbi P13 has been claimed to have super-close vowels /i̧ u̧/ (Odden 1996: 5), but the description of the contrast between /i̧ u̧/ and /i u/ as being "roughly equivalent to the contrast between [ɪ], [ʊ] and [i], [u]" suggests that the vowels likely contrast tongue root position (ATR) rather than tongue height.

A particularly rare phenomenon reported in Hendo C82 involves the class 5 prefix, which is actually the reflex of the Proto-Bantu augment *di- followed by the noun prefix *i̧- (cf. De Blois 1970: 155). Demolin et al. (2002) describe it as an unreleased voiced palatal implosive [ʄ] before a voiceless stop or affricate, e.g., in [ʄpaka] 'moth.' MRI scans indicate that this segment is appropriately viewed as a hyperarticulation of the vowel /i/.

3 CONSONANTS

3.1 Consonant overview

Most Bantu languages are reported as having two series of plosives, voiced and voiceless, and this follows the Proto-Bantu reconstruction of Meeussen (1967). Except in post-nasal environments and sometimes before his reconstructed super-high vowels, the reconstructed voiced plosives most commonly correspond to voiced continuants of one type or another or to implosives in the modern languages. An alternation of some kind is probably to be reconstructed to an early stage, possible even pre-Bantu. Bantu orthographies usually do not indicate these alternations, unless subsequent developments have created a contrast between, say, /b/ and /ɓ/, or /b/ and /β/. This illustrates one instance where the occurrence of cross-linguistically less common phonetic segments may be disguised by notational practices. Aspiration is a contrastive property of voiceless stops (and affricates) in some languages where it is often a reflex of an earlier voiceless prenasalised stop (cf. Kerremans 1980).

Most Bantu languages have a full set of nasals at each place of articulation where a stop or affricate appears, but often intricate (morpho)phonological processes govern nasal/oral alternations and syllabification and other prosodic processes concerning nasals. Most of the languages have relatively limited sets of fricatives of the cross-linguistically common types, although lateral fricatives (and affricates) have developed in or been borrowed into a number of the southern languages, such as Sotho-Tswana S30, Xhosa S41 and Zulu S42. Particularly striking in this connection is the velar ejective lateral affricate [kʟ̝'] of Zulu S42 (cf. Naidoo 2007), which is auditorily reminiscent of a lateral click. There is often only one contrastive liquid, i.e., /l/, /ɾ/ or /r/, though Chaga E60 is among those with more (Davey et al. 1982, Philippson & Montlahuc 2003). High front vowels condition tap allophones of /l/ in Ganda JE15 (Myers 2015) and Tsonga S53 (Bennett & Lee 2015), and of /r/ in the Washili variety of Ngazidja G44a (Patin 2013). The two vocoid approximants /j/ and /w/ occur in many languages, often alternating with high vowels /i u/.

Phonetic studies of labial consonants include the study of plain and prenasalised bilabial trills /ʙ ᵐʙ/ in Medumba, a Narrow Grassfields language, by Olson and Meynadier (2015). Differences in lip posture appear to enhance the contrast between labio-dental /f v/ and labial fricatives /ɸ β/ in Kwangali K33 and in Manyo K332 (Ladefoged 1990). Labial flaps reportedly occur in various Southern Bantu languages, such as Nyanja N31a, Korekore S11, Manyika S13, Ndau S15 and Kalanga S16, and they may contrast with the labio-dental approximant /ʋ/ in the Zezuru S12 variety of Shona (Olson & Hajek 1999).

Although most Bantu languages use only one coronal (typically alveolar) and one dorsal (velar) place of articulation, contrasts between dental and alveolar places are found in several languages, and contrasts between velars and uvulars are found in Kgalagari S311 (Dickens 1987, Monaka 2001, 2005). Tswana S31 has a voiceless uvular affricate and voiceless uvular fricative (Bennett et al. 2016). A voiced pharyngeal fricative /ʕ/ is found in Nyokon A45 (Lovestrand 2011). Consonant gemination has developed through internal processes in languages such as Ganda JE15 (Clements 1986) and by contact with Cushitic languages in Ilwana E701 (Nurse 1994).

In the rest of this section, three of the particular issues of phonetic interest are discussed: the dental/alveolar place contrast, the possible occurrence of articulatorily complex consonants, and the nature of the so-called 'whistling fricatives.' Longer sections of the chapter will be devoted to aspects of laryngeal action in consonants, to the description of clicks and their distribution in Bantu, and to some of the interesting aspects of nasality which occur in these languages.

3.2 Dental vs. alveolar place of articulation

The phonetic realisation of dental and alveolar consonants is dependent on the airstream mechanism. Dental and alveolar implosives and clicks may display constriction patterns that differ from those of corresponding pulmonic stops. Electropalatography (EPG) of Mvita Swahili G42b shows that implosive /ɗ/ has a more retracted occlusion than

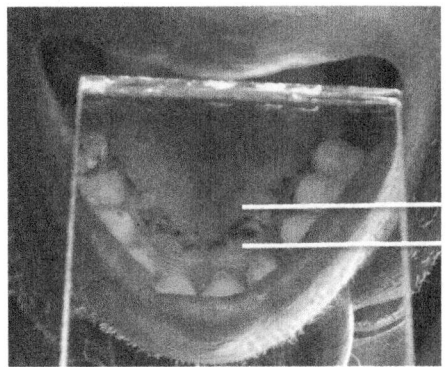 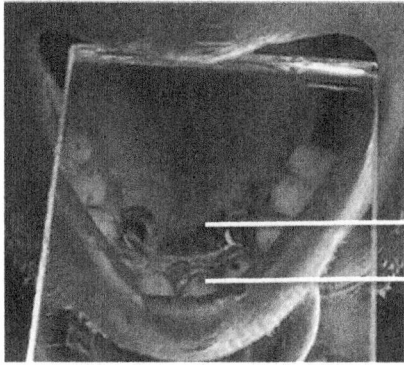

FIGURE 3.12 PALATOGRAM OF [ana] SPOKEN BY A SOGA JE16 SPEAKER. THE WHITE HORIZONTAL LINES INDICATE THE WIDTH OF THE MAXIMUM CONSTRICTION.

Source: Image made available by Gilles-Maurice de Schryver (cf. Nabirye *et al.* 2016: 223).

FIGURE 3.13 PALATOGRAM OF [aɲa] SPOKEN BY A SOGA JE16 SPEAKER. THE WHITE HORIZONTAL LINES INDICATE THE WIDTH OF THE MAXIMUM CONS-TRICTION.

Source: Image made available by Gilles-Maurice de Schryver (cf. Nabirye *et al.* 2016: 223).

pulmonic /t/ and /ɖ/ has a shorter occlusion than /ʈ/ (Hayward *et al.* 1989: 54). The (post: 303) alveolar /ǃ/ clicks in Zulu S42 (Thomas-Vilakati 2010) and Xhosa S41 (Doke 1926: 303) are retracted in comparison to pulmonic alveolar consonants such as /tʰ/ and /s/.

The typical pattern for dental/alveolar contrasts is that the dentals are laminal while the alveolars are apical. Soga JE16 follows this pattern, as shown in the palatograms in Figure 3.12 and Figure 3.13 (Nabirye *et al.* 2016). The width of the constriction for the apical alveolar nasal in Figure 3.12 is narrower than the width of the constriction of the laminal dental in Figure 3.13, as indicated by the positioning of the horizontal white lines superimposed on each photograph.

3.3 Complex or simple consonants?

Doubly articulated labial-velar stops (and nasals) are found almost exclusively in the languages of Africa, but they occur in only relatively few of the Bantu languages, including Londo A11 (Kuperus 1985), "Sawabantu" languages of Guthrie's groups A10–20–30 (Mutaka & Ebobissé 1996–1997), Fang A75 (Medjo Mvé 1997), and Mijikenda E70 (Nurse & Hinnebusch 1993, Kutsch Lojenga 2001) among others. However, from the phonetic point of view, the Bantu languages have fewer articulatorily complex consonants than is sometimes suggested. An interesting process of intensification of secondary articulations into obstruents occurs, inter alia, in Rwanda JD61 (Jouannet 1983) and Shona S10 (Doke 1931a). This process does not result in double articulations that are almost totally overlapped, as in labial-velars, but sequential articulations which are overlapped either not at all or no more than is typical of sequences such as /tk/ or /pk/ in

English words like 'fruitcake' or 'hopkiln.' On the other hand, it does produce rather unusual consonant sequences in onset positions.

Examples of the Rwanda JD61 strengthening of an underlying /u/ or /w/ into a velar stop after a non-homorganic nasal or stop are illustrated by the spectrograms in Figures 3.14–3.16. As these show, the first segment is released before the closure for the second is formed. When the sequence is voiced, as in /mg bg/, a quite marked central vocoid separates the two segments. When the sequence is voiceless, as in /tk/, there is a strong oral release of the first closure. There is no overlap in the closures for the two segments, except optionally in the case of the nasal sequence /mŋ/. Somewhat similar facts have been shown for the Zezuru S12 variety of Shona (Maddieson 1990). However, as was observed long ago by Doke (1931b, 1931a), the phonetic patterns vary quite considerably across the different varieties of Shona S10.

A particularly interesting claim is made by Mathangwane (1999) concerning her pronunciation of parallel forms in Kalanga S16. She suggests that elements like the /pk/ which evolves from earlier or underlying /pw/ are pronounced with almost fully overlapped closures and their duration is similar to that of simple /k/ and /p/ segments, i.e., they are [p͡k, b͡g]. She reports that the labial closure is formed first. This would therefore be an important counter-example to the more common pattern found in labial-velar doubly articulated segments in other languages in which the labial closure is formed very slightly later (10–15 ms) than the velar one. The one spectrogram of a word containing /pk/ published in this study actually shows that the duration of the element is considerably longer than a simple stop, suggesting it contains a sequence of articulations, although no burst is visible for the /p/. Recordings made by the first author of two other female speakers of Kalanga S16, one from Francistown in Botswana and one from Zimbabwe, did not replicate the pattern suggested by Mathangwane. For example, the word meaning

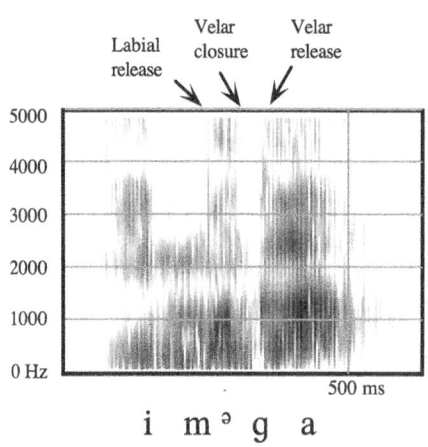

FIGURE 3.14 SPECTROGRAM OF RWANDA JD61 *imwa* [imᵊga] 'DOG' SPOKEN BY A MALE SPEAKER.

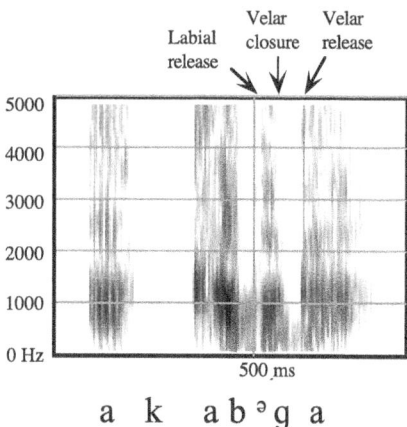

FIGURE 3.15 SPECTROGRAM OF RWANDA JD61 *akabwa* [akabᵊga] 'DOG (DIMINUTIVE)'; SAME SPEAKER AS IN **FIGURE 3.14**.

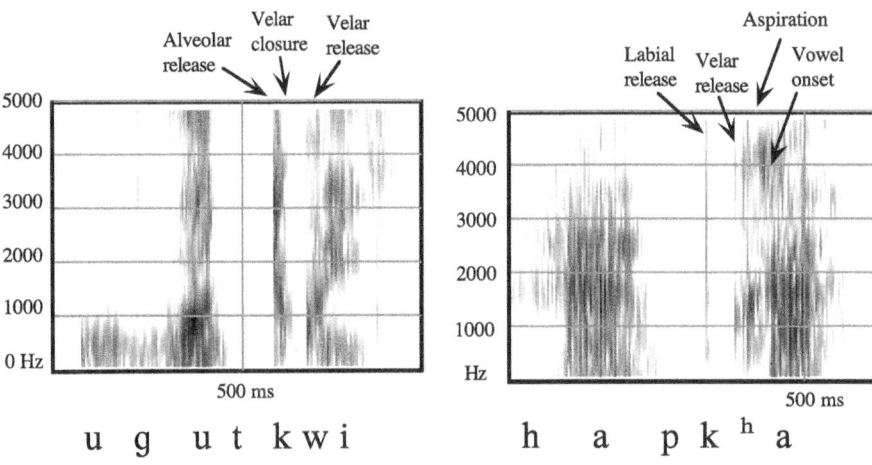

u g u t k w i

h a p kʰ a

FIGURE 3.16 SPECTROGRAM OF RWANDA JD61 *ugutwi* [*ugutkwi*] 'EAR'; SAME SPEAKER AS FIGURE 3.14.

FIGURE 3.17 SPECTROGRAM OF KALANGA S16 [*hapk͡ʰa*] 'AMPIT,' SPOKEN BY A FEMALE SPEAKER FROM ZIMBABWE.

'armpit,' transcribed by Mathangwane as [*h͡apk͡ʰa*], could receive three pronunciations – [*hakʰwa*] with no labial closure, [*hapxa*] with a labial stop followed by a fairly long velar fricative, or [*hapkʰa*] with a sequence of stops with clearly separate releases, as illustrated in Figure 3.17. This third pronunciation was characterised by one of the speakers as being more typical of speakers of 50 or more years of age. Evidently more study of the phonetic and sociolinguistic variation in this area would be of great interest.

3.4 'Whistling' fricatives

Shona S10 and Kalanga S16 are also marked by the occurrence of a type of labialisation co-produced with alveolar fricatives which have led to these segments being named "whistled," or "whistling fricatives" (Doke 1931a, Bladon *et al.* 1987). Unlike "ordinary" labialisation, which involves rounding and protrusion of the lips accompanied by a raising of the tongue back, i.e., a [*w*]-like articulation, this labialisation involves primarily a vertical narrowing of the lips with little or no protrusion and no accompanying tongue back raising. The gesture is also timed differently from "ordinary" labialisation in that it covers the fricative duration rather than being primarily realised as an offglide; hence "whistling fricatives" can themselves be labialised in their release phase. Similar segments are very rare in the world's languages, but do occur in the Dagestanian language Tabasaran (Kodzasov & Muravjeva 1982).

A detailed study of a weakly "whistled" fricative in Tsonga S53 shows that the narrowed lip posture is accompanied by a retroflex lingual gesture and thus may be transcribed with a retroflex fricative symbol [*ʂ*], e.g., [*ʂìɽá*] 'disasters' (Lee-Kim *et al.* 2014). The whistled fricative has more peaked and compact spectra than its non-whistled counterpart, and the fricatives also differ in other acoustic measures. Although lip positions have not been reported for Tshwa S51, the acoustic findings are similar to those in Tsonga

S53 in that the "whistling" fricatives have narrower spectral peak bandwidths and lower spectral peak frequencies when compared to their non-whistled fricative counterparts (Shosted 2006).

In Changana S53, "whistling" fricatives occur with a rounded lip posture (Shosted 2011) rather than the narrowed lip posture seen in Shona S10, Kalanga S16 and Tsonga S53. Since a rounded lip posture can also be seen in non-whistled fricatives, such as in the sequence [usu], the labial constriction alone cannot account for the whistle-like concentration of the frication noise, but it must be due to a particular linguopalatal configuration that is yet undescribed. This study of Changana S53 "whistling" fricatives underscores the fact that the phonetic realisation of a cross-linguistically rare sound may differ from one language to the next. "Whistling" fricatives are very rare cross-linguistically, but they do occur in Mozambican Portuguese (Ashby & Barbosa 2011), clearly due to the influence of Bantu languages.

4 LARYNGEAL ACTION IN CONSONANTS

4.1 Voice onset time

Though most Bantu languages are reported as having voiced and voiceless series of plosives, three-way contrasts in plosives based on Voice Onset Time (VOT) do occur. Engstrand and Lodhi (1985) study one such contrast in Swahili G42 and Monaka (2001, 2005) examines a three-way contrast in Kgalagari S311. Monaka's detailed study combines acoustic data with data about larynx height and vocal fold vibrations obtained using a laryngograph. VOT differs, as expected, between voiced, voiceless unaspirated and aspirated stop categories in Kgalagari S311, and it also varies by place of articulation within each category. She shows that voiceless palatal and velar stops tend to have longer VOT measurements than bilabial, dental or uvular stops (Monaka 2005). Voiced stops tend to be made with a downward movement of the larynx, presumably to help sustain voicing (Monaka 2001). She also uses electropalatography (EPG) to show the susceptibility of stops to coarticulation varies not only by place of articulation, but also according to voice category; aspirated stops are the least susceptible to coarticulation and voiced stops are the most (Monaka 2001). Other studies of coarticulation in Bantu languages have not looked at voicing contrasts (Manuel 1987, Beddor et al. 2002, Malambe 2015), but Dogil and Roux (1996) argue that ejectives and clicks in Xhosa S41 are more resistant to coarticulation than other consonants.

An unusual VOT contrast between partially voiced plosives and fully voiced stops, possibly implosives, has been described in Bekwel A85b (Cheucle 2014: 287) and the Kanincin variety of Ruwund L53 (Demolin 2015: 495).

4.2 Implosives and ejectives

Languages of the North-West, the Eastern coastal area and the South-East often have at least one implosive, most frequently a bilabial, but implosives are generally absent in the languages of the Congo basin and the South-West. Ejective stops and affricates are more rarely found in the Bantu languages, although they occur as variants of the unaspirated voiceless stops in languages of the South, especially in post-nasal contexts. The ejection is generally weak compared to that found in languages of the Afro-Asiatic family, except for Ilwana E701 where the ejectives are in borrowed Cushitic vocabulary, and the ejective

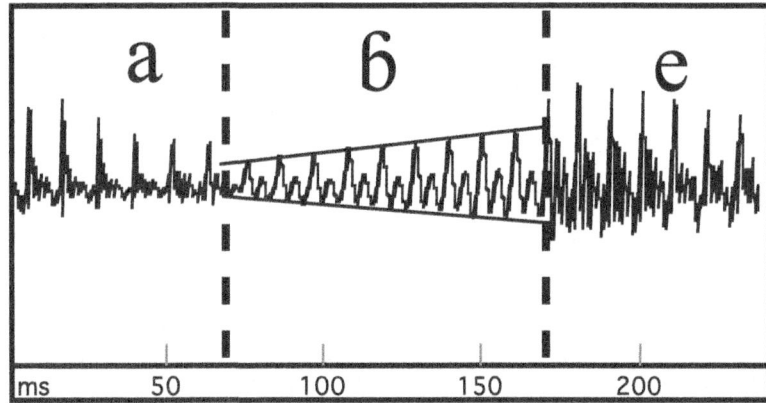

FIGURE 3.18 WAVEFORM OF THE MIDDLE PART OF THE TONGA S62 WORD /ɓàɓé/ 'FATHER,'
ILLUSTRATING THE INCREASING AMPLITUDE OF VOICING DURING THE
IMPLOSIVE.

lateral affricate of Zulu S42 mentioned earlier. In the Ngwato S31c variety of Tswana S31, ejectives are weak and are sometimes lenited, with loss of ejection: /t' k'/ ~ [d͡t k͡x] (Gouskova *et al.* 2011: 2127).

The segments labelled as implosives are sometimes described as if a glottal constriction is characteristic of their production. In Bantu, this is typically not the case; the vocal folds are in the normal position for voicing. Rather, what is critical is that the larynx is lowering during their production, so that the size of the supralaryngeal cavity is being enlarged while the oral closure is maintained. This may have two principal effects. Firstly, it allows the amplitude of vocal fold vibration to increase during the closure, giving a particularly strong percept of voicing at the time of the release. Secondly, it may mean that the intra-oral pressure is relatively low at the time when the closure is released so that at the moment of release the initial airflow is ingressive (Hardcastle & Brasington 1978). The waveform of an intervocalic bilabial implosive in Tonga S62 is shown in Figure 3.18. Dashed vertical lines mark the onset and offset of the bilabial closure. Voicing is continuous through the closure; upper and lower lines have been constructed on the figure linking respectively the positive and negative peaks in the waveform in order to dramatise the growing amplitude of the voicing during the closure.

In Zulu S42, implosive [ɓ] tends to have a shorter closure duration and lower amplitude burst than plosive [b] (Naidoo 2010). In Mpiemo A86c, implosives have a slight rise in F_0 before the onset of a following vowel while voiced plosives have a sharp dip in F_0 (Nagano-Madsen & Thornell 2012).

4.3 Depressor consonants

Another special laryngeal action occurs in the "depressor" consonants which are characteristic of certain Bantu languages of the Eastern and Southern regions. This term was originally applied to consonants which have a particularly salient lowering effect on the pitch of the voice in their neighbourhood (Lanham 1958). It has since sometimes come to be used for any consonant which has any local lowering effect on pitch or, more accurately, on the fundamental frequency of vocal fold vibration, abbreviated F_0, such as an ordinary voiced plosive. It has even been used for those which may simply block a raising or high-tone spreading process. However, the original notion of a depressor consonant is quite different from this expanded use. The most detailed study remains that of Traill *et al.* (1987) on depressor consonants in Zulu S42. This study shows that the F_0 associated with depressors is lower than a low tone, and the lowest pitch is centred on the depressor consonants themselves. At vowel onset, the F_0 difference between High and Low tones after a set of non-depressor consonants is 22 Hz, but a High tone onset after depressor consonants is 44 Hz lower than after the non-depressors and a Low after depressors is 23 Hz lower than after non-depressors. This is the mean across three speakers, two male and one female. Thus a High after a depressor begins considerably lower than a Low elsewhere. Figure 3.19 compares the pitch contours of the Swati S43 words /líhàlà/ 'aloe' and /*líh̤álà*/ 'harrow,' where /̤/ is a diacritic to mark the fact that the consonant is a depressor in the second word. Despite the fact that the lexical tone after the depressor is high (Rycroft 1981), the onset F_0 is about 30 Hz lower than the low tone onset after the non-depressor, and a rapid pitch fall begins during the vowel which precedes the depressor.

Voiceless, voiced, prenasalised and even aspirated stops may all pattern as depressor consonants (Chen & Downing 2011, Cibelli 2015, Lee 2015). Figure 3.19 also illustrates the fact that depression is not necessarily associated with voicing as both /h/ and /h̤/ are voiceless (Downing & Gick 2001, Downing 2009). Equally, voiced segments such as nasals and approximants may contrast in depression (Traill & Jackson 1988, Wright & Shryock 1993, Mathangwane 1998). Since these segments make for easy

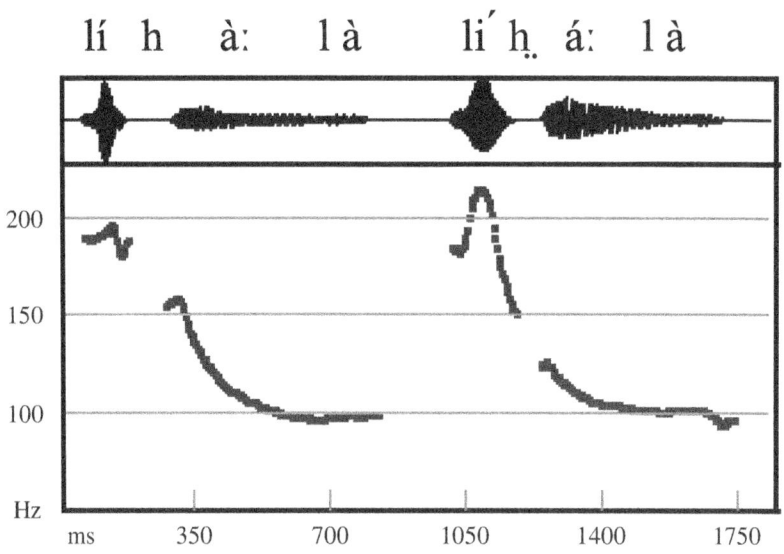

FIGURE 3.19 PITCH CONTOURS ILLUSTRATING EFFECTS OF NON-DEPRESSOR AND DEPRESSOR /*h*/ IN SWATI S43 (MALE SPEAKER).

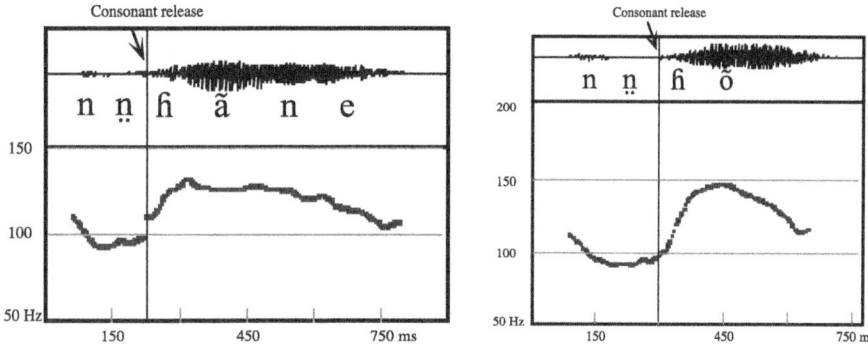

FIGURE 3.20 PITCH EFFECTS OF DEPRESSOR NASALS IN THE GIRYAMA E72A WORDS /nhane/ 'EIGHT' (LEFT PANEL) AND IDEOPHONE /nho/ (RIGHT PANEL). RECORDING COURTESY OF CONSTANCE KUTSCH LOJENGA.

tracking of F_0 through the consonant, the centring of the depression on the consonant can be most easily visualised with them. Two examples from Giryama E72a are illustrated in Figure 3.20. In these cases there is a substantial fall in F_0 from the onset to the middle of the nasal, and pitch begins to rise before the consonant is released; the pitch peak on the vowel is 40Hz (left panel) or 50Hz (right panel) higher than the lowest pitch in the nasal. In these words there is noticeably breathy phonation during part of the consonant and at the vowel onset which is transcribed as [ɦ]. However, breathiness is not an invariable accompaniment of depression as had been proposed by Rycroft (1980). Following Traill et al. (1987), we understand true "depression" to consist of a special laryngeal posture consistent with very low pitch co-produced with the consonant it is associated with. This gesture may become associated with any class of consonants and thus is capable of becoming itself an independent phonological entity deployed for grammatical effect as in the "depression without depressors" described by Traill (1990).

5 CLICKS

Though cross-linguistically rare, clicks are used by millions of people speaking various Bantu languages. They occur in two separate geographical clusters, the South-East (SEB) and the South-West (SWB), as shown in Figure 3.21. Clicks in the South-East cluster were borrowed from Khoe and possibly also from Taa and Kx'a languages into Nguni S40 (Louw 2013, Pakendorf et al. 2017); from Nguni (primarily Zulu S42), they subsequently spread into other SEB languages (Letele 1945, Bailey 1995). Clicks in the South-West cluster were borrowed independently from those in the South-East. In Fwe K402, they were borrowed from Khoe and Ju languages (Bostoen & Sands 2012, Gunnink et al. 2015). In languages of both the South-East and the South-West clusters, clicks can be found in Bantu roots as well as in loanwords. The functional load of clicks varies across languages, as detailed in Pakendorf et al. (2017) and Sands & Gunnink (forthcoming), both in terms of the number of contrastive click consonants, and in terms of the percentage of lexical items which contain clicks.

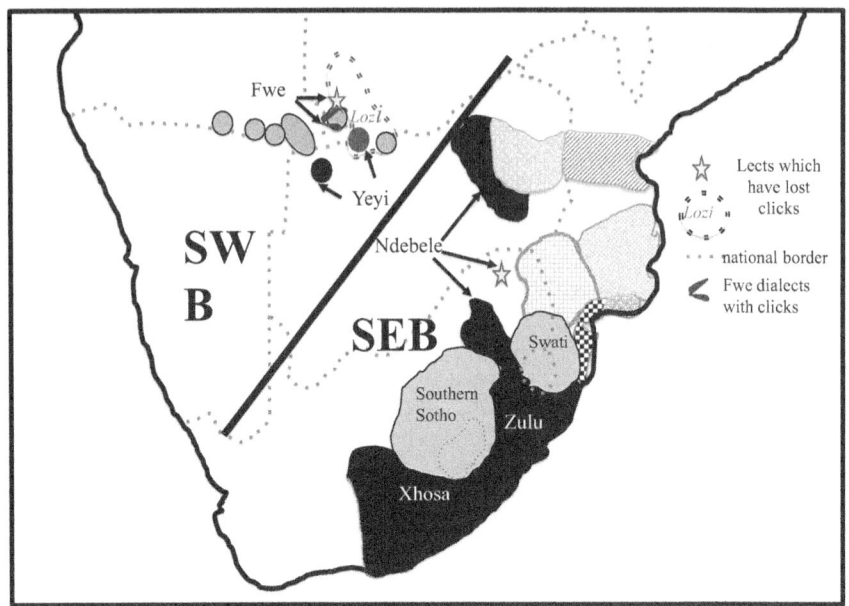

FIGURE 3.21 SOUTH-WEST AND SOUTH-EAST BANTU LANGUAGES WITH CLICKS. LANGUAGE LOCATIONS ARE ESTIMATED FOLLOWING MAHO (2009) AND GIESEKE AND SEIFERT (2007).

In the South-East, the core is formed by the languages of the Nguni group (S40), especially Xhosa S41, Zulu S42, Phuthi S404 and Zimbabwean Ndebele S44, which have between 12 and 15 click consonants; Swati S43 has fewer clicks (Doke 1954, Pakendorf *et al.* 2017). Clicks are found in many words in Southern Sotho S33 (Guma 1971), but only occur in a few sound symbolic words and interjections in Northern Sotho S32 (Poulos & Louwrens 1994). Clicks do not occur in Venda S21 (Ziervogel *et al.* 1981, Poulos 1990). In the Shona S10 group, clicks have only been reported to occur in midlands varieties of Kalanga S16 and in the Ndau S15 variety in Mozambique (Borland 1970, Mkanganwi 1972, Afido *et al.* 1989, Pongweni 1990). In the Tswa-Ronga S50 group, clicks have been reported to occur in Tswa S51, Tsonga S53, Konde S54, Nkuna S53D and Ronga S54 (Passy 1914, Persson 1932, Doke 1954, Baumbach 1974, Afido *et al.* 1989, Sitoe 1996), but their functional load in these lects is not well known. Clicks have also been reported to occur in Chopi S61 (Bailey 1995) and in the Mzimba variety of Tumbuka N21 (Moyo 1995).

In the South-West, the area near where the borders of Namibia, Angola, Botswana and Zambia meet, the largest number of clicks is found in Yeyi R41. Namibian Yeyi is described as having 19 click consonants (Gowlett 1997: 257), while Botswana Yeyi speakers vary, having as few as 12 or as many as 22 distinct click consonants (Fulop *et al.* 2003). Clicks are also found in Manyo (Gciriku) K332, Sambyu K331, Kwangali K33, Mbukushu K333 and Fwe K402 (Baumbach 1997, Möhlig 1997, Gunnink *et al.* 2015). Mbalangwe K401 has clicks, but whether it is a sociolect of Subiya K42 (Maho 1998:

51) or of Yeyi (Baumbach 1997: 307) is unclear. Clicks have been reported to occur in Ikuhane, or Botswanan Subiya (Ndana *et al.* 2017), but they have not been documented in Namibian or Zambian varieties (Baumbach 1997: 311, Jacottet 1896). Click consonants do not occur Herero R31, Umbundu R11, Totela K41 or Lozi K21, nor are they found in languages of the Wambo R20 cluster, such as Kwanyama R21, Mbalanhu R214 and Ndonga R22. Clicks are marginal in Tswana S31 and Kgalagari S311, with the possible exception of the Shetjhauba variety of Kgalagari (Tlale 2005, Lukusa and Monaka 2008).

The separate South-East and South-West groups of Bantu languages with clicks can be seen in the map of Southern Africa in Figure 3.21. Areas in black on the map represent the geographical distribution of languages with large click inventories, and areas in grey represent smaller click inventories. Areas north of Swati S43 and east of Ndebele S44 with grey patterns show the S10, S50 and S61 zones where clicks have been sporadically attested. Like most linguistic maps, this map represents a somewhat fictitious ethnographic idealisation not corresponding precisely with any exact time or population distribution. It is also not possible to definitively state the number of Bantu languages with clicks; clicks may occur in some varieties and not others, as in the case of Fwe K402 (Pakendorf *et al.* 2017: 20, Gunnink forthcoming), and may have even been lost where they were once attested. Tswa S51 may be one such case, as the last attestation was by Persson (1932). It is noteworthy that none of the Bantu languages of East Africa appear to have acquired clicks from the surviving or former languages of this area with clicks (Maddieson *et al.* 1999).

Languages which lost clicks entirely include "Northern" Ndebele of South Africa S408 and Lozi K21 (Ziervogel 1959, Gowlett 1989, Skhosana 2009), though it seems some "Northern" Ndebele S408 speakers are borrowing clicks back from Zulu S42. Some speakers of Southern Ndebele S407 have a reduced click inventory (Schulz & Laine 2016). Click loss is an on-going process in Chopi (Bailey 1995) and in Imusho Fwe (Gunnink forthcoming). Sowetan Zulu S42, too, has a reduced number of click consonants, likely due to contact with Southern Sotho S33 (Gunnink 2014). A rapid reduction in the number of click contrasts occurred more than 100 years ago in the far-flung varieties of Nguni known as Ngoni N12 (Elmslie 1891, Spiss 1904, Doke 1954); Ngoni speakers subsequently shifted from Nguni to languages of the Manda N10 group (Malio 2003). Note that languages of Malawi and Tanzania are not shown on the map in Figure 3.21. Clicks have not been reported for Manda group languages and are unlikely to occur unless efforts to revitalise Malawian Ngoni on a Zulu model prove effective (Kishindo 2002).

The mechanism of producing clicks is now fairly well understood and is illustrated by the sequence of midsagittal real-time MRI in Figure 3.22. A closure in the vocal tract is formed by the back of the tongue contacting the roof of the mouth in the velar or uvular area and a second closure is formed in front of the location of this closure by the tip or blade of the tongue or the lips, as shown at timestep 1. A small quantity of air is entrapped inside the sealed oral cavity. Although not seen in a mid-sagittal diagram, the sides of the tongue are also raised to complete the seal between anterior and dorsal closures. The center portion of the tongue is then lowered while the two main closures are maintained (timesteps 2–3), enlarging the volume of the space between them. In addition, there may be retraction of the tongue tip, dorsum or tongue root for some clicks (Miller 2008, Miller & Finch 2011). Expansion of the closed cavity causes the pressure in the air inside the space to be reduced well below that of the air outside the mouth. Next, the closure at the front and/or side of the mouth is released (timestep 4) and the abrupt equalisation of air pressures inside and outside the mouth results in a sharp acoustic transient. Finally, the

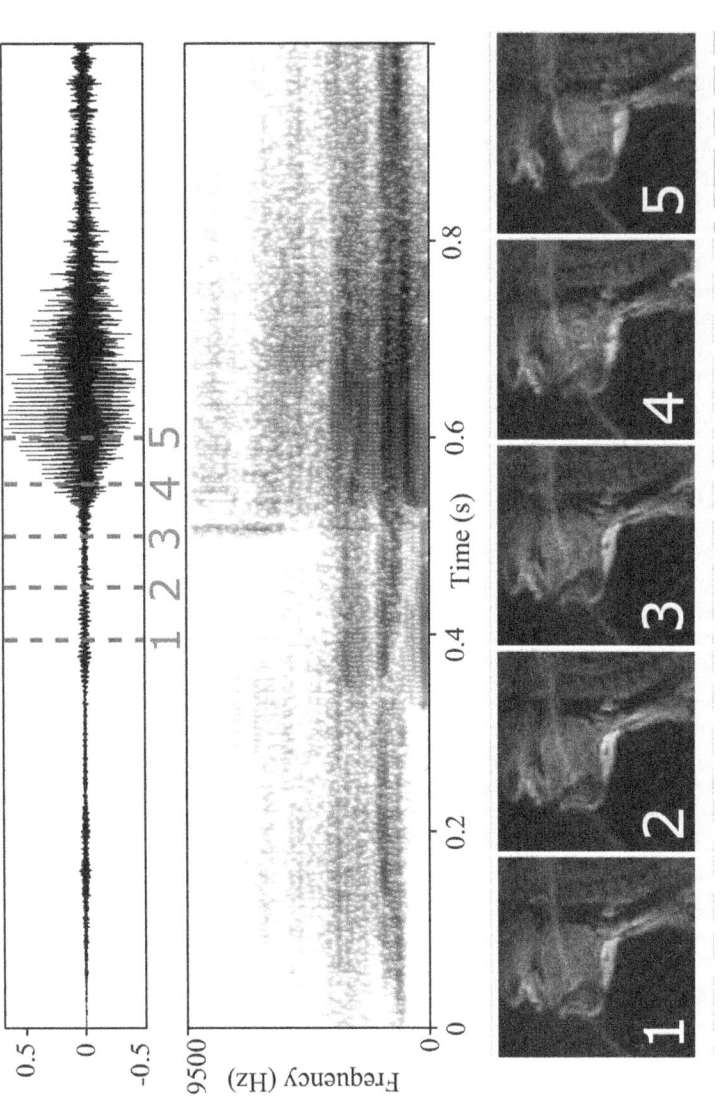

FIGURE 3.22 TIME-ALIGNED AUDIO AND VIDEO DATA OF A SWATI S43 DENTAL CLICK AND FOLLOWING VOWEL IN THE SYLLABLE *ngca*/ŋᶢ!a/. THE TOP AND MIDDLE ROWS SHOW A WAVEFORM AND SPECTROGRAM, RESPECTIVELY. BROKEN VERTICAL LINES INDICATE THE FIVE POINTS IN TIME CORRESPONDING TO THE rtMRI IMAGES SHOWN IN THE BOTTOM ROW. AN ACOUSTIC ARTEFACT OF RECORDING IN THE CYLINDRICAL METALLIC MRI SCANNER BORE IS A SERIES OF ECHOES SPACED AT 53 MS INTERVALS. THE WHITE BOW-SHAPED LINE CROSSING EACH MIDSAGITTAL IMAGE IS ALSO AN ARTIFACT. SEE PROCTOR *ET AL.* (2013), PROCTOR *ET AL.* (FORTHCOMING) FOR A DESCRIPTION OF THE METHODOLOGY USED TO OBTAIN THE IMAGES.

Source: Recording and images made available by Michael Proctor.

back closure is released, and this release may be separately audible. The posture of the vowel following the click is seen in timestep 5.

Clicks in Bantu languages are often made with a back closure that is velar, however uvular constrictions also occur, particularly for post-alveolar [ǃ] and lateral clicks [ǁ] (Miller 2008). Kxʼa, Tuu and Khoe ("Khoisan") languages tend to favour uvular rather than velar constrictions (cf. Miller *et al.* 2007, Miller *et al.* 2009b, Miller 2010, 2016). The velar release of a Xhosa S41 dental click is shown in Figure 3.23, which has a waveform and spectrogram of the word *caca* /ǀàǀà/ 'be clear.' The first unaspirated dental click has a velar burst 17 ms after the anterior click burst. The click in the second syllable has a dorsal release that is closer in time to the release of the anterior click closure. The second click also has a velar closure. This can be seen by the converging F2 and F3 transitions at the end of the first vowel, (as indicated by the arrow), which indicate a velar constriction.

The basic click mechanism does not determine what the larynx is doing while these movements are taking place in the oral cavity, nor whether the velum itself is raised or lowered to block or permit air from the lungs to flow out through the nose. Thus, a click can be accompanied by simple glottal closure, by modal or breathy voicing, by open vocal folds, or by use of the ejective mechanism. Changes in larynx activity can be variously timed in relation to the action in the oral cavity, and to the timing of movements raising and lowering the velum. The possible variations are thus very numerous, and many different categories of individual clicks are found when all the languages which use them are considered (Ladefoged & Maddieson 1996).

In describing clicks, it is customary to talk of the click type and the click accompaniment. The click type refers to the location of the front closure and the manner in which it is released, which may be abrupt or affricated, central or lateral. The accompaniment refers to all the other aspects of the click: laryngeal action and timing, nasal coupling, and the location (uvular or velar) and manner of release (abrupt or affricated) of the back closure. In South-East Bantu languages, three contrastive click types are found, and probably no more than seven accompaniments are used. Zulu S42 and Xhosa S41 have dental /ǀ/, alveolar lateral /ǁ/ and apical post-alveolar /ǃ/ click types. The last of these was often described as "palatal" in older literature. In South-West Bantu languages, Yeyi has these three click types as well as a contrastive laminal post-alveolar type /ǂ/, variously called "alveolar" or "palatal" in different sources. The palatal click type ǂ may be found as a variant of /ǃ/ used in child-directed speech in Zulu and Xhosa (Bradfield 2014: 27). The bilabial click /ʘ/ is not found in Bantu except in paralinguistic utterances, and as a variant pronunciation of a sequence of labial and velar stops, as in Rwanda JD61 (Demolin 2015: 483). Dental and lateral clicks are sometimes called "noisy," "affricated," or "pre-affricated" (Roux 2007), while the (post-)alveolar is described as "abrupt" or "unaffricated." Palatal clicks in Yeyi R41 are somewhat fricated (Fulop *et al.* 2003), though they are typically produced with an "abrupt" or "unaffricated" release in Khoisan languages. Yeyi R41 has eight different accompaniments (Fulop *et al.* 2003), including several contrasts which are not found in other Bantu languages. Yeyi R41 contrasts clicks with a velar fricated and ejective velar fricated release (/ǀˣ/, /ǀˣʼ/) as well as glottalised and ejected clicks (e.g., /ǀˀ/, /ǀʼ/) (Fulop *et al.* 2003). Fwe has four accompaniments including a voiceless nasal accompaniment (Gunnink forthcoming) not known to occur in any other Bantu language. It is possible that phonetic studies of other South-West Bantu click languages will reveal additional click accompaniments.

Thomas-Vilakati's analysis of Zulu click types (Thomas 2000, Thomas-Vilakati 2010), combining insights from acoustic, aerodynamic and electropalatographic techniques, is

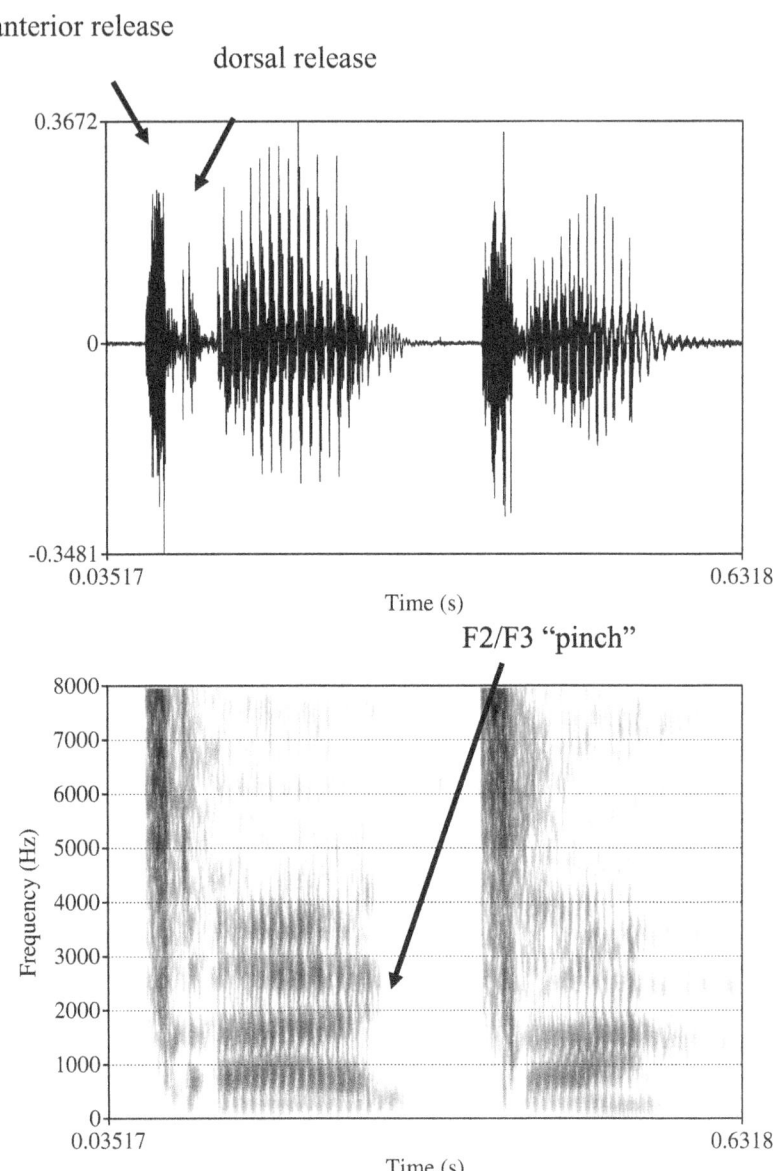

FIGURE 3.23 SPECTROGRAM OF THE XHOSA S41 WORD *CACA* /!a!a/ 'BE CLEAR' SPOKEN BY SIZWE SATYO, A MALE SPEAKER OF XHOSA. THE SMALL ARROWS ON THE WAVEFORM SHOW A DISTINCT ANTERIOR AND DORSAL BURST ON THE FIRST CLICK. THE ARROW IN THE SPECTROGRAM POINTS TO A CONVERGENCE OF F2 AND F3 CHARACTERISTIC OF VELARS.

Source: Recording made by Peter Ladefoged in 1979 and archived at the UCLA Phonetics Lab Archive (Ladefoged 2007).

the most detailed study of click production in a Bantu language to date. Thomas-Vilakati confirms that the velar closure always precedes the front closure; this accounts for the fact that nasals preceding clicks assimilate in place to velar position, and corrects a misobservation by Doke (1926), who believed the front closure was formed first: the velar closure *must* be released after the front closure for the click mechanism to work, but it could in principle be formed later. Dorsal closures for all three click types in Thomas-Vilakati's data are held for about 175 milliseconds, but the front closures show some significant timing differences. The front closure for dental clicks is formed earlier and held longer (about 105 ms) than that for post-alveolar or lateral clicks (about 80 ms). The relative timing and durations of velar and front closures deduced from acoustic and aerodynamic data are graphed in Figure 3.24.

More details on the articulations of clicks are given by electropalatography (EPG). Speakers wear a thin custom-made acrylic insert moulded to the shape of their upper teeth and hard palate in which a number of electrodes are embedded which sense contact between the tongue and the roof of the mouth. In Thomas-Vilakati's study, inserts with 96 electrodes were used, together with software allowing a sweep of the contact patterns to be made every 10 ms. The articulatory contacts can then be examined using stylised displays such as those in Figure 3.29, Figure 3.30 and Figure 3.31, which represent the arc of the teeth and the vault of the palate. Contacted electrodes are shown as black squares and uncontacted ones as grey dots. Figure 3.25 shows the production of a dental click. The first frame, numbered 0, is close to the time that velar closure is first made, as detected from the accompanying acoustic record. Because the insert does not cover the soft palate, this closure cannot be observed on the EPG record at this time. The seal around the inside of the teeth is made by 40 ms later, and as the contact area of the back of the tongue enlarges, the front edge of the velar contact is now visible as a line of contacted electrodes at the bottom of the arc. The closures overlap for 100 ms, until frame 140. During this time, rarefaction is occurring. This figure makes clear that the expansion of the cavity is not solely due to moving the location of the back closure further back. That Zulu dental clicks are produced with a controlled fricated release is also clear from the way the front release initially involves formation of a narrow channel, clearly visible in frame 150.

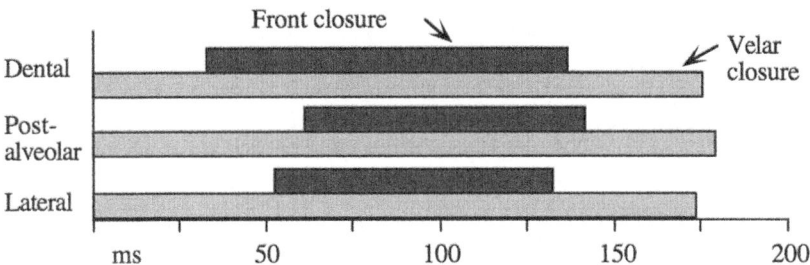

FIGURE 3.24 CLOSURE DURATIONS AND TIMING RELATIONS IN THE THREE CLICK TYPES OF ZULU S42; MEANS FOR VOICELESS CLICKS IN THREE VOWEL ENVIRONMENTS SPOKEN BY THREE SPEAKERS, ADAPTED FROM THOMAS-VILAKATI (2010). FRONT CLOSURE DURATIONS ARE SHOWN AS HEAVILY STIPPLED BARS.

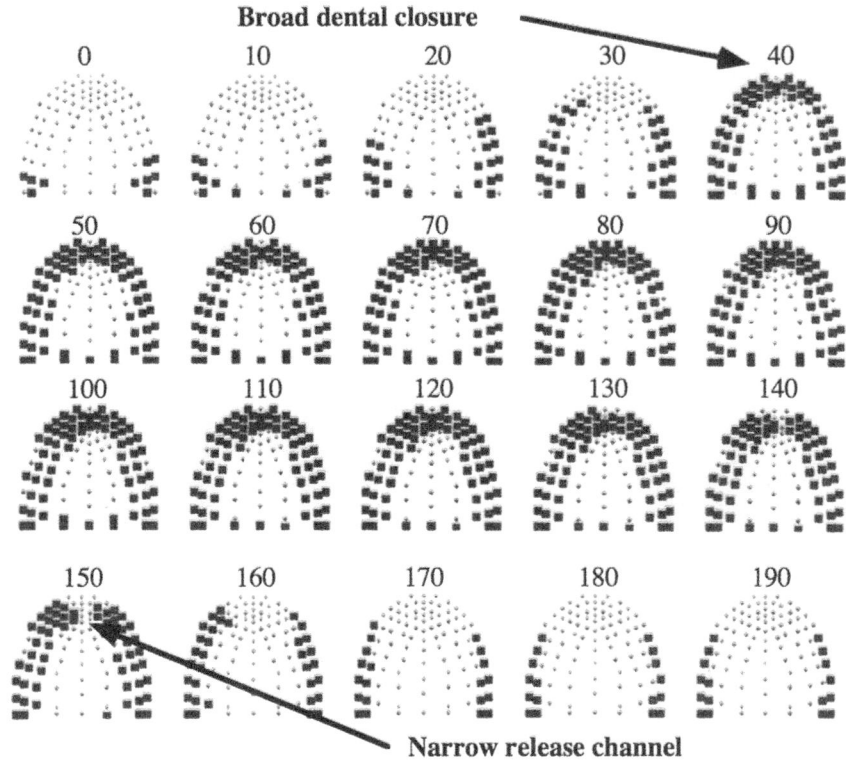

Broad dental closure

Narrow release channel

FIGURE 3.25 EPG FRAMES SHOWING A DENTAL CLICK SPOKEN BY A MALE ZULU S42
SPEAKER.

Source: Thomas-Vilakati 2010.

Production of a lateral click is illustrated in Figure 3.26. In this particular token there is a long lag between the time the velar closure is made and when the front closure is sealed, about 80 ms later. The contact of the front of the tongue is asymmetrical, as the side of the tongue opposite to where the release will be made is braced contra-laterally against the palate. The release of a lateral click is also affricated, occurring initially through a narrow channel quite far back, as shown in frame 170 and continuing in frame 180. In contrast to these two affricated click types, a post-alveolar click is released without affrication. As Figure 3.27 shows, the shift from sealed to open occurs rapidly and completely, here between the two frames numbered 170 and 180. These frames also illustrate the retraction of the tongue tip which occurs just before release of this click type. From frame 150 through to frame 170 the contacted area moves back, so that the configuration at the moment of release is clearly post-alveolar.

During the time period in which the two closures of a click overlap, lowering of the center of the tongue creates a partial vacuum in the cavity between them. Thomas-Vilakati's work provides the first direct measures of how powerful the energy generated by

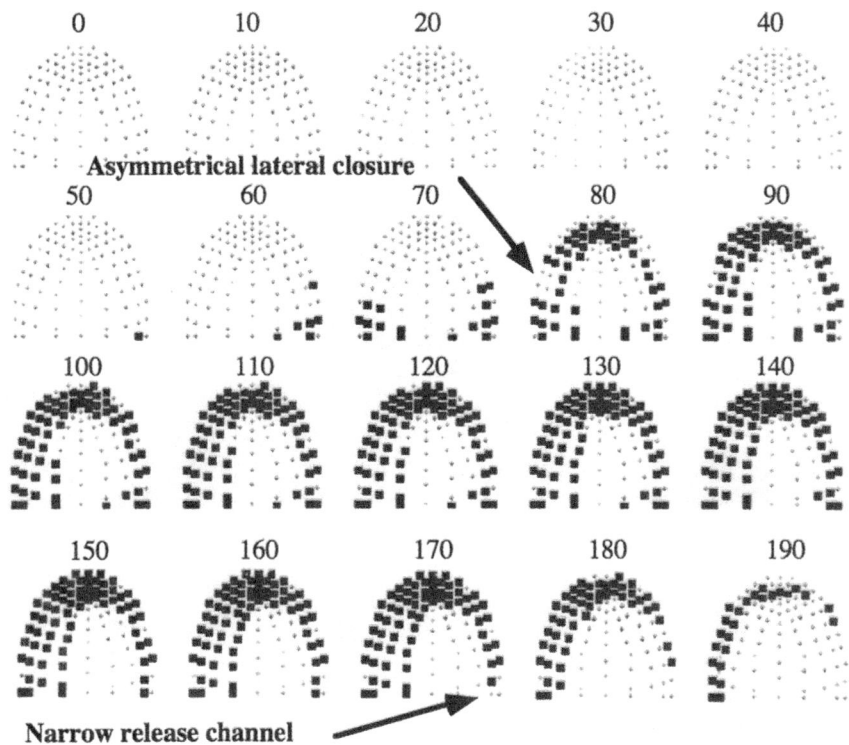

FIGURE 3.26 EPG FRAMES OF A LATERAL CLICK SPOKEN BY A MALE ZULU S42 SPEAKER.

Source: Thomas-Vilakati 2010.

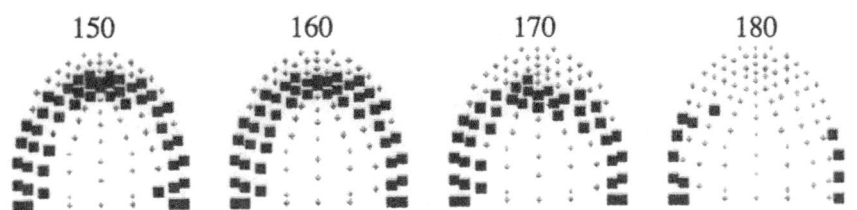

FIGURE 3.27 EPG FRAMES SHOWING THE RELEASING PHASE OF A POST-ALVEOLAR
CLICK SPOKEN BY A MALE ZULU S42 SPEAKER.

Source: Adapted from Thomas-Vilakati 2010.

this gesture is. Air pressure in the oral cavity is measured in relation to the ambient
atmospheric pressure in hectoPascals (hPa, equivalent to the pressure required to support
1 cm of water). For an ordinary pulmonic stop, peak pressure behind the closure ranges
between about 5 and 20 hPa, depending on the loudness of the voice. The peak negative

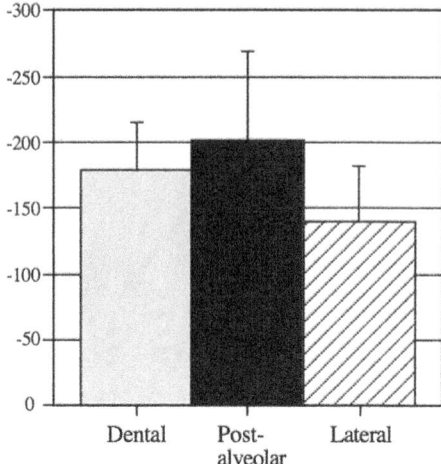

FIGURE 3.28 PEAK NEGATIVE PRESSURE IN THE THREE CLICK TYPES OF ZULU S42
MEANS FOR VOICELESS CLICKS IN THREE VOWEL ENVIRONMENTS SPOKEN
BY THREE SPEAKERS.

Source: Thomas-Vilakati 2010.

pressures reached in clicks are typically -100 hPa or more and may reach over -200, as
shown in Figure 3.28. Post-alveolar clicks have the greatest rarefaction, lateral clicks
the least, perhaps because the contra-lateral bracing of the tongue in the lateral clicks
may constrain the amount of tongue-center lowering that is possible. Thomas-Vilakati's
aerodynamic data also reflect the different dynamics of the affricated and abrupt clicks.
The equalisation of internal and external pressure at release occurs much more quickly in
post-alveolar clicks than for dental and lateral clicks. This can be shown by calculating
the average rate of pressure change over this phase of the click, which is 14.4 hPa/ms for
post-alveolars, 7.9 for dentals, and 4.2 hPa/ms for laterals. Only a small part of this difference can be accounted for by the difference in peak pressre between the click types.

Zulu S42 has four different accompaniments to its three click types: "plain" (voiceless
unaspirated), voiceless aspirated, voiced and voiced nasalised. Because the place of the
dorsal closure is not contrastive, it is not necessary to indicate the (velar in this case)
place before the click type symbol. Our recommended IPA transcription and corresponding Zulu S42 orthographic symbols is given in Table 3.2. There are several ways of
indicating the same click following IPA principles, e.g., /g!, g!, !/ are equivalent ways of
representing a voiced (post-)alveolar click. Similarly, /ŋ!, ⁿ!, ˀ!/ all represent a voiced nasal
(post-)alveolar click. Xhosa S41 has five accompaniments, three of which are the same as
in Zulu S42. The Xhosa S41 "voiced" clicks are breathy or slack voiced (Jessen & Roux
2002) and may even be devoiced (Maphalala et al. 2014, Braver 2017). There is a distinct
breathy/slack voiced nasalised accompaniment; these two series are "depressor consonants." Some speakers of Xhosa S41 produce "plain" clicks with ejection (Jessen 2002).

One of the most striking things about clicks in Bantu is the lack of respect for place
distinctions when few categorical contrasts exist. In Nkuna S53D, Baumbach (1974)
indicates that clicks are indifferently pronounced as dental or post-alveolar. In Mbukushu

TABLE 3.2 IPA SYMBOLS AND ZULU S42 ORTHOGRAPHY FOR CLICKS

	IPA transcription			*Orthography*		
	Dental	*Post-alveolar*	*Lateral*	*Dental*	*Post-alveolar*	*Lateral*
voiceless unaspirated	ǀ	ǃ	ǁ	c	q	x
voiceless aspirated	ǀʰ	ǃʰ	ǁʰ	ch	qh	xh
voiced	ᶢǀ	ᶢǃ	ᶢǁ	gc	gq	gx
voiced nasalised	ᵑǀ	ᵑǃ	ᵑǁ	nc	nq	nx

K333, the one series of clicks is reported to be pronounced "either as dental, palatal or [post-]alveolar sounds" (Fisch 1998). Southern Sotho S33 only has a single click type which may vary in place. Older accounts of Southern Sotho S33 describe both post-alveolar or sub-laminal retroflex articulations (Doke 1923: 713, 1926: 301). In Manyo K332, clicks are "mostly dental, however, with a broad individual variation" (Möhlig 1997). There are four click accompaniments in Fwe K402: voiceless unaspirated, voiced oral, voiced nasal and voiceless nasal, but the language has no contrast for click type or place (Gunnink forthcoming). In the central (Imusho) variety of Fwe, the word 'papyrus' may be pronounced with an unaspirated dental click ([ruǀoma]), as in Figure 3.29, or as an unaspirated alveolar click ([ruǃoma]), as in Figure 3.30, with no difference in meaning. The dorsal constriction of clicks in Fwe is typically velar. In Figure 3.29 and Figure 3.30, the anterior click burst has a higher amplitude than the velar release burst, as is typical for clicks cross-linguistically. In Figure 3.31, however, the dorsal burst has a higher amplitude than the anterior click burst. The click in the word [ruǃoma] 'papyrus' in Figure 3.31 is a very weak click, as indicated by the extended IPA (extIPA) diacritic for a weak articulation, e.g., [ǃ̞], which is similar to the diacritic for an unreleased stop e.g., (c˺), but placed under the consonant rather than after it. The word 'papyrus' may also be articulated with a velar stop in place of the click [rukoma], as seen in Figure 3.32. Because the velar stop burst in the weak click [ǃ̞] is louder than the anterior click burst, it is perhaps not surprising that [k] has come to replace [ǃ̞] for some speakers. The current variation between clicks and velars in Imusho Fwe may eventually lead to the loss of clicks in the variety altogether, as clicks are replaced by velars. Anecdotally, it seems that clicks in other Bantu languages may also vary in amplitude, depending on the individual speaker, stylistic or sociphonetic variables, and prosodic environment. The context-free liberty to vary place of articulation of clicks in some Bantu languages is rarely encountered with other classes of consonants.

6 NASALS AND NASALITY

The special phonetic interest of consonantal nasality in the Bantu languages involves principally the prenasalised segments and the realisation of "voiceless" nasals. In both cases aspects of timing are particularly relevant. Another feature of interest is the presence of a cross-linguistically rare contrast between nasalised and oral glottal approximants (/h̃/ and /h/) found in Kwangali K33 (Ladefoged & Maddieson 1996: 132).

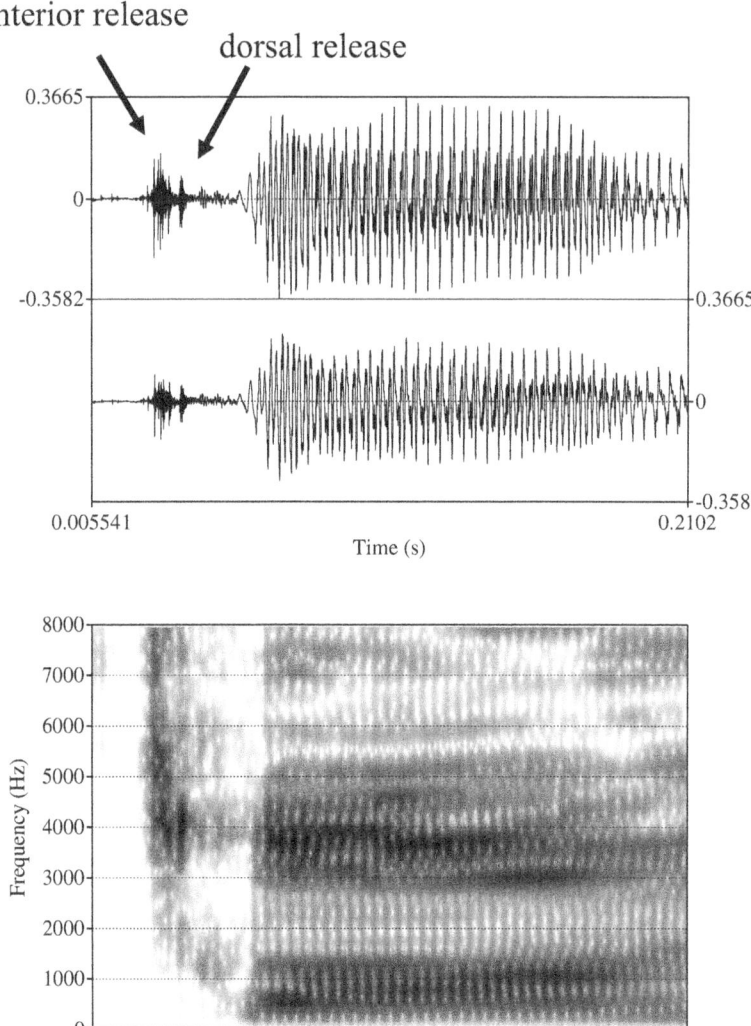

FIGURE 3.29 WAVEFORM AND SPECTROGRAM OF THE MIDDLE SYLLABLE OF THE FWE
K402 WORD [*ru|oma*] 'PAPYRUS,' SPOKEN BY A FEMALE SPEAKER.

Source: Recording made available by Hilde Gunnink.

Detailed studies of timing in prenasalised stops are included in Maddieson (1993), Maddieson and Ladefoged (1993) and Hubbard (1994, 1995). Using data from these sources,
Figure 3.33 compares the durations of nasals and voiced prenasalised stops as well as of
the vowels that precede them in two languages, Ganda JE15 and Sukuma F21. In both languages the oral stop duration in voiced prenasalised stops is very short, so the total segment

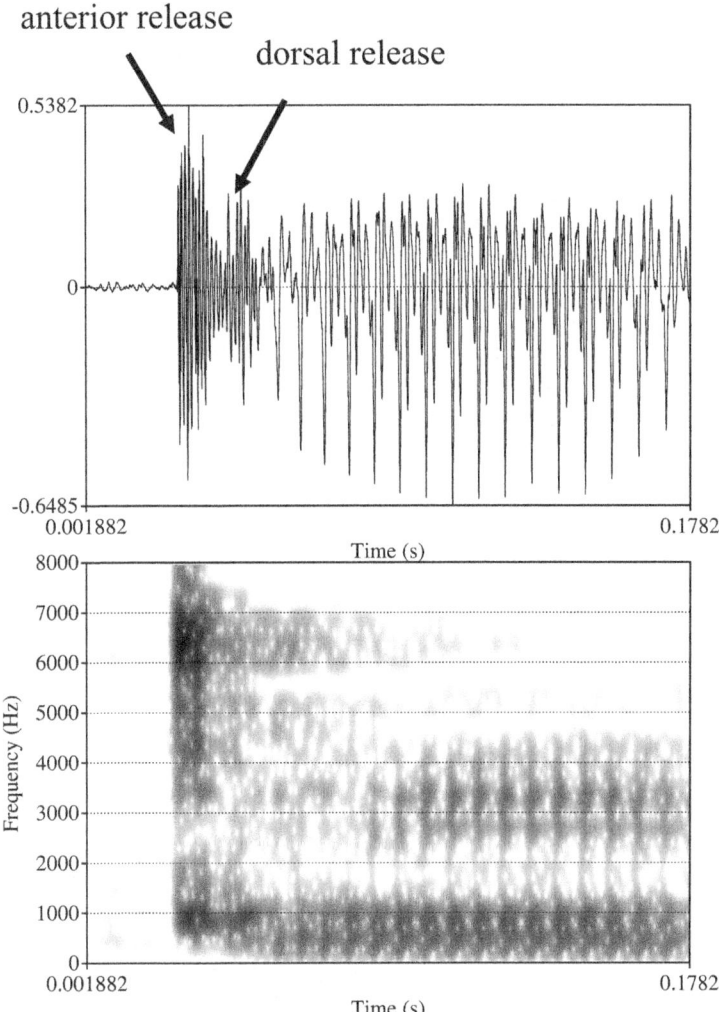

FIGURE 3.30 WAVEFORM AND SPECTROGRAM OF THE MIDDLE SYLLABLE OF THE FWE K402 WORD [ru!oma] 'PAPYRUS,' SPOKEN BY A MALE SPEAKER.

Source: Recording made available by Koen Bostoen.

duration is not so very different from that of a simple nasal. Both languages have contrasts of vowel quantity and compensatory lengthening of vowels before prenasalised stops, but there are interesting differences between the two. Lengthened vowels are much closer in duration to underlying long vowels in Ganda JE15 than they are in Sukuma F21. Sukuma F21 lengthened vowels are almost exactly intermediate between underlying short and long vowels and the nasal portion is quite long. Nyambo JE21 is similar to Sukuma in its pattern.

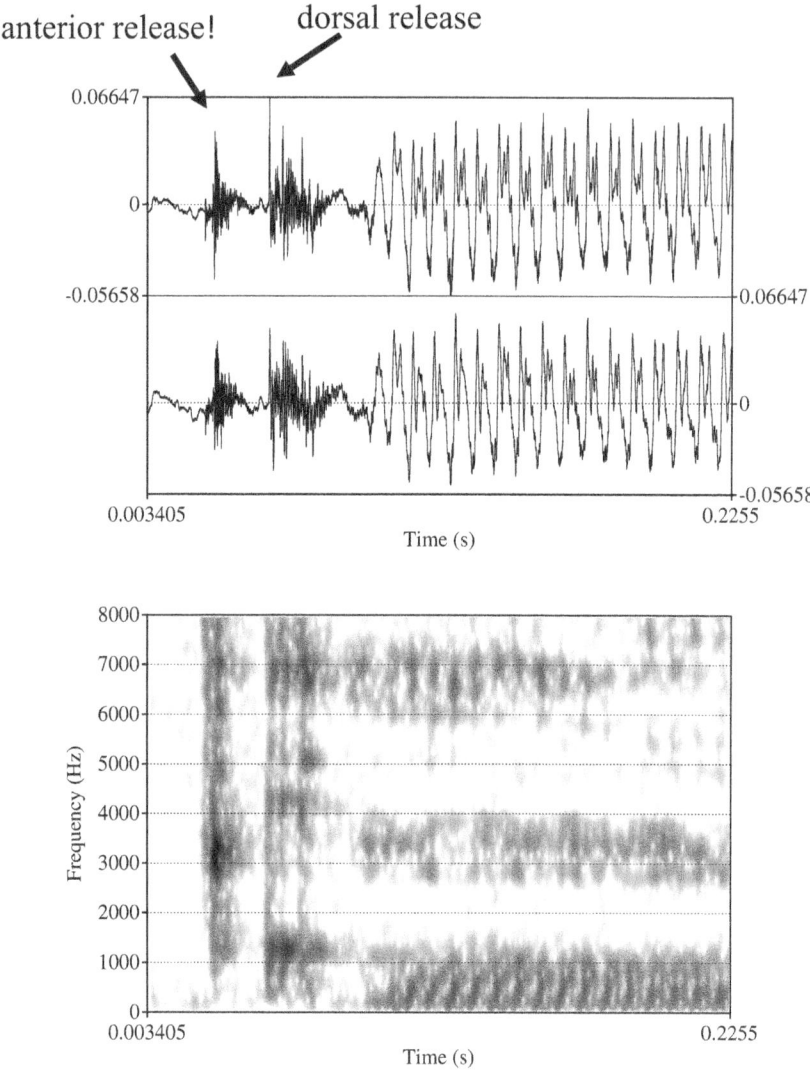

FIGURE 3.31 WAVEFORM AND SPECTROGRAM OF THE MIDDLE SYLLABLE OF THE FWE K402 WORD [*ru!oma*] 'PAPYRUS,' SPOKEN BY A DIFFERENT MALE SPEAKER THAN IN FIGURE 3.30.

Source: Recording made available by Hilde Gunnink.

Hubbard (1994, 1995) suggests that the difference from Ganda is related to the fact that lengthened vowels count in a different way in tone assignment rules in these languages.

Hubbard (1994, 1995) also compared the durations of vowels in three further languages with different patterns. The mean results are given in Table 3.3. Ndendeule N101

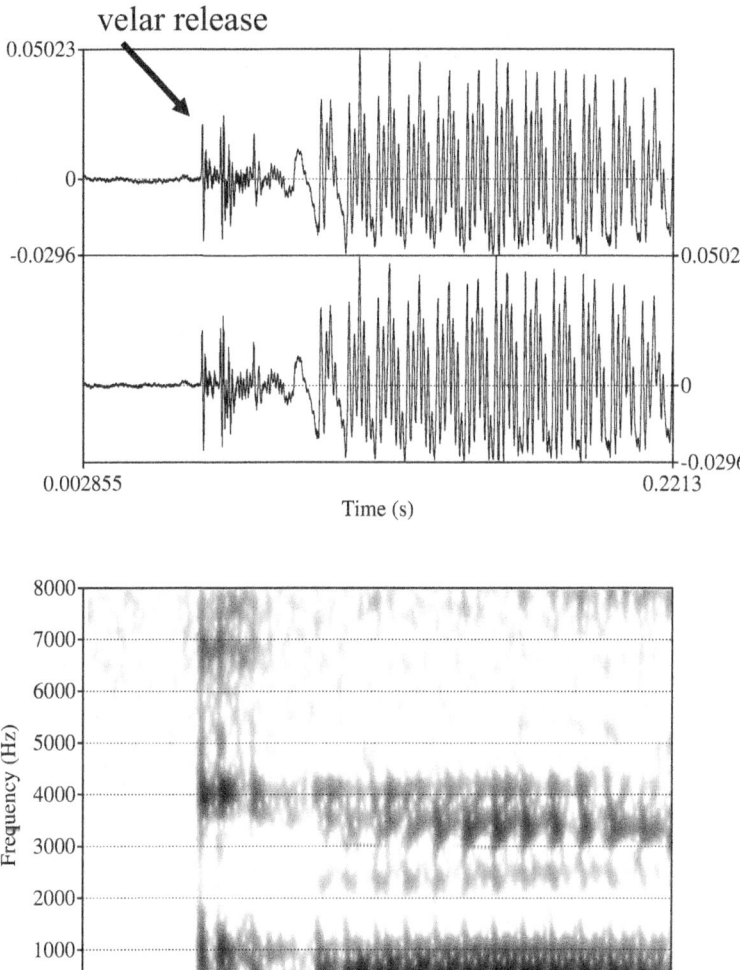

FIGURE 3.32 WAVEFORM AND SPECTROGRAM OF THE MIDDLE SYLLABLE OF THE FWE K402 WORD [*rukoma*] 'PAPYRUS,' SPOKEN BY A DIFFERENT MALE SPEAKER THAN IN FIGURE 3.31 AND FIGURE 3.32.

Source: Recording made available by Hilde Gunnink.

has no long vowels and no lengthening. Yao P21 has a long/short contrast and significant compensatory lengthening so that vowels before prenasalised stops are as long as underlying long vowels and have more than double the duration of short vowels. Tonga M64 has long vowels but does not show any compensatory lengthening before NC. This difference seems to be related to the different origin of long vowels; Yao P21 maintains

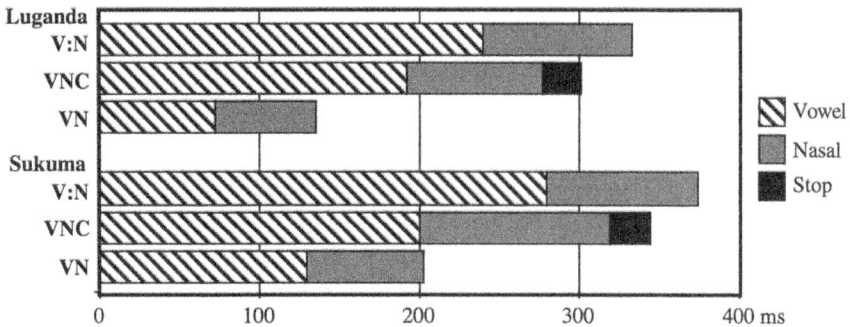

FIGURE 3.33 COMPARISON OF SELECTED VOWEL AND CONSONANTS LENGTHS IN GANDA JE15 AND SUKUMA F21 (SEE TEXT FOR EXPLANATION).

TABLE 3.3 COMPARISON OF MEAN VOWEL DURATIONS IN THREE LANGUAGES, ONE SPEAKER PER LANGUAGE AFTER HUBBARD 1995.

Vowel duration in. . .	Ndendeule N101	Yao P21	Tonga M64
. . . CVC	148	61	100
. . . CVNC	146	130	101
. . . CV:C	–	132	241

Source: After Hubbard 1995

Proto-Bantu vowel length distinctions and adds to them. Tonga M64 does not preserve Proto-Bantu vowel length, but has developed long vowels from intervocalic consonant loss.

Several recent detailed studies have looked at the timing and laryngeal characteristics of stops after nasals in Tswana S31 and Kgalagari S311. These closely related languages have been argued to violate a constraint against voiceless stops after nasals. Post-nasal stops are devoiced in Kgalagari S311 (Solé *et al.* 2010), and in Tswana S31 only for some speakers (Coetzee & Pretorius 2010). There is evidence for post-nasal fortition rather than devoicing in the Ngwato S31c variety (Gouskova *et al.* 2011, Boyer & Zsiga 2013). In this variety, some speakers fail to devoice, and others devoice intervocalically as well as after nasals (Zsiga *et al.* 2006).

In several areas earlier voiceless prenasalised stops have developed into voiceless nasals or related types of segments, including in Sukuma F21 (Maddieson 1991), Pokomo E71, Bondei G24 (Huffman & Hinnebusch 1998), Kalanga S16 (Mathangwane 1998) and Rwanda JD61 (Demolin & Delvaux 2001). Aspects of the original sequencing of nasal + oral and voiced + voiceless portions found in prenasalised stops are sometimes retained and small variations in the timing and magnitude of the different component gestures create quite large variability in the acoustic pattern of these segments as critical alignments are made or missed.

Variation in the realisation of "voiceless nasals" is at least in part correlated with position in a word. Figure 3.34 shows a spectrogram of the Nyamwezi F22 word /ŋ̊apo/
'basket' spoken in isolation. Dotted vertical lines separate the major phonetic components
of the first syllable. The portion marked A, between the first two lines, is phonetically a
voiceless velar nasal [ŋ̊]. The second line marks the time-point at which the velar closure
is released. The fragment marked B has voiceless oral airflow, with resonances similar to
those of the following /a/ vowel. Fragment C is the voiced portion of the vowel /a/. This
type of segment might well be described as an "aspirated voiceless nasal." Figure 3.35
shows a realisation of a medial instance of the same segment in the word /kɔŋ̊á/ 'to suck.'
In this case there is no consonantal nasality. The nasal feature is realised as nasalisation of
the latter part of the vowel /ɔ/ in Fragment C, following an oral portion, B, and the aspiration of the initial stop, A. Fragment D, which is the consonantal part of the /ŋ̊/ is voiceless
but oral, and as often in an [h]-sound, the transition of the formants of the flanking vowels
can be traced through its duration.

In Sukuma F21, the nasal portion of the "voiceless nasals" is often at least partly
voiced or breathy voiced, as described in Maddieson (1991), whereas the parallel segments in Rwanda JD61 are fully voiced (except after voiceless fricatives), but produced with a modified kind of voicing described by Demolin and Delvaux (2001) as
whispery-voice.

7 PROSODIC CHARACTERISTICS

A discussion of Bantu phonetics would not be complete without reference to some of the
studies of the major prosodic characteristics of the languages. Stress in Bantu often falls

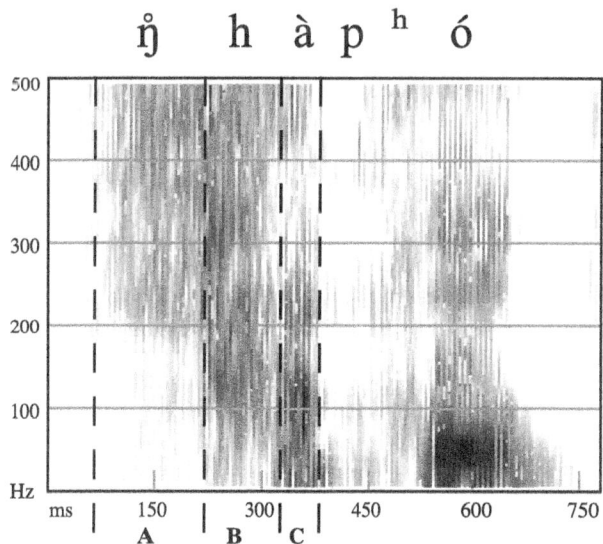

FIGURE 3.34 SPECTROGRAM OF THE NYAMWEZI F22 WORD /ŋ̊apo/ 'BASKET.' SEE TEXT
FOR DISCUSSION OF THE PHONETIC SEGMENTATION.

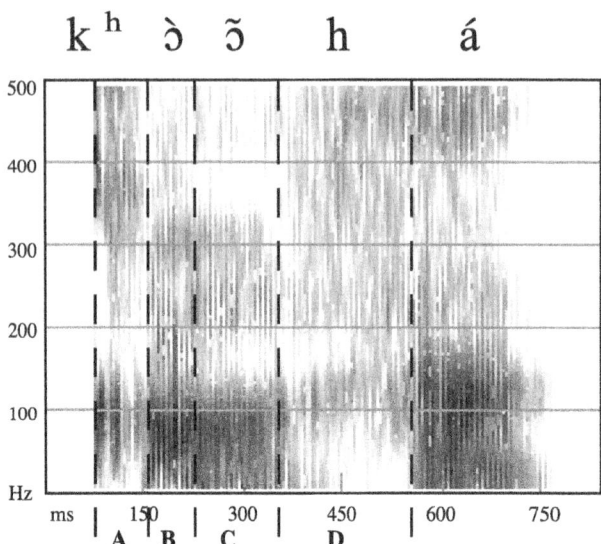

FIGURE 3.35 SPECTROGRAM OF THE NYAMWEZI F22 WORD /kɔ̃j̃á/ 'TO SUCK.' SEE TEXT FOR DISCUSSION OF THE PHONETIC SEGMENTATION.

on the penult, typically with vowel lengthening, but stem-initial prominence also occurs (Downing 2010). A majority of Bantu languages have a tonal distinction of High and Low tones, which often may combine into contour tones. High tone is generally the phonologically marked tone, with Low tone being unmarked (Stevick 1969, Downing 2011) (see also Chapter 5). Languages without tone do occur, e.g., Swahili G42, Mwiini G412, Nyakyusa M31, as do ones with more than two level tones, e.g., Kamba E55 and Oku (Grassfields Bantu) (Downing 2010, Hyman 2014). Aspects of prosody to be covered here include: patterning of tones, phonetic implementation of tones, positional restrictions, intonation, focus prosody and question prosody. Typically, studies of intonation in Bantu languages tend to look at F_0 and duration; measures of intensity and spectral tilt are less often used to identify prosodic cues (Zerbian & Barnard 2008).

The patterning of tones in many Bantu languages resembles that of pitch-accent systems. For instance, the number of High tones which may surface in a word or a stem may be limited to one and prominent peaks tend to occur in a predictable position, often the penult (Downing 2010). In Northern Sotho S32, however, there is speaker variation in the position of the F_0 peak, which may occur "somewhere between the second and the third syllable, counting from the high-tone-bearing, verbstem initial syllable" (Zerbian 2009).

The phonetic shapes of tone sequences can usually be modeled on the basis of the position and height of local H targets, with the Low tones treated as automatically filled valleys between these points. Certain more complex patterns, such as those noted by Hombert (1990) in Fang A75, and by Roux (1995) in Xhosa S41, may require a more elaborate model. High tones tend to fall on the antepenult in Nguni S40 languages such as Xhosa S41, though the penult is stressed/lengthened (Downing 2010). Provisions have to be made for the special effects of depressor consonants on tone in Nguni languages. Depressors also occur in Digo E73 and other Mijikenda E70 group languages and in Kalanga S16 and other Shona S10 group languages (Downing 2010). In Zulu S42, the phonetic effects of depressor consonants on pitch differ from pitch lowering effects caused by implosive consonants (Chen & Downing 2011). In Rwanda JD61, there is anticipatory coarticulation of tone, with the F_0 of a syllable being affected by a High tone in a later syllable (Myers 2003).

The most extensive body of work on the phonetics of tone in a single Bantu language concerns Chewa N31b (Carleton 1996, Myers 1996, Myers & Carleton 1996, Myers 1999a, b). Detailed studies of this type not only illuminate the individual language studied but may provide insights into diachronic issues. For example, in Chewa N31b, as is common cross-linguistically, the High pitch peak is realised at the end of the syllable to which it is associated (Kim 1998, Myers 1999a). This pattern may form the basis for the frequent shifting of a High tone to a later syllable. These studies also address several issues in the relation between intonation and tone. For example, Myers (1999b) shows that syntactically unmarked yes/no questions are characterised by a slower rate of pitch declination than statements. Carleton (1996) demonstrated that units of paragraph length are organised by long-range patterns of tonal declination and resetting.

Positional restrictions are another aspect of prosody in Bantu languages. Tonal contrasts and vowel length contrasts are often restricted to stem-initial syllables (Downing 2010). In Jita JE25, for instance, only the initial syllable of verb roots may contrast in tone (Downing 2011). In Mwiini G412, however, long vowels may surface on the penult or antepenult and only occur word-initially in loanwords (Kisseberth & Abasheikh 2004: xvii). Contour tones may be restricted to heavy syllables. For instance, a contrast between HL and LH contours is restricted to long vowels in Rwanda JD61 (Myers 2003). Stem-initial syllables typically have a greater number of segmental contrasts than found elsewhere (Downing 2010).

Surveys of intonation in Bantu languages include Zerbian and Barnard (2008) and the volume edited by Downing & Rialland (2016a). Both surveys reveal a great deal of variety across Bantu languages. For instance, there are languages with and without down drift, though the former are more common (Downing & Rialland 2016b). Mbochi C25, which does not have downdrift, still has final lowering due to a L% boundary tone (Rialland & Aborobongui 2016). There are different types of downstep attested in some Bantu languages. Downstep due to a floating Low tone is attested in Basaa A43a (Makasso *et al*. 2016). Downstep affects the second of two adjacent High tones in Tswana S31 (Zerbian & Kügler 2015) and Bemba M42 (Kula & Hamann 2016). The interaction of final lowering and downstep in Pare G22 is detailed in Herman (1996). Final lowering is fairly common across Bantu, but is not attested in Basaa A43a (Downing & Rialland 2016b). Final lowering associated with a L% boundary tone at the end of a sentence in Ngazidja G44a is often associated with a devoiced final syllable (Patin 2016). In Tswana S31, declarative sentences are primarily marked by penultimate lengthening and a reduced or devoiced final vowel (Zerbian 2016).

Focus in Bantu is often marked using morphosyntactic means rather than through the use of prosody (Downing & Hyman 2016, Downing & Rialland 2016b). In Bemba M42, however, new information focus is indicated on a subject by its placement in post-verbal position and by pitch raising of the pre-focus constituent (Kula & Hamann 2016). Some North-Western Bantu languages which have stem-initial accent, such as Eton A71, have a focus prosody that causes the lengthening of stem-initial consonants and vowels (Van de Velde & Idiatov 2016). Focus and emphasis are associated with pitch raising in Mwiini G412 (Kisseberth 2016), but this seems to be the exception rather than the rule in Bantu. Chewa N31b and Tumbuka N21, for instance, do not have focus prosody (Downing 2016).

A wide range of means of marking question prosody have been noted for Bantu languages. Rialland's (2007) survey includes seven different prosodic types found in Bantu languages, the most common being the use of register expansion along with the reduction of downdrift. Both falling and rising intonation patterns are found in question prosodies. Final High or rising intonations are found in Ganda JE15, Chewa N31b and Saghala E74b, while final High-Low or High-falling intonations are found in Jita JE25. Mongo-Nkundu C61 has reduction of final lowering, while Zulu S42 and Southern Sotho S33 cancel penultimate lengthening in question prosody. Polar or mid tones are found in Holoholo D28 and Nyanga D43. Zamba C322 and Ganda JE15 raise a final High tone in question prosody. These seven prosodic types do not account for all of the details of the individual languages. For instance, the final High in yes-no questions in Zamba is preceded by a sharp fall (Bokamba 1976: 19). In Bemba M42, polar questions are marked by a final boundary L% on the final syllable, but pitch range expansion is also used (Kula & Hamann 2016).

8 CONCLUSION

Bantu languages provide an opportunity to compare phonetic differences between fairly large numbers of related languages. Phonological theories, phonetic theories, and hypotheses about patterns of sound change can be tested in this real-world laboratory, ensuring the popularity of Bantu languages as subjects of research for years to come. The increasing availability of ultrasound and MRI technologies should lead to future studies examining the effect of prosodic environment on articulation. There is much work that remains to be done on cross-linguistic, intra- and inter-speaker variation of typologically unusual sounds such as clicks and "whistling fricatives." Undoubtedly, studies of intonation and prosody in Bantu languages will continue to increase in number. Corpus studies of Bantu languages are currently few in number (Prinsloo & de Schryver 2001, Niesler *et al.* 2005, Allwood *et al.* 2010), but the increasing availability of such corpora may encourage phonetic studies of natural (unelicited) speech.

REFERENCES

Afido, P., G. Firmino, J. Heins, S. Mbuub & M. Trinta (1989) *Relatório do I Seminário sobre a Padronização da Ortografia de Línguas Moçambicanas*. Maputo: Universidade Eduardo Mondlane.

Allwood, J., H. Hammarström, A. Hendrikse, M. N. Ngcobo, N. Nomdebevana, L. Pretorius & M. van der Merwe (2010) Work on Spoken (Multimodal) Corpora in South Africa. *Proceedings of the Seventh Conference on International Language Resources and Evaluation*, 885–889.

Ashby, S. & S. Barbosa (2011) Bantu Substratum Interference in Mozambican Portuguese Speech Varieties. *Africana Linguistica* 17: 3–31.

Bailey, R. (1995) Issues in the Phonology and Orthography of Chopi (ciCopi S 61). In Traill, A., R. Vossen & M. Biesele (eds.), *The Complete Linguist: Papers in Memory of Patrick J. Dickens*, 135–170. Cologne: Rüdiger Köppe.

Bastin, Y., A. Coupez, E. Mumba & T. C. Schadeberg (eds.) (2002) *Bantu Cologne Reconstructions 3*. Tervuren: Royal Museum for Central Africa (Available online at linguistics.africamuseum.be/BLR3.html, Accessed November 28, 2017).

Batibo, H. M. (1985) *Le kesukuma (langue bantoue de Tanzanie): phonologie, morphologie*. Paris: Centre de Recherches, d'Echanges et de Documentation Universitaire.

Baumbach, E. J. M. (1974) *Introduction to the Speech Sounds and Speech Sound Changes of Tsonga*. Pretoria: J.L. van Schaik.

Baumbach, E. J. M. (1997) Languages of the Eastern Caprivi. In Haacke, W. H. G. & E. D. Elderkin (eds.), *Namibian Languages. Reports and Papers*, 307–450. Cologne: Rüdiger Köppe.

Beddor, P. S., J. D. Harnsberger & S. Lindemann (2002) Language-Specific Patterns of Vowel-to-Vowel Coarticulation: Acoustic Structures and Their Perceptual Correlates. *Journal of Phonetics* 30: 591–627.

Bennett, W. G., M. Diemer, J. Kerford, T. Probert & T. Wesi (2016) Illustrations of the IPA: Setswana (South African). *Journal of the International Phonetic Association* 46(2): 235–246.

Bennett, W. G. & J. S. Lee (2015) A surface constraint in Xitsonga: *Li. *Africana Linguistica* 21: 3–27.

Bladon, A., C. Clark & K. Mickey (1987) Production and Perception of Sibilant Fricatives: Shona Data. *Journal of the International Phonetic Association* 17: 39–65.

Blench, R. (2011) Linguistic Geography or Evidence for Genetic Affiliation? New Proposals for the Phonological Inventory of Proto-Bantu. Paper presented at *West African Phonology Group, London, 28th April, 2011*.

Blench, R. (2015) "The Bantoid Languages." In *Oxford Handbooks Online*. Oxford: Oxford University Press. (Available online at www.oxfordhandbooks.com/view/10.1093/oxfordhb/9780199935345.001.0001/oxfordhb-9780199935345-e-17).

Bokamba, E. G. (1976) *Question Formation in Some Bantu Languages*. Bloomington: Indiana University, PhD.

Borland, C. H. (1970) *Eastern Shona: A Comparative Dialect Study*. Cape Town: University of Cape Town, PhD dissertation.

Bostoen, K. (2008) Bantu Spirantization: Morphologization, Lexicalization and Historical Classification. *Diachronica* 25(3): 299–356.

Bostoen, K. & B. Sands (2012) Clicks in South-Western Bantu Languages: Contact-Induced Vs. Language-Internal Lexical Change. In Brenzinger, M. & A.-M. Fehn (eds.), *Proceedings of the 6th World Congress of African Linguistics Cologne 2009*, 129–140. Cologne: Rüdiger Köppe.

Boyd, V. L. (2015) *The phonological systems of the Mbam languages of Cameroon with a focus on vowels and vowel harmony*. Leiden: Leiden University, PhD dissertation.

Boyer, O. & E. Zsiga (2013) Phonological Devoicing and Phonetic Voicing in Setswana. In Orie, Q. l. & K. W. Sanders (eds.), *Selected Proceedings of the 43rd Annual Conference on African Linguistics*, 82–89. Somerville: Cascadilla Proceedings Project.

Bradfield, J. (2014) Clicks, Concurrency and Khoisan. *Phonology* 31(1): 1–49.

Braver, A. (2017) How Do You Whisper a Click? Acoustic Correlates of Click Voicing in Whispered Speech. Paper presented at the 91st Annual Meeting of the Linguistic Society of America, Austin Texas, 5–8 January 2017.

Carleton, T. (1996) *Phonetics, Phonology and Rhetorical Structuring of Chichewa*. Austin: University of Texas, PhD dissertation.

Chen, Y. & L. J. Downing (2011) All Depressors Are Not Alike: A Comparison of Shanghai Chinese and Zulu. In Frota, S., G. Elordieta & P. Prieto (eds.), *Prosodic Categories: Production, Perception and Comprehension*, 243–265. New York: Springer.

Cheucle, M. (2014) *Etude comparative des langues makaa-njem (bantu A80) : phonologie, morphologie, Lexique. Vers une reconstruction du proto-A80*. Lyon: Université Lyon 2, thèse de doctorat.

Cibelli, E. (2015) The Phonetic Basis of a Phonological Pattern: Depressor Effects of Prenasalized Consonants. In Romero, J. & M. Riera (eds.), *The Phonetics-Phonology Interface: Representations and Methodologies*, 171–192. Amsterdam; Philadelphia: John Benjamins.

Clements, G. N. (1986) Compensatory Lengthening and Consonant Gemination in Luganda. In Wetzels, L. W. & E. Sezer (eds.), *Studies in Compensatory Lengthening*. Dordrecht: Foris Publications.

Coetzee, A. W. & R. Pretorius (2010) Phonetically Grounded Phonology and Sound Change: The Case of Tswana Labial Plosives. *Journal of Phonetics* 38(3): 404–421.

Davey, A., L. Moshi & I. Maddieson (1982) Liquids in Chaga. *UCLA Working Papers in Phonetics* 54: 93–108.

De Blois, K. F. (1970) The Augment in the Bantu languages. *Africana Linguistica* 4: 85–165.

De Wit, G. (2015) *Liko Phonology and Grammar*. Leiden: Leiden University, PhD dissertation.

Demolin, D. (2015) Insights from the Field. In Redford, M. A. (ed.), *Handbook of Speech Production*, 477–504. Malden: Wiley-Blackwell.

Demolin, D. & V. Delvaux (2001) Whispery Voiced Nasal Stops in Rwanda. In Dalsagaard, P., B. Lindberg & H. Benner (eds.), *EUROSPEECH 2001 Scandinavia, 7th European Conference on Speech Communication and Technology, 2nd INTERSPEECH Event, September 3–7, 2001, Aalborg Congress and Culture Centre, Aalborg-Denmark: Proceedings*, 651–654. Bonn: ISCA-Secretariat.

Demolin, D., J.-M. Hombert, P. Ondo & C. Segebarth (1992) Étude du systeme vocalique fang par résonance magnétique. *Pholia* 7: 41–43.

Demolin, D., H. Ngonga-Ke-Mbembe & A. Soquet (2002) Phonetic Characteristics of an Unexploded Palatal Implosive in Hendo. *Journal of the International Phonetic Association* 32(1): 1–15.

Dickens, P. (1987) Qhalaxarzi Consonants. *African Studies* 46(2): 297–305.

Dogil, G. & J. C. Roux (1996) Notes on Unencoded Speech: Clicks and Their Accompaniments in Xhosa. In McCormack, P. (ed.), *Proceedings of the Sixth Australian Conference on Speech Science and Technology*, 55–60. Canberra: Australian Speech Science and Technology Association.

Doke, C. M. (1923) A Dissertation on the Phonetics of the Zulu Language. *Bulletin of the School of Oriental Studies* 2(4): 685–729.

Doke, C. M. (1926) *The Phonetics of the Zulu Language*. Johannesburg: University of Witwatersrand Press.

Doke, C. M. (1931a) *A Comparative Study of Shona Phonetics*. Johannesburg: University of the Witwatersrand Press.

Doke, C. M. (1931b) *Report on the Unification of the Shona Dialects*. Hertford: Stephen Austen and Sons.

Doke, C. M. (1954) *The Southern Bantu Languages*. London: Oxford University Press for the International African Institute (IAI).

Dombrowsky-Hahn, K. (1999) *Phénomènes de contact entre les langues minyanka et bambara (Sud du Mali)*. Cologne: Rüdiger Köppe.

Downing, L. J. (2009) On Pitch Lowering Not Linked to Voicing: Nguni and Shona Group Depressors. *Language Sciences* 31: 179–198.

Downing, L. J. (2010) Accent in African Languages. In van der Hulst, H. G., R. Goedemans & E. van Zanten (eds.), *A Survey of Word Accentual Patterns in the Languages of the World*, 381–427. Berlin: De Gruyter Mouton.

Downing, L. J. (2016) Tone and Intonation in Chichewa and Tumbuka. In Downing, L. J. & A. Rialland (eds.), *Intonation in African Tone Languages*, 365–392. Berlin: De Gruyter Mouton.

Downing, L. J. & B. Gick (2001) Voiceless Tone Depressors in Nambya and Botswana Kalang'a. *Berkeley Linguistics Society* 27: 65–80.

Downing, L. J. & L. Hyman (2016) Information Structure in Bantu Languages. In Féry, C. & S. Ishihara (eds.), *Oxford Handbook of Information Structure*, 790–813. Oxford: Oxford University Press.

Downing, L. J. & A. Rialland (eds.) (2016a) *Intonation in African Tone Languages*. Berlin: De Gruyter Mouton.

Downing, L. J. & A. Rialland (2016b) Introduction. In Downing, L. J. & A. Rialland (eds.), *Intonation in African Tone Languages*, 1–16. Berlin: De Gruyter Mouton.

Downing, L. J. (2011) Bantu Tone. In van Oostendorp, M., C. J. Ewen, E. Hume & K. Rice (eds.), *The Blackwell Companion to Phonology, Chapter 14*. Cambridge; Oxford: Blackwell.

Duke, D. & M. Martin (2012) Introducing Kwasio Pharyngealized Vowels. In Brenzinger, M. & A.-M. Fehn (eds.), *Proceedings of the 6th World Congress of African Linguistics, Cologne, 17–21 August 2009*, 219–224. Cologne: Rüdiger Köppe.

Elmslie, W. A. (1891) *Introductory Grammar of the Ngoni (Zulu) Language, as Spoken in Mombera's Country*. Aberdeen: G. & W. Fraser, 'Belmont' Works.

Engstrand, O. & A. Y. Lodhi (1985) On aspiration in Swahili: Hypotheses, Field Observations and an Instrumental Analysis. *Phonetica* 42: 175–187.

Faytak, M. (2013) Dissimilation by Surface Correspondence in Aghem Velarized Diphthongs. In Kingston, J., C. Moore-Cantwell, J. Pater & R. Staubs (eds.), *Supplemental Proceedings of Phonology 2013*, 1–10. Washington, DC: Linguistic Society of America.

Faytak, M. (2014) Chain Shifts, Strident Vowels, and Expanded Vowel Spaces. Paper presented at *LSA Annual Meeting, January 2–5*, Minneapolis.

Faytak, M. (2015) High Vowel Fricativization as an Areal Feature of the Northern Cameroon Grassfields. Paper presented at *the 8th World Congress of African Linguistics, August 20–24*, Kyoto.

Faytak, M. & J. Merrill (2014) Bantu Spirantization Is a Reflex of Vowel Spirantization. Paper presented at *"Sound Change in Interacting Human Systems," 3rd Biennial Workshop on Sound Change, May 28–31*, University of California, Berkeley.

Fisch, M. (1998) *Thimbukushu Grammar*. Windhoek: Out of Africa Publishers.

Fulop, S. A., P. Ladefoged, F. Liu & R. Voßen (2003) Yeyi Clicks: Acoustic Description and Analysis. *Phonetica* 60(4): 231–260.

Gick, B. (2002) The Use of Ultrasound for Linguistic Phonetic Fieldwork. *Journal of the International Phonetic Association* 32(2): 113–121.

Gick, B., D. Pulleyblank, F. Campbell & N. M. Mutaka (2006) Low Vowels andTtransparency in Kinande Vowel Harmony. *Phonology* 23: 1–20.

Gieseke, S. & M. Seifert (2007) Weiße Geister – Diachrone Stereotype in Nordnamibia und Südangola. Eine Bestandsaufnahme. In Egert, M., F. Heerbaart, K. Kolossa, M. Limanski, M. Mumin, P. A. Rodekuhr, S. Rous, M. Seifert, S. Stankowski & M. Thanassoula (eds.), *Beiträge zur 1. Kölner Afrikawissenschaftlichen Nachwuchstagung (KANT I)*, 1–19. Köln: Institut für Afrikanistik der Universität zu Köln.

Gouskova, M., E. Zsiga & O. T. Boyer (2011) Grounded Constraints and the Consonants of Setswana. *Lingua* 121(15): 2120–2152.

Gowlett, D. F. (1989) The Parentage and Development of Lozi. *Journal of African Languages and Linguistics* 11: 127–149.

Gowlett, D. F. (1997) Aspects of Yeyi Diachronic Phonology. In Haacke, W. H. G. & E. D. Elderkin (eds.), *Namibian Languages. Reports and Papers*, 235–264. Cologne: Rüdiger Köppe.

Greenberg, J. H. 1951. Vowel and Nasal Harmony in Bantu Languages. *Zaïre: revue congolaise* 5(8): 813–820.

Guma, S. M. (1971) *An Outline Structure of Southern Sotho*. Pietermaritzburg: Shuter & Shooter.

Gunnink, H. (2014) The Grammatical Structure of Sowetan Tsotsitaal. *Southern African Linguistics and Applied Language Studies* 32(2): 161–171.

Gunnink, H. (2016) Tone and Vowel Length in Fwe (Bantu, K402). Paper presented at *the Annual Conference of African Linguistics 47*, University of California, Berkeley.

Gunnink, H. (forthcoming) Click Loss and Click Insertion in Fwe. In Sands, B. (ed.), *Handbook of Click Languages*. Leiden: Brill.

Gunnink, H., B. Sands, B. Pakendorf & K. Bostoen (2015) Prehistoric Language Contact in the Kavango-Zambezi Transfrontier Area: Khoisan Influence on Southwestern Bantu Languages. *Journal of African Languages and Linguistics* 36(2): 193–232.

Guthrie, M. (1967) *Comparative Bantu: An Introduction to the Comparative Linguistics and Prehistory of the Bantu languages. Volume 1: The Comparative Linguistics of the Bantu Languages*. London: Gregg International.

Guthrie, M. (1970a) *Comparative Bantu: An Introduction to the Comparative Linguistics and Prehistory of the Bantu languages. Volume 3: A Catalogue of Common Bantu with Commentary*. London: Gregg.

Guthrie, M. (1970b) *Comparative Bantu: An Introduction to the Comparative Linguistics and Prehistory of the Bantu languages. Volume 4: A Catalogue of Common Bantu with Commentary*. London: Gregg.

Guthrie, M. (1971) *Comparative Bantu: An Introduction to the Comparative Linguistics and Prehistory of the Bantu languages. Volume 2: Bantu Prehistory, Inventory and Indexes*. London: Gregg International.

Hamann, S. & N. C. Kula (2015) Illustrations of the IPA: Bemba. *Journal of the International Phonetic Association* 45(1): 61–69.

Hardcastle, W. J. & R. W. P. Brasington (1978) Experimental Study of Implosive and Voiced Egressive Stops in Shona: An Interim Report. *Work in Progress Phonetics Laboratory University of Reading* 2: 66–97.

Hayward, K. M., Y. A. Omar & M. Goesche (1989) Dental and Alveolar Stops in KiMvita Swahili: An Electropalatographic Study. *African Languages and Cultures* 2(1): 51–72.

Herman, R. (1996) Final Lowering in Kipare. *Phonology* 13(2): 171–196.

Hombert, J.-M. (1990) Réalisations tonales et contraines segmentales en fang. *Pholia* 5: 105–111.

Hubbard, K. (1994) *Duration in Moraic Theory*. Berkeley: University of California, PhD dissertation.

Hubbard, K. (1995) Toward a Theory of Phonological and Phonetic Timing: Evidence from Bantu. In Connell, B. & A. Arvaniti (eds.), *Phonology and Phonetic Evidence*, 168–187. Cambridge; New York: Cambridge University Press.

Huffman, M. K. & T. J. Hinnebusch (1998) The Phonetic Nature of "voiceless" Nasals in Pokomo: Implications for Sound Change. *Journal of African Languages and Linguistics* 19(1): 1–19.

Hyman, L. M. (1995) Nasal Consonant Harmony at a Distance: The Case of Yaka. *Studies in African Linguistics* 24(1): 5–30.

Hyman, L. M. (1999) The Historical Interpretation of Vowel Harmony in Bantu. In Hombert, J. M. & L. M. Hyman (eds.), *Bantu Historical Linguistics: Theoretical and Empirical Perspectives*, 235–295. Stanford: CSLI Publications.

Hyman, L. M. (2014) How to Study a Tone Language, with Exemplification from Oku (Grassfields Bantu, Cameroon). *Language Documentation and Conservation* 8: 525–562.

IPA (1999) *Handbook of the International Phonetic Association: A Guide to the Use of the International Phonetic Alphabet*. Cambridge: Cambridge University Press.

Jacottet, E. (1896) Études sur les langues du Haut-Zambèze. Textes originaux, recueillis et traduits en Français et précédés d'une esquisse grammaticale. Première Partie: Grammaires Soubiya et Louyi. Paris: Ernest Leroux.

Jessen, M. (2002) An Acoustic Study of Contrasting Plosives and Click Accompaniments in Xhosa. *Phonetica* 59: 150–179.

Jessen, M. & J. C. Roux (2002) Voice Quality Differences Associated with Stops and Clicks in Xhosa. *Journal of Phonetics* 30: 1–52.

Jouannet, F. (1983) Phonétique et Phonologie: le système consonantique du kinyarwanda. In Jouannet, F. (ed.), *Le kinyarwanda (langue bantu du Rwanda): études linguistiques*, 55–73. Paris: Société des Etudes Linguistiques et Anthropologiques de France avec le concours du Groupe d'Etudes et de Recherches en Linguistique Appliquée, Université Nationale du Rwanda.

Kerremans, R. (1980) Nasale suivie de consonne sourde en proto-bantu. *Africana Linguistica* 8: 159–198.

Kim, S.-A. (1998) Phonetic Assessment of Tone Spreading. In Bergen, B. K., M. C. Plauché & A. C. Bailey (eds.), *Proceedings of the 24th Annual Meeting of the Berkeley Linguistics Society*, 129–138. Berkeley: Berkeley Linguistics Society.

Kishindo, P. J. (2002) "Flogging a dead cow?": The Revival of Malawian Chingoni. *Nordic Journal of African Studies* 11(2): 206–223.

Kisseberth, C. (2016) Chimiini Intonation. In Downing, L. J. & A. Rialland (eds.), *Intonation in African Tone Languages*, 225–284. Berlin: De Gruyter Mouton.

Kisseberth, C. W. & M. I. Abasheikh (2004) The Chimwiini Lexicon Exemplified. Tokyo: ILCAA.

Kodzasov, S. V. & I. A. Muravjeva (1982) Fonetika Tabasaranskogo jazyka. In Zvegintsev, V. (ed.), *Tabasaranskie Etjudy*, 6–16. Moscow: Moscow University.

Kula, N. C. & S. Hamann (2016) Intonation in Bemba. In Downing, L. J. & A. Rialland (eds.), *Intonation in African Tone Languages*, 321–364. Berlin: De Gruyter Mouton.

Kuperus, J. (1985) *The Londo Word: Its Phonological and Morphological Structure*. Tervuren: Royal Museum for Central Africa.

Kutsch Lojenga, C. (2001) The two v's of Giryama. Paper presented at *the 32nd Annual Conference on African Linguistics*. University of California, Berkeley.

Ladefoged, P. (1990) What Do We Symbolize? Thoughts Prompted by Bilabial and Labiodental Fricatives. *Journal of the International Phonetic Association* 20(2): 33–36.

Ladefoged, P. (2000) *A Course in Phonetics*, 4th edition. New York: Harcourt Brace.

Ladefoged, P. (2007) *The UCLA Phonetics Lab Archive*. Los Angeles: UCLA Department of Linguistics (Available online at http://archive.phonetics.ucla.edu/).

Ladefoged, P. & I. Maddieson (1996) *The Sounds of the World's Languages*. Oxford; Cambridge: Blackwell.

Lanham, L. W. (1958) The Tonemes of Xhosa. *African Studies* 17(2): 65–81.

Lee, S. J. (2015) Cumulative Effects in Xitsonga: High-Tone Spreading and Depressor Consonants. *Southern African Linguistics and Applied Language Studies* 33(3): 273–290.

Lee-Kim, S.-I., S. Kawahara & S. J. Lee (2014) The "whistled" Fricative in Xitsonga: Its Articulation and Acoustics. *Phonetica* 71(1): 50–81.

Letele, G. L. (1945) *A Preliminary Study of the Lexicological Influence of the Nguni Languages on Southern Sotho*. Cape Town: University of Cape Town.

Liljencrants, J. & B. Lindblom (1972) Numerical Simulation of Vowel Quality Systems: The Role of Perceptual Contrast. *Language* 48(4): 839–862.

Louw, J. A. (2013) The Impact of Khoesan on Southern Bantu. In Vossen, R. (ed.), *The Khoesan Languages*, 435–444. New York: Routledge.

Lovestrand, J. (2011) *Notes on Nyokon Phonology (Bantu A.45, Cameroon)*. Yaoundé: SIL Cameroon.

Lukusa, S. T. M. & K. C. Monaka (2008) *Shekgalagari Grammar: A Descriptive Analysis of the Language and its Vocabulary*. Cape Town: Centre for Advanced Studies of African Society.

Maddieson, I. (1990) Shona Velarization: Complex Consonants or Complex Onsets? *UCLA Working Papers in Phonetics* 74: 16–34.

Maddieson, I. (1991) Articulatory Phonology and Sukuma "aspirated nasals." In Hubbard, K. (ed.), *Proceedings of the Annual Meeting of the Berkeley Linguistics Society* 17, 145–154. Berkeley: Berkeley Linguistics Society.

Maddieson, I. (1993) Splitting the Mora. *UCLA Working Papers in Phonetics* 83: 9–18.

Maddieson, I. & P. Ladefoged (1993) Phonetics of Partially Nasal Consonants. In Huffman, M. K. & R. A. Krakow (eds.), *Nasals, Nasalization and the Velum*, 251–301. San Diego: Academic Press.

Maddieson, I., P. Ladefoged & B. Sands (1999) Clicks in East African Languages. In Finlayson, R. (ed.), *African Mosaic: Festschrift for J. A. Louw*, 59–91. Pretoria: University of South Africa, UNISA Press.

Maganga, C. & T. C. Schadeberg (1992) *Kinyamwezi: Grammar, Texts, Vocabulary*. Cologne: Rüdiger Köppe.

Maho, J. (1998) *Few People, Many Tongues: The Languages of Namibia*. Windhoek: Gamsberg Macmillan.

System. In Nurse, D. & G. Philippson (eds.), *The Bantu Languages*, 639–651. London; New York: Routledge.

Maho, J. F. (2009) NUGL Online: The Online Version of the New Updated Guthrie List, a Referential Classification of the Bantu Languages (4 Juni 2009) (Available online at http://goto.glocalnet.net/mahopapers/nuglonline.pdf, Accessed December 13, 2010).

Makasso, E.-M., F. Hamlaoui & S. J. Lee (2016) Aspects of the Intonational Phonology of Bàsàá. In Downing, L. J. & A. Rialland (eds.), *Intonation in African Tone Languages*, 167–194. Berlin: De Gruyter Mouton.

Malambe, G. B. (2015) Mid Vowel Assimilation in siSwati. *Southern African Linguistics and Applied Language Studies* 33(3): 261–272.

Manuel, S. Y. (1987) *Acoustic and Perceptual Consequences of Vowel-to-Vowel Coarticulation in Three Bantu Languages*. New Haven: Yale University, PhD dissertation.

Manuel, S. Y. (1990) The Role of Contrast in Limiting Vowel-to-Vowel Coarticulation in Different Languages. *Journal of the Acoustical Society of America* 88: 1286–1298.

Maphalala, Z., M. Pascoe & M. R. Smouse. (2014). Phonological development of first language isiXhosa-speaking children aged 3;0-6;0 years: A descriptive cross-sectional study. Clinical Linguistics and Phonetics, 28(3): 176–194.

Mathangwane, J. T. (1998) Aspirates: Their Development and Depression in Ikalanga. *Journal of African Languages and Linguistics* 19(2): 113–135.

Mathangwane, J. T. (1999) *Ikalanga Phonetics and Phonology: A Synchronic and Diachronic Study*. Stanford: Center for the Study of Language and Information (CSLI), Stanford University.

Medjo Mvé, P. (1997) *Essai sur la phonologie panchronique des parlers fang du Gabon et ses implications historiques*. Lyon: Université Lumière-Lyon2, PhD.

Meeussen, A. E. (1967) Bantu Grammatical Reconstructions. *Africana Linguistica* 3: 79–121.

Meeussen, A. E. (1969) *Bantu Lexical Reconstructions*. Ms. Tervuren: Royal Museum for Central Africa.

Meinhof, C. (1899) *Grundriß einer Lautlehre der Bantusprachen nebst Anleitung zur Aufnahme von Bantusprachen – Anhang : Verzeichnis von Bantuwortstämmen* Leipzig: F.A. Brockhaus.

Miller, A. (2007) Guttural Vowels and Guttural Coarticulation. *Journal of Phonetics* 35: 56–84.

Miller, A. (2008) Click Cavity Formation and Dissolution in IsiXhosa: Viewing Clicks with High-Speed Ultrasound. In Sock, R., S. Fuchs & Y. Laprie (eds.), *Proceedings of the 8th International Seminar on Speech Production*, 137–140. Strasbourg: Institut de Phonetique, available online: issp2008.loria.fr/Proceedings/PDF/issp2008-28.pdf.

Miller, A. (2010) Tongue Body and Tongue Root Shape Differences in Nǀuu Clicks Correlate with Phonotactic Patterns. In Fuchs, S., M. Toda & M. Żygis (eds.), *Turbulent Sounds: An Interdisciplinary Guide*, 245–279. Berlin: De Gruyter Mouton.

Miller, A. (2016) Posterior Lingual Gestures and Tongue Shape in Mangetti Dune !Xung Clicks. *Journal of Phonetics* 55: 119–148.

Miller, A., J. Brugman, B. Sands, L. Namaseb, M. Exter & C. Collins (2009a) Differences in Airstream and Posterior Place of Articulation Among Nǀuu Clicks. *Journal of the International Phonetic Association* 39(2): 129–161.

Miller, A. & K. B. Finch (2011) Corrected High Frame Rate Anchored Ultrasound with Software Alignment. *Journal of Speech, Language and Hearing Research* 54: 471–486.

Miller, A., L. Namaseb & K. Iskarous (2007) Tongue Body Constriction Differences in Click Types. In Cole, J. S. & J. I. Hualde (eds.), *Proceedings of Laboratory Phonology 9*, 643–656. Berlin: De Gruyter Mouton.

Miller, A., A. Scott, B. Sands & S. Shah (2009b) Rarefaction Gestures and Coarticulation in Mangetti Dune !Xung clicks. *Proceedings of the 10th Annual Conference of the International Speech Communication Association (Interspeech 2009)*, 2279–2282. Brighton: Causal Productions.

Mkanganwi, K. G. (1972) The Relationships of Coastal Ndau to the Shona Dialects of the Interior. *African Studies* 31: 111–137.

Möhlig, W. J. G. (1997) A Dialectometrical Analysis of the Main Kavango Languages: Kwangali, Gciriku and Mbukushu. In Haacke, W. H. G. & E. D. Elderkin (eds.), *Namibian Languages. Reports and Papers*, 211–234. Cologne: Rüdiger Köppe.

Monaka, K. C. (2001) *Shekgalagari Stop Consonants: A Phonetic and Phonological Study*. London: University College, University of London, PhD dissertation.

Monaka, K. C. (1999) Shekgalagari Laryngeal Contrasts: The Plosives. *South African Journal of African Languages* 25(4): 243–257.

Moyo, C. T. (1995) Language Contact and Language Change: The Case for chiTumbuka in Northern Malawi. *South African Journal of African Languages* 15(4): 186–191.

Mutaka, N. M. & C. Ebobissé (1996–1997) The Formation of Labial-Velars in Sawabantu: Evidence for Feature Geometry. *Journal of West African Languages* 26(1): 3–14.

Myers, S. (1996) Boundary Tones and the Phonetic Implementation of Tone in Chichewa. *Studies in African Linguistics* 25(1): 29–60.

Myers, S. (1999a) Downdrift and Pitch Range in Chichewa Intonation. *Proceedings of the 14th International Congress of Phonetic Sciences* 3: 1981–1984.

Myers, S. (1999b) Tone Association and F0 Timing in Chichewa. *Studies in African Linguistics* 28(2): 215–239.

Myers, S. (2003) F0 Timing in Kinyarwanda. *Phonetica* 60(2): 71–97.

Myers, S. (2005) Vowel Duration and Neutralization of Vowel Length Contrasts in Kinyarwanda. *Journal of Phonetics* 33(4): 427–446.

Myers, S. (2015) An Acoustic Study of Luganda Liquid Allophones. *Journal of African Languages and Linguistics* 36(1): 67–92.

Myers, S. & T. Carleton (1996) Tonal Transfer in Chichewa. *Phonology* 13: 39–72.

Nabirye, M., G.-M. de Schryver & J. Verhoeven (2016) Illustrations of the IPA: Lusoga (Lutenga). *Journal of the International Phonetic Association* 46(2): 219–228.

Nagano-Madsen, Y. & C. Thornell (2012) Acoustic Properties of Implosives in Bantu Mpiemo. *Fonetik 2012, 15th Swedish Phonetics Conference, May 30–June 1, University of Gothenburg*, 73–76.

Naidoo, S. (2007) A Qualitative and Quantitative Study of Zulu Affricates. *South African Journal of African Languages* 27(3/4): 83–96.

Naidoo, S. (2010) A Re-evaluation of the Zulu Implosive [ɓ]. *South African Journal of African Languages* 30(1): 1–10.

Nchimbi, A. S. A. (1997) Formant Structure of Standard KiSwahili Vowels. Paper presented at *Second World Congress of African Linguistics*, Leipzig.

Ndana, Ndana, Kapule David Mabuta & Andy Chebanne. (2017) Chiikuhane (Subiya) Manual with Orthography. (CASAS Book Series, No. 123). Cape Town: Centre for Advanced Studies of African Society.

Ndinga-Koumba-Binza, H. S. (2011) Interaction of Variables in the Civili Vowel Duration. *The 17th International Congress of Phonetic Sciences (ICPhS XVII), Hong Kong, August 17–21, 2011*, 1458–1461.

Niesler, T., P. Louw & J. C. Roux (2005) Phonetic Analysis of Afrikaans, English, Xhosa and Zulu Using South African Speech Databases. *South African Linguistics and Applied Language Studies* 23(4): 459–474.

Nurse, D. (1994) South Meets North: Ilwana=Bantu+Cushitic on Kenya's Tana River. In Bakker, P. & M. Mous (eds.), *Mixed Languages: 15 Case Studies in Language Intertwining*, 215–224. Amsterdam: Institute for Functional Research into Language and Language Use.

Nurse, D. & T. J. Hinnebusch (1993) *Swahili and Sabaki: A Linguistic History*. Berkeley: University of California Press.

Odden, D. (1996) *The Phonology and Morphology of Kimatuumbi*. Oxford: Clarendon Press.

Olson, K. & Y. Meynadier (2015) On Medumba Bilabial Trills and Vowels. In The Scottish Consortium for ICPhS 2015 (ed.), *Proceedings of the 18th International Congress of Phonetic Sciences. ed. The Scottish Consortium for ICPhS 2015*. Glasgow: University of Glasgow, retrieved from www.icphs2015.info/pdfs/Papers/ICPHS0522.pdf.

Olson, K. S. & J. Hajek (1999) The Phonetic Status of the Labial Flap. Journal of the International Phonetic Association 29(2): 101–114.

Pakendorf, B., H. Gunnink, B. Sands & K. Bostoen (2017) Prehistoric Bantu-Khoisan Language Contact: A Cross-Disciplinary Approach. *Language Dynamics and Change* 7(1): 1–46.

Passy, P. (1914) La langue Thonga. *Miscellanea Phonetica* I: 27–32.

Patin, C. (2013) /r/ in Washili Shingazidja. In Spreafico, L. & A. Vietti (eds.), *Rhotics: New Data and Perspectives*, 173–190. Bolzano: Bozen-Bolzano University Press.

Patin, C. (2016) Tone and Intonation in Shingazidja. In Downing, L. J. & A. Rialland (eds.), *Intonation in African Tone Languages*, 285–320. Berlin: De Gruyter Mouton.

Paulian, C. (1994) Nasales et nasalisation en ŋgùŋgwèl, langue bantu du Congo. *Linguistique africaine* 13: 83–129.

Persson, J. A. (1932) *Outlines of a Tswa Grammar with Practical Exercises*. Cleveland: Central Mission Press.

Philippson, G. & M.-L. Montlahuc (2003) Kilimanjaro Bantu (E60 and E74). In Nurse, D. & G. Philippson (eds.), *The Bantu Languages*, 475–500. London; New York: Routledge.

Pongweni, A. J. C. (1990) *Studies in Shona Phonetics: An Analytical Review*. Harare: University of Zimbabwe.

Poulos, G. (1990) *A Linguistic Analysis of Venda*. Pretoria: Via Afrika.

Poulos, G. & L. J. Louwrens (1994) *A Linguistic Analysis of Northern Sotho*. Pretoria: Via Afrika.

Prinsloo, D. J. & G.-M. de Schryver (2001) Corpus Applications for the African Languages, with Special Reference to Research, Teaching, Learning, and Software. *Southern African Linguistics and Applied Language Studies* 19: 111–131.

Proctor, M., E. Bresch, D. Byrd, K. Nayak & S. Narayanan (2013) Paralinguistic Mechanisms of Production in Human "beatboxing": A Real-Time Magnetic Resonance Imaging Study. *Journal of the Acoustical Society of America* 133(2): 1043–1054.

Proctor, M., Y. Zhu, A. Lammert, A. Tsoutios, B. Sands & S. Narayanan (forthcoming) Studying Clicks Using Real-Time MRI. In Sands, B. (ed.), *Handbook of Click Languages*. Leiden: Brill.

Renaud, P. (1976) *Le bajele: phonologie, morphologie nominale. Volume 1: Phonologie*. Yaoundé: ALCAM, Unité de recherche linguistique et phonétique, Institut des Sciences humaines.

Rialland, A. (2007) Question Prosody: An African Perspective. In Riad, T. & C. Gussenhoven (eds.), *Tones and Tunes. Vol. 1: Typological Studies in Word and Sentence Prosody*, 35–62. Berlin: De Gruyter Mouton.

Rialland, A. & M. E. Aborobongui (2016) How Intonations Interact with Tones in Embosi (Bantu C25), a Two-Tone Language Without Downdrift. In Downing, L. J. & A. Rialland (eds.), *Intonation in African Tone Languages*, 195–222. Berlin: De Gruyter Mouton.

Roux, J. C. (1995) On the Perception and Production of Tone in Xhosa. *South African Journal of African Languages* 15(4): 196–204.

Roux, J. C. (2007) Unresolved Issues in the Representation and Phonetic Description of Click Articulation in Xhosa and Zulu. *Language Matters* 38(1): 8–25.

Roux, J. C. & A. J. Holtzhausen (1989) An Acoustic and Perceptual Analysis of Xhosa Vowels. *South African Journal of African Languages* 9(1): 30–34.

Roux, J. C. & H. S. Ndinga-Koumba-Binza (2011) Perceived Vowel Duration in Civili: Minimal Pairs and the Effect of Post-Vocalic Voicing. *The 17th International Congress of Phonetic Sciences (ICPhS XVII), Hong Kong, August 17–21, 2011,* 1726–1729.

Rueck, M. J., N. Bako, L. Hon, J. Muniru, L. Otronyi & Z. Yoder (2009) *Preliminary Impressions from the Sociolinguistic Survey of the Jar Dialects.* Ms. Jos:

Rycroft, D. K. (1980) *The Depression Feature in Nguni Languages and Its Interaction with Tone.* Grahamstown: Department of African Languages, Rhodes University.

Rycroft, D. K. (1981) *Concise SiSwati Dictionary: SiSwati-English/English-SiSwati.* Pretoria: J.L. van Schaik.

Sands, B. & H. Gunnink. (forth.). Clicks on the fringes of the Kalahari Basin Area. Theory and Description in African Linguistics: Selected Papers from the 47th Annual Conference on African Linguistics. ed. Peter Jenks, Emily Clem & Hannah Sande. Berlin: Language Science Press.

Schadeberg, T. C. (1982) Nasalization in Umbundu. *Journal of African Languages and Linguistics* 4: 109–132.

Schadeberg, T. C. (1995) Spirantization and the 7-to-5 Vowel Merger in Bantu. *Belgian Journal of Linguistics* 9: 71–84.

Schulz, S. & A. Laine (2016) Click Loss, Variation and Acquisition in Two South African Ndebele Varieties. Paper presented at *the 46th Colloquium on African Languages and Linguistics*, Leiden.

Schwartz, J.-L., L.-J. Boë, N. Vallée & C. Abry (1997) The Dispersion-Focalization Theory of Vowel Systems. *Journal of Phonetics* 25: 255–286.

Shosted, R. K. (2006) Just Put Your Lips Together and Blow? Whistled Fricatives in Southern Bantu. In Yehia, H. C., D. Demolin & R. Laboissiere (eds.), *Proceedings of ISSP 2006: 7th International Seminar on Speech Production*, 565–572. Belo Horizonte: CEFALA.

Shosted, R. K. (2011) Articulatory and Acoustic Characteristics of Whistled Fricatives in Changana. In Bokamba, E. G., R. K. Shosted & B. T. Ayalew (eds.), *Selected Proceedings of the 40th Annual Conference on African Linguistics*, 119–129. Somerville: Cascadilla Proceedings Project.

Sitoe, B. (1996) *Dicionário Changana-Português.* Maputo: Instituto Nacional do Desenvolvimento de Educação.

Skhosana, P. B. (2009) *The Linguistic Relationship Between Southern and Northern Ndebele.* Pretoria: University of Pretoria, PhD dissertation.

Solé, M.-J., L. M. Hyman & K. C. Monaka (2010) More on Post-Nasal Devoicing: The Case of Shekgalagari. *Journal of Phonetics* 38(4): 604–615.

Spiss, C. (1904) Kingoni und Kisutu. *Mittheilungen des Seminars für Orientalische Sprachen* 7: 270–414.

Starwalt, C. (2008) *The Acoustic Correlates of ATR Harmony in Seven- and Nine-Vowel African Languages: A Phonetic Inquiry Into Phonological Structure.* Arlington: University of Texas, PhD dissertation.

Stevick, E. W. (1969) Tone in Bantu. *International Journal of American Linguistics* 35(4): 330–341.

Stewart, J. M. (2000) An Explanation of Bantu Vowel Height Harmony in Terms of a Pre-Bantu Nasalized Vowel Lowering. *Journal of African Languages and Linguistics* 21(2): 161–178.

Thomas, K. (2000) Coproduction and Coarticulation of Clicks in IsiZulu: Aerodynamic and Electropalatographic Evidence. In Carstens, V. & F. Parkinson (eds.), *Advances in African Linguistics*, 265–280. Trenton; Asmara: Africa World Press.

Thomas-Vilakati, K. D. (2010) *Coproduction and Coarticulation in IsiZulu Clicks.* Berkeley: University of California Press.

Tlale, O. (2005) *The Phonetics and Phonology of Sengwato, a Dialect of Setswana.* Washington, DC: Georgetown University, PhD dissertation.

Traill, A. (1990) Depression Without Depressors. *South African Journal of African Languages* 10(4): 166–172.

Traill, A. & M. Jackson (1988) Speaker Variation and Phonation Type in Tsonga. *Journal of Phonetics* 16: 385–400.

Traill, A., J. S. M. Khumalo & P. Fridjhon (1987) Depressing Facts About Zulu. *African Studies* 46(2): 255–274.

Van de Velde, M. & D. Idiatov (2016) Stem-Initial Accent and C-Emphasis Prosody in North-Western Bantu. Paper presented at *Special Workshop on Areal Features and Linguistic Reconstruction, 47th Annual Conference on African Linguistics, 23–26 March, 2016*, University of California, Berkeley.

Voll, R. (2012) Tonal Variation in the Tense System of Mundabli, Western Beboid (Bantoid, Cameroon). In Brenzinger, M. & A.-M. Fehn (eds.), *Proceedings of the 6th World Congress of African Linguistics, Cologne, 17–21 August 2009*, 533–543. Cologne: Rüdiger Köppe.

Wissing, D. & W. Pienaar (2014) Evaluating Vowel Normalisation Procedures: A Case Study on Southern Sotho Vowels. *Southern African Linguistics and Applied Language Studies* 32(1): 97–111.

Wright, R. & A. Shryock (1993) The Effects of Implosives on Pitch in SiSwati. *Journal of the International Phonetic Association* 23(1): 16–23.

Zerbian, S. (2009) Phonology and Phonetics of Tone in Northern Sotho, a Southern Bantu language. In Austin, P. K., O. Bond, M. Charette, D. Nathan & P. Sells (eds.), *Proceedings of the Conference on Language Documentation and Linguistic Theory 2*, 313–321. London: SOAS.

Zerbian, S. (2016) Sentence Intonation in Tswana (Sotho-Tswana group). In Downing, L. J. & A. Rialland (eds.), *Intonation in African Tone Languages*, 393–434. Berlin: De Gruyter Mouton.

Zerbian, S. & E. Barnard (2008) Phonetics of Intonation in South African Bantu Languages. *South African Linguistics and Applied Language Studies* 26(2): 235–254.

Zerbian, S. & F. Kügler (2015) Downstep in Tswana (Southern Bantu). In The Scottish Consortium for ICPhS 2015 (ed.), *Proceedings of the 18th International Congress of Phonetic Sciences.* Glasgow: University of Glasgow. (Available online at www.icphs2015.info/pdfs/Papers/ICPHS0291.pdf).

Ziervogel, D. (1959) *A Grammar of Northern Transvaal Ndebele.* Pretoria: van Schaik.

Ziervogel, D., P. J. Wentzel & T. N. Makuya (1981) A Handbook of the Venda Language. Pretoria: University of South Africa.

Zsiga, E. C., M. Gouskova & O. Tlale (2006) On the Status of Voiced Stops in Tswana: Against *ND. *Proceedings of the North Eastern Linguistic Society* 36: 721–734.

Ziervogel, D., P. J. Wentzel & T. N. Makuya (1981) *A Handbook of the Venda Language.* Pretoria: University of South Africa.

Zsiga, E. C., M. Gouskova & O. Tlale (2006) On the Status of Voiced Stops in Tswana: Against *ND. *Proceedings of the North Eastern Linguistic Society* 36: 721–734.

CHAPTER 4

SEGMENTAL PHONOLOGY

Larry Hyman

INTRODUCTION

As in other parts of the grammar, Bantu segmental phonology can be characterised as a theme and variations: despite the large number of languages and great geographic expanse that they cover, the most noteworthy properties concerning Bantu syllable structure, consonant/vowel inventories and phonological processes are robustly attested throughout the Bantu zone. These shared features, although striking, however mask a wide range of differences which are equally, if not more, important in understanding Bantu phonology in general. It is helpful in this regard to consider both the phonological system inherited from Proto-Bantu, as well as the innovations, often areally diffused, which characterise present-day Bantu subgroups and individual languages.

1 PROTO-BANTU

According to most Bantuists, e.g., Meeussen (1967), Proto-Bantu (PB) had the relatively simple consonant and vowel systems in (1):

(1) a   consonants   b   vowels (long and short)

p	t	c	k	i	u
b	d	j	g	ɪ	ʋ
m	n	ɲ		ɛ	ɔ
				a	

Of the two series of oral consonants in (1a), all scholars agree that the voiceless series **p*, **t*, **k* were pronounced as stops. There is, however, disagreement as to whether **b *d *g* should be reconstructed as a parallel voiced stop series or as voiced continuants, i.e., **β *l *γ*, as they are pronounced in many daughter languages today. It also is not clear whether **c* and **j* should be viewed as palatal stops or affricates – or whether they were palatal at all. Many Bantu languages realise **c* as /s/, and some realise **j* as /z/. Realisations of the latter as /y/ or /j/ (i.e., [dʒ]) are, however, probably more common. Although various scholars have occasionally posited additional consonants and series of consonants (e.g., fortis vs. lenis stops) in the proto system, none of these have been demonstrated to the satisfaction of the Bantuist community. On the other hand, much more complex systems have been innovated in the daughter languages as seen in Maddieson (2003) and below (see also chapter 3).

There is, by comparison, more stability in the vowel system, which is reconstructed as in (1b). Most scholars agree that PB had seven distinct vowels (7V). As transcribed in (1b), there would have been an opposition in the high vowels between tense or [+ATR] **i* and **u* and lax or [-ATR] **ɪ* and **ʋ*, which has frequently been transcribed as **i̧ *u̧* vs. *i *u. The mid vowels *ε and *$\supset$ were also lax, although they are often transcribed *e and *o for convenience, which I will generally follow here. Such a system, exemplified from Nande JD42 in (2a) (Mutaka 1995), is widely attested, especially in eastern Bantu, where I have changed the i u transcription of the degree 1 high vowels to be consistent throughout this chapter:

(2) a   -lím-   'exterminate'   -lúm-   'be animated'
        -lìm-   'cultivate'   -lɔ́m-   'bite'
        -lɛ̀m-   'fail to carry sth. heavy'   -tɔ́m-   'put aside'
        -làm-   'heal'

   b   -tìng-   'hook'   -tùng-   'become thin'
        -bèng-   'chase'   -tóng-   'construct'
        -kɛ́ng-   'observe'   -tɔ́ng-   'gather up'
        -tàng-   'flow'

Other Bantu languages, such as Koyo C24 in (2b) (personal notes), have the 7V system /i e ε u o $\supset$ a/, where there is instead a tense/lax or ATR opposition in the mid vowels. Such a system is particularly frequent in Western Bantu. See Stewart (1983, 2000–2001) and Hyman (1999) for further discussion of the reconstructed PB vowel system.

The syllable structures allowed in PB were limited to those in (3).

(3) a   CV, CVV   b   V, N

Most syllables in PB had one of the two shapes in (3a): a single consonant followed by a vowel that was either short (V) or long (VV), e.g., *-$pád$- 'scrape' vs. *-$páad$- 'quarrel.' The syllable shapes in (3b) were most likely limited to prefixes, e.g., *$à$- 'class 1 subject prefix,' *$\dot{N}$- 'class 9 noun prefix.' PB roots with non-identical vowels in sequence have also been reconstructed, e.g., *-$bàij$- 'carve,' *-$gòì$ 'leopard,' but may have involved "weak" intervening consonants, e.g., glides, that dropped out in pre-PB (cf. *-$bí$-ad- 'give birth'). Many vowel sequences, including some identical ones, e.g., *-$tú$-ud- 'rest, put down load,' are analysable as heteromorphemic (cf. *-$tú$-ad- 'carry on the head'), such that Meeussen (1979) questioned whether PB actually had long vowels at all. Others have subsequently arisen through the loss of PB consonants, e.g., *$\dot{N}$-$gòbì$, *$\dot{N}$-$jògù$ > Kamba E55 ŋ-goi 'baby sling,' n-zou 'elephant' (Hinnebusch 1974). In many languages this consonant loss is restricted and results in synchronic alternations between C and Ø. For example, while *-$dìd$- 'cry' is realised li-a (with FV -a) in Swahili G42, the final *d is realised [l] before the applicative suffix -i-, e.g., -lil-i-a 'cry to/for' (from *-$dìd$-$ɪd$-a), since [l] does not normally delete before front vowels (Nurse & Hinnebusch 1993: 100). In contrast with PB *VV sequences, which are tautosyllabic, neo-VV sequences typically remain heterosyllabic. On the other hand, many Bantu languages have lost the inherited V/VV opposition, e.g., *-$dóot$- 'dream' > Tonga M64 -$lót$-, Chewa N31b -$lót$-, Tswana S31 -$lór$-. Onsetless syllables typically have a short vowel, while CVV may be favored in certain alternations. This is the case in roots such as -(y)$ér$- 'sweep' in Ganda JE15, where the initial "unstable-y" is realised in all environments except when preceded by a CV- prefix with which the following vowel fuses, e.g., tw-$èèr$-a 'we sweep' vs. a-$yér$-$à$ 'he sweeps,' n-$jér$-$à$ 'I sweep' (Hyman & Katamba 1999).

The last syllable type, a low tone nasal, is reconstructable in class 9 and 10 noun prefixes, while syllabic nasal reflexes of the first person singular subject and object prefixes derive from earlier *nì-. The nasal of morpheme-internal NC sequences appears never to be itself syllabic, although it frequently conditions length on a preceding vowel, e.g., *-gènd- [-gè:nd-] 'walk' (Clements 1986). If correctly analysed as two segments, NC constitutes the only consonant cluster in PB. This includes heteromorphemic N+N sequences, although these are subsequently degeminated in many Bantu languages. At the same time, new syllabic nasals often derive from the loss of the vowel of *mV- prefixes, especially *mu-CV > m̩-CV > N̩-CV (Bell 1972, Nurse & Hinnebusch 1993: 181–185, Hyman & Ngunga 1997: 139ff, Kadenge 2014, Bostoen & de Schryver 2015, Kadenge 2015).

Some Bantu languages have developed additional syllable structures, typically by the loss of vowels or consonants or through borrowings. Most word-final vowels have been lost in Ruwund L53, whose word-final syllables therefore usually end in a conso-nant, e.g., *N̩-búdà, *dò-kónì > ń-vùl 'rain,' rú-kùɲ 'firewood' (Nash 1992). North-West Bantu languages such as Basaa A43, on the other hand, have not only lost final vowels, e.g., *mò-dómè > n-lóm 'male,' but also create non-final closed syllables by syncopat-ing the medial vowel of CVCVCV stems, e.g., tiɲil 'untie' + a 'passive suffix' → tiɲla (Lemb & de Gastines 1973). Other languages which have closed syllables include the closely related languages Eton A71 (Van de Velde 2008) and Fang A75 (Medjo Mvé 1997), Mbuun B87 (Bostoen & Mundeke 2011), and Nzadi B865 (Crane et al. 2011), the last of which often retains the tones when vowels are deleted, e.g., *mò-(j)ánà > mwàán 'child.' Closed syllables may also be found in incompletely assimilated borrowings, e.g., Swahili G42 m-kristo 'Christian' (Wilson 1985); Yaka H31, mártóo 'hammer' (< French marteau) (Ruttenberg 2000).

Most Bantu languages maintain a close approximation of the PB situation as far as syllable structure is concerned. The open syllable structure is, in fact, reinforced by the well-known Bantu agglutinative morphology. The typical structures of nouns and verbs are schematised in (4).

(4) a   Nouns   AUGMENT   PREFIX   STEM
                CV   CV   CV(V)CV
                V   N   CV(V)
                          V   VCV

   e.g., Bukusu JE31c   kú-mù≠xòò   'arm'   (class 3)
                        ó-mù-xàsì   'woman'   (class 1)
                        é-ǹ-jùxì   'bee'   (class 9)
                        cí-ǹ-jùxì   'bees'   (class 10)
                        lí-ì-bèèlè   'breast' (class 5)

   b   Verbs   PI   SP   NEG   TP   AP   OP   STEM
               CV   CV   CV   CV   CV   CV   CV(V)CV . . .
               V   V   V   N
               N

   e.g., Nande JD42   mó-tù-tétà-yà-mú≠túm-à
                      PI- SP-NEG-AP-OP-SEND-FV
                      'we didn't go and send him'
   (PI = pre-initial morpheme, including augment; SP = subject prefix; NEG = neg-ative, TP = tense prefix, AP = aspect prefix, OP = object prefix, FV = inflectional final vowel)

c   Verb Stems   ROOT   EXTENSION(S)   FV
                 CVC   VC   V
                 CV   V

e.g., Chewa N31b   -lìm-ìts-ìl-àn-à   'cause to cultivate for each other'
                   ROOT-CAUS-APPL-RECP-FV   (causative-applicative-reciprocal)

In the above schemas, V stands for any of the seven PB vowels, while C stands for any of the proto-consonants, including, potentially, NC. The most common shape of each morpheme is given in the first row. As seen in (4a,b), pre-stem morphemes (prefixes) are restricted to the shapes CV-, V- and N-. In (4c), on the other hand, we see that post-root morphemes (suffixes) have the shapes -VC- and -V-. Since most roots begin and end with a C, morpheme concatenation provides almost no potential for consonant clusters, but rather a general alternation of consonant-vowel-consonant etc. The one exception to this occurs when Vs meet across a morpheme boundary. In this case specific rules modify the resulting V+V inputs (see Section 3.2). Other assimilatory and dissimilatory alternations occur when morphemes meet, some of which are restricted to specific domains, e.g., the stem (root+suffixes). Many of these alternations produce output segments beyond those in the above V and C inventories. The most widespread phenomena are treated in the following sections, first for vowels (Section 3), then for consonants (Section 4).

2 VOWEL PHONOLOGY

While PB had the 7V system in (1b), the majority of Bantu languages have merged the degree 1 and 2 vowels to achieve the five-vowel (5V) system in (5a).

(5) a   Swahili G42   b   Budu D332   c   Bafia A53
         i   u   i   u   i   ɨ   u
         ɛ   ɔ   ɪ   ʊ   e   ə   o
             a   e   o   ɛ   ʌ   ɔ
                              ɛ   ɔ   a   ɑ
                                 a

A few languages have gone the other direction and developed the nine-vowel system in (5b). While this system appears to be underlying in Budu D332 (Kutsch Lojenga 1994), other languages, such as Nande JD42 and the Sotho-Tswana S30 languages, derive the eighth and ninth vowels [e] and [o] from the tensing/raising of the degree 3 vowels *ɛ and *ɔ, respectively (Mutaka 1995, Creissels 2005). Finally, some of the languages in zone A, such as Bafia A53 in (5c), have developed "rectangular" vowel systems with back unrounded vowels (Guarisma 1969), which also appear in Grassfields Bantu (Watters 2003). B70 and B80 languages have also developed nasalised and fronted ("umlauted") rounded vowels (Hombert 1986, Bostoen & Koni Muluwa 2014).

3.1 DISTRIBUTIONAL CONSTRAINTS ON UNDERLYING VOWELS

As indicated in Section 2, Bantu phonology is highly sensitive to morphological considerations. Underlying vowel distribution within specific morphological slots and morphological or prosodic domains is thus highly restricted in both seven-vowel and five-vowel languages. Meeussen (1967), for example, allows for the following vowels in each of the indicated positions.

(6) PB vowel reconstructions by position (cf. Hyman 2008, Teil-Dautrey 2008)

	*i	*u	*ɪ	*ʊ	*ɛ	*ɔ	*a
first stem syllable	x	x	x	x	x	x	x
final stem vowel	x	x	x	x	x	x	x
elsewhere	x		x	x			x

The seven vowels of PB contrast in the first and last syllables of a stem, but not in pre-fixes, extensions or stem-internal position, where only four vowels contrast. In a few cases involving reduplication, the vowel *u* appears in the first two syllables of a verb, e.g., *-dùdum-* 'rumble, thunder,' *-pùpum-* 'boil up, boil over.' The root *-tákun-* 'chew,' on the other hand, appears to be exceptional.

Some languages, particularly five-vowel ones, have further restricted this distribution by position within the stem or word. Thus, Punu B43, which has the underlying system /i u ɛ ɔ a/, restricts /ɛ/ and /ɔ/ to stem-initial syllables only (Kwenzi Mikala 1980). In Bobangi C32 (7V), /u/ may not occur in prefixes, nor may any of the rounded vowels /u, o, ɔ/ appear later than the second syllable in stems.

3.2 VOWEL ALTERNATIONS

In addition to underlying constraints on vowel distribution, most Bantu languages severely restrict the sequencing of vowels, particularly within stems. Thus, while Punu B43 allows only /i u a/ in post-root position, /a/ is reduced to schwa in this position, and the expected post-stem sequences [əCi] and [əCu] surface instead as [iCi] and [uCu]. Thus, the histor-ical suffix sequences /-am-il-/ (positional-applicative) and /-am-ul-/ (positional-reversive tr.) are realised [-imin-] and [-umun-]. In addition, a post-stem /a/ ([ə]) assimilates to a FV -i, and both post-stem /a/ and /i/ assimilate to a FV -u (Fontaney 1980). The Punu B43 case demonstrates two general properties of Bantu vowel systems: (i) There are typically more contrasting underlying vowels in the stem-initial syllable, and (ii) vowels in this position may be exempt from reduction and assimilation processes that post-stem vowels undergo. Ruwund L53 (5V) once had the same vowel distribution as Punu B43, disallowing mid vowels from post-stem position. However, it has since undergone con-siderable vowel reduction, among other things by dropping most word-final vowels, e.g., *mò≠(j)ánà > mwáàn* 'child,' *mò≠kádì > mú≠kàj* 'wife' (Nash 1992). While unusual in tolerating word-final closed syllables, which are analysed with a phantom vowel by Nash, Ruwund L53 is perhaps unique in its overall vowel system in (7a).

(7) a   i   u   ii   uu
                    ee   oo
            a   aa

     b   *e > i   e.g. *-dém- > -lím- 'be lame,' *-dèd- > -lìl- 'raise (child)'
         *o > a   e.g. *-bón- > -màn 'see,' *-pót- > -pwàt- 'twist'

As seen in (7a), short /e o/ are missing, since they have reduced, respectively, to the periph-eral vowels [i] and [a], as illustrated in (7b). In this atypical case, vowels in the stem-initial syllable were successfully targeted. In Makonde P23 both /e/ and /o/ optionally reduce to [a] in prepenultimate position (Liphola 2001): /ku-pet-il-a/ -> ku-pet-el-a ~ ku-pat-el-a 'to separate for', /ku-pot-il-a/ -> ku-pot-el-a ~ ku-pat-el-a 'to twist for'.

By far the most widely attested assimilatory process is vowel harmony, particularly vowel height harmony (VHH), as indicated in (8) and (9).

(8) Front height harmony (FHH)
    a   General   : $*_\iota$ → e / /e o/   C ___

    b   Extended : $*_\iota$ → e / /e o a/ C ___

(9) Back height harmony (BHH)
    a   General   :   *ʊ → o / /o/   C ___

    b   Extended :   *ʊ → o / /e o/ C ___

The historical degree-2 vowels (*ɪ *ʊ) harmonise in height with a preceding mid vowel. The process is frequently different with respect to the front vs. back vowel (Bleek 1862). In a wide range of Central and Eastern Bantu languages, degree-2 /ɪ/ lowers after both /e/ and /o/, while degree-2 /ʊ/ lowers only after /o/. Examples of such "asymmetric" FHH vs. BHH are seen from Nyamwezi F22 in (10), based on Maganga and Schadeberg (1992).

(10)   Root + Applicative -ɪl-   Root + Separative -ʊl-
    a βìs-íl-   'hide for/at'   βìs-úl-   'find out'
        gùb-ík-íl- 'put on lid for/at'   gùb-úl-   'take off lid'
        pììnd-íl- 'bend for/at'   pììnd-ùl- 'overturn'
        shòòn-íl- 'gnaw for/at'   shòòn-òl- 'show teeth'
        gàβ-íl-   'divide for/at'   gàβ-òl-   'divide'

    b βòn-èl-   'see for/at'   hòng-ól- 'break off'

    c zèèng-èl- 'build for/at'   zèèng-òl- 'build'

There is no harmony in (10a), where the root vowel is either high or low. In (10b) VHH applies to both -ɪl- and -ʊl- when the root vowel is /o/. However, in (10c), where the root vowel is /e/, -ɪl- lowers to -el-, but -ʊl- remains unchanged. This contrasts with the situation in the JE40 group, e.g., Gusii JE42 (7V), as well as in many North-West Bantu languages. In Mongo-Nkundo C61 (7V), whose vowel system is analysed as /i e ε u o ɔ a/, the back degree-2 vowel also harmonises, and FHH and BHH are thus "symmetric."

(11)   Root + Applicative -el-   Root + Separative -ol-
    a -íy-èl-   'steal for/at'   -ìs-òl-   'uncover'
        -lúk-èl-   'paddle for/at'   -kùnd-òl- 'dig up'
        -ét-èl-   'call for/at'   -bét-òl-   'wake up'
        -tóm-èl-   'send for/at'   -kòmb-òl- 'open'
        -kamb-èl- 'work for/at'   -bák-òl-   'untie'

    b -kɔt-èl-   'cut for/at'   -mɔm-òl-   'unglue'
        -kènd-èl- 'go for/at'   -tɛ́ng-ɔl- 'straighten out'

As indicated in (8b), harmony of the reflex of *ɪ to [e] after /a/ is also attested, particularly in languages towards the Southwest of the Bantu zone, such as Mbundu H21a, Kwangali K33 and Herero R31. While VHH is generally perseverative and limited to the stem domain minus the FV, some languages – particularly those with symmetric FHH/BHH – have extended harmony to the FV and to prefixes. It is important to note that in many 5V Bantu languages, only those /i/ vowels that derive from (degree 2) *ɪ harmonise, while those that derive from *i do not. Other languages have modified this original situation and harmonise the perfective ending *-ìd-è as well (Bastin 1983). In Yaka H31 and certain varieties of Kongo H10, VHH is anticipatory and is triggered primarily by the *-e of perfective *-ìd-è (Hyman 1998, 1999). In addition, some of the same languages that have symmetric VHH have extended the process to prefixes, e.g., Mituku D13 (7V) /tú-mú-lòk-é/ → tó-mó-lòk-é 'let us bewitch him/her (subjunctive)' (Stappers 1973). For more discussion of variations in Bantu VHH, see Leitch (1996), Hyman (1999) and Goes & Bostoen (2019).

Besides height, other features may also participate in vowel harmony. Closely related Konzo JD41 and Nande JD42 have introduced an advanced tongue root (ATR) harmony, whereby /ɪ ʊ ɛ ɔ/ become [i u e o] when followed by /i/ or /u/. They thus have an underlying 7V system (cf. (2)), but introduced two additional, non-contrasting vowels, [e] and [o], by ATR harmony. /a/ may also be phonetically affected by a following /i/ or /u/ (Gick et al. 2006). Clements (1991) provides an overarching framework to treat VHH and ATR harmony in related fashion. Other Bantu languages have innovated rounding harmonies, e.g., perseverative i → u/u ___ in Lengola D12 (Stappers 1971) vs. anticipatory i → u/___ u in Punu B43 (Fontaney 1980). In Maore G44D, regressive rounding harmony even reaches the root vowel: u≠finiki-a 'cover,' u≠funuku-a 'uncover' (Rombi 1983). Finally, it should be noted that many North-Western 7V languages modify /a/ to [ɛ] after /ɛ/ and to [ɔ] after /ɔ/, e.g., Mokpwe A22, Tiene B81, Lingala C36d or, in the case of Bembe H11 (5V), to [e] and [o], respectively.

The above shows the assimilation of one vowel to another across a consonant. When vowels occur in direct sequence, they typically undergo gliding or deletion. Thus, in Ganda JE15, when followed by a non-identical vowel, e.g., the FV -a, the front vowels /i e/ glide to [y], as in (12a), and the back vowels /u o/ glide to [w], as in (12b) (Hyman & Katamba 1999).

(12)	a	/lí-à/	'eat'	[lyáà . . .]	cf.	a≠lí-dd-è	'he has eaten'
		/ké-à/	'dawn'	[kyáà . . .]		lú≠kè-dd-è	'it has dawned'
	b	/gù-à/	'fall'	[gwaa. . .]		a≠gú-dd-è	'he has fallen'
		/mò-à/	'shave'	[mwaa. . .]		a≠mwé-dd-è	'he has shaved'
	c	/bá-à/	'be'	[báà . . .]		a≠bá-dd-è	'he has been'
		/tá-à/	'let go'	[tâa. . .]		a≠tá-dd-è	'he has let go'
						(SM + Root + PFV + FV)	

As seen in the outputs, gliding of /i e/ and /u o/ is accompanied by compensatory lengthening of the following vowel. In Ganda JE15, this length will be realised if word-internal or if the above verb stems are followed by a clitic. Otherwise it, as well as the length obtained from concatenation of /a/ + /a/ in (12c), will undergo final vowel shortening (FVS), e.g., kù≠lyáà=kô 'to eat a little' vs. kù≠lyâ 'to eat.'

The details of vowel coalescence may depend on whether the vowels are tautomorphemic vs. heteromorphemic, and whether the vowel sequence is contained within a word or not. Thus, instead of gliding, the Ganda JE15 mid vowels /e o/ join /a/ in undergoing deletion when followed by a non-identical vowel across a word boundary.

(13) a   mù≠sìbê   +   ò≠mû   →   [mù.sì.bóò.mû]   'one prisoner'
         mù≠wàlâ   +   ò≠mû   →   [mù.wà.lóò.mû]   'one girl'

     b   m≠bògô   +   è≠mû   →   [m.bò.géè.mû]   'one buffalo'
         n≠dígàá   +   è≠mû   →   [n.dì.gèè.mû]   'one sheep'

     c   bà≠sìbê   +   à≠bà-ò   →   [bà.sì.báà.bò]   'those prisoners'
         bà≠kô   +   à≠bà-ò   →   [bà.káà.bò]   'those in-laws'

Deletion, like gliding, is also accompanied by compensatory lengthening. In many cases, the expected glide may not be realised if preceded by a particular consonant or followed by a particular vowel. In Ganda JE15, an expected [w] is not realised when preceded by /f v/, e.g., /fu-a/ → fw-aa → [faa. . .] 'die.' Similarly, an expected [y] is "absorbed" into a preceding /s/, /z/ or palatal consonant, e.g., /se-a/ → sy-aa → [saa. . .] 'grind.' In Ruwund L53, [w] is usually absorbed when preceded by an /m/ or /k/ and followed by /o/, e.g., /kù≠òòš-à/ → [kwòòš] ~ [kòòš] 'to burn' (Nash 1992).

In cases where /a/ is followed by /i/ or /u/ (typically from PB *i and *u), a coalescence process can produce [ee] and [oo], respectively, e.g., Yao P21 /mà≠ísó/ → méésó 'eyes' (Ngunga 1997). This coalescence also occurs in the process of "imbrication" (Section 4.2), whereby the [i] of the perfective suffix is infixed and fuses with a base vowel, e.g., Bemba M42 ísàl-ìl-è → ísàìl-è → íséèl-è 'close (tr.) + perfective' (Hyman 1995a).

With these observations, we now can summarise with examples from Ganda JE15 three of the five sources of vowel length in Bantu: (i) from underlying representations (-lím- 'cultivate,' -líím-(ís-) 'spy on'), (ii) from vowel concatenation, e.g., /bá≠àgàl-à/ → [báàgàlà] 'they want'; (iii) from gliding + compensatory lengthening, e.g., /tú≠àgàl-à/ → [twáàgàlà] 'we want.' A fourth source is the rule of vowel lengthening that occurs before a moraic nasal + consonant, e.g., /kù-ń≠sìb-à/ → [kúúnsìbà] 'to tie me' (cf. Section 4.1.5). A fifth source is penultimate vowel lengthening, which occurs in most Eastern and Southern Bantu languages which have lost the lexical vowel length contrast, e.g., Chewa N31b, t-à≠mèèny-à 'we have hit,' t-à≠mèny-èèl-à 'we have hit for,' t-à-mèny-èl-ààn≠à 'we have hit for each other.' Such lengthening typically applies at the phrase level and may be suspended in certain utterance types, e.g., questions or commands (Hyman 2013).

Besides these lengthening processes, vowel shortening may also apply in one of three contexts. First, there may be final vowel shortening (FVS) with languages varying as to whether this occurs at the end of a word or phrase, or "clitic group," as in Ganda JE15 (Hyman & Katamba 1990). Second, there are a number of languages which restrict long vowels to penultimate or antepenultimate position. Thus, any long vowel that precedes phrase-antepenultimate position will be shortened in Mwiini G412 (Kisseberth & Abasheikh 1974): reeba 'stop,' reeb-er-a 'stop for,' reb-er-an-a 'stop for each other.' Similar observations have been made about the Kongo H10 languages and nearby Yaka H31, which also restricts long vowels to occurring within the stem-initial syllable: zááy-á 'know,' zááy-íl-á 'know+appl,' zááy-is-á 'cause to know'; but: zạy-ákán-á 'be known,' with shortening. In Safwa M25, the shortening process appears to count the two moras of a CVV penultimate syllable: à-gàà≠gúz-y-à 'he can sell,' à-gà≠bùúz-y-à 'he can ask' vs. à-gà≠bùz-y-àág-à 'he may ask' (Voorhoeve n.d.). Finally, a few languages have closed syllable shortening, e.g., before geminate consonants in Ganda JE15: /tú-à-èé≠ˈtt-à/ → [twéttà] 'we killed ourselves.'

In addition to length, vowel height may be sensitive to boundaries. In a number of Great Lakes Bantu languages, historical *ɪ and *ʊ lower to [e] and [o] at the beginning of a constituent. This is particularly noticeable in comparing the augment+prefix sequences across languages, e.g., class 3/4 *u-mu-/i-mi-* in Rwanda JD61 vs. *o-mu-/e-mi-* in Haya JE22 (De Blois 1970). In Nyambo JE21 (personal notes), lowering of /i/ and /u/ occurs only initially in a phrase. As a result, the lowered [o] of the phrase-initial form, ò-mù≠kázì 'woman,' alternates with [ù] in kù-bón'ú-mù≠kázì 'to see a woman,' where the final -à of kù≠bón-à 'to see' has been deleted – in this case, without compensatory lengthening. Besides lowering, root-initial vowels are sometimes deleted initially, e.g., PB *-jíb- > ib-à > Chewa N31b [bà] 'steal.' Although restructured as a prefix, the original [i] appears in the imperative form i-bà 'steal!' (phrase-finally, [ìibà]), where it is needed to fill out the bisyllabic minimality condition on Chewa words.

3 CONSONANT PHONOLOGY

As indicated in Section 2, PB is believed to have had a relatively simple consonant system. In addition, all syllables were open in PB, and syllable onsets mostly consisted of a single consonant. The two possible exceptions to this are nasal+consonant and consonant+glide.

4.1 Nasal+consonant

Besides the consonants in (1a), PB and most present-day languages also have nasal complexes (NC), written *mp, mb, nt, nd, ŋk, ŋg*, etc., and analysed either as clusters of homorganic nasal+consonant or single prenasalised consonants, e.g., *-bómb-* 'mould,' *-gènd-* 'go,' *-táŋg-* 'read.' The class 9/10 nasal prefix *N-* produces equivalent NCs across morphemes, e.g., Tuki A64 m̀≠búá 'dog,' ǹ≠dòànè 'cow,' ɲ̀≠gì 'fly.' The PB 1SG morpheme is also often realised as a homorganic nasal in present-day languages, e.g., Ganda JE15 m̀≠bál-á 'I count,' ǹ≠dúm-à 'I bite,' ŋ̀≠gw-á 'I fall.' While the prefix is an underspecified homorganic *N-* in 9/10, we know from such forms as Yao P21 n-áá≠díp-il-é 'I paid,' where the nasal appears before a vocalic tense marker, that the 1SG prefix has an underlying /n/. In some languages where 9/10 is *N-*, the 1SG prefix has a CV shape, e.g., Swahili G42, Chewa N31b *ni-*, Shona S10 *ndi-*, Nande JD42 *ɲi-* (alternating with *n-*). The 9/10 prefix *N-*, on the other hand, rarely occurs directly before a vowel, since PB roots generally begin with a consonant.

In PB, noun and verb roots did not begin with NC. Root-initial NC has subsequently been introduced in Bantu languages which have lost root-initial *ji or *jɪ, e.g Kalanga S16 *ngín-à* 'enter' (*-jíngɪd-*), *mb-á* 'sing' (*-jímb-*) (Mathangwane 1999). This is true also of the root -*ntù* 'person, thing, entity,' whose reconstruction may ultimately be *-jìntò. In other cases where a stem appears to begin with NC, the nasal may have originally been a prefix, e.g., transferred from 9/10 with the *N-* to another class, which then imposes its own prefix. It is sometimes still possible to analyse such forms as double prefixes, e.g., Chewa N31b chì-m≠bòmbó 7/8 'glutton' (cf. m≠bòmbó 1/2 'greedy person').

While most Bantu languages preserve NC, many have restrictions either on which N+C combinations are possible, or on where within the word structure NC may occur. Thus, Tiene B81 allows NC across morpheme boundaries, e.g., class 9 n≠tàbà 'goat,' but simplifies stem-internal NC, e.g., mù≠òtò 'person' (*-(jì)ntò) (Ellington 1977). In the same language, stem-internal *mb/*nd become [m]/[n] with compensatory lengthening or diphthongisation of the preceding vowel, while *ŋg is deleted with no trace, e.g., -tùùm-à 'cook' (*-tùmb-), -kúón-à 'desire' (*-kónd-), -tú-à 'build' (*-tóng-). On the other hand, Yao P21 deletes a root-initial voiced consonant after the 1SG prefix N-, but not after the 9/10 prefix N-. Thus, /ku-n≠gaadil-a/ → kùù≠ŋáádìlà 'to stare at me' vs. ŋ̀≠gùbò 'cloth' (< PB *N̂≠gòbò) (Ngunga 2000).

Where a consonant C is realised differently (C') after N, it is important to note that this may be due to either of two logical possibilities: C is modified to C' after N; or C' becomes C except after N. The latter situation is frequently found with respect to the weakening of *p to [h] or [w], which is typically blocked after a homorganic nasal, e.g., Nyambo JE21 kú≠h-à 'to give' vs. m̀≠p-à 'give me!.' Through subsequent changes, the relation between C and C' can become quite distant. In Bukusu JE31c, the [h] observed in Nyambo has dropped out, and the preserved labial stop becomes voiced after N, such that the alternation is now úxù≠à 'to give' vs. m̀≠b-à 'give me!.' See Schadeberg (1989), for a case where [p] alternates with [ŋ] in Nyole JE35.

Depending on the nature of the post-nasal consonant, N+C inputs can undergo a variety of processes (cf. Kerremans 1980, Hyman 2001).

3.1.1 Nasal + voiceless stop

Perhaps the most widespread process affecting NC is post-nasal voicing, attested in Nande JD42, Gikuyu E51, Bukusu JE31c and Yao P21. Examples from Yao are illustrated in (14) (Ngunga 2000).

(14) a kù≠pélék-à   'to send'   b kùù-m≠bélèk-à   'to send me'
         kù≠túm-á   'to order'   kùù-n≠dúm-à   'to order me'
         kù≠cápíl-à   'to wash'   kùù-ɲ≠jápìl-à   'to wash for me'
         kù≠kwéél-á   'to climb'   kùù-ŋ≠gwéèl-à   'to climb on me'

Another process affecting voiceless stops is aspiration, e.g., in Chewa N31b, Kongo H10, Pokomo E71 and Swahili G42. The illustration in (15) is from Kongo H10 (Carter 1984).

(15) a /kù-N≠pùn-á/   →   kú-m≠phùn-á   'to deceive me'

      b /kù-N≠tál-à/   →   kú-n≠thàl-à   'to look at me'

      c /kù-N≠kìyílà/   →   kú-ŋ≠khìyíl-à   'to visit me'

The resulting NC^h unit may then undergo nasal effacement ($nt > nt^h > t^h$), as in certain varieties of Swahili G42, or de-stopping ($nt > nt^h > n^h$), as in Rwanda JD61, Rundi JD62, and Shona S10, where *ómò≠(jì)ntò 'person' is realised as (ù)mù≠nhù (cf. Nyamwezi F22 m≠nhò). The resulting N^h may then simplify to N. This is presumably the chain of events that have characterised the southern Tanzanian languages Hehe G62, Pangwa G64 and Kingwa G65, which have $nt > n$ vs. closely related Vwanji G66, which has $nt > n^h$.

3.1.2 Nasal + voiceless fricative

Three different strategies are also commonly seen when a nasal is followed by a voiceless fricative. First, the nasal may simply be effaced, even in languages such as Yao P21, which voice post-nasal voiceless stops, e.g., kù-n≠sóòsà → kùù≠sóòsà 'to look for me.'

The second strategy is seen in Nande JD42, which extends post-nasal voicing to include fricatives, e.g., *ò-lù≠sángá* 'pearl,' pl. *è-n≠zángá*. A third strategy found in languages such as certain varieties of Kongo H10, Yaka H31 and Venda S21 is affrication. As seen in the Kongo forms in (16), post-nasal affrication can also affect voiced fricatives.

(16) Post-nasal affrication in Kongo (Carter 1984)

  a /kù-N≠fíl-à/   →   kú-m≠pfíl-à   'to lead me'
     /kù-N≠síb-à/   →   kú-n≠tsìb-à   'to curse me'

  b /kù-N≠vùn-á/   →   kú-m≠bvùn-á   'to deceive me'
     /kù-N≠zól-à/   →   kú-n≠dzòl-à   'to love me'

In Tuki A64, which has nasal effacement before voiceless consonants, /n+s/ becomes [ts], as expected, but /n+f/ becomes [p], e.g., /à-n≠sèyà-m/ → *à≠tsèyà-m* 'he abuses me,' /à-n≠fùnùnà-m/ → *à≠pùnùna-m* 'he wakes me up' (Hyman & Biloa 1992). This is presumably because [f] comes from earlier *p. Thus, besides conditioning changes which can be characterised as "strengthening" or "fortition," a nasal can block the opposite lenition processes (e.g., *p > f).

3.1.3 Nasal + voiced consonant

As mentioned above and seen in (17a), Yao P21 deletes post-nasal voiced consonants – other than [d], e.g., *kù-n≠dípà* → *kùù-n≠dípà* 'to pay me' (Ngunga 2000).

(17) a /kù-n≠búúcil-a/   →   kùù≠múúcìl-à   'to be angry with me'
         /kù-n≠láp-á/   →   kùù≠náp-à   'to admire me'
         /kù-n≠jíím-à/   →   kùù≠ɲíím-à   'to begrudge me'
         /kù-n≠gónék-à/   →   kùù≠ŋónèk-à   'to make me sleep'

      b /kù-n≠mál-à/   →   kùù≠mál-à   'to finish me'
         /kù-n≠ném-à/   →   kùù≠náp-à   'for me to do incorrectly'
         /kù-n≠ɲál-à/   →   kùù≠ɲál-à   'to cut me in small pieces'
         /kù-n≠ŋáádìl-à/   →   kùù≠ŋáádìl-à   'to play around with me'

This includes nasal consonants in (17b), since many Bantu languages do not tolerate NN sequences. On the other hand, voiced continuants may alternate with stops (or affricates) after a nasal, e.g., Ganda JE15 /n≠láb-à/ → *n≠dáb-à* 'I see,' Tuki A64 /à-n≠ràmà-m/ → *à≠dàmà-m* 'he pulls me.'

Post-nasal voiced consonants may also be nasalised, in which case a geminate nasal is produced. This is most readily observed in the case of Meinhof's Law (Meeussen 1962), known also as the Ganda Law and illustrated from that language in (18).

(18) /n≠bômb-á/   →   m̀≠móòmb-á   'I escape'
      /n≠límb-á/   →   ǹ≠níímb-á   'I lie'
      /n≠júng-á/   →   ɲ̀≠ɲúúng-á   'I join'
      /n≠génd-á/   →   ŋ̀≠ŋéénd-á   'I go'

A nasal+voiced consonant becomes a geminate nasal when the next syllable also begins with a nasal. One motivation for this change is the simplification of NCVNC sequences (cf. Section 4.5). However, many of the languages also apply the process when the voiced NC sequence is followed by a simple nasal consonant, e.g., Ganda JE15 /n≠lím-á/ → n≠ním-á 'I cultivate.' Interestingly, Yao P21, which fails to delete [d] after a nasal in an oral context, will as a result of Meinhof's Law do so if the following syllable is an NC complex, e.g., /kù-n≠dííng-à/ → kù-n≠nííng-à → kùù≠nííng-à 'to try me.'

3.1.4 Other processes

The above seems to indicate that Bantu languages prefer that post-nasal consonants be [+voice] rather than [-voice] and [-continuant] rather than [+continuant]. Voiceless stops tend to become aspirated, and voiced stops tend to become nasalised. While these generalisations reflect the common processes affecting post-nasal consonants, it is important to note that opposing "counter-processes," though less common, are also found. For example, voiced stops are devoiced and variably pronounced as ejectives in Tswana S31 and Southern Sotho S33: bón-á 'see!', m≠pón-á 'see me!', dís-á 'watch,' n≠tís-á 'watch me!' Aspirated stops are deaspirated in Nguni S40 languages, e.g., Ndebele S44 ulu≠thi 'stick,' pl. izin≠ti. Affricates become deaffricated in a number of languages, e.g., Shona S10 bvum-a 'agree, admit,' vs. m≠vum-o 'permission, agreement.' Finally, and perhaps most unusual, nasal consonants are denasalised after another nasal in Punu B43, Bushong C83, Kongo H10, Yaka H31, e.g., Yaka m≠bák-íní 'I carved' (-mák- 'carve'), n≠dúúk-íní 'I smelt' (-núúk- 'smell'). See Hyman (2001) for discussion of the potential theoretical significance of post-nasal processes and counter-processes.

3.1.5 Moricity

In most Bantu languages there is no vowel length opposition before an NC complex. Rather, as seen in many of the cited examples, the preceding vowel is frequently lengthened. The standard interpretation is that this nasal is "moraic," i.e., it contributes a unit of length or "beat," which readily transfers to the preceding vowel (Clements 1986). It also is potentially a tone-bearing unit. This is most transparently seen in languages which allow NN sequences, or when the nasal is syllabic and phrase-initial, e.g., Haya E22 ḿ-bwà '(it's a) dog.' However, even when the nasal loses its syllabicity and compensatorily lengthens the preceding vowel, some languages still treat it as a tone-bearing unit, e.g., Ganda JE15, while others do not, e.g., Haya JE22 and Bemba M42.

3.1.6 New cases of NC

While the preceding subsections characterise the phonology of NC complexes inherited from PB, many Bantu languages have introduced new sequences of N+C. The most common source is the loss of [u] in mu- prefixes, e.g., Swahili G42 m≠thu 'person,' m≠toto 'child' (Nurse & Hinnebusch 1993: 181–185). The resulting syllabic [m] may then undergo homorganic nasal assimilation, as it does in most varieties of Yao P21, e.g., ŋ'≠kúlú 'elder sibling' 1, ǹ'≠sééŋgó 'horn of antelope' 3, ŋ'≠gólógòlò 'in the weasel' 18 (N' = syllabic). The loss of the [u] of mu- prefixes also extends to the 2nd person plural SP and the class 1 OP, but will frequently not take place if followed by a vowel or NC, e.g., Yao P21 mù≠ùsò 'bow of boat' 3, mùù≠ndò 'person' 1, mw-ìì≠gàdsà 'handful' 18 (cf. (d)ì≠gàdsà 'palm of hand' 5). Vowel deletion can also be blocked if the stem is monosyllabic, e.g., *mù≠sì* 'village' 3. The resulting NC may contrast phonetically or phonologically with PB *NC in several ways. First, as the Yao examples illustrate, the nasal from *mò-* is typically syllabic, while the nasal from *NC loses its syllabicity (Hyman & Ngunga 1997). Second, the nasal from *mò-* does not condition the same alternations on the following consonant (e.g., voiceless stops do not become voiced after N'- in Yao). In Tswana S31, where *m+b* is normally realised [*mp*] (cf. Section 4.1.4), *mò-* loses its vowel when followed by stem-initial /b/, which in turn is realised [*m*], e.g., *mò≠bús-i* → *m≠músí* 'governor' (cf. *-bús-á* 'to govern'). Contrast this last example with Matumbi P13 (Odden 1996), where class 9/10 N- does not condition changes on voiced stops (*lù≠góì* 'braided rope,' pl. *N≠góì*), but N'- from *mò-* does (e.g., *mù≠gàálà ~ ɲ≠ɲàálà* 'in the storage place'). A third difference is that N'- does not condition lengthening on the preceding vowel, cf. Haya JE22 /à-kà-ń-bìŋg-à/ → *à-káá-m-bìŋg-à* 'he chased me' vs. /à-kà-mú≠bìŋg-à/ → *à-kà-ḿ≠bìŋg-à* 'he chased him.' Finally, there can be a tonal difference, even in cases where there is no difference in syllabicity. Thus, in Basaa A43, class 3 N- (< *mò-*) is a tone-bearing unit, while class 9 N- is not. The rule of high tone spreading applies in the phrase *púbá m≠bɔ́ndɔ̀* 'white lion' (< *m≠bɔ́ndɔ̀* 'lion' 9), but not in *púbá m≠ɓòmgà* 'white hammer' (< *m≠ɓòmgà* 'hammer' 3), since, although non-syllabic, the *m-* from *mò-* still functions as a tone-bearing unit to which the high tone spreads. See also Bostoen & de Schryver (2015) for similar developments in Kongo H10 varieties and Kadenge (2014, 2015) for the phenomenon in Shona S10 varieties.

4.2 Consonant + high vowel

Besides the post-nasal environment, consonants are frequently realised differently before high vs. non-high vowels. First and foremost is the process of frication, also known as Bantu spirantisation, which affects consonants when they are followed by *i and *u, producing changes such as those schematised in (19) (Schadeberg 1995, Hyman & Merrill 2016).

(19) a *pi > pᴴi > pfi   > fi
         *bi > bᴴi > bvi   > vi

      b   > psi > (tsi) > si
                     > bzi > (dzi) > zi

      c *ti > tᴴi > tsi   > si
         *di > dᴴi > dzi   > zi

      d *ci > cᴴi   > sị
         *ji > jᴴi   > zi

      e *ki > kᴴi > ksi > (tsi) > si
         *gi > gᴴi > gzi > (dzi) > zi

As indicated, these changes are first triggered by the development of "noise" in the release of a consonant before the tense high vowels *i and *u. Indicated as ᴴ in (19), the present-day reflex can, in fact, be aspiration, as in Makhuwa P31 – cf. Kalanga S16 *-thúm-* 'sew' (*-túm-*) vs. *-túm-* 'send' (*-tóm-*). Similarly, /t d/ are aspirated before /i/ in Doko C301 (7V): /ká≠tísá/, /í-dínó/ → [ká≠tʰísá] 'traverse,' [í≠dʰíno] 'tooth.' In most languages, however, Cᴴ is further modified either to an affricate or fricative, as indicated, e.g., Ngom

B22b *kfuɓa* 'chicken' (*-*kúbà*). Such modifications are found in nearly all five-vowel languages (Lengola D12 is a notable exception), as well as in many seven-vowel systems (Schadeberg 1995), although often with irregularities.

As was seen above in (6), **u* was almost entirely restricted to the first and last stem syllables in PB, although it also occurs in stem-internal position in the PB root **-tákun-*, which has reflexes such as Chewa N31b *-tafun-*, Pende L11 *-táfùn-*, Venda S21 *-ṱáfùn-*, Yao P21 *-táwún-* and (with metathesis), Nkore-Kiga JE13/14 and Nyambo JE21 *-fútàn-*. Synchronic alternations are found in languages which use the **-ú* suffix to derive adjectives or nouns from verbs, e.g., Ganda JE15 *-géjj-* 'become fat' → *-gévv-ù* 'fat'; *-lébél-* 'be loose' → *-lèbév-ù* 'loose'; *-támíír-* 'become drunk' → *mù≠tàmíív-ù* 'drunkard.'

*i also most frequently occurred in the first and last stem syllables in PB, but also in noun class prefixes, e.g., class 8 **bì-* > Shona S10 class 8 *zvi-* with a "whistled" labioalveolar [ʐ]. In many Bantu languages, synchronic alternations are conditioned by one or more of the three suffixes reconstructed with **i* (cf. Chapter 6). The first of these is the causative suffix **-i-* which is almost always followed by a vowel. As seen in the Bemba M42 forms in (20), when followed by causative **-i-*, labial /p b/ become [f] and lingual consonants become [s], subsequently modified to [ʃ] by palatalisation, e.g., /-sít-/ → [-ʃít-] 'buy') (Hyman 2003).

(20) a -lèèp- 'be long' → -lèèf-i- 'lengthen' [lèèf-y-à]
       -lùb- 'be lost' → -lùf-i- 'lose' [lùuf-y-à]

   b -fììt- 'be dark' → -fìis-i- 'darken' [fìiʃ-à]
       -cìnd- 'dance' → -cìns-i- 'make dance' [cìnʃ-à]
       -lìl- 'cry' → -lìs-i- 'make cry' [lìʃ-à]
       -bùùk- 'get up (intr)' → -bùùs-i- 'get [s.o.] up' [bùùʃ-à]
       -lúng- 'hunt' → -lúns-i- 'make hunt' [lúnʃ-á]

The second suffix is **-ì*, which derives nouns, often agentives, from verbs, as in Ganda JE15 *ò-mú≠ddùs-ì* 'fugitive' (< -ʹddúk- 'run (away)'), *ò-mù≠léz-í* 'guardian' (< *-lér-* 'raise (child)'). Finally, the bimorphemic perfective suffix **-ìd-è* also frequently conditions frication (Bastin 1983), e.g., Nkore E13 *-réet-* 'bring,' perfective *-réets-ìr-è*; Rundi JD62 *rìr-* 'cry,' perfective *rìz-è* (< *rìr-y-è* < **-dìd-ìd-è*).

While such frications occur frequently throughout the Bantu zone, there is considerable variation (Bostoen 2008, Hyman & Merrill 2016, Downing 2016). Not only can the exact reflexes differ, but so can the environments in which they occur. In Bemba M42, for instance, only causative **-i-* regularly conditions frication. In Yao P21, frication before causative **-i-* is more widespread than before perfective **-ìd-è*, but will almost always occur before the latter if the preceding consonant is either [l] or [k], i.e., the two consonants which occur most frequently in extensions (Ngunga 2000). What this indicates is that frication, although originally an across-the-board phonological process, has acquired morphological restrictions and, in some cases, has been levelled out completely. Thus, the [i] of Bemba M42 *-il-e* not only fails to condition frication, but has been restructured, on analogy with suffixes such as applicative *-il-/-el-*, to undergo vowel height harmony, e.g., *-fik-ìl-è* 'arrive+perfective' vs. *-bèp-èl-è* 'lie+perfective.' The perfective suffix *-il-e* continues to differ from the applicative *-il-* in its ability to fuse or "imbricate" with verb bases that end in a range of consonants, e.g., *-kúngùb-* 'gather' → perfective *kúngwìib-è* vs. applicative *kúngùb-ìl-à* (Bastin 1983, Hyman 1995a).

The vowels *i and *u may have effects on preceding consonants other than frication. While nasals are usually exempt from the effects of degree 1 high vowels, in Ganda JE15, *i palatalises /n/, e.g., ò-mù-sóɲ-í 'tailor' (< -són- 'sew'). A much more frequent phenomenon concerns the realisation of PB *d, which may be preserved as [d] before [i], but realised as [l] or [r] before other vowels. This occurs both in 7V languages, e.g., Duala A24, Tiene B81, Bobangi C32, as well as in 5V languages, e.g., certain varieties of Kongo H10, Lwena K14, Kwezo L13, Manyo K332, Kete L21 and certain varieties of Yao P21. In the 5V languages *di is typically realised as [dzi] or [zi], and *dɪ is realised [di]. Unless preceded by a nasal, *d is realised as [l] or [r] before other vowels. The synchronic situation is considerably obscured in Ruwund L53, where *dɪ is realised [di] and *de is realised as [li] (cf. Section 3.2 above). In other languages the effect is extended to the high back vowel *u, e.g., Tswana S31 (7V). In Kalanga S16 (5V), *du is realised [du], while *dʊ is realised [lu] (Mathangwane 1999). Finally, Chewa N31b exhibits the "hardening" of /l/ to [d] only before glides, e.g., -dy-a 'eat,' -bad-w-a 'be born' (cf. -bal-a 'bear (child)'). What this suggests is that the [d] reali-sation is most preferred before glides, then [i], then [u]. Occurring in both Eastern and Western Bantu – indeed, throughout the zone – there is dialectal evidence in both the Kongo H10 and Sotho-Tswana S30 groups that the [d] was originally pronounced as retroflex [ɖ] before high vowels. See also much of Chaga E60, where *d is realised [r ~ ɻ] before *i and *u, elsewhere as [l] or Ø.

4.3 Consonant + glide

The post-consonant glides [y] and [w] are typically derived from underlying vowels. As a result, consonants often show the same alternations before the glides [y] and [w] as before the corresponding high vowel. Thus, [y] and [w] from *i and *u produce frications, while [y] and [w] from *ɪ and *ʊ or *e and *o typically do not. An exception to this is found in the Mongo C60 group (seven-vowel). In some varieties of Mongo, /t/ is realised [ts] before the high tense vowels /i/ and /u/, while /l/ (< *d) is realised as /j/ [dʒ]. However, all varieties appear to produce the affricate realisations before [y] and [w] – even if they derive from /e/ or /o/: tó≠kàmb-à 'we work,' ló≠kàmb-à 'you pl. work' vs. tsw-án-à 'we see,' jw-án-à 'you (pl.) see' (Hulstaert 1965).

In many Bantu languages, ky/gy develop into alveopalatal affricates. This is seen espe-cially in the different realisations of the class 7 *kɪ- prefix before consonants vs. vowels, e.g., Nyamwezi F22 (7V) kì≠jíìkò 'spoon' vs. c≠èèyò 'broom'; Swahili G42 (5V) ki≠kapu 'basket' vs. c≠ama 'society'; Ha JD66 class 7 ìkì 'this' vs. ìcò 'that (near you).' Other languages front velars with a noticible offglide, i.e., kʸ, gʸ, first before high front vow-els, then before mid-front vowels as well. Thus, Ganda JE15 èkʸì≠kópò 'cup,' ultimately ècì≠kópò. While different patterns of velar palatalisation are found throughout the Bantu zone (Hyman & Moxley 1992), some languages in the Congo basin show analogous developments with respect to alveolar consonants. Although /li/ is realised [di] in both Luba L31a and Pende L11, /ti/ is realised ci [tʃi]: Luba -mác-íl- 'plaster+appl' (-mát-), Pende shíc-ìl- (~ shít-ìl-) 'close+appl' (-shít-).

While a [y] offglide can trigger palatalisation, [w] is responsible for velarisation, e.g., in the Rundi-Rwanda JD60 and Shona S10 groups. Meeussen (1959) summarises the reflexes of labial+glide and coronal+glide complexes in Rundi JD62 as in (21).

(21) a   bw [bg]   b
         fw [fk]   fy [fˢy]
                                 vy [vᶻy]
         mw [mŋ]   my [mɲ]
         tw [tkw]   ty [rtky, r̩ᵗky]
         rw [rgw (gw)]   rg [rgy]
         sw [skw, sᵏw]   sy [sᵏy]
         zw [ᵈzᵍw]
         tsw [tskw, tsᵏw]
         cw [tʃkw, tʃᵏw]
         jw [ᵈʒᵍw]
         shw [ʃkw, ʃᵏw]

                                 nny [ny, nⁱy]

As seen in (21a), *Cw* hardens to *Ck*/*Cg*/*Cŋ*, and the labial offglide is lost ("absorbed")
when the C is labial. (21b) shows that *Cy* undergoes a comparable hardening process.
When the C is velar, one obtains the expected *Cw* sequence. Similar processes occur in
the Shona S10 complex. In Kalanga S16, /l/ becomes [g] before [w], e.g., *-tól-à* 'take,'
-tóg-w-à 'be taken.' Compare also Basaa A43, where the class 4 *mi-* and class 8 *bi-* prefixes are realised *ŋw-* and *gw-* when directly followed by a vowel, e.g., *bì≠tóŋ* 'horns'
vs. *gw≠ɔ́m* 'things' (*y≠ɔ́m* 'thing' 7); cf. the object pronouns *ŋw-ɔ́* (cl. 4) and *gw-ɔ́* (cl. 8)
(Lemb & de Gastines 1973). Finally, geminate *w+w* and *y+y* become, respectively, [ggw]
and [ggy] in Ganda JE15.

(22) a   /pó-à/   →   wo-a   →   ww-aa   →   ggwaa . . .   'become exhausted'

     b   /pí-à/   →   wi-a   →   yy-aa   →   ggyaa . . .   'get burnt'

As seen, both glides derive from **p*, cf. *m≠pw-êdd-è* 'I have become exhausted' and
m≠pî-dd-è 'I have gotten burnt' (~ *n≠jî-dd-è*).

4.4 **i* + Consonant

Consonants may harden not only after nasals or before glides, but also after PB **i*, particularly when this vowel is either word-initial or preceded by a vowel. Thus, Tswana S31
devoices stops not only after N-, but also after reflexive *i-*: *ì≠pón-é* 'see yourself!,' *ì≠tís-é*
'watch yourself!' (*-bón-* 'see,' *-dís-* 'watch'). In Lega D25, PB **t* normally becomes [r],
but is preserved as [t] after the class 5 prefix **i*, e.g., *i-támà* 'cheek, pl. *mà-rámà*. Ganda
JE15, on the other hand, develops geminates from **iC*. Thus, compare *ò-mù≠sájjà* 'man'
and *-bájj-* 'carve' with Haya JE22 *ò-mù≠sáìjà* and *-bàìj-*. This process also produces
root-initial geminates, e.g., *- 'tt-* 'kill' and *- 'bb-* 'steal,' and singular/plural alternations
in classes 5/6 such as *è-g≠gùlù* 'sky,' pl. *à-mà≠gùlù* (cf. Haya JE22 *-ít-*, *-íb-*, *è-ì≠gùlù*).
While class 5 **i-* typically "strengthens" following consonants, it is known also to condition voicing (and implosion), e.g., Zezuru S12 *ɓáŋgá* 'knife' (pl. *mà≠páŋgá*), *ɗàŋgá*
'cattle enclosure' (pl. *mà≠tàŋgá*), *gáŋgá* 'large helmeted guinea-fowl' (pl. *mà≠káŋgá*)
(Hannan 1974).

4.5 Long-distance consonant phonology

In all of the processes discussed thus far, the trigger of the phonological process is adjacent to the targeted consonant. Bantu languages also are known for the ability of a consonant to affect another consonant across a vowel (and beyond). There are several such cases.

The first, Meinhof's Law, was seen already in Section 4.1.3, whereby a nasal+voiced stop is realised NN or N when followed by a second nasal complex (sometimes just N) in the next syllable. Another version of this simplification occurs in Kwanyama R21, where the second nasal+voiced stop loses its prenasalisation (Schadeberg 1987): *N≠gombe > oŋ≠gobe 'cattle,' *N-gandu > oŋ≠gadu 'crocodile.' These dissimilatory changes have the effect of minimising the number of NC complexes in a (prefix+) stem. Another well-known dissimilatory process is Dahl's Law, whereby a voiceless stop becomes voiced if the consonant in the next syllable is also voiceless. This accounts for the initial voiced reflexes in Nyamwezi F22 roots such as -dàkún- 'chew' (*-tákun-), -gùhí 'short' (*-kúpì). Dahl's Law is also responsible for the /t/ or /k/ of prefixes to become voiced and sometimes continuant, e.g., Kuria JE43 /ko≠tema/ → [yo≠tɛm-a] 'to beat.' While there is considerable variation (Davy & Nurse 1982), multiple prefixes may be affected, e.g., Southern Gikuyu E51 /ke-ke-ko≠eta/ → [ye-ye--yw≠eet-a] 'he (cl. 7) called you.' Alternations are also sometimes found stem-internally, as in Rundi JD62 -bád-ìk- 'transplant,' -bád-ùk- 'grow well' vs. -bát-ùr- 'uproot plants to transplant' (Meeussen 1959).

Other long-distance consonant processes are assimilatory in nature. In Bukusu JE31c, an /l/ will assimilate to a preceding [r] across a vowel, e.g., -fúk-ìl- 'stir + APPL' vs. -bír-ìr- 'pass + APPL.' The process is optional when the trigger [r] is separated by an additional syllable, e.g., -rám- 'remain,' -rám-ìl- ~ -rám-ìr- 'remain + APPL.' Several languages in the western Lacustrine area show a process of sibilant harmony which disallows or limits the co-occurrence of alveolar and alveopalatal sibilants, e.g., [s] and [ʃ]. In Rwanda JD61 and Rundi JD62, /s/ becomes alveopalatal across a vowel, when the following consonant becomes (alveo-)palatal as the result of a y-initial suffix, e.g., -soonz- 'be hungry' vs. a-ra≠shoonj-e 'he was hungry' (< -soonz-ye). In Nkore-Kiga JE13/14, the process produces alternations in the opposite (depalatalising) direction, e.g., -shígìsh- 'stir,' ò-mù≠sígìs-ì 'stirrer.' Finally, a third long-distance assimilation involves nasality. A wide range of Bantu languages nasalise [l] or [d] to [n] after an NV(V) syllable (Greenberg 1951), e.g., Bemba M42 -cít-ìl- 'do + APPL' vs. -lìm-ìn- 'cultivate + APPL.' While Suku H32 optionally extends this process across additional syllables, creating extension variations such as -am-ik-il- ~ -am-ik-in-, such long-distance assimilation is obligatory in Kongo H10 – and in nearby Yaka H31, e.g., -ziik-il- 'bury + APPL' vs. -mak-in- 'climb + APPL,' -miituk-in- 'sulk + APPL,' -nutuk-in- 'lean + APPL' (Hyman 1995b).

Other forms of apparent long-distance phonology are highly morpheme-specific. Thus, the passive suffix -w- (*-ʊ-) causes the palatalisation of a preceding labial consonant in the Sotho-Tswana S30 and Nguni S40 groups, e.g., Ndebele S45 -dal-w- 'be created' (-dal-) vs. -bunj-w- 'be moulded' (-bumb-). In Ndebele and other Nguni languages, this process can actually skip syllables, e.g., -funjath-w- 'be clenched' (-fumbath-), -vunjulul-w- 'be uncovered' (-vumbulul-) (Sibanda 2004). In a number of languages where causative *-i-conditions frication, the effect is sometimes seen on non-adjacent consonants. In Bemba M42, the roots -lub- 'be lost' and -lil- 'cry' form the causatives -luf-y- and -lis-y- (> -liš-) and the applicatives -lub-il- and -lil-il-. However, their applicativised causative forms are

-luf-is-y- and *-lis-is-y-*, with frication applying twice. As seen in (23), this is the result of a "cyclic" application of the frication process (Hyman 2003).

(23) ROOT   MORPHOLOGY   PHONOLOGY   MORPHOLOGY   PHONOLOGY

a -lub- → -lub-i- → -luf-i-   → -luf-il-i- → -luf-is-i-
   'be lost'   'lose'   'lose for/at'

b -lil- → -lil-i- → -lis-i-   → -lis-il-i- → -lis-is-i-
   'cry'   'make cry'   'make cry for/at'

In the second morphology stage, applicative *-il-* is "interfixed" between the fricated root and the causative suffix *-i-* (> y before a vowel). Bemba and certain other Bantu languages show the same multiple frications when the *-id-* of perfective *-ìd-è* is interfixed, e.g., *-luf-is-i-e* [*lufife*] 'lose + perfective.' In other languages, frication appears to apply non-cyclically, affecting only the applicative consonant, e.g., Mongo C61 /kál-/ 'dry,' /-kál-i-/ (→ *-káj-à* with FV *-à*) 'make dry,' /-kál-èl-ì-/ → *kál-èj-ì-* 'make dry + APPL' (→ *-kál-èj-à* with FV *-à*). Still others show evidence of cyclicity by "undoing" frication in a fixed (often non-etymological) manner. In Nyamwezi F22 (Maganga & Schadeberg 1992), the verb /-gʊl-/ 'buy' is causativised to /-gʊl-i-/ 'sell,' which undergoes frication to become *-gʊj-i-* (> surface *-gʊj-a* by gliding of *i* to [*y*] and absorption of [*y*] into the preceding alveopalatal affricate). However when *-gʊj-i-* is applicativised, yielding intermediate *-gʊj-ɪl-i-*, the result is *-gʊg-ɪj-i-* (→ *-gʊg-ɪj-a*). The palatal *j* is "undone" as velar [*g*], on analogy with *-og-* 'bathe intr.,' *-oj-i-* 'bathe (s.o.),' *-og-ej-i-* 'bathe + APPL,' not as the [*l*] one would expect if the process were non-cyclic. The most extreme version of this process is seen in languages, such as Nyakyusa M31, which uses a "replacive" [*k*] no matter what the input consonant of an applicativised causative. Thus, *-kees-i-* 'make go by' (the causative of *-keend-* 'go by') is applicativised first to *-kees-el-i-*, which then undergoes frication and de-frication of the *s* to *k*: *-keek-es-i-* 'make go by + APPL' (→ [*-keek-es-y-a*]).

4 FURTHER DISSOLUTION OF THE INHERITED STRUCTURES

The above gives a sketch of some of the phonological properties of syllables, consonants and vowels in Bantu languages. It emphasises languages which have preserved the basic morphological and phonological structure inherited from Proto-Bantu. In order to be complete, it is important to reiterate that quite a few languages in zones A, B and elsewhere have modified this structure significantly, e.g., by allowing closed syllables, developing back unrounded vowels, etc. One striking property of many Northwest Bantu languages is the imposition of prosodic templates on the stem (root + suffixes). Thus, Tiene B81 verb stems can only have the shapes CV, CVV, CVCV, CVVCV and CVCVCV. In the last case the second C must be coronal, while the third must be non-coronal, often requiring infixation, e.g., *-láb-à* 'walk' → *-lásàb-à* 'cause to walk' (Hyman 2010). This stands in stark contrast with Bantu languages to the East and South, where a verb stem can in principle be of any length, cf. Chewa N31b *-mang-its-ir-an-a* 'cause to tie for each other.' By "phonetic erosion" stems have become mostly monosyllabic in Nzadi B865 (Crane *et al.* 2011), e.g., *ò≠kwâ* 'bone' (*-kópà*), *i-kĕl* 'blood' (*-gìdá*) (see also Chapter 14 on Nsong B85d). The tendency to break down the inherited structure is even more pronounced in groups just outside "Narrow Bantu," e.g., Grassfields Bantu (Watters 2003).

REFERENCES

Bastin, Y. (1983) *La finale verbale -ide et l'imbrication en bantou*. Tervuren: Musée royal de l'Afrique centrale.

Bell, A. (1972) The Development of Syllabic Nasals in the Bantu Noun Class Prefixes mu-, mi- and ma-. *Anthropological Linguistics* 14: 29–45.

Bleek, W. H. I. (1862) *A Comparative Grammar of South African Languages. Part I. Phonology*. London: Trübner & Co.

Bostoen, K. (2008) Bantu Spirantization: Morphologization, Lexicalization and Historical Classification. *Diachronica* 25(3): 299–356.

Bostoen, K. & G.-M. de Schryver (2015) Linguistic Innovation, Political Centralization and Economic Integration in the Kongo Kingdom: Reconstructing the Spread of Prefix Reduction. *Diachronica* 32(2): 139–185.

Bostoen, K. & J. Koni Muluwa (2014) Umlaut in the Bantu B80 Languages of the Kwilu (DRC). *Transactions of the Philological Society* 112(2): 209–230.

Bostoen, K. & L. Mundeke (2011) The Causative/Applicative Syncretism in Mbuun (Bantu B87, DRC): Semantic Split or Phonemic Merger? *Journal of African Languages and Linguistics* 32(2): 179–218.

Carter, H. (1984) Geminates, Nasals and Sequence Structure in Kongo. *Oso* 3(1): 101–114.

Clements, G. N. (1986) Compensatory Lengthening and Consonant Gemination in Luganda. In Wetzels, L. W. & E. Sezer (eds.), *Studies in Compensatory Lengthening*. Dordrecht: Foris Publications.

Clements, G. N. (1991) Vowel Height Assimilation in Bantu Languages. *Proceedings of the Annual Meeting of the Berkeley Linguistics Society* 17: 25–64.

Crane, T. M., L. M. Hyman & S. Nsielanga Tukumu (2011) *A Grammar of Nzadi (B.865) A Bantu Language of the Democratic Republic of Congo with an Appendix of Proto-Bantu – Nzadi Sound Correspondences by Clara Cohen*. Berkeley: University of California Publications.

Creissels, D. (2005) L'émergence de systèmes à neuf voyelles en bantou S30. In Bostoen, K. & J. Maniacky (eds.), *Studies in African Comparative Linguistics, with Special Focus on Bantu and Mande*, 191–198. Tervuren: Royal Museum for Central Africa.

Davy, J. I. M. & D. Nurse (1982) The Synchronic Forms of Dahl's Law, a Dissimilation Process in Central and Lacustrine Bantu Languages of Kenya. *Journal of African Languages and Linguistics* 4: 157–195.

De Blois, K. F. (1970) The Augment in the Bantu languages. *Africana Linguistica* 4: 85–165.

Downing, Laura J. (2016) *Explaining the role of the morphological continuum in Bantu spirantisation*. Africana Linguistica 13: 53–78.

Ellington, J. E. (1977) *Aspects of the Tiene Language*. Madison: University of Wisconsin, PhD dissertation.

Fontaney, V. L. (1980) Le verbe. In Nsuka-Nkutsi, F. (ed.), *Eléments de description du punu*, 129–178. Lyon: Centre de Recherches Linguistiques et Sémiologiques (CRLS), Université Lumière-Lyon 2.

Gick, B., D. Pulleyblank, F. Campbell & N. M. Mutaka (2006) Low Vowels and Transparency in Kinande Vowel Harmony. *Phonology* 23: 1–20.

Goes, H. & K. Bostoen (2019) Progressive Vowel Height Harmony in Proto-Kikongo and Proto-Bantu. *Journal of African Languages and Linguistics* 40.

Greenberg, J. H. (1951) Vowel and Nasal Harmony in Bantu Languages. *Zaïre: revue congolaise* 5(8): 813–820.

Guarisma, G. (1969) *Etudes bafia: phonologie, classes d'accord et lexique bafia-français.* Paris: Société des Etudes Linguistiques et Anthropologiques.

Guarisma, G. (2000) *Complexité morphologique, simplicité syntaxique: Le cas du bafia, langue bantoue périphérique* (A50) de l'Ouest du Cameroun. Paris: Peeters.

Hannan, M. (1974) *Standard Shona Dictionary.* Salisbury: Mardon printers Ltd for Rhodesia Literature Bureau.

Hinnebusch, T. J. (1974) Rule Inversion and Restructuring in Kikamba. *Studies in African Linguistics* Supplement 5: 149–167.

Hombert, J.-M. (1986) The Development of Nasalized Vowels in the Teke Language Group. In Bogers, K., H. van der Hulst & M. Mous (eds.), *The Phonological Representation of Suprasegmentals*, 359–379. Berlin: De Gruyter Mouton; Dordrecht: Foris.

Hulstaert, G. (1965) *Grammaire du lomongo. Deuxieme partie, Morphologie.* Tervuren: Musée royal de l'Afrique centrale.

Hyman, L. & F. Katamba (1999) The syllable in Luganda Phonology and Morphology. In van der Hulst, H. & N. Ritter (eds.), *The Syllable: Views and Facts*, 349–416. Berlin: De Gruyter Mouton.

Hyman, L. M. (1995a) Minimality and the Prosodic Morphology of Cibemba Imbrication. *Journal of African Languages and Linguistics* 16(1): 3–39.

Hyman, L. M. (1995b) Nasal Consonant Harmony at a Distance: The Case of Yaka. *Studies in African Linguistics* 24(1): 5–30.

Hyman, L. M. (1998) Positional Prominence and the "prosodic trough" in Yaka. *Phonology* 15(1): 41–75.

Hyman, L. M. (1999) The Historical Interpretation of Vowel Harmony in Bantu. In Hombert, J. M. & L. M. Hyman (eds.), *Bantu Historical Linguistics: Theoretical and Empirical Perspectives*, 235–295. Stanford: CSLI Publications.

Hyman, L. M. (2001) On the Limits of Phonetic Determinism: *NÇ Revisited. In Hume, B. & K. Johnson (eds.), *The Role of Speech Perception Phenomena in Phonology*, 141–185. New York: Academic Press.

Hyman, L. M. (2003) Sound Change, Misanalysis, and Analogy in the Bantu Causative. *Journal of African Languages and Linguistics* 24(1): 55–90.

Hyman, L. M. (2008) Directional Asymmetries in the Morphology and Phonology of Words, with Special Reference to Bantu. *Linguistics* 46(2): 309–350.

Hyman, L. M. (2010) Affixation by Place of Articulation: The Case of Tiene. In Wohlgemuth, J. & M. Cysouw (eds.), *Rara & Rarissima: Documenting the Fringes of Linguistic Diversity*, 145–184. Berlin: De Gruyter Mouton.

Hyman, L. M. (2013) Penultimate Lengthening in Bantu: Analysis and Spread. In Bickel, B., L. A. Grenoble, D. A. Peterson & A. Timberlake (eds.), *Language Typology and Historical Contingency*, 309–330. Amsterdam; Philadelphia: John Benjamins.

Hyman, L. M. & E. Biloa (1992) Transparent Low Tone in Tuki. In Buszard-Welcher, L. A., J. Evans, L. Wee & W. Weigel (eds.), *Proceedings of the Eighteenth Annual Meeting of the Berkeley Linguistics Society: Special Session on the Typology of Tone Languages*, 104–127 (Available online at https://journals.linguisticsociety.org/proceedings/index.php/BLS/).

Hyman, L. M. & F. X. Katamba (1990) Final Vowel Shortening in Luganda. *Studies in African Linguistics* 21(1): 1–59.

Hyman, L. M. & J. Merrill (2016) Morphology, Irregularity and Bantu Frication. In Léonard, J.-L. (ed.), *Actualités des Néogrammariens, Mémoires de la Société de Linguistique de Paris*, 139–157. Paris: Peeters.

Hyman, L. M. & J. Moxley (1992) The Morpheme in Phonological Change: Velar Palatalization in Bantu. *Diachronica* 13(2): 259–282.

Hyman, L. M. & A. Ngunga (1997) Two Kinds of Moraic Nasals in Ciyao. *Studies in African Linguistics* 26(2): 131–163.

Kadenge, M. (2014) Comparing Nasal-Obstruent Clusters Derived from /mu-/ Reduction in Shangwe and Zezuru: An Optimality Theory Analysis. *Language Matters – Studies in the Languages of Africa* 45(1): 40–62.

Kadenge, M. (2015) The Augment and {mu-} Reduction in Bantu: An Optimality Theory Analysis. *South African Journal of African Languages* 35(1): 93–104.

Kerremans, R. (1980) Nasale suivie de consonne sourde en proto-bantu. *Africana Linguistica* 8: 159–198.

Kisseberth, C. W. & M. I. Abasheikh (1974) The Perfect Stem in Chi-Mwiini. *Studies in the Linguistic Sciences* 4(2): 123–138.

Kutsch Lojenga, C. (1994) KiBudu, a Bantu Language with Nine Vowels. *Africana Linguistica* 11: 127–133.

Kwenzi Mikala, T. J. (1980) Esquisse phonologique du punu. In Nsuka-Nkutsi, F. (ed.), *Eléments de description du punu*, 7–18. Lyon: Centre de Recherches Linguistiques et Sémiologiques (CRLS), Université Lumière-Lyon 2.

Leitch, M. F. (1996) *Vowel Harmonies in the Congo Basin: An Optimality Theory Analysis of Variation in the Bantu Zone C*. Vancouver: University of British Columbia, PhD.

Lemb, P. & F. de Gastines (1973) *Dictionnaire basaá-français*. Douala: Collège Libermann.

Liphola, M. (2001) *Aspects of the Phonology and Morphology of Shimakonde*. Ohio State University, PhD dissertation.

Maddieson, I. (2003) The Sounds of the Bantu Languages. In Nurse, D. & G. Philippson (eds.), *The Bantu Languages*, 15–41. London; New York: Routledge.

Maganga, C. & T. C. Schadeberg (1992) *Kinyamwezi: Grammar, Texts, Vocabulary*. Cologne: Rüdiger Köppe.

Mathangwane, J. T. (1999) *Ikalanga Phonetics and Phonology: A Synchronic and Diachronic Study*. Stanford: Center for the Study of Language and Information (CSLI), Stanford University.

Medjo Mvé, P. (1997) *Essai sur la phonologie panchronique des parlers fang du Gabon et ses implications historiques*. Lyon: Université Lumière-Lyon2, PhD.

Meeussen, A. E. (1959) *Essai de grammaire rundi*. Tervuren: Musée royal du Congo belge.

Meeussen, A. E. (1962) Meinhof's Rule in Bantu. *African Language Studies* 3: 25–29.

Meeussen, A. E. (1967) Bantu Grammatical Reconstructions. *Africana Linguistica* 3: 79–121.

Meeussen, A. E. (1979) Vowel Length in Proto-Bantu. *Journal of African Languages and Linguistics* 1(1): 1–8.

Mutaka, N. M. (1995) Vowel Harmony in Kinande. *Journal of West African Languages* 25(2): 42–55.

Nash, J. A. (1992) *Aspects of Ruwund Grammar*. Urbana-Champaign: University of Illinois, PhD dissertation.

Ngunga, A. S. A. (1997) Class 5 Allomorphy in Ciyao. *Studies in African Linguistics* 26(2): 165–192.

Ngunga, A. S. A. (2000) *Phonology and Morphology of the Ciyao Verb*. Stanford: Center for the Study of Language and Information (CSLI), Stanford University.

Nurse, D. & T. J. Hinnebusch (1993) *Swahili and Sabaki: A Linguistic History*. Berkeley: University of California Press.

Odden, D. (1996) *The Phonology and Morphology of Kimatuumbi*. Oxford: Clarendon Press.

Rombi, M.-F. (1983) *Le shimaore: première approche d'un parler de la langue comorienne (Ile de Mayotte, Comores)*. Paris: Société des Etudes Linguistiques et Anthropologiques de France.

Ruttenberg, P. (2000) *Lexique yaka-francais, français-yaka*. Munich: LINCOM.

Schadeberg, T. C. (1987) Silbenanlautgesetze im Bantu. *Afrika und Ubersee* 70: 1–17.

Schadeberg, T. C. (1989) The Velar Nasal in Nyole (E. 35). *Annales Aequatoria* 10: 168–180.

Schadeberg, T. C. (1995) Spirantization and the 7-to-5 Vowel Merger in Bantu. *Belgian Journal of Linguistics* 9: 71–84.

Sibanda, G. (2004) *Verbal Phonology and Morphology of Ndebele*. Berkeley: University of California., PhD dissertation.

Stappers, L. (1971) Esquisse de la langue lengola. *Africana Linguistica* 5: 255–307.

Stappers, L. (1973) *Esquisse de la langue mituku*. Tervuren: Musée royal de l'Afrique centrale.

Stewart, J. M. (1983) The High Unadvanced Vowels of Proto-Tano-Congo. *Journal of West African Languages* 13(1): 19–36.

Stewart, J. M. (2000–2001) Symmetric vs Asymmetric Vowel Height Harmony and e,o vs i,u in Proto-Bantu and Proto-Savannah Bantu. *Journal of West African Languages* 28(2): 45–58.

Teil-Dautrey, G. (2008) Et si le proto-bantu était aussi une langue . . . avec ses contraintes et ses déséquilibres. *Diachronica* 25(1): 54–110.

Van de Velde, M. (2008) *A Grammar of Eton*. Berlin: De Gruyter Mouton.

Voorhoeve, J. (n.d.) *A Grammar of Safwa: Preliminary Draft Based on Previous Research by J. van Sambeek, Checked by C. K. Mwachusa*. Ms.

Watters, J. R. (2003) Grassfields Bantu. In Nurse, D. & G. Philippson (eds.), *The Bantu Languages*, 225–256. London; New York: Routledge.

Wilson, P. (1985) *Simplified Swahili* (New 2nd edition) Nairobi; London: Longman.

CHAPTER 5

TONE

Michael R. Marlo and David Odden

This chapter provides an overview of tone in Bantu languages. It discusses the main representational questions concerning tone in Bantu, i.e., whether L tone is "special" and what the tone-bearing unit is; it surveys major tonal processes found in Bantu languages, such as tone movement and dissimilation; it describes the core tonal properties of nouns, verbs and phrases across a wide range of Bantu languages; and it discusses diachronic developments involving tonal reanalysis.

1 INTRODUCTION

The vast majority of Bantu languages are tonal, and many have complex and divergent tonologies. Even varieties of one (macro-)language, such as Luyia JE32, Shona S10 or Pare G22, may have very different tone systems. Still, there is considerable deep similarity, and much of the complexity and divergence from a simpler proto-language situation can be understood as the historical result of elaborations in morphosyntax and somewhat arbitrary resolution of analytical ambiguities, coupled with recurrent phonologisation of competing phonetic tendencies. What makes Bantu tonology particularly complex is that a given pattern can result from a general phonotactic rule or from one that is restricted but fully productive in a given context, which is sensitive to grammatical properties. Given the richness of syntax and morphology in Bantu, many morphosyntactic properties potentially trigger phonological rules, and morphemes may have many surface variants distinguished by tone, e.g., the eight tonal variants of the noun 'broom' in Shira B41: *duyombulu, duyombúlu, dúyombulu, duyómbulu, dúyómbulu, duyó⁺mbúlu, dúyómbulu* and *dúyó⁺mbúlu* (Blanchon 1999).

Sophisticated and revealing studies of Bantu tone have long existed, including overviews (van Spaandonck 1967, 1971, Hyman 1976, Kisseberth & Odden 2003, Downing 2011, Marlo 2013, Odden & Bickmore 2014), tone-marked descriptive studies arising from the Belgian involvement in the Congo (e.g., by Burssens, Coupez, Stappers, Meeussen); and analyses of tone in specific languages such as Ganda JE15, Kinga G65, Safwa M25, Sukuma F21, and Tonga M64. The advent of autosegmental phonology (Goldsmith 1976) spurred an explosion of theoretically oriented, data-intensive research ranging across a variety of Bantu languages (e.g., by Bickmore, Hyman, Kisseberth, Odden). See Kisseberth and Odden (2003) for examples of such studies, to which we add: Kgalagari S311 (Crane 2009, Hyman & Monaka 2011), Kuria JE43 (Mwita 2008, Marlo *et al.* 2014, 2015), Kwanyama R21 (Halme 2004), Luyia JE32 (Ebarb *et al.* 2014) and Lungu M14 (Bickmore 2007, 2014). A wide range of languages are described and analysed in various collections focusing on Bantu tone (Clements & Goldsmith 1984, Yukawa 1987, Kagaya *et al.* 1989, Kagaya 1992, Hyman & Kisseberth 1998, Blanchon & Creissels 1999, Odden & Bickmore 2014). Marlo and Odden's (2015) bibliography provides a list of data sources on Bantu tone.

We organise this chapter as follows. The first section discusses the main representational questions about Bantu tone concerning the status of L tone and the tone-bearing unit. The second section surveys major tone processes, such as spreading, shifting and dissimilation. The third section investigates the main properties of tones in nouns, verbs and at the phrasal level. TBUs with no tone mark are non-H – whether they are surface specified as L is a matter of debate (see section 2.1). Grave accent means a specifically L toned TBU. We commonly mark the edges of stems with brackets [and].

2 REPRESENTATION OF TONE

Almost all Bantu languages are tonal, but there are languages with just stress like Swahili G42, and there are tonal languages with penult lengthening (cf. Hyman 2013). Languages like Mwiini G412, with contrastive penult and final "prominence," are ambiguous between analyses in terms of tone vs. stress.

Within Narrow Bantu, tone languages have only two robustly contrastive levels, H and L (frequently augmented with downstep). Kamba E55 has amplified this system with extra-H and extra-L tones with limited albeit contrastive distribution, Chaga E60 has developed a restricted extra-H tone, and Bila D311 has developed an extra-L. In Mpiemo A86c (Festen 2008), pre-L pitch-boost changes H to extra-H, but not all L tones trigger this raising, which leads to contrasts like in *lémɔ* 'heart' vs. *lǐma* 'dream' (which may reflect a L/Ø contrast). Reduction of the three surface tones of Mpiemo to two is possible, at the cost of abstractness in the form of "boundary tones" that trigger raising, to account for surface differences like *mé lúɣá kã̌nji mpánjó* 'I subsequently sell a shelter,' *mé bɔ̌ kanji mpánjó* 'I sold a shelter (long ago),' *mḛ̃ rí kã̌nji mpánjó* 'I sell a shelter (habitual),' *mḛ̃ nǎ kã̌nji mpánjó* 'I will sell a shelter.' A number of mostly Western languages like Ntumu A75A and Bafia A53, but also Kuria JE43, have developed a phonetic pre-pausal contrast between falling L (v̀) and non-falling L (v°) or else between L and M, generally from an obscured underlying or historical LL vs. LH contrast. In Ntumu A75A, the distinction is maintained directly only pre-pausally (*ajʷìn*, 'banana,' *aki°* 'egg,' *afʌ́p* 'paper') and is neutralised before another word (*ɑjʷi dáá*, 'one banana,' *aki dáá* 'one egg'), but is indirectly preserved between a H-toned augment and a H possessive pronoun as a downstepped H, as in *á⁺kí dâm* 'my egg,' cf. *ájʷi dâm* 'my banana,' *áfʌ́вʌ́ dâm* 'my paper.' Since L, L° and H contrast in these nouns, the most likely analysis is that /-ki'/ has an underlying floating H.

There are two basic questions about tone representations in a Bantu language. The first is whether the language distinguishes H vs. L or H vs. Ø (no tone); the second is whether the tone-bearing unit is the syllable or the mora.

2.1 The status of L

Significant asymmetries between H and L tones have led analysts since Stevick (1969) to posit privative analysis of tone systems as H vs. Ø (cf. Hyman 2001). Many Bantu languages have a dissimilative process (see Section 3.3) known as Meeussen's Rule whereby HH becomes HL, but Bantu languages do not seem to have a rule where LL becomes LH (or HL). In other words, H is subject to dissimilation, but L is not. Nash (1992–1994: 262–263) claims that such a rule exists in Ruwund L53, applying in a limited domain, but further study will be required to determine if this is an example of L dissimilation. The phonological asymmetry between H and L has been explained representationally in at least two ways: by treating tone in Bantu as epiphenomenal and analysing languages as having 'pitch accent' (Massamba 1982, Goldsmith 1984); or, by analysing the opposition as being between H and Ø, where toneless syllables may receive surface L by a default rule (Pulleyblank 1986), or perhaps never receiving any phonological tone (Myers 1998). The concept of pitch-accent as a legitimate analytic device has been questioned (Pulleyblank 1986, Odden 1999, Hyman 2006, 2009) and has largely given way to underspecification analyses where only H is underlyingly present, or privative analyses where only H is present at any stage.

One line of evidence used in support of underspecification is distributional asymmetry: H tones can appear in fewer places than L tones can. For example, in the verb stem, only the first syllable freely allows H or L; all other syllables are L. A related argument is that inflectional melodic patterns (Section 4.2.3) are composed of H tones which are assigned to locations such as the second stem mora, the penult or the ultima. Such patterns overwhelmingly involve positioning of H tones, not L tones: Mokpwe A22, Fuliiru JD63, and Kamba E55 are among the few languages which have verb tone melodies involving contrastively specified L tones. Bantu languages frequently have rules of H spreading, but not L spreading. Finally, a number of languages such as Haya JE22, Ndanda Yao P21, Bena G63 and various Makonde P23 dialects eschew pre-pausal H (all or nearly all final vowels are L), but we do not find languages prohibiting final L.

These facts can be rationalised if underlying representations only contain H tones, and in verb roots, H only appears on the first syllable (perhaps positioned by an early rule). Melodic inflections involve assignment of a tonal entity to some position, and under the logic that the only tonal entity is H, only H can be in a melody composed solely of tone. The possibility of expulsion ("zeroing") of final H tones follows from the fact that H is a tonal entity which a rule can refer to, and the non-existence of rules expelling L tones in favor of H tones is predicted if L is not a tonal entity and thus cannot be referred to in a rule. Likewise only H can spread, if only H exists as an entity that rules can refer to.

The most compelling theoretical argument is the adjacency argument. Many Bantu languages have tone dissimilation where HH becomes LH or HL – in some languages, this takes place even when L syllables intervene. An example is Tone Loss in Matumbi P13, which deletes phrase-medial word-final H if preceded – at any distance – by another stem H, resulting in alternations like *paníin*[*télekyá*] 'when I cook for him,' *paníin*[*télekya*] *Libulúle* 'when I cook for Libulule,' or *kwíi*[*tyatyákikiyá*] 'to plaster for oneself,' *kwíi*[*tyatyákikiya*] *ɲuúmba* 'to plaster a house for oneself.' Compare *panáa*[*temité*] 'when I chopped (prehodiernal),' *panáa*[*temité*] *mikóongó* 'when I chopped trees' with no deletion of the final H, because in the prehodiernal past there is no preceding H in the stem. If non-H syllables were tonally specified as bearing L rather than being toneless viz. /*kwìi*[*tyàtyákìkìyá*] *ɲùúmbà*/, we would encounter the theoretical problem that the H deleted by Tone Loss can be separated by any number of intervening L tones, contravening generally accepted theoretical principles of locality (cf. Odden 1994) and requiring the inclusion of expressions like "L_0" in the statement of the deletion rule.

Unfortunately, the theoretical desiderata motivating underspecification analyses are just that, i.e., theoretical desiderata. The power of asymmetry arguments is weakened by the fact that the ostensively unspecified member of an opposition actually can be identified, either by directly specifying that the target or trigger must be unspecified for tone, or indirectly by requiring that the rule not delete existing associations (Avery & Rice 1989). Moreover, locality constraints can always be denied their status as linguistic universals. Thus the case for underspecification in Bantu often reduces to the observation that there is little evidence that L is phonologically active.

The most compelling empirical argument for underspecification (and against universal privativity) is the evidence for a three-way lexical contrast between L, Ø and H converging at a two-way surface H/L contrast. Example languages are Fuliiru JD63, Nande JD42, Tiriki JE413, and possibly Tembo JD531, where L and Ø merge into surface L. In Fuliiru JD63 (Van Otterloo 2011, 2014), there is a contrast in CVVC roots between H verbs like /gáaj-/ 'count,' L verbs like /hìiv-/ 'hunt,' and Ø verbs like /biik-/ 'put.' In the infinitive, H is added to the leftmost toneless vowel of the stem and spreads to the right, resulting in úkú[gáájà] (/H/), úkú[hìivà] (/L/), and úkú[bííkà] (/Ø/). In the future, L is assigned to the leftmost toneless vowel of the stem, resulting in túgá[gáàjà] (/H/), 'we will count' túgá[hìivà] (/L/) 'we will hunt,' and túgá[bìikà] (/Ø/) 'we will put.' The special property of Ø roots as contrasted with L roots is that their first syllable becomes entirely L when a L melody is present and entirely H when a H melody is present.

Another kind of evidence for a three-way distinction between H, L and Ø arises from the frequent need to distinguish stem-initial L in verbs from subsequent (possible) L tones. Many languages such as Mokpwe A22 spread word-final melodic H to the left, up to but not including the root-initial vowel, so /nama[kakané]/ becomes nama[kakáné] 'I promised' (cf. noô[kakanɛ] 'I will promise'). One analysis of the limit on spreading is that the root-initial vowel has a specified L but intervening vowels are toneless. This limit could also reflect an explicit restriction against spreading to a root-initial syllable. The matter can be resolved in favor of specified root-initial L in Mokpwe A22, since the same melodic tone assigned to a L CV root yields a rising tone (nama[yɔ̌] 'I laughed'), but the evidence and analysis may be ambiguous in another language lacking this kind of fact.

In Ntumu A75A and some other languages, such as Aghem (Grassfields Bantu), Mokpwe A22, and Kuria JE43, downstep of H results from a preceding floating L, reflecting the traditional view of the representation of downstep (cf. Bird 1966: 135, Clements & Ford 1979: 206, Paster & Kim 2011). In other languages, e.g., Mwanga M22 and Shambaa G23, each discrete H tone is downstepped; L is not necessarily required to trigger the downstep of H (cf. Odden 1982, Schuh 1998, Bickmore 2000).

2.2 The tone-bearing unit

The question of the 'tone-bearing unit' (TBU) remains uncertain in the theoretical and Bantu literature, largely because many criteria are called on, often contradictorily. A widely held view is that tones may associate to syllables or moras, and the choice is a language-specific parameter (cf. Yip 2002: 74).

The main issue relevant to the TBU question is the possible positional contrasts within a phonological constituent. If a language does not have underlying contrasts between level H, rising, or falling tones, as in Jita JE25 or Kerewe JE24, then there is no reason to assume that the mora is the TBU, and the lack of contrast could be taken as evidence for the syllable as TBU. Both languages have rightward shift or spread which moves H by an entire syllable. If the syllable is the TBU in these languages, spreading or shifting would modify the relationship between tones and syllables, regardless of the number of moras in the syllable and movement in these languages would be expressed as a non-iterative operation from syllable to syllable. It appears that attested positional contrasts of H in bimoraic syllables in Bantu are predictable at some level (see Section 3.4), suggesting explanations in terms of impoverished representational power (syllable as TBU or privative tone) in lexical representations. Since contrasts between level H and contour tones frequently do exist in the output, that contrarily suggests that L tones are eventually filled in, or the TBU is the mora, or both.

If a system is based on a privative contrast between H vs. Ø (not H vs. L) and if the TBU is the syllable, then the position of H within bimoraic syllables cannot be lexically contrastive; level H, rising and falling tone would be indistinguishable (simply H). If the TBU is the mora, then on a long syllable a contrast between level H (HH), rise (ØH) and fall (HØ) is unproblematic. Such contours could be diagnostics for the TBU if the status of L could be pinned down. The relevance of underspecification as opposed to privativity is that if L tones are filled in, then even with the syllable as TBU, a derived contrast between HL and LH is also tenable, as long as L is filled in before rules creating the HL/LH contrast apply.

It is typically quite difficult to determine whether the possibility of long contours is due to the TBU being the mora, or due to surface specification of L. Penultimate syllables in Makonde P23, which are always long, contrast level H, L, rise, fall and rise-fall. The first four of those tones are compatible with the mora as TBU or with specified L; the existence of tritonal rise-fall on a bimoraic syllable shows that L must indeed be specified. Such evidence is rarely available.

A related issue is whether processes refer to syllables or moras. One strong form of a mora-reference argument arises in counting languages like Matumbi P13 and Kuria JE43, where tone is assigned by counting moras after a fixed position. In Matumbi P13, the H of the Subjunctive is assigned to the third mora after the SP, thus *ba[telekyáane]* 'they should cook for e.o.,' *ba[tyatyaákiye]* 'they should plaster,' *ba[keengéembe]* 'they should uproot.'

In other languages, a rule may operate in terms of whole syllables. In Tura JE30, H is assigned to all moras of the second stem syllable in the Present Negative, irrespective of the weight of the first two stem syllables.

(1) Present Negative in Tura JE30
  a   s-áa-βá[βukúl-a] tá   'he is not taking them'
      sí-βa-mú[teexéraaŋg-a] tá   'they are not cooking for him'

  b   sí-βa-mú[lexúúl-a] tá   'they are not releasing him'
      sí-βa-mú[liiŋgééraaŋg-a] tá   'they are not watching him'

While this rule makes reference to the syllable in Tura JE30, the evidence of surface contrasts points to the mora as TBU, since there is a contrast between level H, rise and fall on long syllables in comparable contexts, cf. *mu[lííngeré]* 'watch him!,' *alii[niiŋgéér-á]* 'he will watch me (remote future),' *aláa[niiŋgéérá]* 'he will watch me (near future).'

The principles positioning the inflectional H in the Subjunctive in Idakho E411 (Ebarb 2014, Ebarb et al. 2014, Ebarb & Marlo 2015) refer to both syllables and moras. As shown in (2c-d), the melodic H is assigned to the second mora after the initial syllable of the stem. Whether short (2a-b) or long (2c-d), the stem-initial syllable is skipped over.

(2) Subjunctive: 'let him/her V' in Idakho E411
  a   a[xalatʃ-ɛ́]   'cut'   a[βoyoɲán-ɛ]   'go around'

  b   a[sitaátʃ-ɛ]   'accuse'   a[ɲoɲoólits-ɪ] 'tease'

  c   a[saandits-í]   'thank'

  d   a[tsuunzuún-ɪ] 'suck'

Tones sometimes interact with laryngeal features (Bradshaw 1999) – see Cassimjee & Kisseberth (1992), Downing (2011), and Volk (2011) on depressor consonant effects in Mijikenda E72–73 and Southern Bantu – which has occasionally been interpreted as evidence that tone is a segmental feature under the laryngeal node (Clark 1990). Our view is that tone rules may refer to laryngeal features and possibly other types of features, such as vocalic height (Odden 2011), but tones do not associate to the laryngeal node.

3 PROCESSES

Certain processes are commonly observed across Bantu languages, as more extensively discussed in Kisseberth and Odden (2003).

3.1 H-tone spreading/shifting

H in Bantu is quite mobile, meaning that it frequently spreads or shifts. Such movement can be unbounded within a constituent, or bounded, in the latter case being limited to a distance of one or sometimes two syllables or moras, and can operate either to the left or to the right.

In Ikorovere Makhuwa P31G, H spreads one mora to the right. The negative prefix *-hi-* has underlying H, and whatever vowel follows *-hi-* is also H-toned: *u-hí[lím-a]* 'to not cultivate,' *u-hí[máal-a]* 'to not be quiet,' *u-hí-kí[thumel-a]* 'to not buy for me.' On the other hand, Tsonga S53 spreads H rightward in an unbounded fashion to the phrase-penultimate vowel. This can be seen by comparing *h-a[xaviselan-a]* 'we are selling to each other' vs. *v-á-[xávísélán-a]* 'they are selling to each other.' Rightward binary shift is found in Nyamwezi F22 (Maganga & Schadeberg 1992), so /kʊ[tómɪl-a]/ becomes kʊ[tʊmíl-a] 'to send for.' Unbounded rightward shift is found in some languages: to the final in Digo E73, to the penult in Zigula G31, to the antepenult in Nguni S40.

Leftward movement, though much less common, is also found. Lulua L31b (Yukawa 1992a) has unbounded leftward spread whereby /ku[loongesh-á]/ becomes kú[lóóngéshá] 'to teach' and /ku-mu[táandish-á]/ becomes kú-mú[taandishá] 'to threaten him' (lexical H also shifts leftward); Nande JD42, Zambian Tonga M64 and Totela K41 have binary leftward shift. Binary leftward spread as a general phenomenon, i.e., not restricted to the final vowel (see Section 3.2) is rare, but reported for Rundi JD62 and Rwanda JD61 (Philippson 1991, Hyman 2007: 193). Unbounded leftward shift has not been observed. The movement of H toward the penult, from the left or the right, is sometimes interpreted as attraction of H tone to a metrically strong position (Kisseberth 1992, Creissels 1998).

An uncommon process is tone tripling, where H spreads to the following two TBUs, as found in Kanye Tswana S31 (Creissels 1998). The lexical H of a verb stem spreads three syllables to the right, as in the medial context form go[símólólelan-a] . . . 'to begin for one another.' The same pattern appears in other southern Bantu languages, including Shona S10, Venda S21, Kalanga S16 and Tsonga S53, as well as Copperbelt Bemba M42. Double rightward shift of H is found in Sukuma F21 (Richardson 1959), where H can be displaced two TBUs to the right, as in /bákʊkʊ[solel-a]/ → bakʊkó[solel-a] 'they will chose for you.'

A number of language have both spreading and shifting, e.g., Holoholo D28 (Coupez 1955). H shifts from the root-initial vowel to the following vowel in ku[mon-á] 'to see' from /ku[món-a]/, and if there is another syllable, H spreads as in ku[monán-â] 'to see each other' from /ku[mónan-a]/ and ku[tegélél-a] 'to listen' from /ku[tégelel-a]/. Similar patterns of multiple rules of spreading and shifting are found in Kalanga S16, Shona S10, Sukuma F21 and Taita E74.

Tone movement may be restricted by morphosyntactic domain. In Shona S10, the choice between bounded and unbounded spreading is based on morphosyntactic information. There is obligatory unbounded rightward spread of H from a prefix syllable e.g., mú-má-zí-mí-chéro 'it's in the huge fruit trees' from /mú-ma-zi-mi-chero/. Between prefixes and stems, spreading only affects the root-initial syllable, cf. ku[bikira] 'to cook for,' ku-mú[bíkira] 'to cook for him.' Similar bounded/unbounded distinctions based on structural domain are found in Lungu M14 (Bickmore 2007), and one often finds unbounded spreading affecting only the inflectional stem H, e.g., in Mokpwe A22, Kamba E55, Pare G22 and Shona S10.

3.2 Non-finality

'Non-finality' refers to the tendency for the end of certain phonological structures (word, phrase, utterance) to not have H. The most common form of non-finality is failure of H-tone movement to target a final mora. In Makhuwa P31, tone doubling targets a phrase-medial final vowel but not a pre-pausal vowel: u[thúmá] meéle 'to buy millet,' but u[thúma]. Unbounded rightward spreading in Shambaa G23, on the other hand, spreads H up to but not including the word-final syllable, regardless of phrasal position, thus /ání[ghoshoea]/ → ání[ghóshóéa] 'he is making for me,' ání[ghóshóéa] ughoe 'he is making a rope for me.'

A second manifestation is blockage of tone assignment, especially involving melodic H tones. In Ikorovere Makhuwa P31G, H tones are assigned to the first and third stem moras in the infinitive, which spread once to the right (excluding to a pre-pausal syllable), so /u[kʰomaaliha]/ → u[kʰómáálíha], u[kʰómáálíhá] ... 'to strengthen ...,' /u[maaliha]/ → u[máálíha], u[máálíhá] ... 'to make quiet ...' When the third mora is word-final, that H is not assigned, even phrase-medially: /u[tʰumela]/ → u[tʰúméla], u[tʰúméla] ... 'to buy for ...'

A third non-finality effect is deletion of final H. An example found in many Western languages is a verbal alternation known as metatony, where the verb in some tenses has final L pre-pausally, but word-final H utterance medially. Hyman and Lionnet (2012) discuss this phenomenon in Abo A42 and similar languages, where certain tenses, such as the future, have final L in all contexts, e.g., a táa pɔŋɔ, a táa pɔŋɔ bitámbé 'he won't make (shoes),' and some, such as the present, have L pre-pausally but final H (which spreads to the right) before another word, e.g., a tá pɔŋɔ, a tá pɔŋɔ́ bítámbé 'he isn't making shoes.' Underlying /a tá pɔŋɔ́/ undergoes a common pre-pausal lowering rule.

A fourth manifestation of non-finality is for H to shift leftward. In Nkore JE13 (Poletto 1998), final H shifts to the penult pre-pausally, as in /omukamá/ → omukáma 'chief' (cf. omukamá waanje 'my chief'). In contrast, nouns with an underlying penult H retain the H on that syllable both pre-pausally and finally; e.g., enkóko 'chicken,' enkóko yaanje 'my chicken.'

3.3 Obligatory contour principle

Another well-known phenomenon of Bantu tone is the avoidance of successive H tones, known in the theoretical literature as the 'Obligatory Contour Principle' (OCP) (e.g.,

Odden 1986, Myers 1997). The manifestations of the OCP are varied. For example, Tsonga S53 blocks H spread from applying to a target syllable followed by a H. Ordinarily, H at the end of a verb spreads to all toneless syllables in a following noun except for the last. Thus *xi-hlambetwana* 'cooking pot' when combined with *ndzi[vóná]* ... 'I see' results in *ndzi[vóná] xí-hlámbétwána*. However, *ma-tandzá* 'eggs' only permits the H to extend onto its prefix: *ndzi[vóná] má-tandzá*.

Another manifestation of the OCP is deletion of one of the H tones, often called "Meeussen's Rule" after the Belgian linguist who first pointed out the phenomenon. In Nyilamba F31 (Yukawa 1989a), H deletes after H, so compare *ku[dʸágila]* 'to obstruct' retaining its underlying H, versus *kuku-kú[dagila]* 'to obstruct us,' and *naá[dagila]* 'they obstructed' where the root H is deleted after a H. Deletion of H may also target the H to the left (sometimes called "Reverse Meeussen's Rule"). In Nyaturu F32 (Yukawa 1989b), H shifts once to the right so */u[tégheya]/* becomes *u[teghéya]* 'to understand' and */u-vá[ríghitya]/* becomes *u-va[ríghitya]* 'to speak to them.' When a H object prefix precedes a H root, the H of the OP is deleted and */u-vá[tégheya]/* becomes *u-va[teghéya]* 'to understand them.' There are also long-distance versions of H-dissimilation. One of the most common is found in the Great Lakes (zone J) languages, where a grammatical H causes preceding H tones to be deleted. An example is in Kerewe JE24 *ku[káláángila]* 'to fry for,' *a[kalaangíílé]* 'he fried for,' *aki[kalaangíílé]* 'he fried it (cl. 7) for,' *atuki[kalaangíílé]* 'he fried it (cl. 7) for us,' illustrating deletion of H in roots and OPs.

3.4 Avoiding contours

Avoidance of contour tones is well documented in Bantu, and there are no clear cases of contrastive underlying location of association for H within a long syllable, though such contrasts arise by rule. Short contour tones are quite rare and are usually surface phenomena. In Western languages, a complication often arises that vowel length is not contrastive, but is present on contour-toned vowels. Utterance-final (short) syllables with H have falling pitch in various languages, e.g., Tonga S62, Matumbi P13 and Logooli JE41, and in Nguni S40, H is phonetically rising after "depressor" consonants.

Surface long contours arise by direct assignment of tone to a particular target, by movement of H onto a long syllable, or by merger of syllables. When H is assigned to the third mora after the subject marker in the subjunctive in Matumbi P13, rise results if μ3 is on the right side of a long syllable (*u-ki[buúndaye]* 'you should blunt it (cl. 7)'), while fall results if μ3 is on the left side of a long syllable (*u-ki[kaláange]* 'you should fry it (cl. 7)'). Final H tones are retracted to a long penult, thus *u-[kaaté]* becomes *u[kaáte]* 'you should cut,' illustrating contours via tone shift, and glide formation applies to intermediate *u-ka-ki[áke]* to create a surface rising tone in *u-ka-ky[aáke]* 'you should go hunt it (cl. 7).' Similar retraction of final H gives rise to the contrast between fall vs. level H in Haya JE22 and Nkore JE13.

Many languages restrict contours. Quite frequently, expected rising tones systematically become level H, for instance Kamba E55 */ko[étekä]/* becomes *kw[éétekä]* 'to answer,' but falling tones are not changed (*ko[kóolokä]* 'to advance'). Although falling tones are typologically more common than rising tones, Nyamwezi F22, which shifts H once to the right, changes almost all falling tones to level H (*ko[kómʋʋcha]* → *kʋ[komóócha]* 'to make well known,' *bá-lʋ[lola]* → *ba-líí[lola]* 'they are looking'), but does not modify rising tones (*kʋ[léeta]* → *ku[leéta]* 'to bring'). Languages often also eliminate falling tones, Yao P21 being representative. In the infinitive, H is assigned to the first stem vowel and spreads rightwards once, cf. *ku*[*tátánika*] 'to hesitate.' When spreading targets a long syllable, extra spreading eliminates potential fall: *ku*[*sápáángula*] 'to take apart.'

3.5 Plateau

Working at cross-purposes to the OCP is the Plateauing strategy for avoiding HØH (HLH) sequences. Two versions of this can be observed: a single toneless mora between H tones is avoided, or an unbounded number of toneless moras between H tones is avoided. In Bukusu JE31c, alternations are not found when multiple toneless moras intervene between H tones, as in *tʃí-woowáɲi* 'crowned cranes,' but a single toneless mora between H tones is leveled out, e.g., *tʃí-tʃúkúnwe* 'ants' (cf. *tʃukúnwe* 'ant'). Xhosa S41 is a language with unbounded shifting to the antepenult in some contexts, but Plateauing in other contexts. Shifting is observed in *u-ku*[*qononóndiisa*] 'to emphasise,' where the H is underlyingly on the augment. When preceded by *ndi*[*fúna*] 'I want . . .,' the final vowel of the verb elides, placing H immediately in front of the infinitive. Any following toneless syllables are then raised between H tones: *ndi*[*fún* ʼ] *ú-kú*[*qónónóndiisa*]. Unbounded spreading occurs just in the context of an immediately preceding H tone. Other instances of Plateauing are found in Mokpwe A22, and at the phrase and word-level in Ganda JE15, Hunde JD51, and several Luyia JE32 varieties.

4 TONE AND STRUCTURE

The complexity of tonal phonology in Bantu arises from a rich inventory of general processes modifying tone as just outlined, given underlying tones which can be freely placed in many different contexts. A given morpheme may appear in many tonal contexts, and rules apply to the morpheme or not, depending on that context.

4.1 Nouns

Noun tonology in Bantu is relatively simple, because nouns are morphologically simple, with just a noun class prefix, and in some languages, an augment before the stem, and morphemes that combine with noun roots tend to be tonally uniform. Class prefixes are typically toneless, though the augment before the noun class prefix may have a latent H. The most common stem type is monomorphemic and disyllabic, with a length contrast in the first syllable.

Four tone patterns of disyllabic noun stems are reconstructable to Proto-Bantu: *HH, *HL, *LL and *LH. Some modern-day languages maintain the same number of tone distinctions, e.g., Venda S21 (Cassimjee 1986), where a bimoraic stem may take one of four patterns: LL (*mu*[*tuka*] 'youth'), LH (*mu*[*rathú*] 'brother'), HL (*mu*[*sélwa*] 'bride') and HH (*mu*[*sádzí*] 'woman').

There may be more distinctions. Disyllabic stems in Makonde P23 fall into eight patterns, i.e., *shi*[*ndoolo*] 'sp. bean,' *lí*[*maanga*] 'sp. pumpkin,' *li*[*daáda*] 'medal,' *shi*[*ndíili*] 'termite net,' *li*[*doôdo*] 'leg,' *lí*[*njaáno*] 'sp. condiment,' *vá*[*mááka*] 'cats' and *shí*[*mbeêdi*] 'shadow.'

In some languages, nouns are restricted to at most one H, for example Matumbi P13 or Great Lakes languages like Kerewe JE24 and Ganda JE15 (but not Luyia JE32). Extreme reductions in contrast are found in Kuria JE43 and Makhuwa P31. In Kuria JE43, the only contrast is whether CVVCV stems have final H (e[séésé] 'dog') or not (iri[tɔ́ɔ́kɛ] 'banana'). In some dialects of Makhuwa P31, H always appears on the second mora of the nominal word, e.g., i[mátta] 'field,' [nakhúwo] 'maize,' ni[váka] 'spear.'

At the word level, the fact that noun stems are generally monomorphemic means that noun tone tends to remain constant. In Kerewe JE24, en[gólógoombe] 'banana flower' has H on the first two syllables in all contexts. The underlying form of this stem could be /gólogoombe/ or /gólógoombe/ – either way, rules of the language will give the output H.H.L.L. Since tonal alternations generally involve what happens to H tones in various contexts, there is limited potential for alternations arising from noun class prefixation. A noun stem usually remains in the same tonal context throughout the class paradigm, as in Kerewe JE24 omu[sééza] 'man,' aba[sééza] 'men,' i[sééza] 'man (aug.),' ama[sééza] 'men (aug.),' aka[sééza] 'man (dim.),' otu[sééza] 'men (dim.).'

These tendencies are overridden in individual languages. In Shona S10, there are H as well as L noun prefixes – the honorific prefix vá- and sá- 'owner' have H, giving rise to H-spreading alternations (Mi[chero] (name), vá-Mi[chéro] 'Mr. Michero'). Associative prefixes ('of,' 'with') and copular prefixes often cliticise to the noun and cause tone alternations, thus Shona S10 [Mbereko] (name), rá[Mbéreko] 'of Mbereko'; ma[sadza] 'porridges,' má[sádza] 'it is porridges,' with H on the class prefix owing to the copular construction. Matumbi P13 shifts H off of a word-initial syllable, so nouns in cl. 9 are subject to class-related alternations since /N-/ does not provide a syllable, resulting in alternations like lu[bágalo] 'lath,' m[bagálo] 'laths,' ma[kóongoní] 'hartebeests,' n[goongoní] 'hartebeest.'

In languages that have retained it, the augment may contribute a H which is a source of tonal alternation. Some languages reveal the augment H directly, for example Gusii JE42 ó-bo[bé] 'evil,' ó-mo[bere] 'body,' é[mbero] 'speed.' The augment H is often rendered opaque by other processes. In the Nguni S40 languages which retain the augment, the H may undergo shift/spread, depending on the tone of the stem. In Zulu S42, nouns lack the augment as complements to a negative verb: ba[ntu] 'people,' mi[nyango] 'doors,' si[hlalwana] 'small seat.' When the augment is added, H appears on the prefix or first stem syllable in a-bá[ntu], i-mí[nyango], i-si[hlálwana], due to underlying H on the augment, which shifts rightward. Sometimes H appears on the augment only in certain contexts. In Zinza JE23, the augment does not have H in isolation: a-ma[súnunu] 'fresh milk.' However, whenever the initial vowel coalesces with a preceding syllable, the coalesced syllable emerges with the H intact: adding yáá[guliize] yields yáá[guliiz'] ááma[súnunu] 'he has sold fresh milk.'

Another important source of tonal alternations in nouns is phrasal tonology. A citation noun may undergo a pre-pausal rule, e.g., Lowering or Insertion, which can be "undone" in a phrasal form. Mbololo Taita E74 has two citation patterns for disyllabic noun stems: HL (mi[sénge] 'stick') and LL (i[yeɣo] 'tooth,' ir[iso] 'eye'). When combined with a following word, a distinction emerges between underlyingly toneless nouns, such as /i[yeɣo]/, versus nouns with H on the final syllable, like /ir[isó]/ – cf. i[yeɣo] í[baha] 'big tooth,' ir[isó] i[baha] 'big eye.' In Matumbi P13, all nouns have H in their pre-pausal form, but comparison of pre-pausal vs. utterance-medial nouns reveals a contrast between toneless stems (lu[kólogo] 'brewery,' lu[kologo] lúmó 'one brewery') versus H-initial stems (lu[kóngobe] 'wood,' lu[kóngobe] lwaángʊ 'my wood').

4.2 Verbs

While nouns have a fairly simple tone system, verb tone is often very complex. Owing to the rich and productive agglutinative morphology of verbs, any morpheme can usually be freely positioned after H or L and before H or L, and can appear in any number of positions within a word (e.g., initial, peninitial, antepenult, penult), which means that verbs have the maximum potential for paradigmatic tonal alternations. In addition, verbs are often inflected with a tonal melody.

4.2.1 Roots

Regardless of length, verb roots usually contrast H and L, and prediction of the tone shape of a H stem is simplest if one assumes that the H is located on the first root mora. There is usually some context where no other H tones are present, in which case a H verb will realise that H according to the rules of the language. Mbaga Pare G22 has few rules obscuring the realisation of H tones, and H verbs have H on just the first syllable: *ku[chw-á]* 'to cut,' *ku[vón-a]* 'to see,' *ku[bánik-a]* 'to dry,' *ku[finikir-a]* 'to cover.'

Although verbs generally divide along H vs. L lines, some languages have a limited three-way lexical distinction. As noted in Section 2.1, Fuliiru JD63 has a contrast between H, L and Ø CVVC roots. In Shambaa G23, H spreads from the penult to the final syllable so /*ku[kóm-a]*/ becomes *ku[kóm-á]* 'to kill,' but this is exceptionally blocked in a handful of stems (*ku[lál-a]* 'to sleep,' *ku[vyál-a]* 'to give birth'), many of which historically have penult long vowels. Such exceptions, generally a variant of the /H/ class of verbs, constitute a third subclass of roots, failing to undergo some regular spreading or shifting process. Languages with a third root tone class of this type include Herero R31, Kako A93, some varieties of Nguni S40, Nyaturu F32, Kgalagari S311 and Shona S10 (cf. Marlo 2013).

Some languages lack lexical contrasts in verb roots, in which case stem tone is determined by tense-related rules. Such languages are often said to have "predictable" verb tone systems. Languages of this type are scattered throughout Eastern and Central regions, including Ikoma JE45, Lamogi JE16, southwestern Luyia JE32 and languages of southern Tanzania; but also Galwa B11c and Orungu B11b in Gabon. Odden (1989) and Marlo (2013) provide general surveys of this phenomenon.

Some predictable systems developed by acquiring a melodic H in contexts which historically lacked tonal inflection. A core property of proto-Luyia JE32 tone, reflected in conservative systems like Tiriki JE413, is that the Near Future tense lacks a melodic H and preserves the lexical contrast between Ø and H verbs. In that tense, predictable Tura JE30 has a melodic H from the second syllable through the final (and H on -*lá*-).

(3)		Tiriki (conservative)	Tura (predictable)	
	*Ø verbs	a-la[pulux-a]	a-lá[purúx-á]	'he will fly'
	*H verbs	a-la[vúkul-a]	a-lá[βukúl-á]	'he will take'

In some languages e.g., central Luyia (Nyala East JE32F), historically *H roots have been reanalysed as synchronically /L/, while historically *Ø roots have remained /Ø/. In toneless verbs such as *a-lá[kúlíx-á]* 'he will name,' a melodic H extends across the stem, while in /L/ verbs such as *a-lá[xalák-á]* 'he will cut,' the stem-initial L blocks leftward spread of the melodic H. An alternative analysis, which recapitulates the historical developments, is that "toned" roots are /H/ but the root H is deleted/lowered by Reverse Meeussen's Rule, and either through ordering of spreading before deletion or because of the presence of L, the melodic H does not spread to the stem-initial syllable. Other so-called "reversive" languages that may be analysed as having a contrast between /Ø/ and /L/ roots are Musenene Kete L21, Luba L31a, Shi JD53, Sukuma F21 and Tonga M64 (cf. Hyman 2001, 2007, Marlo 2013).

4.2.2 Extensions

Meeussen (1967) reconstructs most verb extensions of Proto-Bantu as *L. These suffixes are typically synchronically toneless and do not affect tone beyond the phonotactic effect of making the stem longer. One exception is Chewa N31b, where applicative -il-, causative -its-, reciprocal -an-, and reversive (tr.) -ul- are Ø, but stative -ik-, intensive -íts-, reversive (intr.) -úk- and, for some speakers, passive -ídw-, contribute a H (Hyman & Mtenje 1999).

Meeussen (1967) reconstructs causative *-í- and passive *-ó- with /H/. While this reconstruction may not be correct (Hyman & Katamba 1990), a number of Bantu languages have tonal effects of the passive and/or causative, including Fuliiru JD63, Ganda JE15, Herero R31, Hehe G62, Holoholo D28, Idakho E411, Lega D25, Makonde P23, Marachi JE342, Mwanga M22, Nande JD42, Nyala West JE18, Shi JD53, and Tiriki JE413. These effects are generally limited only to contexts inflected with a melodic H, sometimes with further restrictions (Hyman & Katamba 1990, Ebarb 2014). The causative and passive suffixes may also indirectly affect tone due to the fact that they typically become glides -y- and -w- in penultimate position (Hyman 1993).

4.2.3 Tonal inflection (melodic H tones)

Throughout Bantu, there are verb inflections for tense, aspect, mood, polarity and clause type with what is usually referred to as grammatical tone, which generally involves assignment of one or more tones – usually H – to a particular mora in the stem. See Odden and Bickmore (2014) for an overview and the associated volume for numerous case studies. The most frequent locations of grammatical H are the final mora or the second stem mora. Kamba E55 has a common pattern in which H spans from the second to the final syllable, as in nétwáa[konéthésyê] 'we made hit' (/kon-/) and nétwáa[táléthésyê] 'we make count' (/tál-/). Mwanga M22 has a grammatical H assigned to the final vowel in the potential tense, so compare the /Ø/ stem tungámú[sakuliil-á] 'we can comb for him/her' and the /H/ stem tungámú[⁺wándúliil-á] 'we can blacksmith for him/her,' which also has rightward spreading of H.

A number of different positions can be specified as targets for grammatical tone, especially in predictable tone languages. In Kuria JE43, primary H can appear on any of the first four moras (H may then spread rightwards), – viz. μ4 in the perfect (n[terekeéye] 'I have cooked for'), μ3 in the subjunctive (n[terekér-ε] 'I should cook for'), μ2 in the recent past (naaga[terékéeye] 'I just cooked for') and μ1 in the completive past (nnaa[téréker-a] 'I cooked'). In Makhuwa P31, the targeted positions are μ1, μ2, μ3 or penult (dialectally), and final; Yao P21 targets μ1, μ2 and final; Matumbi P13 targets μ1, μ2 and μ3. Luyia varieties include other targets such as σ2 and μ2 after σ1.

There is often a special connection between the object prefix and the verb stem in computing where the grammatical H is located, supporting the concept of a "macrostem" above the stem but below the word. In Kuria JE43, tone assignment is based on a count of vowels from the (left) edge of the macrostem, which includes any OPs. Thus in the infinitive, H appears on the first and fourth vowels counting from the OPs, resulting in forms such as *oko[géséra]hó* 'to harvest for at a place,' *oko-mó[géséra]ho* 'to harvest for him at,' *oko-gé-mó[géséra]ho* 'to harvest them for him at.' Makhuwa P31 has patterns targeting the first or second macrostem vowels, as well as the third stem vowel. In Mozambican Yao P21, some melodies are assigned with respect to the stem, others with respect to the macrostem.

The position of the grammatical tone may also depend on the lexical tone of the root, so in Shona S10, H roots realise the grammatical H on the final vowel, as in *havá[tóréséran-á]* 'they didn't make take for each other,' but toneless roots put H on the second stem vowel, as in *havá[bikísíran-a]* 'they didn't make cook for each other.' Similarly, in Nkore JE13 and most Rutara languages, the melodic H appears on the second syllable with toneless roots e.g., *ba[baríran-a]* 'they count for each other.' However, word-finally with H toned stems, e.g., *ba[boneran-á . . .]* 'they see for each other,' H retracts to the penult pre-pausally. Goldsmith (1987) refers to the µ2/FV melody as the 'complex stem tone pattern.'

In Shona S10, melodic H is assigned to all negative and subordinate clause verb tenses (participials, relative clauses, subjunctives). In Matumbi P13, this same class of contexts defines where one finds assignment of tone to µ2 or µ3 instead of µ1. More commonly, the set of tenses selecting a grammatical melody is fairly arbitrary, and must essentially be listed.

Some Luyia varieties have a melodic pattern with no overt melodic H in which other tones are actively lowered. For example, in Nyala West JE18, certain conditional forms surface entirely L, e.g., *xu[paaŋguluule]* 'if we could disarrange,' including when there are one or two object prefixes, e.g., *xu-mu[siindiixire]* 'if we could push him,' *a-muu-m[bukuriire]* 'if he could take him for me,' which normally contribute a H, e.g., *xuri [siindíx-á]* 'we will push,' *xuri-βá[siindíx-á]* 'we will push them.' Surprisingly, the causative/passive H does surface in this context, e.g., *a[kusiis-í-e]* 'if he could sell,' *a[βeker-ú-e]* 'if he could be shaved.'

In languages which have a contrast between two root tones, there is usually some morphosyntactic context, e.g., the infinitive, which lacks tonal inflection. Some languages, such as the "reversive" language Nyala East JE32F have a contrast between two tonal types of roots, but also have tonal inflection in all verb forms. In "predictable" systems, which have eliminated the tonal contrast in verb roots, all verb forms are generally inflected with some melodic H (Marlo 2013), though there are tonally uninflected verb forms in Ikoma JE45 and Mozambican Yao P21.

While melodic inflectional patterns are usually limited to appearing on the verb stem, they occasionally affect larger domains. In Shona S10, the melodic tone pattern is assigned to each verbal prefix as well. A spectacular example of wide-scope melodic patterning is found in Orungu B11b (Maniacky & Ambouroue 2014) where every word from the verb to the end of the intonational phrase is either rendered all-H or all-L, depending on the tense and subject, when the verb is negative: thus /à-á-**H**-lwàn-ày-a ɣó ìdyànjà kwáŋgá òywèrà néŋgénéŋgé/ → à[lwánáɣá] ɣ ídyánjá kwáŋg óywérá néŋgénéŋgé vs. /wá-á-**L**-lwàn-ày-a ɣó ìdyànjà kwáŋgá òywèrà néŋgénéŋgé/ → wá[lwánàɣà] ɣ ìdyànjà kwàŋg òywèrà nèŋgènèŋgè 'that (s/he, they) usually do not stay late at work.'

4.2.4 Prefix tone

Tone on subject prefixes is generally determined by morphosyntactic structure. In Nyala West JE18, SPs are uniformly toneless, but more commonly, the tone of the SP is determined by clause-type and person. In Shona S10, SPs are H in negative and subordinate verb forms (*ha-ndí-chá[bikí]* 'I won't cook,' *ndí-chá[biká]* 'I not being about to cook,' *ra-ndí-ngá[biká]* 'that, which I may cook'), though toneless in a few conditional tenses (*kana ndi-ka[bika]* 'if I had cooked,' *kana a-ka[bika]* 'if he had cooked'). In Matumbi P13, SPs are toneless except in subordinate clause tenses where they are H, and Kerewe JE24 and some other Great Lakes languages have a similar pattern where SPs are generally toneless, but have H in subordinate and negative tenses.

Commonly there is a person-based tonal contrast between H-toned and toneless SPs, where third person SPs have H and first and second person are toneless. In Shona S10, main clause affirmative tenses, SPs follow this pattern, thus *á-chá[bíka]*, *vá-chá[bíka]* '3sg (cl.1)/3pl (cl.2) will cook,' *á-ngá[bíka]*, *vá-ngá[bíka]* '3sg (cl.1)/3pl (cl.2) may cook,' but *ndi-chá[bíka]*, *ti-chá[bíka]*, *u-chá[bíka]*, *mu-chá[bíka]* '1sg/1pl/2sg/2pl will cook,' *ndi-ngá[bíka]*, *ti-ngá[bíka]*, *u-nga[bíka]*, *mu-ngá[bíka]* '1sg/1pl/2sg/ 2pl may cook.' In negative and subordinate tenses, all persons have H. Likewise in the main clause affirmative tenses of Pare G22, first and second person (singular or plural) SPs are toneless and third person SPs have H. However, all SPs are H in negative tenses, and toneless in relative tenses. This same split between toneless non-third person vs. H-toned third person is widespread in Bantu, occurring in Mokpwe A22, Kamba E55, Shona S10 and other languages.

The SPs for cl. 1, 4 and 9 are sometimes toneless when expected to be H, possibly connected to the prefixes for these classes being onsetless vowels. In Kerewe JE24, if the SP is onsetless (1sg, 2sg, cl. 1, 4, 9), its H shifts to the following syllable. In Kamba E55, subordinate SPs are H as a rule, but those for cl. 1, 4, 9 are toneless. In Lungu M14, these prefixes are L, while SPs are generally H, but onsetless 1sg and 2sg prefixes remain H.

Another person-related tonal pattern is that 1sg, 2sg, and cl. 1 are toneless with other persons having H. This pattern is tense-determined and usually co-exists with the more common non-third person toneless vs. third person H pattern. In Somali Zigula (Mushunguli) G311, where tones shift towards the antepenult and all verbs receive a H somewhere resulting in the obscuring of the underlying locus of the tonal distinction, present tense third person SPs are H (*a[hí⁺mídha]*, *wa[hí⁺mídha]* 'cl.1/cl.2 advise') and others are L (*na[himídha]*, *wa[himídha]*, *cha[himídha]*, *mwa[himídha]* '1sg/2sg/1pl/2pl advise), but in the past, 1sg, 2sg and cl. 1 SPs are toneless (*sí[hímidha]*, *kú[hímidha]*, *ká[hímidha]* '1sg/2sg/cl.1 advised'), and other prefixes are H (*chí[hímídha]*, *ḿ[hímídha]*, *wá[hímídha]* '1pl/2pl/cl.3 advised'). Analogous tone distributions are found in Bondei G24 and Kgalagari S311.

Object prefixes tend to be uniform in tonal properties, and usually have H, though may be subject to special tone dissimilation when combined with H roots. In some languages, such as Lungu M14, Nilamba F31, Nyamwezi F22 and Zigula G31, human singular OPs (1sg, 2sg, cl. 1) are L, while other OPs are H. Fuliiru JD63 has the reverse pattern: 1sg, 2sg and cl. 1 are H, while other OPs are L. In Gikuyu E51, 1sg, 2sg and the cl. 1, cl. 4 and cl. 9 OPs are L, while other OPs are H.

Although all post-SP prefixes in Shona S10 receive a melodic H in tenses selecting stem-inflectional H, prefixes marking tense, aspect, mood and negation in Bantu do not tend to have rule-governed tone patterns. See Nurse and Philippson (2006) for commonly encountered tense prefixes across Bantu. It can be difficult to definitively assign a tense prefix a single underlying tone, so for example -*a*- in Fuliiru JD63 is L toned in Conditional/Potential *tw-à-ngà*[*láálíkà*] 'we could/would invite' but H in Conditional Contrary-to-fact Past *tw-á-ngá*[*mwìirì*] 'if we had shaved' (Van Otterloo 2014). Likewise in Kuria JE43, the prefix -*ka*- has H in Habitual Past Focused *nnaa-yá*[*túníre*] 'I used to look for (then),' but L in Hodiernal Past Progressive Anterior Focused *nnaa-ya*[*tuníre*] '(indeed) I have been looking for (today).'

4.3 Tone at the phrasal level

A major demarcative function of tone in Bantu is as an indicator of phrasal syntactic relations. Many languages have phrasal processes whose domains of application are defined in terms of syntactic and semantic relations between words, thus the left edge of the VP defines a limit on tone shift in Tsonga S53 (Kisseberth 1994). Odden (1995) provides an overview. The most typical situation is sandhi that applies only between phrasal heads and their modifiers. For instance, in Zinza JE23, H is deleted from a verb that is followed by a modifier, so compare *aka*[*zína*] 'he sang' versus *aka*[*zina*] *géeta* 'he sang in Geita.' If the following word is not a complement, there is no deletion, cf. *aka*[*zína*] *Bulemo* 'Bulemo sang.' Application of this rule gives a clear indication of the functional relation between verbs and following words.

The kinds of syntactic information which can be reflected in Bantu tone can be complicated. Sikongo H16a deletes H in nouns under complex conditions regarding their relation to the sentence. This yields the alternation between *ma*[*fúnúnúnú*] 'bees' vs. *ma* [*fungununú*] *man*[*tátíkídí*] 'the bees bit me' and [*mweení*] *má*[*fuungununú*] 'I saw bees,' illustrating tone deletion in canonically positioned subjects and objects, as well as objects that are left-dislocated at the sentence level as in *ma*[*fuungununú*] *twa*[*suumbidí*] 'we bought bees,' *ma*[*fuungununú*] *bikuni bya*[*suumbidí*] 'farmers bought bees.' However, when the object is moved into VP-initial focal position, H is unchanged, e.g., *mwaana ma*[*fúngúnúnú*] *ka*[*mwééní*] 'the child saw *bees*' (Odden 1991).

In Luyia JE32 noun phrases, different classes of modifiers trigger different rules. For example, in Tiriki JE413, when a toneless noun like *ma*[*hoondo*] 'pumpkins' combines with an adjective with a H, H spreads unboundedly left (though not onto initial position), e.g., *ma*[*hóóndó*] *máláhi* 'good pumpkins.' When a toneless noun is modified by a demonstrative, possessive, or relative clause, H is inserted onto the penult, e.g., *ma*[*hóóndo*] *kanu* 'these pumpkins,' *ma*[*hóóndo*] *keeru* 'our pumpkins,' *ma*[*hóóndo*] *ka va*[*vukuul-e*] 'the pumpkins that they took.' In other Luyia JE32 varieties, the site of H insertion is different: H is inserted onto the augment in Bukusu JE31c, and onto the final vowel in Wanga JE32a.

In some languages, negative verbs behave differently from corresponding positive forms with respect to tone sandhi. In Tsonga S53, negative verbs spread their H unfettered by restrictions that operate in affirmative verbs. Thus the H-toned subject prefix in *vá*[*xává*] *nyáma* 'they are buying meat' (cf. *ndzi*[*xava*] *nyama* 'I am buying meat') cannot extend onto the final vowel of the nominal complement in this affirmative tense. However, in negative *a-ndzí*[*xáví*] *nyámá* 'I am not buying meat,' the H extends through to the end of the noun (cf. *a-ndzí*[*xáví*] 'I am not buying'). In Makhuwa P31, on the other hand, the initial H-tone of a nominal complement elides after a verb in a class of tense forms connected with focus (Stucky 1979). Thus, *nakhúwo* 'maize' appears with its H tone intact in *ki-nóó[límá] nakhúwo* 'I am cultivating maize,' but without its H tone in *ki-n[límá] nakhuwo* 'it's maize that I am cultivating.' See Hyman (1999) for general discussion of tone and focus in Bantu.

Such categorial restrictions can be general (the general distinction in Tsonga S53 between affirmative and negative forms) or highly specific. The Zinza JE23 rule deleting H in a verb before a modifier does not apply in the present tense: *ni-ba[téeka] maláaya* 'they are cooking bananas.' In Tiriki JE413, there is phrase-medial deletion of the Pattern 2 melodic H in the Present and Recent Past among others, but not in the Indefinite Future or Consecutive.

5 HISTORICAL CHANGE

The current diversity in the tone systems of Bantu languages results from two main historical factors affecting the earlier language. The first is phonologisation of well-known natural phonetic tendencies, especially those that shift pitch to the right, and those which lead to suppression of high pitch in perceptually suboptimal contexts, as outlined in Section 3. The second is grammatical reanalysis, typically based on the fact that the historically original conditions often cannot be reconstructed by children, especially when new rules and constructions are added to the grammar.

The development of melodic tone patterns (Section 4.2.3) throughout Bantu constitutes a case in point. The original system probably involved final suffixes V́ or V̀ with distinctive tone which marked tense and other clausal properties, where the particular vowel quality marking a function is often unclear. Mokpwe A22, though, provides evidence that reflexives specifically had the suffix *-ě*. Bantu languages are replete with tantalising suggestions that tone melody choice was originally related to final vowel selection (cf. Odden & Bickmore 2014). There may have been multiple final suffixes with distinctive tone (Meeussen 1967), subject to elision, so a verb could have inflections of the form / CVCVCV/ versus /CVCVCV-V́/. Owing to an early process of leftward H spreading, the last tone extends to all but the first stem vowel, thus surface CVCVCV vs. CVCV́CV́(-V́). This then gave rise to a general "floating tone" analysis of certain verb tenses, where the final vowel can be realised with various autonomous tones. This situation is most directly seen in Western languages which tend to be of this type, e.g., Yambasa A62 *gu[nyongol-onyo]* 'to tickle' where the final vowel does not have H, versus *amba[nyongólónyó]* from /*amba[nyongolonyo+H]*/, where inflectional H appears on the final vowel and spreads leftwards (Yukawa 1992b).

Subsequent delinking and other phonotactic processes led, through reanalysis, to a wide array of current targets: these outcomes arise from the fact that analytical ambiguity is rampant in many systems. Consider a /Ø/ CVCV verb stem which surfaces CV́CV̀. The HL surface pattern could be the result of assignment of the melodic H to the first mora of the stem, or to the second (third, or fourth) mora of the stem, followed by leftward shift of the melodic H from the final vowel. The same surface pattern could also be the result of assignment to the penultimate vowel, or to the final vowel with pre-pausal leftward shift. Whether the target is oriented toward the left or right edge of the stem can only be determined with data from longer verb stems. Distinguishing "penult" and "final" targets from those where the H is aligned towards the left edge of the stem (e.g., "μ3" or "μ4") requires very long stems to establish the correct analysis. As such verb forms tend to be rare, the conditions are ripe for reanalysis.

Even within very closely related languages, identical surface patterns can be derived differently. Within Luyia JE32, Logooli JE41 and Wanga JE32a both have patterns in cognate forms with H on the first two stem vowels, e.g., of /Ø/ verbs in the Indefinite Future, but they are derived differently. In Logooli JE41, the melodic H is assigned to the second stem vowel and spreads leftward in the stem (/vari[sekérana]/ → vari[sékérana] 'they will laugh for e.o.'), but in Wanga JE32a, the melodic H is assigned to the stem-initial vowel and doubles to the right (/ali[lómaloma]/ → ali[lómáloma] 'he will talk').

When the melodic H spans, on the surface, from the left edge of the stem to the final vowel, it is often not obvious whether the pattern is generated by assigning H to the left edge (the leftmost free vowel, μ2) with subsequent rightward spreading, or, alternatively, to the right edge with leftward spreading. From the perspective of acquisition, "rightmost free vowel" plus "spread leftward" is as good as "leftmost free vowel" plus "spread rightward," and children acquiring the system at a later stage in the development of Bantu could not know of any earlier reason to attribute the addition of H, independent of a segmental morpheme that contributes the H, to the right edge of the verbal form.

A reanalysis in terms of assignment to the leftmost stem vowel with rightward spread took place in Kamba E55, owing in part to the addition of new final tone suffixes, hence one finds H from μ2 to the penult in nétó[alyóóláneete] 'we have been changing e.o.' This form can be understood in terms of the earlier system as reflecting *nétó[alyóólánéété] modified by addition of a final fall and a penult L. However, these additions made a final-vowel target for melodic tone inflection synchronically untenable, and the system was re-cast in terms of a μ2 target with rightward spreading.

Leftward spreading of H can become subject to OCP restrictions which create differences in how the H is positioned on the left. Tachoni JE31E has a pattern in which the melodic H surfaces from σ2 to the penult in /Ø/ verbs (bali[chiichákána] 'they will continue') but σ3 to the penult in /H/ verbs (bali[botooxánáníla] 'they will go around for each other'). These forms derive from assignment of the melodic H to the right edge, followed by leftward-spreading subject to the OCP, and delinking of H from the final vowel. This is one simple OCP-related limit on leftward spreading. An alternative OCP restriction on spreading is to totally block spread if a H precedes, given some domain. Gusii JE42 has a single pattern whose realisation is determined by the distribution of pre-stem H tones. The melodic H generally surfaces from μ2 through the final vowel, e.g., /Ø/ verb to[timókérá] 'we rest for,' /H/ verb to[tákúnérá] 'we chew for.' When a pre-stem H doubles onto the stem-initial vowel, as in tó[tímokeré] 'that we rest for,' or when a pre-stem H fuses with the root-initial H tó[tákuneré] 'that we chew for,' the melodic H surfaces only on the final vowel (Bickmore 1997). A further development, which leads to the common μ2-only pattern of melodic assignment, is phonotactic elimination of multiple-linkages on H tones, effectively yielding a shift in the location of H from the final vowel to μ2. Combined with a Gusii JE42-like ban on leftward spread with H roots, we arrive at the system found in Rutara languages such as Nkore JE13 with final H in /H/ verbs, e.g., ba[shariraná] . . . 'they slice for each other,' and μ2 H in /Ø/ verbs, e.g., ba[rimírana] . . . 'they cultivate for each other.'

Such OCP restrictions could have given rise to the modern-day μ3 and μ4 targets in the predictable-tone system of Kuria JE43. A μ3 target may have arisen from an earlier stage with a lexical tone contrast in verb roots and final H spreading leftward, subject to a one-mora buffer between the lexical H and the melodic H, so that the melodic H commences at μ3, thus earlier *ntoore[hóotooterá] → *ntoore[hóotóótérá]. The lexical tone contrast was lost and the pattern of /H/ verbs was generalised, yielding contemporary *ntoore*[*hootóótéra*] 'we will reassure' and this leads to the current system where the melodic H is directly assigned to μ3. A μ4 target could then arise in such a system if there is rightward doubling of the initial H under some circumstance, so *tora[*hóotooterá*] → *tora[*hóótooterá*], which developed analogously to contemporary *tora*[*hootoótéra*] 'we are about to reassure.'

Reanalysis of phrasal tonal patterns as part of verbal inflection may have played a central role in the development of the synchronically diverse tone systems of Luyia JE32. Central and western Luyia languages have acquired tonal inflection in all tense forms, including Infinitive and Near Future forms which lack a melodic H in Eastern systems like Tiriki JE413 (see Section 4.2.1). Why did these forms acquire a melodic H? In phrase-medial position, a process of H Tone Anticipation (HTA) applies in Tiriki JE413 and other conservative Luyia JE32 varieties, where H from a following word spreads into the verb, e.g., *ala*[*vó⁺yóyáná*] *kálaha* 'he will go around slowly' (cf. *ala*[*vóyoyana*] *vwaangu* 'he will talk quickly'). HTA only affects TAM-forms without a melodic H, e.g., the Near Future and the Infinitive; tense forms inflected with a melodic H (e.g., Hodiernal Perfective and the Remote Past) do not have HTA, e.g., *a*[*voyoyaane*] *kálaha* 'he went around slowly, *ya*[*vóyoyana*] *kálaha* 'he went around slowly.' Speakers may have reinterpreted the phonetic H commonly found on the verb in phrase-medial position in the Near Future and Infinitive as an inherent piece of inflection in those tenses. Due to the fact that melodic H triggers the deletion of lexical tone commonly in Luyia JE32 and throughout Great Lakes languages (see Section 3.3), further changes to root tones resulted once the melodic H became universal throughout the tense system.

6 CONCLUSION

Although a great deal is known about Bantu tone, we have detailed studies of only a fraction of the Bantu languages. Most of the languages for which we do have material seem to exhibit variations that range from subtle differences to what amounts to essentially different systems. We have identified in this chapter many principles and patterns that are pervasive. Doubtless there is still much more to be learnt. The continued documentation of Bantu tone systems remains a critical need – not just for the student of African languages, but also for the general linguist.

The close examination of Bantu tonal systems reveals patterns of extraordinary complexity. The ultimate goal of the study of linguistic systems is to understand how a child learns such systems automatically by virtue of being a human being born into these linguistic communities. There are few other phonological phenomena as complex as Bantu tone, which at the same time are clearly the consequence of the speaker's knowledge of a system of principles rather than the memorisation of specific forms. As a result, the importance of Bantu tone systems for the theoretical study of phonology cannot be overestimated.

7 ACKNOWLEDGEMENTS

We would like to thank our numerous language consultants, and the many funding agencies who have supported our research on Bantu tonology, especially the National Science Foundation, National Endowment for the Humanities, and the Fulbright Foundation. Data with no reference to source derives from our notes, or from the first edition of this chapter in the case of Namwanga and Gusii (thanks to Lee Bickmore), and Bondei, Makhuwa, Tsonga, and Zulu (thanks to Chuck Kisseberth). We further acknowledge Chuck Kisseberth's contributions to the first edition which forms the basis for the present version.

REFERENCES

Avery, P. & K. Rice (1989) Segment Structure and Coronal Underspecification. *Phonology* 6: 179–200.

Bickmore, L. S. (1997) Problems in Constraining High Tone Spread in Ekegusii. *Lingua* 102: 265–290.

Bickmore, L. S. (2000) Downstep and Fusion in Namwanga. *Phonology* 17: 297–333.

Bickmore, L. S. (2007) *Cilungu Phonology*. Stanford, CA: CSLI Publications.

Bickmore, L. S. (2014) Cilungu Tone Melodies: A Descriptive and Comparative Study. *Africana Linguistica* 20: 39–62.

Bird, C. (1966) *Aspects of Bambara Syntax*. Los Angeles: University of California, PhD dissertation.

Blanchon, J. A. (1999) "Tone cases" in Bantu Group B.40. In Blanchon, J. A. & D. Creissels (eds.), *Issues in Bantu Tonology*, 37–82. Cologne: Rüdiger Köppe.

Blanchon, J. A. & D. Creissels (eds.) (1999) *Issues in Bantu Tonology*. Cologne: Rüdiger Köppe.

Bradshaw, M. M. (1999) *A Cross-Linguistic Study of Consonant-Tone Interaction*. Columbus: Ohio State University, PhD dissertation.

Cassimjee, F. (1986) *An Autosegmental Analysis of Venda Tonology*. Urbana-Champaign: University of Illinois, PhD dissertation.

Cassimjee, F. & C. W. Kisseberth (1992) The Tonology of Depressor Consonants: Evidence from Mijikenda and Nguni. In Buszard-Welcher, L. A., J. Evans, L. Wee & W. Weigel (eds.), *Proceedings of the Eighteenth Annual Meeting of the Berkeley Linguistics Society: Special Session on the Typology of Tone Languages*, 26–40.

Clark, M. (1990) *The Tonal System of Igbo*. Dordrecht: Foris.

Clements, G. N. & K. Ford (1979) Kikuyu Tone Shift and Its Synchronic Consequences. *Linguistic Inquiry* 10: 179–210.

Clements, G. N. & J. Goldsmith (eds.) (1984) *Autosegmental Studies in Bantu Tone*. Dordrecht: Foris.

Coupez, A. (1955) *Esquisse de la langue holoholo*. Tervuren: Musée royal du Congo belge.

Crane, T. M. (2009) Tense, Aspect, Mood, and Tone in Shekgalagari. *UC Berkeley Phonology Lab Annual Report* 2009: 224–278.

Creissels, D. (1998) Expansion and Retraction of High Tone Domains in Setswana. In Hyman, L. M. & C. W. Kisseberth (eds.), *Theoretical Aspects of Bantu Tone*, 133–194. Stanford: CSLI Publications.

Downing, L. J. (2011) Bantu Tone. In van Oostendorp, M., C. J. Ewen, E. Hume & K. Rice (eds.), *The Blackwell Companion to Phonology*. Cambridge; Oxford: Blackwell.

Ebarb, K. J. (2014) *Tone and Variation in Idakho and Other Luhya Varieties*. Bloomington: Indiana University, PhD dissertation.

Ebarb, K. J., C. R. Green & M. R. Marlo (2014) Luyia Tone Melodies. *Africana Linguistica* 20: 121–143.

Ebarb, K. J. & M. R. Marlo (2015) Vowel Length (in)sensitivity in Luyia Morphophonology. *Southern African Linguistics and Applied Language Studies* 33(3): 373–390.

Festen, B. (2008) *Verbal Tone in Mpyemo*. Grand Forks: University of North Dakota, MA thesis.

Goldsmith, J. (1976) *Autosegmental Phonology*. Cambridge: MIT, PhD dissertation.

Goldsmith, J. (1984) Tone and Accent in Tonga. In Clements, G. N. & J. Goldsmith (eds.), *Autosegmental Studies in Bantu Tone*, 19–51. Dordrecht: Foris.

Goldsmith, J. (1987) Stem Tone Patterns of the Lacustrine Bantu Languages. In Odden, D. (ed.), *Current Approaches to African Linguistics IV*, 167–178. Berlin: De Gruyter Mouton; Dordrecht: Foris.

Halme, R. (2004) *A Tonal Grammar of Kwanyama*. Cologne: Rüdiger Köppe.

Hyman, L. M. (1976) *Studies in Bantu Tonology*. Los Angeles: University of Southern California.

Hyman, L. M. (1993) Conceptual Issues in the Comparative Study of the Bantu Verb Stem. In Mufwene, S. S. & L. Moshi (eds.), *Topics in African Linguistics*, 3–34. Amsterdam; Philadelphia: John Benjamins.

Hyman, Larry M. (1999) The Interaction Between Focus and Tone in Bantu. In Rebuschi, G. & L. Tuller (eds.), *The Grammar of Focus*, 151–177. Amsterdam; Philadelphia: John Benjamins.

Hyman, L. M. (2001) Privative Tone in Bantu. In Kaji, S. (ed.), *Cross-Linguistic Studies of Tonal Phenomena*, 237–257. Tokyo: ILCAA.

Hyman, L. M. (2006) Word Prosodic Typology. *Phonology* 23: 225–257.

Hyman, L. M. (2007) Universals of Tone Rules: 30 Years Later. In Riad, T. & C. Gussenhoven (eds.), *Tones and Tunes: Studies in Word and Sentence Prosody*, 1–34. Berlin: De Gruyter Mouton.

Hyman, L. M. (2009) How (Not) to Do Phonological Typology: The Case of Pitch-Accent. *Language Sciences* 31: 213–238.

Hyman, L. M. (2013) Penultimate Lengthening in Bantu: Analysis and Spread. In Bickel, B., L. A. Grenoble, D. A. Peterson & A. Timberlake (eds.), *Language Typology and Historical Contingency*, 309–330. Amsterdam; Philadelphia: John Benjamins.

Hyman, L. M. & F. X. Katamba (1990) Spurious High-Tone Extensions in Luganda. *South African Journal of African Languages* 10: 142–158.

Hyman, L. M. & C. W. Kisseberth (eds.) (1998) *Theoretical Aspects of Bantu Tone*. Stanford: CSLI Publications.

Hyman, L. M. & F. Lionnet (2012) Metatony in Abo (Bankon), A42. In Marlo, M. R., N. B. Adams, C. R. Green, M. Morrison & T. M. Purvis (eds.), *Selected Proceedings of the 42nd Annual Conference on African Linguistics*, 1–14. Somerville, MA: Cascadilla Proceedings Project.

Hyman, L. M. & K. C. Monaka (2011) Tonal and Non-Tonal Intonation in Shekgalagari. In Frota, S. n., G. Elordieta & P. Prieto (eds.), *Prosodic Categories: Production, Perception and Comprehension*, 267–290. Dordrecht: Springer.

Hyman, L. M. & A. D. Mtenje (1999) Non-Etymological High Tones in the Chichewa Verb. *Malawian Journal of Linguistics* 1: 121–256.

Kagaya, R. (ed.) (1992) *Studies in Cameroonian and Zairean Languages*. Tokyo: ILCAA.

Kagaya, R., R. M. Besha & Y. Yukawa (eds.) (1989) *Studies in Tanzanian Languages*. Tokyo: ILCAA.

Kisseberth, C. (1992) Metrical Structure in Zigula Tonology. In Gowlett, D. F. (ed.), *African Linguistic Contributions Presented in Honour of Ernst Westphal*, 227–260. Pretoria: Via Afrika Limited.

Kisseberth, C. (1994) On Domains. In Cole, J. & C. Kisseberth (eds.), *Perspectives in Phonology*, 133–166. Stanford: CSLI Publications.

Kisseberth, C. & D. Odden (2003) Tone. In Nurse, D. & G. Philippson (eds.), *The Bantu Languages*, 59–70. London; New York: Routledge.

Maganga, C. & T. C. Schadeberg (1992) *Kinyamwezi: Grammar, Texts, Vocabulary*. Cologne: Rüdiger Köppe.

Maniacky, J. & O. Ambouroue (2014) Melodic Tones in Orungu (Bantu B11b). *Africana Linguistica* 20: 245–261.

Marlo, M. R. (2013) Verb Tone in Bantu Languages: Micro-Typological Patterns and Research Methods. *Africana Linguistica* 19: 137–234.

Marlo, M. R., L. C. Mwita & M. Paster (2014) Kuria Tone Melodies. *Africana Linguistica* 20: 277–294.

Marlo, M. R., L. C. Mwita & M. Paster (2015) Problems in Kuria H Tone Assignment. *Natural Language and Linguistic Theory* 33: 251–265.

Marlo, M. R. & D. Odden (2015). Sources for Bantu Tonology [Data set]. *Zenodo* (Available online at http://doi.org/10.5281/zenodo.35670).

Massamba, D. (1982) *Aspects of Tone and Accent in Ci-Ruri*. Bloomington: Indiana University, PhD dissertation.

Meeussen, A. E. (1967) Bantu Grammatical Reconstructions. *Africana Linguistica* 3: 79–121.

Mwita, L. C. (2008) *Verbal Tone in Kuria*. Los Angeles: University of California, PhD dissertation.

Myers, S. (1997) OCP Effects in Optimality Theory. *Natural Language and Linguistic Theory* 15: 847–892.

Myers, S. (1998) Surface Underspecification of Tone in Chichewa. *Phonology* 15: 367–391.

Nash, J. A. (1992–1994) Underlying Low Tones in Ruwund. *Studies in African Linguistics* 23: 223–278.

Nurse, D. & G. Philippson (2006) Common Tense-Aspect Markers in Bantu. *Journal of African Languages and Linguistics* 27(2): 155–196.

Odden, D. (1982) Tonal Phenomena in Kishambaa. *Studies in African Linguistics* 13: 177–208.

Odden, D. (1986) On the Role of the Obligatory Contour Principle in Phonological Theory. *Language* 62: 353–383.

Odden, D. (1989) Predictable Tone Systems in Bantu. In van der Hulst, H. & N. Smith (eds.), *Autosegmental Studies on Pitch Accent Systems*, 225–251. Dordrecht: Foris.

Odden, D. (1991) The Intersection of Syntax, Semantics and Phonology in Kikongo. In Hubbard, K. (ed.), *Proceedings of the Annual Meeting of the Berkeley Linguistics Society 17*, 188–199. Berkeley: Berkeley Linguistics Society.

Odden, D. (1994) Adjacency Parameters in Phonology. *Language* 70: 289–330.

Odden, D. (1995) Phonology at the Phrasal Level in Bantu. In Katamba, F. (ed.), *Bantu Phonology and Morphology*, 40–68. Munich: Lincom Europa.

Odden, D. (1999) Typological Issues in Tone and Stress in Bantu. In Kaji, S. (ed.), *Proceedings of the Symposium "Cross-Linguistic Studies of Tonal Phenomena: Tonogenesis, Typology, and Related Topics,"* 187–215. Tokyo: ILCAA.

Odden, D. (2011) Features Impinging on Tone. In Goldsmith, J., E. Hume & L. Wetzels (eds.), *Tones and Features: Phonetic and Phonological Perspectives*, 81–107. Berlin: De Gruyter Mouton.

Odden, D. & L. Bickmore (2014) Melodic Tone in Bantu. *Africana Linguistica* 20: 3–13.

Paster, M. & Y. Kim (2011) Downstep in Tiriki. *Linguistic Discovery* 9: 71–104.

Philippson, G. (1991) *Tons et accent dans les langues bantu d'afrique orientale: Étude comparative typologique et diachronique*. Paris: Université Paris V, thèse de doctorat.

Poletto, R. (1998) *Topics in Runyankore Phonology*. Columbus: Ohio State University, PhD dissertation.

Pulleyblank, D. (1986) *Tone in Lexical Phonology*. Dordrecht: Reidel.

Richardson, I. (1959) *The Role of Tone in the Structure of Sukuma*. London: SOAS.

Schuh, R. G. (1998) *A Grammar of Miya*. Berkeley: University of California Press.

Stevick, E. W. (1969) Tone in Bantu. *International Journal of American Linguistics* 35(4): 330–341.

Stucky, S. U. (1979) The Interaction of Tone and Focus in Makua. *Journal of African Languages and Lingusitics* 1: 189–198.

Van Otterloo, K. (2011) *The Fuliiru Language, Volume 1: Phonology, Tone, and Morphological Derivation*. Dallas: SIL International.

Van Otterloo, K. (2014) Tonal Melodies in the Kifuliiru Verbal System. *Africana Linguistica* 20: 385–403.

van Spaandonck, M. (1967) *Morfotonologische analyse in Bantutalen: identificatie van morfotonemen en beschrijving van hun tonologische representatie*. Gent: Rijksuniversiteit Gent, doctoraatsthesis.

van Spaandonck, M. (1971) *L'analyse morphotonologique dans les langues bantoues: identification des morphotonemes et description de leurs representations tonologiques*. Paris: Société pour l'Étude des Langues Africaines.

Volk, E. (2011) *Mijikenda Tonology*. Tel Aviv: Tel Aviv University, PhD dissertation.

Yip, M. J. W. (2002) *Tone*. Cambridge; New York: Cambridge University Press.

Yukawa, Y. (ed.) (1987) *Studies in Zambian Languages*. Tokyo: ILCAA.

Yukawa, Y. (1989a) A Tonological Study of Nilamba Verbs. In Kagaya, R., R. M. Besha & Y. Yukawa (eds.), *Studies in Tanzanian languages*, 405–450. Tokyo: ILCAA.

Yukawa, Y. (1989b) A Tonological Study of Nyaturu Verbs. In Kagaya, R., R. M. Besha & Y. Yukawa (eds.), *Studies in Tanzanian Languages*, 451–480. Tokyo: ILCAA.

Yukawa, Y. (1992a) A Tonological Study of Luba Verbs. In Kagaya, R. (ed.), *Studies in Cameroonian and Zairean Languages*, 303–362. Tokyo: ILCAA.

Yukawa, Y. (1992b) A Tonological Study of Yambasa Verbs. In Kagaya, R. (ed.), *Studies in Cameroonian and Zairean Languages*, 1–46. Tokyo: ILCAA.

CHAPTER 6

WORD FORMATION

Thilo C. Schadeberg and Koen Bostoen

1 INTRODUCTION

Bantu languages are rich in morphology, inflectional as well as derivational. The line between the two kinds of morphological processes is usually easy to draw. The major derivational processes are surveyed in Sections 2 to 5 of this chapter, along with a number of other word formation strategies in Section 6.

General statements refer to "proto-typical" Bantu, which is close to Proto-Bantu (PB) and usually directly reflected in most Bantu languages, except in the extreme North-Western part of the domain. This chapter draws heavily on work done at the Royal Museum for Central Africa in Tervuren, most importantly on "Bantu Grammatical Reconstructions" (Meeussen 1967) and "Bantu Lexical Reconstructions 3" (BLR3) (Bastin *et al.* 2002). Other comparative sources are Büttner (1881); Meinhof (1899, 1904, 1910, 1948); Meinhof and van Warmelo (1932); Doke (1935); Guthrie (1967–1971).

The typical nominal or verbal word form in Bantu consists of a stem preceded by one or more bound morphemes. A nominal stem is usually preceded by a nominal prefix of a given noun class. Preceding a verbal stem, we may find prefixes specifying, amongst other things, the person or noun class of the subject as well as inflectional categories such as time, aspect, negation, etc. The stem-initial position often has special stress-related phonological properties (tone, length, vowel quality). Verb stems can further be divided into a verbal base (B) and a final suffix (F), most often a final vowel. The same analysis into base and final suffix is possible for those nouns that are derived from verbs; most other nominal stems are not amenable to further morphological analysis. In finite verb forms, the final suffix is part of inflectional morphology; in deverbative nouns, the final suffix is a nominalising suffix, as shown in (1). The first four words in (1) all contain the verbal base *-lìm-* 'cultivate'; the last item contains the unanalysable nominal stem *-límá* 'bat.'

(1)   Umbundu R11 (Le Guennec & Valente 1972, Schadeberg 1990)
   tw-à-lìm-á   'we cultivated' $SP_{1Pl.}$-PST-[B-F]$_{stem}$
   óku-lím-à   'to cultivate'   15-[B-F]$_{stem}$
   á-lím-ì   'cultivators'   2-[B-F]$_{stem}$
   á-lím-à   'years'   6-[B-F]$_{stem}$
   á-⁺límá   'bats'   2-Stem

The base is the domain of derivational (lexical) morphology of the verb. The base may be simple or extended. The simple base consists just of a root or radical. The extended base consists of a root followed by one or more "extensions." An extension may be analysable as to form and meaning, in which case we may call it a suffix, or else segmentation may be purely formal, in which case the analysis yields a (formal) radical and an "expansion" or formal suffix. Such formal segmentation can be justified by the parallel phonological behavior of expansions and extensions such as being tonally neutral or having a reduced vowel system. For example, the base *-ták-un- 'chew' is analysed as consisting of the root -tak- and the expansion -un-. A simplex root *-ták- has not been reconstructed (Bastin et al. 2002).

Bantu verbs have a very regular morphological and phonological structure. The R has the structure CVC, where C_2 may be prenasalised or empty. The set of CV-verbs is small, but contains some very basic verbs such as *-dí- 'eat,' *-pá- 'give' and *-kú- 'die.' Almost all verbs longer than CVC can be shown to result from derivational processes, at least from a historical-comparative point of view.

As discussed in Sections 2.1–2.10, one or more extensions may be suffixed to the root. The initial CV of the root may also be reduplicated, as treated in Section 2.11. The verb may equally be derived from a nominal stem, as shown in Section 5. Compounded verbs are very rare in Bantu, but do exist, as discussed in Section 6.3. In Section 6.4, we explain that total reduplication of verbs, a process which is rarely lexicalised, applies to the stem rather than to the base. There are some verbs of the shape *-(j)íCVC- where the initial *(j)í could well be a petrified unidentified morpheme, e.g., *-(j)íbak- 'build,' *-(j)íkʊt- 'be satiated,' *-(j)ípʊd- 'ask.' Most verbs of the shape -CVVC- and -CGVC- can be derived from a -CV- root followed by a -VC- extension. Whether or not the derivational process underlying those verbs with more than one vocalic segment is transparent, all longer verbs obey general structural restrictions in that only the first root vowel is tonally distinctive and exploits the full seven-vowel system.

2 VERB-TO-VERB DERIVATION

The verb extensions reconstructed for PB are listed in (2). They occur in the body of PB reconstructions and their reflexes are found in languages from all zones. For a discussion of possibly cognate verb extensions elsewhere in Niger-Congo, see Hyman (2007).

(2) *-i-/-ici-   causative
     *-ɪd-   applicative
     *-ɪk-   impositive
     *-ɪk-   neuter
     *-am-   positional (stative)
     *-an-   associative (reciprocal)
     *-ag- ~ -ang-   repetitive
     *-ad-   extensive
     *-at-   tentive (contactive)
     *-ʊd- ; -ʊk-   separative tr.; intr. (reversive)
     *-ʊ-/-ibʊ-   passive

The canonical extension has the shape -VC-; the two extensions with the shape -V- have -VCV- allomorphs. Extensions have a reduced five-vowel system; the third-degree vowels *e and *o only occur as phonologically conditioned allophones of the second-degree vowel *ɪ and *ʊ through vowel harmony with mid vowels in the root. The absence of *u may be coincidental, since it does occur in expansions. Nasal harmony affects *d and, very rarely, also velars (Greenberg 1951, Schadeberg 1990: 24–26). Verb extensions of the shape -VC- are tonally neutral or, alternatively, Low. There is some slight evidence that the tone of the two extensions with the shape -V-, i.e., *-i- and *-ʊ-, may have been High (Meeussen 1961).

The element *-a(n)g- behaves tonally as an extension, but it often enters into the inflectional paradigm (with various imperfective meanings such as durative or habitual); it is not discussed further in this chapter (cf. Sebasoni 1967, Nurse & Devos this volume).

Extensions differ widely along the productivity scale, from totally unproductive expansions occurring in just a few verbs to fully productive suffixes.

Several extensions may combine with one root. The last extension determines the syntactic profile of the base. The addition is cyclical in the sense that when the meaning of a base consisting of root + extension$_1$ has developed a specific meaning, this meaning is retained in a further derivation root + extension$_1$ + extension$_2$. The causative extension offers particularly interesting challenges to morphological analysis, since it appears to change its position or has multiple representations (Hyman 2003a, b, Good 2005). In general, more productive extensions follow less productive ones, which is to be expected if verb extensions are added one-by-one.

Bantu verb extensions do not form a neat semantic or syntactic system. The idea that they originate from serial verb constructions (cf. Givón 1971) may be true in some cases, but not in others. The only two etymologies that have been suggested concern reciprocal *-an-, i.e., from the clitic preposition na- 'with, and' (Schladt 1996), and extensive *-ad-, i.e., from *-jàd- 'to spread' (Schadeberg 1994).

The syntactic functions of several extensions are also dealt with in the chapter on clausal morphosyntax by Downing and Marten in this volume.

2.1 Causative

Two Proto-Bantu causative extensions have been reconstructed with an original complementary distribution (Bastin 1986): *-i- after C and *-ici- after V. Leaving aside some exceptional vowel-final verb stems, this means that the long causative extension was used after short roots of the shape -CV-, and the short causative extension was used after the longer stems of the shape -CVC(-VC)-. Later developments have generally extended the range of environments in favor of reflexes of *-ici- which more closely correspond to the canonical shape VC of other extensions. The short causative commonly triggers spirantisation, which is at the origin of some recurrent types of analogical reanalysis across Bantu, among other things because the extension no longer surfaces due to "Y-absorption" into the preceding fricative (Hyman 2003a). As shown in (3), Punu B43 is one such language where the short causative is solely marked by the mutation of the root-final consonant into a fricative or an affricate.

(3)   Punu B43 (Nsuka-Nkutsi 1980: 67)

u-bŏts-a	'to make wet'	<	u-bŏl-a	'to be wet'
u-rŏs-a	'to (make) boil'	<	u-rŏg-a	'to boil'
u-lăs-a	'to show'	<	u-lăb-a	'to see'
u-wĕnz-a	'to let go'	<	u-wĕnd-a	'to go'
u-dónz -a	'to deepen (e.g., a pit)'	<	u-dóng-a	'to be deep'

As illustrated in (3) with examples of the short causative and in (4) with examples of the long causative, causative extensions may be added to transitive as well as to intransitive verbs.

(4)   Mituku D13 (Stappers 1973)
   a   kʊ-súm-ís-a   'to boil (tr.)'   <   kʊ-súm-a   'to boil (intr.)'

   b   kʊ-tút-ís-a   'to make s.o. beat'   <   kʊ-tút-a   'to beat'

   c   kʊ-kúlúm-an-is-a   'gather, assemble'

In both cases a new argument is added to the syntactic frame of the simple verb. This new argument has the syntactic function of subject and the semantic function of agent-causer. Most often, as in (4a), this subject (physically) manipulates the patient-causee who does not act as a volitional entity, even if (s)he is human, viz. direct causation. However, the causative extension may also express indirect causation, as in (4b) and (4c), involving two agentive participants, i.e., 'to make s.o. do sth.' (cf. Shibatani & Pardeshi 2002).

   In the case of transitive verbs, it is usually the agent-subject of the simple verb that ends up as the causee-object of the derived causative verb. Most cases where the object of the simple verb also appears as the object of the derived causative verb occur with lexicalised causatives, such as -guz-a 'to sell (something)' in Shi JD533, which is derived from -gul-a 'to buy' and which contrasts with the productively derived -gul-iis-a 'to cause (someone) to buy' (Polak-Bynon 1975).

   In some languages, particularly in the South-East, the causative extension expresses specialised meanings in addition to its general causative semantics. Two of these are the "adjutive" (< Latin adiutare 'to help') and the "imitative" meanings illustrated in (5a) and (5b) respectively.

(5)   Zulu S42 (Doke 1947: 146–148)
   a   ɓa-yo-fik-a   ɓa-vun-is-e   ɓa-bul-is-e
      SP$_2$-FUI-reach-FV   SP$_2$-harvest-CAUS-SBJV   SP$_2$-thresh-CAUS-SBJV
      'They will come and help to harvest and thresh.'

   b   u-hamb-is-a   o-ko::-nwabu
      SP$_1$-walk-CAUS-FV   AUG-17.11-chameleon
      'He walks like a chameleon.'

The shift from causative to imitative becomes transparent when we paraphrase example (5b) as 'he made a chameleon walk (as an actor).' The adjutive meaning is typically associated with verbs denoting communal activities where helping makes the event possible. It is part of a more comprehensive category known as 'sociative causation,' which is intermediate between direct and indirect causation and most typically concerns a situation whereby a causer makes the causee perform a certain action by simultaneously carrying it out him/herself (cf. Shibatani & Pardeshi 2002).

   In certain Bantu languages, such as Ganda JE15 in (6), causative extensions have become associated with uses, such as the instrumental, which are elsewhere usually associated with the applicative, as shown in (7).

(6) Ganda JE15 (Ashton *et al.* 1954: 344, cited in Wald 1998: 97),
   a tu-lim-y-a   n-kumbi   mu   Bu-ganda
       SP$_{1PL}$-cultivate-CAUS-FV   10-hoe   LOC$_{18}$ 14-Ganda
       'We cultivate with hoes in Buganda.'

   b tu-lim-is-a   n-kumbi   mu   Bu-ganda
       SP$_{1PL}$-cultivate-CAUS-FV   10-hoe   LOC$_{18}$ 14-Ganda
       'We use hoes for cultivating in Buganda.'

(7) Kagulu G12 (Petzell 2008: 134, cited in Marten & Kula 2014: 5)
       ya-ku-dumul-il-a   sigi   mage
       SP$_1$-PRS-cut-APPL-FV   9.rope   14.knife
       '(S)he is cutting the rope with a knife.'

Similar instrumental uses of the causative have been reported, among others, in Shona
S10 (Fortune 1955: 215, cited in Peterson 2007: 66), Rwanda JD61 (Kimenyi 1988: 367–
368, cited in Shibatani & Pardeshi 2002 and Voisin-Nouguier 2003), Nen A44, Ila M63,
Tonga M64, Kwanyama R21 and Zulu S42 (Wald 1998: 97). Such applicative/causative
isomorphism is not uncommon in the world's languages (Peterson 2007: 64–66).

   Another recurring meaning of the causative extension is the intensive (Hyman 2007:
160), where the derived verb behaves syntactically the same as the simple verb, an
atypical use that is also observed with the applicative (Marten 2003). In some languages, formal distinctions between causative and intensive verbs have evolved. For
instance, as shown in (8), Shona S10 expresses the intensive meaning by means of
a reduplicated causative extension (Fortune 1955). A similar use is reported in Zulu
S42 (Doke 1947), whereas elsewhere the double applicative rather fulfills this function
(Hyman 2007: 160).

(8) Shona S10 (Fortune 1955: 217)
       -ɓat-a   'hold' > -ɓat-is-a   'grip'   > -ɓat-is-is-a   'grip strongly'
       -ɗ-a   'love' > -ɗ-is-a   'love much'   > -ɗ-is-is-a   'love very much'
       -bvunz-a 'ask' > -bvunz-is-a 'interrogate' > -bvunz-is-is-a 'shower with
                                                                    questions'

2.2 Applicative

The applicative extension *-ɪd-* is also known in Bantu studies under the terms 'dative,'
'prepositional' and 'directive.' Applicative verbs are typically transitive. The applied object
may fulfill several semantic roles, such as beneficiary including substitutive, maleficiary,
goal, experiencer, recipient, location, instrument, direction/goal, patient/theme, circumstantial and reason/motive (cf. Trithart 1983, Marten & Kula 2014: 4–11, Pacchiarotti 2017).
The beneficiary function is cross-linguistically the most widespread and the most productive (Dammann 1961, Kähler-Meyer 1966). Not all of the other functions are attested in
every single Bantu language. For Lingala C36d, Rapold (1997: 8–9) lists beneficiary (9a),
maleficiary (9b), location (9c) and goal (9d) and considers recipient or addressee (9e) as a
specific kind of benefactive use. Another albeit rare function of the extension in Lingala,
reported by Meeuwis (2013), is 'the expression of use of an instrument' (9f).

(9) Lingala C36d (Rapold 1997, Meeuwis 2013)

  a a-pon-él-ákí   ngáí   elambá
     3SG-choose-APPL-PST 1SG   cloth
     'He chose a cloth for me.' (Meeuwis 2013)

  b ba-yib-él-í   bísó   sukáli
     3PL-steal-APPL-PRS.PRF 1PL   sugar
     'They have stolen some sugar for us.' (Rapold 1997: 8)

  c a-fánd-él-ákí   kíti
     3SG-sit-APPL-PST chair
     'He sat down on a chair.' (Meeuwis 2013)

  d ba-bwák-él-í   yé   libángá
     3PL-throw-APPL-PRS.PRF 3SG   stone
     'They have thrown a stone to him/her.' (Joseph Koni Muluwa pers. comm.)

  e Kengo a-kom-él-í   Marie   mokandá
     Kengo 3SG-write-APPL-PRS.PRF Mary   letter
     'Kengo has written Mary a letter.' (Rapold 1997: 9)

  f na-kát-él-í   yangó   nyama
     1SG-cut-APPL-PRS.PRF 3SG.INAN meat
     'I've used it [mbelí 'knife'] to cut the meat.' (Meeuwis 2013)

Applicative verbs can be derived from about any other verb; when the basic verb is transitive, the objecthood properties of the basic verb's object are typically reduced in the applicative construction. Two salient properties of objects, such as Swahili G42 *barua* 'letter' in (10a), are that they can be referred to by an object concord (10b) and that they can become the subject of a corresponding passive sentence (10c).

(10) Swahili G42 (self-generated)

  a a-li-andik-a   barua
     SP_1-PST-write-FV 9.letter
     'S/he wrote a letter.'

  b a-li-i-andik-a
     SP_1-PST-OP_9-write-FV
     'S/he wrote it.'

  c barua   i-li-andik-w-a   (na-ye)
     9.letter SP_9-PST-write-PASS-FV by-3SG
     'A letter was written (by her/him)'

The "applied object," i.e., the one licensed by the applicative verb 'to write to or for someone,' can also be represented by an object concord (11a) and it can be the subject of a corresponding passive sentence (11b). However, with the applicative verb, the object of the basic verb, i.e., *barua* 'letter,' cannot be referred to by an object concord (11c) nor can it be the subject of a corresponding passive sentence (11d).

(11) Swahili G42 (self-generated)

    a a-li-wa-andik-i-a   wa-zee   barua
        SP_1-PST-OP_2-write-APPL-FV 2-elder   9.letter
        'S/he wrote a letter to/for the elders.'

    b wa-zee wa-li-andik-i-w-a   barua   (na-ye)
        2-elder SP_2-PST-write-APPL-PASS-FV 9.letter by-3SG
        'The elders were written a letter (by her/him).'

    c **   wa-zee   wa-li-i-andik-i-w-a   (na-ye)
             2-elder SP_2-PST-OP_9-write-APPL-PASS-FV by-3SG
             'The elders were written it (by her/him).'

    d **   barua   i-li-wa-andik-i-w-a   wa-zee   (na-ye)
             9.letter SP_9-PST-OP_2-write-APPL-PASS-FV 2-elder by-3SG
             'A letter was written to/for the elders (by her/him).'

Less commonly, the object of the basic verb may keep its object status, as in the Ganda JE15 example in (12), where the object of the basic verb in (12a), i.e., *ente zino* 'these cattle,' is still co-referenced by an object prefix on the derived applicative verb in (12b) and not the 'applied object' *mu-kiraalo* 'into the kraal.' In such constructions, the applicative extension marks the added argument as providing essential new information, which shows that the applicative may play an important pragmatic role in the structuring of information (Hyman & Watters 1984, Creissels 2004, De Kind & Bostoen 2012, Downing & Hyman 2016; Downing & Marten this volume).

(12) Ganda JE15 (Ashton *et al.* 1954: 331)

    a e-n-te   zi-no,   zi-gob-e
        AUG-10-cattle PP_{10}-DEM OP_{10}-chase-IMP
        'These cattle, drive them away'

    b e-n-te   zi-no,   zi-gob-er-e   mu   ki-raalo
        AUG-10-cattle PP_{10}-DEM OP_{10}-chase-APPL-IMP LOC_{18} 7-kraal
        'These cattle, drive them into the kraal!'

Examples as in (12), together with the body of reconstructed applicative verbs, suggest that the primary function of the applicative extension was to tie the non-patient complement closer to the verb. The first such non-patient complements may well have been locative ones (location and/or direction/goal), from which the other roles of the applied object have evolved (Kähler-Meyer 1966, Trithart 1983, Cann & Mabugu 2007, De Kind & Bostoen 2012). For recent comparative studies on the syntax and semantics of applicative morphology in Bantu, see Jerro (2016) and Pacchiarotti (2017).

2.3 Impositive

The impositive extension *-$\imath k$- has been understood as a kind of causative, expressing typically "direct causation," but some "locative" element of meaning has also been observed in several older descriptions (cf. Dammann 1958). Inspection of any list of verbs carrying this extension, either from a particular language or from the body of PB reconstructions, confirms that the more precise meaning is 'to put (sth.) into some position,' hence the term "impositive" < Latin (*im*)*positus* 'put.'

The impositive extension *-ɪk- does not appear to be very productive in any particular language and probably was already highly lexicalised in PB. As shown for Ntandu H16g in (13), it typically commutes with positional *-am- and separative *-ʊd-/-ʊk- (Meeussen 1967). Commutation with zero is less common or archaic.

(13)   Ntandu H16g (Butaye 1909, Polis 1938)

a   -son-ik-a 'write, trace, plot, mark, inscribe' (*-cónɪk- 'draw a line, write')
    <   -son-a (arch.) 'write, trace, plot, mark'   (*-cón- 'draw a line, write')
    cf.   -son-uk-a 'come out when one presses the edges, e.g., wound'
          -son-un-a 'bring out, transcribe'
          -son-am-a 'be written, inscribed'

b   -kob-ik-a 'hide away, hang up (e.g., knife)' (*-kòbɪk- 'hook up')
   **   -kob-a (not attested)   (*-kòb- 'bend, hook up')
    cf.   -kob-uk-a 'fall, sink into; miss'   (*-kòbʊk- 'be unhooked')
          -kob-ul-a 'drop into, miss'   (*-kòbʊd- 'unhook')
          -kob-am-a 'bow, enter'   (*-kòbam- 'be bent, hooked up')

In most languages, the impositive extension is homophonous with the neuter extension; notable exceptions occur in the South-West, e.g., Kwanyama R21, Ndonga R22, Herero R31 where the "assimilating final vowel" suffix of certain tenses is -e after the impositive and applicative extensions, but -a after all other extensions including the homophonous neuter one (Brincker 1886, Halme 2004). In Chewa N31b, the homophonous neuter extension adds a High tone to certain verb forms but the impositive extension does not (Hyman & Mtenje 1999).

2.4 Neuter

The neuter extension *-ɪk- is generally homophonous with the impositive extension (Section 2.3). It is poorly represented in the body of PB reconstructions, but reflexes probably occur in all zones. Reconstructions as well as examples from languages where the neuter extension is least productive show that the extension is best represented with two semantic classes of verbs: verbs of destruction, e.g., 'be breakable/be broken (intr.)' < 'break (tr.),' 'be splittable/be split (intr.)' < 'split (tr.),' and experiencer verbs, e.g., 'be visible' < 'see,' 'be audible' < 'hear.'

In some languages, this extension has gained a high productivity as a detransitivising suffix, which can be combined with a wide range of transitive base verbs. This is a general syntactic function that is closer to passive than to active voice; hence the semantic-syntactic label 'neutro-passive' or shortly 'neuter.' Across Bantu, reflexes of neuter *-ɪk- are productively used in three types of constructions (Dom 2014, Dom et al. 2017). A first group consists of passive-like constructions where the object of the active clause is promoted to subject, whilst its subject is demoted or more commonly omitted. In contrast to canonical passives, such constructions usually do not allow an agent to be demoted to a prepositional phrase, as shown in (14b) and (15b), or to be combined with agent-oriented adverbs or purpose clauses, as illustrated in (16b). Nevertheless, oblique phrases introducing non-agentive initiators, such as instruments, natural forces and motives exemplified in (14a)–(16a), can be licensed. Process or state interpretations may be linked to the aspectual meaning of the particular inflectional category of the verbal form or else may have to be inferred from the context. There are two common subtypes of such passive-like constructions. The first subtype consists of anticausatives, also known as decausatives, illustrated in (14) for Zombo H16h.

(14)   Zombo H16h Anticausative (Fernando 2013: 189ff)
    a   Gyaka ki-uwd-ik-idi   mu   malutelo / mu   tembo / ky-au   mosi.
        7.wall SP_7-break-NEUT-PST LOC_{18} 9.hammer LOC_{18} 7.wind PP_7 -$POSS_{3PL}$ one
        'The wall broke down from a hammer/from the wind/by itself.'

    b. **   Gyaka ki-uwd-ik-idi   kwa   n-kento
             7.wall SP_7-break-NEUT-PST by   1-woman
             Intended: 'The wall was broken down by a woman.'

The second subtype consists of medio-passives, also known as quasi-passives, neutro-passives or statives, illustrated in (15) for Matengo N13 and in (16) for Chewa N31b.

(15)   Matengo N13 – Medio-passive (van der Wal 2015: 86)
    a   Li-ndilíisá   li-hogul-ik-í   nu   ũ-wáai.
        5-window SP_5-open-NEUT-PFV by   14-wind
        'The window has been opened by the wind.'

    b. **   Li-ndilíisá   Li-hogul-ik-í   na   Jóoni.
             5-window SP_5-open-NEUT-PFV by   John
             Intended: 'The window has been opened by John.'

(16)   Chewa N31b – Medio-passive (Dubinsky & Simango 1996: 750–751)
    a   Chi-tseko   chi-na-tsek-ek-a.
        7-food SP_7-PST-cook-NEUT-FV
        'Food was cooked.'

    b **   Chi-tseko   chi-na-tsek-ek-a   mwadala/
            7-food SP_7-PST-cook-NEUT-FV   deliberately
            kuti   a-nthu   a-sa-f-e   ndi   njala
            so.that   2-person SP_2-NEG-die-SBJV from   9.hunger
            Intended: 'Food was cooked deliberately/so that people should not die of starvation.'

Neuter *-ɪk- reflexes functioning as a canonical passive marker are rare, but have been reported, for instance in Tumbuka N21 (Chavula 2016, 2018), where such a verb can by followed by a prepositional agent phrase, as in (17). Other languages are Manda N11 (Bernander 2018) and Tonga N15 (Chavula 2018).

(17)   Tumbuka N21 – Canonical passive (Chavula 2016: 66)
    Kanyiska   wa-ka-temb-ek-a   na   themba.
    Kanyiska SP_1-PST-curse-NEUT-FV by   5.chief
    'Kanyiska was cursed by the chief.'

Notions which Kemmer (1993: 18) categorises as "emotion middle" are also found with neuter *-ɪk- reflexes across Bantu, e.g., *o-nyóóny-ey-a* 'to be upset,' *ó-fwáás-éy-a* 'to be calm' in Cuwabo P34 (Guérois 2015: 272–273) and *-kasir-ik-a* 'be angry, upset,'

-sikit-ik-a 'be sorry, sad,' *-sonon-ek-a* 'be in pain, anxious,' *-tes-ek-a* 'be in pain' in Swahili G42 (Sacleux 1939).

A second group of constructions relying on neuter *-ɩk-* reflexes consists of patient-oriented potentials, as in (18) and (19), where the subject is typically a patient-like participant to which a quality is attributed. Whether the subject is potentially or factually affected by the action expressed by the verb also depends on the aspect and/or the context.

(18) Shangaji P312 – Potential passive (Devos 2008: 10)
      Nakhuwo   ontu   kha-ni-valaz-ey-a
      1a.maize   DEM$_{1A}$ NEG-PRS-grind-NEUT-FV
      'This maize cannot be ground.'

(19) Shona S10 – Facilitative (Hannan 1974: 631)
      MuHarare   muno,   mu-nhu   ha-a-tan-i   ku-ras-ik-a.
      LOC$_{18}$.Harare   DEM$_{18}$ 1-person   NEG-SP$_1$-be.difficult.to.do-NEG   NP$_{15}$-lose-NEUT-FV
      'Here in Harare a person can get lost very easily.'

A third group of constructions principally involves detransitivised perception verbs whose experiencer is demoted or omitted. Either a stimulus participant is promoted to subject or an impersonal construction with an expletive subject concord is used. Two subtypes of this kind are stimulus constructions, as in (20), and evidential constructions, as in (21). Devos (2008: 16–17) reports in Shangaji P312 the use of the negative impersonal verb form *khiñeúweéya* 'it cannot be known,' built on the neuter cognitive verb *-cuw-ey-* 'be known,' in a construction expressing epistemic possibility.

(20) Ganda JE15 – Stimulus (Ashton *et al.* 1954: 446)
      Y-a-lab-ik-a   ng'   a-fumb-a.
      SP$_1$-PST-see-NEUT-FV   like   PP$_1$-cook-FV
      'She appeared to be cooking (like one who cooks).'

(21) Chewa N31b – Evidential (Bresnan & Kanerva 1989: 10)
      Zi-ku-on-ek-a   kutí   ḿ-phunzitsi   w-ánú   á-ma-dy-á   m-bewa.
      SP$_{10}$-PROG-see-NEUT-FV   like   1-teacher   PP$_1$-POSS$_{2SG}$ SP$_1$-HAB-eat-FV   10-mouse
      'It seems that your teacher eats mice.'

2.5 Positional

The PB reconstructions in (22) contain the positional extension *-am-* and are typical for the meaning and the commutation properties of this extension.

(22) *-(j)ég-am-*   'lean against (intr.)'   *-(j)ég-ɩk-*   'lean against (tr.)'
      -(j)ìn-am-   'bend over (intr.)'   *-(j)ìn-ʊk-*   'straighten up (intr.)'
      -cò-am-   'hide (intr.)'   *-cò-ɩk-*   'hide (tr.)'
      -cò-   'set (of sun)'
      -kúk-am-   'kneel'

The common element of meaning is "assuming a position" or, when used in a perfective aspect form, "to be in a position." The positional extension is the intransitive counterpart of the impositive extension with which it often commutes, as it does with the separative extensions discussed below. The label "positional' is preferred over the more widely used one "stative," which is generally used in a grammaticalised, syntactic sense that is not typically associated with this extension. This term covers several of the middle sub-categories distinguished by Kemmer (1993), such as "non-translational motion," "change in body posture," "translational motion," "emotion," "cognition" and "spontaneous events" (Dom *et al.* 2016b, 2017, 2018), as the Bantu lexical reconstructions with *-am-* in (23) indicate (Bastin *et al.* 2002).

(23) a   Change in body posture
           *-bók-am-   'sit sp.'
           *-jìn-am-   'bend over'

      b   Non-translational motion
           *-jác-am-   'open the mouth; yawn'
           *-pàp-am-   'flap wings, flutter'

      c   Translational motion
           *-dòng-am-   'go straight'
           *-búg-am-   'paddle'
           *-tent-am-   'go at the top of'

      d   Emotion
           *-bòng-am-   'be sad'
           *-cʊn-am-   'to be in a bad humour'
           *-dáp-am-   'be greedy in eating'

      e   Cognition
           *-dúd-am-   'forget'

      f   Spontaneous/state
           *-pák-am-   'be jammed, be wedged'
           *-còm-am-   'be inserted, put in'
           *-bòmb-am-   'be wet'

The positional extension has been generalised to form passive verbs, especially in a group of contiguous languages of Guthrie's zone C, such as Lingala C36d, Ngombe C41 and Mongo C61, where this change of function was triggered by the loss of *k* which led to a merger of passive *-ʊ-* and intransitive separative *-ʊk-* (cf. Meeussen 1954). Passive *-am-* is also attested in other languages of the area, where this sound shift did not occur, such as vehicular Kongo H10.

(24) Vehicular Kongo H10 (Bandundu variety) (Bostoen & Mundeke 2011b: 78)
      Di-nkondo   ke-pes-am-a   mu-yimi   na   tata
      5-banana   PRS-give-PASS-FV 1-beggar by   father
      'A banana is given to the beggar by father.'

2.6 Associative (reciprocal)

The most productive use and meaning of the associative extension *-an-*, illustrated in (25), is reciprocal which derives from the wider associative meaning.

(25)   Chewa N31b (Watkins 1937: 81)
       kodí   βa-ntu   βá-múná   βa-ka-kum-an-a   sá-lónj-er-an-a
       Q   2-person   2-man   SP₂-when-meet-ASSOC-FV   NEG.PRF-greet-APPL-ASSOC-FV
       'When men meet do they not greet each other formally?'

Reciprocal verbs require more than one agent, and the agents are at the same time mutual patients of their action. Syntactically, a single plural subject may fulfil these roles, or there may be two noun phrases which then may be either a conjoint subject, as in (26), or one may be the subject and the other one a prepositional phrase, as in (27). The marker of conjunction or preposition generally is the associative preclitic *na-* 'with, and.' When the associative NP follows the verb, agreement may be either with the subject NP or the corresponding plural form, even when the subject is singular, as in (28).

(26)   Zigula G31 (Dammann 1938–1939)
       m-thu   na   m-buya-ye   wa-pang-an-a   dende   dy-a   m-phasi
       1-person   and   1-friend-POSS₃ₛG   SP₂-share-ASSOC-FV   5.leg   PP₅-CON   9-grasshopper
       'A man and his friend share the leg of a grasshopper.' (proverb)

(27)   Zulu S42 (Doke 1947: 145)
       i-ngwe   i-f-an-a   n'ee-kati.
       AUG-9.leopard   SP₉-ressemble-ASSOC-FV   with.AUG-9.cat
       'The leopard resembles ("with") the cat.'

(28)   Mwera P22 (Harries 1950: 73)
       tu-na-lek-an-a   na   m-buya   nj-angu.
       SP₁PL-PRS-leave-ASSOC-FV   with   9-mate   PP₉-POSS₁ₛG
       'I (lit. "we") am parting with my mate.'

In most languages of Angola (zones H, K, R), the reciprocal use of *-an-* is more or less obsolete; its function was first taken over by a compound suffix, for instance *-asan-* or *-angan-*, and subsequently by the reflexive object concord leading to a widespread reflexive-reciprocal polysemy (Bostoen 2010, Dom *et al.* 2016a). A similar tendency can be observed in Mongo C61 and some other languages of zone C where **k* has become zero and **-an-* has taken over the function of neuter **-ik-* (cf. also French '*se quereller*' and German '*sich zanken*' both with reciprocal meaning).

In addition to expressing reciprocity, examples of related but not exactly reciprocal uses are attested from all over Bantu. Bostoen *et al.* (2015) list functions as diverse as "sociative/collective," "natural collective," "natural reciprocal," "chaining," "intensive/extensive," "iterative," "comitative/instrumental," "body action middle," "cognition middle," "spontaneous event middle," "mediopassive" and "potential." A voice construction that largely went unnoticed in Bantu languages, but is commonly associated with *-an-* is the antipassive. This diathesis in which the object of a transitive verb becomes oblique or is omitted developed out of plurality constructions marked by *-an-* through the gradual demotion of the second participant of a co-participative event. In Songye L23, the reciprocal is conveyed through a combination of the reflexive prefix *-i-* and the suffix *-een-*, which would be a merger of *-an-* and applicative *-il-* (Stappers 1964: 27). The simplex suffix *-an-* was previously described as an "alterative" expressing actions directed towards several other people (Stappers 1964: 27), but as the example in (29) shows, the label "antipassive" better covers its actual use. Closely related Luba L31a probably initiated a new cycle, in that *-an-* itself became a middle marker, while the composite suffix *-angan-* manifests reciprocal-antipassive polysemy (Dom *et al.* 2015). The antipassive reading only emerges with extended *-angan-* verbs having a singular subject, as in (30).

(29)   Songye L23 (Bostoen *et al.* 2015: 742)
     bà-mpùlushì   abà-yip-an-a   bi-kile   bu-kùfu
     2-police SP_2-kill-ANTIP-FV   8-much   14-night
     'The police often kill at night.'

(30)   Luba L31a (Dom *et al.* 2015: 355)
     mu-ntu   ù-vwa   mu-ship-angan-a,   bà-vwa   bà-mu-ship-a   p-èndè.
     1-person PP_1-PST NP_1-kill-ANTIP-FV SP_2-PST SP_2-OP_1-kill-FV SP_{17}-POSS$_{3SG}$
     'The person who killed (someone), they also killed him (i.e., he was also killed).'

2.7 Extensive

The extensive extension *-ad-* comes closest to being productive in some Southern languages. In Northern Sotho S32 it expresses "the (active) entering of a passive state where the action expressed by the basic verb potentially or actually affects the subject" (Endemann 1876: 76), e.g., *-bón-al-a* 'to let oneself be seen, to appear' < *-bón-a* 'to see,' cf. *-bón-ɛχ-a* 'to be visible.' The same extension (with loss of its consonant) is probably also contained in the complex extensions described for Lamba M54, but also occurring elsewhere, for example in Umbundu R11: "A derivative form of the verb which indicates that the action is extended in time or space or repeated extensively. The *extensive form* in Lamba is indicated in the intransitive by the suffix *-aaka, -auka* or *-aika*; in the transitive by *-aala, -aula* or *-aila*; in the causative by *-aasya, -ausya* or *-aisya*" (Doke 1938: 204).

Inspection of the sets of verbs containing the extension *-ad-*, either in the body of PB reconstructions or in particular languages, points to a central element of meaning "to be in a spread-out position." Recurring derived meanings are various words for being ill and suffering. The label "extensive" seems to fit this meaning rather well (cf. Schadeberg 1994). Some of the best-attested PB reconstructions follow in (31); others are given in Section 5 on noun-to-verb derivation.

(31)   *-(j)ik-ad-   'sit, dwell, stay'   <   *-(j)ik-   'descend'
     *-tíg-ad-   'remain'   <   *-tíg-   'leave (behind)'
     *-dú-ad-   'wear (clothes)'   cf.   *-dú-ɪk-   'put on (clothes)'
     *-dʊ́-ad-   'be ill'

2.8 Tentive

The tentive extension *-at-* is not known to be productive in any language. Some of the most widespread verbs with this extension are represented by the PB reconstructions in (32).

(32) *-kó-at-   'seize'
      *-dì-at-   'tread on'   variants: *-dìb-at-, *-nì-at-
      *-kúmb-at-   'hold, embrace'   variant: *-kúmb-at-
      *-dàm-at-   'stick to'   variant: *-nàm-at-

The meanings of these four verbs are typical and are taken to contain the common element of "actively making firm contact." The label "tentive" is preferred over "contactive" (Dammann 1962), because it refers to the active element of the contact and is derived from the prototypical verb of this class in Latin: *tenere*/*tentus* 'to hold.'

2.9 Separative

Most descriptions report two separative extensions: transitive *-ʊd-* and intransitive *-ʊk-*. Further morphological segmentation seems to be indicated, but the consonants *d* and *k* do not recur as morphemes with the meanings "transitive/intransitive." The separative extensions occupy an intermediate position on the productivity scale. Separative verbs are quite frequent, but can usually not freely be formed from other verbs. Commutation between the two separative extensions is the norm but exceptions exist, as shown in (33) for Swahili G42.

(33) Patterns of commutation of the separative suffixes in Swahili G42 (-u-/-o- < *-ʊd-)
      with each other: -ond-o-a   'take away'   -ond-ok-a   'leave, go away'
      with zero: -zib-u-   'unblock'   -zib-a   'block'
      with impositive: -fun-u-a   'uncover'   -fun-ik-a   'cover'
      with positional: -in-u-a   'lift up'   -in-am-a   'stoop'
      with neuter: -v-u-a   'take off (clothes)' -v-a-a   'wear (clothes)'
      with none: -pas-u-a   'split'   —
                         -chag-u-a 'choose'   —
                         —   -am-(u)k-a   'wake up'

The most common label for these extensions is "reversive" or "inversive." In more detailed descriptions, we read that only some of the verbs with these extensions have a 'reversive' meaning, the two other most frequently recurring senses are 'intensive' and 'repetitive' (cf. Dammann 1959). The different meanings expressed by the reflex of *-ʊd-* in Mongo C61, as described by Hulstaert (1965), are listed in (34).

(34) Mongo C61 (Hulstaert 1965)
      reversive sense: -bák-ol-   'detach'   -bák-   'fix'
      iterative sense: -nyɔm-ɔl-   'knead again'   -nyɔm- 'knead'
      augmentative sense: -miny-ol-   'smash'   -miny- 'shatter'
      causative sense: -tsíl-ol-   'stretch out (limbs)'   -tsíl- 'agitate '(limbs)'
      multiple senses: -band-ol-   'detach, fix again, fix solidly' -band- 'fetter, gag'
      other senses: -bumb-ol-   'steal'   -bumb- 'find, visit'
                          -lek-ol-   'surpass'   -lek-   'pass'
      "no sense":   -bang-ol-   'begin'   -bang- 'begin'

The synchronic situation described above with respect to frequency, patterns of com-mutation and various senses is also found in the body of PB reconstructions. The label "reversive" is rejected because (i) it fits only a small portion of the data, (ii) it does not explain which member of a given pair, e.g., 'put down'/'pick up,' will have the extension, and (iii) because there are numerous verbs with these extensions for which no plausible non-reversive source can be imagined, e.g., 'choose,' 'comb,' 'jump.' The common ele-ment of meaning which best fits the total data and from which the various senses can be derived is "movement out of some original position." The reversive sense appears only in the specific case where the basic verb incorporates the semantic element 'join' (cf. Schadeberg 1982).

The use of *-ʊd- seems to have become more productive towards that of a general transitiviser in certain languages, such as Mbuun B87, where it phonologically merged with *-ɪd- leading to a causative-applicative syncretism (Bostoen & Mundeke 2011a).

As for *-ʊk-, apart from being "separative," its semantics strongly coincide with that of neuter *-ɪk- suggesting a common origin in the remote past. In Cuwabo P34, for instance, its reflex -uw- conveys different sub-categories of the middle domain, such as change in body posture, translational motion, emotion, spontaneous events (anticaus-ative) and medio-passive, from which it evolved into a canonical passive marker (35a), competing with -iw- (35b), i.e., the reflex of PB passive *-ibʊ- (Guérois & Bostoen 2018).

(35) Cuwabo (P. 34) (Guérois 2015: 269)
     a mí-rí   dhí-ni-ó-j-úw-á   na   nyenyéle
        4-tree   SP$_4$-IPFV.DJ-15-eat-PASS-FV   by   10a.ant
        'The trees are being eaten by the ants.'

     b mw-ááná   o-hí-p-íw-á   na   ńzûwa
        1-child   SP$_1$-PFV.DJ-kill-PASS-FV   by   5.sun
        'The child was killed by the sun.'

2.10 Passive

Two allomorphs of the passive extension have been reconstructed (cf. Stappers 1967): *-ʊ- occurring after C and *-ibʊ- occurring after V. The non-canonical shapes -V- and -VCV- and the conditioning of the allomorphs find their parallel with the causative exten-sion (Section 2.1). Since the passive extension occupies the last position in a sequence of several extensions, the long allomorph not only appears after short radicals of the shape -CV- but also after the causative extension *-i-/-ici-.

Due to the great productivity coupled with syntactic and semantic regularity, few pas-sive verbs have been reconstructed for PB. The one with the largest spread is *-kód-ʊ- 'get drunk,' the corresponding non-passive verb most commonly means 'to be strong,' other attested meanings are 'take,' 'intoxicate' (back formation?), 'choke' and 'work.'

In most of zone A and B there are no clear reflexes of *-(ib)ʊ-; instead, there are suf-fixes of the general shape *-(a)b(e) (the vowels differ from language to language), with a meaning described as passive, neuter or middle voice, which is also part of the composite passive suffix -Vban in the Bantu A70 languages (Bostoen & Nzang-Bie 2010). Since Basaa A43 attests both passive *-ib(ʊ)- and reflexive *-ab(e)-, and some other languages of the area, such as Ngumba A81, Myene B11 and Tsogo B31, also attest *-ʊ-, we reject doubts expressed by Guthrie (1967) and Heine (1973) as to the presence of the passive extension(s) in PB.

There are several other areas where the passive extension is used sparingly or not at all. One of these areas covers parts of zones B, H, K and L, but generally a few synchronically underived verb stems attesting *-ʋ- have survived. In the greater part of zone C the passive extension has been lost due to a phonologically triggered merger with *-ʋk- (Meeussen 1954). Where the passive extension is not commonly used, other extensions (*-am-, *-an-) have filled the gap, or other types of constructions are used (cf. Bostoen & Mundeke 2011b). Luba L31a and Mbundu H21 use a "passive participle" (Chatelain 1964: 83–85, Willems 1970); more widespread is the use of an active verb with a class 2 subject concord (cf. Kula & Marten 2010). Both strategies are illustrated for Luba L31a, i.e., the passive participle in (36a) and the so-called *ba*-passive in (36b) as well as in (30) above.

(36)   Luba L31a (Willems 1970: 160–162)

a   n-dí   mw-ípík-íl-á   n-kundé   léélo   kudi   máámu
     SP_{1sg}-be NP_1-cook-APPL-FV NP_{4n}-bean   today   by   mother
     'Beans are cooked for me today by my mother'
     (lit. 'I am being cooked for beans today there where is my mother').

b   ba-kwac-ilé   m-bují   kudi   n-kashaama
     SP_2-take-PST NP_{1n}-goat   by NP_{1n}-leopard
     'The goat was seized by the leopard'
     (lit. 'They seized the goat there where is leopard').

2.11 Partial reduplication

There are a good number of verbs of the shape -CVCVC- where the first two CV-sequences are identical and which contain "repetitive motion" or "oscillation" as a recurring element of meaning, e.g., -babat-a 'tap (as in metal working),' -gugun-a 'gnaw, nibble, nag,' -kokot-a 'pull by tuggIng' In Swahili G42 (Sacleux 1939).

Some of these verbs are derived by partial reduplication, i.e., -CV.CVC-. Where C_2 of the R is NC, the nasal may be reduplicated as in Bemba M42, e.g., -sansant-a 'disperse, scatter, spread,' -fyomfyont-a 'suck' (Anonymous 1954) or may not be reduplicated as in Rwanda JD61, e.g., -ríriimb-a 'sing,' -ruruumb-a 'be obsessed by gluttony' (Coupez et al. 2005).

In many cases, the actual path of derivation is difficult to trace. Other possible sources are -CVC-VC- or derivation from a (reduplicated) noun (Section 5.2) or ideophone (Section 5.3). Compare the set of reconstructions in (37), where the noun has a much wider distribution, also including North-Western languages from zones A and B, than the (extended) verbs. In a BLR2 comment, Coupez et al. (1998) ascribe possible ideophonic qualities to the root consonants, i.e., "*Réflexes p: iconique dans certaines langues (+ suffixe).*"

(37)   *-pàpá   'wing'   (zones A B J L S)
       *-pàp-   'flap wings, flutter'   (zones G H J M N P R S)
       *-pàpad-   'flap wings, flutter'   (zones H J L)
       *-pàpam-   'flap wings, flutter'   (zones H R S)

3 VERB-TO-NOUN DERIVATION

The derivation of nouns from verbs involves several productive processes, some of which are so widespread that they have been reconstructed for PB. The process involves two parts: the derivation of a nominal stem from a verbal base by the addition of a final suffix and the assignment of the derived nominal stem to a nominal class or gender (noun class pair).

3.1 Infinitives

From a morphological point of view, infinitives are nouns by virtue of having a nominal prefix, but they also have verbal characteristics such as the possibility to include an object concord as well as a limited range of inflectional morphemes in pre-stem position, e.g., motional -*ka*- and negative markers.

The infinitive stem has the final suffix *-*a* and is thus the same as the default stem used in verbal inflection. In some languages, the infinitive has a conjoint form with final *-*á* linking it syntactically to a following complement. The negative infinitive sometimes has a different final suffix, which it shares with certain negative tenses.

Infinitives are generally assigned to class 15, whose nominal prefix is *kò-, less commonly also to class 5 with nominal prefix *ì-. Some languages attest both. Other noun classes (9, 14) are also sometimes employed (cf. Forges 1983, Hadermann 1999). The use of the prefix *kà- after verbs of motion may derive from the motional -*ka*- rather than from class 12. The homophony between the nominal prefixes of class 15 and the locative class 17 raises the question whether the infinitive with *kò- derives from the locative. Whatever the answer may be, the infinitive has in many languages instigated the removal of the original small set of non-verbal nouns from class 15: *kò-bókò 'arm,' *kò-gòdò 'leg,' *kò-túì 'ear,' *ko-jápà 'armpit.'

3.2 Agent nouns

The primary kind of agent nouns are derived from verbs by adding the final suffix *-*ì* to the base; stems are generally assigned to class 1/2, since typical agents are human, but other classes may also be used, as shown in (38). Although this final is a PB high front vowel, it unevenly triggers spirantisation across Bantu (Bostoen 2005, 2008).

(38) *mʊ-(j)íb-ì   (cl. 1/2)   'thief'   <   *-(j)íb-   'steal'
      *mʊ-kó-ì   (cl. 1/2)   'in-law'   <   *-kó-   'pay bridewealth'
      *mʊ-(j)éd-ì   (cl. 3/4, 15) 'moon, moonlight'   <   *-(j)éd-   'shine, be clear'
      *m-bón-ì   (cl. 9/10)   'pupil (of eye)'   <   *-bón-   'see'

The final suffix is *-*á*, rather than *-*ì*, when it follows a causative or passive extension, as in (39). The same suffix is used to form complemented agent nouns, as discussed in Section 6.3. The derivation from passive verbs shows that we are dealing with a subject noun rather than with an agent noun.

(39) Nyamwezi F22 (Maganga & Schadeberg 1992)
      m-βʊój-á   'the one who asks'   <   -βóʊj- 'ask'   (CAUS)   <   -βóʊl-   'reveal'
      m-toól-w-á   'bride'   <   -tóol-w-   (PASS)   <   -tóol-   'marry'

3.3 Actions, results, instruments

Deverbative stems with a final suffix *-ò are very common and appear in all classes except 1/2. Such nouns refer to the action itself, the result of the action, the place or the instrument, often with applicative *-ìd-. Unlike the formation of infinitives and agent nouns, the derivation of nouns in -ò is generally not fully productive and the meaning of such nouns is not entirely predictable. For instance, the derived noun *m-púk-ò (cl. 9/10) 'mole' in (40), actually refers to a non-human agent, but this meaning may be the outcome of a metonymic shift from the mole-hill, i.e., the visible result of the activity of this animal that is not easily seen.

(40) *mʊ-dòng-ò   (cl. 3/4)   'line'   < *-dòng-   'arrange, pack'
      *ì-tém-ò   (cl. 5/6)   'hoe'   < *-tém-   'cut, cut down'
      *ì-dì-ìd-ò   (cl. 5)   'place or thing for eating' < *-dì-ìd-   'eat + APPL'
      *kì-dìd-ò   (cl. 7/8)   'mourning'   < *-dìd-   'cry, weep'
      *m-púk-ò   (cl. 9/10) 'mole'   < *-púk-   'dig, fling up (earth)'
      *dʊ̀-gènd-ò   (cl. 11)   'journey'   < *-gènd-   'walk, travel'
      *bʊ̀-dìmb-ò   (cl. 14)   'birdlime'   < *-dìmb-   'stick to sth.'

3.4 Adjectives

Stems derived from verbs of quality with the final suffix *-ú function as adjectives, i.e., they take a nominal prefix in agreement with the syntactic head noun. Further derivation leads to nouns of quality in class 14 (Section 4.3) and sometimes to nouns denoting persons in class 1/2. This kind of formation is productive in some languages such as Nyamwezi F22 (Maganga & Schadeberg 1992). In many more languages, such as Swahili G42 in (41), only isolated examples are found in the lexicon. A few such stems have been reconstructed. The derived adjective *-bod-ú 'rotten' is commonly used with meat (class 9 whose nominal prefix is represented by prenasalisation) or a fruit (typically class 5), the two phonological environments where *b has been retained in Swahili G42. Elsewhere, *b has been lost in Swahili before rounded vowels, including after the class 1/2 (human) with nominal prefixes *mʊ-/*ba- which provided the other common pattern of usage for this adjective. The semantic split into 'rotten' vs. 'evil' and the largely parallel phonological split -bovu vs. -ovu has led to the complete split into two separate lexical items.

(41) Swahili G42 (Sacleux 1939)
      -bovu   rotten   < *-bòd- (class 5 with nominal prefix *i-)
      -ovu   bad, evil   < *-bòd- (class 1, other classes)
      -kavu   dry   < *-kàd-
      -gumu   hard   < *-gùm- (cf. *-júm- 'be dry')
      -refu   long   < *-dài-p-
      m-fumu   chief [loanword?]   < *mʊ̀-kúm-ú < *-kúm- 'be honored'

3.5 Other final suffixes

Practically all vowels, with different tones, are attested as final suffixes in deverbative nouns, with a wide range of meanings. Nouns with final suffix -a are particularly frequent. Their exact path of derivation is not always traceable (Section 5.2). The examples are too varied to allow firm reconstructions, but a good candidate is the final suffix *-ꞌè (HL rather than "polar" tone) deriving nouns or participles often referring to a state (cf. Bastin 1989). The best such reconstruction, *-dóm-è 'male, husband' < *-dóm- 'bite,' does not fit this meaning. It rather appears to be an agent noun, a metaphoric reference to the male sex.

(42) *kì-kunt-í   (cl. 7)   'fist'   < *-kunt-   'knock'
      *mò-dumb-í   (cl. 3)   'continuous rain'   < *-dumb-   'rain'
      *-kád-ì   (adj.)   'bitter, sharp'   < *-kád-   'be bitter, be sharp'
      *n-gob-é   (cl. 9)   'hook'   < *-gòb-   'bend, crook'
      *m-pét-é   (cl. 9)   'ring'   < *-pèt-   'bend, curve' (tone!?)
      *mò-kél-è   (cl. 3)   'salt'   < *-kél-   'filter'
      *kì-pàc-è   (cl. 7)   'splinter'   < *-pàc-   'split'
      *m-bíd-á   (cl. 9)   'call'   < *-bíd-   'call'
      *kì-dìm-ìd-à   (cl. 7)   'pleiades'   < *-dìm-   'cultivate'
      *dò-kómb-ó   (cl. 11)   'broom'   < *-kómb-   'sweep'
      *-kúd-ú   (adj.)   'old'   < *-kúd-   'grow'
      *-(j)éd-ʊ   (adj.)   'white'   < *-(j)éd-   'shine, be clear'

Some languages have developed new productive derivational processes. For example, Nyamwezi F22 has two kinds of manner nouns (Maganga & Schadeberg 1992), shown in (43), and Luba L31a has active and passive participles (Burssens 1946: 56–57), illustrated in (44).

(43) Nyamwezi F22 (Maganga & Schadeberg 1992)
      -zeeng- 'build > kì-zeeng-elé   class 7   (type of construction)
                       > kà-zeeng-elé   class 12   (manner of building)

(44) Luba L31a (Burssens 1946: 56–57)
      -kwàt- 'take, seize' > mêmé mú-kwàt-é   'me having seized. . .'
                             > ndí/mvwá mú-kwàt-á   'I am/was seized.'

4 NOUN-TO-NOUN DERIVATION

Nouns are derived from nouns by shifting them from one class (gender) to another (cf. Bastin 1985). Nouns are derived from adjectives by assigning them to a specific class. It is this derivational or "autonomous" use of noun class assignment which most clearly shows (some of) the semantic content of Bantu nominal classes. In addition to the widespread and productive processes listed in Sections 4.1 to 4.6, there are sporadic, non-recurring cases in the body of PB reconstructions, as listed in (45).

(45) *mò-ntò   (cl. 1/2)   'person, someone'   >   *kì-ntò   (cl. 7/8) 'thing, something'
      *mò-jómà   (cl. 1/2)   'person, someone'   <   *kì-jómà   (cl. 7/8) 'thing, something'
      *mò-tí   (cl. 3/4)   'tree'   >   *kì-tí   (cl. 7/8) 'stool; stick'
      *njíkì   (cl. 9/10) 'bee'   <> *bò-jíkì   (cl. 14) 'honey'

The direction of such sporadic noun-to-noun derivations is not always easy to determine. The series of derivation by shifts of noun class in (46) has been analysed by Grégoire (1976).

(46) *kì-bánjá   (cl. 7/8)   'clearing prepared for building'
      *m-bánjá   (cl. 9/10)   'chief's village'
      *dò-bánjá   (cl. 11/10)   'dwelling-place, courtyard, family; meeting, affair, law-court, guilt'
      *ì-bánjá   (cl. 5)   'debt'

4.1 Locatives

Locative nouns are formed by adding one of the locative nominal prefixes to a whole noun (nominal prefix + stem). Three locative classes have a general geographical distribution; they indicate spaces, which are described relative to the entity referred to by the basic noun. A fourth locative nominal prefix, i.e., *ɩ-, has a more sporadic spread and limited use (cf. Grégoire 1975).

(47) Chewa N31b (Watkins 1937)
      class 16: *pà-   space adjacent to X   pa-phiri   'on the mountain'
      class 17: *kò-   space anywhere in the area of X   ku-mudzi   'at the village'
      class 18: *mò-   space inside X   mu-mphika   'in the cooking pot'

In *pà-ntò 16, *kò-ntò 17, *mò-ntò 18 'place' (and in parallel forms with *-jómà) the locative nominal prefix directly precedes the stem, which is the rule for adjectival stems but atypical for nouns.

A locative suffix *-ini is used, often together with the locative nominal prefixes in languages of the extreme east, i.e., from Guthrie's zones E, G, P, S (see also Schadeberg 2003b, Bostoen & Grégoire 2007).

Locative nominal prefixes are not directly prefixed to nouns referring to people. Locative nouns behave syntactically much like ordinary nouns, not as prepositional phrases. They control agreement within the nominal phrase as well as on the verb, and a phrase such as 'hair on the head' or 'head hair' has to be rendered as 'hair of on the head,' e.g., Lamba M54 imisisi yā pa-mutwi (Doke 1938: 167).

4.2 Diminutives and augmentatives

Diminutives are widely formed by assigning nouns to classes 12/13 (kà-/tò-), and in the North-West also to classes 19/13 (pì-/tò-). Classes 12 and 13 have very few nouns inherently assigned to them, and these have no obvious semantic feature of "smallness," e.g., *kà-nòà 'mouth,' *kà-já 'home,' *tò-bíi 'excrements,' *tò-dó 'sleep.'

Augmentatives are widely formed by assigning nouns to classes 5/6 (ì-/mà-) or 7/8 (kì-/bì-). In some Eastern languages, there is also an augmentative class gò-. In Nyamwezi F22, animals shifted to the "augmentative" gender 5/6 are presented as being individuals (Maganga & Schadeberg 1992). Augmentative and diminutive classes often have affective values; common pairings are "small is beautiful" and "big is ugly and dangerous," but the inverse relationship also occurs. Such secondary meanings may explain the occurrence of shifts between augmentative and diminutive meanings.

In the formation of diminutives and augmentatives, the derived nominal prefix is sometimes placed before the inherent nominal prefix rather than substituted for it, especially with classes 9/10 where the (historical, underlying) identity of the stem-initial C may be obscured by the sound changes of prenasalisation (nominal prefixes *N-).

4.3 Qualities

Nouns of quality are freely derived from adjectives and from nouns denoting kinds of people (generally classes 1/2) by placing the stem into class 14 (*bò-). Underived nouns belonging to this class have no obvious semantic feature of "quality" or "abstractness," e.g., *bò-játò 'boat,' *bò-táà 'bow,' *bò-jògà 'mushroom,' *bò-tíkò 'night.' Some reconstructible lexicalisations attest the ancientness of this derivation, e.g., *bò-dòg-ì 'witchcraft' < *mò-dòg-ì (cl. 1/2) 'witch' < *-dòg- 'bewitch' and *bò-tí 'medicine, witchcraft' < *mò-tí (cl. 3/4) 'tree.'

4.4 Peoples, countries, languages

Many languages freely assign ethnonymic stems to different classes (genders), deriving words for the specific people, the country, and the language or any behaviour that is attributed to the particular ethnic group. People are generally in classes 1/2, the country in class 14, and the language in class 11 or 7 (also referring to manner). Some languages may also assign some ethnonyms to other classes, e.g., Gogo G11 va-gogo class 1/2 (people) vs. ci-gogo class 7 (language) vs. u-gogo class 14 (territory) (Rugemalira 2009); Ganda JE 15 (a)ba-ganda class 1/2 (people), (o)lu-ganda class 11 (language), (o)bu-ganda class 14 (territory) (Ashton et al. 1954); Umbundu R11 óví-mbúndu class 7/8 (people), ú-mbúndu class 14–3 (language) (Schadeberg 1990).

4.5 Trees and fruits

Trees and their fruits are often referred to by words of the same root in different classes. Trees are in class 3/4 and their fruits in classes 5/6, sometimes also in class 9 or 14. Economically highly relevant plants and their fruits may be referred to by unrelated words.

(48) Koti P311 (Schadeberg & Mucanheia 2000)

n-ráràṅca	3/4	'orange tree'	>	n-ráràṅca	5/6	'orange'
n-khunazi	3/4	'Ziziphus sp.'	>	khunazi	9/10	'(its fruit)'
n-lapa	3/4	'baobab tree'	>	o-lapa	14	'(its fruit)'
n-nazi	3/4	'coconut palm'	>	naazi	9/10	'coconut (fully ripe)'
			cf.	n-xala	5/6	'half-ripe, for drinking'
m-oopo	3/4	'banana plant'	cf.	laazu	5/6	'banana'

4.6 Number

If shift of noun class is the formal characteristic of noun-to-noun derivation, the formation of plurals (or singulars) may also be considered as a derivational process in Bantu.

Many languages also form collectives and plurals of non-count nouns with the meaning 'kinds of X'; the most common class for this kind of derivation is class 6 (*ma-).

(49) Swahili G42 (Schadeberg 1992)
      simba/simba   (cl. 9/10)   'lion/lions'   ma-simba   'a pride of lions'
      u-gonjwa   (cl. 11~14)   'illness; disease'   ma-gonjwa   '(kinds of) diseases'

Number ("singular" and "plural") is at best a marginal category in Bantu grammar to which reference is hardly ever needed; "third person singular" and "third person plural" have no common exponents due to the presence of specific noun class agreement markers.

5 NOUN-TO-VERB DERIVATION

5.1 Verbs from adjectives

Noun-to-verb derivation is not a productive process in Bantu languages. The only clearly attested regular process involves the suffix *-p- which is added to adjectival stems to derive inchoative verbs of quality: 'to become (be) X.' When added to monosyllabic roots the vowel always appears as long; the reconstructed inherent length of such adjectival roots should be considered as doubtful.

(50) *-kée-p-   <   *-kéè   'small'   *-néne-p-   <   *-nénè   'big'
      *-dài-p-   <   *-dàì   'long'   *-jípí-p-   <   *-jípí   'short'
      *-bîi-p-   <   *-bîì   'bad'   *-jàngʋ-p-   <   *-jàngò   'quick'
      *-kádi-p- <   *-kádì   'bitter, sharp'

5.2 Verbs from nouns or ambivalent derivation

There are pairs of obviously related nouns and verbs where the derivational path is difficult to trace. Most of these pairs consist of a nominal stem -CVCV and a verbal stem -CVCVC-a. In some cases, both stems may be derived from a no longer existing root -CVC-, but often a derivation of a verb by adding a C to the nominal stem seems more likely, e.g., Swahili G42 ki-pofu 'blind person' (< *-póku) vs. -pofua, -pofuka, -pofusha 'to blind, be blind, make blind' (Sacleux 1939). The noun occurs in various classes and is attested from all zones (A to S). The verbs are probably independent innovations in Swahili and a few other Eastern languages. The verbs cannot be analysed as -CVC-VC- since there are no extensions of the shape *-ul-/-uk- (first-degree vowel, no vowel harmony). The noun semantically fits the derived adjectives in *-ú (Section 3.4), but the noun has a final Low tone, and there is no trace of a (verbal) root -pók-.

BLR3 contains many comparable sets, e.g., *-kókó 'crust' vs. *-kókot- 'scrape, crunch' vs. *-kókod- 'remove crust' (Bastin et al. 2002). Again, the final High tone of the noun and the unattested expansion *-ot- point to a denominal derivation. The form with *-od- semantically and formally fits the separative extension. The consonantal suffix is problematic in that most PB consonants are attested (*p *t *k *b *l *m *n). In some cases, it seems that the final vowel of the nominal stem is replaced or deleted, often resulting in a form with the appearance of containing a semantically fitting extension. In view of this, a verb such as *-jót- 'warm oneself' could be a back formation from *mò-jótò 'fire.'

(51) *bʊ̀-jógà   'fear'   > *-jógop-   'fear'
     *dʊ̀-jìgì   'door'   > *-jìgad-   'shut'; *-jìgʊd- 'open'
     *mʊ̀-túè   'head'   > *-tʊ́-ad-   'carry on head'; also: *-tʊ́-ɩk-, *-tʊ́-ʊd-, *-tʊ́-ʊk-
     *tʊ-dó   'sleep'   > *-dóot-   'dream' > *-dóótó, *-dóótò, *-dóótà, *-dóótì 'dream'

5.3 Ideophones and derivation

Basic ideophones have the shape CV, and verbs are occasionally derived from them in the same way as from adjectives and nouns (Sections 5.1 and 5.2). Derivation of nouns and adjectives is also possible by assigning the ideophonic stem to a noun class. The ideophonic origins given in parentheses are assumed rather than attested; note that the chicken has been introduced long after the PB period.

(52) (ko-ko)   >   *nkókó, *nkʊ́kʊ́ 'chicken'; > *-kókʊd- 'cackle'
     (di)   >   *-dìdì 'cold'; > *-dìdim-, *-dìdid- 'be cold, shiver'
     *tú   'blunt'   >   *-túùp-, *-túùn-, *-túùmp-
     *pì   'black'   >   *-pìip-, *-pìit-; *-pínd- (tone?); *-pìnd- (V?)
                     >   *-pì (adj.); *-pìipí 'darkness,' *mʊ̀-pìiti 'very dark person'
     *pè   'white'   >   *-pémb-; *-pémbé, *-pémbá 'white clay'

The derivations from *pí and *pè with voiced prenasalised C₂ are doubtful. As an alternative one could assume the inverse derivational path. The first syllable of a verb stem or the first two in the case of trisyllabic stems may be used as an ideophone, which then has to be redefined as an adverbial part of speech giving a vivid description of the event without necessarily involving sound symbolism. This procedure is extremely common in Zulu S42 and well attested in other southern languages.

(53) Zulu S42 (Doke & Vilakazi 1958)
       -mony-ul-a, -mony-uk-a   'pull out'   >   mónyu
       -volokohlel-a   'trample down' >   volokohlo
But: -ngqongqoz-a   'knock'   <   ngqo

Grammars written in the school of Doke analyse all such verbs as derived from the ideophone (Doke 1931: "radical"), which is unlikely in view of the fact that some of these forms include VC-sequences that are analysable as extensions, and the corresponding verbs but not the ideophones exist elsewhere in Bantu, e.g., *-móny-ʊd-, -móny-ʊk- 'break off.' Doke, like other early Bantuists, may have believed that ideophones represent the earliest form of words from which other categories have evolved.

6   OTHER WORD FORMATION STRATEGIES

Compound nouns of the structure A+B generally refer to a "B-like kind of A," i.e., the first part is the head of the compound. This is also true of the more petrified types of compounds (Sections 6.1 and 6.2). Most types of compounds are restricted to fully lexicalised forms; regional productive processes are the formation of names (Section 6.1) and diminutives (Section 6.2), compounds with *mw-ana* and *mw-ene* as well as agent nouns with complements (Section 6.3). Compound verbs are extremely rare (Section 6.3), but reduplication is much more common with verbs than with nouns (Section 6.4).

6.1 Pre-stem elements

The elements *na-* or *nya-* and *ca-* or *ci-* are often prefixed to a noun or a nominal stem. These forms are derived from the words for 'mother' (**nyàngó*, **nyòkó*, **nìnà ~ jìnà* 'my, your, his/her mother') and 'father' (**cángó*, **có*, **cé ~ jìcé* 'my, your, his/her father'). The resulting nouns have no nominal prefix and are generally assigned to class 1a. They refer to more complex kinship relations, but are also used as personal names and ethnonyms, and as name-like common nouns often referring to small animals and plants. The formation of personal names is productive in some languages such that one element is used to form proper names of women and the other proper names of men, e.g., Umbundu R11 *ná-ngómbe* (feminine proper noun) vs. *sá-ngómbe* (masculine proper noun), both derived from *óngómbe* 'cattle' (Schadeberg 1990).

The interpretation of such nouns is not necessarily "mother/father of X"; the relationship may be the inverse one: "someone whose mother/father is X," e.g., in Nyamwezi F22, someone born during a famine may be called *Nya-nzala* (< **n-jàdà* 'hunger'), and the *Nya-mwezi* are 'people from the west' (*mwezi* 'moon – west') (Maganga & Schadeberg 1992).

The element *ka-* is usually synchronically analysable as a class 12 nominal prefix; historically it seems to have a second origin, i.e., the Niger-Congo "person marker" (Greenberg 1963: 152), which also survives in Bantu as class 1 verbal prefix (Güldemann 1996, Bostoen & Mundeke 2012). This origin is manifest in its use to form names and name-like common nouns, which have no semantic link with the mainly diminutive class 12, and formally by the frequent absence of the augment and/or assignment of such nouns to class 1a, which is another typical feature of names.

(54) Kwanyama R21 (Tobias & Turvey 1954)
      ka-lunga   'God'
      ka-dina   'namesake'   <   *ì-jínà   'name'

6.2 Post-stem elements

Five nominal stems (rather than full nouns) often occur as quasi-suffixes: **-ntò* 'person,' **-lómè* 'husband, male,' **-kádí* 'wife, woman, female,' **-jánà* 'child' and **-kóló* 'adult, old, big.' The form **-kádí* also appears as *kái* > *ké*, which may make it difficult to distinguish it from reflexes of **-ké* 'small.' Some of these stems may even combine with each other, e.g., **mò-kái-ntò* 'woman,' **mò-kái-kóló* 'old woman,' **mò-kádí-jánà* 'girl, daughter-in-law' or **mò-jána-kádí* 'woman.'

The stem for 'child' has become a diminutive suffix, most productively used in the south (Güldemann 1999, Gibson *et al.* 2018: 371–373), e.g., Zulu S42 *intatshana* 'a small mountain' < *intaba* 'mountain'; *izinsukwana* 'a few days < *izinsuku* 'days' (Doke & Vilakazi 1958).

6.3 Compounds

N+N is not a common syntactic phrase (except with names and titles), but most languages have compounds consisting of two nouns. The second noun may sometimes shed its nominal prefix. In some but not all cases, such nouns may originate from a genitival or connective construction, e.g., **m-bódó-ko-túì* 'ear' < **m-bódó* (an animal?) + **kò-túì* 'ear'; **n-kódò-tímà* 'breastbone' < **n-kódo* 'heart, breastbone' + **mò-tímà* 'heart.'

Two kinds of N+N compounds have gained complete productivity in some languages. In Swahili G42, *-enye* + X (< **mò-jéné* 'owner, chief'), with a nominal prefix of class 1 or a pronominal prefix of any other class, refers to someone or something that owns or has X, e.g., *miti yenye matunda* 'trees with fruits,' while *mw-ana* + X (< **mò-jánà* 'child' – the relational term) refers to a person who is characterised by X, e.g., *mwana-siasa* 'politician' < *siasa* 'politics' (Schadeberg 1992).

N+A compounds may look like phrases and are therefore known as exocentric phrasal compounds. Their status as compound words manifests itself by their meaning and by (external) agreement, which is class 1/2 in the Swahili G42 example *kalamu-mbili* 'secretary bird' < *kalamu* 10 'pens' + *-wili* 'two.'

Swahili G42 also has two adjectives + ideophone compounds, i.e., *-eupe* 'white' < **-elu* < **-jéd-ò* 'white' + **pe* 'white' and *-eusi* 'black' < **-elu* < **-jíd-ò* 'black' + **pi* 'black.' Paradigm levelling accounts for the initial mid vowel in 'black,' possibly after the derived adjectives were made more distinct by adding monosyllabic ideophones for 'black' and 'white.' PB verbs for 'be white, shine' and 'be dark' are very similar.

Connective phrases (N+PP-a-N) may become lexicalised as nouns. The generic head nouns **mò-ntò* 'someone' and **kì-ntò* 'something' are often omitted, e.g., Koti P311 *etthú y'oóca* 'food' < *e-tthú ya o-c-a* 'something of to eat'; (*ntthú*) *w'oohóóni* 'blind (person)' < (*n-tthú*) *wa o-hí-on-i* '(someone) of not to see' (Schadeberg & Mucanheia 2000).

A productive type of compound has the form [NP-B-*a*]+N, as in **mò-dìnd-à* + *mì-nòè* 'ring finger' (lit. 'watcher + fingers'). Such nouns are basically agent nouns with complements, though not necessarily objects in the traditional syntactic sense as shown in (55).

(55)   Swahili G42
      mw-end-a   kimya   m-l-a   nyama
      1-walk-FV   silently   1-eat-FV   9.meat
      'the one (lion) who walks silently will eat the meat'

In some examples, the noun seems to be the subject of the preceding verb, e.g., Ganda JE15 *e-bugwa-njuba* 'west' < *-gò-* 'fall' + *ì-jóbà* 'sun' (Snoxall 1967), Umbundu R11 *úkú≠-á-mbámbi* 'tree sp.' < *-ku≠-* 'gnaw' + *o-mbambi* 'gazelle' (Schadeberg 1990). Note that these exocentric compounds as a whole refer neither directly nor metaphorically or metonymically to the apparent agent, in our examples the falling sun and the gnawing gazelle. A more consistent analysis therefore is to regard the syntactic source or correlate of such compounds as sentences with locative subjects (e.g., "there-falls the sun").

Derivation of verbs by incorporation of nouns is extremely rare; a widespread exceptional case is the verb 'to sit down' < **-(j)ìk-ad-a* + *ncí* 'sit + ground' (cf. Botne 1993), e.g., Makhuwa P31 *o-khala-thi, o-kilathi* > *o-kilathiha* (CAUS), *o-kilathihiwa* (CAUS-PASS) (Frizzi 1982).

6.4 Reduplication

Reduplication occurs with all kinds of words. Reduplicated nouns and adjectives are usually lexicalised, and generally only the stem is reduplicated. Pronouns and confirmative demonstratives are frequently reduplicated. In all kinds of non-verbal words, reduplication is particularly frequent with monosyllabic stems, which indicates a rhythmic preference for polysyllabic stems. Semantically, reduplication often appears to indicate smallness and/or repetition or iteration.

(56) *maamá   'mother'
      *-kéèké   'small'   < *-ké   'small'
      *-nàìnàì   'eight'   < *-nàì   'four'
      *mò-dìdì   'root'   < *mò-dì   'root'
      *kì-díóngʊdí(óngʊ) 'vertigo' < *-díóngò 'vertigo' (◇ *-díʊng-ʊd- 'turn around')

Reduplication of verbs is a productive process, indicating repetition often coupled with low intensity. Reduplication affects the verb stem; languages differ as to how they treat the reduplicand. In the most complete case, the whole stem with its final suffix and tonal profile is reduplicated; more commonly, the first part (the reduplicand) carries the lexical tone on its first mora and has the unmarked final F -*a*, while the second part carries the inflectional final and the tonal profile is spread over the whole reduplicated stem. Sometimes, the reduplicand is limited to two syllables, in which way it approaches the process of partial reduplication (Section 2.11). Reduplication of monosyllabic stems requires multiple application or the intercalation of the infinitive nominal prefix or some other "stabilizing" element, e.g., Nyamwezi F22 *kʊ-lyaá-lyaa-lya* < *kʊ-lí-a* 'to eat' (Maganga & Schadeberg 1992), Swahili G42 *ku-la-ku-la* < *ku-l-a* 'to eat' (Schadeberg 1992), Ndebele S44 *uku-dla-yi-dla* < *uku-dl-a* 'to eat' (Sibanda 2004).

ACKNOWLEDGEMENTS

The second author revised an earlier version of this chapter called "Derivation" (Schadeberg 2003a). The first author gave his consent to this update and participated in the final revisions following the peer review process. The second author wishes to thank Sebastian Dom for his feedback and for the fruitful discussions on verbal derivation in Bantu, which they have had over the past years. Thanks also go to Mark Van de Velde, Kyle Jerro and Heidi Goes for their useful comments. The usual disclaimers apply.

REFERENCES

Anonymous (1954) *The White Fathers' Bemba-English Dictionary*. London: Longmans, Green & Co. for the Northern Rhodesia & Nyasaland Joint Publications Bureau.

Ashton, E. O., E. M. K. Mulira, E. G. M. Ndawula & A. N. Tucker (1954) *A Luganda grammar*. London; New York; Toronto: Longmans, Green.

Bastin, Y. (1985) *Les relations sémantiques dans les langues bantoues*. Bruxelles, Belgique: Académie royale des sciences d'outre-mer.

Bastin, Y. (1986) Les suffixes causatifs dans les langues bantoues. *Africana Linguistica* 10: 56–145.

Bastin, Y. (1989) Les déverbatifs bantous en -e. *Journal of African Languages and Linguistics* 11: 151–174.

Bastin, Y., A. Coupez, E. Mumba & T. C. Schadeberg (eds.) (2002) *Bantu Lexical Reconstructions 3*. Tervuren: Royal Museum for Central Africa (Available online at linguistics.africamuseum.be/BLR3.html, Accessed November 28, 2017).

Bernander, R. (2018) The neuter in Manda, with a focus on its reinterpretation as passive. *Southern African Linguistics and Applied Language Studies* 36(3): 175–196.

Bostoen, K. (2005) Comparative Notes on Bantu Agent Noun Spirantization. In Bostoen, K. & J. Maniacky (eds.), *Studies in African Comparative Linguistics, with Special Focus on Bantu and Mande*, 231–258. Tervuren: Royal Museum for Central Africa.

Bostoen, K. (2008) Bantu Spirantization: Morphologization, Lexicalization and Historical Classification. *Diachronica* 25(3): 299–356.

Bostoen, K. (2010) Reflexive-Reciprocal Polysemy in South-Western Bantu: Typology and Origins. Paper presented at *Syntax of the World's Languages IV Conference*, Lyon.

Bostoen, K., S. Dom & G. Segerer (2015) The Antipassive in Bantu. *Linguistics* 53(4): 731–772.

Bostoen, K. & C. Grégoire (2007) La question bantoue: bilan et perspectives. *Mémoires de la Société de Linguistique de Paris – Nouvelle Série* 15: 73–91.

Bostoen, K. & L. Mundeke (2011a) The Causative/Applicative Syncretism in Mbuun (Bantu B87, DRC): Semantic Split or Phonemic Merger? *Journal of African Languages and Linguistics* 32(2): 179–218.

Bostoen, K. & L. Mundeke (2011b) Passiveness and Inversion in Mbuun (Bantu B87, DRC). *Studies in Language* 21(1): 72–111.

Bostoen, K. & L. Mundeke (2012) Subject Marking and Focus in Mbuun (Bantu, B87). *Southern African Linguistics and Applied Language Studies* 30(2): 139–154.

Bostoen, K. & Y. Nzang-Bie (2010) On How "middle" Plus "associative/reciprocal" Became "passive" in the Bantu A70 Languages. *Linguistics* 48(6): 1255–1307.

Botne, R. D. O. (1993) Noun Incorporation Into Verbs: The Curious Case of "ground" in Bantu. *Journal of African Languages and Linguistics* 14: 187–199.

Bresnan, J. & J. M. Kanerva (1989) Locative Inversion in Chichewa – a Case-Study of Factorization in Grammar. *Linguistic Inquiry* 20(1): 1–50.

Brincker, H. (1886) *Wörterbuch und kurzgefasste Grammatik des Otji-Hérero mit Beifügung verwandter Ausdrücke und Formen des Oshi-Ndonga – Otj-Ambo*. Leipzig: republished 1964 Ridgewood Gregg Press.

Burssens, A. F. S. (1946) *Manuel de Tshiluba (Kasayi, Congo Belge)*. Anvers: De Sikkel.

Butaye, R. (1909) *Dictionnaire kikongo-français, français-kikongo*. Roulers: Jules De Meester.

Büttner, C. G. (1881) Kurze Anleitung für Forschungsreisende zum Studium der Bantu-Sprachen. *Zeitschrift der Gesellschaft für Erdkunde zu Berlin* 16: 1–26.

Cann, R. & P. Mabugu (2007) Constructional Polysemy: The Applicative Construction in ChiShona. *Metalinguistica* 19: 221–245.

Chatelain, H. (1964) *Grammatica elementar do Kimbundu ou lingua de Angola*. New Jersey: The Gregg Press.

Chavula, J. J. (2016) *Verbal Derivation and Valency in Citumbuka*. Utrecht: LOT.

Chavula, J. J. (2018) The polysemy of the neuter extension -ik in Citumbuka (N21) and Citonga (N15). *Southern African Linguistics and Applied Language Studies* 36(3): 197–209.

Coupez, A., Y. Bastin & E. Mumba (eds.) (1998) *Reconstructions lexicales bantoues 2/Bantu lexical reconstructions 2 (electronic database)*. Tervuren: Musée royal de l'Afrique centrale.

Coupez, A., T. Kamanzi, S. Bizimana, G. Samatama, G. Rwabukumba & C. Ntazinda (2005) *Dictionnaire rwanda-français et français-rwanda/Inkoranya y'ikinyarwaanda mu kinyarwaanda nó mu gifaraansá*. Tervuren: Musée royal de l'Afrique centrale.

Creissels, D. (2004) Non-Canonical Applicatives and Focalization in Tswana. Paper presented at *Syntax of the World's Languages*, Leipzig (Available online at http://email. eva.mpg.de/~cschmidt/SWL1/handouts/Creissels.pdf).

Dammann, E. (1938–1939) Sprichwörter der Zigula. *Zeitschrift für Eingeborenensprachen* 29: 71–76.

Dammann, E. (1958) Die sogenannten Kausativa auf -eka in Bantusprachen. *Afrika und Ubersee* 42: 173–178.

Dammann, E. (1959) Inversiva und Repetitiva in Bantusprachen. *Afrika und Ubersee* 43: 116–127.

Dammann, E. (1961) Das Applikativum in den Bantusprachen. *Zeitschrift der Deutschen Morgenländischen Gesellschaft* 36: 160–169.

Dammann, E. (1962) Kontaktiva in Bantusprachen. *Afrika und Ubersee* 46: 118–126.

De Kind, J. & K. Bostoen (2012) The Applicative in Cilubà Grammar and Discourse: A Semantic Goal Analysis. *Southern African linguistics and applied language studies* 30(1): 101–124.

Devos, M. (2008) The Expression of Modality in Shangaci. *Africana Linguistica* 14: 3–35.

Doke, C. M. (1931) *Text Book of Zulu Grammar*. Johannesburg: Witwatersrand University Press.

Doke, C. M. (1935) *Bantu Linguistic Terminology*. London; New York: Longmans, Green.

Doke, C. M. (1938) *Text Book of Lamba Grammar*. Johannesburg: Witwatersrand University Press.

Doke, C. M. (1947) *Text-Book of Zulu Grammar*. Johannesburg: Longmans, Green & Co.

Doke, C. M. & B. W. Vilakazi (1958) *Zulu-English Vocabulary*. Johannesburg: Witwatersrand University Press.

Dom, S. (2014) *The Neuter in Bantu: A Systemic Functional Analysis*. Ghent: Ghent University, MA thesis.

Dom, S., H. Goes, G.-M. de Schryver & K. Bostoen (2016a) Multiple Reciprocity Marking in the Kikongo Language Cluster: Functional Distribution and Origins. Paper presented at *the 6th International Conference on Bantu Languages*, Helsinki.

Dom, S., L. Kulikov & K. Bostoen (2016b) Middle Voice in Bantu. Paper presented at *the 6th International Conference on Bantu Languages*, Helsinki.

Dom, S., L. Kulikov & K. Bostoen (2017) The Middle as a Voice Category in Bantu: Setting the Stage for Further Research. *Lingua Posnaniensis* 59(2): 129–149.

Dom, S., L. Kulikov & K. Bostoen (2018) Introduction – Valency-decreasing derivations and quasi-middles in Bantu: A typological perspective. *Southern African Linguistics and Applied Language Studies* 36(3): 165–173.

Dom, S., G. Segerer & K. Bostoen (2015) Antipassive/Reciprocal Polysemy in Cilubà (Bantu, L31a): A Plurality of Relations Analysis. *Studies in Language* 39(2): 354–385.

Downing, L. J. & L. Hyman (2016) Information Structure in Bantu Languages. In Féry, C. & S. Ishihara (eds.), *Oxford Handbook of Information Structure*, 790–813. Oxford: Oxford University Press.

Dubinsky, S. & S. R. Simango (1996) Passive and Stative in Chichewa: Evidence for Modular Distinctions in Grammar. *Language* 72(4): 749–781.

Endemann, K. (1876) *Versuch einer Grammatik des Sotho*. Berlin,: W. Hertz.

Fernando, M. (2013) *The Causative and Anticausative Alternation in Kikongo (Kizombo)*. Stellenbosch: Stellenbosch University, PhD dissertation.

Forges, G. (1983) La classe de l'infinitif en bantou. *Africana Linguistica* 9: 257–264.

Fortune, G. (1955) *An Analytical Grammar of Shona*. London; Cape Town; New York: Longmans, Green and Co.

Frizzi, G. (1982) *Dicionário de emakhuwa-português e português-emakhuwa*. Lichinga (Moçambique).

Givón, T. (1971) On the verbal origin of the Bantu verb suffixes. *Studies in African Linguistics* 2(2): 145–164.

Gibson, H., R. Guérois & L. Marten (2017) Patterns and Developments in the Marking of Diminutives in Bantu. *Nordic Journal of African Studies* 26(4): 344–383.

Good, J. C. (2005) Reconstructing Morpheme Order in Bantu: The Case of Causativization and Applicativization. *Diachronica* 22: 55–109.

Greenberg, J. H. (1951) Vowel and Nasal Harmony in Bantu Languages. *Zaïre: revue congolaise* 5(8): 813–820.

Greenberg, J. H. (1963) *The Languages of Africa*. Bloomington: Indiana University.

Grégoire, C. (1975) *Les locatifs en bantu*. Tervuren: Musée royal de l'Afrique centrale.

Grégoire, C. (1976) Le champ sémantique du thème bantu *-bánjá. *African Languages/ Langues Africaines* 2: 1–12.

Guérois, R. (2015) *A Grammar of Cuwabo (Mozambique, Bantu P34)*. Lyon: University of Lyon 2, PhD dissertation.

Guérois, R. & K. Bostoen (2018) On the origins of passive allomorphy in Cuwabo (Bantu P34). *Southern African Linguistics and Applied Language Studies* 36(3): 211–233.

Güldemann, T. (1996) *Verbalmorphologie and Nebenprädikationen im Bantu. Eine Studie zur funktional motivierten Genese eines konjugationalen Subsystem*. Bochum: Universitätsverlag Dr. N. Brockmeyer.

Güldemann, T. (1999) Head-Initial Meets Head-Final: Nominal Suffixes in Eastern and Southern Bantu from a Historical Perspective. *Studies in African Linguistics* 28(1): 49–91.

Guthrie, M. (1967) *Comparative Bantu: An Introduction to the Comparative Linguistics and Prehistory of the Bantu Languages. Volume 1: The Comparative Linguistics of the Bantu Languages*. London: Gregg International.

Guthrie, M. (1967–1971) *Comparative Bantu: An Introduction to the Comparative Linguistics and Prehistory of the Bantu languages (4 volumes)*. London: Gregg International.

Hadermann, P. (1999) Les formes nomino-verbales de classes 5 et 15 dans les langues bantoues du Nord-Ouest. In Hombert, J.-M. & L. M. Hyman (eds.), *Bantu Historical Linguistics: Theoretical and Empirical Perspectives*, 431–471. Stanford: CSLI.

Halme, R. (2004) *A Tonal Grammar of Kwanyama*. Cologne: Rüdiger Köppe.

Hannan, M. (1974) *Standard Shona Dictionary*. Salisbury: Mardon printers Ltd for Rhodesia Literature Bureau.

Harries, L. (1950) *A Grammar of Mwera, a Bantu Language of the Eastern Zone, Spoken in the South-Eastern Area of Tanganyika Territory*. Johannesburg: Witwatersrand University Press.

Heine, B. (1973) Zur genetischen Gliederung der Bantu-Sprachen. *Afrika und Übersee* 56(3): 164–185.

Hulstaert, G. (1965) *Grammaire du lomongo. Deuxieme partie, Morphologie*. Tervuren: Musée royal de l'Afrique centrale.

Hyman, L. & J. R. Watters (1984) Auxiliary Focus. *Studies in African Linguistics* 15(3): 233–273.

Hyman, L. M. (2003a) Sound Change, Misanalysis, and Analogy in the Bantu Causative. *Journal of African Languages and Linguistics* 24(1): 55–90.

Hyman, L. M. (2003b) Suffix Ordering in Bantu: A Morphocentric Approach. In Booij, G. & J. van Marle (eds.), *Yearbook of Morphology 2002*, 245–281. Dordrecht: Kluwer Academic Publishers.

Hyman, L. M. (2007) Niger-Congo Verb Extensions: Overview and Discussion. In Payne, D. L. & J. Peña (eds.), *Selected Proceedings of the 37th Annual Conference on African Linguistics*, 149–163. Somerville: Cascadilla Press.

Hyman, L. M. & A. D. Mtenje (1999) Non-Etymological High Tones in the Chichewa Verb. *Malawian Journal of Linguistics* 1: 121–256.

Jerro, K. J. (2016) *The Syntax and Semantics of Applicative Morphology in Bantu*. Austin: University of Texas, PhD dissertation.

Kähler-Meyer, E. (1966) Die örtliche Funktion der Applikativendung in Bantusprachen. *Neue Afrikanistische Studien* 5.

Kemmer, S. (1993) *The Middle Voice*. Amsterdam; Philadelphia: John Benjamins.

Kimenyi, A. (1988) Passives in Kinyarwanda. In Shibatani, M. (ed.), *Passive and Voice*, 355–386. Amsterdam; Philadelphia: John Benjamins.

Kula, N. & L. Marten (2010) Argument Structure and Agency in Bemba Passives. In Legère, K. & C. Thornell (eds.), *Bantu Languages: Analyses, Description and Theory*, 115–130. Cologne: Rüdiger Köppe Verlag.

Le Guennec, G. & J. F. Valente (1972) *Dicionário português-umbundu*. Luanda: Instituto de Investigação Científica de Angola.

Maganga, C. & T. C. Schadeberg (1992) *Kinyamwezi: Grammar, Texts, Vocabulary*. Cologne: Rüdiger Köppe.

Marten, L. (2003) The Dynamics of Bantu Applied Verbs: An Analysis at the syntax-Pragmatics Interface. In Lébikaza, K. K. (ed.), *Actes du 3e congrès mondial de linguistique africaine (Lomé 2000)*. Cologne: Rüdiger Köppe.

Marten, L. & N. C. Kula (2014) Benefactive and Substitutive Applicatives in Bemba. *Journal of African Languages and Linguistics* 35(1): 1–44.

Meeussen, A. E. (1954) Werkwoordafleiding in Mongo en Oerbantoe. *Aequatoria* 17: 81–86.

Meeussen, A. E. (1961) Le ton des extensions verbales en bantou. *Orbis* 10(424–427).

Meeussen, A. E. (1967) Bantu Grammatical Reconstructions. *Africana Linguistica* 3: 79–121.

Meeuwis, M. (2013) Lingala Structure Dataset. In Michaelis, S. M., P. Maurer, M. Haspelmath & M. Huber (eds.), *Atlas of Pidgin and Creole Language Structures Online*. Leipzig: Max Planck Institute for Evolutionary Anthropology (Available online at http://apics-online.info/contributions/60, Accessed August 10, 2016).

Meinhof, C. (1899) *Grundriß einer Lautlehre der Bantusprachen nebst Anleitung zur Aufnahme von Bantusprachen – Anhang : Verzeichnis von Bantuwortstämmen* Leipzig: F.A. Brockhaus.

Meinhof, C. (1904) Einige Bantuwortstämme. *Mitteilungen des Seminars für Orientalische Sprachen* 7: 127–149.

Meinhof, C. (1910) *Grundriß einer Lautlehre der Bantusprachen nebst Anleitung zur Aufnahme von Bantusprachen – Anhang: Verzeichnis von Bantuwortstämmen*. Berlin: Dietrich Reimer (Ernst Vohsen).

Meinhof, C. (1948) *Grundzüge einer vergleichenden Grammatik der Bantusprachen.* Hamburg: Eckardt & Messtorff.

Meinhof, C. & N. J. van Warmelo (eds.) (1932) *Introduction to the Phonology of the Bantu Languages Being the English Version of "Grundriss Einer Lautlehre Der Bantusprachen" by Carl Meinhof, Translated, Revised and Enlarged in Collaboration with the Author and Dr. Alice Werner by N. J. Van Warmelo.* Dietrich Reimer/Ernst Vohsen, published under the auspices of The International Institute of African Languages and Cultures, The Carnegie Corporation of New York, and The Witwatersrand Council of Education, Johannesburg.

Nsuka-Nkutsi, F. (ed.) (1980) *Eléments de description du punu.* Lyon: Centre de Recherches Linguistiques et Sémiologiques, Université Lumière Lyon 2.

Pacchiarotti, S. (2017) *Bantu Applicative Construction Types Involving *-Id: Form, Functions and Diachrony.* Eugene: University of Oregon, PhD dissertation.

Peterson, D. A. (2007) *Applicative Constructions.* Oxford; New York: Oxford University Press.

Petzell, M. (2008) *The Kagulu Language of Tanzania: Grammar, Texts and Vocabulary.* Cologne: Rüdiger Köppe.

Polak-Bynon, L. (1975) *A Shi Grammar. Surface Structures and Generative Phonology of a Bantu language.* Tervuren: Royal Museum for Central Africa.

Polis, C. (1938) *Lexique kikongo – français (manuscrit en neuf fascicules, dupliqué par stencil).* Ms. Leuven:

Rapold, C. (1997) *The Applicative Construction in Lingala.* Leiden: Universiteit Leiden, M.A.

Rugemalira, J. M. (2009) *Kamusi Ya Kigogo-Kiswahili-Kiingereza Kiingereza-Kiswahili-Kigogo.* Dar Es Salaam: Languages of Tanzania Project, University of Dar es Salaam.

Sacleux, C. (1939) *Dictionnaire swahili-français.* Paris,: Institut d'ethnologie.

Schadeberg, T. C. (1982) Les suffixes verbaux séparatifs en bantou. *Sprache und Geschichte in Afrika* 4: 55–66.

Schadeberg, T. C. (1990) *A Sketch of Umbundu.* Cologne: Rüdiger Köppe.

Schadeberg, T. C. (1992) *A Sketch of Swahili Morphology.* Cologne: Rüdiger Köppe.

Schadeberg, T. C. (1994) Die Extensive Extension im Bantu. In Geider, T. & R. Kastenholz (eds.), *Sprachen und Sprachzeugnisse in Afrika: eine Sammlung philologischer Beiträge Wilhelm J. G. Möhlig zum 60. Geburtstag zugeeignet.* Cologne: Rüdiger Köppe.

Schadeberg, T. C. (2003a) Derivation. In Nurse, D. & G. Philippson (eds.), *The Bantu Languages*, 71–89. London; New York: Routledge.

Schadeberg, T. C. (2003b) Historical Linguistics. In Nurse, D. & G. Philippson (eds.), *The Bantu Languages*, 143–163. London; New York: Routledge.

Schadeberg, T. C. & F. U. Mucanheia (2000) *Ekoti: The Maka or Swahili Language of Angoche.* Cologne: Rüdiger Köppe.

Schladt, M. (1996) *Reciprocals in Bantu: A Case of Grammaticalization.* Cologne: University of Cologne, PhD dissertation.

Sebasoni, S. (1967) La préfinale du verbe bantou. *Africana Linguistica* 3: 122–135.

Shibatani, M. & P. Pardeshi (2002) The Causative Continuum. In Shibatani, M. (ed.), *The Grammar of Causation and Interpersonal Manipulation*, 85–126. Amsterdam; Philadelphia: John Benjamins.

Sibanda, G. (2004) *Verbal Phonology and Morphology of Ndebele*. Berkeley: University of California., PhD dissertation.

Snoxall, R. A. (1967) *Luganda-English Dictionary*. Oxford: Clarendon P.

Stappers, L. (1964) *Morfologie van het Songye*. Tervuren: Koninklijk Museum voor Midden-Afrika.

Stappers, L. (1967) Het passief suffix -u- in de Bantoetalen. *Africana Linguistica 3*, 139–145.

Stappers, L. (1973) *Esquisse de la langue mituku*. Tervuren: Musée royal de l'Afrique centrale.

Tobias, G. W. R. & B. H. C. Turvey (1954) *English-Kwanyama Dictionary*. Johannesburg: Witwatersrand University Press.

Trithart, M. L. (1983) *The Applied Suffix and Transitivity: A Historical Study in Bantu*. Los Angeles: UCLA, PhD.

van der Wal, J. (2015) A Note on the (Non-Existing) Passive in Matengo. *Linguistique et langues africaines* 1: 81–98.

Voisin-Nouguier, S. (2003) Un syncrétisme causatif/applicatif en wolof? In Sauzet, P. & A. Zribi-Hertz (eds.), *Typologie des langues d'Afrique & universaux de la grammaire. Volume 2: Benue-Kwa Wolof*, 185–203. Paris: L'Harmattan.

Wald, B. (1998) Issues in the North/South Syntactic Split of East Bantu. In Maddieson, I. & T. J. Hinnebusch (eds.), *Language History and Linguistic Description in Africa*, 95–106. Trenton: Africa World Press.

Watkins, M. H. (1937) *A Grammar of Chichewa: A Bantu Language of British Central Africa*. Philadelphia: Linguistic Society of America.

Willems, E. (1970) *Le Tshiluba du Kasayi*. Luluabourg: Mission de Scheut.

CHAPTER 7

ASPECT, TENSE AND MOOD

Derek Nurse and Maud Devos

1 INTRODUCTION

This chapter differs in two main ways from the corresponding chapter in the first edition (Nurse 2003). First, it has a section on mood/modality, not included in the first edition. Second, the sections on aspect and tense (henceforth AT) are revised in light of Nurse (2008). We also refer to other authors who offer alternative theoretical approaches, even if they often lack the broad geographical scope of this chapter.

We focus here on the typology of aspect, tense and mood in Bantu (henceforth ATM), more specifically on the semantics of these ATM categories, how they interact, the systems they form, the range of variation they manifest and their expression both segmental and tonal.

Several features make Bantu and this chapter interesting: the multiple time divisions (Section 6), the attempt to quantify aspect in Bantu (Section 5), the encoding of these categories (Section 3), which is often different from how other languages in the world encode, the intensity and rapidity of grammaticalisation in Bantu ATM, and the first formal view of mood and modality in Bantu (Section 7).

Valid generalisations about AT rely on having a geographically inclusive and typologically representative base. The base used here is that in Nurse (2008: 4), comprising more than 210 languages that are thought to be reasonably discrete and representative of the whole area. It includes at least one language from each of Guthrie's zones and groups (Guthrie 1971, Maho 2009). A schematic analysis of the languages is available online: www.faculty.mun.ca/dnurse/Tabantu/. The 210 breaks down into 100 core languages, used for detailed examination and statistical statements, and 110 peripheral languages. The amount of detail depends on the source. The description of any language involves three main variables: author's knowledge of how the language, and especially AT, works; how much data is presented; and the theoretical approach. The data presented tends to support the approach. Readers coming to a description with a different theoretical stance and looking for different data will not find all they seek. Many general grammars and even some specialised analyses use quite disparate terminology for the same phenomena, and the quantity of good and relevant data presented for tense and especially aspect is often limited. Although a firm line between "good" and "less good" data is hard to draw, we would say there was "good" data for 20–25 languages in the AT database, and "less good" for the rest. Generalisations are only as good as the data they rest on. There has been a recent upswing in higher quality descriptive work. We hope this will continue with new studies also being better theoretically informed. Section 7 on Mood and Modality uses a slightly modified version of the Nurse (2008) 100 languages sample (cf. Devos & Van Olmen 2013: 55–56).

2 VERB STRUCTURES, CATEGORIES EXPRESSED IN THE VERB

The verbal word has a similar structure in most Bantu languages. Meeussen (1967) reconstructs it for Proto-Bantu with a slightly more elaborate schema than in (1).

(1)   Initial – Subject – Negative – T(A) – Object ≠ Root – Extension(s) – Final – Suffix

Slots to the left and right of the Root and Extension(s) in (1) involve inflection. The Initial expresses only two categories common to many Bantu languages, negative and relative, but individual languages express a range of other categories at Initial, because this is a slot where new material often becomes grammaticalised. Extension involves a small, closed set of often valency-changing categories, of which causative, applicative, stative, reciprocal, separative and passive are the most common (cf. Chapter 6). Final also includes a small, closed set originally having to do with mood and aspect, but now including negation and tense in some languages. Across Bantu, Suffix includes only a marker of imperative plural, although, as the Initial, newly grammaticalised material can become attached here. The other labels are self-explanatory. Sections 4, 5 and 6 below deal with the slots important to the expression of tense and aspect: Initial, T(A) and Final. In nearly all structures of this type, tense is encoded to the left, aspect to the right. Aspectual categories are often more closely linked to the root than tense.

Many Bantu languages also have verb compounds or multi-verb constructions, treated by some analysts as biclausal. In such structures, the first word is usually an inflected auxiliary which is followed by the infinitive of the main verb, as in (2a). However, certain languages also have multi-verb constructions in which the lexical verb is inflected, as in (2b).

(2)   a   Ngindo P14 (Nurse 2003: 91)
          tu-ø-tenda   ku-hemer-a
          SP_{1PL}-PRS-do NP_{15}-buy-INF
          'We are buying.'

      b   Kimbu F24 (Derek Nurse field notes)
          xʊ-xa-lɪ   xʊ-xʊ-gʊl-a
          SP_{1PL}-PER-be SP_{1PL}-PROG-buy-FV
          'We are still buying.'

Some languages even have verbal compounds that have three words or have been analysed this way, as in (3).

(3)   a   Hehe G62 (Nurse & Makombe 1979: 117)
          saa   tu-ø-va   tu-ø-gus-ile
          FUT (< come?) SP_{1PL}-PRS-be SP_{1PL}-PRS-buy-PRF
          'We will have bought.'

      b   Sukuma F21 (Nurse 2008: 30)
          d-áá-lí   dǒ-ṭaalí   dʊ́-líí-gʊ́l-a
          SP_{1PL}-PST_4-be SP_{1PL}-PER SP_{1PL}-PROG-buy-FV
          'We were still buying.'

Structures of the three types are common across the Bantu area. In the North-West, there may be a structural spectrum, from languages having the one-word structure just described, through languages where this structure is loosening, to languages where some or all of the pre-stem material is not phonologically bound (Beavon 1991, Ernst 1991). A corollary is that grammaticalised material forms into what is known as serial verbs. Because of this typological discrepancy between a few North-Western languages and the rest, they are sometimes exceptions to the generalisations following. We deal with the widespread rather than the narrow.

Tense, aspect and mood are not the only categories expressed at Initial, T(A), Final and Suffix. Lack of space prevents proper discussion of other categories. Instead we list them and mention works giving an overview, where we are aware of them. The categories are negation, at the Initial and/or Negative slots above, or less commonly via a separate word (Kamba Muzenga 1981, Güldemann 1996, Devos & van der Auwera 2013); relatives, marked at Initial, Suffix, tonally or via a separate word (Nsuka-Nkutsi 1982); focus and assertion, indicated at Initial, T(A) or tonally (Dalgish 1979, Hyman & Watters 1984, Wald 1997); degrees of certainty in affirmation, marked variously; and conditionals (Saloné 1979, Parker 1991) (see also Section 7).

3 HOW TENSE AND ASPECT ARE ENCODED

Most tense and aspect encoding in Bantu languages involves a combination of three main components: inflection of the verb, tone, and the use of verbs additional to and commonly preceding the main lexical verb. These additional verbs occur as auxiliaries in most Bantu languages or as serial verbs in a few North-Western languages.

The simplest *way* of encoding tense and aspect is via a single segmental or tonal marker in a single word. The Venda S21 example in (4) illustrates the segmental approach; the tones here play little role.

(4)   Venda S21 (Ziervogel *et al.* 1981)
      r-ó-rém-á   vs.   ri-do-rém-á
      SP$_{1PL}$-PST-chop-FV   SP$_{1PL}$-FUT-chop-FV
      'We chopped.'   'We will chop.'

Each form in (4) is a word with four morphemes, but only the pre-stem morpheme is significant for AT: *do* marks future, *o* marks past. Contrary to most Eastern Bantu languages, the fourth morpheme, the FV, plays no role in the marking of AT here. As Swahili G42, Venda has three possible FVs, one associated with general negation, one with subjunctive, leaving -*a* as the default case, associated with any or all tenses and aspects, thus semantically neutral for AT. All AT marking thus occurs at the pre-stem position.

Since Bantu languages typically have tone or pitch-accent, most use tone in a major way in indicating verbal categories. First, lexical stems typically fall into one of a small number of underlying tone classes. Then, on the left of the stem the (tense) marker may carry its own tone, and on the right, the (aspect) Final likewise. Additionally, total verb forms may carry an imposed tonal melody for individual tenses and aspects (cf. Odden & Bickmore 2014). There may also be floating and other kinds of tones, which tend to fuse with the kinds of tones described. Finally, a number of phonetic processes link these underlying tones to each other and to surface realisations, so that underlying and surface

tones are rarely the same. Surface tones normally therefore carry grammatical information. We exemplify only three of the myriad tonal possibilities that result from these
variables. The three are Lingala C36d (Dzokanga 1995), Haya JE22 (Muzale 1998) and
Koonzime A842 (Beavon 1991).

Lingala has a quite simple tonal system: the surface tones are also the underlying
tones. Therefore, the differences between the three forms are the tone differences at Subject and FV.

(5) Lingala C36d (cf. Meeuwis 1995: 100ff)
    a-pés-aka   vs. a-pés-áká   vs. á-pés-aka
    SP$_{3SG}$-give-HAB   SP$_{3SG}$-give-REM.PST   SP$_{3SG}$.SBJV-give-INT
    'S/he regularly gives.'   'S/he gave.'   '. . . (that) s/he (may) really give.'

Haya is a restricted tone language. The subject *tu-*, stem *-gur-*, NEG *ti-* and PROG *ni-*
are underlyingly L, while FV *-á* is H. There are several possible explanations for the H
on *-á*, but the tone of FVs cannot be predicted on general grounds (so other final *-a* in
Haya are low). None of the three words has a tense morpheme at T(A), because none is
marked for time reference. There is a regular process whereby final H is thrown back to
the penultimate before pause. For a more complete analysis of this in Haya, see Hyman
and Byarushengo (1984).

(6) Haya JE22 (Nurse 2010)
    [tu-ø-gúr-a] /tu-ø-gur-á/ vs. [ti-tu-ø-gúr-a]   vs. [ni-tu-ø-gúr-a]
    SP$_{1PL}$-PRS-buy-FV   NEG-SP$_{1PL}$-PRS-buy-FV   PROG-SP$_{1PL}$-prs-buy-FV
    'We buy.'   'We don't buy.'   'We are buying.'

Koonzime is more complicated tonally. Its PST$_1$ is only marked tonally and in four places.
As analysed by Beavon (1991), *be* 'they,' *fumo* 'build' and *mi-mbɛr* '4-house' in (7) have
tones as indicated; the LH on 'house' is realised as rising. PFV *si* is inherently toneless.
PST$_1$ is indicated by the four underlying tones, which surface in different ways. The first
L fuses with the preceding H, giving H. The second H surfaces on the preceding toneless
PFV. The third L appears on the second syllable of the root, while the fourth replaces the
following L.

(7) Koonzime A842   (Beavon 1991)
    be-   si-   fumo   #   mi- mbɛr
    H   L   H   L   L   H   #   L   LH
    SP$_2$ PST$_1$ PFV   PST$_1$ build   PST$_1$ PST$_1$ #   NP$_4$ house
    'They built houses.'

A common pattern has prefixes and finals combine to show AT. Typically, the pre-stem mor-
pheme(s) mark tense, while the finals indicate aspect, as in the Gikuyu E51 example in (8).

(8) Gikuyu E51 (Johnson 1980)
    tw-a-hanyok-irɛ   vs. tw-a-hanyok-etɛ   vs. tw-a-hanyok-aga
    SP$_{1PL}$-PST$_3$-run-PFV   SP$_{1PL}$-PST$_3$-run-PRF   SP$_{1PL}$-PST$_3$-run-IPFV
    'We ran.'   'We had run.'   'We were running.'

This type of pattern, with its multiplicity of encoded categories and use of post-stem structure, is probably older, while systems such as that of Swahili G42, because of their fewer encoded categories, greater use of auxiliaries and diminished use of post-stem structure, are probably newer.

A major source of innovation is the constant emergence of forms based on auxiliaries or modals and a main verb, either in the infinitive or in an inflected form, as in the Mwera P22 example in (9) (Harries 1950). This is the remotest degree of futurity. A Mwera speaker would have little trouble analysing the form as deriving from *ci-tu-ø-ji-e ku-uma* 'we-will-come to-buy,' by simple deletion of infinitival *ku*. One could easily imagine the next stage to be simplification of the vowel sequence [*ieu*]. The previous two stages are also pretty transparent. All future formation in Mwera involves the subjunctive -*e*: (non-factual) subjunctive was semantically extended to (non-factual) future. Initial *ci*- points to an even earlier stage as it is a grammaticalised form of 'say,' used in all Mwera future tenses and common in the languages of the area (Botne 1998).

(9)   Mwera P22 (Harries 1950)
      ci-tu-ji-e-uma
      FUT₃-SP₁ₚₗ-come-FUT-buy
      'We will buy.' (Far Future)

Another transparent case is the Haya JE22 Present Progressive Negative presented in (10). Stage 1 represents an assumed past stage, where auxiliary and main verb were separate. At Stage 2, they have fused in a form commonly heard today. Stages 3–6, dialectal or personal forms also heard today, derive from Stage 2 by r-deletion, not otherwise a regular Haya feature, and by other tonal and segmental adjustments. The forms with falling tone are still to be analysed as compounds, while those with level H are one word. This kind of grammaticalisation is a constant process. All the languages examined in detail show signs of recent or on-going grammaticalisation.

(10) Haya JE22
      Stage 1 (historical)   ti-tú-ri   ku-gur-a
                                        NEG-SP₁ₚₗ-be   NP₁₅-buy-INF
      Stage 2 (today)   ti-tú-riku-gur-a
                                        NEG-SP₁ₚₗ-PROG-buy
                                        'We aren't buying.'
      Stages 3–6 (today, personal or dialectal)   [titwíikugura, titúùkugura,
                                        titúúkugura, titwííkugura]

A possible stage, intermediate between the Eastern/Southern and North-Western patterns, is where successive operations of grammaticalisation result in a complex string of pre-stem morphemes, up to four in number, unlike the more usual single morpheme. This mostly occurs in two geographical areas: one in northeast Tanzania, the other in a few languages on the eastern side of Lake Victoria (JE40).

(11) a   Chaga E60 (Vunjo E622C dialect)   (Moshi 1994: 142)
          à-lé-màà-èndà-írzérz-a
          SP₁ₛ₉-PST₂-PFV(< finish)-IPFV(< go)-speak-FV
          'S/he had already spoken.'

b   Shambaa G23   (Besha 1989: 64)
    ní-tà-zà-nà-hè-mù-ítáng-à
    SP$_{1sg}$-PRS-ANT-FIRST-WHEN-OP$_1$-call-FV
    'I might call her.'

c   Gusii JE42   (Whiteley 1960: 37)
    ba-tá-á-kó-raa-ná-gó-sang-er-er-ek-an-a
    SP$_2$-NEG-PST-PROG-SEQ-CNT-PROG-meet-APPL-APPL-STAT-ASSOC-FV
    'They should perhaps first meet together.'

This section has outlined five strategies for encoding AT: purely segmental, purely tonal, a combination of segments and tones, the use of two word structures with auxiliaries and a string of morphemes at T(A). For purposes of exposition they have been presented discretely, but in practice a combination of these often carries the grammatical message.

4 ASPECT AND TENSE CATEGORIES

The terms tense (abbreviated T) and aspect (A) have been used hitherto without definition. Tense is classically defined as the grammaticalised representation of (that is, verbal inflection for) location in time, and aspect as the representation of the internal temporal constituency of a situation (cf. Comrie 1976). Tense establishes the time framework, and aspect then sets out how the situation is distributed within the time framework.

Aspects can be identified in many Bantu languages by considering the two word structures mentioned in (1) and illustrated in the first example in (2a). The auxiliary verb of such structures may be inflected for several categories, the most prominent being tense. The auxiliary itself carries limited semantic material, its main function being as a placeholder to carry the categories that set the temporal scene. Aspect is marked in the second verb. Since encoded aspects are few, only a few morphemes typically occur in the T(A) position in the second verb. Thus, a good method of identifying aspect is to examine the morphology of the second verb. The first verb is often omitted, leaving only aspect marked on the verb, because the time framework is usually known to the participants in daily discourse.

Affirmative, negative and relative forms are three interlocking dimensions of the same system, but are not necessarily mapped directly on to one another. Affirmatives tend to show the largest set of tense-aspect contrasts. Negatives and relatives usually have a reduced set of contrasts. The theoretical framework underlying Sections 3 and 4 rests on certain assumptions:

(i)   AT categories are not just a list to be learnt or memorised. While there are notable exceptions, many treatments of Bantu languages have tended to treat individual AT morphemes as self-standing, which are listed, have labels attached and meanings given, with little or no reference to the other members of the system. Consequently, it is often unrightfully claimed that aspect is hard to distinguish from tense.

(ii) AT categories form an interlocking system, in which most aspects co-occur with most tenses.

(iii) Systems and categories are cognitively based and do not necessarily directly reflect the events or objects of this world. They reflect human organisation and categorisation of these objects and events. These categories have a strong cognitive component, which is why the categories themselves tend to be relatively stable over time, and tend to re-occur across languages.

(iv) A discrete AT form has a specific and unique range of meaning, different from that of other forms in the language.

(v) A form derives its basic meaning by contrast with other forms within the verbal paradigm. Since each form and meaning is so derived, while there can be some overlap between forms, there is rarely total overlap, because that would make a form redundant. Some Swahili speakers would claim that the verb forms in (12) are always or often semantically identical, as are the ones in (13). There are two possibilities: either the many speakers are wrong, because they have overlooked certain subtle semantic differences which they have trouble articulating, or they are right, in which case one member of each pair can probably look forward to a short life, as language does not usually tolerate absolute redundancy for long, or one member will undergo semantic change.

(12) Swahili G42 (Nurse 2008: 13)

tu-na-sem-a	vs.	tw-a-sem-a
SP$_{1PL}$-PROG-say-FV		SP$_{1PL}$-PRS-say-FV
'We are talking.'		'We talk.'

(13) Swahili G42 (Nurse 2008: 13)

tu-li-ku-w-a	tu-ki-zungumz-a vs.	tu-li-ku-w-a	tu-na-zungumz-a
SP$_{1PL}$-PST-SM-be-FV	SP$_{1PL}$-SIT-chat-FV	SP$_{1PL}$-PST-SM-be-FV	SP$_{1PL}$-PROG-chat-FV
'We were chatting.'			

(vi) The meaning of a form is flexible and can be modified in use and discourse. Therefore, an AT system is not inflexible or unchanging.

(vii) Aspect is more fundamental than tense, which explains the order AT adopted here, rather than the commoner TA. Many languages in the world have aspectual, but no tense, contrasts. Bantu AT systems developed out of early Niger-Congo, for which only aspect is assumed. Every finite verb form has aspect.

(viii) A verb form can have only one tense but more than one aspect. It can have but a single tense, as an event cannot normally take place at two different times. Multiple aspects are possible because an event can be viewed and represented in more than one way simultaneously, as for instance in English 'they might have been eating.' However, verb compounds can encode two tenses, one representing the relationship between speech time and first event time, the second that between first event time and a second event. Views differ on whether such structures should be considered biclausal.

(ix) Regardless of their morphological exponence, tenses and especially aspects have certain common semantic features in most languages. While these may not be quite universal, they are certainly widespread (Comrie 1976, 1985, Dahl 1985, Bybee *et al.* 1994).

Finally, as this is not a theoretical treatise but an account of Bantu, there is a necessary balance between theoretical finesse and making the aunt accessible to readers. Other theoretical positions are defended in, for instance, Aksenova (1997), Bochnac and Matthewson

(2015), Botne and Kershner (2008), Güldemann (1996, 2003), Hewson (2012, 2016b, 2016a).

5 ASPECT

The same few aspect categories occur constantly across Bantu, with relatively little variation, while tenses vary more. Even when their morphological exponents are destroyed or recycled, aspects are often maintained in a new guise.

Six aspects are widespread in Bantu, in the related Niger-Congo languages and worldwide: PERFECTIVE (PFV) (a similar category is also referred to by others as FACTATIVE, PERFORMATIVE, even AORIST, cf. Welmers (1973: 346), Faraclas *et al.* (2007)), contrasting with IMPERFECTIVE (IPFV), PERFECT (PRF, also ANTERIOR or RETROSPECTIVE), PROGRESSIVE (PROG), PERSISTIVE (PER) and HABITUAL (HAB)/ITERATIVE (ITR). The advantage of reducing aspects to a few major categories and labels is that they become more transparent. The disadvantage is that the picture may become oversimplified and ignore other subtleties and less common aspectual categories. Few languages show clear evidence of all six categories together. In (14) and (15), the basic aspects of Basaa A43 and Ndendeule N101, in conjunction with the different tense morphemes with which they can be combined, are illustrated.

(14) Basáa A43 (Nurse 2010)

    a   PFV

        a   ń-'jɛ́   a   ń-lɔ

        3SG PRS-eat   3SG PRS-come

        'He eats.'   'He comes.'

    b   IPFV

        a   ḿ-ɓɛl-ɛ́k

        3SG PRS-plant-IPFV

        'He is/will be planting (today).'

    c   PRF

        a   má   'jɛ́   a   má   lɔ

        3SG PRF   eat   3SG PRF   come

        'He has already eaten.'   'He has already come.'

    d   PROG

        a   yé   'jɛ́   a   yé   lɔ

        3SG PROG   eat   3SG PROG   come

        'He is eating.'   'He is coming.'

    e   HAB

        a   ḿ-'ɓéná   'jɛ́

        3SG PRS-HAB   eat

        'He often eats.'

    f   PER

        a   ngí   jêk   a   ngí   lɔ̂k

        3SG PER   eat.IPFV   3SG PER   come.IPFV

        'He is still eating.'   'He is still coming.'

(15) Ndendeule N101 (Nurse 2010)

    a PST-PFV
       ti-aki-telek-a
       SP_{1PL}-PST-cook-FV
       'We cooked (before today).'

    b PRS-PFV
       ti-ø-telek-a
       SP1PL-PRS-cook-FV
       'We cook, are cooking.'

    c FUT-PFV
       ca-ti-telek-a
       FUT-SP_{1PL}-cook-FV
       'We will cook.'

    d PST-IPFV
       ti-a-hip-á
       SP_{1PL}-PST-smoke-IPFV
       'We were smoking.'

    e PRS-IPFV
       ti-ø-hip-it-á
       SP_{1PL}-PRS-smoke-IPFV
       'We are smoking.'

    f PST-PRF
       ti-a-hik-ite
       SP_{1PL}-PST-arrive-PRF
       'We had arrived.'

    g PRS-PRF
       ti-ø-hip-ite
       SP_{1PL}-PRS-smoke-PRF
       'We have smoked.'

    h PST-HAB
       ti-aki-hipahipa
       SP_{1PL}-PST-smoke.HAB
       'We used to smoke.'

    i PRS-HAB
       ti-ø-hipahipa
       SP_{1PL}-PRS-smoke.HAB
       'We smoke regularly.'

    j FUT-HAB
       ca-ti-hipahipa
       FUT-SP_{1PL}-smoke.HAB
       'We will smoke regularly.'

PERFECTIVE (PFV) presents an event as an undifferentiated and time-bounded whole without regard to the internal constituency of the event. It takes an exterior view of the event as a whole. It typically answers questions such as "When did you X?," to which the answer could be "We X-ed this morning (yesterday, last week)." The X could take a longer or a shorter time. Thus "wrote" as in "wrote the number 2" or "wrote a long letter" or "wrote a book" refers to time periods of different lengths, but presents the writing as a single past act. Dahl (1985) lists questions that typically produce PFV responses along with questions that produce other kinds of responses. Although PFV can have future reference, they more often co-occur with past, because past events are better known and more easily defined than future ones. All languages examined had PFV forms, often restricted to past. They are typically expressed by one-word forms, inflected at T(A). Every language in the Nurse (2008) database is analysed as having a PFV.

By contrast, IMPERFECTIVE (IPFV) has to do with the internal constituency of events. IPFV verbs usually represent events characterising a longer period, so they are not punctual. They often represent backgrounded events for events foregrounded by the use of PFV. IPFV is used in two senses, to contrast with PFV or to represent an unbounded situation that lasts over a period of time. A typical imperfective represents a situation as part complete, part incomplete. Some languages have a single category to express imperfectivity. Others have an IPFV coordinate with PROG, HAB, PER and others. Yet others have no single imperfective category but only distinct categories to represent PROG and HAB, and sometimes CONTINUOUS and PER. All languages examined had an imperfective in one or other of these senses. The available data sometimes gets in the way of a clear analysis of individual languages. Often we had to rely on data which is not exemplified but merely translated (often inadequately) into English or other languages. For instance, in several cases, a single form might be translated as 'we were buying,' which could be progressive or imperfective, and also as habitual 'we used to buy' – is it progressive, imperfective, habitual or do the terms need revising? Similarly, some writers have several "present tenses," which all translate as English 'we buy, we are buying,' none really a present or a tense. While most languages have some overlapping forms, three or four semantically identical "presents" is implausible. Given this, it is currently impossible to state IPFV categories common to all Bantu languages with certainty. The best that is possible is describing common situations. All the languages examined have a contrast between a PFV and at least one IPFV category in Comrie's sense. As illustrated in (16), the PFV is unmarked or morphologically simpler than the IPFV. Thus:

(16) PFV vs. IPFV in number of Bantu languages
   a   Kako A93 (Ernst 1991)
       á-tă-kwá (PFV)   á-tă-bɛ́   kɛ́-kwà (IPFV)
       SP₁-FUT-leave   SP₁-FUT-be   PROG-come
       'She will leave.'   'She will be leaving/is about to leave.'

   b   Punu B43 (Jean Blanchon, personal communication)
       á-tsí-lá:mb-ə̀ (PFV)   à-lá:mb-ə̀   à-tsì-là:mb-a:ng-ə̀ (IPFV)
       SP₁-PST-cook-FV   SP₁-cook-FV
       'She cooked.'   'She was cooking.'

   c   Leke C14 (Vanhoudt 1987)
       bá-támbuz-í (PFV)   bá-bé-ak-í   mô-támbud-é (PFV)
       SP₂-walk-PST₁   SP₂-be-IPFV-PST₁   PROG-walk-FV
       'They walked.'   'They were walking.'

   d   Lega D25 (Meeussen 1971)
       tu-ka-bulut-a (PFV)   tu-ka-bulut-ag-a (IPFV)
       SP₁ₚₗ FUT-pull-FV   sp₁ₚₗ-fut-pull-ipfv-fv
       'We will pull.'   'We will be pulling, pull regularly.'

   e   Rangi F33 (Oliver Stegen personal communication)
       tw-a:-sék-iré (PFV)   kʊ-séka   tw-á-rɪ (IPFV)
       SP₁ₚₗ-PST-laugh-PFV   NP₁₅-laugh-INF   SP₁ₚₗ-PST-be
       'We laughed.'   'We were laughing.'

                                tw-á:-rɪ   tw-á-sék-a
                                SP₁ₚₗ-PST-be   SP₁ₚₗ-PST-laugh-FV

   f   Pogoro G51 (Nurse 2010)
       tu-sek-iti (PFV)   tu-wer-iti   tu-anku-sek-a (IPFV)
       SP₁ₚₗ-laugh-PFV   SP₁ₚₗ-be-PFV   SP₁ₚₗ-PROG-laugh-FV
       'We laughed.'   'We were laughing.'

   g   Nyakyusa M31
       a-tʊ-kʊ-ʊla (PFV)   a-tʊ-kʊ-j-a   pa-ku-ʊla (IPFV)
       FUT-SP₁ₚₗ-NP₁₅-buy   FUT-SP₁ₚₗ-NP₁₅-come-FV   NP₁₆-NP₁₅-buy-INF
       'We will buy.'   'We will be buying.'

As the data in (16) suggests, IPFV may be expressed by inflection, usually at Final, or by using auxiliaries (tones not marked where not reliably known). The suffix -a(n)g- is often associated with IPFV. A pre-stem zero form, unmarked for aspect or tense, and suggesting that an action characterises a huge, unspecified period, occurs in many languages. It is used when the distinction between IPFV, PROG or HAB is not important, relevant or known. All database languages have an IPFV.

PRF denotes a situation that started in the past but continues into the present, or the continuing present relevance of a previous situation. It focuses on the result phase, subsequent to the situation. In Bajuni G41, for instance, the PRF marker is *-ndo-*, e.g., *i-ndo-vunda* 'it is rotten' (*-vunda* is a stative verb) vs. *Masudi ndo-andoka* 'Masudi has gone out' (*-andoka* is a dynamic verb). 80% of the languages examined seem to have a PRF. This figure should be treated with care, partly because it is not always possible to distinguish PRF and (near) past. PRF is primarily expressed by reflexes of the Proto-Bantu final **-ide*, i.e., *-ile, -ele, -ire, -ie*, etc. (cf. Bastin 1983), which suggests this is a category with a long Bantu history. From the definition and the facts it becomes clear that PRF easily shades over into stative or past (in French up to the late 18th century 'I saw' was rendered by *je vis* PST, today replaced in speech by *j'ai vu* PRF). It becomes past by loosening, and then losing, the requirement that it have present relevance. Our analysis suggests that since the kinds of past event that have present relevance are often recent events, PRF first becomes near or middle past, then perhaps associated with all pasts. Many languages have reflexes of **-ide* associated with some form of past. On the other hand, it can become purely stative by losing the requirement that some past event have led to the present state. There are fewer examples of languages that have gone or are going in this direction and there is then a possibility that it will become completely associated with a small set of stative verbs, such as 'lie,' 'sleep,' 'sit,' 'be,' 'stand' and 'know,' and be no longer productive, as in Shona S10, Yeyi R41 and some Chaga E60 varieties. Our data suggests that where **-ide* has been replaced as a marker of PRF, it is most often by grammaticalised forms of verbs meaning 'finish.'

PROG is a focused imperfective, which narrows attention to the temporal space around the time of reference or speaking, and for a short preceding period, so contrasting with general IPFV. As such, it is incompatible with stative verbs, whose emphasis is permanent state, e.g., *I am knowing that $2 + 2 = 4$. The class of stative verbs has a fairly common core across languages, but there is some cross-linguistic variation. As noted above, PROG is often hard to distinguish in the available data from similar categories, such as CONTINUOUS. Whereas in English 'we were fishing' is both PROG ('. . . when you saw us') and CONTINUOUS ('. . . all week'), many Bantu languages may contrast these two, so when the data source gives a single translation into English, without context, it is impossible to know which category is intended. So, although we cannot state with certainty that all Bantu languages have this category, it appears to be very common. 63% of the database languages were analysed as having a PROG. It is so often and widely renewed that it is hard to say whether older Bantu might have expressed it by a single inflected form. Nowadays it is mainly expressed by grammaticalised forms that visibly derive from 'be' or 'have' plus locative or verbal noun ('be in, be at, be with, have, etc.') (cf. Bastin 1989a, Heine *et al.* 1993). In most cases LOC and INF (verbal noun) are represented by *ku-*. Proto-Bantu had a class 15, mostly the INF, marked by *ku-*, and a class 17, marked by *ku-*. In most languages, the two are segmentally, and in most instances tonally, identical. It is therefore likely that in many languages in the database both could be labelled identically. Grammaticalisation theory suggests that most are in fact LOC (Heine *et al.* 1993). Note what is called the INF in English has the shape *to X* (*to go*), where *to* is a LOC. Bantu has several verbs that translate 'be' and the one that figures most often in PROG is the one that indicates temporary being

(*-li, -ri*): 'be in/at a particular place' becomes 'be at a particular time or in a particular situation (or both).' So present PROG forms that involve *-li/-ri* 'be' plus *-ku-/-i-* 'verbal noun,' often with an overt locative, are common.

(17) Lwena K14 (Yukawa 1987a)

    a ngu-li   (na)   ku-tángis-a

        SP$_{1sg}$-be   (with)   NP$_{15}$-teach-INF

        'I am (with) teaching.'

    b ngu-ná-p-ú   (na)   ku-tángis-a

        SP$_{1sg}$-PST-be-PST (with)   NP$_{15}$-teach-INF

        'I was (with) teaching.'

(18) Luba L33 (Beckett 1951)

    ngi-di   mu-ku-dim-a

    SP$_{1sg}$-be   NP$_{18}$-NP$_{15}$-cultivate-INF

    'I am (in) cultivating.'

(19) Umbundu R11 (Schadeberg 1990)

    a ha-ví-lí   l(a)   ó-ku-kolà

        NEG-SP$_8$-be   with   AUG-NP$_{15}$-grow-INF

        'They are not growing.'

    b ó-kasí   l(a)   ó-ku-yw-á

        SP$_1$-be   with   AUG-NP$_{15}$-bathe-INF

        'She is bathing.'

    c nd-a-kála   l(a)   ó-ku-túng-a

        SP$_{1sg}$-PST-be   with   AUG-NP$_{15}$-build-INF

        'I was building.'

PERSISTIVE (PER), occurring in at least 50% of the core languages, affirms that a situation has held continuously since an explicit or implicit point in the past up to the time of speaking ('be still Xing'). SITUATIVE (SIT), occurring only in 11% of the core languages and only in Eastern Bantu, suggests that a situation is open-ended and could continue for a long time. Mentioned in Doke (1935), it was taken up again of late in Hewson (2007). PER and SIT are discussed in Nurse (2008: 145–149) and PER also in Nurse (2003: 99).

HABITUAL (HAB) also occurs widely, i.e., in 43% of the database languages. It represents a situation "characteristic of an extended period of time, so extended that the situation is viewed as a characteristic feature of a whole period" (Comrie 1976: 27). We consider ITERATIVE (ITR) here as representing a situation that is repeated, an incomplete series of complete events. This definition differs from that of authors such as Bybee *et al.* (1994: 160) and Bhat (1999: 53ff) who reserve ITR for mono-occasional repetition in contrast to repetition that happened on different occasions signaled by categories like the FREQUENTATIVE and/or HAB. While the distinction between HAB and ITR is easy enough to grasp objectively, in practice it is less clear. Some sources describe as ITR situations that others label HAB. Here we use the single label HAB, but readers should keep in mind that this may include ITR. It is very often associated with reflexes of *-a(n)g-* after the root. If limited in time reference, it is mostly to past or timeless situations, for pragmatic reasons. HAB can merge semantically with other types of HAB, apparently the common cause for its disappearance.

(20) Mbala H41 (Ndolo 1972)
      ga-loomb-aang-a   ga-gu-loomb-aang-a
      SP₁-request-HAB-PST   SP₁-HOD.PST-request-HAB-FV
      'She used to request.'   'She used to request (recently).'

Keeping in mind the caveats mentioned, aspects in the 100 core languages are distributed as follows: PFV 100%, PRF 83%, PROG 63%, IPFV 60%, PER 50%, HAB 43%.

6 TENSE

Tenses locate a situation in time relative to a reference point, which is most often – but not always – the present or time of speech. Since tenses reflect not the world but our categorisation of the world, different languages may divide the timeline up differently, resulting in a different number of tenses. In principle, a timeline can be cut at many points. Many Bantu languages – but notably not some of the widely used ones such as Swahili G42 or Shona S10 – are known for their multiple time divisions, leading analysts to comment that Bantu had the richest set of contrasts in their sample. Numbers of tense contrasts in the 100 core languages are as shown in (21).

(21)

No. tenses	No. languages with _____ pasts	No. languages with _____ futures	No. languages where total pasts = total futures
0	0	9	0
1	17	47	15
2	41	25	17
3	31	16	9
4	10	1	1
5	1	2	0

These figures should be seen as grossly and not absolutely accurate, because of uncertainty in the analysis of some languages. Since such languages are few, the overall picture is clear enough and can be taken as being typical of Bantu. It can also be seen that this table makes no mention of "present tenses." When grammars claim that a language has several presents, that is, several identical ways of referring to the same fleeting moment, we have to be sceptical. In our experience, these "presents" are usually aspects, that is, they are different representations of the time within the event. Often forms referring to "present" are not marked for time, or only minimally. Why mark for time what is obvious to the participants? Present is the default case after past and future.

Future is problematic in a different way. Past time and past tenses are generally easier to characterise than non-pasts. Because the future hasn't happened yet, and especially if it is only a short distance away, it can be an extension of the "present" and thus often represented by PROG or HAB. This is a natural semantic extension of "present." Such forms have not been counted as discrete futures in what follows. Such extensions of the present are often coeval with formally discrete near futures, differing only by factors such as degree of certainty or firmness of intent, or not at all. Firmness of intent and degree of certainty often appear in characterisations of futures. Some languages are even described as having, for instance, a near future, a far future and an uncertain future. Just because future is not only concerned with stating firm facts, it relies quite heavily on the use of modal or volitional verbs, on auxiliary verbs, of which "come" and "go" are the commonest in our sample, and on the subjunctive. Future reference is more often renewed than past, and the morphemes involved range along the grammaticalisation path from full auxiliary to post-subject inflectional morpheme with classic CV shape. Younger and older speakers of the same language sometimes differ in their choice of future alternatives.

Moreover, it should be recognised with Botne (2012: 541–542) that "[a]lthough the hodiernal/hesternal or hodiernal/prehodiernal divisions (or their future equivalents) appear to be robust crosslinguistically, perhaps even prevalent [. . .], there is evidence to suggest that this is not always the correct characterisation of the distinction. Rather, a more appropriate characterisation in many cases may be one of currently relevant [. . .] versus adjoining time units" and that "remoteness needs to be conceptualised not only in terms of prehodiernal intervals and Currently Relevant vs. Adjoining time units, but also in terms of separate, dissociated worlds or domains (Botne & Kershner 2008)" (Botne 2012: 548). In certain languages, remoteness distinctions such as hodiernal and hesternal, although used in grammars, are to be considered as secondary interpretations rather than fundamental TA categories.

With these caveats in mind, certain generalisations are possible about the data. With few exceptions, nearly all Bantu languages examined have either one, two, three or four past tenses. Those with two (41%) and three (31%) are the most numerous, followed by those with one (17%) and finally with four (10%). Multiple pasts are normal, whereas multiple futures are much less common, so it can be said that most Bantu languages have multiple past tenses, whereas under half (44%) have multiple futures. In those with three, near past normally refers to the events of today, i.e., "hodiernal," starting this morning (it is our impression that most "Bantu" days start at sunrise but this needs wider study); middle past refers either just to the events of yesterday, i.e., "hesternal," or yesterday and some few days before; and far past to events prior to those of the middle past. In languages with four pasts, the fourth most often refers to events that occurred immediately before the time of speaking/present. In those with two pasts, the near past refers either to the events of today, or today and yesterday, with the further past referring to prior events. In languages with one past, reference to recent events is also often made via the PRF. For examples of languages with two, three and four pasts, see the sketches of Rwanda JD61, Ganda JE15 and Bemba M42, respectively, in Nurse (2010). For a single past, see (4) above.

With few exceptions, all languages had one, two or three discrete future tenses. Those with one (47%) and two (25%) are the largest groups, followed by those with three (16%). With a doubtful exception (Mwera P22), the number of futures never exceeded that of pasts. General temporal divisions ("hodiernal," "crastinal") mirror those of the past, except that the semantic extension of the "present" often overlaps with the nearest future. Swahili G42, Zulu S42, Mande A46, for which see Nurse (2010), and Logooli JE41 in (22) exemplify languages with one, two, three and four futures, respectively.

(22) Logooli JE41 (Lwane 1976, Nurse 2003: 101–102)

    a kʊ-ra-gʊr-a   b na-kʊ-gʊr-ɪ
       SP$_{1PL}$-FUT-buy-FV   FUT-SP$_{1PL}$-buy-SBJV
       'We will buy.' (Near Future)   'We will buy.' (Middle Future)

c   kʊ-rika-gʊr-a
  SP$_{1PL}$-FUT-buy-FV
  'We will buy.' (Far Future)

d   kʊ-rı-gʊr-a
  SP$_{1PL}$-FUT-buy-FV
  'We will buy.' (Uncertain future)

The exceptions to the statement that most languages have between one and four discrete pasts are Mwiini G412 (Somali coast), Ilwana E701 (northeast Kenya), Comorian G44, some languages in the North-East and some Zone S (southern Africa) languages. The first three have collapsed PRF and PST and use the former PRF final to encode the combined single category. In the first two this seems to have come about under heavy influence from neighbouring Cushitic communities. It would be interesting to see similar cases from other parts of the Bantu area. The exceptions to the claim about the number of futures are some G30 languages from Tanzania, plus Yeyi R41, which have a single non-past category.

Widespread inherited markers of past outnumber those of future, although whether this reflects the original situation or results from the frequent renewal of future markers is unclear. For past reference, most languages surveyed had reflexes of -a(a)-, -ile, and -ka-, in that order of frequency. -a(a)- is so written because both long and short forms occur, often tonally contrastive, semantically different, and sometimes in languages which have otherwise neutralised the inherited vowel length distinction. This merits further examination. The Final -ile probably originally represented PRF but has become a past, or the past, marker in a significant number of languages (cf. Botne 2010, Crane 2012).

It may surprise readers that so many Bantu languages have multiple past reference, yet apparently involving so few morphemes. Any overview of Bantu tense and aspect encoding would conclude that there is in fact no contradiction. With two or three pre-stem morphemes, two Finals (-ile, neutral -a), two contrastive tones and occasional new grammaticalised morphemes, multiple tense contrasts can be achieved. The overall data strongly suggests constant flux, and constant systemic and semantic permutation.

Although -laa- and -ka- seem to be inherited markers of futurity, both appear to be geographically limited, especially -laa-, which occurs only in eastern Africa, although a morpheme with similar meaning and shape occurs in some Grassfields Bantu. The constant renewal of future reference via grammaticalisation should be kept in mind here. Reflexes of the former subjunctive -e occur in most contemporary languages and are often associated in some way with future reference (cf. Section 7). A zero pre-stem marker is the commonest indicator of 'present'/non-past/IPFV.

The literature does not much discuss whether time reference is flexible or fixed. Since neither our own data nor that in the source grammars is complete on this issue, we rely on impression rather than hard figures. The impression is based on detailed examination of a few languages, on discussions with speakers of some of those languages, and on a visible discrepancy between analytical statements and actual textual use in some descriptions. While some representations of time are quite concrete ("We saw him on January 23 at 2pm"), many are not. Some variability is factual. Thus, if your language has three pasts and you refer to something you do several times a day, every day, the three pasts will be as described above (today: yesterday and maybe a few days earlier: prior to that). But if you refer to planting, which in large parts of Africa occurs annually, then the near past can refer to the most recent planting, which could be several months earlier, the middle past to last year's planting, the far past to planting seasons before that. If you refer to acts of God, or the origins of your ethnic group, far past would have a different referent again. Some variability can also be subjective, in how the speaker sees or represents the facts. Events can be moved nearer or further using appropriate tenses (cf. Besha 1989: 288–300). It is our impression that this flexibility of reference occurs widely across Bantu.

Another category, probably best regarded as a tense and widespread in Bantu is the CONSECUTIVE (CONS). It is not discussed here, but treated in Nurse (2003: 101–102) and Nurse (2008: 120–123).

7 MOOD AND MODALITY

Mood and modality have not been the most popular topics in Bantu studies. Doke (1935: 146–147) identifies eight moods: imperative, infinitive, indicative, subjunctive, participial, potential, conditional and contingent. This classificatory system has been applied in many descriptive grammars of zone S Bantu languages (cf. also Güldemann 1997: 78), languages which have seen the highest number of studies dedicated to mood so far (e.g., Fourie 1989, Louwrens 1990, Crane 2009), although zone J languages are gaining ground (e.g., Bostoen *et al.* 2012, Mberamihigo 2012, Kawalya *et al.* 2014, Mberamihigo 2014, Mberamihigo *et al.* 2016). Doke (1935: 146) defines "modal" as well, but it refers to a very broad category including aspect. Following Meinhof (1906: 61), he characterises verbal suffixes as being "modal" in contrast to prefixes which are temporal. The idea that modality is mainly marked through suffixes, and especially the subjunctive final vowel *-e*, is prevalent within Bantu literature (cf. Nurse 2003: 91, Nurse *et al.* 2010). Such is the case in broader typological studies on mood and modality where Bantu languages are typically classified with languages like Spanish that are well known for their indicative/ subjunctive system, and not with languages like English known for their modal auxiliaries (Palmer 2001: 107–108, de Haan 2006: 33).

We adopt here a definition of mood and modality which allows for the inclusion of at least some of the moods traditionally distinguished in Bantu grammars, but which also includes modal notions that were hardly investigated in Bantu studies until very recently. Following Hengeveld (2004: 1190), we define mood as the grammatical expression of a large semantic area that can be subdivided in illocution and modality. Illocution is concerned with types of speech act and modality refers to expressions that can be characterised in terms of possibility and necessity. For the definition and delimitation of the category of modality, we mainly follow van der Auwera and Plungian (1998).

Cross-linguistically modal meanings tend to be expressed by a wide range of morphological, syntactic and lexical categories (de Haan 2006: 32–41 and confirmed by Schicho (1995: 127) for Swahili G42). For reasons of space, this section focuses on verbal forms dedicated to the expression of different types of illocution and modality.

7.1 Illocution

Under illocution, we subsume optatives and directive speech acts, i.e., second-person imperatives, cohortatives, jussives and prohibitives. We do not include interrogatives here as they tend to be expressed by question words, particles or intonation and not by verbal elements. The above illocutionary use categories are typically expressed by only two form categories: the Imperative and the Subjunctive. As stated in the above, we cannot consider the full range of expressions of illocution and modality. Non-verbal means of conveying illocution and modality, like the optative conjunctive in (23), are therefore not discussed below. Similarly, the use of dedicated TA-markers for illocutionary purposes, like the directive use of the future tense in (24), though interesting, also goes beyond the scope of this overview chapter.

(23) Zulu S42 (Doke & Vilakazi 1948: 5, Fourie 1989: 53)
     akwaba   i-zulu   li-zo-dum-a
     would.that NP$_5$-weather   SP$_5$-FUT-thunder-FV
     'One could wish that it would thunder.'

(24) Swahili G42 (Schicho 1995: 155)
     kesho   u-ta-andik-a   sentensi N-zima   kwa   u-fasaha,
     tomorrow   SP$_{2SG}$-FUT-write-FV NP$_{10}$.sentence NP$_{10}$-wholefor   NP$_{11}$-correct
     mara   kumi.   U-na-elew-a?
     NP$_{10}$.time ten   SP$_{2SG}$-PRS-understand-FV
     'Write the whole sentence correctly ten times by tomorrow. Do you understand?'

Both the Imperative and the Subjunctive are widely attested in Bantu languages. What is more, PB reconstructions have been suggested for them (Meeussen 1962, 1969, 2014). The Bantu Imperative consists of the verbal base with a final suffix -*a*. As reconstructed by Meeussen (1962: 74, 1967: 112, 2014: 35), the verb root has its own lexical tone and the final suffix has a high tone, while the extensions have a polar tone, i.e., a tone opposite to the verb root's lexical tone. North-Binja D24 has retained this tonal pattern.

(25) North-Binja D24 (Meeussen 1962: 70, 2014: 31)
     lindík-á 'arrange!'   síndik-á 'push!'

Devos and Van Olmen (2013: 10) show that 97 out of a sample of 100 Bantu languages, an adapted version of Nurse's 2008 sample, use the morphologically specialised Imperative for the singular. Plural Imperatives typically include a plural addressee suffix, which is reconstructed as **Vni* for PB and possibly derives from a second person plural pronoun (Van de Velde & van der Auwera 2010: 137).

(26) Orungu B11b (Ambouroue 2007: 244)
     ỳòl-à-ánì 'buy (pl.)!'

The Imperative is thus a well-established and stable verb form throughout Bantu. However, it may be modified, i.e., reinforced or mitigated, a processes called "illocutionary modification" by Hengeveld (2004: 1192). In some exceptional cases, the modified forms are neutralised and become the regular Imperative form; for a preliminary overview, see Devos and Van Olmen (2013: 10–15). Imperative forms which obligatorily include a reflex of the imperfective extension **-ag-/*-ang-* are cases in point. Meeussen (1962: 70, 2014: 30–31) identifies three such languages, one of which is Kele C55 as shown in (27).

(27) Kele C55 (Carrington 1943: 205)
     kel-áká 'do!'

Interestingly, an origin in a more polite and a more emphatic Imperative are equally conceivable. In the first scenario, the development goes from a plural addressee marker, through an honorific/politeness marker to an imperative marker (Van de Velde & van der Auwera 2010: 135–136, Devos & Van Olmen 2013: 10–13). In the second scenario, an originally more emphatic form linked to the intensive or imperfective sense of the pluractional neutralises to become the regular Imperative (Van de Velde & van der Auwera 2010: 136, footnote 19).

The reconstructed Proto-Bantu Subjunctive has a high-toned subject marker and a final high tone suffix -e while all syllables in between are low (Meeussen 1962: 74, 1967: 112, 2014: 35). Lega D25 has retained this tonal pattern.

(28) Lega D25 (Meeussen 1962: 60, 2014: 20)
      túbuluté 'Let's pull.'   túkụbulé 'Let's pour.'

All languages in the Devos and Van Olmen (2013) sample have a form that can be considered a Subjunctive. It is important to note that the final suffix -e is not a sufficient characteristic of Bantu Subjunctives. First, most languages in Zone B, many in Zone A, and several in Zone C, D and H do not have the Subjunctive final suffix -e. They do have forms with final -a, which are reminiscent of Subjunctives in -e from a tonal and functional perspective (Nurse & Philippson 2006: 179). Next, a final suffix -e does not automatically imply a Subjunctive form (Güldemann 1996: 153–158, Van de Velde & van der Auwera 2010: 134).

Apart from expressing the wish of the speaker as an optative, as in (29), the Subjunctive is often described as a polite equivalent of the Imperative (see also Nurse 2008: 28).

(29) Shangaji P312
      t'   ii-fííy-e   y-iíta   ki-ń-ráfun-e   manttúwi   ń-wiíxi
      COP SP_9-arrive-SBJV NP_9-rain.season SP_{1SG}-OP_1-chew-SBJV NP_{1A}.peanut NP_1-raw
      'May the end of the rainy season arrive, so I can eat raw peanuts.'

However, it also stands in a double complementary distribution with the Imperative. First, it fills the gaps in the "imperative/hortative paradigm" (Xrakovskij 2001). Whereas the morphologically specialised Imperative is dedicated to the expression of second person imperatives, the Subjunctive is used for cohortatives (28) and jussives (30). In addition, the negative Subjunctive serves as a prohibitive in many languages (31).

(30) Swahili G42 (Sacleux 1909: 238)
      haya!   ni-anz-e   kazi   y-angu   sasa
      come.on SP_{1SG}-begin-SBJV   work PP_9-POSS$_{1SG}$ now
      'Come on! Let me start my work now!'

(31) Shambaa G23 (Besha 1989: 71)
      ú-she-dik-é
      SP_{2SG}-NEG-cook-SBJV
      'Do not cook!'

Second, the Subjunctive very frequently serves as an "unmarked" second person imperative in the case of plural directives and, especially, directives with object prefixes. As is argued in Devos and Van Olmen (2013: 17–22), its use in the latter context is probably at the origin of hybrid subjunctive-imperative forms, like the one in (32), which has the Subjunctive final suffix -*e*, but lacks a subject prefix, as is typical for the Imperative. In Subiya K42, the hybrid form ended up replacing the deeply entrenched Bantu Imperative (33).

(32) Ha JD66 (Harjula 2004: 88–89)
      mu-bwiir-e
      OP₁-tell-SBJV
      'Tell him!'

(33) Subiya K42 (Nurse 2010)
      nyw-ĕ   menzi
      drink-SBJV   NP₆.water
      'Drink water!'

Although many Bantu languages thus make use of two forms for a series of speech acts, some have a more differentiated set of illocutionary forms. Lega D25 with separate "Hortative" and "Optative" forms is a case in point (34).

(34) Lega D25 (Botne 2003: 443–444)
      a   kangúl-á   į-swá   'Clear the field!' [Imperative]
          clear-IMP   NP₅-field

      b   tú-kangul-é   į-swá   'Let's clear the field!' [Subjunctive]
          SP₁ₚₗ-clear-SBJV   NP₅-field

      c   tw-a-kangul-a   į-swá   'We should clear the field!' [Hortative]
          SP₁ₚₗ-FUT-clear-FV   NP₅-field

      d   tw-a-nŭ-kangul-ag-a   į-swá   'If only we could clear the field!' [Optative]
          SP₁ₚₗ-REM.PST-SEQ-clear-FV   NP₅-field

Likewise, many Bantu languages express prohibition through a form different from the negative Subjunctive. Devos and Van Olmen (2013: 24–43) show that 29 languages of their sample make use of the negative Subjunctive, either exclusively or along with other strategies, for the expression of prohibition. The other strategies include – in order of decreasing frequency – periphrastic constructions, dedicated prohibitive markers, negative infinitives, negative indicatives and negative imperatives.

7.2 Modality

Following van der Auwera and Plungian (1998), we explain modality as the expression of possibility and necessity in four domains: 1) participant-internal (or dynamic) modality, 2) participant-external modality, 3) deontic modality and 4) epistemic modality. Illustrations of all types are given in (35) to (42). Category 3 is seen as subset of category 2 by van der Auwera and Plungian (1998).

(35) Eton A71 (Van de Velde 2008: 341)
      à-nè   kwàm   d-âm   lé-sê
      SP₁-can(<exist)   INF.do   NP₅-thing   PP₅-every
      'He's capable of everything.' (participant-internal possibility)

(36) Changana S53 (E. Nhampoca personal communication)

ndzi-síndzís-ék-á	ku-y-a	mahósí	svósvi
SP$_{1sg}$-force-STAT-PRS	NP$_{15}$-go-INF	behind	now

'I need to go to the toilet now.' (participant-internal necessity)

(37) Rundi JD62 (Bostoen *et al.* 2012: 16)

n-ra-shóbor-a	ku-siinziir-a	ha-ri-hó	i-ki-tǎnda
SP$_{1sg}$-PRS.DJ-can-FV	NP$_{15}$-sleep-FV	SP$_{16}$-be-LOC$_{16}$	AUG$_7$-NP$_7$-bed

'I can sleep (because) there is a bed.' (non-deontic participant-external possibility)

(38) Shangaji P312 (M. Devos field notes)

oóntú	oówél-el-e	ma-xaála	ee-vulúw-á
DEM$_1$	SP$_1$.climb-APPL-PFV	NP$_6$-unripe.coconut	SP$_1$.SUBS-fall-FV
m-mú-náazi		e-n-ón-á	o-m-óóngolééla
NP$_{18}$-NP$_3$-coconut.tree		SP$_9$-PRS-must(<see)-FV	NP$_{15}$-OP$_1$-massage-INF
n'	o-ń-khánttel-a-khánttéela		
and	NP$_{15}$-OP$_1$-rub-INF-RED		

'This one climbed after young coconuts and fell out of the coconut tree. S/He has to be massaged and rubbed with oil.' (non-deontic participant-external necessity)

(39) Zulu S42 (Poulos & Msimang 1998: 284)

u-nga-ngen-a
SP$_{2sg}$-POT-enter-FV
'You may enter.' (deontic possibility)

(40) Ngoni N12 (Ngonyani 2013: 7)

n-dut-e
SP$_{1sg}$-go-SBJV
'I should go.' (deontic necessity)

(41) Fuliiru JD63 (Van Otterloo 2011: 258)

à-àngà-b-à	y-é-w-à-yàbíír-á	yìzó	fwárángà
SP$_1$-POT-be-FV	PP$_1$-FOC-SP$_1$-PST$_1$-take-FV	DEM$_{10}$	money

'He might be the one who took that money.' (epistemic possibility)

(42) Mongo C61 (Hulstaert 1965: 370)

b-ïfo-tsingol-a	ɔlɔ́kɔ
SP$_2$-must-explain-FV	tomorrow

'They will certainly explain tomorrow.' (epistemic necessity)

In the remainder of this section we look at inflected verb forms, as in (39)-(41), and auxiliary constructions, as in (35), (37), (38) and (42), expressing modal notions. For reasons of space, we do not discuss other means of expressing modality, as there are modal adverbs, such as *ngirango* 'maybe' in (43), the modal use of derivational affixes, such as neuter -*ik*- in (44), and modal lexical verbs, such as -*sindziseka* 'be necessary, need' in (35). In Changana S53, more or less the same meaning can be expressed by at least two other lexical modal verbs, i.e., -*kombela* 'ask, beg, be necessary' and -*fanela* 'be appropriate, be necessary.' For an inventory and discussion of lexical modal verbs in Rundi JD62 and Swahili G42, see Mberamihigo (2014) and Schicho (1995: 140–146) respectively.

(43) Rundi JD62 (Mberamihigo *et al.* 2016: 252)
      tweebwé   tu-ri   a-ba-áarimú   tu-ø-ígiish-a   mu
      we SP_{1PL}-be AUG_2-NP_2-professor SP_{1PL}- PRS-teach-IPFV LOC_{18}
      i-ø-shuúre   ø-kaminúuza   i   Bujuumbura
      AUG_5-NP_5.school NP_9.university LOC_{19} Bujumbura
      ngirango   mu-ra-mar-ye   ku-ha-úmv-a
      maybe SP_{2PL}-DJ-finish-PFV NP_{15}-OP_{16}-hear-INF
      'We, we are professors, we teach at the University of Bujumbura, maybe you have already heard about it.' (epistemic possibility)

(44) Swahili G42 (Seidl & Dimitriadis 2003: 254, cited in Dom 2014)
      ki-tanda   hiki   ki-na-lal-ik-a   vizuri
      NP_7-bed DEM_7 SP_7-PRS-sleep-NEUT-FV   well
      'This bed sleeps well.' (participant-internal possibility)

7.2.1 Modality through verbal inflection

For reasons of space, we only discuss the most recurrent affirmative inflections here. As suggested by the examples in (35)–(42), the Subjunctive and the Potential/Conditional, including a reflex of *ngá*, labelled 'potential' by Guthrie (1971: 145) and 'conditional' by Meeussen (1967: 109), are the most recurrent inflectional means of expressing modality in Bantu languages. The Subjunctive is widespread in Bantu and, as mentioned before, has been reconstructed for Proto-Bantu. Still, as far as we know, there is no comparative study of its meaning and use throughout the Bantu domain. It is generally accepted that the Subjunctive has subordinate as well as main clause uses. It is not clear, however, whether it is in origin a subordinate tense which acquired independent uses through "insubordination" (cf. Evans 2007) or whether it started out as a kind of optative which developed harmonic subordinate uses. In grammars of individual languages, one either gets a list of different uses or a rather vague encompassing description often including a reference to "non-factuality." Some efforts have been made to arrive at a coherent description of the Subjunctive in individual languages. Givón (1971, 1994), referring mainly to Bemba M42, argues that the Subjunctive is a dependent tense dedicated to the expression of "coercion." Ngonyani (2013) claims that the Subjunctive in Kisi G67, Ndendeule N101 and Ngoni N12 relates non-facts expressed through deontic modality (weaker manipulation) and epistemic possibility. Leonard (1980) identifies "high likelihood of occurrence" as the unitary meaning of the Subjunctive in Swahili G42. It is our impression that the Subjunctive is mainly used for the expression of participant-external necessity implying weaker, often subjective, obligation, as in (38). Examples of Subjunctives relating permissive meanings, i.e., participant-external possibility, are mainly found in questions, as in (45).

(45) Shangaji P312 (M. Devos field notes)
      ki-viír-e   n-nyúumpa   ttíimpho
      SP_{1SG}-pass-SBJV NP_{18}-NP_9.house   EMPH.DEM_{18}
      'May I enter into the house itself?'

The expression of epistemic possibility through the use of a Subjunctive, as reported by Ngonyani (2013), appears to be rare. He describes the use of a Subjunctive after an adverb expressing 'maybe, probably' in Ndendeule N101, but in Swahili G42, for example, the adverb *huenda* 'maybe' is not followed by an affirmative Subjunctive (Devos & de Schryver 2013).

(46) Ndendeule N101 (Ngonyani 2013: 11)
      pangi   a-hik-e
      probably   SP$_1$-arrive-SBJV
      'Perchance she/he might come.'

(47) Swahili G42 (Devos & de Schryver 2013)
      huenda   ma-tatizo   y-ao   ya-ta-pungua   a-ki-anza   ku-pata
      maybe   NP$_6$-problem PP$_6$-POSS$_2$ SP$_6$-FUT-diminish SP$_1$-sit-begin NP$_{15}$-get
      m-shahara
      NP$_3$-wages
      'Maybe their problems will grow less once he starts to earn wages.'

Non-modal extensions according to the model of van der Auwera and Plungian (1998) occur in several languages. The best-known one is the future. Typically, it is not the Subjunctive final alone but rather the combination with future T(A) markers which expresses the future meaning. Different types of future, including Near Futures, as in (48), Remote Futures, as in (9) and Imminent futures, as in (49), can take final -*e*. Modal overtones are obvious with uncertain Near Futures of Great Lakes Bantu languages (Nurse & Muzale 1999: 528–530) and with Remote Futures which refer to events that will take place at an imprecise future time. A modal sense is less clear with Imminent Futures.

(48) Ruri JE253 (Nurse & Muzale 1999: 530)
      ci-la-gur-e
      SP$_{1PL}$-FUT-buy-SBJV
      'We will buy (near future, not certain).'

(49) Kaonde L42 (Foster 1960: 32)
      ke   tú-j-e
      IMM   SP$_{1PL}$-come-SBJV
      'We are about to eat.'

Subjunctives are also recurrently attested in the protasis of conditional clauses, be they what Timberlake (2007: 322) calls "general," as in (50), "potential," as in (51), or "counterfactual," as in (52). In Lingala C36d, the suffix -*aka* is an alternative for the suffix -*a* of the Subjunctive mostly denoting speaker's insistence. In the protasis of counterfactual conditional sentences, it appears to be the preferred alternative (Meeuwis 2010: 118, 189).

(50) Sukuma F21 (Batibo 1985: 275)
      a-laa-bok-ag-e   w-aa-gw-á
      SP$_1$-FUT-get.up-IPFV-SBJV   SP$_1$-PST-fall-FV
      'Each time he got up, he fell.'

(51) Totela K41 (Crane 2011: 333)
      ési   tù-ly-é   àhúlù,   tù-lékùt-à
      if   SP$_{1PL}$-eat-SBJV   a.lot   SP$_{1PL}$-NONCMPL.become.full-FV
      'If we eat a lot, we('ll) get full.'

(52) Lingala C36d (Meeuwis 2010: 189)
      sókí   ná-kéb-aka   té   ndé   na-síl-ís-ákí   mbóngo   nyónso
      if   SP$_{1SG}$.SBJV-beware-FV   NEG   so   SP$_{1SG}$-end-CAUS-PST$_1$ money   all
      'If I hadn't been careful, I would have lost all my money.'

Another non-modal extension is the use of the Subjunctive in utterances expressing that
something is expected to happen, but has not for the moment. In Nkoya L62, for example,
shílóló followed by the Negative Subjunctive expresses 'not yet.'

(53) Nkoya L62 (Yukawa 1987b: 154)
      shílóló   ni-ta-lŭsh-e
      yet   SP$_{1SG}$-NEG-explain-SBJV
      'I have not explained yet.'

So far, the feature linking the illocutionary (cf. Section 7.1), modal and extra-modal uses
of the Subjunctive quite straightforwardly is that they all refer to events that are not
realised yet, which brings to mind the notion of "irrealis." However, other more factual
extra-modal uses are attested as well. The Subjunctive, or at least the subjunctive final,
can mark Narratives, as in (54), Past Imperfectives, as in (55) and subordinate clauses of
anterior or simultaneous taxis, as in (56).

(54) Bena G63 (Morrison 2011: 278)
      a-gon-áge
      SP$_1$-sleep-NARR
      'Then s/he slept.'

(55) Ndengeleko P11 (Ström 2013: 256)
      aándík-age   balua
      SP$_1$.write-PST.IPFV   NP$_{10}$.letter
      'He was writing letters.'

(56) Makwe G402 (Devos 2008b: 337–338)
      n-ní-uúm-a   panda   ti-ka-úum-e   panda   ti-ní-wíin-a
      SP$_{1SG}$-PFV-leave-DJ   NP$_{16}$.out   SP$_{1PL}$-CONS-leave-SBJV   NP$_{16}$.out   SP$_{1PL}$-PFV-dance-dj
      'I went outside. . . . When going outside/having gone outside, we danced.'

Güldemann (1996: 154–158), examining the latter use in some M60 languages, argues
convincingly that not the subjunctive final but rather an anterior final -*e* is involved, such
as the second part of the Perfective final **-ide* (Voeltz 1980) or the so-called deverbative nominal suffix -*e* (Bastin 1989b). The use of the Subjunctive after *tangu* 'since' in
Swahili (G42) might be another case in point, e.g., *tangu nifike* 'since I arrived.' Miehe

(1979: 191–192) suggests that rather than a Subjunctive, a remnant of an Old Perfect is involved. Still, more detailed investigations are definitely needed, seeing that the phenomenon is not restricted to M60 languages. More examples are found in Mongo C61 (Hulstaert 1965: 701–702), Kagulu G12 (Petzell 2008: 115) and Kaonde L42 (Wright 2007: 34). Exactly the same tail-head sequence, as in (56), is in other eastern Bantu languages marked by the "narrative/imperfective" (cf. Nurse & Hinnebusch 1993: 365) marker -ka- and a final -a. Fuliiru JD63 is a case in point (Van Otterloo 2011: 230). An affinity between narrative and subjunctive tenses has been recognised in the literature; see, for example, Carlson (1992) for a number of African languages and Leonard (1980) and Seidel (2008: 326–328) for Swahili G42 and Yeyi R41 respectively. Further in-depth research and cross-linguistic comparisons will make clear whether and how all these uses are connected and whether a vague term like "irrealis" could be significant for Bantu (Bybee et al. 1994: 236–240, van der Auwera & Schalley 2004: 92–93, de Haan 2006: 41–45).

The Potential/Conditional is less widespread than the Subjunctive. It occurs in 29% of the core languages (Nurse 2008: 251). Nurse and Philippson (2006: 193–194) and Nurse (2008: 251–252) argue that the conditional and potential meanings are transparently derived from a particle *nga* expressing 'like, as' for which they propose a grammaticalisation path going from a particle to a clitic, to a prefix, and eventually to a T(A) marker. Whether all these stages are necessary or whether these represent all the necessary stages is as yet unclear. Moreover, a number of north-western Bantu languages have -*nga*- forms with past or future meanings, as in Ewondo A72 *mə-ngá-dí* (SP_{1SG}-PST_3-eat) 'I ate' (Nurse 2010), but these forms might be unrelated (Nurse 2008: 252, footnote 240). Northern Swahili dialects show interesting variation with regard to *nga*. It can be used as a particle (57a), as a defective verb (57b-c), as a concessive marker (57d) and as a marker of epistemic possibility and necessity with the copula -*li* (57e-f). The latter use possibly gave rise to its use in counterfactual conditional clauses marked by -*ngali*- (57g).

(57) Swahili G42 (Sacleux 1939: 677, Miehe 1979: 238–247)

a	*si kwamba nga maskini*	'It is not as if they were poor.'
b	ni-nga watu sao	'I resemble those people.'
c	a-ki-nga mwenye wazimu	'He was like a maniac.'
d	ni-nga-w-a na nyingi thweka	'Although I had many burdens. . .'
e	n-na-zani a-nga-li nyumba-ni	'I think he is at home.'
f	a-nga-li m-zima	'He must be well.'
g	u-ngali-ku-dy-a dyana u-ngali-m-kut-a	'If you would have come yesterday, you would have met him'

Both the desyntactisation of *nga* and its concomitant semantic evolution thus need further investigation. When used as a T(A) marker *nga* can express possibility in all four modal domains. Examples of deontic and epistemic possibility have been presented in (39) and (41) respectively. The remaining types are shown in (58) and (59). The Potential/Conditional is also attested in the protasis of conditional clauses with or without a conditional conjunction (60). In languages such as Kuria JE43, Zulu S42 and Northern Sotho S32, the Potential/Conditional is used to express all the modal meanings as well as the conditional meaning. For other languages only the modal, only the conditional or a combination of some modal meanings and the conditional have been reported. In Venda S21, where *nga-* may convey participant-internal possibility (58), this meaning is preferably expressed through the use of the auxiliary *-kon-* 'be able to' (Poulos 1990: 277). The apparent loss of ground of the Potential in favour of auxiliary constructions for the expression of participant-internal possibility is attested in several zone S Bantu languages.

(58)   Venda S21 (Poulos 1990: 277)
    a-nga-bambel-a   naa?
    SP$_1$-POT-swim-FV   inter
    Can he swim? (participant-internal possibility)

(59)   Safwa M25 (Voorhoeve n.d.: 45)
    ba-ga-hw-iíb-il-a   pásiku
    SP$_2$-POT-OP$_{2sg}$-steal-appl-FV   at.night
    'They can steal from you at night.' [non-deontic participant-external possibility]

(60)   Kagulu G12 (Petzell 2008: 115)
    u-ng'hej-a   ko-ni-fik-a
    SP$_{2sg}$-COND.come-FV   SP$_{2sg}$.FUT-OP$_{1sg}$-find-FV
    'If you come, you will find me.'

More research is needed to establish whether this variation in meaning and use reflects a cross-linguistically well-attested evolution from a modal marker of possibility to a conditional marker (van der Auwera & Plungian 1998: 91–93). Exactly this path is hypothesised for the potential/conditional marker *-oo-* in Rundi JD62, whose etymology is still unknown and whose distribution is restricted to a couple of very closely related languages (Mberamihigo 2014: 197–261). However, Kawalya *et al.* (2018) claim a reverse pathway for the Ganda JE15 verbal prefix *-andi-* which they consider to be an "irrealis" marker. It would have been primarily a conditional marker that only subsequently developed different modal meanings.

7.2.2 Modal auxiliary constructions

In line with Anderson (2006: 5), we define modal auxiliaries as elements "that in combination with a lexical verb form a mono-clausal verb phrase with some degree of (lexical) semantic bleaching that performs" a modal function. In his list of common source-target semantics, Anderson (2011: 51–52) only provides one modal auxiliary derived from a source meaning 'come.' Nevertheless, grammars of Bantu languages make clear that auxiliaries play an important role in the expression of modality. In the examples (35)–(42), four out of the eight modal types involve the use of modal auxiliaries. What is more, modal auxiliaries can convey the other types as well, as shown in (61)–(64). Whenever an auxiliary has a known lexical meaning, we add the lexical semantics to the grammatical gloss. The etymology of Kongo ya Leta H10A *lenda* in (62) is not straightforward, but this auxiliary is widespread withing the Kikongo Language Cluster (see for instance Dom 2013: 47, 82, 96, De Kind 2014: 97, De Kind *et al.* 2015: 147). Bentley (1887: 694) translates *-lenda* as 'to possess, to have power' and reports it as used for the expression of participant-internal and possibly also deontic possibility.

(61) Shangaji P312 (Devos 2008a: 19)
      ki-ná   y'   oo-ráfúun-a
      SP₁ₛ₉-need(<have)   CONN₉   NP₁₅-chew-INF
      'I need to eat.' (participant-internal necessity)

(62) Kongo ya Leta H10A (Fehderau 1962: 86)
      yandi   lenda   (ku)-sal-a   sesepi
      he   can   (NP₁₅)-work-FV   now
      'He may work now.' (deontic possibility)

(63) Saamia JE34 (Botne et al. 2006: 62)
      o-xóy-eré   o-many'-é   óó-ler-á   ó-mw-ana
      SP₂ₛ₉-ought-PFV SP₂ₛ₉-know-SBJV   NP₁₅-raise-FV   AUG₁-NP₁-child
      'You ought to know to take care for a child.' (deontic necessity)

(64) Kanyok L32 (Mukash Kalel 1982: 287)
      w-aa-zàb   w-àà-tùm
      SP₂ₛ₉-CONS-might(<play)   SP₁-CONS-send
      'He might send.' (epistemic possibility)

Grammars often give very little information on the etymology of the auxiliary, the modal domains covered by it and the possible extra-modal uses. Recently, innovative research has been carried out on auxiliaries in Rundi JD62 and Ganda JE15, identifying all modal and non-modal uses of a given auxiliary through corpus-driven research and tracing the semantic evolution with the help of a diachronic corpus. The auxiliary -shobor- in (37) is a case in point. It is used in Rundi JD62 for the expression of possibility in the whole modal domain (Bostoen *et al.* 2012, Mberamihigo 2014: 80–92). The modal uses of the cognate form -sóból- in Ganda JE15 are less diversified, covering mainly participant-internal and external possibility. The diachronic corpus analysis suggests an increasing "subjectification" in the use of the auxiliary -sóból- with the participant-external meanings gaining ground over time on the lexical and participant-internal meanings (Kawalya *et al.* 2014). Following Traugott (2003), subjectification refers to an increasing tendency to express the speaker's subjective perspective on the situation.

7.3 Summary

In Section 7 we have tried to give a more diversified picture of mood and modality in Bantu languages. The Imperative and the Subjunctive are the main grammatical means of expressing basic illocutions. The Subjunctive is also involved in the expression of modality but covers only a limited range of modal meanings mostly centered on participant-external and deontic modality. The Potential/Conditional is known to express possibility in all four modal domains and modal auxiliaries do not only fill the remaining gaps but may denote any modal type. Many lacunae in our knowledge of these means of expressing illocution and modality have been indicated in the main text and call for further dedicated research.

REFERENCES

Aksenova, I. S. (1997) *Kategorii vida, vremeni i naklonenija v jazykax bantu [Categories of aspect, tense and mood in Bantu languages]*. Moscow: Nauka.

Ambouroue, O. (2007) *Eléments de description de l'orungu, langue bantu du Gabon (B11b)*. Bruxelles: Université Libre de Bruxelles, thèse de doctorat.

Anderson, G. D. S. (2006) *Auxiliary Verb Constructions*. Oxford: Oxford University Press.

Anderson, G. D. S. (2011) Auxiliary Verb Constructions in the Languages of Africa. *Studies in African Linguistics* 1–2: 1–409.

Bastin, Y. (1983) *La finale verbale -ide et l'imbrication en bantou*. Tervuren: Musée royal de l'Afrique centrale.

Bastin, Y. (1989a) El prefijo locativo de la clase 18 y la expresión del progresivo presente en Bantu. *Estudios africanos* 4: 35–55; 61–86.

Bastin, Y. (1989b) Les déverbatifs bantous en -e. *Journal of African Languages and Linguistics* 11: 151–174.

Batibo, H. M. (1985) *Le kesukuma (langue bantoue de Tanzanie): phonologie, morphologie*. Paris: Centre de Recherches, d'Echanges et de Documentation Universitaire.

Beavon, K. (1991) Koozime Verbal Structure. In Anderson, S. C. & B. Comrie (eds.), *Tense and Aspect in Eight Languages of Cameroon*, 47–104. Dallas: Summer Institute of Linguistics; University of Texas at Arlington.

Beckett, H. W. (1951) *Handbook of Kiluba*. Mulongo: Garenganze Evangelical Mission.

Bentley, W. H. (1887) *Dictionary and Grammar of the Kongo Language as Spoken at San Salvador, the Ancient Capital of the Old Kongo Empire, West Africa*. London: Baptist Missionary Society and Trübner & Co.

Besha, R. M. (1989) *A Study of Tense and Aspect in Kishambala*. Berlin: Dietrich Reimer.

Bhat, D. N. S. (1999) *The Prominence of Tense, Aspect, and Mood*. Amsterdam; Philadelphia: John Benjamins.

Bochnac, M. R. & L. Matthewson (eds.) (2015) *Methodologies in Semantic Fieldwork*. Oxford: Oxford University Press.

Bostoen, K., F. Mberamihigo & G.-M. de Schryver (2012) Grammaticalization and Subjectification in the Semantic Domain of Possibility in Kirundi. *Africana Linguistica* 18: 5–40.

Botne, R. (1998) The Evolution of Future Tenses from Serial "Say" Constructions in Central Eastern Bantu. *Diachronica* 15(2): 207–230.

Botne, R. (2010) Perfectives and Perfects and Pasts, Oh My!: On the Semantics of -ILE in Bantu. *Africana Linguistica* 16: 31–64.

Botne, R. (2012) Remoteness Distinctions. In Binnick, R. I. (ed.), *The Oxford Handbook of Tense and Aspect*, 536–562. Oxford: Oxford University Press.

Botne, R. & T. L. Kershner (2008) Tense and Cognitive Systems: On the Organization of Tense/Aspect Systems in Bantu Languages and Beyond. *Cognitive Linguistics* 19: 145–218.

Botne, R., H. Ochwada & M. Marlo (2006) *A Grammatical Sketch of the Lusaamia Verb*. Cologne: Rüdiger Köppe.

Botne, R. D. O. (2003) Lega. In Nurse, D. & G. Philippson (eds.), *The Bantu Languages*, 422–449. London; New York: Routledge.

Bybee, J. L., R. D. Perkins & W. Pagliuca (1994) *The Evolution of Grammar: Tense, Aspect, and Modality in the Languages of the World*. Chicago: University of Chicago Press.

Carlson, R. J. (1992) Narrative, Subjunctive and Finiteness. *Journal of African Languages and Linguistics* 13(1): 59–85.

Carrington, J. F. (1943) The Tonal Structure of Kele (Lokele). *African Studies* 2: 193–209.

Comrie, B. (1976) *Aspect: An Introduction to the Study of Verbal Aspect and Related Problems*. Cambridge: Cambridge University Press.

Comrie, B. (1985) *Tense*. Cambridge: Cambridge University Press.

Crane, T. M. (2009) Tense, Aspect, Mood, and Tone in Shekgalagari. *UC Berkeley Phonology Lab Annual Report* 2009: 224–278.

Crane, T. M. (2011) *Beyond Time: Temporal and Extra-temporal Functions of Tense and Aspect Marking in Totela, a Bantu Language of Zambia*. Berkeley: University of California, PhD dissertation.

Crane, T. M. (2012) -ile and the Pragmatic Pathways of the Resultative in Bantu Botatwe. *Africana Linguistica* 18: 41–96.

Dahl, Ö. (1985) *Tense and Aspect Systems*. Oxford: Blackwell.

Dalgish, G. M. (1979) The Syntax and Semantics of the Morpheme ni in Kivunjo (Chaga). *Studies in African Linguistics* 10(1): 47–63.

de Haan, F. (2006) Typological Approaches to Modality. In Frawley, W., E. Eschenroeder, S. Mills & T. Nguyen (eds.), *The Expression of Modality*, 27–69. Berlin: De Gruyter Mouton.

De Kind, J. (2014) Pre-Verbal Focus in Kisikongo. *ZAS Papers in Linguistics* 57: 95–122.

De Kind, J., S. Dom, G.-M. de Schryver & K. Bostoen (2015) Event-Centrality and the Pragmatics-Semantics Interface in Kikongo: From Predication Focus to Progressive Aspect and Vice Versa. *Folia Linguistica Historica* 36: 113–163.

Devos, M. (2008a) The Expression of Modality in Shangaci. *Africana Linguistica* 14: 3–35.

Devos, M. (2008b) *A Grammar of Makwe*. Munich: LINCOM.

Devos, M. & G.-M. de Schryver (2013) From Habitually Going to Maybe: Grammaticalization (Lexicalization?) of an Epistemic Sentence Adverb in Swahili. Paper presented at *the 21st International Conference on Historical Linguistics*, University of Oslo, Norway.

Devos, M. & J. van der Auwera (2013) Jespersen Cycles in Bantu: Double and Triple Negation. *Journal of African Languages and Linguistics* 34(2): 205–274.

Devos, M. & D. Van Olmen (2013) Describing and Explaining the Variation of Bantu Imperatives and Prohibitives. *Studies in Language* 37(1): 1–57.

Doke, C. M. (1935) *Bantu Linguistic Terminology*. London; New York: Longmans, Green.

Doke, C. M. & B. W. Vilakazi (1948) *Zulu-English Dictionary*. Johannesburg: Witwatersrand University Press.

Dom, S. (2013) *Tijd en aspect in Kikongo (Bantoe H16): Een comparatieve benadering van het Kimbata, Kimbeko, Kinkanu, Cizali, Ciwoyo en Kisolongo*. Gent: Universiteit Gent, MA thesis.

Dom, S. (2014) *The Neuter in Bantu: A Systemic Functional Analysis*. Ghent: Ghent University, MA thesis.

Dzokanga, A. (1995) *Grammaire pratique du lingala illustré*. Paris: INALCO.

Ernst, U. (1991) Temps et aspects en kako. In Anderson, S. C. & B. Comrie (eds.), *Tense and Aspect in Eight Languages of Cameroon*, 17–45. Dallas: Summer Institute of Linguistics; University of Texas at Arlington.

Evans, N. (2007) Insubordination and Its Uses. In Nikolaeva, I. (ed.), *Finiteness: Theoretical and Empirical Foundations*, 366–431. New York: Oxford University Press.

Faraclas, N. G., Y. Rivera-Castillo & D. E. Walicek (2007) No Exception to the Rule: The Tense-Aspect-Modality System of Papiamentu Reconsidered. In Huber, M. & V. Velupillai (eds.), *Synchronic and Diachronic Perspectives on Contact Languages*, 257–278. Amsterdam; Philadelphia: John Benjamins.

Fehderau, W. H. (1962) *Descriptive Grammar of the Kituba Language: A Dialectal Survey*. Ms. Léopoldville.

Foster, C. S. (1960) *Lessons in Kikaonde*. Tervuren: Manuscript Consultable in the Linguistics Library of the Royal Museum for Central Africa.

Fourie, D. J. (1989) Modality in Isizulu. *Logos* 9(1): 45–56.

Givón, T. (1971) Dependent Modals, Performatives, Factivity, Bantu Subjunctives and What Not. *Studies in African Linguistics* 2: 61–81.

Givón, T. (1994) The Pragmatics of De-transitive Voice: Functional and Typological Aspects of Inversion. In Givón, T. (ed.), *Voice and Inversion*, 3–46. Amsterdam; Philadelphia: John Benjamins.

Güldemann, T. (1996) *Verbalmorphologie and Nebenprädikationen im Bantu. Eine Studie zur funktional motivierten Genese eines konjugationalen Subsystem*. Bochum: Universitätsverlag Dr. N. Brockmeyer.

Güldemann, T. (1997) Prosodic Subordination as a Strategy for Complex Sentence Construction in Shona: Bantu Moods Revisited. In Herbert, R. K. (ed.), *African Linguistics at the Crossroads: Papers from Kwaluseni*, 75–98. Cologne: Rüdiger Köppe.

Güldemann, T. (2003) Present Progressive vis-à-vis Predication Focus in Bantu. A Verbal Category between Semantics and Pragmatics. *Studies in Language* 27(2): 323–360.

Guthrie, M. (1971) *Comparative Bantu: An Introduction to the Comparative Linguistics and Prehistory of the Bantu Languages. Volume 2: Bantu Prehistory, Inventory and Indexes*. London: Gregg International.

Harjula, L. (2004) *The Ha Language of Tanzania: Grammar, Texts and Vocabulary*. Cologne: Rüdiger Köppe.

Harries, L. (1950) *A Grammar of Mwera, a Bantu Language of the Eastern Zone, Spoken in the South-Eastern Area of Tanganyika Territory*. Johannesburg: Witwatersrand University Press.

Heine, B., T. Güldemann, C. Kilian-Hatz, D. A. Lessau, H. Roberg, M. Schladt & T. Stolz (1993) *Conceptual Shift: A Lexicon of Grammaticalization Processes in African Languages*. Cologne: University of Cologne.

Hengeveld, K. (2004) Illocution, Mood and modality. In Booij, G., C. Lehmann & J. Mugdan (eds.), *Morphology. An International Handbook on Inflection and Word-Formation (Volume 2)*, 1190–1201. Berlin: De Gruyter Mouton.

Hewson, J. (2007) L'aspect situatif. In Bres, J. & T. Ponchon (eds.), *Actes du XIe Colloque de l'Association Internationale de la Psychomécanique du Langage*, 255–260. Limoges: Lambert-Lucas.

Hewson, J. (2012) Tense. In Binnick, R. I. (ed.), *The Oxford Handbook of Tense and Aspect*, 507–535. Oxford: Oxford University Press.

Hewson, J. (2016a) Schematic Diagrams. In Nurse, D., S. Rose & J. Hewson (eds.), *Tense and Aspect in Niger-Congo*, 299–321. Tervuren: Royal Museum for Central Africa.

Hewson, J. (2016b) Tense and Aspect: Discussion and Conclusions. In Nurse, D., S. Rose & J. Hewson (eds.), *Tense and Aspect in Niger-Congo*, 288–298. Tervuren: Royal Museum for Central Africa.

Hulstaert, G. (1965) *Grammaire du lomongo. Deuxieme partie, Morphologie*. Tervuren: Musée royal de l'Afrique centrale.

Hyman, L. & J. R. Watters (1984) Auxiliary Focus. *Studies in African Linguistics* 15(3): 233–273.

Hyman, L. M. & E. R. Byarushengo (1984) A Model of Haya Tonology. In Clements, G. N. & J. A. Goldsmith (eds.), *Autosegmental Studies in Bantu tone*, 53–104. Berlin: De Gruyter Mouton; Dordrecht: Foris Publications.

Johnson, M. R. (1980) A Semantic Description of Temporal Reference in the Kikuyu Verb. *Studies in African Linguistics* 11: 269–320.

Kamba Muzenga, J.-G. (1981) *Les formes verbales négatives dans les langues bantoues*. Tervuren: Musée royal de l'Afrique centrale.

Kawalya, D., K. Bostoen & G.-M. de Schryver (2014) Diachronic Semantics of the Modal Verb -sóból- in Luganda: A Corpus-Driven Approach. *International Journal of Corpus Linguistics* 19(1): 60–93.

Kawalya, D., G.-M. de Schryver & K. Bostoen (2018) From Conditionality to Modality in Luganda (Bantu, JE15): A Synchronic and Diachronic Corpus Analysis of the Irrealis Marker -andi-. *Journal of Pragmatics*, 127: 84–106.

Leonard, R. A. (1980) Swahili e, ka and nge as Signals of Meaning. *Studies in African Linguistics* 11(2): 209–226.

Louwrens, L. J. (1990) Mood and Modality in Northern Sotho. *South African Journal of African Languages* 10(1): 10–17.

Lwane, B. G. (1976) *Teachers' Handbook Lulogooli/Kitabu cho Mwegetsi*. Nairobi: East African Literature Bureau.

Maho, J. F. (2009) NUGL Online: The Online Version of the New Updated Guthrie List, a Referential Classification of the Bantu Languages (Juni 4, 2009) (Available online at http://goto.glocalnet.net/mahopapers/nuglonline.pdf, Accessed December 13, 2010).

Mberamihigo, F. (2012) Grammaticalisation et evolution sémantique des adverbs épistémiques en kirundi. Paper presented at *42nd Colloquium on African Languages and Linguistics*, Leiden University, the Netherlands.

Mberamihigo, F. (2014) *L'expression de la modalité en kirundi. Exploitation d'un corpus électronique*. Bruxelles/Gand: Université libre de Bruxelles (ULB), Université de Gand (UGent), thèse de doctorat.

Mberamihigo, F., G.-M. de Schryver & K. Bostoen (2016) Entre verbe et adverbe: Usages et évolutions du marqueur épistémique umeengo/umeenga en kirundi. *Journal of African Languages and Linguistics* 37(2): 247–286.

Meeussen, A. E. (1962) De tonen van subjunktief en imperatief en het bantoe. *Africana Linguistica* 1: 57–74.

Meeussen, A. E. (1967) Bantu Grammatical Reconstructions. *Africana Linguistica* 3: 79–121.

Meeussen, A. E. (1969) *Bantu Lexical Reconstructions*. Ms. Tervuren: Royal Museum for Central Africa.

Meeussen, A. E. (1971) *Éléments de grammaire lega*. Tervuren: Musée royal de l'Afrique centrale.

Meeussen, A. E. (2014) Tones of the Subjunctive and the Imperative in Bantu. *Africana Linguistica* 20: 15–38.

Meeuwis, M. (1995) The Lingala Tenses: A Reappraisal. *Afrikanistische Arbeitspapiere* 43: 97–118.

Meeuwis, M. (2010) *A Grammatical Overview of Lingála* Munich: LINCOM.

Meinhof, C. (1906) *Grundzüge einer vergleichenden Grammatik der Bantusprachen*. Berlin: Reimer.

Miehe, G. (1979) *Die Sprache der älteren Swahili-Dichtung*. Berlin: Dietrich Reimer.

Morrison, M. E. (2011) *A Reference Grammar of Bena*. Houston: Rice University, PhD dissertation.

Moshi, L. (1994) Time Reference in KiVunjo-Chaga. *Journal of African Languages and Linguistics* 15: 127–159.

Mukash Kalel, T. (1982) *Le kanyok, langue bantoue du Zaïre: phonologie, morphologie, syntagmatique*. Paris: Université Sorbonne nouvelle, Paris III, PhD.

Muzale, H. R. T. (1998) *A Reconstruction of the Proto-Rutaran Tense/Aspect System*. St John's: Memorial University of Newfoundland, PhD dissertation.

Ndolo, P. (1972) *Essai sur la tonalité et la flexion verbale du gimbala*. Tervuren: Musée royal de l'Afrique centrale.

Ngonyani, D. S. (2013) The Subjunctive Mood in Kikisi, Kindendeule and Chingoni. *Journal of Linguistic and Language in Education* 7(1): 1–19.

Nsuka-Nkutsi, F. (1982) *Les structures fondamentales du relatif dans les langues bantoues*. Tervuren: Musée royal de l'Afrique centrale.

Nurse, D. (2003) Aspect and Tense in Bantu Languages. In Nurse, D. & G. Philippson (eds.), *The Bantu Languages*, 90–102. London New York Routledge.

Nurse, D. (2008) *Tense and Aspect in Bantu*. Oxford: Oxford University Press.

Nurse, D. (2010) *Bantu Tense and Aspect Systems*. St. John's: Memorial University of Newfoundland. (Available online at www.faculty.mun.ca/dnurse/Tabantu/).

Nurse, D. & T. J. Hinnebusch (1993) *Swahili and Sabaki: A Linguistic History*. Berkeley: University of California Press.

Nurse, D. & I. A. M. Makombe (1979) Note on Hehe. *African Languages/Langues Africaines* 5(1): 114–118, 141–145.

Nurse, D. & H. R. T. Muzale (1999) Tense and Aspect in Great Lakes Bantu Languages. In Hombert, J. M. & L. M. Hyman (eds.), *Bantu Historical Linguistics: Theoretical and Empirical Perspectives*, 517–544. Stanford: CSLI Publications.

Nurse, D. & G. Philippson (2006) Common Tense-Aspect Markers in Bantu. *Journal of African Languages and Linguistics* 27(2): 155–196.

Nurse, D., S. Rose, & J. Hewson (2010) *Verbal Categories in Niger-Congo*. St. John's: Memorial University of Newfoundland. (Available online at www.africamuseum.be/sites/default/files/media/docs/research/publications/rmca/online/documents-social-sciences-humanities/tense-aspect-niger-congo.pdf).

Odden, D. & L. Bickmore (2014) Melodic Tone in Bantu. *Africana Linguistica* 20: 3–13.

Palmer, F. R. (2001) *Mood and Modality*. Cambridge; New York: Cambridge University Press.

Parker, E. (1991) Conditionals in Mundani. In Anderson, S. C. & B. Comrie (eds.), *Tense and Aspect in Eight Languages of Cameroon*, 165–187. Dallas: Summer Institute of Linguistics; University of Texas at Arlington.

Petzell, M. (2008) *The Kagulu Language of Tanzania: Grammar, Texts and Vocabulary*. Cologne: Rüdiger Köppe.

Poulos, G. (1990) *A Linguistic Analysis of Venda*. Pretoria: Via Afrika.

Poulos, G. & C. T. Msimang (1998) *A Linguistic Analysis of Zulu*. Cape Town: Via Afrika.

Sacleux, C. (1909) *Grammaire swahilie*. Paris: Procure des PP. du Saint- Esprit.
Sacleux, C. (1939) *Dictionnaire swahili-français*. Paris,: Institut d'ethnologie.
Saloné, S. B. (1979) Typology of Conditionals and Conditionals in Haya. *Studies in African Linguistics* 10(1): 65–80.
Schadeberg, T. C. (1990) *A Sketch of Umbundu*. Cologne: Rüdiger Köppe.
Schicho, W. (1995) Modalität und Sprecherintention. In Miehe, G. & W. J. G. Möhlig (eds.), *Swahili-Handbuch*, 125–166. Cologne: Rüdiger Köppe.
Seidel, F. (2008) *A Grammar of Yeyi. A Bantu Language of Southern Africa*. Cologne: Rüdiger Köppe.
Seidl, A. & A. Dimitriadis (2003) Statives and Reciprocal Morphology in Swahili. In Sauzet, P. & A. Zribi-Hertz (eds.), *Typologie des langues d'Afrique & universaux de la grammaire. Volume 1: Approches transversales, domaine bantou*, 239–284. Paris: L'Harmattan.
Ström, E.-M. (2013) *The Ndengeleko Language of Tanzania*. Gothenburg: University of Gothenburg, PhD dissertation.
Timberlake, A. (2007) Aspect, Tense, Mood. In Shopen, T. (ed.), *Clause Structure, Language Typology and Syntactic Description, Volume 3: Grammatical Categories*, 315–333. Cambridge: Cambridge University Press.
Traugott, E. C. (2003) From Subjectification to Intersubjectification. In Hickey, R. (ed.), *Motives for Language Change*, 124–139. Cambridge; New York: Cambridge University Press.
Van de Velde, M. (2008) *A Grammar of Eton*. Berlin: De Gruyter Mouton.
Van de Velde, M. & J. van der Auwera (2010) Le marqueur de l'allocutif pluriel dans les langues bantu. In Floricic, F. (ed.), *Essais de typologie et de linguistique générale. Mélanges offerts à Denis Creissels*, 119–141. Lyon: ENS Editions.
van der Auwera, J. & V. A. Plungian (1998) Modality's Semantic Map. *Linguistic Typology* 2: 79–124.
van der Auwera, J. & E. Schalley (2004) From Optative and Subjunctive to Irrealis. In Brisard, F., M. Meeuwis & B. Vandenabeele (eds.), *Seduction, Community, Speech: A Festschrift for Herman Parret*, 87–96. Amsterdam; Philadelphia: John Benjamins.
Van Otterloo, R. (2011) *The Kifuliiru Language, Volume 2: A Descriptive Grammar*. Dallas: SIL International.
Vanhoudt, B. (1987) *Eléments de description du leke, langue bantoue de zone C*. Tervuren: Musée royal de l'Afrique centrale.
Voeltz, E. F. K. (1980) The Etymology of the Bantu Perfect. In Bouquiaux, L. (ed.), *L'expansion bantoue: actes du colloque international du Centre National de la Recherche Scientifique, Viviers 4–16 avril 1977*, 487–492. Paris: Société des Etudes Linguistiques et Anthropologiques.
Voorhoeve, J. (n.d.) *A Grammar of Safwa: Preliminary Draft Based on Previous Research by J. van Sambeek, Checked by C. K. Mwachusa*. Ms.
Wald, B. (1997) The 0 Tense Marker in the Decline of the Swahili Auxiliary Focus System. *Afrikanistische Arbeitspapiere* 51: 55–82.
Welmers, W. E. (1973) *African Language Structures*. University of California Press.
Whiteley, W. H. (1960) *The Tense System of Gusii*. Kampala: East African Institute of Social Research.
Wright, J. L. (2007) *An Outline of Kikaonde Grammar*. Lusaka: Bookworld Publishers – UNZA Press.

Xrakovskij, V. S. (ed.) (2001) *Typology of Imperative Constructions*. Munich: LINCOM.

Yukawa, Y. (1987a) A Tonological Study of Luvale Verbs. In Yukawa, Y. (ed.), *Studies in Zambian Languages*, 1–34. Tokyo: ILCAA.

Yukawa, Y. (1987b) A Tonological Study of Nkoya Verbs. In Yukawa, Y. (ed.), *Studies in Zambian Languages*, 129–184. Tokyo: ILCAA.

Ziervogel, D., P. J. Wentzel & T. N. Makuya (1981) *A Handbook of the Venda Language*. Pretoria: University of South Africa.

CHAPTER 8

NOMINAL MORPHOLOGY AND SYNTAX

Mark Van de Velde

1 INTRODUCTION

One of the quintessential typological properties of the Bantu languages is their pervasive system of noun classes and noun class agreement. This is undoubtedly the aspect of their grammatical structure that is most discussed in the literature, if only because every grammar sketch of a Bantu language contains a section on noun classes. The most complete discussion can be found in Maho (1999). In contrast, the structure of the noun phrase has received little attention, which cannot be attributed to a lack of interesting features. This chapter briefly introduces aspects of the noun and noun phrase that are well studied, such as the noun class system and the different types of adnominal modifiers, and provides them with a reference. It also aims at filling some gaps in the literature. Section 2.2. provides a first systematic overview of the types of semantic agreement that can be found in the Bantu languages. Section 3 is entirely dedicated to the augment, a pervasive element in the grammar of the Bantu languages, of which the last comparative study (De Blois 1970) was in need of an update. The typologically unusual word order patterns that can be found in the Bantu languages receive a first comparative analysis and diachronic explanation in Section 5.

2 NOUNS AND NOUN CLASSES

2.1 The form and structure of nouns

Bantu nouns minimally consist of a stem, which is usually preceded by a class prefix, itself sometimes preceded by one or more prefixes or proclitics, which can be inflectional or derivational. Noun stems can contain a derivational suffix (cf. Schadeberg & Bostoen, this volume). Reconstructed Proto-Bantu noun stems tend to have a CVCV-structure. In many contemporary languages, this is still the canonical stem form. However, in some North-Western languages, such as Nzadi B865 (Crane *et al.* 2011) and Nsong B85d (cf. Koni Muluwa & Bostoen this volume), the last vowel or syllable of noun stems tends to have dropped, giving rise to a monosyllabic pattern as the most frequent stem type. The prefix + stem template is so strong in the Bantu languages, that monomorphemic nouns may behave prosodically as if they consisted of a prefix and a stem, and that class prefixes that were historically incorporated in stems can be reactivated in different ways.

As for tone, augments and derivational prefixes added to full nouns, i.e., class prefix plus stem, are reconstructed high, except in class 1 and 9, whereas class prefixes are reconstructed low (Meeussen 1967). Proto-Bantu disyllabic stems can have any of the four logically possible tone patterns, viz. *HH, *HL, *LL and *LH (Bastin *et al.* 2002).

Here too, many changes have taken place in individual languages, such as the appearance of H class prefixes in a subset of Punu B43 nouns (Blanchon 1997) or the emergence of two extra tone patterns on disyllabic noun stems in Nande JD42 (Kenstowicz 2008). In a number of Western Bantu languages, the tone pattern of nouns depends on their syntactic position, a phenomenon known as "tone case" in Bantu studies (see for instance Schadeberg 1986, Kavari et al. 2012).

2.2 The noun class system

Noun classes can be defined as sets of nouns that trigger the same agreement pattern. Noun class assignment is typically coded by means of a nominal class prefix in the Bantu languages. Bantuists use numbers, rather than labels such as "feminine" or "neuter," to refer to individual noun classes. Numbers are assigned to classes in individual languages on the basis of cognacy. Odd numbers are used for classes that contain singular nouns and even numbers for plural classes, with some exceptions, most notably class 12 (SG) and 13 (PL). Singular-plural class pairings are usually called "genders." The classic Swahili G42 examples in (1) show the noun class prefixes and agreement prefixes of the noun -*kapu* 'basket' in the singular (class 7) and plural (class 8).

(1)   Swahili G42 (Welmers 1973: 171)
    a   Ki-kapu   ki-kubwa   ki-moja   ki-li-anguk-a.
       7-basket NP_7-big NP_7-one SP_7-PST-fall-FV
       'One large basket fell.'

    b   Vi-kapu   vi-kubwa   vi-tatu   vi-li-anguk-a.
       8-basket NP_8-big NP_8-three SP_8-PST-fall-FV
       'Three large baskets fell.'

Meeussen (1967: 97) reconstructed nineteen noun classes in Proto-Bantu, summarised in Table 8.1. According to Maho (1999: 50–55), contemporary Bantu languages have between zero and 19 classes, with an average of about 15 (without counting the locative classes 16, 17 and 18). The most common genders are 1/2, 3/4, 5/6, 7/8, 9/10, 11/6, 11/10, 12/13 and 14/6 (Maho 1999: 54). Classes from number 12 (included) upward generally contain less nouns than the other classes. The distribution of nouns over the noun classes is little studied, but appears to be quite variable across Bantu. Class 1, for instance, tends to be very small in (North-)Western languages and larger in the East and South. Class 7, in contrast, is usually much larger in the West than in the East.[1]

    Languages that have fully or nearly lost their noun class system are very rare among the Bantu languages, an observation that has sometimes been used as an – obviously imperfect – criterion for distinguishing Narrow Bantu from related language groups, where typologically diverse reduced class systems are common (see Good 2012 for an overview of most of Niger-Congo). An example of a language classified as Narrow Bantu that has lost its noun class system is Polri A92a (Wega Simeu 2012). A small set of about five nouns with a human denotation and a vowel-initial stem has a different prefix in the singular ($m\overset{\scriptstyle\grave{}}{V}$-) as in (2a) and the plural (*bò-*) as in (2b). Plural formation for all other nouns, such as *sámá* 'sheep' in (2c), involves the use of the plural marker *bè* in NP-initial position, as in (2d). There is no agreement in number or class anywhere. Plural NPs

TABLE 8.1 THE PROTO-BANTU NOUN CLASS SYSTEM AS RECONSTRUCTED BY MEEUSSEN (1967: 97)

	Nom	*Num*	*Pron*	*Subj*	*Obj*
1	mò-	(ò?)	jò-	ó-, á-	mò-
2	bà-	bá-	bá-	bá-	bá-
3	mò-	(ó?)	gó-	gó-	gó-
4	mì-	(í-?)	gí-	gí-	gí-
5	ì-	dí-	dí-	dí-	dí-
6	mà-	(á-?)	gá-	gá-	gá-
7	kì-	kí-	kí-	kí-	kí-
8	bì-	bí-	bí-	bí-	bí-
9	n-	(ì-)	jì-	jì-	jì-
10	n-	í-	jí-	jí-	jí-
11	dò-	dó-	dó-	dó-	dó-
12	kà-	ká-	ká-	ká-	ká-
13	tò-	tó-	tó-	tó-	tó-
14	bò-	bó-	bó-	bó-	bó-
15	kò-	kó-	kó-	kó-	kó-
16	pà-	pá-	pá-	pá-	pá-
17	kò-	kó-	kó-	kó-	kó-
18	mò-	mó-	mó-	mó-	mó-
19	pì-	pí-	pí-	pí-	pí-

with a prenominal modifier have to start with the plural morpheme, also when their head noun has plural marking, as shown in (2b).

(2)   Polri A92a (Wega Simeu 2012)

    a   jíkɛ́   mw-ǎn   jà   wàŋgɔ̀-ím
       other   SG-child   DEM   come-HOD.PST
       'That other child has come.'

    b   bɛ̀   jíkɛ́   bw-ǎn   jà   wàŋgɔ̀-ím
       PL   other   PL-child   DEM   come-HOD.PST
       'Those other children have come.'

    c   jíkɛ́   sámá   jà   wàŋgɔ̀-ím
       other   sheep   DEM   come-HOD.PST
       'That other sheep has come.'

    d   bɛ̀   jíkɛ́   sámá   jà   wàŋgɔ̀-ím
       PL   other   sheep   DEM   come-HOD.PST
       'Those other sheep have come.'

Most Bantu languages that lack noun classes, such as Polri A92a or Bira D32, are in contact with non-Benue-Congo languages, and contact-induced change has been suggested as an explanation for the loss of classes in these cases (see for instance Dimmendaal 2011: 196). However, Nzadi B865 shows that the very strong reduction of noun classes can also be due to purely language-internal changes (Crane *et al.* 2011).

2.2.1 Counting noun classes

Establishing the number of noun classes in a language is not as straightforward as it may seem. Different authors have used different criteria, depending on their goals. One option is to distinguish noun classes (sets of nouns that trigger the same agreement pattern) from morphological classes (sets of nouns with identical number marking). The advantages of this approach are that it provides the most accurate synchronic analysis of the nominal morphosyntax, and that it allows for straightforward cross-linguistic comparison with other systems of grammatical gender. The disadvantage of this approach is that it is not very compatible with the comparative concepts "class 1," "class 2," etc. developed in the Bantuist linguistic tradition, so that adopting it complicates Bantu-internal comparison. The reason for this is that there are many instances of noun classes whose agreement patterns have merged in the history of individual languages, whereas their nominal class markers have remained unchanged. Therefore, most Bantuists prefer to follow the tradition of splitting up sets of nouns that trigger the same agreement pattern into two or more classes if they have different nominal prefixes AND if they participate in different singular-plural pairings, or if one set contains singular nouns and the other plural nouns. For instance, Makwe G402 has two sets of nouns that trigger the same agreement pattern in the singular: one with a prefix *mu-* and plurals of class 4; and one with a prefix *u-* ~ *lu-* and plurals in class 4, 6, 10, 10a or 10b. Following Bantuist practice, Devos (2008: 43) treats these as two separate classes, 3 and 11 respectively, which reflects the historical origin of the difference in nominal prefix.

Bantuists also have the somewhat inconsistent habit of distinguishing between a class 15 and a class 17, even though these two classes tend to have exactly the same nominal prefix and agreement pattern. Class 17 is one of the locative classes, together with 16 and 18. Its class marker can be added as a pre-prefix to a full noun to derive a locative noun, which can trigger agreement according to the locative class or its original noun class, depending on the language and/or the construction. Class 15, on the other hand, is the class of infinitives in most Bantu languages, except in the north of the domain, where infinitives are often of class 5, sometimes also other classes (cf. Forges 1983, Hadermann 1999). Class 15 also contains a small set of canonical nouns, mostly body part terms. Six have been reconstructed in Proto-Bantu with a plural in class 6: **-bókò* 'arm,' **-gòdò* 'leg,' **-tóì* 'ear,' **-dúì* 'knee' and **-jápà* 'armpit' (Doneux 1967). The arguments for splitting agreement class 15–17 into two distinct noun classes, when explicitly mentioned, are rather diverse and usually not compatible with any working definition of noun classes.

When sets of singular nouns trigger the same agreement pattern and have their plural in the same class, but differ in the shape of their nominal prefix in ways that are not phonologically predictable, they tend to be divided into subclasses.[2] Bantu subclasses are typically labelled by means of a letter after the class number. For example, Devos (2008) distinguishes in Makwe G402 between class 10 (prefix *ji-*), class 10a (prefix ∅-) and class 10b (prefix *jiN-*). Such subclasses and their labels are typically language-specific, or even description-specific, with one notable exception: class 1a, which can be found throughout Bantu. The so-called class 1a was first systematically described by Doke (1927) as a set of nouns that lack a prefix in the singular and usually trigger class 1 agreement. It typically contains proper names, some kinship terms, personified animals and borrowings. Class 1a is radically different from the other subclasses. Its lack of a nominal prefix in most languages is not due to prefix loss, but goes back to Proto-Bantu at least. It can be explained by pointing out that proper names and certain suppletive kin terms tend to lack a determiner in the languages of the world, or an augment in the Bantu languages (see Section 3), because they are inherently determined. Names for personified animals function as proper names and borrowings are easily attracted to any class without an overt class marker. The plurals of class 1a nouns are often marked by an element *ba(a)* or *(b)ɔ* in which case they trigger the same agreement pattern as nouns of class 2 and are treated as belonging to subclass 2a, when the vowel of the marker is *a*, or 2b, when the plural marker has a back vowel. In Eton A71, the class 2b marker *bɔ́* has the phonological characteristics of a separate word and I have analysed it as a number word (Van de Velde 2006a), an analysis that may be extended to other Bantu languages. Class 2a/2b markers often express associativity. Associative constructions are used to refer to collectives that consist of an identified referent and a number of associate referents. The relation between the identified referent and its associates can be metonymic, i.e., based on contiguity, or metaphoric, i.e., based on similarities. The former type is typically reserved for human referents in the languages of the world (3a), but in some Bantu languages it can also be used with place names (4a). Examples of metaphoric associativity, also called "similative plurals" (Daniel & Moravcsik 2005), can be found in different parts of the Bantu-speaking area, as shown in (3b) and (4b).

(3)   Mongo C61 (Hulstaert 1965: 145)
    a   baa Byeka
        'Byeka and his family/pupils/followers. . .'

    b   baa mésá
        'tables and similar things; tables, for instance; tables and the like'

(4)   Xhosa S41 (Hendrikse 1990: 391), *oo* is the class 2b marker
    a   oomaRhini < oo+ amaRhini
        'Grahamstown and environment'

    b   ookulamba < oo + ukulamba
        'hunger and similar feelings; hunger and the like'

2.2.2 Semantics

There is widespread agreement among Bantuists that the gender assignment of a noun cannot generally be predicted on the basis of its meaning, but that the noun class systems of Bantu languages are not devoid of semantic regularities either. Disagreement exists about whether or not the gender assignment of all or most nouns can be shown to be semantically motivated, if not in contemporary languages, then at least in a proto-stage.

One area in which gender assignment could in principle be clearly shown to be semantically motivated is the integration of borrowings into the noun class system (cf. Mous, this volume). Unfortunately, there is no thorough comparative study on this subject. In my experience, gender assignment on semantic grounds is relatively exceptional and mostly restricted to language names and some nouns for human beings. Most often, borrowings are assigned to classes on formal grounds. Either they are inserted into classes that have no, or no clearly recognisable, prefix, such as class 9, 5 or 1a, or the initial segments of their stem are reinterpreted as an existing class prefix. Note in this respect that for comparative-historical reasons class 9 is often analysed as having a nasal prefix *N*-, but that in many languages the plural (class 10) has the same initial nasal, which therefore does not commute with anything else, and is strictly speaking not morphologically separable from the stem, unless, perhaps, in some deverbal nouns, where it can be analysed as a derivational affix.

The remainder of this section will be devoted to a number of uncontroversial semantic characteristics of Bantu class systems, starting with gender assignment and then discussing meaning in grammar. The human/non-human distinction will be a recurrent theme. Due to space limitations, this section cannot do justice to the extensive literature on the subject. A more elaborate introduction with many references can be found in Maho (1999: 63–99).

2.2.2.1 Gender assignment

The first semantic regularity that naturally comes to mind is the semantic homogeneity of gender 1/2 in many languages, which tends to be restricted to nouns with human reference. However, it is quite common to find terms for humans in other classes, especially if they are used to refer to persons with unusual characteristics. Second, words for liquids are typically found in class 6. Other semantic sub-regularities in gender assignment that have been pointed out in the literature are not Bantu-wide, or they tend to show many more exceptions. Language names, for instance, tend to be of class 7 or of class 11 and nouns for abstract concepts are in class 14 in many languages.

Many patterns of nominal derivation have a target gender, in which they create semantically coherent subgroups. Agentive nouns assigned to gender 1/2 are a typical example (cf. Schadeberg & Bostoen, this volume). Special cases are evaluative and locative denominal derivation. When used derivationally, locative class markers are normally additive, i.e., added to the inherent class marker, whereas diminutive and augmentative prefixes can be either substitutive, i.e., replacing the inherent class marker, or additive, with a general preference for the former. Typical diminutive classes are 12 and 19, sometimes 5. Class 7 is the most recurrent augmentative class. Other combinations of classes and evaluative meanings can be found too. For instance, class 5 can have an augmentative meaning in some languages and class 7 a diminutive one.

The grammaticalisation of evaluative morphemes can lead to the emergence of new class markers. In Eton A71, a diminutive can be formed of any noun by means of the proclitic $mɔ́^H=$ (plural $bɔ́^H=$), historically derived from the noun m-$ɔ̀ŋɔ́$ (cl.1)/b-$ɔ̀ŋɔ́$ (cl.2) 'child(ren).' These procliticised diminutives trigger agreement pattern 1 in the singular and 2 in the plural, as predicted by their etymology. Leitch (2003: 400) analyses Babole C101 as having a diminutive gender 19/20, with a prefix $mwá$- in the singular and $báná$- in the plural. These too are derived from the gender 1/2 noun 'child,' but they can be analysed as the markers of a separate noun class, because they trigger a unique agreement pattern. These agreement patterns correspond to those of class 19 and 13 respectively in other Bantu languages, typical diminutive classes. So at one point in the history of Babole, the noun for 'child' must have been used as the head of a genitive phrase to express a diminutive meaning, in which case it triggered semantic agreement.

2.2.2.2 Meaning and grammar

We speak of semantic agreement, as opposed to syntactic agreement, when the choice of an agreement pattern depends on aspects of the meaning of the controller rather than on its morphological class defined by the nominal prefix (Corbett 1991). Here, the link between noun classes and semantics is more straightforward than in gender assignment.

The best-known example is animate agreement, as found especially in zone K and among the coastal languages of zone G and E (Wald 1975, Maho 1999: 124), where animate nouns trigger agreement pattern 1 in the singular and 2 in the plural, whatever the morphological class to which they belong, as shown in (5b), where the Swahili class 7 noun *kiboko* 'hippo' triggers class 1 agreement on both the demonstrative and the verbal object prefix, in contrast to *kisu* 'knife' which triggers syntactic class 7 agreement.

(5)   Swahili G42 (Wald 1975: 241–242)
   a   ki-le   ki-su,   ni-li-ki-on-a.
       PP_7-dem   7-knife SP_{1SG}-PST-OP_7-see-FV
       'That knife, I saw it.'

   b   Yu-le   ki-boko,   ni-li-mw-ona.
       PP_1-DEM   7-hippo SP_{1SG}-PST-OP_1-see-FV
       'That hippo, I saw it.'

There is some variation between languages concerning the obligatoriness of its application and the agreement targets that are involved. For the latter, Corbett's (1979) agreement hierarchy (attributive < predicate < relative pronoun < personal pronoun) makes the right predictions, viz. the higher (i.e., the more to the right) an agreement target is on the hierarchy, the more likely it is to have semantic agreement. In fact, the Bantu languages might require a refinement of the hierarchy, since adnominal modifiers can be split into possessive pronouns versus the others. Possessive pronouns are at the bottom of the hierarchy, i.e., most likely to agree according to the morphological class of the head noun (more on this in Section 5).

   The notion of animate agreement can only make sense when it is opposed to syntactic agreement. It should not be confounded with the situation found in the few Bantu languages that have a radically restructured class system in which animacy determines gender assignment. An interesting example is Kako A93, which has the typologically unusual characteristic of distinguishing two genders in the plural, but none in the singular (Ernst 1992: 34). Kako does not have nominal prefixes on singular nouns. Plural nouns take the prefix *ɓè-* (6a-b) if they are animate and *mè* if they are inanimate (6c-d). These prefixes also mark agreement in gender within the noun phrase (6e-f).

(6)   Kako A93 (Ernst 1992)
   a   mbam 'the man'/ɓè-mbam 'the men'
   b   mbiyè 'the dog'/ɓè-mbiyè 'the dogs'
   c   gwàlɔ 'the hoe'/mè-gwàlɔ 'the hoes'
   d   tù 'the house'/mè-tù 'the houses'
   e   ɓè-ŋgo ɓa-ka 'these pigs'
   f   mè-kandɔ ma-ka 'these clothes'

The main variety of Lingala C36d, also called "Kinshasa Lingala" (Meeuwis 2010: 37), has retained a typical Bantu system of morphological classes: 11 noun prefixes define 15 classes plus two subclasses, if one counts in a traditional Bantuist way, taking into account singular-plural pairings of nouns.The nominal prefixes have inflectional uses (number marking) and derivational uses. However, the morphological class of a noun is completely irrelevant for the agreement it triggers. Agreement in gender is restricted to subject markers on verbs and defines two genders: Animate and Inanimate. Animate controllers trigger a subject prefix *a-* in the singular and *ba-* in the plural; inanimate controllers trigger the subject marker *e-* in the singular and the plural. Therefore, despite its full set of nominal prefixes, the grammatical gender of Lingala C36d is typologically more similar to that of Kako A93 or English than to that of the majority of Bantu languages.

TABLE 8.2 MORPHOLOGICAL CLASSES
IN LINGALA C36D

	SG	*PL*
1/2 1a	mo- ∅-	ba-
3/4	mo-	mi-
5/6	li-	ma-
7/8 7a	e- ki-	bi-
9/10	∅-	ba-
11	lo-	
14	bo-	
15	ko-	

Another kind of semantic agreement could be called "superclassing," in that it involves a hierarchical organisation of the agreement patterns of a language along semantic lines. This is described for a number of multi-class stems in Luba L31a, such as the non-selective interrogative stem *-nyi*, the indefinite stem *-ntu* and the locative adverbials (van den Eynde & Mufuta 1994). Thus, wherever one can use the indefinites *mu-ntu* 'somebody' (class 1), *ci-ntu* 'something' (class 7) or *ka-ntu* 'a small person or thing' (class 12), the use of the more general term *ci-ntu* 'some entity' (class 7) is felicitous as well. Likewise, wherever one can use *ku-ntu* 'some direction' (class 17) or *mu-ntu* 'some place inside' (class 18), one can also always use *pa-ntu* 'somewhere' (class 16), which can have the more specific meaning of 'some place outside' too. This shows class 7 and 16 to be at the top of a semantic hierarchy of agreement patterns for entities and locations respectively. The locative hierarchy is illustrated by the use of adverbials in (7) (van den Eynde & Mufuta 1994: 102). In examples (7a) and (7b) the same message can be expressed in two ways. The initial demonstrative can either agree in noun class with the locative noun, or it can take the default locative form of class 16. These options are not available in (7c), because the default locative class is identical to the one for the specific meaning of super-position. Examples (7d-g) are ungrammatical because the demonstrative fails to agree with the noun, whether by syntactic agreement or by superclassing.

(7)   Luba L31a (van den Eynde & Mufuta 1994)

a   apa   tu-di   ku-n-zubu   or   eku   tu-di   ku-n-zubu
     DEM$_{16}$ SP$_{1PL}$-be   17-9-house   DEM$_{17}$ SP$_{1PL}$-be   17-9-house
     'Here, we are towards the house.'

b   apa   tu-di   mu-n-zubu   or   emu   tu-di   mu-n-zubu
     DEM$_{16}$ SP$_{1PL}$-be   18-9-house   DEM$_{18}$ SP$_{1PL}$-be   18-9-house
     'Here, we are in the house.'

c   apa   tu-di   pa-n-zubu
     DEM$_{16}$ SP$_{1PL}$-be   16-9-house
     'Here, we are on the house.'

  d   *emu tu-di ku-n-zubu

  e   *emu tu-di pa-n-zubu

  f   *eku tu-di mu-n-zubu

  g   *eku tu-di pa-n-zubu

A similar semantically based hierarchical organisation of the class system is at work in contexts of enforced agreement, e.g., where conjoined NPs in subject position have to agree with the verb, a situation also known as concord resolution (cf. Marten & Downing, this volume, Section 3.4).

A special case of enforced agreement can be found with proper name and suppletive kin term controllers (Van de Velde 2006b, 2009). In some languages they trigger a default agreement pattern, e.g., class 9 for place names in Makwe G402 (Devos 2008: 61). This default pattern may be mixed, as in Orungu B11b, where names trigger class 9 agreement within the nominal constituent and class 1 agreement on verbs and pronouns (Van de Velde & Ambouroué 2011). Similarly, in Kagulu G12, suppletive kin terms trigger agreement pattern 5 (sg) or 10 (pl) on possessive pronouns and 1/2 on other agreement targets (Petzell 2008: 56). In other languages, such as Gikuyu E51 and Rundi JD62, as shown in (8), proper names and suppletive kin terms trigger the same agreement pattern as the basic level term that expresses their categorical presuppositional meaning (Van de Velde 2009).

(8)   Rundi JD62 (Van de Velde 2009)
  a   *u-ru-kara* 'black' (cl. 11); *u-muu-ntu* 'person' (cl.1); *i-m-bwá* 'dog' (cl.9)

  b   Rukara   a-rikó   a-ra-fuungur-a
      Rukara   SP₁-prog   SP₁-DJ-eat-IPFV
      'Rukara (a person) is eating.'

  c   Rukara   i-rikó   i-ra-ry-á
      Rukara   SP₉-PROG   SP₉-DI-eat-IPFV
      'Rukara (a dog) is eating.'

  d   *u-muu-ntu* 'person' (cl. 1); *i-nká* 'cow' (cl. 9)

  e   nyina   a-ra-ryam-ye
      mother   SP₁-DJ-sleep-PFV
      'His/her mother is sleeping.' (person)

  f   nyina   i-ra-ryam-ye
      mother   SP₁-DJ-sleep-PFV
      'His/her mother is sleeping.' (cow)

The so-called class 1a, the mixed agreement patterns of Orungu B11b and Kagulu G12 and the semantic proper name agreement of Rundi JD62 and Gikuyu E51 are basically the same phenomenon, which can be dealt with coherently by analysing proper names and grammatically similar common nouns as being classless and the agreement they trigger as a kind of enforced agreement (Van de Velde 2006b).

So far, we have seen three types of semantic agreement, viz. animate agreement, superclassing and agreement determined by the categorical sense of proper name controllers. A fourth type could be called "evaluative agreement." It can be observed in at least two contexts. The first is found in those languages with animate agreement, such as Swahili

G42 and Ndengeleko P11 (Ström 2013: 190), where nouns with animate reference oblig-atorily trigger agreement patterns 1 or 2 on all or most targets, except when they are derived diminutive or augmentative nouns. In that case, as illustrated in (9), they trigger the agreement pattern of the diminutive or augmentative class.

(9)   Ndengeleko P11 (Ström 2013: 163, 195)

a   m-bésa   a-úu
10-hare NP_2-white
'White hares'

b   ka-pésa   ka-úu
12-hare NP_{12}-white
'Little white hare'

When the lexical class of a noun with animate reference is the same as one of the eval-uative classes, as in Swahili G42 (10), agreement alone signals whether the noun is derived (Gregersen 1967: 19). Adherents of modular approaches to grammar may find this an interesting example of denominal derivation signalled by the syntax rather than the morphology.

(10)   Swahili G42

a   Yu-le   ki-pofu,   ni-li-mw-on-a
PP_1-DEM   7-blind SP_{1sg}-PST-OP_1-see-FV
'That blind man, I saw him.'

b   Ki-le   ki-pofu,   ni-li-ki-on-a
PP_7-DEM   7-blind   1SG-PST-OP_7-see-FV
'That tiny blind man, I saw him.'

Agreement in examples like (8b) and (8e) is usually analysed as syntactic agreement, e.g., in Wald (1975: 273). However, it is impossible to account for it without making reference to the semantic properties of the head noun. The fact that evaluative agreement outranks animate agreement can be ascribed to the conspiracy of formal and semantic conditions in evaluative agreement.

The second context in which evaluative agreement can be observed provides indepen-dent evidence for its analysis as a type of semantic agreement. It is found in languages such as Rundi JD62, where the agreement pattern of proper name and suppletive kin term controllers is determined by their presuppositional meaning. Since emotive connotations are typically part of the meaning of names (Van Langendonck 2007: 83), it is not sur-prising that they can trigger evaluative agreement of diminutive class 12 or augmentative class 7, as shown in (11a-c). This option is not available for common noun controllers, which always trigger syntactic agreement, as shown in (11d-e).

(11)   Rundi JD62 (Van de Velde 2009)

a   Taama   a-ra-aje
Taama SP_1-DJ-come.PFV
'Taama arrives'

b   Taama   ki-ra-aje
Taama SP_7-DJ-come.PFV
'(big/horrible) Taama arrives.'

   c   Taama   ka-ra-aje
       Taama   SP$_{12}$-DJ-come.PFV
       '(little/dear) Taama arrives.'

   d   u-mu-ganwa   a-ra-aje
       AUG-1-prince   SP$_1$-DJ-come.PFV
       'The prince arrives.'

   e   *u-mu-ganwa   ka-ra-aje
       AUG-1-prince   SP$_{12}$-DJ-come.PFV
       'The (little/dear) prince arrives.'

The fifth and last type of semantic agreement is locative agreement. It can be argued to exist in some of the languages that derive locative nouns from other nouns by means of a suffix. These languages are spoken in zones E, G, F, P and S, i.e., the Eastern part of the Bantu area from North to South. The most common locative suffix is a form similar to -*ini* (Grégoire 1975). There is a small group of languages of zone G in which locative nouns are derived from non-locative nouns uniquely by means of a suffix, i.e., without also adding a locative class prefix, and in which locative nouns can trigger agreement of class 16, 17 or 18 according to the type of location the speaker wishes to express, as the Bondei G24 examples in (12) show. Even if one chooses to analyse the locative suffix as a formally unusual class marker, the choice between three different agreement patterns can only be analysed in terms of semantic agreement.

(12) Bondei G24 (Grégoire 1975: 192)
      a nyumba-ni mw-ako (18-POSS$_{2SG}$)   'in your house'
      b nyumba-ni ha-kwe (16- POSS$_{3SG}$) 'close to his house'
      c nyumba-ni kw-etu   (17-POSS$_{1PL}$)   'at our house'

To conclude, although lexical gender assignment has received most attention in studies on the semantic basis of Bantu noun class systems, evidence for semantic regularities is stronger in the grammar, i.e., derivation and agreement.

3 THE AUGMENT

The term "augment" (also "initial vowel" or "pre-prefix") is used by Bantuists to refer to an element that precedes the class prefix of nouns and some adnominal or nominalised modifiers and that changes neither the class assignment of the noun, nor its lexical meaning, nor the syntactic positions it can occupy. In Meeussen's (1967) tentative Proto-Bantu reconstruction, the augment is a separate word that is formally identical to a pronominal prefix. Since the augment is pervasive in Bantu and since it hasn't been the object of a comparative study since De Blois (1970), I will go into considerable detail discussing the shape (Section 3.1), use (Section 3.2) and origin (Section 3.3) of the augment.

3.1 Shape

Contemporary languages can be typologised according to the shape of their augment (cf. De Blois 1970: for a comparative overview). The reconstructed CV-shape, identical to the reconstructed shape of the pronominal prefix, can be found in a very restricted set of Eastern languages of zones J and G. Interestingly, the augment in these languages often only has a CV-shape in classes 1, 3, 4, 6 and 10, i.e., the nasal classes, in which the segmental form of the pronominal prefix differs from that of the nominal prefix. In other classes, the form of the augment is reduced to a vowel identical to that of the class prefix. A variation on this pattern can be found in Swati S43, for example, where the nasal classes have a V-augment and the other classes have no segmental augment (Ziervogel & Mabuza 1976). Given its distribution, the reduction of the augment to a single vowel may be explainable as partial haplology (see Patin *et al.* this volume for a synchronic case of allomorphy of the augment of class 7 that can be interpreted along the same lines). In most languages that have an augment, its shape is *V-* in all classes. Here, languages differ according to whether they have a "full" paradigm of three augment vowels, i.e., the three vowels that can be found in pronominal prefixes, or whether the paradigm has been partly or fully reduced. An example of the former is Rundi JD62 in (13), where we find *i-* in class 4, 5, 7, 8, 9 and 10, *u-* in class 1, 3, 11, 13, 14, 15 and 18, and *a-* in class 2, 6, 12 and 16 (Meeussen 1959: 62). Note that *u* and *i* are the reflexes of PB **ʋ* and **ɪ* in Rundi and that the augment vowels correspond to those of the pronominal prefixes reconstructed for their respective classes (see Table 8.1).

(13)   Rundi JD62 (Meeussen 1959: 71)
      ù-mùù-ntù / à-bàà-ntù   'person/s'   class 1/2
      ì-kìì-ntù / ì-bìì-ntù   'thing/s'   class 7/8

In partly reduced paradigms, the reflex of **a* has been replaced by that of **ɪ* or **ʋ*. Fully reduced paradigms have the same augment vowel in all classes. De Blois (1970) cites Mbundu H21 (*o-* in all classes), Hunde JD51 (everywhere *a-*) and Tonga M64 (always *i-*) as examples of the latter type, among many others. It is not always clear in such cases whether we are dealing with an augment according to the definition given at the beginning of this section or with a different morpheme altogether, as discussed for Eton A71, Fang A75 and Basaa A43a below. Some languages, scattered over the Bantu domain, do not have an augment, but show traces of its previous existence, which De Blois (1970: 107) calls "latent augments." Typically, these traces can be found in the tone and/or vowel quality of grammatical morphemes that end in -*a*, such as the Final Vowel of verb forms, the stem -*a* of the connective relator, and the comitative-instrumental preposition **nà*. The Shona S10 examples in (14b, d & f) show the effect of the latent augment on the vowel of the similative marker *sá*. There is no context in the language in which the nouns in (14a, c, e and g) appear with an initial vowel. The word for 'tree,' for instance, is never realised as *úmùtí*. Example (14f) shows that changes in the quality of the similative vowel cannot be ascribed to assimilation and (14h) shows that we do not find an altered similative vowel in front of a noun of class 1a, i.e., in front of nouns that should never have had an augment.

(14)   Shona S10 (Mudzingwa & Kadenge 2013: 89)
     a   mù-tí   'tree' (cl 3)
     b   só=mù-tí   'like a tree'
     c   ʧì-tótà   'locust' (cl 7)
     d   sé=ʧì-tótà 'like a locust'
     e   gòdò   'bone' (cl 5)
     f   sé=gòdò   'like a bone'
     g   sèkúrú   'uncle' (cl 1a)
     h   sá=sèkúrú 'like uncle'

Finally, there are many Bantu languages that do not have an augment or traces of it, especially in the North-West of the Bantu area.

3.2 Use

The definition of the augment provided at the beginning of Section 3 is necessarily restricted to formal characteristics. The use of the augment is too language-specific to allow for a crosslinguistically valid functional definition. In many languages it depends on an intricate set of conditions, including propositional act function (Sections 3.2.1 and 3.2.2), syntactic context (Section 3.2.3), discourse-referential properties (Section 3.2.4) and stylistics (Section 3.2.5).

In the majority of Bantu languages that have an augment, the default form of nouns is the augmented form. It is therefore usually more insightful and economic to provide conditions for its absence than for its presence (Greenberg 1978). The reason for this is that nouns are prototypically used to refer (Croft 2001: 88) and that the augment typically codes the propositional act of referring. It is therefore often absent in contexts where nouns are not used to refer and in those where nouns are inherently referential. The following is a comparative overview of the most recurrent conditions on the presence versus absence of the augment on nominal heads. These conditions are sometimes in competition and they are here presented in decreasing order of dominance.

3.2.1 *Absence of the augment on nouns that are not used to refer*

Nouns are not referring in vocatives, and also not in some cases when they are used for modification or predication. There are many augment languages where nouns used as vocatives lack an augment. In Zulu S42, for instance, "the general rule for the formation of vocatives from nouns is as follows: Elide the initial vowel of the noun-prefix" (Doke 1997: 280): *umuntu > muntu* (O person!).

Descriptions of individual languages are sometimes contradictory, as is the case for Ganda JE15, which has been described as lacking an augment in the vocative at least since Torrend (1891). However, vocatives do have an initial vowel augment according to Hyman and Katamba (1993), unless if followed by a second person pronoun. This may be due to dialectal differences and/or to language change.

Another context in which nouns are not used to refer is in attributive nominal predication. Predicate nouns lack the augment in Bemba M42, if they are not used to identify the subject.

(15) Bemba M42 (Givón 2001: 123)

    a   uyu   u-muu-ntu   muu-puupu
        DEM$_1$ AUG-1-person   1 -thief
        'This person is a thief.'

    b   uyu   u-muu-ntu   u-muu-puupu
        DEM$_1$ AUG-1-person   AUG-1-thief
        'This person is the/a thief (I told you about).'

A reflex of the absence of the augment in this use can be found in the phenomenon of predicative lowering in Makhuwa P31, a change in the tone pattern of nouns when they are used predicatively (Van der Wal 2006).

Finally, in some cases the absence of an augment could be attributed to the fact that the noun is used to modify, rather than to refer. De Blois (1970: 127) lists a number of eastern Bantu languages in which nouns lose their augment in adverbial use, i.e., when they modify a verb. In Gusii JE42, for instance, the adverb *botuko* 'at night' is derived from the class 14 noun *obotuko* 'night' by dropping the augment. Likewise, nouns tend to lack the augment when they are in the second, non-referring position of compounds, as in the Zulu S42 example in (16).

(16)   Zulu S42 (Buell 2009, citing von Staden 1973)
       u-m-lindimasango   'gatekeeper'   <   u-m-lindi   +   a-ma-sango
       AUG-1-gatekeeper   AUG-1-guard   AUG-6-gate

Thus far, we saw examples of the absence of an augment where nouns are not used to refer. The flip side of the coin is that the augment can be present on word classes other than nouns, when elements that usually fulfil the propositional act of modifying are nominalised and used to refer. Examples can be found in Nande JD42 (Valinande 1984), where adjectives, possessive pronouns and other adnominal modifiers can function as NPs if they are preceded by an augment (17).

(17)   Nande JD42 (Valinande 1984: 642, 709, 714)
       a   ò-mò-kìrá ɣw-áː-yɔ   'his tail' (class 9 possessor)
       b   ɔ́-ɣw-áː-yɔ̀   'his one'
       c   ò-mò-tí mù-kúhí   'the short tree'
       d   ò-mù-kúhí   'the short one'

3.2.2 Absence of the augment on nouns that are inherently referential

Proper names tend to lack an augment for a reason that is exactly the opposite of the non-referentiality discussed in Section 3.2.1. Unlike common nouns, proper names are inherently referential. Their referential status never needs to be coded, which is also the reason why they lack a noun class prefix, i.e., why they tend to belong to the so-called class 1a. There is a small number of languages in zones G, M and S, in which the nouns of this category do have an augment, but lack a class prefix. In languages such as Rundi JD62, proper names are productively derived by omitting the augment of a common noun (Meeussen 1959, Van de Velde 2009). The nominal prefix of the source noun is morphologically integrated into the name stem. It is often pointed out that personified animals in fairy tales have the same behaviour as proper names. In fact, they are names. Another category of nouns that behave like names, but that can be distinguished from them, are kinship terms, especially the suppletive ones, which are also inherently referring in any given discourse context, due to the deictic element in their definition. Suppletive kinship terms such as *kìtááwê* 'his father' and *bàzé* 'my husband' never take an initial vowel augment in Ganda JE15 (Hyman & Katamba 1993: 222). When borrowings are integrated into class 1a on formal grounds, i.e., due to their lack of a recognisable class marker, they are often fully assimilated to proper names, also in their incompatibility with the augment.

3.2.3 Absence of the augment in certain syntactic environments

Together with the propositional act function, the syntactic context is the most widespread conditioning factor for the presence or absence of the augment. Usually, it determines whether the use of the augment is prohibited, obligatory or syntactically optional, but only for those nouns that are not inherently augmentless for reasons specified in Sections 3.2.1 and 3.2.2. Hence, the propositional act function generally outranks the syntactic context as a conditioning factor. On clause level, the augment is often absent or contrastive, i.e., syntactically optional, on the object of a negative verb form. In Ganda JE15, the augment is likewise absent on an object noun when it is under focus as in (18b) (Hyman & Katamba 1993).

(18) Ganda JE15 (Hyman & Katamba 1993: 228)
    a y-à-gúl-ìr-à   à-bá-ànà   è-bí-tábó
       SP$_1$-PST-buy-APPL-FV   AUG-2-child AUG-8-book
       'He bought the children books.'

    b y-à-gúl-ìr-à   bá-ànà   è-bí-tábó
       SP$_1$-PST-buy-APPL-FV 2-child   AUG-8-book
       'He bought THE CHILDREN books.'

On the level of the noun phrase, head nouns often lose their augment in the presence of (certain types of) preposed modifiers. The augment can then either function as a phrasal affix and appear on the prenominal modifier (19) or be left out (20).

(19) Nande JD42 (Valinande 1984: 819)
    a ɔ̀-mò-tí   'the tree'
    b ɔ̀-yò-ntì mò-tí 'the other tree'

(20) Zulu S42 (Buell 2009)
    a i-ndawo leyo   'that place'
    b le ndawo   'this place'

Locative prefixes are also often incompatible with the augment. In many languages they are not preceded by the augment of their own locative class, nor are they followed by the augment of the noun to which they attach, but exceptions to both types of absence of the augment occur (De Blois 1970: 117–119, Grégoire 1975: 156–169).

A few postnominal modifiers have been found to be incompatible with the augment too, at least in parts of the Bantu domain, such as the interrogative 'how many' (Grégoire & Janssens 1999: 417).

A completely different type of syntactic conditioning within the noun phrase is attested in Basaa A43a and in the A70 languages, where the non-augmented form is the default form of nouns and an augment appears if and only if the noun is restrictively modified by certain adnominal modifiers. The augmented form can therefore be analysed as a construct form of the noun (Van de Velde 2017). The set of modifiers that require the augmented/construct form is language specific, but always includes postnominal demonstratives. Note, however, that the form of the augment in these languages is an invariant vowel *i-* or *é-*, sometimes reduced to a floating high tone, which makes it hard to determine whether this form is cognate with the forms identified as the augment in other Bantu languages.

3.2.4 Absence of the augment with indefinite or non-specific nouns

In many Bantu languages, an augmented form of the noun can contrast with a non-augmented form, at least in some syntactic contexts, sometimes including nouns in isolation.

The augment is usually described as expressing definiteness or specificity in such cases. Unfortunately, evidence for this tends to be restricted to isolated examples accompanied by a translation equivalent in French or English that contains a definite article for the Bantu form with an augment and an indefinite article for that without an augment (21). An oft-cited example of a language in which the augment is claimed to express definiteness is Zamba C322 (Bokamba 1971, Kamanda Kola 1994). Other examples include Ngazidja G44a (cf. Patin et al, this volume) and Nande JD42 (Valinande 1984: 432). These three languages have in common that the use of the augment on nouns modified by a postnominal demonstrative is obligatory.

(21)   Zamba C322 (Kamanda Kola 1994: 402)
    a   bá-bà- áná   bà-nd'   ó-dǎn-á
       AUG_2-2-child SP_2-PROG   INF-play-FV
       'The children are playing.'

    b   bà-áná   bà-nd'   ó-dǎn-á
       2-child SP_2-PROG   INF-play-FV
       'Children are playing.'

Patin (2017) does provide evidence for the use of the augment as a marker of definiteness in Ngazidja, in contexts where its use is syntactically optional. Similar evidence may exist for languages such as Zamba and Nande. Note that in none of these languages the augment is entirely dedicated to the expression of definiteness. In Zamba C322, for instance, it also functions as a nominaliser of modifiers and as an intensifier on nouns (including infinitives).

In the absence of clear evidence, some scepticism is warranted by the fact that relying on translation equivalents in European languages in order to establish whether the augment expresses definiteness can be treacherous. An example of this can be found in Fang Ntumu A75A, where the augment has been claimed to be a marker of definiteness (Ondo Mebiame 2001). However, Fang being an A70 language, the conditioning for the appearance of the augment is basically syntactic (see my claim in Section 3.2.3). The augment is triggered by the presence of certain types of modifiers that can be functionally characterised as anchoring (or localising) (Rijkhoff 2002: 173–212).[3]

Anchoring modifiers are those modifiers that allow the hearer to identify the referent of the noun phrase in the world of discourse by locating it in space or by linking it to an already identified entity. Modifiers that allow both an anchoring and a non-anchoring use are responsible for the impression that the augment marks definiteness, since the presence of an anchoring modifier changes the discourse-referential properties of a nominal constituent. The only concrete indication for calling the augment a definiteness marker in Fang Ntumu A75A is that the modifier -fɔ́ translates as 'the other' when the head noun takes an augment and as 'another' in the absence of an augment. However, it is clear that the presence of the augment changes the type of modification from non-selective to selective (and therefore anchoring), which merely implies a definite interpretation of the head noun. This analysis is confirmed by word order in the nominal constituent, in that anchoring modifiers obligatorily follow other types of modifiers in the A70 languages. Adnominal -fɔ́ in Fang Ntumu A75A can precede or follow a cardinal number in the absence of an augment (22a-b), but it has to follow when the head noun is augmented (22c-d) (Van de Velde 2017).

(22) Fang Ntumu A75A (Van de Velde 2017)

    a   màtá mə́⁺béɲ mə́⁺fə́
        mə̀-tá   mə́-běɲ   mə̀-fə́
        6-pile   PP$_6$-two   PP$_6$-other
        'Two other piles.'

    b   màtá mə́⁺fə́ mə́⁺béɲ
        mə̀-tá   mə̀-fə́   mə́-běɲ
        6-pile   PP$_6$-other   PP$_6$-two

    c   mə́tá mə́⁺béɲ mə́⁺fə́
        ᴴ-mə̀-tá   mə́-běɲ   mə̀-fə́
        AUG-6-pile   PP$_6$-two   PP$_6$-other
        'The two other piles.'

    d   *mə́tá mə́⁺fə́ mə́⁺béɲ

The role of the augment as a marker of specificity (or referentiality) in languages in which augmentless nouns can contrast with augmented nouns has been shown more convincingly, most notably for Bemba M42 (Givón 2001: 453). Nouns that are in the scope of a realis proposition are necessarily specific/referring, and therefore must take the augment in Bemba.

(23) Bemba M42 (Givón 2001)

    a   a-a-som-ene   i-ci-tabo
        SP$_1$-PST-read-PFV   AUG-7-book
        'She read a/the book.'

    b   *a-a-som-ene ci-tabo
        *'She read a book (not a specific one).'

In other contexts, where indefinite nouns are not necessarily referring, the absence of the augment signals non-referentiality.

(24) Bemba M42 (Givón 2001)

    a   a-a-fwaay-ile   u-ku-soma   i-ci-tabo
        SP$_1$-PST-want-PFV   AUG-15-read   AUG-7-book
        'She wanted to read a/the book' (a specific book)

    b   a-a-fwaay-ile   uku-soma   ci-tabo
        SP$_1$-PST-want-PFV   AUG-15-read   7-book
        'She wanted a book to read (any book).'

    c   ta-a-a-som-ene   i-ci-tabo
        NEG-SP$_1$-PST-read-PFV   AUG-7-book
        'He didn't read the book.'
        *'He didn't read a book.'

    d   ta-a-a-somene   ci-tabo
        NEG-SP$_1$-PST-read-PFV   7-book
        'He didn't read any book.'

The absence of the augment to signal non-referentiality in these examples is related to the cases of lack of an augment discussed in Sections 3.2.1 and 3.2.2. Moreover, the syntactic conditionings on clause level discussed in Section 3.2.3 must result from differences in frequency between a referential and a non-referential reading in contexts where languages such as Bemba M42 allow the presence of the augment to contrast with its absence.

3.2.5 Stylistic and prosodic conditionings

The use of the augment has been reported to be subject to stylistic conditions in some languages. Thus, according to Carter (1963, cited via Greenberg 1978: 254), speakers of Tonga M64 find the excessive use of the augment undignified, and there are considerable differences in the frequency of augment use according to text style. Likewise, a comparative study of eight languages of zone G in which the augment is on its way out or has lost its traditional type of distribution, has shown that the use of the augment depends on register and is sometimes used by communities to differentiate their speech from that of neighbouring groups (Aunio & Petzell 2013).

Finally, Bantu languages tend to "care about" prosodic well-formedness of words, affixes and stems, and there are numerous examples of the presence of an augment solely motivated by a necessity to comply to minimality constraint on words, for instance. Thus, Gisu JE31 nouns have an augment in the classes where the pronominal prefix differs segmentally from the nominal prefix, i.e., classes 3, 4, 6 and 10 (see Section 3.1 for this prosodic conditioning). In the other classes, an augment appears only before monosyllabic or vowel-initial stems (De Blois 1970: 94), i.e., *u-mu-ndu/ba-ba-ndu* 'person/people' (cl. 1/2) vs. *ba-kana* 'girls' (cl. 2).

3.3 Origin

It is undeniable that augments or traces of them can be found throughout the Bantu area, also in the North-West (Grégoire & Janssens 1999). Meeussen (1967: 99) tentatively reconstructs it as a prenominal demonstrative form, "a separate word, identical in form with the pronominal prefix, and used as a weak demonstrative, or rather anaphoric, in affirmative, non-predicative constructions, with definite meaning." Greenberg (1978: 254) characterises the Proto-Bantu augment as (probably) a "stage II" article. Stage II refers to a phase in a continuous evolution of markers that start out as demonstratives (stage zero) and become definite articles, i.e., pragmatically conditioned markers of definiteness (stage I). The presence of these markers subsequently becomes grammatically conditioned. They tend to become obligatory with common nouns, except in a number of language-specific grammatical contexts (stage II). The fact that the article has lost its functional load at stage II can either lead to its generalisation, so that it becomes a mere nominal marker (stage III), or to its erosion and disappearance. Thus, both Meeussen (1967) and Greenberg (1978) reconstruct the Proto-Bantu augment at a rather advanced stage of grammaticalisation, assuming that syntactic restrictions on its distribution already existed before the Bantu languages dispersed. However, it is very difficult to determine whether Proto-Bantu had an augment and, if it did, at what stage of the Greenberg cycle it has to be situated. The reason is that there is ample evidence for the existence of multiple cycles of augment creation and loss in Bantu and beyond. Some of this evidence is presented in the remainder of this section.

For instance, we saw that the augment in Zamba C322 is claimed to express definiteness, i.e., that at least in some of its uses it functions as a Stage I article, which puts it earlier in the cycle of grammaticalisation than the proposed reconstructed augment. Moreover, Zamba is the only language in its region and genealogical subgroup that has a fully functional augment (Kamanda Kola 1994). The other languages of zone C, such as Doko C301 (Twilingiyimana 1984), show at most some traces. The Zamba C322 augment should be interpreted as an innovation, rather than an exceptional retention. Formal evidence for this is provided by the tone of the nominal prefix in Zamba, which Kamanda Kola (1993) analyses as underlyingly high. Since nominal prefixes are low in Proto-Bantu and the contemporary languages surrounding Zamba and since the augment has a high tone, the most straightforward explanation for the high tone on Zamba nominal prefixes is that it is the reflex of an augment that generalised into a Stage III article. Consequently, the current segmental augment must be a renewal. There are other Bantu languages in which the basic tone pattern of nouns includes a generalised historical augment High. An interesting example is Yombi H16c, where a tonal distinction between definite and indefinite nouns has been innovated by the creation of a new indefinite form, derived from the form of nominal predicates (Blanchon 1998).

Indications for the renewal of augments can also be found in languages with two paradigms of augments, one of which typically has a CV-shape and the other a V-shape. This can be found in Safwa M25, Kinga G65 and Nyakyusa M31, where the CV-augment is claimed to have a special "emphatic" use (De Blois 1970: 98, confirmed for Nyakyusa by Bastian Persohn pers. comm.)

(25) Nyakyusa M31 (Bastian Persohn, pers. comm.)
ʊ-mu-ndʊ 'the person'
ʊ-mu-ndʊ ʊ-jʊ 'this person'
jʊ-mu-ndʊ 'the very person'

The Zamba C322 augment can also be used as an intensifier (26b), and preposed demonstratives have been characterised as "emphatic" in Nkore JE13, Kanyok L32, Bemba M42 and Bolia C35b, as opposed to their postposed counterparts (Van de Velde 2005). In fact, it is not always straightforward to know whether we are dealing with an augment or rather with a prenominal demonstrative.

(26) Zamba C322 (Kamanda Kola 1994: 403)
a ândó wà Mádángá
'He is from (the village of) Madanga.'

b ândó wà Mǎdángá (má-mádángá)
'He is from Madanga itself, from the very village of Madanga.'

The existence of augment-like morphemes in Wide Bantu or Bantoid languages has been adduced as evidence for the antiquity of the augment in Bantu, but it can just as well show that Niger-Congo languages with elaborate noun class systems are very prone to develop augment-like categories and that this strong tendency constantly feeds cycles of grammaticalisation within Narrow Bantu as well. To give only one example, the Atlantic language Keerak has a morpheme that Segerer (2015: 107) calls "minimal determiner." It is an enclitic =aC (where C is identical to the consonant of the noun class prefix), which appears in most syntactic contexts, including in isolation. In the few contexts where the minimal determiner is optional, e.g., before an adjective, its presence may express definiteness.

4 ADNOMINAL MODIFIERS

Adnominal modifiers normally agree in noun class with their head noun. They are traditionally classified according to their paradigm of agreement markers. Meeussen (1967: 97) reconstructed the five paradigms provided in Table 8.1: nominal prefixes, enumerative prefixes, pronominal prefixes, subject prefixes and object prefixes, the first four of which are used to mark agreement within the noun phrase. Contemporary languages may distinguish either more or less different paradigms than those reconstructed in Proto-Bantu. This section provides a brief overview of some types of adnominal modifiers.

4.1 Demonstratives

Bantu languages tend to be rich in demonstratives, both in types and in frequency of use. They typically have three or four series of demonstratives, which, in their exophoric use, express different degrees of spatial distance from the speaker and/or the hearer.

(27) Nande JD42 (Valinande 1984: 787–792)
a ɔ̀-mʊ̀-sìkáà ɔ̀-nɔ̀   'this girl (within reach of the speaker)'
b ɔ̀-mʊ̀-sìkáà ɔ̀-lìà   'that girl (far from speaker and hearer)'
c ɔ̀-mʊ̀-sìkáà ò-ìù   'this girl (close to the speaker, but out of their reach)'
d ɔ̀-mʊ̀-sìkáà ɔ̀-ìʊ̀-ɔ̀   'that girl (close to the hearer)'

Formally, demonstratives minimally contain the pronominal prefix, sometimes accompanied by a stem and a full or partial repetition of the agreement marker. Reconstructed demonstrative stems include *-nóò [close to speaker] and *-díà [far from speaker and hearer] (Meeussen 1967: 107). Recent research on demonstratives in Bantu, most notably by Nicolle (2012, 2014), has focused on their discourse functions, an important aspect of their use that is hardly mentioned in grammatical descriptions. In a comparative corpus study of ten Eastern Bantu languages, Nicolle (2014) identified four different discourse level functions of demonstratives, viz. indicating the so-called activation status of participants (distal demonstratives reactivate participants after a period of inactivity in languages such as Jita JE25); indicating agentivity (agentive participants are modified by distal demonstratives, others by referential demonstratives); "text structuring" (distinguishing between different types of participants, delimiting episodes) and thematic development (indicating important developments in the narrative).

4.2 Connectives (genitives)

Nominal possessors are introduced by a relator that typically consists of a pronominal prefix and a stem -a, the tone of which is identical to that of the prefix. Bantuists call this relator the "connective," sometimes also "connexive" or "associative" marker.

(28) Nyamwezi F22 (Maganga & Schadeberg 1992: 89)
 m̀-zuna   w-aa-m̀-kúma
1-younger_sister PP_1-CON-1-woman
'the younger sister of the woman'

Van de Velde (2013) provides a typology of connective constructions in the Bantu family, which shows, among other things, that the expression of linguistic possession is only one of their multiple functions. In fact, possessive examples such as (28) may be relatively rare in discourse in North-Western Bantu languages, since many of them prefer external possessor constructions, in which the possessor is expressed as an argument of the verb. In Orungu B11b, for instance, there are almost no restrictions on external possessors and it is not unusual to find multiple external possessors in one clause (29b).

(29) Orungu B11b (Odette Ambouroue p.c.)

  a áfè wádyóní ìgúgè ɲ ínàgò yâmì

     á-fè   wá-á-dyón-ì   ì-gûgè   ɲ-á   ì-nâgò   ìy-âmì

     2-burglar.DTP SP$_2$-PRF-break-PRF 5-door.DTP PP$_5$-CON AU$_9$-9.house.DTP PP$_9$-POSS$_{1SG}$

     'The burglars have broken down the door of my house.'

  b áfè wádyóní myɛ́ nág ìgúgè

     á-fè   wá-á-dyón-ì   myɛ́   nâgò   ì-gûge

     2-burglar.DTP SP$_2$-PRF-break-PRF 1SG 9.house.DTP 5-door.DTP

     'The burglars have broken down the door of my house.' (lit. 'have broken me the house the door')

4.3 Possessive pronouns

Possessive pronouns consist of a stem and a pronominal prefix that agrees with the possessee. In most Bantu languages, the form of the stem depends on the person and number of the possessor, but in some languages it agrees with the possessor in noun class as well, giving rise to huge paradigms. These languages can be found in different regions, e.g., Herero R31 (Möhlig *et al.* 2002: 60), Binza C321 (Van Leynseele 1977: 35) and Ha JD66 (Harjula 2004: 69) and Nande JD42 (Valinande 1984). Possessive pronouns for non-human possessors are connective constructions with a pronominal possessor.

(30) Herero R31 (Möhlig *et al.* 2002: 60)

  a ò-mù-tí   n-ò-ví-yàò   vy-á-⁺w-ó

     AUG-3-tree   and-AUG-8-leave PP$_8$-CON-PP$_3$-PPR

     'the tree and its leaves'

  b ò-tjì-kúnìnò   n-ò-mí-tí   vy-á-⁺ty-ó

     AUG-7-garden and-AUG-4-tree PP$_4$-CON-PP$_7$-PPR

     'the garden and its trees'

The Bantu languages typically have a number of suppletive kinship terms, in which a kinship term and a possessive pronoun are merged. A comparative-historical study of the form of possessive pronouns can be found in Kamba Muzenga (2003).

4.4 Numbers

Cardinal numbers from 'one' to 'five' usually agree with the noun they modify, sometimes using a dedicated set of numeral agreement markers. From 'six' upwards they tend to be uninflected (Stappers 1965, Meeussen 1967). Ordinal numbers are usually expressed by means of a connective construction with a cardinal number in the modifier position (Polak-Bynon 1965).

(31)   Makwe G402 (Devos 2008: 133)
       sìíkù   y-a-táànò
       9.day PP_9-CON-five
       'the fifth day'

4.5 Relative clauses

Some parameters of variation in the structure of relative clauses across Bantu languages are briefly discussed by Downing & Marten (this volume). Two additional parameters of variation related to the agreement prefix of the relative verb are especially relevant for this section: its paradigm and its controller (Nsuka-Nkutsi 1982). Relative verbs can take an agreement prefix of the verbal (subject) paradigm or of the pronominal one. In non-subject relatives, the controller of agreement can be either the subject of the relative verb or the relativised noun. There is a partial correlation between these two parameters, in that relative verbs that always agree with their subject never take a pronominal prefix. The reverse correlation is strong, but not without exceptions: the great majority of relative verbs that agree with the relativised noun take a pronominal prefix rather than a verbal prefix. A straightforward explanation for this correlation is that the pronominal prefix in relative verbs originates in a demonstrative that was used to nominalise verb forms.

4.6 Adjectives

Bantu adjectives are defined as agreeing words that take an adjectival prefix, a paradigm that is fully or nearly identical to that of the nominal prefixes. The contents of this Bantu-specific word category differs from language to language. In some languages, such as Ngazidja G44a (cf. Patin et al. this volume), it also contains numbers. Most Bantu languages have a limited set of adjectives (e.g., eight in Rwanda JD61), some have none at all (e.g., Eton A71, Mongo C61). A small number of qualifying adjectives can be reconstructed into Proto-Bantu: *-bíì 'bad,' *-bícì 'unripe,' *-dàì 'long,' *-dìtò 'heavy,' *-nénè 'big,' *-páì 'new' and *-tádí 'long' (Meeussen 1967, Baka 2000, Bastin et al. 2002, Baka 2005). In languages with an open class of adjectives, often most are derived from verbs, usually change-of-state verbs. The most common productive deverbal derivation types involve the suffixes *-é (Bastin 1989) and *-ú (Schadeberg 2003: 81).

The scarcity of adjectives in the majority of languages is compensated for by alternative strategies for nominal qualification. A common strategy is to use a connective construction, in which the modifying noun can be a property denoting (32) or entity denoting (33) noun or the infinitive of a change-of-state verb (34).

(32)   Ha JD66 (Harjula 2004: 135)
       umutaama wíkigongwe
       u-mu-taama   u-a   i-ki-gongwe
       AUG-1-old.man PP_1-CON   AUG-7-kindness
       'a sympathetic old man'

(33) Mongo C61 (Hulstaert 1966: 247)
      e-kútu   ě-a=n-dɔsɔ́
      7-calabash   PP$_7$-CON=10-pore
      'a porous calabash'

(34) Digo E73 (Nicolle 2013)
      ng'ombe   z-a   ku-ond-a
      10.cow   PP$_{10}$-CON   INF-become.thin-FV
      'thin cows'

In some North-Western Bantu languages, the qualifying noun is construed as the mor-phosyntactic head of the connective construction, in a construction type that instantiates possessee-like qualifier constructions. Such constructions are typical in a large area of northern Sub-Saharan Africa.

(35) Eton A71 (Van de Velde 2008: 214)
      ìvèvèz ḿpég í⁺té kù
      ì-vàvèz   ᴴ=ɴ̀-pɛ́g   í-ᴸtɛ́   ᴸ-kù
      7-light   PP$_7$.CON=3-bag   SP$_7$-PRS   INF-fall
      'The light bag falls.'

Some Southern Bantu languages (S30–40) have a word class traditionally called "rela-tives," which has the functions and distributional potential of adjectives. These words have a stem preceded by an agreement prefix of the paradigm of subject markers on verbs. As most other adnominal modifiers in these languages, they are obligatorily pre-ceded by a linker in attributive use. Creissels (2014), who prefers the term "new adjec-tives," argues that they emerged as a new word class from the use of nouns as descriptive predicates. Example (36) contrasts an old type adjective (36a) with a new adjective (36b) in Tswana S31.

(36) Tswana S31 (Creissels 2014)
      a   mò-sádì   jó   mò-léèlé
          1-woman   LNK$_1$ NP$_1$-tall
          'A tall woman.'

      b   mò-sádì   jó   ú-bòtɬʰálí
          1-woman   LNK$_1$ SP$_1$-clever
          'A clever woman.'

Other strategies for adnominal qualification include the use of relative clauses and participles.

5 THE STRUCTURE OF THE NOUN PHRASE

The Bantu languages show a lot of variation between them in the word order patterns of the noun phrase. Some of the patterns that can be found in Bantu, such as N DEM ADJ NUM or N NUM DEM ADJ, are typologically very unusual or have never been attested in the typological literature (see for instance Greenberg 1963, Hawkins 1983, Rijkhoff 2002,

Cinque 2005), and (Wide) Bantu languages have had a heavy impact on the way universals of word order in the NP have been formulated. Thus, Greenberg's universal 20 states that demonstratives, numbers and adjectives occur in this order (i.e., DEM NUM ADJ) when they all precede and in the same or in the exact opposite order if they all follow (Greenberg 1963). The fact that Greenberg allows the order DEM NUM ADJ among postnominal modifiers is apparently due to his knowledge of Gikuyu E51, which is not part of his 30-language sample. Hawkins (1983) later reformulated Universal 20, based on a sample of more than 300 languages, stating that nothing can be predicted about the mutual ordering of postnominal demonstratives, numbers and adjectives. The Wide Bantu languages Aghem and Noni, which allow N ADJ DEM NUM and N DEM NUM ADJ respectively, are responsible for this further weakening of Greenberg's universal 20.

In light of this prominence of the Bantu languages in the general literature, it is somewhat surprising that there exists no comparative study of the structure of the noun phrase in Bantu. The following subsections are a first attempt. Comparing the word order in the NP of individual Bantu languages to typological findings is complicated by three factors. First, the three adnominal modifiers to which typologists have restricted their studies exclude the possessive pronoun, the position of which is noteworthy in Bantu. Second, it is not a priori clear which kind of definition for the different types of adnominal modifiers would produce the most insightful results in a comparative study, a functional one or a formal one. For instance, according to Rijkhoff's (2002) (functional) layered model of the NP, we expect numbers to be serialised differently from qualifiers, whatever the language-specific grammatical category to which they belong. In many Bantu languages, numbers from 'one' to 'five' are analysed as Adjectives, since they take the same paradigm of agreement markers as qualifying adjectives. The two are therefore often not distinguished in schematic representations of word order in the NP. The third complicating factor in comparing word order in Bantu NPs to what is found in the typological literature is that many Bantu languages allow for considerable syntactic freedom within the NP.

5.1 Common word order patterns

Most adnominal modifiers are postnominal, but demonstratives (37a), various quantifiers (37b) and/or the modifier 'other' (37c) are prenominal in many languages. This is also sometimes the case for possessive pronouns, for which prenominal position is usually (or always?) syntactically optional and pragmatically marked, e.g., in Songye L23 (Stappers 1964) and Makaa A83 (38b).

(37) Ha JD66 (Harjula 2004: 75, 131)

    a izo   súka   zi-bíri
       DEM_{10} 10.hoe PP_{10}-two
       'those two hoes'

    b burú   mu-ntu
       each   1-person
       'each person'

    c u-wú-ndi   mú-si
       AUG-PP_3-other 3-day
       'another day'

(38) Makaa A83 (Heath 1998: 3)

   a me   angane   dug   boog   j-am
      1SG   PROG.NEG see   7.hoe PP_7-$POSS_{1SG}$
      'I don't see my hoe.'

   b me ke   gule   ne   j-am   boog, wo   ke   ne   gwoo   boog
      1SG go   hoe   with PP_7-$POSS_{1SG}$ 7.hoe 2SG   go   with $PP_7.POSS_{2SG}$ 7.hoe
      'I am going to hoe with MY hoe, you go with YOUR hoe.'

Among postnominal modifiers, the position of the possessive pronoun is remarkable, as in many languages it is obligatorily adjacent to the noun, i.e., preceding any other postnominal modifiers. Again, this is typologically highly unusual, since possessive pronouns are typically anchoring modifiers, which are nearly universally placed further away from the head noun than classifying, qualifying and quantifying modifiers (Rijkhoff 2008). The same is true for postposed demonstratives, which in some languages are serialised immediately after the noun, or after the possessive pronoun, preceding other modifiers. Some examples of the order of demonstratives, possessive pronouns, numbers and adjectives are: Digo E73 DEM N POSS NUM ADJ (Nicolle 2013), Nande JD42 N POSS NUM ADJ DEM (Valinande 1984: 633), Nkore-Kiga JE13–14 N POSS DEM ADJ NUM (Taylor 1985: 55 who adds that "this order is not rigidly adhered to"), and Orungu B11b N POSS ADJ DEM (Odette Ambouroue pers. comm., see below for the position of NUM).

The preference for the immediate postnominal position of possessive pronouns is so strong in some Bantu languages, that it gives rise to external possession within the noun phrase, as in the Mbugwe F34 example in (39) (cf. Wilhelmson, this volume). Semantically, the possessive pronoun modifies the noun 'body' in the connective construction, but formally it follows the head noun of the NP, with which it also agrees.

(39) Mbugwe F34 (Wilhelmson, this volume)

     n-yèèŋgɔ   jì-ááné   jì-ɔ́ɔ́nsè   jì-á   mò-vèrè
     10-joints PP_{10}-$POSS_{1SG}$ PP_{10}-all PP_{10}-CON 3-body
     'All the joints of my body.'

The core position that the possessive pronoun occupies in the NP structure of many Bantu languages, is also reflected in its agreement properties. As was said in Section 2.2.2.2, possessive pronouns are at the top of the agreement hierarchy in some Bantu languages, since they are the only agreement targets that can have syntactic agreement with animate controllers. Strangely, syntactic agreement is restricted to certain noun classes. In Swahili G42, human controllers trigger animate agreement on all agreement targets, except on the possessive pronoun, if the head noun has a class 9/10 prefix.

(40) Swahili G42 (Van de Velde 2006b: 195)

     N-dugu   y-angu   a-me-anguk-a
     9-brother PP_9-$POSS_{1SG}$ SP_1-PFV-fall-FV
     'My brother has fallen.'

Likewise, in Fe'Fe' (Wide Bantu), possessive pronouns are the only agreement target for which the full range of noun classes are differentiated (Hyman 1972).

5.2 Word order variation in the NP

In many Bantu languages, the mutual ordering of some or all postnominal modifiers is syntactically free, represented by curly brackets around the interchangeable modifiers in the schemes that follow: Basaa A43a N {POSS, NUM, ADJ} DEM (DEM and POSS can both or either be prenominal as well) (Hyman 2003); Ha JD66 DEM QUANT N {POSS, ADJ, NUM} {REL, PART} (Harjula 2004); Eton A71 N {POSS, NUM, CON} REL DEM (Van de Velde 2008: 227). The position of cardinal numbers is syntactically free in Orungu B11b, and in their presence the otherwise rigid order of DEM and ADJ becomes free as well. The possessive pronoun remains in immediate postnominal position (Odette Ambouroue, pers. comm.): N POSS {NUM, ADJ, DEM}.

The possible semantic and information structural implications of alternative word orders in the noun phrase have received little attention in the literature. For Rwanda JD61, Wilkins & Kimenyi (1975) claim that the order of adnominal modifiers is determined by a generality hierarchy, such that the modifiers that contribute most to the identification of the NP's referent are placed at the back. This hierarchy can explain both the default order in Rwanda and departures from it. In the following example, the relative clause can optionally precede the adjective only if it expresses a habitual activity, i.e., a more or less permanent quality.

(41)   Rwanda JD61 (Wilkins & Kimenyi 1975: 160)

    a   u-mu-gabo   mu-re-mu-re   u-ririimb-a   n'   u-mu-txwaare
       AUG-1-man NP_1-tall-NP_1-tall PP_1-sing-FV   is   AUG-1-chief
       'The tall man who sings is the chief.'

    b   u-mu-gabo   u-ririimb-a   mu-re-mu-re   n'   u-mu-txwaare
       AUG-1-man PP_1-sing-FV NP_1-tall-NP_1-tall   is   AUG-1-chief
       'The TALL man who sings, is the chief.'

5.3   Towards a historical explanation for the uncommon word order patterns in Bantu NPs

A possible explanation for the typologically unusual word order patterns in Bantu is that in many languages the noun phrase is or was not fully integral, and that the relation between a head noun and some of its semantic modifiers is or was appositional. There are several types of indications for the validity of this hypothesis. First, as we saw in Section 3.2.1, the augment is often used to nominalise modifiers. The fact that several types of adnominal modifiers can take the augment in some Bantu languages is an indication that these modifiers are or were nominalised and in an appositional relation with the head noun. In Bemba M42, for instance, modifiers are non-restrictive – and therefore arguably in loose apposition – when they take an augment, and restrictive when they don't (Givón 1974: 132, Kasonde 2009: 167).

(42)   Bemba M42 (Givón 1974: 132, Kasonde 2009: 167)

    a   a-ba-ntu   ba-suma
       AUG-2-person NP_2-good
       'the good people'

    b   a-ba-ntu   a-ba-suma
       AUG-2-person   AUG-NP_2-good
       'the people, who were (all) good'; 'the people, the good ones'

c a-ba-ana   ba-andi
   AUG-2-child   PP₂-POSS₁ₛ₉
   'my children'

d a-ba-ana   a-ba-andi
   AUG-2-child   AUG-PP₂-POSS₁ₛ₉
   'the children, those that are mine'

A number can only be inserted between the noun and the adjective if the latter is augmented/apposed (Kasonde 2009).

(43) Bemba M42 (Kasonde 2009)
   a à-báá-ntù   bà-bìlì   á-bà-kúlú
      AUG-2-person   NP₂-two   AUG-NP₂-big
      'the two men, the big ones'

   b *à-báá-ntù   bà-bìlì   bà-kúlú

The use of appositive structures involving a nominalised modifier is not necessarily reserved for non-restrictive modification. It can also distinguish the anchoring use of modifiers from other uses. In Babole C101, for instance, the connective relator has a short and a long form (Leitch 2003). The long form consists of the short form, prefixed by an augment. It is used when the connective construction expresses linguistic possession, i.e., when it has an anchoring function. Connective modifiers with a classifying or qualifying function take the short, non-augmented form of the connective relator. The same distinction exists in Zamba C322 (Kamanda Kola 1994).[4]

(44) Zamba C322 (Kamanda Kola 1994: 405)
   a ímúntɔ̀dù mwä mwäsì   'the navel of a woman'

   b imúntɔ̀dù ímwä mwäsì   'the woman's navel'

The creation of such appositive structures may be instigated by the tendency, noted for Rwanda JD61 in Section 5.2, to put the modifiers that contribute most to the identification of the NP's referent at the right edge of the phrase.

Second, there is morphosyntactic evidence for analysing certain modifiers as extraphrasal. In Ganda JE15, for instance, an adnominal modifier can be "exbraciated." It then follows any other modifiers and takes the augment, also if the noun phrase from which it is extracted follows a negative verb form, which does not allow for the augment (Hyman & Katamba 1993).

(45) Ganda JE15 (Hyman & Katamba 1993)
   a tè-y-à-láb-à   bì-tábó   bì-néné   bì-nó
      NEG-SP₁-PST-see-FV 8-book   NP₈-big   PP₈-DEM
      'He didn't see these big books.'

   b tè-y-à-láb-à   bì-tábó   bì-nò   è-bì-nènè
      NEG-SP₁-PST-see-FV 8-book   PP₈-DEM   AUG-NP₈-big
      'He didn't see these big books.'

In Nen A44, which has verb-final clausal syntax, part of a discontinuous object NP can even end up after the verb, marking contrastive focus on the extracted modifier (Mous 2003).

(46) Nen A44 (Mous 2003: 345)

mè-ná	ìmìtə̀	yè	mʷə̀nífí	índí	mè-ŋéŋ	ò	hè-lɔ́bátɔ̀
1SG-HOD.PST	9.calabash	CON₉	6.water	give.HOD.PST	NP₉-big	LOC	19-child

'I gave the BIG water calabash to the child.'

Recall that the augment is a renewal of agreement morphology, that cycles of augment creation and disappearance are likely to have occurred and that eventually all agreement markers within the noun phrase must have started as demonstratives used to nominalise adnominal modifiers.[5] Consequently, the structure of the noun phrase in contemporary languages can be the result of cycles of appositional extraposition of modifiers followed by subsequent reintegration, giving rise to multiple layers of modifiers around a nuclear NP. This scenario should be tested in a thorough comparative study, which should pay special attention to prosodic clues for the syntactically layered nature of Bantu NPs. In loose apposition, an anchor is typically separated from the appositive by means of a prosodic boundary, as O'Connor and Patin (2015) have demonstrated for Ngazidja G44a. Differences between types of modifiers in prosodic bonding with their head in the Bantu languages could be reflexes of such boundaries. In Binza C321, for instance, there is high tone dissimilation with the last syllable of a nuclear NP (N (ADJ)) on a following demonstrative or possessive pronoun, but not on a following quantifier or connective (Van Leynseele 1977). In Makwe G402, adnominal modifiers can be classified into conjoint, disjoint and conjoint/disjoint, depending on whether they must, cannot or may form a phonological phrase with the immediately preceding head of the NP (Devos 2008: 377–382). Finally, nouns that end in two high tones become high-low before a pause in Tswana S31. They acquire the same final tone pattern when preceding certain types of adnominal modifiers. Since the final HL pattern is phonologically unpredictable in these cases, Creissels (2009: 79) rightly analyses it as a construct form of the noun.

NOTES

1 These observations are based on data from a sample of 15 languages extracted from the RefLex database (Segerer & Flavier 2011–2016). The sample is restricted to sources of at least a thousand lexical entries, created by specialists of the language (i.e., excluding lexical surveys) and counts exclude the even-numbered classes and everything up from class 13 (included). Percentages of nouns belonging to class 1 in the (North-)West include: 2% (Eton A71), 1% (Ngom/Koya B22), 5% (Viya B301) and 2% (Yombe H16c). More to the East we find: 14% (Lega D25), 13% (Nyamwezi F22), 16% (Ndamba G52), 15% (Makonde P23) and 11% (Tswana S31). The North-Western A80 languages are a notable exception to this trend with an average of 20% of nouns belonging to class 1 in three languages (Kol A832; Njyem A84 and Mekaa A83). In two of these languages, class 9 is marginal, which may or may not be a coincidence. For class 7, differences in size between western and eastern languages are less dramatic. Typical figures are 34% (Njyem A84) versus 16% (Makonde P23).

2 These subclasses are not to be confused with the notion of "subgenders" in the typological literature, which involve minimally distinct agreement patterns (Corbett 1991: 163).

3 Restrictive relative clauses always require the augment in these languages, whatever the type of modification they provide. This may be due to the historical or current presence of a demonstrative relativiser.

4 Both authors, Leitch (2003) and Kamanda Kola (1994), describe the distinction in terms of definiteness, but again this analysis seems to be more inspired by the French translation equivalents of these examples than by the Babole C101 and Zamba C322 data. Note in this respect that, if the augment were a definiteness marker in (44b), we would have expected it to agree with and be prefixed to the dependent noun 'woman.' Instead, it is prefixed to the connective relator, and it agrees with the head noun 'navel.'

5 The linkers that are obligatorily used with a subset of adnominal modifiers in Southern Bantu languages, such as Tswana S31 (Creissels 2006: 75, 2014) are another instance of the same general phenomenon. They originate in demonstratives and can be used to turn sequences of proclitic subject marker plus descriptive predicate into adnominal modifiers, arguably via a stage of nominalisation.

REFERENCES

Aunio, L. & M. Petzell (2013) The Impact of Sociolinguistic Factors on the Usage of the Pre-Prefix in Some G Languages. Paper presented at *Fifth International Conference on the Bantu Languages, June 12–15*, Paris.

Baka, J. (2000) *L'adjectif en bantu*. Bruxelles: Université libre de Bruxelles, thèse de doctorat.

Baka, J. (2005) Pourquoi une forme de type *-pái "nouveau" en protobantou. In Bostoen, K. & J. Maniacky (eds.), *Studies in African Comparative Linguistics, with Special Focus on Bantu and Mande*, 139–146. Tervuren: Royal Museum for Central Africa.

Bastin, Y. (1989) Les déverbatifs bantous en -e. *Journal of African Languages and Linguistics* 11: 151–174.

Bastin, Y., A. Coupez, E. Mumba & T. C. Schadeberg (eds.) (2002) *Bantu Lexical Reconstructions 3*. Tervuren: Royal Museum for Central Africa (Available online at linguistics.africamuseum.be/BLR3.html, Accessed November 28, 2017).

Blanchon, J. A. (1997) Les formes nominales de citation a prefixe haut en pounou (bantu B43). *Journal of African Languages and Linguistics* 18(2): 129–138.

Blanchon, J. A. (1998) Semantic/Pragmatic Conditions on the Tonology of the Kongo Noun-Phrase: A Diachronic Hypothesis. In Hyman, L. M. & C. W. Kisseberth (eds.), *Theoretical Aspects of Bantu Tone*, 1–32. Stanford: Center for the Study of Language and Information, Stanford University.

Bokamba, G. D. (1971) Specificity and Definiteness in Dzamba. *Studies in African Linguistics* 2: 217–237.

Buell, L. (2009) The Distribution of the Nguni Augment: A Review. Paper presented at *Bantu Augment Workshop, Leiden University, June 17*, Leiden.

Carter, H. (1963) Coding, Style and the Initial Vowel in North Rhodesian Tonga: A Psycholinguistic Study. *African Language Studies* 4: 1–42.

Cinque, G. (2005) Deriving Greenberg's Universal 20 and Its Exceptions. *Linguistic Inquiry* 36(3): 315–332.

Corbett, G. G. (1979) The Agreement Hierarchy. *Journal of Linguistics* 15: 203–224.

Corbett, G. G. (1991) *Gender*. Cambridge: Cambridge University Press.

Crane, T. M., L. M. Hyman & S. Nsielanga Tukumu (2011) *A Grammar of Nzadi (B.865) A Bantu Language of the Democratic Republic of Congo with an Appendix of Proto-Bantu – Nzadi Sound Correspondences by Clara Cohen*. Berkeley: University of California Publications.

Creissels, D. (2006) *Syntaxe générale, une introduction typologique. 2. La phrase*. Paris: Lavoisier.

Creissels, D. (2009) Construct Forms of Nouns in African Languages. In Austin, P. K., O. Bond, M. Charette, D. Nathan & P. Sells (eds.), *Proceedings of Conference on Language Documentation & Linguistic Theory 2*, 73–82. London: SOAS.

Creissels, D. (2014) The "new adjectives" of Tswana. In Simone, R. & F. Masini (eds.), *Word Classes: Nature, Typology and Representations*, 75–94. Amsterdam; Philadelphia: John Benjamins.

Croft, W. (2001) *Radical Construction Grammar: Syntactic Theory in Typological Perspective*. Oxford: Oxford University Press.

Daniel, M. & E. A. Moravcsik (2005) Associative Plurals. In Dryer, M. S., M. Haspelmath, D. Gil & B. Comrie (eds.), *The World Atlas of Language Structures*, 150–153. Oxford: Oxford University Press.

De Blois, K. F. (1970) The Augment in the Bantu Languages. *Africana Linguistica* 4: 85–165.

Devos, M. (2008) *A Grammar of Makwe*. Munich: LINCOM.

Dimmendaal, G. (2011) *Historical Linguistics and the Comparative Study of African Languages*. Amsterdam; Philadelphia: John Benjamins.

Doke, C. M. (1927) The Significance of Class 1a of Bantu Nouns. In Boas, F., O. H. Dempwolff, G. Panconcelli-Calzia, A. Werner & D. Westermann (eds.), *Festschrift Meinhof: Sprachwissenschaftliche und andere Studien*, 196–203. Hamburg: : Kommissionsverlag von L. Friederichsen.

Doke, C. M. (1997) *Textbook of Zulu Grammar. Sixth Edition*. Cape Town: Maskew Miller Longman.

Doneux, J. L. (1967) Données sur la classe 15 nominale en bantou. *Africana Linguistica* 3: 1–22.

Ernst, U. (1992) *Esquisse grammaticale du kako*. Yaoundé: Summer Institute of Linguistics.

Forges, G. (1983) La classe de l'infinitif en bantou. *Africana Linguistica* 9: 257–264.

Givón, T. (1974) Syntactic Change in Lake Bantu: A Rejoinder. *Studies in African Linguistics* 5(1): 117–139.

Givón, T. (2001) *Syntax: An Introduction*. Amsterdam; Philadelphia: John Benjamins.

Good, J. (2012) How to Become a "Kwa" Noun. *Morphology* 22(2): 293–335.

Greenberg, J. H. (1963) Some Universals of Grammar with Particular Reference to the Order of Meaningful Elements. In Greenberg, J. H. (ed.), *Universals of Language*. Cambridge: M.I.T. Press.

Greenberg, J. H. (1978) How Does a Language Acquire Gender Markers. In Greenberg, J. H., C. A. Ferguson & A. Moravcsik (eds.), *Universals of Human Language, Volume 3 Word Structure*, 47–82. Stanford: Stanford University Press.

Gregersen, E. (1967) *Prefix and Pronoun in Bantu*. Baltimore: Published at the Waverly Press by Indiana University, Bloomington.

Grégoire, C. (1975) *Les locatifs en bantu*. Tervuren: Musée royal de l'Afrique centrale.

Grégoire, C. & B. Janssens (1999) L'augment en bantou du nord-ouest. In Hombert, J.-M. & L. Hyman (eds.), *Bantu Historical Linguistics: Theoretical and Empirical Perspectives*, 413–429. Stanford: CSLI.

Hadermann, P. (1999) Les formes nomino-verbales de classes 5 et 15 dans les langues bantoues du Nord-Ouest. In Hombert, J.-M. & L. M. Hyman (eds.), *Bantu Historical Linguistics: Theoretical and Empirical Perspectives*, 431–471. Stanford: CSLI.

Harjula, L. (2004) *The Ha Language of Tanzania: Grammar, Texts and Vocabulary.* Cologne: Rüdiger Köppe.

Hawkins, J. A. (1983) *Word Order Universals*. New York: Academic Press.

Heath, T. (1998) *A Preliminary Grammar Sketch of the Mekaa Noun and Verb Morphology.* Yaoundé: SIL Cameroon.

Hendrikse, A. P. (1990) Number as a Categorizing Parameter in Southern Bantu: An Exploration in Cognitive Grammar. *South African Journal of African Languages* 10(4): 384–400.

Hulstaert, G. (1965) *Grammaire du lomongo. Deuxieme partie, Morphologie.* Tervuren: Musée royal de l'Afrique centrale.

Hulstaert, G. (1966) *Grammaire du lomongo. Troisième partie, Syntaxe.* Tervuren: Musée royal de l'Afrique centrale.

Hyman, L. M. (1972) A Phonological Study of Fe'fe'-Bamileke. *Studies in African Linguistics*. Supplement 4.

Hyman, L. M. (2003) Basaá (A43). In Nurse, D. & G. Philippson (eds.), *The Bantu Languages*, 257–282. London; New York: Routledge.

Hyman, L. M. & F. X. Katamba (1993) The Augment in Luganda: Syntax or Pragmatics? In Mchombo, S. A. (ed.), *Theoretical Aspects of Bantu Grammar*, 209–256. Stanford: CSLI.

Kamanda Kola, R. (1993) A propos de la tonalité en zamba. *Afrikanistische Arbeitspapiere* 33: 83–103.

Kamanda Kola, R. (1994) Notes sur l'augment en zamba. *Annales Aequatoria* 15: 399–409.

Kamba Muzenga, J.-G. (2003) *Substitutifs et possessifs en bantou.* Paris: Peeters.

Kasonde, A. R. M. (2009) *Phonologie et morphologie de la langue bemba.* Munich: LINCOM.

Kavari, J. U., L. Marten & J. Van der Wal (2012) Tone Cases in Otjiherero: Head-Complement Relations, Linear Order, and Information Structure. *Africana Linguistica* 18: 315–353.

Kenstowicz, M. (2008) On the Origin of Tonal Classes in Kinande Noun Stems. *Studies in African Linguistics* 37(2): 115–151.

Leitch, M. (2003) Babole (C101). In Nurse, D. & G. Philippson (eds.), *The Bantu Languages*, 392–421. London New York: Routledge.

Maganga, C. & T. C. Schadeberg (1992) *Kinyamwezi: Grammar, Texts, Vocabulary.* Cologne: Rüdiger Köppe.

Maho, J. F. (1999) *A Comparative Study of Bantu Noun Classes.* Göteborg: Acta Universitatis Gothoburgensis.

Meeussen, A. E. (1959) *Essai de grammaire rundi.* Tervuren: Musée royal du Congo belge.

Meeussen, A. E. (1967) Bantu Grammatical Reconstructions. *Africana Linguistica* 3: 79–121.

Meeuwis, M. (2010) *A Grammatical Overview of Lingála.* Munich: LINCOM.

Möhlig, W. J. G., L. Marten & J. U. Kavari (2002) *A Grammatical Sketch of Herero.* Cologne: Rüdiger Köppe.

Mous, M. (2003) Nen (A44). In Nurse, D. & G. Philippson (eds.), *The Bantu Languages*, 283–306. London; New York: Routledge.

Mudzingwa, C. & M. Kadenge (2013) An Analysis of the Ghost Augment in ChiShona. *South African Journal of African Languages* 33(1): 87–93.

Nicolle, S. (2012) Semantic-Pragmatic Change in Bantu -No Demonstrative Forms. *Africana Linguistica* 18: 193–234.

Nicolle, S. (2013) *A Grammar of Digo: A Bantu language of Kenya and Tanzania*. Dallas: SIL International.

Nicolle, S. (2014) Discourse Functions of Demonstratives in Eastern Bantu Narrative Texts. *Studies in African Linguistics* 43(2): 125–144.

Nsuka-Nkutsi, F. (1982) *Les structures fondamentales du relatif dans les langues bantoues*. Tervuren: Musée royal de l'Afrique centrale.

O'Connor, K. M. & C. Patin (2015) The Syntax and Prosody of Apposition in Shingazidja. *Phonology* 32(1): 111–145.

Ondo Mebiame, P. (2001) L'augment en fang ntumu. *Revue gabonaise des sciences du langage* 2: 60–77.

Patin, C. (2017) The Augment in Shingazidja. Paper presented at *the SOAS-UGent Workshop on Parametric Approaches to Morphosyntactic Variation in Bantu, 31 March 2017*, Ghent University.

Petzell, M. (2008) *The Kagulu Language of Tanzania: Grammar, Texts and Vocabulary*. Cologne: Rüdiger Köppe.

Polak-Bynon, L. (1965) L'expression des ordinaux dans les langues bantoues. *Africana Linguistica* 2: 127–159.

Rijkhoff, J. (2002) *The Noun Phrase*. Oxford: Oxford University Press.

Rijkhoff, J. (2008) Descriptive and Discourse-Referential Modifiers in a Layered Model of the Noun Phrase. *Linguistics* 46(4): 789–829.

Schadeberg, T. C. (1986) Tone Cases in Umbundu. *Africana Linguistica* 10: 423–447.

Schadeberg, T. C. (2003) Derivation. In Nurse, D. & G. Philippson (eds.), *The Bantu Languages*, 71–89. London; New York: Routledge.

Segerer, G. (2015) Les classes nominales en keerak. In Creissels, D. & K. Pozdniakov (eds.), *Les classes nominales dans les langues atlantiques*, 103–146. Cologne: Rüdiger Köppe.

Segerer, G. & S. Flavier (2011–2016) *RefLex: Reference Lexicon of Africa, Version 1.1*. Paris, Lyon: CNRS (Available online at http://reflex.cnrs.fr/).

Stappers, L. (1964) *Morfologie van het Songye*. Tervuren: Koninklijk Museum voor Midden-Afrika.

Stappers, L. (1965) Het hoofdtelwoord in de Bantoe-talen. *Africana Linguistica* 2: 177–199.

Ström, E.-M. (2013) *The Ndengeleko Language of Tanzania*. Gothenburg: University of Gothenburg, PhD dissertation.

Taylor, C. (1985) *Nkore-Kiga*. London: Croom Helm.

Torrend, J. (1891) *A Comparative Grammar of the South-African Bantu Languages, Comprising Those of Zanzibar, Mozambique, the Zambezi, Kafirland, Benguela, Angola, the Congo, the Ogowe, the Cameroons, the Lake Regions*. London: Kegan Paul, Trench, Trübner & Co.

Twilingiyimana, C. (1984) *Éléments de description du doko*. Tervuren: Musée royal de l'Afrique centrale.

Valinande, N. K. (1984) *The Structure of Kinande*. Washington, DC: Georgetown University, PhD dissertation.

Van de Velde, M. (2005) The Order of Noun and Demonstrative in Bantu. In Bostoen, K. & J. Maniacky (eds.), *Studies in African Comparative Linguistics, with Special Focus on Bantu and Mande*, 425–441. Tervuren: Royal Museum for Central Africa.

Van de Velde, M. (2006a) The Alleged Class 2a Prefix bɔ in Eton, a Plural Word. Paper presented at *31st Annual Meeting of the Berkeley Linguistics Society, February 2005*, Berkeley (Available online at https://halshs.archives-ouvertes.fr/halshs-00603470).

Van de Velde, M. (2006b) Multifunctional Agreement Patterns in Bantu and the Possibility of Genderless Nouns. *Linguistic Typology* 10: 183–221.

Van de Velde, M. (2008) *A grammar of Eton*. Berlin: De Gruyter Mouton.

Van de Velde, M. (2009) Agreement as a Grammatical Criterion for Proper Name Status in Kirundi. *Onoma* 44: 219–241.

Van de Velde, M. (2013) The Bantu Connective Construction. In Carlier, A. & J.-C. Verstraete (eds.), *The Genitive*, 217–252. Amsterdam; Philadelphia: John Benjamins.

Van de Velde, M. (2017) The Augment as a Construct Form Marker in Eton Relative Clause Constructions. In Atindogbe, G. & R. Grollemund (eds.), *Relative Clauses in Cameroonian Languages: Structure, Function and Semantism*, 47–66. Berlin: De Gruyter Mouton.

Van de Velde, M. & O. Ambouroué (2011) The Grammar of Orungu Proper Names. *Journal of African Languages and Linguistics* 32: 113–141.

Van den Eynde, K. & P. Mufuta (1994) Le système de référence primaire du ciluba (luba-Kasai): analyse en traits. *International Journal of Lexicography* 7: 90–105.

Van der Wal, J. (2006) Predicative Tone Lowering in Makhuwa. *Linguistics in the Netherlands* 23: 224–236.

Van Langendonck, W. (2007) *Theory and Typology of Proper Names*. Berlin: De Gruyter Mouton.

Van Leynseele, H. (1977) *An Outline of Libinza Grammar*. Leiden: Rijksuniversiteit Leiden, MA thesis.

von Staden, P. M. S. (1973) The Initial Vowel of the Noun in Zulu. *African Studies* 32(3): 163–181.

Wald, B. (1975) Animate Concord in Northeast Coastal Bantu: Its Linguistic and Social Implications as a Case of Grammatical Convergence. *Studies in African Linguistics* 6(3): 267–314.

Wega Simcu, A. (2012) *Grammaire descriptive du pólrì. Eléments de phonologie, de morphologie et de syntaxe*. Yaounde: University of Yaounde I, thèse de doctorat.

Welmers, W. E. (1973) *African Language Structures*. Berkeley; Los Angeles; London: University of California Press.

Wilkins, W. & A. Kimenyi (1975) Strategies in Constructing a Definite Description: Some Evidence from Kinyarwanda. *Studies in African Linguistics* 6(2): 151–170.

Ziervogel, D. & E. J. Mabuza (1976) *A Grammar of the Swati Language: Siswati*. Pretoria: J.L. Van Schaik.

CHAPTER 9

CLAUSAL MORPHOSYNTAX AND INFORMATION STRUCTURE

Laura J. Downing and Lutz Marten

1 INTRODUCTION

Research in the morphology and syntax of Bantu languages has a long tradition, dating back to the 19th century. The earliest studies tended to focus on morphology, and in particular on the noun class system and verbal morphology. Considerable analytical work is contained in early descriptive grammars. The first comparative work includes Bleek (1862) and Meinhof (1906), the companion volume to his comparative phonology (Meinhof 1899). Both comparative grammar works provide detailed discussions of Bantu noun classes and lay the foundation of the widely adopted Bantu noun class numbering system. Meeussen (1967) remains the most comprehensive account of Proto-Bantu morphosyntax to date.

More recently, one important strand of research has led to extensive comparative studies for particular aspects of Bantu grammar – for example, noun classes (Maho 1999), relative clauses (Nsuka-Nkutsi 1982, Henderson 2006), subordination (Güldemann 1996), or tense-aspect marking (Nurse 2008). Since the 1960s, another important strand of research has become the formal study of Bantu languages, addressing a variety of topics from different theoretical perspectives. Bantu languages have played a key role in theoretical studies in areas such as valency changing, word order, and agreement (see, for example, papers in Mchombo 1993), as well as information structure and the interface between prosody and syntax. With some 350–400 quite closely related languages, the Bantu family provides a wealth of examples for the study of fine-grained micro-variation within overall typological similarity. Analysis of this variation with respect to processes and constraints relating to language change, language contact and language universals is a central current research issue, whose systematic investigation is just beginning (cf. Marten *et al.* 2007).

In this chapter, we survey the following main themes in Bantu clausal morphosyntax: word order (Section 2), agreement in the phrase and the clause (Section 3), valency-changing morphosyntax (Section 4), non-verbal predication (Section 5), complex clause structure and dependent clauses (Section 6) and the phonology-syntax interface (Section 7).

2 WORD ORDER

2.1 Syntactic constraints on word order

The basic word order in the vast majority of Bantu languages is SVO (Heine 1976). However, there is considerable flexibility as well as cross-linguistic variation resulting from different syntactic constraints, interaction of word order and agreement morphology, as well as information structure effects, which are discussed in Section 2.2. For Chewa N31b, for example, Bresnan and Mchombo (1987) show that when an object prefix (OP) occurs on the verb in inflected transitive clauses, all permutations of subject, verb and object are possible.

(1)   Chewa N31b (Bresnan & Mchombo 1987: 744–745)
      a   SVO   N-jûchi   zi-ná-wá-lum-a   a-lenje
                10-bee SP_{10}-PST-OP_2-bite-FV   2-hunter
                'The bees bit them, the hunters'
      b   VOS   Zináwáluma alenje njûchi
      c   OVS   Alenje zináwáluma njûchi
      d   VSO   Zináwáluma njûchi alenje
      e   SOV   Njûchi alenje zináwáluma
      f   OSV   Alenje njûchi zináwáluma

Data like these raise questions about the status of subject and object markers and the agreeing lexical NPs, for example whether one or both of the NPs in (1b-f) are dislocated (cf. Section 2.2). Furthermore, the notions of subject – in particular as opposed to topic (Section 3.2) – and object (Schadeberg 1995, Thwala 2006, Van de Velde 2008b, Nicolle 2013: 95–99) have been critically discussed with respect to Bantu. Of specific relevance in this context are inversion constructions where word order, thematic structure, information structure, argument structure and agreement interact (Section 3.3).

Within the verb phrase, a typical word order restriction in double object constructions is for the recipient-benefactive-causee argument to precede the theme argument, whether the verb phrase is headed by a predicate like 'give' licensing two objects (2), or by a derived verb such as an applicative (3) or causative (4).

(2)   Luguru G35 (Marten & Ramadhani 2001: 266)
      Chibua   ko-w-eng'-a   iwa-na   ipfi-tabu
      Chibua SP_1.PRS-OP_2-give-FV   2-children   8-books
      'Chibua is giving children books'

(3)   Chewa N31b (Mchombo & Firmino 1999: 217)
      A-lenje   a-ku-phík-ír-a   a-nyaní   zí-túmbúwa
      2-hunters SP_2-PRS-cook-APPL-FV   2-baboons   8-pancakes
      'The hunters are cooking (for) the baboons some pancakes'

(4)   Ndamba G52 (Edelsten & Lijongwa 2010: 99)
      Simba   a-ka-m-somol-es-ile   n-dembo   lw-imbo
      9.lion SP_1-PST-OP_1-sing-CAUS-PFV   9-elephant   11-song
      'The lion caused the elephant to sing a song'

Double object constructions are often analysed as being either "symmetrical" or "asymmetrical" (e.g., Bresnan & Moshi 1990, Alsina & Mchombo 1993, Rugemalira 1993), depending on whether both objects behave identically with respect to syntactic tests such as word order, passivisation and object marking, or whether the two objects behave differently. For example, in benefactive applicatives in Chewa N31b the order benefactive-theme object is grammatical (3), while the opposite order is ungrammatical (5). Chewa N31b is thus asymmetrical according to this syntactic test. In contrast, in Ha JD66, both orders of objects are acceptable (6).

(5)   Chewa N31b (Mchombo & Firmino 1999: 217)
      *A-lenje   a-ku-phík-ír-a   zí-túmbúwa   a-nyǎni
      2-hunters   SP$_2$-PRS-cook-APPL-FV   8-pancakes   2-baboons

(6)   Ha JD66 (Harjula 2004: 148)
      a   Ya-a-mú-haa-ye   umu-káaté   umw-áana
          SP$_1$-PST-OP$_1$-give-PFV   3-bread   1-child
          'S/he gave bread to the child'

      b   Ya-a-mú-haa-ye   umw-áana   umu-káaté
          SP$_1$-PST-OP$_1$-give-PFV   1-child   3-bread
          'S/he gave bread to the child'

Symmetry can be conditioned by factors such as animacy, information structure or the thematic roles of the two objects. For example, in Chaga E60, benefactive applicatives are asymmetrical, whereas instrumental applicatives are symmetrical, i.e., the order of the two objects is free.

(7)   Vunjo-Chaga E622C (Moshi 1998: 148)
      a   M-solro   n-á-lé-wé-í-á   kí-shú   nyáma
          1-man   FOC-SP$_1$-PST-slice-APPL-FV   7-knife   9.meat
          'The man sliced with a knife the meat.'

      b   M-solro   n-á-lé-wé-í-á   nyáma   kí-shú
          1-man   FOC-SP$_1$-PST-slice-APPL-FV   9.meat   7-knife
          'The man sliced the meat with a knife.'

A similar difference exists between benefactive and motive/reason applicatives in Swahili G42 (Buell 2005: 190). Van der Wal (2017) notes a cross-linguistic implicational hierarchy of symmetry, namely, that if a Bantu language has symmetrical causatives, then applicatives and lexical ditransitives will also be symmetrical; and if applicatives are symmetrical, lexical ditransitives will be, too. It should be noted, however, that for languages like Swahili G42, a high degree of variation in the data has been reported (e.g., Bentley 1994, Ngonyani 1996, Marten et al. 2007, Riedel 2009), which appears to be related to dialectal and speaker variation (Murrell 2012). Clearly, more detailed studies of variation and usage patterns are needed.

It has been argued in work since Hawkinson and Hyman (1974) that asymmetry in Bantu double object constructions is related to the intrinsic topicality of object NPs. Highly topical objects are most likely to be licensed to occur in immediate post-verbal (IAV) position in VPs under broad focus. Intrinsic topicality is defined in terms of the semantic hierarchies in (8).

(8)   Topicality hierarchies (Hyman & Duranti 1982: 224)
      a   Benefactive > Recipient > Patient > Instrument
      b   1st > 2nd > 3rd human > 3rd animal > 3rd inanimate
      c   definite > indefinite

The Hyman and Duranti (1982) analysis of Southern Sotho S32 illustrates the role of the topicality hierarchy – and especially the human factor – in accounting for asymmetries in the morphosyntactic properties of ditransitive objects. In the canonical role-animacy mapping (benefactive = human, patient = inanimate), either object can be in IAV position and be passivised (object symmetry). In a non-canonical case, however, object asymmetry is observed. We can see this in (9), where a non-human purpose object of an applicative co-occurs with a human patient object. Note that the non-human object cannot occur in IAV position (9b), and it cannot be passivised (9d).

(9)   Southern Sotho S32 (Hyman & Duranti 1982: 225–226)
    a   Ke-bítselítsé   baná   mo-kéte.
        SP$_{1SG}$-call.for.PST   2.child   3-feast
        'I called children for the feast.'

    b   *Ke-bítselítsé   mo-kété   baná
        SP$_{1SG}$-call.for.PST 3-feast   2.child
        'I called for the feast the children.'

    c   Baná   bá-bítselítsoé   mo-kéte
        2.child   SP$_{2}$-call.for.PASS.PST   3-feast
        'Children were called for the feast.'

    d   *Mo-kété   ó-bítselítsoé   baná
        3-feast   SP$_{3}$-call.for.PASS.PST   2.child
        'Feast was called for the children.'

The role of such syntacticised hierarchies of intrinsic topicality in accounting for object asymmetries has also been demonstrated for other Bantu languages (cf. Hawkinson & Hyman 1974, Duranti & Byarushengo 1977, Rugemalira 1991, 1993, Riedel 2009). Another aspect of double (and indeed multiple) object constructions is the relation between the order of overt object NPs and the order of object markers, which often, although not always, stand in a mirror relation to each other:

(10)   Tswana S31 (Cole 1955: 431–432)
    a   Ke-tla-kwal-êl-êl-a   ngwana   ba-sadi   lo-kwalô
        SP$_{1SG}$-FUT-write-APPL-APPL-FV   1.child   2-parents   11-letter
        'I will write a letter to the parents on behalf of the child.'

    b   Ke-tla-lo-ba-mo-kwal-êl-êl-a
        SP$_{1SG}$-FUT-OP$_{11}$-OP$_{2}$-OP$_{1}$-write-APPL-APPL-FV
        'I will write it to them on his/her behalf.'

There are different formal and functional explanations for this pattern, e.g., order restrictions related to topicality hierarchies (8) or constraints on the order of functional heads. One diachronic explanation assumes that the pre-verbal distribution of object markers reflects an older (pre-Bantu) language stage with head-final word order (cf. Nurse 2008: 62 and references therein for discussion).

Characteristics of head-finality are also found in other domains of Bantu syntax, although restricted to a small set of Bantu languages. In Nen A44, basic word order is SOV (11a). However, when the object is focused, it is placed after the verb (11b).

(11)   Nen A44 (Mous 1997: 126)

   a   Àná   mòné   índì.
      SP_1.PST   money   give
      'S/he gave money.'

   b   Àná   índì   á   mòné.
      SP_1.PST   give   FOC   money
      'S/he gave MONEY.'

Potential traces of head-final word order are also found in complex verb forms in a number of North-Eastern Bantu languages. In Rangi F33, for example, in the future tense, the lexical verb – *rína* in (12) – precedes the inflected auxiliary verb – *úrɨ* in (12) – contrary to the expected Aux-V order.

(12)   Rangi F33 (Gibson 2012: 17)
      Weéwe   rín-a   ú-rɨ   ɨ-hɨ   mi-rɨ́ɨ́nga
      2SG   open-FV SP_{2SG}-FUT PP_4-DEM   4-beehive
      'You will open these beehives.'

Rangi F33 and some other languages with V-Aux order are in a language contact situation with South Cushitic head-final languages, so that the atypical head-final order may result from structural transfer. However, head-final order seems to interact with information structure, as Rangi V-Aux order reverts to Aux-V in a number of syntactic contexts, including argument focus and questions; see Gibson (2012) for a detailed analysis of the Rangi case and Güldemann (2007) for more general discussion of this point. Indeed, while there certainly are syntactic and semantic constraints on Bantu word order, a complete analysis must also take into account information structure discussed further in Section 2.2.

2.2   Effect of information structure on word order in statements and questions

In many Bantu languages, information structure (IS) considerations influence word order: focused (new or contrastive) information and given (old, topical) information can occur (or tend to occur) in some positions in the clause or verb phrase but not in others. For a recent survey of this topic, see Downing and Hyman (2016). A common restriction is that focused subjects cannot occur in canonical pre-verbal position (Morimoto 2000, Zerbian 2006, van der Wal 2009). The incompatibility of subject and focus can be illustrated with content question/answer pairs, which have inherent new information focus. As Zerbian (2006) shows, in Northern Sotho S32, the focused subject must be either clefted or post-verbal.

(13)   Northern Sotho S32 (Zerbian 2006: 69–72)

   a   Ké   mang   (yo)   a-nyaka-ng   ngaka?   [cleft]
      COP   who DEM_1 SP_1-look.for-REL   9.doctor
      'Who is looking for the doctor?'

   b   Go-fihla   mang?   [post-verbal]
      SP_{17}-arrive   who
      'Who is arriving?'

   c   Ké   mo-kgalabje   (yo)   a-thlokomela-ng   ngaka.   [cleft]
      COP   1-old.man DEM_1 SP_1-look.for-REL   9.doctor
      'An old man is looking for the doctor.'

d *Mang o-nyaka   ngaka?   [*pre-verbal]
   who SP_1-look.for 9.doctor
   Intended: 'Who is looking for the doctor?'

Focused objects and adjuncts often occur in a restricted position in the verb phrase, typically in the Immediately After the Verb (IAV) position. IAV focus was first demonstrated systematically for Aghem (Narrow Grassfields) (Watters 1979, Hyman 2010).

(14) Aghem (Narrow Grassfields, Ring Group) (Hyman 2010: 96–98)
   a tí-bvú   tì-bìghà   mô   zì   kí-bɛ́   ꜜnɛ́
      10-dog NP_{10}-two PST_1 eat 7-fufu today
      'The two dogs ate fufu today.'

   b tí-bvú   tì-bìghà   mô   zì   nɛ́   ꜜbɛ́   ꜜkɔ́
      10-dog NP_{10}-two PST_1 eat today fufu DET_7
      'The two dogs ate fufu TODAY.'

   c tí-bvú   tì-bìghà   mô   bɛ́   ꜜkí   zí   nɛ́
      10-dog NP_{10}-two PST_1 fufu DET_7 eat today
      'The two dogs ate fufu TODAY.' (pre-posing 'fufu' places TODAY in IAV)

   d à   mɔ̂   zì   tí-bvú   tì-bìghà   bɛ́   ꜜkɔ́   nɛ́
      EXPL.SM PST_1 eat 10-dog NP_{10}-two fufu DET_7 today
      'The TWO DOGS ate fufu today' (EXPL.SM = expletive subject marker)

The broad focus word order S-AUX-V-O-X is shown in (14a). In (14b), the focused temporal adjunct 'today' is in IAV position, and the backgrounded object 'fufu' appears after it. The IAV effect is achieved in (14c) by placing defocused 'fufu' in pre-verbal position, thereby isolating 'today' as the focused IAV element. (14d) parallels (13b), above: the logical subject may be focused by placing it in IAV position, and an expletive is the grammatical subject. It is not only explicitly focused constituents that occur in IAV position, but also WH elements, which are inherently focused. As a result of their proximity to the verb, they often develop into enclitics, as in Ganda JE15.

(15) Ganda JE15 (Downing & Hyman 2016)
   a bá-wá=ání   è-kì-kópò   'Who do they give the cup to?'
      SP_2-give=who   AUG-7-cup

   b bá-wá=kí   ò-mw-áànà   'What do they give the child?'
      SP_2-see=what   AUG-1-child

   c bá- lábá=ddí   ò-mw-áànà   'When do they see the child?'
      SP_2-see=when   AUG-1-child

   d bá-lábá=wá   ò-mw-áànà   'Where do they see the child?'
      SP_2-see=where AUG-1-child

IAV focus has been demonstrated for a number of Bantu languages (cf. Downing 2012, Downing & Hyman 2016, Gibson et al. 2017). While many Bantu languages exploit the IAV position to mark focus, an almost inverse effect is found in Zone B and H languages, where a focused object occurs immediately before the verb (IBV) (e.g., Hadermann 1996). For example, Bostoen and Mundeke (2011b: 75–76) show that although Mbuun B87 has the unmarked SVO word order, as in (16a), the IBV position is used for focus-marking, as in the Q/A exchange in (16b-c). (16d) shows that topical information occurs initially, as expected in a SVO language (cf. also Bostoen & Mundeke 2012).

(16)   Mbuun B87 (Bostoen & Mundeke 2011b: 75–76)

a	ɔ-káár	o-á-súm	ki-te	'The woman buys a chair.'
	1-woman	SP_1-PRS-buy	7-chair	

b	ɔ-káár	nké	ká-wó-kon?	'What did the woman plant?'
	1-woman	what	SP_1-PST-plant	

c	ɔ-káár	a-sáŋ	ká-wó-kon	'The woman planted millet.'
	1-woman	6-millet	SP_1-PST-plant	

d	a-sáŋ	maam	o-á-(á-)kon	'Millet, mother plants it.' ~
	6-millet	mother	SP_1-PRS-(OP_6)-plant	'Millet is planted by mother.'

IBV focus is described in other B and H languages in work like De Kind (2014) and Koni Muluwa and Bostoen (2014). They show that it frequently correlates with OV and not with canonical VO word order. However, Nen A44, a Bantu SOV language, does not exploit the IBV for focus (Mous 1997), as shown in (11b) above. Obviously, much more work is needed on the syntax of focus in the Bantu languages in these zones.

A final common effect of IS on word order is that the requirement for focused elements to occur in IAV position often leads to displacement (or "dislocation") of non-focused nominal phrases from their canonical positions. The displaced nominal phrases, if they are objects, require a co-referential (resumptive) object marker in many Bantu languages. This is illustrated in (17) with data from Zulu S42; the class 6 object prefix is resumptive here.

(17)   Zulu S42 (Cheng & Downing 2009)

a	(*si-thwéle*	*ámá-thánga*	*ngó-bhasikíídi*)	(= "broad focus")
	SP_{1PL}-carry	6-pumpkin	with.1a-basket	

'We carry (the) pumpkins with a basket'

b	Q	(*u-wa-thwéle*	*ngáan'*)	(amá-thaanga)
		SP_{2SG}-OP_6-carry	how	6-pumpkin

'How are you carrying the pumpkins?'

c	A	(*ámá-thaanga*)	(*si-wa-thwéle*	*ngó-bhasikíídi*)
		6-pumpkin	SP_{1PL}-OP_6-carry	with.1a-basket

'We are carrying the pumpkins with a basket.'

While both left and right dislocation are widely attested in Bantu languages, asymmetries in the information structure status of left vs. right dislocation have been reported for some languages. Both contrastive topics and aboutness topics must be pre-verbal in Zulu S42, as Cheng and Downing (2009) show, while right dislocations like (17b) are afterthoughts, rather than aboutness topics. Zerbian (2006) finds for Northern Sotho S32 that right dislocations are almost non-existent in her natural language corpus and proposes that this is due to their status as afterthoughts. However, Tenenbaum (1977) demonstrates that in Haya JE22 right dislocations can be used to connote surprise, emphasis or contrast. She proposes that this follows from the fact that utterance-initial and utterance-final positions are both perceptually salient. Marten (2007) develops a formal analysis of the interaction of context and word order choices in left-right asymmetries in Bantu, which notes the connection to interpretive restrictions of agreement markers (cf. Sections 3.1 and 3.2). Asymmetries in the information structure of left vs. right dislocations require further investigation.

In short, as work since Stucky (1981) has observed, word order is rather flexible in many Bantu languages, since it is conditioned by both information structure and syntax.

2.3 Word order, information structure and verb morphology: the conjoint-disjoint alternation

A particularly complex effect of the interaction between information structure and word order is the so-called conjoint-disjoint alternation in verb inflection. The alternation has been described in older literature in various ways: e.g., "long vs. short," "strong vs. weak bond," "dependent vs. independent." The terminology "conjoint-disjoint" is adopted from Meeussen's (1959) description of Kirundi JD62. One verb form (typically the marked form) can occur in clause-final position and often implies predicate focus ("disjoint"), as in (18a), while the other form (typically the unmarked form) cannot occur clause-finally and often implies term focus on the following phrase ("conjoint"), as in (18b). In (18a) *la-* is the Bemba M42 disjoint marker.

(18) Bemba M42 (Sharman 1956: 40, Kula 2017: 276)
    a Bushé mu-la-peep-a?
       Q   SP$_{2PL}$-DJ-smoke-FV   (disjoint)
       'Do you smoke?'

    b Ee   tu-peep-a   sekelééti. (conjoint)
       Yes   SP$_{1PL}$-smoke-FV cigarettes
       'Yes, we smoke cigarettes (and not a pipe, for instance).'

In most languages, the conjoint-disjoint alternation is marked segmentally, as in (18a), at least in some instances, but tonal marking has been argued for in Tswana (Creissels 1996, 2017). Conjoint-disjoint marking is typically only found with a restricted set of tenses, and since the formal marking of the conjoint-disjoint alternation resembles tense-aspect marking (e.g., by verbal prefixes or tonal alternation), the alternation has sometimes been analysed as part of the TAM system. However, because the function of the alternation is unrelated to tense and aspect, it is nowadays seen as an independent inflection category, called "junctivity" in Buell (2015). The conjoint-disjoint alternation is mainly found in Bantu and in a few other Niger-Congo languages, but has not been described outside of the phylum. Important recent work focusing on the conjoint-disjoint alternation is found in, for example, Creissels (1996) and Buell (2006), with the most comprehensive discussion provided in the collective volume edited by van der Wal and Hyman (2017). Typological and theoretical work on the conjoint-disjoint alternation has focused mainly on the distinct roles of morphosyntax and information structure in the analysis of the construction, and on the diachronic and functional contexts in which it is embedded.

With respect to the first question, there appears to be cross-Bantu variation. While Buell (2006) shows that in Zulu S42, a conjoint form is required even if the verb phrase-internal constituent in IAV is not in focus, van der Wal (2009) shows that in Makhuwa P31 every constituent following a conjoint form is in focus. Such contrast between constituency-based and directly focus-based conjoint-disjoint alternations even occurs between very closely related varieties like Rwanda JD61 (Ngoboka & Zeller 2017) and Rundi JD62 (Nshemezimana & Bostoen 2017) respectively. Disjoint forms are typically characterised structurally by the absence of any following constituents in the same domain, and in terms of information structure they often also encode predicate focus (cf. Güldemann 2003).

With respect to the second question, functional and diachronic relations have been noted between the conjoint-disjoint alternation and two other grammatical phenomena found in Bantu – metatony and tonal cases – where prosodic marking, morphosyntactic domainhood and/or information structure interact (cf. Hyman 2017). In the process of metatony, found for example in Songye L23, a final low tone of a verb form is realised as High if an object follows (Stappers 1964 cited by Dimmendaal 1995: 32 and Schadeberg 1995: 176). The prosodic marking of nouns in languages with so-called tone cases, found for example in Herero R31, is also to some extent sensitive to distributional restrictions which seem quite similar to those relevant for the conjoint-disjoint alternation, in particular to conjoint verb forms (Kavari et al. 2012). There may well be a historic link between all three of these phenomena, as Hyman (2017) speculates, but much more research is needed to establish this.

3 INFLUENCE OF SYNTAX ON MORPHOLOGY

3.1 Object markers between agreement-like and pronominal-anaphoric function

Since at least Bresnan and Mchombo (1987), considerable discussion of Bantu agreement has centered on the difference between the anaphoric pronoun-like function and the agreement-like function of (subject and) object markers. Creissels (2005: 44–45) proposes three diachronic stages of development in the function of object (and other pronominal) markers from purely anaphoric to purely agreement function. At Stage I the object marker has a *purely anaphoric* function, as it cannot occur within the same clause as the corresponding, co-referential object NP. It can occur with a dislocated object NP and it can represent on its own a co-referential NP. At Stage II the object marker acquires an *agreement* function, as it obligatorily occurs even if there is a corresponding, co-referential object NP within the same clause, although the object marker can also occur on its own. At Stage III the object marker has a *purely agreement* function and cannot by itself represent a co-referential NP.

Bresnan and Mchombo (1987) analyse Chewa N31b as the canonical example of a language where object markers are anaphoric. However, subsequent work has demonstrated that object marking in this language is much more like Swahili G42: human objects are marked whatever their position; non-human are unmarked, whatever their position (Bentley 1994, Henderson 2006, Riedel 2009, Downing (2018)). For this reason, we use Zulu S42 as the canonical language to illustrate purely anaphoric use of object marking – Stage I – in (19). While dislocated objects must be co-indexed with an object marker, in situ post-verbal objects, i.e., those within the same clause (or verb phrase constituent) as the verb, cannot be. Prosodic evidence, i.e., penultimate lengthening before a prosodic phrase boundary, shows that *ínkukhu* 'chicken' is dislocated in (19a-b), where object marking through a class 9 object prefix is obligatory, but not in (19c) where object marking is disallowed. Similarly, (19d) is ungrammatical, because the object marker refers to an object that has not been dislocated.

(19) Zulu S42 (Cheng & Downing 2009)

   a Q ((Ú-siipho)   (ú-yí-phékéla   BAANI)   ín-kuukhu)
      1-Sipho   SP$_1$-OP$_9$-cook.for   1.who   9-chicken
      'Who is Sipho cooking the chicken for?'

   b A ((Ú-síph'   ú-yí-phékél'   ízí-VAKÁASH')   ín-kuukhu)
      1-Sipho   SP$_1$-OP$_9$-cook.for   8-visitor   9-chicken
      'Sipho is cooking the chicken for the visitors.'

   c *Ú-síph'   ú-yí-phékél'   ízí-VAKÁSH'   ín-kuukhu
      1-Sipho   SP$_1$-OP$_9$-cook.for   8-visitor   9-chicken

   d *Ú-síph'   ú-zí-phékél'   ízí-VAKÁASH'   ín-kuukhu
      1-Sipho   SP$_1$-OP$_8$-cook.for   8-visitor   9-chicken

The data show that in Zulu S42 a lexical object and a co-referential object marker cannot co-occur, and so the object marker always functions as an incorporated pronoun (20).

(20) Zulu object marker as incorporated pronoun

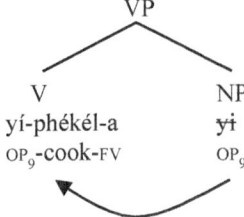

Bantu languages with grammatical agreement-like object marking (Stage II) include, for example, Swahili G42. Swahili object markers can serve a pronominal function, resuming objects mentioned earlier in the discourse, but they also serve a grammatical agreement function: object marking is (almost) obligatory with overt human objects (especially definite ones), as in (21a), and common with definite non-human objects, as in (21c).

(21) Swahili G42 (Riedel 2009)

   a Ni-li-mw-ona   mw-ana-we
      SP$_{1SG}$-PST-OP$_1$-see   1-child-POSS$_{3SG}$
      'I saw his child.'

   b *Ni-li-ona mwanawe

   c Ni-li-zi-ona   picha   hizo
      SP$_{1SG}$-PST-OP$_{10}$-see   10.picture   DEM$_{10}$
      'I saw those pictures.'

Riedel (2009: 42, 46) affirms that the object marker in these examples occurs even though the co-referential object noun phrase is in its base position, not dislocated. This means that class 1/2 object markers in Swahili can be analysed as agreement markers (22).

(22) Swahili object marker as agreement marker

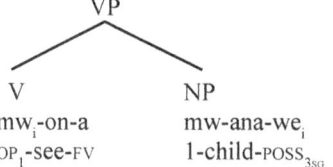

Recent detailed studies – either of individual languages, e.g., Chewa N31b (Downing 2018), Haya JE22 and Shambaa G23 (Riedel 2009), Bukusu JE31c (Diercks 2010), Manyika S13 (Bax & Diercks 2012), Zulu S42 (Van der Spuy 1993, Buell 2005, Adams 2010, Halpert 2012, Zeller 2012b) or from a comparative perspective (e.g., Morimoto 2002, Henderson 2006, Ngonyani 2006, Marten & Kula 2012, Marlo 2014, 2015, van der Wal 2015) – demonstrate the extensive cross-Bantu variation in object marking and show a great deal of variation as to whether the markers are obligatory or optional. This work shows that agreement-like object markers tend strongly to co-occur with human or animate objects or with definite objects (Duranti 1979, Bentley 1994, Morimoto 2002, Riedel 2009, Marten & Kula 2012). The generalisation that emerges is that Stage II agreement in Bantu languages tends to show what has been called Differential Object Marking (DOM) (Bossong 1985, Aissen 2003). It is conditioned by the Topicality Hierarchies in (8), which play a central role in defining other object properties in Bantu languages, as we saw in Section 2.1.

In light of the high degree of variation found with Bantu object marking, recent studies call into question the dichotomy between agreement-like and pronominal status of object markers. Riedel (2009), for example, proposes that all Bantu object markers are agreement markers, while Bax and Diercks (2012) propose an analysis of Manyika S13 object markers as clitics. An alternative view is to postulate that a specific object marking system can be located along a cline from agreement marker to anaphoric, incorporated pronoun (Creissels 2005, Zeller 2012b, 2014).

There is still surprisingly little systematic, detailed information about the distribution of object markers available for most Bantu languages. More micro-variation studies, taking DOM properties into account as a parameter of variation in all contexts where object marking can occur – for instance in situ, in relative clauses, dislocated, multiple object marking – are still needed, as are detailed textual and corpus studies, investigating pragmatic and usage-based aspects of object markers (cf. Seidl & Dimitriadis 1997, Poeta 2016).

3.2 Subject agreement or topic agreement?

According to Bresnan and Mchombo (1987) subject markers in Bantu languages like Chewa N31b are ambiguous between agreement marker and incorporated pronoun (cf. Section 3.1). As agreement markers, they are part of the verbal inflection, and agree with a lexical subject. However, in the absence of a lexical subject, the subject marker functions like a pronoun that is incorporated into the verb – thus itself fulfilling the role of the subject. One consequence of this analysis is that when the subject marker functions as incorporated pronoun, any co-indexed overt lexical NP has to be analysed as a dislocated, extra-clausal syntactic topic NP. In SVO languages, topical information typically occurs sentence-initially (pre-verbally), while focused elements occur postverbally (Morimoto 2000, Güldemann 2007), and work since, at least, Givón (1976) has shown that there is a (diachronic) relation between subjects and topics. Under this view, pre-verbal subjects are in a canonical topic position, and there is a body of work on various Bantu languages demonstrating the topic-like properties of subjects (e.g., Bresnan & Kanerva 1989, Morimoto 2000, Zerbian 2006, van der Wal 2009, Marten 2011, Cheng & Downing 2014). First, subject concord on the verb is obligatory in most Bantu languages, whether there is an overt subject noun phrase in the clause or not. Further, as shown in Section 2.3, the pre-verbal subject position is incompatible with focus in many Bantu languages. Focused subjects must either be clefted or be post-verbal, while both topics and non-focused subjects occur pre-verbally. Finally, Demuth (1990) has argued that Southern Sotho speaking children's relatively early acquisition of the passive construction correlates with the fact that Southern Sotho S33 subjects must be discourse topics. The passive is acquired early, as it is a canonical topicalisation construction. These generalisations fall out if subjects are analysed as topics (e.g., Morimoto 2000, Zerbian 2006).

However, there are also arguments against confounding subjects and topics. Most generative analyses have, in fact, rejected the Bresnan and Mchombo (1987) proposal that the subject marker may function as an incorporated pronoun. For one thing, subject marking is typically obligatory in inflected verb forms, and obligatoriness is more consistent with morphological agreement than with syntactic incorporation. Furthermore, in complex verb phrases, several subject markers can occur in agreement with one subject.

(23) Swati S43 (Thwala 2006: 357)
      Ba-fana   ba-ngahle   ba-phidze   ba-bheme   i-sangu
      2-boys SP_2-might SP_2-repeat SP_2-smoke   9-pot
      'The boys might smoke pot again'

If it is assumed that incorporated pronouns are referentially unrestricted, the obligatory co-reference in (23) provides evidence for analysing the subject marker as an agreement marker, agreeing with the overt subject (Buell 2005, Thwala 2006: 357, Henderson 2007, Diercks 2010, Carstens 2011, but see Gibson and Marten 2016 for an alternative view). Furthermore, work like van der Wal (2009) and Halpert (2012) has shown that there is a syntactic distinction between subject and topic in many Bantu languages. As Zerbian (2006) shows, subjects can have all-new focus in thetic sentences, and they can have semantic properties incompatible with topic. In short, even though topic properties of subjects are striking and can have significant syntactic consequences in many Bantu languages, it would be a mistake to simply equate subject with syntactic topic in any Bantu language in the absence of detailed semantic and syntactic study.

3.3 Inversion constructions

Inversion constructions, in which the logical subject or agent follows the verb, i.e., is inverted, provide a noteworthy example of how agreement is conditioned both by syntax and information structure. Next to the most well-known example, locative inversion (24), Marten and van der Wal (2014) show that Bantu languages have a number of related inversion constructions, including semantic locative inversion (25), instrument inversion

(26), patient inversion (also called subject-object reversal) (27), default agreement inversion (28) and agreeing inversion (29).

(24) Herero R31 (Möhlig *et al.* 2002: 102)
Mò-n-djúwó mw-á-hìtí é-rùngà.
18-9-house SP_{18}-PST-enter 5-thief
'The thief entered the house.'

(25) Zulu S42 (Buell 2007: 108)
Lezi zi-ndlu zi-hlala aba-ntu aba-dala.
DEM_{10} 10-houses SP_{10}-live 2-people NP_2-old
'Old people live in these houses.'

(26) Zulu S42 (Zeller 2012a: 134)
Isi-punu si-dl-a u-John.
7-spoon SP_7-eat-FV 1a-John
'John is using the spoon to eat.'

(27) Swahili G42 (Whiteley & Mganga 1969: 113)
W-imbo u-ta-imb-a wa-tu mia.
11-song SP_{11}-FUT-sing-FV 2-people hundred
'A hundred people will sing the song.'

(28) Tswana S31 (Creissels 1996: 113)
Gó-tsàmá-ílé Mphó.
SP_{17}-go-PRF.CJ Mpho
'There has gone Mpho.'

(29) Manyo K332 (Möhlig 1967: 242)
βa-jĭmb-ire βa-mắti li-dína.
SP_2-sing-PRF 2-boy 5-song
'The boys had sung a song.'

All these inversion constructions share the following properties: inversion of the agent and the optional fronting of a non-agent, a particular information structure (most commonly, focus on the post-verbal agent), close "bonding" between verb and post-verbal NP (e.g., through phonological phrasing, absence of augment, conjoint verb form, or complement tone pattern), and the absence of object marking.

The prominence of inversion constructions in Bantu grammar has prompted a range of formal and comparative analyses. Locative inversion (e.g., Bresnan & Kanerva 1989, Demuth & Mmusi 1997, Marten 2006, Khumalo 2010, Diercks 2011, Guérois 2014) and patient inversion (e.g., Bokamba 1976, Givón 1979, Kimenyi 1980, Bokamba 1985, Russell 1985, Ndayiragije 1999, Morimoto 2000, Henderson 2006, Morimoto 2006, Henderson 2011, Hamlaoui & Makasso 2015) have been analysed most thoroughly, while semantic locative inversion (Buell 2007) and instrument inversion (Zeller 2013) have only been addressed more recently. The comparative study of 46 Bantu languages by Marten and Van der Wal (2014) shows that formal locative inversion, default agreement inversion and agreeing inversion are widespread, that there is near-complementarity between formal and semantic locative inversion and that the presence of instrument inversion or patient inversion implies the presence of locative inversion. The main conceptual questions which have been addressed in the theoretical discourse surrounding inversion constructions are related to the similarities and differences between the different inversion types (and passives) (cf. Bostoen & Mundeke 2011b, Marten & Gibson 2015) and the extent to which inversion constructions can be seen as involving a change of grammatical function, the influence of subject vs. topic agreement (cf. Section 3.2) on inversion constructions, the status of the post-verbal NP, and the exact relation between information structure and inversion constructions.

Thematic and semantic constraints are also relevant for analysing inversion constructions. For example, thematic constraints on the predicate are a factor conditioning cross-Bantu variation in locative inversion (e.g., Khumalo 2010) and thematic constraints can play a role in the distribution of different inversion constructions (Marten & Gibson 2015). The availability of patient inversion in Rwanda JD61 is related to animacy (Kimenyi 1980, Morimoto 2000). Whiteley and Mganga (1969) refer to patient inversion in Swahili as "counter-experiental": the coding of the agent as non-subject gives rise to readings of surprise or unexpectedness. Despite the extensive literature on Bantu inversion constructions, more detailed description and analysis, in particular of the semantic and pragmatic aspects of the construction, are still needed.

3.4 Agreement with conjoined NPs

Since Bantu languages typically have (near-)obligatory subject marking of the inflected verb, in addition to different object marking possibilities, the question of how subject and object agreement works with conjoined NPs is a natural one, and there is a long tradition of research from different perspectives. Early work such as Corbett and Mtenje (1987) develops rule-based analyses of agreement resolution. A number of subsequent works from a generative perspective have aimed to derive different agreement patterns from syntactic configurations, in particular locality, for example Riedel (2009) for Haya JE22 and Shambaa G23 and Simango (2012) for Nsenga N41. Marten (2005) provides a Dynamic Syntax analysis which is particularly concerned with the effect of linear order in Swahili G42 conjunct agreement, while De Vos and Mitchley (2012) develop an Optimality Theory analysis of agreement with conjoined NPs in Southern Sotho S33, using constraints relating to number resolution, noun class resolution and animacy. A summary of different strategies for agreement with conjoined NPs is given in (30). Agreement resolution involves morphological (30a), semantic (30b-c), phonological (30d), syntactic (30e-f), as well as pragmatic (30g) considerations.

(30) Strategies for agreement with conjoined NPs
    a Appropriate corresponding (typically plural) class agreement
    b Default class for animates/humans (class 2)
    c Default class for non-animates/non-humans (typically class 8 or class 10)
    d Phonological resolution
    e Partial agreement with the second conjunct (typically with a preceding conjoined NP)
    f Partial agreement with the first conjunct (typically with a following conjoined NP)
    g Partial agreement with the pragmatically prominent conjunct
    h Avoidance of agreement with conjoined NPs

The work of Simango (2012) on Nsenga N41 illustrates a number of the strategies: the use of an appropriate plural class for animate conjuncts (31a), an appropriate class for non-animates of the same class (31b), and the use of default class 8 for two non-animate conjuncts from different classes (31c).

(31) Nsenga N41 (Simango 2012: 175–179)

    a Newo   na   Khuzwayo   t-e-lal-a   pa-mphasa
       1SG   and   Khuzwayo SP_{1PL}-PST-sleep-FV 16-mat
       'I and Khuzwayo slept on the mat.'

    b Tu-mbwili na   tu-temo   tw-a-soŵ-a
       13-hoe   and   13-axe SP_{13}-PST-miss-FV
       'Hoes and axes are missing.'

    c Ci-muti na   li-tepo v-a-phy-a   na   mu-lilo
       7-tree   and   5-leaf SP_8-PST-burn-FV by   3-fire
       'The tree and the leaves were burnt by the fire'

Phonological resolution is described for Xhosa S41 (Voeltz 1971). Since the subject marker of class 8 and class 10 is *zi-*, and class 11 nouns regularly take a class 10 plural, any conjunction of nouns of classes 7/8, 9/10 and 11/10 can be resolved by using the subject marker *zi-*.

Pragmatic aspects are relevant to agreement with conjoined NPs as well. In Luguru G35, non-human conjoint NPs can show default class 8 agreement (32a), but also partial agreement with any of the conjuncts, if the agreed-with conjunct is focused/emphasised (indicated by italics in the translation) (Marten & Ramadhani 2001).

(32) Luguru G35 (Marten & Ramadhani 2001: 272)

    a Ghumu-biki ne-chi-kapu   pfi-ghu-li-w-a
       3-tree   and-7-basket SP_8-buy-PASS-FV
       'The tree and the basket were bought'

    b Chi-ti na   ghumu-biki chi-ghul-iw-a
       7-chair and   3-tree SP_7-buy-PASS-FV
       'The chair and the tree were bought'

    c Chi-ti na   ghumu-biki u-ghul-iw-a
       7-chair and   3-tree SP_3-buy-PASS-FV
       'The chair and *the tree* were bought'

Another factor related to partial agreement is word order. Typically, a *pre-verbal* conjoint NP can only show plural (full) or default agreement, while a *post-verbal* conjoint NP can show plural (full), default, or first-conjunct (partial) agreement. This system is found, for example, in Haya JE22, Luguru G35, Shambaa G23 and Swahili G42, and is indifferent to whether the relevant conjoined NP functions as subject or object (Marten 2000, 2005, Riedel 2009).

(33) Swahili G42 (Marten 2005: 541–543)

    a Amina a-li-wa-on-a   Haroub na   Nayla
       Amina SP_1-PST-OP_2-see-FV Haroub and Nayla
       'Amina saw Haroub and Nayla.'

b   Amina   a-li-mw-on-a   Haroub   na   Nayla
Amina   SP$_1$-PST-OP$_1$-see-FV   Haroub   and   Nayla
'Amina saw Haroub and Nayla.'

c   Haroub   na   Nayla,   Amina   a-li-wa-on-a
Haroub   and   Nayla   Amina   SP$_1$-PST-OP$_2$-see-FV
'Haroub and Nayla, Amina saw them.'

d   *Haroub   na   Nayla,   Amina   a-li-mw-on-a
Haroub   and   Nayla   Amina   SP$_1$-PST-OP$_1$-see-FV

Data from agreement with conjoined NPs will continue to pose a challenge to theories of agreement and conjunction. Empirically, most work so far focuses on "and/with" conjunction, and on conjoined NPs with only two conjuncts (but see Marten 2000 on more complex conjoined NPs in Swahili). Future work should go beyond this and explore a wider set of data from a wider range of languages.

4 INFLUENCE OF MORPHOLOGY ON SYNTAX

The morphological process of verbal derivation, which is dealt with in length in Chapter 6, interacts closely with clausal syntax and semantics. Verbal derivational suffixes, or "extensions" are widespread in Bantu and include causative, applicative, reciprocal, passive, stative/neutro-passive and reversive/separative. The extensions change the meaning of the base verb and often interact unpredictably with its lexical semantic meaning. In addition to their semantic contribution, many extensions also interact in intricate ways with the valency of the base verb. Guthrie (1962) divides Bantu verbal extensions into three groups: 1) valency increasing (applicative, causative), 2) valency-decreasing (passive, stative/neutro-passive, reciprocal), and 3) valency-neutral (reversive/separative). More recent research shows that there are exceptions to this classification, in particular with respect to pragmatic functions of extensions.

Applicatives are a central research topic in Bantu morphosyntax, and are well-described and analysed. Applicatives typically introduce a new, applied object into the set of the verb's arguments. Different thematic roles can be expressed by the applied object with some variation among different languages. Typical roles include beneficiary (34) and its subtype substitution (35), goal/direction/recipient (36), instrument (37), location (38) and motive/reason (39).

(34)   Vunjo-Chaga E622 (Moshi 1998: 146–148)
Lémúnyí   n-á-lé-úlr-í-á   máná   sházru
Lemunyi   FOC-SP$_1$-PST-buy-APPL-FV   1.child   10.shoes
'Lemunyi bought the child shoes.'

(35)   Bemba M42 (Marten & Kula 2014: 3)
Ábá-icé   bá-ká-send-el-a=kó   im-fúmu   ubu-ta
2-children   SP$_2$-FUT-carry-APPL-FV=LOC$_{17}$ 9-chief   14-bow
'The children will carry the bow on behalf of (instead of) the chief.'
or 'The children will carry the bow for the chief there.'

(36) Ngoni N12 (Ngonyani & Githinji 2006: 34)
      Kuku   a-ku-va-pelek-el-a   va-jukulu   v-aki   chi-viga.
      grandpa SP_1-PRS-OP_2-send-APPL-FV   2-grandchild PP_2-his   7-pot
      'Grandpa is taking the pot to his grandchildren.'

(37) Kagulu G12 (Petzell 2008: 134)
      Ya-ku-dumul-il-a   sigi   mage
      SP_1-PRS-cut-APPL-FV   9.rope   14.knife
      'S/he is cutting the rope with a knife.'

(38) Chewa N31b (Alsina & Mchombo 1993: 42)
      A-lēnje   a-ku-pá-lúk-ir-á   mí-kêka   (pa-m-chēnga)
      2-hunters SP_2-PRS-OP_{16}-weave-APPL-FV   4-mats   16-3-sand
      'The hunters are weaving mats there (on the beach).'

(39) Zulu S42 (Buell 2005: 189)
      Ngi-lahl-el-a   i-mali   u-doti.
      SP_{1SG}-dispose-APPL-FV   9-money   1-trash
      'I'm taking out the trash for money.'

Several semantic analyses of Bantu applicatives have proposed that the underlying meaning of the extension is locative or "direction towards" and that other uses are semantic extensions of this basic meaning (Dammann 1961, Kähler-Meyer 1966, De Kind & Bostoen 2012, Marten & Kula 2014). Formal approaches have mainly concentrated on explaining how the additional argument is licensed, and what the exact status of the different arguments in applicative constructions is. Specific proposals include lexical licensing in LFG (Alsina & Mchombo 1990, Bresnan & Moshi 1990, Alsina & Mchombo 1993: 42, Murrell 2012), incorporation (Baker 1988), the postulation of a syntactic applicative head (Marantz 1993), and the differentiation of two such applicative heads, one "high" and one "low" (Pylkkänen 2008).

Comparatively little attention has as yet been paid to pragmatic and usage aspects of applicatives, although a correlation of applicatives with focus has sometimes been noted (Marten 2003, Creissels 2005, Bostoen & Mundeke 2011a). For example, in (40), even though the verb forms have applicative morphology, there is no corresponding change in valency. Rather, the applicative indicates focus on the predicate (40b) or motive phrase (41).

(40) Swahili G42 (Marten 2003: 215)
      a Juma   a-li-va-a   kanzu
         Juma SP_1-PST-wear-FV   9.man's.long.white.robe
         'Juma was wearing a kanzu'

      b Juma   a-li-val-i-a   nguo   rasmi
         Juma SP_1-PST-wear-APPL-FV   9.clothes   official
         'Juma was *dressed up* in official/formal clothes'

(41) Mbuun B87 (Bostoen & Mundeke 2011a: 192)
      ó-á-mó-dzwíllé   óngira   n-dzim
      SP_1-PRS -OP_1-kill.APPL   for   9-money
      'He kills her *for the money*.'

Causatives typically introduce an additional "causer" argument, which is encoded as subject, while the original subject is demoted to object and assumes the role of "causee" in addition to "agent."

(42) Luguru G35 (Marten & Ramadhani 2001: 267)
Wa-nzehe wa-mw-ambik-its-a Chuma ipfi-dyo
2-elders SP$_2$.PST-OP$_1$-cook-CAUS-FV Chuma 8-food
'The elders made Chuma cook food.'

Across Bantu, different morphological forms of causative markers can be distinguished (Bastin 1986) and are sometimes analysed as resulting in different interpretations, such as direct vs. indirect causation (Simango 1999, 2009). Instrument arguments are introduced by causative morphology in some languages, such as Ganda JE15 in (43), but by applicatives in others, such as Swahili G42 in (44).

(43) Ganda JE15 (Wald 1997: 222)
Tu-lim-is-a n-kumbi
SP$_{1PL}$-farm-CAUS-FV 9-hoe
'We are farming with hoes.'

(44) Swahili G42 (Wald 1997: 222)
Tu-na-lim-i-a ma-jembe
SP$_{1PL}$-PRS-farm-APPL-FV 6-hoe
'We are farming with hoes.'

Wald (1997) proposes to locate this difference along a geographic division between North-Eastern Bantu languages (applicatives) and South-Western Bantu languages (causatives), although he notes that e.g. Ganda JE15 complicates the picture.

The main valency-reducing suffixes in Bantu are passives, statives (or neutro-passives) and reciprocals. In reciprocals, agent and patient typically form a plural subject, and the erstwhile transitive verb becomes intransitive. Alternatively, one participant in the action can be expressed by an oblique, introduced by a preposition such as *na* in Ndamba G52.

(45) Ndamba G52 (Edelsten & Lijongwa 2010: 100)
Simba a-ku-tov-an-a na ndembo
9.lion SP$_1$-PRS-hit-RECP-FV with 9.elephant
'The lion is fighting (lit. hit each other) with the elephant'

The reciprocal also can be used to form antipassives, as shown in recent work by Bostoen *et al.* (2015) and Dom *et al.* (2015).

Both passives and statives typically promote a non-agent expression to subject and demote the agent. In passives, the agent can be encoded as oblique (46a), while it normally remains unexpressed in statives (46b).

(46) Chewa N31b (Simango 2009: 122, 124)
a N-dodo i-dza-thyol-edw-a ndi Shuko
9-stick SP$_9$-FUT-break-PASS-FV by Shuko
'A stick will be broken by Shuko.'

b   N-dodo   i-dza-thyok-a   (* ndi Shuko)
     9-stick   SP$_9$-FUT-break.STAT-FV   by Shuko
     'A stick will break.'

While many Bantu languages, like Chewa N31b, have a passive suffix reflecting the reconstructed Proto-Bantu passive allomorphs *-ʊ- and *-ibʊ- (Stappers 1967, Schadeberg
2003: 72), some Bantu languages have lost a dedicated passive construction, e.g., Mbuun
B87 (Bostoen & Mundeke 2011b), and several others have innovated passive constructions. Newly developed passive constructions are formed in different ways, e.g., by adapt-
ing other extensions, like -am-, as in many zone C languages (Schadeberg 2003: 78–79),
or -ik- as in Tumbuka N21 (Chavula 2016), or by developing innovative constructions,
such as a participle passive construction as in Luba L31a (Willems 1970: 160–162), or a
(former) impersonal construction, as in Bemba M42 exemplified in (47).

(47)   Bemba M42 (Kula & Marten 2010: 119)
        Umw-áána   bá-alí-mu-ít-a   ku   mu-mbúlu
        1-child   SP$_2$-PST-OP$_1$-call-FV   by   3-wild.dog
        'The child was called by the wild dog.'

There is also variation in the encoding of the oblique agent phrase: e.g., with a comitative
preposition, a copula form, a locative or without any overt marker (Fleisch 2005).

     Several extensions can be used in the same verb form, although there are idiosyn-
cratic and language-specific restrictions on combinations, as well as some that result in
non-compositional meanings. The order and interpretation of multiple extensions has
been the cause of much debate. In some languages, the order of extensions seems to
reflect their semantic scope, or, in syntactic terms, the order of their corresponding syntactic operations, following the "mirror principle" (Baker 1985, Alsina 1999). However,
there are a number of counterexamples, and an alternative view is to see the order of
extensions as being constrained by a morphological template, imposing the order Causative-Applicative-Reciprocal-Passive (CARP) (Hyman 2003). In Chewa N31b, for example, Hyman (2003) shows that only the order causative-applicative is allowed and is used
also in cases where an applicative modifies the base verb first, and the whole event is
then causativised. In (48), 'the spoon' is the applicative instrument, not the causer of the
stirring:

(48)   Chewa N31b (Hyman 2003: 248)
        A-lenjé   a-ku-tákás-its-il-a   m-kází   m-thíko
        2-hunters   SP$_2$-PROG-stir-CAUS-APPL-FV   1-woman   3-spoon
        'The hunters are making the woman stir with a spoon'

Other areas of interest have been the interaction of passives and applicatives, in particular the effect of passives on object marking (Woolford 1995, Zeller 2012b), the combination of stative and reciprocal resulting in potential meaning (Seidl & Dimitriadis
2003), and the interaction of extensions with inflectional suffixes. In Ndamba G52, for
example, the habitual and completive aspect suffixes precede the passive (Edelsten &
Lijongwa 2010), even though the derivational extensions are expected to precede
inflectional suffixes.

5 NON-VERBAL PREDICATION

Bantu languages display a wide array of non-verbal clauses. Nominal and adjectival predication, existence, location and possession can all be expressed without a lexical verbal head. Different kinds of copulas can be used in these constructions, as well as prosodic marking of the predicate. The interaction of different formal means of encoding non-verbal predication with the specific syntactic configurations in which they are used often results in differences in interpretation and appropriateness in different pragmatic contexts.

Across Bantu, the copula takes different forms, including an invariant, uninflected copula, as in (49), which may be cliticised to the following post-copula complement, as in (50). Inflected copulas include: forms taking a subject concord (51), possessives built from the subject concord and the comitative preposition *na* (52), and locatives, often based on the subject concord and a locative clitic (53).

(49)   Nen A44 (Dugast 1971: 348)
       Mɛ   lɛ   mù-ɛs
       1SG   COP NP_1-good
       'I am fine.'

(50)   Yeyi R41 (Seidel 2008: 415)
       Muraliswani   ndi=mu-teriki
       Muraliswani   COP=1-cook
       'Muraliswani is the cook.'

(51)   Zulu S42 (Zeller 2013: 1120)
       U-Thandi   u-ng-u-m-fundi.
       AUG-1a.Thandi SP_1-COP-AUG-1-student
       'Thandi is a student.'

(52)   Herero R31 (Möhlig & Kavari 2008: 214)
       Ú-ná   omú-tímá   omu-wá.
       SP_1-with   3-heart NP_3-good
       'S/he has a good heart.'

(53)   Swahili G42 (Marten 2013: 56)
       Yeye   yu-ko   U-kerewe   mimi   ni-ko   U-sukuma.
       3SG SP_1-be($<LOC_{17}$)   11-kerewe   1SG SP_{1SG}-be($<LOC_{17}$)   11-sukuma
       'He is in Ukerewe; I am in Usukuma.'

In addition to segmentally expressed copulas, non-verbal predication can be expressed purely prosodically. In Swahili, for example, the predicative adjective receives a "final tonic" (Maw 1969) and the whole expression receives sentence intonation.

(54)   Swahili G42 (Maw 1969: 42)
       Ku-tenda | ku-zuri
       15-act NP_{15}-good
       'Acting is good.'

In many Bantu languages, nouns and adjectives can be used predicatively through the use of specific tonal inflection, either by the use of special (predicative) High tone associated with the nominal prefix or the augment, or by the lowering of lexical High tones. The example from Shona S10 in (55) shows how nominal predication is expressed by a High tone on the noun class prefix.

(55)   Shona S10 (Welmers 1973: 323)
       a   mù-nhù   b   mú-nhù
           1-person   1-person
           'person'   'It is a person.'

In Herero R31, a High tone is found on the augment as part of a system of nominal inflection (Möhlig & Kavari 2008, Kavari *et al.* 2012).

(56)   Herero R31(Möhlig & Kavari 2008: 122, Kavari *et al.* 2012: 318)
       a   òtjì-hávérò   b   ótjì-hávérò
           7-chair   7-chair
           'chair'   'It's a chair.'

       c   òzò-ngòmbè   òzò-néné   d   òzò-ngòmbè   ózó-néné
           10-cow NP_{10}-big   10-cow NP_{10}-big
           'big cows'   'The cows are big.'

In contrast, in Zulu S42, a depressor Low tone on the augment is used to indicate predication (Doke 1961: 215ff).

(57)   Zulu S42 (Doke 1961: 215)
       a   í-m-buzi   b   ì-m-buzi
           AUG-9-goat   AUG-9-goat
           'goat'   'It is a goat.'

A particular (diachronic) variant of this system is found in Cuwabo P34 and Makhuwa P31 where a process of "predicate lowering" deletes the lexical High tones of predicatively used nouns and adjectives (in some cases with the exception of the final lexical High tone).

(58)   Cuwabo P34 (Guérois 2015: 126)
       a   mu-sáno   /   mú-yaná   b   mu-sano   /   mu-yaná
           1-queen   1-woman   1-queen   1-woman
           'queen'   'woman'   'It is a queen.'   'It is a woman.'

(59)   Makhuwa P31 (van der Wal 2009: 120)
       a   n-thálí   mw-áńkhaáni   b   n-thálí   mw-ankhaáni
           3-tree NP_3-small   3-tree NP_3-small
           'the small tree'   'The tree is small.'

Simple copula constructions are typically restricted to present or timeless contexts, and not inflected for tense – although they can often be relativised (60), negated (61) or receive some aspectual marking (62).

(60) Zulu S42 (Buell & de Dreu 2013: 431)
i-zingane   ezi-se-zincane.
AUG-10.children   REL$_{10}$-PER-10.small
'children who are still small'

(61) Digo E73 (Nicolle 2013: 287)
Mino   si   dza   a-tu   a-njina
1SG   NEG.COP   like   2-people   NP$_2$-other
'I am not like other people.'

(62) Zulu S42 (Buell & de Dreu 2013: 427)
Ngi-se-ngu-m-fundisi.
SP$_{1SG}$-PER-COP-1-teacher
'I am still a teacher.'

In non-present contexts an alternative construction involving some (possibly historically fossilised) form of the verbs 'be' or 'become' that can be inflected for tense is typically used.

(63) Zulu S42 (Buell & de Dreu 2013: 450)
Ngi-zo-ba   se-Thekw-ini.
SP$_{1SG}$-FUT-be   LOC-Durban-LOC
'I will be in Durban.'

(64) Ndamba G52 (Edelsten & Lijongwa 2010: 94)
A-ka-v-ele   m-henja   w-ako.
SP$_1$-PST-be-PFV   1-guest   PP$_1$-POSS$_{2SG}$
'S/he was your guest.'

(65) Swahili G42 (Marten 2013: 48)
Ku-li-ku-w-a   na   ma-endeleo   sana
SP$_{17}$-PST-INF-be-FV   with   6-development   much
'There was a lot of development.'

Bantu non-verbal predication has received little attention in the formal literature so far (Buell & de Dreu 2013, Marten 2013: 48, Zeller 2013).

6 COMPLEX CLAUSE STRUCTURE AND DEPENDENT CLAUSES

Studies in Bantu morphosyntax have historically focused more on clausal syntax and predicate-argument relations than on complex sentences. Yet, the research that has been undertaken shows the theoretical and comparative interest this area provides. In the next sections, we briefly survey properties of relative clauses and of other dependent clauses.

6.1 Relative clauses

The comprehensive study of Nsuka-Nkutsi (1982) reveals considerable variation in the form of relative constructions across Bantu (cf. Zeller 2004, Henderson 2006, Downing *et al.* 2010 for more recent surveys). We illustrate here with a few chosen languages the following parameters of variation: (1) the form and position of the relative morpheme; (2) word order within the relative clause; and (3) whether a non-subject relative head is resumed within the relative clause or leaves a gap.

According to the survey of Nsuka-Nkutsi (1982), the two most common forms of the relative morpheme are derived from demonstrative pronouns, often emphatic demonstratives. The relative pronoun can be a free form, identical to one set of demonstratives, or a bound, reduced form. As Downing and Mtenje (2011) show, Chewa N31b has both types of relative morpheme: -*méné*, a free form which occurs at the beginning of the relative clause, and the bound enclitic -*o*, which occurs at the end of the relative clause. The likely historical source of the relative pronoun, -*méné*, is the homophonous emphatic demonstrative, while the -*o* relative enclitic is homophonous with the remote demonstrative, both illustrated in *nyumbá zi-ménee-zo* 'those very houses.'

(66) Chewa N31b (Downing & Mtenje 2011)

   a   M-balá   i-méné   í-ná-bá   n-dalámá   z-angáa-yo   i-ku-tháawa

       9-thief PP_9-REL SP_9-PST_1-steal 10-money PP_{10}-my-REL_9 SP_9-PROG-run.away

       'The thief who stole my money is running away.'

   b   A-lendó   a-méné   á-ná-wa-bweretsérá   m-phátsoo-wo   a-koondwa

       2-visitor PP_2-REL SP_2-PST_2-OP_2-bring.APPL 10-gift-REL_2 SP_2.PRF-be.happy

       'The visitors who they brought the gifts for are happy.'

Both relative morphemes in (66a-b) show agreement with the head of the relative. Note that sentence (66b) is an object relative and the head is resumed as an object marker on the relative verb, i.e., the class 2 object prefix. Both relativisers can be omitted in Chewa N31b. In languages like Eton A71 (Van de Velde 2008a), relative clauses are typically not introduced by a relativiser.

In languages which do usually use a relativiser, it is very common, in Nsuka Nkutsi's survey, for the free relative morpheme to occur at the beginning of the relative clause, while cognates of the -*o* relative occur as an enclitic to the (auxiliary) verb or the relative clause, as in Chewa N31b. However, Southern Bantu languages (Zeller 2004) and Ganda JE15, for example, illustrate other possibilities. In Ganda JE15, as Hyman and Katamba (2010) show, the -*e* relativiser, required with non-subject relatives, occurs as a proclitic immediately before the relative verb (67a), even though it is considered related to the -*o* relativiser by Nsuka-Nkutsi (1982). Subject relatives (67b) do not have an overt relativiser.

(67) Ganda JE15 (Hyman & Katamba 2010: 88)

   a   tw-áá-gùl-à   è-bì-kópò   wàlúsìmbì   by-è   y-à-léètá

       SP_{1PL}-PST-buy-FV   AUG-8-cup   Walusimbi PP_8-REL SP_1-PST-bring

       'We bought the cups that Walusimbi brought.'

   b   à-mànyí   bá-kázì   bá-á-láb-á   bí-kópò

       SP_1-know   2-woman SP_2-PST-see-FV   8-cup

       'S/he knows the women who saw the cups.'

The third most widely distributed form of the relative morpheme in Nsuka-Nkutsi's survey is the connective or associative morpheme -*a*, which across Bantu commonly connects noun phrases to indicate possession or other forms of modification (cf. Van de Velde 2013 for recent discussion). This type of relative is found, for example, in Kalanga S16 and other dialects of Shona S10 (Güldemann 1997, Demuth & Harford 1999, Letsholo 2009).

(68) Kalanga S16 (Letsholo 2009)
    a isípá   Néo   cha-á-ká-pá   Nchídzi
       7.soap   Neo   REL$_7$-SP$_1$-PST-give Nchidzi
       'The soap that Neo gave Nchidzi . . .'

    b N-lume   bo-Néo   wa-bá-ká-bóna   wá-énda
       1-man   2a-Neo   REL$_1$-SP$_2$-PST-see SP$_1$-leave
       'The man whom Neo and others saw left.'

    c ngwanáná   wa-ká-ízela   (cf. ngwanáná wá-ka-ízela 'The girl is sleeping')
       1.girl   REL$_1$-PST-sleep
       'The girl who is sleeping . . .'

Note in (68c) that the relative morpheme and subject prefix can fuse in the case of subject relatives, so that only tone distinguishes a relative from a non-relative construction. Tone is, in fact, a widely distributed relative marker in Bantu languages.

In all the data so far, word order in the relative clause is the same as in main clauses, in both subject and object relatives. However, in some languages, we find subject inversion in the relative clause with object relatives. Widely discussed examples of this come from the Shona S10 variety discussed in Demuth and Harford (1999) and from Zamba C322 (Bokamba 1976). Note in the Zamba example, that, as is typical for subject inversion (Section 3.3), the pre-verbal noun phrase – not the logical subject – conditions subject marking on the relative verb.

(69) Shona S10 (Demuth & Harford 1999)
    m-batya   dza-va-ka-soner-a   va-kadzi   mw-enga
    10-clothes REL$_{10}$-SP$_2$-TAM-sew.APPL 2-woman 1-bride
    'Clothes which the women sewed for the bride . . .'

(70) Zamba C322 (Bokamba 1976)
    izi-bata   i-zi-eza-áki   o-Poso   ba-butu   loome
    5-duck   REL$_5$-SP$_5$-give-PST 1-Poso 2-guest today
    'The duck that Poso gave the guests today . . .'

Work like Givón (1972) and Demuth and Harford (1999) has proposed that there is a link between subject inversion in relatives and the form of the relative morpheme. Subject inversion should be obligatory in languages where the relative morpheme is a bound proclitic on the verb, while main clause word order should be found only in languages with free relative complementisers. As Henderson (2006) and Letsholo (2006) point out, however, one easily finds counterexamples. In the Ganda JE15 (67) and Kalanga S16 (68) data cited above, for instance, we find bound proclitic relativisers but no subject inversion.

The last parameter of variation we take up also involves object relatives. As work like Henderson (2006), Ngonyani (2006) and Marten and Kula (2012) show, in some

Bantu languages, a resumptive object marker is found on the relative verb, while other languages tolerate a "gap." This variation is illustrated in the sample of languages in this section. We find a resumptive pronoun in Chewa N31b (66b), though only with human objects. We do not find an object marker on the relative verb in the object relatives in the other languages. The Marten & Kula (2012) object marking survey identifies three types of languages: (1) object marking is obligatory in the relative clause, e.g., Tswana S31 or Zulu S42; (2) object marking is optional, e.g., Swahili G42; (3) object marking is disallowed, e.g., Bemba M42. In this area and in others related to the syntactic structure of relatives, we see the need for more detailed research on more languages.

6.2 Issues in complex clause constructions

Complex clauses other than relative clauses are less well studied. In this section we survey a few current research topics related to Bantu complex clauses.

Güldemann (1996) shows that, across Bantu, distinctions between main and dependent clauses are often reflected in verbal morphology. For example, two different forms of negation are often employed in main and dependent clauses, the former typically preceding the subject marker, the latter following it.

(71) Nkore-Kiga JE13-14 (Taylor 1985: 59)
    a Ti-ba-giire   Mbarara
        NEG-SP_2-go.PST Mbarara
        'They didn't go to Mbarara'

    b N-ta-riku-mu-reeb-a
        SP_{1SG}-NEG-PRS.PROG-SP_1-see-FV
        '(if) I didn't see him' [literally: 'I not seeing him']

A related phenomenon is the presence of different (class 1) subject marker forms in main and subordinated clauses (and other contexts), sometimes called "anti-agreement" (e.g., Schneider-Zioga 2007, Henderson 2013).

The licensing of arguments – in particular subjects – in complex sentences is another current research topic. In Bukusu JE31c (72a), the subject of the lower clause, *Sammy*, is licensed by the inflected verb. However, in (72b), the predicate is an uninflected infinitive, which in many syntactic theories is not able to license an overt subject.

(72) Bukusu JE31c (Diercks 2012: 260)
    a Ka-nyal-ikhana   mbo   Sammy   a-la-khila   ku-mw-inyawe okwo
        SP_6-be.possible-STAT that   Sammy SP_1-FUT-win AUG-3-game DEM_3
        'It is possible that Sammy will win the game.'

    b Ka-nyal-ikhana   Sammy   khu-khila   ku-mw-inyawe o-kwo
        SP_6-be.possible-STAT Sammy   15-win.INF   AUG-3-game DEM_3
        'It is possible for Sammy to win the game.'

Similarly, in Herero R31 (73b-c), the lower clause subjects precede the complementiser *kutja*. While in (73b), *òvánátjè* 'children' could be analysed as the object of the matrix verb *ívángà* 'want,' this is unlikely in (73c) as the verb is a stative form.

(73) Herero R31 (Kavari *et al.* 2012: 328–330)

a Ò-mìtìrì   í-váng-à   kùtjá   òvà-nátjè   vé-tjáng-é   òm-bàpírà
9-teacher SP_9.HAB-want-FV   that   2-child SP_2-write-SBJV   10-letter
'The teacher wants that the children should write a letter.'

b Ò-mìtìrì   í-váng-à   òvá-nátjè   kùtjá   vé-tjáng-é   òm-bàpírà
9-teacher SP_9.HAB-want-FV   2-child   that SP_2-write-SBJV   10-letter
'The teacher wants that the children should write a letter.'

c P-á-mún-ík-á   òvá-éndá   kùtjá   vè-rí   mò-n-gándà
SP_{16}-PST-seen-NEUT-FV   2-guests   that SP_2.HAB-be   18-9-home
'It was seen that the guests are at home/in the house.'

In (73c) *òváéndá* 'children' is semantically, and presumably syntactically, the subject of the lower clause ('are at home'). However, the fact that it appears before the complementiser *kùtjá* seems to show that it is not part of that clause, and so it is not clear what the underlying syntactic licensing relations are.

A similar problem obtains in so-called "hyper-raising" constructions in Saamia JE34 (Carstens 2011), where the subject appears to be raised out of a finite clause. In (74), both verbs are inflected for subject agreement.

(74) Saamia JE34 (Carstens 2011: 725)
Efula   yi-bonekhana   i-na-kwa   muchiri
9.rain SP_9-appear SP_9-FUT-fall   tomorrow
'It seems that it will rain tomorrow.'(Lit: 'Rain seems will fall tomorrow.')

Examples such as these raise fundamental questions about the nature of subjects in Bantu and their grammatical, thematic and pragmatic status (Diercks 2012). Investigation of these questions and work on other aspects of complex syntactic phenomena in Bantu is just beginning and is likely to yield significant new results in the years to come.

7 PHONOLOGY-SYNTAX INTERFACE

Work on Bantu phrasal phonology since Byarushengo *et al.* (1976) on Haya JE22 and Kisseberth and Abasheikh (1974) on Mwiini G412 has established that High tone realisation and other prosodies are often conditioned by syntactic context. For example, in Haya JE22, all pre-verbal nouns have what Byarushengo *et al.* (1976) term a phrase-medial tone pattern (75a-b). This pattern contrasts with a phrase-final pattern: compare the tone of 'snuff' and 'Kakulu' in (75b) vs. (75c). Parentheses indicate the boundaries of domains relevant to High tone realisation.

(75) Haya JE22 (Byarushengo *et al.* 1976)
a *Canonical order* – one assertion = one tone domain
(aba-kázi   ba-bon'   ómw-áána)
2-woman SP_2-see   1-child
'The women see the child.'

b *Left dislocation* – one assertion = one tone domain
(aba-kázi   Kakúlw   óbu-goló   ba-bu-mú-ha)
2-woman   Kakulu   14-snuff SP_2-OP_{14}-OP_1-give
'The women, Kakulu, the snuff, they GIVE it to him.'

c   *Right dislocation* – assertion ends with the verb; right dislocations in separate
domains
(a-bu-bón')   óbu-góló)   Kakûlu)
SP_1-OP_{14}-see   14-snuff   Kakulu
'He SEES it, the snuff, Kakulu.'

In Mwiini G412, the cue to phrasing is the (potential) occurrence of a long vowel and
obligatory accent marked with an acute accent. In (76), a prosodic phrase break follows
every noun phrase. In the Mwiini G412 data, underlined coronal consonants are [dental].
See Kisseberth and Abasheikh (1974), Abasheikh (1978), Kisseberth (2010), for more
detailed discussion of Mwiini G412 phonology and syntax.

(76)   Mwiini G412 (Kisseberth 2010); '/' separates prosodic phrases
    a   sultani úyu/sulile m-loza mw-aanáwe/mú-ke
        'This sultan/wanted to marry his son/to a woman.'

    b   Hamádi/mw-andikilile mw-áana/xáti/ka Núuru
        'Hamadi/wrote for the child/a letter/to Nuuru.'

    c   mu-nthu ofeto x-fakatá/na-x-pumúla
        'The man who is tired from running/is resting now.'

    d   n-uzize chi-buku ch-a Nuurú/m-bozelo mw-aaná
        'I sold the book that Nuuru/stole from the child.'

As we can see from these two languages, one parameter of cross-Bantu variation in prosodic phrasing is the size of the domains. In Haya JE22, phrases can include an entire
"assertion" or clause (Byarushengo *et al.* 1976). In Mwiini G412, phrase boundaries align
with XPs (noun phrases or other lexical phrases) (Kisseberth 2010). Work since the 1970s
confirms this basic dichotomy (cf. Inkelas & Zec 1990). Cheng and Downing (2016)
provide a recent overview.

   Another parameter of variation lies in the phrasing of pre-verbal topics, as recent surveys like Downing (2011) and Zerbian (2007) show. In languages where phrase boundaries align with right clause edges, one would expect pre-verbal noun phrases to phrase
with the following clause. This is exactly what one finds in the Haya JE22 data in (75a-
b). However, in languages like Chewa N31b, pre-verbal topics and subjects are phrased
separately from what follows, even though one has not reached the edge of the clause.
Parentheses indicate phrasal domain boundaries, motivated by the distribution of long
penult vowels.

(77)   Chewa N31b (Cheng & Downing 2016)
    a   (M-fúumu)   (i-na-pátsá   mw-aná   zó-óváala)
        9-chief SP_9-PST_2-give   1-child   10-clothes
        'The chief gave the child clothes.'

    b   (a-leenje)   (zi-ná-wá-luuma)   njúuchi
        2-hunter SP_{10}-PST_2-OP_2-bite   10.bees
        'The hunters, they bit them, the bees [did].'

While one important function of Bantu prosodic phrasing is to mark the right edges of
constituents like clause, XP and topic, Odden (1995) demonstrates that prosody can be

conditioned by quite specific morphosyntactic information. For example, negative vs. affirmative main verbs in Tsonga S53 condition different phrasal tone realisations. In the affirmative, High tone spreads to the penult of the word following the verb; in contrast a High tone spreads to the final syllable of the phrase from a negative verb.

(78) Tsonga S53 Kisseberth (1994: 148)
    a Affirmative
        vá-xávélá   mú-nhu   tin-guuvu
        SP_2-buy.for   1-person   10-clothes
        'They are buying someone clothes.'

    b Negative
        a-vá-xav'élí   xí-kóxá   nyáámá
        NEG-SP_2-buy   7-old.woman   9.meat
        'They are not buying meat for the old woman'

Finally, as work since Bickmore (1990) shows, modified nouns often phrase differently from non-modified nouns in the same position (cf. Cheng & Downing 2016 for a recent survey). For example, in Nyambo JE21, prosodic domains condition the phrasal process of High tone deletion (HTD): the rightmost High tone of a word is deleted if the following word within the phrase has a High tone. Examples (79a) and (79c) show that the entire assertion can be the domain of HTD. However, in (79b) and (79d), we see that a branching (modified) noun phrase (underlined) interrupts the HTD domain.

(79) Nyambo JE21 (Bickmore 1990: 14–15)
    a (ba-kuru   bá-ka-júna).
        2-mature.ones SP_2-PST-help
        'The mature ones helped.'

    b (<u>Aba-kozi ba-kúru</u>)   (bá-ka-júna)
        2-worker NP_2-mature SP_2-PST-help
        'The mature workers helped.'

    c (Nejákworech'   ába-koz'   ém-bwa)
        SP_1.FUT.show   2-worker   9-dog
        'S/he will show the workers the dog.

    d (Nejákworech'   <u>ómukama w'   ába-kózi</u>)   (ém-bwa)
        SP_1.FUT.show   1-chief CON_1 2-worker   9-dog
        'S/he will show the chief of the workers the dog.'

Their robust phrasal prosody explains why Bantu languages are bound to continue to play an important role in the development of theories of the phonology-syntax interface.

8 CONCLUSION

Research on Bantu morphosyntax has made a strong contribution to theoretical linguistics, linguistic typology and historical and comparative linguistics. Core themes in the study of Bantu morphosyntax are related to the interaction of morphology, syntax and information structure, with specific focus on agreement, word order and valency. Information structure, in particular, has become a major research interest, highlighting the interaction of morphosyntactic structure, pragmatics and contextual information, on the one hand, with prosody and information packaging, on the other. In terms of wider cross-linguistic typology, Bantu languages not only provide abundant and complex examples for some widely attested grammatical phenomena, such as those associated with pronominal subject and object marking or valency-changing morphology, but also present unique constructions not found in this form outside of the family: for example, the conjoint-disjoint alternation or the rich array of inversion constructions.

Future research directions in Bantu morphosyntax are likely to build on the strong tradition of interaction between formal, theoretical and descriptive linguistics in Bantu. Significant new developments will come as more sophisticated research questions and more fine-grained, detailed analyses become available for more Bantu languages on a wider range of constructions. Work on comparative and historical linguistics, grammaticalisation, micro-variation and language contact is likely to remain highly relevant for Bantu linguistics and will open further opportunities for interdisciplinary work involving, for example, historical, archaeological and sociological research. Finally, research in all aspects of Bantu morphosyntax will benefit from the increasing availability of different kinds of data through open access corpora, which not only allow the study of patterns of use, but also enable data and analyses to be critically assessed by a wider community of practitioners and researchers.

REFERENCES

Abasheikh, M. I. (1978) *The Grammar of Chimwi:ni Causatives*. Urbana-Champaign: University of Illinois,

Adams, N. (2010) *The Zulu Ditransitive Verb Phrase*. Chicago: University of Chicago, PhD dissertation.

Aissen, J. (2003) Differential Object Marking: Iconicity vs. Economy. *Natural Language & Linguistic Theory* 21: 435–483.

Alsina, A. (1999) Where's the Mirror Principle? *The Linguistic Review* 16: 1–42.

Alsina, A. & S. Mchombo (1990) The Syntax of Applicatives in Chicheŵa: Problems for a Theta Theoretic Asymmetry. *Natural Language & Linguistic Theory* 8(4): 493–506.

Alsina, A. & S. A. Mchombo (1993) Object Asymmetries and the Chichewa Applicative Construction. In Mchombo, S. A. (ed.), *Theoretical Aspects of Bantu Grammar*, 17–45. Stanford: CSLI.

Baker, M. (1985) The Mirror Principle and Morphosyntactic Explanation. *Linguistic Inquiry* 16: 373–415.

Baker, M. C. (1988) *Incorporation. A Theory of Grammatical Function Changing*. Chicago: University of Chicago Press.

Bastin, Y. (1986) Les suffixes causatifs dans les langues bantoues. *Africana Linguistica* 10: 56–145.

Bax, A. & M. Diercks (2012) Information Structure Constraints on Object Marking in Manyika. *Southern African Linguistics and Applied Language Studies* 30(2): 185–202.

Bentley, M. (1994) *The Syntactic Effects of Animacy in Bantu Languages*. Bloomington: Indiana University, PhD dissertation.

Bickmore, L. S. (1990) Branching Nodes and Prosodic Categories: Evidence from Kinyambo. In Inkelas, S. & D. Zec (eds.), *The Phonology-Syntax Connection*, 1–17. Chicago: University of Chicago Press.

Bleek, W. H. I. (1862) *A Comparative Grammar of South African Languages. Part I. Phonology*. London: Trübner & Co.

Bokamba, E. G. (1976) *Question Formation in Some Bantu Languages*. Bloomington: Indiana University, PhD dissertation.

Bokamba, E. G. (1985) Verbal Agreement as a Noncyclic Rule in Bantu. In Goyvaerts, D. L. (ed.), *African Linguistics: Essays in Memory of M.W.K. Semikenke*, 9–54. Amsterdam; Philadelphia: John Benjamins.

Bossong, G. (1985) *Empirische Universalienforschung. Differentielle Objektmarkierung in den neuiranischen Sprachen*. Tübingen: Narr.

Bostoen, K., S. Dom & G. Segerer (2015) The Antipassive in Bantu. *Linguistics* 53(4): 731–772.

Bostoen, K. & L. Mundeke (2011a) The Causative/Applicative Syncretism in Mbuun (Bantu B87, DRC): Semantic Split or Phonemic Merger? *Journal of African Languages and Linguistics* 32(2): 179–218.

Bostoen, K. & L. Mundeke (2011b) Passiveness and Inversion in Mbuun (Bantu B87, DRC). *Studies in Language* 21(1): 72–111.

Bostoen, K. & L. Mundeke (2012) Subject Marking and Focus in Mbuun (Bantu, B87). *Southern African Linguistics and Applied Language Studies* 30(2): 139–154.

Bresnan, J. & J. M. Kanerva (1989) Locative Inversion in Chichewa – a Case-Study of Factorization in Grammar. *Linguistic Inquiry* 20(1): 1–50.

Bresnan, J. & S. Mchombo (1987) Topic, Pronoun, and Agreement in Chichêwa. *Language* 63: 741–782.

Bresnan, J. & L. Moshi (1990) Object Asymmetries in Comparative Bantu Syntax. *Linguistic Inquiry* 21(2): 147–185.

Buell, L. (2005) *Issues in Zulu Verbal Morphosyntax*. Los Angeles: University of California, PhD dissertation.

Buell, L. (2006) The Zulu Conjoint/Disjoint Verb Alternation: Focus or Constituency? *ZAS Papers in Linguistics* 43: 9–30.

Buell, L. (2007) Semantic and Formal Locatives: Implications for the Bantu Locative Inversion Typology. *SOAS Working Papers in Linguistics* 15: 105–120.

Buell, L. C. (2015) The Bantu Languages. In Kiss, T. & A. Alexiadou (eds.), *Syntax Theory and Analysis, Volume 3*, 1622–1657. Berlin: De Gruyter Mouton.

Buell, L. C. & M. de Dreu (2013) Subject Raising in Zulu and the Nature of PredP. *The Linguistic Review* 30(3): 423–466.

Byarushengo, E. R., L. M. Hyman & S. Tenenbaum (1976) Tone, Accent and Assertion in Haya. In Hyman, L. M. (ed.), *Studies in Bantu Tonology*, 183–205. Los Angeles: Department of Linguistics, University of Southern California.

Carstens, V. (2011) Hyperactivity and Hyperagreement in Bantu. *Lingua* 121: 721–741.

Chavula, J. J. (2016) *Verbal Derivation and Valency in Citumbuka*. Utrecht: LOT.

Cheng, L. L.-S. & L. J. Downing (2009) Where's the Topic in Zulu? *The Linguistic Review* 26: 207–238.

Cheng, L. L.-S. & L. J. Downing (2014) Indefinite Subjects in Durban Zulu. *ZAS Papers in Linguistics* 57: 5–25.

Cheng, L. L.-S. & L. J. Downing (2016) Phasal Syntax = Cyclic Phonology? *Syntax* 19(2): 159–191.

Cole, D. T. (1955) *An Introduction to Tswana Grammar*. London; New York: Longmans, Green.

Corbett, G. G. & A. Mtenje (1987) Gender Agreement in Chichewa. *Studies in African Linguistics* 181: 1–38.

Creissels, D. (1996) Conjunctive and Disjunctive Verb Forms in Setswana. *South African Journal of African Languages* 16(4): 109–115.

Creissels, D. (2005) A Typology of Subject and Object Markers in African Languages. In Voeltz, E. F. K. (ed.), *Studies in African Linguistic Typology*, 43–70. Amsterdam; Philadelphia: John Benjamins.

Creissels, D. (2017) The Conjoint/Disjoint Distinction in the Tonal Morphology of Tswana. In van der Wal, J. & L. M. Hyman (eds.), *The Conjoint/Disjoint Alternation in Bantu*, 200–238. Berlin: De Gruyter Mouton.

Dammann, E. (1961) Das Applikativum in den Bantusprachen. *Zeitschrift der Deutschen Morgenländischen Gesellschaft* 36: 160–169.

De Kind, J. (2014) Pre-Verbal Focus in Kisikongo. *ZAS Papers in Linguistics* 57: 95–122.

De Kind, J. & K. Bostoen (2012) The Applicative in Cilubà Grammar and Discourse: A Semantic Goal Analysis. *Southern African Linguistics and Applied Language Studies* 30(1): 101–124.

De Vos, M. & H. Mitchley (2012) Subject Marking and Pre-Verbal Coordination in Sesotho: A Perspective from Optimality Theory. *Southern African Linguistics and Applied Language Studies* 30: 155–170.

Demuth, K. (1990) Subject, Topic and Sesotho Passive. *Journal of Child Language* 17: 67–84.

Demuth, K. & C. Harford (1999) Verb Raising and Subject Inversion in Comparative Bantu. *Journal of African Languages and Linguistics* 20(1): 41–61.

Demuth, K. & S. Mmusi (1997) Presentational Focus and Thematic Structure in Comparative Bantu. *Journal of African Languages and Linguistics* 18: 1–19.

Diercks, M. (2010) *Agreement with Subjects in Lubukusu* Georgetown: Georgetown University, PhD dissertation.

Diercks, M. (2011) The Morphosyntax of Lubukusu Locative Inversion and the Parameterization of Agree. *Lingua* 121: 702–720.

Diercks, M. (2012) Parameterizing Case: Evidence from Bantu. *Syntax* 15(3): 253–286.

Dimmendaal, G. (1995) Metatony in Benue-Congo: Some Further Evidence for an Original Augment. In Emenanjo, E. N. & O.-M. Ndimele (eds.), *Issues in African Languages and Linguistics: Essays in Honour of Kay Williamson*, 30–38. Aba: National Institute for Nigerian Languages.

Doke, C. M. (1961) *Textbook of Zulu Grammar*. Cape Town: Longmans.

Dom, S., G. Segerer & K. Bostoen (2015) Antipassive/Reciprocal Polysemy in Cilubà (Bantu, L31a): A Plurality of Relations Analysis. *Studies in Language* 39(2): 354–385.

Downing, L. J. (2011) The Prosody of "dislocation" in Selected Bantu Languages. *Lingua* 121(5): 772–786.

Downing, L. J. (2012) On the (Non-)congruence of Focus and Prominence in Tumbuka. In Marlo, M. R., N. B. Adams, C. R. Green, M. Morrison & T. M. Purvis (eds.),

Selected Proceedings of the 42nd Annual Conference on African Linguistics, 122–133. Somerville: Cascadilla Proceedings Project.

Downing, L. J. (2018) Differential Object Marking in Chichewa. In Seržant, I. A. & A. Witzlach-Makarevich (eds.), *The Diachronic Typology of Differential Argument Marking*. Berlin: Language Science Press, 41–67.

Downing, L. J. & L. Hyman (2016) Information Structure in Bantu Languages. In Féry, C. & S. Ishihara (eds.), *Oxford Handbook of Information Structure*, 790–813. Oxford: Oxford University Press.

Downing, L. J. & A. Mtenje (2011) Prosodic Phrasing of Chichewa Relative Clauses. *Journal of African Languages and Linguistics* 32: 65–111.

Downing, L. J., A. Rialland, J.-M. Beltzung, S. Manus, C. Patin & K. Riedel (eds.) (2010) *Papers from the Workshop on Bantu Relative Clauses*. Berlin: ZAS Papers in Linguistics 53.

Dugast, I. (1971) *Grammaire du tùnen*. Paris,: Klincksieck.

Duranti, A. (1979) Object Clitic Pronouns in Bantu and the Topicality Hierarchy. *Studies in African Linguistics* 10(1): 31–45.

Duranti, A. & E. R. Byarushengo (1977) On the Notion of "direct object." In Duranti, A., E. R. Byarushengo & L. M. Hyman (eds.), *Haya Grammatical Structure: Phonology, Grammar, Discourse*, 45–71. Los Angeles: Department of Linguistics, University of Southern California.

Edelsten, P. & C. Lijongwa (2010) *A Grammatical Sketch of Chindamba, a Bantu Language (G52) of Tanzania*. Cologne: Rüdiger Köppe.

Fleisch, A. (2005) Agent Phrases in Bantu Passives. In Voeltz, E. F. K. (ed.), *Studies in African Linguistic Typology*, 93–111. Amsterdam; Philadelphia: John Benjamins.

Gibson, H. (2012) *Auxiliary Placement in Rangi: A Dynamic Syntax Perspective*. London: SOAS, University of London, PhD dissertation.

Gibson, H., A. Koumbarou, L. Marten & J. van der Wal (2017) Locating the Bantu Conjoint/Disjoint Alternation in a Typology of Focus Marking. In van der Wal, J. & L. M. Hyman (eds.), *The Conjoint/Disjoint Alternation in Bantu*, 61–99. Berlin: De Gruyter Mouton.

Gibson, H. & L. Marten (2016) Variation and Grammaticalisation in Bantu Complex Verbal Constructions: The Dynamics of Information Growth in Swahili, Rangi and siSwati. In Nash, L. & P. Samvelian (eds.), *Approaches to Complex Predicates*, 70–109. Leiden: Brill.

Givón, T. (1972) Pronoun Attraction and Subject Postposing in Bantu. In Peranteau, P. M., J. N. Levi & G. C. Phares (eds.), *The Chicago Which Hunt: Papers from the Relative Clause Festival*, 190–197. Chicago: Chicago Linguistic Society.

Givón, T. (1976) Topic, Pronoun, and Grammatical Agreement. In Li, C. N. (ed.), *Subject and Topic*, 147–188. New York: Academic Press.

Givón, T. (1979) *On Understanding Grammar*. New York: Academic Press.

Guérois, R. (2014) Locative Inversion in Cuwabo. *ZAS Papers in Linguistics* 57: 49–71.

Guérois, R. (2015) *A Grammar of Cuwabo (Mozambique, Bantu P34)*. Lyon: University of Lyon 2, PhD dissertation.

Güldemann, T. (1996) *Verbalmorphologie and Nebenprädikationen im Bantu. Eine Studie zur funktional motivierten Genese eines konjugationalen Subsystem*. Bochum: Universitätsverlag Dr. N. Brockmeyer.

Güldemann, T. (1997) Prosodic Subordination as a Strategy for Complex Sentence Construction in Shona: Bantu Moods Revisited. In Herbert, R. K. (ed.), *African Linguistics at the Crossroads: Papers from Kwaluseni*, 75–98. Cologne: Rüdiger Köppe.

Güldemann, T. (2003) Present Progressive vis-à-vis Predication Focus in Bantu. A Verbal Category Between Semantics and Pragmatics. *Studies in Language* 27(2): 323–360.

Güldemann, T. (2007) Preverbal Objects and Information Structure in Benue-Congo. In Aboh, E. O., K. Hartmann & M. Zimmermann (eds.), *Focus Strategies in African Languages: The Interaction of Focus and Grammar in Niger-Congo and Afro-Asiatic*, 83–112. Berlin De Gruyter Mouton.

Guthrie, M. (1962) The Status of Radical Extensions in Bantu Languages. *Journal of African Languages* 1(3): 202–220.

Hadermann, P. (1996) Grammaticalisation de la structure "infinitif + verbe conjugué" dans quelques langues bantoues. *Studies in African Linguistics* 25(2): 155–196.

Halpert, C. (2012) *Argument Licensing and Agreement in Zulu*. Cambridge: MIT, PhD dissertation.

Hamlaoui, F. & E.-M. Makasso (2015) Focus Marking and the Unavailability of Inversion Structures in the Bantu Language Bàsàá (A43). *Lingua* 154: 35–64.

Harjula, L. (2004) *The Ha Language of Tanzania: Grammar, Texts and Vocabulary*. Cologne: Rüdiger Köppe.

Hawkinson, A. & L. M. Hyman (1974) Hierarchies of Natural Topics in Shona. *Studies in African Linguistics* 5: 147–170.

Heine, B. (1976) *A Typology of African Languages Based on the Order of Meaningful Elements*. Berlin: Dietrich Reimer.

Henderson, B. M. (2006) *The Syntax and Typology of Bantu Relative Clauses*. Urbana-Champaign: University of Illinois, PhD dissertation.

Henderson, B. M. (2007) Multiple Agreement and Inversion in Bantu. *Syntax* 9: 275–289.

Henderson, B. M. (2011) Agreement, Locality, and OVS in Bantu. *Lingua* 121: 742–753.

Henderson, B. M. (2013) Agreement and Person in Anti-agreement. *Natural Language and Linguistic Theory* 31: 453–481.

Hyman, L. M. (2003) Suffix Ordering in Bantu: A Morphocentric Approach. In Booij, G. & J. van Marle (eds.), *Yearbook of Morphology 2002*, 245–281. Dordrecht: Kluwer Academic Publishers.

Hyman, L. M. (2010) Focus Marking in Aghem: Syntax or Semantics? In Fiedler, I. & A. Schwarz (eds.), *The Expression of Information Structure :A Documentation of Its Diversity Across Africa*, 95–116. Amsterdam; Philadelphia: John Benjamins.

Hyman, L. M. (2017) Disentangling Conjoint, Disjoint, Metatony, Tone Cases, Augments, Prosody, and Focus in Bantu. In van der Wal, J. & L. M. Hyman (eds.), *The Conjoint/Disjoint Alternation in Bantu*, 100–121. Berlin: De Gruyter Mouton.

Hyman, L. M. & A. Duranti (1982) On the Object Relation in Bantu. In Thompson, S. A. & P. J. Hopper (eds.), *Studies in Transitivity*, 217–239. New York: Academic Press.

Hyman, L. M. & F. X. Katamba (2010) Tone, Syntax and Prosodic Domains in Luganda. *ZAS Papers in Linguistics* 53: 69–98.

Inkelas, S. & D. Zec (1990) *The Phonology-Syntax Connection*. Chicago: CSLI.

Kähler-Meyer, E. (1966) Die örtliche Funktion der Applikativendung in Bantusprachen. *Neue Afrikanistische Studien* 5.

Kavari, J. U., L. Marten & J. van der Wal (2012) Tone Cases in Otjiherero: Head-Complement Relations, Linear Order, and Information Structure. *Africana Linguistica* 18: 315–353.

Khumalo, L. (2010) Passive, Locative Inversion in Ndebele and the Unaccusative Hypothesis. *South African Journal of African Languages* 30(1): 22–34.

Kimenyi, A. (1980) *A Relational Grammar of Kinyarwanda*. Berkeley: University of California Press.

Kisseberth, C. (1994) On Domains. In Cole, J. & C. Kisseberth (eds.), *Perspectives in Phonology*, 133–166. Stanford: CSLI Publications.

Kisseberth, C. W. (2010) Phrasing and Relative Clauses in Chimwiini. *ZAS Papers in Linguistics* 53 109–144.

Kisseberth, C. W. & M. I. Abasheikh (1974) Vowel Length in Chi-Mwi:ni – A Case Study of the Role of Grammar in Phonology. In Bruck, A., R. Fox & La Galy, M. (eds.), *Papers from 10th Regional Meeting of the Chicago Linguistic Society: Papers from the Parasession on Natural Phonology*, 193–209. Chicago: Chicago Linguistic Society.

Koni Muluwa, J. & K. Bostoen (2014) The Immediate Before the Verb Focus Position in Nsong (Bantu B85d, DR Congo): A Corpus-Based Exploration. *ZAS Papers in Linguistics* 57: 123–135.

Kula, N. & L. Marten (2010) Argument Structure and Agency in Bemba Passives. In Legère, K. & C. Thornell (eds.), *Bantu Languages: Analyses, Description and Theory*, 115–130. Cologne: Rüdiger Köppe.

Kula, N. C. (2017) The Conjoint/Disjoint Alternation and Phonological Phrasing in Bemba. In van der Wal, J. & L. M. Hyman (eds.), *The Conjoint/Disjoint Alternation in Bantu*, 258–293. Berlin: De Gruyter Mouton.

Letsholo, R. (2006) WH Constructions in Ikalanga: A Remnant Movement Analysis. In Arasanyin, O. F. & M. A. Pemberton (eds.), *Selected Proceedings of the 36th Annual Conference on African Linguistics*, 258–270. Somerville: Cascadilla Proceedings Project.

Letsholo, R. (2009) The "forgotten" Structure of Ikalanga Relatives. *Studies in African Linguistics* 38(2): 131–154.

Maho, J. F. (1999) *A Comparative Study of Bantu Noun Classes*. Göteborg: Acta Universitatis Gothoburgensis.

Marantz, A. (1993) Implications of Asymmetries in Double Object Constructions. In Mchombo, S. A. (ed.), *Theoretical Aspects of Bantu Grammar*, 113–150. Stanford: CSLI.

Marlo, M. R. (2014) Exceptional Patterns of Object Marking in Bantu. *Studies in African Linguistics* 43: 85–123.

Marlo, M. R. (2015) On the Number of Object Markers in Bantu Languages. *Journal of African Languages and Linguistics* 36: 1–65.

Marten, L. (2000) Agreement with Conjoined Noun Phrases in Swahili. *Afrikanistische Arbeitspapiere* 64: 75–96.

Marten, L. (2003) The Dynamics of Bantu Applied Verbs: An Analysis at the Syntax-Pragmatics Interface. In Lébikaza, K. K. (ed.), *Actes du 3e congrès mondial de linguistique africaine (Lomé 2000)*. Cologne: Rüdiger Köppe.

Marten, L. (2005) The Dynamics of Agreement and Conjunction. *Lingua* 115: 527–547.

Marten, L. (2006) Locative Inversion in Otjiherero: More on Morphosyntactic Variation in Bantu. *ZAS Papers in Linguistics* 43: 97–122.

Marten, L. (2007) Focus Strategies and the Incremental Development of Semantic Representations: Evidence from Bantu. In Aboh, E. O., K. Hartmann & M. Zimmermann (eds.), *Focus Strategies in African Languages: The Interaction of*

Focus and Grammar in Niger-Congo and Afro-Asiatic, 113–138. Berlin De Gruyter Mouton.

Marten, L. (2011) Information Structure and Agreement: Subjects and Subject Agreement in Swahili and Herero. *Lingua* 121(5): 787–804.

Marten, L. (2013) Structure and Interpretation in Swahili Existential Constructions. *Italian Journal of Linguistics/Rivista di Linguistica* 25(1): 45–73.

Marten, L. & H. Gibson (2015) Structure Building and Thematic Constraints in Bantu Inversion Constructions. *Journal of Linguistics* 52(3): 565–607.

Marten, L. & N. C. Kula (2012) Object Marking and Morphosyntactic Variation in Bantu. *Southern African Linguistics and Applied Language Studies* 30: 237–253.

Marten, L., N. C. Kula & N. Thwala (2007) Parameters of Morphosyntactic Variation in Bantu. *Transactions of the Philological Society* 105(3): 253–338.

Marten, L. & N. C. Kula (2014) Benefactive and Substitutive Applicatives in Bemba. *Journal of African Languages and Linguistics* 35(1): 1–44.

Marten, L. & D. Ramadhani (2001) An Overview of Object Marking in Kiluguru. *SOAS Working Papers in Linguistics and Phonetics* 11: 259–275.

Marten, L. & J. van der Wal (2014) A Typology of Bantu Subject Inversion. *Linguistic Variation* 14(2): 318–368.

Maw, J. (1969) *Sentences in Swahili: A Study of Their Internal Relationships*. London: SOAS.

Mchombo, S. A. (ed.) (1993) *Theoretical Aspects of Bantu Grammar*. Stanford: CSLI.

Mchombo, S. A. & G. Firmino (1999) Double Object Constructions in Chichewa and Gitonga: A Comparative Analysis. *Linguistic Analysis* 29: 214–233.

Meeussen, A. E. (1959) *Essai de grammaire rundi*. Tervuren: Musée royal du Congo belge.

Meeussen, A. E. (1967) Bantu Grammatical Reconstructions. *Africana Linguistica* 3: 79–121.

Meinhof, C. (1899) *Grundriß einer Lautlehre der Bantusprachen nebst Anleitung zur Aufnahme von Bantusprachen – Anhang : Verzeichnis von Bantuwortstämmen*. Leipzig: F.A. Brockhaus.

Meinhof, C. (1906) *Grundzüge einer vergleichenden Grammatik der Bantusprachen*. Berlin: Reimer.

Möhlig, W. J. G. (1967) *Die Sprache der Dciriku. Phonologie, Prosodologie und Morphologie*. Köln: Universität zu Köln, PhD dissertation.

Möhlig, W. J. G. & J. U. Kavari (2008) *Reference Grammar of Herero (Otjiherero)*. Cologne: Rüdiger Köppe.

Möhlig, W. J. G., L. Marten & J. U. Kavari (2002) *A Grammatical Sketch of Herero*. Cologne: Rüdiger Köppe.

Morimoto, Y. (2000) *Discourse Configurationality in Bantu Morphosyntax*. Stanford: Stanford University, PhD dissertation.

Morimoto, Y. (2002) Prominence Mismatches and Differential Object Marking in Bantu. In Butt, M. & T. H. King (eds.), *Proceedings of the LFG02 Conference, National Technical University of Athens*, 292–314. Stanford: CSLI Publications.

Morimoto, Y. (2006) Agreement Properties and Word Order in Comparative Bantu. *ZAS Papers in Linguistics* 43: 161–187.

Moshi, L. (1998) Word Order in Multiple Object Constructions in KiVunjo-Chaga. *Journal of African Languages and Linguistics* 19(2): 137–152.

Mous, M. (1997) The Position of the Object in Tunen. In Déchaine, R.-M. & V. Manfredi (eds.), *Object Positions in Benue-Kwa: Papers from a Workshop at Leiden University, June 1994*, 123–137. The Hague: Holland Academic Graphics.

Murrell, P. (2012) The Applicative Construction and Object Symmetry as a Parameter of Variation in Kiswahili and Maragoli. *Southern African Linguistics and Applied Language Studies* 30(2): 255–275.

Ndayiragije, J. (1999) Checking Economy. *Linguistic Inquiry* 30(3): 399–444.

Ngoboka, J. P. & J. Zeller (2017) The Conjoint/Disjoint Alternation in Kinyarwanda. In van der Wal, J. & L. M. Hyman (eds.), *The Conjoint/Disjoint Alternation in Bantu*, 350–389. Berlin: De Gruyter Mouton.

Ngonyani, D. S. (1996) *The Morphosyntax of Applicatives (Bantu, Ndendeule, Swahili)*. Los Angeles: University of California,

Ngonyani, D. S. (2006) Resumptive Pronominal Clitics in Bantu Languages. In Arasanyin, O. F. & M. A. Pemberton (eds.), *Selected Proceedings of the 36th Annual Conference on African Linguistics*, 51–59. Somerville: Cascadilla Proceedings Project.

Ngonyani, D. S. & P. Githinji (2006) The Asymmetric Nature of Bantu Applicative Constructions. *Lingua* 116: 31–63.

Nicolle, S. (2013) *A Grammar of Digo: A Bantu Language of Kenya and Tanzania*. Dallas: SIL International.

Nshemezimana, E. & K. Bostoen (2017) The Conjoint/Disjoint Alternation in Kirundi (JD62): A Case for Its Abolition. In van der Wal, J. & L. M. Hyman (eds.), *The Conjoint/ Disjoint Alternation in Bantu*, 390–425. Berlin: De Gruyter Mouton.

Nsuka-Nkutsi, F. (1982) *Les structures fondamentales du relatif dans les langues bantoues*. Tervuren: Musée royal de l'Afrique centrale.

Nurse, D. (2008) *Tense and Aspect in Bantu*. Oxford: Oxford University Press.

Odden, D. (1995) Phonology at the Phrasal Level in Bantu. In Katamba, F. (ed.), *Bantu Phonology and Morphology*, 40–68. Munich: Lincom Europa.

Petzell, M. (2008) *The Kagulu Language of Tanzania: Grammar, Texts and Vocabulary*. Cologne: Rüdiger Köppe.

Poeta, T. (2016) *Object Marking in Makhuwa and Swahili Discourse*. London: SOAS, University of London, PhD dissertation.

Pylkkänen, L. (2008) *Introducing Arguments*. Cambridge: MIT Press.

Riedel, K. (2009) *The Syntax of Object Marking in Sambaa. A Comparative Bantu Perspective*. Utrecht: LOT.

Rugemalira, J. (1991) What Is a Symmetrical Language? Multiple Object Constructions in Bantu. *Proceedings of the Annual Meeting of the Berkeley Linguistics Society* 17: 200–209.

Rugemalira, J. (1993) Bantu Multiple Object Constructions. *Linguistic Analysis* 23: 226–252.

Russell, J. (1985) Swahili Quasi-Passives: The Question of Context. In Goyvaerts, D. L. (ed.), *African Linguistics: Essays in Memory of M.W.K. Semikenke*, 477–490. Amsterdam, Philadelphia: John Benjamins.

Schadeberg, T. (1995) Object Diagnostics in Bantu. In Emenanjo, E. N. & O.-M. Ndimele (eds.), *Issues in African Languages and Linguistics: Essays in Honour of Kay Williamson*, 173–180. Aba: National Institute for Nigerian Languages.

Schadeberg, T. C. (2003) Derivation. In Nurse, D. & G. Philippson (eds.), *The Bantu Languages*, 71–89. London; New York: Routledge.

Schneider-Zioga, P. (2007) Anti-Agreement, Anti-Locality and Minimality: The Syntax of Dislocated Subjects. *Natural Language and Linguistic Theory* 25: 403–446.

Seidel, F. (2008) *A Grammar of Yeyi. A Bantu Language of Southern Africa*. Cologne: Rüdiger Köppe.

Seidl, A. & A. Dimitriadis (1997) The Discourse Function of Object Marking in Swahili. *Papers from the Regional Meeting of the Chicago Linguistic Society* 33(1): 373–388.

Seidl, A. & A. Dimitriadis (2003) Statives and Reciprocal Morphology in Swahili. In Sauzet, P. & A. Zribi-Hertz (eds.), *Typologie des langues d'Afrique & universaux de la grammaire. Volume 1: Approches transversales, domaine bantou*, 239–284. Paris: L'Harmattan.

Sharman, J. C. (1956) The Tabulation of Tenses in a Bantu Language (Bemba: Northern Rhodesia). *Africa* 16: 29–46.

Simango, S. R. (1999) Lexical and Syntactic Causatives in Bantu. *Linguistic Analysis* 29: 69–86.

Simango, S. R. (2009) Causative Disguised as Stative: The Affix -ik/-ek in ciCewa. *Southern African Linguistics and Applied Language Studies* 27(2): 121–134.

Simango, S. R. (2012) Subject Marking, Coordination and Noun Classes in ciNsenga. *Southern African Linguistics and Applied Language Studies* 30: 171–183.

Stappers, L. (1964) *Morfologie van het Songye*. Tervuren: Koninklijk Museum voor Midden-Afrika.

Stappers, L. (1967) Het passief suffix -u- in de Bantoe-talen. *Africana Linguistica* 3: 137–145.

Stucky, S. (1981) *Word Order Variation in Makua: A Phrase Structure Grammar Analysis*. Urbana-Champaign: University of Illinois, PhD dissertation.

Taylor, C. (1985) *Nkore-Kiga*. London: Croom Helm.

Tenenbaum, S. (1977) Left- and Right-Dislocation. In Duranti, A., E. R. Byarushengo & L. M. Hyman (eds.), *Haya Grammatical Structure: Phonology, Grammar, Discourse*, 161–170. Los Angeles: Department of Linguistics, University of Southern California.

Thwala, N. (2006) On the Subject-Predicate Relation and Subject Agreement in SiSwati. *Southern African Linguistics and Applied Language Studies* 24(3): 331–359.

Van de Velde, M. (2008a) *A Grammar of Eton*. Berlin: De Gruyter Mouton.

Van de Velde, M. (2008b) The Syntax of Verb Complements and the Loss of the Applicative in Eton (A71). In Legère, K. & C. Thornell (eds.), *Bantu Languages: Analyses, Description and Theory*, 281–293. Cologne: Rüdiger Köppe.

Van de Velde, M. (2013) The Bantu Connective Construction. In Carlier, A. & J.-C. Verstraete (eds.), *The Genitive*, 217–252. Amsterdam; Philadelphia: John Benjamins.

Van der Spuy, A. (1993) Dislocated Noun Phrases in Nguni. *Lingua* 90: 335–355.

van der Wal, J. (2009) *Word Order and Information Structure in Makhuwa-Enahara*. Utrecht: LOT.

van der Wal, J. (2015) Object Clitics in Comparative Bantu Syntax. Paper presented at *BLS 41, 7–8 February 2015*, UC Berkeley.

van der Wal, J. (2017) Flexibility in symmetry: An implicational relation in Bantu double object constructions. In M. Sheehan & L. R. Bailey (eds.), *Order and structure in syntax II: Subjecthood and argument structure*. Berlin: Language Science Press, 115–152.

van der Wal, J. & L. M. Hyman (eds.) (2017) *The Conjoint/Disjoint Alternation in Bantu*. Berlin: De Gruyter Mouton.

Voeltz, E. F. K. (1971) Surface Constraints and Agreement Resolution: Some Evidence from Xhosa. *Studies in African Linguistics* 21: 37–60.

Wald, B. (1997) Instrumental Objects in the History of Topicality and Transitivity in Bantu. In Déchaine, R.-M. & V. Manfredi (eds.), *Object Positions in Benue-Kwa*, 221–253. The Hague: Holland Academic Graphics.

Watters, J. R. (1979) Focus in Aghem: A Study of Its Formal Correlates and Typology. In Hyman, L. M. (ed.), *Aghem Grammatical Structure*, 137–197. Los Angeles: Department of Linguistics, University of Southern California.

Welmers, W. E. (1973) *African Language Structures*. Berkeley, London: University of California Press.

Whiteley, W. H. & J. D. Mganga (1969) Focus and Entailment: Further Problems of Transitivity in Swahili. *African Language Review* 8: 108–125.

Willems, E. (1970) *Le Tshiluba du Kasayi*. Luluabourg: Mission de Scheut.

Woolford, E. (1995) Why Passive Can Block Object Marking. In Akinlabi, A. (ed.), *Theoretical Approaches to African Linguistics*, 199–215. Trenton: Africa World Press.

Zeller, J. (2004) Relative Clause Formation in the Bantu Languages of South Africa. *Southern African Linguistics and Applied Language Studies* 22(1/2): 75–93.

Zeller, J. (2012a) Instrument Inversion in Zulu. In Marlo, M. R., N. B. Adams, C. R. Green, M. Morrison & T. M. Purvis (eds.), *Selected Proceedings of the 42nd Annual Conference on African Linguistics*, 134–148. Somerville: Cascadilla Proceedings Project.

Zeller, J. (2012b) Object Marking in isiZulu. *Southern African Linguistics and Applied Language Studies* 30(2): 219–235.

Zeller, J. (2013) Locative Inversion in Bantu and Predication. *Linguistics* 51(6): 1107–1146.

Zeller, J. (2014) Three Types of Object Marking in Bantu. *Linguistische Berichte* 239: 347–367.

Zerbian, S. (2006) *Expression of Information Structure in the Bantu Language Northern Sotho*. Berlin: Humboldt-Universität zu Berlin, PhD dissertation.

Zerbian, S. (2007) Phonological Phrasing in Northern Sotho (Bantu). *The Linguistic Review* 24: 233–262.

CHAPTER 10

RECONSTRUCTING PROTO-BANTU

Koen Bostoen

1 INTRODUCTION

Since the early days of historical linguistics, the Comparative Method (CM) has been the core tool for examining linguistic prehistory, both in Bantu and other families. It is an upstream approach that reconstructs proto-languages from cognate morphemes in related languages (Nurse 1997: 361, Weiss 2014: 127). In the absence of historical language records, which allow for the study of distinct stages in the evolution of one single language, the CM predominantly relies on linguistic evidence with limited time depth. The oldest Bantu material is early 17th century Kongo H16 (Doke 1959). Thanks to a catechism from 1624 (Cardoso 1624, Bontinck & Ndembe Nsasi 1978), a dictionary from 1652 (Van Gheel 1652, De Kind *et al.* 2012) and a grammar from 1659 (Brusciotto 1659, Guinness 1882), language change in Kongo can be studied on the basis of empirical evidence spanning almost 400 years (cf. Bostoen & de Schryver 2015, 2018a, b). This is quite unique in Bantu studies. Having a catechism from the 1640s (Pacconio & do Couto 1642) and a grammar from the late 1690s (Dias 1697, Angenot *et al.* 2011, Rosa 2013, Fernandes 2015), Kongo's southern neighbour Mbundu H21 is the only other Bantu language documented over a comparable time span. The earliest Swahili G42 texts in Arabic script are not older than the mid-18th century (Knappert 1971: 5). For the large majority of the Bantu languages, documentation did not start before the late 19th century.

The 19th century is also the period when Bantu historical linguistics emerges as a scientific discipline. It was the German missionary Wilhelm Bleek who presented Bantu as a family and coined its name (Bleek 1858, 1862, 1869). His countryman Carl Meinhof laid the foundations for the reconstruction of *Ur-Bantu*, known today as Proto-Bantu (PB), the putative common ancestor (Meinhof 1899). Benefiting from a classical training in comparative Indo-European philology, Meinhof primarily relied on the CM, as did his successors in Bantu reconstruction studies, such as Meeussen (1967, 1980), Guthrie (1967, 1970a, b, 1971), Coupez *et al.* (1998), and Bastin *et al.* (2002). PB, with its estimated time depth of 4,000 to 5,000 years (Vansina 1995, Blench 2006), has thus been reconstructed from language data that are younger than 150 years.

Despite the general lack of ancient language data, the CM has been quite effective for the reconstruction of PB and subsequent ancestor languages, especially compared to other families within the Niger-Congo phylum or to Niger-Congo as a whole. This relative success can be attributed to several factors.

First, the CM is not a method for generating hypotheses of genealogical relatedness, but rather serves to confirm or reject them (Weiss 2014: 128). Such a hypothesis has been agreed upon for Bantu ever since Bleek (1862) and was further reinforced thanks to the CM. For other branches of Niger-Congo or for other African phyla, such a widely accepted relatedness hypothesis has been less straightforward or emerged much later, among other things because they usually count much less languages spoken in fewer countries, which means that much fewer studies were/are being made of them.

Second, the success of the CM greatly depends on the identifiability of cognates, i.e., morphemes having identical or similar forms and meanings across related languages and going back to a common ancestral source. These morphemes can be lexemes or grammemes. Thanks to their relatively close genealogical relatedness, identifying cognates between Bantu languages turns out to be fairly easy. The fact that Bantu roots and stems are reasonably long (CVC, CVCV) further facilitates comparison and reconstruction, unlike for languages that have a stronger tendency towards CV roots and stems. This is no doubt the reason why the first real attempt to apply the CM to Bantu was quite successful. Through the comparison of a restricted set of languages from different parts of the domain, Meinhof (1899) could immediately establish several hundreds of series of both lexical and grammatical cognates. Table 10.1 presents six comparative series of cognate lexemes (Meinhof & van Warmelo 1932: 221–227). Even languages separated by several thousands of kilometres and belonging to different subgroups share many forms that manifest obvious similarities.

TABLE 10.1 COMPARATIVE SERIES OF COGNATES IDENTIFIED BY MEINHOF AND VAN WARMELO (1932: 221–227)

Duala[1] A24	Swahili G42	Kongo[2] H16b	Herero R31	N. Sotho S32	PB[3]	Gloss
-lalo	-tatu	-tatu	-tatu	-rarọ[4]	*-tátὸ	'three'
bw-ele	m-ti	n-ti	omu-ti	mo-re	*-tí	'tree'
mu-lema	m-tima	n-tima	omu-tima	/	*-tímà	'heart'
-loma	-tuma	-tuma	-tuma	-roma	*-tóm-	'send'
-loŋga	-tuŋga	-tuŋga	-tuŋga	-roka	*-tóng-	'sew, build'
-loa	-tuk-ana	-tuka	-tuk-ana	-roχ-aka	*-tók-	'curse, abuse'

Third, the diachronic effectiveness of the CM is directly proportional to the amount of synchronic data available and the degree of variation between languages. The number of Bantu languages is high (Hammerström this volume; Maho 2009) and their study has exponentially increased ever since Meinhof's pioneering research (Maho 2008). Descriptions of poorly known Bantu languages continue to become available (cf. part II of this volume). This cumulative descriptive research adds to the documentation of variation and allows for even more effective applications of the CM. It also urgently calls for a systematic update of research in both Bantu lexical and grammatical reconstruction. The most recent reference works in these two domains are respectively 17 years (Bastin *et al.* 2002) and 52 years old (Meeussen 1967). Providing such an update is not the objective of this chapter. For reasons of time and space, this is impossible to achieve for PB lexicon and grammar within the limits of a book chapter, since we deal here with much more open paradigms than for PB phonology, whose reconstruction is extensively discussed in Section 2. The rest of this chapter is organised as follows. Section 3 discusses how one proceeds from phonological to lexical reconstruction through the CM. In Section 4, issues in PB grammatical reconstructions are treated. The principles on which scholars have relied to reconstruct PB have also been relied on to reconstruct ulterior ancestor languages, but such attempts are only occasionally referred to here. If the CM primarily serves to detect shared retentions, it obviously also allows to identify shared innovations, a crucial device for genealogical classification. The relationship between Bantu reconstruction and classification is discussed in Section 5.

2 PHONOLOGY

The CM proceeds according to a number of more or less standardised steps (Nurse 1997: 361–362, Weiss 2014: 128–129). The reconstruction of PB phonology from comparative series as in Table 10.1 is only possible after the establishment of regular sound correspondences, i.e., phonological similarities between related languages that are recurrent, systematic and without unexplainable exceptions. The late-19th-century Neogrammarians put great emphasis on the regularity of sound changes in order to distinguish historically significant similarities from those resulting from historical accident. Historical-comparative analysis should exclude random resemblances, as for instance between French *bateau* 'boat' and *bwato* 'canoe' (< PB *-(j)átò*), found in many Bantu languages, such as Zambian Ila M63 (Smith 1964: 21), to name but one. It should also disregard correspondences due to sound symbolism, the direct non-arbitrary linkage between sound and meaning, such as the verbs *kukokola* in Swahili G42, *kakelen* in Dutch and *caqueter* in French, all referring to the *cackling* of chickens (Dimmendaal 2011: 7). Regular sound correspondences also help to distinguish between linguistic items inherited through regular intergenerational transmission only and those acquired through contact (cf. Mous this volume). They are established thanks to the comparison of numerous sets of cognates as in Table 10.1. The segments of the morphemes compared in Table 10.1 can be organised in several correspondence sets. The examples have been chosen so as to highlight one sound correspondence that is recurrent in all cognates. Stem-initially, each language has the same consonant that consistently corresponds to a phonetically identical or similar consonant in the other languages: Duala *l* ≈ Swahili *t* ≈ Kongo *t* ≈ Herero *t* ≈ N. Sotho *r*. The recurrence of *a* can also be established from the comparative data in Table 10.1, which bears evidence of two more vocalic correspondence sets, i.e., Duala *e* ≈ Swahili *i* ≈ Kongo *i* ≈ Herero *i* ≈ N. Sotho *e* and Duala *o* ≈ Swahili *u* ≈ Kongo *u* ≈ Herero *u* ≈ N. Sotho *o*.

Having identified several regular sound correspondence sets, one can start to reconstruct proto-sounds from shared retentions observed in the comparative data set. To tell apart retentions from innovations, one can rely on the principles of frequency, economy and directionality (Campbell 1998: 114–122). In a large sample of Bantu languages, the most frequent sound in a regular correspondence set tends to be the most plausible candidate for reconstruction. This is certainly the case for **a* as proto-phoneme for *a* ≈ *a* ≈ *a* ≈ *a* ≈ *a* in Table 10.1, but also for **t* as proto-phoneme for *l* ≈ *t* ≈ *t* ≈ *t* ≈ *r*. This consonant is attested in a majority of languages spoken throughout the Bantu domain (Guthrie 1967: 72). It is also the most economical option for this correspondence set, since it implies only two shifts for the five languages involved, while every other option would require at least four. Moreover, in terms of directionality, **t > l* and **t > r* are two common types of lenition in the world's languages, respectively known as lambdacism and rhotacism (Dimmendaal 2011: 24, 29)[5]. By means of such simple and straightforward principles, the fathers of Bantu historical linguistics succeeded in the reconstruction of the systems of PB simple consonants presented in Table 10.2.

TABLE 10.2 SIMPLE CONSONANT SYSTEMS RECONSTRUCTED FOR PB

*k	*ɣ	(*ŋ)	*y	*p	*t	*k	*c	*y
*t	*l	*n	*w	*b	*d	*g	*j	
*p	*ʋ	*m		*m	*n	*ŋ	*ny	
*ḵ	*ỿ							
*ṭ	*ḽ							
(Meinhof & van Warmelo 1932: 33)				(Guthrie 1967: 52)				

*m	*n	*ɲ	
*b	*d	*j	*g
*p	*t	*c	*k
(Meeussen 1967: 83)			

As one can see from Table 10.2, the application of straightforward principles did not lead to uniform outcomes: only *p, *t, *c, *k, *j, *m, *n occur in all three systems. Meinhof and van Warmelo (1932: 30–32) consider *k/*t and *ɣ/*l "secondary palatalized forms" of respectively *k/*t and *ɣ/*l and mostly correspond to palatal *c and *j in the systems of Guthrie (1967) and Meeussen (1967, 1969). With regard to PB voiced consonants, Meinhof and van Warmelo (1932: 31) primarily reconstruct fricatives instead of plosives, but admit that these manifest "a certain tendency to become plosive." In Meinhof and van Warmelo (1932) ʋ stands for β. The reconstruction of *b and *g, as Guthrie (1967) and Meeussen (1967) proposed, is more plausible, not only because these plosives are more frequent among present-day Bantu languages, but also because the lenitions *b > β and *g > ɣ are much commoner sound shifts than the reverse fortifications. As for *d, both Guthrie (1962: 13) and Meeussen (1967: 83) acknowledge that one could also use *l to represent this proto-phoneme. The liquid is the commonest present-day reflex and stands in allophonic variation to d as witnessed in the widespread Bantu morphophonological rules $l \rightarrow d/N\_\_$ and $l \rightarrow d\_\_i/u$. This allophony is also attested elsewhere in Niger-Congo (Stewart 1993). The same distribution can be observed morpheme-internally as a diachronic sound change (Schadeberg 2003: 146; Hyman this volume). Meeussen (1967: 83) also concedes that instead of *c one might just as well use the symbol *s and *z or *y instead of *j. While *c is indeed most often reflected by s or ʃ in present-day Bantu languages, the conflation of current-day z and y/Ø into *j – a convention also adopted in Bantu Lexical Reconstructions (Bastin et al. 2002) – is not justified. Meinhof and van Warmelo (1932: 33) and Guthrie (1967: 52) did reconstruct both *j and *y. The reconstruction of this contrast is required. Certain stems, such as *-jòngó[6] 'cooking pot' in (1a), have a stem-initial consonant in certain languages but not in others. Others, such as *-(j)átò 'canoe' in (1b), are always vowel-initial and should actually be reconstructed with stem-initial *y if one wishes to respect the common Bantu CVCV noun stem template or otherwise without stem-initial consonant (Creissels 1999, Bostoen 2005: 179–198, Bulkens 2009). If one admits the existence of vowel-initial noun stems in PB, it is enough to reconstruct just *j and not *y.

(1) a *-jòngó 'cooking pot' > Mbochi C25 *n-zúngù*, Bembe H11 *n-dzúúngù*,
Lega D25 *ka-zongó*, Kamba E55 *kasungu*,
Nyika E72 *ka-dzungu*, Shambaa G23 *ki-zùngú*,
Subiya K42 *chi-zungu* vs. Embu E52 *ki-ʊngʊ*,
Nilamba F31 *ky-ungú*, Ndamba G52 *cì-yúngù*,
Mbunga P15 *ki-yungu* (Bostoen 2005: 179–198)

b *-(j)átò 'canoe' > Lundu A11 *bw-ádò*, Leke C14 *lw-átò*, Mbochi
C25 *bw-áre*, Lega D25 *bw-ărɔ*, Holoholo
D28 *bw-àtó*, Pokomo E71 *bw-aho*, Pogoro
G51 *bw-ato*, Bembe H11 *bw-aatu*, Holu L12b
bw-áàtù, Chewa N31b *ly-atô*, Kwanyama R21
bw-átò, Tsonga S53 *bw-àtsò* (cf. Botne 1994,
Bulkens 2009: 30, Kouarata 2014, 2016)

Finally, the PB reconstruction of the nasals *n and *$ŋ$ outside NC clusters is a matter of debate. Only a very restricted number of PB roots require the reconstruction of a palatal nasal; *-*ɲàmà* 'animal' and *-*ɲó-* 'drink' are two of them, but *$ɲ$ might result here from contact between *ni* and a following vowel at an earlier stage and thus actually represent *ny. Proto-Benue-Congo *$ŋ$ was probably not conserved in PB (Schadeberg 2003: 147). The velar nasal does often occur before vowels in present-day Bantu languages, as Meinhof and van Warmelo (1932: 32) observe, but generally as a phonological innovation. It may be the reflex of *$ŋg$ as the outcome of Meinhof's rule (Meinhof 1912–1913, Meeussen 1962, Dammann 1972, Johnson 1979; Hyman this volume). This cluster reduction exists in two variations, i.e., (a) $NC_{[+voice]}VNC_{[+voice]}V > (N)NVNC_{[+voice]}V$ and (b) $NC_{[+voice]}VNV > (N)NVNV$, which are illustrated in (2).

(2) a Ganda JE15 òlùlimì vs. ènnimì 'tongue(s)'
   òlùgendo vs. èŋŋendo 'journey(s)'
   ùbûmba vs. mmûmba 'you/I mould'
   ùgènda vs. ŋŋènda 'you/I travel'

b Lenje M61 *-gòmbè > ŋombe 'cow'
   *-gàndá > ŋanda 'house'
   *-gòmà > ŋoma 'drum'
   *-dòngó > nongo 'pot'

c Lamba M54 lulembo vs. inembo 'tattoo, incision'
   *-gòmbè > ŋombe 'cow'
   BUT
   *-gòmà > ngoma 'drum'

d Umbundu R11 -laman- vs. onamani 'jump vs. jumper'
   ulima vs. nima 'you/I cultivate'
   BUT
   -ling- vs. ondinga 'work vs. worker'
   ulanda vs. ndanda 'you/I buy'

In some languages, such as Ganda (2a) and Lenje (2b), both are attested. In others, such as Lamba (2c) and Umbundu (2d), only one of both is operational. In Ganda, it gives rise to geminated nasals, but elsewhere NC is reduced to simple N. It is observed both as a diachronic sound shift, especially in nouns belonging to cl. 9/10, and as a morpho-phonological alternation, i.e., in nouns belonging to cl. 11/10 and in verb conjugations involving a 1SG subject or object marker. Meinhof's rule – either lexicalised or produc-tive – is widespread in Bantu (Nurse & Philippson 2003). Meeussen (1967: 85) therefore considers it for reconstruction in PB. However, if it were really productive in the ancestor, its traces in the lexicon of present-day Bantu languages should be more pervasive. It is at most an inherited ancestral tendency.

The examples in (2) furthermore show the occurrence of NC clusters or nasal com-pounds that can be reconstructed to PB. As summarised in Table 10.3, each of the PB consonants has an equivalent preceded by a non-syllabic homorganic nasal. Some few exceptions notwithstanding, such as *-ntò 'person, thing,' ultimately probably a short-ening of *-jìntò, stem-internal prenasalised consonants only occur in C_2 position. At the beginning of the stem, NC is commonly heteromorphemic, the homorganic nasal being either a cl. 9/10 noun prefix or a 1SG marker. Lexical reconstructions with $NC_{[-voice]}$ are significantly less frequent than with $NC_{[+voice]}$ (Meeussen 1967: 83). Except in some rare languages, such as Ganda JE15, geminated nasals are not found in Bantu. The reconstruc-tion of *mm, *nn, *ɲɲ and *ŋŋ in PB by Meeussen (1967: 83) is debatable.

TABLE 10.3 SIMPLE CONSONANT SYSTEMS RECONSTRUCTED FOR PB

*mb	*nd	*ɲj	*ŋg
*mp	*nt	*ɲc	*ŋk
(Guthrie 1967: 52, Meeussen 1967: 83)			

As for vocalic reconstruction, the regular correspondence between close vowels in Swahili G42, Kongo H16b and Herero R31, and mid-close vowels in Duala A24 and N. Sotho S32, as identified in Table 10.1, is suggestive of the PB seven-vowel system and the subsequent seven > five vowel merger it underwent in the first group of languages. Further evidence is presented in Table 10.4.

TABLE 10.4 COMPARATIVE SERIES OF COGNATES IDENTIFIED BY MEINHOF AND VAN WARMELO (1932: 224–227)

Duala	Swahili	Kongo	Herero	N. Sotho	PB	Gloss
-lulɛ	-fua	-fula	-ṱura	-rula	*-túd-	'hammer'[7]
-lunda[8]	-fundiʃa	-funda	-ṱunda	-ruta	*-túnd-	'teach, punish'
/	u-siku	fuku	ou-ṱuku	bo-ʃeɣo	*-tíkò/túkò	'night'
/	-sia	-sa	-ṣewa (-ṣia)	-ʃiɣa	*-tíg-	'leave behind'
mu-sima	ki-sima	sima	omu-ṣema	mo-ʃima	*-timà	'well, pool'
mu-singa	u-singa	n-singa	oru-ṣinga	le-ʃika	*-tíngà	'thread, string'
/	ma-sika	ma-sika	/	ma-riɣa/ ma-reɣa	*-tíkà	'cold season'

Close vowels in Swahili, Kongo and Herero do not correspond here with mid-close vowels in Duala and N. Sotho, but mostly with close vowels: Duala *i* ≈ Swahili *i* ≈ Kongo *i* ≈ Herero *i* ≈ N. Sotho *i* and Duala *u* ≈ Swahili *u* ≈ Kongo *u* ≈ Herero *u* ≈ N. Sotho *u*. Only the N-Sotho reflex for 'night' is irregular in that respect. In contrast to Duala and N. Sotho, Swahili, Kongo and Herero only have three contrastive degrees of vowel aperture. Their highest degree of aperture corresponds to two aperture degrees in Duala and N. Sotho: close and mid-close. This regular correspondence is most economically accounted for as Meinhof and his successors did, by the reconstruction of seven vowels in PB, as shown in Table 10.5.

TABLE 10.5 VOWEL SYSTEMS RECONSTRUCTED FOR PB

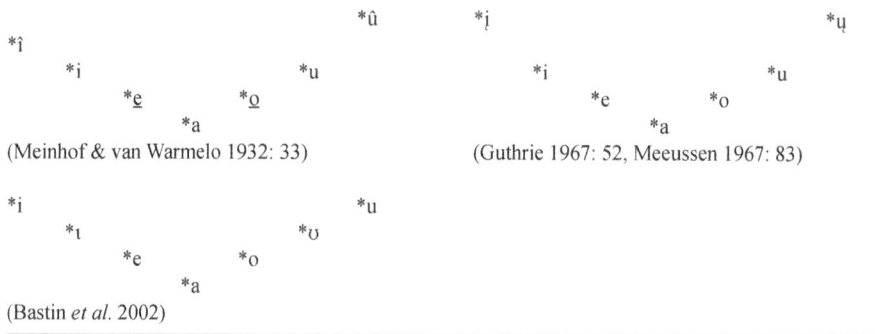

(Meinhof & van Warmelo 1932: 33)   (Guthrie 1967: 52, Meeussen 1967: 83)

(Bastin *et al.* 2002)

While Duala and N. Sotho retained the four contrastive degrees of vowel aperture[9], Swahili, Kongo and Herero fused the two highest degrees. This seven > five-vowel merger did not lead to homonymic clash, because consonants preceding PB first-degree vowels underwent a mutation commonly known as Bantu Spirantisation (Schadeberg 1995, Labroussi 1999, Bostoen 2008). In contrast to the examples in Table 10.1 where **t* is conserved before a non-close PB vowel, it is reflected here by *f* before **u* and by *s* before **i* in Swahili and Kongo. Labio-dental and alveolar fricatives are the commonest outcome of Bantu Spirantisation. The *ɟ* and *ʂ* reflexes of Herero, which Möhlig (2009: 127) defines as "addental,"[10] result from a rather uncommon dentalisation which other South-West Bantu languages also have as output of Bantu Spirantisation (Janson 2007: 98). Whatever the phonetic outcome of Bantu Spirantisation may be, it usually generates a new set of consonants. As a consequence, the original vocalic opposition is transphonologised in five-vowel languages to a consonantal opposition (Hyman 2003: 56). In seven-vowel languages, Bantu Spirantisation is much less pervasive: if present, then often only very irregularly, as is the case in Duala and N. Sotho. In Table 10.4, both languages manifest the same unconditioned reflex of **t* before **u*, as they do before any other vowel in Table 10.1. The fricative occurring before **i* is not consistent, cf. N. Sotho *mareya* 'winter' or Duala **-tínà* 'root, trunk' > *tína*, **-tíndí* 'heel' > *tíndí*, **-dóótì* 'dream' > *ndɔti*. This non-fricativised *t* is still different from the unconditioned reflex of **t*, i.e., *l* as in Table 10.1.

It was Bantu Spirantisation that made Meinhof (1899: 18) consider the PB first-degree vowels as phonetically odd. He therefore characterises them as *schwer* ('heavy'), currently also known as "superclose" (cf. Maddieson 2003: 19–20), and marks them with a circumflex. Guthrie (1967: 52) and Meeussen (1967: 83) sustain this view by marking the first-degree vowels with a cedilla. Bastin *et al.* (2002) break with that convention by using **ɪ* and **ʊ* for the second-degree vowels and noting the first-degree vowels as unmarked *i* and *u*. In doing so, they adhere to the view that PB did not have "superclose" vowels, but simply a contrast between tense and lax high vowels, i.e., with or without Advanced Tongue Root [±ATR] (Schadeberg 1995: 73–75, Hyman 1999: 247). Phonetically speaking, the highest vowels in present-day seven-vowel languages are indeed generally [*i*] and [*u*]. Second-degree vowels are cross-linguistically more variable. Certain seven-vowel languages have exactly the same system as in Bastin *et al.* (2002), other seven-vowel languages have mid vowels [*e*] and [*o*] instead of [*ɪ*] and [*ʊ*]. The latter are universally much rarer than their mid counterparts (Maddieson 1984). It is therefore more plausible to assume that the shift from *ɪ/ʊ* to *e/o* occurred recurrently in Bantu rather than the other way around. Moreover, *ɪ* and *ʊ* occur elsewhere in Niger-Congo (Stewart 2000). Despite their typological rareness, Proto-Bantu may thus well have inherited them from an earlier Niger-Congo stage. Then again, as Maddieson (2003: 19–20) argues, certain Narrow Bantu languages do have first-degree vowels which could be considered "superclose," because they manifest "an unusually narrow constriction, nearly consonantal in character." Connell (2007) claims that PB high vowels actually were fricative vowels, as they occur in several Wide Bantu languages, among others in cognates of Narrow Bantu words reconstructed with first-degree vowels (see also Faytak 2015). It still needs to be established, however, whether these fricative vowels are not a secondary development. The release of a stop into a high vocoid involves a brief period of turbulence. This noise is believed to have triggered Bantu Spirantisation (Hyman 2003), as it is at the origin of assibilation elsewhere in the world (Kim 2001, Hall *et al.* 2006). It could well be that in certain Benue-Congo languages it became associated with the following vowel giving rise to fricative vowels.

Greenberg (1948) was the first to reconstruct tone in PB, i.e., a system with two units – high and low – [11]accompanying either a vowel or a syllabic nasal (Meeussen 1967: 84). As illustrated in Table 10.6, regular correspondences are also observed on a supra-segmental level. While Mongo C61 reflects the PB tone system directly, Luba L31a developed a reverse tone system (Hulstaert 1941, Van Spaandonck 1971, Maddieson 1976).

TABLE 10.6 REGULAR TONE CORRESPONDENCES BETWEEN MONGO AND LUBA (HULSTAERT 1941)

Mongo	Luba	BLR	Gloss
n-jàlà	n-zálá	*-jàdà	'hunger'
m-bólókó	m-bùlùkù	*-bódókó	'antelope'
bò-lómè	mú-lùmé	*-dómè	'male, man'
lì-kálà	di-kalá	*-kádà	'charcoal'
lì-itá	dí-vítà	*-bìtá	'war'

Phonological reconstruction obviously goes beyond the reconstitution of individual segments and tones. It is also about reconstructing consonant and vowel sequences, syllables and morphophonological rules. Vowel length in PB was only reconstructed relatively late (Meeussen 1954a, 1979). Sequences of both identical and different vowels, which do not emerge from the combination of morphemes, only occur in the initial syllable of the root. Preceding NC and following CG sequences, vowel length is automatic and never contrastive. Reconstructed vocabulary generally consists of sequences of *(N)CV syllables. In verbs and derived nouns, syllable and morpheme structure rarely coincide, since reconstructed roots most commonly have CVC structure, while many PB suffixes have V or VC shape (Schadeberg 2003: 147), e.g., *-báb-ʋd-a 'to singe, burn,' *-báb-ʋk-a 'to be singed, burned' vs. [*bá.bʋ.da], [*bá.bʋ.ka]; *kì-bímb-ò 'swelling' (< *-bímb- 'swell') vs. [*kì.bí.mbò].

The commonest morphophonological rules, which in all likelihood can be reconstructed to PB, are the homorganicity of the non-syllabic nasal of the cl. 9/10 noun prefixes and the 1SG prefixes and the vowel harmony displayed by PB suffixes with a second-degree vowel when following a root having a third-degree vowel.

The characteristic morphophonological behaviour of the non-syllabic nasal is illustrated in (3) with data from the Cabinda Kikongo variety Cisundi H10 (Futi 2012: 49–50). It regressively assimilates the point of articulation of the stem-initial consonant when the latter is voiced as in (3a-c). The assimilation is mutual in (3b), in that the 1SG object prefix triggers post-nasal hardening or strengthening. In (3c), on the contrary, it triggers progressive dissimilation. The geminate nasal resulting from the agglutination of the 1SG object prefix is turned here into a homorganic NC cluster. This dissimilation rule, which has the reverse effect of Meinhof's rule (Meeussen 1962: 29, Herbert 1986: 227), is typical for Kongo H10 varieties. Before voiceless consonants, the non-syllabic nasal disappears, as is common in Bantu (Kerremans 1980), but not without triggering either aspiration (3d) or affrication (3e). As illustrated in (3f), none of these morphophonological rules, often observed in Bantu with non-syllabic cl. 9/10 and 1SG nasals, occurs in association with the syncopated syllabic nasals of classes 1 and 3 (*mʋ-). As in Swahili G42, but in contrast to most other Kongo H10 varieties, this nasal is not homorganic in Cabinda Cisundi (Bostoen & de Schryver 2015).

(3)	a	N-gaánga	→	ŋgaánga	'soothsayer'
		N-diindi	→	ndiindi	'float'
		N-zíizi	→	nzíizi	'fly'
		N-baambi	→	mbaambi	'salamander'
	b	N-longa	→	ndonga	'teach me'
	c	N-nona	→	ndona	'insert me'
		N-mona	→	mbona	'see me'
	d	N-tala	→	thala	'look at me'
		N-vaana (*-páan-)	→	phaana	'give me'
	e	N-soola	→	tsoola	'choose me'
		N-futa	→	pfuta	'pay me'
	f	mu-vondi	→	m'vondi	'killer'
		mu-loonje	→	m'loonje	'teacher'
		mu-moni	→	m'moni	'the who sees'
		mu-tima	→	m'tima	'heart'
		mu-ti	→	m'ti	'tree'

PB noun class prefixes and verb extensions have four instead of seven vowels. None has been reconstructed with the first-degree vowel *u or with the third-degree vowels *e and *o. However, many present-day languages do attest reflexes of the PB mid vowels in these affixes through the application of progressive and/or regressive harmony with the root vowel. Especially the vowel harmony to which extensions reconstructed with a second-degree vowel, such as the applicative *-ɪd-*, the neuter *-ɪk-*, the impositive *-ɪk-* and the separative *-ʊd-/*-ʊk-*, are subject is so widespread that it has been reconstructed to Proto-Bantu (Meeussen 1967: 84). Since the vowel of these extensions is most often only lowered when the root has a mid-open vowel, i.e., the reflex of a PB third-degree vowel, but not when it has a low vowel, it is considered to be vowel height harmony. Following the terminology of Hyman (1999), it can either be "symmetric," as in (4), or "asymmetric," as in (5). The 17th-century South-Kongo H16a data in (4), presented in the original spelling by Van Gheel (1652), show that the vowel of both the applicative and the separative are lowered whether the root has *e* or *o*.

(4)   Applicative   Separative
      cu-ssamb-il-a   'to climb'   cu-amb-ul-a   'to abandon'
      cü-bhel-el-a   'to turn white'   cu-lebh-ol-a   'to insult'
      cu-bhobh-el-a   'to intercede'   cu-somb-ol-a   'to liberate'

In Luba L31a, however, the separative only harmonises after the back mid vowel, but not after the front mid vowel, while the applicative harmonises after both (Kabuta 2008).

(5)   Applicative   Separative
      -samb-il-a   'pray'   -amb-ul-a   'pick up, transport'
      -ep-el-a   'avoid'   -lèp-ul-a   'exhaust'
      -pot-el-a   'annoy'   -somb-ol-a   'take vengeance'

Meeussen (1967: 84) reconstructs the asymmetric type to PB, since it is far more common than the symmetric type which is limited to the North-Western part of the domain (see map in Hyman 1999: 239). Neither type is restricted to any given subgroup though. Alternative analyses are therefore conceivable, such as the one proposed by Hyman (1999), i.e., that the extensions reconstructed with a second-degree front vowel should actually have been reconstructed with a third-degree front vowel. Nevertheless, North-Western languages tend to be more liberal in the application of vowel harmony (Leitch 1996, Kutsch Lojenga 2008). Meeussen was therefore probably right in proposing the asymmetric type as the most archaic. If North-Western languages could extend vowel harmony to prefix vowels and/or low suffix vowels, it is plausible that they also innovated the assimilation of *ʊ* in extensions to the height of *e* in the root. Goes & Bostoen (2019) argue that symmetric and asymmetric vowel harmony can be reconstructed in neither Proto-Bantu nor Proto-Kikongo.

3 LEXICON

Once the phonology of a proto-language is reconstructed, one can return to the initial database of lexical cognates and reconstruct a proto-form by "triangulating backwards" from each of the comparative series used to establish regular sound correspondences (cf. Nurse 2008: 228). This close interconnection between phonological and lexical reconstruction explains why both enterprises have been done simultaneously since the early days of Bantu historical linguistics. Meinhof's pioneering *Grundriß einer Lautlehre der Bantusprachen* already included an inventory of PB stems (Meinhof 1899). The online database *Bantu Lexical Reconstructions* 3 (BLR3) is the most recent reference tool reflecting the output of more than one century of research within that specific branch of Bantu historical linguistics (Bastin *et al.* 2002). For a history of Bantu lexical reconstruction and a detailed discussion of BLR3, we refer the reader to Schadeberg (2002) and Bostoen and Bastin (2016).

BLR3 contains slightly less than 10,000 lexical reconstructions. In contrast to what the introductory note to BLR3 specifies, these are NOT all PB reconstructions. Crucial in this respect are the reliability judgements in BLR2 (Coupez *et al.* 1998), the offline forerunner to BLR3, i.e.:

- "0" = a previously proposed but rejected proto-form (731 items);
- "1" = a well-established reconstruction having a Bantu-wide distribution and thus going back to PB (310 items);
- "1a" = a derived form whose root is a well-established PB reconstruction (514 items);
- "1(a)" = a derived form which can itself be reconstructed to PB (20 items);
- "2/2a" = a cluster of variable reconstructions which cover together the whole Bantu domain and have each a large distribution (17 items);
- "3" = a well-established regional proto-form (593 items);
- "3a" = a derived form whose root is a well-established regional proto-form (519 items);
- "4" = a problematic reconstruction (1243 items);
- "5" = reconstructions corresponding to a comparative series whose attestations are few and/or spatially restricted (5397 items).

Hence, not all reconstructions have the same time depth or equal reliability. This important distinction is somewhat camouflaged in the online BLR3 due to a replacement of the reliability codes by a system of colour marks.

To go by the BLR2 codes "1" and "1(a)," only 330 out of a total of 9821 reconstructions actually have a firm PB status, i.e., only ca. 3.4 %. Based on a preliminary assessment, about 110 reconstructions coded as "1a" should actually be coded as "1(a)," which would result in ca. 440 PB reconstructions, i.e., ca. 4.5%. The reconstructions coded as "2/2a" are so-called "osculant" reconstructions. For Guthrie (1962, 1967) "osculance" refers to the state of affairs in which a presumption of common origin arises between two or more comparative series, either because they have the same meaning but slightly different forms, or identical forms but different meanings. Thorough diachronic semantic research often leads to solving osculant series of the latter kind. A better study of complex sound correspondences also sometimes solves osculance of the first type (Bostoen 2001, Ricquier & Bostoen 2008). Nevertheless, phonologically osculant reconstructions that cannot be reduced to one single proto-form because they are at odds with regular sound changes remain numerous in BLR3, especially among those coded as "4" in BLR2. A systematic study of "2/2a" and "4" reconstructions would certainly help to increase PB vocabulary. A further increase of PB reconstructions in a follow-up of BLR3 would be possible if one more systematically incorporated the comparative lexical research output realised by both historians and linguists over the past decades (e.g., Vansina 1990, Philippson & Bahuchet 1994–1995, Mouguiama-Daouda 1995, Schoenbrun 1997, Ehret 1998, Bulkens 1999, Klieman 2003, Vansina 2004, Bostoen 2005, Stephens 2007, de Luna 2008, Bostoen *et al.* 2013, Ricquier 2013, Bostoen 2014, Koni Muluwa 2014).

A considerable part of the reconstructed PB vocabulary belongs to the so-called "basic vocabulary" that is commonly used for quantitative approaches to language classification (cf. Philippson & Grollemund this volume). Table 10.7 is a list of PB forms that could be reconstructed for the 92 concepts which Bastin *et al.* (1999) retained and adapted from the original Swadesh-100 list.[12] The targeted concepts are italicised. Other English translations correspond to other meanings that are frequently expressed by present-day reflexes of these proto-forms. On the basis of Narrow Bantu evidence, no PB reconstruction could be established for the concepts *bark, blood, cloud, good, hair* (as opposed to

TABLE 10.7 PB RECONSTRUCTIONS FOR THE TERVUREN-92 LIST OF BASIC CONCEPTS (BASTIN *ET AL.* 1999)

1	ncè	all	1	41	jíjɪ	know	1a
2	bókò	arm; hand; front paw	1		jíjɪb	know	1a
3	bú	soil, ashes; dust	1	42	jánì	leaf; grass	4
4		bark		43	gòdò	leg; hind leg	1
5	dà	abdomen, intestines; [belly]	1	44	dáad	lie down, sleep; spend night	1
	bùmò	abdomen; pregnancy [belly]	3	45	tímà	heart; liver	1
6	nénè	big	1	46	dàì	long	4
7	jònì	bird	4	47		louse	
8	dóm	bite	1	48	dómè	male; husband [man]	1
9	jídò	black	1(a)	49	jíngí	many	4
10		blood		50	nyàmà	animal; meat	1
11	kúpà	bone	1		tító	animal, meat	1a
12	béèdè	breast, udder	4	51	jédì	moon; month	1(a)
13	pí	be burnt; be hot; be cooked; be ripe; be red	1	52		mountain	
14		cloud		53	nòà	mouth; lip	4
15	pépò	wind; cold	1	54	jádà	finger/toe-nail; claw	4
	pód	be cold, cool down; be quiet	1	55	jínà	name	4
16	jìj	come	1	56	kòtì	(nape of) neck; occiput	1
17	kú	die	1		kíngó	neck; nape; voice	1
18	bòà	dog	1	57	píà	new	4
19	nyó	drink	1	58	tíkò	night	4
20	jóm	be dry	1	59	bòmbó	forehead; (bridge of) nose	1
21	tòì	ear	1		jòdò	nose, nostril	4
22	dí	eat	1	60	mòì	one	4
23	gì	egg	4	61	jìdà	path	4
24	jícò	eyɛ	1	62	ntò	person	1
25	kútà	fat; oil	1	63	búdà	rain	1
26	cádá	feather	3	64	kódà	red colour/substance	4
27	píà	fire	1(a)	65	tínà	root, base of tree trunk; banana-tree	1
	jótò	fire, fireplace	1a		dì	root, fibre	1
28	cúí	fish	4	66		round	
29	gòdɔk	fly; run fast	1	67	céké	sand; grains; dregs; chaff	4
30	jíjad	be full	1(a)		cèngà	sand; sandy ground	4
	jíjɔd	become full	1(a)	68	gàmb	speak; answer	1
31	pá	give	1		tì	say; quote	1
32		good		69	bón	see; find; acquire; undergo	1
33	cí	ground; country; underneath	1	70	bótò	seed	1a
34	búì	white hair	1	71	jìkad	dwell; be; sit; stay	1
35	tóè	head	1	72	kóbá	skin; strap; girdle	1
36	jígu	hear	1	73	dáad	lie down, sleep	1
37	tímà	heart; liver	1	74	kéè	little, small	3
38	jìgá	horn	4	75	jíkì	smoke	4
39	bòd	break, smash; kill	1	76	jìm	stand; stop, intr.	4
40	dúí	knee	4	77		star	

(Continued)

TABLE 10.7 (Continued)

78	bòè	stone	1	86	gènd	walk; travel; go (away)	1
	tádè	stone	1	87		warm (weather)	
79	jóbà	sun	4	88	jíji	water	1
80	jóg	bathe; wash; swim	4	89	í	what ?	4
82	dímì	tongue; language; flame	1	90		white	
83	jínò	tooth	1	91		who?	
	gègò	molar tooth; tooth	1	92	kádí	woman, wife	
84	tí	tree; stick	1	(93	gì	go	1)
85	bìdí	two	4				
	bàdí	two	2				

Source: Bastin *et al.* 1999

white hair), *louse, mountain, round, star, warm, white* and *who*. As one can judge from the BLR2 reliability codes, 26 proto-forms have a distribution that is wide enough to justify a PB reconstruction but correspond to comparative series which are not fully regular ("4"). Three proto-forms considered regional ("3") have a distribution that actually justifies reconstruction in PB: **-bùmò* 'belly' (A B H K L M R), **-cádá* 'feather' (A B C D H J K L), **-kéè* 'little, small' (B C D H J L M). See Schadeberg (2003: 160–163), for a similar list including distribution according to Guthrie's zones.

BLR3 also contains several PB reconstructions referring to concepts belonging to other basic vocabulary lists, such as Swadesh-200 (Hedinger & Ekandjoum 1980) and Jakarta-Leipzig-100 (Holman *et al.* 2011). Table 10.8 presents, in alphabetical order, only those which were encoded "1" or "1(a)" in BLR2 and do not occur in the Tervuren-92 list. The targeted concepts are underlined. In contrast to Table 10. 7, concepts for which no well-established PB form exists are not listed.

Apart from these relatively universal concepts, PB vocabulary can be reconstructed for much specific natural and cultural vocabulary, which is more telling with regard to the natural environment and lifestyle of early Bantu speech communities. In order to do so, it is important to study not only phonological correspondences between Bantu languages, but also semantic variation. Diachronic semantics and semantic reconstruction have received considerably less attention within Bantu historical linguistics (Fleisch 2008). The lack of insight into semantic change has even caused a serious inflation of reconstructions (Schadeberg 2002: 183). Instead of reconstructing an etymological meaning, as Grégoire (1976) did for the highly polysemic stem **-bánjá*, i.e., 'land prepared for building, uncovered or cleared land,' cognate forms with divergent meanings often ended up as homonymous proto-forms. Moreover, as can be observed in Table 10.7 and Table 10.8, English translation equivalents in BLR3 rather represent the present-day cross-linguistic polysemy of an etymon than a truly reconstructed meaning. More systematic diachronic semantic reconstruction is required to obtain a better understanding of Proto-Bantu vocabulary.

4 GRAMMAR

As lexemes can be reconstructed backwards from cognates between related languages, so can grammemes, especially in Bantu where languages tend to be particularly rich in

TABLE 10.8 PB RECONSTRUCTIONS FOR CONCEPTS BELONGING TO THE SWADESH-200 AND/OR JAKARTA-LEIPZIG LISTS

bá	dwell, be, become [live]	kád	be bitter; be sour; be sharp; be fierce
bàd	count	kám	squeeze; wring
bímb	swell	kèd	cut
bíì	bad	kín	dance; play; gambol
bìng	chase; chase away; go after [hunt]	kóm	hit with hammer; beat; kill
bòd	be rotten	kòd	pull
cèk	laugh; joke	kódó	adult; important person; master; elder
cèng	cut spp.		brother (sister); old
cìngʊ	rub, smear	kónì	firewood
dèm	be heavy; be honoured	pád	scrape, scratch
dìbà	pool, pond; deep water; well [lake]	pèep	blow (as wind); winnow; smoke; breathe
díng	turn round, tr. intr.; wind round, wrap up	pèèpè	wind
dìd	weep; shout; wail	píc	hide; cover
dìmb	trap by birdlime; stick to (something), intr; be firm, intr., [hunt]	pin	press, squeeze (especially with the fingers)
dùmb	smell intr	píp	suck; extract juice
dùt	pull	pùk	dig up, fling up (earth)
dò	fight	tá	throw away; throw; lose; put; trap; play
dòd	be bitter; be sharp		game; do
dók	vomit	tààtá	father, my father
dòng	join by tying	tái	spittle
dòng	be straight, right; put straight, right; be fitting; adjust	tákò	buttock, rear part; back
gàdʊk	turn, intr. ; go back; come back	tím	dig
gàng	tie up	tínd	push, push back
gìd	act, do; say; have	tító	forest, thicket
gò	fall	tí	fear
gòndà	forest, garden, luxuriant vegetation	tín	cut
jákà	year; cultivation season; harvest	tó	stamp; pound; bite [grind, crush]
jánà	child	tú	spit; fix the price
ját	split, tr.; separate	túm	stab
jícì	day, daytime	túm	sew, plait
jíjì	river; waterhole	tóád	carry on the head; carry; bring; carry away; be chief; include
jìmb	sing; dance	tóng	put through; thread on string; plait; sew; tie up; build; close (in)
jókà	snake; intestinal worm		

(bound) grammatical morphology. Pioneering research in Bantu phonological and lexical reconstruction was therefore combined with attempts in morphological reconstruction. Meinhof (1899: 200–204) includes some tables with *Urbantu* noun prefixes and verb suffixes. *Bantu Grammatical Reconstructions* is the last global sketch of PB grammar in which Meeussen (1967) synthesises the knowledge acquired so far and makes several new proposals. Ever since then, new reconstruction research has been carried out on specific components of Bantu grammar, such as noun (class) morphology (Grégoire 1975, Miehe 1991, Maho 1999), inflectional verb morphology (Bastin 1983, Polak-Bynon

1983, 1986, Miehe 1992, Güldemann 1996, Botne 1999, Güldemann 1999, Nurse 2008), derivational verb morphology (Schadeberg 1982, Bastin 1986, Schadeberg 1994) and independent pronouns (Voorhoeve 1980, Weier 1985, Kamba Muzenga 2003).

An updated PB grammar comparable to Meeussen (1967) would be welcome but beyond the scope of this chapter section which will be restricted to some important issues involving Bantu grammatical reconstruction. Most of these issues are particularly relevant for verbal prefixes with the exception of subject and object prefixes. Well-established PB reconstructions exist for noun class and agreement prefixes (Van de Velde this volume), nominal and verbal derivation affixes (Schadeberg & Bostoen this volume) and different types of unbound grammemes (cf. supra). Much more uncertainty exists, however, about the elements preceding the macrostem in the inflected verb template in (6), which Nurse (2008: 40) adapted from the one proposed for PB by Meeussen (1967: 108–111).

(6)   Pre-SP + SP + NEG$_2$ + TA + $\underline{OP + root + extension + FV}$ + post-FV
                                        macrostem

One concern is that the identification of cognate grammatical morphemes is often more challenging than that of cognate lexemes. Especially inflectional verb prefixes are usually not longer than one syllable and sometimes consist of just one segment or even a supra-segment. A good case in point to illustrate the intricacies of reconstructing pre-stem verbal morphology are -a- markers of different length and with different tones expressing a range of different tense/aspect-related values. Nurse (2008: 237) notes at least one -a- marker of some kind in 84% of his sample languages, in 78% of which it has some past reference and in 27% another reference, such as present, future, imperfective, disjunctive focus and anterior. Some of these markers are true cognates, possibly corresponding to either -a- 'recent' or -á- 'preterite' which Meeussen (1967: 109) tentatively reconstructs for PB. The latter proto-grammeme may be at the origin of many present-day -a- markers with (remote) past reference. The original value of Meeussen's -a- 'recent' is maybe rather aspectual, i.e., 'anterior' or 'perfect,' which may have shifted in certain languages to 'present' or '(near) past,' thus giving rise to part of the cross-linguistic polysemy (cf. Nurse 2008: 237–238). However, -a- may also have obtained its 'present' or other 'non-past' meaning through a parallel semantic pathway, e.g., via predicate-centred focus -a- to which Nurse (2008: 212–213, 240) attributes a certain time depth, possibly PB. Then again, still other present-day -a- markers are to be considered homonyms resulting from phonemic merger through sound change. This is particularly likely for those expressing values that are less common among -a- markers, such as 'future' or 'disjunctive focus.' In Nurse's sample, 17% of the languages have some kind of -laa- future marker and 5% a -la-'disjunctive, present' marker (Nurse 2008: 238). Given that the intervocalic loss of liquids is attested across Bantu (Guthrie 1967: 73), and occurs maybe even more frequently in grammemes than in lexemes, it is not unlikely that certain -a- markers are rather cognate to consonant-initial markers than to other -a- markers. The establishment of cognacy among grammemes is further complicated by the fact that sound change as observed in the lexicon does not always manifest the same regularity in grammatical paradigms. It can either be blocked through analogy and morphological embedding (Hyman 2003, Bostoen 2008, Garrett 2014) or occur irregularly through the process of phonetic reduction or erosion, one of the mechanisms of change commonly associated with grammaticalisation (Traugott 2011: 29).

Grammaticalisation as a diachronic change has greatly contributed to the morphological complexity of many Bantu languages through "desyntacticization" (Klausenburger 2000: 142), i.e., the "univerbation of a loose syntactic structure into a morphologically complex word whereby the constituent order becomes petrified in morphotactic regularity" (Güldemann 2003: 183). It is primarily a formal evolution along the "cline of grammaticality" content word > grammatical word > clitic > inflectional affix (Hopper & Traugott 2003: 7). "Today's morphology is yesterday's syntax," as Givón (1971: 413) typifies this process, especially holds for Bantu verbal morphology where the process has been most active and can still be observed at work today (Güldemann 2003: 183–187). Bound morphemes filling the different slots of the finite verb template in (6) are often found either as lexical items or as independent grammemes elsewhere in Benue-Congo or beyond, as shown in the Tikar (South-Bantoid) examples in (7) cited by Güldemann (2011: 125).

(7) a wù   sh-ê   mùn   . . .
         2SG.SUBJ say-IRR 1SG
         'If you had told me. . .' (Stanley 1991: 71)

     b à   tǎ   ǹshe   shɛ
        3SG.SUBJ IPFV   luggage   carry
        'He carries the luggage.' (Stanley 1991: 103)

     c à   dunmi   nun   ji   fɛ
        3SG.SUBJ   begin.PFV 3SG.OBJ food   give
        'He has begun to give him food.' (Stanley 1991: 133)

The subject pronouns in (7) formally resemble the subject prefixes which have been reconstructed for PB and occur in most Narrow Bantu languages as bound morphemes in the initial slot of the inflected verb. Objects are post-verbal, as in (7a), but occur in between the auxiliary and the main verb, as in (7b) and (7c), since Tikar has both S-V-O and S-AUX-O-V clause orders. PB object prefixes, which occur in most Narrow Bantu languages as bound morphemes immediately preceding the root, are thought to have "desyntacticised" from free object pronouns, such as *nun* in (7c). The fact that the subject marker and the aspectual morpheme are separated from the verb by the direct nominal object in (7b) confirms that unlike in most Narrow Bantu languages, they are independent pronouns and not prefixes. The same holds for the object pronoun as *nun* in (7c), which is separated from the main verb by the direct object noun.

Narrow Bantu polarity, focus and TAM markers occurring in between SP and OP in the template in (6) originate from auxiliaries, such as *tǎ* in (7b) and *dunmi* in (7c). Since even the most agglutinative Bantu languages still express certain grammatical categories through auxiliaries, the innovation of bound verb morphology through univerbation and phonetic reduction is continuous and highly cyclic. Synchronic pre-stem verb morphology generally reflects successive diachronic layers of grammaticalisation.

This is much less so for verb morphology following the root. As can be seen in (7a) and (7c), even the rather analytical Tikar conveys certain TAM categories through suffixes, such as irrealis *-ê* which is a likely cognate of PB subjunctive *-e* (Meeussen 1967: 112; Nurse & Devos this volume) and the final perfective *-i* which is possibly cognate to PB resultative/ anterior **-ide* (Botne 2010, Crane 2012) and/or the anterior *-i* widely attested in Bantu.

As for Bantu verb prefixes, they commonly grammaticalise from widespread lexical verbs of PB origin, such as *-màd- 'finish' > past/perfective/anterior (Nurse 2008: 305), *-gì- 'go' > (near) future, *-jìj- 'come' > (remote) future (Botne 1999: 497) and different 'be' verbs leading to a variety of different meanings, e.g., *-bá- 'dwell, be, become,' *-dì- 'be' and *-jìkad- 'dwell; be; sit; stay' (Nurse 2008: 60). Hence, it is extremely hard to tell apart verb prefixes that arise through convergent grammaticalisation from the regularly inherited ones, all the more because phonetic erosion often results in the same output. A good case in point is the plethora of Bantu -ka- markers. Besides the motional (aka itive), inceptive and future -ka- prefixes, which Meeussen (1967: 109) tentatively proposes for PB (see also Botne 1999), many languages share homophonous markers of more recent origin that derive from the auxiliary use of verbs such as *-jìkad- 'dwell; be; sit; stay' and *-jìk- 'come/go down' (Nurse 2008: 240). In order to obtain more robust verb prefix reconstructions, a stepwise bottom-up approach may therefore prove necessary. While other components of Bantu grammar could be fairly well reconstructed through Bantu-wide comparison, the reconstruction of pre-stem verb morphology may benefit from a focus on lower-level subgroups, as for instance Nurse and Hinnebusch (1993) did for Sabaki or Nurse and Muzale (1999) for Great Lakes Bantu.

Finally, grammaticalisation is of key importance to the question whether the inflectional verb template in (6) should really be reconstructed to PB, as Meeussen (1967: 108–111) did, and if so, where this PB should be situated in the Bantu genealogical tree. Especially Güldemann (2011) and Hyman (2011) have brought this issue to the forefront. Although it is impossible to do justice here to their article-length arguments, the debate essentially boils down to two diametrically opposed visions on the morphological typology of Proto-Bantu, and by extension Proto-Niger-Congo.

Hyman defends the traditional view that the highly agglutinative verb morphology reconstructed for PB is not only archaic within Bantu, but also within Niger-Congo. Güldemann contests this interpretation on account of grammaticalisation theory. He claims that analytical languages like Tikar are both predominant and more conservative within Niger-Congo. Languages licensing multiple verbal prefixation and suffixation, as most Narrow Bantu languages and certain Atlantic languages do, are rather innovative. In his understanding, Proto-Bantu and Proto-Niger-Congo had no inflectional verb prefixes and only a moderately agglutinative system of derivational verb suffixes.

It is thanks to the parallel desyntacticisation of free morphemes that certain daughter languages acquired more complex verb morphology. It is true that even within Narrow Bantu certain subgroups are distinctively more analytical than others. As Hyman (2004) argues himself, North-West Bantu languages are very "Kwa-like" (or Tikar-like) in that they have much stricter maximum size constraints on the verb form than most other Bantu languages do. However, on account of attested cases of "debonding" across Niger-Congo, Hyman (2011) conceives this as an erosion of the ancestral agglutinative system leading to the rise of new analytical structures, a process that is likely to have happened cyclically given the huge time depth that allows things to change back and forth. Furthermore, grammaticalisation can cause loss of morphology, i.e., transparent structures like in (7b) are recent and not evidence for an earlier structure.

On the other hand, taking into account that even Western Bantu languages further south are generally less permissive in terms of affix stacking than their Eastern relatives, it is not that heretic from a Bantu-internal perspective to assume that verb morphology gradually became more agglutinative, especially in the pre-stem domain. Meeussen (1967) explicitly acknowledges that he relied exclusively on Eastern languages for certain grammatical reconstructions. His conception of PB grammar more generally may have been biased towards East Bantu, given that he had mainly done descriptive research on Eastern languages, such as Luba L31a (Meeussen 1943), Bangubangu D27 (Meeussen 1954b) and Rundi JD62 (Meeussen 1959), which are strongly agglutinative.

Thanks to different lexicostatistical and phylogenetic approaches to basic vocabulary that have produced largely similar results, we now have a basic idea of the Bantu family tree. To put it schematically, it consists of the following subgroups: Mbam-Bubi (A40-A60) straddling Narrow and Wide Bantu; North-West (rest of zone A + B10–30); Congo Basin (mainly zone C); West-Coastal (mainly B40–80, zone H); South-West (zones K and R + parts of zone L); East (remainder apart from some hard-to-classify zone D languages in North-East Congo) (Vansina 1995, Bastin *et al.* 1999, Bostoen *et al.* 2015, Grollemund *et al.* 2015). Given that most of these subgroups are quite different in terms of morphosyntax and morphological typology, it might also be beneficial to reconstruct first their respective proto-languages in order to acquire a more balanced understanding of PB grammar.

5 RECONSTRUCTION AND CLASSIFICATION

It has become clear from the above that linguistic reconstruction does not happen independently from linguistic classification. An elementary understanding of the phylogeny of a language family is needed to make correct reconstructions (Weiss 2014: 142). Comparative research aiming at reconstructing Proto-Bantu has traditionally been restricted to Narrow Bantu languages, i.e., those conventionally included in Guthrie's referential classification (Hammerström this volume; Maho 2009), to the detriment of Wide Bantu languages, aka Bantoid. Comparative work on the latter languages, such as Hyman (1980, 2010), did not lead yet to any revisions of Proto-Bantu. Nonetheless, several studies have emphasised that Narrow and Wide Bantu languages form a continuum rather than two clearly distinct branches of the Benue-Congo's South-Bantoid branch (Piron 1998, Bastin & Piron 1999, Grollemund 2012). Even within Narrow Bantu, linguistic reconstruction has predominantly relied on so-called "Savannah" languages, for which there used to be more and better descriptions. We know today that the diversity among so-called "Forest" languages, which are increasingly better described, is much higher despite their more limited distribution. The reconstruction of PB lexicon and of certain aspects of PB grammar did not suffer much from this bias, but other parts possibly did as discussed above.

The establishment of a good phylogeny also hinges on correct reconstructions (Weiss 2014: 142). The CM is not only applicable to linguistic reconstruction, but also to linguistic classification through the principle of shared innovations. Especially if an innovative feature is unlikely to have arisen through parallel evolution, languages that have it in common are thought to have inherited it from a more recent common ancestor and to be more closely related among each other than to other relatives. Such use of shared innovations for linguistic sub-grouping obviously only works if one succeeds in distinguishing them from shared retentions that serve for linguistic reconstruction. Hence, new visions on what is archaic and innovative, especially in Bantu grammar, may also lead to new ideas on internal Bantu classification.

While traditional lexicostatistical and new phylogenetic studies of basic vocabulary have helped us to identify the main Bantu subgroups (cf. supra), we still do not have a solid idea of how they relate to each other. Nurse and Philippson (2003) is one of the few and still the most recent attempt to achieve an internal classification of the entire Bantu family based on non-lexical innovations. It is known that the application of the CM on a global Bantu level is hampered by the fact that Bantu languages have not only diverged from each other in the course of their history, but also transferred features among each other through multilingualism and intensive and protracted contacts inducing convergence (cf. Schadeberg 2003: 156–160 for a discussion of the problem of cross-cutting innovations). Here, again, more systematic bottom-up research in reconstruction and classification may be useful, since the CM has already proven to be more efficient for sub-grouping when applied to lower-level Bantu subgroups.

ACKNOWLEDGEMENTS

I wish to express my profound gratitude to Thilo C. Schadeberg and Gerrit Dimmendaal, who once showed me the way to African Historical Linguistics when I was still an undergraduate at Ghent University. Thilo also trusted me the rewriting of his historical linguistic chapter from the first edition of this volume (Schadeberg 2003). The current chapter is strongly indebted to Claire Grégoire, Yvonne Bastin and the late Baudouin Janssens from the Royal Museum for Central Africa in Tervuren, who not only further trained me in Historical Bantu Linguistics at Brussels University (ULB), but from whose unpublished courses – especially those of Yvonne and Baudouin – I have also copied some examples with their kind permission. I wish to thank them for their intellectual generosity. Finally, I also thank Sebastian Dom (Ghent University), Larry Hyman (UC Berkeley), Roland Kiessling (Hamburg University) and Mark Van de Velde (LLACAN – CNRS) for their comments on an earlier version of this chapter. The usual disclaimers apply.

NOTES

1   For reasons of uniformity, the examples of Meinhof and van Warmelo (1932) are not rendered here in the original spelling, but in IPA spelling.

2   Kongo was not part of the original sample of Bantu languages which Meinhof used for his *Grundriß* (Meinhof 1899). Central-Kongo H16b data was added to the revised English version (Meinhof & van Warmelo 1932). Neither Makonde P23, which was already used in the first edition, nor Zulu S42, which was added to the revised edition, are included in Table 1 for space constraints.

3   For reasons of uniformity, the original toneless Ur-Bantu reconstructions are not given here, but the PB reconstructions as found in Bastin *et al.* (2002).

4   According to Meinhof and van Warmelo (1932: 4), "[. . .] o and u denote the open, o and u the close pronunciation of o and u."

5   Rhotacism more commonly refers to the shift of coronal fricatives to /r/ (Campbell 1998: 36, Dimmendaal 2011: 29).

6   The hyphen after an asterix marking a lexical reconstruction indicates that what follows is either a noun stem or verb root. In the former case, a noun class prefix needs to be added to make it a full-fledged noun; in the latter case, verbal morphology to obtain a finite or infinite verb form.

7   This root has become the common Bantu verb to refer to the art of forging, but this meaning cannot be reconstructed to PB (de Maret & Nsuka-Nkutsi 1977). In Herero R31, it refers to crushing and pounding.

8 This potential Duala A24 reflex is not included in Meinhof and van Warmelo (1932: 227). It is found in Helmlinger (1972: 255) with the meaning 'leave in anger,' which is deviant, as is the meaning 'destroy, spoil' of the Herero R31 reflex.

9 The S30 languages recently turned the inherited Proto-Bantu seven-vowel system into a nine-vowel system through the phonologisation of raised allophones of *e [ɛ] and *o [ɔ] that was triggered by the loss of the conditioning high vowel (Gowlett 2003: 611, Creissels 2005, 2007: 33). A similar evolution led to the reshuffle of a classical reduced five-vowel system into a new seven-vowel system in Hungan H42 (Bostoen & Koni Muluwa 2011).

10 The lowering of the high vowel to e observed in the Herero R31 words for 'leave behind' and 'well' seems to be unconditioned and thus irregular.

11 Stevick (1969) proposes an alternative H vs. Ø tone system (see also Odden & Marlo this volume).

12 Bastin et al. (1999) dropped *I, we, thou; not; this, that; green, yellow* from the Swadesh-100 list and renamed *earth > ground; foot > leg; hand > arm; claw > nail* (de Schryver et al. 2015).

REFERENCES

Angenot, J.-P., C. B. Kempf & V. Kukanda (2011) Arte da Língua de Angola de Pedro Dias (1697) sob o prisma da Dialetologia Kimbundu. *Papia* 21(2): 231–252.

Bastin, Y. (1983) *La finale verbale -ide et l'imbrication en bantou.* Tervuren: Musée royal de l'Afrique centrale.

Bastin, Y. (1986) Les suffixes causatifs dans les langues bantoues. *Africana Linguistica* 10: 56–145.

Bastin, Y., A. Coupez & M. Mann (1999) *Continuity and Divergence in the Bantu Languages: Perspectives from a Lexicostatistic Study.* Tervuren: Royal Museum for Central Africa.

Bastin, Y., A. Coupez, E. Mumba & T. C. Schadeberg (eds.) (2002) *Bantu Lexical Reconstructions 3.* Tervuren: Royal Museum for Central Africa (Available online at linguistics.africamuseum.be/BLR3.html, Accessed November 28, 2017).

Bastin, Y. & P. Piron (1999) Classifications lexicostatistiques: bantou, bantou et bantoïde. De l'intérêt des 'groupes flottants'. In Hombert, J. M. & L. M. Hyman (eds.), *Bantu Historical Linguistics: Theoretical and Empirical Perspectives,* 149–164. Stanford: CSLI Publications.

Bleek, W. H. I. (1858) *The Library of His Excellency Sir George Grey, K.C.B. Philology. Vol. I – Part I. South Africa (Within the Limits of British Influence).* London: Trübner & Co.

Bleek, W. H. I. (1862) *A Comparative Grammar of South African Languages. Part I. Phonology.* London: Trübner & Co.

Bleek, W. H. I. (1869) *A Comparative Grammar of South African Languages. Part II. The Concord.* London: Trübner & Co.

Blench, R. (2006) *Archaeology, Language and the African Past.* Lanham: Altamira Press.

Bontinck, F. & D. Ndembe Nsasi (1978) *Le catéchisme kikongo de 1624. Ré-édition critique.* Bruxelles: Académie Royale des Sciences d'Outre-Mer.

Bostoen, K. (2001) Osculance in Bantu Reconstructions: A Case Study of the Pair °-kádang-/°-káng- ("fry", "roast") and Its Historical Implications. *Studies in African Linguistics* 30: 121–146.

Bostoen, K. (2005) *Des mots et des pots en bantou. Une approche linguistique de l'histoire de la céramique en Afrique*. Frankfurt: Peter Lang.

Bostoen, K. (2008) Bantu Spirantization: Morphologization, Lexicalization and Historical Classification. *Diachronica* 25(3): 299–356.

Bostoen, K. (2014) Wild Trees in the Subsistence Economy of Early Bantu Speech Communities: A Historical-Linguistic Approach. In Stevens, C. J., S. Nixon, M. A. Murray & D. Q. Fuller (eds.), *Archaeology of African Plant Use*, 129–140. Walnut Creek: Left Coast Press.

Bostoen, K. & Y. Bastin (2016) "Bantu Lexical Reconstruction." In *Oxford Handbooks Online*. Oxford: Oxford University Press. (Available online at www. oxfordhandbooks.com/view/10.1093/oxfordhb/9780199935345.001.0001/ oxfordhb-9780199935345-e-36&gt).

Bostoen, K., B. Clist, C. Doumenge, R. Grollemund, J.-M. Hombert, J. Koni Muluwa & J. Maley (2015) Middle to Late Holocene Paleoclimatic Change and the Early Bantu Expansion in the Rain Forests of West Central-Africa. *Current Anthropology* 56(3): 354–384.

Bostoen, K. & G.-M. de Schryver (2015) Linguistic Innovation, Political Centralization and Economic Integration in the Kongo Kingdom: Reconstructing the Spread of Prefix Reduction. *Diachronica* 32(2): 139–185.

Bostoen, K. & G.-M. de Schryver (2018a) Seventeenth-Century Kikongo Is Not the Ancestor of Present-Day Kikongo. In Bostoen, K. & I. Brinkman (eds.), *The Kongo Kingdom: The Origins, Dynamics and Cosmopolitan Culture of an African Polity*, 60–102. Cambridge: Cambridge University Press.

Bostoen, K. & G.-M. de Schryver (2018b). Langues et évolution linguistique dans le royaume et l'aire kongo. In Clist, B., P. de Maret & K. Bostoen (eds.), *Une archéologie des provinces septentrionales du royaume Kongo*, 51–55. Oxford: Archaeopress.

Bostoen, K., R. Grollemund & J. Koni Muluwa (2013) Climate-induced Vegetation Dynamics and the Bantu Expansion: Evidence from Bantu Names for Pioneer Trees (Elaeis guineensis, Canarium schweinfurthii and Musanga cecropioides). *Comptes Rendus Geoscience* 345: 336–349.

Bostoen, K. & J. Koni Muluwa (2011) Vowel Split in Hungan (Bantu H42, Kwilu, DRC): A Contact-Induced Language-Internal Change. *Journal of Historical Linguistics* 1(2): 247–268.

Botne, R. (1999) Future and distal -ka-'s: Proto-Bantu or nascent form(s). In Hombert, J.-M. & L. Hyman (eds.), *Bantu Historical Linguistics: Theoretical and Empirical Perspectives*, 473–515. Stanford: CSLI.

Botne, R. (2010) Perfectives and Perfects and Pasts, Oh My!: On the Semantics of -ILE in Bantu. *Africana Linguistica* 16: 31–64.

Botne, R. D. O. (1994) *A Lega and English Dictionary with an Index to Proto-Bantu Roots*. Cologne: Rüdiger Köppe.

Brusciotto, G. (1659) *Regulae quaedam pro difficillimi Congensium idiomatis faciliori captu, ad grammaticae normam redactae. A F. Hyacintho Brusciotto a Vetralla Concionatore Capuccino Regni Congi Apostolicae Missionis Praefecto*. Rome: Sacra Congregatio de Propaganda Fide.

Bulkens, A. (1999) Linguistic Indicators for the Use of Calabashes in the Bantu World. *Afrikanistische Arbeitspapiere* 57: 79–104.

Bulkens, A. (2009) Quelques thèmes pour "pirogue" dans les langues bantoues. *Africana Linguistica* 15: 27–58.

Campbell, L. (1998) *Historical Linguistics: An Introduction*. Edinburgh: Edinburgh University Press.

Cardoso, M. (1624) *Doutrina christãa [. . .] De novo traduzida na lingua do Reyno de Congo*. Lisboa: Geraldo da Vinha.

Connell, B. (2007) Mambila Fricative Vowels and Bantu Spirantization. *Africana Linguistica* 13: 7–31.

Coupez, A., Y. Bastin & E. Mumba (eds.) (1998) *Reconstructions lexicales bantoues 2 / Bantu lexical reconstructions 2 (electronic database)*. Tervuren: Musée royal de l'Afrique centrale.

Crane, T. M. (2012) -ile and the Pragmatic Pathways of the Resultative in Bantu Botatwe. *Africana Linguistica* 18: 41–96.

Creissels, D. (1999) Remarks on the Sound Correspondences Between Proto-Bantu and Tswana (S.31), with Particular Attention to Problems Involving *j (or *y), *i and Sequences *NC. In Hombert, J. M. & L. M. Hyman (eds.), *Bantu Historical Linguistics: Theoretical and Empirical Perspectives*, 297–334. Stanford: CSLI Publications.

Creissels, D. (2005) L'émergence de systèmes à neuf voyelles en bantou S30. In Bostoen, K. & J. Maniacky (eds.), *Studies in African Comparative Linguistics, with Special Focus on Bantu and Mande*, 191–198. Tervuren: Royal Museum for Central Africa.

Creissels, D. (2007) L'influence des voyelles sur les évolutions des consonnes en tswana (S31). *Africana Linguistica* 13: 33–52.

Dammann, E. (1972) Das Meinhofsche Gesetz. *Afrika und Übersee* 55: 242–244.

De Kind, J., G.-M. de Schryver & K. Bostoen (2012) Pushing Back the Origin of Bantu Lexicography: The Vocabularium Congense of 1652, 1928, 2012. *Lexikos* 22: 159–194.

de Luna, K. M. (2008) *Collecting Food, Cultivating Persons: Wild Resource Use in Central African Political Culture, c. 1000 B.C.E. to c. 1900 C.E.* Evanston: Northwestern University, PhD dissertation.

de Maret, P. & F. Nsuka-Nkutsi (1977) History of Bantu Metallurgy: Some Linguistic Aspects. *History in Africa* 4: 43–56.

de Schryver, G.-M., R. Grollemund, S. Branford & K. Bostoen (2015) Introducing a State-of-the-Art Phylogenetic Classification of the Kikongo Language Cluster. *Africana Linguistica* 21: 87–162.

Dias, P. (1697) *Arte da Lingua de Angola, Oeferecida a Virgem Senhora N. do Rosario, Mãy, & Senhora dos mesmos Pretos*. Lisboa: Miguel Deslandes.

Dimmendaal, G. (2011) *Historical Linguistics and the Comparative Study of African Languages*. Amsterdam; Philadelphia: John Benjamins.

Doke, C. M. (1959) Early Bantu Literature – The Age of Brusciotto. *African Studies* 18(2): 49–67.

Ehret, C. (1998) *An African Classical Age: Eastern and Southern Africa in World History, 1000 B.C. to A.D. 400*. Charlottesville; Oxford: University Press of Virginia; James Currey.

Faytak, M. (2015) High Vowel Fricativization as an Areal Feature of the Northern Cameroon Grassfields. Paper presented at *the 8th World Congress of African Linguistics, August 20–24*, Kyoto.

Fernandes, G. (2015) The First Known Grammar of (Kahenda-Mbaka) Kimbundu (Lisbon 1697) and Álvares' ars minor (Lisbon 1573). *Africana Linguistica* 21: 213–232.

Fleisch, A. (2008) The Reconstruction of Lexical Semantics in Bantu. *Sprache und Geschichte in Afrika* 19: 67–106.

Futi, J. M. (2012) *Essai de morphologie lexicale du cisuundi du Cabinda (Angola)*. Paris: L'Harmattan.

Garrett, A. (2014) Sound Change. In Bowern, C. & B. Evans (eds.), *The Routledge Handbook of Historical Linguistics*, 227–248. London: Routledge.

Givón, T. (1971) Historical Syntax and Synchronic Morphology: An Archaeologist's Field Trip. *Chicago Linguistic Society* 7: 394–415.

Goes, H. & K. Bostoen (2019) Progressive Vowel Height Harmony in Proto-Kikongo and Proto-Bantu. *Journal of African Languages and Linguistics* 40.

Gowlett, D. F. (2003) Zone S. In Nurse, D. & G. Philippson (eds.), *The Bantu Languages*, 609–638. London New York Routledge.

Greenberg, J. H. (1948) The Tonal System of Proto-Bantu. *Word* 4(3): 196–208.

Grégoire, C. (1975) *Les locatifs en bantu*. Tervuren: Musée royal de l'Afrique centrale.

Grégoire, C. (1976) Le champ sémantique du thème bantu *-bánjá. *African Languages/ Langues Africaines* 2: 1–12.

Grollemund, R. (2012) *Nouvelles approches en classification: Application aux langues bantu du Nord-Ouest*. Lyon: Université Lumière Lyon 2, thèse de doctorat.

Grollemund, R., S. Branford, K. Bostoen, A. Meade, C. Venditti & M. Pagel (2015) Bantu Expansion Shows That Habitat Alters the Route and Pace of Human Dispersals. *Proceedings of the National Academy of Sciences of the United States of America* 112(43): 13296–13301.

Guinness, H. G. (1882) *Grammar of the Congo Language as Spoken Two undred Years Ago, Translated from the Latin of Brusciotto*. London: Hodder & Stoughton.

Güldemann, T. (1996) *Verbalmorphologie and Nebenprädikationen im Bantu. Eine Studie zur funktional motivierten Genese eines konjugationalen Subsystem*. Bochum: Universitätsverlag Dr. N. Brockmeyer.

Güldemann, T. (1999) The Genesis of Verbal Negation in Bantu and Its Dependency on Functional Features of Clause Types. In Hombert, J. M. & L. M. Hyman (eds.), *Bantu Historical Linguistics: Theoretical and Empirical Perspectives*, 545–587. Stanford: CSLI Publications.

Güldemann, T. (2003) Grammaticalization. In Nurse, D. & G. Philippson (eds.), *The Bantu Languages*, 182–194. London; New York: Routledge.

Güldemann, T. (2011) Proto-Bantu and Proto-Niger-Congo: Macro-areal Typology and Linguistic Reconstruction. In Hieda, O., C. König & H. Nakagawa (eds.), *Geographical Typology and Linguistic Areas with Special Reference to Africa*, 109–141. Amsterdam; Philadelphia: John Benjamins.

Guthrie, M. (1962) A Two-Stage Method of Comparative Bantu Study. *African Language Studies* 3: 1–24.

Guthrie, M. (1967) *Comparative Bantu: An Introduction to the Comparative Linguistics and Prehistory of the Bantu Languages. Volume 1: The Comparative Linguistics of the Bantu Languages*. London: Gregg International.

Guthrie, M. (1970a) *Comparative Bantu: An Introduction to the Comparative Linguistics and Prehistory of the Bantu Languages. Volume 3: A Catalogue of Common Bantu with Commentary*. London: Gregg.

Guthrie, M. (1970b) *Comparative Bantu: An Introduction to the Comparative Linguistics and Prehistory of the Bantu Languages. Volume 4: A Catalogue of Common Bantu with Commentary*. London: Gregg.

Guthrie, M. (1971) *Comparative Bantu: An Introduction to the Comparative Linguistics and Prehistory of the Bantu Languages. Volume 2: Bantu Prehistory, Inventory and Indexes*. London: Gregg International.

Hall, T., S. Hamann & M. Zygis (2006) The Phonetic Motivation for Phonological Stop Assibilation. *Journal of the International Phonetic Association* 36(1): 59–81.

Hedinger, R. & J. Ekandjoum (1980) *Swadesh 200 Word List of 27 Varieties from the Mbo Cluster or the Manenguba Languages. Phonetic Transcriptions*. Yaoundé: SIL Cameroon.

Helmlinger, P. (1972) *Dictionnaire duala-français, suivi d'un lexique français-duala*. Paris: Klincksieck.

Herbert, R. K. (1986) *Language Universals, Markedness Theory, and Natural Phonetic Processes*. Berlin: De Gruyter Mouton.

Holman, E. W., C. H. Brown, S. Wichmann, A. Müller, V. Velupillai, H. Hammarström, S. Sauppe, H. Jung, D. Bakker, P. Brown, O. Belyaev, M. Urban, R. Mailhammer, J.-M. List & D. Egorov (2011) Automated Dating of the World's Language Families Based on Lexical Similarity. *Current Anthropology* 52(6): 841–875.

Hopper, P. J. & E. C. Traugott (2003) *Grammaticalization* (2nd edition). Cambridge; New York: Cambridge University Press.

Hulstaert, G. (1941) Tonetiek van Lomongo en Tshiluba. *Aequatoria* 4(3): 56–58.

Hyman, L. M. (ed.) (1980) *Noun Classes in the Grassfield Bantu Borderland*. Los Angeles: University of Southern California.

Hyman, L. M. (1999) The Historical Interpretation of Vowel Harmony in Bantu. In Hombert, J. M. & L. M. Hyman (eds.), *Bantu Historical Linguistics: Theoretical and Empirical Perspectives*, 235–295. Stanford: CSLI Publications.

Hyman, L. M. (2003) Sound Change, Misanalysis, and Analogy in the Bantu Causative. *Journal of African Languages and Linguistics* 24(1): 55–90.

Hyman, L. M. (2004) How to Become a Kwa Verb. *Journal of West African Languages* 30(2): 69–88.

Hyman, L. M. (2010) Pronoun Systems in Grassfields Bantu. In Pozdniakov, K., V. Vydrin & A. Zheltov (eds.), *Personal Pronouns in Niger-Congo Languages*, 48–51. Saint-Petersburg: Saint-Petersburg State University.

Hyman, L. M. (2011) The Macro-Sudan Belt and Niger-Congo Reconstruction. *Language Dynamics and Change* 1: 1–47.

Janson, T. (2007) Bantu Spirantisation as an Areal Change. *Africana Linguistica* 13: 79–116.

Johnson, M. R. (1979) The Natural History of Meinhof's Law in Bantu. *Studies in African Linguistics* 10(3): 261–271.

Kabuta, N. S. (2008) *Nkòngamyakù Cilubà – Mfwàlànsa*. Ghent Research Centre of African Languages and Literatures, Ghent University.

Kamba Muzenga, J.-G. (2003) *Substitutifs et possessifs en bantou*. Paris: Peeters.

Kerremans, R. (1980) Nasale suivie de consonne sourde en proto-bantu. *Africana Linguistica* 8: 159–198.

Kim, H. (2001) A Phonetically Based Account of Phonological Stop Assibilation. *Phonology* 18: 81–108.

Klausenburger, J. (2000) *Grammaticalization: Studies in Latin and Romance Morphosyntax*. Amsterdam; Philadelphia: John Benjamins.

Klieman, K. A. (2003) *"The Pygmies Were Our Compass": Bantu and Batwa in the History of West Central Africa, Early Times to C. 1900 C.E*. Portsmouth, NH: Heinemann.

Knappert, J. (1971) *Swahili Islamic Poetry*. Leiden: Brill.

Koni Muluwa, J. (2014) *Noms et usages de plantes, animaux et champignons chez les Mbuun, Mpiin, Ngong, Nsong et Hungan en RD Congo*. Tervuren: Musée royal de l'Afrique centrale.

Kouarata, G. N. (2014) *Variations de formes dans la langue mbochi (Bantu C25)*. Lyon: Université Lumière Lyon 2, thèse de doctorat.

Kouarata, G. N. (2016) *Dictionnaire beembe-français*. Saint-Denis: Edilivre.

Kutsch Lojenga, C. (2008) Nine Vowels and ATR Vowel Harmony in Lika, a Bantu Language in DR Congo. *Africana Linguistica* 14: 63–84.

Labroussi, C. (1999) Vowel Systems and Spirantization in Southwest Tanzania. In Hombert, J.-M. & L. M. Hyman (eds.), *Bantu Historical Linguistics: Theoretical and Empirical Perspectives*, 335–377. Stanford, CA: CSLI.

Leitch, M. F. (1996) *Vowel Harmonies in the Congo Basin: An Optimality Theory Analysis of Variation in the Bantu Zone C*. Vancouver: University of British Columbia, PhD.

Maddieson, I. (1976) Tone Reversal in Ciluba: A New Theory. In Hyman, L. M. (ed.), *Studies in Bantu Tonology*, 143–165. Los Angeles: Department of Linguistics, University of Southern California.

Maddieson, I. (1984) *Patterns of Sounds*. Cambridge: Cambridge University Press.

Maddieson, I. (2003) The sounds of the Bantu Languages. In Nurse, D. & G. Philippson (eds.), *The Bantu Languages*, 15–41. London; New York: Routledge.

Maho, J. F. (1999) *A Comparative Study of Bantu Noun Classes*. Göteborg: Acta Universitatis Gothoburgensis.

Maho, J. F. (2008) *The Bantu Bibliography*. Cologne: Rüdiger Köppe.

Maho, J. F. (2009) NUGL Online: The Online Version of the New Updated Guthrie List, a Referential Classification of the Bantu Languages (4 Juni 2009) (Available online at http://goto.glocalnet.net/mahopapers/nuglonline.pdf, Accessed December 13, 2010).

Meeussen, A. E. (1943) Syntaxis van het Tshiluba (Kasayi) [deel II]. *Kongo-Overzee* 9(4–5): 113–159.

Meeussen, A. E. (1954a) Klinkerlengte in het Oerbantoe. *Kongo-Overzee* 20: 423–431.

Meeussen, A. E. (1954b) *Linguistische schets van het Bangubangu*. Tervuren: Koninklijk Museum van Belgisch-Kongo.

Meeussen, A. E. (1959) *Essai de grammaire rundi*. Tervuren: Musée royal du Congo belge.

Meeussen, A. E. (1962) Meinhof's Rule in Bantu. *African Language Studies* 3: 25–29.

Meeussen, A. E. (1967) Bantu Grammatical Reconstructions. *Africana Linguistica* 3: 79–121.

Meeussen, A. E. (1969) *Bantu Lexical Reconstructions*. Ms. Tervuren: Royal Museum for Central Africa.

Meeussen, A. E. (1979) Vowel Length in Proto-Bantu. *Journal of African Languages and Linguistics* 1(1): 1–8.

Meeussen, A. E. (1980) *Bantu Lexical Reconstructions (Reprint)*. Tervuren: Royal Museum for Central Africa.

Meinhof, C. (1899) *Grundriß einer Lautlehre der Bantusprachen nebst Anleitung zur Aufnahme von Bantusprachen – Anhang : Verzeichnis von Bantuwortstämmen* Leipzig: F.A. Brockhaus.

Meinhof, C. (1912–1913) Dissimilation der Nasalverbindungen im Bantu. *Zeitschrift für Kolonialsprachen* 3: 272–278.

Meinhof, C. & N. J. van Warmelo (eds.) (1932) *Introduction to the Phonology of the Bantu Languages Being the English Version of "Grundriss Einer Lautlehre Der Bantusprachen" by Carl Meinhof, Translated, Revised and Enlarged in Collaboration with the Author and Dr. Alice Werner by N. J. Van Warmelo*. Dietrich Reimer/Ernst Vohsen, published under the auspices of The International Institute of African Languages and Cultures, The Carnegie Corporation of New York, and The Witwatersrand Council of Education, Johannesburg.

Miehe, G. (1991) *Die Präfixnasale im Benue-Congo und im Kwa: Versuch einer Widerlegung der Hypothese von der Nasalinnovation des Bantu.* Berlin: Dietrich Reimer.

Miehe, G. (1992) Zur Herkunft von Tempus- und Aspekt-morphemen im Bantu. In Müller, E. W. & A.-M. Brandstetter (eds.), *Forschungen in Zaïre: in memoriam Erika Sulzmann (7.1.0911–17.6.1989),* 289–310. Münster: LIT.

Möhlig, W. J. G. (2009) The Language History of Herero as a Source of Ethnohistorical Interpretations. In Bollig, M. & J.-B. Gewald (eds.), *People, Cattle and Land. Transformations of a Pastoral Society in Southwestern Africa,* 119–146. Cologne: Rüdiger Köppe.

Mouguiama-Daouda, P. (1995) *Les dénominations ethnoichtyologiques chez les bantous du Gabon: étude linguistique historique.* Lyon: Université Lyon 2, thèse de doctorat.

Nurse, D. (1997) The Contributions of Linguistics to the Study of History in Africa. *Journal of African History* 38(3): 359–391.

Nurse, D. (2008) *Tense and Aspect in Bantu.* Oxford: Oxford University Press.

Nurse, D. & T. J. Hinnebusch (1993) *Swahili and Sabaki: A Linguistic History.* Berkeley: University of California Press.

Nurse, D. & H. R. T. Muzale (1999) Tense and Aspect in Great Lakes Bantu Languages. In Hombert, J. M. & L. M. Hyman (eds.), *Bantu Historical Linguistics: Theoretical and Empirical Perspectives,* 517–544. Stanford: CSLI Publications.

Nurse, D. & G. Philippson (2003) Towards a Historical Classification of the Bantu Languages. In Nurse, D. & G. Philippson (eds.), *The Bantu Languages,* 164–181. London; New York: Routledge.

Pacconio, F. & A. do Couto (1642) *Gentio de Angola sufficientemente instruido nos mysterios de nossa sancta Fé. Obra posthuma, composta pello Padre Francisco Pacconio da Companhia de Iesu. Redusida a methodo mais breve & accomodado á capacidade dos sogeitos, que se instruem pello Padre Antonio do Couto da mesma Companhia.* Lisboa: Domingos Lopes Rosa.

Philippson, G. & S. Bahuchet (1994–1995) Cultivated Crops and Bantu Migrations in Central and Eastern Africa: A Linguistic Approach. *Azania* 29–30: 103–120.

Piron, P. (1998) The Internal Classification of the Bantoid Language Group, with Special Focus on the Relations between Bantu, Southern Bantoid and Northern Bantoid languages. In:. In Maddieson, I. & T. J. Hinnebusch (eds.), *Language History and Linguistic Description in Africa,* 65–74. Trenton: Africa World Press.

Polak-Bynon, L. (1983) L'infixe réfléchi en bantou. *Africana Linguistica* 9: 271–304.

Polak-Bynon, L. (1986) Les infixes ("préfixes objets") du bantou et leur réconstruction. *Africana Linguistica* 10: 365–422, 449–454.

Ricquier, B. (2013) *Porridge Deconstructed: A Comparative Linguistic Approach to the History of Staple Starch Food Preparations in Bantuphone Africa.* Bruxelles: Université Libre de Bruxelles, PhD dissertation.

Ricquier, B. & K. Bostoen (2008) Resolving Phonological Variability in Bantu Lexical Reconstructions: The Case of "to bake in ashes." *Africana Linguistica* 14: 109–150.

Rosa, M. C. (ed.) (2013) *Uma língua africana no Brasil colônia de Seiscentos: O quimbundo ou língua de Angola na Arte de Pedro Dias, S.J.* Rio de Janeiro: 7 Letras.

Schadeberg, T. C. (1982) Les suffixes verbaux séparatifs en bantou. *Sprache und Geschichte in Afrika* 4: 55–66.

Schadeberg, T. C. (1994) Die extensive Extension im Bantu. In Geider, T. & R. Kastenholz (eds.), *Sprachen und Sprachzeugnisse in Afrika: eine Sammlung philologischer Beiträge Wilhelm J. G. Möhlig zum 60. Geburtstag zugeeignet.* Cologne: Rüdiger Köppe.

Schadeberg, T. C. (1995) Spirantization and the 7-to-5 Vowel Merger in Bantu. *Belgian Journal of Linguistics* 9: 71–84.

Schadeberg, T. C. (2002) Progress in Bantu Lexical Reconstruction. *Journal of African Languages and Linguistics* 23(2): 183–195.

Schadeberg, T. C. (2003) Historical Linguistics. In Nurse, D. & G. Philippson (eds.), *The Bantu Languages*, 143–163. London; New York: Routledge.

Schoenbrun, D. L. (1997) *The Historical Reconstruction of Great Lakes Bantu: Etymologies and Distributions*. Cologne: Rüdiger Köppe.

Smith, E. W. (1964) *A Handbook of the Ila Language (Commonly Called the Seshukulumbwe), Spoken in North-Western Rhodesia, South-Central Africa: Comprising Grammar, Exercises, Specimens of Ila Tales, and Vocabularies*. Farnborough: Gregg Press.

Stanley, C. (1991) *Description phonologique et morphosyntaxique de la langue tikar*. Yaoundé: Summer Institute of Linguistics.

Stephens, R. (2007) *A History of Motherhood, Food Procurement and Politics in East-Central Uganda to the Nineteenth Century*. Evanston, IL: Northwestern University, PhD dissertation.

Stevick, E. W. (1969) Tone in Bantu. *International Journal of American Linguistics* 35(4): 330–341.

Stewart, J. M. (1993) The Second Tano Consonant Shift and Its Likeness to Grimm's Law. *Journal of West African Languages* 23(1): 3–39.

Stewart, J. M. (2000) An Explanation of Bantu Vowel Height Harmony in Terms of a Pre-Bantu Nasalized Vowel Lowering. *Journal of African Languages and Linguistics* 21(2): 161–178.

Traugott, E. (2011) Grammaticalization and Mechanisms of Change. In Heine, B. & H. Narrog (eds.), *The Oxford Handbook of Grammaticalization*, 19–30. Oxford: Oxford University Press.

Van Gheel, J. (1652) *Vocabularium Latinum, Hispanicum, e Congense. Ad Usum Missionariorû transmittendorû ad Regni Congo Missiones*. Rome: National Central Library, Fundo Minori 1896, MS Varia 274.

Van Spaandonck, M. (1971) On the So-Called Reversing Tonal System in Ciluba: A Case of Restructuring. *Studies in African Linguistics* 2(2): 131–144.

Vansina, J. (1990) *Paths in the Rainforest: Toward a History of Political Tradition in Equatorial Africa*. Madison: University of Wisconsin Press.

Vansina, J. (1995) New Linguistic Evidence and the Bantu Expansion. *Journal of African History* 36(2): 173–195.

Vansina, J. (2004) *How Societies Are Born: Governance in West Central Africa Before 1600*. Charlottesville: University of Virginia Press.

Voorhoeve, J. (1980) Le pronom logophorique et son importance pour la reconstruction du proto-bantou (PB). *Sprache und Geschichte in Afrika* 2: 173–187.

Weier, H.-I. (1985) *Basisdemonstrativa im Bantu*. Hamburg: Helmut Buske.

Weiss, M. (2014) The Comparative Method. In Bowern, C. & B. Evans (eds.), *The Routledge Handbook of Historical Linguistics*, 127–146. London: Routledge.

CHAPTER 11

CLASSIFYING BANTU LANGUAGES

Gérard Philippson and Rebecca Grollemund

1 INTRODUCTION

For decades, the problem of the Bantu Expansion has raised many questions. Since the 19th century, numerous studies (linguistic, archaeological and, more recently, genetic) have been conducted in order to better understand the origin of the Bantu-speaking peoples and their history. In that perspective, the classification of Bantu languages has been a major preoccupation.

The Bantu languages belong to the Niger-Congo phylum, according to the well-known classification of Greenberg (1963). Bantu constitutes a sub-branch of Benue-Congo, also sometimes called "Benue-Kwa" (Elugbe & Williamson 1977), itself a sub-branch of Niger-Congo, yet Bantu is the group with the largest number of speakers within that phylum.

The classification of the Bantu languages can thus be approached from two angles: 1) the definition of Bantu languages as opposed to related languages not included under that label, viz. the external classification of Bantu and 2) classifying the c. 600 Bantu varieties into coherent subgroups of variable depth, viz. the internal classification of Bantu. This chapter will mainly be dealing with the second question, but we nevertheless address the first one, though not in detail.

In Section 2, we discuss the limits of Bantu and its position among its closest relatives. In Section 3, we discuss the various attempts at classification of the Bantu languages, according to referential, lexicostatistical and phylogenetic methods. Finally, in Section 4, a conclusion will briefly point to paths which might be followed by future research.

2 EXTERNAL CLASSIFICATION OF THE BANTU LANGUAGES

From quite an early period on, it was noticed that many languages in Cameroon and Nigeria – and even further afield, e.g., Temne spoken in Sierra Leone – exhibited a noticeable resemblance to the core Bantu languages spoken in central, eastern and southern Africa. Those languages were then generally known under the name of "Semi-Bantu" (Johnston 1919) – the term itself appearing in 1891 according to Williamson (1989: 6). However, Westermann (1927) made a much more detailed study of the "Western Sudanic" languages and came up with a more precise classification of those languages closest to Bantu which he gathered under the label "Benue-Cross River." It is within this group – renamed "Central Branch" (Greenberg 1949), then "Benue-Congo" – that Greenberg put Bantu, and his solidly argued classification has been generally accepted, although the exact boundaries of Benue-Congo have been redefined several times. The next stage was then to work out the precise boundaries of Bantu and on this point no unanimity has been achieved.

The languages closest to Bantu within Benue-Congo are called Bantoid. There are about 150 such languages, among which are the Tivoid, Ekoid, Jarawan and Grassfields languages. Guthrie (1948), whose classification is discussed in detail below, provided an inventory of what he considered to be Bantu languages. His criteria can be summarised as follows: (1) a system of nominal classes marked by prefixes, organised into singular/ plural pairs and determining agreement on dependent elements and (2) common lexicon with regular correspondences. He remarked that certain languages did have the first criterion, but not the second, and so he did not acknowledge them as Bantu.

But more recent classificatory attempts have found it difficult to draw a precise line between some of the North-Western-most languages retained as Bantu by Guthrie (1948, 1971) and others that he excluded. In certain classifications (Bennett & Sterk 1977, Lewis 2009), some of these languages appear closer to some non-Bantu languages than to the rest of Bantu; in other classifications (Piron 1995); their position varies according to the method of computation. In the following, we will only deal with Guthrie's "Narrow Bantu" as opposed to "Wide Bantu," which would include, for instance, Grassfields or Jarawan languages. However, the matter is in no way settled.

3 INTERNAL CLASSIFICATIONS OF THE BANTU LANGUAGES

We can roughly distinguish two periods with regard to internal Bantu classification, i.e., before and after the turn of the 21st century.

3.1 Before the 21st century

Since the 19th century, linguists have proposed several classifications based on different techniques and often leading to different results: referential classifications, lexicostatistical classifications and Comparative Method-based classifications relying on linguistic innovations.

The question only began to be treated seriously in the mid-20th century. Before that date, the various proposals put forward were based on very insufficient data and mostly unspecified criteria; a summary can be found in Doke (1945, 1954) and Cole (1961). Here, we will briefly review referential classifications before discussing in more detail the several lexicostatistical classifications. We will also give some space to discussing several other approaches, like those of Möhlig (1977, 1978, 1981, 1984–1985) and Ehret (1999).

3.1.1 *Referential classifications*

Referential classifications classify the Bantu languages according to some linguistic and geographical criteria with mostly practical aims. By 1945, when a sizeable amount of data from Bantu languages had been collected, some sort of sub-grouping had to be established, if only to identify and locate individual languages.

Doke (1945) proposed a classification of the Bantu languages which would serve as a framework for his survey of the existing literature on those languages. It was thus not meant as a complete classification of the field. He divided the Bantu-speaking area into several "zones" mainly on a geographical basis (North-Western, East-Central, etc.) and "groups" whose languages were so similar as to be almost inter-intelligible, because of "possessing common salient phonetic and grammatical features" (Doke 1945: 1, see also Cole 1959: 198).

Van Bulck (1952) proposed another referential classification whose criteria were not made explicit. They seem to have been defined as much on ethnological as linguistic grounds.

Guthrie (1948: 23ff) states that there are three possible methods for classifying languages on a linguistic, i.e., non-geographical, basis: (1) the "historical" method, which he deems impossible for want of historical documents; 2) the "empirical" method, to which he devotes considerable attention, and which would consist in the mapping of isoglosses capturing various significant "differentia," namely lexical, grammatical, phonological, phonetic and tonal; he concludes, however, that since isoglosses constantly criss-cross each other the method cannot be used consistently, but has to be used with "some modification"; this is what he calls 3) the "practical" method where the empirical method is modified by "some element of arbitrariness." Guthrie (1948: 27) adds the (perhaps tongue-in-cheek) comment "It is presumably because others have not made this admission that no satisfactory classification has yet been achieved." The practical method will thus consist in taking as its point of departure "the *individual* language and move outward from it" (italics ours); this initial language will be grouped with others similar to it until it is decided that "we have moved into another group" (ibid.). This is where the arbitrariness lies, since most characteristic features will not coincide strictly with any group's limits.

In fact, Guthrie aims here at establishing a referential classification – which would make it easy to manipulate data from the several hundred Bantu languages – by using a unique alpha-numeric code for each language: a letter represents a "zone," itself divided into "groups." It is not clear to what extent Guthrie's use of these two terms was influenced by Doke, since he does not refer to the latter. Due to the coding format (X.00, e.g., Gikuyu E.51, Kamba E.55) each zone could include at most nine groups and each group nine languages. Guthrie sometimes tried to go around that constraint by introducing a "dialect" format (X.00x, e.g., Unguja G42d), which allowed him a few more members. On the other hand, the number of zones was sufficiently high (16 at first, later reduced to 15: A, B, C, D, E, F, G, H, K, L, M, N, P, R, S) to prevent the crowding of groups, only one zone (A) reaching the full complement of nine groups. The two smallest zones (F and P) are comprised of only three groups each. This format entirely obscures the fact that some zones are extremely heterogeneous in terms of the genealogical relatedness of their members, e.g., zones A and B, whereas others are much more coherent, e.g., zone S. The consequences this has for a genetic classification will be dealt with below. Although the main use of Guthrie's classification is that it helps approximately to localise both geographically and linguistically a language in relation to its neighbours, some attempts have been made to "improve" it by correcting parts of it, with not always felicitous results. One modification which has found wide acceptance, however, is the creation by the "Tervuren school" of a zone J, which groups parts of Guthrie's zones D and E, along with other genealogically inspired revisions which were not retained (cf. Bastin 1978). Nowadays, in accordance with Maho (2003, 2009), the languages in question are most often quoted with a double letter where the second one indicates the zone the language used to belong to, e.g., Rundi JD62, Zinza JE23 etc. (see also Chapter 2, this volume).

Although Guthrie claimed his classification to be solely referential, he in fact made use of it for historical purposes. By comparing contemporary Bantu languages, he recognised a large number of cognate roots which he termed C(ommon) B(antu). He studied the distribution of these roots throughout the Bantu domain and found that they clustered in two regions, which he termed Eastern Bantu and Western Bantu. He then claimed that the area with the greatest number of CB roots, which was situated south of the rainforest, must have been the center of origin of the Bantu languages. However, the method was vitiated at its basis by the fact that to establish one CB root (or C(omparative) S(eries)), the item must be exemplified with regular sound correspondences in at least three different zones. Since the zones were established in the referential perspective mentioned above, there was no guarantee that each of them would represent a taxon of equivalent ranking, and in fact zones A, B and C were very much more diverse internally than the others, which Guthrie would not consider. As a result, the largest numbers of CS did not really belong to CB in Guthrie's sense, but only to a subset corresponding to Guthrie's Eastern Bantu, plus the southern part of his Western Bantu. For a detailed critique of Guthrie's comparative method, see Möhlig (1976) and Flight (1980).

Finally, the last referential classification was established by Maho, initially for the first edition of this volume (Maho 2003) followed by several updates, the last in Maho (2009). He mostly kept Guthrie's original coding intact and expanded it, by attributing to new languages numbers not used by Guthrie or introducing three-digit codes, e.g., Mwani G403. For an overview of the present state of the referential classification, see Chapter 2 of this volume.

3.1.2 Lexicostatistical studies

In principle, lexicostatistics could refer to any method applying statistics to a given lexical corpus. In fact it generally refers to the type of study developed by Swadesh (1952), relying on standard word lists of so-called "basic vocabulary." It is closely linked with "glottochronology," which is an attempt to provide absolute dating for various stages of language evolution, although the two should by no means be necessarily taken as synonymous (McMahon & McMahon 2005). Swadesh proposed 200 and 100-word lists which were supposed to represent the most stable concepts found across the world's languages. From these lists, the relatedness between two languages was calculated on the basis of the percentage of words shared. The 100-word list was eventually preferred because it offered less scope for including loanwords, thus giving a better view of genetic relationships. Then glottochronology was applied to determine the time depth of the separation between the two languages, it being held by Swadesh and his followers that the rate of change was stable for basic vocabulary across the world's languages. This assumption having been subject to strong criticism (Bergsland & Vogt 1962, Dyen et al. 1967, McMahon & McMahon 2005), it was generally (but not entirely) abandoned in the 1970s.

Strict cognacy between lexical items can only be determined by the Comparative Method, and this was indeed always claimed to be applied in the earlier studies, which mostly dealt with already well-reconstructed families such as Indo-European, Algonquian, Turkic, etc. However, a tendency to skip the comparative stage and base the cognacy judgments on look-alikes also occurred when lexicostatistics was applied to less well-known languages (cf. the very apt discussion in Fairbanks 1955), especially when trying to determine distant relationships. Swadesh (1954) thought he could "push" the method to assess Amerindian linguistic relationships, which date back to the first peopling of the continent, whereas the probability of finding enough real cognates to establish correspondences for such ancient relationships is minute indeed. Even in case the correspondences would not be too hard to establish, the sheer amount of data might dissuade the researchers, especially since lexicostatistical studies can provide a shortcut to getting a general idea of the internal structure of the group. So for instance, most lexicostatistical studies of Bantu languages presented below did not make use – or only made use in part – of the Comparative Method. This is also true for the newer phylogenetic methods dealt with in Section 3.2.

Moreover, lexicostatistics does not take into account complex linguistic phenomena, such as borrowings, language contact or convergence, which by all accounts are very frequent in Bantu languages (Nurse 1997, Nurse & Philippson 2003b), nor does this method differentiate properly between retentions and innovations. We will see below that some phylogenetic methods have a somewhat better claim to be able to achieve this. Lexicostatistics has thus been rejected by some linguists or used only as a preliminary to other methods, such as qualitative lexical analysis (Ehret 1999) or diachronic phonology (Hinnebusch et al. 1981, Möhlig 1981).

Although lexicostatistics had its heyday in the 1950s and 1960s, only a few attempts were made then to apply it to Bantu languages (Coupez 1956, Meeussen 1956, Olmsted 1957). Guthrie (1967: 96–110) did try to apply quantitative methods to his Bantu corpus. The results, although not very significant due to the small number of his test languages, i.e., 28, seemed to him to confirm his theory about a "Bantu nucleus" south of the rainforest (Guthrie 1962).

Henrici (1973) proposed a lexicostatistical classification of Bantu languages based on the same 28 Bantu test languages and on the 2235 CB roots used by Guthrie. He measured the degree of language similarity by comparing pairs of languages and studied the linguistic distances between each pair. With the matrix of distance thus generated, he was able to propose a tree, and his main results consisted in showing a first group composed of the A languages (A24 and A74), followed by a branch composed of the B and C languages (B75, C32 + C71) (Henrici 1973: 97). Contrary to Guthrie, he did not distinguish a Western/Eastern division, and he proposed to locate the proto-Bantu nucleus in Cameroon, whereas Guthrie, as mentioned above, placed it in the savannah south-east of the rainforest.

The same year, Heine (1973) proposed a lexicostatistical classification of the Bantu languages that would be revised in Heine et al. (1977). Heine (1973) was based on 100 basic vocabulary items documented from the available literature for 137 languages. The results obtained showed a tree-like representation with eight groups coordinated at the same level: (i) Tiv (ii) Ekoi-Branch, (iii) Bobe A31, (iv) Sanaga-Branch (A and B20 languages), (v) Kande-Mpongwe-Branch (B10–30 languages), (vi) Aruwimi-Branch (C40 and D30 languages), (vii) Ituri-Branch (other D languages) and (viii) Congo-Branch (all the other languages). According to Heine (1973), Narrow Bantu started at branch number 3 with Bobe A31. The ensuing work by Heine et al. (1977) is at the origin of "the East-out-of-the-West" model (cf. Wiesmüller 1997, Pakendorf et al. 2011), which postulates on glottochronological grounds that from the proto-Bantu nucleus in Cameroon, there was a first split that might have occurred between 5,000 and 3,500 years ago; 1,000 years later, the Bantu languages expanded south through the equatorial rainforest. From a second center of dispersion called "Congo nucleus" or "Western nucleus" located in the Congo Basin, there was a second main migration wave that moved to the east and formed the "East Highland nucleus." This third center of dispersion would have given birth to

Eastern Bantu languages. In this model, Eastern Bantu is a sub-branch within a diverse Western Bantu.

Bastin *et al.* (1999) – henceforth CDBL – represents a momentous achievement in Bantu lexicostatistics. It is the culminating point of a series of lexicostatistical publications by the "Tervuren school" (Coupez *et al.* 1975, Bastin *et al.* 1983). Besides being the most complete statement of Bantu lexicostatistical relationships, it has also been the source of data for various phylogenetic contributions discussed in detail below. For this reason, it is necessary to raise a number of questions arising from the elaboration and collation of its database on top of the more general evaluation of lexicostatistics discussed above.

CDBL has sometimes been claimed (not by its authors) to offer a comparative word list for 542 Bantu languages. Since the total number of Bantu languages has been put more or less vaguely at around 600 (Nurse & Philippson 2003a), this would mean the near-complete covering of the field with a few insignificant gaps left. In fact, it is not so. Although there is no complete fit between "ethnic" labels and individual languages, a rough survey of the data, available online at www.africamuseum.be/research/human-sciences/cultsoc/lexico-1/, shows that between 20% and 25% of the lists provide very slightly different variants of other lists, either under the same "language" name or a different one, so that the actual number of individual "languages" or well-defined "dialects" is more on the order of 450, as pointed out by Vansina (1995). While this is still a not insignificant figure, there are serious geographical gaps in the database, specifically – and paradoxically – for Tanzania and Northern Mozambique, for which data are rather easily available. In Northern Tanzania, the absence of any of the East Lake languages (Guthrie's E42 to E45), of which there exist around 12 varieties, some of them fairly well documented, and of the crucial Sonjo E46, invalidates any attempt at analysing the relationship between the Lake languages and the Central Kenya group, which consequently appears as an isolate within Eastern Bantu in the various trees. Furthermore, Chaga E60, which has been known since Nurse (1979) and Philippson (1984) to consist of four different languages themselves fairly diverse internally, is only represented by two rather similar varieties. In Central and Southern Tanzania, apart from Nilamba F31, Guthrie's F30 languages (Rimi F32, Rangi F33, Mbugwe F34, plus Kimbu F24 which is genealogically more closely related to the F30 than to the other F20 languages) do not appear at all. Guthrie's G50 and G60 (about 10 varieties) are represented only by Kinga G65. Of the western Rufiji languages (Guthrie's N10 and P10), only Chimpoto N14 is included and the very important and diversified Makonde P23 is left out. This last area is certainly not minor, *pace* Vansina (1995: 180, footnote 110). Although this considerable gap probably would not influence the deeper splits in the various trees, it definitely contributes to muddling up the Eastern Bantu-internal situation.

Furthermore, CDBL relies on data selected on the basis of the Swadesh 100-word list. It was decided not to retain the two color terms, i.e., 'green' and 'yellow,' nor some items "that belonged more to grammar than lexis" (Bastin *et al.* 1999: 5), i.e., pronouns, demonstratives, and 'not.' One then wonders why they kept 'who?' and 'what?' which belong to the same category and are notoriously difficult to compare. They were thus left with only 92 items, but it should be mentioned that even some of those almost never yield satisfactory results in Bantu comparisons, e.g., 'walk' which Bastin *et al.* (1999: vi) belatedly suggested should be replaced by 'go,' 'round,' 'warm.' Others are ambiguous, e.g., 'human skin' (Bantu languages normally only have words for 'animal skin,' otherwise just 'body'), 'cold weather' (does it mean 'cold season'?), 'dry,' 'full' (verbs in most languages), 'say' ('utter' or 'tell'?), 'animal fat' ('oil' would have been better), etc. Note that Swadesh (1952: 157) himself had suggested replacing the original 'speak' with 'say' – this was definitely a bad move with regard to Bantu. Finally, many languages have gaps in the data so that the number of items is in fact much less than 92, e.g., Mituku D13 has only 74 entries, Buyu D55 has 75, Kami G36 has 83, Nyanja N31a has 85, etc. Moreover, as noted by Currie *et al.* (2013), the published database only includes 90 items due to a page-setting error. As a consequence, the important items 'name' and 'neck' are missing. Currie *et al.* (2013) thus had to make their calculations over those 90 items only. Whether this was also the case for the other phylogenetics studies using the same database is not known. They might have had access to the unpublished original data. Anyhow, to quote Embleton (1981: 113) "comparison of results for N=200 with those for N=100 shows that accuracy [. . .] is considerably decreased by using a 100-word list," and she concludes that a 200-word list should be used whenever possible. An anonymous reviewer disputed this assertion, pointing to Holman *et al.* (2008) where it is supposedly demonstrated that a word list of only 40 items gives results, which are just as accurate. Pending an experiment on this basis which would certainly be worthwhile, we still prefer to side with Embleton in view of the great internal homogeneity of Bantu (see also Heggarty *et al.* 2010). 40 items might suffice to determine whether a language is Bantu or not, but would not help much in sub-grouping. For their study of about 80 languages of Uganda, Kenya and northern Tanzania, Nurse and Philippson (1980) used a 400-word list; Möhlig (1974) used a 600-word list for Central Kenya.

Although it could not be claimed that CDBL is *the* definitive Bantu classification, it nevertheless represents a very considerable achievement. Furthermore, the statistical methodology developed by Mann (1999) led to the computation of a number of what he called "similarity" trees, since he rightly considered them not as genetic, but "phenetic," i.e., based on resemblances and not innovations. While the authors did not consider any of the trees as "truer," four groupings emerged consistently, whatever fluctuations might be undergone by marginal languages: (i) Mbam-Bubi (A44, A60, A31); (2) North-Western (all remaining A + B10–20–30); (3) Central-Western (remaining B + C, H, K, R and a few L); (4) Eastern (all others). The tables showing the computations and the details of all resulting trees can be most useful (Bastin *et al.* 1999: 126–222).

3.1.3 Other classificatory approaches

A very original approach is that of Möhlig (1977, 1978, 1981, 1984–1985). Rejecting the possibility of reconstructing the Proto-Bantu language, he studied the phonological correspondences among Bantu languages to reconstruct various stages or "strata" in their evolution. This quite intricate model is based on the assumption that each contemporary language consists of a mixture of several of those "strata," each of them having contributed specific phonological developments, e.g., the seven > five vowel shift is a characteristic of Savanna stratum IV, the voicing of $NC_{[-voice]}$ belongs to Savanna stratum I, the loss of contrastive vowel length while retaining seven vowels belongs to Rainforest strata etc. While Möhlig does not deny the ultimate existence of some "Proto-Bantu" language, he claims that each of these strata corresponds to a real ancestral language – presumably itself somehow derived from an ancestral Proto-Bantu – spoken at some point in time. This seems to us rather unlikely, since most of the phonological developments in question are natural and can easily arise in the course of language evolution. Note that we say "most," since some of them are much more marked and could then be considered diagnostic of shared ancestry. The very characteristic sound shift $NC_{[+voiced]} > C_{[-voiced]}$ is a very good case in point. This sound change, known as "nasal strengthening" (Schadeberg 1999: 386), is a shared innovation between Makhuwa P31 and the Southern Bantu languages of Sotho-Tswana S30 group (Janson 1991–1992). However as "ideal types" for a typology of synchronic Bantu phonological systems from which diachronic inferences can be drawn, Möhlig's strata can certainly prove useful, especially since his model explicitly takes contact into account. Needless to say, the "trees" resulting from this approach are very network-like.

Ehret (1999) proposed a new classification of what he calls "Savanna Bantu," i.e., the Bantu languages classified in Guthrie's zones E, F, G, K, L, M, N, P, R, S, plus the two languages or dialect clusters Lega D25 and Mbundu H21. For the other Bantu languages, he refers to the more inclusive lexicostatistical classification of Klieman (1997), where Ehret's Savanna Bantu appears as a "probable" branch of Sangha-Kwa, which also includes B, C and H languages. Ehret's classification is based on the study of shared lexical innovations, which are according to him more reliable than a lexicostatistical study, which does not take borrowings into account. According to Ehret's results, Savanna Bantu splits into two groups, Western Savanna and Eastern Savanna. Western Savanna is comprised of six non-hierarchical groups: Luyana-South-West Bantu, Lwena, Lunda, Pende, Kimbundu and Umbundu. Eastern Savanna has a much richer internal structure with Lega D25, Luban L30, Botatwe M60, Sabi M40–50 and finally Mashariki, grouping all the languages of Eastern and Southern Africa, which is itself again subdivided in two branches, i.e., Kaskazi and Kusi, and many sub-branches. Note that the terms "Mashariki," "Kaskazi" and "Kusi" are from Swahili G42 and mean "East," "North" and "South" respectively.

Ehret's work is welcome, since it is qualitative and puts to the fore the study of lexical innovations, most important for sub-groupings. It cannot, however, be accepted as it stands. First of all, it depends almost entirely for its database on the data from Guthrie (1967, 1970a, 1970b, 1971), which are not 100% reliable. Second, one gets the feeling that his analysis of innovations depends too much on his *a priori* classification: if an item considered by him as defining subgroup X appears in one or two languages of subgroup Y, the latter are authoritatively stated to have borrowed them, which should be demonstrated by other criteria than just distribution, as is done for smaller subgroups by, for example, Nurse and Hinnebusch (1993), Hinnebusch (1999) and Nurse (2000).

In her preliminary attempt to compute non-lexical criteria, Bastin (1980, 1983) chose 52 phonological and morphological criteria for 80 Bantu languages, the sample being determined by the availability of descriptive grammars. Although the study did not permit a complete classification of the domain, it pointed out a number of interesting innovations, such as the shape of locative markers or progressive nasal assimilation.

This approach was taken up again by Nurse and Philippson (2003b). In order to establish their classification, they worked on linguistic innovations and a list of 30 phonological and morphological criteria for more than 80 Bantu languages. The criteria, such as the verb suffix -*i* "past/anterior," the verb suffix -*a(n)g-a* "imperfective," Dahl's Law, Meinhof's Law, lenition of **p*, etc., were similar but not identical to the list of Bastin (1980, 1983). Although the method could not be consistently applied, Nurse and Philippson (2003b) proposed several levels of groupings. However, they could not reach a final classification of the whole domain due to the (apparent) lack of diagnostic innovations. It is nevertheless noteworthy that they came up with no "Eastern" Bantu, which is especially challenging, since all the lexical computations done so far agree on the existence of such a node! On the other hand, they proposed as being reasonably well-supported: (1) Western Bantu comprised of A, B, C, H40, K40, L10, L30, L40, parts of M60, D10–20-D30 and some H and K languages, based on two innovations, and itself incorporating (2) Forest Bantu (A, B, C, H and some D10–20–30 languages), (3) West Central Bantu (K10, K30–40, L20, L60, R20–30–40: R10?, parts of D10–20–30 and parts of H based on three innovations); and (4) Northeastern Savanna (that included E50, E60/E74a, F21–22, F33–34, J and G60)[1].

Although the classification of Nurse and Philippson (2003b) remains somewhat impressionistic, due to the unsystematic application of the criteria, the question of the use of morphosyntactic criteria for sub-grouping is attracting renewed attention at present, and several projects are under way and expected to give new results.

3.2 From the turn of the 21st century onwards

Since the beginning of the current century, phylogenetic methods borrowed from biology have been employed to classify languages. Numerous analogies between biological evolution and language evolution established by Pagel (2000), Atkinson and Gray (2005) and Pagel (2009), among others, have encouraged scholars to apply phylogenetic tools to the classification of languages. Languages, like species, evolve. As in biology, the similarities observed between taxa cannot be fortuitous, but might reveal a potential common ancestor.

The analogy between language evolution and species evolution was first proposed by Darwin (1859) in the *Origin of Species*. Ever since, the analogy has been developed. Pagel (2000), for example, has pointed out the fact that linguistic evolution and gene-sequence evolution share many features, such as vertical transmission. The differences observed between two or more organisms or languages can be explained by the length of time since they diverged from a common ancestor. Linguistic evolution and biological evolution are both composed of element units: nucleotide, codons, genes for genetics; and words, morphology and syntactic constructions for languages. Moreover, they both present mechanisms for replicating these units: transcription for genetics and learning and language acquisition for languages. Finally, there are horizontal transmissions of genetic information in the same way as words are borrowed horizontally between languages.

The methods used in phylogenetic reconstructions are of two main types, which can be labelled "phenetic" and "cladistic" respectively. The units compared are *taxa* in zoological or botanical systematics, which can be equated with languages or language families in linguistics. The elements of comparison are *characters*, i.e., elements susceptible to take certain values, e.g., presence or absence of wings in vertebrates, number of appendices among arthropods, etc., each value being termed a *state*. In linguistics, a character might be a unit of meaning, e.g., 'arm,' and the various states attested in Bantu languages would be (to simplify) reflexes of CB *-bókò, or *-kónò[2], or else the character might be phonological, e.g., "Bantu Spirantization" being either present or absent, morphological, e.g., presence or absence of cl. 12, etc.

The data can be presented either in the form of *trees*, with an identified vertex (*root*) and splits (often binary) at each node; each node is supposed to represent an intermediate ancestral language, and the whole tree straightforwardly represents the phylogeny of the entire group. However, this only works in cases where there has been little contact between the languages, which is a fairly unlikely assumption, certainly for Bantu languages. In order to better evidence contacts and lateral influences, a second method of presentation is preferable, namely *networks*, where links are shown between languages not necessarily sharing a unique common ancestor. In such cases, the phylogeny is by no means easy to establish.

Phenetic methods compare taxa by measuring the distance between them, computed over the number of shared character states, without distinguishing between innovations and retentions. They are thus quite similar to classical linguistic lexicostatistics discussed above. The best-known of these methods are UPGMA (Unweighted Pair-Group Method using Arithmetic Averages) (Sneath & Sokal 1973), Neighbour-joining (Saitou & Nei 1987) and NeighbourNet (Bryant & Moulton 2004).

Cladistic methods endeavor to distinguish between ancestral and derived states. A representative one is *parsimony*, implying that "trees with minimum changes are preferred" (Edwards & Cavalli-Sforza 1963). Taxa that share derived characters are grouped together. This method will produce various trees and the best possible tree (optimal tree), i.e., the most parsimonious, will be chosen. Parsimony methods are based on the idea that the most probable evolutionary pathway is the one that requires the smallest number of changes from some ancestral state.

Another family of methods is based on *maximum likelihood*, i.e., which parameters are the most likely to produce the observed data. One of the methods most frequently employed is *Bayesian inference*, which calculates the likelihood of given parameters by applying random fluctuations to the parameter set; this randomisation process is repeated a very large number of times (*Markov Chain Monte Carlo* or MCMC), producing a huge number of trees of which the most likely in terms of the data will be selected, normally more than one so that further methods must be used to find a consensus tree. We cannot do more than give the barest outline of these methods here. Interested readers should consult the relevant literature, perhaps starting with McMahon and McMahon (2005). Slightly more technical, but clear and up-to-date, is Dunn (2014). Several phylogenetic classifications of Bantu languages using those different methods have been proposed: Holden (2002), Holden *et al.* (2005), Holden and Gray (2006), Rexová *et al.* (2006), Currie *et al.* (2013), Bostoen *et al.* (2015), Grollemund *et al.* (2015), de Schryver *et al.* (2015). Note that all these studies, apart from the last three, wholly depend on data tabulated in CDBL. Bostoen *et al.* (2015) only cover the North-Western part of the domain, using data from Grollemund (2012), whereas de Schryver *et al.* (2015) focus on the Kikongo Language Cluster.

Holden (2002) proposed a phylogenetic classification of 73 Bantu languages using two Bantoid languages (Ejagham and Tiv) as outgroups. The languages were selected from those also listed in the Ethnographic Atlas of Murdock (1967), so as to enable further computation of cultural traits concerning the relationship of matrilineality and cattle husbandry (cf. Holden & Mace 2003). This probably explains the rather strange sample, with extremely few North-Western languages included, whereas some very closely related ones are selected, e.g., Lamba M54 and Lala M52 or Southern Sotho S33 and Tswana S31. In order to establish her classification, she applied to the data a maximum parsimony method. Her results show the divergence of six North-Western languages first. Then her tree shows a binary split opposing roughly a Western to an Eastern group; however these do not correspond to Guthrie's West (without North-West: zones B, C, H, K, L, R)/East (zones D, E, F, G, M, N, P, S) split; specifically her L languages are in the Eastern group, whereas several D languages are in Western. Due to the very small and skewed sample, not much confidence can be placed on this classification, which has in any case been superseded by the more complete ones to follow. In her conclusion, the author tentatively linked the Bantu expansion to the spread of farming, basing herself on archaeological dates – which admittedly provide at the very best only circumstantial evidence for farming (ceramics, iron-working, etc.).

Holden and Gray (2006) proposed a new phylogenetic classification of 93 Bantu languages, plus Ejagham and Tiv again. Although counting 20 more taxa than Holden (2002), the selection was still weak on North-Western languages – in any case, the languages were selected as in Holden (2002) from those also appearing in Murdock (1967) and with the same aims. This time, instead of Maximum Parsimony, the authors used a Bayesian method to construct a majority rule tree. The results showed the successive divergence of four main groups: West Bantu (zones A, B, C, H and some D), South-West Bantu (K and R), Central Bantu (some D, L, M40 and M50) and East Bantu (all the rest). Some of the nodes were very poorly supported, particularly East Bantu, comprising Eastern and Southern African languages. But in spite of the good support enjoyed by most of the other nodes, the authors pointed out that other published and unpublished analyses gave very different groupings and that most of the groups in their studies appeared in others as paraphyletic, i.e., not forming a meaningful genetic unit; so that they concluded "there is significant uncertainty regarding the shape of Bantu history that we cannot currently resolve" (Holden & Gray 2006: 22). They suggest, in apparent agreement with us, that "A 200-word vocabulary list would probably be preferable" (Holden & Gray 2006: 24). Also presented in the article is a network representation generated by the NeighbourNet method, and therefore distance-based. The network confirmed the previous groupings observed in the tree-like representation with Western, Southwest, Central and Eastern Bantu groups. According to Holden and Gray (2006), there would have been a rapid radiation among West Bantu languages, whereas some extensive borrowings would have been the case for the Eastern Bantu languages, where huge reticulations appear. It should be noted that the authors are impressed by their not finding any deep branching among their Western Bantu group. This might appear surprising since Western Bantu is close to the root of the tree, but this is an artifact of their not having included any of the A44–46, A50 and A60 languages (the Mbam languages). Furthermore, the details of the classification are unconvincing due to the small number of languages involved, e.g., it is out of the question that Yao P21 and Nyakyusa M31 should form a clade with the Ruvu G20 and G30 languages, as indicated in Holden and Gray (2006: 21, Figure 2.2). In fact, the authors themselves mention that Yao "has conflicting affiliations with both East and Southeast Bantu languages" (Holden & Gray 2006: 22). This is probably due to CDBL leaving out most of the N10 and P languages, where Yao's closest relatives are to be found. This small sample is clearly the source of the classification error, which does not appear in Currie *et al.* (2013) where the *entire* CDBL database is entered in the computation. Russell Gray (personal communication) readily admits this shortcoming.

Approximately contemporaneous with Holden and Gray (2006), but independent of it was Rexová *et al.* (2006). This work is the first phylogenetic study to try to include lexical as well as non-lexical data, i.e., the 19 phonological and 33 morphosyntactic criteria already treated statistically by Bastin (1980, 1983). The drawback of this choice is that the only word lists that could serve for the computation were those of languages for which grammatical data existed, which cut down the number to 87, although it must be acknowledged that – no doubt due to Bastin's expertise – the coverage was rather better than that of the two previous studies, since Nen A44 and Kpa (Bafia) A53 were this time included and appeared as expected at the root of the tree. No Bantoid languages were selected. A maximum parsimony method and a Bayesian method were followed. Whether

due to a more judicious sampling of the languages or to the inclusion of non-lexical characters – which appears unlikely since the tree based on the latter does not differ much
from the one based on lexical characters – they were able to remove the uncertainty about
primary branching within Bantu and confirmed the East-out-of-the-West model. They
refer approvingly to the classification and expansion model of Ehret (1998), which is all
the more significant since Bastin has long been known as a supporter of the East-next-to-
the-West model. The results show a Bantu expansion in five steps: with an initial radiation
from Cameroon (A languages), followed by an expansion through the equatorial rainforest (C and D languages); the third step corresponds to the main radiation in the south-east
part of the Democratic Republic of the Congo (D, L and M languages), followed by a
spread westward (K, R, H and B languages)[3] with finally a radiation south-eastward in the
Eastern area (J, F, E, G, N, P and S languages). The main merit of the study is to have set
out the main branches, which were by and large confirmed by further studies. At the level
of sub-grouping, the sample is much too small to give convincing results.

Two other studies (de Filippo *et al.* 2012, Li *et al.* 2014) did not propose new trees, but
did consider linguistic data in conjunction with genetic data and endeavored to establish
routes of expansion. Space considerations prevent us from further considering their contributions here.

The next true phylogenetic classification of Bantu languages is the one proposed by
Currie *et al.* (2013). This classification is based on the entire sample of languages of
Bastin *et al.* (1999). To build their tree, they used a Bayesian MCMC method. In order
to better understand the Bantu migrations, they proposed to reconstruct the locations of
ancestral Bantu societies and mapped the results. The expansion scenario resembles the
one proposed by Rexová *et al.* (2006) and the results show that from the Proto-Bantu
nucleus, the Bantu languages expanded south-eastward through the rainforest. From the
present-day DRC, a branch went south, whereas another branch moved east toward the
Great Lakes region.

The latest phylogenetic classification of the entire Bantu region is the one established
by Grollemund *et al.* (2015). With Bostoen *et al.* (2015), and de Schryver *et al.* (2015),
it is the first phylogenetic classification not entirely based on CDBL. Grollemund *et
al.* (2015) worked on 424 languages including 409 Bantu and 15 Bantoid languages
(Tivoid, Grassfields and Jarawan languages) and relied on a 100-word list, but not
strictly identical with the Swadesh-100 list. For each word, cognate sets were identified
on the basis of resemblances in shape rather than through a strict application of the comparative method. Finally, a likelihood model of lexical evolution that allows different
rates of evolution for the words studied and Bayesian inference of phylogeny (using
MCMC) was applied to the data. The Grollemund *et al.* (2015) classification divides
the Bantu area into five subgroups: North-Western, Central-Western, West-Western,
South-Western and Eastern. The Bantu ancestral homeland is located through a method
that reconstructs the probable location of each of the nodes of the phylogenetic tree.
The results suggest that the Bantu migration was coupled with the evolution of the
Central African rainforest, and the several climate-induced changes that occurred in
Central Africa during the Middle to Late Holocene, i.e., around 4000 BP and around
2500 BP (Vincens *et al.* 1994, Vincens *et al.* 1998, Maley 2001, Ngomanda *et al.* 2009,
Maley & Willis 2010). These phases of climate-induced forest crisis had an impact on
the Bantu expansion, which appears to be characterised by preferences for following
familiar savannah habitats. Ever since the pioneering work of Schwartz (1992), the
idea that a climate-induced rainforest crisis during the third millennium BP facilitated

the rapid spread of Bantu speech communities throughout Central Africa has gained growing agreement (Maley 2001, Oslisly 2001, Neumann *et al.* 2012a, Neumann *et al.* 2012b, Bostoen *et al.* 2013, Bostoen *et al.* 2015).

An important problem remains, however, for iron-working populations, presumably Bantu-speaking, are found in the East African Great Lakes region around the middle of the first millennium BC, i.e., more or less contemporaneously with the drying up of the rainforest; since a certain time must be assumed for these populations to have reached the Great Lakes area, the proposed chronology encounters difficulties here, as rightly pointed out by Ehret (2015). This is an area for further research. One should of course consider the apt reminder of Vansina (1980: 317): "if every specialist stays within his own discipline and does not worry unduly about the effects his findings might have on the question of Bantu expansion, less distortion will result."

4 CONCLUSION

The first comment that should be made about the various quantitative classifications presented above is that apart from Bastin (1980, 1983) and Rexová *et al.* (2006), they were all based exclusively on lexical data and a fairly restricted list at that, mostly basic vocabulary à la Swadesh. There is an obvious need to expand the number and types of character used. No study such as for instance that reported in Ringe *et al.* (2002) and Nakhleh *et al.* (2005), which used for Indo-European a lexical database of around 330 items (Swadesh-200 list plus meanings of cultural importance for ancient Indo-European languages), 22 phonological characters and 15 morphological characters, has ever been attempted for the whole of Bantu. It must be pointed out, however, that their database contained only 24 languages, whereas a Bantu database would have to be much larger to achieve reasonable representativity. We had considered including in this chapter a phylogenetic comparison of a number of phonological correspondences in East African Bantu languages. Considerations of space as well as methodological issues led us to postpone this project that we will pursue independently, preferably extending it to the whole Bantu field.

The study of more specialised cultural vocabulary can also be expected to bring results, certainly for sub-grouping (Grollemund & Hombert 2012), but even for classificatory issues of Bantu as a whole as explored by Bostoen (2005).

The second lesson to be learnt is that from a purely classificatory point of view, the various trees published over the last 15 years or so by and large agree in their results: the Mbam languages (Guthrie's A44-A46 + A60 minus A63) are the first to branch off, often but not always with Guthrie's A50. Bobe A31 also appears as an early branching off, but its apparent proximity to the previous group is perhaps an artifact – a qualitative overview of the respective vocabularies reveals no close relationship; whether other types of data as mentioned above would bring out this relationship remains to be tested. Then a series of successive branches of forest languages, then a considerable residuum corresponding rather well with Heine's Kongo-Zweig and Ehret's Savanna. So at this level of generality, phylogenetic studies only seem to confirm and quantitatively support earlier findings. But this is only part of the story, since the tree-like structure clearly appears only at a macro-level. A real advance introduced by phylogenetic methods is the network representation that allows visualising the amount of contact at work at the level of sub-grouping, thus pointing out those areas where more detailed and local studies will hopefully be undertaken (Nakhleh *et al.* 2005, Heggarty *et al.* 2010).

ACKNOWLEDGEMENTS

We want to thank Koen Bostoen, Brigitte Pakendorf and two anonymous reviewers whose sometimes-severe criticism helped us to better focus our contribution and hopefully make it more accessible. All remaining mistakes are obviously our own.

NOTES

1 A remark is in order here: since Guthrie's classification – even as completed by Maho – is purely referential and has no direct relationship to genetic groupings, the latter should never be mentioned under the zone labels, but under new names univocally referring to them, preferably geographic, since ethnic labels are liable to arouse some susceptibilities. This is the practice initiated by Christopher Ehret and his students, and followed by Nurse and Philippson in their publications. So "Sabaki" = Guthrie's E71-E73 + G40; "Rufiji-Ruvuma" = N13-N14 + P11-P14 + P20; "Kilombero" = G50 + P15; "Luhya" = E18 + E30 + E41, etc. These are fairly widely accepted by researchers working on East Africa, but the practice has unfortunately hardly caught on elsewhere.

2 We are specifying reflexes of CB, since ideally the comparison should be made only after a *complete* series of correspondences have been established with the Comparative Method. To the best of our knowledge, this has never been *consistently* applied to overall Bantu classifications, although Nurse and Philippson (1980) did it for 125 languages and dialects of Kenya, Uganda and most of Tanzania. The correspondence tables have been published in Nurse (1979) to which were added 17 languages of southern Tanzania (Nurse 1988) – contra Petzell and Hammarström (2013: 133, footnote 139).

3 This of course only applies to the three B languages in their sample, i.e., two B70 and one B80. It is well known, and Bastin herself has commented on the fact (cf. Bastin & Piron 1999), that B10 and part at least of B20 languages are much closer to A languages.

REFERENCES

Atkinson, Q. D. & R. D. Gray (2005) Curious Parallels and Curious Connections – Phylogenetic Thinking in Biology and Historical Linguistics. *Systematic Biology* 54(4): 513–526.

Bastin, Y. (1978) Les langues bantoues. In Barreteau, D. (ed.), *Inventaire des études linguistiques sur les pays d'Afrique noire d'expression française et de Madagascar*, 123–185. Paris: Conseil international de la langue française.

Bastin, Y. (1980) Statistique grammaticale et innovations en bantou. In Bouquiaux, L. (ed.), *L'expansion bantoue. Actes du Colloque International du CNRS, Viviers (France), 4–16 avril 1977*, 387–400. Paris: Société des Etudes Linguistiques et Anthropologiques.

Bastin, Y. (1983) Essai de classification de quatre-vingts langues bantoues par la statistique grammaticale. *Africana Linguistica* 9: 11–108.

Bastin, Y., A. Coupez & B. de Halleux (1983) Classification lexicostatistique des langues bantoues (214 relevés). *Bulletin des séances de l'Académie royale des sciences d'Outre-Mer* 37(2): 173–199.

Bastin, Y., A. Coupez & M. Mann (1999) *Continuity and Divergence in the Bantu Languages: Perspectives from a Lexicostatistic Study*. Tervuren: Royal Museum for Central Africa.

Bastin, Y. & P. Piron (1999) Classifications lexicostatistiques: bantou, bantou et bantoïde. De l'intérêt des "groupes flottants." In Hombert, J. M. & L. M. Hyman (eds.), *Bantu Historical Linguistics: Theoretical and Empirical Perspectives*, 149–164. Stanford: CSLI Publications.

Bennett, P. R. & J. P. Sterk (1977) South Central Niger-Congo: A Reclassification. *Studies in African Linguistics* 8(3): 241–273.

Bergsland, K. & H. Vogt (1962) On the Validity of Glottochronology. *Current Anthropology* 3(2): 115–153.

Bostoen, K. (2005) *Des mots et des pots en bantou. Une approche linguistique de l'histoire de la céramique en Afrique*. Frankfurt: Peter Lang.

Bostoen, K., B. Clist, C. Doumenge, R. Grollemund, J.-M. Hombert, J. Koni Muluwa & J. Maley (2015) Middle to Late Holocene Paleoclimatic Change and the Early Bantu Expansion in the Rain Forests of West Central-Africa. *Current Anthropology* 56(3): 354–384.

Bostoen, K., R. Grollemund & J. Koni Muluwa (2013) Climate-Induced Vegetation Dynamics and the Bantu Expansion: Evidence from Bantu Names for Pioneer Trees (Elaeis guineensis, Canarium schweinfurthii and Musanga cecropioides). *Comptes Rendus Geoscience* 345: 336–349.

Bryant, D. & V. Moulton (2004) Neighbor-net: An Agglomerative Method for the Construction of Phylogenetic Networks. *Molecular Biology and Evolution* 21(2): 255–265.

Cole, D. T. (1959) Doke's Classification of Bantu Languages. *African Studies* 18: 197–213.

Cole, D. T. (1961) *Contributions to the History of Bantu Linguistics*. Witwatersrand: Witwatersrand University Press.

Coupez, A. (1956) Application de la lexicostatistique au mongo et au rwanda. *Aequatoria* 19: 85–87.

Coupez, A., E. Evrard & J. Vansina (1975) Classification d'un échantillon de langues bantoues d'après la lexicostatistique. *Africana Linguistica* 6: 131–158.

Currie, T. E., A. Meade, M. Guillon & R. Mace (2013) Cultural Phylogeography of the Bantu Languages of Sub-Saharan Africa. *Proceedings of the Royal Society B (Biological Sciences)* 280(1762): 1–8.

Darwin, C. (1859) *On the Origin of Species*. London: John Murray.

de Filippo, C., K. Bostoen, M. Stoneking & B. Pakendorf (2012) Bringing Together Linguistic and Genetic Evidence to Test the Bantu Expansion. *Proceedings of the Royal Society B (Biological Sciences)* 279(1741): 3256–3263.

de Schryver, G.-M., R. Grollemund, S. Branford & K. Bostoen (2015) Introducing a State-of-the-Art Phylogenetic Classification of the Kikongo Language Cluster. *Africana Linguistica* 21: 87–162.

Doke, C. M. (1945) *Bantu: Modern Grammatical, Phonetical, and Lexicological Studies*. London: Percy Lund, Humphries & Co. for the International Institute of African Languages and Cultures.

Doke, C. M. (1954) *The Southern Bantu Languages*. London: Oxford University Press for the International African Institute (IAI).

Dunn, M. (2014) Language Phylogenies. In Bowern, C. & B. Evans (eds.), *The Routledge Handbook of Historical Linguistics*, 190–211. London: Routledge.

Dyen, I., A. T. James & J. W. T. Cole (1967) Language Divergence and Estimated Word Retention Rate. *Language* 43(1): 150–171.

Edwards, A. W. F. & L. L. Cavalli-Sforza (1963) The Reconstruction of Evolution. *Annals of Human Genetic* 27: 105–106.

Ehret, C. (1998) *An African Classical Age: Eastern and Southern Africa in World History, 1000 B.C. to A.D. 400.* Charlottesville; Oxford: University Press of Virginia; James Currey.

Ehret, C. (1999) Subclassifying Bantu: The Evidence of Stem Morpheme Innovations. In Hombert, J. M. & L. M. Hyman (eds.), *Bantu Historical Linguistics: Theoretical and Empirical Perspectives*, 43–147. Stanford: CSLI Publications.

Ehret, C. (2015) Bantu history: Big advance, although with a chronological contradiction. *PNAS* 112(44): 13428–13429.

Elugbe, B. & K. Williamson (1977) Reconstructing Nasals in Proto-Benue-Kwa. In Juilland, A. (ed.), *Linguistic studies offered to Joseph H. Greenberg on the occasion of his sixtieth birthday.* Saratoga: Anma Libri & Co.

Embleton, S. (1981) Lexicostatistical Tree Reconstruction Incorporating Borrowing. *Toronto Working Papers in Linguistics* 2(0): 93–122.

Fairbanks, G. H. (1955) A Note on Glottochronology. *International Journal of American Linguistics* 21(2): 116–120.

Flight, C. (1980) Malcolm Guthrie and the Reconstruction of Bantu Prehistory. *History in Africa* 7: 81–118.

Greenberg, J. H. (1949) Studies in African Linguistic Classification: III. The Position of Bantu. *Southwestern Journal of Anthropology* 5(4): 309–317.

Greenberg, J. H. (1963) *The Languages of Africa.* Bloomington: Indiana University.

Grollemund, R. (2012) *Nouvelles approches en classification: Application aux langues bantu du Nord-Ouest.* Lyon: Université Lumière Lyon 2, thèse de doctorat.

Grollemund, R., S. Branford, K. Bostoen, A. Meade, C. Venditti & M. Pagel (2015) Bantu Expansion Shows That Habitat Alters the Route and Pace of Human Dispersals. *Proceedings of the National Academy of Sciences of the United States of America* 112(43): 13296–13301.

Grollemund, R. & J.-M. Hombert (2012) Use of Plant Names for the Classification of the Bantu Languages of Gabon. In Connell, B. & N. Rolle (eds.), *Selected Proceedings of the 40th Annual Conference on African Linguistics*, 150–163. Toronto: Cascadilla Proceedings Project.

Guthrie, M. (1948) *Bantu Word Division: A New Study of an Old Problem.* London; New York: Published for the International African Institute by the Oxford University Press.

Guthrie, M. (1962) Bantu Origins: A Tentative New Hypothesis. *Journal of African Languages* 1(1): 9–21.

Guthrie, M. (1967) *Comparative Bantu: An Introduction to the Comparative Linguistics and Prehistory of the Bantu Languages. Volume 1: The Comparative Linguistics of the Bantu Languages.* London: Gregg International.

Guthrie, M. (1970a) *Comparative Bantu: An Introduction to the Comparative Linguistics and Prehistory of the Bantu Languages. Volume 3: A Catalogue of Common Bantu with Commentary.* London: Gregg.

Guthrie, M. (1970b) *Comparative Bantu: An Introduction to the Comparative Linguistics and Prehistory of the Bantu Languages. Volume 4: A Catalogue of Common Bantu with Commentary.* London: Gregg.

Guthrie, M. (1971) *Comparative Bantu: An Introduction to the Comparative Linguistics and Prehistory of the Bantu Languages. Volume 2: Bantu Prehistory, Inventory and Indexes*. London: Gregg International.

Heggarty, P., W. Maguire & A. McMahon (2010) Splits or Waves? Trees or Webs? How Divergence Measures and Network Analysis Can Unravel Language Histories. *Philosophical Transactions of the Royal Society B: Biological Sciences* 365(1559): 3829–3843.

Heine, B. (1973) Zur genetischen Gliederung der Bantu-Sprachen. *Afrika und Übersee* 56(3): 164–185.

Heine, B., H. Hoff & R. Voßen (1977) Neuere Ergebnisse zur Territorialgeschichte der Bantu. In Möhlig, W. J. G., F. Rottland & B. Heine (eds.), *Zur Sprachgeschichte und Ethnohistorie in Afrika*, 52–72. Berlin: Dietrich Reimer.

Henrici, A. (1973) Numerical Classification of Bantu Languages. *African Language Studies* 14: 82–104.

Hinnebusch, T. J. (1999) Contact and Lexicostatistics in Comparative Bantu Studies. In Hombert, J. M. & L. M. Hyman (eds.), *Bantu Historical Linguistics: Theoretical and Empirical Perspectives*, 173–205. Stanford: CSLI Publications.

Hinnebusch, T. J., D. Nurse & M. J. Mould (1981) *Studies in the Classification of Eastern Bantu Languages*. Hamburg: Helmut Buske.

Holden, C. J. (2002) Bantu Language Trees Reflect the Spread of Farming Across Sub-Saharan Africa: A Maximum-Parsimony Analysis. *Proceedings of the Royal Society of London Series B-Biological Sciences* 269(1493): 793–799.

Holden, C. J. & R. D. Gray (2006) Rapid Radiation, Borrowing and Dialect Continua in the Bantu Languages. In Forster, P. & C. Renfrew (eds.), *Phylogenetic Methods and the Prehistory of Languages*, 19–31. Cambridge: The MacDonald Institute for Archaeological Research.

Holden, C. J. & R. Mace (2003) Spread of Cattle Led to the Loss of Matrilineal Descent in Africa: A Coevolutionary Analysis. *Proceedings of the Royal Society B-Biological Sciences* 270(1532): 2425–2433.

Holden, C. J., A. Meade & M. Pagel (2005) Comparison of Maximum Parsimony and Bayesian Bantu Language Trees. In Mace, R., C. J. Holden & S. Shennan (eds.), *The Evolution of Cultural Diversity: A Phylogenetic Approach*, 53–65. London: UCL Press.

Holman, E. W., S. Wichmann, C. H. Brown, V. Velupillai, A. Müller, P. Brown & D. Bakker (2008) Explorations in Automated Language Classification. *Folia Linguistica* 42: 331–354.

Janson, T. (1991–1992) Southern Bantu and Makua. *Sprache und Geschichte in Afrika* 12–13: 63–106.

Johnston, H. H. (1919) *A Comparative Study of the Bantu and Semi-Bantu Languages*. Oxford: Clarendon Press.

Klieman, K. A. (1997) *Hunters and Farmers of the Western Equatorial Rainforest: Economy and Society, 3000 BC to AD 1880*. Los Angeles: University of California, PhD dissertation.

Lewis, M. P. (2009) *Ethnologue: Languages of the World*, 16th edition. (Available online at www.ethnologue.com/).

Li, S., C. Schlebusch & M. Jakobsson (2014) Genetic Variation Reveals Large-Scale Population Expansion and Migration During the Expansion of Bantu-Speaking Peoples. *Proceedings of the Royal Society B (Biological Sciences)* 281: 20141448.

Maho, J. F. (2003) A Classification of the Bantu Languages: An Update of Guthrie's Referential System. In Nurse, D. & G. Philippson (eds.), *The Bantu Languages*, 639–651. London; New York: Routledge.

Maho, J. F. (2009) NUGL Online: The Online Version of the New Updated Guthrie List, a Referential Classification of the Bantu Languages (June 4, 2009) (Available online at http://goto.glocalnet.net/mahopapers/nuglonline.pdf, Accessed December 13, 2010).

Maley, J. (2001) La destruction catastrophique des forêts d'Afrique centrale survenue il y a environ 2500 ans exerce encore une influence majeure sur la répartition actuelle des formations végétales. *Syst. & Geogr. Plants* 71: 777–796.

Maley, J. & K. J. Willis (2010) Did a Savanna Corridor Open Up Across the Central African Forests 2500 Years Ago? *Letter CoForChange* 2: 5p.

Mann, M. (1999) A Note on Historical and Geographical Relations Among the Bantu Languages. In Hombert, J. M. & L. M. Hyman (eds.), *Bantu Historical Linguistics: Theoretical and Empirical Perspectives*, 165–171. Stanford: CSLI Publications.

McMahon, A. & R. McMahon (2005) *Language Classification by Numbers*. Oxford: Oxford University Press.

Meeussen, A. E. (1956) Lexico-statistiek van het bantoe: Bobangi en Zulu. *Kongo-Overzee* 22(1): 86–89.

Möhlig, W. J. G. (1974) *Die Stellung der Bergdialekte im Osten des Mt. Kenya. Ein Beitrag zur Sprachgliederung im Bantu*. Berlin: Dietrich Reimer.

Möhlig, W. J. G. (1976) Guthries Beitrag zur Bantuistik aus heutiger Sicht. *Anthropos: internationale Zeitschrift für Völker- und Sprachenkunde* 71: 673–715.

Möhlig, W. J. G. (1977) Zur frühen Siedlungsgeschichte der Savannen-Bantu aus lauthistorischer Sicht. In Möhlig, W., F. Rottland & B. Heine (eds.), *Zur Sprachgeschichte und Ethnohistorie in Afrika*, 166–193. Berlin: D. Reimer.

Möhlig, W. J. G. (1978) Versuch einer historischen Gleiderung der nordöstlichen Bantusprachen auf lautvergleichender Grundlage. *Afrika und Übersee* 51(3–4): 175–189.

Möhlig, W. J. G. (1981) Stratification in the History of the Bantu Languages. *Sprache und Geschichte in Afrika*: 251–317.

Möhlig, W. J. G. (1984–1985) The Swahili Dialects of Kenya in Relation to Mijikenda and to the Bantu Idioms of the Tana Valley. *Sprache und Geschichte in Afrika* 6: 253–308.

Murdock, G. P. (1967) *Ethnographic Atlas*. Pittsburgh: University of Pittsburgh Press.

Nakhleh, L., D. Ringe & T. Warnow (2005) Perfect Phylogenetic Networks: A New Methodology for Reconstructing the Evolutionary History of Natural Languages. *Language, Journal of the Linguistic Society of America* 81(2): 382–420.

Neumann, K., K. Bostoen, A. Höhn, S. Kahlheber, A. Ngomanda & B. Tchiengué (2012a) First Farmers in the Central African Rainforest: A View from Southern Cameroon. *Quaternary International* 249(6): 53–62.

Neumann, K., M. K. H. Eggert, R. Oslisly, B. Clist, T. Denham, P. de Maret, S. Ozainne, E. Hildebrand, K. Bostoen, U. Salzmann, D. Schwartz, B. Eichhorn, B. Tchiengué & A. Höhn (2012b) Comment on "Intensifying Weathering and Land Use in Iron Age Central Africa." *Science* 337(6098): 1040.

Ngomanda, A., A. Chepstow-Lusty, M. v. Makaya, C. Favier, P. Schevin, J. Maley, M. Fontugne, R. Oslisly & D. Jolly (2009) Western Equatorial African Forest-Savanna Mosaics: A Legacy of Late Holocene Climatic Change? *Climate of the Past* 5: 647–659.

Nurse, D. (1979) *Classification of the Chaga Dialects*. Hamburg: Helmut Buske.

Nurse, D. (1988) The Diachronic Background to the Language Communities of Southwestern Tanzania. *Sprache und Geschichte in Afrika* 9: 15–115.

Nurse, D. (1997) The Contributions of Linguistics to the Study of History in Africa. *Journal of African History* 38(3): 359–391.

Nurse, D. (2000) *Inheritance, Contact and Change in Two East African Languages.* Cologne: Rüdiger Köppe.

Nurse, D. & T. J. Hinnebusch (1993) *Swahili and Sabaki: A Linguistic History.* Berkeley: University of California Press.

Nurse, D. & G. Philippson (1980) The Bantu Languages of East Africa: A Lexicostatistical Survey. In Polomé, E. C. & C. P. Hill (eds.), *Language in Tanzania*, 26–67. London: Oxford University Press for the International African Institute.

Nurse, D. & G. Philippson (2003a) Introduction. In Nurse, D. & G. Philippson (eds.), *The Bantu Languages*, 1–13. London; New York: Routledge.

Nurse, D. & G. Philippson (2003b) Towards a Historical Classification of the Bantu Languages. In Nurse, D. & G. Philippson (eds.), *The Bantu Languages*, 164–181. London; New York: Routledge.

Olmsted, D. L. (1957) Three Tests of Glottochronological Theory. *American Anthropologist* 59(5): 839–842.

Oslisly, R. (2001) The History of the Human Settlement in the Middle Ogoué Valley Gabon. Implications for the Environment. In Weber, W., L. J. T. White, A. Vedder & L. Naughton-Treves (eds.), *African Rain Forest Ecology and Conservation. An Interdisciplinary Perspective*, 101–118. New Haven: Yale University Press.

Pagel, M. (2000) Maximum-Likelihood Models for Glottochronology and for Reconstructing Linguistic Phylogenies. In Renfrew, C., A. P. McMahon & L. Trask (eds.), *Time Depth in Historical Linguistics*, 189–207. Cambridge: McDonald Institute for Archaeological Research.

Pagel, M. (2009) Human Language as a Culturally Transmitted Replicator. *Nature Reviews Genetics* 10: 405–415.

Pakendorf, B., K. Bostoen & C. de Filippo (2011) Molecular Perspectives on the Bantu Expansion: A Synthesis. *Language Dynamics and Change* 1: 50–88.

Petzell, M. & H. Hammarström (2013) Grammatical and Lexical Comparison of the Greater Ruvu Bantu Languages. *Nordic Journal of African Studies* 22(3): 129–157.

Philippson, G. (1984) *"Gens des bananeraies": Contribution linguistique à l'histoire culturelle des Chaga du Kilimanjaro.* Paris: Editions Recherches sur les Civilisations.

Piron, P. (1995) Identification lexicostatistique des groupes bantoïdes stables. *Journal of West African Languages* 25(2): 3–39.

Rexová, K., Y. Bastin & D. Frynta (2006) Cladistic Analysis of Bantu Languages: A New Tree Based on Combined Lexical and Grammatical Data. *Naturwissenschaften* 93(4): 189–194.

Ringe, D., T. Warnow & A. Taylor (2002) Indo-European and Computational Cladistics. *Transactions of the Philological Society* 100(1): 59–129.

Saitou, N. & M. Nei (1987) The Neighbor-Joining Method: A New Method for Reconstructing Phylogenetic Trees. *Molecular Biology and Evolution* 4(4): 406–425.

Schadeberg, T. C. (1999) Kathupa's Law in Makhuwa. In Hombert, J. M. & L. M. Hyman (eds.), *Bantu Historical Linguistics: Theoretical and Empirical Perspectives*, 379–394. Stanford: CSLI Publications.

Schwartz, D. (1992) Assèchement climatique vers 3000 B.P. et expansion bantu en Afrique centrale atlantique: quelques réflexions. *Bulletin de la Societé Géologique de France* 163(3): 353–361.

Sneath, P. H. A. & R. R. Sokal (1973) *Numerical Taxonomy – The Principles and Practice of Numerical Classification*. San Francisco: W. H. Freeman.

Swadesh, M. (1952) Lexico-Statistic Dating of Prehistoric Ethnic Contacts: With Special Reference to North American Indians and Eskimos. *Proceedings of the American Philosophical Society* 96(4): 452–463.

Swadesh, M. (1954) Time Depths of American Linguistic Groupings. *American Anthropologist* 56(3): 361–377.

Van Bulck, G. (1952) Langues bantou. In Meillet, A. & M. Cohen (eds.), *Les langues du monde, Volume 2*, 891–892. Paris: CNRS.

Vansina, J. (1980) Bantu in the Crystal Ball 2. *History in Africa* 7: 293–325.

Vansina, J. (1995) New Linguistic Evidence and the Bantu Expansion. *Journal of African History* 36(2): 173–195.

Vincens, A., G. Buchet, H. Elenga, M. Fournier, L. Martin, C. de Namur, D. Schwartz, M. Servant & D. Wirrmann (1994) Changement majeur de la végétation du lac Sinnda (vallée du Niari, Sud Congo) consécutif à un l'assèchement climatique holocène supérieur: apport de la palynologie. *Comptes Rendus de l'Academie des Sciences, Series IIA, Earth and Planetary Science* 318: 1521–1526.

Vincens, A., D. Schwartz, J. Bertaux, H. Elenga & C. de Namur (1998) Late Holocene Climatic Changes in Western Equatorial Africa Inferred from Pollen from Lake Sinnda, Southern Congo. *Quaternary Research* 50: 34–45.

Westermann, D. (1927) *Die Westlichen Sudansprachen und ihre Beziehungen zum Bantu*. Berlin: de Gruyter & Co.

Wiesmüller, B. (1997) Möglichkeiten der interdisziplinären Zusammenarbeit von Archäologie und Linguistik am Beispiel der frühen Eisenzeit in Afrika. In Klein-Arendt, R. (ed.), *Traditionelles Eisenhandwerk in Afrika: geschichtliche Rolle und wirtschaftliche Bedeutung aus multidisziplinärer Sicht*, 55–90. Cologne: Heinrich-Barth-Institut für Archäologie und Geschichte Afrikas.

Williamson, K. (1989) Niger-Congo Overview. In Bendor-Samuel, J. & R. L. Hartell (eds.), *The Niger-Congo Languages: A Classification and Description of Africa's Largest Language Family*, 3–45. Lanham: University Press of America.

CHAPTER 12

LANGUAGE CONTACT

Maarten Mous

1 INTRODUCTION

The Bantu languages cover a large part of Africa and have spread thousands of years ago from Cameroon to the southern tip of the continent. Over a long period of time, this large group of relatively closely related languages have been in contact with each other and with languages of different descent. It is likely that they have absorbed elements of pre-existing languages in their initial spread, but there is little linguistic evidence for this process. This chapter concentrates on contact-induced change. The first edition of the current volume (Nurse & Philippson 2003) contained a chapter on contact languages in the Bantu area (Mufwene 2003). To complement that chapter, less attention is paid to contact languages (Section 3) and more to contact-induced change (Section 4). First, a short overview of where Bantu languages are in contact with non-Bantu languages, and the methodological issues in reconstructing history from contact-induced language change, are presented (Section 2).

2 BANTU LANGUAGES IN CONTACT

2.1 Panorama of Bantu in contact with non-Bantu

The Bantu languages are in contact with several other African language families, mainly at the edges of the Bantu-speaking area. In Cameroon, Bantu is in contact with its closest relatives from the Bantoid branch of Benue-Congo, but this contact is relatively difficult to observe and to examine. There is no direct contact between (Narrow) Bantu and the Chadic languages spoken in northern Cameroon. In the northern Congo, there is direct contact with the Ubangi and Central Sudanic languages, and this has led to changes in the phonologies of those Bantu languages including the introduction of labial-velar consonants. The Bantu languages that are in close contact with Central Sudanic in north-eastern Congo show various uncommon Bantu features and many no longer have a functioning noun class system: Komo D23, Bira D32, Bila D311, Bhele D31, Humu-Amba D22 (Kutsch Lojenga 2003: 452). It is likely that language contact played a role in this development, but the details are still unclear. In East Africa, there are and have been contacts to various degrees with Nilotic and Cushitic languages. Dimmendaal (1995, 2001, 2011: 188–196) discusses several cases of structural influence of Bantu on Nilotic, but not the other way around. The lexicon of many East African Bantu languages shows influence of early contacts with Cushitic languages whose speech communities often have vanished (Nurse 1988, Nurse & Hinnebusch 1993, Ehret 2001).

An interesting case of remarkable lack of Cushitic influence is provided by the story of the Mushunguli, whose language variety is known as Mushungulu G311. The Mushunguli kept their Zigula G31 language from Tanzania, when they were transported in the course of the 19th century to the Juba valley in Somalia to work there in the banana plantations (Declich 2002). Some reunited with their forefathers as refugees of the civil war in Somalia, and their Zigula G31 seems to be indistinguishable; others settled in the USA. In 2004, I briefly visited Handeni in Zigula area. This town is near the former Mkuyu refugee camp with many Mushunguli refugees. I interviewed some of these refugees. They claim that there is no difference between the way they speak Zigula and the Zigula of those who never left the area; the Zigula speakers present agreed to this. Noting down some expressions, I had the same impression when comparing this to my Zigula field notes from 1994. A comparison of the speech of the different communities of Mushunguli refugees could establish whether there was Somali influence and how fast that disappeared in the Zigula environment. Current studies on the speech of Mushunguli in the USA reveal remarkably little Somali influence (Hout 2012, Williams 2012, Barlew 2013, Tse 2015a, b, Odden n.d.). Nevertheless, Tse (2013) reports on an interesting case of transfer and subsequent loss of a Somali sound: k^h in Ziglua was reinterpreted as a uvular stop or fricative, q or χ, in Mushunguli, but once the Mushunguli refugees were back in Zigula area, this uvular was quickly replaced by h.

Malagasy has been in contact with Bantu languages, but its impact on Bantu seems limited. It is likely that Malagasy and Comorian have influenced each other in lexicon and phonology (Adelaar 2012).

In Southern Africa, Bantu languages have been and still are in contact with the various languages from the different Khoisan families. The salient feature of clicks in the inventory of the Nguni S40 languages is due to Khoisan influence and they entered the languages partly through the manipulation of words by sound replacement in surrogate languages, such as the famous in-law register *hlonipha* with subsequent normalisation into the non-marked vocabulary (Van Rooyen 1968, Herbert 1990b, 1990a). In the area of upstream Zambezi too, a number of Bantu languages contain clicks due to borrowing from Khoisan languages (Gunnink *et al.* 2015, Pakendorf *et al.* 2017); the enrichment of their phonology with clicks enables these languages to use them for sound symbolic functions, such as mimicry of noise (Bostoen & Sands 2012: 129).

The Bantu languages have also been in contact with various European languages, especially from the colonial times onwards. The influence of Afrikaans on the lexicon of Southern Bantu languages is well documented. Owen Lloyd (1955), quoted in Branford and Claughton (2002: 203), provides some 300 Afrikaans words in Xhosa S41 in the expected areas of religion, law, army and dress. A number of these are no longer in use. Similar observations can be made for the influence of English, French, Portuguese, German and Spanish on the lexicon of Bantu languages, and Latin and Arabic, specifically in religious vocabulary. Many instances of influence from non-African languages on Bantu languages will be discussed in Section 4.

2.2 Methodological issues

It requires careful analysis to detect and prove contact-induced change among closely related languages, such as the Bantu languages. The borrowed forms will not stick out as being different from inherited forms and careful examination of the regularity of sound change is needed. Transfer from one Bantu language to another can be difficult to detect when adjustments to structures of the receiving language are automatically made, especially when the genetic relatedness of forms is recognised by speakers.

Mouguiama-Daouda (2005) provides detailed instructions on how to go about recognising borrowing and how to distinguish inherent irregularities in phonological reconstruction from those as a consequence of external influence. He evokes the concept of a set of virtual reconstructions for cultural vocabulary by reversing established sound changes and thus establishing chronological order of transfer of certain roots.

Once that is done carefully, it is possible to distinguish layers of historical stages of language contact. A number of such studies have been dedicated to the Bantu languages in Southern Africa that have been influenced by the languages belonging the different Khoisan language families. The article of Möhlig (2007) on the history of Manyo K332 speakers, on the basis of a careful analysis of the various changes in the language through contact, is exemplary. He shows that Manyo developed terminology for aquatic animals reflecting acquired knowledge from an earlier hunting community. Manyo also shows evidence of having absorbed a different language containing clicks while its speakers shifted from a hunting economy to a fishing economy. Gunnink *et al.* (2015) is a similar study for an adjacent area, in which the authors show different intense contact periods with Khoisan Ju, followed by a period of even more intense contact involving bilingualism with Khoisan Khwe, leading to structural influences as well. The outstanding nature of clicks rendered them useful for expressing identity, which played a role in the Khoisan influence as is suggested by the way the transferred words are integrated into the noun class system with their appropriate noun class prefix (see Section 4.3 below). Careful studies like these ones can reveal separable strata of language contact in the language's history referred to as stratification or stratigraphy (Möhlig 1981, Nurse & Masele 2003).

As a result of the Bantu Expansion, Bantu languages are now spoken in a huge part of Africa. It is likely that there were other languages in that area before the arrival of the Bantu languages. Linguistic evidence for such earlier speech communities is mainly available in the border areas. In addition, it has been proposed that remnants of earlier pygmy languages can be found in common specialised non-Bantu vocabulary for flora and fauna (Bahuchet 1992, 2012). Methodologically, it is questionable to assume substrate influence from an unattested language group.

A further complication for the recognition of Bantu-internal borrowing is the rampant purism in dictionaries. Mafela (2010: 695) provides insightful detailed examples:

[T]he word *mukhukhu*, which appears to have originated among the Sotho people, specifically the Basotho ba Leboa. This word has been adopted by all South African indigenous languages. It refers to a building made of corrugated iron, or a temporary shelter; the door is so low that when people want to enter, they have to crouch (*khukhuna*). This structure occurs mostly in urban areas where people of all indigenous language groups coexist. As the word is in popular usage, one would expect to find *mukhukhu* in all African language dictionaries. However, it does not form part of the lexicon in any of the checked Tshivenda (S21) dictionaries. Instead, the word *mushasha* has been included. Although, according to Van Warmelo (1989), both *mukhukhu* and *mushasha* have the same meaning, *mushasha* is a temporary shelter of branches and grass, meant for travellers, whereas *mukhukhu* is a temporary shelter made of corrugated iron. This suggests that *mukhukhu* should be part of the words in an indigenous African language dictionary. In addition, the meaning of *mukhukhu* can also be linked to a type of dance practised by almost all indigenous language speakers in this country. This dance is performed by members of the Zion Christian Church, a church not restricted to a particular ethnic group, but one which embraces all South

Africans as well as foreigners from neighbouring countries. This word has been accepted into the spoken indigenous languages in South Africa. It is therefore surprising to find that it is not included in the dictionaries of some of these languages.

Such purism comes from research based on elicitation and meta-conversation on language use, which strengthens awareness of origin of the terms used and the strong wish to provide the "real" term. Ngonyani (2003: 5), for example, remarks that "speakers provided examples that show very little influence of Kiswahili. The sample text, however, illustrates heavy influence of Kiswahili in everyday Chingoni."

In recognising language change we need the abstraction of "language," but in order to understand the various routes of influence we need to pay attention to the complexity of the linguistic landscape for an individual and a community. This involves much more than multilingualism. We have to realise that several kinds of special-purpose "languages" exist in most communities, including the in-law languages that have been mentioned for South Africa, also attested in East Africa in the father-in-law respect register of the Nyakyusa M31 (Kolbusa 2000). During initiation too, such surrogate languages were learnt (Storch 2011: 75–80). We can add to this linguistic landscape the youth languages that pop up in many African cities and in rural areas too (cf. Kiessling & Mous 2004, Nassenstein & Hollington 2015), as discussed in Section 3.1 below.

3 THE EFFECTS OF LANGUAGE CONTACT

The extreme effects of language contact leading to languages or language areas that are characterised by these effects are briefly discussed in this section. I first discuss individual languages, which are designated by the rather vague term "contact languages" that is intended to be neutral (Section 3.1). I refer the reader to Mufwene (2003) for an overview of "contact languages" in the Bantu domain. Here I add some contact languages that are only briefly discussed. Convergence areas are considered subsequently (Section 3.2). I use the term "convergence" for the result of extensive contact involving mutual interference between all languages spoken in a given area and leading to a greater similarity between them. The term convergence is used here to cover a combination of contact-induced changes and as a characterisation of the end result, rather than the individual separate processes themselves having led to this situation (Section 4).

3.1 Contact languages

There is quite a long list of languages that are candidates to be included in this section. I only devote short sections to the mixed language Ma'á/Mbugu, to Lingala as an instance of a language originating from trade, to Fanagalo as an example of a language that thrived through explicit language policy for communication, and finally to the youth languages. Other interesting contact languages include Nubi, the Arabic creole with Bantu lexis originating in a community of Sudanese soldiers (Heine 1982, Owens 1985, 1991, Kaye & Tosco 1993, Luffin 2005, Wellens 2005, de Smedt 2011). Pidgin A70, based on Ewondo A72 and related Bantu languages of the A70 group in Cameroon used by truck drivers, appears in many lists of pidgins and creoles, but all go back to Alexandre (1963, 1964), who provides little linguistic information. Another contact variety that received its own name and due attention is Town Bemba (Spitulnik 1998). Kituba or Kongo ya Leta H10A is a variety of Kongo that is used for wider communication (Fehderau 1967, Mufwene 1988, Samarin 1990–1991, Mufwene 1994, 2013, Samarin 2013).

3.1.1 Ma'á/Mbugu

The most well-known instance of Cushitic-Bantu contact language is the famous case of the mixed language Ma'á/Mbugu G221. I have argued elsewhere that the mixed language is in fact a parallel lexicon to a variety of Pare G22, also known as Asu, which arose partly through deliberate creation after a regretted near-complete language shift from a Cushitic language to that Pare dialect (Mous 2003). The speakers of the mixed language actually speak two languages or, better, one language with two lexicons. For most of the lexicon they have a separate word form that feels different enough from the original Pare and that is taken from a variety of non-Bantu languages or from Pare, but then manipulated, for example, by truncation and addition of é. This double lexicon shares one identical meaning and morphosyntactic value (noun class, verb extension) for each of its members. The grammar is completely shared, and Pare. Speakers choose between using all roots from the normal or all from the extra (mixed) lexicon. Natural texts in mixed Ma'á/Mbugu have more than 95% of their roots (including many function words) from a source that is deviant from Pare, while Pare provides all grammatical affixes. The core of this deviant lexicon is the leftover of the original Cushitic language that was put at a par to the new dominant Pare language. Thomason (1997) has instead argued for explanation by extreme and prolonged Bantu influence on a Cushitic language. Given that the extensive morphological nominal and verbal system is completely Pare, I consider Ma'á/Mbugu to be a Bantu language after language shift and an unsuccessful attempt at returning to the original language, or at least one that reflects their different identity and culture as "cattle people." The difference between her analysis and mine depends, in my view, on what to consider language shift or not.

3.1.2 Lingala

The development of new trade routes attracts many people from different linguistic backgrounds who shift to the language of the first traders on the route. Massive numbers of second-language speakers leave their influence on that language. Lingala C30B must have started this way (Samarin 1990–1991, Motingea Mangulu 1996, Motingea Mangulu & Mwamakasa Bonzoi 2008, Samarin 2013) and has subsequently become the language of the cities, of the music, and of the western half of the DRC. The success of Lingala as a contact language resulted in formal instruction in the language and in attempts to engineer the language by re-bantuising the reduced noun class system (Meeuwis 2009). In other contact languages, any instruction was aimed at simplification rather than at complexification. This is the case for Fanagalo S40A.

3.1.3 Fanagalo

Fanagalo is a master-servant pidgin of southern Africa with a strong Zulu S42 component in its lexicon. The few grammatical morphemes are Zulu, but it also shows English/Afrikaans syntactic features and some lexical items from those Western languages too (Mesthrie 1989, Kaltenbrunner 1996). Adendorff (2002) sees an origin in early missionary master-servant talk, but this is at odds with the fact that missionary attitudes towards Zulu were to learn the language proficiently (Gilmour 2006: 131, 133). The language was used by the middlemen on the plantations in Natal around 1860 and in the mines near Kimberley and Johannesburg from the 1870s onwards, but also in households. The language is Zulu-based, but it is clearly a stripped-down version of Zulu. It is not an attempt to speak Zulu, but one to use enough Zulu to be understood by Zulu speakers while resorting to a minimum of grammatical affixes: -*ile* past tense, -*iw*- passive, -*is*- causative and *ma*- nominal plural. Word order copies English and Afrikaans with modifiers preceding the head and question words placed sentence-initially. For subject marking independent pronouns (Zulu forms) are used as in English and Afrikaans rather than prefixes on the verb as in Zulu. Nouns are often preceded by an invariant article *lo* but interestingly also where a definite article in English or Afrikaans is not expected, for example, *mina dlala lo futbol manje* /1SG play ART football now/ 'I am now playing football' (Kaltenbrunner 1996: 77). Fanagalo has no agreement. The pidgin is conventionalised and the various instruction books and classes in the mines entail prescriptive standardisation, which is rare for a pidgin. This is not foreigner talk that is copied by the "masters," but a choice to use a simple form of Zulu for communication in a context where the "masters" want to keep their language for themselves and impose communication in a simplified version of the other's language that they can learn. The English- and Afrikaans-speaking "masters" had neither the intention nor the capacity to learn proper Zulu. It does not represent the ideal of the mineworkers. Yet the language has gained popularity on both sides and is still used, mainly by former miners.

A very similar situation existed in the use of Swahili G42 in the East African Army, the King's African Riffles (KAR pronounced as [kea]), where a simplified form of Swahili was used and instructed (Mutonya & Parsons 2004). Such simplified Swahili was used by colonial white settlers in Kenya at the time (Vitale 1980).

3.1.4 Sheng, tsotsitaals, youth languages

Urbanisation gave rise to particular youth slang varieties of urban languages. These youth languages (or stylects) show a perplexing ingenuity and innovative power, expressing an anti-identity, but often developing into vehicles for identity linked to urbanity and modernity. The phenomenon has developed into a booming research domain (cf. Kiessling & Mous 2004, Nassenstein & Hollington 2015, Hurst 2017). The Swahili-based Sheng from Nairobi and Kenya has been discussed in many PhD dissertations and articles, and the various forms of what Mesthrie (2008) calls "*tsotsitaals*" thrive in South Africa (Hurst & Mesthrie 2013, Gunnink 2014, Hurst & Buthelezi 2014, Mesthrie 2014).

3.2 Convergence

There are regions in the Bantu area where there has been intense internal contact leading to convergence zones. One such area is the coast of Mozambique where Swahili G42 used to be the dominant language and culture leaving a string of varieties that have some semblance of Swahili but are heavily influenced by the various languages of the hinterland, which were dominant for many of the bilinguals. Schadeberg (1994, 1996) shows how new lexical forms of Mwani G403, a variety of Swahili, calque the (cognitive) structure of the other, dominant language of the community, Makonde P23. For example, Swahili *uso* 'face' is replaced by *kumaso*, literally 'at the eyes' which is a calque from Makonde *ku-meho* 'at the eyes,' although it is not a straightforward calque because in Swahili *uso* is not 'eyes'; in this respect too, the speakers follow the Makonde model.

There are many more instances in the Bantu area of intense Bantu-internal contact. A fascinating topic for research is to compare the fates of the languages of the various Nguni groups that moved over large distances after the internal Zulu power struggle known as *mfecane* in the first half of the 19th century. The people who now speak Ngoni N12 in Tanzania speak a variety of a local Bantu language, but with a complex situation regarding ethnic identity: "people sought to identify themselves with the 'True Ngoni,' they labelled their language Ngoni and the same time they were resisting Ngoni domination in a rather subtle way. Indeed this is the present-day Tanzanian Ngoni" (Ngonyani 2001: 344). In the initial stages, the original Nguni of the invaders was influenced by the languages of the youth and women whom they assimilated, but it was still close to Zulu S42 and remained the court language of the ones in power. In Malawi, some Tumbuka N21 speakers claim that what they speak is Ngoni N121 to express their wish to associate with the Ngoni (Kishindo 2002: 206–207, 211–212). The various outcomes of the *mfecane* spread of Nguni peoples all involve issues of ethnic identity and labelling (e.g., language naming). In some cases, the migration led to a new Nguni variety with lesser or stronger effects of contact-induced change such as Ndebele S407. In the same historical period of upheaval, Sotho warriors conquered the territories of Luyana K31 speech communities, which gave rise to Lozi K21, a Sotho S30 language heavily influenced by the conquered Luyana (Gowlett 1989). In other cases, it led to a new language after language shift but recognisably different from the language shifted to, for example Tanzanian Ngoni and Mozambican Ngoni. In still other cases, only the magic name Ngoni remained after complete language shift, as it happened in Malawi. The different stages of Tanzanian Ngoni have been recorded to some extent in several successive publications (Elmslie 1891, Spiss 1904, Ebner 1951 [1930], Stirniman 1963, Ebner 1987 [1959], Ngonyani 2003). The initial phonological adaptation of Zulu sounds (laterals in these examples) by second language learners was followed by the lexical replacement when other speakers adapt Ngoni identity, e.g., *indhlu* 'house' becomes *nslu* in Old-Ngoni and is replaced by *nyumba* in New-Ngoni; *-hle* 'good' is adapted to *-she* in Old-Ngoni and replaced by *-bwina* in New Ngoni (Ebner 1987 [1959]: 156–160, Ngonyani 2003). The various linguistic outcomes of the Nguni migrations show instances of conversion and sociolinguistic factors of identity, adaptation and resistance to powerful immigrants.

In current days, convergence continues in various places. In Uganda there is the official rise of a "new language," RunyaKitara JE10A, which combines closely related varieties and is proclaimed to be one language combining Nkore JE13, Nyoro JE11, Tooro JE12 and Kiga JE14. It provides an emblem which represents a numerically strong unit, the largest in the country, but its linguistic standard is still abstract at the time of Bernsten (1998). The new unit RunyaKitara is recognised by young speakers, though. Nassenstein and Baraka Bose (2015) claim that the youth link their "Oruyáyé" to RunyaKitara rather than to the subgroups.

4 CONTACT-INDUCED CHANGE IN VARIOUS LINGUISTIC DOMAINS

The approximately 500 Bantu languages form an ideal research area for comparative research into contact-induced change. Such research is still largely to be done. In the following sections I hope to show that a comparative approach to phonological change through loanwords, to the adaptation of borrowed nouns into a gender system such as the Bantu noun class system and to the integration of foreign verbs into an agglutinating (Bantu) language, are promising areas for such studies. I have singled out tone from the phonology section because of its unique potential for understanding contact influence in the prosodic domain.

4.1 Loan phonology

When foreign words are borrowed, there is usually some adaptation in pronunciation to the phonological system of the receiving language and sometimes this adaptation is only partial and foreign sounds enter the system of the languages. Bostoen and Donzo (2013) show that Ngombe C41 acquired its labial-velar stops through contact with Ubangi languages. They argue that these must have come into the language via loanwords but the sounds subsequently spread to inherited lexicon. Similar processes must have occurred in other Bantu languages, as many of them in the area have labial-velars in their inventory. Even some loans from Ubangi that do not have a labial-velar in the Ubangi source acquired a labial-velar in Ngombe, and labial-velar stops also occur in roots inherited from Proto-Bantu. Moreover, labial-velar stops are quite frequent in sound-symbolic words, which often allow for sounds that are not part of the core phonological inventory. The same is true for labial-velars in Liko D201, which is influenced by the Central Sudanic language of the powerful Mangbetu neighbours. The labial part of labial-velars is realised as a trill, which most likely happened under Mangbetu influence. Cross-linguistically, the labial trill is a rare and highly marked sound, but it is prominent in Mangbetu. Liko speakers must have mastered Mangbetu well enough to keep this sound in their Mangbetu loans and to expand it to some ideophones (De Wit 2015: 39). The more common alveolar trill is also restricted to ideophones and to one loan from Mangbetu *átràbá* 'a small potter's tool'; in other loans *r* is adapted to *l* (De Wit 2015: 38). Afrikaans had influence on Phuthi S404, a Nguni language in the Sotho S30 area, resulting in a loan phoneme *p* where *p* is otherwise adapted to *b*. Recurrent adaptations in Africa are the unrounding of French round front vowels [y] to [i] and of the mid vowel [œ] to [ɛ] (e.g., Kwenzi-Mikala 1989: 160–161); the replacement of English [θ] by [f] (Tucker 1947: 230); of Arabic uvulars to velars (Kharusi 1995: 106).

Very common among Bantu languages is the adaptation of syllable structure to the CV pattern, for example *cricket* > *ikhilikithi* in Xhosa S41 (Drame 2000: 239). On the other hand, exceptions to the CV pattern arise through not fully adapted loans: "Due to contact with Afroasiatic languages word-internal clusters which are clearly not onset clusters are found in some eastern Bantu languages (Mushunguli *barshi* 'pillow,' Mwiini G412 *kuβla* 'to kill,' Swahili *ɣafla* 'suddenly')" (Odden 2015). Batibo (1996) shows that Swahili G42 is more tolerant to consonant clusters than Tswana S31 is. He also shows that the most common strategy to impose a CV syllable structure to loans is by vowel epenthesis. The added vowel is often assimilated to the preceding vowel or to the following one if there is no preceding one; an alternative is that the epenthetic vowel *i* or *ɪ* is assimilated to labial after a labial consonant. This is the rule for added final vowels in Swahili and for initial clusters in Tswana, except that right-to-left assimilation occurs if C_2 is *r* or *l*. The adaptation rules are not completely predictive. For example, in a number of loans the added final vowel in Tswana is not assimilated but is *a*, as in *oura* 'hour' from English 'hour' and *gouta* 'gold' from Afrikaans 'goud.'

4.2 Borrowing and tone

Most Bantu languages are tone languages and thus it is important to study what happens with tone in situations of contact. It is often hard to prove that language contact is the actual cause of changes in a tone system. Kaji (2010) shows how Tooro JE12 has lost tone as a last small step in a series of simplifications of the tone system in which the phonetic distinction is smaller than the phonological one. Tooro is spoken close to the area where non-tonal Nilotic languages are spoken, but there is no evidence that contact played a role in the internal simplification processes in a range of related languages in the area (Kaji 2010: 99). Both Kinga G65 (Schadeberg 1973) and Safwa M25 (Voorhoeve 1973) have restricted tone systems and are spoken in roughly the same area, and are not into contact with non-tone languages other than Swahili G42. The distribution of this reduction seems areal but, like in the Tooro case, it has not been shown that it is due to language contact, let alone how such change through contact took place (see also Philippson 1991).

It has both been claimed that prosody is easily borrowed (Matras 2009: 231), and that tone is easily lost as a consequence of language contact (Salmons 1992). Loss of tone in contact languages can be observed in the lack of tone in most of the Atlantic creoles and in Kongo ya Leta H10A (Mufwene 2003). For Swahili G42, it is sometimes claimed that it lost tone because of the large number of second-language speakers and its presumed *koine* character (e.g., Mugane 2015), even though most of the second-language speakers must have spoken tone languages. However, the alternative of internal development is equally feasible, because the predecessors of Swahili, Proto-Sabaki and Proto-Seuta, had already developed stages of pitch-accent systems and some dialects of Comorian G44, not a lingua franca to the same extent as Swahili, have a stress system (Philippson 1993).

The Bantu languages are by and large tonal and thus an ideal research area to study what happens with tone in contact. However, relatively few such studies have been conducted. The claims for loss of tone or acquisition of extra tones are often based solely on observations of areal distribution and hence on indirect evidence. To my knowledge there has been no comparative study of how (Bantu) tone languages deal with tone and stress of their loans; which strategies are used to interpret stress; what consequences massive borrowing has on the tonal system; what happens in loans from a tone language; whether tones in loans are only taken as surface manifestations or adapted to the receiving tonology; whether tone rules of the donor language can influence those of the receiving language, etc.

Bantu languages tend to borrow words from languages with various stress systems (English, French, Swahili). Reinterpretation of stress by a High tone or a Fall is intuitively the most sensible way for a tone language to deal with stress in the source word, but this pattern is not always followed. A comprehensive comparative study could determine which factors play a role: tone properties of the receiving language; segmental properties and/or properties of the stress realisation in the donor language. The accent is interpreted as High or Falling tone in Luba L31a (Kabuta 1998), but there are deviations from this pattern. In Ma'á, lexemes from Swahili are sometimes reinterpreted with a penultimate High, but more often as Low throughout (Mous 2003: 221–298).

Borrowing words into a tone language has repercussions for distributional properties of tone. Kabuta (1998) remarks that loanwords proportionally increase the number of low tones in the lexicon of Luba L31a. Odden (2015) observes that in Shona S10 the HL pattern occurs in loanwords such as [tʃikóro] 'school' while otherwise HL is rare due to the historical development in Shona of Proto-Bantu HL becoming HH; and he also reports that Matumbi P13 has no long level H tones except in the Swahili loan *sáána* 'very.' Datooga loans in Ikoma JE45 introduce new tone patterns that go against the essence of the native tonal system. Aunio (2014: 14) shows that

all H pattern has been introduced to Ikoma through Datooga loanwords. The tonal pattern matches Datooga Nominative case tone which is used in post- verbal subject position (Kiessling 2007). This new tonal pattern in Ikoma has been applied to noun stems of non-Datooga origin as well and it has affected the typology of Ikoma nominal tone system. Without this newly introduced tonal pattern Ikoma tone system could be defined as a reduced tonal system, where each morpheme can have at most one prominent syllable, but the introduced tonal pattern also introduced multiply linked H tones which are not allowed in other contexts in Ikoma.

In these cases, the surface High tone or stress of the donor language is taken as such in the receiving language with minor or major repercussions for the tone system of the receiving language.

One wonders if the phonology of the receiving language can treat the tone of the loanwords as if the donor language tone is underlying. Such an analysis is proposed by Batibo and Rottland (2001: 21), suggesting that Sukuma F21 High tones in Datooga loans are two syllables (or mora) away from their original position in line with the Sukuma rule that High tones surface two syllables away from the syllable to which they are underlyingly associated; thus *déeda* becomes *dɪɪdá* 'old cow,' and *aseéeta* becomes *sɪɪtá* 'God.' This would imply that Datooga loans are reinterpreted as having an abstract underlying tone rather than surface tone and therefore they are affected by the Sukuma rightward H-displacement rule, or they were transferred before Sukuma developed the H-tone shift. Other examples are *βulugé* 'light yellow,' *mʊʊlí* 'grey/beige,' *sanagó* 'black and white tail tip,' *fʊʊlí* 'eland-coloured,' *nhʊligá* 'giraffe-coloured,' *lɪɪlá* 'red cow' and *mʊʊgá* 'weaned calf'; but there are also exceptions such as *gidaléela* 'red,' *máala* 'white-faced, *hʊómʊ* 'dotted black/brown' and *saámu* 'brown/black.' The data presented are limited and not fully analysed though. Another possible analysis is that this final H-tone is due to a reinterpretation of the Datooga word plus a definite suffix and word-final reduction to whisper, much like Iraqw loans *digée* 'name for cow bought with donkey' from Datooga *digéetị* 'donkey' and similar loans.

An example of contact influence on a tone rule comes from the Nguni S40 influence on Sotho S30. The survey of Sotho tone systems by Zerbian (2006) shows that in most Sotho dialects, High tone spreads either just one syllable to the right or to the penultimate: Southern Sotho S33 *go khúrúmeleetsa* 'to cover for' vs. a North-Western dialect of Northern Sotho S32 *go khúrúmééétša* 'to cover for.' However, in the Setswapo dialect of Northern Sotho, Monareng (1992) shows that a High tone in verbs of more than 3 syllables spreads to the antepenultimate, *go khúrúméleetša* 'to cover for.' Spread – or shift – to the antepenult is typical of neighbouring Nguni S40 Bantu languages like Zulu S42 and Ndebele S407 (Rycroft 1983).

4.3 Borrowing into the noun class system

When gender languages borrow nouns from other languages (with or without gender), the question arises what gender will be assigned to the loan. Gender assignment to loanwords usually follows language-internal criteria of mainly a semantic and morphological nature. Stolz (2008: 404–405) shows that Romance borrowings into German may follow the gender of a near-synonym, have natural gender or a gender associated with a lexical field, e.g., borrowed fruit items are feminine. The gender can also be determined by analogy with a suffix that is associated with a particular gender. In addition, there is a default category of neuter gender. Gender in German is covert, while Bantu noun classes are overt due to their noun class prefixes.

4.3.1 Formal principles underlying gender assignment

Gender assignment to lexical borrowings may happen according to a number of formal principles in the Bantu languages:

- Initial syllables of the source word are recognised as noun class prefixes and the loan is placed in the noun class having that same segmental sequence as an allomorph, for example *ki-mau/vi-mau* (class 7/8) 'tunic(s)' in Swahili G42 from Portuguese *quimão* (Schadeberg 2009: 91).
- In case of loans from other Bantu languages, this principle results in a situation in which loans end up in the same noun class as in the source language as long as there is noun class prefix equivalence for the speakers, e.g., Swahili G43 *m(u)hanga/mihanga* (class 3/4) 'aardvark' from Shambaa G23 *mhanga* (Schadeberg 2009: 89) or Zigula G31 *mohanga* (Sacleux 1939: 622). This strategy of gender copy is cross-linguistically rare, but not for overt noun class prefixes of closely related Bantu languages with easily recognisable correspondences.
- Loans are put in the noun class that has zero as one of the allomorphs as a prefix. This principle can leave some room for choice. In Swahili G42, both class 9 and class 5 have zero as an allomorph for noun class prefix and both are often used for loans, e.g., *samaki/samaki* (class 9/10) 'fish' from Arabic *samak* vs. *kaburi/ma-kaburi* 'grave(s)' (class 5/6) from Arabic *qabr* (Schadeberg 2009: 90). Whiteley (1967) gives for Swahili some percentages for English loans per class in one speaker: 46% in 9/10, 43% in 5/6 and 5% ambivalent (remainder in other classes).

These principles are mentioned by several authors, for example Fleisch (2000: 61) for Lucazi K13:

> Loans from foreign languages tend to be incorporated into this [class 9] irrespective of their meaning. [Note: the class that does not necessarily show overt marking of the noun class by a prefix, i.e., class 9.] Loans that do have noun class prefixes do so because they are from other Bantu languages or the beginning of the loan is reinterpreted as prefix, e.g., *ka-valu* (12) 'horse' in LuCazi from Portuguese *cavalo*.

If the first syllable of the source word coincides with a noun class prefix in the receiving language, there is no necessity to reinterpret it as such. For example, in Punu B43, there are cases such as *bisəku* 'biscuit,' *bilətə* 'bulletin' and *biru* 'bureau' that are not reinterpreted as belonging to class 8 which has the noun class prefix *bi-*, *masu* 'builder' that is not class 6 which has the noun class prefix *ma-* and *bwatɔ* 'box' that not class 14 which has the noun class prefix *bu-*. Reinterpretation of the initial syllable happens for syllables that are not exactly identical, for example the initial nasal mid vowel in French *enveloppe* is also reinterpreted as the class 3 prefix *mu-* in *mù-ßílɔ̀pɔ̀* (Kwenzi-Mikala 1989).

How close does the match need to be to reinterpret an initial syllable as a noun class prefix? Nurse (2000: 133) observes that borrowed nouns from Bantu languages are usually put into the same class, but in Ilwana E701, several Swahili G42 class 7/8 nouns with vowel-initial roots end up in class 9/10. The noun class 7 prefix in Swahili is *ki-* for consonant-initial roots, but shifts to palatal *ch-* before vowel-initial roots. Hence, those loans are not recognised as containing a CV- noun class prefix. The source word is considered to have a zero prefix and interpreted as a class 9 noun, which has zero as an allomorph. We can assume that Ilwana speakers are aware of the allomorphy in Swahili, even though Ilwana does not have the same allomorphy for the class 7 noun class prefix.

Where we just saw that allomorphs in the source language are not automatically recognised as such in the transfer from one Bantu language to another, in other instances noun class membership is recognised for source words that do have a slightly different noun class prefix in the source language compared to the receiving language, and the speakers recognise the equivalence. Loans from Swahili G42 in other Bantu languages of East Africa show this: for example, the recognition and adaptation of the Swahili prefix *m-* as *mũ-* in Gikuyu E51 *mũ-cumaa* 'candle' from *m-shumaa*; *mũ-thitarĩ* 'line, row' from *m-stari*, as *mu-* in Ganda JE15 *mù-zigiti* 'mosque' from *m-sikiti* and as *uN-* in Nyakyusa M31 *un-súmali* 'nail' from *m-sumari*. Even in cases where the link between the prefixes is only historical, speakers may make the connection, as is the case with the recognition and adaptation of Swahili class 11 *u-* as *lu-* in Lwena K14 *lu-pàwa* 'spoon' from *u-pawa* 'ladle' and of Swahili class 11/14 *u-* as class 14 *bw-* in Ganda JE15 *bw-îno* 'ink' from *w-ino* 'ink' (Baldi 2011).

These examples of transfer of nouns from the Swahili G42 *u-* class to other languages are of interest because of the fact that classes 11 *lu-* and 14 *bu-* have merged in Swahili. There are more Swahili loanwords with *u-* into the historical correct Ganda JE15 class 14, e.g., *bu-pumbaavu* 'foolishness' from Swahili *u-pumbavu*, which is derived from the verb *-pumbaa* 'be speechless' and *bu-kaafiiri* 'paganism,' parallel to *omu-kaafiiri* 'a pagan,' from Swahili *u-kafiri* derived from *kafiri* (class 5/6). In these additional examples from Murphy and Kiggundu (1972), we can argue, however, that the class 14 prefix is a result of Ganda-internal word formation processes. The Swahili loans in Ganda class 11 are either the result of reinterpretation of the initial syllable as in *lu-kusa* 'permission' from Swahili *ruhusa* (class 9/10) or Ganda has added the class 11 *lu-* prefix to a Swahili root as in *lu-balaza* 'porch, veranda' from Swahili *baraza* (class 9/10). This latter process of prefix addition is productive for language names in the loan lexicon, e.g., *lu-falansa* 'French,' *lu-latini* 'Latin.' The appropriate association with class 11 and 14 in Ganda for loans from Swahili, which does not distinguish these classes, can largely be accounted for by Ganda-internal word formation processes. Nevertheless, in certain loans, there is no reinterpretation of the initial syllable in Ganda, e.g., *lupiiya* (class 9/10) 'rupee' from Swahili, *bulangeti* (class 9/10) 'blanket' from English.

One wonders what the limits of phonetic deviance for such interpretation are, because it seems to go beyond the language-internal rules of phonological adaptation, and it is based on establishing morphological equivalence. Another interesting question is whether we can make generalisations about when such initials are automatically interpreted as class prefixes and when they are not. Examples of the reinterpretation of the initial syllable as prefix usually involve class 7 *ki-* and there seems to be a strong tendency for reinterpretation of the first syllable when it involves *ki*. Schadeberg (2009) shows that all source words that start in *ki* are interpreted as class 7 *ki-*, e.g., *ki-tabu* 'book' from Arabic *kitāb*, *ki-biriti* 'box of matches' from Arabic *kibrīt* or from Hindi, *ki-mau* 'tunic type shirt' from Portuguese *quimão* and *ki-diri* 'bush squirrel' from a Bantu source, possibly Zigula G31. Others reinterpreted as *ki-* are *ki-tani* 'linen' from Arabic *kattān* 'linen' or from Persian, *ki-azi* 'potato' from Malagasy, Malay *kəladi*. Few words starting in *ch* in the source, the pre-vowel allomorph of the noun class prefix *ki-*, are recognised as class 7 such as *ch-epeo* (class 7/8) 'hat with brim' from Portuguese *chapéu* 'hat.' Most others follow the default strategy and are treated as class 9, e.g., *chai* 'tea,' *chizi* 'cheese' from English, *cheti* 'certificate' from Persian *čiṭṭhī* 'note, letter' or from Hindi. The loanword *chenza* 'tangerine' (class 5/6) from Chinese *chenzə* 'kind of orange' follows the semantic criterion of assigning fruits to class 5. Reinterpretation of initial *ki* as noun class prefix is not completely automatic in Swahili as Whiteley (1967: 170) notes that some loans in *ki-* in 7/8 such as *kisi* 'kiss,' *kilabu* 'club' are hybrids; they can also take 9/10 agreement. All class 7/8 Swahili loanwords in the World Loan Word Database (Haspelmath & Tadmor 2009) have a source that starts in kV; most source words that start in kV get this syllable reinterpreted as class 7 *ki-*, but not all. Maybe the syllabic pattern is a factor as Swahili *kilo* is in class 9/10 and not in 7. The initial syllable tends to be reinterpreted as a syllable if is a prototypical syllable CV and if what remains is interpretable as a typical root, i.e., CVCV.

An initial syllable *ma* is rarely interpreted as a noun class prefix *ma-* in Swahili. The only examples in Schadeberg (2009) are from Arabic: *ma-radi* (6) 'disease' from *marad* 'disease, illness,' and *majira* (6 or 9/10) from *maḡran* 'course (of water, events).' Most source words that start in *ma* follow the default strategy of assignment to class 9/10: *mashine* (9/10) 'machine,' *manjano* (9) 'yellow' probably from Hindi-Urdu *manǧa* 'turmeric,' and the Arabic loans *madhabahu* 'altar' (9/10) from *maḏbaḥ* 'altar, abattoir,' *mahakama* (9/10) 'court' from *maḥkama(t)* 'court,' *maisha* (9) 'life' from *maʿīša(t)* 'life,' *maiti* (9/10) 'corpse' from *mait, mayyit, maita(t)* 'dead, corpse,' *malkia* 'queen' (9/10) from *malika; malakīya(t)* 'queen; monarchy,' *mashua* (9/10) 'kind of boat' from *mašawa* 'kind of (fishing) boat,' *mara* 'time' (9/10) from *marra(t)* 'time (occasion).' In an early Cushitic loan in Eastern Bantu, **-díbà* (class 5/6) 'milk' (Bastin *et al.* 2002) from South Cushitic **'ilibaa* (Kiessling & Mous 2003), the class 6 noun prefix was added, i.e., *maziwa* 'milk' in Swahili.

Loans may be borrowed in Swahili G42 class 5 if they start in *ji* in the source, e.g., *jini* (class 5) 'demon' from Arabic *ǧinn* 'demon.' The initial syllable does not function as a prefix in Swahili, though, since its plural will be *ma-jini*. The rationale behind this is that although most Swahili words that start in *ji* in class 5 have a zero prefix (including these loans), some core members of this class have an allomorph *ji-* for this class prefix, such as the inherited *ji-no* 'tooth,' plural *m-eno* 'teeth.' Another example is *jinai* (class 5) 'crime' from Arabic *ǧināya(t)* 'crime.'

Other CV- syllables are only occasionally reinterpreted as noun class prefixes: for example, class 4 in *mi-la* 'custom' from Arabic *milla(t)* 'nation, religious community,' and *mu-* for class 3 in *mu-da* 'time' from Arabic *mudda(t)* 'period of time.' (Semi)vowel-initial segments are rarely recognised as class prefixes; they are in *u-mri* (11) 'age' from Arabic *ʕumr* 'period of life, age,' and *w-akati* (11) 'time' from Arabic *waqt* 'time' and *w-asiwasi* (11) 'anxiety' from Arabic *waswās* 'doubt,' but not in *wiki* (9) 'week' from English and *w-ali* '(cooked) rice' from a predecessor of Malagasy *váry* '(cooked) rice.' Notice that all these words fit in the *u/w-* class due to their abstract semantics.

The noun class attribution of loans in Swahili G42 is similar to that in other Bantu languages. In Punu B43, the reinterpretation of an initial syllable as noun class prefix is frequent for loans that denote non-countables. The French article *du* which is standardly used with such non-countables is phonologically adapted to *di* and reinterpreted as a class 5 prefix *di-*: *dí-pə̀* 'bread,' *dí-ßə̀* 'wine,' *dí-là* 'milk.' Note that these are all monosyllables, which might be relevant, because the prefix has a high tone and presumably that is only possible with vowel-initial roots and subsequent vowel merger. However, it is puzzling why this prefix would be borrowed as having a high tone, as the French words have the stress on the final syllable. Clearly semantics play a role here, as the regular plural of class 5, class 6, is commonly used as a general plural marker. An initial *ma* in French loans in Punu is in fact commonly reinterpreted as class 6 prefix *ma-*, *mà-sînə̀* 'machine,' not only for liquids or mass nouns. Sometimes the loan is given plural interpretation as in *mà-dâyì* from *médaille* and in the singular the plural prefix is replaced, *dì-dâyì*. Occasionally other initial syllables are reinterpreted as noun class markers, as in class 11 *du-* in *dú-mătù*, plural *bá-mătù* 'tomato(s)' (Kwenzi-Mikala 1989). Loans tend to use the singular form of the source language and borrow it as a singular. Exceptions do occur as in the 'medal' example above and in Swahili G42 *m-sumari* 'nail' from Persian *mismār* with a singular by back formation.

Over time an initial syllable in a loan may be reinterpreted in the history of the language after the transfer. Arabic *maqass* 'scissors (plural)' was borrowed as *ma-kasi* in Swahili, which Sacleux (1939) has as collective (class 6), but it is now *m-kasi* (class 3/4) 'scissors.' When newer loanwords follow the default class strategy, older loans may seemingly follow the reinterpretation strategy.

In conclusion, the principle of reinterpretation of the initial syllable is particularly popular for *ki* to class 7 in Swahili. This is also the only singular class prefix of CV- shape in Swahili. Such reinterpretation is never predictive. The match has to be quite exact and it is rare that allomorphs are reinterpreted as noun class prefixes. Initial syllables that consist of a vowel only are seldom reinterpreted as a noun class marker. It is also rare that syllables that resemble the prefixes of plural classes are reinterpreted as prefixes. This reflects a general tendency for singulars to be borrowed, unless there is semantic motivation to borrow a plural or as a plural. All these observations are (strong) tendencies but not predictive. Exceptions occur and semantics can play a role too.

4.3.2 Semantic principles underlying gender assignment

We now turn to the issue of whether semantics plays a role in noun class assignment for loans. Semantic criteria especially play a strong role when the loans involve trees and their fruits, as has been discussed by Contini-Morava and Kilarski (2013: 271). Fruits and trees tend to be borrowed into the appropriate noun class and add a noun class prefix.

Similar regularities are documented for Spanish, and Romance more generally, by Pountain (2005: 332), who points out that in Spanish the tree/fruit distinction is productively applied to loanwords such as *aceituna*/*aceituno* 'olive'/'olive-tree,' from Arabic. Pountain (2005: 330) argues that "gender can be used in nouns which inherently refer to inanimates to encode semantic contrasts relating to features other than sex precisely because there is no risk of such nouns having an animate reading." The tree/fruit distinction is a common exploitation of noun class/gender differentiation, found in genetically unrelated languages. This may also reflect a shape distinction, as trees/plants may be classified with long and thin things, while fruits may be classified with round or three-dimensional things. In Swahili G42, for example, names of plants are usually in the class pair 3/4, whereas names of the associated fruits are usually in the class pair 5/6, as in *m-boga* 'gourd plant' (class 3) vs. *Ø-boga* 'gourd' (class 5). Similarly to the example given above from Spanish, this pattern is pervasive enough to be extended to many loanwords designating plants, even if the source word did not begin with *m-*, as in *m-papai* 'papaya plant' (class 3) vs. *Ø-papai* 'papaya' (class 5) from Hindi *papai*. Similar cases are Shona S10 *mu-firinya* 'cassava' from Portuguese *farinha* 'flour' (Knappert 1999: 206).

Adding the appropriate noun class prefix is rare otherwise. In the case of trees, this may also be considered as language-internal derivation from the fruit, which was borrowed in class 5 (with zero as allomorph noun class prefix) combining semantic principles and the default strategy. Surprisingly, sometimes loan nouns do appear in the appropriate noun class with an extra CV- prefix. This seems to be the regular strategy in Phuthi S404, according to Coetser (1999) who mentions the following loans from Afrikaans: *i-treni* 'train,' *li-plase* 'place' from *plaas*, *ti-tabule* 'potato' from *ertappel*, *li-haletere* 'dumbbell' from *halter*, *i-zaca* 'sugar' from *suiker*, *bu-khetso* 'choice' from *keuse*, *i-tloloko* 'clock' from *klok*, *mu-diakone* 'deacon' from *diaken*. Mojela (2010) reports the addition of the class 5 prefix for loanwords in Northern Sotho S32 as a minor strategy *le-botlelo* 'bottle,' *le-polanka* 'plank' from Afrikaans *plank*, *le-jakane* 'Christian' from Afrikaans *diaken* 'deacon,' *le-famolele* or *le-famolebe* 'person who went away for a long time' from Afrikaans *vanmelewe* 'long ago.' I have suggested that such cases of addition of noun class prefixes are used when loans are put on par with existing nouns as a kind of 'paralexification' (Mous 2001). Bostoen and Donzo (2013: 458) observe for Ngombe C41 that a number of loans have added the appropriate noun class prefix suggesting such a historical process of paralexification, e.g., *li-kpóka* (class 5/6) 'ulcer' from Ngbandi *kpoká*. The instances in Phuthi S404 do not qualify for paralexification, although the language is a hybrid result of intense language contact. Phuthi is basically a Swati S43 variety spoken in a Northern Sotho S32 and Xhosa S41 environment, for which Donnelly (1991: 114) noticed high levels of morphological isomorphism in a process of language attrition.

Assignment of loans to noun classes in Bantu differs from the general principles that Stolz (2008: 404–405) established for German, partly because the Bantu noun class system is overt and classes have prefixes. In German, gender may be determined by analogy with a suffix that is associated with a particular gender. The equivalent in Bantu languages is reinterpretation of the initial syllable as prefix, which is a stronger factor, because the system is overt and because the marking is more salient at the beginning of the word. The phonological and morphological principles behind this strategy should be examined comparatively. Gender associated with a lexical field, such as borrowed fruit items, can be observed in Bantu, as it is in Romance languages, and probably reflects a universal tendency involving round objects in classification. While in German the loan may follow the gender of a near-synonym, this does not seem to happen often in Bantu or it has not received attention yet. Some instances have been reported of loans following the noun class of the original word in replacive borrowing.

Bantu languages that are used as contact languages and that developed a simplified variant such as up-country Swahili, Lingala and Pidgin A70 reduce the noun class system in similar ways: breakdown of external agreement on the verb into animate (agreement of class pair 1/2) versus inanimate (sometimes no number distinction for inanimate), and similar developments in the internal agreement within the noun phrase coupled with a general loss of agreement and an increase of modifiers that are invariant (Alexander 1967).

Occasionally it may seem that a noun class marker is borrowed. Both Budu D332 and Liko D201 have a number of nouns beginning in *nV* reflecting a Mangbetu prefix *ne-*, *nɛ-* or *na-*. The Mangbetu prefix is a definite prefix that can be added to any noun in the singular. Hence, most nouns in their citation form start in this prefix. Frieke-Kapper (2007: 57) presents a list of approximately 75 Budu nouns that start in *ne-*, *nɛ-* or *na-*. However, none of them can be linked directly to a Mangbetu root. Many of them denote peripheral lexicon of flora and fauna. This suggests another possible etymology: the Bantu proclitic *na-* or *nya-* 'mother-' (cf. Meeussen 1967: 95), which is used in naming and hence in words for rarely used words for specific flora and fauna. It is cross-linguistically very common to find special lexical innovation in these domains, often reflecting name-giving practices. In Liko too, a number of words in class 1a with zero prefix begin in *ne-*, *nɛ-* or *na-*, but some also in *si-* which is the male equivalent to *n(y)a-* in Bantu naming. These are partly loans from Mangbetu, as the source words in Mangbetu can be identified, but most others cannot be shown to be of Mangbetu origin, although speakers claim they are. Some of them are shared with Budu (De Wit 2015: 212–213). The *nV* initial syllable in loanwords is not treated as a prefix, since these words do not form a new noun class but fall in the prefix-less 1a class.

A new class can arise in the receiving language if nouns are not adapted to the cognate noun class, but copied with their agreement behaviour and class prefix. A case in point is the Mbugu G221 noun class "14.2," whose prefix is *u-*. It hosts loans from Swahili G42 and Shambaa G23 which have *u-* as the reflex of the Proto-Bantu class 14 prefix *$bʊ$-; next to Mbugu's regular reflex "14.1," *vu-*. Both 14.1 and 14.2 have their own agreement system (Mous 2003: 171).

4.4 Loan verbs

The Bantu languages show little evidence for the cross-linguistically common rule that the source words for verbs are taken as nouns when borrowed (Moravcsik 1975, Wohlgemut 2009). Verbs borrowed in Bantu rarely show denominal verbalising morphology. Such denominal verbalising morphology does exist in the form of applicative or causative suffixes. Some applicatives are added to Arabic loan in Swahili G42: *-ashiria* 'signal' from Arabic *ʔaʃʃir* 'to mark, record,' and *-hudhuria* 'be present at' from *ʔuhdˤur* 'to be present, be in the presence, to attend, participate, arrive, appear, to be settled' (Kharusi 1995: 235, 242). In Luba L31a, loan verbs are adapted with a causative suffix in some instances: *ku-been-esh-a* 'to bless' from French *benir* and *ku-sopw-esh-a* 'to warn' from Dutch *pas op* 'be careful!' (Kabuta 1998).

Borrowing verbs from non-Bantu sources poses challenges to Bantu morphology. One issue is which verb form is taken from the source. We see a number of different strategies in the Bantu languages, and the strategy may vary within one language for the different source languages. In Rundi JD62, verbs borrowed from English are taken in the participle form ending in *ing*, hence *gu-traaⁿsiretiiⁿg-a* /INF-translate-FV/ 'to translate' (Ntihirageza 2009: 72). But for verbs from French, Rundi takes the root without any suffixes, e.g., *gu-saaⁿt-a* /INF-smell-FV/ 'to smell' (Ntihirageza 2009: 75). Native rules only apply after the English stem – not the root – is imported into the Kirundi structure. Morphophonological rules, such as consonant prenasalisation and compensatory lengthening, apply to the English stem (root+-ing) rather than the root: *ku-move-iing-a*

[*kumuviiŋga*] INF-move+ing-FV, *ya-pop-iing-ye* [*yapopiinze*] 3sg.NEAR.PST-pop+ing-PRF
'S/he popped in.' (Ntihirageza 2009: 72). In Maore G44D, the French infinitive is taken
as the unit and the final vowel *e*, in which most French infinitives end, is interpreted as a
non-inflectional root vowel, e.g., *a-tso-barje* 's/he will take the ferry' from *barger* (Alnet
2009: 127). Luba L31a too borrows French infinitives as they are with an invariable
final vowel (Kabuta 1998: 50): -*pròpòzê* 'propose' from *proposer*, -*kònsèvwâr* 'conceive'
from *concevoir*, -*dèfìnîr* 'define' from *définir*, -*dìdèbrùyê* 'to manage, to get on' from *se
debrouiller* (including the reflexive prefix -*dì*- which is a calque of the French *se* pronoun), -*kòmprandrè* 'understand' from *comprendre*. However, it also has apparent French
loan verbs with an inflectional final vowel *a*, e.g., *ku-batiiza* 'baptise' from *baptiser* and
ku-penta 'paint' from *peindre*. It is also conceivable that these are older Portuguese loans
form *baptizar* and *pintar* respectively. Swahili G42 verbs from Arabic are based either on
the imperative form or on the past tense form (Kharusi 1995: 202, 208). In code-switch-
ing between Ndali M301 and Swahili G42, the infinitival Swahili prefix can be retained
in conjugated forms, e.g., *a-kusoma leka* |s/he-to.read a.lot| 'S/he reads a lot' (Swilla
2008: 237). There is little to say about which source form will be taken for borrowing
verbs except that the same base form is used for all borrowed verbs from the same source
language.

The issue that is most crucially linked to the imposition of the Bantu verb structure is
how to deal with the final vowel. Phonological adaptation in borrowing generally requires
a final open syllable of the borrowed verb form. The default final vowel for the Bantu
verb is *a*. This final vowel is prone to changes to express TAM, polarity, etc. If the source
final vowel is received as being *a*, this vowel is interpreted as this inflectional for the
cognate Bantu verbs as in the Mbugu G221 examples above.

Some borrowed Swahili G42 verbs from Arabic origin end in an inflectional *a*, either
because it is borrowed from the past tense form in Arabic, e.g., *rash.a* 'smear, whitewash'
from *raʃʃa* (past) 'splash,' or because it ends in *a* and an *ain ʕ* in Arabic as in Swahili
hada.a 'deceive, cheat' from Arabic *ʔixdaʕ > ʔaxdaʕ* (Kharusi 1995: 348). Other Arabic loan verbs in *a* are -*shiba* 'be satisfied, sated' from *ʔiʃbaʕ* 'to be or become sated,
satisfied, full, to have enough'; *kab.a* 'press, squeeze' from *qabba* 'to chop off, cut off,
straighten up, rise, stand on end,' -*tawal.a* 'rule' from *tawalla* 'to occupy, be entrusted, to
assume the responsibility, seize control, come to power, to refrain,' and -*tawadh.a* 'perform ablution' from *tawadydya* with the same meaning (Kharusi 1995: 331).

The majority of Arabic loan verbs in Swahili end in an invariant *i* which assimilates
to *u* after labial consonants, e.g., -*ishi* 'live,' -*rudi* 'return,' -*furahi* 'be glad,' -*hesabu* 'to
count,' -*dumu* 'to last.' There are a few Arabic loan verbs that end in *u* but not because of
labial assimilation to the preceding consonant, but to the preceding vowel plus alveolar
consonant -*busu* 'kiss,' -*thubutu* 'dare,' -*abudu* 'worship,' -*dhuru* 'damage.' Loan verbs
from English, such as -*kisi* 'kiss,' also take final *i* (Schadeberg 2009), but in general
English verbs are used without any final vowel (Myers-Scotton 2000: 204). Myers-Scotton (2000) also shows that English verbs in Zulu G42 either have no final vowel or add
the inflectional vowel *a* while in Sotho S30 and Shona S10 the inflectional vowel seems
to be added to English verb roots. The different behaviour of the final vowel is difficult
to explain. Languages have different strategies. Possibly the strategy not to add a final
vowel represents a situation of little adaptation and proficient bilingualism, while adding
a final vowel is a subsequent adaptation or one in which the contact situation requires
adaptation to the receiving language. A final vowel *a* is interpreted as an inflectional
default vowel but other final vowels are seen as part of the stem. As soon as a loan

verb undergoes derivation within the receiving language, it also has an inflectional final vowel *a*.

Another strategy that is common in accommodating foreign verbs in a Bantu language is the use of a light verb construction. In Shona S10 and Chewa N31b, this light verb is 'do' *-ita* and *-chita* respectively. In Shona the light verb is followed by the English participle in *ing*, e.g., *-ita binding* 'to bind,' while in Chewa by the bare form of the English verb, e.g., *-chit-a think.about* 'to think about' (Myers-Scotton 2000: 207–208). In Swahili G42, the light verb construction is with 'to hit,' *ku-piga* X, where X is a nominal instrument or an ideophone and originates in loan translations from Arabic using *dʲaraba* 'to strike.' Kharusi (1995: 192) provides examples such as Swahili *-piga miθali* from Arabic *dʲaraba maθalan* 'to give an example' and *-piga mstari* 'to draw a line' from Arabic *dʲaraba xatʲtʲan*.

There is little available evidence for the Bantu-internal borrowing of verbs. Verbs are less often borrowed than nouns in general, and verbs borrowed from related Bantu languages may not always be recognisable as such. Cognate Bantu verb stems are added to the target lexicon without adaptation. The WOLD files for Swahili G42 do not contain Swahili loan verbs from a Bantu source (Haspelmath & Tadmor 2009, Schadeberg 2009). The Pare variety Normal Mbugu G221 in the Usambara mountains borrows verbs from Shambaa G23 and these verbs are borrowed without adaptation. The Normal Mbugu loans are identical to their Shambaa verbs: *-ambal-a* 'borrow (cow),' *-zul-a* 'to skin,' *-shul-a* 'to smell, fart,' *-vughut-a* 'to forge,' *-umbuk-a* 'to have leprosy.' These verbs have identical form and meaning in Normal Mbugu and Shambaa, with one slight difference in meaning: 'to forge' in Mbugu has a more general meaning in Shambaa 'to generate wind' (LangHeinrich 1921, Mous 1994).

4.5 Morphological borrowing and structural transfer

The transfer of grammatical affixes in form and function is in general rare in language contact. Occasionally this can be observed in situations of profound language contact. One such Bantu example is Ilwana E701, a Sabaki Bantu language spoken at the Tana river, which has been heavily influenced by the Orma variety of Oromo (Nurse 2000). Its phonology is reshaped by the influx of Cushitic lexicon: consonant clusters consisting of a liquid plus obstruent are allowed and glottalised consonants are retained as such in loans. In the nominal system, Ilwana has introduced plural suffixes to distinguish between singular class 9 and plural class 10, which are otherwise homophonous. These suffixes, i.e., *-eena*, – *imʊ* and – *era*, are productive and not restricted to Orma loans. Interestingly, they cannot be identified as Orma plural suffixes and although they conspicuously look like Cushitic suffixes no Cushitic source has been identified (Nurse 2000: 131–132). The tense/aspect system of Ilwana has been restructured under Orma influence, but with inherited Bantu formatives. Ilwana's lexicon is heavily influenced by a number of languages, with Swahili G42 and Orma being the main contributors. For a large part of the lexicon (43% out of 2200 lexemes), no source has been identified. Using the detailed reconstruction of Sabaki by Nurse and Hinnebusch (1993), Nurse (2000) can establish that these words are not inherited. From the history of Ilwana, it is clear that there was a strong influence of Orma in the pre-colonial period and probably a shift to Orma was under way, but arrested due to changes in the socio-political power relations.

4.6 Prepositions and conjunctions

Independent function words can more easily be borrowed than dependent grammatical morphemes. Conjunctions, in particular, being initial and disjoint in the clause are often borrowed, both cross-linguistically (Matras 1998) and in Bantu specifically. In Swahili G42, the conjunctions *bali* 'however, on the contrary,' *ama* 'yet, however,' *ila* 'except, unless, but, otherwise,' *hata* 'until, since, so that,' *kama* 'as,' *wala* 'not' are all borrowed from Arabic (Kharusi 1995: 222–223). Prepositions can also be borrowed. Bantu languages are often relatively poor in inherited prepositions and have a multifunctional preposition *na* and not many others, but those languages that are in contact with Swahili nearly all borrow *mpaka* 'until' as a preposition from Swahili. In Swahili, this is a grammaticalisation of the word for 'border, boundary' (Mous 2016).

4.7 Syntax

In the domain of syntax, predominantly contact with non-Bantu languages occurs or is at least most visible. External influence on syntax is often paralleled with language-internal evolutions. Cushitic influence on Rangi F33 and Mbugwe F34 in Verb-Aux order has been reported (Mous 2000, Gibson 2012), but internal developments played a role in this development too. Nilamba F31 and Nyaturu F32 developed a verbal system with a pre-verbal clitic complex, apparently following the model of the Cushitic languages of the area, but with inherited Bantu material (Kiessling *et al.* 2008). Gibson and Wilhelmsen (2015) show that Rangi F33 and Mbugwe F34 developed a double negative with *toko* as final element as an internal development following the path of a Jespersen cycle but with external material. They relate *toko* to the surrounding Cushitic languages Alagwa and Burunge. Another potential source is Maasai *tokol* 'completely' (Mol 1996).

5 PROSPECTS

Language contact is one of the reasons why the internal classification of the Bantu languages is still a major challenge (Schadeberg 2003: 143). To study language contact between Bantu languages is a challenging activity, because the close relationships and the fact that speakers can often recognise lexicon and language structure lead to adjustments that render the effect of language contact difficult to discern. With careful study, it is possible to disentangle inherited lexicon and grammar from contact-induced items. Once such research advances, it should be possible to arrive at a more convincing internal classification, as well as an understanding of the histories of the languages concerned and their people, using insights from our general knowledge about language contact, but specifically also using an enhanced insight into how Bantu languages tend to behave in contact situations regarding the typical Bantu features, such as tone and noun classes. What we need is comparative studies of language contact in the Bantu domain. A lot can easily be done with the influence of the European languages of wider communication on Bantu, provided that the base studies are analytical in outlook and comprehensive, which until now is only the case for a selected number of languages. The massive number of Bantu languages makes them a unique area for developing and testing general theories.

REFERENCES

Adelaar, A. (2012) Malagasy Phonological History and Bantu Influence. *Oceanic Linguistics* 51(1): 123–159.

Adendorff, R. (2002) Fanakalo: A Pidgin in South Africa. In Mesthrie, R. (ed.), *Language in South Africa*, 179–198. Cambridge: Cambridge University Press.

Alexander, P. (1967) Note sur la réduction du systèmes des classes dans les langues véhiculaires à fonds bantu. In Manessy, G. (ed.), *La classification nominale dans les langues négro-africaines: actes du colloque international CNRS, Aix-en-Provence, 3–7 juillet 1967* 277–290. Paris: Éditions du Centre National de la Recherche Scientifique.

Alexandre, P. (1963) Aperçu sommaire sur le pidgin A70 du Cameroun. *Cahiers d'Etudes Africaines* 12: 577–582.

Alexandre, P. (1964) Aperçu sommaire sur le pidgin A70 du Cameroun. *Colloque sur le multilingualisme/Symposium on multilingualism: the second meeting of the Inter-African Committee on Linguistics, Brazzaville, 16–21 August 1962*, 251–256. London; Lagos; Nairobi: Commission pour Coopération Technique en Afrique (CCTA); Conseil Scientifique pour l'Afrique (CSA).

Alnet, A. J. (2009) *The Clause Structure of the Shimaore Dialect of Comorian (Bantu)*. Urbana-Champaign: University of Illinois, PhD dissertation.

Aunio, L. (2014) Contact Leading to a More Complex Tonal System: The Case of Ikoma and Datooga in North-West Tanzania. Paper presented at *The 6th international Conference on Tone and Intonation in Europe (TIE 6), 10–12 September*, Utrecht. (Available online at www.linguistics.fi/contact/Book_of_abstracts_10.7.2014.pdf).

Bahuchet, S. (1992) *Dans la forêt d'Afrique centrale: les Pygmées Aka et Baka*. Paris: Peeters, Société des Etudes Linguistiques et Anthropologiques.

Bahuchet, S. (2012) Changing Language, Remaining Pygmy. *Human Biology* 84: 11–43.

Baldi, S. (2011) Swahili: A Donor Language. *Lingua Posnaniensis* 53(1): 7–24.

Barlew, J. (2013) Point of View in Mushunguli Locatives. In Orie, O. l. O. l. & K. W. Sanders (eds.), *Selected Proceedings of the 43rd Annual Conference on African Linguistics*, 115–129. Somerville: Cascadilla Proceedings Project.

Bastin, Y., A. Coupez, E. Mumba & T. C. Schadeberg (eds.) (2002) *Bantu Lexical Reconstructions 3*. Tervuren: Royal Museum for Central Africa. (Available online at linguistics.africamuseum.be/BLR3.html, Accessed November 28, 2017).

Batibo, H. M. (1996) Loanword Clusters Nativization Rules in Tswana and Swahili: A Comparative Study. *South African Journal of African Languages* 16(2): 33–41.

Batibo, H. M. & F. Rottland (2001) The Adoption of Datooga Loanwords in Sukuma and Its Historical Implications. *Sprache und Geschichte in Afrika* 16/17: 9–50.

Bernsten, J. G. (1998) Runyakitara: Uganda's "new" Language. *Journal of Multilingual and Multicultural Development* 19(2): 93–107.

Bostoen, K. & J.-P. Donzo (2013) Bantu-Ubangi Language Contact and the Origin of Labial-Velar Stops in Lingombe (Bantu, C41, DRC). *Diachronica* 30(4): 435–468.

Bostoen, K. & B. Sands (2012) Clicks in South-Western Bantu Languages: Contact-Induced Vs. Language-Internal Lexical Change. In Brenzinger, M. & A.-M. Fehn (eds.), *Proceedings of the 6th World Congress of African Linguistics Cologne 2009*, 129–140. Cologne: Rüdiger Köppe.

Branford, W. & J. S. Claughton (2002) Mutual Lexical Borrowings Among Some Languages of Southern Africa: Xhosa, Afrikaans and English. In Mesthrie, R. (ed.), *Language in South Africa*, 199–215. Cambridge: Cambridge University Press.

Coetser, A. (1999) Die invloed van Afrikaans op Phuthi. *South African Journal of Linguistics* 17(2–3): 205–212.

Contini-Morava, E. & M. Kilarski (2013) Functions of Nominal Classification. *Language Sciences* 40: 263–299.

de Smedt, J. (2011) *The Nubis of Kibera: A Social History of the Nubians and Kibera Slums*. Leiden: Leiden University, PhD dissertation.

De Wit, G. (2015) *Liko Phonology and Grammar*. Leiden: Leiden University, PhD dissertation.

Declich, F. (2002) *I Bantu della Somalia: etnogenesi e rituali mviko*. Milano: F. Angeli.

Dimmendaal, G. (1995) The Role of Bilingualism in Nilotic Sound Change. *Belgian Journal of Linguistics* 9: 85–109.

Dimmendaal, G. (2001) Language Shift and Morphological Convergence in the Nilotic Area. *Sprache und Geschichte in Afrika* 16/17: 83–124.

Dimmendaal, G. (2011) *Historical Linguistics and the Comparative Study of African Languages*. Amsterdam; Philadelphia: John Benjamins.

Donnelly, S. S. (1991) *Phonology and Morphology of the Noun in Yeeyi*. Cape Town: University of Cape Town,

Drame, A. (2000) Foreign Words as Problems in Standardisation/Lexicography: English and Afrikaans Loan-Words in isiXhosa. *Lexikos* 10: 231–241.

Ebner, E. (1951 [1930]) *Grammatik des Neu-Kingoni*. Peramiho: Mission Magagura.

Ebner, E. (1987 [1959]) *History of the Wangoni and Their Origin in the South African Bantu Tribes*. Ndanda: Benedictine Publications.

Ehret, C. (2001) *An African Classical Age: Eastern and Southern Africa in World History, 1000 B.C. to A.D. 400*. Charlottesville: The University of Virginia Press.

Elmslie, W. A. (1891) *Introductory Grammar of the Ngoni (Zulu) Language, as Spoken in Mombera's Country*. Aberdeen: G. & W. Fraser, 'Belmont' Works.

Fehderau, H. W. (1967) *The Origin and Development of Kituba, Lingua Franca Kikongo*. Kisangani: Université libre du Congo.

Fleisch, A. (2000) *Lucazi Grammar: A Morphosemantic Analysis*. Cologne: Rüdiger Köppe.

Frieke-Kapper, C. (2007) *The Creative Use of Genre Features: Continuity and Change in Patterns of Language Use in Budu, a Bantu Language of Congo (Kinshasa)*. Amsterdam: Vrije Universiteit Amsterdam, PhD dissertation.

Gibson, H. (2012) *Auxiliary Placement in Rangi: A Dynamic Syntax Perspective*. London: SOAS, University of London, PhD dissertation.

Gibson, H. & V. Wilhelmsen (2015) Cycles of Negation in Rangi and Mbugwe. *Africana Linguistica* 21: 233–257.

Gilmour, R. (2006) *Grammars of Colonialism: Representing Languages in Colonial South Africa*. Basingstoke: Palgrave Macmillan.

Gowlett, D. F. (1989) The Parentage and Development of Lozi. *Journal of African Languages and Linguistics* 11: 127–149.

Gunnink, H. (2014) The Grammatical Structure of Sowetan Tsotsitaal. *Southern African linguistics and applied language studies* 32(2): 161–171.

Gunnink, H., B. Sands, B. Pakendorf & K. Bostoen (2015) Prehistoric Language Contact in the Kavango-Zambezi Transfrontier Area: Khoisan Influence on Southwestern Bantu Languages. *Journal of African Languages and Linguistics* 36(2): 193–232.

Haspelmath, M. & U. Tadmor (eds.) (2009) *World Loanword Database*. Leipzig: Max Planck Institute for Evolutionary Anthropology (Available online at http://wold.clld.org).

Heine, B. (1982) *The Nubi Language of Kibera – An Arabic Creole*. Berlin: Dietrich Reimer.

Herbert, R. K. (1990a) The Relative Markedness of Click Sounds: Evidence from Language Change, Acquisition and Avoidance. *Anthropological Linguistics* 32(1/2): 120–138.

Herbert, R. K. (1990b) The Sociohistory of Clicks in Southern Bantu. *Anthropological Linguistics* 32(3–4): 295–315.

Hout, K. (2012) *The Vocalic Phonology of Mushunguli*. Lawrence: University of Kansas MA thesis.

Hurst, E. (2017) "African (Urban) Youth Languages." In *Oxford Research Encyclopedia of Linguistics*. Oxford: Oxford University Press (Available online at http://linguistics.oxfordre.com/view/10.1093/acrefore/9780199384655.001.0001/acrefore-9780199384655-e-157).

Hurst, E. & M. Buthelezi (2014) A Visual and Linguistic Comparison of Features of Durban and Cape Town Tsotsitaal. *Southern African Linguistics & Applied Language Studies* 32(2): 185–197.

Hurst, E. & R. Mesthrie (2013) "When you hang out with the guys they keep you in style": The Case for Considering Style in Descriptions of South African tsotsitaals. *Language Matters* 44(1): 3–20.

Kabuta, N. S. (1998) Loanwords in Cilubà. *Lexikos* 8: 37–64.

Kaji, S. (2010) A Comparative Study of Tone of West Ugandan Bantu Languages, with Particular Focus on the Tone Loss in Tooro. *ZAS Papers in Linguistics* 53: 99–107.

Kaltenbrunner, S. (1996) *Fanakalo: Dokumentation einer Pidginsprache*. Vienna: Afro-Pub.

Kaye, A. & M. Tosco (1993) Early East African Pidgin Arabic. *Sprache und Geschichte in Afrika* 14: 269–305.

Kharusi, N. S. (1995) *The Linguistic Analysis of Arabic Loan-Words in Swahili*. Washington D.C.: Georgetown University, PhD dissertation.

Kiessling, R. (2007) The "marked nominative" in Datooga. *Journal of African Languages and Linguistics* 28(2): 149–191.

Kiessling, R. & M. Mous (2003) *The Lexical Reconstruction of West Rift (Southern Cushitic)*. Cologne: Rüdiger Köppe.

Kiessling, R. & M. Mous (2004) Urban Youth Languages in Africa. *Anhropological linguistics* 46(3): 303–341.

Kiessling, R., M. Mous & D. Nurse (2008) The Rift Valley Area of Central Tanzania as a Linguistic Contact Zone. In Heine, B. & D. Nurse (eds.), *A Linguistic Geography of Africa*, 186–227. Cambridge; New York: Cambridge University Press.

Kishindo, P. J. (2002) "Flogging a dead cow?": The revival of Malawian Chingoni. *Nordic Journal of African Studies* 11(2): 206–223.

Knappert, J. (1999) Loanwords in African Languages. In Finlayson, R. (ed.), *African Mosaic (Festschrift for J.A. Louw)*, 203–220. Pretoria: Unisa Press, University of South Africa.

Kolbusa, S. (2000) *Schwiegermeidung bei den Nyakyusa*. Bayreuth: Universität Bayreuth, MA thesis.

Kutsch Lojenga, C. (2003) Bila (D32). In Nurse, D. & G. Philippson (eds.), *The Bantu Languages*, 450–474. London New York: Routledge.

Kwenzi-Mikala, J. T. (1989) Contribution à l'analyse des emprunts nominaux du yipunu au français. *Pholia* 4: 157–170.

LangHeinrich, F. (1921) *Schambala-Wörterbuch*. Hamburg: Friederichsen, De Gruyter & Co.

Luffin, X. (2005) *Un créole arabe: le kinubi de Mombasa.* Munich: LINCOM.
Mafela, M. J. (2010) Borrowing and Dictionary Compilation: The Case of the Indigenous South African Languages. *Lexikos* 20: 691–699.
Matras, Y. (1998) Utterance Modifiers and Universals of Grammatical Borrowing. *Linguistics* 36: 281–331.
Matras, Y. (2009) *Language Contact.* Cambridge: Cambridge University Press.
Meeussen, A. E. (1967) Bantu Grammatical Reconstructions. *Africana Linguistica* 3: 79–121.
Meeuwis, M. (2009) Involvement in Language: The Role of the Congregatio Immaculati Cordis Mariae in the history of Lingala. *The Catholic Historical Review* 95(2): 240–260.
Mesthrie, R. (1989) The Origins of Fanagalo. *Journal of Pidgin and Creole Languages* 4(2): 211–240.
Mesthrie, R. (2008) 'I've Been Speaking Tsotsitaal All My Life Without Knowing It': Towards a Unified Account of Tsotsitaals in South Africa. In Meyerhoff, M. & N. Nagy (eds.), *Social Lives in Language*, 95–109. Amsterdam; Philadelphia: John Benjamins.
Mesthrie, R. (2014) English Tsotsitaals? – An Analysis of Two Written Texts in Surfspeak and South African Indian English Slang. *Southern African Linguistics & Applied Language Studies* 32(2): 173–183.
Möhlig, W. J. G. (1981) Stratification in the History of the Bantu Languages. *Sprache und Geschichte in Afrika*: 251–317.
Möhlig, W. J. G. (2007) Linguistic Evidence of Cultural Change. The Case of the Rumanyo Speaking People in Northern Namibia. *Sprache und Geschichte in Afrika* 18: 131–154.
Mojela, V. M. (2010) Borrowing and Loan Words: The Lemmatizing of Newly Acquired Lexical Items in Sesotho sa Leboa. *Lexikos* 20: 700–707.
Mol, F. (1996) *Maasai Language and Culture Dictionary.* Lemek: Maasai Centre.
Monareng, W. M. (1992) *A Domain-Based Approach to Northern Sotho Tonology: A Setswapo Dialect.* Urbana-Champaign: University of Illinois, PhD dissertation.
Moravcsik, F. (1975) Borrowed Verbs. *Wiener Linguistische Gazette* 8: 3–30.
Motingea Mangulu, A. (1996) Pool Malebo Lingala. *Afrikanistische Arbeitspapiere* 46: 55–117.
Motingea Mangulu, A. & Mwamakasa Bonzoi (2008) Aux sources du lingála: cas du Mbenga de Mankanza. *African Study Monographs* 38: 1–93.
Mouguiama-Daouda, P. (2005) Phonological Irregularitites, Reconstruction and Cultural Vocabulary: The Names of Fish in the Bantu Languages of the Northwest (Gabon). *Diachronica* 22(1): 59–107.
Mous, M. (1994) Ma'a or Mbugu. In Bakker, P. & M. Mous (eds.), *Mixed Languages: 15 Case Studies in Language Intertwining*, 175–201. Amsterdam: Instituut voor Functioneel Onderzoel van Taal en Taalgebruik.
Mous, M. (2000) Counter-Universal Rise of Infinitive-Auxiliary Order in Mbugwe (Tanzania, Bantu F 34). In Vossen, R., A. Mietzner & A. Meissner (eds.), *Mehr als nur Worte: Afrikanistische Beitrage zum 65. Geburtstag von Franz Rottland*, 469–481. Cologne: Rüdiger Köppe.
Mous, M. (2001) Paralexification in Language Intertwining. In Smith, N. & T. Veenstra (eds.), *Creolization and Contact*, 113–123. Amsterdam; Philadelphia: John Benjamins Publishing Company.
Mous, M. (2003) *The Making of a Mixed Language : The Case of Ma'a/Mbugu.* Amsterdam; Philadelphia: John Benjamins.

Mous, M. (2016) Transfer of Swahili "until" in Contact with East African Languages. Paper presented at *the 29th Swahili Colloquium, May 6–8*, Bayreuth.

Mufwene, S. S. (1988) Formal Evidence of Pidginization/Creolization in Kituba. *Journal of African Languages and Linguistics* 10(1): 33–51.

Mufwene, S. S. (1994) Restructuring, Feature Selection and Markedness: From Kimanyanga to Kituba. In Moore, K. E., D. A. Peterson & C. Wentum (eds.), *Proceedings of the Annual Meeting of the Berkeley Linguistics Society, February 18–21, 1994. Special Session on Historical Issues in African Linguistics*, 67–90. Berkeley: Berkeley Linguistics Society.

Mufwene, S. S. (2003) Contact Languages in the Bantu Area. In Nurse, D. & G. Philippson (eds.), *The Bantu Languages*, 195–208. London; New York: Routledge.

Mufwene, S. S. (2013) Kikongo-Kituba. In Michaelis, S. M., P. Maurer, M. Haspelmath & M. Huber (eds.), *The Survey of Pidgin and Creole Languages. Volume 3: Contact Languages Based on Languages from Africa, Asia, Australia, and the Americas.*, 3–13. Oxford: Oxford University Press.

Mugane, J. M. (2015) *The Story of Swahili*. Athens: Ohio University Press.

Murphy, J. D. & C. Kiggundu (1972) *Luganda – EnglishDdictionary*. Washington, DC: Consortium Press for the Catholic University of America Press.

Mutonya, M. & T. H. Parsons (2004) KiKAR: A Swahili Variety in Kenya's Colonial Army. *Journal of African Languages and Linguistics* 25(2): 111–125.

Myers-Scotton, C. (2000) Comparing Verbs in Swahili/English Codeswitching with Other Data Sets. In Kahigi, K., Y. Kihore & M. Mous (eds.), *Lugha za Tanzania/ Languages of Tanzania: Studies Dedicated to the Memory of Prof. Clement Maganga*, 203–214. Leiden: CNWS.

Nassenstein, N. & P. Baraka Bose (2015) Oruyáyé: Youth Language as Regional Identity in Western Uganda. Paper presented at *2nd Conference on African Urban Youth Languages*, Kenyatta University, Nairobi, Kenya, 11–13 December 2015.

Nassenstein, N. & A. Hollington (eds.) (2015) *Youth Language Practices in Africa and Beyond*. Berlin: De Gruyter Mouton.

Ngonyani, D. S. (2001) The evolution of Tanzanian Ngoni. *Sprache und Geschichte in Afrika* 16/17: 321–354.

Ngonyani, D. S. (2003) *A Grammar of Chingoni*. Munich: LINCOM.

Ntihirageza, J. (2009) Context Sensitive Morphology: Adaptations of English and French Verbs into Kirundi. *Concentric: Studies in Linguistics* 35: 65–86.

Nurse, D. (1988) Extinct Southern Cushitic Communities in East Africa. In Bechhaus-Gerst, M. & F. Serzisko (eds.), *Cushitic-Omotic: Papers from the International Symposium on Cushitic and Omotic Languages. St. Augustin, Jan. 6–9 1986*, 93–104. Hamburg: Helmut Buske.

Nurse, D. (2000) *Inheritance, Contact and Change in Two East African Languages*. Cologne: Rüdiger Köppe.

Nurse, D. & T. J. Hinnebusch (1993) *Swahili and Sabaki: A Linguistic History*. Berkeley: University of California Press.

Nurse, D. & B. F. Y. P. Masele (2003) Stratigraphy and Prehistory: Bantu Zone F. In Andersen, H. (ed.), *Language Contacts in Prehistory: Studies in Stratigraphy*, 115–134. Amsterdam; Philadelphia: John Benjamins.

Nurse, D. & G. Philippson (eds.) (2003) *The Bantu languages*. London New York: Routledge.

Odden, D. (2015) "Bantu Phonology." In *Oxford Handbooks Online*. Oxford: Oxford University Press. (Available online at www.oxfordhandbooks.com/view/10.1093/oxfordhb/9780199935345.001.0001/oxfordhb-9780199935345-e-59).

Odden, D. (n.d.) The Mushunguli – Chizigua language of Somalia. (Available online at www.ling.ohio-state.edu/~odden.1/mushunguli/).

Owen Lloyd, G. (1955) A Study of Some Xhosa Words of Afrikaans Origin. *South African outlook* 85: 90–93.

Owens, J. (1985) The origins of East African Nubi. *Anthropological Linguistics* 27: 229–271.

Owens, J. (1991) Nubi, Genetic Linguistics, and Language Classification. *Anthropological Linguistics* 33: 1–30.

Pakendorf, B., H. Gunnink, B. Sands & K. Bostoen (2017) Prehistoric Bantu-Khoisan Language Contact: A Cross-Disciplinary Approach. *Language Dynamics and Change* 7(1): 1–46.

Philippson, G. (1991) *Tons et accent dans les langues bantu d'afrique orientale: Étude comparative typologique et diachronique*. Paris: Université Paris V, thèse de doctorat.

Philippson, G. (1993) Tone (and stress) in Bantu. In Nurse, D. & T. J. Hinnebusch (eds.), *Swahili and Sabaki: A Linguistic History*. Berkeley: University of California Press.

Pountain, C. J. (2005) Gender Without Sex: The Semantic Exploitation of the Masculine/Feminine Opposition in the History of Spanish. In Wright, R. & P. Ricketts (eds.), *Studies on Ibero-Romance Linguistics Dedicated to Ralph Penny*, 329–348. Newark: Juan de la Cuesta.

Rycroft, D. K. (1983) Tone-patterns in Zimbabwean Ndebele. *Bulletin of the School of Oriental and African Studies* 46(1): 77–135.

Sacleux, C. (1939) *Dictionnaire swahili-français*. Paris,: Institut d'ethnologie.

Salmons, J. (1992) *Accentual Change and Language Contact: Comparative Survey and a Case Study of Early Northern Europe*. Stanford: Stanford University Press.

Samarin, W. J. (1990–1991) The Origins of Kituba and Lingala. *Journal of African Languages and Linguistics* 12(1): 47–77.

Samarin, W. J. (2013) Versions of Kituba's Origin: Historiography and Theory. *Journal of African Languages and Linguistics* 34(1): 111–181.

Schadeberg, T. C. (1973) Kinga: A Restricted Tone System. *Studies in African Linguistics* 4(1): 23–47.

Schadeberg, T. C. (1994) KiMwani at the Southern Fringe of KiSwahili. In Bakker, P. & M. Mous (eds.), *Mixed Languages: 15 Case Studies in Language Intertwining*, 239–244. Amsterdam: Instituut voor Functioneel Onderzoel van Taal en Taalgebruik.

Schadeberg, T. C. (1996) *De Swahili-talen van Mozambique*. Amsterdam: Koninklijke Nederlandse Academie voor Wetenschappen.

Schadeberg, T. C. (2003) Historical Linguistics. In Nurse, D. & G. Philippson (eds.), *The Bantu Languages*, 143–163. London; New York: Routledge.

Schadeberg, T. C. (2009) Loanwords in Swahili. In Haspelmath, M. & U. Tadmor (eds.), *Loanwords in the World's Languages: A Comparative Handbook*, 76–102. Berlin: De Gruyter Mouton.

Spiss, C. (1904) Kingoni und Kisutu. *Mittheilungen des Seminars für Orientalische Sprachen* 7: 270–414.

Spitulnik, D. (1998) The Language of the City: Town Bemba as Urban Hybridity. *Journal of Linguistic Anthropology* 8(2): 30–59.

Stirniman, H. (1963) *Nguni und Ngoni: eine kulturgeschichtliche Studie*. Vienna: Österreichische Ethnologische Gesellschaft.

Stolz, C. (2008) Loan Word Gender: A Case of Romancisation in Standard German and Related Enclave Varieties. In Stolz, T., D. Bakker & R. Salas Palomo (eds.), *Aspects of Language Contact: New Theoretical, Methodological and Empirical Findings with Special Focus on Romancisation Processes*, 399–440. Berlin: De Gruyter Mouton.

Storch, A. (2011) *Secret Manipulations: Language and Context in Africa*. Oxford: Oxford University Press.

Swilla, I. (2008) Signs of Language Shift in Chindali and the Impact of Swahili. *Language Matters* 39(2): 230–241.

Thomason, S. G. (1997) Ma'a (Mbugu). In Thomason, S. G. (ed.), *Contact Languages*, 469–488. Amsterdam; Philadelphia: John Benjamins.

Tse, H. (2013) Methodological Considerations in the Study of Sociophonetic Variation in an Underdocumented Minority Language: Somali Bantu Kizigua as a Case Study. *Proceedings of Journées d'Études Toulousaines (JéTou 2013), University of Toulouse*: 129–139 (Available online at http://jetou2013.free.fr/documents/JeTou2013-Actes-p2129-2139-Tse.pdf).

Tse, H. (2015a) The Diachronic Emergence of Retroflexion in Somali Bantu Kizigua: Internal Motivation or Contact-Induced Change? In Kramer, R., E. C. Zsiga & O. T. Boyer (eds.), *Selected Proceedings of the 44th Annual Conference on African Linguistics*, 277–289. Somerville: Cascadilla Proceedings Project.

Tse, H. (2015b) The Role of Shift-Induced Interference in the Development of a Typologically Rare Phonological Contrast in Somali Bantu Kizigua. *TIPA Travaux Interdisciplinaires sur la Parole et le Langage* 31 (Available online at http://tipa.revues.org/1426).

Tucker, A. N. (1947) Foreign Sounds in Swahili. *Bulletin of the School of Oriental and African Studies* 12(1): 214–232.

Van Rooyen, C. S. (1968) A Few Observations on the Hlonipha Language of Zulu Women. *Limi* 5: 35–42.

Van Warmelo, N. J. (1989) *Venda Dictionary: Tshivenda – English*. Pretoria: J.L. van Schaik.

Vitale, A. J. (1980) Kisetla: Linguistic and Sociolinguistic Aspects of a Pidgin Swahili of Kenya. *Anthropological Linguistics* 22(2): 47–65.

Voorhoeve, J. (1973) Safwa as a Restricted Tone System. *Studies in African Linguistics* 4(1): 1–22.

Wellens, I. (2005) *An Arabic Creole in Africa: The Nubi language of Uganda*. Leiden: Brill.

Whiteley, W. H. (1967) Swahili Nominal Classes and English Loan-words: A Preliminary Survey. In Manessy, G. (ed.), *La classification nominale dans les langues négro-africaines: actes du colloque international CNRS, Aix-en-Provence, 3–7 juillet 1967* 157–174. Paris: Éditions du Centre National de la Recherche Scientifique.

Williams, S. T. (2012) *Gender Conflict Resolution in Mushunguli*. Columbus: Ohio State University, MA thesis.

Wohlgemut, J. (2009) *A Typology of Verbal Borrowings*. Berlin: De Gruyter Mouton.

Zerbian, S. (2006) High Tone Spread in Northern Sotho varieties. In Mugane, J. M., J. P. Hutchison & D. A. Worman (eds.), *Selected Proceedings of the 35th Annual Conference on African Linguistics – African Languages and Linguistics in Broad Perspective*, 147–157. Somerville: Cascadilla Proceedings Project.

PART 2

CHAPTER 13

KWAKUM A91

Elisabeth Njantcho Kouagang and Mark Van de Velde

1 INTRODUCTION[1]

Kwakum (A91, ISO 639-3 kwu, glottocode kwak1266) is a cluster of Bantu language varieties spoken in the East Province of Cameroon. Ethnologue distinguishes four dialects: Baki, Betɛn, Til and Kwakum, which is the focus of this description. There is a high degree of intelligibility between Kwakum and Til, whereas dialectal variation is stronger between Kwakum, Baki and Betɛn. The Kwakum variety discussed in this chapter is spoken in the Doume sub-division. It is referred to by its speakers as *Kwàkúm*. The speakers' folk etymology for the language name is *kwày né kúm* 'generosity and fame/prosperity.' According to a 2013 census by Elecam, the Kwàkúm dialect has about 7,000 native speakers, whose villages are spread around the Doume sub-division (Grand Sibita, Petit Sibita, Grand Paki, Petit Paki, Mendim, Loumbou, Kempong, Kobila, etc.).

Kwakum communities share borders with communities speaking Makaa A83, Pol A92, Gbaya (Gbaya-Ngbaka-Manza) and Baka (Ubangi). Moreover, Kwakum-speaking settlements typically have immigrant communities speaking Kako A93 and/or languages from the Grassfields area and northern Cameroon. In this multilingual setting, Kwakum is mainly used in informal or traditional settings (family, market, cultural ceremonies, etc.), while communication in formal contexts (administration, school, etc.) is mostly in French. In addition to these two languages, many Kwakum speakers also speak Ewondo A72, Kako A93 and/or Gbaya.

The data used for the grammatical analyses in this chapter were collected by the first author between 2013 and 2017 from Kwakum speakers living in Grand Sibita. We rely on elicited materials as well as on a corpus of recorded and transcribed spontaneous speech (narrative and procedural texts, speeches, etc.). The only existing descriptive work on the language is the grammar sketch written by Belliard (2005, 2007), whose focus was an ethno-musicological study of the Kwakum people.

Kwakum is in many ways typologically unusual for a Bantu language. In order to characterise the structure of the language within the available space limits, we had to concentrate heavily on the morphology and the tone system. The morphological sections do contain quite a lot of syntactic information. A more thorough analysis of the syntax and segmental phonology of Kwakum is provided in the doctoral dissertation of Njantcho Kouagang (2018).

2 PHONOLOGY

2.1 Vowels

Kwakum has a seven-vowel system with contrastive vowel length. The (mid-)open vowels are much more frequent than the (mid-)close vowels in stems.

TABLE 13.1 VOWEL PHONEMES

	Front	Back
close	i iː	u uː
mid-close	e eː	o oː
mid-open	ε εː	ɔ ɔː
open	a aː	

The (near-)minimal pairs in (1) illustrate that vowel length is contrastive.

(1) a   /i/ vs. /iː/   dʒì° 'be'   dʒì: 'excrement'
     b   /u/ vs. /uː/   ì-bú⁺ʃὲ 'ash'   bùːʃὲ 'gather'
     c   /e/ vs. /eː/   kὲ 'what'   ì-kὲː° 'egg'
     d   /o/ vs. /oː/   lò 'grow'   lòː 'shoot'
     e   /ε/ vs. /εː/   ì-ʃὲ 'raffia'   ì-ʃέː 'sand'
     f   /ɔ/ vs. /ɔː/   ɲɔ́ 'snake'   ɲɔ́ː 'hip bone'
     g   /a/ vs. /aː/   bà 'cut up'   bàː 'kola nut'

2.2 Consonants

Kwakum has 28 consonant phonemes, including a series of aspirated stops and a series of prenasalised stops. The aspirated stops are reflexes of historical NC clusters, of which the initial N tends to be a class 9/10 prefix that has dropped (2).

(2) *n-pígò (9/10) 'kidney'   >   pʰĭ⁺kí 'kidney'
     *n-bèdì (9) 'front'   >   pʰὲl° 'in front'
     *n-kákà (9/10) 'pangolin'   >   kʰà° 'pangolin'
     *n-gáŋgà (1/2, 9/10) 'medicine man' >   kʰàː 'medicine man'

TABLE 13.2 CONSONANT PHONEMES

		Labial		Alveolar		Palatal		Velar		Labial- velar	
oral stops	- aspirated	p	b	t	d	ʧ	dʒ	k	g	k͡p	g͡b
	+aspirated	pʰ		tʰ		ʧʰ		kʰ			
fricatives	- aspirated	f	(v)			ʃ					
	+aspirated					ʃʰ					
nasals		m		n		ɲ		ŋ			
prenasalised stops		mb		nd		ɲdʒ		ŋg		ŋ͡mg͡b	
approximants				l		y				w	

In our lexical database, the voiced labio-dental fricative is rare. When it occurs before a front vowel, it is in free variation with its voiceless counterpart /f/: *ì-fítlɔ̀* ~ *ì-vítlɔ̀* 'darkness,' *kì-vèklà* ~ *kì-fèklà* 'statuette.' The prenasalised labial-velar stop also has a marginal distribution, as it is found in two stems only: *ŋmgbéŋgá* 'fishing net' and *ŋmgbàŋ* 'crow.' The borrowing *hámà* 'hammer' is the only word that contains an [h], which we did not include in the phoneme inventory. The voiced velar stop /g/ occurs either in front of /w/ (36 occurrences) or a (mid-)close back vowel (eight occurrences) (3). Furthermore, /g/ is found before an open vowel in one stem (*-ŋgàànɔ̀* 'refuse, deny,' *ŋgàtì* 'refusal') and before a /t/ in *-dùgtàà* 'get tired.'

(3) a   gwɔ́ɔ́⁺mbɔ́ 'chase'
     b   gùʃlɛ̀ 'pluck'

The lateral approximant /l/ is optionally realised as a trill [r] in nine words in our lexical database. In each case, it is preceded by an alveolar obstruent /nd/ or /t/ and followed by a mid-open vowel: *-fítlɔ̀* [~fítrɔ̀] 'night,' *-fìndlè* [~fìndrè] 'knock down,' *tʃéndlè* [~tʃéndrè] 'candle' and *-tándlè* [~tándrè] 'attach,' or a low vowel *-dʒàndlàà* [~dʒàndràà] 'move (millipede)' and *-ʃàndlàà* [~ʃàndràà] 'urine.'

Voiceless stops are in free variation with affricates whenever they occur before /i/ (*ì-dí* ~ *ì-dʒí* 'bait,' *tìlà* ~ *tʃìlà* 'lion'). Likewise, there is a free variation between alveolar and palatal nasals before /i/ (*nìkɔ̀* ~ *ɲìkɔ̀* 'bend').

2.3 Morphophonology

Nasals assimilate to a following consonant in place of articulation if a morpheme boundary separates the two consonants. In (4), assimilation takes place after the deletion of the last vowel of *mɔ̀ɔ̀nɔ́* 'child.'

(4)   mɔ̀ɔ̀m pʰàâm
       mɔ̀-ɔ̀nɔ́   ᴴ-pʰàâm
       1-child con-1.man
       'boy' or 'the man's child'

(5)   àpʰĭ ŋmgbɔ́⁺ndɔ́
       à-pʰiᴴ   ᴴ-n-gbóndɔ́ᴸ
       2-dog   prs-prs-bark
       'Dogs are barking.'

In non-prepausal context, there is a high tendency for some words ending in /i/, /u/, /ɛ/ and /ɔ/ to drop their final vowel (6a). Once deletion takes place, the epenthetic vowels *i~i* or *u* are optionally inserted after obstruents. Their choice is determined by the roundness feature of the preceding vowel: *i~i* occurs after an unrounded vowel (6b), while *u* is required when preceded by a rounded vowel or the glide /w/ (6c). Both can occur after /(m)b/ in free variation, whatever the roundness of the vowel that precedes it (6d).

(6)   a |pʰĭkíᴸ bùláàwὲᴴ| → pʰĭk bùláàwὲ° 'many kidneys'
            |à-yéklὲ bùláàwὲᴴ| → àyékl̀ bùláàwὲ° 'many teachers'
            |ì-kààmɔ̀ bùláàwὲᴴ| → ìkààm bùláàwὲ° 'to love very much'

b |ì-tàǎkɔ̀ bùláàwὲᴴ| → ìtààk(ì) bùláàwὲ° 'to take much'

c |ǹ-bɔ̀ɔ̀ʃɔ̀ bùláàwὲᴴ| → m̀bɔ̀ɔ̀ʃ(ù) bùláàwὲ° 'much misfortune'
|pʰyààwɔ́ bùláàwὲᴴ| → pʰyààw(ù) búláàwὲ° 'much blood'

d |ì-dàámbɔ́ᴸ bùláàwὲᴴ| → ìdàámbí bùláàwὲ°~ ìdàámbú bùláàwὲ° 'to cook much'

This morphophonological process is conditioned by various parameters, such as syllable structure, grammatical category and tone of the final vowel. For instance, the deletion of final /ɔ/ can occur in nominal stems with an initial heavy syllable like tʰààlɔ̀ 'grandchild' (7), but /ɔ/ cannot drop if the noun stem has an initial light syllable, as in tʃìlɔ̀ 'gorilla' (8). Verbs, in contrast, allow the deletion of final /ɔ/ irrespective of their syllable structure (9).

(7) tʰààl dʒì tɛ́
    tʰààlɔ̀   dʒìᴴ   tὲᴴ
    1.grandchild be   there
    'The grandchild is over there.'

(8) tʃìlɔ̀ dʒì tɛ́   (*tʃil dʒì tɛ́)
    tʃìlɔ̀   dʒìᴴ   tὲᴴ
    1.gorilla   be   there
    'The gorilla is over there.'

(9) áɲtʃil tàmbyὲ
    ᴴ-à-n-tʃìlɔ̀   tàmbyὲ
    PRS-3SG-PRS-write good
    'He writes well.'

2.4 Phonotactics

The syllable nucleus can be a vowel, a nasal or the lateral /l/. Only syllables with a vocalic nucleus can have a coda. Syllables with a nucleus /l/ can have an onset, as in à.yὲ.kĺ.kòò 'he taught.' There is no straightforward way to identify syllable boundaries where two consonants succeed each other. We decided to assume a complex onset only in those cases where a succession of consonants can occur in utterance-initial position. Defined this way, complex onsets always consist of a consonant followed by /w/ or /y/. If the glide is /w/, the initial consonant has to be velar.[2] The syllable types attested in the Kwakum lexicon are V, N, CV(:), CGV(:), CV(:)C, CGV(:)C and CL. The rule of non-prepausal vowel deletion described in Section 2.3 gives rise to complex codas with successions of obstruents, as in /bǎkʃɛ́/ → [bǎkʃ] 'keep!', /pùʃkɔ́/ → [pùʃk] 'error' or /fóktὲ/ → [fókt] 'listen carefully.'

Only four words of a total of 1900 in our lexical database have a closed syllable with a long vowel, viz. gbù:ŋlὲ 'plough, turn the soil,' ì-tǎːn 'five,' mù:ŋlὲ 'uproot' and pʰààm 'man, male person.'

As is typical in the North-Western Bantu languages, the distribution of consonant phonemes over stems is heavily skewed. Table 13.3 shows that the occurrence of half of the consonants is restricted to the onset of stem-initial syllables (O_1). Restrictions on the possible occurrence of consonants become stronger when we move to onsets of non-stem initial syllables (O_2), non word-final codas (C_{NWF}) and word-final codas (C_{WF}). Note that whenever the succession of symbols <kp> occurs outside of stem initial position, it is the orthographic representation of a succession of /k/ and /p/, as found in a small number of reduplicated stems, such as *kì-pèkpékì* 'end.'

TABLE 13.3 PHONOTACTIC DISTRIBUTION OF CONSONANT PHONEMES

	O_1	O_2	C_{mvf}	C_{vf}
Cʰ, k͡p, g͡b, f, v, b, d, g, dʒ, ɲ	+	–	–	–
ʧ, ɲdʒ, ŋg	+	+	–	–
p, t, k, ʃ, mb, nd	+	+	+	–
m, n, ŋ, l, y, w	+	+	+	+

Out of a total of 911 occurrences, there are three exceptions to the generalisation that voiced stops are restricted to stem-initial onsets: *fìlbà* 'cooking pot,' *tèndbò* 'spider' and *-dùgtàà* 'become tired.' The first two of these can be respectively explained as a borrowing (< English *silver*) and a historical compound, cf. **-tanda* 'spider, spider's web' and PB **-bòbì* 'spider' (Bastin *et al.* 2002). Whenever the voiceless affricate /ʧ/ occurs outside of O_1 position, it is followed by the vowel /ɛ/, for a reason that we do not know. The distribution of the velar nasal /ŋ/ is exceptional, because it does not occur in O_1 position, except in the verb stem *ŋwèŋlè* 'persist.'

In reduplicated stems, the base forms a prosodic stem: *pè~bèlà* 'seed' (cf. *-bèlɔ* 'plant, sow'), *pà~bám* 'disapproval' (cf. *ì-bàá⁺mɔ́* 'reprimand'). The prosodic status of the reduplicant is somewhat ambiguous. If the base starts in a voiced oral stop, it is devoiced in the reduplicant, arguably because voiced stops are restricted to stem-initial position. However, labial-velar stops are allowed in O_1 position of the reduplicant, as in *kpà~kpá⁺tí* 'scissors,' although they too are normally restricted to stem-initial position.[3]

The mid-close vowels are absent from final open syllables of polysyllabic stems. The few exceptions to this generalisation mostly involve borrowings, reduplicated stems or the Past 3 suffix *-kòò*.

2.5 Tone

Syllables can be realised with a low (à), a high (á), a falling (â) a rising (ǎ), or a down-stepped high (⁺á) tone. In utterance-final position, low tones can be realised low-falling (à) or level low (à°) (10). In certain contexts, utterance-final downstepped high tones can be alternatively realised as level lows and vice versa. In our current analysis, the tone-bearing unit (TBU) is the mora and there is a three-way underlying opposition between low, high and zero. Underlying tones can be floating or attached to a TBU. In underlying representations, floating low and high tones are respectively represented by the superscript letters ᴸ and ᴴ (10a) and toneless TBUs are represented by means of the absence of a tone mark (10b). Example (10) also illustrates the fact that the level realisation of utterance final low tones is due to a following floating high.

(10) a |ŋgwɔ̀ɔ̀ᴴ| → ŋgwɔ̀ɔ̀° 'brain' (level low realisation)

　　 b |ʃòo| → ʃòò 'fish sp.' (low-falling realisation)

　　 c |bùpà| → bùpà 'animal' (low-falling realisation)

On top of lexical tones, Kwakum also has a low boundary tone ᴸ﹪ that can be optionally inserted at the beginning of every utterance.

A floating tone attaches to the first TBU to its right, whose underlying tone it delinks (11a). If it finds a succession of two TBUs of which the second is toneless, it links to both of them (11b).

(11)   a   ᴴ-kù → kúᴸ → [kú] 'of the hole'
       b   ᴸ﹪tʃóo w-ɛ́ → tʃòoᴴ wɛ́ → [tʃòò wɛ́] 'that iron'

Kwakum has a rule of rightward tone spreading across word boundaries. Spreading tones behave the same as floating tones in the way they attach (12).

(12)   a   pú kòndù → pú kɔ́ndù 'the girl's misfortune'
       b   kòndù w-ɛ́ mòò mè-ʃé → kòndù wὲ mɔ́ò mὲʃé 'that girl who came'

Tone spreading also takes place from prefixes to stems (13a), except in words that are in utterance final position (13b). Example (13b) also illustrates the fact that tones link to a following TBU if it is structurally toneless, which is why kìléwɔ́ surfaces with two high tones in isolation.

(13)   a   kìlèwɔ̀ wáàmbɔ́
           kì-léwɔ   w-ààmbɔ́
           7-baby   PP₁-1SG.POSS
           'my baby'

       b   kìléwɔ́ 'baby'

There are three contexts in which the attachment of a tone to a following TBU is blocked. The first is that a low tone cannot delink a following high if the latter is the last tone of an utterance. Compare the tone on the demonstrative wɛ́ in example (12b) to that in (14).

(14)   kòndù w-ɛ́ → kòndù wɛ́ 'that girl'   *kòndù wɛ̀°

The second is the failure of a low tone to attach to a following high tone if the latter is immediately followed by a low tone within the same word.⁴ In (15), spreading of the final low tone of tààkɔ is blocked by the floating low that follows the linked high of fénᴸ 'handles.' The same floating low prevents the linked high from spreading to the right, so that the initial low of bùláàwὲ° 'many, much, lots' is preserved.

(15) |ì-tààkɔ̀ fénᴸ bùláàwɛᴴ| → ìtààk(ì) fén bùláàwὲ° 'to take many handles'

The third context in which tonal attachment is blocked is the mirror image of the second one: a floating or spreading high cannot attach to a following TBU with a low tone when the latter is itself followed by a high tone. However, an extra condition for blocking high tone attachment is that the LH contour must occur in a verb stem (16a) or it must occur within one syllable and the high part should not be floating (16b).

(16)   a |ndóm dʒɔ̀ʃɛ́ᴸ kòò yὲᴴ| → ndóm dʒɔ̀ʃɛ́ kòò yὲ° 'the husband hid it'
       b |ᴴ-kɔ̀ɔ́ndὲ| → kɔ̀ɔ́ndὲ 'of the fish sp.'

Finally, there are some morphemes, including agreement prefixes, that have fixed tones and are (optionally?) impervious to tone spreading.

Kwakum also has a number of rules of downstepping. First, downstepping takes place on the second high in a succession of two high tones in case of an intervening floating low. This floating low may be the result of delinking due to an incoming high from the left. In (17), the floating high tone of the connective relator attaches to the first TBU of *mòtú* 'head,' of which it delinks the low tone, which subsequently attaches to the following high. Instead of delinking this high, it combines with it to create a downstepped high.

(17) |ᴴ-mòtú| → mó⁺tú 'of the head'

In (18), the high of *fén* 'handle' cannot spread, because it is blocked by the final floating low of this noun. Since this low is trapped in between two highs, it creates downstep.

(18) |fénᴸ wɛ́| → fén ⁺wɛ́ 'that handle'

No downstep formation takes place in (19) for reasons that we have already explained. The connective high links to the first TBU of *ʃùkɛ́* 'mouse,' of which it delinks the low tone. This delinked low cannot attach to the following high, because it is itself followed by a low within the same word.

(19) |ᴴ-ʃùkɛ́ᴸ| → ʃúkɛ́ 'of the mouse sp.'

Downstepping also takes place where two TBUs that are linked to a high tone meet. When this happens across a morpheme boundary, downstep is always optionally possible, as shown by the second downstep in (20), the one on ⁺*dú*.[5] In contrast, within a morpheme two adjacent high TBUs lead to downstep of the second high only in pre-pausal position (21a). We have never heard clear cases of downstep of the high part of a falling tone.

(20) áŋ⁺gwí ⁺dúŋɛ́ŋ
      ᴵᴵ-â-n-gwí   dùŋɛŋ
      PRS-3SG-PRS-die   morning
      'It dies in the morning.'

(21) a |ʃúlyɛ́ᴸ| → ʃú⁺lyɛ́ 'smoke'
      b |ʃúlyɛ́ᴸ nè bèetàà| → ʃúlyɛ́ nè bèètàà 'smoke and fire'

The need for positing underlyingly toneless TBUs is demonstrated in three different ways in examples (22–23). First, the high tone nouns in (22) appear without downstep on the second syllable in pre-pausal position. This can be easily formalised by assigning an underlying High tone to their first TBU only, which surfaces on the next TBU as well.

(22) a |ʃólɔᴸ| → ʃólɔ́ 'bench' (*ʃó⁺lɔ́)
      b |dʒíki| → dʒíkí 'river' (*dʒí⁺kí)

Second, when the low tone of *nè* 'with' spreads onto the following TBU in (23a-b) and deletes its high tone, it links to all unattached TBUs. This is why the low links to one TBU in (23a) versus two TBUs in (23b). Third, in (23c), we see that the high of the first

TBU of *fóló* 'bench' is protected by a its final floating low, meaning that there can be no intervening tone between them.

(23) a |nὲ ʃúlyɛ́ᴸ bùláàwɛᴴ| → nὲ ʃùlyɛ́ bùláàwὲ° 'with much smoke'
   b |nὲ dʒíki bùláàwɛᴴ| → nὲ dʒìkì búláàwὲ° 'with many rivers'
   c |nὲ ʃólɔᴸ bùláàwɛᴴ| → nὲ ʃólɔ bùláàwὲ° 'with many benches'

Finally, two tonal phenomena that cannot be represented in the basic description provided so far need to be mentioned. The first is that a number of syllables with an underlying low tone become rising when a high tone attaches to them (versus the expected high or falling pattern). This is the case of the past tense prefix *àà-* (Section 6.3.1), the first syllable of the stem of certain possessive pronouns (Section 4.4) and that of a number of nouns (24).

(24) base noun   after a connective ᴴ
   pʰòŋgò 'maize'   pʰǒŋgò 'of the maize'
   tʰà[à]lɔ̀ 'grandchild'   tʰàálɔ̀ 'of the grandchild'
   ʃɔ̀kù 'elephant'   ʃɔ̌kù 'of the elephant'

The second concerns a number of morphemes that change the final low tone of a preceding word to a high. This cannot be represented by means of an initial floating ᴴ, because that would be supposed to attach to the right, rather than to the left. We will mark these morphemes with an initial upward arrow in their underlying representation. Examples are the complementiser ⁺*nàá*ᴸ, interrogative ⁺*fὲ* 'where' and the near-speaker demonstrative ⁺*nɛ́*ᴸ.

(25) |ᴴ-à-n-kὲὲ ⁺nàáᴸ| → áŋkèé nàá 'He says that. . .'

3 NOUNS

3.1 The syllable structure of noun stems

The majority of noun stems are disyllabic: 63%. Two-thirds of the disyllabic stems have a CV.CV pattern. 28% of noun stems are monosyllabic, two-thirds of which have a CVC pattern. 7% are trisyllabic, mostly of the CV.CV.CV type. These figures only contain consonant-initial stems. The remaining 2% of noun stems begin in the vowel *a*, which adds a syllable to the above patterns, giving rise to di-, tri- and quadrisyllabic stems.

3.2 Nominal classification

Kwakum has eight morphological classes, defined as sets of nouns that have the same nominal prefix. There are seven prefixes, *mò-*, *gwò-*, *kì-*, *ì-*, *à-*, *ǹ-*, *n-* and the lack of a prefix, symbolised as Ø-. Three of these forms, *mò-*, *n-* and Ø-, are used exclusively to mark singular nouns. Three others, *gwò-*, *ǹ-* and *à-*, are restricted to marking plural nouns. The prefixes *kì-* and *ì-* mark the singular of some nouns and the plural of others. Figure 13.1 is a somewhat simplified presentation of the singular-plural pairings in morphological classes.

   Figure 13.1 excludes a small number of marginal patterns, such as *kì-/ǹ-* found only in the noun *kìbámbú / mbámbú* 'board/s.' Equally excluded from Figure 13.1 are the five

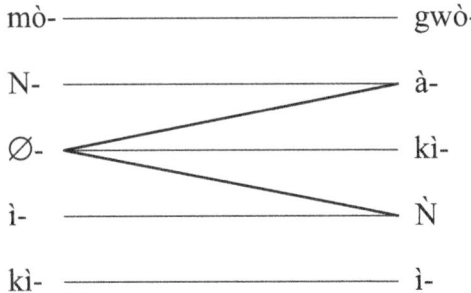

FIGURE 13.1 KWAKUM MORPHOLOGICAL CLASSES.

nouns in our database with irregular singular-plural pairings, two of which are reflexes of PB class 5/6 nouns with a vowel-initial stem: *dʒ-íˉʃí / m-íˉʃí* 'eye/s' and *díˉnɔ́ (~ì-díˉnɔ́) / m̀-míˉnɔ́ (~n̄-díˉnɔ́ ~ kì-díˉnɔ́)* 'name/s.' We did include pattern *mò-/gwò-* despite its very low number of members, because the three nouns it contains are important: *mò-mɔ́* 'person,' *m-ɔ̀ɔ̀nɔ́* 'child' and *mó-myáá* 'woman.'

The morphological class system shows a high degree of variability, in that many nouns can have alternative class prefixes, mostly in the plural, but sometimes also in the singular. During elicitation, speakers often say they do not know the plural of nouns that lack a class prefix in the singular, or they accept alternative plural forms (26).

(26) a   Ø-mɔ̀ ~ ì-mɔ̀ 'shed' / m̀-mɔ̀ ~ à-mɔ̀ 'sheds'
      b   Ø-kótú 'bag' / à-kótú ~ kì-kótú 'bags'

6% of the nouns in our database begin in a syllabic nasal *n̄* in the singular. This nasal is always preserved in the plural, therefore does not commute and should not be analysed as a morphological class marker according to our definition, even though historically it is certainly a class marker and synchronically it can be analysed as a derivational prefix in nouns that have a derivational relation with a verb (27). We will have more to say on this set of nouns when we discuss gender assignment.

(27)   Ø-n̄làkʃàà / à-nlàkʃàà 'question/s' < làkʃɛ̀ 'ask'

Finally, plural markers can be stacked (i.e., two plural markers) or additive (i.e., plural marker added to the singular marker). We find the optional stacked plural marking *à-n̄-* in ten out of 155 nouns that have a singular in *ì-* and a plural in *n̄-* (28).

(28)   ì-ʃúm / n̄-ʃúm ~ à-n̄-ʃúm 'bracelet/s'

Stacking is more generalised among younger speakers, who add an *à* before all other plural markers. Additive plural marking can be found in nine out of 136 nouns with a singular prefix *kì-* and a plural *ì-*. These can have an alternative plural in *à-*, which is then added to the singular prefix *kì-* (29). Both phenomena point to a certain tendency for *à-* generalising as a plural marking.

(29) kì-bèkɔ̀ / ì-bèkɔ̀ ~ à-kì-bèkɔ̀ 'shoulder/s'

Kwakum has five noun classes, defined as sets of nouns that trigger the same agreement pattern. Agreement in noun class is restricted to a relatively small set of adnominal modifiers, almost each of which has a separate paradigm of agreement markers, given in Table 13.4. These modifiers are possessive pronouns (I), demonstratives (I-V) and the connective relator (VI). For ease of reference, we have given the Kwakum noun classes a Bantu-style number, chosen somewhat arbitrarily (but see below for a partial justification). The agreement prefixes of the six paradigms will all be glossed as PP in this chapter. Nouns are assigned to noun classes on the basis of their morphological class and number.

TABLE 13.4 PARADIGMS OF NOUN CLASS AGREEMENT MARKERS

Noun class #	Morphological class & number	I	II	III	IV	V	VI
1	mò-, n-, Ø-	w-	Ø-	yí-	Ø-	yí-	Ø-
2	à-, gwò-, ì- (PL)	y-	yí-	yí-	yí-	yí-	Ø- ~ yì-
5	ì- (SG)	ly-	Ø- ~ lí-	lí- ~ yí-	Ø-	yí-	Ø⁻
6	ǹ(-)	m-	mí-	mí-	mí-	mí-	Ø- ǹ-
7	kì-	tʃ-	tʃí-	tʃí-	tʃí-	tʃí-	Ø- ~ kì-

Class 6 has a subclass for the plural of five nouns, viz. ì-tɔ́ / ǹ-tɔ́ 'ear/s,' ì-lɔ́ / ǹ-lɔ́ 'intestine/s,' ǹ-ʃéé 'work,' ì-tú / ǹ-tú 'day/s' and mbɔ́ / m̀-mbɔ́ 'hand/s,' defined by the agreement prefix ǹ- in paradigm VI (connectives). The brackets around the hyphen in ǹ(-) in the second column of Table 13.4 are meant to show that nouns that begin in a non-commuting syllabic nasal consonant in the singular tend to be assigned to class 6, as are the plural nouns that begin in ǹ-, where this nasal can be straightforwardly analysed as a class prefix, because it commutes with a different singular prefix. This could be used as an extra argument to recognise ǹ- as a class marker in singular nouns anyway. However, a minority of 13% of singular nouns that start in ǹ trigger agreement of class 1, as if they had no prefix. Singular ǹ- is therefore somewhere in between a canonical class prefix and the initial syllable of a possibly derived stem.

Some of the Kwakum class markers are easily identified as cognate to the class markers of languages with a more typical Bantu noun class system. Among the morphological class markers, mò- corresponds to PB class 1, ì- (SG) to PB class 5, kì- (SG) to PB class 7 and ì- (PL) to PB class 8. Many instances of Kwakum ǹ- correspond to PB class 6, as ǹ- marks the plural of nouns that take ì- in the singular and the near totality of nouns for liquids take ǹ-. The à- prefix may be a reflex of the PB class 2 prefix bà- that generalised to become a default plural marker. The form gwò- that currently marks the plural of nouns that have the prefix mò- in the singular must be an innovation.

3.3 Derivation

Slightly more than 7% of the nouns in our lexical database are clearly in a derivational relation with a verb. In many instances, nouns are derived from verbs by the addition of a class prefix to the verb stem. This prefix is most often ì- (30), sometimes also kì- (30b) or ǹ- (30c). Some nouns are derived from verbs through reduplication of the verb stem (31).

(30) a ì-fɔ́ɔ́⁺lɔ́ 'peace, coolness' < fɔ́ɔ́⁺lɔ́ 'be cool'
     b kì-lɔ́⁺ɔ́ 'depth' < lɔ́⁺ɔ́ 'be deep'
     c ǹ-dʒòòlàà 'bath' < dʒòòlàà 'bathe'

(31) a pɛ̀~bɛ̀là 'seed' < bɛ̀lɔ̀ 'plant'
     b pà~bám 'disapproval' < bàá⁺mɔ́ 'reprimand'
     c kà~kàn° 'story' < kàɲɛ̀ 'tell a story'

In the other types of derivational relations, there is no straightforward way to decide which is derived from which. Sometimes the addition of a prefix is accompanied by a change in the stem, which may be segmental (32a), tonal (32b) or both (32c). Sometimes, the noun does not have a prefix and the related verb and singular noun do not differ (33a) or only in the shape of their stem (33b).

(32) a ì-dʒɔ̀ʃù 'hiding place' – dʒɔ̀ʃɛ̀ 'hide'
     b ì-yépyɛ́ 'religion' – yé⁺pyɛ́ 'believe'
     c ì-byàndʒí 'disobedience' – byàndʒɛ̀ 'disobey'

(33) a yéklɛ̀ 'teacher' – yéklɛ̀ 'teach'
     b tʃɛ̀w 'game' – tʃɛ̀wɔ̀ 'play'

The only more or less regular derivational pattern we were able to identify thus far is the derivation of instrument nouns involving the prefix kì- and the suffix -kà, on top of other formal changes.

(34) a kì-pètkà 'plug, lid < pètɔ̀ 'close'
     b kì-dʒìwùkà 'key' < dʒítlɛ̀ 'open'
     c kì-tʃálkà 'sharp weapon' < tʃàá⁺lɔ́ 'be sharp, be fast'

4 PRONOUNS, ADNOMINAL MODIFIERS AND NOUN PHRASE STRUCTURE

4.1 Personal pronouns

Table 13.5 provides the paradigms of person markers in Kwakum. We distinguish between free forms (independent pronouns) and bound forms, analysing a person marker as an affix when its occurrence is restricted to immediate pre- or postverbal position. Some person/number combinations lack a subject and/or object affix and are therefore always represented by means of an independent pronoun. Kwakum does not have object prefixes.

The first- and second-person singular object suffixes are restricted to verbs that lose their final vowel in non-prepausal position, i.e., they attach only to verb forms that end in a consonant (35a). Other verbs are followed by an independent pronoun (35b).

TABLE 13.5 PERSON MARKERS

	Free forms	Subject prefixes	Object suffixes
1sg	ɲì	ǹ-	-ɛɛ
2sg	gwɛ̀	ɔ̀-	-ɔɔ
3sg	dʒɛ̀ᴴ	à-	-yɛ̀ᴴ
1pl	ʃɛ̀ᴴ		
2pl	díné	nɛ̀ᴴ-	
3pl	dʒàᴴ	yɛ̀ᴴ-	-yàᴴ

(35) a   ám⁺bíwɛ́ɛ̀
          ᴴ-à-n-bíwɔ́ᴸ-ɛɛ
          prs-3sg-prs-beat-1sg
          'He beats me.'

      b   ámbèè ɲì
          ᴴ-à-n-bèè   ɲì
          prs-3sg-prs-follow 1sg
          'He follows me.'

The second-person singular subject prefix ɔ̀- is in free variation with an independent pronoun (36).

(36) a   ɔ́m⁺bíwɛ́ɛ̀
          ᴴ-ɔ̀-n-bíwɔ́ᴸ-ɛɛ
          prs-2sg-prs-beat-1sg
          'You beat me.'

      b   gwɛ́ ⁺mbíwɛ́ɛ̀
          ᴴ-gwɛ̀   n-bíwɔ́ᴸ-ɛɛ
          prs-2sg   prs-beat-1sg
          'You beat me.'

In the presence of a nominal subject, the use of an additional subject marker is optional in the plural and ungrammatical in the singular (37). However, when the nominal subject is in focus, the verb always takes a subject marker.

(37) a   pʰàâm ɲ́ʃè°
          pʰàâm   ᴴ-n-ʃèᴴ
          1.man   prs-prs-come
          'The man comes.'

      b   àpʰàâm (yɛ́-)ɲ́ʃè°
          à-pʰàâm   ᴴ-yɛ̀ᴴ-n-ʃèᴴ
          2-man   prs-3pl-prs-come
          'The men come.'

Kwakum also has a set of four dual number pronouns, presented in Table 13.6, which can take any position in the clause: subject (38a), primary object (38b), secondary object or prepositional complement (38c).

TABLE 13.6 DUAL NUMBER PRONOUNS

1sg+2sg	díʃɔɔ̀ᴴ
1sg+3sg	díʃèɛ̀ᴴ
2sg+3sg	díɲèɛ̀ᴴ
3sg+3sg	yáɲèɛ̀ᴴ

(38) a   díʃɔɔ̀ᴴ mɛ́⁺dʒí támbyɛ̀
    díʃɔɔ̀ᴴ   ᴴ-mɛ̀-dʒí   tàmbyɛ̀
    1sg+2sg pst2-pst2-eat.pst2 good
    'We (you and I) ate well.'

   b   pʰàâm mɛ́⁺dʒɛ́ɛ́ díʃɔɔ̀°
    pʰàâm   ᴴ-mɛ̀-dʒɛ́ɛ́   díʃɔɔ̀ᴴ
    1.man   pst2-pst2-see.pst2 1sg+2sg
    'The man saw us (you and me).'

   c   pʰàâm mɛ́⁺ʃɛ́ɛ́ nɛ́ díʃɔɔ̀°
    pʰàâm   ᴴ-mɛ̀-ʃɛ́ɛ́   nɛ̀   díʃɔɔ̀ᴴ
    1.man   pst2-pst2-work.pst2 with 1sg+2sg
    'The man worked with us (you and me).'

Where a dual pronoun can be used, the corresponding plural pronoun can normally be used too (39b), as well as the two relevant singular pronouns linked by the preposition *nɛ̀* 'and, with' (39c).[6]

(39) a   pʰàâm mɛ́⁺dʒɛ́ɛ́ díʃɔɔ̀°
    pʰàâm   ᴴ-mɛ̀-dʒɛ́ɛ́   díʃɔɔ̀ᴴ
    1.man   pst2-pst2-see.pst2 1sg+2sg
    'The man saw us (you and me).'

   b   pʰàâm mɛ́⁺dʒɛ́ɛ́ ʃɛ́
    pʰàâm   ᴴ-mɛ̀-dʒɛ̀ɛ́   ʃɛ̀ᴴ
    1.man   pst2-pst2-see.pst2 1pl
    'The man saw us.'

   c   pʰàâm mɛ́⁺dʒɛ́ɛ́ ɲí nɛ̀ gwɛ̀
    1.pʰàâm   ᴴ-mɛ̀-dʒɛ́ɛ́   ɲì   nɛ̀   gwɛ̀
    1.man   pst2-pst2-see.pst2 1sg with 2sg
    'The man saw you and me.'

4.2 Connectives

The connective construction is used to link two, typically nominal, constituents. Because the dependency relations between these constituents are ambiguous in Kwakum, we will designate them by means of the neutral terms *first relatum* (R1) and *second relatum* (R2). The connective relator in Kwakum is an optional floating high tone prefixed to R2, as in (40), where the connective construction is used to express a possessive relation. The connective relator optionally agrees in gender with the head noun (41). Only two classes have a connective agreement prefix, viz. class 2 and class 7. Surprisingly, these agreement prefixes, respectively *yì-* and *kì-*, have a low tone.

(40) kàkàn mɔ́ɔ̀nɔ́
      kàkànᴴ   ᴴ-mɔ̀-ɔ̀nɔ́
      1.story   CON-1-child
      'the story of the child'

(41) kìdʒìmá kìfyâl
      kì-dʒìmá   kì-ᴴ-fyàl
      7-beautiful PP₇-CON-daughter_in_law
      'a beautiful daughter-in-law'

Some of the adnominal modifiers discussed in the remainder of this section involve a connective relator.

4.3 The nominaliser-linker mòò/gwòòm

Kwakum has a marker *mòò* (PL *gwòòm*) that originates in the noun *mòmɔ́* 'person' and that is used as a nominaliser or a linker used to introduce adnominal modifiers. It is used to nominalise adnominal demonstratives, e.g., *mòó ⁺né* 'this one' (see Section 4.5 for demonstratives). Moreover, it can head R2 in connective constructions that express a possessive relation (optionally) (42) or an ordinal number (obligatorily) (43). The element that follows *mòò* takes the connective relator ᴴ-, optionally if it is a number (43), obligatorily elsewhere, including in the expression of 'first' and 'last' (44). *Mòò* can itself optionally be linked to the preceding element by the connective prefix ᴴ-.

(42) bùpà mòò mɔ́ɔ̀nɔ́
      bùpà   mòò   ᴴ-mɔ̀-ɔ̀nɔ́
      1.animal   NLNK.SG   CON-1-child
      'the child's animal'

(43) a   mòò íbá⁺á ~ mòò ìbá⁺á 'the second one'
      b   mɔ́ɔ̀n mòò íbá⁺á   ~mɔ́ɔ̀n mòò ìbá⁺a
          mɔ̀-ɔ̀nɔ́   mòò   ᴴ-ì-báàᴴ   mɔ̀-ɔ̀nɔ́   mòò   ì-báàᴴ
          1-child   NLNK.SG   CON-2-two   1-child   NMLZ.SG   2-two
          'the second child'

(44) mɔ́ɔ̀n móò pʰɛ̌⁺í
      mɔ̀-ɔ̀nɔ́   mòò   ᴴ-pʰɛ̀lᴴ
      1-child   NLNK.SG   CON-1.front
      'the first child'

The nominaliser-linker *mòò* is also used to introduce relative clauses (see Section 7.3). Its current distribution strongly suggests that *mòò* was initially grammaticalised as a nominaliser of adnominal modifiers that were used in apposition to their head noun and subsequently reintegrated in the noun phrase, leading to the further evolution of *mòò* from nominaliser to linker in some of its uses. This is a common scenario in the Bantu languages, responsible for the typologically unusual word order patterns in the noun phrase structures of the family (see Section 5.3 in Van de Velde, this volume). In contrast, the origin of the Kwakum nominaliser-linker in a noun for 'person' is a departure from the much more common demonstrative origin of this element.

4.4 Possessive pronouns

Possessive pronouns follow the noun and take an agreement prefix of paradigm I (see Table 13.4). There are segmental and tonal differences between the stem of possessive pronouns that agree with class 1 controllers and the stem of those that agree with nouns from the other classes. Since these differences cannot be described by means of the synchronic rules of the language, we treat them as being suppletive. Segmentally, the stems for first- and second-person plural possessors have vowels that are identical to those of the agreement prefix, viz. /u/ in class 1 and /i/ elsewhere. Tonally, the 1SG possessor stem of the classes 2–7 has a rising tone on its first syllable, the high part of which must historically originate in the high tone of their prefix.

TABLE 13.7 POSSESSIVE PRONOUNS

	1SG	*2SG*	*3SG*	*1PL*	*2PL*	*3PL*
class 1	wààmbɔ́	ɔ̀ɔ̀^{H}	ὲὲ^H	ùʃú	ùn^H	àa^H
other classes	-ààmbɔ́^{L}	-ɔ́ɔ	-ὲὲ^L	-íʃí^L	ín^L	áa

As shown in example (45), singular nouns can trigger agreement of their own class (45a) or of class 1 (45b), in free variation. Note that a high tone that links to the 1SG possessor pronoun of class 1 links to its first TBU (45b).

(45) a   kìlὲwɔ̀ tʃàá⁺mbɔ́
          kì-léwɔ   tʃ-ààmbɔ́^{L}
          7-baby   PP$_7$-1SG.POSS

      b   kìlὲwɔ̀ wáàmbɔ́
          kì-léwɔ   w-ààmbɔ́
          7-baby   PP$_1$-1SG.POSS
          'my baby'

Two nouns have an alternative inalienable possessive construction, in which the possessive modifier is merged with their stem. The inalienable construction is the most frequently used.

(46) a   kòtʃ-έὲ° ~ kòkù w-έὲ° 'his/her maternal uncle'
      b   ɲótʃ-ὲὲ° ~ ɲótú wὲὲ° 'his/her body'

In order to express dual possessives, the noun is followed by a connective relator and a dual independent pronoun (47).

(47) gwɔ̀ɔ̀n (yì)díʃɔ̀ɔ̀°
      gwɔ̀-ɔ̀nɔ́   yì-^{H}-díʃɔ̀ɔ̀^{H}
      2-child   PP$_2$-CON-1SG+2SG
      'our children'

Independent possessive pronouns agree with their possessee in number, but not in noun class. In the singular they take the prefix *gú-*, which may be a reflex of the Proto-Bantu class 17 marker, and the stem of the adnominal possessive pronouns of class 1. In the plural, their prefix is *ʤ-* and their stem that of the adnominal form of classes 2–7.

TABLE 13.8 INDEPENDENT POSSESSIVE PRONOUNS

	SG possessee	*PL possessee*
1SG	gwáàmbɔ́	ʤàámbɔ́ᴸ
2SG	gwɔ́ɔ́	ʤɔ́ɔ
3SG	gwéɛ́	ʤéɛᴸ
1PL	gúʃû	ʤíʃîᴸ
2PL	gúń	ʤíńᴸ
3PL	gwáá	ʤáa

4.5 Demonstratives

Kwakum has five series of demonstratives. One is used to identify referents as being close to the speaker (*-né*), two are used to identify referents as being close to the hearer (*-ɔ́⁺ɔ́* and *-έ*), one for items far from speaker and hearer (*-kέ*) and one for anaphoric use (*-έ*). The difference in use between the two near-listener demonstratives is not clear yet. In elicited utterances they are interchangeable. The roman numbers in the headers of Table 13.9 refer to the paradigms of agreement prefixes provided in Table 13.4.

TABLE 13.9 DEMONSTRATIVES

	Near speaker (II)	*Near listener (III)*	*Near listener (IV)*	*Far (V)*	*Anaphoric (I)*
1	⁺néᴸ	y-ɔ́ɔ́ᴸ	kéᴸ	yí-kéᴸ	(w-)έ
2	yí-néᴸ	y-ɔ́ɔ́ᴸ	y-έᴸ	yí-kéᴸ	y-έ
5	li-néᴸ	ly-ɔ́ɔ́ᴸ ~ y-ɔ́ɔ́ᴸ	kéᴸ	yí-kéᴸ	ly-έ
6	mí-néᴸ	m-ɔ́⁺ɔ́ᴸ	m-έᴸ	mí-kéᴸ	m-έ
7	tʃí-néᴸ	tʃ-ɔ́ɔ́ᴸ	tʃ-έᴸ	tʃí-kéᴸ	tʃ-έ

4.6 Quantifiers

Numbers do not agree in class with the noun they quantify. Numbers from one to five have the formal characteristics of nouns and are assigned to morphological class *ki-* (SG) / *i-* (PL). Numbers from six to ten do not have a class assignment.

In cardinal adnominal use, numbers are postposed to the noun they modify (49–50). In this construction, the numeral nouns 2–5 are always in their plural form, but the quantified head noun can take either its singular or plural form (48), unless if they belong to morphological class *mò-/gwò-*, in which case they have to be plural.

TABLE 13.10 ADNOMINAL NUMBERS

1	mótù^H	6	tówo
2	ì-báà^H	7	tàmbályὲ^H
3	ì-tátí	8	ʃál
4	ì-néὲ^H	9	bùyɛ́
5	ì-tàán	10	káamɔ

(48) a   ìtóó ìbá⁺á
      ì-tóó^L   ì-báà^H
      5-house   2-two

    b   ǹtóó ìbá⁺á
      ǹ-tóó^L   ì-báà^H
      6-house   2-two
      'two houses'

(49) kìfyὲtì tówó
      kì-fyὲtí   tówo
      7-tree   six
      'six trees'

The nominal prefixes of numbers two to five have in common with agreement prefixes that they can optionally prevent a preceding high from attaching.

(50) kìfyὲtì ìbá⁺á   ~   kìfyὲtì íbá⁺á
      kì-fyὲtí   ì-báà^H
      7-tree   2-two
      'two trees'

Numbers from one to five can also be the first relatum in a connective construction, in which case the resulting NP has a definite interpretation. The stem of number two is -bàlá in this construction (51).

(51) ìbàlá kífyὲtí
      ì-bàlá   ^H-kì-fyὲtí
      2-two   CON-7-tree
      'the two trees'

The numbers two to five can be used in the singular or in the plural, in free variation (52). When such an NP is extraposed in a topic construction, the resumptive subject pronoun is always in the plural, showing semantic agreement with the subject NP (the semantic head of which can be a singular noun) (53). The quantified noun too can be either in its singular or plural form, except, again if it belongs to morphological class mò-/gwò-.

(52) ɲàádʒɛ́ɛ́ íbàlá yífyâl ~ kíbàlá kífyâl
      ɲì-^H-àà-dʒɛ́ɛ́   ì-bàlá   yì-^H-fyàl
      1SG-PST-PST2-see.PST2   2-two   PP₂-CON-1.daughter_in_law
      'I saw the two daughters-in-law.'

(53) kìbàlá kìpʰàâm wὲ yέɲʃὲ°
      kì-bàlá   kì-ᴴ-pʰàâm   w-έ   ᴴ-yὲᴴ-n-ʃὲᴴ
      7-two   PP₇-CON-1.man   PP₁-ANA   PRS-3PL-PRS-come
      'As for those two men, they are coming.'

The number one is used to mean 'the only,' with a prefix *kì-* in the singular and *ì-* in the plural.

(54) a   kìmɔ̀t kífyâl
           kì-mɔ́tùᴴ   kì-ᴴ-fyàl
           7-one   PP₇-CON-1.daughter_in_law
           'the only daughter-in-law'

      b   ìmɔ̀t yìgwɔ́ɔ̀nɔ́
           ì-mɔ́tùᴴ   yì-ᴴ-gwɔ̀-ɔ̀nɔ́
           2-one   PP₂-CON-2-child
           'the only children'

There is a set of four quantifiers that have the same grammatical behaviour as the cardinal numbers, viz. *kì-pyàpyá* 'very little, very few,' *kì-ɲàá⁺ʃí* 'little, few,' *kòŋàá⁺ʃí* 'little' and *ʧìndí* 'all.' The first two belong to class *kì-/ì-*, while the other two have no class prefix, but trigger agreement of class 2. The special tonal behaviour of numbers two to five (optional blocking of ᴴ-attachment) does not apply to these quantifiers.

(55) a   ɲ̀ʧìk kípyàpyá
           ǹ-ʧìkí   kì-pyàpyá
           6-water   7-little
           'a little water'

      b   ìtàw ìpyàpyá
           ì-tàw   ì-pyàpyá
           5-sheep   5-few
           'few sheep'

(56) a   àbùpà ʧìndí
           à-bùpà   ʧìndíᴸ
           2-animal   2.all
           'all the animals'

      b   tʃìndí yìbúpà
           tʃìndíᴸ   yì-ᴴ-à-bùpà
           2.all   PP₂-CON-2-animal
           'all the animals'

(57) gwɔ̀ɔ̀n kóŋàá⁺ʃí
      gwɔ̀-ɔ̀nɔ́   kòŋàáʃíᴸ
      2-child   little
      'few children'

(58) pʰàâm mέdʒí kípyàpyá kìpŏŋgò
      pʰàâm   ᴴ-mὲ-dʒí   kì-pyàpyáᴸ   kì-ᴴ-pòŋgò
      1.man   PST2-PST2-eat.PST2   7-few   PP₇-CON-maize
      'The man ate a small quantity of maize.'

When *ʧìndí* is used prenominally to mean 'every,' it is not linked to the quantified noun by means of a connective relator (59).

(59) tʃĩndʒí bùpà dʒì nɛ́ kùl ə̀ndʒí
     tʃĩndʒí^L   bùpà   dʒì^H nɛ̀   kùl   à-n-dʒí^L
     2.every   1.animal COP with strength 3SG-CONS-eat.CONS
     'Every animal can eat.'

4.7 Qualifiers

Nominal qualification is particularly interesting in Kwakum, due to the wide range of constructions in this domain. From a lexical perspective, Kwakum qualifiers can be divided into those whose use is restricted to qualification (dedicated qualifiers), and those who are also used to refer (referring-qualifying nouns). The former group can be divided into a set of 12 underived qualifiers and an open class of derived qualifiers. Qualifiers are derived from verbs by means of one of three suffixes, shown in Table 13.11.

TABLE 13.11 LEXICAL TYPES OF QUALIFIERS, WITH AN EXAMPLE OF EACH

referring-qualifying		ʃómtu 'wise, wisdom'	
dedicated to qualification	underived	tàmbyɛ̀ 'good'	
	derived	-áá	dèt-áa 'hard'
		-áàwè^H	dèt-áàwè^H 'hard'
		-ɛ́ŋ/-áŋ	dèt-ɛ́ŋ 'hard'

The suffix *-áàwè^H* most probably originates in a possessive form consisting of the Proto-Bantu connective stem *a* and a third-person pronominal form. It is an instance of the possessee-like qualifiers that are common in Northern Sub-Saharan Africa.

Qualifiers are similar to quantifiers in that they can be used in a variety of constructions, as illustrated for *dètáàwè^H* 'hard,' which is simply preposed in (60a), R2 in a connective construction in (60b) and R1 in a connective construction in (60c).

(60) a dètáàwè kídʒà
       dèt-áàwè^H   kì-dʒà
       be.hard-ADJ   7-chair
       'a hard chair'

     b kídʒà kìdétáàwɛ̀°
       kì-dʒà   kì-^H-dèt-áàwè^H
       7-chair   PP₇-CON-be.hard-ADJ
       'a hard chair'

     c dètáàwè yìkífyètí
       dèt-áàwè^H   yì-^H-kì-fyètí
       be.hard-ADJ   PP₂-CON-7-tree
       'hard trees'

The underived qualifier *tàmbyɛ̀* 'good' is simply postposed to the semantic head in the singular (61a) and R2 in a connective construction in the plural (61b). It cannot occur in front of the semantic head. There is no space here for a full description of the morphosyntactic behaviour of qualifiers.

(61) a kɔɔ̀ndɛ̀   tàmbyɛ̀
       kɔɔ̀ndɛ̀   tàmbyɛ̀
       1.plantain   in.good.condition
       'a plantain in good condition'

   b àkɔɔ̀ndɛ̀ yìtámbyɛ̀
       à-kɔɔ̀ndɛ̀   yì-^{H}-tàmbyɛ̀
       2-plantain PP_2-CON-in.good.condition
       'plantains in good condition'

The underived qualifier *tàmbyɛ̀* and the ones derived by means of the suffix *-áàwɛ̀^{H}* can also be used as adverbs, whereas those derived by means of the suffix *-ɛ́ŋ/-áŋ* can be used as secondary predicates.

4.8 Agreement and word order in the noun phrase

Agreement in complex noun phrases is often determined by proximity, rather than syntactic structure or semantic scope. The demonstrative in (62), for instance, modifies the noun *ǹtóó* 'houses,' but agrees in noun class and number with the quantifier *kìpyàpyá* 'few, little.'

(62) ǹtóó kìpyàpyá $^+$tʃɛ́
      ǹ-tóó^L kì-pyàpyá^L tʃĭ-ɛ́
      6-house   7-little PP_7-ANA
      'those (aforementioned) few houses'

Elsewhere, speakers volunteer agreement with either the head noun (63–64a), or the immediately preceding nominal form (63–64b).

(63) a ǹtóó ìbáà mí$^+$kɛ́
        ǹ-tóó^L ì-báà^H mí-kɛ́^{L}
        6-house   2-two PP_6-DEM
        'those two houses'

    b ǹtóó ìbáà yí$^+$kɛ́
        ǹ-tóó^L ì-báà^H yí-kɛ́^{L}
        6-house   2-two PP_2-DEM
        'those two houses'

(64) a kìbàlá kìntóó màá$^+$mbɔ́
        kì-bàlá   kì-^{H}-ǹ-tóó^L m-àámbɔ́^{L}
        7-two PP_7-CON-6-house PP_6-1SG.POSS
        'my two houses'

    b kìbàlá kìntóó tʃ-àá$^+$mbɔ́
        kì-bàlá   kì-^{H}-ǹ-tóó^L tʃ-àámbɔ́^{L}
        7-two PP_7-CON-6-house PP_6-1SG.POSS
        'my two houses'

Class 1 serves as a default agreement class for all singular controllers, which can option-ally be used instead of the controller's lexically determined agreement class. Occasionally, we have found examples of this default agreement with a plural controller too, mostly in noun phrases that contain a numeral.

Word order in complex noun phrases is schematised in (65). QUAL and QUANT are for-mal-functional notions, used for the quantifiers (including numbers) and qualifiers that are juxtaposed to the noun they modify, i.e., excluding the use of these lexemes in con-nective constructions.

(65) HN − {QUAL, QUANT, POSS, ANA, CON} − REL − DEM

The abbreviation *HN* stands for the head noun from a morphosyntactic point of view: the nominal element that is not used to modify any other element in the noun phrase. Needless to say, this is not necessarily the semantic head, which happens to come at the very end of the NP in (66).

(66) ìbàlá yìdétáá tʃìndí wɛ̀ɛ̀ dʒó⁺wɔ́
      ì-bàlá   yì-^{H}-dèt-áa ^{H}-tʃìndí^L w-ɛ̀ɛ̀^{H} ^{H}-dʒòwɔ́
      2-two   PP$_2$-CON-be.hard-ADJ   CON-all   PP$_1$-3SG.POSS   CON-day
      'all his two difficult days'

5 ADPOSITIONS

Kwakum has six prepositions and three "ambipositions." The latter are postposed with nominal complements and preposed with pronominal complements.

(67) a   prepositions
           nɛ̀ 'with, by'
           pɔ̌mbú 'for'
           pʰɛ̀l^H 'in front of'
           ʃìmɔ́ ~ ʃìm 'behind'
           lémɛ́ lémɛ́ 'between'
           pákláá 'among'

      b   ambipositions
           ʃì^H 'under'
           kóó^L ~ kól^L 'on'[7]
           téé 'in'

Except for the comitative-instrumental-agentive preposition *nɛ̀*, they all originate from nouns in a connective construction, which is synchronically evidenced by the fact that they take possessive pronouns as pronominal complements, rather than personal pronouns. On their pronominal complements, they take the agreement pattern of the nouns from which they orig-inate, which is in all cases class 1. However, the pronominal complements of the prepositions *pɔ̌mbú* 'for,' *pʰɛ̀l^H* 'in front of,' *kól^L* 'on' and *ʃìmɔ́* 'behind' can alternatively have the *y*- prefix of class 2, in which case they do still have the tone pattern of possessives of class 1 (68).

(68) a   pɔ̌mb yɛ́⁺ɛ́ ~ pɔ̌mb wɛ́⁺ɛ́ 'for him'
      b   ʃìm yɛ́⁺ɛ́ ~ ʃìmɔ́ wɛ́⁺ɛ́ 'behind him'

In order to disambiguate between the adpositional and the nominal use of these lexemes, the nominaliser-linker *mòò* can be used in the nominal use.

(69) pʰɛ̀l (móò) tóó
      pʰɛ̀l[H]   mòò   [H]-ì-tóó[L]
      1.front   NLNK   CON-5-house
      'the front of the house'

The complements of all Kwakum adpositions are accessible to relativisation, but only *nè* can be stranded. The other adpositions require a resumptive pronoun.

6 VERBS

6.1 The structure of verb stems

Table 13.12 summarises the most frequent syllable structures of underived verb roots. As can be seen, 90% of the verb stems are disyllabic, which is unusual for a Bantu language.

TABLE 13.12 THE MOST COMMON SYLLABLE SCHEMES OF UNDERIVED VERB STEMS

CV scheme	%	example
CV.CV	25	bè.nɔ̀ 'deny'
CVC.CV	21	yék.lè 'teach'
CVV.CV	17	bà.àndɔ̀ 'peel'
CV.CVV	10	dò.wáà 'call'
CVC.CVV	5	lùk.làà 'buzz'
CV.CCV	3	fì.⁺myɛ́ 'wipe'
CVV	5	bèè 'follow'

Ten verb stems in our database of 614 verbs are trisyllabic. There is only one verb stem that ends in a consonant, viz. *kèn* 'go.' There are strong phonotactic constraints on the last vowel of verb stems: /ɛ, ɔ, aa/ in disyllabic stems and /ɛ, aa/ in trisyllabic stems.

6.2 Derivation

Kwakum has four verb-to-verb derivational suffixes. One is valency increasing (causative *-fɛ̀*), two are valency reducing (*-yɛ̀* and reciprocal *-àà*) and one can be either (*-lè*). These suffixes almost always attach to a base with CVC-shape, usually obtained by the addition of /y/ to CV roots[8] and /ŋ/ to CVV roots[9] or by deletion of the last vowel of CV(V)CV roots and the shortening of the vowel of their initial syllable if that happens to be long.[10] CVCV roots with an alveolar second consonant drop their last syllable and lengthen their first vowel in front of the suffix *-fɛ̀*. We have no examples of verbs derived from a trisyllabic root.

Verbs derived from roots that are not entirely low have a fixed tone pattern determined by the derivational suffix, as illustrated in (70). Entirely low roots remain low when a derivational affix is added.

(70) a *-ʃê*: HL.L
    líí^L 'get black'   >   lîŋ-ʃè 'blacken'
    bélɔ́^{L} 'be cooked'   >   béè-ʃè 'cook'
    dʒáálɔ́^{L} 'give birth'   >   dʒáà-ʃè 'deliver'

    b *-lè*: H.L
    béʃɛ́^{L} 'rise'   >   béʃ-lè 'lift'

    c *-àà*: L.HL
    fɛ́ 'give'   >   fêy-áà 'receive'
    dʒɛ́ɛ́^{L} 'see'   >   dʒèn-áà 'see each other'
    bíwɔ́^{L} 'beat'   >   bìw-áà 'beat each other'

    d *-yè* : H.H^L
    dʒàálɔ́^{L} 'give birth'   >   dʒál-yɛ́^{L} 'be born'

There is only one verb in which the suffix *-ʃê* is not clearly causative, viz. *ʃèk-ʃê* 'shake' (<*ʃèkɔ̀* 'sieve'). Some typical causative examples are provided in (71).

(71) lîŋ-ʃè 'darken'   <   líí 'get dark'
    pûp-ʃè 'clean'   <   pú$^+$pɔ́ 'recover'
    ɲîŋ-ʃè 'let in'   <   ɲîŋlè 'enter'
    dáà-ʃè 'put to bed (sp.)'   <   dàá$^+$lɔ́ 'sleep'
    dèè-ʃè 'strengthen'   <   dètɔ̀ 'be strong'

The suffix *-lè* has a causative meaning in one verb, namely *béʃ-lè* 'lift' (<*bé$^+$ʃê* 'rise'). In *dít-lè* 'open' (<*dí$^+$tɔ́* 'close'), it encodes reversal. Its function is decausative in *tʃĭk-lè* 'stay' (<*tʃĭ$^+$kɔ́* 'abandon, leave something') and perhaps intensive or repetitive in *bòm-lè* 'hammer' (<*bòmɔ̀* 'break').

(72) ámbéʃì bùpà
    ^{H}-à-n-béʃɛ́^{L}-lè   bùpà
    PRS-3SG-PRS-rise-CAUS   animal
    'He lifts up the animal.'

The suffix *-àà* can be used to derive reciprocal, reflexive, passive and/or decausative verbs. Some examples are provided in (73–74). In addition, in two verbs, *-àà* serves to express collective actions (*dùlyáà* 'shout together'< *dú$^+$lɔ́* 'shout' and *kàw-àà* 'share among many people <*kààwɔ̀* 'share'). The reciprocal meaning is often reinforced by the phrases *yè né dʒè°* 'each other' (SG, i.e., the subject and one other participant) and *yà né dʒà°* 'each other' (PL, i.e., the subject and more than one other participant), while a pronoun preceded by a floating high tone and followed by *tʃĭtʃé* 'oneself' is used to specify a reflexive meaning. Example (74) shows that passive constructions with a verb derived by *-àà* can have an agentive complement flagged by means of the preposition *nè*.

(73) a Reciprocal
    dʒèn-áà 'see each other'   <   dʒɛ́ɛ́ 'see'
    kàm-áà 'love each other'   <   kààmɔ̀ 'love'

    b Reflexive
    kèŋ-àà 'shave' (intr.)   <   kèè 'shave' (tr.)
    fiʃ-áà 'cover oneself'   <   fĭ$^+$ʃɔ́ 'cover'

c   Passive voice
   bày-àà 'be operated'   <   bà 'operate'
   ʃàl-áà 'be done'   <   ʃáá 'do'
   kèk-áà 'be circumcised'   <   ké⁺kɔ́ 'circumcise'
   màŋ-àà 'be gathered'   <   màà 'gather (tr.)'

d   Decausative
   màŋ-àà 'gather (intr.)'   <   màà 'gather (tr.)'
   bànd-áà 'bend (intr.)'   <   bàá⁺ndɔ́ 'bend (tr.)'
   pàɲdʒ-àà 'scatter (intr.)' <   *pàɲdʒè* 'separate'

(74) àdɔ́ɔ́k ḿmàŋàà nὲ pʰàâm
     à-dɔ́ɔ́kɔ́ᴸ   ᴴ-n-màŋ-àà   nὲ   pʰàâm
     2-mango   PRS-PRS-gather-PASS   with   1.man
     'The mangoes are gathered by the man'

Our lexical database contains four verbs derived by *-yὲ*.

(75) ʃá⁺n-yέ 'split (intr.)'   <   ʃàá⁺nɔ́ 'split (tr.)'
     pém-⁺yέ 'change (intr.)'   <   pémlè 'change (tr.)'
     dʒál-⁺yέ 'be born'   <   dʒàá⁺lɔ́ 'give birth'
     bòm-yὲ 'burst (intr.)'   <   *bòmɔ́* 'burst (tr.)'

We have found only one example of stacked derivational suffixes, viz. the verb *bɔ́mʃáà* 'sell' (<*bɔ́ɔ́⁺mɔ́* 'buy'), which is formed by means of the suffixes *-ʃὲ* and *-àà*.
     Verbal derivation is not productive in Kwakum. Causation can be expressed by means of a complex sentence that has *ʃáá* 'do' as its main verb. The lexical verb occurs in a complement clause introduced by the complementiser *⁺nàá·* (76).

(76) àfὲέ ʃáá nàá ɔ̀ʃùm ìtóó
     à-fὲέ ᴸ   ʃáá ᴸ   ⁺nàá ᴸ   ɔ̀-ʃùmɔ́   ì-tóó ᴸ
     3SG-FUT1   do   COMP   2SG-build.SBJV   5-house
     'He will make you build a house.'

Periphrastic passives are productively formed with a copula and a participial form of the verb derived by means of the suffix *-έŋ/-áŋ* (see Section 4.6, Table 13.11). As with derived passive verbs, the agent is introduced by the preposition *nὲ*.

(77) ìtóó dʒì ʃú⁺méŋ né pʰàâm
     ì-tóó ᴸ   dʒì ᴴ   ʃùm-έŋ   nὲ   pʰàâm
     5-house   COP   build-ADJ   by   man
     'The house is built by the man.'

There is no applicative derivation. Beneficiaries are dependent marked by means of the preposition *pɔ̌mbú*.

(78) áɲʃùm ìtóó pɔ̌mb pʰàâm wέ
     ᴴ-à-n-ʃùmɔ́   ì-tóó ᴸ   pɔ̌mbú   pʰàâm   w-έ
     PRS-3SG-PRS-work   5-house   for   man   PP₁-ANA
     'He is building a house for the man.'

6.3 Inflection

6.3.1 Tense and aspect in the indicative mood

In the indicative mood, Kwakum distinguishes 14 tense-aspect combinations, summarised in Table 13.13. It has four past tenses, a present and three future tenses, as well as a distinction between perfective versus imperfective aspect. Past 1 and 2 are not distinguished in the imperfective. The present tense shows a formal distinction between a synthetic and a periphrastic verb form that corresponds to the perfective-imperfective distinction in other tenses. The periphrastic form corresponding to the imperfective in the other tenses is used for verb focus.

TABLE 13.13 TEMPORAL AND ASPECTUAL DISTINCTIONS

Perfective	Temporal reference	Imperfective
Past 4	remote	Past 4
Past 3	yesterday	Past 3
Past 2	today	Past 2
Past 1	immediate	
Present	now	Present
Future 1	immediate/today	Future 1
Future 2	tomorrow	Future 2
Future 3	remote	Future 3

More research is needed on the exact use of the different past and future tenses. Basically, Past 4 is a remote past, while Past 3 is typically used for events that took place the day before utterance time. Past 2 and Past 1 are today's pasts, with Past 1 being used for events that took place right before the time of utterance and Past 2 for earlier events. The distinctions in the future are similar.

Table 13.14 gives an overview of the verb forms that express the eight tenses in the perfective. By way of an example, the last column gives the surface representation of the inflected verb *ì-dàá⁺mbɔ́* 'to cook' in utterance final position, with a third-person singular subject prefix *à-*. Verb forms that can only be used with nominal subjects are illustrated with the noun *pʰàâm* 'man' in subject position.

TABLE 13.14 KWAKUM TENSES (PERFECTIVE FORMS)

PST4	(SM-)STEM-mɛ	àdámb(ì)mɛ́
PST3	(SM-)STEM kòò	àdàmb(ì) kóò
PST2	ᴴ-mɛ̀-STEM	pʰàâm médáámbɛ́
	SM-ᴴ-àà-STEM	àádáámbɛ́
PST1	mɛ̀-STEM	pʰàâm mɛ̀dáámbɔ́ ~ mɛ̀dáámbɛ́
	SM-àà-STEM	ààdáámbɔ́ ~ ààdáámbɛ́
PRS	ᴴ-(SM)-n-stem	ándàá⁺mbɔ́
FUT1	(SM-)fɛ̀ɛ́ᴸ-STEM	àfɛ̀ɛ́dàá⁺mbɔ́
	ʰ-SM-stem	ádàá⁺mbɔ́
FUT2	(SM-)ʃɔ̀ɔ̀-ᴴ-STEM	àʃɔ̀ɔ̀dàá⁺mbɔ́
FUT3	(SM-)ʃɔ̀ɔ̀-ŋgɛ̀-ᴴ-STEM	àʃɔ̀ɔ̀ŋgɛ̀dàá⁺mbɔ́

The schemes in Table 13.14 show that tense is expressed by means of prefixes, suffixes and postverbal particles or clitics (in PST3) and that tense prefixes can be tonal morphemes. Past 2 and Past 1 have different tense prefixes depending on whether their subject is nominal or pronominal. An interesting feature of the initial floating high tone in the Present and Future 1 is that it is realised on the subject whenever it is pronominal, whether it is a prefix or an independent pronoun, as shown in (79), repeated from (36).

(79) a   ómⁿ⁺bíwéὲ
          ᴴ-ɔ̀-n-bíwɔ́ᴸ-ɛɛ
          PRS-2SG-PRS-beat-1SG
          'You beat me.'

      b   gwέ ⁺mbíwέὲ
          ᴴ-gwὲ   n-bíwɔ́ᴸ-ɛɛ
          PRS-2SG   PRS-beat-1SG
          'You beat me.'

Table 13.14 also shows that the future has as many formal distinctions as the past. Since we were not able to find any difference in the use of the two hodiernal future tense forms, we treat them as allostructs of Future 1. More research might identify a functional difference between both verb forms.

Finally, the forms in the third column of Table 13.14 show that the verb stem in the past tenses differs tonally and segmentally from the basic allomorph dàá⁺mbɔ́ found in the other tenses and in the infinitive. Every verb stem has three or four non-basic allomorphs. Their shape is predictable and determined by tense and mood, by the position of the verb in the utterance (final versus non-final), as well as by the tone, the syllable structure and the final segment of the basic allomorph. A full description of the patterns of allomorphy is provided in the doctoral dissertation of the first author (Njantcho 2018). In Table 13.14, the use of a non-basic allomorph of the verb stem is signalled by curly underlining.

Tense suffixes are treated as part of the verb stem with respect to the application of fixed tone schemes in non-basic stem allomorphs. This observation justifies the distinction in morphological status of the post-stem tense marker between the Past 4 suffix -mɛ and the Past 3 unbound morpheme (or clitic?) kòò. The difference in their behaviour is illustrated in the verb forms in Table 13.15. In the second column, F is short for (utterance) final and NF for non-final. The third column contains the tone scheme of the non-basic allomorph of the verb stem. The examples in the last column have a 3SG subject prefix à-. The verb bà means 'cut up' and dʒɔ̀fὲ means 'hide.'

TABLE 13.15 THE DIFFERENCE IN MORPHOLOGICAL BONDING BETWEEN -Mɛ AND KÒÒ

Stem	Conjugation	Tone scheme	Example
bà	PST4 F	LH	àbàmέ
	PST4 NF	Lᴴ	àbàmë̈ᴴ
	PST3	Lᴴ	àbà kóò
dʒɔ̀fὲ	PST4	LHH	àdʒɔ̀fὲ⁺mέ
	PST3	LHᴸ	àdʒɔ̀fὲ kòò

The verb 'be,' whether used as a copula or to express existence or location, has a suppletive paradigm, with the stem ʤì^H (most probably cognate with ʤìlɔ 'stay, live') in the Present indicative and bé in the other verb forms, including non-indicative moods.

Imperfective verb forms are construed by means of an auxiliary and the stem of the lexical verb, as shown in Table 13.16. The last column is again an example with the verb dàá⁺mbɔ 'cook.'

TABLE 13.16 KWAKUM TENSES: IMPERFECTIVE FORMS

PST4	(SM-)ʤì^H/yì^H-mɛ STEM	àʤìmé dàá⁺mbɔ
PST3	(SM-)ʤì^H/yì^H kòò STEM	àʤì kóò dàá⁺mbɔ
PST1/2	^{H}-mè-yì^H STEM	pʰàâm méyì dàá⁺mbɔ
	SM-^{H}-àà-yì^H STEM	àáyì dàá⁺mbɔ
PRS	(SM-)yɔkù^H STEM	àyɔkù dàá⁺mbɔ
	(SM-)ʤì^H/yì^H STEM	àyì dàá⁺mbɔ
FUT1	(SM-)fɛ̀ɛ́^{L}-bé SM-ʤì^H/yì^H STEM	àfɛ̀ɛ́⁺bé àʤì dàá⁺mbɔ
FUT2	(SM-)ʃɔ̀ɔ̀-^{H}-bé SM-ʤì^H/yì^H STEM	àʃɔ̀ɔ̀bé àʤì dàá⁺mbɔ
FUT3	(SM-)ʃɔ̀ɔ̀-ŋgɛ̀-^{H}-bé SM-ʤì^H/yì^H STEM	àʃɔ̀ɔ̀ŋgɛ̀bé àʤì dàá⁺mbɔ

Interestingly, past imperfective forms involve the Present tense stem ʤì^H of the auxiliary 'be' with past tense morphology. In some TA forms this stem has the reduced form yì^H, suggesting that it is prosodically evolving towards affix status. Imperfective forms in the future tenses involve a succession of two 'be' auxiliaries. The first one is inflected for tense and may take a pronominal or nominal subject. The second one is the Present tense stem preceded by a subject pronoun.

6.3.2 Non-indicative moods

The three non-indicative moods of Kwakum – Imperative, Subjunctive and Consecutive – all involve non-basic allomorphs of verb stems. The Imperative singular has no extra marking, except with CV-stems, where it takes the suffix -ké^L (81). The suffix of the Imperative plural is -kín^L (82). Two verbs have a suppletive Imperative form (83).

(80) Imperative verb forms
    a   CV-ké^L (SG)
    b   STEM (SG)
    c   STEM-kín^L (PL)

(81) bàké bùpà
      bà-ké^L bùpà
      cut_up.IMP-2SG.IMP   1.meat
      'Cut up (SG) the meat!'

(82) bàkín bùpà
      bà-kín^L bùpà
      cut_up.IMP-2PL.IMP   1.meat
      'Cut up (PL) the meat!'

(83) a   ʤìᴴ 'eat' → ʤìké ~ ʤòké 'Eat!'
      b   ʃèᴴ 'come' → ʃɔ́ɔ́⁺kɔ́ 'Come!'

The Subjunctive mood is used in subordinate clauses and to express hortative modality (85). It is formed by means of a non-basic allomorph of the verb stem and a subject prefix (84).

(84) the Subjunctive verb form
      (SM-)STEM

(85) ʃèbáà ìtàá⁺mbɔ́
      ʃèᴴ-báà   ì-tàámbɔ́ᴸ
      1PL-set.SBJV   5-trap
      'Let us set the trap.'

The Consecutive mood is formed by means of the prefix *n-* and a non-basic allomorph of the verb stem (86).

(86) the Consecutive verb form
      (SM-)n-STEM

It has a wide array of uses. In its consecutive use it functions as a relative tense form indicating that an event takes place after a previously mentioned event. Second, it can function as a general present, typically in proverbs (87). Third, it can be used in either or both the apodosis and the protasis of conditional clauses. Finally, it can be used in subordinate clauses instead of the Subjunctive.

(87) mòmyàà mbé tʰà kìbàpù
      ᴸ‰mó-myaa   N-béᴸ   tʰà   kì-bàpù
      1-woman   CONS-COP.CONS   CMP   7-horsefly
      'A woman is like a horsefly.'

6.3.3 Negation

The negative form of the verb *ʤìᴴ* 'be' is *fètéé~téé* (88).

(88) pèŋgè fètéé tʃàk bɔ́ɔ́⁺ʃéŋ
      pèŋgè   fètééᴸ   tʃàkí   bɔ̀ɔʃ-éŋ
      1.money   NEG.COP   1.thing   be.bad-ADJ
      'Money is not a bad thing.'

In indicative verb forms, the negative marker is *wéé*. It originates in a third-person singular possessive pronoun, a path of grammaticalisation that is not unusual in the Bantu languages (Devos & van der Auwera 2013). When the subject is a first-person pronoun, the 1SG possessive pronoun *wàá⁺mbɔ́* can be used as well, and some speakers use the 2SG possessive pronoun *wɔ́ɔ́* with 2SG subjects. In the present and future tenses, the negative marker is in between the Tense marker(s) and the stem (89). In the past tenses it follows the stem of the conjugated verb immediately (90).

(89) àfɛ́ɛ́wɛ́ɛ́dʒì búpà
    à-fɛ̀ɛ́ᴸ-wɛ́ɛ́ᴸ-dʒìᴴ   bùpà
    3sɢ-ꜰᴜᴛ1-ɴᴇɢ-eat   1.meat
    'He will not eat meat.'

(90) àádʒìwɛ́ɛ́ bùpà
    à-ᴴ-àà-dʒìᴴ-wɛ́ɛ́ᴸ   bùpà
    3sɢ-ᴘsᴛ2-ᴘsᴛ2-eat.ᴘsᴛ2-ɴᴇɢ 1.meat
    'She did not eat meat.'

In non-indicative moods, negation is marked by means of the marker *bĕk* (91).

(91) bĕkdʒì búpà
    bĕkᴸ-dʒìᴴ   bùpà
    ɴᴇɢ-eat.2sɢ.ɪᴍᴘ 1.meat
    'Don't eat meat!'

7 CLAUSAL SYNTAX

7.1 Simple clauses

As in most Bantu languages, the subject precedes the verb and the object follows it. In clauses with two unmarked complements the Goal precedes the Theme (92), an order that can be optionally reversed if and only if the Theme is pronominal and the Goal a noun with human reference (93). Both the Theme and the Goal are accessible to relativisation, but only the Theme is accessible to passivisation.

(92) pʰàâm ḿ⁺fɛ́ mɔ́ɔ̀n á⁺máŋgòlò
    pʰàám̀   ᴴ-n-fɛ́   mɔ̀-ɔ̀nɔ́   à-máŋgòlò
    1.man   ᴘʀs-ᴘʀs-give 1-child   2-mango
    'The man is giving the child mangoes.'

(93) pʰàâm ḿ⁺fɛ́ já mɔ́ɔ̀nɔ́
    pʰàám̀   ᴴ-n-fɛ́   jàᴴ   mɔ̀-ɔ̀nɔ́
    1.man   ᴘʀs-ᴘʀs-give 3ᴘʟ   1-child
    'The man is giving them to the child.'

7.2 Questions

Polar questions are marked intonationally by a rising tone on the last syllable of the sentence. Information questions are formed by replacing the questioned constituent by an interrogative, such as *kè* 'what, why' or *tà* 'who,' which remain in situ (94). Alternatively, interrogatives can occur in clause-initial position in a focus construction.

(94) gwɛ́ ndàámbú kè?
    ᴴ-gwɛ̀   n-dàámbɔ́ᴸ   kè
    ᴘʀs-2sɢ   ᴘʀs-cook   what
    'What are you cooking?'

7.3 Subordinate clauses

Subordinate clauses are optionally marked by means of the clause-final subordinator *jí*. Relative clauses can additionally be marked by means of a connective relator, which may or may not be preceded by the nominaliser-linker *mòò/gwòòm* (95a). The connective relator that introduces relative clauses is not an agreement target. Alternatively, relative clauses with a plural head noun can be introduced by the copula *dʒî^H* (95b). Since both the subordinator and the relativiser are optional, relative clauses can be totally unmarked (96). All positions in the clause are accessible to relativisation.

(95) a àpʰàâm (gwóòm) bá kóò bùpà (yí)

	à-pʰàâm	ᴴ-gwòòm	ᴴ-bàᴴ		kòò	bùpà	yíᴸ
	2-man	CON-NLNK.PL	CON-cut_up.PST3	PST3	1.animal	SUB

b àpʰàâm dʒì bá kóò bùpà (yí)

	à-pʰàâm	dʒìᴴ	bàᴴ		kòò	bùpà	yíᴸ
	2-man	COP	cut_up.PST3	PST3	1.animal	SUB

'the men who cut up the meat'

(96) àpʰàâm bà kóò bùpà

	à-pʰàâm	bàᴴ		kòò	bùpà
	2-man	cut_up.PST3	PST3	1.animal

'the men who cut up the meat' or 'The men cut up the meat.'

Complement clauses are introduced by the complementiser ⁺*nàá^L*, which has an initial *í* when introducing a clause functioning as subject.

(97) ínàá pùlò mɛ́⁺nó búláàwě (yí) ńdɛ̌lkáà ɲì

	ínàáᴸ	pùlɔ́	ᴴ-mɛ̀-nɔ́	bùl-áàwèᴴ		yíᴸ	n-dɛ̌lkáà	ɲì
	COMP	1.rain	PST2-PST2-fall.PST2	be_numerous-ADJ	SUB	PRS-surpass	1SG

'It surprises me that it rained so abundantly.'

NOTES

1 Research for this chapter was carried out as part of the research projects BantuTyp (granted by USPC) and LC2 Areal phenomena in Northern sub-Saharan Africa of the Labex EFL (ANR-10-LABX-0083). We wish to thank Simon Charles Ndengue Ndengue, Paul Nargaba and Nicolas Noël Sibita, our main Kwakum consultants, as well as Larry Hyman and Koen Bostoen for their useful comments.

2 The only exception in our lexical database is the verb stem *ʃwíjè°* 'leave.'

3 A possible explanation for this is that the voiceless labial-velar stop /kp/ is not restricted to stem-initial position, but that it simply has not been found elsewhere in our lexicon due to its low lexical frequency.

4 The only exception attested so far is the numeral *mótù^H* 'one,' of which the high can be replaced by a preceding low.

5 The first downstep in (20), on ⁺*gwí*, is due to the low tone of the 3SG subject prefix *à-* that was delinked from its TBU by the preceding floating ᴴ.

6 There are some co-occurrence restrictions in topic constructions between anteposed topical pronouns and resumptive pronouns. Both cannot be dual pronouns, for instance.

7   The form *kól*ᴸ is obligatory in prepositional use, i.e., with a pronominal complement.
8   The only examples we have in our lexical database consist of CV-verbs derived by means of -*àà*. We don't know whether /j/ will also occur in coda position if the verb is derived with other suffixes.
9   Some exceptions include *fáá* 'do' and *féé*ᴸ 'work' where the inserted consonant is /l/ (e.g., *fáá*ᴸ 'do'> *fàláà*, 'be done'; *féé*ᴸ 'work' > *fèláà* 'be processed') and *dʒéé* 'see' where it is /n/ (*dʒéé*ᴸ 'see' > *dʒènáà* 'see each other'). These "inserted" consonants are most probably retentions of root consonants that have eroded in other contexts.
10 The only exception in our database is *dʒòòlàà* 'bathe (intr.)' (cf. *dʒòòlè* 'bathe (tr.)') where shortening of the initial long vowel fails to occur.

REFERENCES

Bastin, Y., A. Coupez, E. Mumba & T. C. Schadeberg (eds.) (2002) *Bantu Lexical Reconstructions 3*. Tervuren: Royal Museum for Central Africa (Available online database at: linguistics.africamuseum.be/BLR3.html, Accessed November 28, 2017).

Belliard, F. (2005) *Instruments, chants et performances musicales chez les Kwakum de l'arrondissement de Doume (Est-Cameroun). Etude ethnolinguistique de la conception musicale d'une population de langue bantoue A91*. Paris: Université Paris 7; LLACAN, thèse de doctorat.

Belliard, F. (2007) *Parlons kwàkúm, langue bantu de l'Est Cameroun : langue et culture*. Paris: L'Harmattan.

Devos, M. & J. van der Auwera (2013) Jespersen Cycles in Bantu: Double and Triple Negation. *Journal of African Languages and Linguistics* 34(2): 205–274.

Njantcho Kouagang, E. (2018) *A Grammar of Kwakum*. Paris: INALCO, PhD dissertation.

CHAPTER 14

NSONG B85d

Joseph Koni Muluwa and Koen Bostoen

1 INTRODUCTION

Nsong [è-nsɔ̀ŋ] (ISO 639-3 soo) is spoken in the vicinity of Kikwit (5°2'S, 18°48'E), Kwilu Province, DR Congo, by some 7,080 people (Koni Muluwa & Bostoen 2008). Nsong scores low on all nine UNESCO criteria for assessing language vitality and can therefore be considered severely endangered. Guthrie (1971) classified 'Ntsuo' B85d as a dialect of Yans B85 (Maho 2003). It is also known as Itsong, Tsong and Kisongo (Hammarström *et al.* 2016, Lewis *et al.* 2016). The variety described here is spoken in the Kipuka sector (Bulungu territory). Our main consultants originate from the villages of Kabamba, Kisala, Kafumba and Bulumbu. The Nsong of de Beaucorps (1941) live further north of Kikwit, i.e., in the Gobari, Kwilu-Kimbata, Luniungu and Nko sectors (Bulungu territory). Apart from our earlier work (Koni Muluwa 2006, Koni Muluwa & Bostoen 2007, Koni Muluwa 2010, Koni Muluwa & Bostoen 2010, 2012, Bostoen & Koni Muluwa 2014, Koni Muluwa 2014, Koni Muluwa & Bostoen 2014, 2015), Iliku Mimpya (1979) provides the only description of Nsong. Nsong is distinct from Songo H24 (Angola).

This sketch is based on Nsong data which the first author collected as part of his MA and PhD research (Koni Muluwa 2006, 2010) and a text corpus which he collected, transcribed and annotated in 2013 and 2014 as part of a documentation project supervised by the second author and funded by the DoBeS program of the Volkswagen Foundation (2012–2015). This corpus consists of oral discourse, currently counts 48,022 tokens and 11,973 types and comprises different text genres, such as political speeches, historical traditions, folk music, tales, proverbs, hunting language, ceremonial discourse and popular biological knowledge.

2 PHONOLOGY

2.1 Vowels

Nsong has seven basic vowel phonemes, viz. *i ɪ ɛ a ɔ ʊ u*, whose phonemic status is apparent from the minimal pairs in (1).

(1)	mpíl	'manner'	vs.	mpíl	'viper'	vs.	mpέl 'squirrel sp.'
	mpúr	'fallow'	vs.	mpʊ́r	'poor'		
	kɔ̀bʊ́ʊ́m	'to make potttery'	vs.	kɔ̀bɔ́ɔ́m	'to leave'		
	ndìts	'serval'	vs.	ndèts	'beard'		
	ɛ̀lɛ̀l	'madness'	vs.	ɛ̀làl	'orange'		
	kɔ̀sál	'to do, work'	vs.	kɔ̀sɔ́l	'to choose'		

Along with its closest relatives, Nsong distinguishes itself from the majority of Western Bantu seven-vowel languages in having *ɩ* and *ʊ* as second-degree vowels (Koni Muluwa & Bostoen 2008: 5, 2012: 359). Vove B305 is another case in point (van der Veen 1991). The vowels *e* and *o*, more common as second-degree vowels elsewhere in western Bantu, also occur in Nsong, though at best as marginal phonemes. They are only attested in a very limited set of stems listed in (2). We have found only one minimal-pair involving *e* in the stem. As discussed below, *e* and *o* occur in noun class prefixes as allophones of *ɛ* and *ɔ*.

(2) ébwèts   'poultry'
     èwùmè   'diarrhoea'
     kódyé   'to blow'
     kóbyó   'to belch'
     ngyôm   'fruit bat'
     ntèn   '*Garcinia kola* nuts'   vs.   ntín   'speed'   vs.   ntùn   'emissary'

Nsong also has several diphthongs, both on-glides and off-glides. The latter always end in *y*. As is commonly the case in Bantu, certain diphthongs are the outcome of contact between two vowels that belong to different morphemes or lost an intermediate consonant, e.g., *mó-its* → *mwîts* 'light,' **-dòkì* > *ndɔ̀y* 'namesake,' or reflect historical vowel clusters within the stem, e.g., **-búà* > *m̀bwà* 'dog.' Others, as those in (3), originate in a monophthong that broke into a vowel cluster through diphthongisation (Koni Muluwa & Bostoen 2012) and umlaut (Bostoen & Koni Muluwa 2014), which are regular diachronic sound shifts characteristic for Nsong and related B80 languages (Daeleman 1977, Rottland 1977).

(3) mókyέŋ   'mongoose'   (< *-kéngè)
     nswìŋ   'moon'   (< *-cóngé)
     épwìm   '*Erythrophleum guineense*'   (< *-pùmí)
     móswèm   'eel'   (< *-còmbì)

Vowel length is phonologically distinctive in Nsong (4a), except in CGV and VNC contexts where lengthening is automatic and not noted (4b). Long vowels also occur word-finally.

(4) a mátà   'bows'   vs.   mátàà   'buttocks'
        mpù   'hat(s)'   vs.   mpùù   'rat'
        mbɔ́m   'python'   vs.   mbɔ́ɔ́m   'nose'
        mbím   'border'   vs.   mbíím   'craftsman'

     b kɔ̀fwέr   'to suck'   [kɔ̀fwέːr]
        lótέndz 'vein, tendon'   [lótέːndz]

2.2 Consonants

Nsong has 24 simple consonant phonemes and 16 prenasalised ones, as listed in Table 14.1.

The phonemic status of the consonants in Table 14.1 is apparent from the minimal pairs in (5).

TABLE 14.1 NSONG CONSONANT SYSTEM

	Bilabial		Labio-dental		Alveolar		Palatal	Velar	Glottal
Plosive	p	b			t	d		k	
Nasal		m				n	ɲ	ŋ	
Trill					r				
Fricative			f	v	s	z	ʃ 3		h
Approximant							y	w	
Lateral approxim.					l				
Affricate			pf	bv	ts	dz	(c)	(j)	
Prenasalised	mp	mb			nt	nd		ŋk ŋg	
			mf	mv	ns	nz	nʃ	n3	
			mpf	mbv	nts	ndz			

(5) kɔ́tyà   'to descend'   vs.   kɔ́dyà   'to eat'
     mán   'wine'   vs.   máŋ   'branches'
     kɔ̀ŋésél   'to show teeth'   vs.   kɔ̀ŋésél   'to lower'
     kɔ́wɔ́r   'to warm up'   vs.   kɔ́wɔ́l   'to take'
     kólii   'to be exhausted'   vs.   kódii   'to wander'
     kɔ̀sɛ́n   'to take'   vs.   kɔ̀ʃén   'to carry, take away, bring'
     kɔ̀ʃím   'to catch'   vs.   kɔ̀ʒím   'to extinguish'
     lɔ̀háŋ   'oath'   vs.   lɔ̀páŋ   'enclosure'
     kɔ́pfà   'to die'   vs.   kɔ́pà   'to give'
     étsín   'taboo'   vs.   étín   'short'
     mbvù   'blat sp. (pl.)'   vs.   mvù   'raw'
     ndzál   'fingernails'   vs.   nzál   'hunger'

The palatal affricates c and j, represented in parentheses in Table 14.1, are rare and occur only in free allophonic alternation with their alveolar equivalents. They also freely alternate with ʃ and 3 before i. The bilabial affricate bv is also marginal and most often – though not always – in free allophonic variation with v. The alveolars l and d respect to a great extent the complementary distribution that is common in Bantu, namely l → d /N__ (N = homorganic nasal) and l → d /__i. However, l is also found today in front of close i, especially because many present-day speakers freely alternate close and near-close vowels under the influence of vehicular Kongo or Kongo ya Leta H10A, which only has five vowels. For instance, kɔ̀líís 'to show' with a long near-close front vowel is often heard as kòlíís with a long close front vowel and thus in phonemic contrast with kòdíís 'to feed.' The fricatives s/z and ʃ/3 are also partially in complementary distribution, namely s/z → ʃ/3 /__i, but they do contrast phonemically in other contexts. All consonants occur in stem-initial and stem-medial positions, but many consonants are excluded from stem-final position. These are predominantly but not exclusively voiced: d, ɲ, v, z, 3, h, w, pf, bv, dz, mv, mbv, ŋg. Stem-finally ŋk is only heard in loanwords, such as flàŋk 'franc.'

2.3 Morphophonological rules

2.3.1 Vowel hiatus resolution

When two vowels belonging to different morphemes enter in direct contact, the most commonly observed vowel hiatus resolution is glide formation. If the first vowel is a front vowel, it is turned into a palatal glide. If it is a back vowel, it becomes a labial-velar glide.

(6) mɛ́-ànd   'dances'   →   myând
     mɛ́-ɔ̀r   'stars'   →   myɔ̂r
     mɔ́-àn   'child'   →   mwân
     mɔ́-ìb   'thief'   →   mwîb

Two identical non-front vowels merge into a long vowel, as in (7), where it carries a falling tone. Contour tones always occur on long vowels.

(7) mɔ́-ɔ̀r   'star'   →   mɔ̂r
     bá-àn   'children' →   bân

A prefix with *a* undergoes total assimilation and merges with the stem.

(8) má-ɛ̀ts   'oil'   →   mɛ̂ts
     má-ɛ̀n   'ground'   →   mɛ̂n
     bá-ìb   'thiefs'   →   bîb

2.3.2 Regressive ATR vowel harmony

Nsong prefix vowels are subject to regressive vowel harmony. Noun class prefixes manifest an alternation between half-close and half-open mid vowels. This assimilation with root vowels is based on the distinction between vowels pronounced with or without an advanced tongue root. Roots having a [-ATR] vowel, viz. ɪ ɛ a ɔ ʊ, trigger a prefix with a [-ATR] vowel, while roots having a [+ATR] vowel, viz. i u, always occur with a prefix having a [+ATR] vowel.

(9) [-ATR]   [+ATR]
     ɛ́-tì   'spring'   vs.   é-tìì   'orphan'
     ɛ́-tɔ́r   'cloud'   vs.   é-tùr   'full moon'
     ɛ́-tɔ̀l   'crop'   vs.   é-ʃìn   'squirrel'
     mɔ́-lɛ́ts   'sorcerer' vs.   mó-dím   'husband'
     kɔ̀-wɔ́l   'to take'   vs.   kò-bil   'to explain'
     lɔ́-jwɛ́m   'bulrush'   vs.   ló-pin   'finger'
     ɛ́-tàm   'cheek'
     mɔ́-ban   'valley'
     lɔ́-zal   'claw'

This alternation between half-close and half-open mid vowels is also observed in verbal concord prefixes, as shown in (10) with the class 1 object prefix.

(10)   akímɔ́mɔ́n   vs.   kàmòcwíl
       à-kí-mɔ́-mɔ́n   kà-mò-cúl-í
       SP₂ₛ𝗀-FUT-OP₁-see   SP₁-OP₁-cut-ANT
       'You will see him.'   'He cut him.'

2.3.3 Progressive ATR vowel harmony

Progressive vowel harmony is observed between the root and extensions having a front
or back vowel as opposed to the central vowel *a*. Root and extension vowels also agree in
terms of ATR, but it does not involve an alternation between half-close and half-open mid
vowels. The system is more complex in that extensions exhibit alternation between close,
near-close and half-open vowels, as shown in (11) with examples of extensions having a
front vowel, viz. the applicative (*-il/-ɪl/-ɛl*) and the causative (*-is/-ɪs/-ɛs*). Roots with central and back vowels combine this progressive assimilation with a shift of the root vowel
known as umlaut (Section 2.3.4). Harmony with close vowels is shown in (11a), with a
near-close vowel in (11b), and with half-open ones in (11c).

(11) a   kò-kír   'to make'   kò-kír-íl   'to make for'   kò-kír-ís   'to make make'
          kò-sùùm   'to buy'   kò-swím-ín   'to buy for'   kò-swím-ís   'to make buy'

      b   kɔ̀-wɔ́m   'to dry'   kɔ̀-wɪ́m-ɪ́n   'to dry for'   kɔ̀-wɪ́m-ɪ́s   'to make dry'

      c   kɔ̀-lɔ́b   'to fish'   kɔ̀-lwɛ́b-ɛ́l 'to fish for'
          kɔ̀-bwár   'to wear'   kɔ̀-bwɛ́r-ɛ́l 'to wear for'   kɔ̀-bwɛ́r-ɛ́s 'to make wear'

The reflexes of the separative extensions *-ʊd-* (12a) and *-ʊk-* (12b), which have a
back vowel, exhibit the same three-way alternation following roots with a back vowel.
Additional assimilation to the place of articulation occurs with roots having a central or
front vowel.

(12) a   kò-wúndúl   'to rest'
          kò-wʊ́sól   'to wake up (tr.)'
          kɔ́-wɔ́bɔ́l   'to skin, bark'   cf.   *-jòbʊd-   'skin (fruit, animal)'
          kɔ́-páb-ál   'to tear'   cf.   *-papʊd-   'tear', -páb-á 'be torn'
          kɔ́-káŋál   'to open, reveal'   cf.   *-kaŋʊd-   'open (tr.)'
          kɔ̀-lwɛ́l-ɛ́l   'to forgive'   cf.   kɔ̀-lwɛ́l   'to punish'
          kò-dzwíb-íl   'to open'   cf.   kò-dzwíb-í 'to close'

      b   kó-fùrú   'to return, go back' cf.   *-bútʊk-   'come or go back'
          kɔ-wʊ́sʊ́   'to wake up (intr.)'
          kɔ́-tɔ́ɔ́mɔ́   'to jump'   cf.   *-tómbʊk-   'rise up, jump'
          kɔ́-pábá   'to be torn'

2.3.4   Umlaut

Not only affix vowels, but also root vowels, mutate in Nsong under the influence of a
(underlying) front vowel in the following syllable. This diachronically regular regressive
assimilation known as "umlaut" is also active as a synchronic morphophonological rule
in Nsong (Bostoen & Koni Muluwa 2014). It leads to the raising and fronting of *a* to *ɛ*
and to the fronting of *ɔ* to (*w*)*ɛ*, *ʊ* to (*w*)*ɪ and u to* (*w*)*i*. It can be triggered by an extension

having a front vowel, as in (11), and/or TAM suffixes consisting of a front vowel that most often does not appear on the surface, such as the anterior (13a-b) and the subjunctive/imperative (13c-d).

(13) a bátwìɲɛ́
        bà-tùŋ-í
        SP$_2$-build-ANT
        'They have built.'

     b bèɲém
        bè-ɲàm-í
        SP$_{1PL}$-install-ANT
        'We have installed.'

     c bèmwɛ́nɛ́
        bè-mɔ́n-í
        SP$_{1PL}$-see-SBJV
        'Let us see.'

     d mòkwíl
        mo-kúl-í
        OP$_3$-remove-IMP
        'Remove it.'

2.3.5 Nasal harmony

As several examples above show, for instance in (11), extensions ending in *l* undergo nasal harmony if the verb root ends in a bilabial or alveolar nasal. Nasal harmony occurs neither when the root ends in a velar nasal or a NC cluster nor when the nasal does not occur in root-final position, e.g., -*nwíl* 'drink for.'

2.3.6 Contact between nasal and oral sounds

The major morphophonological shifts involving non-syllabic homorganic nasals are the velarisation and unpacking of the homorganic nasal *N* to *ŋg* in front of verb roots starting with a glide (14), and the strengthening of *l* to *d* (15). Meinhof's rule (NC$_{[+voiced]}$VN(C$_{[+voiced]}$) → NVN(C$_{[+voiced]}$) or Kongo rule (NN → NC$_{[+voiced]}$) are not observed synchronically.

(14) N-wɔ́b   →   ŋgwɔ́b   'I take a bath'
      N-yɛ́ndɛ́k   →   ŋgyɛ́ndɛ́k   'I begin'

(15) kɔ̀-N-láb-í   →   kɔ̀ndɛ́b   'don't follow me'
      N-lund-í   →   ndwínd   'let me add'
      lɔ́làm   →   ńdàm   'lizard(s) sp.'

2.3.7 Spirantisation

Spirantisation is prolific morpheme-internally as a diachronic shift, but across the board it is only observed in agentive derivations, which are synchronically not productive. The underlying high front vowel underwent apocope and never appears on the surface.

(16) kɔ́bǜl   'to be sick'   vs.   mɔ́bǜts   'sick person'
      kɔ́lɔ̀ɔ̀   'to bewitch'   vs.   mɔ́lɛ́ts   'witch'

2.3.8 Palatalisation

Synchronically, palatalisation is commonly observed when the anterior suffix -*i* is affixed to a verb stem ending in *s*.

(17) è-s-í   →   èʃí   'I have put.'
      SP$_{1SG}$-put-ANT
      à-pf-ìs-í   →   àpfíʃí 'Someone of his family has died.'
      SP$_1$-die-CAUS-ANT

2.3.9 Imbrication

Imbrication is a common Bantu type of morpheme merger whereby certain verbal suffixes merge or whereby a suffix is infixed into certain verb roots (Bastin 1983). Given a maximality constraint of two syllables per verb base (cf. Hyman 2004: 79), such infixation is observed in Nsong when a derived verb takes more than one extension. In (18), a causative suffix followed by an applicative suffix is imbricated into the root.

(18) a   bábwɛ́sɛ́l épùs
          bá-bwár-ɛ́s-ɛ́l-ɛ́   é-pùs
          SP$_2$-dress$_{(intr.)}$-CAUS-APPL-IPFV   7-tissue
          'They dress (him) with a tissue.'

      b   bàlàwɛ́ ɔ̰́wɛ́sɛ́l màts
          ba-la-wɛ́-í   ɔ̰́-wɔ́b-ɛ́s-ɛ́l   màts
          SP$_2$-CONS-go-ANT   OP$_1$-bath-CAUS-APPL   6.water
          'They subsequently washed her with water.'

Final vowel suffixes triggering umlaut accompanied by diphthongisation could also be considered "imbricated" (Bostoen & Koni Muluwa 2014: 217).

2.4 Syllable structure

Due to diachronic final vowel loss, an areal feature that Nsong shares with several closely and more distantly related languages (Daeleman 1977, Rottland 1977, Mukash Kalel 1982, Nash 1992, Grégoire 2003, Bostoen & Koni Muluwa 2011, Bostoen & Mundeke 2011a, Koni Muluwa & Bostoen 2012, Bostoen & Koni Muluwa 2014), closed syllables occur as an innovation of the inherited Proto-Bantu pattern in which only open syllables are admitted. Nsong thus has two syllable types, namely CV and CVC. C can be either an oral, nasal or prenasalised consonant; V either a short or long monophthong or a diphthong. Syllables consisting of nothing but a vocalic nucleus only occur in word-initial position. Syllabic homorganic nasals occur in Nsong as prefixes of noun classes 1n, 9 and 10 (Section 3.1). While these nasal prefixes are syllabic and tone-bearing word-initially in isolation, e.g., ŋ́.gɔ̀ 'leopard' and m̀.pù 'hat,' they are syllabified as CV.NCV and no longer bear a tone when preceded by a plural prefix, e.g., *bá.ŋgɔ̀* 'leopards' and *má. mpù* 'hats,' or when preceded in context by a word with a final vowel. Words tend to end in a closed syllable, especially if they are not monosyllabic. However, polysyllabic words with a final vowel do exist, commonly verbs with a vocalic extension, as seen for instance in (9) and (12). Word-final suffixes usually do not surface, but may appear for euphonic reasons that are not well understood yet as certain examples in (13) show. Complex onsets are rare, but not inexistent, e.g., *kɔ́.tsà.tslèŋ* 'to stumble.'

2.5 Tone

As the minimal pairs in (19) indicate, tone is phonologically distinctive in terms of lexical meaning. Nsong has two fundamental level tones, viz. high (H) marked with an acute accent (´), and low (L) marked with a grave accent (`). A syllable can be level H or L or involve a fall from H to L (ˆ) or a rise from L to H (ˇ). Contour tones always occur on long vowels. Synchronically, they do not necessarily derive from the combination of level tones. They often result from the contraction of the noun prefix and the stem into one syllable, but they are also observed stem-internally, especially falling.

(19)   èdzín   'tooth'   édzìn   'name'
      lɔ̀bá   'arrow'   lɔ́bà   'palm nut'
      ébɔ́ŋ   'horn'   ébɔ̀ŋ   'knee'
      kólùù   'to vomit'   kòlúú   'to give a name'
      mwǐʃ   'pestle'   mwîʃ   'smoke'
      ŋgwĕn   'crocodile'   ŋgwên   'his/her mother'
      tăr   'three'   târ   'father'
      mbɔ̂m   'nose'   mbɔ́m   'python'
      nzɔ̂   'elephant'   ńzɔ̀   'house'

Based on a test sample of 220 basic vocabulary items, the basic tone patterns of nouns (including infinitives) in isolation are all H (47%) and H followed by L (44%), which is realised as F (falling) on monosyllabic nouns. All L (5%) and L followed by H (4%) are rather exceptional. Pre-pausally, the same tone scheme surfaces as in isolation (20).

(20) mókèts   'wife'   àyìʃén mókèts   'He came to marry a wife.'
      mbáá   'fire'   bŏtìn mbáá   'Those who fled the fire.'
      ńdìb   'bell'   nsés kàbwìrí ńdìb   'Blue duiker brought a bell.'
      éʃìn   'squirrel' bàjwí éʃìn   'They killed a squirrel.'

The surface realisation of nominal tone patterns in context is generally not the same as in isolation, especially not at the beginning of certain noun phrases (21) or in pre-verbal subject position (22). An in-depth tonal analysis is needed to account for these tone shifts.

(21) kán   'mouth'   kàn lí   'this mouth'   kàn lámwés   'one mouth'
      kɔ̂   'arm'   kɔ̌ lí   'this arm'   kɔ̌ lámwés   'one arm'
      bwàl   'village'   bwál bá   'this village'   bwàl bámwés   'one village'
      mwĕm   'medicine'   mwèm wú   'this medicine'   mwèm mwés   'one medicine'
      ébás   'insect sp.'   èbàs kí   'this insect sp.'   èbàs kémwés   'one insect sp.
      mɔ́tì   'tree'   mɔ̀tí wú   'this tree'   mɔ̀tí mwés   'one tree'

bɔ́kwên   'hunting net'   bɔ̀kwɛ̌n bá   'this net'   bɔ̀kwɛ̌n bámwɛ́s   'one net'
éʃíkíl   'stool'   èʃikìl kí   'this stool'   èʃikìl kémwɛ́s   'one stool'
ékwìtsì   'heel'   èkwítsí kí   'this heel'   èkwítsí kémwɛ́s   'one heel'
èsáŋéŋ   'island'   èsàŋèŋ kí   'this island'   èsàŋèŋ kémwɛ́s   'one island'

(22) éʃìn   'squirrel'   èʃíín kèwíŋ mwébɔ̀ŋ   'A squirrel has entered the knee.'
      ébɔ̀ŋ   'knee'   èbɔ́ŋ èndwíndz   'My knee hurts.'
      mbáá   'fire'   mbàà èjímí   'The fire was extinguished.'
      myɛ́ts   'rivers'   myɛ̀ts àmí ntɔ̀ à myá 'These rivers (have) their sources.'

Tone can also be grammatically distinctive, as illustrated in (23). More in-depth study is needed to understand the contribution of melodic tone within the TAM system.

(23) bɔ̀wàbíl lɛ̀ nkí?   'How do you call (it)?'
      bɔ̀wábíl lɛ̀ nkí?   'How did you call (it)?'

3 NOUNS AND NOUN PHRASES

3.1 Noun class system

Nsong has maintained 16 of the 18 basic noun classes reconstructed for Proto-Bantu (Meeussen 1967: 97–98) with only minor reshuffling of the concord patterns and the noun class pairings involved in plural formation. Classes 12 and 13 were lost. Some of its ancient members are still found in the lexicon, but they were integrated in other noun classes. The basic forms of Nsong noun class and concord prefixes, which are subject to the morphophonological rules of vowel hiatus resolution (Section 2.3.1) and regressive vowel harmony (Section 2.3.2), are listed in Table 14.2.

TABLE 14.2 NOUN CLASS AND CONCORD PREFIXES IN NSONG

Class	Noun Prefix (NP)	Adnominal Concord Prefix (ACP)	Subject Prefix (SP)	Object Prefix (OP)
1	mɔ́-	mɔ́-/ú-	a-, ɔ-, ka-	-mɔ-
1a	ø-	mɔ́-/ú-	a-, ɔ-, ka-	-mɔ-
1n	ń-	mɔ́-/ú-	a-, ɔ-, ka-	-mɔ-
2	bá-	bá-	ba-, a-	-ba-
3	mɔ́-	wú-	a-	-mɔ-
4	mɛ́	mí-	mi-, mɛ-	-mɛ-
5	ɛ́-	lá-	la-	-lɔ-
6	má-	má-	ma-, mo-	-ma-
7	kɛ́- / ɛ́-	kí-	ki-, kɛ-	-kɛ-
8	bɛ́- / ɛ́ɛ́-	bí-	bɛ-	-bɛ-
9	ń-	yí-, ɛ́-	yi-, ɛ-	-yi-
10	ń-	yí-, ɛ́-	yi-, ɛ-	-yi-
11	lɔ́-	lí-, la-	lɔ-	-lɔ-
14	bɔ́-	bá-	ba-, ɔ-	-ba-
15	kɔ́-	kú-	kɔ-, ka-	-ka-
16	pa-	pa-, pɔ-, pu-	×	×
17	kɔ-	ɔ-, ku	×	×
18	mɔ-	mɔ-/ɔ̨	×	×

As discussed in Koni Muluwa and Bostoen (2008: 7), certain prefixes underwent phonological reduction in the variety spoken north of the Lokelenge River: (1) the noun prefixes of classes 1, 3, 4, 6 and 18 (all NV-) are realised as nasalised vowels; (2) those of classes 2, 7 and 14 lost their initial consonant; (3) the prefix of class 8 also became vocalic, but with a long vowel. The Lokelenge River constitutes a natural barrier for this regiolectal variation, though not an impermeable one in that reduced prefixes can also be heard in Nsong-speaking villages situated south of it.

We use three criteria to define noun classes in Nsong: the shape of the noun prefix, the agreement patterns it triggers and the noun class with which it forms a regular pair to mark the singular and plural of nouns. Their semantic content is not discretely defined, but certain tendencies can be distinguished. As is commonly the case in Bantu, the Nsong noun class system has four main functions: nominal classification, number, agreement and derivation. The former three functions are dealt with in this section. Nominal derivation is discussed in the following section.

Out of the 16 noun classes, 12 enter in nine recurrent noun class pairings, which are also common elsewhere in Bantu, to distinguish between the singular and plural of nouns: 1/2, 3/4, 5/6, 7/8, 9/10, 9/6, 11/10, 11/6 and 14/6. Some of the classes include specific subsets that do require some additional discussion.

Class 1 incorporates two distinct subsets: class 1a and class 1n. As elsewhere in Bantu, subclass 1a consists of kinship terms, such as *mâm* 'mother,' *târ* 'father, *yêy* 'older sibling' and *kwér* 'sibling-in-law.' They are prefix-less, but agree in subclass 1 and take class 2 prefix *bà-* in the plural. Nouns of subclass 1n also have an additive plural prefix *bà-*. We distinguish them from class 1a, because they all start with a homorganic nasal that is syllabic and tone-bearing in the singular, e.g., *m̂pfùm/bá-mpfùm* 'chief(s),' *ŋgáŋ/bà-ŋgáŋ* 'healer(s),' *ǹzɔ́ɔ́/bà-nzɔ́ɔ́* 'elephant(s),' *ǹtáám/bà-ntáám* 'lion(s).' It also includes certain kin terms beginning with a homorganic syllabic nasal, such as *m̀pèŋ* 'relative, parent, sibling,' *ŋ̀gwâs* 'mother's brother,' *ǹzèl* 'sibling-in-law.' In contrast to subclass 1a nouns, all subclass 1n nouns are ancient class 9 nouns that were reclassified on grounds of their animacy. Similar animacy-motivated reclassification is observed with certain ancient class 5 nouns, which maintained the class 5 prefix in the singular, but agree in class 1 and take their plural in class 2, e.g., *è-bàál/bà-bàál* 'male person(s),' *è-kúb/bà-kúb* 'adult(s).' The concord patterns of these reclassified nouns are illustrated in (24).

(24) a   N-bwá   mó-kwíndz
         1n-dog   ACP₁-naked
         'naked dog'

      b   bà-ŋkím   bà   N-bálá
         2-monkey   CON₂   9-forest
         'forest monkeys'

      c   è-bààl   wú
         5-male   PROX.DEM₁
         'this male person'

Class 5 has *ɛ-* as noun prefix, but incorporates two distinct subsets of nouns that deviate in that regard. The first subset consists of ancient members of class 13, which is extinct in Nsong. These nouns begin with *t*, which is reminiscent of NP₁₃ *tò-*: *tìb/mátìb* 'excreta' (< *tò-bìì*), *tɔ̀l* 'sleep' (< *tò-dò*), *twî* 'flying ant.' However, they currently agree in class 5, as shown in (25a), and take their plural in class 6, if they have one. The second subset consists of two nouns referring to body parts, namely *kù-ùl/mì-ìl* 'leg(s)' and *kɔ́-ɔ̀/mí-ɔ̀* 'arm(s).' They used to belong to class 15, of which they have conserved the noun class prefix, but nowadays they take their concord in class 5, as illustrated in (25b). In the plural, both nouns belong to and agree in class 4, as seen in (25c), and not class 6, as is often the case elsewhere in Bantu.

(25) a   Ø-tìb   lá   mɔ́-àn
          5-excreta   CON₅   1-child
          'excreta of the child'

      b   kɔ́-ɔ̀   lá   nzɛ́n
          (1)5-arm   CON₅   him
          'his arm'

      c   mì-ìl   myán   my-ɛ̀lɛ̀
          4-leg   MED.DEM₄ ACP₄-two
          'those two legs'

Class 7 has *kè-* as canonical noun prefix, but incorporates a subset of nouns that have *kà-* as noun prefix, but also take their plural in class 8, i.e., *ká-kàngàl/bé-kàngàl* 'kind of swallow,' *kà-mbɔ́lɔ́ŋ/bè-mbɔ́lɔ́ŋ* 'plant species (*Mondia whitei*),' *kà-yɔ́ŋ/bè-yɔ́ŋ* 'plant species (*Biophytum talbotii*),' *kà-dyámb/bè-dyámb* 'plant species (*Chenopodium ambrosioides*).' Most of them are plant and animal names, which used to belong to class 12, but were reanalysed as class 7 nouns in terms of plural formation and agreement. As shown in (26), they no longer take their concord in class 12, which therefore does not exist anymore.

(26) kà-dyámb   kí   kì-lí   ké   mɛ̀
      7-*Chenopodium.ambrosioides*   PROX.DEM₇ SP₇-be   CON₇ me
      'This *Chenopodium ambrosioides* is mine.'

Class 9 nouns commonly have an indistinct plural form in class 10, but some have their plural in class 6 with an additive noun prefix *mà-*, e.g., *mbɔ̂m/mà-mbɔ̂m* 'noise(s),' *ńdàà/ mándàà* 'language(s), palaver(s),' *ńtàts/mántàts* 'bed(s),' *ḿpù/mámpù* 'hat(s).'

Classes 5 and 11 are still distinct in terms of nominal prefix and plural formation, but they merged their concord prefixes.

(27) a   lɔ́-béŋ   lí   lé-mvù   lá-bwì   ɛ̀wɛ́   è-wú-lá-bùù
          11-fruit   PROX.DEM₁₁   ACP₁₁-raw   SP₁₁-fall.ANT   I   SP₁ₛ₉-MOT-OP₁₁-collect
          'This raw fruit has fallen, I am going to collect it.'

      b   è-yɔ́r   lí   lé-pfin   lá-pfĭ   è-bà-fú-lá-lɔ́ɔ́n   lɔ́ɔ́
          5-potter's.kiln PROX.DEM₅ ACP₅-new   SP₅-die.ANT   SP₁ₛ₉-NEG-REP-OP₅-repair NEG
          'This new potter's kiln is broken, I can repair it no more.'

3.2 Nominal derivation

3.2.1 Noun-to-noun derivation

Noun class alternation is a common device to derive nouns from other nouns. While certain derivational noun prefixes are substitutive, others are additive. Reduplication is another process involved in noun-to-noun derivation either in combination with noun prefix alternation or independently.

Diminutives occur in classes 7/8 and their derivation involves reduplication. As shown in (28), only C(G)V can be fully reduplicated. Of CVC only the first consonant and vowel are copied. When the noun stem consists of only a consonant, the prefix is copied. When the copied element is *Ca*, the low central vowel shifts to a high front vowel.

(28) mɔ́-cwì   'head'   →   ὲ-cwícwì   'small head'
      ḿ-bwà   'dog'   →   è-mbúmbwà   'little dog'
      mɔ̀-kánd   'woman'   →   ὲ-kíkánd   'small woman'
      lɔ́-lím   'tongue'   →   ὲ-lílím   'little tongue'
      mú-nd   'man'   →   è-múmúnd   'short man'
      bá-nd   'men'   →   bè-bíbánd   'short men'
      kú-úl   'leg'   →   ὲ-kúkúúl   'little leg'
      mí-íl   'legs'   →   bè-mímíl   'little legs'

Augmentatives are derived through the reduplication of the entire word without any noun prefix alternation (29). The augmentative value often acquires an emphatic reading, real or figuratively big. Sometimes the insertion of an epenthetic vowel *a* is observed. This may indicate that this derivation strategy is in origin a connective construction, i.e., *ngɔ̀m à ngɔ̀m* 'a drum of a drum' → 'big drum,' whose connective element tends to be dropped. The fact that the second nasal in an augmentative like *ńdàà-ńdàà* remains tone-bearing and thus syllabic suggests that this is rather a sequence of two separate words than a truly reduplicated word. In the latter case, the nasal would become the onset of the next syllable and its high tone would be realised on the preceding vowel, as is for instance observed with the diminutive *èmbúmbwà* 'little dog' in (28).

(29) mú-nd   'man'   →   múnd-múnd   'big man, real man'
      bá-nd   'men'   →   bánd-bánd   'big men, real man'
      mɔ̀-kánd   'woman'   →   mɔ̀kánd-mɔ̀kánd   'big woman, real woman'
      ńdàà   'problem'   →   ńdàà-ńdàà   'big problem, real problem'
      bɔ́ɔ́   'mushroom' →   bɔ́ɔ́ á bɔ́ɔ́   'big mushroom, real mushroom'
      ngɔ̀m   'drum'   →   ngɔ̀m à ngɔ̀m   'big drum, real drum'

A common prefix alternation is observed between names for peoples (classes 1/2) and language names (class 7).

(30) mɔ̀-nsɔ̀ŋ   'Nsong person'   vs.   ὲ-nsɔ̀ŋ   'Nsong language'
      mɔ̀-mpíín   'Mpiin person'   vs.   ὲ-mpíín   'Mpiin B863 language'

Similarly, plant and tree names generally – though not exclusively – occur in classes 3/4, while their fruits are designated with the same noun stem preceded by the noun prefixes of classes 5/6 or 11/10 (see also Koni Muluwa 2010, Koni Muluwa 2014).

(31) mɔ́-bèts   'kola tree'   vs.   ὲ́-bèts   'kola nut'
      mɔ́-bìl   '*Canarium schweinfurthii*'   vs.   lɔ́-bìl   'fruit of *C. schweinfurthii*'
      ὲ́-bà   'oil palm'   vs.   lɔ́-bà   'palm nut'

Class 14 incorporates numerous nouns designating abstract concepts, which have a corresponding equivalent as part of another noun class.

(32) bɔ̀-kár   'imbecility'   vs.   kɛ̀-kár   'imbecile'
      bò-kwíndz   'nudity'   vs.   mò-kwíndz   'naked person'

Since Nsong does not have locative nouns consisting of only a locative noun class prefix
and a noun stem, the locative classes 16 ('upon,' 'on,' 'at'), 17 ('on,' 'from,' 'towards')
and 18 ('in') can be considered as exclusively derivational. Their prefix is always additive and generally does not trigger agreement on adnominal modifiers except on demonstratives (not obligatorily) and relative verbs (33).

(33) a   pɔ̀bwálá pɔ̀kɔ̀zwél
          pɔ̀-bɔ̀-álá   pɔ̀-kɔ̀-zɔ́l-í
          16-14-village   ACP₁₆-SP₁-love-ANT
          'at the village that he loves'

      b   kwiáŋ sɛ̀ kùbɛ̀kwɛ́ kùmàlúm kwâ
          kwìáŋ   sɛ̀   kù-bɛ̀-kɔ́-í   ù-mà-lúm   kwâ
          MED.DEM₁₇ also   ACP₁₇-SP₁ₚₗ-go-ANT   17-6-graveyard there
          'there as well where we have gone to the graveyard'

      c   mùmàlɔ̀ŋ àmán mànsɔ́ mùbàwúbɛ̀písɛ́ mpɛ̀s
          mù-mà-lɔ̀ŋ   àmán   mà-nsɔ́   mù-bà-wú-bɛ̀-p-ís-í   N-pɛ̀s
          18-6-lesson END.DEM₆   ACP₆-all ACP₁₈-SP₂-MOT-OP₁ₚₗ-give-CAUS-ANT   9-pain
          'in all those lessons in which they were going to cause us pain'

3.2.2 Verb-to-noun derivation

Due to regular final vowel loss (Section 2.4), the suffixes *-i, -o, -a*, and *-e*, through which
nouns are commonly derived from verbs in Bantu (Bastin 1985: 10, see also Chapter 6
of this volume), no longer appear on the surface in Nsong. Only the agentive suffix *-i* left
a lexicalised trace through its effect on the root vowel (umlaut, Section 2.3.4) and/or the
root-final consonant (spirantisation, Section 2.3.7).

(34) kɔ́-bìil   'to be sick, ill'   >   mɔ́-bìits   'sick person'   (spir)
      kɔ̀-lɔ́ɔ́   'to bewitch'   >   mɔ̀-léts   'sorcerer'   (spir + umlaut)
      kɔ́-búúm   'to make pottery'   >   m-bíím   'craftsman'   (umlaut)

Otherwise deverbative nominal derivation is only marked through noun prefix
alternation.

(35) a   Agents (cl. 1/2)
          kó-wib   'to steal'   >   mw-îb/b-îb   'thief'
          kò-nún   'to grow old'   >   mó-nún/bá-nún   'senile person'

      b   Results (cl. 7/8, 11)
          kɔ̀-kɛ̀n   'to remember'   >   lɔ́-kɛ̀n   'memory'
          kɔ́-pfà   'to die'   >   lɔ́-pfà/má-pfà   'mourning'

      c   Actions (cl. 6, 9/10)
          kò-tín   'to flee, run'   >   n-tín   'speed'
          kɔ̀-kín   'to dance'   >   mà-kín   'dance'

d   Tools, instruments (cl. 7/8, 9/10)

kɔ̀-kɔ́ɔ́m   'to brush'   >   kɛ̀-kɔ́ɔ́m/bɛ̀-kɔ́ɔ́m   'brush'
kɔ̀-lɔ́b   'to fish'   >   n-dɔ́b/n-dɔ́b   'fishhook'

3.3 Adnominal modifiers

The most frequent word order within the noun phrase is noun-determiner. Adjectives, numerals, possessives and demonstratives generally follow the noun. Some exceptions notwithstanding, adnominal modifiers agree with the head noun through an "adnominal concord prefix" (ACP). The contrast reconstructed for Proto-Bantu between nominal, numeral and pronominal concord prefixes (Meeussen 1967: 97) is not observed in Nsong.

3.3.1   Adjectives

As do many Bantu languages, Nsong has a restricted number of adjectives which manifest noun class agreement with the head noun e.g., è-bwél é-ŋkyè 'little pepper.'

(36)   -bì   'bad'   -bím   'whole'
        -bîts   'sick'   -bɔ̂   'red'
        -bɔ̀ŋ   'good, beautiful'   -bwâs   'empty, mere'
        -mvû   'unripe'   -nɛ́n   'big, fat'
        -ŋkyɛ̀   'little, small, young'   -pfin   'new'
        -pir   'black'   -pìì   'white'
        -tír   'thin'   -tín   'short'
        -vùl   'many'

Apart from these adjectives, qualification of nouns happens through connective constructions in which the second noun modifies the first noun, e.g., mùnd wù kɔ̀bɔ̀ŋ/bànd bà kɔ̀bɔ̀ŋ 'good person(s)' (lit. 'person(s) of goodness'), mbíts à mêts/bàmbíts à mêts 'fat animal(ʒ)' (lit. 'animal(ʒ) of fat'), mùnd à bwél/bùnd bù bwél 'tall person(s)' (lit. 'person(s) of tallness'). Another qualifying strategy is the use of relative constructions.

(37)   mɛ̀-tí   mɛ̀-ŋgú-bɔ́l
        4-tree   ACP₄-RES-rot
        'rotten trees' (lit. 'trees that have rotten')

3.3.2 Numerals

In counting, the numbers from one to ten occur invariably as mwés 'one,' yɔ̌l 'two,' ètár 'three,' énà 'four,' ètéén 'five,' èʃéém 'six,' nsámwên 'seven,' énán 'eight,' éwà 'nine,' èkwím 'ten.'

The cardinal numbers between 'one' and 'nine' with the exception of 'seven' behave as variable adjectives, e.g., ètíí kèmwès 'one orphan,' lòpìn lɔmwés 'one finger.' The numeral 'two' manifests vowel assimilation with concord prefixes having a front vowel, e.g., bàbààl bɔ́ɔ́l 'two men,' mɛ̀tí myéél 'two trees,' nzwɛ̀m ywéél 'two rushes.' The numeral 'seven' is invariable, e.g., bân nsámwân 'seven children,' màyúmú nsámwân 'seven ancient villages,' bèlúm nsámwân 'seven days.' The numeral èkwím 'ten' is a noun that belongs to class 5 and forms its plural in class 6. It is used after the noun. Both *èkwím* and its plural *màkwím* are followed by the invariable numerals 'one' to 'nine' to form compound numerals, e.g., *bàkánd èkwím è mwés* '11 women.' Decimals are formed by multiples of *èkwím* 'ten,' e.g., *màkwím móól* '20,' *màkwím mátár* '30,' etc. Higher basic numbers include *ńkàm* '100' and *fúún* '1000' (pl. *mà-fúún*).

To form ordinal numbers, the stem of the cardinal number is preceded by a connective (Section 3.3.3), e.g., *èkánd ké yóól* 'the second clan,' *mwán wá tár* 'the third child,' *màyùm má téén* 'the fifth ancient village.' Only for 'first' a specific stem is used, i.e., *tóm*, as *bákùb bà tóm* 'the first ancestors.'

3.3.3 Possessive constructions

Nsong has two possessive constructions, viz. nominal and pronominal, which are structurally identical in that the modifier following the head noun is introduced by a morpheme commonly known as a connective. In a nominal possessive construction, the connective precedes a noun (38).

(38) a   mɔ̀-ɛ̀ndèl   wá   N-fùm
          3-back CON_3 1n-chief
          'the back of the chief'

      b   bɔ̀-àl   á   N-fùm
          14-village CON_{14} 1-chief
          'the chief's village'

      c   mɔ̀-kéts   á   N-fùm
          1-wife CON_1 1n-chief
          'the wife of the chief'

      d   bé-kyèl   bé   N-kàà
          8-shell CON_8 1n-pangolin
          'the shells of the pangolin'

In a pronominal one, it is followed by a personal pronoun (39). The personal pronouns, aka 'substitutives,' that figure invariably in pronominal connective constructions are the following: *mɛ* 'I,' *nzé* 'you (sg),' *nzên* 'he, she,' *bús* 'we,' *bé* 'you (pl),' *bâ* 'they.'

(39) a   lò-zú   là   nzén
          11-peanut CON_{11} him
          'his peanut'

      b   é-kwítsì   á   mè
          7-heel CON_7 me
          'my heel'

      c   kè-mpfúm   ké   bús
          7-power CON_7 us
          'our power'

The connective is observed under three distinct forms: the PP-*à* form reconstructed for Proto-Bantu (Meeussen 1967: 106) as in (38a) and (39a), the linker *a* without ACP as in (38b+c) and (39b) and the ACP alone as in (38d) and (39c). As far as we can judge, these three types of connectives are free allomorphs, which occur both in nominal and pronominal constructions. In a nominal possessive construction, the modifying noun basically refers to a possessor (38), but it can also refer to a quality (40a), an origin (40b) or an agent (40c).

(40) a bɔ̀-kár   bò   mù-nd
         14-stupidity CON_{14} 1-person
         'the human stupidity'

      b mɔ̀-pɛ̀ndɛ̀   wà   Kɛlɛmb
         1-Pende CON_1 Kilembe
         'a Pende person from Kilembe village'

      c kɔ̀-tswá   ká   bà-kánd   bà-ɔ̌l
         15-pound CON_{15} 2-woman ACP_2-two
         'pounding of/by women'

Possessive constructions are commonly adnominal, but they can also occur without head noun (26).

3.3.4 Demonstratives

Nsong has three exophoric demonstratives and one endophoric demonstrative. To refer to an element that is external to the text, the degrees of deixis are "close to speaker" (proximal) (41a), "close to the hearer" (medial) (41b) and "remote to both speaker and hearer" (distal) (41c). The proximal demonstrative takes the form of the ACP. The medial and distal demonstratives consist of that same ACP combined with the stems -ìán and -ìáán respectively. The final nasal is sometimes realised as velar instead of alveolar (33). The structure of the endophoric demonstrative is $a+ACP+n$ (41d). Its initial vowel can be dropped, but is mostly present. Its reference is usually anaphoric (referring to a unit earlier mentioned in the discourse), but can also be cataphoric (referring to a unit later in the text).

(41) a m-bǎ   yí
         9-fire $PROX.DEM_9$
         'this fire'

      b mɔ̀-ɛ́ndɛ́l   wyán
         3-back $MED.DEM_3$
         'that back'

      c è-kyɛ́   lyáán
         5-egg $DIST.DEM_5$
         'that egg'

      d mɛ̀-sɔ́r   àmín
         4-forest $END.DEM_4$
         'those (aforementioned) forests'

Locative demonstratives are only employed independently (33b). The other demonstratives are also used independently, as in (42), where the deixis is temporal rather than spatial.

(42) kì   kì-lí   è-lúm   kὲ   má-ndàà
PROX.DEM$_7$ SP$_7$-be 7-day   CON$_7$ 6-palaver
'Today is palaver day.' (Lit. 'This is the day of the palaver.')

All demonstratives have an emphatic equivalent that is formed by prefixing *ACP-á* to the basic form.

(43) a   bá-sɔ̀ɔ̀   bábá
         2-slave   EMPH.PROX.DEM$_2$
         'these very slaves'

     b   ὲ-bòŋ   lályán
         5-pitcher   EMPH.MED.DEM$_5$
         'that very pitcher'

     c   kὲ-ŋgɔ́mb   kyákyáán
         7-okra   EMPH.DIST.DEM$_7$
         'that very okra'

     d   mò-nún   wáwún
         4-old man   EMPH.END.DEM$_1$
         'that really old man (in question)'

3.3.5 Interrogatives

Nsong interrogatives are *ná?* 'Who? What?,' *kwέ?* 'How many?, *nkí?* 'What?,' How much?,' *ná étàŋ?* 'When?,' *pén? kwín? mwín?* 'Where?,' *mòngyúr á nkì?* 'Why?,' *èbwín?* 'How?'. The first two are also used as adnominal modifier: *ná* precedes the head noun (44a-c), while *kwέ* follows it (44e-f). In terms of agreement, *ná* and *kwέ* only take an ACP$_2$ when they modify plural animate beings, as in (44b) and (44d), which is optional (44e). The interrogative *ná* not only modifies nouns referring to human being (44a-b), but it is also used for non-humans (44c), which is typologically rare but also attested in zone C (Idiatov 2009).

(44) a   ná   mɔ́-náŋ
         which   1-rich.person
         'which rich person?'

     b   bà-ná   bá-sám
         ACP$_2$-which 2-friend
         'which friends?'

     c   ná   bɔ́-ɔ́   àbá
         which   14-mushroom PROX.DEM$_{14}$
         'which mushroom is this?'

     d   bà-nzέnz   bà-kwέ
         2-guest   ACP$_2$-how.many
         'how many guests?'

     e   bà-àn   kwέ
         2-child   how.many
         'how many children?'

f   mέ-tènz   kwέ
    4-small.hammer   how.many
    'how many (small) hammers?'

4 VERBS

4.1 Verb roots and bases

The canonical shape of a verb root is CVC (*-pùr-* 'dig,' *-mìn-* 'swallow,' *-lìm-* 'cultivate,' *-náŋ-* 'sit down,' *-wíb-* 'steal') or a variant, such as CVVC (*-bàdl-* 'speak,' *-bɔ́ɔ́m-* 'leave), CGVC (*-fwér-* 'suck,' *-kyέl-* 'cut') and CVNC (*-dzúndz-* 'sting'). Less common forms are C (*-t-* 'throw, launch,' *-pf-* 'die'), CV (*-dì-* 'eat,' *-tì-* 'go down, *-nù-* 'drink,' *-cù-* 'pound') and CVV (*-lìì-* 'be tired,' *-lɔ̀ɔ̀-* 'bewitch'). C and CV roots originate in Proto-Bantu CV roots and are the only ones to still take the common Bantu Final Vowel *-*a* to form a stem, e.g., *-pfà* 'to die,' *-dyà* 'to eat,' *-cwà* 'to pound.' CVV roots originate in Proto-Bantu CVC roots that lost their final consonant. Polysyllabic verb bases, such as *-pábál* 'tear,' *-wɔ́bɔ́l* 'strip,' and *-wósó* 'wake up,' which do not have a matching monosyllabic verb root, are historically derived or expanded. Their CVCV(C) shape corresponds to that of synchronically derived verb bases. Both derived or expanded bases are subject to a maximality constraint of two syllables.

4.2 Verbal derivation suffixes

The repertoire of verbal derivation suffixes is reduced in Nsong. The language lost several extensions which have been reconstructed for Proto-Bantu and tend to be common elsewhere in Bantu, such as passive *-*ʊ-/-ibʊ-*, neuter *-*ɪk-*, impositive *-*ɪk-* and positional *-*am-*. Moreover, stacking of verbal derivational suffixes is exceptional in Nsong. A derived verb rarely takes more than one extension. In the rare case of two extensions, imbrication (Section 2.3.9) commonly takes places, cf. (18). Most extensions are subject to progressive vowel harmony (Section 2.3.3).

4.2.1 Applicative

Nsong productively uses the reflex of the Proto-Bantu applicative *-*ɪd-* to add an object to the verb's argument structure which would otherwise be an oblique argument. The applied object can be a beneficiary (45), a maleficiary (46), a recipient (47), a reason (48) or an instrument (18). The applicative extension may also be lexicalised (49). The base verb *-band-* is not attested.

(45) èkírílέ múnd ńzɔ̀
      e-kúr-íl-έ   mú-nd   Ń-zɔ̀
      SP$_{1sg}$-make-APPL-IPFV   1-person   9-house
      'I am building a house for someone.'

(46) nsέs òyí bὲbwérὲl ńdìb
      N-sέs   ò-yí   bɛ-bwár-έl   N-dìb
      1n-blue.duiker SP$_1$-be   OP$_{2pl}$-dress-APPL   9-bell
      'The blue duiker is wearing a bell against you (as a weapon).'

(47) mùnd wàwélèl mósúr
mù-nd   ò-à-wél-ɛl-í   mó-súr
1-person   ACP$_1$-SP$_1$-sell-APPL-ANT   3-forest
'the person to whom he sold the forest. . .'

(48) kɔ̀ndébèl mwând
kɔ̀-N-láb-èl   mɔ́-ànd
NEG-OP$_{1SG}$-follow-APPL.IMP   3-dance
'Don't follow me for the dance.'

(49) bábèndèl ébwánkɔ̀
bá-bànd.èl-í   ébwánkɔ̀
SP$_2$-begin.APPL-ANT   morning
'They have begun in the morning.'

4.2.2 Causative

Nsong productively uses the reflex of the Proto-Bantu causative *-ici- to add a causee
object to the verb's argument structure. It typically expresses "direct causation" (Shibatani & Pardeshi 2002: 89). The causee usually is patientive, as in (18) and (50), where
none of the object participants acts as a volitional entity.

(50) bàládís mâm àwún
bà-lá-dí-ís   mâm   awún
SP$_2$-SUBS-eat-CAUS   1a.mother   END.DEM$_1$
'And then that mother is fed.'

The subject of the derived causative verb, which is an agentive causer, affects both objects
physically. However, the causee can also be an experiencer. In (51), the causative verb,
derived from -wòò 'hear,' is commonly used as a speech verb.

(51) kán mówísí "nzè ɔ́kwɛ́"
kán   mó-wòò-is-í   nzè   ɔ́-kwɛ́-ɛ́
mouth   OP$_1$-hear-CAUS-ANT   you   SP$_{2SG}$-go-IPFV
'Mouth said to him "you, you go."'

Causative -is- also serves to express indirect causation involving an agentive causee
(52).

(52) bâ bɔ̀fwɛ́t kókírís bèsál àbín
bâ   bɔ̀-fwɛ́t   kó-kír-ís   bè-sál   àbín
they   SP$_2$-must   15-do-CAUS   8-work   END.DEM$_8$
'They must make carry out those works.'

It may also have an "adjutive" or "sociative" causative meaning in the sense of helping
someone to do something (53).

(53) mùnd sɛ̀ àbwìtísí nzɛ̂n wù
      mù-nd   sɛ̀   à-bùt-ís-í   nzɛ̂n   wù
      NP₁-person also   SP₁-give.birth-CAUS-ANT   her   REL₁
      'Also the person who helped her to give birth. . .'

4.2.3 Reciprocal

Nsong has two extensions involved in the expression of reciprocity: -ɛn-/-ɛŋ- and
-aŋgan-. The short marker occurs with either an alveolar or a velar nasal and is in all
likelihood the reflex of Proto-Bantu *-àn-. It is typically used to express naturally reciprocal events, namely actions or states in which the relationship among two participants
is usually or necessarily mutual (Kemmer 1993: 17), such as -fɛ́n-ɛ́ŋ 'resemble,' -kís-ɛ́ŋ
'agree,' -mwɛ́n-ɛ́ŋ (-mɔ́n-ɛ́ŋ) 'meet' and -bùn-ɛ́ŋ 'assemble.' Two more examples follow
in (54) and (55).

(54) é pàpàn bɛ́kɛ̀bɛ̀ŋ
      é   pàpàn   bɛ́-kàb-ɛ̀ŋ-í
      COP   EMPH.MED.DEM₁₆   SP₁ₚₗ-divide-RECP-ANT
      'It is from there that we separated (from each other).'

(55) mpìl àfùʃɛ̀nɛ̀ŋ lɔ́ɔ́
      N-pil   à-fù-ʃɛn-ɛ̀ŋ   lɔ́ɔ́
      9-way   SP₂-ITR-take-RECP   NEG
      'No way they marry (each other) again.'

As illustrated in (56), the short marker is also found with meanings that are not strictly
reciprocal, but rather belong to the broader "middle" domain as defined by Kemmer
(1993). Given that the base verb -pánz- is not attested, -ɛ́n- is to be considered lexicalised
here. Similar lexicalised middle verbs are -ŋɛ́tsɛ́ŋ 'be bright' and -tsàtslɛ̀ŋ 'stumble.'

(56) àmbìts àpɛ́nzɛ́n
      à-N-bìts   à-pánzɛ́n-í
      2-1n-fish   SP₂-disperse-ANT
      'The fish dispersed.'

The long marker seems to refer primarily to events that are not naturally mutual (57). It
possibly is the more productive reciprocal suffix. However, the distribution between the
long and short marker needs more in-depth study. Moreover, being polysyllabic, -aŋgan-
provides for one of the few contexts in which the maximality constraint of two syllables per verb base is violated (cf. Hyman 2004 for similar exceptions in other Western
languages).

(57) bàwábìlángáné
      bà-wá-bìl-ángán-ɛ́
      SP₂-PST.HAB-call-RECP-IPFV
      'They used to call each other.'

4.2.4 Separative

The reflexes of the Proto-Bantu separative extensions *-ʊk- (intransitive) and *-ʊd- (transitive) are fully lexicalised in Nsong. Their current-day forms are respectively -ù- and -ùl- and their different allomorphs are shown in (12). They generally occur with expanded verbs that no longer have a corresponding base verb, although exceptions exist, e.g., -lwél-él 'forgive' vs. -lwél 'punish.' Their synchronic morpheme status is furthermore established on the co-existence of certain intransitive and transitive equivalents, such as -páb-á 'be torn' vs. -páb-ál 'tear.' Due to progressive vowel assimilation turning back vowels into front vowels (Section 2.3.3), it is difficult to distinguish possible reflexes of Proto-Bantu neuter *-ɪk- from *-ʊk-. However, since -é- extensions are only observed with roots having a front vowel, e.g., -wémé 'calm down (intr.),' -tsélé 'ripen, be ripe,' we assume them to be reflexes of *-ʊk- and neuter *-ɪk- to be entirely extinct in Nsong. On the other hand, the final front vowel of transitive verbs, such as -dzwíbí 'close, shut' and -wéné 'spread out (e.g., maize),' is possibly a reflex of Proto-Bantu impositive *-ɪk-.

4.2.5 Passive

Nsong has no suffix to express a canonical passive. One possible alternative is the impersonal 3PL construction, illustrated in (50), (58) and (91). It is also attested in closely related Mbuun B87, which also lost the Proto-Bantu passive *-ʊ-/-ibʊ- (Bostoen & Mundeke 2011b).

(58) èdzwín lá bwàl bɔ̀wàpúr lwétìn.
      è-dzwín   lá   bɔ̀-àl   bɔ̀-wà-púr   lwétìn
      7-hole   CON₅   14-village   SP₂-HAB-dig   ACP₅.short
      'The hole of the village is usually dug shallow'
      (Lit. 'The hole of the village, they dig shallow.')

4.3 Tense, aspect and mood affixes

No coherent analysis of the TAM system in Nsong is offered in this section. Such an enterprise would need a dedicated in-depth study, which has not been undertaken yet. We limit ourselves here to an overview of the most important affixes involved in the expression of TAM according to their position in the conjugated verb form. Three verb slots are relevant for marking TAM distinctions in Nsong: verb-initially, verb-finally and verb-medially, i.e., between the subject prefix and the root.

4.3.1 Verb-final TAM markers

Many conjugated forms are unmarked verb-finally. The default -à, which normally marks the verb's right edge in Bantu, is no longer present due to final vowel loss. Nevertheless, three important aspectual and modal distinctions are still marked at the end of the verb by front vowel suffixes, which are often segmentally indistinguishable in surface realisation. We distinguish them here as imperfective -é, anterior -í and subjunctive -í.

4.3.1.1 Imperfective -ɛ́

Nsong uses a final suffix -ɛ́ to mark imperfective situations. In contrast to anterior -i and subjunctive -i, this marker does not trigger umlaut and it always surfaces as a half-open front vowel, as in (59), where it refers to three on-going events taking place at the time of utterance.

(59) ɛ̀kwɛ́ɛ́, ɛ̀wímɛ́ mwând. mwɛ́nsɔ́ŋ ɛ̀bàlɛ̀.

ɛ̀-kwɛ́-ɛ́	ɛ̀-wím-ɛ́	mɔ́-ànd	mɔ́-ɛ́-nsɔ́ŋ	ɛ̀-bàl-ɛ́
SP₁ₛɢ-walk-IPFV	SP₁ₛɢ-sing-IPFV	3-song	18-7-Nsong	SP₁ₛɢ-speak-IPFV

'I walk and sing a song. In Nsong I speak.'

In the absence of any other TAM marker, verbs marked by -ɛ́ at the end canonically designate situations with present time reference (59). In (60), the present is generic.

(60) bóʒíngɛ́ sɛ̀ mwámù.

bó-ʒíng-ɛ́	sɛ̀	mwámù
SP₂-live-IPFV	also	here

'They also live here.'

It is also observed in conjunction with other imperfective markers, such as the auxiliary -yì 'be' in (61), which marks progressive aspect and whose main verb takes the final suffix -ɛ́.

(61) ngyɛ̀dyɛ́ bùù; òyì ònwɛ́ màn.

N-yì	ɛ̀-dì-ɛ́	bùù	ò-yì	ò-nɔ̀-ɛ́	màn
SP₁ₛg-PROG	SP₁ₛg-eat-IPFV	14.fufu	SP₂ₛg-PROG	SP₂ₛg-drink-IPFV	6.palm-wine

'I'm eating cassava bread; you're drinking palm wine.'

It also occurs in combination with the generic present/habitual marker -wà- (62), as well as with its high-toned past equivalent -wá- (63).

(62) bɔ̀ɔ̀ àbá bɔ̀wàkálɛ́ bwír.

bɔ̀-ɔ̀	àbá	bɔ̀-wà-kál-ɛ́	bɔ́-ír
14-mushroom	PROX.DEM₁₄	SP₁₄-HAB-be-IPFV	ACP₁₄-hard

'This mushroom is (usually) hard.'

(63) bàkúb bàwábílɛ́ ŋgwáánkùb mwɛ̀nsɔ̀ŋ.

bà-kúb	bà-wá-bíl-ɛ́	N-gwáánkùb	mɔ̀-ɛ̀-nsɔ̀ŋ
2-ancestor	SP₂-PST.HAB-call-IPFV	9-mushroom.sp	18-7-Nsong

'The ancestors used to call (it) ŋgwá:nkub in Nsong.'

4.3.1.2 Anterior -í

To mark situations that are completed, Nsong uses a final suffix -i, which triggers umlaut (Section 2.3.4) and surfaces then either as -ɛ́ or as zero, in contrast to the imperfective marker, which always surfaces as -ɛ́ and never triggers umlaut. The ending -i is most often observed in situations that can be characterised as "anterior" or "perfect" in the strict sense of the word, namely referring to a state of affairs that is completed but still has relevance at the time of utterance. The child in (64) died the day the utterance was produced.

(64) ndùùr èyí éʃímí mwǎn; àpfí.

    N-dùùr   é-ʃím-í   mɔ̀-án   à-pf-í

    9-disease.sp SP_9-catch-ANT   1-child SP_1-die-ANT

    'The disease called *nduur* has caught the child, it has died.'

The present reference time can be made explicit through the use of the time adverbial *nôbù* 'today,' as in (65), where it is understood more generally as 'at the present time.'

(65) nôbù àwú bàmɛ̀mpɛ́ bàyí.

    nôbù   àwú   bà-mɛ̀-mpɛ́   bà-yà-í

    today   PROX.DEM_3 2-4-priest SP_2-come-ANT

    'Today the priests have come.'

The past situation, which still has present relevance, may have started long before (66).

(66) bɛ̀ɛ̀sɛ̀s, àkwí mùmàmvúl màkwím mɔ́ɔ́l.

    bɛ̀-ɛ̀-sàs-í   à-kú-í   mɔ̀-mà-N-vúl   mà-kwím   mɔ́ɔ́l

    SP_{1PL}-OP_{2SG}-raise-ant SP_{2SG}-reach-ant   18-6-9-year ACP_6-ten ACP_6.two

    'We have raised you, you have reached twenty years.'

Apart from prototypical anterior situations, the final -*i* suffix is also found in contexts that lack present relevance. In (67), it refers to the remote past of the ancestors.

(67) mwɛ́tàŋ làkɛ́lɛ́ bákùb, bàkɛ́lɛ́ èyí bànkír bɔ́ɔ́l.

    mɔ̀-ɛ́-tàŋ   là-kàl-í   bá-kùb,   bà-kàl-í   èyí   bà-N-kír   bɔ́ɔ́l

    NP_{18}-NP_5-time ACP_5-be-ANT NP_2-elder SP_2-be-ANT   with NP_2-NP_{1n}-fetish ACP_2.two

    'In the era when our ancestors lived, they had two idols.'

At this stage, these anterior verb forms with present and past time reference seem identical, at least from a purely segmental point of view. Further research needs to clarify whether they have distinct tone patterns. The anterior -*i* may also co-occur with other TA-affixes, such as the pre-initial past marker *lí-* in (68).

(68) bètsìn líbɛ́kɛ̀lɛ̀

    bè-tsìn   lí-bɛ́-kal-í

    8-taboo   PST-SP_8-be-ANT

    'Taboos existed.'

4.3.1.3 Subjunctive and imperative -*í*

A final front vowel -*i* also marks the subjunctive mood in Nsong. Subjunctive verbs can be used subordinately, as in (69), where it expresses purpose. In (70), it is used in the main clause with a hortative meaning.

(69) láám búú, bús kédí.

    láám   búú   bús   ké-di-í

    cook.IMP   14.fufu   we SP_{1PL}-eat-SBJV

    'Cook cassava bread so that we (can) eat.'

(70)   bὲ nsɔ́, bɔ̀kέl mbǔ mwέs.

bὲ   nsɔ́   bɔ̀-kál-í   N-bǔ   mwέs
you   all   SP₂ₚₗ-be-SBJV   9-place   one
'You all, be together!'

As seen in (69), the singular imperative consists of nothing but the verb base, which is always high-toned. As shown in (13b), the imperative also ends in -*í*, when it takes an object prefix. In the plural, the imperative is additionally marked by the "plural allocutive" clitic -*éŋ* (cf. Van de Velde & van der Auwera 2010), e.g., *sɔ̀l-éŋ* 'Choose! (pl.),' *lɔ́n-éŋ* 'Repair! (pl.).' When the root ends in a vowel, this marker is realised as -*néŋ*, e.g., *nwénéŋ* 'Drink! (pl.),' *dí-néŋ* 'Eat! (pl.).' Plural subjunctive forms optionally take the same clitic, e.g., *kέ-sέm-éŋ* = *kέ-sέm* 'let us pray.'

Nsong does not distinguish between a negative subjunctive and a negative imperative; it has one single prohibitive form, which is marked by pre-initial negative *ko-*, which is low-toned, and final -*í*, e.g., *kòdíí* 'Do not eat!,' *kònwí* 'Do not drink!.' This form can be used to address one person or several people. The clitic -*éŋ* does not occur in the negative.

4.3.2   Verb-initial TAM markers

Nsong has only one pre-initial tense marker, namely the past marker *lí-*, which seems always to combine with other TA-markers. It is observed with "anterior" -*í* to refer to events that took place in a distant past (68). This remote past marker is also observed in conjunction with past habitual -*wá-* and imperfective -*é*, as in (71), where the speaker was asked whether people also divorced in ancestral times.

(71)   líbáwákàbὲŋὲ. Màwâl límáwápfὲ.

lí-bá-wá-kàb-ὲŋ-έ   mà-wâl   lí-má-wá-pfà-έ
PST-SP₂-HAB-divide-RECP-IPFV   6-marriage   PST-SP₆-HAB-die-IPFV
'They used to divorce. Marriages used to break up.'

It is also found with other verb-medial aspect markers, such as the 'already' completive -*fúlá-* in (72) to refer to events that occurred in a more recent past. We therefore consider it to be a general past marker rather than a dedicated remote past marker.

(72)   mὲ límfúlábáál mὲ lὲ yí mbàkàsὲs.

mὲ   lí-N-fúlá-báál   mὲ   lὲ   yí   N-bàkàsὲs
me   PST-SP₁ₛɢ-CMPL-say   me   that   PROX.DEM₉   9-mushroom.sp.
'I already told that this is the *mbakasɛs* mushroom.'

4.3.3   Verb-medial TAM markers

4.3.3.1   Future -kí-

Nsong has one future marker -*kí-*. There are no temporal subdivisions in the future.

(73)   mósúú àkíwɔ̀ɔ̀

mósúú   à-kí-wɔ̀ɔ̀
tomorrow   SP₂ₛɢ-FUT-hear
'Tomorrow you will hear.'

(74) bùsú kèkíbáál; kèkímwènéŋ
    bùsú kè-kí-báál   kè-kí-mòn-éŋ
    me   SP₁ₚₗ-FUT-say   SP₁ₚₗ-FUT-see-RECP
    'We will talk. (One day) we will meet.'

4.3.3.2 Iterative -fú-

The marker -fú- expresses iterative aspect, used to express a single repetition 'again,' both with affirmative (75) and negative verbs (76). The variant forms -fí- (77) and -fwí- (78) only occur in negative forms.

(75) èfúdyà
    è-fú-dyà
    SP₁ₛɢ-ITR-eat
    'I eat again.'

(76) ncìk èbáfúkál, bàmbú bátwí
    N-cìk   è-bá-fú-kál   bà-N-bú   bá-tú-í
    9-taboo   SP₉-NEG-ITR-be   2-1n-twin   SP₂-go.out-ANT
    'The taboo is no more, the twins have gone out.'

(77) kòfíbénd lwáŋ
    kò-fí-bénd-í   ló-áŋ
    NEG-ITR-pull-IMP   11-branch
    'Do not pull the branch again.'

(78) bànkír à bámbùt bòbáfwíyíl lóó.
    bà-N-kír   à   bá-N-bùt   bɔ-bá-fwí-yíl   lóó
    2-1n-fetish   CON₂   2-1n-forefather   SP₂-NEG-ITR-exist   NEG
    'The fetishes of the forefathers no longer exist.'

4.3.3.3   Generic present/habitual -wà-

Nsong has a generic present/habitual marker -wà- (62) and a high-toned past equivalent -wá- (63), which generally combines with imperfective -é. Especially with generic present reference, it is also observed without this ending, as in the proverbs in (58) and (79).

(79) èpàŋ kèwàkál yí èkúl
    è-pàŋ   kè-wà-kál   yí   è-kúl
    7-adze   SP₇-HAB-be   with   7-corner
    'The adze has a corner.' (Each object has a proper place to be stored away)

4.3.3.4 Habitual -kà-

Apart from -wà-, Nsong has another verb-medial marker to refer to a habit or a general truth, namely -kà-, which also often occurs in proverbs, as in (80). It generally does not appear with the imperfective ending -é.

(80) mbwá mìl ménà àkàlàb njìl mwés
      N-bwá   m-ìl   mé-nà   à-kà-làb   N-jìl   mwés
      1n-dog   4-leg   ACP₄-four   SP₁-HAB-follow   9-path   one
      'A four-legged dog usually follows one way.'

This habitual marker also has a longer variant -kìká-.

(81) bàpéndé bókìkákír yì mbwím
      bà-péndé   bó-kìká-kír   yì   N-bwím
      2-Pende   SP₂-HAB-make   with   9-rattan.
      'The Pende people usually produce (baskets) with rattan.'

It also occurs with -fú- 'again,' as in (82), where it is explained that people nowadays tend to accuse each other much more in court than in the past. In (83), -fú- and -kà- contract to -fwá-.

(82) bófúkàdìlàngàn
      bó-fú-kà-dì-ìl-àngàn
      SP₂-REP-HAB-eat-APPL-RECP
      'People usually eat each other again.'

(83) bèbùfwámwénéŋ kèbèn kɔbɔŋ lɔɔ
      bè-bù-fwá-mɔn-éŋ   kèbèn   kɔbɔŋ   lɔɔ
      SP₁ₚₗ-REP-HAB-see-RECP   really   well   NEG
      'As a matter of fact we no longer meet very often.'

Its negative equivalent is -búká-, as in the proverb in (84), which may contract to -bwá- (85).

(84) lòpín lɔmwés lòbúkáʃím ntsìn lɔɔ
      lò-pín   lɔ-mwés   lò-bú-ká-ʃím   N-tsìn   lɔɔ
      11-finger   ACP₁₁-one   SP₁₁-NEG-HAB-catch   1n-path   NEG
      'One finger usually does not catch a louse.'

(85) mè mbwázɔl lɔɔ
      mè   N-bwá-zɔl   lɔɔ
      I   SP₁ₛ₉-NEG.HAB-love   NEG
      'I do not like (it).'

4.3.3.5 Priorative -bá-

The priorative marker -bá- (possibly from *-bád 'begin') refers to events that happen "first" before another takes place.

(86) ébálínd bɔɔ bá.
      é-bá-línd   bɔ-ɔ   bá
      SP₁ₛ₉-PRIOR-look.at   14-mushroom   PROX.DEM₁₄
      'I first look at these mushrooms.'

4.3.3.6 Resultative -ŋgú-, -ŋgí- and -mú-

Nsong has several prefixes conveying perfective aspect, such as resultative *-ŋgú-* (87), which highlights a state as the outcome of a completed process and has a variant *-ŋgí-* (88).

(87) bùsú bèŋgúkúl.
      bùsú   bè-ŋgú-kúl
      we   SP$_{1PL}$-RES-grow
      'We are grown up (adult).'

(88) nyèŋ-mɔ́kɔ́ɔ́ yí èŋgíbáŋá
      N-yèŋ-mɔ́kɔ́ɔ́   yí   è-ŋgí-báŋá
      9-mushroom.sp.   PROX.DEM$_9$ SP$_9$-RES-blush
      'This mushroom species is red.'

A similar marker both in form and meaning is *-mú-*.

(89) kèzúngú kàmúfwà . . . mὲ kàmbíím ɛmúwɔl kémfum
      kèzúngú   kà-mú-fwà   mὲ   kambíím   è-mú-wɔl   ké-mfùm
      Kizunga   SP$_1$-RES-die   I   Kamembo   SP$_{1SG}$-RES-take   7-power
      'Kizungu is dead . . . me Kambembo I am in power now.'

4.3.3.7 Completive -fúlá-

Another perfective marker is completive *-fúlá-* meaning 'already' (90). In spontaneous speech, this specific notion is commonly expressed by the French adverb *déjà* (91).

(90) ndáá èfúlábwà
      N-dáá   è-fúlá-bwà
      9-problem   SP$_9$-CMPL-fall
      'The problem has already occurred.'

(91) mansáb déjà bákéŋ mɔ́njɔ́
      mà-N-sáb   déjà   bá-káŋ-í   mɔ́-N-jɔ́
      6-9-key   already   SP$_2$-close-ANT   18-9-house
      'They had already locked (closed with key) the house/The house was already locked.'

4.3.3.8 Subsecutive -lá-

The prefix *-lá-* is a relative tense marker. It indicates that an event takes places after another. The reference time may be past (92), present (93) or future (94). It is normally preceded by the subject prefix, but may also occur verb-initially, since the subject prefix tends to be dropped in front of *-lá-*, especially with 3SG/PL of classes 1 and 2.

(92) kàlwîm bɔ́tà mɔ̀ŋgyèŋ, bànd lámpànzéŋ
      kà-lùm-í   bɔ́-tà   mɔ̀-N-yèŋ   bà-nd   Ø-lá-mpànzéŋ
      SP$_1$-fire-ANT   14-bow/gun   18-9-sky   2-person   SP$_2$-SUBS-spread.RECP
      'He fired in the air and then the people spread.'

(93) kácúl, lásà pàlɔ́kà, látùl, bàláyɛ́ndɛ́k kɔ́dyà

ká-cúl   Ø-lá-sà   pà-lɔ́-kà   Ø-lá-tùl   bà-lá-yɛ́ndɛ́k   kɔ́-dyà
SP₁-cut   SP₁-SUBS-put   16-11-leaf   SP₁-SUBS-put.down   SP₂-SUBS-start   15-eat

'She cuts (the cassava bread), she subsequently puts it on a leaf and puts it down and then they start to eat.'

(94) akímɔ́n bèláŋkwíkíl

à-kí-mɔ́n   bè-lá-N-kwíkíl
SP₂ₛ₉-FUT-see   SP₂ₚₗ-SUBS-OP₁ₛ₉-believe

'You will see and then you will believe me.'

4.4 Negation

Nsong marks negation in two different verb slots, either in front of or following the subject prefix. The pre-initial negation marker is *kɔ̀-* with specific allomorphs for 1SG (*kè-*), 2SG (*kwà-* or *kɔ̀-*) and 3SG of class 1 (*kà-*). It is mainly observed in conjunction with perfective verb-medial aspect markers.

(95) àtá kîm kèŋgúbál lɔ́ɔ́.

àtá   kí-ìm   kè-ŋgú-bál   lɔ́ɔ́
even   7-thing   NEG₁ₛ₉-RES-say   NEG

'I haven't said even a thing.'

Negation is more commonly marked behind the subject prefix. Post-initially, four different allomorphs are observed, namely *-bá-*, *-bú-*, *-bwá-* and *-búká-*, of which the functional distribution across conjugational paradigms needs more in-depth study.

(96) ɛ̀bákír ŋkír lɔ́ɔ́.

ɛ̀-bá-kír   N-kír   lɔ́ɔ́
SP₁ₛ₉-NEG-make   9-fetish   NEG

'I will not produce a fetish.'

(97) mwǎn áwún, mbúmɔ́mɔ́n lɔ́ɔ́, mbúmɔ́ʃíím lɔ́ɔ́

mɔ̀-án   áwún   N-bú-mɔ́-mɔ́n   lɔ́ɔ́   N-bú-mó-ʃíím   lɔ́ɔ́
1-child   END.DEM₁   SP₁ₛ₉-NEG-OP₁-see   NEG   SP₁ₛ₉-NEG-OP₁-catch   NEG

'That child, I have not seen it, I have not caught it.'

(98) bàbwákírís lɔ́ɔ́

bà-bwá-kír-ís   lɔ́ɔ́
SP₂-NEG-do-CAUS   NEG

'They had not made (them).'

(99) mètí mènsɔ́ bèbúkákíríl lɔ́ɔ́

mè-tí   mè-nsɔ́   bè-búká-kír-íl   lɔ́ɔ́
4-tree   ACP₄-tree   SP₁ₚₗ-NEG-do-APPL   NEG

'We do not use all trees.'

As the examples above show, Nsong has double negation marking. Besides a verbal negative morpheme, it also uses the post-verbal negation marker *lɔ́ɔ́*, which tends to be clause-final and is also used as negative answer particle. It sometimes occurs as the only sentence negation marker (117).

4.5 Grammatical relations

4.5.1 Subject marking

Each conjugated verb, except for the imperative, usually has a subject prefix that either signals agreement with a clause-internal lexical subject, which is generally pre-verbal, or functions as an anaphoric pronoun referring to a subject participant that is known from the context. The subject prefixes of the different noun classes are given in Table 14.2 above; those of the discourse participants are listed in (100). Each of the person markers has several allomorphs, which are partially conditioned by vowel harmony and by regiolectal variation, and which correlate to a certain extent with the different conjugations of the verb. As for 1SG, for instance, the alternation between *N-* and *è-/-ɛ̀-* is conjugationally determined. As for 1PL, the alternation does not so clearly pattern with distinctive verb conjugations, but might once have been regiolectal. It is striking that the subject prefix paradigm manifests (partial) 2SG/PL and 3SG/PL syncretism. Their exact reference is contextually determined and often disambiguated by the use of personal pronouns, aka "substitutives" (Section 3.3.3). Mbuun B87 has a similar allomorphemic 3SG variation involving *kà-* (Bostoen & Mundeke 2012).

(100) 1SG *è-/ɛ̀-*   or *N-*   1PL *bè-/bɛ̀-* or *kè-/kɛ̀-*
      2SG *ò-/ɔ̀-*   or *à-*   2PL *bè-/bɛ̀-* or *bò-/bɔ̀-*
      3SG *à-*   or *ò-/ɔ̀- + kà-*   3PL *bà-/à-*   or *bò-/bɔ̀-*

Another striking feature is the possibility to omit the subject prefix in the presence of a lexical subject. This phenomenon mainly occurs with verbs carrying an object prefix.

(101) mɔ́sɛ́s wú pá, èbún bùsú mɔ́sà pá?
      mɔ̀-sɛ́s   wú   pá   èbún   bùsú   mɔ́-sà   pá
      3-string   PROX.DEM$_3$ PROX.DEM$_{16}$ how   we   OP$_2$-put   PROX.DEM$_{16}$
      'This string here, how will we put it here?'

4.5.2 Object marking

An object can only be co-referenced on the verb if it is clause-external. The object prefixes of the different noun classes are given in Table 14.2 above. Those of the discourse participants and the 3SG/PL subject prefixes of classes 1 and 2 are listed in (102).

(102) 1SG *-N-*   1PL *-bè-/-bɛ̀-* or *-è-/-ɛ̀-*
      2SG *-ò-/-ɔ̀-*   or *-à-*   2PL *-bò-/-bɔ̀-* or *-bè-/-bɛ̀-*
      3SG *-mò-/-mɔ̀-* or *-ɔ̀-* or *-ɔ̣̀-* 3PL *-bà-*   or *-à-*

A verb commonly does not take more than one object prefix, even not a ditransitive verb. The aforementioned patientive object is usually left implicit.

(103) kàtwís mán làmɔ́pà.
     kà-tús-í   mán   Ø-là-mɔ́-pà
     SP_2-take.out-ANT 6.wine SP_1-SUBS-OP_1-give
     'He took out the wine and then gave (it) to him.'

However, double object prefixes are grammatical and can easily be elicited.

(104) wɔ́l mwɛ̀m mumɔ́pɛ̀
     wɔ́l   mɔ̀-ɛ̀m   mù-mɔ́-pɛ̀
     take.IMP   3-medicine OP_3-OP_1-give.IMP
     'Take the medicine and give it to him.'

Objects can also be pronominalised by post-verbal personal pronouns, especially in relative clauses (105) and in infinitives (106).

(105) ɔ́sásɛ́ nzên. . .
     ɔ́-sás-ɛ́   nzên
     ACP_1-raise-IPFV   him
     'the one who raises him. . .'

(106) nzɛ́n òyí é bîm bì kɔ́pá mɛ̀ . . .
     nzɛ́n ò-yí   é   bí-m   bì   kɔ́-pá   mɛ̀
     he ACP_1-be with 8-thing REL_8 15-give me
     'He who has things to give me. . .'

5 CLAUSE TYPES

5.1 Word order variation in the main clause

The canonical word order in Nsong is SVO. Variations in this linear order commonly occur for information structural purposes. While SVO is mainly associated with unmarked topic-comment structures, narrowly focused arguments occur in the Immediate Before the Verb position. This focus site can host arguments under the scope of either identification or information focus, not only core arguments, but also adjuncts. Object focus triggers (S)OV word order. If the lexical subject is overtly expressed, which is rare in object focus constructions, the focused object is positioned in between the topical subject and the verb. Subject focus triggers OSV word order. In contrast to topical subjects in object focus constructions, topical objects in subject focus constructions are often overtly expressed. If so, they cannot occur postverbally. They need to occur in the clause-initial topic position. This focus site remains empty in topic-comment structures, where the subject is topical and thus clause-initial and the post-verbal object is part of the focused verb phrase. Subjects that are neither narrowly focused nor topical obligatorily appear behind the verb, as evidenced by thetic utterances. The mono-clausal OSV word order associated with subject focus is far more prominent in our Nsong text corpus than (S)OV word order associated with object focus. More generally, bi-clausal cleft constructions seem to be more frequent argument focus strategies. For more details, the reader is referred to Koni Muluwa and Bostoen (2014).

5.2 Relative clauses

5.2.1 Direct relative

In direct relative constructions, the relativised NP is also the subject of the relative clause and triggers subject agreement on the relative verb. The relative clause can be introduced by a relativiser (107). It is a resumptive pronoun following the relativised noun phrase with which it is co-referential. This pronoun, which is segmentally identical to the proximal demonstrative but low-toned, is not obligatory (108). Another co-referential relativiser may follow after the relative verb, either before or after its object if present, but it is not obligatory.

(107)   bámfùm bà bàfúlábùr bà máyàs. . .

    bá-N-fum   bà   bà-fúlá-bùr   bà   má-yàs

    2-1n-chief   REL$_2$ SP$_2$-CMPL-generate   REL$_2$ 6-twin

    'Chiefs who have already brought forth twins. . .'

(108)   kìm kètóm kèwàpá mé kyês (kì) . . .

    kì-m   kè-tóm   kè-wà-pá   mé   ké-ès   (kì)

    7-thing   ACP$_7$-first   SP$_7$-HAB-give   me   7-joy   REL$_7$

    'The first thing that makes me happy. . .'

5.2.2 Indirect relative

In indirect relative clauses, the relativised NP is not the subject of the relative clause, but its object, as in (109) and (110), or an adjunct (111). Such constructions are obligatorily marked by a circumposition consisting of a low-toned relativiser, which agrees in class with the head noun and which both introduces and closes the relative clause. An adverbial may still follow the post-verbal relativiser (111). The subject prefix of the relative verb, if present, does not manifest concordance with the relativised noun phrase, but with the subject of the relative clause, which is always post-verbal if expressed lexically (110). Also, pronominal subjects tend to be resumed after the relative verb, as in (109) and (111).

(109)   màyúmú àmán mà àngúfúl nzé mà . . .

    mà-yúmú   àmán   mà   à-ngú-fúl   nzé   mà

    6-abandoned.village   END.DEM$_6$ REL$_6$ SP$_{2SG}$-RES-ask.for   you   REL$_6$

    'these former villages for which you have asked. . .'

(110)   ndáá yì éwís bákùb yì . . .

    N-dáá   yì   é-wís.í   bá-kùb   yì

    9-matter   REL$_9$ OP$_{2SG}$-say.ANT   2-ancestor   REL$_9$

    'the matter the ancestors told you. . .'

(111)   kèsál àkí kì bényéré bús kì nŏbù

    kèsál   àkí   kì   bé-nyár-í   bús   kì   nŏbù

    7.Kisala   PROX.DEM$_7$ REL$_7$ SP$_{1PL}$-walk-ANT   we   REL$_7$ today

    'This village of Kisala to which we walked today. . .'

5.2.3 Adverbial clauses

Subordinate adverbial clauses are generally relative clauses and frequently involve the locative markers of classes 16, 17 and 18. Temporal clauses rely on either an initial class 16 marker (112) or on an adverbial noun phrase introduced by the noun prefix of class 18 (113). The class 18 noun prefix is commonly used in temporal noun phrases, such as *mwὲbwánkɔ́ɔ́* 'in the morning,' *mɔ̀máʃíí* 'in afternoon,' *mὸmpìmb* 'by night.'

(112) pà àkέl málìm myáŋ mándέl mà . . .

      pà   à-kál-í   má-lìm   myáŋ   má-ndέl   mà

      NP_{16} SP_2-be-ANT   6-uncle $MED.DEM_6$ ACP_6-old REL_6

      'When those old uncles lived. . .'

(113) mwὲtáŋ là àmúpfá mfúm lì . . .

      mɔ̀-ὲ-táŋ   là   à-mú-pfá   N-fúm   lì

      18-5-time REL_5 SP_1-RES-die   1n-chief REL_5

      'When a leader has died. . .'

A reason clause is either an indirect relative with class 18 relativisers, as in (114), or as in (115), it is introduced by the conjunction *ntɔ́*, which is on its way out in favour of Kongo *sámbù*, which is used to express both reason and purpose and even has a nativised Nsong equivalent, namely *sâm*.

(114) mὸdím mù kapfî mu. . .

      mɔ̀-dím   mù   kà-pf-í   mù

      1-husband REL_{18} SP_1-die-ANT REL_{18}

      'Because the husband has died. . .'

(115) bààn á ŋkέnzὲl bí bὲbí kɔ̀ŋgwὲ ntɔ́ ŋgwên kàŋgútsɔ́tsɔ́

      bà-àn   á   N-kέnzὲl   bí   bὲ-bà-í   kɔ̀ŋgwὲ   ntɔ́   N-gwên

      2-child   CON   1n-goat $PROX.DEM_8$ SP_8-be-ANT   behind   because   1n-mother

      ka-ŋgú-tsɔ́tsɔ́

      SP_1-RES-go.fast

      'These baby goats stayed behind because their mother had gone fast.'

Class 18 is also involved in the expression of purpose (116).

(116) lá pá mùkòkúú kòkír bèlúm kwέ?

      lá   pá   mɔ̀-kò-kúú   kò-kír   bè-lúm   kwέ

      $PROX.DEM_5$ $PROX.DEM_{16}$ 18-15-remove   15-do   8-day   how.many

      'For this to be removed, how many days would be needed?'

      A purpose clause can also be introduced by the conjunction *nà* 'so that.'

(117) bὲὲpέ mwὲm pá nà òfúbúr lɔ́ɔ́

      bà-ὲ-pέ   mɔ̀-ὲm   pá   nà   ò-fú-búr   lɔ́ɔ́

      SP_2-OP_{2SG}-give.PRS   3-medicine $PROX.DEM_{16}$ so.that SP_{2SG}-REP-give.birth   NEG

      'You are given the medicine here so that you do not give birth again.'

A condition clause is marked by the conjunction *lὲ* 'if.'

(118) bùsú lὲ bὲbέέl ɛ́nsɔ́ŋ, bǎ ɛ̀nsɔ̀ŋ ὲ bús bókùwáwὸὸ
      bùsú lὲ bὲ-báál-í   ɛ́-nsɔ́ŋ   bǎ   ɛ̀-nsɔ̀ŋ   ὲ   bús   bó-kὲ-wá-wὸὸ
      we   if SP_{1PL}-speak-ANT   7-Nsong   they   7-Nsong CON_7 us SP_2-OP$_7$-HAB-hear
      'If we speak Nsong, our Nsong, they will understand it.'

In an unreality conditional clause, *lὲ* is preceded by *áfù*.

(119) ɔ̀ʃɛ́ɛ́ áfù lɛ́ mὲ ŋgyìlì mɔ́sàm á nzɛ́.
      ɔ̀-ʃɛ́-ɛ́   áfù   lɛ́   mὲ   N-yìlì   mɔ́-sàm   á   nzɛ́
      SP_{2SG}-laugh-IPFV   as   if   I SP_{1SG}-be.PRS   1-friend   CON   you
      'You laugh as if I am your friend.'

ACKNOWLEDGEMENTS

We wish to thank Sebastian Dom, Heidi Goes, Hilde Gunnink, Lolke van der Veen and Mark Van de Velde for their helpful feedback on an earlier version of this chapter.

REFERENCES

Bastin, Y. (1983) *La finale verbale -ide et l'imbrication en bantou*. Tervuren: Musée royal de l'Afrique centrale.

Bastin, Y. (1985) *Les relations sémantiques dans les langues bantoues*. Bruxelles: Académie royale des sciences d'outre-mer.

Bostoen, K. & J. Koni Muluwa (2011) Vowel Split in Hungan (Bantu H42, Kwilu, DRC): A Contact-Induced Language-Internal Change. *Journal of Historical Linguistics* 1(2): 247–268.

Bostoen, K. & J. Koni Muluwa (2014) Umlaut in the Bantu B80 Languages of the Kwilu (DRC). *Transactions of the Philological Society* 112(2): 209–230.

Bostoen, K. & L. Mundeke (2011a) The Causative/Applicative Syncretism in Mbuun (Bantu B87, DRC): Semantic Split or Phonemic Merger? *Journal of African Languages and Linguistics* 32(2): 179–218.

Bostoen, K. & L. Mundeke (2011b) Passiveness and Inversion in Mbuun (Bantu B87, DRC). *Studies in Language* 21(1): 72–111.

Bostoen, K. & L. Mundeke (2012) Subject Marking and Focus in Mbuun (Bantu, B87). *Southern African Linguistics and Applied Language Studies* 30(2): 139–154.

Daeleman, J. (1977) A Comparison of Some Zone B Languages in Bantu. *Africana Linguistica* 7: 93–144.

de Beaucorps, R. (1941) *Les Basongo de la Luniungu et de la Gobari*. Bruxelles: Institut Royal Colonial Belge.

Grégoire, C. (2003) The Bantu Languages of the Forest. In Nurse, D. & G. Philippson (eds.), *The Bantu Languages*, 349–370. London New York: Routledge.

Guthrie, M. (1971) *Comparative Bantu: An Introduction to the Comparative Linguistics and Prehistory of the Bantu Languages. Volume 2: Bantu Prehistory, Inventory and Indexes*. London: Gregg International.

Hammarström, H., R. Forkel, M. Haspelmath & S. Bank (eds.) (2016) *Glottolog 2.7*. Jena: Max Planck Institute for the Science of Human History (Available online at http://glottolog.org, Accessed March 11, 2016).

Hyman, L. M. (2004) How to Become a Kwa Verb. *Journal of West African Languages* 30(2): 69–88.

Idiatov, D. (2009) A Bantu Path Towards Lack of Differentiation Between "who?" and "what?." *Africana Linguistica* 15: 59–76.

Iliku Mimpya, D. (1979) *Esquisse grammaticale de la langue tsong: phonologie et morphologie*. Lubumbashi: Université de Lubumbashi, mémoire de licence.

Kemmer, S. (1993) *The Middle Voice*. Amsterdam; Philadelphia: John Benjamins.

Koni Muluwa, J. (2006) *Phytonymes et zoonymes en nsong (RD Congo), une étude linguistique de la faune et de la flore*. Bruxelles: Université libre de Bruxelles, mémoire de DEA.

Koni Muluwa, J. (2010) *Plantes, animaux et champignons en langues bantu. Etude comparée de phytonymes, zoonymes et myconymes en nsong, ngong, mpiin, mbuun et hungan (Bandundu, RD Congo)*. Bruxelles: Université libre de Bruxelles, thèse de doctorat.

Koni Muluwa, J. (2014) *Noms et usages de plantes, animaux et champignons chez les Mbuun, Mpiin, Ngong, Nsong et Hungan en RD Congo*. Tervuren: Musée royal de l'Afrique centrale.

Koni Muluwa, J. & K. Bostoen (2007) Un recueil de proverbes nsong (RD Congo, bantu B85d). *Annales Aequatoria* 28: 521–578.

Koni Muluwa, J. & K. Bostoen (2008) *Noms et usages de plantes utiles chez les Nsong (RD Congo, bantu B85F)*. Göteborg: University of Gothenburg, Department of Oriental and African Languages.

Koni Muluwa, J. & K. Bostoen (2010) Les plantes et l'invisible chez les Mbuun, Mpiin et Nsong (Bandundu, RD Congo): une approche ethnolinguistique. *Sprache und Geschichte in Afrika* 21: 95–122.

Koni Muluwa, J. & K. Bostoen (2012) La diphtongaison dans les langues bantu B70–80 (Bandundu, RDC): typologie et classification historique. *Africana Linguistica* 18: 355–386.

Koni Muluwa, J. & K. Bostoen (2014) The Immediate Before the Verb Focus Position in Nsong (Bantu B85d, DR Congo): A corpus-based exploration. *ZAS Papers in Linguistics* 57: 123–135.

Koni Muluwa, J. & K. Bostoen (2015) *Lexique comparé des langues bantu du Kwilu (République démocratique du Congo)*. Cologne: Rüdiger Köppe.

Lewis, M. P., G. F. Simons & C. D. Fennig (eds.) (2016) *Ethnologue: Languages of the World, Nineteenth edition*. Dallas: SIL International. (Available online at www.ethnologue.com).

Maho, J. F. (2003) A Classification of the Bantu Languages: An Update of Guthrie's Referential System. In Nurse, D. & G. Philippson (eds.), *The Bantu Languages*, 639–651. London; New York: Routledge.

Meeussen, A. E. (1967) Bantu Grammatical Reconstructions. *Africana Linguistica* 3: 79–121.

Mukash Kalel, T. (1982) *Le kanyok, langue bantoue du Zaïre: phonologie, morphologie, syntagmatique*. Paris: Université Sorbonne nouvelle, Paris III, PhD dissertation.

Nash, J. A. (1992) *Aspects of Ruwund Grammar*. Urbana-Champaign: University of Illinois, PhD dissertation.

Rottland, F. (1977) Reflexes of Proto-Bantu Phonemes in Yanzi (B85). *Africana Linguistica* 7: 375–396.

Shibatani, M. & P. Pardeshi (2002) The Causative Continuum. In Shibatani, M. (ed.), *The Grammar of Causation and Interpersonal Manipulation*, 85–126. Amsterdam; Philadelphia: John Benjamins.

Van de Velde, M. & J. van der Auwera (2010) Le marqueur de l'allocutif pluriel dans les langues bantu. In Floricic, F. (ed.), *Essais de typologie et de linguistique générale. Mélanges offerts à Denis Creissels*, 119–141. Lyon: ENS Editions.

van der Veen, L. J. (1991) *Étude comparée des parlers du groupe okani (B.30 Gabon)*. Lyon: Université Lumière Lyon 2, thèse de doctorat nouveau régime.

CHAPTER 15

PAGIBETE C401

JeDene Reeder

1 INTRODUCTION

The approximately 30,000 speakers of Pagibete C401 live primarily in Businga Territory, Nord-Ubangi District of the Equateur Province of north-western Democratic Republic of the Congo. They live in three non-contiguous areas, as well as in Businga, the territorial capital. This chapter describes the Monveda dialect, which has the largest number of speakers. These speakers live north of the Dua River in 12 villages stretched out along a 96 kilometre (60 mile)-long trail passing through the forest. The Mongbapele dialect is spoken by Pagibete living along the Businga-Lisala road south of the Dua in seven villages spread out over a 14 kilometre (nearly nine mile)-long section of the road. The final dialect area is found on the road from Butu to Bokonzi in two large villages.

Language use is vigorous, although most adults also speak Lingala. Lingala is the dominant language in most of the church denominations of the area and is used during the first three years of primary school, after which French becomes the medium of instruction. Marriage with women from other tribes is common; approximately one in five wives in the Monveda area comes from a neighbouring group, with Ngombe C41 being the most common. Social pressure on these wives to learn Pagibete is quite strong. Pagibete women who marry outside the ethnic group often teach their language to their children who then use it when visiting maternal relatives. The main trading partners of the Monveda and Mongbapele Pagibete are Ngombe. At the weekly markets, Ngombe speakers are more likely to use Pagibete than Pagibete speakers are to use Ngombe. Other neighbouring language groups are the Ngbandi and the Mbandja.

Virtually all the data for this chapter were gathered over a two-year period from July 1994 through August 1996, one month of which was in the town of Businga while living with a Pagibete family, and nine months of which was spent living in the Monveda dialect area. The remaining time was spent in Gemena, where language lessons and analysis continued with the help of Pagibete living in town. The author and a Pagibete colleague, Nzongo Ngole Roger, also did a one-week dialect survey in the Mongbapele dialect area. Accuracy of all entries in the lexical database was confirmed by at least three other speakers. Some additional data have been added by Nzongo more recently via email.

2 PHONOLOGY

Pagibete has 40 phonemes, including 15 prenasalised consonants. Unusually among Bantu languages, although not among languages of the Ubangi including other Bantu borderland languages, plosive and implosive consonants contrast.

2.1 Vowels

Pagibete has an inventory of seven vowels, presented in Table 15.1.

TABLE 15.1 THE VOWEL PHONEMES OF PAGIBETE

i		u
e		o
ɛ		ɔ
	a	

Vowel length is not phonemic. Vowels become slightly lengthened when they carry a contour tone. Thus, double vowels are used in this article to show surface contour tones. The exception is the word *tíí* 'until,' the only word in the corpus where the vowel is definitely lengthened on a single tone. Contour tones are discussed in Section 2.3.

2.1.1 Vowel harmony processes

In seven-vowel systems, height harmony of mid vowels frequently occurs within morphemes, if not across morpheme boundaries. Although this process exists in Pagibete, it is only a tendency. Numerous examples of words where height harmony does not exist even within morphemes have been confirmed with multiple speakers, for example (1–3).

(1)   mɔ́nè
       6.sun
(2)   zɔ̀zɛ̀
       9.itch
(3)   té-yèkɛ́
       7-slippery

Pagibete has two vowel harmony processes that operate in verb bases, defined as root + (extension +) tense/mood suffix: vowel rounding and vowel raising. Rounding is triggered by /ɔ/ in the root of the verb and spreads right. It is, however, blocked by close and mid-close vowels. The resulting phonetic output in (4–7) is given in square brackets.

(4)   /àɓɔ́tà/   →   [àɓɔ́tɔ̀] 'she will give birth'
(5)   /ɓáNdɔ́pagí/ →   [ɓá^{n}dɔ́pɔ́gí] 'they trod on it repeatedly'
(6)   /àtɔ́tɛlɛké/ →   [àtɔ́tɔ́lɔ́ké] 'it should be pierced'
(7)   /Ndɔ̀pàgiséyè/ →   [^{n}dɔ̀pɔ̀gìséyè] 'to cause one to tread on repeatedly'

Raising in verb bases is triggered by /i/ and raises preceding mid-close vowels to close (8). This process does not apply to mid-open vowels (9).

(8)   a   /lék-/ 'set a snare'   b   /sòs-/ 'wash'
            [lékéyè] 'to set a snare'   [sòséyè] 'to wash'
            [àlékà] 's/he will set a snare']   [àsòsà] 's/he will wash'
            [àlíkí] 's/he set a snare'   [àsùsí] 's/he washed'
                                             [^{n}sùsísɛ́] 'to cause to wash it'

(9)   a   /ɓɛ̀m-/ 'look around'   b   /ɓɔ́t-/   'give birth'
         [ɓɛ̀méyè] 'to look around'   [ɓɔ́téyè] 'to give birth'
         [àɓɛ̀mɛ̀] 's/he will look around'   [àɓɔ́tɔ̀] 'she will give birth'
         [àɓɛ̀mí] 's/he looked around'   [àɓɔ́tí] 'she gave birth'
         [ɓɛ̀míséyè] 'to cause to look around'   [ɓɔ́tíséyè] 'to cause to give birth'

2.1.2 Adjacent vowels

Within morpheme boundaries, the following vowel sequences occur in Pagibete, with an
optional glide interposed: /ea, eo, ɛu, eu, ia, iɛ, ua/. In addition, /óa/ and /óɛ/ exist in one
verb in the database (*bóéyè* 'to go far') and its conjugations (e.g., *mábóà* 'you(PL.) go
far'), but never with a glide.

Across morpheme boundaries, Pagibete has two vowel deletion rules. When the phoneme /a/ precedes a mid-front vowel across a morpheme boundary, the former is deleted.
Note that for example (10), the final vowel of *èkyá* acquires a contour tone. This was a
consistent phenomenon for this phrase, possibly due to a phrase-level floating tone.

(10) nèkyáà
      nà   è-kyá
      with   5-morning
      'in the morning'

(11) ɓèbògó
      ɓà-èbògó
      2-wild.duck
      'wild duck'

(12) kwɛ́yè
      kwá-ɛ́yè
      die-INF
      'to die'

An /ɔ/ immediately followed by another vowel across a morpheme boundary deletes that
vowel.

(13) a   kɔ́yè
          kɔ́-ɛ́yè
          cough-INF
          'to cough'

      b   ákɔ̀
          á-kɔ́-à
          SP$_{3sg}$\PRS-cough-PRS
          's/he coughs'

      c   àkɔ́
          à-kɔ́-i
          SP$_{3sg}$\PST-cough-PST
          'he coughed'

2.2 Consonants

Table 15.2 presents the definitively identified consonant phonemes of Pagibete. As no unambiguous CVN syllables exist (e.g., none are found in word-final position), it seems that all nasals that precede a consonant are prenasalised. While nearly the entire inventory of consonants can be prenasalised, all word-internal prenasalised consonants are voiced unless reduplication or a probable historic morpheme break exists (e.g., *mankodo* 'snail').

TABLE 15.2 THE CONSONANT PHONEMES OF PAGIBETE

	Bilabial	*Labio-dental*	*Alveolar*	*Palatal*	*Velar*	*Labio-velar*
Plosives	p, b		t, d		k, g	k͡p, g͡b
Implosives	ɓ		ɗ			
Fricatives		f, v	s, z			
Nasals	m		n			
Glides				y		w
Liquids			l			
Prenasalised series	^{m}p, ^{m}b		^{n}t, ^{n}d		$^\eta$k, $^\eta$g	$^{\eta m}$k͡p, $^{\eta m}$g͡b
	^{m}ɓ		^{n}ɗ			
		$^\eta$v	^{n}s, ^{n}z			
			$^\eta$y			$^\eta$w

In Pagibete, nasal morphemes assimilate to the following point of articulation and change that following consonant to a prenasalised one (5). When nasal morphemes precede a liquid consonant, that consonant is deleted (14).

(14) a   nókò   ɓàlókò
         N-lókò   ɓà-lókò
         'male'   'males'

      b   nú^{m}bá   lú^{m}bɛ́yè
         N-lú^{m}b-á   lú^{m}b-ɛ́yè
         'bury him!'   'to bury'

2.3 Tone

Pagibete has two phonemic tones, high and low. Tones are only found on vowels. Both lexical tone and grammatical tone are present. Changes in the lexical tones for citation forms have been observed on sentence or phrase level. While a few observations follow (Section 2.3.3), this is an area for further research.

2.3.1 Lexical tone

Lexical tone in Pagibete carries a very light functional load. With over 1,000 nouns and verbs in the data corpus, merely 21 pairs are contrastive only for tone. However it should be noted that one area of dialectal difference is lexical tone. See Table 15.3 for several minimal pairs.

TABLE 15.3 LEXICAL TONE: SELECTED MINIMAL PAIRS

L	màk-ɛ́yè	lɔ̀k-ɛ́yè	è-kpòkpó-kè	0-èsàkà
	throw-INF	vomit-INF	7-knife.sheath-7	9-basket.type
	'to throw'	'to vomit'	'knife sheath'	'basket type'
H	mák-ɛ́yè	lɔ́k-ɛ́yè	è-kpókpó-kè	è-sàká
	call-INF	rain-INF	7-cranium-7	5-rainwater
	'to call'	'to rain'	'cranium'	'rainwater, puddle'

2.3.2 Grammatical tone

Grammatical tone, in contrast to lexical tone, carries a much heavier functional load. Tone patterns distinguish the near demonstrative from the connective agreement morpheme (see Tables 15.7 and 15.10). Additionally, tone is used to distinguish between most tense/moods. Contrastive tones occur on both the subject prefix and the tense/mood vowel of verbs. Table 15.4 presents the tonal melodies for conjugated verbs for three CVC verbs. The tense/mood vowel may be final, as in Table 15.4, but both aspect and adverbial suffixes (see Section 4.2) may occur after it; this is why "FV" does not appear in any glosses of Pagibete. There are two verb classes in Pagibete, which affect the vowel of the subjunctive, but not the tone.

TABLE 15.4 VERB PARADIGMS IN 3sG SHOWING GRAMMATICAL TONE CONTRAST

Tense/mood expressed	gènéyè 'to go'	pùpéyè 'to go out'	mákéyè 'to call'
Present	ágènà	ápùpà	ámákà
Future	àgènà	àpùpà	àmákà
Narrative	àgèná	àpùpá	àmáká
Past	àginí	àpùpí	àmákí
Subjunctive	ágèné	ápùpí	ámáké

Tonological variants for the suffixes of imperative and narrative also exist. For narrative, one change is tied to the conjunction suffix /-to/ 'when.' It appears that the H of the narrative suffix is moved to the conjunction and deleted from the mood suffix.

As for other Bantu languages, extensions take the tone of the following (non-extension) morpheme. In Pagibete, this is the tense/mood morpheme, which may or may not be word final.

2.3.3 Discourse-level tone

One area where tonal changes have been observed are for phonological words that in isolation have contour tones. In discourse, for nouns and some particles the second tone frequently disappears (15). On the other hand, verb roots that have a single vowel in isolation may acquire a contour tone in discourse (16). Contour tones thus appear to have a discourse function. This is an area for further study.

(15) *èdámè* léngàɓú ɓáɓànánágí sókè
 è-dàámè   lé-ngà-ɓú   ɓá-ɓà[n]-an-ag-í   sók-è
 5-friendship PP$_5$-POSS-3PL SP$_{3PL}$\PST-love-RECP-REP-PST surpass-INF
 'Their friendship was very strong.'

(16) *àwáàmá* àpàná àdè kóngí.
 à-wám-á   à-pàn-á   à-dè   kó-ngí
 SP$_{3SG}$\NARR-stop-NARR SP$_{3SG}$\NARR-return-NARR SP$_{3SG}$\NARR-arrive\NARR 17-village
 'He stopped (talking), returned and arrived in the village.'

2.4 Syllable structure

Pagibete manifests only open syllable structures. Possible syllable types in Pagibete include V, CV, and CV_1V_2. The maximum syllable onset allowable is C, where C may or may not be prenasalised. All consonants and all vowels can occur in a CV syllable, with the restrictions noted in Section 2.2 for prenasalised consonants. All vowels except /u/ and /ɔ/ can occur in a V syllable, which is always word-initial except in one insect name. Examples are presented in Table 15.5.

TABLE 15.5 SAMPLE SYLLABLE TYPES

V-CV	*CV-CVV-CVV*	*V-CV-CV-CV*
àtà	máléàléà	èdibákè
even.though	1a.cicada	è-dibá-kè
		7-skin-7

3 NOUNS AND NOUN PHRASES

Pagibete is somewhat atypical of Bantu languages in its affixation of noun class markers. Plural noun classes have prefixes, as do the singular noun classes 1, 3, 5, 15 and 15a. However, only classes 9, 10, and 17 never have suffixes. Pagibete is also atypical in not having the verb's subject prefix agree with the class of the subject.

3.1 Noun classes

Pagibete has 16 noun classes. With reference to the standard Bantu noun class number-ing, Pagibete has the following: 1, 1a, 1x, 2, 3, 4, 5, 6, 6a, 7, 8, 9, 10, 15, 15a, 17. The label *1x* is used instead of *1b* because neither the form nor the semantic content (animals) agrees with the usual class 1b (Welmers 1973: 165–166). Classes are distinguished from each other by one or a combination of the following factors: affix shape (see Table 15.6), agreement pattern (see Table 15.7), and the pairing of one class with another as singular and plural. For example, classes 2 and 8 have many of the same affixes and agreement patterns. Where these are identical, class membership is assigned solely on the basis of the class membership of the stem in its singular form. Class 1x and class 7 have the same

TABLE 15.6 PAGIBETE NOUN CLASSES

Class	Affix(es)	Examples		
1	N- N- -yè N- -ɛ́	N-bɛ́tù 1-firefly	-ká-yè 1-female.person-1	N-bikà-ɛ́ 1-visitor-1
1a	ø- ø- -yè ø- -ɛ́	èbògó 1a.wild.duck	mázeyá-yè 1a.woman.in.post-birth.confinement-1a	gbùká-ɛ́ 1a.large.forest.rat-1a
1x	ø- -kè	èkòpí-kè 1x.leopard-1x		
2	ɓà- ɓà- -ɓà	ɓà-bɛ́tù 2-firefly ɓà-èbògó 2-wild.duck ɓà-èkòpí 2-leopard	ɓà-ká-ɓà 2-female.person-2 ɓà-mázàyèyá-ɓà 2-woman.in.post-birth.confinement-2	ɓà-bìká 2-visitor ɓà-gbùká 2-large.forest.rat
3	(N)- (N)- -yè (N)- -ɛ́	N-pèmú 3-doorway	N-mòó-yè 3-head-3	nsálá-ɛ́ work-3
4	N- -mè	N-pèmú-mè 4-doorway-4	N-mòó-mè 4-head-4	nsálá-mè work-4
5	è- è- -lè	è-bògó 5-rock	è-bá-lè 5-liver-5	
6	mà- mà- -mà	mà-bògó 6-rock mà-ɓɔ́kɔ̀ 6-arm	mà-bá-mà 6-liver-6	mà-kwá 6-die 'funerals, wakes'
6a	mà-	mà-káná 6a-alcoholic.drink		
7	ø- -kè	èkólógbɔ́-kè knee-7	ɛ̀gbɔ̀kèlí-kè handle-7	
8	ɓà- ø ɓà- -ɓè	ɓà-èkólógbɔ́ 8-knee	ɓà-ègbɔ̀kèlí-ɓè 8-handlè-8	
9	ø	bìtí 9.night	èkèlè 9.spoon	
10	ɓà-	ɓà-bìtí 10-night	ɓà-èkèlè 10-spoon	
15a	ò- ò- -kò	ò-ɓɔ́kɔ̀ 15a-arm	ò-kwá-kò 15a-die-15a 'death, funeral, wake'	
15	ó- -kò	ó-pàsá-kò 15-improve-15 'improving oneself'		
17	kó-	kó-pèmú 17-doorway 'in the doorway'	kó-ɓɔ́kɔ̀ 17-arm 'on the arm'	

TABLE 15.7 CLASS AGREEMENT MORPHEMES

Class	Adj.	Con./Rel.	Num.	Possessives		Verbals	
				Pron.	Nom.	Id.	Proc.
1, 1a, 1x	mó-	mó-	ó-	mó-	mó-	-nò, -mò	-yò
2	ɓá-	ɓá-	ɓé-	ɓá-	ɓá-	-ɓà	-ɓò
3	ó-	wé-	ó-	ó-	wé-	-wè	-wɔ̀
4	má-	má-	mí-	má-	má-	-mà	
5	lé-	lé-	lé-	lé-	lé-	-lè	-lò
6, 6a	má-	má-	mí-	má-	má-		-mò
7	té-	té-	té-	té-	té-	-tè	-tɔ̀
8	ɓá-	ɓá-	ɓé-	ɓá-	ɓá-		-ɓò
9	é-	yé-	é-	é-	yé-	-yè	-yò
10	é-	yé-	é-	é-	yé-	-yè	-yò
15, 15a	ó-	wé-	ó-	ó-	wé-		-wɔ̀
17	ó-			kó-			

suffix, but can be distinguished by the agreement pattern. Some class 3 members lose the initial N when other class prefixes are added, while others do not. This is signaled by parenthesis in Table 15.6. The class 17 prefix replaces all other class affixes as a general rule. Plurals are found in even-numbered classes.

A certain amount of residue is present in the data. Four items remain (*lémbò/mámbò* 'song,' *líɓá/máɓá* 'water,' *línà/mánà* 'name' and *lísò/mísò* 'eye') which are in class 5–6, but retain the class 5 prefix of Proto-Bantu *li-* (Welmers 1973: 165). No rule can be formulated that adequately accounts for the identification of all four stems when the appropriate class prefix is added.

Noun class agreement occurs with affixes, such as those for adjectives, or with independent morphemes, such as connectives/relatives. The commonality of form across a class is clearly seen when these morphemes are placed next to each other, as in Table 15.7.

The shift in the class 3 agreement marker for adjectives, numerals and possessives from *wé-* to *ó-* seems to have occurred in the last half of the 20th century, as de Boeck (1949: 835–840) gives *wé-* as the prefix used in these instances. The verbals, discussed in Sections 4.3.1.4 and 4.3.1.5, participate in the class agreement system. Gaps in Table 15.7 represent gaps in the data.

3.2 Nouns

The pairings commonly occurring in Pagibete are the following: 1/2, 1a/2, 1x/2, 3/4, 5/6, 7/8, 9/10, 15a/6. In addition, 3/10, 5/10, 7/10 and 15a/10 have numerous exemplars; 9/6 and 5/2 also are found, but far more rarely. Several of these more unusual pairings are also documented in Ngombe C41 (Kadima 1969: 89), the language spoken immediately to the south of the Pagibete.

A number of nouns are found in only one class. These occur in classes 1, 2 or 8, 3, 6a, 9 or 10, 15 and 17. Placement in classes 1, 3, and 15 is based upon the agreement pattern. Since the concord marker of 2 and 8, *ɓa*, is identical, as is that of 9 and 10, *ye*, tentative assignment to those classes is based mainly on the semantic content.

Nominal derivation from verbs (17) is much less productive in Pagibete than in other Bantu languages. The only class consistently derived from verbs is class 15 (18).

(17) a   kwɛ́yè
          kwá-ɛ́yè
          die-INF
          'to die'

      b   ò-kwá-kò
          15-die-15
          'death, funeral, wake'

      c   ŋkwénɛ̀
          N-kwa-én-ɛ̀
          1-die-APPL?-1
          'orphan'

(18) a   ódàsɛ́gɛ́kò
          ó-dàs-ɛg-(ɛ́)kò
          15-gather-REPA-15
          'unity'

      b   ósòsákò
          ó-sòs-(á)kò
          15-wash-15
          'washing oneself'

Examples of noun-to-noun derivation (19–20) exist also.

(19) a   mbúwá-yè
          rain-1a
          'rain'

      b   ò-mbúwá-kò
          15-rain-15
          'rainy season'

(20) a   ŋgbàyɛ́
          N-gbàyá-ɛ́
          1-chief-1
          'chief'

      b   è-gbàyá
          5-chief
          'chiefdom'

3.3 Pronouns and nominal modifiers

Pagibete has both independent and dependent pronouns. Nominal modifiers agree in class with the head noun of the phrase.

3.3.1 Pronouns

Pagibete personal pronouns may occur both in subject and object positions. The third-person forms provided in Table 15.8 apply for animate (21) and inanimate (22) across all classes.

TABLE 15.8 INDEPENDENT PRONOUNS

Person	Singular	Plural
1	ɛ̀mɛ̀	ɓàsù
2	òwɛ̀	ɓànù
3	yí	ɓú

(21) yí   ɓéè   ɓɔ̀   "Gùndìyà   kó-ngí?"
     3SG   that thus how.goes.it 17-village ?
     'He (the father) said thus, "How is it going in the village?" '

(22) nɛ̀-gɛ̀n-á   nà   yí   mùngá
     1SG\PRS-go-PRS   with 3SG 9.hunting.with.dog
     'I go hunting with it (a dogbell).'

A pronoun used for things that functions as a focaliser has the form of the connective/relative morpheme with (23) or without (68b) a HL contour tone. In (23), *yéè* refers to the class 9 noun *kwáyè*, 'problem.'

(23) yéè   wánè   kó-ɓɔ̀kɔ̀   gò   à-gɛ̀n-á   à-sí-à.
     9.FOC   here   17-arm   also SP₃ₛ₉\NARR-go-NARR   SP₃ₛ₉\PRS-be.finished-PRS
     'The one here on my arm also is finished.'

3.3.2 Possessives

Possession is expressed in various ways. Pronominal possessives (24) are composed of an agreement marker for the possessed (see Table 15.7), the morpheme *-ngá-*, and a suffix indicating the possessor. See Table 15.9 for possessor suffixes, which only agree in person and number, not in class.

(24) a N-gbàyá-ɛ́   mó-ngá-sù
        1-chief-1   PP₁-POSS-1PL
        'our chief'

TABLE 15.9 POSSESSOR SUFFIXES

Person	Singular	Plural
1	-mɛ̀	-sù
2	-kò	-nù
3	-kè	-ɓù

b nsàlá-mὲ   má-ngá-kὲ
   work-4 PP_4-POSS-3SG
   'his works'

These possessor suffixes are also obligatorily affixed directly to several kinship terms, as well as optionally by other relational terms, without the use of a possessive morpheme. In the latter case, these suffixes replace the final syllable of the noun (25). In the plural of obligatory kinship terms, the possessor suffix is added after the class 2 suffix (26). *bàbá* 'father' becomes *àbá-* when a possessor suffix is attached (44).

(25) a dámò
     1a.friend

    b dá-kò
       1a.friend-2SG
       'your friend'

    c ɓà-dá-mὲ
       2-friend-1SG
       'my friends'

(26) a N-kálá-mὲ
     1-wife-1SG
     'my wife'

    b ɓà-kálá-ɓá-mὲ
       2-wife-2-1SG
       'my wives'

When the possessor is named or referred to by a noun other than a kinship term, Pagibete has two ways of expressing possession, both involving a particle between the possessed and the possessor. The first strategy applies only when the possessor is animate. In this strategy, the first particle is an agreement morpheme prefixed to the possessive morpheme *-ka* (27). As for possessive pronouns, the prefix agrees with the head noun.

(27) a ⁿgúyè
     ngùúyè   yé-kà   Nyɔ́mbò
     9.power NP_9-POSS 1.God
     'the power of God'

    b ngí   wé-kà   ɓà-kómbòzó
       3.village NP_3-POSS 2-chimpanzee
       'village of the chimpanzees'

A connective particle is used when the possessor is inanimate. This morpheme agrees with the head noun and has either a high (28) or a HL contour tone (38).

(28) ɓà-gbàyá ɓá   ngí
    2-elder CON_2 3.village
    'elders of (the) village'

3.3.3 Demonstratives

Pagibete has a series of four demonstratives, all independent words that agree with the head noun. They roughly correspond in meaning to 'this,' 'that,' 'this which was just mentioned' (reference near), and 'that which had been referred to already' (reference far). The near demonstrative ('this') is composed of the agreement morpheme with a falling tone, while the other demonstratives consist of the consonant of the agreement marker plus a demonstrative stem. See Table 15.10. For use of the near, far, and reference near demonstratives in context, see (42, 45, 46, & 52).

TABLE 15.10 DEMONSTRATIVES

Class no.	Near	Far	Ref. near	Ref. far
1, 1a, 1x	móò	yáà	mɔ́ɔ̀	móní
2	ɓáà	ɓíyá	ɓɔ́ɔ̀	ɓóní
3	wéè	yáà	wɔ́ɔ̀	wóní
4	máà	míyá	mɔ́ɔ̀	móní
5	léè	líyá	lɔ́ɔ̀	lóní
6, 6a	máà	míyá	mɔ́ɔ̀	móní
7	téè	tíyà	tɔ́ɔ̀	tóní
8	ɓáà	ɓíyá	ɓɔ́ɔ̀	ɓóní
9	yéè	yáà	yɔ́ɔ̀	yóní
10	yéè	yáà	yɔ́ɔ̀	yóní
15, 15a	wéè	wíyá	wɔ́ɔ̀	wóní

3.3.4 Adjectives

Adjectives may be divided into three groups. The first group is derived from verbs and takes an agreement marker (29). The second group is not derived from verbs and also takes an agreement marker (30). Both of these classes include a few members whose stem changes in the plural, either from reduplication of the first syllable of the root with a high tone added to the reduplicated syllable, or from a change of the second consonant (31).

(29) a   míkí   mó-ɓùɓé
          1a.child NP_1-be.clean
          'a clean child'

      b   è-ɓèmbé   lé-ɓùɓé
          5-house NP_5-be.clean
          'a clean house'

(30) a   nkéngànó   ò-kùɗù
          3.walk NP_3-short
          'a short walk'

      b   è-ɓógò   lé-kùkù
          5-plantain NP_5-unripe
          'an unripe plantain'

(31) a è-ɓèmbé   má-kὲdὲ
        5-house   NP₅-small
        'a small house'

     b mà-ɓèmbé   má-kὲkὲ
        6-house   NP₆-small
        'small houses'

The majority of adjectives that do not agree with the head noun appears to be formed by the addition of the prefix ɓé- 'that' to a stem (32-33). However, none of these stems exists in any other form. Additionally, these adjectives often have a reduplicated syllable (33).

(32) è-sàsó   ɓédòkìlò
     5-metal.pot deep
     'deep metal pot'

(33) dɔ̀ótὲ   ɓésὲdὲsὲdὲ
     7.dirt   slippery
     'slippery ground'

3.3.5 Enumeratives

The traditional Pagibete number system has numbers for one to five; after five, they used to use an addition system, except for ten màbókò, 'hands.' Today, while they continue to use their words for one through five, and sometimes that for ten, they commonly use the Lingala numbers for six through ten. Numbers over ten may use a mix of the two languages. When used in counting, numbers one through four have the form é- + stem.

(34) é-mòtí 'one'
     é-bàlé 'two'
     é-sálò 'three'
     é-kwàngànè 'four'

When used as adjectives, these numbers have the form agreement + stem (35). 'Five,' òɓúmòtí, has an invariable form, which appears to be composed of the class 15 prefix ò-, plus the 3PL affix ɓú, plus the stem for 'one' -mòtí ('they ones,' which may refer to the five digits on a hand). The Lingala loanwords generally remain invariable in form, as they do in the dialect of Lingala spoken in the area.

(35) a è-ɓèmbé   lé-mòtí
        5-house   EP₅-one
        'one house'

     b mà-ɓèmbé   mí-sálò
        6-house   EP₆-three
        'three houses'

The word meaning 'how many,' -éngà, like numbers, is also composed of an agreement marker plus a stem.

(36) ò-lì-èk-í   ná   ɓà-míkí   ɓá-ɛ́ngà?
SP₂ₛ𝒈\PRS-be-N-PRS   CJ   2-child   EP₂-how.many
'How many children do you have?'

3.3.6 Quantifiers

No single morphological statement can be made about Pagibete quantifiers. Two are invariable in form; *zù* 'all,' and *mánkàkà* 'alone.' The former is borrowed from their non-Bantu neighbours to the north, the Ngbandi.

Other quantifiers behave as adjectives, agreeing in class with the head noun.

(37) ɓà-sɔ̀kɔ̀   ɓá-ngúyè
2-caterpillar   NP₂-many
'many caterpillars'

3.3.7 Conjunctions

Noun phrases are conjoined with other noun phrases by one of two forms. The first is *nà*. In this context, *nà* generally signals accompaniment and may be translated 'and' or 'with.' It is usually an independent morpheme, but elision will occur if the initial vowel of the following word is /e/ or /o/.

(38) è-tɔ́   léè-gá   nà-ètɔ́   à-lì-èk-í   lé-kà   nvà-ɛ́   nà   ɓúsù.
5-story   CON₅-PROG   CJ-story   SP₃ₛ𝒈\PRS-be-N-PRS   NP₅-POSS   dog-1a   CJ   1a.cat
'This story is a story about Dog and Cat.'

The second, used exclusively to associate animate beings, is *ɓúnà*, recognisable as a compound morpheme, *ɓú* '3PL' + *nà* 'CJ.' Again, this conjunction may be translated either 'and' or 'with.'

(39) tinè   yá-pínè   ndí-yò, má-wén-à-gá   ɓétè ɓúsù   ɓúnà   mbàɓú
9.meaning 9-CONT   PROC-9   SP₂ₚₗ\PRS-see-PRS-PROG   that   1a.cat   CJ   1a.rat
ɓá-ɓà[n]-an-ag-è-píné   ká.
SP₃ₚₗ\PRS-love-RECP-REPI-PRS-CONT   neg
'The (on-going) meaning is this, we see that Cat and Rat do not love each other. . .'

3.3.8 Locatives

Locative phrases may involve the class 17 (locative) prefix (40), or a preposition (41) which may contain the locative prefix (42). In all cases the locative precedes the noun.

(40) ɓá-gèn-á   kó-pá, . . .
SP₃ₚₗ\NARR-go-NARR   17-forest
'They went to the forest, . . .'

(41) àtàⁿgàtó
à-tàng-á-tó   sòpòní   lí-ɓá,
SP₃ₛ𝒈\NARR-walk-NARR-when   in   5-water
'When he walked in the water, . . .'

(42) ɓà-míkí   ɓá-lì-à   kó-gùná   ngwàngwà
      2-child   sp₃ₚₗ\prs-sit-prs   17-on   1a.grass
      'The children sit on the grass.'

3.3.9 Relative clauses

Relative clauses occur after the noun which they modify. They begin with a connective/relative morpheme that agrees with the relativised noun. No differentiation exists for subject (43) and object (44) relativisation. If the dependent clause is inflected for aspect, the aspect marker will be attached to the relative pronoun instead of the verb (43).

(43) lí-ɓá   [lé-píné   ɓá-sòs-ag-á   nà   ɓà-sánì]
      5-water   REL₅-CONT   sp₃ₚₗ\NARR-wash-REPI-NARR   CON   10-dish
      'The water [that they (always use to) wash dishes] . . .'

(44) míkí   móò   à-wɔ́k-à   nzòmbí   [wé   àbá-kè   yí
      1a.child   1.NDEM   sp₃ₛ\NARR-hear-NARR   3.word   REL₃   1a.father-3sg   3sg
      à-N-kpítí   yí.]
      sp₃ₛ\PST-OM-speak-PST   3sg
      'This son listened to the advice [that his father gave him].'

3.4 Word order and agreement in the noun phrase

From the corpus of texts, it appears that possessive pronouns occur in the closest proximity to the head noun, as seen in (45). Relative clauses occur after any other modifiers (46).

(45) a nsálá-mè   má-ngá-kè   má   è-gbàyá   kó-ngí
        work-4   PP₄-POSS-3sg   CON₄   5-chiefdom   17-village
        'his works of the chiefdom in the village'

     b è-zègé-lè   lé-ngá-mè   léè
        5-illness-5   PP₅-POSS-1sg   NDEM₅
        'this illness of mine'

(46) míkí   mó   ó-mòtí   [mó   à-N-bís-í   àbá-kè   kó-pà]
      1a.child   CON₁   EP₁-one   REL₁   sp₃ₛ\PST-OM-put-PST   1a.father-3sg   17-forest
      'The one son [who had put his father in the forest] . . ."

To date, all collected texts that contain complex noun phrases involve either a possessive or relative clause (e.g., 45 & 46), or have merely two of the same type of word modifying the noun (47). In the latter case, the successive demonstratives play the same role (and have the same form) as the pronominal focus marker noted in (23).

(47) è-mbá-lè   líyá   lɔ́ɔ̀
      5-money-5   FDEM₅   RNDEM₅
      'that money just referred to'

4 VERBS AND CLAUSE STRUCTURE

Pagibete verbs include both prefixes and suffixes. Unlike most Bantu languages, while the bound subject and object pronouns are prefixed to the verb root, the entire TAM system is generally located after the verb root except for tonal modifications on the subject prefix (see Table 15.4). The vowel carrying the tense/mood (in conjunction with the tone on the subject marker) may be final, but any aspect morphemes will occur after it. Finally, as was discussed in Section 3.3.8, the aspect marker can be suffixed to the relative pronoun in a dependent clause.

4.1 Verbs

Verbal roots in Pagibete cannot stand alone. At the very least they must have a suffix. Thus, the minimal form is a base (root + suffix). This is the form of both the infinitive and the 2SG imperative, as well as the Identificational Copula (see 4.3.1.4) and Process Verbal (see 4.3.1.5).

The verb (other than the identificational and process morphemes) can consist of quite a few more elements. Thus, a more complete picture of the Pagibete verb is this: (negative/adverb) + (subject prefix) + (object marker) + root + (extension1) + (extension2) + tense/mood + (aspect) + (adverb). However, no verb in the data corpus to date includes all of these possibilities; six morphemes seem to be the maximum permitted, and even that is quite rare.

4.1.1 Syllable structure

Verbal roots in Pagibete can have one of five forms: C, CV, CVC, CVCV and CVCVC. It should be noted that this last is questionable, as the second syllable has no independent tone, but acts as extensions do. However, several CVCVC verbs are irreducible to a CVC root. Note that an *o* or *ɔ* preceding an *έ* deletes that vowel, but attaches the high tone to the resulting vowel.

TABLE 15.11 VERB SYLLABLE STRUCTURE

C- *root*	CV- *root*	CVC- *root*	CVCV- *root*
p-έyè	mɔ̀-έyè	gèn-έyè	sàpò-έyè
give-INF	[mɔ́yè]	go-INF	[sàpóyè]
	kill-INF		divert-INF

CVCVC- *root*
dɔ̄ngɔ̀t-έyè
stretch.to.end-INF
'to stretch to the end (but still be inadequate)'

Although in general Bantu languages have a prefix for the infinitive marker, in Pagibete, this is not the case. Verbs have the suffix, -έ(yè).

4.1.2 Extensions

Pagibete has seven well-attested extensions to accomplish verbal derivation. These are as given in Table 15.12. The posited underlying form is that which surfaces when the tense/

mood vowel is *a*. Most have at least two allophones, based on the vowel harmony rules given in 2.1.1.

TABLE 15.12 PAGIBETE EXTENSIONS

Label	Underlying form	Example of the extension with a verb
Repetitive, animate object	-ɛg-	bòm-ɛg- 'beat' (an animate being, such as a dog)
Repetitive, inanimate object	-ag-	bòm-ag- 'beat' (an inanimate object, such as a drum)
Neuter	-ek-	wàl-ek- 'split open' (as a fruit's skin does in hot water)
Causative	-is-	bib-is- 'strengthen, raise'
Associative	-ɛn-	kíl-ɛn- 'run together' (as do rivers)
Stative	-an-	tíg-an- 'remain'
Applicative	-ɛl-	v-ɛl- 'marry' (literally, 'to take to')

In addition, there appear to be two frozen extensions, *-it-* and *-em-*. These usually appear in irreducible verb bases, and so their semantic component is unclear. A contrast was given by native speakers, however, between *-it-* and *-is-* for the verb *bíng* 'roll,' which must have one of these two extensions attached.

(48) a àbíngísàgá pípà.
          à-bíng-is-à-gá   pípà
          SP$_{3sg}$\PRS-roll-CAUS-PRS-PROG   9.barrel
          'S/he's rolling the barrel (on one edge).'

      b àbíngítàgá pípà.
          à-bíng-it-à-gá   pípà
          SP$_{3sg}$\PRS-roll-?-PRS-PROG   9.barrel
          'S/he's rolling the barrel (on its side).'

Pagibete does not have a reflexive extension. To express the reflexive, the verb root is nominalised and becomes a gerund in class 15. Because of the syllable constraints of the language, an epenthetic vowel must be inserted between a C- or CVC- root and the suffix *-kò* (Table 15.13).

TABLE 15.13 INFINITIVES OF REFLEXIVE VERBS

C reflexive	CVC reflexive
o-p-áko	o-kís-áko
[òpákò]	[òkísáko]
15-give-15	15-lower-15
'giving oneself'	'humbling oneself'

Most extensions may co-occur with at least one other. For some, an ordering restriction applies; for others, it does not. The repetitives, whether for an animate object (REPA) or for an inanimate object (REPI), must be the final extension in the base (50). The associative follows all except the causative; it does not co-occur with either of the repetitives (51).

(49) ɓàv̀ɛlɛ̀nɛ̀gbá
      ɓà-v-ɛl-ɛn-ɛ̀-gbá
      SP$_{3PL}$\FUT-take-APPL-ASSOC-FUT-REM.FUT
      'They will marry.'

(50) àwátɛ́lɛ́gé nà kìngò ɛ́ngákɛ̀ zù
      à-wát-ɛl-ɛg-é   na   kìngò   ɛ́-ngá-kɛ̀   zù
      SP$_{1SG}$\NARR-twist-APPL-REPA-NARR   CJ   neck   PP$_9$-POSS-3SG   all
      'He (the snake) wrapped himself completely around his (the chief's) neck.'

(51) màdɔ̀gɔ̀pí ngímɛ̀ nà ngímɛ̀ mádásékánà?
      mà-dɔ̀-ɛg-ɔ̀-pí   ngí-mɛ̀   nà ngí-mɛ̀   má-dás-ek-an-à
      SP$_{2PL}$\FUT-come.from-REPA-FUT-HAB village-4 CJ village-4 SP$_{2PL}$\PRS-gather-N-ASSOC-PRS
      'Will you come from many countries to join together?'

4.2 TAM

The TAM markers of Pagibete surround the verb root plus extensions. The subject marker carries the initial information regarding tense/mood. The first slot after the extension is the one which takes the second part of either a basic tense morpheme (past, present or future), or the morpheme for the imperative, subjunctive or narrative mood. The mood morphemes never co-occur with the basic tense morphemes. The singular and 1PL subject prefixes carry the tone of the tense/mood, but 2PL and 3PL subject prefixes are always high.

When the conjugated verb is in isolation, the present and future are marked with the same post-root vowel à. The two tenses are differentiated by the tone on the subject prefix; present takes a high tone, and future a low tone. The post-root vowel of the past tense is í and it co-occurs with a low tone on the subject prefix. The subjunctive is marked with a high tone on the subject prefix and é or í in the tense/mood slot, according to verb class. The narrative takes a low tone on the subject prefix, and á in the tense/mood slot. The second person imperative has no subject prefix; the post-root vowel is á. All but the past tense have one or more phonological variants caused by vowel harmony processes in CVC- or CVCVC- roots or after extensions. This is summarised in Table 15.14; sample verb paradigms are found in Table 15.4.

TABLE 15.14 TENSE/MODE MARKING

Tense/mode	Tone on SG, 1PL prefix	Vowel and tone in tense/ mode slot	Phonological variants for vowel of tense/mode slot
Present	high	à	ɛ̀, ɔ̀, è
Future	low	à	ɛ̀, ɔ̀, è
Narrative	low	à	ɛ̀, ɔ̀, è
Past	low	í	(none)
Subjunctive	high	é	í
Imperative	n.a.	á	ɛ́, ɔ́

For C-, CV-, and CVCV- roots, the same allomorphs exist for present, future, narrative and imperative. However, the past tense for these roots has the allomorphs $\acute{\varepsilon}$, $\acute{u}$ and $\acute{\jmath}$. $\acute{\varepsilon}$ occurs when the root has an (underlying) ε. $\acute{u}$ occurs when the root has a non-low back vowel, and $\acute{\jmath}$ occurs when the root has an (underlying) $\jmath$.

The TAM system of Pagibete distinguishes degree of distance in the past or future. The six markers of temporal remoteness (see Table 15.15) occur in the same slot as the aspect markers habitual, present progressive and continuous, and so are considered as aspects. If any of these morphemes are in the main verbal phrase, they will occur immediately after the tense slot. Co-occurrence restrictions are fairly severe, however. Only three of the nine may occur with more than one tense slot morpheme. The future aspects, for example, must co-occur with the future tense (52). The immediate past may co-occur with either the present tense, which indicates that the action has occurred within the last two or three hours, or with the past tense, which indicates that it happened that same day, but longer ago than if it is used with the present tense (59). Remote past may co-occur with either the past tense or the narrative mood marker (53 & Table 15.15). These morphemes may also occur on the connective/relative marker in relative clauses (see Section 3.3.8), and with the subordinating conjunction *gòndé* 'since, because.'

(52) à-súng-à-lɔ́   ndéngé   tìnà?
      SP$_{3SG}$\FUT-end-FUT-IMM.FUT   9.sort   how
      'How will this end?'

(53) ɓá-gèn-á-ndé   kó-pà
      SP$_{3PL}$\NARR-go-NARR-REM.PST   17-forest
      'They used to go to the forest [for hunting].'

TABLE 15.15 TENSE AND ASPECT CO-OCCURRENCE RESTRICTIONS

	-ndè	-yókò	-yáká	-gá	-lɔ́	-púmá	-gbá	-pí	-píné
	Remote past	Mid past	Immediate past	Progressive	Immediate future	Mid future	Remote future	Hab-itual	Continual
Past	x	x	x						
Present			x						x
Future				x	x	x	x	x	
Narrative	x							x	

Subjunctive and imperative moods do not allow the addition of aspect markers under most conditions. They do, however, allow the suffixation of adverbial morphemes that indicate plurality (on imperatives) or emphasis, which then permit the addition of an aspect marker (55).

(54) pùp-á-nì!
      go.out-IMP-PL
      'Go out!'

(55) ò-bí-sè-gbá   gòtó   ká
     SP₂ₛ₉\SUBJ-come\SUBJ-EMPH-REM.FUT   again   NEG
     'You should never come again.'

The progressive aspect co-occurs only with the future tense. In order to express the idea of past progressive, one of two other constructions is used. The first requires a past tense 'be' verb followed by followed by the locative *ko*, which in turn is followed by another verb.

(56) báliàgí kó ɓágèná kóbònà ndé ɓáwòná kpòtò yáà.
     ɓá-lì-ag-í   kó   ɓá-gèn-á   kó-bònà   ndé
     SP₃ₚₗ\PST-be-REPI-PST   LOC   SP₃ₚₗ\NARR-go-NARR   17-market   then
     ɓá-wòn-á   kpòtò   yáà
     SP₃ₚₗ\NARR-see-NARR   10.mushroom   FDEM₁₀
     'They were there going to market when they saw those mushrooms.'

The second construction requires a past tense 'be' verb with the conjunction *né* or *ná* 'in order to' before an infinitive verb.

(57) tè-bìy-à-gá   ná   gót-έyè   mbásò   yè
     SP₁ₚₗ\FUT-come-FUT-PROG   CJ   cut-INF   tree.10 NDEM₁₀
     'We were coming to cut these trees (and are in place and cutting).'

The exception to the general rule of subjunctive moods not allowing aspect markers is the perfective marker *-ní* (58). In questions, although it is still affixed to the subjunctive, it carries other connotations, but this needs to be further explored.

(58) á-gɛn-á-ní   go   kó-ngi
     SP₃ₛ₉\SUBJ-go-SUBJ-PRF   again   17-village
     'He had already gone again to the village.'

The imperfective aspect, however, is expressed with a special construction using a verbal morpheme *ndóò* or *ndíyò*. This morpheme may be independent or, in the form of *ndò-*, take a suffix which agrees with the subject of the clause (see Table 15.7). When a verb follows, it may be in the present tense or narrative mood. See Section 4.3.1.5 for further discussion of this construction.

(59) ndíyò   tè-bìy-á   ná   gót-έyè   mbáso   yè
     PROC   SP₁ₚₗ\NARR-come-NARR   CJ   cut-INF   tree.10 NDEM₁₀
     'We are on the way to cut these trees.'

4.3 Clause structure and types

Pagibete, like other Bantu languages (Heine 1980), is an SVO language. Pagibete has both verbal and verbless clause types. The verbal clauses are distinguished further by the type of verb required.

4.3.1 Verbal clauses

Verbal clauses generally have SVO order. The syntax of specific clause types is discussed in the following subsections, with consideration of what is allowed in each slot.

4.3.1.1 Intransitive clauses

The basic form of intransitive clauses is a verb word, having a subject marker and the verb (60). Descriptive clauses may have an adjective after the verb (61). For most intransitive clauses, the verb may be an active verb or a stative verb. In the descriptive clause, however, the copular verb /-lì-/ 'be' is used.

(60) nὲ-yɔ̀y-í-yákà
      SP$_{1SG}$\PST-be.tired-PST-NPST
      'I was tired (this morning).'

(61) mbìyáyè léὲ àlìkí lédùwè
      mbìyáyè   léὲ   à-lì-ek-í   lé-dùwè
      5.oil.palm.tree   NDEM$_5$ SP$_{3SG}$\PRS-be-N-PRS   NP$_5$-alive
      'This oil palm tree is alive.'

4.3.1.2 Transitive clauses

The basic form of a transitive clause is SVO. Four types of transitive verbs exist: the first never takes an object marker (62), the second takes an object marker only if the object is animate (63), the third takes an object marker regardless of animacy (64), and the fourth always takes both an object marker and a nominal object (65). When object markers are used, the object is not always overtly stated as a clause-level constituent (64). The object marker is a nasal morpheme and only marks existence of an object; it does not agree with the object in class.

(62) nὲ-lék-à-gá   ò-kpìà-kò
      SP$_{1SG}$\FUT-set.up-FUT-PROG   15a-net-15a
      'I am setting up a net.'

(63) a   àbìbìsí òɓɔ́kɔ
          à-bìb-is-í   ò-ɓɔ́kɔ
          SP$_{3SG}$\PST-raise-CAUS-PST   15a-arm
          'He strengthened the arm.'

      b   àmbìbìsí míkí
          à-N-bìb-is-í   míkí
          SP$_{3SG}$\PST-OM-raise-CAUS-PST   1a.child
          'She raised the child.'

(64) à-N-dùl-í-yáká
      SP$_{3SG}$\PST-OM-hit-PST-IMM.PST
      'He hit him.'

(65) nè-N-zéng-à-gá   kpɔ́tɔ̀
    SP$_{1SG}$\FUT-OM-slice-FUT-PROG   9.greens
    'I'm slicing greens.'

4.3.1.3 Ditransitive clauses

In ditransitive clauses, two objects are found after the verb. The verb is one whose root is inherently ditransitive, or one which includes the applicative extension (66). The primary direct object may not surface except as an object marker before the verb base (67).

(66) tènkónélègá ɓàpasteur tíkò
    tè-N-kón-ɛl-è-gá   ɓà-pasteur   tíkò
    SP$_{1PL}$\FUT-OM-plant-APPL-FUT-PROG   2-pastor   9.field
    'We are going to plant the pastor's family a field.'

(67) à-gèn-à-pí   á-N-p-á   nyé   pá   kó
    SP$_{3SG}$\PRS-go-PRS-HAB   SP$_{3SG}$\NARR-OM-give-NARR   3.food   only   LOC
    'He went regularly to give him food there.'

4.3.1.4 Identificational clauses

Identificational clauses involve the use of the copula *ndé* (68). If a noun does not follow it, a class agreement marker (see table 15.7) is suffixed (69). *Ndé-* has two possible phonological variants, *ndó-* (69a) and *ndí-*. The syntax is noun phrase + copula (+ noun phrase).

(68) lí-ná   lé-ngá-kè   ndé   Ɓàbú
    5-name   PP$_5$-POSS-3SG   COP   Ɓàbú
    'Her name was Ɓabú.'

(69) a   nvà-ɛ́   mó-ngá-mè   ndó-nò
        dog-1a   PP$_1$-POSS-1SG   COP-1.IDS
        'This is my dog.'

      b   lé   lɛ́-kùkù   ndé-lè
        5.FOC   NP$_5$-unripe   COP-5.IDS
        'The unripe one is this one.'

When an independent pronoun is used as the subject, the syntax is simply pronoun + copula.

(70) yí   ndó-mò
    3SG   COP-IDS$_1$
    'This is he.'

4.3.1.5 Process clauses

Process clauses are how Pagibete expresses that the effect of an action or state is ongoing. The Process particle may stand in place of the verb, or may co-occur with a verb. In either case, it agrees with the subject. The root of the Process particle is *ndí-*. However, it has an allomorph, *ndú-*, when the agreement marker is *-wɔ̀*. The known agreement suffixes for each class are given in Table 15.7.

(71) a   kó-gɛ̀bɛ́   è-pwá-lè   ndí-lò   kó-gɛ̀bɛ̀
       17-little.finger   5-wound-5   PROC-5   17-little.finger
       'The wound is on the little finger (with all its attendant effects).'

   b   ɓa-mɛmɛ   ndí-ɓo   ɓá-n-dɔp-ɔ́g-í   biliki
       2-goat   PROC-2   SP₃ₚₗ-OM-stamp-REPA-PST   2.brick
       'The goats walked all over the bricks (with on-going results).'

Another Process particle, *ndóò or ndó*, does not agree in class with any noun in the sentence. Motingea Mangulu (2002) labelled *ndó* assertive for the language of the Bokala-Nkole, but his translation does not fit the Pagibete meaning, although the function is similar.

(72) ngòlù   ndó   à-gbùng-á   dɔ́ótè
     1a.pig PROC   SP₃ₛₒ\NARR-break.up-NARR   7.dirt
     'The pig breaks up the dirt.'

4.3.2 Non-verbal clauses

Pagibete has two types of clauses which do not require a verb or copula of any kind. These are descriptive clauses and equative clauses.

While descriptive clauses may take the form of an intransitive clause (see Section 4.3.1.1), a second form is often heard. This form consists of a noun phrase and one or more descriptors. The descriptor agrees with the subject noun.

(73) àbá-kò   ó-kyàkà
     1a.father-2sg   1a-tall
     'Your father (is) tall.'

The equative clause involves placing a noun phrase or pronoun next to a noun. As stated in Section 4.3.1.4, this type of clause indicates membership of the subject in the set given in the predicate nominal.

(74) yí   mángèná
     3sg   1a.owner
     'She (is) the owner.'

REFERENCES

de Boeck, L. B. (1949) Note sur la langue des Apakabete. *Bulletin de l'Institut Royal Colonial Belge* 20: 834–843.

Heine, B. (1980) Language Typology and Linguistic Reconstruction: The Niger-Congo Case. *Journal of African Languages and Linguistics* 2(2): 95–112.

Kadima, M. (1969) *Le système des classes en bantou*. Leuven: Vander.

Motingea Mangulu, A. (2002) Notes sur le parler des Bokala-Nkole (Territoire de Yahuma – RDC). *Afrikanistische Arbeitspapiere* 70: 5–53.

Welmers, W. E. (1973) *African Language Structures*. Berkeley: University of California Press.

CHAPTER 16

ZIMBA D26

Constance Kutsch Lojenga

1 INTRODUCTION

Zimba [zmb], called Binja by the native speakers and also known as Binja-Sud, is a Bantu language spoken in the Democratic Republic of the Congo classified as D26 (Guthrie 1971). The language is spoken in the Maniema province, in the Mulu and Maringa chiefdoms of the Kasongo territory. There are minor differences in speech between the inhabitants of these two chiefdoms.

Surrounding and closely related languages are first of all Songola, also known under the names Songoora or Binja-Nord D24. Two other names of related languages were mentioned by Zimba speakers: Kwange and Mamba, which are not cited as separate languages in the Ethnologue. Others may consider these as dialects of Zimba. Further research will have to reveal how close the relationship is between these two and Zimba.

The Ethnologue quotes 120,000 speakers as census figures from 1994 (Lewis *et al.* 2016). Now, 20 years later, the census figures of 2015 present a total of 293,184 inhabitants. The speakers estimate the number of Zimba in the diaspora (in Kindu, the capital of Maniema, and in other big cities in Congo, and worldwide) at least at 200,000, which means that the total may be around half a million, though not all those living in the diaspora are capable of speaking their original language any more, which is a big concern for those who do speak the language. The language is used vigorously in the home area and in Kindu.

Children learn to speak both Swahili and Zimba. Zimba is spoken at home as well as in many other traditional situations, like village meetings, and also in church services. The language of wider communication in the area is Swahili, which is also the language used for teaching in the first few years of the primary schools, followed by French.

The data presented in this chapter result primarily from a month-long language development workshop which I directed in the village of Kipaka in the Maringa chiefdom in June 2011. The workshop was organised by Malendela Kate-ke-Sombe, a native speaker of the language, and counted between 15 and 20 people, who actively participated, day after day, in collecting vocabulary (at present over 2,000 lexical items), checking vowels and consonants, and who contributed to the choice of pictures for an alphabet poster and a small booklet for teaching the vowels, going on from there to researching various parts of the nominal and verbal morphology, writing proverbs, riddles and folktales. Other Zimba speakers who have contributed to the data in this chapter are Mbutu Feruzi Kabemba and Sakodi Ndanga.

Documentation on the Zimba language can be divided into recent works and older documents. Recent works include Motingea Mangulu (1996), Mukubingwa (1997) and Kabungama Yuka (2006). Of these three authors, only Kabungama Yuka is a native speaker of the language. In all three, the seven-vowel system is well represented, but without paying attention to the details of the vowel-harmony and vowel-height-harmony systems. The present study, therefore, intends to introduce quite a lot of detail in this domain. Then there are some older documents on the language. I unearthed an anonymous, undated and untitled typed manuscript of 66 pages at the parish of the White Fathers in Kipaka. There is also an undated typed manuscript by R. Hennin (II + 119 pp.) conserved in the RMCA (Tervuren). Both wrote the language with five vowels, and have ignored the vowel-harmony system. These two documents use an older numbering system for the noun classes than is customary at present. There are also two handwritten undated manuscripts containing notes from Meeussen and Sebasoni and from Meeussen alone, written in a seven-vowel system, represented with the traditional vowel symbols of those days: į, i, e, a, o, u, ų. The vowel symbols are confusing for a language like Zimba which has a clear vowel-harmony contrast in the mid vowels (e/ɛ and o/ɔ) rather than in the high vowels (*i/ɪ and *u/ʊ), like various Bantu Zone D languages further north in Congo.

There are three main dialects: Kisémbɔ́mbɔ, Kisémolo, and Kɛsɔlɛ́. Like Kabungama Yuka's dissertation, this sketch is based on the Kisémolo variety, which is geographically in the centre, and which is considered "common" Binja (Kabungama Yuka 2006: 3). Particularly the vowel-height assimilation in prefixes (see 2.1.3) may only be a feature of the Kisémolo variety, unknown in the other dialects.

In addition to presenting an overview of the phonology and morphology of Zimba, the present sketch introduces not only the vowel inventory, but particularly the vowel-harmony system of the language in a descriptive way. Quite a bit of attention is also paid to the function of tone. Both in the lexicon and in the grammar, tone plays an important role. Finally, the double system of diminutives and augmentatives is an interesting phenomenon, a system which, by my knowledge, only Meeussen (1953) and Kabungama Yuka (2006) have briefly alluded to.

2 PHONOLOGY

In this section, I present the regular and symmetrical seven-vowel system (Section 2.1). There will be some discussion on vowel harmony, based on the feature RTR[1] in roots first of all. This type of vowel harmony as well as vowel-height assimilation are frequently found phenomena in the language, and details and examples of morphophonological changes are presented in the chapter about morphology, both nominal and verbal. The consonants are presented in Section 2.2. The language has a two-tone system (Section 2.3). The function of tone in the grammar is presented in the various subsections dealing with the morphological and syntactic structures in the language. The syllable structures of the language are treated in Section 2.4.

2.1 Vowels

2.1.1 Vowel contrasts

Zimba has a symmetrical vowel system consisting of seven contrastive vowels. The vowel system is typical for many Bantu Zone C languages and several languages classified in the Bantu Zone D (Leitch 1996: 9, Motingea Mangulu 1996: 85). The language has no contrast between short and long vowels. Any seemingly long vowels must be considered as two adjacent syllables of short vowels (see 2.4).

TABLE 16.1 THE ZIMBA VOWEL SYSTEM

	Front	Central	Back
high	i		u
mid	e		o
mid [RTR]	ɛ		ɔ
low		a	

The following three sets of examples prove the contrastivity of these seven vowels:[2]

* nouns with disyllabic roots in which $V_1 = V_2$
* nouns with monosyllabic roots
* monosyllabic verb roots

TABLE 16.2 CONTRAST BETWEEN THE SEVEN VOWELS

	Nouns $V_1 = V_2$		Monosyllabic noun roots		Monosyllabic verb roots	
i	ì-bìmbì	wave	mì-chì	roots	kù-líng-à	to deceive
e	lò-lémè	tongue	mè-té	trees	kò-bès-à	to bring
ɛ	lò-ményɛ́	open space	mɛ̀-tɛ̀	muscles	kɔ̀-tɛ́m-à	to farm
a	kè-wàwá	wing	kà-yá	fire	kò-kàk-à	to tie up
ɔ	n-gɔ́kɔ́	chicken	ì-bɔ̌	oil palm	kɔ̀-tɔ̀t-à	to choose
o	n-gòsò	parrot	mò-ndò	person	kò-chòk-à	to flee
u	ì-túmù	spear	ì-lú	knee	kù-tút-à	to pound

2.1.2 Static Vowel Harmony: within roots

The RTR contrast in the mid vowels gives rise to some distributional restrictions: within roots, the non-RTR mid vowels /e/ and /o/ are incompatible with the RTR mid vowels /ɛ/ and /ɔ/, whereas both sets can co-occur with the other three vowels, /i/, /a/, /u/. Some (near-) minimal pairs between /e/ and /ɛ/ are the following:

(1) ì-féfé   'snake, sp.'   ì-fɛ́fɛ́   'knot in a rope'
    kò-békà   'to keep, conserve'   kɔ̀-bɛ́kà   'to cut into pieces'
    kè-kwĕkwĕ 'firefly'   kà-kwɛ̆kwɛ̆ 'turkey, sp.'

Some (near-)minimal pairs between /o/ and /ɔ/ are the following:

(2) m-bóbó   'jackal, sp.'   m-bɔ̀bɔ̀   'locust, sp.'
    n-gókó   'grandchild'   n-gɔ́kɔ́   'chicken'
    n-góá   'tick (insect)'   n-gɔ́á   'snail, sp.'

The following are some minimal pairs between combinations of /e/ and /o/ on the one hand, and /ɛ/ and /ɔ/ on the other hand.

(3) lò-kòmbé   'drum'   lò-kòmbɛ̀   'vine, sp.'
    m-bèó   'speed (PL)'   m-bɛ̀ɔ́   'cold (N)'

2.1.3 Dynamic vowel harmony: across morpheme boundaries

Dynamic vowel harmony relates to vowel-harmony processes which occur across morpheme boundaries within words. In Zimba (and most likely in all other languages with a seven-vowel harmony system with two contrastive sets of mid vowels in Bantu Zones C and D), the root vowels never change. The vowel-harmony processes therefore take place in the prefixes and suffixes. The trigger which determines the vowels of the prefixes and suffixes is the adjacent vowel to the right (for prefixes) or to the left (for the suffixes), which may be a root or another prefix or suffix.

In Zimba suffixes, there is RTR harmony: the default vowels /e/ and /o/ change to the RTR vowels /ɛ/ and /ɔ/ when preceded by /ɛ/ or /ɔ/. Following any other root vowel (obviously /e/ and /o/, but also /a/, /i/, and /u/), they retain their default quality.

In the prefixes, however, a system of vowel-height harmony has developed out of the RTR harmony: the default vowels /e/ and /o/ change to the RTR vowels /ɛ/ and /ɔ/ when preceding the vowels /ɛ/ or /ɔ/, but they also change to /i/ and /u/ when preceding a high vowel /i/ or /u/. Preceding the root vowel /a/, the default form of the prefix is used: /e/ or /o/. Any affix with an underlying vowel /a/, /i/, or /u/ does not participate in vowel-harmony processes.

The first example shows vowel-height harmony in the prefixes. The noun class 7 prefix has three phonologically conditioned allomorphs: *kè-* ~ *kɛ̀-* ~ *kì-;* the class 15 infinitive prefix has three phonologically conditioned allomorphs: *kò-* ~ *kɔ̀-* ~ *kù-:*

(4)   kè-bàmbà   'chair'   kò-bèsà   'to bring'
       kɛ̀-kɔ́yɔ̀   'eyelid'   kɔ̀-témà   'to farm'
       kì-lúbá   'basket'   kù-língà   'to deceive'

The following example shows RTR harmony in suffixes: the Subjunctive has a verb-final vowel *-é*. This vowel changes to *-ɛ́* following the RTR vowels /ɛ/ or /ɔ/, and remains *-é* following all other vowels. The 3sɢ forms are given.

TABLE 16.3 VOWEL HARMONY IN SUFFIXES

Root vowel	Infinitive		Subjunctive 3sɢ	
i	kù-chìb-à	to close	ú-chìb-é	he should close
e	kò-két-à	to do	ó-kèt-é	he should do
ɛ	kɔ̀-mèn-à	to swallow	ɔ́-mèn-ɛ́	he should swallow
a	kò-kàmb-à	to cook	ó-kàmb-é	he should cook
ɔ	kɔ̀-lɔ̀l-à	to look	ɔ́-lɔ̀l-ɛ́	he should look
o	kò-kónd-à	to love	ó-kònd-é	he should love
u	kù-tút-à	to push	ú-tùt-é	he should pound

The following examples of the verb kɔ̀-lɔ̀l-à 'to look' (with the RTR root vowel /ɔ/) show that a prefix vowel which is subject to vowel harmony, i.e., which contains the vowel /e/ or /o/ underlyingly, bases the choice of its surface realisation exclusively on the adjacent vowel to the right. The prefix of the Perfective Present is *-só-*, and it will change to *-sɔ́-* preceding /ɔ/; the reflexive prefix is an invariable *-i-*, and causes a prefix to its left to assimilate for height: no longer *-só-*, but *-sú-*. The 3sɢ subject prefix also changes to *ú-*.

(5) ɔ́-sɔ́-lɔ̀l-à   'he looks'
     ú-sù-í-lɔ̀l-à   'he looks at himself'

The same verb will now be used with the Reciprocal extension -*an*-. Without this extension, the verb-final vowel of the Subjunctive (underlyingly -*é*) is realised -*ɛ́*, to harmonise
with the root vowel /ɔ/. However, with the extension -*àn*-, following the vowel /a/, it
changes to -*é*.

(6) tɔ́-lɔ̀l-ɛ́   'we should look'
     tɔ́-lɔ̀l-àn-é   'we should look at each other'

2.2 Consonants

Zimba has a relatively small inventory of contrastive consonants, displayed in the following table, presented in orthographic symbols, accompanied by the IPA symbols where
different.

TABLE 16.4 THE ZIMBA CONSONANT SYSTEM

		Bilabial	Alveolar	Palatal	Velar
voiceless	stops	p	t	ch [tʃ]	k
voiced	stop	b			
prenas.	stops	mb [ᵐb]	nd [ⁿd]	nj [ⁿdʒ]	ng [ᵑg]
voiceless	fricatives	f	s		
nasal	sonorants	m	n	ny [ɲ]	
oral	sonorants	w	l	y [j]	

- Except the voiced bilabial stop /b/, there are no other voiced stops.
- In some cases, /n/ and /ny/ are interchangeable: e.g., *mw-ènì* ~ *mw-ènyì* 'visitor,' and
  kà-nwà ~ *kà-nywà* 'mouth'
- The consonant /h/, as mentioned in Motingea Mangulu (1996: 87, 89) does not exist
  in my data. For each of his examples with /h/, I posit a vowel-initial root.

TABLE 16.5 H OR Ø

Motingea Mangulu (1996)	Kutsch Lojenga	
mɔ̀hémbɛ́	mɔ̀émbɛ́	'nose'
mòhéndé	mòéndé	'leg'

2.3 Tone

Zimba has a basic two-tone system: High (H) and Low (L). A High tone is marked by an
acute accent /á/; a Low tone is marked by a grave accent /à/. L and H can be combined
on one (short) syllable, with a resulting Rising (LH) or a Falling (HL) contour tone, /ǎ/
and /â/ respectively. In many cases, those contour tones are the result of the combination
of two tones on one syllable after a morphophonological process in which the vowels of

two adjacent syllables have merged to become one short syllable. There are also some instances of a LHL combination on one short syllable, which is always the result of a morphophonological process (see the case of L-tone prefixes with the adjective *-âchi* 'good' in 3.5). Other Rising or Falling contour tones are realised on two adjacent identical vowels, which represent two syllables. They are therefore interpreted as a sequence of two unlike tones.

One tonal process frequently attested in Zimba TAM paradigms is Meeussen's Rule, in which a High tone changes to Low. The tonal environments in which Meeussen's Rule operates have not yet been precisely identified. It is clear that sequences of H tones resulting from at least more than one underlying H tone are possible (see the Subjunctive negative of the H-tone verb, Table 16.40 in 5.3.3.2). Two subsequent underlying H tones also seem possible. A clear environment in which H -> L is attested when the H tone is both preceded and followed by another H tone. In that case, the H tone in the middle changes to L (see the Subjunctive affirmative for H-tone verbs, Table 16.39 in 5.3.3.2). There are also instances in which two H tones following a H tone will change to L (see the Remote Past 3SG/PL (CL.1/2) of the H-tone verb, Table 16.43 in 5.3.3.5).

Verbs are divided into two tone classes: H-tone verbs and L-tone verbs. In the infinitive, the H tone of H-tone verbs is realised on the first root vowel, and L-tone verbs carry a L tone on the first root vowel. The following are some examples of lexical minimal pairs between verb infinitives:

(7) L-tone verbs   H-tone verbs
     kòtàndà   'to climb'   kòtándà   'to close'
     kɔ̀pὲpà   'to winnow'   kɔ̀pέpà   'to cut'
     kɔ̀sὲà   'to hunt'   kɔ̀séà   'to vomit'

On nouns with disyllabic roots, all four tonal melodies consisting of combinations of L and H tones are found. The prefixes all carry a L tone. The following are examples of class 5 nouns.

(8) H.H   ì-sébέ   'joke'   H.L   ì-sɔ́kì   'saliva'
     L.L   ì-sòmbà   'boy'   L.H   ì-sìkí   'dowry'

The nominal system counts an extremely high number of tonal minimal pairs, in part due to the extensively used double system of diminutives and augmentatives (see Section 3.3.1). The following lexical minimal triplet may serve as example:

(9) L.L.L   mòlàngà   'spear shaft'
     L.L.H   mòlàngá   'fish trap'
     L.H.H   mòlángá   'discussion'

Besides the heavy functional load of tone in the lexicon, tone also has a contrastive function in various domains of the grammar: in the pronominal system (3.12) and in subject and object prefixes (5.3.2). In the verbal inflectional system, it differentiates between immediate and remote past, both in affirmative and in negative forms (5.3.3.4, 5.3.3.5, and 5.3.3.7) as well as between several affirmative and negative verb forms in the present continuous (5.3.3.8). Finally, tone may be the sole means of distinguishing relative clauses from regular main clauses (3.10).

2.4 Syllable Structure

Zimba has only short, open syllables: with a consonantal onset, CV, or with a vocalic onset, V, in word-initial and word-medial position.

At first sight, it looks as if the language has many long vowels. However, these must be considered as juxtaposed syllables consisting of two identical vowels. This may happen root-internally or between prefix and root.

TABLE 16.6 JUXTAPOSED SYLLABLES WITH IDENTICAL VOWELS

Two identical vowels root-internally		*Two identical vowels across morpheme boundaries*	
ì-káá	crab	lò-ómbé	vapour
lò-bèèkèsè	cockroach	bà-ámbì	cooks (n, PL)
lù-búú	monkey, sp.	mè-èmò	works (n, PL)

Different morphophonological processes may take place between prefixes and roots. On the one hand, there is glide formation and elision; on the other, there are also many examples of heterosyllabification, whereby the two vowels are juxtaposed without any further modification. It is not predictable which words will display glide formation and which words choose heterosyllabification. There are a few examples of free variation between the two.

An example of glide formation in the singular and elision in the plural, whereby the L tone of the prefix and the H tone of the root merge into a LH rising contour.

(10) mò-ánà   ->   mw-ǎnà   'child'
      bà-ánà   ->   b-ǎnà   'children'

Heterosyllabification is exemplified in the following example:

(11) mò-ànì   ->   mò-ànì   'counselor'
      bà-ànì   ->   bà-ànì   'counselors'

In the following example, both possibilities are found in free variation.

(12) mw-ǎkà / my-ǎkà   'year(s)'
      mò-ákà / mè-ákà   'year(s)'

With verbs, no glide formation takes place; only heterosyllabification:

(13) ko-émb-a   'to sing'
      ko-amb-a   'to destroy (tr)'

Some onsetless syllables are the result of historical consonant loss. Quite often, people vary in their pronunciation of a word with or without the lateral consonant /l/.

(14) mè-lòngé   ~   mè-òngé   'sunlight'
      lò-lémè   ~   lò-émè   'tongue'
      -kàtàà   ~   -kàtàlà   'big' (ADJ)

3 NOMINAL MORPHOLOGY

3.1 Introduction

Zimba has a noun-class system consisting of 19 noun classes. Class 15 contains all verb infinitives, and only one noun: kɔ̀bɔ́kɔ̀ / màbɔ́kɔ̀ 'arm(s)' 15/6. There are no nouns in the locative classes, 16–18,[3] but the original class markers are found in some locative adverbs, and they may be used as pre-prefix with any noun, marking a location. Vowel harmony applies, as with all other prefixes. Locative agreement marking does exist but is not frequently used.

There are four series of prefixes: nominal prefixes (NP), adjectival prefixes (AP), pronominal prefixes (PP) and subject prefixes (SM). The adjectival prefixes are only used for a very small set of qualifying adjectives. They differ from the nominal prefixes only in classes 5, 9 and 10 (see 3.5). The pronominal prefixes are used for a few quantifiers (3.5), numerals (3.6), demonstratives (3.7), the connective construction (3.8) possessives (3.9) and relative clauses (3.10). The subject prefixes (5.3.2) are segmentally the same as the pronominal prefixes, but tonally different in classes 1, 4, 9, and 18, where the pronominal prefix has a L tone, and the subject prefix a H tone.

The series of prefixes are presented in their underlying forms in Table 16.7.

TABLE 16.7 ZIMBA CLASS PREFIXES

Noun class	Noun-class px (NP)	Adjectival px (AP)	Pronominal px (PP)	Subject px (SM)
1	mò-	mò-	ò-	ó-
2	bà-	bà-	bá-	bá-
3	mò-	mò-	ó-	ó-
4	mè-	mè-	è-	é-
5	ì-	lè	lé-	lé-
6	mà-	mà-	má-	má-
7	kè-	kè-	ké-	ké-
8	bì-	bì-	bí-	bí-
9/9a	N-/Ø-	mè-	è-	é-
10/10a	N-/Ø-	chi-	chì-	chí-
11	lò-	lò-	ló-	ló-
12	kà-	kà-	ká-	ká-
13	tò-	tò-	tó-	tó-
14	bò-	bò-	bó-	bó-
15	kò-	kò-	kó-	kó-
16	à-	à-	á-	á-
17	kò-	kò-	kó-	kó-
18	mò-	mò-	mò-	mó-
19	sì-	sì-	sí-	sí-

The class 5 prefix *i*- has an allomorph *ch*- before vowel-initial roots. Class 5a is introduced for a small number of mostly loanwords which take *li*- as their basic prefix. Classes 9 and 10 consist of homorganic prenasalisation, thereby voicing any underlyingly voiceless obstruent, resulting in a voiced prenasalised obstruent. Classes 9a and 10a contain a number of loanwords, which take a zero prefix before voiceless consonants, the lateral /l/, and some vowel-initial stems. In addition, the language has a constraint against monosyllabic nouns. Whenever this occurs, the vowel *i*- is added to the prefix *N*- or *Ø*-, resulting in a prosodically conditioned allomorph *iN*- or *i*-, in order to make the word disyllabic.

3.2 Nominal prefixes

This section contains an overview of the noun-class system with all its prefixes.

TABLE 16.8 ZIMBA NOUN CLASS PREFIXES

Class	Prefixes	Noun	English gloss
1	mò-	mò-ndò	person
2	bà-	bà-ndò	people
3	mò-	mò-té	tree
4	mè-	mè-té	trees
5	ì-	i-bòbè	spider
5a	lì-	lì-kèmbé	finger piano
6	mà-	mà-bòbè	spiders
7	kè-	kè-kàngà	mat
8	bì-	bi-kàngà	mats
9	N-	n-gémà	monkey
9a	Ø-	Ø-sóngé	moon
10	N-	n-gémà	monkeys
10a	Ø-	Ø-fîsì	small lions
11	lò-	lò-chókè	bee
12	kà-	kà-péndé	spear
13	tò-	tò-péndè	spears
14	bò-	bò-làmbà	traditional dance
15	kò-	kò-bàkà	to come
16	à-	à-mù-kíndá	at the door
17	kò-	kò-mà-síò	in the fields
18	mò	mò-n-dábò	inside the house
19	sì-	sì-nyɔ̀nyì	small bird

Table 16.9 contains a few examples of the prosodically conditioned allomorphs *-iN-* and *i-*, used with monosyllabic roots in classes 9 and 10.

TABLE 16.9 PROSODICALLY CONDITIONED CLASS 9/10 ALLOMORPHS

cl 9	cl 10	
ì-m-bwá	ì-m-bwá	dog
ì-Ø-sɛ́	ì-Ø-sɛ́	wild goat
cl 11	cl 10	
lò-pwé	ì-m-bwé	white hair
lò-swá	ì-Ø-swá	flying termite

Class pairings are very much standard as in other Bantu languages: 1/2, 3/4, 5(a)/6, 7/8, 9(a)/10(a), 11/10(a), 12/13, 14/(6), 15/(6), and 19/13. Class 19 is exclusively used for the diminutive derivation (3.3.1).

3.3 Diminutives and augmentatives

Zimba has a rich system of class-changing derivation for diminutives and augmentatives. For each of these, there are two strategies to form the singular with a common plural form for these two different diminutive and augmentative singular classes.

3.3.1 Diminutives

There are two degrees of diminutives in Zimba. First-degree diminutives are formed in class 12, *kà-*, and are used for small items; second-degree diminutives are formed in class 19, *sì-*, and are used to mark nouns as extra small, a fact also noted by Meeussen (1953: 389) and Kabungama Yuka (2006). There is only one plural form for both: class 13, *to-*. The meaning of these diminutives can be neutral and only refer to size, small and very small, or, according to the context, there may be negative connotations, especially with the class 19 diminutives. The language already has quite a lot of nouns which use the class 12/13 prefixes in their basic form. These have only one option for their diminutives, namely the class 19 prefix, and they have no separate form for the plural diminutive.

TABLE 16.10 DIMINUTIVE FORMATION

Classes	Noun SG	Noun PL	Dim SG 1	Dim SG 2	Dim PL	Gloss
			cl 12	cl 19	cl 13	
1/2	mò-káchí	bà-káchí	kà-káchí	sì-káchí	tò-káchí	woman
3/4	mò-té	mè-té	kà-té	sì-té	tò-té	tree
5/6	i-túmù	mà-túmù	kà-túmù	sì-túmu	tù-túmù	lance
7/8	kè-kòmbé	bi-kòmbé	kà-kòmbé	sì-kòmbé	tò-kòmbé	drum
9/10	n-dòsó	n-dòsó	kà-tòsó	sì-tɔsó	tò-tòsó	banana
11/10	lò-lémè	n-démè	kà-lémè	sì-lémè	tò-lémè	tongue
12/13	kà-lyá	tò-lyá	–	sì-lyá	–	village
14/6	bɔ̀-lɔ̀lèɔ̀	mà-lɔ̀lèɔ̀	kà-lɔ̀lèɔ̀	sì-lɔ̀lèɔ̀	tɔ̀-lɔ̀lèɔ̀	window
15/6	kɔ̀-bókɔ̀	mà-bókɔ̀	kà-bókɔ̀	sì-bókɔ	tò-bókɔ̀	arm

3.3.2 Augmentatives

Like for the diminutives, there are also two forms for the augmentatives, marking two degrees of augmentation, which may refer to size, but more often have a negative connotation or indicate an abnormal situation. The first augmentative is found in class 7, *ke-*, with its plural in class 8, *bi-*. The second degree of augmentative is formed in class 11, *lo-*. If a plural form is needed, it will use the class 8 prefix *bi-* as well. For class 7/8 nouns, there is only one form for the augmentative, the one in class 11; nouns of class 11 can only make augmentatives with the prefixes of classes 7/8.

TABLE 16.11 AUGMENTATIVE FORMATION

Classes	Noun SG	Noun PL	Aug SG 1	Aug SG 2	Aug PL	Gloss
			cl 7	cl 11	cl 8	
1/2	mw-ănà	b-ănà	ky-ănà	lw-ănà	by-ănà	child
3/4	mò-kɔ̀kɔ̀	mè-kɔ̀kɔ̀	kè-kɔ̀kɔ̀	lò-kɔ̀kɔ̀	bi-kɔ̀kɔ̀	sheep
5/6	i-túmù	mà-túmù	ki-túmù	lù-túmù	bi-túmù	lance
7/8	kè-kòmbó	bi-kòmbó	–	lò-kòmbó	–	hat
9/10	n-dòsó	n-dòsó	kè-tòsó	lɔ̀-tòsó	bi-tòsó	banana
11/10	lò-lémè	n-démè	kè-lémè	–	bi-lémè	tongue
12/13	kà-lyá	tò-lyá	kè-lyá	lò-lyá	bi-lyá	village
14/6	bɔ̀-lɔ̀lèɔ̀	mà-lɔ̀lèɔ̀	kè-lɔ̀lèɔ̀	lò-lɔ̀lèɔ̀	bi-lɔ̀lèɔ̀	window
15/6	kɔ̀-bókɔ̀	mà-bókɔ̀	kè-bókɔ̀	lò-bókɔ̀	bi-bókɔ̀	arm

3.4 Deverbative nouns

In the following subsections, two types of deverbative nouns are treated: agentive nouns, in classes 1/2, and abstract nouns, in class 14.

3.4.1 Agentive nouns

Agentive nouns are formed by a verb root preceded by the class 1/2 prefixes, and followed by a derivational suffix -ì. No changes are attested in the final consonant of the verb. The tonal melody of the verb, H or L, is retained on the agentive noun. Semantically, an agentive noun can mark an incidental or a continued or habitual action, like a profession.

TABLE 16.12 AGENTIVE NOUNS

verb	gloss	Agentive N sg	Agentive N pl	gloss
kò-àn-à	to advise	mò-àn-ì	bà-àn-ì	adviser
kò-émb-à	to sing	mò-émb-ì	bà-émb-ì	singer
kɔ̀-tɛ́m-à	to cultivate	mɔ̀-tɛ́m-ì	bà-tɛ́m-ì	farmer

3.4.2 Abstract nouns

Abstract nouns can be derived from verbs by a productive process of creating class 14 nouns, also with a derivational suffix -ì. The verb root keeps its original H or L tone.

TABLE 16.13 ABSTRACT NOUNS

kù-líng-à	to deceive	bù-líng-ì	deceit
kò-kwách-à	to help	bò-kwách-ì	help
kɔ̀-lɔ́b-à	to fish	bɔ̀-lɔ́b-ì	fishing

3.5 Adjectives

Only eight adjectives have been found in a corpus of over 2,000 words: four qualifiers and four quantifiers. The qualifying adjectives show agreement with the head noun by the use of an adjectival prefix. The quantifying adjectives take agreement markers from the set of pronominal prefixes. Other adjectival concepts are expressed by connective constructions (those marking colours, and a number of concepts which are derived from stative verbs).

TABLE 16.14 INVENTORY OF ADJECTIVES

Qualifying adjectives		Quantifying adjectives	
-kàtàà	big	-sɛ̀	each, all
-kɛ́ɛ̀kɛ́	few	-kìnɛ́	other
-chîniní	small	-àbùá	many
-âchí	good	-mɔ́	a certain

Table 16.15 illustrates all four qualifying adjectives in the three classes (5, 9, and 10) in which the adjectival prefix differs from the nominal prefix. The class 9 adjectival prefix is identical to that of class 4.

TABLE 16.15 ADJECTIVES WITH AP

Class	Noun	-kàtàà 'big'	-kéèké 'few'	-chìníní 'small'	-âchí 'good'	Gloss
5	ì-làngì	lè-kàtàà	lè-kéèké	lì-chìníní	ly'-âchí	guinea fowl
9	n-gɔ́kɔ́	mè-kàtàà	mè-kéèké	mì-chìníní	my'-âchí	chicken
10	n-gɔ́kɔ́	chì-kàtàà	chì-kéèké	chì-chìníní	ch'-âchí	chicken

In the example below, the pronominal prefixes are used to show agreement with the nouns. The prefixes for classes 1 and 3 are segmentally the same but tonally different. The prefixes for classes 4 and 9 are segmentally and tonally identical.

TABLE 16.16 ADJECTIVES WITH PP

Class	Noun	-sè 'each, all'	-kìnɛ́ 'other'	-àbùá 'many'	
1	mò-ndò	ɔ̌-sè	ù-kìnɛ́		person
3	mò-té	ɔ́-sè	ú-kínɛ́		tree
4	mè-té	ɛ̌-sè	i-kìnɛ́	y-àbùá	trees
9	n-gɔ́kɔ́	ɛ̌-sè	i-kìnɛ́		chicken

3.6 Numerals

The numerals 1 to 5 take pronominal prefixes, agreeing with the class of the head noun of the noun phrase. In their citation forms, they take the prefixes of classes 12/13. The numerals from 6 to 9 are invariable and must be considered as nouns.

TABLE 16.17 NUMERALS 1–9

1	kú mà	6	mòtóbá (cl.3)
2	tó-béé	7	mòsàmbò (cl.3) or mòtóbá nà mwènjì
3	tó-sátò	8	mòàndà (cl.3) or kìnănɛ̌ (cl.7)
4	tó-nàchì	9	kètémá (cl.7) or ìbwá (cl.5)
5	tó-tánù		

The numerals 10, 100, and 1000 are nouns and have singular and plural forms.

TABLE 16.18 NUMERALS 10, 100, 1000

	singular	plural	
10	ì-kúmì	mà-kúmì	classes 5/6
100	lò-kámà	n-gámà	classes 11/10
1000	kà-nùnù	tù-nùnù	classes 12/13

Table 16.19 shows the nouns of different classes followed by the numerals 1 to 5 with their agreement prefixes, including RTR harmony. Height harmony is not relevant, since none of these five numerals has a high vowel as first root vowel.

TABLE 16.19 NUMERAL AGREEMENT

Noun		-mɔ̀ '1'	-béé '2'	-sátò '3'	-nàchì '4'	-tánù '5'	Gloss
1	mò-ndò	ɔ́-mɔ̀					person
2	bà-ndò		bá-béé	bá-sátò	bá-nàchì	bá-tánù	people
3	mò-té	ɔ́-mɔ̀					tree
4	mè-té		è-bèé	è-sátò	è-nàchì	è-tánù	trees
5	ì-làngì	lέ-mɔ̀					guinea fowl
10	n-dɔ̀sɔ́		chí-béé	chí-sátò	chí-nàchì	chí-tánù	bananas

3.7 Demonstratives

Zimba has two main sets of demonstratives which relate to distance: close and far. The final vowel of the set which points to far-away participants can be lengthened to point to something very far away, still visible, but nearly out of sight. The third set in the table below contains referential demonstratives: they refer to a participant who is absent, but which both speaker and hearer know and are referring to.

The demonstrative root for the first set of demonstratives has two different forms, based on whether the pronominal prefix consists of a CV-syllable or a V-syllable. In the first case, it consists of the noun-class prefix (L tone), which is preceded by the pronominal prefix, e.g., *bá-bà* (cl. 2), *ké-kè* (cl. 7) etc. However, when the concord marker consists of a V-syllable (classes 1, 3, 4 and 9), the demonstrative root is *-nà*, preceded by the pronominal prefix as class concord: *ŏnà* (cl. 1, 3), *ĕnà* (cl. 4, 9). In spite of the fact that the pronominal prefixes for classes 1 and 3 are tonally different, their demonstratives are tonally identical. There is as yet no explanation for this.

The demonstrative root for the second set of demonstratives is *-á* when preceded by a pronominal prefix with a CV-syllable, and *-lyá* when preceded by a pronominal prefix consisting of a vowel only. However, the Kenyambombo speech variety has all forms with the

TABLE 16.20 DEMONSTRATIVES

	Noun	Gloss	DEM set 1 close-by	DEM set 2 far away	DEM set 3 referential
1	mò-ndò	person	ŏ-nà	ŏ-lyá	ɔ̌-y-ɔ̀
2	bà-ndò	people	bá-bà	bâ-á	bá-b-ɔ̀
3	mɔ̀-kɔ̀kɔ́	sheep	ŏ-nà	ŏ-lyá	ɔ̌-y-ɔ̀
4	mè-kɔ̀kɔ́	sheep (PL)	ĕ-nà	ĕ-lyá	ɛ̌-y-ɔ̀
5	ì-làngì	guinea fowl	lé-lè	lê-á	lέ-ly-ɔ̀
6	mà-làngì	guinea fowl (PL)	má-mà	mâ-á	má-m-ɔ̀
7	kè-kàngà	mat	ké-kè	kê-á	ké-ky-ɔ̀
8	bì-kàngà	mats	bí-bì	bî-á	bí-by-ɔ̀
9	n-jèá	road	ĕ-nà	ĕ-lyá	ɛ̌-y-ɔ̀
10	nyɔ̀nyì	bird	chí-chì	chî-á	chí-ch-ɔ̀
11	lò-bàò	knife	ló-lò	lô-á	lɔ̀-l-ɔ̀
12	kà-mbàsà	scorpion	ká-kà	kâ-á	ká-k-ɔ̀
13	tò-mbàsà	scorpions	tó-tò	tô-á	tɔ́-t-ɔ̀
14	bò-ótà	honey	bó-bò	bô-á	bɔ́-b-ɔ̀
15	kò-bɔ́kɔ̀	arm	kó-kò	kô-á	kɔ́-k-ɔ̀
19	sì-nyɔ̀nyì	small bird	sí-sì	sî-á	sí-sy-ɔ̀

demonstrative root *-lyá*. The vowel of the root *-(ly)á* can be lengthened: *-lyáá* to mark further distance, pointing to something which is so far away that it is hardly visible any more.

Finally, the set of referential demonstratives refers to a participant which is not visible, but both speaker and hearer know what or whom they are referring to. It is similar to the first set with the addition of a final *-ɔ* and the application of morphophonological processes, including vowel harmony. In addition, in classes 1, 3, 4 and 9, the demonstrative root consists of *-y-*.

3.8 Connective construction

The connective construction consists of two nouns linked by a connective marker. Underlyingly, the connective marker consists of the default mid vowel *-é*, with an [RTR] allomorph *-ɛ́*. The choice between the two is determined by the vowels of N$_2$. When N$_2$ contains [RTR] vowels /ɛ, ɔ/, the allomorph *-ɛ́* is used; in all other cases, the allomorph *-é* is chosen. This is different from the vowel harmony as it operates with the nominal prefixes and therefore a reason not to consider the connective marker a prefix, but rather a proclitic. The connective marker is preceded by a pronominal prefix, agreeing with the class of N$_1$.

Meeussen (1953: 389) notes that in a number of languages in the area, there is a "latent" front vowel, something which I would interpret as an augment or preprefix. Synchronically, Zimba does not have any trace of augments on nouns. However, Meeussen's thought sounds very plausible, i.e., that all nouns had a high-toned pre-prefix vowel *í-*, or *é-*. This would account for the vowel, *e-* (or *ɛ-* preceding RTR nouns) being a merger or coalesced form of *a + i* or *a + e*, as well as for the H-tone element, which combines with the L tone of the agreement marker in classes 1, 4, 9, and 18.

In the following examples, N$_1$ is the same in both columns, hence also the agreement markers: *mwǎnà* 'child,' *ìbwè* 'stone' and *ndɔ̀sɔ́* 'bananas.' For N$_2$, I have chosen ngòsò 'parrot,' with non-RTR mid vowels, and ngɔ́kɔ́ 'chicken' with the RTR mid vowels, to show the alternation in the vowel of the connective marker.

TABLE 16.21 CONNECTIVE CONSTRUCTION

class		of	parrot	class		of	chicken
1	mwǎnà	wě	ngòsò	1	mwǎnà	wě	ngɔ́kɔ́
5	ìbwè	lyé	ngòsò	5	ìbwè	lyɛ́	ngɔ́kɔ́
10	ndɔ̀sɔ́	ché	ngòsò	10	ndɔ̀sɔ́	chɛ́	ngɔ́kɔ́

3.9 Possessive pronouns

Possessive pronouns are formed by a possessive root, preceded by an agreement marker which consists of the pronominal prefix. The possessive pronoun roots for participants and classes 1 and 2 are the following:

TABLE 16.22 POSSESSIVE PRONOUN ROOTS

	singular	plural
1	-ánɛ́	-ító
2	-àò	-ínú
3	-ákɛ́	-ábɔ́

Table 16.23 presents some examples of nouns followed by possessive pronouns.

TABLE 16.23 POSSESSIVE PRONOUNS

		-áné	-àò / -áò	-ákɛ́	-ító	-ínú	-ábɔ́
		'my'	'your'	'his/her'	'our'	'your PL'	'their'
1	mw-ǎnà	w-ǎné	w-àò	w-ǎkɛ́	w-ǐtó	w-ǐnú	w-ǎbɔ́
3	mò-té	w-áné	w-áò	w-ákɛ́	w-ító	w-ínú	w-ábɔ́
5	i-làngì	ly-áné	ly-àò	ly-ákɛ́	l-ító	l-ínú	ly-ábɔ́

Two kinship terms have possessive pronouns incorporated in the noun, which does not exist without a possessive pronoun. The possessive roots are the same as the ones presented in the table above.

TABLE 16.24 KINSHIP TERMS WITH POSSESSIVE PRONOUNS

is-á 'father of'		in-á 'mother of'	
is-áné	my father	in-áné	my mother
ìs-áò	your father	ìn-áò	your mother
ìs-ákɛ́	his/her father	in-ákɛ́	his/her mother
ìs-ító	our father	in-ító	our mother
is-ínú	your (PL) father	in-ínú	your (PL) mother
ìs-ábɔ́	their father	in-ábɔ́	their mother

Possessive marking for possessives of the other noun classes is expressed by a construction of PP (referring to the possessed item), followed by the possessive marker -á-, followed by the PP (referring to the class of the possessor) followed by -ó (15–16).

(15) mò-té   ně   bɛ̀-sɛ̀sí   by-á-w-ó
     3-tree   and   8-leaves PP_8-POSS-PP_3-ó
     'the tree with its leaves'

(16) tò-péndé ně   mè-té   y-ǎ-tw-ó
     13-spears and 4-trees PP_4-POSS-PP_{13}-ó
     'the spears and/with their (wooden) shafts'

3.10 Relative clauses

Since relative clauses are modifiers to the head of the noun phrase, they are treated at this point in the chapter. The following remarks specifically concern subject relative clauses, which take the pronominal prefix as subject to the verb in order to mark agreement with the head noun. In a regular main clause, however, the subject prefix is used in the same position. This situation leads to a tonal contrast between the subject prefix in the matrix clause and the pronominal prefix in the relative clause in classes 1, 4, 9 and 18, as shown in the two sets of examples below.

(17) mw-ǎnà   ó-kách-éè
     1-child SM_1-play-RP
     'the child played.'

(18) mw-ănà   ò-kách-éè
      1-child   PP₁-play-RP
      'the child that played . . .'

(19) n-gémà   é-sò-ké-à
      9-monkey   SM₉-PFV.PRS-fall-FV
      'the monkey falls.'

(20) n-gémà   è-só-ké-à
      9-monkey   PP₉-PFV.PRS-fall-FV
      'the monkey that falls . . .'

In most classes, the pronominal prefix and the subject prefix are tonally identical. In those cases, the prefix to a main clause and a relative clause are identical, but there may be other tonal alternations in the verb form, which have to do with the application of Meeussen's Rule.

(21) mà-é   má-sò-ké-à
      6-egg   SM₆-PFV.PRS-fall-FV
      'The eggs fall.'

(22) mà-é   má-só-kè-à
      6-egg   PP₆-PFV.PRS-fall-FV
      'the eggs which fall . . .'

3.11 Word order in the noun phrase

Word order in the noun phrase seems to be quite flexible. Numerals, adjectives, possessives and connectives generally follow the noun, but they can be used in free ordering internally, with some differences in meaning, as shown in the following examples.

(23) mà-ábò   má-tánù   mà-kàtàà   or   mà-ábò   mà-kàtàà   mà-tánù
      6-house   PP₆-five   AP₆-big   6-house   AP₆-big   PP₆-five
      'five big houses'

All modifiers except regular adjectives can also precede the noun, particularly to mark contrastive emphasis, as in the second of each of the following sets of examples.

(24) mà-ábò   má-tánù
      6-house   PP₆-five
      'five houses'

(25) má-tánù   mà-ábò
      PP₆-five   6-house
      'five houses (not six, or seven)'

(26) n-dábò   y-ănɛ́
      9-house   PP₉-my
      'my house'

(27) y-ăné   n-dábò
PP$_9$-my   9-house
'my house (and not yours)'

(28) mà-ábò   má-mà
6-house   PP$_6$-DEM1
'these houses'

(29) má-mà   mà-ábò
PP$_6$-DEM1   6-house
'these houses (and not those)'

(30) n-dábò   y-ě   mú-chó   w-ăné
9-house   PP$_9$-CON   1-uncle   PP$_1$-my
'the house of my uncle'

(31) y-ě   mú-chó   w-ăné   n-dábò
PP$_9$-CON   1-uncle   PP$_1$-my   9-house
'the house of my uncle (and not somebody else's)'

3.12 Pronouns

The full set of independent pronouns, which may serve to replace nouns of all different classes, is presented in this section. The second and third person singular are differentiated by tone alone. There are also pronominal forms for the other noun classes. They are formed in a regular way by an initial vowel *à-* followed by the pronominal prefix and the vowel *-ó*. The independent pronominal forms for the locatives are not provided, since there are no regular nouns in these classes. If they do exist, they will be very rare.

TABLE 16.25 INDEPENDENT PRONOUNS

1SG	mèé	I
2SG	àwè	you
cl.1	àwέ	he/she
1PL	bàé	we
2PL	bănú	you
cl.2	àbɔ́	they
cl.3	àwó	it
cl.4	àyó	they
cl.5	àlyó	it
cl.6	àmó	they
cl.7	àkyó	it
cl.8	àbyó	they
cl.9	àyó	it
cl.10	àchó	they
cl.11	àlwó	it
cl.12	àkó	it
cl.13	àtwó	they
cl.14	àbwó	it
cl.15	àkwó	it
cl.19	àsyó	it

4 PREPOSITION

Zimba has a general preposition *nà*, used with comitative (32) as well as instrumental meaning (33). It is realised as *nà* when followed by a proper name (including the animals in a folktale), an inalienably possessed kinship term or an independent pronoun. On the other hand, it surfaces as *ně ~ ně* when followed by a common noun. Both the vowel and the LH contour tone are unusual. This could be seen synchronically as a separate preposition, but historically as the same preposition *na* in which there is vowel coalescence with the latent front-vowel augment, as proposed by Meeussen (1953) (see also Section 3.8 on the connective construction). We could therefore posit that this preposition consists of a merger of the preposition *nà* with a high-toned pre-prefix vowel *$\ast i$-, or *$\ast é$-, which no longer surfaces preceding nouns.

(32)   Kìkúù nà Lwèlèmè 'Turtle and Monitor Lizard'
        mwǎnà nà ìsáké 'the child and/with his father'

(33)   ně lòbàò 'with a knife'
        ně kèkómbò 'with a broom'

5 VERBAL MORPHOLOGY

5.1 Introduction

In this section, the Infinitive is presented first, as basis for the following subsections. In Section 5.2, verbal derivational categories are introduced and Section 5.3 focusses on the inflectional morphology.

Infinitives consist of a verb stem, preceded by the class 15 prefix *kò-* and followed by the verb-final suffix *-à*. The verb stem generally consists of a monosyllabic root (one vowel, optionally preceded and followed by a C), which, in turn, may be followed by one or more verbal extensions. The root carries the lexical tone belonging to the verb, H or L. The prefix and the verb-final vowel carry a L tone.

TABLE 16.26 VERB INFINITIVES

-CVC-a	
kò-kónd-à	to love
kò-mèn-à	to swallow

Infinitives can take an object prefix following the infinitive prefix and preceding the verb stem, as in the following examples (for a list of object prefixes, see 5.3.3.2 below).

TABLE 16.27 VERB INFINITIVES WITH OBJECT PREFIX

Verb infinitive	With object prefix	Meaning
kò-kónd-à	kò-<u>mò</u>-kónd-à	'to love him/her'
kò-lòl-à	kò-<u>bá</u>-lòl-à	'to see them'
kù-súkù-à	kù-<u>n</u>-júkù-à	'to push me'

5.2 Derivational morphology

5.2.1 Introduction

Verbal derivations are formed by a suffix following the verb stem. In this section, I present an overview of the morphology of the following verbal derivational categories: Passive, Applicative, Causative, Stative Passive, Reciprocal, Reversive and Habitual. It is likely that the language has several more derivational categories, but these are less productive and it is therefore harder to find suitable examples. In addition, I will only present the main strategy for forming these derivations. For most of them, there are different allomorphs in certain environments, as well as exceptional cases. These are not shown in the sections below.

5.2.2 Passive

The passive is regularly formed by the extension -w following the verb stem.

TABLE 16.28 PASSIVE DERIVATION

kɔ̀-mèn-à	to swallow	kɔ̀-mèn-w-à	to be swallowed
kù-líng-à	to deceive	kù-líng-w-à	to be deceived

5.2.3 Applicative

The applicative extension consists of the morpheme -è ~ -èl. When the verb stem ends in a consonant, the simple vowel allomorph -è is used; when the verb stem ends in a vowel, there is a strong tendency to use the allomorphs with the consonant /l/, so as to avoid sequences of three vowels.

TABLE 16.29 APPLICATIVE DERIVATION

Root vowel	Basic verb		Extension – e ~ -ɛ ~ -el ~ -ɛl	
o	kò-òl-à	to buy	kò-mò-òl-è-à	to buy for him/her
ɛ	kɔ̀-tém-à	to cultivate	kɔ̀-mɔ̀-tém-è-à	to cultivate for him/her
a	kò-byá-à	to plant	kò-mò-byá-èl-à	to plant for him/her
ɛ	kɔ̀-bɛ́-à	to break	kɔ̀-mɔ̀-bɛ́-èl-à	to break for him/her

5.2.4 Causative

The causative is formed with the extension -èky.

TABLE 16.30 CAUSATIVE DERIVATION

Root vowel	Basic verb		Causative	
e	kò-bék-à	to keep	kò-bék-èky-à	to cause to keep
ɔ	kɔ̀-kɔ́mb-à	to sweep	kɔ̀-kɔ́mb-ɛ̀ky-à	to cause to sweep

5.2.5 Stative passive

For the stative-passive extension, the morpheme -èk is used. Stative-passive verbs are intransitive.

TABLE 16.31 STATIVE-PASSIVE DERIVATION

Root vowel	Verb	Gloss	Stative passive	Gloss
o	kò-bót-à	to give birth	kò-bót-èk-à	to be born
ɛ	kɔ̀-bɛ́nj-à	to break	kɔ̀-bɛ́nj-èk-à	to be broken

5.2.6 Reciprocal

The reciprocal extension has only one form, -àn, which is used after all vowels in consonant-final and vowel-final verb stems.

TABLE 16.32 RECIPROCAL DERIVATION

Root vowel	Verb	Gloss	Reciprocal	Gloss
e	kò-chéb-à	to know	kò-chéb-àn-à	to know e.o.
a	kò-àmb-à	to pay	kò-àmb-àn-à	to pay e.o.

5.2.7 Reversive

The Reversive extension consists of the morpheme -ò.

TABLE 16.33 REVERSIVE DERIVATION

Root vowel	Verb	Gloss	Reversive	Gloss
a	kò-kàk-à	to tie	kò-kàk-ò-à	to untie
i	kù-chib-à	to close	kù-chib-ò-à	to open
ɔ	kɔ̀-tɔ̀t-à	to humidify	kɔ̀-tɔ̀t-ɔ̀-à	to dehumidify

5.2.8 Habitual

The Habitual is formed by the invariable extension -ay. This extension is highly productive, marking that the action is often repeated, regularly or habitually done. The Habitual is considered derivational rather than inflectional, not only because the suffix is found in the same position as all other derivational suffixes, but particularly since verbs with the Habitual extension can be inflected in nearly all TAM paradigms. The only exception is that the Perfective Present is semantically incompatible with the Habitual extension, since an action which is/was finished at the present moment cannot also indicate an action which is repeated or often done. Other TA paradigms marking Past can all be found in combination with the Habitual extension, since the action may have been regularly done in the past, but was finished or completed at some point in the past already. Verbs with the Habitual extension may also be inflected for Subjunctive, Present Continuous and Future.

TABLE 16.34 HABITUAL DERIVATION

Root vowel	Verb	Gloss	Habitual	Gloss
a	kò-kách-à	to play, dance	kò-kách-ày-à	to play habitually
ɛ	kɔ̀-sɛ́k-à	to laugh	kɔ̀-sɛ́k-ày-à	to laugh habitually

5.3 Inflectional morphology

5.3.1 Introduction

This section focusses on the TAM marking of verbs (5.3.3). Prior to that, there is a brief section on pronominal reference (5.3.2), giving an overview of subject, object and reflexive prefixes, which are morphemes preceding the verb stem. The section on TAM marking begins with the Irrealis Mood: Imperative and Subjunctive. Following that, I present the Perfective Aspect with four tense distinctions, and the Imperfective Aspect with three tense distinctions. The present description shows both affirmative and negative forms. A number of paradigms are differentiated by tone alone, both in the affirmative and negative, or between affirmative and negative forms.

5.3.2 Subject, object and reflexive prefixes

Table 16.35 introduces the subject and object prefixes, as well as the reflexive prefix. All prefixes containing the vowels /i/, /u/ and /a/ are invariable; those with the vowels /e/ or /o/ have allomorphs according to the vowel-harmony system, determined by the first vowel of the following syllable. The final allomorph listed in each case is used when the prefix precedes a vowel-initial morpheme.

When the prefix ends in a glide, /w/ or /y/, its tone is realised on the following vowel.

The second- and third-person singular subject prefixes are differentiated by tone alone (ò- and ó-), as was the case with the independent pronouns as well. The class 1 and second-person plural object prefixes are also differentiated by tone alone (mò- and mó-).

Object prefixes preceding the verb stem and nominal objects are mutually exclusive within a clause, both for animate and inanimate nouns. This means that an object first needs to be introduced by a noun, and only in a following clause or sentence can it be referred to by an object prefix.

(34) nàlɔ́lá mwǎnà 'I saw a/the child (distant past)'
       nà*mɔ̀lɔ́lá* 'I saw him (distant past)'
       but: *nà*mɔ̀lɔ́lá* mwǎnà

5.3.3 Tense/Aspect/Mood

In this section, I present the main verb paradigms representing tense, aspect and mood. Further study in the language will undoubtedly reveal more verb paradigms, but the present set will give a representative sample of the basic TAM distinctions in the language.

The first basic distinction is between realis and irrealis mood. Irrealis mood is represented by the Subjunctive paradigm and the imperative. Realis mood can be subdivided into the perfective and the imperfective aspect. The perfective aspect contains the

TABLE 16.35 SUBJECT, OBJECT AND REFLEXIVE PREFIXES

	Subject prefixes	*Object prefixes*
1SG	nà- ~ n-	-N-
2SG	ò- ~ w'-	-kò-
3SG/cl.1	ó- ~ w'-	-mò-
reflexive prefix		-í-
1PL	tò- ~ tw'-	-tó-
2PL	mò- ~ mw'-	-mó-
3PL/cl.2	bá- ~ b' -	-bá-
cl.3	ó- ~ w'-	-mó
cl.4	é- ~ y'-	-mé-
cl.5	lé- ~ ly'-	-lé-
cl.6	má- ~ m'-	-má-
cl.7	ké- ~ ky'-	-ké-
cl.8	bí- ~ by'-	-bí-
cl.9	é- ~ y'-	-mé-
cl.10	chí- ~ ch'-	-chí-
cl.11	ló- ~ lw'-	-ló-
cl.12	ká- ~ k'-	-ká-
cl.13	tó- ~ tw'-	-tó-
cl.14	bó- ~ bw'-	-bó-
cl.15	kó- ~ kw'-	-kó-
cl.19	sí- ~ sy'-	-sí-

following four paradigms: Present, Immediate Past, Recent Past and Remote Past. The imperfective aspect contains three paradigms: Present Continuous, Near Future and Distant Future. These paradigms are presented in the following sections, with their affirmative and negative forms.

5.3.3.1 List of TAM marking

Tables 16.36 and 16.37 introduce the TAM markers for the different paradigms. This concerns the marker used before the stem or before the object prefix and then the verb-final vowel. The first table in this section shows the TAM marking for the affirmative paradigms; the second table shows the markers for the negative paradigms. In the Perfective aspect, three affirmative paradigms all use one and the same negative paradigm. In fact, there are only two negative paradigms in the Perfective: one as counterpart of the Immediate Past and the Recent Past, and the other one as counterpart of the Remote Past. In the same way as the two affirmative paradigms – Immediate Past and Remote Past – are tonally contrastive, the two negative Perfective paradigms are also tonally contrastive. All other affirmative paradigms, Subjunctive and the three paradigms of the Imperfective Aspect, have a negative counterpart.

The negative marker used in most TA paradigms is *nde- ~ nd-*, occurring in initial position preceding the subject prefix. The two allomorphs are used before a following consonant or vowel respectively. There is a suppletive morpheme *si- ~ s-* for the first-person singular. Some paradigms have no negative prefix, but a suffix (Subjunctive), or negation may be marked by tone (Present Continuous, plural forms).

TABLE 16.36 TAM MORPHEMES IN AFFIRMATIVE PARADIGMS

	TAM marker	Verb stem	Verb-final vowel
Imperative	-		-á
Subjunctive	H tone on sm		-è
Perfective Present	-(s)ó-		-à
Immediate Past	-á-		-à
Recent Past	-		-éè
Remote Past	-à-		-á
Present Continuous	-tò-		-à
Near Future	-tá-		-é
Distant Future	-kà- ~ -kǎ-		-à

TABLE 16.37 TAM MORPHEMES IN NEGATIVE PARADIGMS

	TAM marker	Verb stem	Verb-final
Subjunctive	H tone on sm		-yá
Immediate and Recent Past (and Present Perfective)	-á-		-à
Remote Past	-á-		-á
Present Continuous (and Present Perfective)	-tó-		-à
Near Future	-kà-		-é
Distant Future	-kà-		-à

5.3.3.2 Imperative and subjunctive

The Imperative and Subjunctive are to a certain degree intertwined, since there is no form for the second-person singular Subjunctive. Instead, the Imperative singular is used to fill the gap. There is, however, an Imperative plural form as well as a second-person plural in the Subjunctive paradigm without a real difference in meaning.

The following are the structures of the Imperative singular and plural. Singular Imperatives are formed by the stem of the verb, followed by a H-toned verb-final vowel -á. Plural Imperatives add an additional plural addressee morpheme -nì. With H-tone verbs, a L tone is inserted between the H tone of the root and the H tone of the suffix, which results in a HL contour tone on the root vowel. L-tone verbs retain a L tone on the root without any other tonal changes.

The Subjunctive is the only paradigm which has all H-tone subject prefixes. This H tone must be considered a marker for the Subjunctive. The paradigm has no segmental prefix; the verb-final consists of a H-tone -é.

There is neutralisation of tone on the verb root between L-tone verbs and H-tone verbs in all forms of the Subjunctive affirmative paradigm, caused by the application of Meeussen's Rule.

The negative Subjunctive forms are presented in Table 16.40. All subject prefixes carry a H tone, as in the affirmative paradigm. The fused forms of subject prefix and negative marker are sí-, ndó- and ndá- for first, second and third person singular respectively. There is no trace of a segmental negative morpheme in the plural subject prefixes. However, there is a verb-final morpheme -yá, which is only used for the negative Subjunctive.

TABLE 16.38 IMPERATIVE

	-CVC-a	
SG	kônd-á	love!
PL	kônd-á-nì	you (PL) love!
SG	mèn-á	swallow!
PL	mèn-á-nì	you (PL) swallow!

TABLE 16.39 SUBJUNCTIVE AFFIRMATIVE

	L-tone verb kɔ̀-lɔ̀l-à	To see	H-tone verb kò-két-à	To do
1SG	ná-lɔ̀l-ɛ́	I should see	ná-kèt-é	I should do
2SG	lɔ̀l-á	see, look!	kêt-á	do!
3SG/cl.1	ɔ́-lɔ̀l-ɛ́	he should see	ó-kèt-é	he should do
1PL	tɔ́-lɔ̀l-ɛ́	we should see	tó-kèt-é	we should do
2PL	mɔ́-lɔ̀l-ɛ́	you should see	mó-kèt-é	you should do
3PL/cl.2	bá-lɔ̀l-ɛ́	they should see	bá-kèt-é	they should do

TABLE 16.40 SUBJUNCTIVE NEGATIVE

	L-tone verb		H-tone verb	
1SG	sí-lɔ̀lá-yá	I should not look	sí-kétá-yá	I should not do
2SG	ndɔ́-lɔ̀lá-yá	you should not look	ndó-kétá-yá	you should not do
3SG/cl.1	ndá-lɔ̀lá-yá	he should not look	ndá-kétá-yá	he should not do
1PL	tɔ́-lɔ̀lá-yá	we should not look	tó-kétá-yá	we should not do
2PL	mɔ́-lɔ̀lá-yá	you should not look	mó-kétá-yá	you should not do
3PL/cl.2	bá-lɔ̀lá-yá	they should not look	bá-kétá-yá	they should not do

5.3.3.3 Perfective present

In addition to the regular series of subject prefixes (L tone for the participants; H tone on the class 1 and class 2 prefixes), the TA marker is -(s)ó-.[4] The Perfective Present is used for an action whose completed status is relevant right at the moment of speaking (as opposed to the Present Continuous, which refers to an on-going process). Tables 16.41 and 16.42 give a paradigm with a L-tone verb first, followed by one with a H-tone verb.

Both the negative of the Immediate and Recent Past (5.3.3.7) and the negative of the Present Continuous (5.3.3.8) can be used to negate the Perfective Present.

TABLE 16.41 PERFECTIVE PRESENT AFFIRMATIVE

	L-tone verb		H-tone verb	
	kɔ̀-sèk-à	To laugh	kɔ̀-bék-à	To cut (meat)
1SG	nà-sɔ́-sèk-à	I laugh	nà-sɔ̂-bék-à	I cut
2SG	ɔ̀-sɔ́-sèk-à	you laugh	ɔ̀-sɔ̂-bék-à	you cut
3SG/cl.1	ɔ́-sɔ́-sèk-à	he/she laughs	ɔ́-sɔ̀-bék-à	he/she cuts
1PL	tɔ̀-sɔ́-sèk-à	we laugh	tɔ̀-sɔ̂-bék-à	we cut
2PL	mɔ̀-sɔ́-sèk-à	you laugh	mɔ̀-sɔ̂-bék-à	you cut
3PL/cl.2	bá-sɔ́-sèk-à	they laugh	bá-sɔ̂-bék-à	they cut

5.3.3.4 Immediate past

The Immediate Past has a H-tone -á- as TA marker and a L-tone verb-final -à. L-tone subject prefixes in 1SG/PL and 2SG/PL and H-tone TA-markers are fused, and surface with a LH rising tone. The Immediate Past describes something as having taken place very recently.

TABLE 16.42 IMMEDIATE PAST AFFIRMATIVE

	L-tone verb	'just now'	H-tone verb	'just now'
	kɔ̀-sèk-à	to laugh	kɔ̀-bɛ́k-à	to cut (e.g., meat)
1SG	nǎ-sèk-à	I laughed	nǎ-bɛ́k-à	I cut
2SG	wǎ-sèk-à	you laughed	wǎ-bɛ́k-à	you cut
3SG/cl.1	wá-sèk-à	he laughed	wá-bɛ́k-à	he cut
1PL	twǎ-sèk-à	we laughed	twǎ-bɛ́k-à	we cut
2PL	mwǎ-sèk-à	you laughed	mwǎ-bɛ́k-à	you cut
3PL/cl.2	bá-sèk-à	they laughed	bá-bɛ́k-à	they cut

5.3.3.5 Remote past

The Remote Past has a L-tone -à- as TA marker and a H-tone verb-final -á. Tonally, this is exactly the opposite of the Immediate Past. The Remote Past describes something as having taken place quite a while ago.

TABLE 16.43 REMOTE PAST AFFIRMATIVE

	L-tone verb	'a long time ago'	H-tone verb	'a long time ago'
	kɔ̀-sèk-à	to laugh	kɔ̀-bɛ́k-à	to cut (e.g., meat)
1SG	nà-sèk-á	I laughed	nà-bɛ́k-á	I cut
2SG	wà-sèk-á	you laughed	wà-bɛ́k-á	you cut
3SG/cl.1	wá-sèk-á	he laughed	wá-bèk-à	he cut
1PL	twà-sèk-á	we laughed	twà-bɛ́k-á	we cut
2PL	mwà-sèk-á	you laughed	mwà-bɛ́k-á	you cut
3PL/cl.2	bá-sèk-á	they laughed	bá-bèk-à	they cut

In the paradigm of the H-tone verb kɔ̀-bɛ́k-à 'to cut,' we find an example of Meeussen's Rule: the 3SG and 3PL forms wábèkà and bábèkà contain L tones after a H tone on the subject prefix.

5.3.3.6 Recent past

In addition to the Immediate and Remote Past forms, there is another paradigm, which points to a Recent Past. Further study will have to reveal if this past form has different usages than the Immediate and the Remote Past. The time of action is definitely a bit further back than the Immediate Past.

The regular subject prefixes are used, without a specific TA marker. The verb-final is -èè. This very much points to the verb-final -à followed by a Perfective morpheme -ile. The /l/ was dropped, as in so many environments in this language, and the vowels -à and -i followed by -è then coalesced into -èè.

TABLE 16.44 RECENT PAST AFFIRMATIVE

	L-tone verb	*'recently'*	*H-tone verb*	*'recently'*
	kɔ̀-lɔ̀l-à	to look	kù-líng-à	to deceive
1SG	nà-lɔ̀l-ɛ́ɛ̀	I looked	nà-líng-éè	I deceived
2SG	ɔ̀-lɔ̀l-ɛ́ɛ̀	you looked	ù-líng-éè	you deceived
3SG/cl.1	ɔ́-lɔ̀l-ɛ́ɛ̀	he looked	ú-líng-éè	he deceived
1PL	tɔ̀-lɔ̀l-ɛ́ɛ̀	we looked	tù-líng-éè	we deceived
2PL	mɔ̀-lɔ̀l-ɛ́ɛ̀	you looked	mù-líng-éè	you deceived
3PL/cl.2	bá-lɔ̀l-ɛ́ɛ̀	they looked	bá-líng-éè	they deceived

5.3.3.7 Negative perfective paradigms

There are two negative Perfective paradigms, one for the Immediate Past and the Recent Past together and another one to negate to Remote Past. The two negative paradigms are tonally contrastive. The TA marker is a H-tone -*á*- for both; the final vowel is a L-tone -*à* in the first paradigm, for the non-Remote Past, and a H-tone -*á* for the Remote Past. The negative paradigms for Immediate and Recent Past together can also be used to negate the Perfective Present.

TABLE 16.45 IMMEDIATE AND RECENT PAST NEGATIVE; PERFECTIVE PRESENT NEGATIVE

	L-tone verb	*'just now, recently'*	*H-tone verb*	*'just now, recently'*
	kɔ̀-sèk-à	to laugh	kɔ̀-bék-à	to cut
1SG	sá-sèk-à	I did not laugh	sá-bɛ́k-à	I did not cut
2SG	ndwá-sèk-à	you did not laugh	ndwá-bɛ́k-à	you did not cut
3SG/cl.1	ndá-sèk-à	he did not laugh	ndá-bɛ́k-à	he did not cut
1PL	ndè-twá-sèk-à	we did not laugh	ndè-twá-bɛ́k-à	we did not cut
2PL	ndè-mwá-sèk-à	you did not laugh	ndè-mwá-bɛ́k-à	you did not cut
3PL/cl.2	ndè-bá-sèk-à	they did not laugh	ndè-bá-bɛ́k-à	they did not cut

Table 16.46 shows the negative paradigm for the Remote Past. In the H-tone verb *kɔ̀-bék-à* 'to cut,' the underlying form of the negative would be *sá-bék-á*; however, Meeussen's Rule converts the last two H tones into L tones, with the surface result *sá-bèk-à*.

TABLE 16.46 REMOTE PAST NEGATIVE

	L-tone verb	*'long ago'*	*H-tone verb*	*'long ago'*
	kɔ̀-sèk-à	to laugh	kɔ̀-bék-à	to cut
1SG	sá-sèk-á	I have not laughed	sá-bèk-à	I have not cut
2SG	ndwá-sèk-á	you have not laughed	ndwá-bèk-à	you have not cut
3SG/cl.1	ndá-sèk-á	he has not laughed	ndá-bèk-à	he has not cut
1PL	ndè-twá-sèk-á	we have not laughed	ndè-twá-bèk-à	we have not cut
2PL	ndè-mwá-sèk-á	you have not laughed	ndè-mwá-bèk-à	you have not cut
3PL/cl.2	ndè-bá-sèk-á	they have not laughed	ndè-bá-bèk-à	they have not cut

5.3.3.8 Present continuous

In this section, both the affirmative and negative paradigms are presented. The TA marker is a L-tone *-tò-* in the affirmative paradigm. The verb-final is a L-tone *-à*. The Present Continuous marks an action as ongoing in the present.

TABLE 16.47 PRESENT CONTINUOUS AFFIRMATIVE

	L-tone verb		H-tone verb	
	kɔ̀-lɔ̀l-à	to see, look	kù-súkù-à	to push
1SG	nà-tɔ̀-lɔ̀l-à	I am looking	nà-tù-súkù-à	I am pushing
2SG	ɔ̀-tɔ̀-lɔ̀l-à	you are looking	ù-tù-súkù-à	you are pushing
3SG/cl.1	ɔ́-tɔ̀-lɔ̀l-à	he is looking	ú-tù-súkù-à	he is pushing
1PL	tɔ̀-tɔ̀-lɔ̀l-à	we are looking	tù-tù-súkù-à	we are pushing
2PL	mɔ̀-tɔ̀-lɔ̀l-à	you are looking	mù-tù-súkù-à	you are pushing
3PL/cl.2	bá-tɔ̀-lɔ̀l-à	they are looking	bá-tù-súkù-à	they are pushing

The negative of the Present Continuous takes a H-tone *-tó-* morpheme, with a verb-final L-tone *-à*. Overt negative subject markers exist for the singular forms only; in the plural negative, the subject prefixes have a LH rising tone, which leads to some tonal contrasts between affirmative and negative in the plural forms.

TABLE 16.48 PRESENT CONTINUOUS NEGATIVE; PERFECTIVE PRESENT NEGATIVE

	L-tone verb		H-tone verb	
1SG	sí-tó-lɔ̀l-à	I am not looking	sí-tú-sùkù-à	I am not pushing
2SG	ndɔ́-tó-lɔ̀l-à	you are not looking	ndú-tú-sùkù-à	you are not pushing
3SG/cl.1	ndá-tó-lɔ̀l-à	he is not looking	ndá-tú-sùkù-à	he is not pushing
1PL	tɔ̌-tó-lɔ̀l-à	we are not looking	tǔ-tú-sùkù-à	we are not pushing
2PL	mɔ̌-tó-lɔ̀l-à	you are not looking	mǔ-tú-sùkù-à	you are not pushing
3PL/cl.2	bǎ-tó-lɔ̀l-à	they are not looking	bǎ-tú-sùkù-à	they are not pushing

Apart from the regular tonal contrast between 2SG and 3SG forms in the affirmative paradigms: *ɔ̀-tɔ̀-lɔ̀l-à* 'you are looking' and *ɔ́-tɔ̀-lɔ̀l-à* 'he is looking,' there is exclusive tonal contrast between the affirmative and negative forms in the plural.

The tonally minimal verb forms in Table 16.49 between affirmative and negative are attested:

TABLE 16.49 PRESENT CONTINUOUS AFFIRMATIVE AND NEGATIVE

tɔ̀-tɔ̀-lɔ̀l-à	we are looking	tɔ̌-tó-lɔ̀l-à	we are not looking
mɔ̀-tɔ̀-lɔ̀l-à	you are looking	mɔ̌-tó-lɔ̀l-à	you are not looking
bá-tɔ̀-lɔ̀l-à	they are looking	bǎ-tó-lɔ̀l-à	they are not looking

5.3.3.9 Near Future

The Near Future affirmative (Table 16.50) is formed by a H-tone -*tá*- TA prefix and a verb-final H-tone -*é*. Various morphophonological and tonological processes give the following result with the verbs:

TABLE 16.50 NEAR FUTURE AFFIRMATIVE

	L-tone verb	'soon'	H-tone verb	'soon'
	kò-kàmb-à	to do, cook	kù-tút-à	to pound
1SG	n-dá-kàmb-é	I will cook	n-dâ-tút-é	I will pound
2SG	ò-tá-kàmb-é	you will cook	ò-tâ-tút-é	you will pound
3SG/cl.1	ó-tá-kàmb-é	he will cook	ó-tà-tút-é	he will pound
1PL	tò-tá-kàmb-é	we will cook	tò-tâ-tút-é	we will pound
2PL	mò-tá-kàmb-é	you will cook	mò-tâ-tút-é	you will pound
3PL/cl.2	bá-tá-kàmb-é	they will cook	bá-tà-tút-é	they will pound

The Near Future negative (Table 16.51) is formed by a L-tone TA prefix -*kà*- and the same verb-final H-tone -*é* as in the affirmative, exemplified in the following paradigms.

TABLE 16.51 NEAR FUTURE NEGATIVE

	L-tone verb	'soon'	H-tone verb	'soon'
1SG	sí-kà-kàmb-é	I will not cook	sí-kà-tút-é	I will not pound
2SG	ndó-kà-kàmb-é	you will not cook	ndó-kà-tút-é	you will not pound
3SG/cl.1	ndá-kà-kàmb-é	he will not cook	ndá-kà-tút-é	he will not pound
1PL	tǒ-kà-kàmb-é	we will not cook	tǒ-kà-tút-é	we will not pound
2PL	mǒ-kà-kàmb-é	you will not cook	mǒ-kà-tút-é	you will not pound
3PL/cl.2	bǎ-kà-kàmb-é	they will not cook	bǎ-kà-tút-é	they will not pound

5.3.3.10 Distant future

The Distant Future affirmative is marked by a TA marker -*kà*- ~ -*kǎ*- and a verb-final L-tone -*à*. The verbs used are the same as in the Near Future:

TABLE 16.52 DISTANT FUTURE AFFIRMATIVE

	L-tone verb	'later'	H-tone verb	'later'
1SG	n-gǎ-kàmb-à	I will cook	n-gǎ-tút-à	I will pound
2SG	ò-kǎ-kàmb-à	you will cook	ò-ká-tút-à	you will pound
3SG/cl.1	ó-kà-kàmb-à	he will cook	ó-kà-tùt-à	he will pound
1PL	tò-kǎ-kàmb-à	we will cook	tò-ká-tút-à	we will pound
2PL	mò-kǎ-kàmb-à	you will cook	mò-ká-tút-à	you will pound
3PL/cl.2	bá-kà-kàmb-à	they will cook	bá-kà-tùt-à	they will pound

In the negative, following the negative marker, there are two TA prefixes: -*á*- and -*kà*-. whereby the -*á*- is fused with the negative or the subject prefixes. The -*kà*- is likely the same as the one used for the Near Future. The final vowel is a L-tone -*à*.

TABLE 16.53 DISTANT FUTURE NEGATIVE

	L-tone verb	'later'	H-tone verb	'later'
1SG	sá-kà-kàmb-à	I will not cook	sá-kà-tùt-à	I will not pound
2SG	ndwá-kà-kàmb-à	you will not cook	ndwá-kà-tùt-à	you will not pound
3SG/cl.1	ndá-kà-kàmb-à	he will not cook	ndá-kà-tùt-à	he will not pound
1PL	ndè-twá-kà-kàmb-à	we will not cook	ndè-twá-kà-tùt-à	we will not pound
2PL	ndè-mwá-kà-kàmb-à	you will not cook	ndè-mwá-kà-tùt-à	you will not pound
3PL/cl.2	ndè-bá-kà-kàmb-à	they will not cook	nde-bá-kà-tùt-à	they will not pound

NOTES

1 Leitch (1996) has devoted his entire dissertation on vowel-harmony systems of Bantu Zone C languages. He has introduced the feature RTR to describe these systems, and gives an explanation on why he chooses the feature RTR rather than ATR as the active (dominant) harmonic feature in the system (Leitch 1996: 75).
2 Prefixes and suffixes are separated from the root by a hyphen marking the morpheme break where necessary. A nasal separated from the following consonant does not mean that it is syllabic: the consonant is prenasalised; however, the morpheme boundary falls between the prenasalisation and the root-initial obstruent.
3 The locative classes are only represented in this overall table of agreement markers. Since it is exceedingly difficult to find examples using these three prefixes, they are omitted from all other tables in this chapter.
4 The TA marker for the Perfective Present varies between -só- and -ó-. This variation is regional and individual. For the sake of clarity, the forms with the /s/ will be used in this chapter.

REFERENCES

Guthrie, M. (1971) *Comparative Bantu: An Introduction to the Comparative Linguistics and Prehistory of the Bantu languages. Volume 2: Bantu Prehistory, Inventory and Indexes.* London: Gregg International.
Kabungama Yuka, E. (2006) *Fonctions syntaxiques dans la phrase binja.* Kisangani: Université de Kisangani, thèse de doctorat.
Leitch, M. F. (1996) *Vowel Harmonies in the Congo Basin: An Optimality Theory Analysis of Variation in the Bantu Zone C.* Vancouver: University of British Columbia, PhD.
Lewis, M. P., G. F. Simons & C. D. Fennig (eds.) (2016) *Ethnologue: Languages of the World, Nineteenth edition.* Dallas: SIL International. Online version: www.ethnologue.com.
Meeussen, A. E. (1953) De talen van Maniema (Belgisch-Congo). *Kongo-Overzee* 19: 385–391.
Meeussen, A. E. (n.d.) *Notes sur les préfixes binja N. et S.* Ms. Tervuren: Musée Royal de l'Afrique Centrale, 6pp.
Meeussen, A. E. & S. Sebasoni (n.d.) *Notes binja-sud.* Ms. Tervuren: Musée Royal de l'Afrique Centrale, 91pp.
Motingea Mangulu, A. (1996) Esquisse du kibinja-sud: langue bantoue de la frontière C-D. *Journal of Asian and African Studies* 52: 81–123.
Mukubingwa, Z. (1997) Morphèmes verbaux du binja-sud. *Afrikanistische Arbeitspapiere* 50: 35–52.

CHAPTER 17

THE MARA LANGUAGES JE40

Lotta Aunio, Holly Robinson, Tim Roth, Oliver Stegen and John B. Walker

1 INTRODUCTION

The JE40 languages are a cluster of Lacustrine Bantu languages spoken on the eastern shore and hinterland of Lake Victoria, both in Kenya and in Tanzania. In the *Ethnologue* (Lewis *et al.* 2016), Idakho JE411-Isukha JE412-Tiriki JE413 – considered three distinct languages by Maho (2009) –, Suba E403, and the extinct Singa are listed for Kenya; Gusii JE42, Logooli JE41 and Kuria JE43 for both Kenya and Tanzania; and Ikizu-Sizaki – two languages in Maho (2009) –, Ikoma-Nata-Isenye (three ethnicities speaking closely related language varieties), Kabwa, Ngoreme, Suba-Simbiti (not to be confused with Kenyan Suba, hence in the following just Simbiti) and Zanaki for Tanzania. Although Bastin (2003), in her chapter on Bantu J, showed Gusii, Suba, Kuria, Ngoreme, Zanaki and Nata as JE40 languages (Isukha, Idakho, Tiriki and Logooli were counted under JE30), the references cited by her only deal with Gusii and Kuria. In the meantime, research on Tanzanian JE40 (henceforth, the Mara languages) has intensified (e.g., Aunio 2010, Higgins 2012, Rundell 2012, Aunio 2013, Gambarage 2013, Gray 2013, Osa-Gómez del Campo 2013, Sadlier-Brown 2013, Walker 2013, Roth 2014, Aunio 2015, Gambarage *et al.* 2017). Unless otherwise indicated, all data comes from the authors' own fieldwork or from fieldwork by SIL Tanzania. This chapter endeavours to complement Bastin's chapter on Zone J by means of giving a descriptive and comparative overview of six only recently described Mara language varieties, viz. Ikizu, Ikoma, Kabwa, Ngoreme, Simbiti and Zanaki.

TABLE 17.1 BASIC INFORMATION ON THE TARGET JE40 LANGUAGES

Ikizu JE402/[ikz]	2 ethnicities speaking two varieties: (1) Ikizu, (2) Sizaki
	Distribution: Bunda and Musoma districts, Tanzania
	Number of speakers: 55,000 in 2005
Ikoma JE45/[ntk]	3 ethnicities speaking three varieties: (1) Ikoma, (2) Nata, (3) Isenye
	Distribution: Serengeti district, Tanzania
	Number of speakers: total of 36,000 in 2005, (1) 15,000, (2) 11,500, (3) 9,500
Kabwa JE405/[cwa]	Dialects: none
	Distribution: Bukabwa ward in Musoma rural district, Tanzania
	Number of speakers: 14,000 in 2011
Ngoreme JE401/[ngq]	Dialects: (1) Mwibara, (2) Rogoro
	Distribution: Serengeti district, Tanzania
	Number of speakers: 55,000 in 2005
Simbiti JE431/[ssc]	8 ethnicities speaking eight varieties: (1) Simbiti proper, (2) Kiroba, (3) Hacha, (4) Kine, (5) Surwa, (6) Sweta, (7) Iregi, (8) Rieri
	Distribution: Tarime and Musoma districts, Tanzania
	Number of speakers: 113,000 in 2011
Zanaki JE44/[zak]	Dialects: (1) Northeast, (2) West, (3) South
	Distribution: Musoma rural district, Tanzania
	Number of speakers: 100,000 in 2005

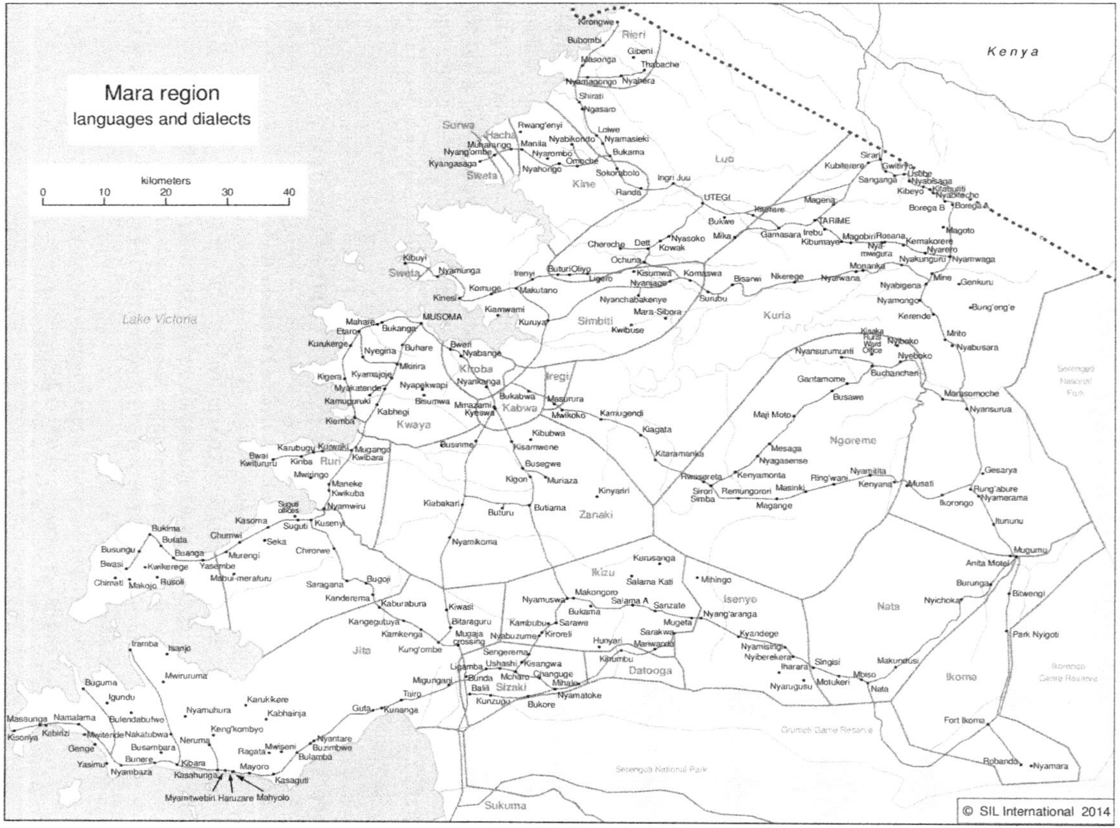

Mara region
languages and dialects

MAP 17.1

Five of the six areas where these languages are spoken share borders with each other, the exception being Ikoma – however, the closely related Nata variety borders on the Ngoreme area (Map 17.1). Due to Lake Victoria in the west and the Serengeti in the east, the Mara languages do not have many non-JE40 neighbours, these being confined to the Suguti languages Kwaya JE251, Ruri JE253 and Jita JE25 in the west and south-west, Sukuma F21 in the extreme southwest, and the Nilotic languages Datooga and Luo in the south and in the north respectively. This description of the internal Mara language differences will provide a basis for further research into Mara language history and language contact. It will also facilitate a more precise placement of JE40 within Bantu J than Bastin (2003) was able to provide with her representation of JE40 through the Kenyan JE40 languages.

A high degree of linguistic diversity in the Mara languages is found at all levels, from phonology and morphology to syntax and information structure. This will be evidenced by the descriptions in the following sections.

2 PHONOLOGY

The vowel systems of the Mara languages contain some unusual features for Bantu, and these are highlighted in Section 2.1. This is followed by a brief section (2.2) on the consonants, which are fairly standard for Bantu. The phonology chapter concludes with a detailed discussion of the diverse array of tonal phenomena (Section 2.3), both lexical and grammatical, across the six languages.

2.1 Vowels

All six languages exhibit contrastive vowel length with length neutralisation through lengthening after labialisation and palatalisation and before prenasalisation. Long vowels are also found when vowels become adjacent across morpheme boundaries. Four of the languages have seven-vowel /i e ɛ a ɔ o u/ systems, the exceptions being Kabwa with a reduction to five vowels (as in the neighbouring Suguti languages and in Swahili G42) and Zanaki, which is between a seven-vowel [+ATR]-dominant /i ɪ ɛ a ɔ ʊ u/ system and a Nilotic-type ten-vowel system with complete [ATR] symmetry. Like Casali (2008), we assume a categorical difference between [+ATR]-dominant versus [-ATR]-dominant vowel harmony systems. The dominance of [+ATR] in Zanaki is evidenced by leftward [+ATR] spread from suffixes like agentive -i or causative -ja, as shown in (1).

(1)   Zanaki
    ɔkʊβwɛɛma 'to hunt'   →   omuβweemi 'hunter'
    ɔkwɔɔma 'to dry (INTR)'   →   okwoomjɜ 'to dry (TR)'
    ɔkʊrɪŋaana 'to be flat'   →   okuriŋɜɜnjɜ 'to flatten'

Some Zanaki speakers apparently extend the canonical seven-vowel [+ATR]-dominant system by reinterpreting the [ɜ, e, o] allophones of /a, ɛ, ɔ/ as phonemes in their own right, thus arriving at a two-by-five [ATR]-symmetric system, with /ɪ ɛ a ɔ ʊ/ as [-ATR] and /i e ɜ o u/ as [+ATR] vowels. In Table 17.2, some contrastive Zanaki words are given which cannot be explained by [+ATR] spread as shown above.

TABLE 17.2 ZANAKI VOWEL SYSTEM

	/i/ vs /ɪ/	/e/ vs /ɛ/	/ɜ/ vs /a/	/o/ vs /ɔ/	/u/ vs /ʊ/
+ATR	eriirigo 'cattle pen'	okukweerɜ 'marry'	omwɜɜkɜ 'year'	wujo 'DEM.REF'	enzugu 'elephant'
-ATR	ɛrɪrɪhɔ 'payment'	ɔkʊhwɛɛra 'finish'	ɔmwaana 'child'	wʊnɔ 'DEM.PROX'	ɛnzʊgʊ 'groundnut'

A comparative overview of the vowel systems of all six languages is given in Table 17.3 (expanded from Higgins 2012: 269). Simbiti and Ngoreme have different vowel inventories in nouns and verbs: while there are seven vowel phonemes in nouns, there are only five vowels in basic verb roots. In both languages, however, the two additional phonemes are present in derived verb stems as a result of vowel harmony. While the five-vowel language Kabwa has only height harmony in verbal extensions, other languages have height harmony extending leftward through stems and prefixes, resulting in as many as three prefix vowel variants. Those languages with [ATR] harmony exhibit two different prefix variants. In all languages but Kabwa, there are examples of both leftward and rightward spreading harmony.

TABLE 17.3 VOWEL SYSTEM OVERVIEW

	CWA	SSC	NGQ	NTK	IKZ	ZAK
Inventory	/i ɛ a ɔ u/	/i e ɛ a ɔ o u/ (/ i ɛ a ɔ u/ in verb roots)	/i e ɛ a ɔ o u/ (/ i ɛ a ɔ u/ in verb roots)	/i e ɛ a ɔ o u/	/i e ɛ a ɔ o u/	/i e ɜ o u/ ɪ ɛ a ɔ ʊ/
Primary Harmony	n.a.	height (/a/ blocks)	[-ATR] (also [+ATR], height)	[-ATR] (also [+ATR])	height (/a/ blocks)	[+ATR]
Prefix variants	always ɔmu-	umu-: i u omo-: e o ɔmɔ-: ɛ a ɔ	omo-: i e o u omu-: ɛ a ɔ	omo-: i e o u omu-: ɛ a ɔ	umu-: i u omo-: e a o ɔmɔ-: ɛ ɔ	omu-: i e ɜ o u ɔmʊ-: ɪ ɛ a ɔ ʊ
Root vowel raising	n.a.	Caus [-i] etc. /ɛ ɔ/ → [e o]	Caus [-i] etc. /ɛ ɔ/ → [e o]	Caus [-i] etc. /ɛ/ → [e]	Caus [-i] etc. /e o/ → [i u] /ɛ ɔ/ → [e o]	Caus [-i] etc. -ATR → +ATR
Applicative variants with root vowels	-ir-: i a u -er-: ɛ ɔ	always -ɛr- but lowering verb root /i u/ → [ɛ ɔ]	always -ɛr- but lowering verb root /i u/ → [ɛ ɔ]	-er-: e o -ɛr-: i ɛ a ɔ u	-er-: i e a o u -ɛr-: ɛ ɔ	-ir-: i ɜ u -ir-: i a ʊ -er-: e o -ɛr-: ɛ ɔ
Subjunctive	always -ɛ	always -ɛ	always -ɛ	-eː i e ɛ a o u -ɛː ɔ	-eː i e a o u -ɛː ɛ ɔ	-eː i e ɜ o u -ɛː ɪ ɛ a ɔ ʊ

Individual languages may be viewed as intermediate speech varieties of others, e.g., Ngoreme is like Ikoma in nominal prefixes but like Simbiti in verbal suffixes. The range of differences is staggering, e.g., vowel quality of Applicative is determined by the verb root in Ikoma, Ikizu and Zanaki, but it determines the verb root vowel in Simbiti and Ngoreme (Table 17.4), like in Kuria (Chacha & Odden 1998).

The height harmony which occurs in Ikizu and Simbiti is very pervasive within these languages, certainly more so than the more familiar Applicative height harmony found in many Bantu languages. To use Ikizu as an example, within stems and suffixes, leftward-spreading harmony involves a first degree vowel (*i, u*) to the right raising leftward vowels by one degree. That is, third degree vowels (*ɛ, ɔ*) are raised to second degree

TABLE 17.4   VARIATION IN APPLICATIVE VOWEL HARMONY

Root vowel	CWA	SSC	NGQ	NTK	IKZ	ZAK
	/-ìr-/	/-ɛr/	/-ɛr/	/-ɛr/	/-er/	/-ɪr/
/i u/	kù-símb-ìr-à 'to dig for'	ɔ̀kɔ̀-téɣ-ɛ̀r-à 'to leave for' (cf. uku-tíɣ-a)	ɣù-téɣ-ɛ̀r-à 'to leave for' (cf. ɣo-tíɣ-a)	ɣò-tʃíβ-ɛ̀r-à 'to weed for'	kù-βís-ér-á 'to hide for'	oku-riβ-ir-ɜ 'to plug for'
/ɪ ʊ/	-	-	-	-	-	ɔkw-ʊt-ɪr-a 'to pour for'
/e o/	-	-	-	kò-réh-ɛ̀r-à 'to pay for'	kò-βéék-ér-á 'to bury for'	oku-kweer-er-ɜ 'to marry for'
/ɛ ɔ/	kù-kɔ́r-ɛ̀r-à 'to do for'	ɔ̀kɔ̀-tém-ɛ̀r-à 'to hit for'	ɣù-sɛ́k-ɛ̀r-à 'to laugh at'	ɣù-kɔ́r-ɛ̀r-à 'to do for'	kɔ̀-rɛ́ɛ́t-ér-á 'to bring for'	ɔkʊ-kɛɛz-ɛr-a 'to be late for'
/a ɜ/	kù-már-ìr-à 'to finish for'	ɔ̀kɔ̀-hák-ɛ̀r-à 'to paint for'	ɣù-sáám-ɛ̀r-à 'to move to'	ɣù-ɣáβ-ɛ̀r-à 'to inherit for'	kò-már-ér-á 'to finish for'	ɔkʊ-haat-ɪr-a 'to leave for' oku-mwɜɜj-ir-ɜ 'to shout for'

vowels (*e, o*), and second degree vowels are raised to first degree vowels. The process applies iteratively unless interrupted by the low vowel (*a*), which is opaque.

(2) Ikizu vowel raising in stems

		Vowel raising	Unraised Infinitive	Nominalised form	
(a)		2nd degree to 1st degree	òkò-rém-à 'to farm' òkò-gór-à 'to buy'	ùmù-rìm-ì 'farmer' rì-gúr-í 'price'	
(b)		3rd degree to 2nd degree	ɔ̀kɔ̀-rɔ́g-à 'to bewitch' ɔ̀kɔ̀-gɛ́ɛ́nd-à 'to travel'	òmò-ròg-ì 'sorcerer' òmò-gèènd-ì 'traveler'	cf. ɔ̀βɔ̀-rɔ̀gɔ̀ 'witchcraft' cf. ɔ̀rɔ̀-gèèndɔ̀ 'journey'
(c)		4th degree with no raising	òkò-sáβ-à 'to pray, beg'	òmò-sàβ-ì 'beggar'	

A different form of leftward-spreading harmony occurs across the prefix-stem boundary. Most Ikizu prefixes have underlying second degree vowels (e.g., /omo-/), and these vowels can be either raised (*umu-*) or lowered (*ɔmɔ-*) depending on the first stem vowel. See (3), with examples from noun classes 1, 7 and 9. This leftward harmony is also iterative, unless interrupted by /a/.

(3) Ikizu prefix harmony

Stem vowel a	Stem vowels ɛ, ɔ	Stem vowels e, o	Stem vowels i, u
òmò-rámú 'brother-in-law'	ɔ̀mɔ́-hɔ́tʃá 'servant'	òmò-kékóró 'old woman'	ùmú-βísá 'enemy'
èké-sági 'cloth'	èkɛ́-kɔ́βá 'lip'	èkè-gòrò 'mountain'	ikì-hùɲò 'stopper'
è-gàβò 'inheritance'	è-kɔ́rɔ́ 'heart'	èm-bézé 'wild pig'	in-gúβú 'antelope'

Finally, rightward spreading occurs from verb roots to suffixes if the suffix has an underlying second degree vowel. See examples of Ikizu Applicative vowel harmony in Table 17.4.
Simbiti height harmony has many similarities to Ikizu height harmony. For example, prefix vowels are realised at three heights, depending on the stem vowels: ɔ̀mɔ̀-βɔ́hὲ

'prisoner,' òrò-rémè 'tongue' and ùmù-kúúŋgù 'old lady.' One interesting difference is in the treatment of the low vowel. Stem-initial *a* harmonises with the second degree vowels in Ikizu (see examples in (3)), but *a* harmonises with the third degree vowels in Simbiti: ɔmɔ-ɣááŋgɔ 'porridge stirrer,' ɔrɔ-βáɣɔ 'fence,' ɔmɔ-βásò 'sunshine.' Vowel raising in stems is also similar in Simbiti (e.g., òmò-rémì 'farmer' from ɔkɔ-rém-à 'to farm'), with the exception that there are only five underlying vowels in Simbiti verb stems, so there are no cases of second degree vowels being raised to first degree vowels.

A number of unusual phenomena are found in the vowel harmony system of Ikoma, as argued by Higgins (2012). Whereas [-ATR] is marked and triggers vowel harmony, the nominal prefixes preceding [-ATR] roots actually undergo a process of dissimilation. As illustrated in (4) with class 7 nouns, mid [+ATR] prefixes alternate to high-vowel prefixes before [-ATR] stems.

(4)   Ikoma prefix dissimilation
      [+ATR] stems: èkè-mìrà 'mucus'   [-ATR] stems: èkì-héénɔ 'fever'
                      èkè-mèrò 'throat'   èɣì-kɔɔmbè 'shoulder blade'
                      èɣè-sóβè 'shell'   èɣì-sàré 'twin'
                      èɣè-kúúndì 'fist'

Ikoma also stands out against the other Mara languages in that root vowel raising caused by causative suffixes only pertains to /ɛ/ and not to both /ɛ/ and /ɔ/, as in the other languages. By contrast, the [-ATR] allomorph of the subjunctive suffix is triggered only by the back mid vowel /ɔ/.

At least two of these vowel systems appear to be in a current state of transition (for example, Zanaki's ambivalence between a seven-vowel and a ten-vowel system). In Ngoreme, the reduction of a seven-vowel to a five-vowel system is evidenced not only by its implementation in verb roots but also by the breakdown of consistent harmony patterns in noun prefixes. For example, the plural diminutive of òmó-tè 'tree' is èβí-tè, not the expected *eβe-te. Similarly, class 10 nouns have the high-vowel prefix t∫ì- with both [-ATR] and [+ATR] mid-vowel stems, with t∫e- only before the high vowels *i* and *u*. These examples point toward a recategorisation of all mid vowels as [-ATR].

2.2 Consonants

The consonant systems of the Mara languages are shown below in Table 17.5. The bilabial, alveolar, palatal and velar places of articulation are fully represented in the nasals and the plosives/fricatives.

All of the languages have the bilabial fricative /β/ and Simbiti, Ngoreme and Ikoma have the velar fricative /ɣ/. In addition, the relevant corpus languages have the following related phonological rule: /β, ɣ/→ [b, g]/[+nas] __.

All of the languages have undergone *p-lenition (*p > h), although Kabwa, Simbiti and Ikizu maintain /p/ in loanwords, whereas Zanaki and Ikoma have *p > β* in loanwords. Kabwa and Simbiti only have /d/ in loanwords, as *d has gone to /r/ in the majority of reflexes in each of the corpus languages. In the context of a preceding nasal, however, *d is retained as [d], as expressed in the phonological rule: /r/→ [d]/[+nas] __. Kabwa /t∫/ and Simbiti /f, dʒ/ also only occur in loanwords.

TABLE 17.5 CONSONANT SYSTEM OVERVIEW

	CWA	*SSC*	*NGQ*	*NTK*	*IKZ*	*ZAK*
Plosive	/p, t, k/	/p, t, k/	/t, k/	/t, k/	/p, t, k/	/t, k/
	/d, g/	/d/			/b, d, g/	/g/
	/tʃ, dʒ/	/ dʒ/	/tʃ/	/ tʃ/	/tʃ, dʒ/	
Fricative	/s, ʃ, h/	/f, s, ʃ, h/	/s, ʃ, h/	/s, ʃ, h/	/s, ʃ, h/	/s, h/
	/β/	/β, ɣ/	/β, ɣ/	/β, ɣ/	/β, z/	/β, z/
Trill		/r/	/r/			
Flap	/ɾ/	/ɾ/	/ɾ/	/ɾ/	/ɾ/	/ɾ/
Glide	/w, j/	/w, j/	/w, j/	/w, j/	/w, j/	/w, j/
Nasal	/m, n, ɲ, ŋ/	/m, n, ɲ, ŋ/	/m, n, ɲ, ŋ/	/m, n, ɲ, ŋ/	/m, n, ɲ, ŋ/	/m, n, ɲ, ŋ/
Nasal compounds	+	+	+	+	+	+
Palatalisation	+	+	+	+	+	+
Labialisation	+	+	+	+	+	+

Both Simbiti and Ngoreme attest the trill /r/, in phonological opposition to the tap both root-internally and at certain morpheme boundaries (normally a verb root ending in /r/ coupled with the applicative or perfective suffixes). Root-internally, the trill results from the combination of two taps: compare Ikoma *èrì-háɾáɾá* 'grasshopper' and *yù-kóɾɔ̀r-à* 'to cough' with the Simbiti examples with the trill below in (5). The vowel is lengthened before the trill. Examples of root-internal minimal pairs in Simbiti are shown in (5), while (6) demonstrates the contrast at morpheme boundaries in Ngoreme involving the Applicative and Perfective suffixes.

(5)   Simbiti
    ɾì-háɾà   'spot'   ɔ̀-kɔ̀-kóɾ-à   'to do, mend'
    ɾì-hááɾà   'grasshopper'   ɔ̀-kɔ̀-kóór-à   'to cough'

(6)   Ngoreme
    ɣù-kóɾ-à   'to do' (INF)
    ɣù-kóór-à   'to do for' (APPL)
    n-á-kòòr-è   'he/she has done' (PRF)

Table 17.6 demonstrates the variety of non-nasal reflexes in the Mara languages as compared to Proto-Bantu. The majority of the differences in the consonant systems of the corpus languages occurs among the reflexes of *c and *ɟ.

In contrast to many other Great Lakes languages, the Mara languages generally do not display Bantu spirantisation processes.

Dahl's Law is a process of voicing dissimilation in which "a voiceless stop becomes voiced if the consonant in the next syllable is also voiceless" (Hyman 2003: 56). Dahl's Law is attested in Ikoma, Ngoreme and Simbiti, but not in the other languages. In each of the attesting languages the voiced counterpart for /k/ is the fricative /ɣ/ (see Table 17.5). Dahl's Law occurs both root/stem-internally and across morpheme boundaries. In Simbiti, the underlying /k/ in the prefixes of classes 7, 12 and 15 is in free variation with /ɣ/, while this is not the case in Ikoma and Ngoreme. However, Ikoma (along with related

TABLE 17.6 JE40 REFLEXES OF PROTO-BANTU CONSONANTS

Proto-Bantu	CWA	SSC	NGQ	NTK	IKZ	ZAK
*p	/h/	/h/	/h/	/h/	/h/	/h/
*b	/β/	/β/	/β/	/β/	/b, β/	/β/
*t	/t/	/t/	/t/	/t/	/t/	/t/
*d	/ɾ/	/ɾ/	/ɾ/	/ɾ/	/ (d), ɾ/	/ɾ/
*c	/s, (ʃ) /	/s, ʃ/	/s, ʃ/	/s, ʃ/	/s, (ʃ) /	/s/
*ɟ (root-initial)	Ø	Ø	Ø	Ø	Ø	Ø
*ɟ (elsewhere)	/dʒ/	/ʃ/	/tʃ/	/tʃ/	/z/	/z/
*k	/k/	/k/	/k/	/k/	/k, tʃ/	/k/
*g	/g/	/ɣ/	/ɣ/	/ɣ/	/g/	/g/

Isenye) has /ɣ/ in prefixes not just before voiceless consonants but also before /ɣ/. This is an active process that may be an assimilation distinct from Dahl's Law.

2.3 Tone and syllable structure

The Mara languages have previously been reported as accentual or "two type" languages with fairly reduced or non-existent lexical tonal contrasts (Bastin 2003). However, a closer look at the Mara languages shows that many have more than two lexical tonal contrasts and make use of melodic tones in their verbal systems. On the other hand, there is considerable difference in how prevalent tones are, even within single languages. For example, in Zanaki (the variant described in this chapter) some speakers make use of regular penultimate stress – possibly under Swahili G42 influence –while others have several contrasting lexical and grammatical tone patterns. This variation is not directly dialectal, or based on age, but more research is needed to describe the variables involved and the degrees of variation.

The underlying contrast is between H and Ø (toneless) since only H tones are active in the tonal rules. There is at most one lexical H tone per morpheme with the exception of Kabwa double H pattern (see below).

Simbiti resembles Kuria in that the TBU is the mora and moras are counted in assigning melodic tones (Marlo *et al.* 2015). Ikizu also assigns tones to moras but counts syllables. In all the other corpus languages the TBU is the syllable and syllables are counted in assigning melodic tones. None of the languages have underlying contrasts between level H and contour tones, except Ikizu in some restricted contexts. H tones are attracted to heavy syllables in Ikoma (e.g., final melodic H shifted to long penult) and Ngoreme (e.g., H tone assignment in the Infinitive, see (16) below). Syllables with long vowels, both contrastive and derived, are considered heavy by the tonal rules despite the fact that, at least in Ikoma, the derived long vowels are phonetically shorter than the contrastive long vowels (Higgins 2012).

There is a rather direct relationship between the underlying position of tones and their surface realisations. Utterance-final Hs are allowed in Ikoma and Ngoreme. In Simbiti and Ikizu, melodic Hs are shifted from the utterance-final syllable to the penult, and in Ikizu, the lexical tone spread does not target the utterance-final syllable. In Kabwa, utterance-final Hs are deleted (see explanation following Table 17.8). Throughout this chapter, all the tones are the ones found in phrase-medial position, unless stated otherwise.

Tone spread is attested in Ikizu, Kabwa and Simbiti. In Ikizu, for example, the penultimate and antepenultimate lexical Hs spread right to the rest of the nominal stem (see Table 17.7). In Simbiti (7), tone spread takes place syllable-internally when a melodic H assigned to either the first or the second vowel of a bimoraic syllable spreads to the other vowel of the syllable as well, thus avoiding rising and falling tones.

(7) a Simbiti
βa-tó-βerek-éeje   →   βàtóβèrèkééjè
SP₂-OP₁ₚₗ-call-APPL.PFV
'they have called us'

b βa-ra-tó-haɣaáʃ-er-a   →   βàràtóhàɣááʃèrà
SP₂-PRS-OP₁ₚₗ-build-APPL-FV
'they are building for us'

Since morphemes can have at most one H, with the exception of Kabwa double H pattern, and tone spread and shift are not common, there are not many contexts with two Hs on adjacent TBUs. In Ikoma, two TBUs with Hs can become adjacent within a word with the class 2a prefix and in some verb forms. When that happens, the second H is deleted by Meeussen's Rule. Across word boundaries, the second H is downstepped, not deleted (Aunio 2010).

(8) a Ikoma
βáámààmè   →   βáámààmè
2a-maternal_uncle
'maternal uncles'

b èkì-βàrí kíírè   →   èkìβàrí ꞌkíírè
7-flower   DEM.DIST₇
'that flower'

Although Ikoma allows word-final Hs, word- and phrase-final HØH sequences are not allowed. In the Perfective there is a melodic H on the first and the last syllable of the macrostem, as in nàà[ré-rèmìrí 's/he farmed it.' However, the final H is deleted when there would only be one toneless syllable between the two Hs, as in nàà[rémìrì 's/he farmed.' Final H deletion also applies to nouns with a class 2a prefix (βáá-nɔɔrá 'agama lizards' → βáá-nɔɔrà). With these, it is also possible to form a H plateau instead of deleting the final H (βáá-nɔɔrá).

In Ikoma, there is an interesting tonal restriction in the connective phrases: a maximum of two H tones are allowed within the phrase (Aunio 2010). The Connective marker (9) has a H tone which is not realised when both nouns involved have a H tone.

(9) a Ikoma
àmà-ɣòrò   ɣá   tʃàà-kà
6-leg   CON₆ 10-lion
'legs of lions'

b àmà-ɣòrò   ɣá   βà-témì
6-leg   CON₆ 2-king
'legs of kings'

c àβà-témì   βá   tʃààn-tʃèrà
2-king   CON₂ 10-path
'kings of paths'

d àβà-témì   βà   tʃàà-sùná
2-king   CON₂ 10-mosquito
'kings of mosquitos'

The nominal prefixes and pre-prefixes are underlyingly toneless; some exceptions to this are the Ikoma and Ngoreme class 2a prefix βáá- and the Kabwa class 5 prefix éri-. Nominal lexical tone is attested in all six languages to some extent. Ikizu, Ikoma and Kabwa have the most complex nominal tone systems. Ikizu has two tonal patterns for monosyllabic noun stems and three patterns for longer stems (Table 17.7), with H spreading to the end of the stem. Kabwa has n+1 patterns, i.e., two for monosyllabic, three for bisyllabic and four for trisyllabic noun stems (cf. Table 17.8 and following explanation). Ikoma has two patterns for monosyllabic stems (toneless and H), three for bisyllabic stems (ØØ, HØ, ØH), and n+2 patterns for longer stems, e.g 3+2 patterns (ØØØ, HØØ, ØHØ, ØØH, HHH) for trisyllabic stems (Table 17.9).

TABLE 17.7 IKIZU NOMINAL TONES

Ø	Penult H	Antepenult H
èè-kà 'lion'	íí-swé 'fish'	
ùmù-rìmì 'farmer'	òmò-rámú 'brother-in-law'	ùmú-βísá 'enemy'
ùmù-nììβì 'rich person'	òmò-géégé 'stupid person'	òró-téété 'spine'
ìkì-rìβòkò 'crossing'	èkè-tènéká 'piece'	òrò-kérégé 'hill'
èkè-tààŋgòrà 'bed'	èkè-rèrééná 'baby'	rì-góróóβá 'afternoon'

TABLE 17.8 KABWA NOMINAL TONES

final H	Penult H	Double H	Antepenult H
òβù-sú 'nest'	òβú-sù 'face'		
èŋ-gìgé 'locust'	èŋ-gégè 'tilapia'	èŋ-gíhó 'kidney'	
èkì-rèèsá 'chin'	èkì-βééβè 'sadness'	èm-bíídʒí 'wild pig'	
èn-dìmìrá 'Pleiades'	[accidental gap]	èŋ-gúrúβè 'pig'	è-góròhè 'billy-goat'
èkj-ààβìrá 'mercy'	èn-sòróórò 'termite'	èn-sááríírù 'hawk'	èn-tááŋànà 'steer'

TABLE 17.9 IKOMA NOMINAL TONES

Ø	H on one stem syllable		Multiply linked H	
à-kà 'lion'	à-ká 'home'			
òmò-rèmì 'farmer'	òmò-kérà 'tail'	èkì-βàrí 'flower'		
	òrù-βéérè 'millet'	èkè-hùùhé 'butterfly'		
à-βùɣùsì 'upper arm'	èɣè-tííŋgìrò 'heel'	à-ɲàɣárà 'lizard'	èkè-hùrèró 'pot'	àm-búsúrú 'seed'
sòòkòrò	èɣì-táámòònì	òmù-sùméénò 'saw'	òmò-sòkààní	èrì-βúrúúŋgé 'egg'
'grandparent'	'scorpion'	à-kòòróótò 'hoof'	'compromise'	à-mátóóɲí 'vulture'
mìʃèèŋgè 'aunt'	èɣè-súrùùrà 'a kind of tree'			èrì-βóónóónó 'lump'

Kabwa nouns in Table 17.8 are written with underlying tone, as the tonal contrasts of different tone groups are neutralised in certain contexts. For example, an utterance-final H-deletion rule causes both nouns with final H and nouns with double H to be pronounced low in isolation, thus èŋgìgè 'locust' and èŋgìhò 'kidney.' Only in non-final position, for example with following demonstrative, is the tonal contrast realised, thus èŋgìgé jínù 'this locust' versus èŋgíhó jínù 'this kidney.' With following demonstratives, however, the tonal contrast between nouns with double H and nouns with penultimate H is neutralised due to H-spreading, thus èŋgégé jínù 'this tilapia' and èŋgíhó jínù 'this kidney,' whereas these two contrast in isolation, viz. èŋgégè 'tilapia' and èŋgìhò 'kidney.' Equivalent neutralisations apply to trisyllabic noun stems.

In Ngoreme and Zanaki, the lexical tones are much less stable and many speakers have shifted to penultimate stress (Zanaki) or weight-determined tone placement (Ngoreme). However, some Zanaki speakers still make use of a fairly complex tonal system, possibly with four patterns for bisyllabic noun stems. In Ngoreme, some speakers have three contrastive tonal patterns for bisyllabic and longer stems. Simbiti nouns all have a H on the first syllable of the stem, except for a handful of CVVCV-shape nouns that have a H on the noun class prefix in addition to the stem-initial H (10).

(10) Simbiti
    èkè-ɣésò   'knife'
    èkè-tɔɔ́ʃɔ   'hare'   vs.   èké-mɔɔ́rì   'calf'

In some languages (e.g., Ngoreme, Kabwa), historically toneless stems are realised with the all-H pattern, but other languages (e.g., Ikizu, Ikoma) have retained toneless stems. For example, in Ngoreme the three patterns for trisyllabic noun stems are HHH, HØØ and ØHØ, while in Ikizu the three underlying patterns are ØØØ, HØØ and ØHØ. However, Ikizu toneless patterns are realised with a stem-initial H when the noun is pronounced in isolation. Ikoma is the only one that has a contrast between toneless stems and stems with the all-H pattern. In the all-H pattern, the H is linked to all stem syllables and it functions as one H tone; adjacent H tones are not allowed in Ikoma. Further proof for multiple linking is found, for example, in contexts in which the multiply linked H tone is the second H in final HØH sequences: in these contexts, a H plateau is formed by pronouncing the toneless noun class prefix as H between two Hs.

(11) Ikoma
    àmà-ɣòrò   ɣá   tʃààm-bérété   →   àmàɣòrò   ɣá   tʃáámbérété
    6-leg   CON$_6$ 10-goat
    'legs of goats'

It is worth mentioning that the varieties closest to Ikoma (Nata and Isenye) do not share the nominal tone system of Ikoma (Aunio 2015). The Nata nominal tone system is a stress system in which every noun can have only one stressed syllable, realised as a H tone, and toneless nouns do not exist, as analysed by Anghelescu (2013) and confirmed with our own field data. There are three lexical tone types that are assigned 1) to the second syllable of a word unless the first syllable of the word is heavy, in which case the high tone is associated to the heavy word-initial syllable; 2) to the third syllable of the word; and 3) to the last syllable of the word. In Isenye, there are only two-tone types for all noun stems: either there is a H tone on the first syllable of the noun stem or all the stem syllables are H.

All six languages lack lexical tonal contrast in the verb stem. In Ikoma (12), Kabwa (13), and Simbiti (14), all verb stems in the Infinitive have a H on the first syllable of the macrostem (the left edge of the macrostem marked with "["); in Simbiti the H is retracted to the prefix with monosyllabic macrostems. Zanaki has regular penultimate stress in all Infinitive forms (15).

(12) Ikoma
    kò-[rómà 'to bite'   kù-[rááɣèrà 'to eat'
    kù[βá-ròmà 'to bite them'   kù[βá-rààɣèrì 'to feed them'

(13) Kabwa
    òkù-[rúmà 'to bite'   òkw-[ímùkà 'to leave'
    òkù[tú-rúmà 'to bite us'   òkù-[tígítà 'to be late'

(14) Simbiti
    ùkú-[ɲwà 'to drink' ùkù-[tíɣà 'to leave' ɔ̀kɔ̀-[mááhà 'to see' ɔ̀kɔ̀-[βérèkèrà 'to call'
    ɔ̀kɔ́[tɔ́-βèrèkèrà 'to call us'

(15) Zanaki
    ɔkʊ['βɔha 'to close' oku['sɜkjɜ 'to try' oku[sɜ'kirjɜ 'to help' ɔkʊ[βɔhɛ'rɛra 'to tether'

In Ngoreme, the location of the Infinitive H is conditioned by syllable weight (Aunio 2017). This is seen most clearly with trisyllabic stems in which the H is realised on the (second) heavy syllable and on the final syllable if there are no heavy syllables in the stem (16). Vowel-initial stems have the H on the second or the last syllable (17). In quadrisyllabic stems there are only two syllables in which the tone can be realised: the first syllable of the stem when there are no heavy syllables or the first syllable is the only heavy syllable, and the second syllable of the stem with all other stem shapes (18). Monosyllabic stems have the H on the only stem syllable (19), and bisyllabic stems have two-tone patterns, stems with short vowels have the H on the first syllable while stems with long vowels have a final H which is often anticipated on the previous syllable, creating a surface rising tone (20). Ngoreme verb tones are assigned to the stem, that is, their location is not affected by the presence of an object prefix (21).

(16) Ngoreme
    kù-ráágèrà 'to eat'   kò-mìmííntà 'to suck' kù-βààndéérì 'to squeeze' ɣù-sèβɔ́ká 'to sprout'
    kò-hííŋgìrà 'to push' kù-hèhéénà 'to pant' ɣù-sààsáámà 'to pray' kò-hìɣàmá 'to kneel'

(17) Ngoreme
    gw-ììkàrá 'to sit' kw-ììhííɲà 'to bow'

(18) Ngoreme
    kù-βérèkèrà 'to call'   kù-βéétʃèɣèrà 'to ruminate'
    ɣù-tʃàβáátʃɔrì 'to peel (away from oneself)' kù-hɔ̀ɔ̀jéérànà 'to copulate' kù-hèrékèèrà 'to escort'

(19) Ngoreme
    ɣò-tú 'to pick' kò-ɲó 'to drink'

(20) Ngoreme
    kò-rúɣà 'to cook'   ɣù-tɔ́mà 'to send'
    kù-ɣèéndá 'to walk' kù-βɔ́ɔ́ká 'to wake up'

(21) Ngoreme
    kò[tò-hííŋgìrà 'to push us'

As with nouns, Ikizu assigns the Infinitive H in relation to the end of the stem: the H is assigned to the antepenultimate syllable of the macrostem. When the macrostem has fewer than three syllables, the Infinitive H is realised macrostem-initially.

(22) Ikizu

ùkù[ɾjá 'to eat'	ùkù[kí-ɾjà 'to eat it'
òkò[ɾómà 'to bite'	òkò[mó-ròmà 'to bite him/her'
ɔ̀kɔ̀[βέɾèkà 'to carry on back'	ɔ̀kɔ̀[mɔ̀-βέɾèkà 'to carry him/her on back'
ùkù[ɲáhàarà 'to hurt	ùkù[mù-ɲáhàarà 'to hurt him/her'

Ikoma has 5 melodic H tones that can be assigned to 1) the first syllable of the macrostem (*mbàayò[kú-rèètèrà* 'they bring/will bring you'); 2) the second syllable of the macrostem (*βà[kàrárè* 'let them get angry'); 3) to the final syllable (*mbààŋgà[ràayìiré* 'they would eat'); 4) to both the first syllable of the macrostem and the final syllable (*mbàà[tú-ɲààkìrí* 'they disturbed us'); and 5) the initial syllable of the verb, i.e., the subject prefix (*βá[ɲàhàrèkà* 'when they had been hurt'). The melodic Hs are syllable weight sensitive: heavy syllables attract melodic Hs in some verb forms, and for example in the Narrative, the melodic H fails to be realised if the penultimate syllable is not heavy (Aunio 2013).

The Simbiti grammatical tone system resembles that of Kuria (Marlo *et al.* 2014), but Simbiti does not have the iterative spread. Simbiti has four melodic tones that are assigned to 1) the third mora of the macrostem; 2) the first and the fourth mora of the macrostem; 3) the first mora of the macrostem; and 4) the fourth mora of the macrostem.

Ngoreme has one melodic tone pattern: H assigned to the first syllable of the stem. In other forms, the TAM prefix has a H tone, but toneless forms are not found. Ikizu assigns melodic tones to the first or the second syllable of the macrostem, as well as to the penult. Kabwa assigns melodic Hs to the first syllable of the macrostem or to the whole macrostem.

Ikoma, Simbiti, Ngoreme, Kabwa and Ikizu also have verb forms that have a H on the TAM prefix (Ikoma *βàráá[βùyà* 'when they say'; Kabwa *βàrá[gùrà* 'they are buying'). Ikoma and Kabwa also have toneless verb forms (Ikoma *mbà[ràayìirè* 'they have eaten'; Kabwa *βàà[gùrà* 'we usually buy').

3 NOUNS AND NOUN PHRASES

While the Mara noun class systems are canonical, considerable differences are found in the shape of noun class prefixes due to different vowel harmony processes and sound changes. There are also considerable differences in constituent order within the noun phrase.

3.1 Noun classes and agreement

All of the Mara languages have canonical noun class systems, with a range of 17 to 20 of the roughly 24 classes reconstructed for Proto-Bantu (Katamba 2003: 104ff). All of the languages have retained their augments (or pre-prefixes) in the normal contexts as well as the class 5 prefix *li-/ri-* (which has been lost in many other languages). Most prefixes and augments are toneless.

Only Ikizu and Zanaki have maintained class 13 for plural diminutives. Kabwa attests class 22, and Ngoreme, Ikoma and Ikizu have class 19. All of the languages have at least scattered remnants of class 23 (e.g., demonstratives, possessives and the lexeme 'home').

One of the more typologically unusual features is the augment vowel *a-* in the Ikoma class 9/10 *a(N)-/tʃa(N)-* prefixes, which is not attested in any of the other Mara languages, except for Nata.

Table 17.10 lists one variant of the noun class prefixes that are found in each of the languages; however, as was noted in Table 17.3, all of the languages except for Kabwa have several vowel alternations for most of the class prefixes and augments.

TABLE 17.10 NOUN CLASS PREFIXES

Class	CWA		SSC		NGQ		NTK		IKZ		ZAK	
1	ɔ-	mu-	o-	mo-	o-	mo-	o-	mo-	o-	mo-	ɔ-	mu-
1a		Ø		Ø		Ø		Ø		Ø		Ø
2	a-	βa-	a-	βa-	a-	βa-	a-	βa-	a-	βa-	a-	βa-
2a		βa-		βa-		βaa-		βáá-		βa-		βa-
3	ɔ-	mu-	o-	mo-	o-	mo-	o-	mo-	o-	mo-	ɔ-	mu-
4	ɛ-	mi-	e-	me-	e-	me-	e-	me-	e-	me-	ɛ-	mɪ-
5	é-	ɾi-	(e-)	ɾi-	e-	ɾi-	e-	ɾi-	(i-)	ɾi-	ɛ-	ɾɪ-
6	a	ma-	a-	ma-	a-	ma-	a-	ma-	a-	ma-	a-	ma-
7	ɛ-	ki-	e-	ke-	e-	ke-	e-	ke-	e-	ke-	ɛ-	kɪ-
8	ɛ-	βi-	e-	βe-	e-	βe-	e-	βe-	e-	βe-	ɛ-	βɪ-
9	ɛ-	(N)-	e-	(N)-	e-	(N)-	a-	(N)-	e-	(N)-	ɛ-	(N)-
10	ɛgi-	(N)-	se-	(N)-	tʃe-	(N)-	tʃa-	(N)-	ze-	(N)-	ɛzɪ-	(N)-
11	ɔ-	ɾu-	o-	ɾo-	o-	ɾo-	o-	ɾo-	o-	ɾo-	ɔ-	ɾu-
12	a-	ka-	a-	ka-	a-	ka-	a-	ka-	a-	ka-	a-	ka-
13									o-	to-	ɔ-	tu-
14	ɔ-	βu-	o-	βo-	o-	βo-	o-	βo-	o-	βo-	ɔ-	βu-
15	ɔ-	ku-	o-	ko-	o-	ko-	o-	ko-	o-	ko-	ɔ-	ku-
16	a-	ha-	a-	ha-	a-	ha-	a-	ha-	a-	ha-	a-	ha-
17		ku-		ko-		ko-		-		ko-		ku-
18		mu-		mo-		mo-		mo-		mo-		mu-
19					e-	hi-	e-	hi-	e-	he-	ɛ-	hɪ-
20		gu-	o-	ɣo-	o-	ɣo-	o-	ɣo-	o-	go-	ɔ-	gu-
22		ga-										
23		i-		e-(y-)						i-/e-		

"-" indicates that the class is in use in the relevant language but the prefix is not attested.

All of the corpus languages have the following expected standard class pairings: 1/2, 1a/2a, 3/4, 5/6, 7/8, 9/10, 9a/10a, 11/10, 14/6, 15/6. The non-standard class pairings are as follows: Kabwa, Simbiti and Zanaki have 12/8, whereas Ikoma, Ngoreme and Ikizu attest 12/19 (in Ikizu as an alternative for 12/13). Zanaki, Ikoma, Ikizu and Ngoreme have 11/14. Zanaki, Ikoma, Simbiti and Ngoreme have 9/6. Zanaki, Simbiti and Ngoreme have 11/6. Ikizu and Zanaki have 20/6. Most of the corpus languages have additional non-standard pairings: Zanaki (9a/6, 5/14, 5/4); Kabwa (20/22); Ikizu (5/10, 11/8, 12/13); Simbiti (7/6); Ngoreme (14/8, 14/10).

In terms of the animate/inanimate split and resulting concord agreement, animate nouns take agreement with their noun class.

Locative markers on nouns are prefixed before the lexical noun class prefixes without the augments, and any modifiers agree with the lexical noun class. The secondary prefixes are subject to prefix vowel harmony in all languages except Ikoma. In Ikoma, only class 18 is used as a secondary (23).

TABLE 17.11 SUBJECT AND OBJECT PREFIXES

Class	CWA SP	CWA OP	SSC SP	SSC OP	NGQ SP	NGQ OP	NTK SP	NTK OP	IKZ SP	IKZ OP	ZAK SP	ZAK OP
1SG	n-/ni-	n-	n-	n-	ne-/n-	n-	ne-/n-	n-	ne-/n-	n-	ni-	nɪ-
2SG	o-	ku-	o-	ko-	o-	ko-	o-	ko-	o-	ko-	ɔ-	kʊ-
1	a-	mu-	a-	mo-	a-	mo-	a-	mo-	a-	mo-	a-	mʊ-
1PL	tu-	tu-	to-	to-	to-	to-	to-	to-	to-	to-	tʊ-	tʊ-
2PL	mu-	βa-	mo-	βa-	mo-	βa-	βa-	βa-	mo-	βa-	mʊ-	βa-
2	βa-	βa-	βa-	βa-	βa-	βa-	βa-	βa-	βa-	βa-	βa-	βa-
3	gu-	gu-	yo-	yo-	yo-	yo-	o-	wo-	go-	go-	gʊ-	gʊ-
4	gi-	gi-	ye-	ye-	ye-	ye-	ye-	ye-	ge-	ge-	gɪ-	gɪ-
5	ri-	ri-	re-	re-	re-	re-	re-	re-	re-	re-	rɪ-	rɪ-
6	ga-	ga-	ya-	ya-	ya-	ya-	ya-	ya-	ga-	ga-	ga-	ga-
7	ki-	ki-	ke-	ke-	ke-	ke-	ke-	ke-	ke-	ke-	kɪ-	kɪ-
8	βi-	βi-	βe-	βe-	βe-	βe-	βe-	βe-	βe-	βe-	βɪ-	βɪ-
9	e-	ji-	e-	ye-	e-	je-	e-	je-	e-	e-/je-	ɛ-	I-
10	gi-	gi-	se-	se-	tʃe-	tʃe-	tʃe-	tʃe-	ze-	ze-	zɪ-	zɪ-
11	ru-	ru-	ro-	ro-	ro-	ro-	ro-	ro-	ro-	ro-	rʊ-	rʊ-
12	ka-	ka-	ka-	ka-	ka-	ka-	ka-	ka-	ka-	ka-	ka-	ka-
13									to-	to-	tʊ-	tʊ-
14	βu-	βu-	βo-	βo-	βo-	βo-	βo-	βo-	βo-	βo-	βʊ-	βʊ-
15	ku-	ku-	ko-	ko-	ko-	ko-	ko-	ko-	ko-	ko-	kʊ-	kʊ-
16	ha-	ha-	ha-	ha-	ha-	ha-	ha-	ha-	ha-	ha-	ha-	ha-
17	-	-	-	-	-	-	-	-	-	-	-	kʊ-
18	-	-	-	-	-	-	-	-	-	-	-	mʊ-
19					he-	he-	he-	he-	he-	he-	hɪ-	hɪ-
20	gu-	gu-	yo-	yo-	yo-	yo-	yo-	yo-	go-	go-	gʊ-	gʊ-
22	ga-	ga-										
23	-	-	-	-					-	-		

"-" indicates that class is in use in the relevant language but the prefix is not attested.

(23) Ikoma
　　　mò-βì-táárò　　èβéé-ndè
　　　18-8-river　　　PP_8-other
　　　'in the other rivers'

(24) Kabwa
　　　kù-βì-kùkù　　βí-jó　　　βjá　　Kwíírààŋgì
　　　17-8-hill　　　PP_8-DEM.REF　CON$_8$　Kwirangi
　　　'in those hills of Kwirangi'

Ikoma has two lexicalised uses of secondary locative prefixes: the prefix o- (possibly from class 17 ko-), which can be used with kinship terms meaning 'to,' and the prefix a- (possibly from class 16 ha-) meaning 'at,' which is only used with a small set of nouns (25).

(25) Ikoma

tààtà 'father'   mò-tààtà 'inside father's room'   ò-tààtà 'to father'
[1a]father   18-[1a]father   17?-[1a]father

òmò-ɣóóndò 'farm'   mò-mò-ɣóóndò 'inside the farm'   à-mò-ɣóóndò 'at the farm'
3-farm   18-3-farm   16?-3-farm

3.2 Noun modifiers and substitutives

Modifiers (i.e., adjectives, numerals and pronominal forms) mostly follow the head noun in the Mara languages with the most common order being N-POSS-DEM-ADJ-NUM (Ikoma, Kabwa, Zanaki). Zanaki allows for demonstratives to precede the noun while in Ikizu demonstratives occur before the noun in the standard word order. In Ikoma, numerals are rather movable and can even precede the noun. In Simbiti, the standard order is N-POSS-ADJ-NUM-DEM, but generally, the ordering is quite flexible for the sake of what appears to be emphasis on the post-N element.

Adjectives take the nominal prefix (NP) without the augment. The enumerative prefix (EP) is identical to the pronominal prefix (PP) in most classes, except in vowel qualities. In Simbiti, Ngoreme and Ikoma the enumerative prefix has a fixed vowel quality that does not harmonise with the numeral stem vowels like the other prefixes. In Ikizu, only class 10 enumerative prefix *i-* is exceptionally high and does not harmonise. There is variation in the presence of the consonant *g-* in classes 3, 4 and 6; these prefixes are reconstructed either with or without the consonant, as in Meeussen (1967: 97). Likewise, the class 10 enumerative prefix is canonical in that it is a single vowel in all six languages. The interrogative *-reeŋge* (Ngoreme, Ikoma)/*-reeŋga* (Simbiti)/*-riiŋga* (Ikizu) 'how many' also takes the enumerative prefix.

In Ngoreme, the numerals *-mwé* '1,' *-βérè* '2,' *-tátò* '3,' *-né* '4,' *-táánó* '5,' *-sáánsàβà* '6,' and *-nááné* '8' are inflected while *mòhúúŋgàtè* '7' and *kééndà* '9' are invariables. In the other languages, the numerals are similar in form although there is some variation in the quality of the vowels. In Ikoma, where word-final labialised syllables are not allowed, numeral 1 is *-mù*. The numerals '7,' '8,' '9' and '10' are invariables in Kabwa (*mùhúúŋgáti, mùnááné, kèéndá, èkúmí*) and Simbiti (*mùhùùŋgátè, mònáánè, kééndà, ìkómì*).

The nasals in numerals *-né* '4' and *-nááné* '8' are palatalised with the class 10 enumerative prefix in all languages. With the class 10 prefix the stem initial *t* is fricativised to *s* or *ʃ* in *-tátò* '3' and *-táánó* '5' in all the languages except Kabwa where *-sátú* '3' and *-táánú* '5' are found with all prefixes. In Simbiti, the class 10 fricativisation is optional with *-tátó* '3.'

For '10,' a variant of *ìkómì* is used. For multiples of ten, a variant of *miroongo* is used, e.g., *mìróóŋgó èβèrè* '20.' Again, vowel qualities change from one language to another but otherwise the systems are identical.

Pronominal forms include demonstratives, connectives, independent and possessive pronouns, quantifiers, specifiers and interrogatives. These are invariable forms (independent pronouns for persons 1, 2, and 3) or take the pronominal prefix. In Simbiti, the pronominal prefix has the augment with most pronominal forms in all classes except class 23.

All six languages have the same set of demonstratives: the Proximal Demonstrative *-no/-nɔ/-nu/-no*, the Distal Demonstrative *-rja/-Vre/-Vra*, and the Referential Demonstrative *-jo/-jɔ*. In Ikoma, the demonstratives are toneless when utterance-initial but have an initial H tone utterance medially.

TABLE 17.12 PRONOMINAL PREFIXES (PP)

Class	CWA	SSC	NGQ	NTK	IKZ	ZAK
1	wa-	u-	o-/wo-	o-	o-/wo-	wa-
2	βa-	(a-) βa-	βa-	βa-	βa-	βa-
3	gu-	(u-) ɣo-	ɣo-	o-	go-	gʊ-
4	gi-	(i-) ɣe-	ɣe-	ɣe-	ge-/dʒ-	gɪ-
5	ɾi-	(i-) ɾe-	ɾe-	ɾe-	ɾe-	ɾɪ-
6	ga-	(a-) ɣa-	ɣa-	ɣa-	ga-	ga-
7	ki-	(i-) ke-	ke-	ke-	ke-/tʃ-	kɪ-
8	βi-	(i-) βe-	βe-	βe-	βe-	βɪ-
9	ja-	i-	e-/je-	e-	e-/je-	ja-
10	gi-	(i) se-	tʃ-/tʃe-	tʃe-	z(e)-	zɪ-/za-
11	ɾu-	(u-) ɾo-	ɾo-	ɾo-	ɾo-	ɾʊ-
12	ka-	(a-) ka-	ka-	ka-	ka-	ka-
13					to-	tʊ-
14	βu-	(u-) βo-	βo-	βo-	βo-	βʊ-
15	ku-	(u-) ko-	ko-	ko-	ko-	kʊ-
16	ha-	(a-) ha-	ha-	ha-	ha-	ha-
17	ku-	-	ko-	ko-	-	kʊ-
18	mu-	-	mo-	mo-	-	mʊ-
19			he-	he-	he-	hɪ-
20	gu-	(u-) ɣo-	ɣo-	ɣo-	go-	gʊ-
22	ga-					
23	(o)-wa-	(a)-ha-/wa-			(i)wa-	

"-" indicates that class is in use in the relevant language but the prefix is not attested.

(26) Ikoma
Class 1:	uno	uɾe	ujo
Class 2:	βano	βaaɾe	βajo
Class 9:	ino	jiiɾe	ijo
Class 10:	tʃino	tʃiiɾe	tʃijo

The Connective stem in Simbiti, Kabwa, Ikizu and Zanaki is simply -a, which is elided before an augment vowel. In Ikoma, rather than using the a-stem, the Connective is identical to the pronominal prefix (see Table 17.12). The vowel of the Connective causes the following pre-prefix to elide. In Ngoreme, some of the Connectives have the a-stem (e.g., class 1 wa, class 9 ja, class 10 tʃa) while others are identical to the pronominal prefix.

(27) a òmóónà   ó   mùɣènì
         omo-óna   ó   omu-ɣeni
         1-child   CON₁   1-visitor
         'child of visitor'

      b òmóónà   ó   βàɣènì
         omo-óna   ó   aβa-ɣeni
         1-child   CON₁   2-visitor
         'child of visitors'

(28) a   Ngoreme
          òmóónà   wò   mùɣénì
          omo-óna   wa   omu-ɣéni
          1-child   CON₁   1-visitor
          'child of visitor'

      b   òmóónà   wà   βàɣénì
          omo-ónà   wa   aβa-ɣéni
          1-child   CON₁   2-visitor
          'child of visitors'

      c   òβò-rítò   βò   táátà
          14-weight   CON₁₄   father
          'weight of father'

      d   òβò-rítò   βwὲ   ŋ-gókɔ̀
          oβo-ríto   βo   ɛŋ-gókɔ
          14-weight   CON₁₄   chicken
          'weight of chicken'

Independent pronouns occur in singular and plural forms (Table 17.13), with Ikoma and Ngoreme exhibiting the unusual 2SG form *ije*. Possessive pronouns are typically formed with pronominal stems, preceded by the Connective and the pronominal prefix. In Ikoma and Ngoreme, where the Connective -*a* is mostly missing, the vowel quality of the possessive is the quality of the pronominal prefix except in 1st person plural -*ito*/-*itu*. In Ikoma and Ngoreme, the tones of the independent and possessive pronouns vary depending possibly on the syntactic position.

In addition to the forms in Table 17.13, Ikizu and Zanaki also have a possessive stem -*aku* that can refer to both singular and plural possessors, mostly nouns in classes other than 1/2. In Kabwa, the possessive stems listed in Table 17.13 are only used with classes 1 and 2. In all the other classes, the possessive is inflected according to the noun class of both the possessed and the possessor.

(29) Kabwa
      a   èrì-gúhá   rj-áá-j-ò   àmà-gúhá gá-á-j-ò
          5-bone PP₅-CON-PP₉-POSS   6-bone PP₆-CON-PP₉-POSS
          'its (cl. 9) bone'   'its (cl. 9) bones'

      b   èrì-gúhá   rj-áá-gj-ò   àmà-gúhá gá-á-gj-ò
          5-bone PP₅-CON-PP₁₀-POSS   6-bone PP₆-CON-PP₁₀-POSS
          'their (cl. 10) bone'   'their (cl. 10) bones'

TABLE 17.13 INDEPENDENT AND POSSESSIVE PRONOUNS

Language/ Person	CWA	SSC	NGQ	NTK	IKZ	ZAK
1SG	ɔni/-ani	oni/-anɛ	eni/-Vne	eni/-Vne	iɲe/-ane	eɲe/-ɜni
2SG	uwɛ/-ahɔ	uwɛ/-ahɔ	ije/-Vtʃo	ije/-Vtʃo	awe/-azo	ɜwe/-ɜzo
3SG	wɛɛki/-ajɛ	wɛ/-ajɛ	we/-Vtʃe	iwe/-Vtʃe	ɛwɛ/-aze	ewe/-ɜje
1PL	βɛɛtu/-ɛtu	βeeto/-eto	itu/-itu	itu/-ito	itwe/-ito	etwe/-etu
2PL	βɛɛɲu/-ɛɲu	βeeɲu/-eɲu	iɲu/-Vɲu	iɲu/-Vɲu	imwe/-aɲu	emwe/-eɲu
3PL	βɛɛki/-aβɔ	βɔ/-aβɔ	aβu, βuu/-Vβo	iβo/-Vβo	eβo/-aβo	eβo/-ɜβu

The quantifier -*ose* 'all,' specifiers -*ene* 'self' and -*nde* 'other' and interrogative -*Vhe* 'which' are found in all the Mara languages in slightly different shapes. These all take the pronominal prefix. All of the languages also use the augment vowel with -*εnε* 'self' and -*nde* 'other.'

4 VERBS AND CLAUSE STRUCTURE

This section covers verbal extensions/derivation (4.1), TAM and negatives (4.2), and clause types and discourse/information structure (4.3). While the extensions are fairly typical and straightforward, the TAM systems in the Mara languages are surprisingly diverse. The final section provides a detailed overview of the various clause types and discourse features in the Mara languages including topic, focus and pragmatic concerns.

4.1 Verbal extensions

As with many other Bantu languages, the Mara languages make use of various derivational extensions, which are realised immediately after the verb root. The basic forms of the most productive extensions are shown in Table 17.14, which excludes possible variants based on vowel harmonisation. A more detailed treatment of the vowel harmonisation effects, especially when Causative and Applicative extensions are added to the verb root, is found in Section 2.1 and Tables 17.3 and 17.4.

TABLE 17.14 PRODUCTIVE VERBAL EXTENSIONS

	CWA	SSC	NGQ	NTK	IKZ	ZAK
Applicative	-iɾ-	-εɾ-	-εɾ-	-εɾ-	-eɾ-	-ɪɾ-
Causative	-i-	-i-	-i	-i	-i-	-i-
Reciprocal	-an-	-an-	-an-	-an-	-an-	-an-
Inversive/Reversive/Separative	(-uɾ-)	-ɔɾ-	-ɔɾ-	-oɾ-	-oɾ-	-ʊɾ-
Passive	-u-	-w-	-u	-u	-u-	-ʊ-
Stative	-ik-	(-εk-)	(-εk-)	-εk-	-ek-	-ɪk-

It is difficult to determine the underlying quality of the vowel of Simbiti Passive, or even whether it is underlyingly a vowel or a glide, since it always surfaces as a labialisation on a preceding consonant, and it does not cause vowel raising as the causative extension does. The passive is left as -*w*- in the chart, following Chacha and Odden (1998) in their analysis of closely related Kuria.

The Inversive extension is a less productive suffix than the Applicative in the Mara languages, and it tends to display asymmetric patterns of vowel harmony.

For the Stative extension, the vowel in the root controls the harmony of the suffix in Kabwa, Ikoma, Ikizu and Zanaki. In Ngoreme and Simbiti, the Stative extension is not productive. Simbiti has a frozen -*εk*- form in several verb stems, indicating that a suffix-controlled vowel harmony pattern once occurred. The Stative meaning in Simbiti and Ngoreme is currently conveyed through the use of either an Anterior aspect marker or a Reflexive morpheme.

Ngoreme (30) and Ikoma (31) have atypical Passive and Causative forms. In these forms, the final vowel -*a* drops out and the passive (-*u*) or causative (-*i*) fill the final slot on the verb stem; labialisation and palatalisation are only allowed root-initially in these languages.

(30) Ngoreme Passive
      kù-ɾɛ́m-à   kù-ɾɛ́m-ù
      INF-farm-FV   INF-farm-PASS
      'to farm'   'to be farmed'

(31) Ikoma Causative
      kò-ɾɛ́m-à   kò-ɾɛ́m-ì
      INF-farm-FV   INF-farm-CAUS
      'to farm'   'to cause to farm'

Multiple verbal extensions on a single verb root are possible in the Mara languages, but they have yet to be studied extensively. There are some interesting phonological phenomena that occur due to vowel harmonisation effects when a derivational extension is strung together with the Anterior suffix. For example, when the Anterior (-iɾi) and Passive (-ʊ-) suffixes co-occur in Zanaki, the result (-iɾwe) combines the two morphemes together in an unpredictable way (32). When these same suffixes co-occur in Ikoma, the combination of -iɾi and -u is realised as -uɾu (or -βuɾu on monosyllabic roots; 33).

(32) Zanaki Anterior Passive
      βa-tem-iɾwe
      SP$_2$-hit-ANT.PASS
      'they have been hit'

(33) Ikoma Anterior Passive
      tʃè-ɾèè-βùɾù
      SP$_{10}$-eat-ANT.PASS
      'it (meat) has been eaten'

4.2 TAM(P)

The Mara languages use the interaction of tense, aspect (both lexical and grammatical), and mood/modality (TAM) as grammatical categories that encode temporal relations. The focus in this section is on the grammatical categories of tense and aspect. As in other areas of their grammars, a striking amount of variety exists among the TAM systems of the Mara languages. The systems in Kabwa and Simbiti are more traditional Bantu systems in that they have remoteness distinctions and a wide range of TA combinations. In contrast, the systems found in the remaining corpus languages are organised with aspect as their central defining feature and make a much more limited use of tense. The interplay between Aktionsart (lexical aspect) and grammatical aspect to convey temporal relations is striking. All languages also make use of a variety of periphrastic constructions, not all of which have been listed in the tables below.

4.2.1 Kabwa and Simbiti

The TA systems of Kabwa and Simbiti (as well as closely related Kuria; Cammenga 2004) resemble more traditional Bantu systems by having many tense distinctions. An overview of the TA forms in these languages is found in Tables 17.15 and 17.16. In these tables PST$_1$ refers to the most recent ("hodiernal" or earlier today) past and is also used as ANT; PST$_2$ is the "hesternal" (or yesterday) past and PST$_3$ is "prehesternal" (before

TABLES 17.15 & 16 KABWA AND SIMBITI TA SYSTEMS

	CWA	SSC
PAST		
PST$_1$	SBJ-VB-írí	[5] (n)-SBJ-VB-ire
PST$_2$	SBJ-aa-VB-írí	
PST$_3$	[10] SBJ-aa-VB-a	[6] (n)-SBJ-aa-VB-ire
PRESENT	SBJ-aa-VB-a	[5] SBJ-ra-VB-a/(SBJ-Ø-VB-a)
FUTURE		
FUT$_1$	[10] (n)-SBJ-ráá-VB-ɛ	(n)-SBJ-ráá-VB-ɛ
FUT$_2$	[1] SBJ-aka-VB-ɛ	[7] (n)-SBJ-aka-VB-ɛ
FUT$_3$		[6] (n)-SBJ-ri-VB-a
FUT$_4$	[4] SBJ-ri-VB-a	
NAR/CSC	[1] SBJ-ka-VB-a	[8] SBJ-ka-VB-a
PERSISTIVE	SBJ-kja-VB-a	[6] n-ko-VB-a SBJ-ke-re/ ([5] SBJ-kee-VB-a)
CONDITIONAL	[10] SBJ-ráá-VB-ɛ	SBJ-ráá-VB-ɛ
HYPOTHETICAL	[1] SBJ-kaa-VB-írí	[5] n-SBJ-kaa-VB-ire

	CWA HAB/IPFV	PROG	SSC HAB/IPFV	PROG
PAST				
PST$_1$	-	SBJ-aa-rí [1] SBJ-ra-VB-a	-	(n)-SBJ-á-rɛ ko-VB-a
PST$_2$	SBJ-aa-VB-aŋga	SBJ-aa-rí-ŋga [1] SBJ-ra-VB-a	[6] (n)-SBJ-aa-VB-aŋga	(n)-SBJ-aa-rɛ [6] ko-VB-a
PST$_3$				(n)-SBJ-aa-rɛ-ŋga [6] ko-VB-a
PRESENT	SBJ-ra-VB-aŋga	[10] SBJ-rá-VB-a	[5] SBJ-ra-VB-aŋga	n-ko-VB-a [9] SBJ-rɛ
FUTURE				
FUT$_1$	[10] SBJ-ráá-VB-ɛŋga	SBJ-ráá-β-ɛŋga [1] SBJ-ra-VB-a	(n)-SBJ-ráá-VB-ɛŋga	(n)-SBJ-ráá-β-ɛŋga [5] SBJ-ra-VB-a
FUT$_2$	[10] SBJ-aka-VB-ɛŋga	SBJ-aka-β-ɛŋga [1] SBJ-ra-VB-a	[7] (n)-SBJ-aka-VB-ɛŋga	(n)-SBJ-aka-β-ɛŋga [5] SBJ-ra-VB-a
FUT$_3$			[6] (n)-SBJ-ri-VB-aŋga	(n)-SBJ-ri-β-áŋga [5] SBJ-ra-VB-a
FUT$_4$	[11] SBJ-ri-VB-áŋga	SBJ-ri-β-áŋga [1] SBJ-ra-VB-a		

The superscript numbers in the Tables refer to tone melodies:
H on σ1 of the macrostem (stem in Ngoreme): 1
H on μ1+μ4 of the macrostem: 5
H on μ3 of the macrostem: 6
H on μ1 of the macrostem: 7
H on μ4 of the macrostem: 8
H on SBJ: 9
H plateau on the whole macrostem, excluding the final vowel: 10
H on antepenult: 11
The forms without an indication of tonal melody are realised with the prefix/suffix tones only and as all L when there are no H tones on the prefixes/suffixes.

yesterday). FUT$_1$ refers to the immediate/imminent future, FUT$_2$ is used for the "hodiernal" (later today) future, FUT$_3$ is the category for the "crastinal" (tomorrow) future, and FUT$_4$ represents the "postcrastinal" (after tomorrow) category.

Looking just at reference to past tense, the Kabwa and Simbiti systems are very similar. Although Kabwa has three distinct forms while Simbiti has two, Simbiti also has a third TAM form with past reference, *SBJ-aa-VB-a*, which has been referred to as an "Untimed Past Anterior" because it does not have a specific temporal reference but merely indicates that the action occurred at some time in the past (Cammenga 2004: 287, Walker 2013: 62–63). This form aligns quite nicely with the PST$_4$ (PreHesternal or "Before Yesterday" Past tense) in Kabwa, though the Simbiti form does not carry a specific temporal reference with it. The future tense systems of the two languages are also quite similar.

In several complex verb forms, Simbiti (as well as Kuria JE43 and Gusii JE42) exhibits a pattern of auxiliary inversion which is atypical for Eastern Bantu languages (Walker 2013: 101), although a similar pattern of inversion with different functions has been observed in Mbugwe F34 (Mous 2000) and in Rangi F33 (Gibson 2013). For Simbiti, auxiliary inversion occurs in the Persistive and complex Progressive forms where the main verb appears before the auxiliary (34). Interestingly, the Progressive forms in Ngoreme (see Table 17.17) also display this type of auxiliary inversion.

(34) Simbiti Auxiliary Inversion (adapted from Walker 2013: 101)
    ùrù-síkò   ùrù-ɣímà   ŋ-kù-tùk-à   tó-rè
    11-day   11-whole   FOC-INF-dig-FV   SP$_{1PL}$-COP
    'We are digging all day long.'

Another interesting feature of Simbiti verbs is the capability to include focus-marking (N-) on almost any verb form in a main clause where it indicates emphasis on the action itself (Walker 2013: 128–129). In example (35), the focus is on the verb in the first sentence, and on the pre-verbal temporal adverb in the second.

(35) Simbiti Focus-marking (adapted from Walker 2013: 128–129)
    n-tw-àà-ʃùmááʃ-ìrè   nà-wè̀   mɔ̀-ɔ́kà
    FOC-SP$_{1PL}$-PST$_2$/PST$_3$-speak-ANT with-3SG 3-year
    'We spoke with him last year.'

    ŋ-kàrɛ́   tw-àà-ʃùmááʃ-ìrè   nà-wè̀
    FOC-long.time.ago   SP$_{1PL}$-PST$_2$/PST$_3$-speak-ANT with-3SG
    'A long time ago, we spoke with him.'

Although both Ikoma and Ngoreme also have verb forms with the nasal, it appears to be a lexicalised feature of specific verb forms rather than an active synchronic strategy for indicating emphasis.

4.2.2 Ngoreme and Ikoma

The Ikoma and Ngoreme TA systems rely less on tense distinctions, and have many more aspectual distinctions than Kabwa and Simbiti. The tense contrasts in Ikoma and Ngoreme (along with Ikizu and Zanaki in 4.2.3 below) are clearly appendages to the aspectual system (cf. Nurse 2008: 68). Tense contrasts are limited to the narrative *-ka-/-Vka-* and periphrastic

TABLE 17.17   IKOMA AND NGOREME TA SYSTEMS

	NGQ	*NTK*
Perfective	⁹ (n)-SBJ-VB-iɾe	n-SBJ-VB-iɾi
Retrospective	¹ (n)-SBJ-a-VB-iɾe	³ n-SBJ-V-VB-iɾi
Imperfective	¹ SBJ-ɾa-VB-a	¹ (n)-SBJ-Vko-VB-a
Progressive	¹ n-ko-VB-a SBJ-V-ní	¹ SBJ-ɾa-VB-a
	¹ n-ko-VB-a SBJ-áɾe	n-SBJ-V-ɾé ¹ SBJ-ɾa-VB-a
Habitual	n-SBJ-háá-VB-a	n-SBJ-háá-VB-a
	n-SBJ-ɾáá-VB-a	
Persistive	SBJ-kééɾe ¹ SBJ-ɾa-VB-a	(n)-SBJ-keeɾé ¹ SBJ-ɾa-VB-a
	SBJ-kééɾe ko-VB-a	(n)-SBJ-keeɾé ¹ ko-VB-a
	SBJ-kééɾe ¹ (n)-aŋga-SBJ-VB-e (?)	(n)-SBJ-keeɾé ³ n-SBJ-V-
	SBJ-kééɾe ⁹ (n)-SBJ-VB-iɾe	VB-iɾi
Conditional	¹ SBJ-ko-VB-a	SBJ-ɾáá-VB-a
Inceptive	SBJ-Vká-VB-a	² SBJ-Vká-VB-a

H on σ1 of the macrostem (stem in Ngoreme): 1
H on final σ: 2
H on σ1 of the macrostem + final H: 3
H on SBJ: 9
The forms without an indication of tonal melody are realised with the prefix/suffix tones only and as all L when there are no H tones on the prefixes/suffixes.

constructions with auxiliaries derived primarily from copulas, movement verbs and "to finish, be finished" verbs (Bybee *et al.* 1994: 55), which are all relatively recent according to the "grammaticalization cycle" of compounding and inflection (Nurse 2008: 170). These systems in Ikoma and Ngoreme mainly use the interaction of Aktionsart (lexical aspect) and grammatical aspect to impart temporal meaning. For example, consider that the Imperfectives can have both present and future readings, as in examples (36–39) below.

Osa-Gómez del Campo (2013) analyses closely related Nata in like manner, using insights from Kershner (2002) in conceptualising Bantu lexical aspect in terms of a split between punctives (statives and achievements) and duratives (activities and accomplishments). The categories in Table 17.17 also reflect terminological influence from Hewson (2012: 525), e.g., Retrospective, a "Perfective [. . .] viewed from a later position." As can be seen from the table, the fundamental contrast is perfective:imperfective, with a host of additional imperfective categories in combination with a general Imperfective (*-ɾa-* in Ngoreme and *-Vko-* in Ikoma).

The primary difference between the Ngoreme and Ikoma patterns occurs between the general Imperfective and Progressive forms. The Imperfective and Progressive morphology is inverted across these systems (Imperfective *-ɾa-* in Ngoreme and *-Vko-* in Ikoma, while Progressive *-ko-* in Ngoreme and *-ɾa-* in Ikoma). Consider for instance the Imperfective futurates in examples (36) and (37).

(36)   Ikoma Imperfectives with future readings
　　　 m-bà-àkò-hík-à　　　　n-tò-òγò-sííkèr-à
　　　 FOC-SP$_2$-IPFV-arrive-FV　FOC-SP$_{1PL}$-IPFV-speak-FV
　　　 'they will arrive'　　　 'we will speak'

(37) Ngoreme Imperfectives with future readings
βà-ɾà-hík-à   tù-ɾà-ɣáámbàn-à
SP₂-IPFV-arrive-FV   SP₁ₚₗ-IPFV-speak-FV
'they will arrive'   'we will speak'

These patterns reflect the general cline: focus > progressive > general present > future or non-past (Nurse 2008: 294). Progressives commonly originate from locatives, which explains the -ko- formative, while -ra- is presumed to have originated as a focus marker (Nurse 2008: 139, 294). Bybee et al. (1994: 140) further note that progressives can develop into more general imperfectives. Also note the auxiliary inversion that occurs in the Ngoreme Progressive construction.

Another key difference between Ngoreme and Ikoma is in how the Inceptive -Vká- form is utilised. Ikoma makes use of the Inceptive as a pseudo-present depending on whether the verb in question is punctive (stative, achievement) or durative (activity, accomplishment).[1] If punctive, Ikoma often uses the Inceptive in such cases; if durative, Imperfective -Vko-.

(38) Ikoma punctive Inceptives with present readings
à-ká-hìk-à   à-ká-mèɲ-à
SP₁-INC-arrive-FV   SP₁-INC-know-FV
's/he arrives, is arriving' 's/he knows'

(39) Ikoma durative Imperfectives with present readings
n-tò-òɣò-túk-à   n-à-àɣù-kóɾ-à
FOC-SP₁ₚₗ-IPFV-dig-FV   foc-sp₁-ipfv-build-fv
'we dig, are digging'   's/he builds, is building'

This is in contrast to Ngoreme which for pseudo-presents tends to use the Perfective for punctives, and either the Progressive n-ko-VB-a SBJ-ni construction or the more general Imperfective -ra- formative for duratives.

TABLE 17.18 IKIZU AND ZANAKI TA SYSTEMS

	IKZ	ZAK
Perfective	[1] SBJ-VB-iɾe	SBJ-VB-iɾɪ
Retrospective	[4] SBJ-aa-VB-iɾe	SBJ-a-VB-iɾɪ
Imperfective	[1] SBJ-ɾa-VB-a	SBJ-ɾa-VB-a
Habitual	SBJ-háá-VB-a	SBJ-haa-VB-a
Persistive	SBJ-kee-ɾé SBJ-ɾa-VB-a	SBJ-ka-β-a SBJ-kyaa-VB-a SBJ-ki-VB-a
Conditional	SBJ-ráá-VB-e	SBJ-ɾaa-VB-e

H on σ1 of the macrostem (stem in Ngoreme): 1
H on penult σ: 4
The forms without an indication of tonal melody are realised with the prefix/suffix tones only and as all L when there are no H tones on the prefixes/suffixes.

(40) Ngoreme punctive Perfectives with present readings
á-hìk-ìrè   n-á-màɲ-ìrè
SP₁-arrive-PFV   FOC-SP₁-know-PFV
's/he arrives, is arriving' 's/he knows'

(41) Ngoreme durative Progressive and Imperfective with present readings
n-gù-kɔ́ɾ-à à-ní   tù-ɾà-túk-à
FOC-PROG-build-FV SP$_1$-COP   SP$_{1PL}$-IPFV-dig-FV
's/he builds, is building'   'we dig, are digging'

4.2.3 Ikizu and Zanaki

The Ikizu and Zanaki TA systems (Table 17.18) resemble the systems in Ikoma/Ngoreme, but make use of even fewer aspectual categories (no Inceptive and only additional Progressive/Imperfective -ku-/-ko- category in relative clause affirmative verbs).

Much like Ikoma/Ngoreme, the majority of tense contrasts are limited to the narrative -ka- and periphrastic constructions. However, both Ikizu and Zanaki make use of the -ka- formative as both a Narrative *and* Remote Past tense (42; not included in Table 17.18). The remainder of the Mara languages only use this -ka- in narrative function.

(42) Zanaki Remote Past -ka- (adapted from Walker 2013: 61)
tʊ-ka-gaamb-an-a   na-we   ɛkaɾɛ
SP$_{1PL}$-PST-speak-RECP-FV   with-3SG   long.time.ago
'We spoke with him a long time ago.'

Furthermore, what sets Ikizu apart from its close neighbour Zanaki is the use of the *SBJ-aa-VB-a* construction (43; not included in Table 17.18), which resembles the Present tense form in Kabwa. The function of this construction in Ikizu needs more study.

(43) Ikizu Present (adapted from Walker 2013: 91)
tw-àà-j-á   dàrèsàráámù
SP$_{1PL}$-PRS-go-FV   Dar-es-Salaam
'We go to Dar es Salaam.'

4.2.4 Negatives

Negatives in the majority of the Mara languages are formed with -ta- between the subject marker and TA slots, often with a H tone prefix. Simbiti is the exception as it uses a combination of two elements, namely pre-verbal ti-/te-/tɛ- and post-verbal *he*, for the majority of its negative forms (apart from the use of -ta- in the Hypothetical). There are several other instances of a two-element negation strategy in Eastern Bantu, as reported for closely related Kuria (Cammenga 2004: 247, Mwita 2008: 178–180), as well as Rangi and Mbugwe (Gibson 2013), along with many others (Devos & van der Auwera 2013).

(44) Simbiti
tɛ̀-βà-àkɔ̀-βét-à hè   tɛ̀-β-àà-βét-ìrè hè
NEG-SP$_2$-PRS-burn-FV NEG   NEG-SP$_2$-PST-burn-PFV NEG
'they do not burn'   'they did not burn'

Although the post-verbal *he* does not need to be adjacent to the verb, as in *tì-tù-ù-hík-ɛ̀ bɔ̀ŋgó hè* 'we will (FUT$_1$) not arrive early,' it is considered part of the melodic tone domain when it is adjacent to the verb stem. For example, in FUT₃, the melodic H is assigned to the third mora of the macrostem, and retracted to the penult if it should fall on the final syllable. In shorter macrostems, the melodic H is not realised at all. However, the melodic H is realised on the stem-final syllable when the enclitic follows the stem.

(45)   Simbiti

a   m-bà-rè-βèrék-à   vs.   tè-βà-rè-βèrèk-á hè
    FOC-SP₂-FUT₃-carry-FV   NEG-SP₂-FUT₃-carry-FV NEG
    'they will carry'   'they will not carry'

b   m-bà-rè-βèt-à   vs.   tè-βà-rè-βèt-á hè
    FOC-SP₂-FUT₃-burn-FV   NEG-SP₂-FUT₃-burn-FV NEG
    'they will burn'   'they will not burn'

4.3 Clause types, information structure, and related discourse features

The Mara languages conform to the standard Bantu SVO order in pragmatically neutral main clauses.[2] Similarly, common phenomena like object left-dislocation (46) and subject right-dislocation have been observed. Despite these general similarities, the differences at clause and discourse level, as described in the following subsections, are striking.

(46)   Simbiti (Masatu 2015; tones added)
à-kà-ɣààmb-á íɣáà   sè-ŋɔɔ́mbέ sέ-nɔ́   sé-réé-ŋgé   kà   há-nɔɔ́
SP₁-NAR-say-FV COMP   10-cow   PP₁₀-DEM.PROX SP₁₀-be-HAB home PP₁₆-DEM.PROX
mùùʃ-έ   ɔ̀kɔ̀-háán-à βà-nɔ̀   βà-rèè-ŋgé jèèkà há-nɔ̀ɔ́   nà
2PL.COME-SBJV INF-give-FV PP₂-DEM.PROX SP₂-be-HAB home   PP₁₆-DEM.PROX   and
ɔ̀ɔ̀rà   ùwà   mɔ̀-má-ɣììŋgà
DEM.DIST₁   CON₁   18-6-island
'He said, "These cows that are at home here, come and give them to the people at home here and to the one in the islands."'

4.3.1 Dependent clauses

Dependent clauses frequently occur as the second part in tail-head linkage in which "information from one clause is repeated at the start of the following clause, and the main verb is repeated" (Nicolle 2015: 11; see also Guillaume 2011). While all six languages use the Proximal Demonstrative of class 16 hanɔ (the final vowel varies across languages) to introduce such temporal head clauses (47), Ikizu and Kabwa more frequently employ different TAM-forms without explicit connectives to indicate tail-head linkage (48).

(47)   Ikoma
à-k(à)   é-ɣà-j(έ)   é-kà-mò-ʃóómi.   Hà-nò   é-mò-ʃòòm-ìrì . . .
9-lion   SP₉-NAR-go SP₉-NAR-OP₁-spy_on PP₁₆-DEM.PROX   SP₉-OP₁-spy_on-PFV
'And the lion went and spied on him. When it (the lion) had spied on him. . .'

(48)   Ikizu
ì-kù-βúmár-á   ᴴhá-jò   àβ-ííβúrí   βí   kí-sùsù   βà-ká-áz-à.
SP₉-PROG-sleep-FV PP₁₆-DEM.REF 2-parent CON₂ 7-Hare SP₂-NAR-come-FV
'While he [Lion] was sleeping there, Hare's parents came.'

In relative structures as well, a variety of strategies are found. More common are either a -*ku*- or -*ko*- prefix in the verbal complex, or the combination of a demonstrative and the Anterior/Perfect -*ire* suffix. Instead of -*ku*-, Kabwa uses -*ra*- as relative prefix (also Progressive aspect in Ikoma, Ikizu, Ngoreme and Zanaki, cf. 4.2.2). While Ikoma uses the demonstrative -*nɔ* in combination with an initial melodic H on the relative verb form (Aunio 2013: 311–312), both Simbiti and Ngoreme display a relative pronoun -*nɔ* which is tonally distinct from the Proximal Demonstrative (49).

(49) Simbiti
    ùmù-múrá   ó-mwé   ɔ̀-nɔ̀   jàà-ɾéésj-ààŋg-à   sì-túɣɔ̀   wɔ́ɔ̀nswὲ
    1-young_man   EP$_1$-one   PP$_1$-REL   1.PST-eat.CAUS-HAB-FV   10-cattle   with.3SG
    'One young man who used to herd the cattle with him. . .'

Rundell (2012: 26–27, 54–55) also mentions a marked use of the relative marker *niɣɔ* in Ngoreme as point of departure of manner (50). Table 17.19 summarises the main strategies for dependent clauses.

(50) Ngoreme
    . . . à-ɾà-sɔ́h-á   ɲùùmbà.   nì=ɣò   á-sɔ́h-íɾέ   ɲúúmbá . . .
    SP$_1$-PROG-enter-FV   [9]house   COP=manner.REL   SP$_1$-enter-PFV   [9]house
    '. . . she goes inside. Having gone inside in this way . . .'

TABLE 17.19 DEPENDENT CLAUSE STRATEGIES

	CWA	SSC	NGQ	NTK	IKZ	ZAK
Dependent clause as temporal head	TAM > hanu	hanɔ/βɔːnɔ	hanɔ/niɣo	hano	TAM > hano	hanɔ
Relatives	-ɾa- (prefix)	PRON -nɔ > -ko- (prefix)	PRON -nɔ	DEM -nɔ + initial H > -Vko- (prefix)	-ko- (prefix)	-ku- (prefix)

4.3.2 Participant reference, topic and focus

Two salient features in participant reference are the introduction of participants and the usage of demonstratives for participant tracking (Nicolle 2014). At the beginning of a text, major participants are typically introduced in thetic sentences (Nicolle 2015: 55–58), which usually take the form of a locative verb form of 'to be' and a post-verbal subject noun phrase (51).

(51) Zanaki
    ɛ-kaɾɛ   hɜ-joo   j-a-ɾɪ   a-ɾɪ-hʊ   ɔmʊ-ʊtʊ
    9-old_times   PP$_{16}$-DEM.REF   SP$_1$-PST-be   SP$_1$-be-LOC   1-man
    'Once upon a time there was a man . . .'

Exceptions to this formula are found in Ngoreme, which frequently has the subject noun phrase of the introduced participant pre-verbally in situ, and in Simbiti which exhibits a clause-final past tense auxiliary in pre-clitic copula constructions (52).

(52) Simbiti
mù-síímbètè nà mɔ́-hààʃà m=bá-ánà àβà ɛ́n-dá é-mwé βà-à-ɾè
1-Simbiti and 1-Hasha COP=2-child CON$_2$ 9-stomach EP$_9$-one SP$_2$-PST-be
'The Simbiti and the Hasha were children of the same mother.' (lit. 'of one stomach')

Concerning participant tracking, whereas all six languages use the Distal Demonstrative for reactivation of participants, the functions of Proximal and Referential Demonstrative usage are more disparate across languages. For example, the Referential Demonstrative is used for salience in Kabwa and Ikizu, versus its use for demoting a participant's salience in Zanaki. Additionally, the function of identification is covered by the Proximal Demonstrative in Simbiti and Ikizu but by the Distal Demonstrative in Zanaki. Table 17.20 shows the predominant features of each language in the domains of participant reference, topic and focus.

TABLE 17.20 FEATURES OF INFORMATION STRUCTURE

	CWA	SSC	NGQ	NTK	IKZ	ZAK
Intro of major participants	-(β)aangahɔ/ -anga aɾihɔ; post-V	-aɾeengahɔ; post-V but also final -aare	-aɾeho; SUBJ in situ > post-V	-aɾeho; post-V	-aɾɛhɔ; post-V	jaaɾɪ aɾɪhɔ; post-V
DEM usage a) PROX b) REF c) DIST	a) focus b) salient c) reactivate	a) ident b) peak c) reactivate	a) intro b) info c) reactivate	a) intro b) info b/c) reactivate	a) ident b) salient c) reactivate	a) emph b) non-sal c) ident b/c) reactivate
Topic a) object marking b) switch	a) internal b) SP	a) internal b) full NP; DEM.DIST	a) zero b) SP	a) internal b) SP	a) internal b) 'other' PRO	a) internal b) full NP; DEM.DIST
Focus	DEM.PROX + cleft	COP ni- + cleft	COP n- + cleft; postposed	postposed; COP n- + cleft	COP ni- + cleft	COP nɪ + cleft

When constituents are left-dislocated for topicalisation purposes, most Mara languages follow the common Bantu characteristics of a) not repeating a pronoun in situ for the left-dislocated noun phrase, and b) for objects, obligatorily marking the object on the main verb. However, Ngoreme never features such an obligatory object marker throughout the entire data set, as illustrated in (53).

(53) Ngoreme
ɛ̀-ŋànà ɛ̀-jɔ́ tɛ́ɛ́-ŋ-kwɛ̀ɛ̀β-à
9-matter PP$_9$-DEM.REF NEG-SP$_{1sg}$-forget-FV
'This matter, I will not forget.'

Also, Rundell (2012: 52) reports a case in Kabwa where a pronoun in situ duplicates for its left-dislocated subject noun (54).

(54) Kabwa (Rundell 2012; tones added)
βɔ́ɔ́nɔ̀ ɔ̀mù-kárì ɾùù-ndì wɔ́ɔ́nsɛ̀ à-ɲɔ̀ɔ̀ɾ-ɛ́ɾí ɛ̀βj-ɔ̀ɔ̀kùɾjà
now 1-woman 11-one 1.PRO 1-acquire-PFV 8-food
'Now the woman, one day, she also had got some food. . .'

Switch topic is usually marked either in the subject prefix on the main verb form (in
Kabwa, Ngoreme and Ikoma) or with a full noun phrase including a distal demonstrative
(in Simbiti and Zanaki). The marking differences between the languages are probably
pragmatically motivated. A third strategy of switch topic marking makes use of additive
and different-set pronouns (Rundell 2012), including nominal modifiers meaning 'other.'
This is preferred in Ikizu as shown in (55).

(55) Ikizu
àβá-ánà àβáá-ndé βà-kà-βóɾéɾ-á wììsè
2-child PP$_2$-other SP$_2$-NAR-tell-FV father.POSS$_{3PL}$
'The other children told their father . . .'

Finally, the main focus construction observed in the six language data sets is identificational articulation, realised as a copula plus cleft sentence (56).

(56) Zanaki
nɪ m3-gin3 j-a-ɾɪ kʊ-ɾɔɾa
COP 6-stone SP$_1$-PST-be INF-see
'It was stones which he was seeing.'

4.3.3 Questions

Yes/No-questions are differentiated from their corresponding statements by intonation or
different tone assignment only; i.e., no specific yes/no-question markers have been found
in the database. For a detailed discussion of intonation versus tone assignment in yes/
no-questions in the neighbouring Jita JE25 language, see Downing (1989).

Information questions differ across the Mara languages whether the question word
is positioned at the beginning of the question or left in situ. In Simbiti and Ikoma, the
question word is in initial position (Simbiti *Hài ákùjá?*, Ikoma *Hàjì àɣàjè?* 'Where is he
going?'), whereas in the other languages, the question word is in-situ (Kabwa *Àrágjá hà?*,
Ngoreme *Àràgí hàì?*, Ikizu *Àrádʒà hàjì?*, and Zanaki *Àràgjà hàjì?* 'Where is he going?').

In cleft-construction questions, the clause consisting of copula and question word can
be either left- or right-dislocated:

(57) Ngoreme
nè=újé á-ɣó-súk-ìrè tʃèèn-sùkò tʃí-nó?
COP=who SP$_1$-OP$_{2SG}$-braid-PFV 10-braid PP$_{10}$-DEM.PROX
'Who is it that braided you this braid?'

(58) Ngoreme
káná kí-nó ɣé-ɣó-tú-ɲééɾ-á tʃíí-ɲámá tʃí-nò ŋ=gètóké?
now PP$_7$-REL.PROX SP$_7$-PROG-OP$_{1PL}$-soften-FV 10-meat PP$_{10}$-DEM.PROX COP=what
'Now, that which is softening for us this meat is what?'

When it comes to the functions of questions, the Mara languages also differ considerably. For example, whereas rhetorical questions are hardly ever found in Kabwa, they abound in Simbiti where they are almost exclusively used for rebuke (59). In Ikoma, by contrast, rhetorical questions are mainly used to affirm the opposite of the corresponding statement (60).

(59) Simbiti

mò-ká-áhó	nà	àβá-ánà	àβá-áhò	nàwè	ò-ráá-βá-téɣ-éɾ-éè?
1-wife-POSS$_{2SG}$	and	2-child	PP$_2$-POSS$_{2SG}$	who	SP$_{2SG}$-FUT-OP$_2$-leave-APPL-FV

'Your wife and children, to whom will you leave them?' (Context: addressee is contemplating suicide, and speaker is rebuking him by implying that he should not leave them in the first place.)

(60) Ikoma

ìβééɾ(e)	ámá-ŋáná	ɣá-jò	nìβù	ò-ɣò-ɣá-kòɾ-á?
now	6-matter	PP$_6$-DEM.REF	how	SP$_{2SG}$-FUT-OP$_6$-do-FV

'Now, these matters, how will you do them?' (Implied: it is impossible to do them.)

5 CONCLUSION

Despite both the close geographical proximity and the close historical and genetic relationship of the Mara languages, a wide array of divergent linguistic phenomena appears to have developed, ranging from vowel harmony and tonal complexity to tense-aspect systems and clause structure. This shows that even closely related languages in geographical and historical proximity can be diverse. While language contact with Nilotic languages has possibly played a role in the diversity of the Mara languages, it is not a sufficient explanation by itself. The role of dissimilatory processes in the development of divergent language identities merits further investigation.

NOTES

1 This is the opposite of the situation Osa-Gómez del Campo (2013: 147–148) describes for Nata. It is possible that in the closely related language varieties of Ikoma, Isenye, Nata and Ngoreme that similar morphology applies to different event types, yet still using the same fundamental punctive/durative split. This is an area for further research.
2 In this section, tone diacritics mark surface *pitch* distinctions, not necessarily tones. At this point of analysis it is not possible to distinguish intonation and sentence-level tone processes in longer chunks of text.

REFERENCES

Anghelescu, A. (2013) Morphophonology and Tone in Nata. *UBC Working Papers in Linguistics* 34: 89–103.
Aunio, L. (2010) Ikoma Nominal Tone. *Africana Linguistica* 16: 3–30.
Aunio, L. (2013) Ikoma Verbal Tone. *Nordic Journal of African Studies* 22(4): 274–321.

Aunio, L. (2015) A Typological Perspective on Bantu Nominal Tone: The Case of Ikoma-Nata-Isenye in Western Tanzania. *Southern African linguistics and applied language studies* 33(3): 359–371.

Aunio, L. (2017) Syllable Weight and Tone in Mara Bantu Languages. In Newman, P. (ed.), *Syllable Weight in African Languages*, 191–214. Amsterdam: John Benjamins.

Bastin, Y. (2003) The Interlacustrine Zone: Zone J. In Nurse, D. & G. Philippson (eds.), *The Bantu Languages*, 501–528. London; New York: Routledge.

Bybee, J. L., R. D. Perkins & W. Pagliuca (1994) *The Evolution of Grammar: Tense, Aspect, and Modality in the Languages of the World*. Chicago: University of Chicago Press.

Cammenga, J. (2004) *Igikuria Phonology and Morphology: A Bantu Language of South-West Kenya and North-West Tanzania*. Cologne: Rüdiger Köppe.

Casali, R. (2008) ATR Harmony in African Languages. *Language and Linguistics Compass* 2(3): 496–549.

Chacha, N. C. & D. Odden (1998) The Phonology of Vocalic Height in Kikuria. *Studies in African Linguistics* 27(2): 129–158.

Devos, M. & J. van der Auwera (2013) Jespersen Cycles in Bantu: Double and Triple Negation. *Journal of African Languages and Linguistics* 34(2): 205–274.

Downing, L. J. (1989) The Interaction of Tone and Intonation in Jita Yes/No Questions. *Studies in the Linguistic Sciences* 19: 91–113.

Gambarage, J. J. (2013) Nominal ATR Harmony in Nata: An Assessment of Root Faithfulness. In Luo, S. (ed.), *Proceedings of the 2013 Annual Conference of the Canadian Linguistic Association*, 1–15. Victoria: University of Victoria, Canadian Linguistic Association. (Available online at homes.chass.utoronto.ca/~cla-acl/actes2013/Gambarage-2013.pdf).

Gambarage, J. J., A. Anghelescu, S. Burton, J. Dunham, E. Guntly, H. Keupdjio, Z. W.-M. Lam, A. Osa-Gómez del Campo, D. Pulleyblank, D. Si, Y. Yoshino & R.-M. Déchaine (2017) The Nata Documentation Project: An Overview. In Kandybowicz, J. & H. Torrence (eds.), *Africa's Endangered Languages*. Oxford: Oxford University Press.

Gibson, H. (2013) Auxiliary Placement in Rangi: A Case of Contact-Induced Change? *SOAS Working Papers in Linguistics* 16: 153–165.

Gray, H. (2013) *Locatives in Ikizu*. Leiden: Leiden University, MA thesis.

Guillaume, A. (2011) Subordinate Clauses, Switch-Reference, and Tail-Head Linkage in Cavineña Narratives. In Van Gijn, R., K. Haude & P. Muysken (eds.), *Subordination in Native South American Languages*, 109–139. Amsterdam: John Benjamins.

Hewson, J. (2012) Tense. In Binnick, R. I. (ed.), *The Oxford Handbook of Tense and Aspect*, 507–535. Oxford: Oxford University Press.

Higgins, H. A. (2012) *Ikoma Vowel Harmony: Phonetics and Phonology*. Ms. (www.sil.org/resources/publications/entry/51528).

Hyman, L. M. (2003) Segmental Phonology. In Nurse, D. & G. Philippson (eds.), *The Bantu Languages*, 42–58. London; New York: Routledge.

Katamba, F. (2003) Bantu Nominal Morphology. In Nurse, D. & G. Philippson (eds.), *The Bantu Languages*, 103–120. London; New York: Routledge.

Kershner, T. L. (2002) *The Verb in Chisukwa: Aspect, Tense, and Time*. Bloomington: Indiana University, PhD dissertation.

Lewis, M. P., G. F. Simons & C. D. Fennig (eds.) (2016) *Ethnologue: Languages of the World, Nineteenth edition*. Dallas: SIL International. Online version: www.ethnologue.com.

Maho, J. F. (2009) NUGL Online: The Online Version of the New Updated Guthrie List, a Referential Classification of the Bantu Languages (4 Juni 2009) (Available online at http://goto.glocalnet.net/mahopapers/nuglonline.pdf, Accessed December 13, 2010).

Marlo, M. R., L. C. Mwita & M. Paster (2014) Kuria Tone Melodies. *Africana Linguistica* 20: 277–294.

Marlo, M. R., L. C. Mwita & M. Paster (2015) Problems in Kuria H Tone Assignment. *Natural Language and Linguistic Theory* 33: 251–265.

Masatu, A. (2015) Suba-Simbiti Narrative Discourse. *SIL Language and Culture Documentation and Description* 31. (Available online at www.sil.org/resources/publications/entry/61345).

Meeussen, A. E. (1967) Bantu Grammatical Reconstructions. *Africana Linguistica* 3: 79–121.

Mous, M. (2000) Counter-Universal Rise of Infinitive-Auxiliary Order in Mbugwe (Tanzania, Bantu F 34). In Vossen, R., A. Mietzner & A. Meissner (eds.), *Mehr als nur Worte: Afrikanistische Beitrage zum 65. Geburtstag von Franz Rottland*, 469–481. Cologne: Rüdiger Köppe.

Mwita, L. C. (2008) *Verbal Tone in Kuria*. Los Angeles: University of California, PhD dissertation.

Nicolle, S. (2014) Discourse Functions of Demonstratives in Eastern Bantu Narrative Texts. *Studies in African Linguistics* 43(2): 125–144.

Nicolle, S. (2015) A Comparative Study of Eastern Bantu Narrative Texts. *SIL Electronic Working Papers* 3: www.sil.org/resources/publications/entry/61479.

Nurse, D. (2008) *Tense and Aspect in Bantu*. Oxford: Oxford University Press.

Osa-Gómez del Campo, A. (2013) Future Expressions in Nata, a Bantu language. *UBC Working Papers in Linguistics* 38: 141–152.

Roth, T. (2014) The Case of Ngoreme: Distinguishing Contact from Genetic Inheritance in Mara Bantu. Paper presented at *the "Language Contact: The State of the Art" Conference*, Helsinki.

Rundell, O. (2012) *A Comparative Discourse Analysis of Some Bantu Languages in Tanzania*. London: Middlesex University, MA thesis.

Sadlier-Brown, E. (2013) The Nata Applied Double Object Construction. In Luo, S. (ed.), *Proceedings of the 2013 Annual Conference of the Canadian Linguistic Association*. Victoria: University of Victoria, Canadian Linguistic Association, homes.chass.utoronto.ca/~cla-acl/actes2013/Sadlier-Brown-2013.pdf.

Walker, J. B. (2013) *Comparative Tense and Aspect in the Mara Bantu Languages: Towards a Linguistic History*. Langley: Trinity Western, MA thesis.

CHAPTER 18

MBUGWE F34

Vera Wilhelmsen

1 INTRODUCTION

Mbugwe F34 has approximately 37,000 speakers (Mradi wa Lugha za Tanzania 2009). It is mainly spoken in villages located in the area stretching from the south end of Lake Manyara to the town of Babati in the Manyara region in northern Tanzania. In the Mbugwe villages the children speak Mbugwe before they learn Swahili G42, which is the language of instruction in schools. Mbugwe is spoken by all generations in the villages. Young adults, however, usually have a stronger Swahili influence on their Mbugwe than older speakers, and code mixing is widespread in all generations. The language has no official status and no standard orthography yet, but there is an on-going SIL project, which is in the process of establishing an orthography. No dialects have been identified for Mbugwe.

According to Kiessling *et al.* (2008), Mbugwe is part of a Sprachbund consisting of the Bantu languages Kimbu F24, Nilamba F31, Nyaturu F32 and Rangi F33, as well as languages from other language families, such as Iraqw, Gorowa, Alagwa and Burunge (Cushitic), Maasai and Datooga (Nilotic), Sandawe (Khoe-Kwadi according to Güldemann 2014: 35–36) and Hadza (isolate). Mbugwe is closely related to Rangi F33, with a lexical similarity of 72% according to Nurse and Masele (2003: 121). The genealogical relationship between Rangi and Mbugwe and the other languages in the F zone is uncertain (cf. Nurse & Masele 2003).

The data in this chapter is mostly based on the author's own fieldwork, but useful references are the Mbugwe grammar sketch by Mous (2004) and a database (FLEx) which was provided by the SIL team working on Mbugwe (Larsen *et al.* 2011). For a more developed analysis, see Wilhelmsen (2018). The fieldwork consisted mostly of elicitation, including elicited narratives. The two main language consultants were Naomi Richards, born 1967, and Colman Chuchu, born 1949, both born and raised in the Mbugwe area.

2 PHONOLOGY

The phonemes of Mbugwe are given in a broad phonetic transcription in this section, with the graphemes used in this work in arrow brackets. The graphemes are based on the proposed orthography for Mbugwe, which is influenced to a large extent by Swahili. The underlying forms and phonemes are written with slanted brackets in the text and tables in this section, and surface forms are written without any brackets or with square brackets in cases where a more detailed phonetic transcription is employed. Underlying forms are only given when they differ from the surface form.

2.1 Vowels

The vowel phonemes of Mbugwe are found in Table 18.1. There are seven different vowel qualities, and length is distinctive.

In addition to underlying distinctive vowel length, there are three additional sources of surface vowel length: vowel concatenation, compensatory lengthening and conditional lengthening. If two identical vowels become adjacent across morpheme boundaries, they form a long vowel, as in *á-à-kúy-á* (SP$_{3SG}$-PST-die-FV) 'he/she died.'

Compensatory lengthening occurs when a vowel is glided before another vowel. The front vowels /i/ and /e/ are glided to /y/ before any other vowel, and the back vowels /o/ and /u/ are glided to /w/. The vowel following the glide is then lengthened. Examples are *wèèvà* /o-ev-a/ (INF-forget-FV) 'to forget' and *yààngì* /i-ángi/ 'bird of prey (sp.)' (cl. 5). Conditional lengthening happens before prenasalised consonants (see section on consonants below).

There is limited vowel height harmony in Mbugwe, and it is found for instance in the applicative verbal extension *-er*, which is realised *-ɛr* after /ɛ/, such as in *ò-fèèng'-èr-à* (INF-run-APPL-FV) 'run (somewhere)' and *ò-fét-ér-à* (INF-go-APPL-FV) 'to go (somewhere).'

TABLE 18.1 THE VOWELS OF MBUGWE

	Front		*Central*		*Back*	
Close	i	iː <ii>			u	uː <uu>
Close-Mid	e	eː <ee>			o	oː <oo>
Open-Mid	ɛ	ɛː <ɛɛ>			ɔ	ɔː <ɔɔ>
Open			a	aː <aa>		

2.2 Consonants

In Table 18.2 the consonant phonemes of Mbugwe are given.

In addition to the phonemes in Table 18.2, all the stops may be prenasalised. The nasal is always assimilated for place with the consonant. The nasal also assimilates to the voicing of the stop, so that the nasal becomes voiceless in [ᵐp], [ⁿt], [ᶮc] and [ᵑk]. In addition to the stops, there is also the prenasalised /s/, [ⁿs]. Younger speakers of Mbugwe tend to drop the voiceless nasal completely utterance initially. All of the prenasalised consonants occur root medially, and /nt/, /nk/, /nj/, /ng/ and /ns/ occur root-initially, as well.

When a nasal and another consonant, except for /w/, /y/ and /h/, become adjacent across morpheme boundaries, prenasalised consonants are formed. That is, when a nasal precedes a /p/ or /f/ across a morpheme boundary, it is realised as [ᵐp], and if a nasal precedes a /b/ or /v/ across a morpheme boundary, it is realised as [ᵐb]. If a nasal precedes a /l/ or /r/ across a morpheme boundary, the result is [ⁿd], as in the noun *ndòsèkà* /n-loseka/ (cl. 9/10) from the verb *lòsèk-à* 'speak.' The vowel preceding the NC cluster across morpheme boundaries is always lengthened (see Section 2.1).

2.3 Syllable shape

The most common syllable shape is CV, followed by CVV and V. VV also occurs. The C here includes NC clusters. Glides may occur alone or after another C in onset, and /y/

TABLE 18.2 THE CONSONANTS OF MBUGWE

	Labial	*Alveolar*	*Palatal*	*Velar*	*Glottal*
Voiceless stop	p	t	c <ch>	k	
Voiced stop	b	d	ɟ < j>	g	
Nasals	m	n	ɲ <ny>	ŋ <ng'>	
Voiceless fricative	f	s	ʃ <sh>		h
Voiced fricative	v				
Liquid trill		r			
Lateral liquid		l			
Glide	w		j <y>		

Adapted from Mous (2004)

occurs in the coda, as well. Syllabic nasals are rare in Mbugwe, but there are a few examples of geminate /n/, where the nasal is syllabic, as in *ǹ-nò* 'toe' cl. 9/10. There is also an allomorph of the class 1 and 3 prefix *mo-* which is realised as *m̩-*.

2.4 Tone

Mbugwe has a high (H) underlying tone, and a default surface L tone, which is analysed here as underlyingly zero (Ø). The mora is the Tone-Bearing Unit (TBU). This is based on the fact that there are contour tones occurring on long vowels only. A sequence of HØ or ØH tones on the two moras of a long vowel results in a falling or rising tone (see for example Table 18.3). Another indication that the mora is the TBU is the High Tone Spread (HTS) rule: the underlying H tone spreads one mora to the right, creating a falling tone on a following long vowel, as in *ò-vá̱-láàn-à* (INF-OP$_{3PL}$-bid.farewell-FV) 'to bid them farewell' (more on HTS below). Underlyingly H vowels are underlined in transcriptions of surface realisation of tone in this section. The target of the Melodic High tone (see Section 2.4.3.) is underlined with double lines in this section.

2.4.1 Lexical tone on noun and verb roots

Each mora of a noun root may carry a lexical H tone. Bimoraic nouns have four possible tone patterns, whether they are monosyllabic or disyllabic, as seen in Table 18.3.

Mbugwe verb roots fall into two-tone classes lexically: those with a H tone which is associated with the initial syllable of the verb root, and the ones which are toneless throughout the root. Examples of minimal pairs for verbs are found in Table 18.4.

In addition to the verb roots having lexical tone, verbal prefixes also have underlying tone. The paradigms for the subject prefixes (SP), object prefixes (OP) and various tense, aspect and mood (TAM) prefixes will be given in Section 4 on verbs, but the tone of personal SPs warrants some comments here. The underlying tones of the personal SPs are toneless for 1st and 2nd person, and H for the 3rd person. The underlying tones surface in most TAM-forms, but in others, the tones of the personal SPs are neutralised and they are either all H or all toneless. The TAM-forms where the tone of the SP is neutralised are listed in Table 18.5 (see also Section 4.3 for TAM-forms).

TABLE 18.3 BIMORAIC NOUN ROOTS

Noun	Translation
dúṵ	'darkness' (cl. 9/10)
kè-dákɔ̰́	'duck' (cl. 7)
vὲὲ̀	'shoulder' (cl. 5)
kè-màkà	'thing' (cl. 7)
n-chḭi	'(a kind of) small rat' (cl. 9/10)
bá̰sà	'twin'(cl. 5)
yùṵ	'ash' (cl. 5)
mò-sìyá̰	'fiancée (female)' (cl. 1)

TABLE 18.4 MINIMAL PAIRS FOR VERBS (INF-ROOT-FV)

H verb root	Translation	Ø verb root	Translation
ò-vá̰l-à	'to start'	ò-vàl-à	'to count'
ò-sá̰á̰l-à	'to pray'	ò-sààl-à	'to spread'
ò-lá̰á̰l-à	'to sleep'	ò-lààl-à	'to fall down'
ò-rɛ̰́r-à	'to awake, leave'	ò-rɛ́r-à	'to raise a child'

TABLE 18.5 TAM-FORMS WITH TONALLY NEUTRALISED SP

TAM	Form
Future Perfective	SP(H)-jɛ́-ROOT-a
Present Imperfective	SP(H)-kɛ́ɛ́-/kɛ́ɛ́n-/kɛ́ɛ́ndé-ROOT-a
Past Imperfective	SP(ø)-kee-/keen-/keende-ROOT-a
Subjunctive Perfective	SP(H)-ROOT-ɛ̰́
Subjunctive Imperfective	SP(H)-ɛ̰́ɛ̰́/ɛ̰́ɛ̰́nd-ROOT-a
Subjunctive Consecutive	SP(H)-á̰-ROOT-a
Conditional	SP(ø)-kɛ́ɛ́-ROOT-á̰
Irrealis Recent Past	SP(ø)-ká̰á̰-ROOT-iyɛ
Irrealis Far Past	SP(ø)-ká̰á̰-ROOT-á̰

Verbal suffixes are underlyingly toneless, but the ultimate or penultimate vowel of the verb stem (verb root plus suffixes) may be H due to the Melodic High tone which occurs in certain TAM-forms (see Section 2.4.3).

2.4.2 Tonal processes

In Mbugwe, there is bounded H tone spread, that is, an underlying H tone spreads one mora to the right. This is illustrated by verb stems with three syllables or more with a lexical H tone in the infinitive (Table 18.6). The Infinitive marker *o-* is toneless, as are the applicative extension *-er* and the FV *-a*.

TABLE 18.6 HTS IN H VERB ROOTS IN THE INFINITIVE

Infinitive	Gloss	Translation
ò-vɛ́kér-à	INF-dress-FV	'to dress somebody'
ò-sɛ́ɛ́rɛ́r-à	INF-help-FV	'to help'
ò-fwɛ́ɛ́r-ɛ́r-y-à	INF-rest/breathe-APPL-CAUS-FV	'to dismiss'
ò-tɔ́mám-ɛ̀r-à	INF-work-APPL-FV	'to serve'

As seen in Table 18.6, the H lexical tone of the initial syllable of the verb root spreads one mora to the right. In the verb roots with a long vowel on the initial syllable, such as *ò-sɛ́ɛ́rɛ́r-à* (INF-help-FV) 'to help' and *ò-fwɛ́ɛ́r-ɛ́r-y-à* (INF-breathe-APPL-CAUS-FV) 'to dismiss,' the H tone is realised not only on both moras of the long vowel, but it also spreads to the following mora. This shows that the whole initial syllable is underlyingly H in these verbs, and not just the initial mora. The H tone does not spread in the bisyllabic verb stem *ò-mút-à* (INF-beat-FV) 'to beat.' This is due to a restriction against HTS to an utterance final mora. If the verb is not utterance final, however, as in *ò-mút-á mònɔ̀* (INF-beat-FV a.lot) 'to beat a lot,' the H tone spreads to the final *-a*.

Another restriction for HTS is against spreading if the target for the spreading is followed by a H tone, as in the noun phrase *mò-té mò-nɛ́nɛ̀* (3-tree NP₃-big) 'big tree,' where the spreading of the first H tone is blocked by a following H mora. The second H tone does not spread to the utterance final mora, due to the HTS restriction above. HTS does occur, however, if the following adjective is toneless, as in for instance *mò-té mó-lèy* (3-tree NP₃-tall) 'tall tree.' The HTS rule, incorporating the restrictions above, can be formulated as follows: an underlying H tone spreads one mora to the right, if the target mora is followed by a toneless mora.

In Mbugwe there is also downdrift and final lowering of H tones. As these processes are predictable and well documented for Bantu languages, they will not be described here. There is another tonal process which will be described in more detail, due to the relative rarity of the phenomenon, even if the process is predictable and not contrastive in Mbugwe. An underlyingly H mora which follows another H mora and precedes a toneless mora is upstepped, which means it is pronounced at a higher pitch than the previous H mora. HTS may happen after the upstep, and the mora to which the H tone spreads is realised on the same level as the upstepped H mora. An utterance-final H mora is not upstepped.

Upstep may be illustrated by the infinitive verb *ò-vá-⁺tɔ́mám-ɛ̀r-à* (INF-OP₃ₚₗ-work-APPL-FV) 'to serve them' (Figure 18.1). Here it is evident that the root-initial syllable *tɔ́-* is

TABLE 18.7 UPSTEP IN H VERBS IN THE INFINITIVE WITH H OP

Infinitive	Gloss	Translation
ò-vá-f-á	INF-OP₃ₚₗ-give-FV	'to give them'
ò-vá-⁺mút-à	INF- OP₃ₚₗ-beat-FV	'to beat them'
ò-vá-⁺sóóch-a	INF- OP₃ₚₗ-hate-FV	'to hate them'
ò-vá-⁺tɔ́mám-ɛ̀r-à	INF- OP₃ₚₗ-work-APPL-FV	'to serve them'

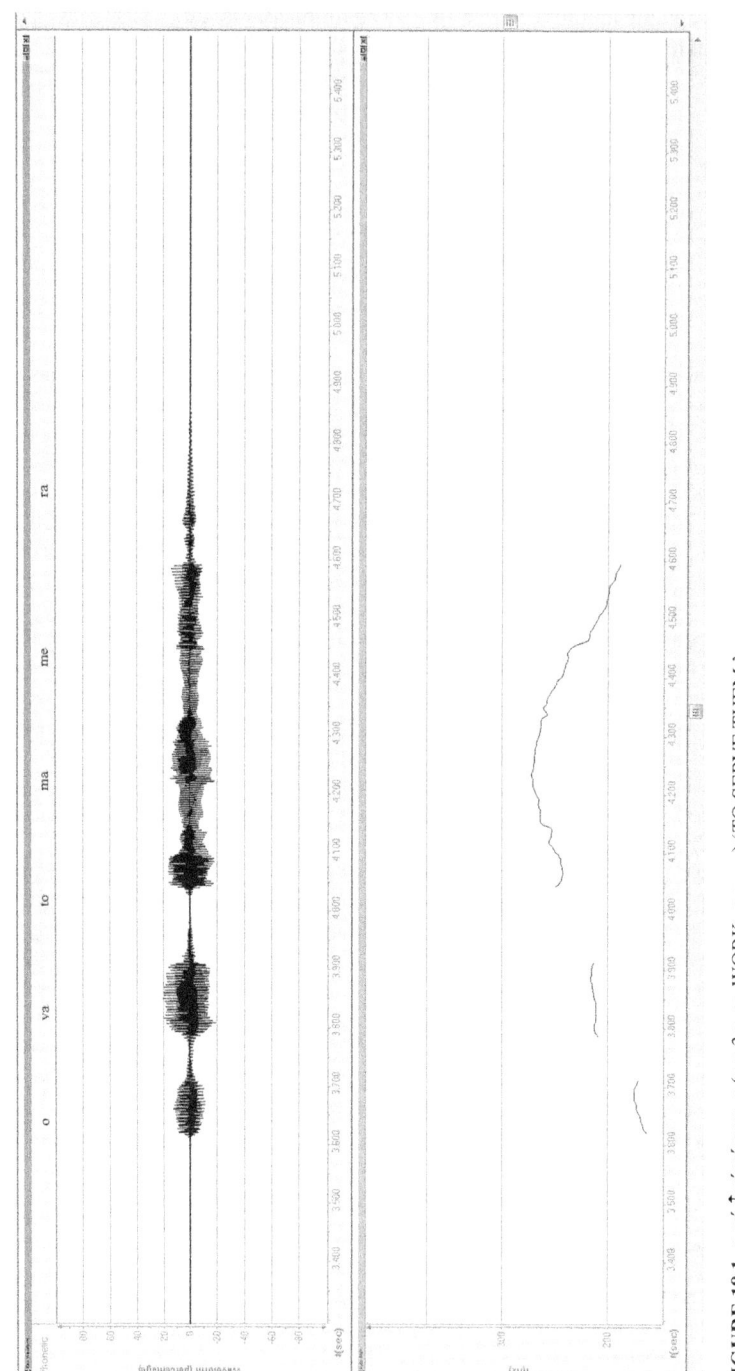

FIGURE 18.1 *o-vá-↑ tōmám-er-a* (INF-3PL.OP-WORK-APPL-FV) 'TO SERVE THEM.'

upstepped in relation to the H OP *vá-* (3PL), and that the following syllable (*-má-*) is also realised with a higher pitch.

As seen in Table 18.7, upstep occurs in all verbs with a lexical H tone in the infinitive with the H OP, except for the monosyllabic verb *ò-vá-f-á* (INF-3PL.OP-give-FV), 'to give them.' This is because upstep does not happen utterance finally.

For an analysis of upstep in a Bantu language which is spoken not far from Mbugwe, see the analysis of upstep in Nyaturu F32 by Hyman (1992). Like final lowering and downdrift, upstep is predictable and will therefore not be marked in other sections of this study.

2.4.3 Melodic H tones on verbs

In addition to lexical tones on noun and verb roots, there are Melodic H tones, which are assigned to a certain syllable of the verb stem in order to mark various TAM categories. See Odden & Bickmore (2014), for an introduction to Melodic H tones in Bantu. In this section, the inflectional verb stem, which is the domain of the Melodic H, is marked with square brackets. Only the main patterns of affirmative verbs are described here. The assumed target of the Melodic H is underlined with double lines. An overview is found in Table 18.8. For information on the various TAM-forms, see Section 4.3.

The Conditional, Irrealis Far Past and Subjunctive Perfective verbs without a lexical H tone have a Melodic H tone on the final mora of the verb stem.

TABLE 18.8 OVERVIEW OF TAM-FORMS WITH MH

Pattern	TAM form
1a) MH on the ultimate	Conditional Ø verbs
	Irrealis Far Past Ø verbs
	Subjunctive Perfective Ø verbs
1b) MH on ultimate with left spread till σ2	Conditional H verbs
	Irrealis Far Past H verbs
	Subjunctive Perfective H verbs
2) MH on σ2-ultimate	Far Past Perfective Verbs
3a) MH on penultimate	Imperative singular Ø Verbs
3b) MH on penultimate with left spread till σ2	Imperative singular H Verbs
	Hodiernal Perfective H verbs
4) MH on σ2-penultimate	Hodiernal Perfective Ø verbs
	Imperative plural Verbs

TABLE 18.9 TAM-FORMS WITH MH ON THE ULTIMATE

TAM form	Form and gloss	Translation
Conditional Ø verb	kò-kéé-[tákànèr-á] SP$_{1PL}$-COND-persuade-FV	'if we persuade'
Irrealis Far Past Ø verb	kò-káá-[tákànèr-á] SP$_{1PL}$-IRR-persuade-FV	'if we would have persuaded'
Subjunctive Perfective Ø verb	kó-[tákànèr-ɛ́] SP$_{1PL}$-persuade-SBJV	'we should persuade'

TABLE 18.10 TAM-FORMS WITH MH ON THE ULTIMATE WITH LEFT UNBOUNDED SPREAD

TAM form	Form and gloss	Translation
Conditional H verb	kò-kéé-[tómám-ér-á] SP$_{1PL}$-COND-work-APPL-FV	'if we serve'
Irrealis Far Past H verb	kò-káá-[tómám-ér-á] SP$_{1PL}$-IRR-work-APPL-FV	'if we would have served'
Subjunctive Perfective H verb	kó-vá-[tómám-ér-ɛ́] SP$_{1PL}$-OP$_{3PL}$-work-APPL-SBJV	'we should serve them'

TABLE 18.11 FAR PAST PERFECTIVE VERBS

Far Past Perfective	Underlying and gloss	Translation
kwà-à-[takánér-á]	/ko-a-takaner-a/ SP$_{1PL}$-PST-persuade-FV	'we persuaded'
kwà-à-[tómám-ér-á]	/ko-a-tómam-er-a/ SP$_{1PL}$-PST-work-APPL-FV	'we served'

TABLE 18.12 IMPERATIVE VERBS WITH NO LEXICAL TONE

Imperative	Gloss	Translation
[làán-à]	bid.farewell-FV	'bid farewell!'
[sààkér-à]	search-FV	'search!'
[tàkànér-à]	persuade-FV	'persuade!'

The verbs with a H lexical tone in the Conditional, the Far Past Irrealis and the Subjunctive Perfective have an all H stem (Table 18.10). The Melodic H tone is assigned to the final mora of the verb stem, and then spreads to the left until the H tone of the initial syllable of the verb roots with a H lexical tone.

In the Far Past Perfective, the whole verb from the second syllable and onwards is H. The first syllable is H in verbs with a lexical H tone, and toneless for lexically toneless verbs (Table 18.11).

The Imperative verbs without a lexical tone have a Melodic H on the penultimate mora of the verb stem. The same pattern is found on Hodiernal verbs with no lexical tone, but in these verbs the whole penultimate syllable is H (Table 18.12 and 18.13).

The verbs with a H lexical tone in the Imperative singular form display a pattern where the Melodic H tone links to the penultimate syllable of the verb stem and then spreads to the left, until the first syllable of the verb stem (which already is H) (Table 18.14).

In the Hodiernal Perfective verbs with a H in the lexical root, the Melodic H, which is anchored to the penultimate syllable of the verb stem, spreads to the left, until the first syllable of the verb stem (Table 18.15).

TABLE 18.13 HODIERNAL PERFECTIVE VERBS WITH NO LEXICAL TONE

Hodiernal Perfective	Underlying and gloss	Translation
vá-[rèf-íyè]	/vá-ref-i̱yɛ/ SP₃ₚₗ-pay-PFV	'they have paid'
vá-[láàn-íyɛ]	/vá-laan-i̱yɛ/ SP₃ₚₗ-bid.farewell- PFV	'they have bid farewell'
vá-[sáàk-ɛ́ɛ́ɛ̀]	/vá-saaker-i̱yɛ/ SP₃ₚₗ-search- PFV	'they have searched for'
vá-[tákān-íryè]	/vá-takaner-i̱yɛ/ SP₃ₚₗ-persuade-PFV	'they have persuaded'

TABLE 18.14 IMPERATIVE SINGULAR VERBS WITH A H LEXICAL TONE

Imperative	Gloss	Translation
[vɛ́k⁺ér-à]	dress-FV	'dress someone!'
[sɛ́ɛ́r⁺ɛ́r-à]	help-FV	'help!'
[tɔ́mám-⁺ér-à]	work-APPL-FV	'serve!'

TABLE 18.15 HODIERNAL PERFECTIVE VERB WITH A H LEXICAL TONE

Hodiernal Perfective	Underlying/gloss	Translation
kò-[tómám-⁺ɛ́ɛ́ɛ̀]	/ko-tɔ́mam-er-i̱yɛ/ SP₁ₚₗ-work-APPL-PFV	'we have served'

TABLE 18.16 HODIERNAL PERFECTIVE VERB WITH NO LEXICAL TONE

Hodiernal	Underlying/gloss	Translation
kò-[tàkán-⁺íryè]	/ko-takaner-i̱yɛ/ SP₁ₚₗ-persuade-PFV	'we have persuaded'

In the Hodiernal Perfective verbs with no lexical tone, the whole verb stem after the initial syllable until the penultimate is H (Table 18.16).

The Imperative plural verbs have a suffix -i after the FV, which results in the ending -ey (-a-i → -ey). The Melodic H is realised on every mora of the verb stem after the first syllable of the verb stem, until the final – é, which is the penultimate mora underlyingly (Table 18.17 and 18.18).

TABLE 18.17 IMPERATIVE PLURAL VERBS WITHOUT A LEXICAL TONE

Imperative plural	Underlying/gloss	Translation
[rèf-ɛ́y]	/ref-a̱-i/ pay-FV-PL	'you all pay!'
[lààn-ɛ́y]	/laan-a̱-i/ bid-farewell-IMP-PL	'you all bid farewell!'
[sààkɛ́r-ɛ́y]	/saakɛ̱r-a̱-i/ search-FV-PL	'you all search for!'
[tàkán-ɛ́r-ɛ́y]	/taka̱nɛ̱r-a̱-i/ persuade-FV-PL	'you all persuade!'

TABLE 18.18 IMPERATIVE PLURAL VERBS WITH A H LEXICAL TONE

Imperative plural	Underlying/gloss	Translation
[f-ɛ́y]	/f-á̱-i/ give-FV-PL	'you all give!'
[mút-ɛ́y]	/mút-a̱-i/ beat-FV-PL	'you all beat!'
[sóóch-ɛ́y]	/sóóch-a̱-i/ hate-FV-PL	'you all hate'
[sɛ́ɛ́rɛ̱r-ɛ́y]	/sɛ́ɛ́r-ɛ̱r-a̱-i/ help-FV-PL	'you all help!'
[tómám-ɛ̱r-ɛ́y]	/tóma̱m-er-a̱-i/ work-APPL-FV-PL	'you all serve!'

3 NOMINALS

3.1 Noun classes

The noun classes and nominal agreement markers are found in Table 18.19. The singular-plural pairs are listed in the far right column. No augment is found in Mbugwe. All of the nominal prefixes are toneless.

The agreement markers for adjectives are the same as the noun class markers (NP) for each class, except for 1a, where the adjective prefix is *mo-*, as in class 1. The pronominal prefix (PP) is used for the connective markers, possessive pronouns, quantifiers and demonstratives, and the numeral prefix (EP) is used for numerals and the interrogative word *reengá* 'how many.' As seen in Table 18.22, PP and EP are different from the NP in many classes, but PP and EP differ from each other only in classes 9 and 10. Sometimes, the EPs *e-* and *i-* are used for certain modifiers that normally take the PP. This is an indication that the two sets of nominal agreement prefixes might be collapsing into one group.

The noun classes of Mbugwe are quite typical for Bantu, but some unusual pairings and classes will be discussed here. The pair 12/19 for singular and plural diminutives is unique to the F30 languages, according to Maho (1999: 187). Class 19 has two different allomorphs in the PP paradigm: *fi-* and *sha-*. They appear to be in free variation, at least for some nouns. The reason for this variation is unknown, but Mous (2004: 4) suggests that there is allophonic variation, and refers to the sound changes from PB *py > fy > sh*

TABLE 18.19 NOMINAL CONCORD

Classes	NP	PP	EP	SG/PL
1	mo-	o-	o-	1/2
1a	Ø-	o-	o-	1a/2
2	va-	va-	va-	1/2
3	mo-	o-	o-	3/4
4	me-	e-	e-	3/4
5	Ø-/ri-/i-	re-	re-	5/6
6	ma-	a-	a-	5/6
7	ke-	ke-	ke-	7/8
8	vi-	vi-	vi-	7/8
9	N-/nj(i)	ji/e-	e-	9/10
10	N-/nj(i)	ji-/i-	i-	9/10
11	lo-	lo-	lo-	11/10
12	ka-	ka-	ka-	12/19
14	o-	o-	o-	14/6
15	ko-	ko-	ko-	15/6
15a	o-	-	-	-
16	fa-	fa-	fa-	-
17	ko-	ko-	ko-	-
19	fi-	fi-/sha-	fi-	12/19

TABLE 18.20 DIMINUTIVES (BASED ON *LARSEN ET AL.* 2011)

Class	Noun	Cl. 12	Cl. 19
1/2	mo-díma	ka-díma	fi-díma
	'fool'	'small fool'	'small fools'
3/4	mo-tííkó	ka-mo-tííkó	sha-me-tííkó/ fi-me-tííkó
	'wooden spoon'	'small wooden spoon'	'small wooden spoons'
5/6	lʊʊla	ka-lóóla	sha-loola
	'(kind of) bird'	'small (kind of) bird'	'small (kind of bird)'
7/8	ke-ráátó	ka-ke-ráátó	sha-vi-ráátó
	'shoe'	'small shoe'	'small shoes'
9/10	m-pérá	ka-pérá	fi-pérá
	'rhinoceros'	'small rhinoceros.'	'small rhinoceroses'
11/6	lo-dí	ka-lo-dí	fi-dí
	'rope'	'small rope	'small ropes'
15/6	ko-olo	ka-ko-olo	sho-olo
	'leg or foot'	'small leg or foot.'	'small feet or legs'

which have happened in Mbugwe according to Dempwolff (1915–1916: 12). The variation in the prefixes *fy-* and *sha-* may therefore represent different stages in the historical development. The prefix **pi-* is reconstructed for class 19 in PB (Meeussen 1967: 98). This does not explain the vowel change from /i/ to /a/, however, and it is unknown where the /a/ came from.

When the diminutive prefix of class 12/19 is added, the original noun class prefix is retained for classes 3/4, 7/8, 11 and 15 (see Table 18.20). For classes 1/2, 5/6 and 9/10, the original noun class prefix is deleted, and only the class 12/19 prefix is added to the noun stem. Class 11 and 15 have plurals in class 6, and here also the original prefix is dropped.

The infinitive prefix *o-* is different from the prefix of the nouns in class 15, which is *ko-*, hence the subclass 15a for the infinitive of the verb (Mous 2004: 15). However, the adjectival concord prefix is *ko-* (Mous 2004: 17).

Class 15 consists of two nouns, and they are both reconstructed to class 15 in PB also: *ko-olo* 'foot, leg' and *ko-tó* 'ear.' Their concord prefix is *ko-* (NP, PP and EP).

3.2 Nominal derivation

The only productive derivational nominal suffix is the locative suffix *-i*, which can be attached to any noun, denoting the location of the noun, i.e., 'at/on NOUN.' For instance, *túúmbéy* /*túmbé-i*/ 'on the chair' is formed from *túúmbé* /*túmbé*/ 'chair.' Some of the forms are lexicalised, such as *lòtàángéy* /*lo-tangé-i*/ 'toilet' from *lòtàángé* /*lo-tangé*/ 'place far away' (cl. 11).

3.3 Personal pronouns

Personal pronouns may replace the noun in subject or object position. In addition to the regular forms, there are partially reduplicated forms for first and second person. It has not been established yet whether there is a semantic difference between the short forms and the partially reduplicated forms (see Table 18.21.)

TABLE 18.21 PERSONAL PRONOUNS

Person	Pronoun	Long form
1SG	nɛ́ɛ	nɛɛ́~nɛ
2SG	wɛ́ɛ	wɛɛ́~wɛ
3SG	wɛɛ́	-
1PL	síyɛ	sií~siyɛ
2PL	nyɛɛ	nyɛɛ́~nyɛ
3PL	vɔɔ́	-

3.4 Nominal modifiers

The nominal modifiers in this section are organised according to the prefixes: first adjectives are introduced, and they take the NP. Then the adnominals taking the PP are introduced: the connective marker, possessive pronouns, quantifiers and demonstratives. Then follow the modifiers which take the EP, which are numerals and the interrogative word *reengá* 'how many.'

3.4.1 Adjectives

Only 37 adjectives have been found in the data. The adjectives are here organised according to Dixon's (1982: 16) lists of seven semantic types of adjectives. The seventh type, speed, is not represented in the database. Some of the adjectives have antonyms, and they are listed on the same row in Table 18.22. The adjectives take agreement with the NP, and

TABLE 18.22 ADJECTIVE SEMANTIC CLASSES

Semantic class	Adjective	Translation	Adjective	Translation
1. DIMENSION	-íma	'whole, complete'		
	-dídi	'small'	-nέnɛ	'big, important, many'
	-kúfé	'short'	-ley	'tall'
	-séséré	'narrow'	-aaré	'wide'
2. PHYSICAL PROPERTY	-túúfú		-fóómbú	
		'dull'		'sharp'
	-vése	'unripe'	-vérú	'ripe'
	-hálí	'dirty'	-já	'good, clean'
	-fafú	'hard, difficult, expensive'	-ɔ́lɔ	'easy, sharp'
	-uundú	'rotting, smelling'		
	-óómo	'dry'		
	-fɔ́lɔ, -mpéfɔ	'cold'		
	-ritɔ	'heavy'		
3. COLOUR	-έro	'white, light colour'	-íro	'black, dark'
4. HUMAN PROPENSITY	-lomé	'male'	-ká	'female'
	-tarí	'careful, sly'		
	-sɛu	'silent'		
	-vira	'lazy'		
5. AGE	-feyá	'new'	-kɔ́lɔ́	'old'
6. VALUE	-já	'good, clean'	-vé	'bad'
	-enú	'expensive'		
	-kέ, -rimo	'scarce'		

cach mora my take a H tone. Like Nyamwezi F22 (Nurse 2008: 81), Mbugwe appears to have a suffix -ú, which turns verbs into adjectives (for instance -fafú 'hard, difficult, expensive' from fafya 'harden,' -uundú 'rotten' from uunda 'rot' and -vérú 'ripe' from vérya 'ripen'). It is not productive in Mbugwe.

3.4.2 The connective marker

The connective marker -a/-á connects various constituents in the noun phrase. It can be used to connect nouns with other nouns, or with adjectives, ordinal numerals, interrogatives or possessives. The prefix is the PP, and the connective marker has a H tone in all noun classes except for class 1.

3.4.3 The possessive pronouns

The possessive pronouns occur after the noun, and agree with the noun, taking the PP. 3PL has two variants: -aavɔ́/-áávɔ́ and -aaɔ́/-ááɔ́. They are in free variation. The tone of the possessives is ØØH for class 1, 4 and 9, and HHH for the rest of the classes. The paradigm for the possessive pronouns is given in Table 18.23.

TABLE 18.23 POSSESSIVE PRONOUNS

1SG	2SG	3SG	1PL	2PL	3PL ~	3PL
-aanέ/ -ááné	-aakɔ́/ -áákɔ́	-aachwέ/ -ááchwέ	-ɛytó/ -ɛ́ytó	-aanyú/ -áányú	-aavɔ́/ -áávɔ́	-aaɔ́/ -ááɔ́

3.4.4 Quantifiers

The quantifiers are ɔ́ɔnsɛ/ɔ́ɔnsɛ 'all/whole,' ɔ-ɔ́ɔnsɛ/ɔ-ɔ́ɔnsɛ 'any,' and ngé 'other.' The word ɔ́ɔnsɛ/ɔ́ɔnsɛ means 'whole' in the singular noun classes, and 'all' in the plural classes. The rising initial tone is found in class 1, 4, 9 and 14 for both ɔ́ɔnsɛ 'all/whole' and ɔ-ɔ́ɔnsɛ 'any' (Table 18.24). For ɔ-ɔ́ɔnsɛ 'any' class 2 also has an initial rising tone: v-ɔ́-v-ɔ́ɔnsɛ.

TABLE 18.24 QUANTIFIERS

Class	All/whole	Any	Other
1	ɔ́ɔnsɛ	w-ɛɛ́-ɔ́ɔnsɛ	o-ngé
2	v-ɔ́ɔnsɛ	v-ɔ́-v-ɔ́ɔnsɛ	vá-ngé
3	ɔ́ɔnsɛ	w-ɔ-w-ɔ́ɔnsɛ	o-ngé
4	y-ɔ́ɔnsɛ	y-ɔ-y-ɔ́ɔnsɛ	ye-ngé
5	r-ɔ́ɔnsɛ	r-ɔ-r-ɔ́ɔnsɛ	ré-ngé
6	ɔ́ɔnsɛ	ɔ-ɔ́ɔnsɛ	á-ngé
7	ch-ɔ́ɔnsɛ	ch-ɔ-ch-ɔ́ɔnsɛ	ché-ngé
8	vy-ɔ́ɔnsɛ	vy-ɔ-vy-ɔ́ɔnsɛ	ví-ngé
9	y-ɔ́ɔnsɛ	y-ɔ-y-ɔ́ɔnsɛ	ye-ngé
10	j-ɔ́ɔnsɛ	j-ɔ-j-ɔ́ɔnsɛ	jí-ngé
11	l-ɔ́ɔnsɛ	l-ɔ-l-ɔ́ɔnsɛ	lɔ́-ngé
12	k-ɔ́ɔnsɛ	k-ɔ-k-ɔ́ɔnsɛ	ká-ngé
14	ɔ́ɔnsɛ	ɔ-ɔ́ɔnsɛ	ó-ngé
15	k-ɔ́ɔnsɛ	k-ɔ-kɔ́ɔnsɛ	kó-ngé
16	f-ɔ́ɔnsɛ	f-ɔ-f-ɔ́ɔnsɛ	fá-ngé
17	k-ɔ́ɔnsɛ	k-ɔ-k-ɔ́ɔnsɛ	kó-ngé
19	fy-ɔ́ɔnsɛ	fy-ɔ-fy-ɔ́ɔnsɛ	fí-ngé

2.4.5 Demonstratives

Mbugwe has three different sets of demonstratives: proximate, distant and anaphoric. The proximate form can be described as the PP prefix (forming a root) with the vowel of the PP copied as a prefix, which is also what is reconstructed for PB: *V-PP (Meeussen 1967: 107). The distant demonstrative is formed by the PP prefix and rá, which is reconstructed as *PP-día. The anaphoric demonstrative is reconstructed as *V-PP-o, that is, the proximate form with the addition of an -o suffix. In Mbugwe also, it consists of a vowel prefix (copied from the PP), the consonant of the PP, if any, and the suffix -ɔ. The tone varies for the proximate and the anaphoric demonstrative, depending on whether it occurs before the noun, in which case it is HH, or after the noun, in which case it is HØ. Both forms are given in Table 18.25.

TABLE 18.25 DEMONSTRATIVES

Class	Proximate DEM-N /N-DEM	Anaphoric DEM-N /N-DEM	Distant
1	óó/ óo	óɔ́/óɔ	o-rá
2	ává/áva	ávɔ́/ávɔ	vá-rá
3	óó/óo	óɔ́/óɔ	ó-rá
4	éé/ée	éɔ́/éɔ	é-rá
5	éré/ére	érɔ́/ érɔ	ré-rá
6	áá/áa	áɔ́/áɔ	á-rá
7	éché/éche	échɔ́/echɔ	ké-rá
8	íví/ívi	ívyɔ́/ívyɔ	ví-rá
9	éé/ée	éɔ́/éɔ	e-rá
10	íjí/ íji	íjɔ́/íjɔ	ji-rá
11	ól ó/ólo	ólɔ́/ólɔ	ló-rá
12	áká/áka	ákɔ́/ákɔ	ká-rá
14	óó/óo	óɔ́/óɔ	ó-rá
15	ókó/óko	ókɔ́/ókɔ	kó-rá
16	áfá/áfa	áfɔ́/áfɔ	fá-rá
17	ókó/óko	ókɔ́/ ókɔ	kó-rá
19	ífí/ífi	ífyɔ́/ífyɔ	fí-rá

3.4.5 Numerals

Mbugwe speakers usually use the Swahili numbers. Only older speakers use the Mbugwe numbers over 5. The numerals in Table 18.28 are elicited from Colman Chuchu. Numerals 1 (*moontí*), 7, 9 and 10 do not take the EP, but are modifying nouns that function as numerals (Mous 2004: 20–21). They are juxtaposed the noun, as in *mweérí kɛɛndá* /*mo-eéri kɛɛndá*/ 'nine months.'

The variation and inconsistencies in the numeral system, for example the variation of numbers 1, 4 and 8, may reflect the mix of an older and a newer system, with clear borrowings from Swahili in certain forms. The lexemes as well as noun class system and agreement system may be a mix of archaic and news forms. The variety used in counting is listed first in Table 18.26.

There is no morphological ordinal form of the numerals. Ordinal numbers are formed by the noun plus the connective -*a*, followed by the cardinal number. One exception is -*a mbéére* 'first,' which also means 'in front of' (*mbéére* means 'front').

4 VERBS

4.1 Verb structure

Below are the slots which have been identified for the Mbugwe verb, adapted from Nurse (2008: 40).

(NEG1) + SP + (NEG2) + (TAM ()) + (OP) + root + (extension ()) + FV + (post-FV)

The NEG1 position refers to the negative prefix *te-*. It is used for the negative in the indicative, except for the first-person singular, which uses the NEG2 marker (see below). It is also used in the Irrealis mood, but not in the Subjunctive or Prohibitive. No negative

TABLE 18.26 NUMERALS

	1–20		21–1000
1	moontí, -mɔ́	21	me-rɔɔngɔ e-veeré na i-mɔ́
2	-veeré	22	me-rɔɔngɔ e-veeré na i-veeré
3	-saató	23	me-rɔɔngɔ e-veeré na i-sááto
4	-ínya, -ínye	24	me-rɔɔngɔ e-veeré na í-ínye
5	-sáánɔ	25	me-rɔɔngɔ e-veeré na i-sáánɔ
6	-áánsáto	26	me-rɔɔngɔ e-veeré na i-sáánsáto
7	fúnkáte	27	me-rɔɔngɔ e-veeré na mo-fúnkáte
8	-náná, nyáánye	28	me-rɔɔngɔ e-veeré i-nyáánye
9	kɛɛndá	29	me-rɔɔngɔ e-veeré na i-ré kɛɛndá
10	kómi (cl. 10), me-rɔɔngɔ (cl.4)	30	me-rɔɔngɔ e-táto
11	kómi na i-mɔ́	40	me-rɔɔngɔ e-nne
12	kómi na i-veeré	50	me-rɔɔngɔ e-táánɔ
13	kómi na i-sááto	60	me-rɔɔngɔ e-táántáto
14	kómi na í-ínye	70	me-rɔɔngɔ mo-fúnkáte
15	kómi na i-sáánɔ	80	me-rɔɔngɔ e-nááne
16	kómi na i-sáánsáto	90	me-rɔɔngɔ kɛɛndá
17	kómi na i-ré mo-fúnkáte	100	gana re-mɔ́
18	kómi na i-nyáánye	200	ma-gana a-veeré
19	kómi na i-ré kɛɛndá	1000	ma-gana na ma-gana
20	me-rɔɔngɔ e-veeré		

TABLE 18.27 SP FOR NOUN CLASS 1–19

Noun class

	Third person	Second person	First person
1	ó-/á	o-	N-
2	vá-	mo-	ko-
3	o-		
4	ya-		
5	re-		
6	a-		
7	ke-		
8	vi-		
9	e-		
10	ji-		
11	lo-		
12	ka-		
14	o-		
15	ko-		
16	fa-		
18	ko-		
19	fi-		

TABLE 18.28 TAM PREFIXES

Prefix	Full form	TAM form
á-	SP-á-ROOT-iyɛ	Hesternal Perfective
á-	SP(H)-á-ROOT-a	Consecutive Subjunctive (see also Past Progressive 1)
a-	SP-a-ROOT-á	Far past Perfective
jé-	SP(H)-jé-ROOT-a	Future Perfective
kéé-/kéén-/kééndé-	SP(H)-kéé-/kéén-/kééndé-ROOT-a	Present Imperfective
kee-/keen-/keende-	SP(H)-kee-/keen-/keende- ROOT-a	Past Imperfective
áándá-	SP-áándá-ROOT-a	Habitual 1
jéé-	SP-jéé-ROOT-a	Habitual 2
re-	SP-á-re-ROOT-a	Past Progressive 1
áysé-	SP-áyse-ROOT-a	Past Progressive 2
kéré-	SP-kéré-ROOT-a	Persistive
ká-	SP-ká-ROOT-a	Consecutive
jé-	SP-jé-ROOT-a	Situative
éé-/éénd-	SP(H)-éé-/éénd-ROOT-a	Imperfective Subjunctive
kéé-	SP(H)-kéé-ROOT-á	Conditional
káá-	SP(ø)-káá-ROOT-á	Far past Irrealis

of the Conditional has been observed, probably due to a gap in the data. For more on negation in Mbugwe, see Gibson and Wilhelmsen (2015).

The SP slot is the Subject Prefix, which is obligatory in all indicative verb forms. The SP for all classes and persons, including underlying tone, are given in Table 18.27. Only noun classes 1 and 2 (3SG and 3PL) have a H tone.

There are three negative markers which occur in the NEG2 slot: *sí-, káysé-* and *jé-*. *sí-* is the negative 1SG prefix, which occurs in all indicative negative forms of the 1SG. *káysé-* is the negative subjunctive. The pronunciation *káysé-* is in free variation with *késé-* (cf. Mous 2004: 7); *jé-* is used in non-past negation. It occurs together with *sí-* if the subject is 1SG (see (6) above).

TAM is the slot for tense, aspect and mood markers, and this slot is usually, but not obligatorily filled. The TAM slot allows for up to two TAM markers, but usually there is only one. An example of two TAM markers can be found in the Past Progressive 1: it is made up of the prefix *á-*, which refers to the past, and *re-*, which marks the Progressive, grammaticalised from a locative copula. Table 18.28 provides an overview of all the TAM prefixes.

A Ventive also occurs in the TAM slot. The form is *ja-*, transparently grammaticalised from the verb *j-a*, 'come.' It has only been observed with the consecutive *ká-* in the data (*ká-ja-*) (see (3) above).

The object prefix (OP) is not obligatory, but it is often present if the object is animate. Only one object can be marked on Mbugwe verbs. As in several other F languages (Stegen 2002: 139), the reflexive and reciprocal have merged, and are positioned in the OP slot as *é-* (see example 20 below). The form of all the OPs is found in Table 18.29. Only class 1 prefixes (1SG, 2SG and 3SG) do not have a H tone.

TABLE 18.29 OP IN CLASS 1–19

Noun class	Third person	Second person	First person	Reflexive/reciprocal
1	mo-	ko-	N-	é-
2	vá-	mó-	kó-	
3	ó-			
4	é-			
5	ré-			
6	á-			
7	ké-			
8	ví-			
9	é-			
10	jí-			
11	ló-			
12	ká-			
14	ó-			
15	kó-			
16	fá-			
18	kó-			
19	fí-			

The productive extensions are -er (applicative), -y (causative) and w- (passive). The causative and applicative are often lexicalised, and are then analysed as part of the verb root. The applicative increases the valency with one, adding a receiver or beneficiary of the verb, whereas the passive decreases the valency and lifts the object to subject position. The causative changes the meaning of the verb into 'cause someone to VERB,' and adds an object to an intransitive verb. The extensions may also be combined, for instance applicative + causative, or applicative + passive.

The Final Vowel (FV) is usually -a, which is the default and usually marks a verb in the indicative mood. Certain subjunctive forms have the FV -ɛ. A third FV is -iyɛ, which occurs in the Hodiernal and in the Hesternal, which are also indicative.

There is only one post-FV suffix: the Imperative Plural. Here, the suffix -i is added to the final vowel. Nurse (2008: 39) notes that the F zone is the only Bantu zone which has -ɪ (sic.) instead of the more common -ni suffix for the plural imperative.

4.2 Periphrastic verb forms

The periphrastic verb construction in Mbugwe has attracted attention from researchers (see, for instance, Mous 2000) as it is typologically unexpected: the lexical verb comes first, in the infinitive, and the auxiliary comes after, with an SP marker. The OP may occur after the infinitive prefix of the lexical verb. Six auxiliaries have been found in this construction in the textual data for this study. All of them have a counterpart in a simple form, where the auxiliary occurs as a prefix (Table 18.30). For an explanation of the TAM categories, see Section 4.3.

In these TAM-forms, the periphrastic is the default construction. Below are examples of the periphrastic Future Perfective (1), Present Imperfective (2) and Habitual 1 (3)

TABLE 18.30 PERIPHRASTIC VERBS

TAM	*Periphrastic*	*Simple verb*
Present Imperfective	o-ROOT-a SP-kɛ́ɛ́ndɛ́	SP(H)-kéé(n)-ROOT-a
		SP(H)-kééndé-ROOT-a
Past Progressive 1	o-ROOT-a SP-á-re	SP-á-re-ROOT-a
Past Progressive 2	o-ROOT-a SP-áyse	SP-áyse-ROOT-a
Habitual 1	o-ROOT-a SP-áándá	SP-áándá-ROOT-a
Habitual 2	o-ROOT-a SP-jéénde	SP-jéé-ROOT-a
Future Perfective	o-ROOT-a SP-je	SP(H)-jé-ROOT-a

(1) CC traditional story 1.2
   òsírá kòjè nà vàánà
   o-sír-a   ko-je   na   va-ána
   INF-finish-FV   SP$_{1PL}$-FUT   and   2-child
   'We are going to die, and the children too.'

(2) EK life story, 1.1
   nè **òsáákà nkɛ́ɛ́ndɛ́** òlóósá mìkàlɔ yààné kwɛ́ɛ̀ndà nàyáálwá
   ne   o-sáák-a   n-kɛ́ɛ́ndɛ́
   FOC   INF-want-FV   SP$_{1SG}$-PRS

   o-lóós-a   me-ikalɔ   e-aané   kwɛɛnda
   INF-tell-FV   4-life   PP$_4$-1SG.POSS   since

   n-a-yáál-w-á.
   SP$_{1SG}$-PST-give.birth-PASS-FV
   'I want to tell about my life since I was born.'

(3) CC life story 1.40
   nà **òmòtúúmbà náándá** ìjóvà báá éènsíkò
   na   o-mo-túmb-a   n-áándá
   FOC   INF-OP$_{3SG}$-follow-FV   SP$_{1SG}$-HAB1

   i-jóva   báá éénsíko
   5-god   even today
   'I follow God, even today.'

The simple forms are used in certain syntactic conditions, specifically, interrogative (4), relative (5) and subordinate clauses (6). More research is needed in order to determine whether the simple form occurs in other syntactic contexts.

(4) CC traditional story 4.5
   áfá **ókéésííta** chákòrà
   áfá   ó-kéé-síít-a   ke-ákora
   DEM.PROX$_{16}$ SM$_{2SG}$-PRS-deny-FV   7-food
   'Why do you refuse food?'

(5)   CC traditional story 1.25
      nà chákòrà **áárètwárèrwà** nè chákòrà chá mpúúmbà

na	ke-ákora	á-á-re-twál-er-w-a		ne
and	7-food	SP$_{3SG}$-PST-PROG1-bring-APPL-PASS-FV		COP

ch-ákora	ke-á	m-púmba.
7-food	PP$_7$-CON	9/10-chaffs.of.grain

'And the food she was brought was chaffs of grain.'

(6)   CC traditional story 7.23
      áfá **ááysètótèkà** nkó méè kónɔ vákámònólà njéérè vàkàmɔɔ̀yà vákámòfɔ́tɔ̀là
      màkútà vákámòvékèrà ngɔ̀ mpèyá

áfá	á-áyse-tótek-a	n-kó
DEM.PROX$_{16}$	SP$_{3SG}$-PROG2-bring-FV	9/10-firewood

meé	kónɔ	vá-ká-mo-nól-a	n-jéére
then	here	SP$_{3PL}$-CONS-OP$_{3SG}$-cut-FV	10-hair

vá-ká-mo-ɔ́y-a	vá-ká-mo-fɔ́tɔl-a
SP$_{3PL}$-CONS-OP$_{3SG}$-bath-FV	SP$_{3PL}$-CONS-OP$_{3SG}$-apply-FV

ma-kúta	vá-ká-mo-vék-er-a		n-gɔ	m-feyá
6-oil	SP$_{3PL}$-CONS-OP$_{3SG}$-dress-APPL-FV		9/10-clothing	NP$_{9/10}$-new

'While she was collecting firewood, they cut her hair, bathed her and applied
lotion, and dressed her in new clothes.'

The alternation between periphrastic and simple verbs in Mbugwe coincides with a similar phenomenon in Rangi F33. In Rangi, the alternation (which is analysed as an alternation between pre- and post-verbal auxiliary placement) is found in wh-clauses, negative constructions, relatives, cleft constructions and subordinate clauses (Gibson 2012). The origin of the INF-AUX construction is still unclear, but it has been proposed that it is due to language contact (Mous 2000, Gibson 2012), or that it is a language-internal grammaticallisation process motivated by considerations of information structure (Mous 2000, Gibson 2012, Gibson & Marten 2015).

4.3 Tense, aspect and mood

In this section, the various TAM-forms of the Mbugwe verbs will be presented. TAM is marked by a combination of TAM prefixes, suffixes (FV) and tone on the verb in Mbugwe. It is not always straightforward to decide what function a single morpheme marks, but each morpheme is given a gloss according to the present analysis.

There are three past tenses in the perfective aspect in Mbugwe. They are called Hodiernal, which refers to event which have taken place earlier than the day of speaking, but also functions as an anterior, Hesternal, which refers to events that took place the day before the time of utterance, and Far Past, which took place before yesterday. Future time reference also occurs in the perfective aspect, but there are no degrees of remoteness in future reference. The perfective forms (Section 4.3.1) contrast with the imperfective ones (4.3.2). In the imperfective aspect, there is only one degree of past time reference.

In addition, present time reference usually occurs in the imperfective aspect. Future time reference is also possible in the imperfective aspect. The following imperfective categories have separate verb forms in Mbugwe: the Imperfective, the Habitual, the Progressive and the Persistive. In addition to these categories, there are two TAM-forms which are called relative tenses by Nurse (2008): the Consecutive and the Situative (4.3.3. and 4.3.4). The moods that are found in Mbugwe are the Indicative, the Subjunctive (4.3.4), the Imperative (4.3.6), the Conditional (4.3.7) and the Irrealis (4.3.8).

4.3.1 Perfective forms

As mentioned above, there are three past tenses and one future tense in the perfective aspect. In addition, the most common Subjunctive form has perfective aspect (see Section 3.3.5). In Table 18.31, the Melodic H is indicated with a subscript MH and the position of the tone.

TABLE 18.31 PERFECTIVE FORMS

TAM	Simple form	Periphrastic form	Negative
Hodiernal Perfective	SP-ROOT-íyɛ $_{\text{MH(2)-PU}}$		te-SP-ROOT-íyɛ $_{\text{MH(2)-PU}}$
Hesternal Perfective	SP-á-ROOT-iyɛ		te-SP-á-ROOT-iyɛ
Far Past Perfective	SP-a-ROOT-á$_{\text{MH2-U}}$		te-SP-a-ROOT-á$_{\text{MH2-U}}$
Future Perfective	SP(H)-jé-ROOT-a	INF-ROOT-a SP-je	te-SP-jé-ROOT-a

When an event has occurred earlier on the day of speaking, the Hodiernal Perfective is used. The Hodiernal is also used to express anterior aspect. For a discussion of this usage see Wilhelmsen (2011).

The Hesternal Perfective is used primarily for events that happened the day before the time of utterance. Sometimes, however, the form is used more subjectively. For instance, the forms in (7) were given with the Hesternal in elicitation by both language consultants, initially.

(7) CC LOT #158a
nárémìyɛ̀ yòòndà ráànɛ́ wíkì éè yálɔ́ɔ̀kìyɛ̀ nà ìjɔ́
n-á-rem-iyɛ   i-onda   r-áánɛ́   wíki
SP$_{\text{1SG}}$-PST-cultivate-PFV 5-field PP$_5$-1SG.POSS 9/10.week

ée   y-á-lɔɔk-iyɛ   na   ìjɔ́
DEM.PROX$_9$ SP$_9$-PST-pass-PFV   CONN   DEM.ANA$_{10}$
'I cultivated my farm last week.'

Colman Chuchu used the Hesternal for cultivating last month also, but for 'last year' both language consultants used the Far Past. This indicates that the three degrees of past time reference may be used to refer to various time frames: the closest, the one before that, and the most remote one. For a discussion of the subjective use of the various past tenses, where also Mbugwe is mentioned, see Botne (2014).

The Far Past is used for anything that happened before yesterday. In a narrative, the first verb is usually in the Far Past Perfective, and then the following verbs are in the Consecutive, the Past Imperfective or another imperfective form.

The Future Perfective is periphrastic in neutral main clauses.

4.3.2 Imperfective forms

Four different aspects are subsumed in what is here referred to as imperfective forms. The forms are Imperfective, Habitual, Progressive and Persistive. They are grouped together as there is some semantic overlap between the categories. Common to these forms is that the focus is on the continuing event, with no reference to the beginning or end of it (Comrie 1976: 16). All of the imperfective forms except for the Persistive are periphrastic by default (Table 18.32).

TABLE 18.32 IMPERFECTIVE FORMS

TAM	Simple form	Periphrastic form	Negative
Present Imperfective	sp(H)-kéé-/kéén-/kééndé-root-a	INF-root-a sp-kééndé	te-sp-jé-root-a
Past Imperfective	sp(L)-kee-/keen-/keende-root-a		te-sp-kee-root-a
Habitual 1	sp-áándá-root-a	INF-root-a sp-áánda	te-sp-áándá-root -a
Habitual 2	sp-jéé-root-a	INF-root-a sp-jéénde	
Past Progressive 1	sp-á-re-root-a	INF-root-a sp-á-re	te-sp-á-re-root-a
Past Progressive 2	sp-áyse-root-a	INF-root-a sp-áyse	
Persistive	sp-kéré-root-a		

The most common form used to refer to present time is the Present Imperfective. It is the neutral way of expressing a present event. The fact that Present Imperfective and Future share a common negative form points to the presence of a non-past or 'general' negative in Mbugwe" (see also Mous 2004: 7).

The Past Imperfective form emphasises the duration of the event, but a habitual or progressive reading is also possible in some cases.

The Habitual aspect has two forms. There are only two examples of *jéénde/jéé-* in the data. For this reason, no semantic difference between the forms has been possible to establish, and they are therefore simply labelled Habitual 1 and Habitual 2.

Two progressive forms are found, and both refer to past time. The most common form is what is here called Past Progressive 1, with the prefixes *á-re-*. The TAM prefix *á-* is the same as the past tense prefix in the Hesternal form (and *a-* with no H tone in the Far Past tense). The other past progressive form, with the prefix *áyse-*, is rarely found in the data, and is only used by the older language consultant. It is referred to as Past Progressive 2. Again, no temporal or other semantic difference has been possible to establish between them.

In (8) and (9), which both refer to a situation where the progressive typically is used for the first verb, the younger language consultant used the Past Progressive 1, and the elderly language consultant gave the Past Progressive 2 form.

(8)   NR LOT #182
       òrèmà náré yòòndà rááné áfà áàfìká
       o-rem-a   n-á-re   i-onda
       INF-cultivate-FV   SP$_{1sg}$-PST-PROG1   5-field

       re-ááné   áfa   á-a-fìk-á
       PP$_5$-1SG.POSS   DEM.PROX$_{16}$ SP$_{3sg}$-PST-arrive-FV
       'I was cultivating my farm when he arrived.'

(9)   CC LOT #182
       nè **òrèmà náysé** yòòndà rááné áfá áàfìká
       ne   o-rem-a   n-áyse   i-onda
       FOC   INF-cultivate-FV   SP$_{1sg}$-PROG2   5-field

       r-ááné   áfá   á-a-fìk-á
       PP$_5$-1SG.POSS   DEM.PROX$_{16}$ SP$_{3sg}$-PST-arrive-FV
       'I was cultivating my farm when he arrived.'

The Persistive is analysed as diachronically made up of two prefixes, the prefixes *ké-* and *ré-*. *re-* is known from the Past Progressive 1, although it has a different tone, and *ké-* is probably a reflex of persistive PB *kí-*. They are analysed together synchronically, however, as the TAM prefix *kéré-*.

4.3.3 Consecutive

The Consecutive form is used for most verbs in a narrative, except for the first, which sets the time frame for the story. Each verb which is marked with the Consecutive refers to an event which happened after the previous verb in the narrative. The Consecutive TAM prefix is *ká-*, and the FV is *-a*.

The consecutive does not always refer to the past time, but is also used with imperatives (10).

(10)   NR LOT #183b
        fétá **ókárémà** yòòndà rááká
        fét-á   ó-ká-rem-a   i-onda   r-ááká
        GO-IMP   SP$_{2sg}$-CONS-cultivate-FV   5-field   PP$_5$-2SG.POSS
        'Go and cultivate your farm.'

4.3.4 Situative

The Situative is similar to the Consecutive in that the time reference of the second verb depends on the first verb, but here, the events do not happen one after the other, but at the same time, and the second verb is subordinated to the first (11).

(11)   CC story 1.14
vààfétá **vájóórékèryà** nyòòmbá yá kwá wààlàvɔ̀
vá-a-fét-á   vá-jé-órekery-a   nyombá
SP$_{3PL}$-PST-GO-FV   SP$_{3PL}$-SIT-ask-FV   9/10.house
e-á   ko-á   waalavɔ
PP$_9$-CON   PP$_{17}$-CON   their.sibling
'They walked around asking about the house of their relative.'

4.3.5 Subjunctive mood

The indicative-subjunctive dichotomy is fundamental when it comes to moods in Mbugwe. The subjunctive serves a wide range of functions in Mbugwe, expressing both propositional modality and event modality (Palmer 2001: 111). They often occur in subordinate clauses. In addition, the subjunctive sometimes functions as an imperative or jussive.

There are three different subjunctive forms in Mbugwe: the Subjunctive Perfective, the Subjunctive Imperfective and the Subjunctive Consecutive (Table 18.33).

TABLE 18.33 SUBJUNCTIVE FORMS

TAM	Simple Form	Negative
Subjunctive Perfective	SP(H)-ROOT-ɛ́ $_{MH(2-)U}$	SP-káysé/késé-ROOT-a
Subjunctive Imperfective	SP(H)-éé/éénd-ROOT-a	
Subjunctive Consecutive	SP(H)-á-ROOT-a	

The Subjunctive Perfective expresses the subjunctive mood in a perfective sense, that is, as a whole, completed event. It may occur with SP, SP and OP, or just OP. Below is an example of a verb with just an OP, and no SP (12). This is not grammatical in the indicative.

(12)   NR Song, 1.10
nányù njémwéérà **áìkàlɛ́** nèshòpárkì
nányu   n-jé-mo-wéér-a   á-ikal-ɛ́   neshopárki.
who   SP$_{1SG}$-OP$_{3SG}$-FUT-tell-FV   OP$_{3SG}$-stay-SBJV   national.park
'Who will I tell to stay in the national park?'

The Subjunctive Imperfective refers to an action which is ongoing or continuous, without referring to the beginning or end of it. In (13), the first Subjunctive Imperfective is the main verb, and it functions as a polite command (imperative), with an OP only. The second Subjunctive Imperfective functions as a purposive, and has an SP only.

(13)   NR Song 1.6
éékórìrà kwéésíítà òjìshà mpɔ̀ɔ̀ngɔ́
éé-kó-rir-a   kó-éé-síít-a
SBJV-IPFV-OP$_{1PL}$-protect-FV   SP$_{1PL}$-SBJV.IPFV-refuse-FV

o-jish-a   m-pɔɔngɔ́
IND-do-FV   9/10-thing
'Protect us so that we refuse to do things.'

There is a separate form called Subjunctive Consecutive. It is distinguished from the Far
Past Perfective by tone only. There is no Melodic H, but a H tone on the TAM prefix. In
(14), it functions as a jussive.

(14)   CC life story, 1.29
       èbú **kwáívà** nyááfù
       ebú   kó-á-iv-a   n-yááfu
       GO.IMP   SP$_{1PL}$-CONS.SBJV-steal-FV   9/10-net
       'Let us go and steal nets.'

The Negative Subjunctive form is *káysé-/késé-*, which occurs after the SP, but before any
OP (15). No aspectual variation has been observed with the Negative Subjunctive.

(15)   LOT CC #179
       òkáysérémà yòòndà rááné
       o-káysé-rem-a   i-onda   r-ááné
       SP$_{2SG}$-NEG.SBJV-cultivate-FV   5-field   PP$_5$-1SG.POSS
       'Do not cultivate my farm.'

4.3.6 Imperative mood

The imperative may occur with an OP, and occasionally with an SP (Table 18.34).

TABLE 18.34 IMPERATIVE FORMS

TAM	Simple Form	Negative
Imperative Singular	(SP)(OM)-ROOT-á $_{MH(2-)PU}$	a-ré SP-ROOT-a
Imperative Plural	(SP)(OM)-ROOT-éy $_{MH2-PU}$	a-ré SP-ROOT-a

There are several suppletive forms of the Imperative. For example, the verb *render-a*
'wait' has the Imperative form *nendera*, and *fét-a* 'go' has the imperative form *ebú*. A polite
form of *óch-a* 'take' in the Imperative is *asé*. The Plural Imperative has the suffix *-i*.

The Imperative cannot be negated with the regular negative prefix *te-*, but there is a
form with a prohibitive reading (16). The form is *aré* SP-(OP)-ROOT-*a*. More research
is needed in order to determine the full function and origin of this form.

(16)   NR Song 2.1
       síré nà málì síré ná ng'ɔɔmbè **àré mònsóóchà**
       sí-ré   na   máli   sí-ré
       NEG.1SG-COP.LOC with   wealth   NEG.1SG-COP.LOC
       na   ng'ɔɔmbɛ aré   mo-n-sóóch-a
       with   9/10.cow PROH   SP$_{2PL}$-OP$_{1SG}$-hate-FV
       'I do not have riches, I do not have cows, do not hate me.'

4.3.7 Conditional mood

In conditional clauses, the first, conditional verb has the prefix *kéé-*, and the FV is *-a*.
There is also a Melodic H tone on the ultimate of the verb stem (see Section 2.4.3).

The *kéé-* prefix may also function as an indicator of time, translated as 'when' instead of 'if.' A few examples of this are found in the data. The two sentences 'If you come, you will find me' (LOT #247a) and 'When you come, you will find me' (LOT #250) are identical in Mbugwe (17).

(17)   NR LOT #247a, 250
    òkééjá ònkúúndyá òjè
    o-kéé-j-á   o-n-kúndy-a   o-je
    SP_{2sg}-COND-COME-FV   INF-OP_{1sg}-meet-FV SP_{2sg}-FUT
    'When/if you come, you will find me.'

4.3.8 Irrealis mood

The Irrealis (counterfactual) mood refers to a hypothetical situation that is known not to be true or possible. It can combine with the suffixes *-iyɛ* and *-a* in order to form Recent and Far Past Irrealis (Table 18.35). There is a Melodic H tone on the ultimate mora of the Far Past Irrealis, see Section 2.4.3.

TABLE 18.35 IRREALIS FORMS

TAM	Simple Form	Negative
Irrealis Recent Past	SP(ø)-káá-ROOT-iyɛ	te-SP-káá-ROOT-iyɛ
Irrealis Far Past	SP(ø)-káá-ROOT-á $_{MH(2-)U}$	te-SP-káá-ROOT-á $_{MH(2-)U}$

The Irrealis Recent Past refers to a situation that is hypothesised to have happened not too long ago, for instance today or yesterday, as in (18). It may also include the time of utterance.

(18)   Adapted from Dahl's questionnaire (106)
    àkáátɔ́ɔ́yɛ̀ mpíyà nàyjɔ́ **ákáámòòrèrèyɛ̀** sàwádì mwàìrétù
    á-káá-tɔ́ɔ́-iyɛ   m-píya   nayjɔ́
    SP_{3sg}-IRR-receive-PFV 9-money   yesterday

    á-káá-mo-ol-er-iyɛ   sawadi   mo-irétu
    SP_{3sg}-IRR-OP_{3sg}-buy-APPL-PFV gift   1-girl
    'If the boy had gotten the money yesterday he would have bought a present for the girl'

The Irrealis Far past refers to something which is hypothesised to have happened before yesterday, and does not include the time of speaking (19).

(19)   NR Song 1.5
    nà **ákááfétá** ááùnà kwá Nùùmbè
    na   a-káá-fét-á   a-a-un-a
    and SP_{3sg}-IRR-go-FV SP_{3sg}-PST-HURT-FV

    ko-á   Numbe
    PP_7-CON   Numbe
    'If he would have gone he would have hurt Numbe'

5 SYNTAX

Syntax is yet to be researched in Mbugwe, but in this section some basic syntactic observations will be presented.

5.1 Basic word order

The basic word order in Mbugwe is SVO, although the subject often is dropped, and only
marked on the verb. Subject agreement is obligatory in the indicative when the subject
is present (20).

(20) CC LOT #243
Màsànjà **wéékèríyɛ̀** nà lwéɛmbɛ̀ mònwè
Masanja   á-é-kɛr-íyɛ   na   lo-ɛ́mbɛ   mo-nwe.
Masanja   SP₃ₛ₉-REFL-cut-PFV   with   14-razor   3-finger
'Masanja has cut his finger with a razor'

Only one object may be marked on the verb in Mbugwe. If there is more than one object
in a clause, there has to be a separate noun phrase in order to express the second object. In
(21), the indirect object, being animate, is marked on the verb. The indirect object comes
before the direct object in the clause.

(21) LOT questionnaire, CC #237
mòòntòmòká **ókééválúvèrà** vàánà vááchwè chákɔ́rà
mo-ntomoká   ó-kéé-vá-luv-er-a   va-ána
1-woman   SP₃ₛ₉-PRS-OP₃ₚₗ-cook-APPL-FV   2-child

va-ááchwɛ́   ke-ákɔra
pp₂-3sg.poss   7-food
'The woman is cooking food for her children'

5.2 Word order in the noun phrase

Nominal modifiers usually occur after the noun, as illustrated in (22).

(22) EK's life story 1.3
nyɛ̀ɛ̀ngɔ̀ jáánɛ́ jɔ́ɔ̀nsɛ̀ já móvèrè jàlwàálá.
n-yɛɛngɔ   ji-áánɛ́   ji-ɔ́ɔ́nsɛ   ji-á   mo-vere
10-joints   PP₁₀-1SG.POSS   PP₁₀-all   PP₁₀-CON   3-body
ji-a-lwaal-á
SP₁₀-PST-hurt-FV
'. . . all the joints of my body hurt' (Lit. ' . . . all my joints of the body hurt.')

The possessive pronoun always occurs right after the noun (22). If there are several modifiers in the noun phrase, the noun is usually first, and the adjective is last (23). As seen in
(23), numerals come before demonstratives and adjectives.

(23) EK's life story, 1.23
neé nkájarérá **vaáná vaveeré ává vadídí** namwɛ́ɛ́nɛ́
neé   n-ká-ja-rɛ́r-a   va-ána
then   SP$_{1\text{SG}}$-CONS-VENT-raise-FV   2-child

va-veeré   ává   va-dídí   namw-ɛ́ɛ́nɛ́.
EP$_2$-two   DEM.PROX$_2$ NP$_2$-small   1SG-self
'Then I was taking care of these two small children on my own.'

The only modifier that comes before the noun is the demonstrative, as in (24).

(24) CC traditional story 1.12
áfá áakúyá nyináávɔ́ vákáva váchááyɛ́ **vará vaáná** vɛɛnɛ
áfá á-a-kúy-á nyináávɔ́   vá-ká-va
dem.prox$_{17}$ SP$_{3\text{sg}}$-pst-die-pfv   their.mother   sp$_{3\text{pl}}$-cons-cop

vá-cháál-íyɛ   va-rá   va-ána   va-ɛɛné
SP$_{3\text{PL}}$-left.behind-PFV   PP$_2$-DEM.DIST   2-child   NP$_2$-self.
'When their mother died, those children had been left by themselves.'

Sometimes, however, for reasons yet to be discovered, the demonstrative follows the noun it modifies, as in (25).

(25) CC traditional story 7.12
chákóra éche nɛ́ɛ nsíjékekwéérya **chákóra éché** nachéévá káley
ke-ákora   éche   nɛ́ɛ
7-food   DEM.PROX$_7$ 1SG

n-sí-jé-ke-kwééry-a   ke-ákora
SP$_{1\text{SG}}$-1SG.NEG-NPST.NEG-OP$_7$-be.able-FV   7-food

éche   n-a-ke-év-á   kaley
DEM.PROX$_7$ SP$_{1\text{SG}}$-PST-OM$_7$-forget-FV   long.time.ago
'I can't eat this food because I have forgotten it a long time ago.'

5.3 Relative clauses

The relative is not morphologically marked on the verb, but a relative clause takes a phrasal tone. For TAM-forms that are periphrastic by default, the simple form will be used in a relative clause.

5.4 Interrogatives

Yes/no-questions are marked by an extra H tone utterance-finally in Mbugwe. Content questions are formed with interrogative pronouns, which occur clause-initially, as in (26). Examples of interrogative pronouns are *-á kee* 'what, why,' *nányu* 'who,' *ndee* 'when,' *kɔ́ɔ* 'where' and *jore* 'how.'

(26) CC story 7.20
      nè **wá kèè** òkéémòséérérà òrìrà óò
      ne   o-á   kee   o-kéé-mo-séérɛr-a
      COP   PP₁₄-CON   what   SP₂ₛ₉-PRS-OP₃ₛ₉-help-FV

      o-rir-a   óo
      INF-protect-FV   DEM.PROX₁
      'Why are you helping her to protect this one?'

5.5 Focus markers

There is a possible focus marker *ne/na* in Mbugwe (27). Two similar discourse markers are described for Rangi, and it is there considered a variant of the copula *ní* (Mbugwe *ne*) (Gibson 2012: 92–95). For an example, see (27).

(27) From LOT questionnaire Mzee Colman #182
      nè òrèmà náysé yòòndà rááné áfá ààfiká
      ne   o-rem-a   n-áyse   i-onda
      FOC   INF-cultivate-FV   SP₁ₛ₉-PROG2   5-field

      r-ááné   áfá   á-a-fik-á
      PP₅-1SG.POSS   DEM.PROX₁₆   SP₃ₛ₉-PST-arrive-FV
      'I was cultivating my farm when he arrived.'

More research is needed in order to determine the function and nature of the focus markers in Mbugwe. In general, syntax is a ripe area for future research in Mbugwe, as this short section illustrates. Topics for future research include but are not limited to information structure, clause structure and argument structure.

REFERENCES

Botne, R. (2014) Resultatives, Remoteness, and Innovation in Eastern and Southern Bantu T/A Systems. *Nordic Journal of African Studies* 23: 16–30.

Comrie, B. (1976) *Aspect: An Introduction to the Study of Verbal Aspect and Related Problems.* Cambridge: Cambridge University Press.

Dempwolff, O. (1915–1916) Beiträge zur Kenntnis der Sprachen in Deutsch-Ostafrika. 7. Buwe, 8. Irangi. *Zeitschrift für Kolonialsprachen* 6: 1–27, 102–103.

Dixon, R. M. W. (1982) *Where Have All the Adjectives Gone? And Other Essays in Semantics and Syntax.* Berlin: De Gruyter Mouton.

Gibson, H. (2012) *Auxiliary Placement in Rangi: A Dynamic Syntax perspective.* London: SOAS, University of London, PhD dissertation.

Gibson, H. & L. Marten (2015) *Contact and Change in Bantu: The Case of Rangi.* London: SOAS, University of London, Unpublished manuscript.

Gibson, H. & V. Wilhelmsen (2015) Cycles of Negation in Rangi and Mbugwe. *Africana Linguistica* 21: 233–257.

Güldemann, T. (2014) 'Khoisan' Linguistic Classification Today. In Güldemann, T. & A.-M. Fehn (eds.), *Beyond Khoisan: Historical Relations in the Kalahari Basin*, 1–40. Amsterdam; Philadelphia: John Benjamins.

Hyman, L. M. (1992) Register Tones and Tonal Geometry. In van der Hulst, H. & K. Snider (eds.), *The Phonology of Tone: The Representation of Tonal Register*, 75–108. Berlin: De Gruyter Mouton.

Kiessling, R., M. Mous & D. Nurse (2008) The Rift Valley Area of Central Tanzania as a Linguistic Contact Zone. In Heine, B. & D. Nurse (eds.), *A Linguistic Geography of Africa*, 186–227. Cambridge; New York: Cambridge University Press.

Larsen, J., V. Larsen & O. Stegen (2011) *Mbugwe Data*. Ms.

Maho, J. F. (1999) *A Comparative Study of Bantu Noun Classes*. Göteborg: Acta Universitatis Gothoburgensis.

Meeussen, A. E. (1967) Bantu Grammatical Reconstructions. *Africana Linguistica* 3: 79–121.

Mous, M. (2000) Counter-Universal Rise of Infinitive-Auxiliary Order in Mbugwe (Tanzania, Bantu F 34). In Vossen, R., A. Mietzner & A. Meissner (eds.), *Mehr als nur Worte: Afrikanistische Beitrage zum 65. Geburtstag von Franz Rottland*, 469–481. Cologne: Rüdiger Köppe.

Mous, M. (2004) *A Grammatical Sketch of Mbugwe, Bantu F34, Tanzania*. Cologne: Rüdiger Köppe.

Mradi wa Lugha za Tanzania (2009) *Atlasi ya Lugha za Tanzania*. Dar es Salaam: Chuo Kikuu cha Dar es Salaam.

Nurse, D. (2008) *Tense and Aspect in Bantu*. Oxford: Oxford University Press.

Nurse, D. & B. F. Y. P. Masele (2003) Stratigraphy and Prehistory: Bantu Zone F. In Andersen; H. (ed.), *Language Contacts in Prehistory: Studies in Stratigraphy*, 115–134. Amsterdam; Philadelphia: John Benjamins.

Odden, D. & L. Bickmore (2014) Melodic Tone in Bantu. *Africana Linguistica* 20: 3–13.

Palmer, F. R. (2001) *Mood and Modality*. Cambridge; New York: Cambridge University Press.

Stegen, O. (2002) Derivational Processes in Rangi. *Studies in African Linguistics* 31(1/2): 129–153.

Wilhelmsen, V. (2011) Grammaticalization of Tense and Aspect in Mbugwe: A Preliminary Investigation. In Svantesson, J.-O., N. Burenhult, A. Holmer, A. Karlsson & H. Lundström (eds.), *Language Documentation and Description (Volume 10)*, 247–264. London: SOAS.

Wilhelmsen, V. (2018) *A Linguistic Description of Mbugwe with Focus on Tone and Verbal Morphology*. PhD dissertation. Uppsala: Acta Universitatis Upsaliensis.

CHAPTER 19

KAMI G36

Malin Petzell and Lotta Aunio

1 INTRODUCTION

Kami is an endangered, under-described Eastern Bantu language, classified as G36 in the referential classification of Guthrie (1971).[1] It is reported to be spoken by 5,518 people in the Morogoro region of Tanzania (Mradi wa Lugha za Tanzania 2009). This figure indicates the total number of persons who consider themselves to speak Kami, but it does not say anything about the competence of those speakers. The number of fluent speakers left is significantly lower, as was established during field trips to the area (in 2008, 2009, 2014, and 2016). We found no children or adolescents speaking the language, which means that the language is threatened with extinction. Our youngest informant was in his thirties, and he could only understand Kami, not speak it. Swahili G42, the national language of Tanzania, is gaining ground at the expense of Kami, and is the only language (apart from English) allowed in education, media, parliament and church. That said, Swahili is not the major threat to Kami – the regional language Luguru G35 is. Luguru is the major language in the Morogoro region, with 403,602 speakers (Mradi wa Lugha za Tanzania 2009). In the smaller Morogoro district, where most Kami speakers live, the Luguru speakers amount to 73.5% while the Kami speakers amount to only 1.3%, and in the entire Morogoro region, the Kami speakers constitute only 0.3% (Petzell 2012b: 19). This means that most Kami speakers are in fact trilingual in Kami, Luguru and Swahili. Kami is linguistically similar to its neighbours Kutu G37, Kwere G32 and Zaramo G33 (Petzell & Hammarström 2013). There are no regional varieties of Kami. The area where most Kami speakers live is called Mikese and is situated to the east of Morogoro town. Speakers can be found in Mkunga Mhola, Dete, and (Lukonde) Koo.

All the data in this chapter were collected in collaboration with a total of 11 Kami mother tongue speakers. We had some difficulties finding these speakers since the language is so marginalised. Our database consists of elicited sentences, words and stories, as well as recordings of spontaneous speech. However, our data is limited and there are many gaps. For instance, information on word order, double objects and syntactic matters generally requires further work.

Kami exhibits a great deal of variation, much of which can be traced to influence from neighbouring languages and Swahili. We do not oppose the claim that language obsolescence may lead to simplification, but in our case, the contrary seems to be true. A factor that most likely plays a role in the abundance of forms is the fact that the endangered language is influenced by more than two languages. As an endangered language declines, "we expect it to be flooded with an influx of patterns and forms from the dominant language" (Aikhenvald 2012: 77).

There are only a small number of written records of Kami. A few secondary sources briefly mention its existence, such as Johnston (1919: 141) and Guthrie (1948, 1971).

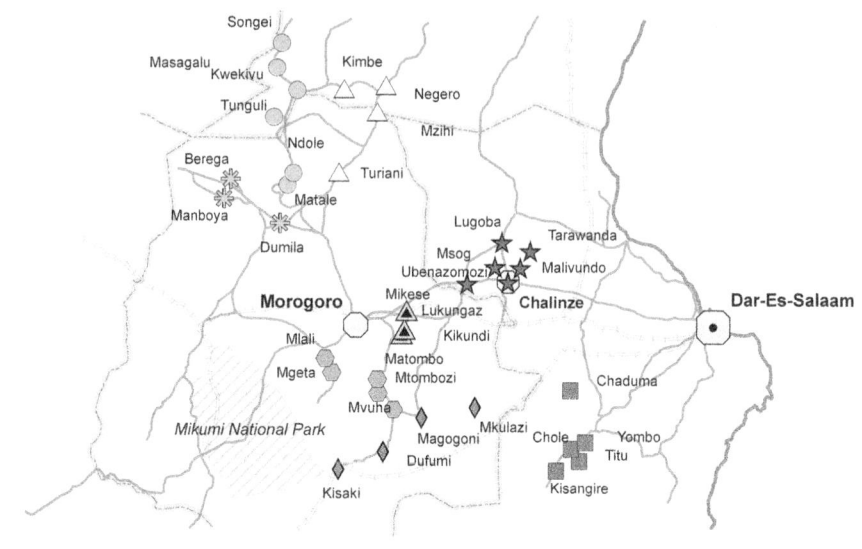

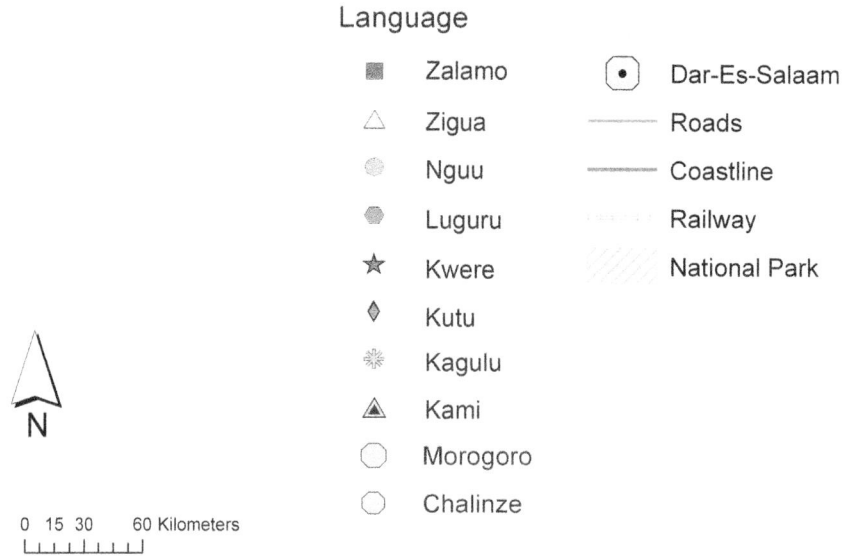

MAP 19.1 MAP OF KAMI AND NEIGHBOURING LANGUAGES

(Map created by Malin Petzell and Ulf Ernstsson)

Their data in turn stem from a brief and dated account of Kami written over 100 years ago (Velten 1900). In addition, Kami appears in Nurse and Philippson (1975, 1980), in Petzell (2012a, 2012b) and in Petzell and Hammarström (2013).

2 PHONOLOGY

Since there is no standardised orthography for Kami, the writing system used here is based on how the speakers write their language, which in turn follows the Swahili orthography (i.e., the language they are all taught in school). The following sections are a brief overview of the Kami sound system, in which the threatened status of the language is manifested by a remarkable amount of phonetic variation.

2.1 Vowels

Kami has a five-vowel system /a e i o u/ with no vowel length contrast. Of these, /u/ and /o/ are sometimes in free variation, most notably in the prefixes of classes 15 and 17 (*ku-lima*/*ko-lima* 'to cultivate' and *ku-m-gunda*/*ko-m-gunda* 'at the field').

While the data available does not allow for a comprehensive analysis of Kami vowel hiatus resolution, some observations can be listed. One frequent resolution is when /a/ and /i/ meet, they coalesce into /e/ as seen in the underlying *ka-iz-a* 's/he came' which is realised as *keza* 's/he came.' Another involves the close vowels /i/ and /u/ which are often glided across morpheme boundaries before another vowel. This happens, for example, with noun class prefixes, pronominal prefixes, and some object prefixes (1–5). As seen in example (3), the noun class 8 prefix, and also the pronominal prefix, alternates with the prefix form in which the /i/ is deleted instead of glided. Attested consonant + glide sequences are listed in Table 19.1.

(1)   mwana
      mu-ana
      1-child
      'child'

TABLE 19.1 KAMI CONSONANT PHONEMES WITH THE CORRESPONDING GRAPHEMES

Plosives/affricates	p \<p\>	t \<t\>	tʃ \<ch\>	k \<k\>
	b	d \<d\>	dʒ \<j\>	g \<g\>
Fricatives	f \<f\>	s \<s\>		h \<h\>
	v \<v\>	z \<z\>		
Approximants	w \<w\>	l/ɾ \<l, r\>	j \<y\>	
Nasals	m \<m\>	n \<n\>	ɲ \<ny\>	ŋ \<ng'\>
NC combinations	mp, mb, mf, mv	nt, nd, nz		nk, ng
	\<mh/m, mb, mf, mv\>	\<nh/n, nd, nz\>		\<ng'h/ng', ng\>
Cy combinations	my, vy	dy/ly \<dy\>		
	\<my, vy/v\>			
Cw combinations	bw, mbw, mw	tw, dw, ndw, sw, lw		kw, gw, ŋgw, ŋw
	\<bw, mbw, mw\>	\<tw, dw, ndw, sw, lw\>		\<kw, gw, ngw, ng'w\>

(2)   lwifi
     lu-ifi
     11-door
     'door'

(3)   vyana/vana vya/va mgeni
     vi-ana   vi-a   m-geni
     8-child PP_8-CON   1-guest
     'little children of guest'

(4)   mbiki   wangu/wetu   mi-biki   yangu/yetu
     m-biki   u-angu/u-etu   mi-biki   i-angu/i-etu
     3-tree PP_3-$POSS_{1SG}$/PP_3-$POSS_{1PL}$ 4-tree PP_4-$POSS_{1SG}$/PP_4-$POSS_{1PL}$
     'my/our tree'   'my/our trees'

(5)   niwona
     ni-u-on-a
     SP_{1SG}-OP_{11}-see-FV
     'I saw it'

The vowel of the nominal prefix of classes 4, 7 and 10 *mi-*, *chi-* and *zi-* and of the pronominal prefixes of class 7 and 10 *chi-* and *zi-* is deleted, rather than glided (6). Object prefixes of classes 4, 7 and 10 retain the vowel /i/ even before another vowel, and so does the 1SG subject prefix *ni-* (7).

(6)   chana cha mgeni
     chi-ana   chi-a   m-geni
     7-child PP_7-CON   1-guest
     'little child of guest'

(7)   nichiona
     ni-chi-on-a
     SP_{1SG}-OP_7-see-FV
     'I saw it'

When two /a/ phonemes are adjacent, one /a/ is deleted:

(8)   wana
     wa-ana
     2-child
     'children'

In addition, there are many vowel combinations in which both vowels are retained across morpheme boundaries (9–11). Some of these combinations also trigger glide formation or vowel elision as presented above, but it is not known whether this is context- or speaker-specific, or free variation.

(9)   kuikadzi
      ku-i-kadzi
      17-9-work
      'at work'

(10) koibaka
      ko-i-bak-a
      SP₁.NPST-REFL/RECP-smear-FV
      's/he is smearing her/himself'

(11) niwaona
      ni-wa-on-a
      SP₁ₛ₉-OP₂-see-FV
      'I saw them'

With some vowel-initial verb roots, the vowel of the infinitive prefix is either glided or deleted. For example, 'to throw' is either *kw-asa* or *k-asa*. Similar alternations are seen in other roots as well: *ko-kwandik-a* 's/he is writing' vs. *kw-andik-a* 'to write' and *ko-mw-andikila* 's/he is writing to'; *kw-ingil-a* 'to enter' vs. *ko-kwingil-a* 's/he is entering.' With some verbs the alternating consonant is *y*: *ku-yol-a* 'to be rotten' vs. *g-ol-a* 'they (class 6) are rotten; *k-em-a* 's/he stood up' vs. *ku-yemil-a* 's/he is standing.'

2.2 Consonants

Some of the variation in the Kami consonant inventory can be traced to Swahili influence, and some to neighbouring languages such as Luguru G35. There is variation between speakers but also within the idiolect of individual speakers.

The voiceless plosives /p t k/ are aspirated, /t/ often more so than the others. Voiced plosives /b d g/ are slightly implosive, except when preceded by a nasal. /tʃ/ and /dʒ/ are affricates.

The phoneme status of the fricatives is difficult to establish conclusively. When Gérard Philippson worked with Kami in the mid-1970s, he found an apparently active contrast between *fi* and *pfi* (Gérard Philippson pers. comm.). In our data, /pf/ appears to have merged with /f/: *-fika* 'arrive' < *-pik- and *-fimba* 'swell' < *-bimb-, while Philippson found *-fika* and *-pfimba*. While in the data from the 1970s, class 8 prefix is mostly *vi-*; in our data it is *vi-* or *fi-*. Therefore, the sound [f], in addition to being the realisation of the phoneme /f/, is also found as an allophone of /v/. Moreover, [pf] also occurs as an allophone of /v/ in our data, including in class 8 prefixes.

The alveolar fricative /z/ shows wide-ranging variation. The most common realisations are [z] and [ts], as for example in *mw-ezi/mw-etsi* 'moon'; [nz] and [s] can also be heard. For instance, the word 'water' can be *ma-nzi, ma-zi* or *ma-tsi* and the word 'string' can be *lu-sabi, lu-tsabi* or *lu-zabi*.

/l/ is a lateral approximant [l] or a flap [ɾ]. The allophone [d] of /l/ is attested after nasals and as a free variant in nominal and inflectional prefixes of class 5 ([li] or [di]).

Voiced plosives can be preceded by nasals both within stems and across morpheme boundaries. With some speakers, a trill release is found with /nd/ ([ndr]) and even with /mb/ ([mbr]). The sequence /nl/ is realised as [nd].

When the voiceless consonants /p t k/ are preceded by a nasal, only one of the sounds is pronounced. Most often these combinations are pronounced as nasals, with voicing ranging from voiceless [m̥ ŋ̊ ŋ̊] to fully voiced nasals [m n ŋ], for example in *mhene*/*mene* 'goat' and *n-kulu* → *ng'hulu*/*ng'ulu* 'big.' Historically, the same sound change is seen in words such as *mu-nhu*/*mu-nu* 'person' < **ntò*. In our examples, these sounds are written according to actual pronunciation, i.e., as a plain nasal when the nasal is voiced and with the nasal followed by *h* when the nasal is voiceless.

Of the fricatives, /f v z/ can be preceded by a nasal across morpheme boundaries (12–14).

(12) nimfika
      ni-m-fik-a
      SP$_{1SG}$-OP$_1$-meet-FV
      'I met her/him'

(13) komvaza
      ko-m-vaz-a
      SP$_1$.NPST-OP$_1$-dress-FV
      'S/he dresses her/him.'

(14) nzabi
      n-zabi
      9/10-string
      'string(s)'

/pj/ is only found in *-pya* 'new,' most likely as a recent loan from Swahili. /sh/ ([ʃ]) appears only in some Swahili loanwords, such as *ishirini* '20,' but Swahili [ʃ] is usually pronounced as [s] in Kami.

2.3 Tone and syllable structure

Kami is a language with regular penultimate stress and no tonal contrast, either lexical or grammatical. In addition to the canonical syllable shapes (CV, NCV, CGV, NCGV), onsetless syllables (V) are allowed word-initially, word-medially and word-finally.

3 NOUNS AND NOUN PHRASES

Kami is a fairly typical Bantu language, with a total of 16 noun classes. The unstable status of the language is seen in the variation in noun class prefixes and agreement as well as in the singular-plural pairs.

3.1 Noun classes and agreement

The 16 noun classes in Kami include three locative classes. Class 15 contains only the nominal forms of verbs. Most classes have an overt class prefix, as for example in *chi-goda*/*vi-goda* 'chair(s).' Nouns consisting only of a root are found in classes 5, 9 and 10. In class 5, some speakers use the nominal prefix *li-*/*di-* only with monosyllabic

TABLE 19.2 KAMI NOUN CLASS PREFIXES

	NP	PP	EP	SP	OP
1				ni-	ni-
	m-	yu-	yu-	u-/ku-	ku-
				ka-/va-/ya-	m-/mw-
2				ti-	ti-
	wa-	wa-	wa-	m-	wa-
				wa-	wa-
3	m-	u-	u-	u-	u-
4	mi-/i-	i-/mi-	mi-	i-	i-
5	Ø//li-/di-	li-/di-	li-/di-	li-/di-	li-/di-
6	ma-	ya-/ga-	ma-	ya-/ga-	ga-
7	chi-	chi-	chi-	chi-	chi-
8	vi-/pfi-/fi-	vi-/pfi-/fi-	vi-	vi-/pfi-/fi-	vi-/pfi-/fi-
9	N/Ø	i-	i-	i-	i-
10	N/Ø	zi-	N-	zi-	zi-
11	lu-	lu-	lu-	lu-	u-
14	u-	u-	u-	u-	u-
15	ku-				
16	ha-	ha-	ha-	ha-	ha-
17	ku-	ku-		ku-	
18	mu-/m-	mu-/m-		mu-/m-	

noun stems, such as *di-bwa* 'dog' and *di-twi* 'head,' whereas other speakers also use it with longer stems, such as *di-tuka* 'car,' *di-taya* 'jaw,' *di-figa* 'cooking stone,' *di-buyu* 'gourd,' *di-fagilo* 'broom' and with some names of animals, such as *di-nguluwe* 'pig' and *di-simba* 'lion.' In classes 9 and 10, the nasal prefix is realised as *n*- or *m*- (with assimilation to place of articulation) with noun stems that have a voiced plosive or fricative stem-initially. Vowel initial stems have the *ny*- prefix, while stems with initial nasals and voiceless consonants do not carry prefixes.

3.2 Augment

There are few occurrences of the augment in present-day Kami, most of which can probably be ascribed to influence of Luguru G35, where the pre-prefix is frequently used (Petzell & Kühl 2017). What is more, there are barely any records of an augment ever existing in Kami; a handful of tokens in class 10 only are listed in Velten (1900) and there is nothing in Johnston (1919). That said, today the pronominal prefix *zi-/tsi-* in addition to the nasal nominal prefix for class 10 (*tsi-n-guo* 'cloths') can be heard, which is most likely a reinterpretation of the augment. The same is occasionally found with some class 6 nouns that are used with both *ga-* (pronominal prefix) and *ma-* (nominal prefix) prefixes (*ga-ma-finga* 'eggs').

When used, the augment is usually used for definite distant things as in *i-ng'anda i-la* 'that house.' If the augment is used on a common word in a generic sense, such as *mu-lume* 'a husband,' the meaning becomes 'your husband,' *i-mu-lume*. When the augment is in the form of a single vowel, it is always *i-*, regardless of the quality of the vowel in the following noun class prefix:

(15) . . . yu-ja   i-mu-ke   ka-long-a
     PP₁-DEM.DIST   AUG-1-wife   SP₁-talk-FV
     '. . . this wife said.'

3.3 Singular/plural pairings

Singular/plural pairings are canonical, that is, classes 2, 4, 6, 8 and 10 function as plural classes for classes 1, 3, 5, 7 and 9, respectively. Both classes 11 and 14 have plurals in class 10. In addition, classes 4 and 6 function as the plural class for some class 11 and 14 nouns (*lw-ifi*/*mi-lifi* 'door(s),' *u-tamu*/*mi-tamu*/*ma-tamu* 'sickness(es)'). Class 14 and 15 nouns referring to abstract concepts do not have plurals. Classes 11 and 14 are on occasion used interchangeably; for example, 'bead' is either *u-salu* or *lu-salu*. With some words, there is considerable variation in noun class assignment. For example, 'tree' is assigned to class 3 (*m-biki*), 5 (*biki*) or 11 (*lu-biki*) in the singular, but the plural for all of these is in class 4 (*mi-biki* 'trees'). Moreover, some animals in class 5 such as *zoka* 'snake' take their plural in class 4 instead of class 6: *mi-zoka* 'snakes.' The Swahili class and agreement prefix *ki-* in class 7 is often used instead of *chi-*.

TABLE 19.3 KAMI NOUN CLASS PAIRINGS

1/2	mu-ganga – wa-ganga	'doctor – doctors'
3/4	mw-ezi – mi-ezi	'moon – moons'
5/6	finga – ma-finga	'egg – eggs'
7/8	chi-goda – vi-goda	'chair – chairs'
9/10	n-dege – n-dege	'bird – birds'
	mene – mene	'goat – goats'
	ny-oka – ny-oka	'snake – snakes'
11/10	lu-zabi – n-zabi	'string – stings'
14/10	u-salu – salu	'bead – beads'

3.4 Agreement in the noun phrase

Adjectives take prefixes that are identical to noun class prefixes (NP), and numerals take the enumerative prefixes (EP). Subject prefixes (SP) are identical to pronominal prefixes (PP) except in class 1 where a variety of prefixes is found (*ka-*/*va-*/*a-*/*ya-*). Class 14 agreement markers are identical with class 3 agreement markers. Class 11 sometimes uses class 3 agreement markers (*Lw-ifi u-no m-dodo* 'This door is small').

Some of the kinship terms in class 1a take the pronominal prefix of class 5 on possessives in the singular and class 10 on possessives in the plural: *lumbu dy-angu* 'my sibling of opposite sex' and *wa-lamu ts-angu* 'my relatives by marriage.' This is not uncommon among the neighbouring languages (Petzell 2008: 56). In addition to kinship terms, Kami has some animals in class 1a (and their plural in 2a). These nouns do not have noun class prefixes, but take class 1 and 2 concords:

(16) buga   ya-ih-ile   /   buga   wa-ih-ile
     [1a]rabbit SP₁-be_bad-PFV /   [2a]rabbit SP₂-be_bad-PFV
     'bad rabbit /bad rabbits'

(17) buga   wa-no   wo-kimbil-a
     [2a]rabbit PP$_2$-DEM.PROX   SP$_2$.NPST-run-FV
     'These rabbits run.'

The (classes 1 and 2) agreement markers are used to some extent with animate nouns in other classes than 1 and 2. Adjectives that modify animate nouns usually take class 1 and 2 prefixes (*mene m-kulu* 'big goat,' *mene wa-kulu* 'big goats'), but with pronouns there is more variation, i.e., sometimes animate agreement is used (*vi-jana wa-no* 'these young-sters,' *mene yu-no* 'this goat'), whereas at other times inherent class agreement is used (*mene i-no* 'this goat'). Every now and then, non-animate agreement markers in class 9 are used even with class 1 nouns (*m-bwanga i-dya* 'that youngster' instead of *m-bwanga yu-dya*; *mu-nhu i-no* 'this person' instead of *mu-nhu yu-no*) or with animate nouns found in other classes (*ki-jana i-no* 'this youngster'; *kijana* is a loan from Swahili). Agreement marking of animate nouns on the verb usually follows classes 1 and 2 (18), but the ani-mals in class 5 commonly take the inherent class concord (19).

(18) Mene   w-a   m-geni   wa-danganyik-a
     [10]goat PP$_2$-CON 1-visitor SP$_2$-die-FV
     'The goats of the visitor have died.'

(19) di-bwa   di-nog-ile
     5-dog   SP$_5$-be_good-PFV
     'good dog, lit. a dog that has become good'

Class 11 nouns are occasionally used with a plural meaning together with the class 10 pronominal prefix:

(20) lu-zabi   zi-no   lu-tali
     11-string PP$_{10}$-DEM.PROX NP$_{11}$-long
     'these strings are long'

Classes 7 and 8 function as diminutive classes (21–22). Animate concords are not used with diminutives. Class 12 is intermittently used instead of class 7, possibly as a conse-quence of colloquial Swahili influence (23).

(21) Ch-ana ki-no   ki-nog-a
     7-child   PP$_7$-DEM.PROX SP$_7$-be_good-FV
     'This small baby is beautiful/good.'

(22) Vi-mene vi-no   vi-nog-a
     8-goat   PP$_8$-DEM.PROX SP$_8$-be_good-FV
     'These small goats are beautiful/good.'

(23) ka-mene ka-dodo
     12-goat   NP$_{12}$-small
     'small goat'

The augmentative can be formed by substituting the nominal prefix with the class 5 pre-fix for the singular, and classes 4 or 6 for the plural, or adding the prefix in front of the existing noun class prefix (24), but like the diminutive, the forms are not commonly used and had to be elicited.

(24) di-chi-nu   kulu   /   mi-chi-nu   mi-kulu
      5-7-thing   [5]big   /   4-7-thing   NP₄-big
      'big thing/big things'

There is only one locative noun in Kami, *hanu* 'place,' in class 16, although the Swahili *mahali* 'place' can be heard. Classes 17 and 18 contain no nouns at all, but other nouns can take their locative prefixes. In most cases, the three locative prefixes are added in front of the inherent noun class prefixes rather than replacing them, which results in double prefixes (25–27). Additionally, all three locative classes display agreement such as subject and object marking as well as inflection on pronouns. The modifiers of the locativised nouns take agreement in the locative class (28).

(25) vi-nu   v-a   ku-i-kadzi
      8-thing   PP₈-CON   17-9-work
      'things of work'

(26) ko-m-gunda
      17-3-field
      'at the field'

(27) m-li-tuka
      18-5-car
      'in the car'

(28) m-ng'anda   mw-ako
      18-house   PP₁₈-POSS₂SG
      'in your house'

3.5 Word order and nominal modifiers

Nominal modifiers, i.e., adjectives, numerals and pronominal forms (demonstratives, connectives/associatives, independent and possessive pronouns, quantifiers, specifiers and interrogatives) follow the head noun (29). An exception to this is the occasional use of demonstratives preceding the head noun to mark emphasis or definiteness (30). The relative order of the modifiers seems to be rather free.

(29) mw-ana   yu-dya
      1-child   PP₁-DEM.DIST
      'that child'

(30) hi-no   mu-nu
      PP₉-DEM.PROX   1-person
      'this/the person'

TABLE 19.4 KAMI DEMONSTRATIVES

	Near -no	*Far -(d)ya/-ja*	*Referential a-* . . . *-o*
1	(a)yuno/(h)ino	(a)yudya, (a)ija	ayo, iyo
2	wano	wadya/waja	ao
3	uno	udya/uja	ao
4	(h)ino	idya/ija	ayo
5	dino	lidya/dija	ajo
6	gano	yadya/yaja	ago
7	chino/kino	chidya/kija	(h)acho
8	vino/fyino	vidya/vija	afo
9	(h)ino	idya/ija	ayo
10	zino	zidya/zija	a(t)so
11	luno	ludya/luja	alo
14	uno	udya/uja	ao
15			
16	hano	hadya/haja	aho
17		kudya/kuja	ako/kuko
18			

Note: A gap in the table means that the item is absent in our data.

We found only ten inflected adjectives in our data: *-dodo* 'small or young,' *-kulu* 'big,' *-guhi* 'short,' *-tali* 'tall,' *-sisili* 'narrow or thin,' *-bovu* 'rotten,' *-bisi* 'raw, unripe,' *-duhu/ dung'hu* 'red,' *-zelu* 'white,' and *-titu* 'black.' The qualities 'bad' and 'good' are expressed by verbs (*kw-ih-a* 'to be bad' and *ku-nog-a* 'to be good'), which is cross-linguistically unusual – if there is an adjective class in a language, it normally includes these items (Dixon 1982: 56). In addition to inflected adjectives, Kami has adjectives of Arabic origin (through Swahili), which are uninflected, such as *hodari* 'diligent' and *maalum* 'special.'

The numerals *-mwe* '1,' *-bili* '2,' *-datu* '3,' *-ne* '4,' *-tano* '5,' and *-nane* '8' take the EP, while *sita* '6,' *saba* '7,' and *tisa* '9' are invariables and Swahili loans. Swahili *moja* is also used for '1,' and *nane* '8' is sometimes used without an inflectional prefix. Numbers from ten onwards, likewise, are the same as those in Swahili, e.g., *kumi* '10,' *ishirini* '20,' and *themanini* or *semanini* '80.' The question word *-ingahi* 'how many' takes the pronominal prefix:

(31) Ku-na   ng'ombe   wa-ingahi   /   zi-ingahi?
SP_{2SG}-be_with   [9/10]cow PP_2-how_many / PP_{10}-how_many
'How many cows do you have?'

Kami has three sets of demonstratives: the proximal demonstrative PP-*no*, the distal demonstrative PP-*dya*, and the referential demonstrative *a*-PP-*o*, which refers to something that has been previously mentioned or that is of common knowledge. Distal demonstratives can be used to emphasise the distance further by raising the voice to a falsetto towards the end and by prolonging the last syllable (*mu-nu yu-dyaaa* 'that person far away'). The Kami demonstratives vary the most in comparison to the other G-languages of the area (Petzell & Hammarström 2013: 143). This is probably yet another sign of influence from neighbouring languages where the form of the demonstratives may differ.

TABLE 19.5 KAMI INDEPENDENT/POSSESSIVE PRONOUNS

	SG		*PL*	
1	mie, nie, niye	-angu	tie	-etu
2	weye	-ago, -ako	mweye	-enu
3	yeye	-age, -ake	wao	-ao

Connectives/associatives are formed with the canonical -*a* stem and pronominal prefix:

(32) chi-goda   ch-a   m-geni
     7-chair   PP₇-CON   1-visitor
     'chair of visitor'

(33) vi-goda   v-a   m-geni
     8-chair   PP₈-CON   1-visitor
     'chairs of visitor'

(34) ng'anda   y-a   m-geni
     [9/10]house PP₉-CON   1-visitor
     'house of visitor'

(35) ng'anda   z-a   m-geni
     [9/10]house PP₁₀-CON   1-visitor
     'houses of visitor'

The quantifiers -*ose* 'all' and -*o* -*ose* 'any' (36) and the specifiers -*ngi* 'another' take the pronominal prefix (37–38).

(36) m-beyu   y-o   y-ose
     9/10-seed   PP₉-REF   PP₉-all
     'all seed'

(37) mu-nu   yu-ngi
     1-person   PP₁-another
     'another person'

(38) chi-nu   chi-ngi
     7-thing   PP₇-another
     'another thing'

4 VERBAL DERIVATION

There are only three productive verbal extensions in Kami, namely the Applicative, the Causative and the Passive. There are, however, many non-productive or frozen forms. These include the Impositive, the Reversive and the Neuter. The extensions typically commute with other extensions. This can be seen in the verb *ku-lum-a* 'to bite.' If an Applicative extension is added, the meaning changes: *ku-lum-il-a* 'to get hurt.' The

Causative can be added instead, and then the verb becomes transitive: *ku-lum-iz-a* 'to wound.' As seen in these examples, even with productive extensions, the meaning of the derived form is not always fully predictable. Verbal extensions are virtually never combined. The few examples that appear to be a concatenation of extensions are all made up of a frozen extension plus a verb and a productive extension. Example (39) shows this "double extension" – the frozen Impositive formative *-ik-* (not segmented), followed by the Applicative formative *-il-*.

(39) ka-wa-gubik-il-a   wa-ng'onyo   na   nongo
     SP$_1$-OP$_2$-cover-APPL-FV   2-insect   and   [9/10]pot
     'She covered the pot to keep out insects.'

4.1 Applicative

The most frequently occurring derivational extension is the Applicative. It has two allomorphs, *-il-* and *-el-*, which are determined by vowel harmony. A preceding mid vowel in the verb root is followed by *e* in the extension, whereas all other vowels take *i* in the extension. The Applicative extension encodes actions that are directed to, with or for a person or object, for instance by introducing an instrument that does the action, or a beneficiary of the action. The Applicative extension requires either a following complement, or a complement prefix on the verb (or both).

(40) ku-kan-a
     INF-cut-FV
     'to cut'

(41) Ko-kan-il-a   lu-zabi   mw-ele
     SP$_1$.NPST-cut-APPL-FV   11-rope   3-knife
     'He is cutting the rope with a knife.'

(42) ku-tend-a   kasi
     INF-do-FV   [9/10]work
     'to work'

(43) Ko-tu-tend-el-a   kasi.
     SP$_1$.NPST-OP$_{1PL}$-do-APPL-FV   [9/10]work
     'S/he is working for us.'

(44) ku-gw-a
     INF-fall-FV
     'to fall'

(45) Ka-ni-gw-il-a   m-di-twi
     SP$_1$.PST-OP$_{1SG}$-fall-APPL-FV   18-5-head
     'It fell on my head.'

It is not uncommon for the Applicative verb forms to be lexicalised, while the underived counterpart without the Applicative extension does not exist in the language.

These so-called Pseudo-Applicatives are exemplified with *ku-kimbil-a* 'to run away' and *ku-wingil-a* 'to enter.'

4.2 Causative

The Causative extension does not occur frequently in Kami, although it is productive. As the name suggests, this extension adds an argument to the verb which causes an action to happen, as in *ku-mem-ez-a* 'to fill,' from *ku-mem-a* 'to be full' and *ku-lek-es-a* 'to make (somebody) stop,' from *ku-lek-a* 'to cease.' Vowel harmony determines the vowel of the extension, as in the case of the Applicative (46).

(46) Ni-mw-ang'-iz-a   ma-zi   m-bwanga
      SP_{1SG}-OP_1-drink-CAUS-FV   6-water   1-boy
      'I made the boy drink water.'

4.3 Passive

The Passive morpheme in Kami is realised as *-igw-* (47). When the Passive co-occurs with the Imperfective post-root formative *-ag-*, the two are merged into *-agw-* (48). This hypothetical shorter formative for the Passive (*-w-*) is not otherwise attested in Kami.

(47) Mi-savu   i-kan-igw-a   na   m-kalizaji
      4-grass SP_4-cut-PASS-FV   and   1-guard
      'The grass was cut by the guard.'

(48) Chupa   i-sol-ag-w-a   (na   Babu)
      [9/10]bottle SP_9-take-HAB-PASS-FV   and   Babu
      'The bottle was (usually) taken (by Babu).'

4.4 Non-productive/frozen extensions

The Impositive (*ku-gub-ik-a* 'to cover') and the Reversive (*ku-gub-ul-a* 'to uncover') extensions are not productive in Kami. The reconstructed extensions cited in our text follow Meeussen (1967: 92). An unproductive formative can give rise to frozen forms, i.e., forms that do not have a basic form and that may have gone through a semantic shift and become lexicalised. These frozen forms can in turn take productive extensions such as the Applicative (see example 38 above).

An extension that appears to be productive in neighbouring languages but not in Kami is the Neuter (*ku-dah-ik-a* 'to be possible,' *ku-ben-ek-a* 'to brake'). As opposed to the Passive, the Neuter extension removes the actor from the clause. As seen in (49), the Passive alters only the thematic roles while the verb's argument structure remains unaffected – the oblique agent is still present implicitly (even though the agent phrase is not obligatory). In contrast, no agent or 'by' phrase can be added to the Neuter in example (50).

(49) Lw-ifi   lu-fugul-igw-a   na   Said
      11-door SP_{11}-open-PASS-FV   and   Said
      'The door was opened by Said.'

(50) Lw-ifi   lu-fug-uk-a   (*na Said)
       11-door   SP₁₁-open-NEUT-FV
       'The door opened.'

5 VERBAL INFLECTION

The following sections will examine the inflectional morphology of the verb phrase, including TAM, morphological negation, reduplication and periphrastic verbs. To facilitate an overview of all slots in the verbal complex and what morphemes they can take, a slot matrix is presented below.

TABLE 19.6 THE VERBAL SLOTS

1	2	3	4	5	6	7	8	9
NEG	SP	TAM	OP	ROOT	EXT	TAM	PASS	FV
NEG	SP+TAM		OP	ROOT	EXT	TAM+PASS		FV

The first slot can hold an adverbial marker or negators. Slots 2 (SP) and 3 (TAM) may be merged if there is a portmanteau morpheme. Slot 4 can hold the object prefix or Reciprocal/Reflexive marker. The only compulsory slots to fill are slots 5 (the root) and 9 (the FV). However, slot 2 (the SP) is usually filled, but the subject prefix often merges either with the negator or with the TAM markers, which makes it difficult to tell what slot the resulting morpheme actually occupies. The same goes for the post-root TAM marker -ag- and the Passive -igw-. The formatives have merged to -agw-, so we can only speculate what the original order was. The majority of the slots can be filled concurrently; the restriction lies in which morphemes may be combined with which. Table 19.7 shows the slot matrix including example clauses.

TABLE 19.7 THE VERBAL SLOTS EXEMPLIFIED

NEG	SP	TAM	OP	ROOT	EXT	TAM	PASS	FV	Gloss
				lim				a	Cultivate (sg)!
	ti			lim				e	Let us cultivate!
		ni		long	il		igw	a	I was (already) told.
	i			sol		agw		a	It (cl.9) was taken (habitually).
		two		himb		ag		a	We are usually digging (all day).
		ko	i	lol				a	S/he looks at herself.
	ka		m	long	el			a	S/he said to her/him.

5.1 Infinitive

The infinitive is created by adding the nominal prefix of class 15 to the verb stem, as in ku-lim-a 'to cultivate.' The ku- may also be added to the macrostem, i.e., the verb stem including an optional object prefix and an optional verbal extension (51). There is no negative infinitive in Kami.

(51) ku-m-lams-a   mu-nu
      INF-OP₁-wake_up-FV   1-person
      'to wake up a person'

5.2 Imperative

The imperative consists of a verbal stem which includes the FV -a (52). The post-final plural marker -ni is used for plural addressees (53). To give a polite command, the FV -e is used. The imperative can be formed with an object prefix (54).

(52) Di(y)-a!
      eat-FV
      'Eat!'

(53) Himb-e-ni   vi-bogwa!
      dig-SBJV-PL   8-potato
      'Dig up (PL) the potatoes!'

(54) M-kem-e!
      OP₁-call-SBJV
      'Please call (him/her)!'

A negative command is created by using a form of the verb -lek-a 'cease', a reflex of the verb reconstructed as *-dèk- 'let go, cease, allow' (Bastin et al. 2002), followed by an infinitive complement (55–56) or a verb in the subjunctive (57). The verb form seke (57) has gone through phonological changes that merit further investigation. Addressing plural addressees can take many different forms (58).

(55) Lek-a   ku-lil-a!
      cease-FV   INF-cry-FV
      'Do not cry!'

(56) U-lek-e   ku-ng'w-a!
      SP₂ₛ₉-cease-SBJV   INF-drink-FV
      'Don't drink!'

(57) Sek-e   u-tend-e   fi-no!
      cease-SBJV   SP₂ₛ₉-do-SBJV   PP₈-DEM.PROX
      'Do not do that!'

(58) Lek-e-ni   /   m-lek-e   /   m-lek-e-ni   ku-ng'w-a!
      CEASE-SBJV-PL   /   SP₂ₚ-cease-SBJV   /   SP₂ₚ-cease-SBJV-PL   INF-drink-FV
      'Do not drink (pl.)!'

The subjunctive is also used to express exhortatives, both in affirmative and negative forms:

(59) Tu-was-e!
      SP₁ₚ-sleep-SBJV
      'Let us sleep!'

(60) Ti-lek-e   k-uk-a!
      SP_{1PL}-cease-SBJV   INF-go-FV
      'Let us not go!'

5.3 Subject and object prefixes

The person or class of a noun is encoded on the verb by subject prefixes (SP) and/or object prefixes (OP). In the indicative, the subject prefix is obligatory for all finite verb forms. The markers are identical with the pronominal prefix of the nouns in all classes but classes 1 and 2 (see Table 19.2).

The 1PL marker *tu-* occurs as *ti-* on the second verb in periphrastic verb forms (60). The same variation goes for class 1, which makes use of *ka-* (or *ko-* in the Non-Past) for simple verbs, and *ya-* on the second verb in periphrastic verbs (and intermittently elsewhere too). Apart from the variation mentioned above, there are two different prefixes for the first person plural, regardless of the verb form. This variation appears to be speaker-related. Two speakers consistently use *chi-* (*cho-* for the Non-Past) and three use *tu-* (with *ti-* usually in periphrastic forms and *to-* for Non-Past). The usage of *ti-* in non-periphrastic constructions is reported for the mid-1970s as well (Gérard Philippson pers. comm.). The variation could partly be due to the neighbouring languages, since the speakers who chose *chi-* lived closer to a Kwere-speaking area and the others lived closer to a Luguru-speaking area, where the prefix is *tu-* or *ti-*. These subject prefixes are not attested in the Proto-Bantu reconstructions, neither for Kami nor for the neighbouring languages, but Guthrie (1970: 240) reports a "mixed situation."

(61) To-lond-a   ti-lim-e   m-gunda   w-angu
      SP_{1PL}.NPST-want-FV   SP_{1PL}-cultivate-SBJV   3-farm   PP₃-POSS_{1SG}
      'We want to cultivate my farm.'

5.4 Reflexive and reciprocal

Both the Reflexive (62) and the Reciprocal (63) are in the form of a single vowel *-i-* that sits in the object prefix slot.

(62) Wa-nu   o-i-pend-a
      2-person   SP_{2PL}.NPST-REFL/RECP-like-FV
      'People like each other.'

(63) M-dele   ko-i-bak-a   ma-futa
      1-girl   SP₁.NPST-REFL/RECP-smear-FV   6-oil
      'The girl is smearing herself with oil.'

5.5 TAM(N)

There are surprisingly few tense, aspect and mood (TAM) markers in Kami, considering how multifaceted verbal morphology can be in other Bantu languages. Morphologically, there is basically a Past and a Non-Past. Additionally, there is a conditional marker *-ng'(h)a-* (64).

(64) U-ng'a-iz-a   ko-ni-fik-a
      SP$_{2SG}$-COND-come-FV   SP$_{2SG}$.NPST-OP$_{1SG}$-arrive-FV
      'If you come you will find me.'

5.5.1 Non-past

The morpheme -o- replaces the vowel of the SP in both the Present and the Future in the affirmative, i.e., it is a Non-Past (65–67).

(65) Mweye   mo-wa-pend-a   i-wa-an-enu
      PRO$_{2PL}$ SP$_{2PL}$.NPST-OP$_2$-like-FV   AUG-2-child-POSS$_{2PL}$
      '[As for] you (pl), you love your children.'

(66) Lelo   to-gend-a   Dar es Salaam.
      today   SP$_{1PL}$.NPST-go-FV   Dar es Salaam
      'Today we are going to Dar es Salaam.'

(67) Mwakani   no-bwel-a
      next_year   SP$_{1SG}$.NPST-return-FV
      'I will come back next year.'

5.5.2 Past

All past forms in both the Perfective and the Imperfective aspect, as well as the Anterior, carry zero marking. There is only a subject prefix preceding the root (and an optional object prefix) (68–70). Note that even for the Anterior 'have already' there is zero marking and no additional adverbials can occur (71).

(68) Tu-himb-a   simo
      SP$_{1PL}$-dig-FV   [9/10]hole
      'We (have) dug a hole.'

(69) Tu-dy-a   lelo
      SP$_{1PL}$-eat-FV   today
      'We ate/have eaten today.'

(70) Ni-long-a   na-yo   mi-tondo   i-no.
      SP$_{1SG}$-discuss-FV   with-REF$_1$ 4-morning   PP$_4$-DEM.PROX
      'I spoke with her/him this morning.'

(71) Tu-long-a   na-yo
      SP$_{1PL}$-discuss-FV   with-REF$_1$
      'We have (already) spoken with her/him.'

For stative verbs, the Anterior can be used even when referring to the present. These so-called Inchoative verbs express a change in the condition or state of the subject (72–73).

(72) Ka-neneh-a
     SP$_1$-become_fat-FV
     'S/he is healthy (lit. s/he has become fat)'

(73) Niye   chi-kami ni-many-a   ch-a   ku-lomb-el-a   mu-nyu
     PRO$_{1SG}$ 7-Kami   SP$_{1SG}$-know-FV PP$_7$-CON INF-ask-APPL-FV 3-salt
     '(As for) me, Kami I (have come to) know enough of it to ask for salt i.e., I manage
     to make myself understood.'

5.5.3 Habitual and progressive

The Habitual (74) and the Progressive (75) are represented by -ag- in the post-root TAM
slot. It occurs in complementary distribution with -ile. The morpheme -ag- can be combined
with the past as well as the non-past subject prefixes, depending on the context. However,
there are semantic restrictions. For instance, an anterior reading of a verb cannot take -ag-.

(74) Chi-gend-ag-a   chila   mara Dar es Salaam
     SP$_{1PL}$-go-IPFV-FV every   time   Dar es Salaam
     'We go to Dar es Salaam frequently.'

(75) Ukaye   ko-som-ag-a   mw-anafunzi
     in_the_house SP$_1$.NPST-read-IPFV-FV 1-student
     'In the house a student is (in the process of) reading.'

When -ag- co-occurs with the Passive, the two morphemes are amalgamated to -agw-.
The realisation of this merged formative could be the result of a bimorphemic quality of
-ag- (Nurse 2008: 37), but it still does not explain the shortened passive formative -w-
(see Section 4.3).

5.5.4 Conditional

The Conditional is made up of the marker -ng'(h)a- (see 64 above). In the past, it takes
the post-root morpheme -ile:

(76) Kama   u-ng'a-iz-ile. . .
     if   SP$_{2SG}$-COND-come-PFV
     'If you came. . .'

In our data, there is one occurrence of the Conditional marker preceded by the formative
na-: Tu-na-ng'-end-a samba. . . . 'If we depart now. . .' Philippson reports a morpheme
na-, but used for the Persistive (Gérard Philippson pers. comm.).

5.5.5 Negation

The negative morpheme in Kami is ha- (77), apart from in 1SG and 2SG where it is merged
with the subject prefix to become si- (78) and hu- (79) respectively. In the Past Negative,
-ile is used (79).

(77) Ha-m-lim-a   m-gunda   w-enu   wiki   i-kw-iz-a
      NEG-SP$_{2PL}$-cultivate-FV   3-farm   PP$_3$-POSS$_{2PL}$ [9/10]week   SP$_9$-15-come-FV
      'You (pl) will not cultivate your farm next week.'

(78) Si-hand-a   m-beyu   y-a   aina   y-o   y-ose
      SP$_{1SG}$.NEG-plant-FV   9/10-seed   PP$_9$-CON   [9/10]type   PP$_9$-REF   PP$_9$-all
      'I will not plant any seeds.'

(79) Hu-lim-ile   m-gunda   w-ako   jana
      SP$_{2SG}$.NEG-cultivate-PFV   3-farm   PP$_3$-POSS$_{2SG}$ yesterday
      'You did not cultivate your farm yesterday.'

In addition to morphological negation, there is a negative word, *bule*, meaning 'nothing at all' or 'not yet' (80–81). Another example can be seen in a commonly occurring greeting (82).

(80) Si-lond-a   bule
      SP$_{1SG}$.NEG-speak-FV   NEG
      'I do not speak at all.'

(81) I-ja   ha-som-ile   bule
      PP$_9$-DEM.DIST   SP$_1$.NEG-read-PFV   NEG
      'S/he has not studied yet.'

(82) Wa-ana   ha-wa-lumw-e   bule
      2-child   NEG-SP$_2$-be_bitten-PFV   NEG
      'The children are not ill/hurting/bitten at all (i.e., they are well).'

Bule stands alone in these examples but it can also take subject prefixes (see Section 6.6). The word *bule* is found in the neighbouring languages Kutu, Kwere and Zalamo, both used as an independent word and taking inflection.

5.5.6 The function of reduplication

In Kami, reduplication carries a diversity of meanings, such as aimlessly doing something (83), repeating the action (84) or intensifying the action (85). It is usually only the stem (i.e., the verb root plus the FV) that is reduplicated, but the entire verb may occasionally be reduplicated.

(83) Ko-lim-a-lim-a
      SP$_1$.NPST-cultivate-FV-cultivate-FV
      'S/he cultivates a little here and there'

(84) Two-himb-a   two-himb-a
      SP$_{1PL}$.NPST-dig-FV   SP$_{1PL}$.NPST-dig-FV
      'We are digging over and over.'

(85) I-mw-ana   ka-ben-a-ben-a   lu-balati
AUG-1-child   SP₁-break-FV-break-FV   11-stick
'The child broke up the stick completely.'

5.5.7 Persistive

The Persistive consists of an auxiliary and lexical verb. The lexical verb may carry inflection as well as the auxiliary, although the lexical verb is usually in the infinitive. The defective verb *ng'ali* functions as an auxiliary. It conveys both the notion of 'still' (86) and 'not yet' (87). It takes only a subject prefix and no TAM markers, while the following lexical verb may be in the infinitive or inflected with subject prefixes (including SP + TAM) (88).

(86) Di-tunda   di-ng'ali   dyo-di-igw-a
5-fruit   SP₅-be_still   SP₅.NPST-eat-PASS-FV
'The fruit is still eatable.'

(87) Ni-ng'ali   ku-maliz-a
SP₁ₛ₉-be_still INF-finish-FV
'I'm not yet finished.'

(88) Yeye   ya-ng'ali   yo-hand-a   m-beyu   goya
PRO₁   SP₁-be_still   SP₁.NPST-plant-FV   9/10-seed   carefully
'S/he was planting every seed carefully.'

5.5.8 The past in periphrastic constructions

The verb *ku-kal-a* 'to be, remain, sit,' reconstructed as *-jikad- 'dwell; be; sit; stay' (Bastin *et al*. 2002; Nurse & Philippson 2006: 166), is used in periphrastic constructions for the Past, such as Imperfect (89) and Pluperfect (90). *Ku-kal-a* can be followed by *na*, thus meaning 'to have' (91).

(89) Mw-ana   ka-kal-a   yo-lumw-a   m-tamu
1-child   SP₁-be-FV   SP₁.NPST-bitten-FV   NP₁-sick
'The child was ill/hurt/bitten.'

(90) To-kal-a   tu-lim-a
SP₁ₚₗ.NPST-be-FV   SP₁ₚₗ-cultivate-FV
'We (had) cultivated.'

(91) Mu-nu   yu-no   ka-kal-a   na   mw-ehe
1-person   PP₁-DEM.PROX   SP₁-be-FV   and   1-wife
'This person had a wife.'

5.5.9 Modality in periphrastic constructions

Modality is usually expressed periphrastically. Below are examples of permission (92) and possibility (93).

(92) No-wez-a   ku-fik-a?
      SP$_{1SG}$.NPST-can-FV   INF-arrive-FV
      'Can/may I get (there)?'

(93) Cho/two-dah-a   ku-himb-a simo
      SP$_{1PL}$.NPST-can-FV   INF-dig-FV [9/10]hole
      'We could/might dig a hole.'

6 SYNTAX

The following sections on syntax will illustrate and discuss basic clause structures, agreement including object agreement and object word order, copulas and predication, and finally different clause types.

6.1 Basic clause structure

Like the majority of the Bantu languages, the canonical word order in Kami is SVO. There is obligatory subject-verb agreement in Kami, even when a subject noun phrase is present:

(94) Mu-nu   i-no   ka-long-a.
      1-person   PP$_9$-DEM.PROX   SP$_1$-discuss-FV
      'This person (s/he) spoke.'

6.2 Noun class resolutions

There are three possible resolutions when a coordinated noun phrase consists of nouns belonging to various classes. If an animate noun is involved, the subject prefix is always in class 2. Otherwise, class 8 may be used as a default subject prefix, since the noun for general 'things' is in this class:

(95) Di-goda   na   peni   fi-gul-igw-a
      5-chair   and   [9/10]pen   SP$_8$-buy-PASS-FV
      'The chair and the pen were bought.'

A third way to resolve conflicting noun classes is to use the subject prefix of the noun closest to the verb:

(96) Mi-biki   na   ng'anda   i-gw-a
      4-tree   and   [9/10]house   SP$_9$-fall-FV
      'The trees and the house fell.'

6.3 Impersonal constructions

Class 9 is used as a dummy subject prefix in impersonal constructions (97). Class 9 agreement is also often used in constructions regarding the weather (98).

(97) I-nog-a
      SP_9-be_good-FV
      'It is good.'

(98) Yo-tony-a   m-vula
      SP_9.NPST-rain-FV   9/10-rain
      'It is raining.'

6.4 Inversion

In locative inversion constructions, the locative is the grammatical subject, which is in the subject position, while the inverted, or logical, subject is in the object position after the verb. Any of the locative subject markers (from classes 16, 17 or 18) can be used (99). The inherent agreement marker may also be used in a similar construction (100), but then it is "agreeing inversion" (cf. Marten & van der Wal 2014), plus, in this case, fronting of the locative (100).

(99) Mw-i-biki   mu-kal-a   ga-ma-nyani
      18-4-tree SP_{18}-sit-FV   AUG-6-baboon
      'In the tree sit baboons.'

(100) Mw-i-biki   ga-kal-a   ga-ma-nyani
      18-4-tree SP_6-sit-FV   AUG-6-baboon
      'In the tree sit baboons.'

6.5 Object agreement and object word order

Although there may be several objects in the phrase, Kami can never take multiple object prefixes (101). If there is more than one object, a separate noun phrase is required.

(101) M-lume   ka-iz-a   ukaye   ka-ti-ing'-a   barua
      1-man SP_1-come-FV   home SP_1-OP_{1PL}-give-FV   [9/10]letter
      'The man came to our house and gave us a letter.'

If both objects are animate, it is always the beneficiary, in this case the receiver, which is encoded on the verb:

(102) No-mw-ing'-a   wa-ana   m-hehe
      SP_{1SG}.NPST-OP_1-give-FV   2-child   1-woman
      'I am giving the woman children.'

The nominal object that agrees with the verb (the beneficiary) has to follow any other objects (103–104) (although it precedes temporal adjuncts). This is the opposite of some other Eastern Bantu languages where the indirect object immediately follows the verb, such as Shambaa G23 (Riedel 2009: 79). Even though this position is usually used for given information (topic), we cannot assume this based on the Kami data available, since there are no counterexamples (i.e., no instances of the beneficiary preceding the direct object).

(103) I-mu-lume   ka-mw-ing'-a   chi-tabu   i-m-ke   (jana)
       AUG-1-man   SP₁-OP₁-give-FV   7-book   AUG-1-woman   yesterday
       'The man gave the woman a book (yesterday).'

(104) Ali   ko-mw-ing'-a   sendi   John
       Ali   SP₁-OP₁-give-FV money   John
       'Ali gives John money.'

6.6 Copulas and predication

Non-verbal predicates are juxtaposed to their subject in non-past affirmative utterances
(105). In the past and when negated, inflected forms of the verb *ku-w-a* 'to be' or the
verb *ku-kal-a* 'to be, remain, sit' must be used (106). There is no uninflected copula in
the present tense.

(105) Siso   di-no   kulu
       [5]eye   PP₅-DEM.PROX   [5]big
       'This eye is big.'

(106) Si-w-ile   mw-alimu
       SP₁ₛɢ.NEG-be-PFV   1-teacher
       'I was not a teacher.'

There are some locational constructions consisting of a demonstrative that denotes location, and the subject prefix:

(107) Ali   ka-ha-no
       Ali   SP₁-PP₁₆-DEM.PROX
       'Ali is here.'

(108) Cha-ha-ja
       SP₇-PP₁₆-DEM.DIST
       'It is there.'

The negative copula *bule* can take subject prefixes and thus form a complete phrase
(109). It cannot take TAM markers nor any verbal extensions. However, it appears that it
can take an object prefix adjacent to the subject prefix (110). It may also be inflected with
a locative subject prefix, thus creating a negative existential (111).

(109) Wa-bule   heshima
       SP₂-NEG.COP   [9/10]respect
       'They have no respect/They are without respect.'

(110) Ya-ha-bule!
       SP₁-OP₁₆-NEG.COP
       'Go away (be not here)!'

(111) Sweden   ha-bule   tangawizi
       Sweden   SP$_{16}$-NEG.COP   [9/10]ginger
       'There is no ginger in Sweden.'

6.7 Clause types

This section discusses the declarative, interrogative, imperative (see Section 5.2 above) and exclamatory clause types.

6.7.1 Interrogative

Yes/No-questions are differentiated from their corresponding statements by intonation only (pitch rises towards the end of the last word and drops on the last syllable, which often has a lengthened vowel), while content questions are formed with the help of invariable interrogative pronouns or with inflected interrogative pronouns. Question words include *kwani* 'where,' *choni/yachi* 'what/which,' *nani* 'who,' and *lini* 'when.' The question word *habali* 'why' is a loanword from Swahili *habari* 'news' that has gone through a semantic shift (112).

(112) Habali   ko-ni-to-a?
       why   SP$_1$.NPST-OP$_{1sg}$-hit-FV
       'Why is s/he hitting me?'

6.7.2 Exclamatory clauses and greetings

Exclamatory clauses are most often used to express surprise (113–114). As in many places in Tanzania, kinship terms are used in greetings even though they may be directed at strangers. The term 'sister' was used between two women of the same age who did not know each other (115).

(113) Anhaaa!
       'Aha/I see!'

(114) Ka-ha-bule,   ko-ka-ko!
       SP$_1$-OP$_{16}$-NEG.COP   SP$_1$.NPST-PP$_{17}$-DEM.REF
       'S/he is not here, s/he is there!'

(115) Dada!
       'Sister!'

6.7.3 Complement clauses

Clausal complements (arguments) may be in the indicative (116), the subjunctive (117) or the infinitive (118).

(116)   Wa-zungu   wo-long-a   mwo-hanik-a   mi-gunda
        2-European   SP₂.NPST-say-FV   SP₂ₚₗ.NPST-destroy-FV   4-farm
        'The Europeans said you are destroying the farms.'

(117)   No-lond-a   ni-lim-e   m-gunda   w-angu.
        SP₁ₛ₉.NPST-want-FV   SP₁ₛ₉-cultivate-SBJV   3-farm   PP₃-POSS₁ₛ₉
        'I want to cultivate my farm.'

(118)   I-mu-ke   yu-no   ya-pend-ile   ku-hand-a   m-beyu
        AUG-1-wife   PP₁-DEM.PROX   SP₁-like-PFV   INF-plant-FV   9/10-seed
        'This (particular) wife liked to sow seeds.'

6.7.4 Relative clauses

There are two types of relative clauses, depending on whether it is the subject or the object that is relativised. When the subject is relativised, the relative verb is usually marked by the formative -ile in the past (119), but in the non-past it is often not marked. There is no difference between the clauses 'the book which is readable' and 'the book is readable'; both are *chi-tabu cho-som-ek-a*. For class 1, the subject prefix is *ya-* instead of *ka-* in the relative, where *ya-ib-ile* 's/he who stole' can be contrasted with the non-relative *ka-ib-a* 's/he stole.' The object relative has the verb in the indicative and an object prefix. The object prefix is placed after the subject prefix, i.e., in the regular object prefix slot (120).

(119)   Mu-geni   ya-ib-ile   ng'ombe   ka-kimbil-a
        1-stranger   SP₁-steal-PFV   [9/10]cow   SP₁-run-FV
        'The stranger who stole the cows ran away.'

(120)   chi-nu   chi-no   wa-chi-sol-a   wa-uz-a
        7-thing   PP₇-DEM.PROX   SP₂-OP₇-take-FV   SP₂-sell-FV
        'This thing which they took, they sold.'

Another way of marking a relative clause is to use the verb *ku-kal-a* 'to be, remain, sit' (121–122). The corresponding non-relative to example 121 is *chi-nu ha-chi-no* 'the thing is here.'

(121)   chi-nu   chi-kal-ile   ha-no
        7-thing   SP₇-be-PFV   PP₁₆-DEM.PROX
        'the thing which is/sits here'

(122)   Chi-tabu   chi-kal-a   ki-som-igw-a . . .
        7-book   SP₇-be-FV   SP₇-read-PASS-FV
        'The book, which is readable. . .'

6.7.5 Adverbial clauses of time

There are two ways of forming dependent clauses of time with the sense of 'when' or 'as': the most common is using *(h)a-* (class 16) in slot 3 (123), or the formative *vi-* (class 8) in slot 1 usually together with *-ile* (124). These forms cannot be used with Non-Past formatives containing *-o-*:

(123) Wa-ha-to-a   ngoma. . .
       SP$_2$-16-play-FV   [9/10]drum
       'When they play the drums. . .'

(124) Vi-ni-gend-ile   u-tali   m-kulu. . .
       8-SP$_{1sg}$-go-PFV 14-distance   NP$_{14}$-big
       'When I had walked a long distance. . .'

There is another formative relating to time (*za* or *tsa*) that can only appear independently and immediately before the verb (125). The formative seems to bear a referential meaning, similar to 'at a specific time.' Similarly, the influence of *tsa* can be illustrated with the sentences *nikala* 'I sat' and *tsanikala* 'I sat at that point in time' (which also means I am not sitting anymore). There is a similar phenomenon (*zaa*) in neighbouring Luguru G35 which is used when restricting the event to one determinate moment in the past (Mkude 1974: 95). Since it only occurs a few times in our data, it is difficult to say if it really is a Kami morpheme or simply used due to Luguru influence.

(125) Za   ni-gend-a   nzila
       PART   SP$_{1sg}$-go-FV [9/10]path
       '(At that time/because) I walked on the path.'

7 CONCLUSION

Kami is a fairly typical Bantu language, although there are some areas that stand out. For instance, the verbal morphology is heavily reduced, with only two tense markers (Past and Non-Past), one modal and few aspectual markers. The lack of tones is unusual for Bantu, although not particularly so in this area, as there is also no tone in Luguru (Mkude 1974) nor in Kagulu (Petzell 2008). Another interesting feature is the striking variety of allophones representing the same phoneme. Some of these sounds can be traced to Swahili influence and some to local languages such as Luguru. There is variation between speakers, but also within the idiolect of individual speakers. Finally, adjectives are few in number and other parts of speech such as nouns or verbs are used in adjectival constructions. This in itself is not unusual. What is atypical, however, is the lack of the widespread adjectives 'good' and 'bad.' Their function is instead filled by ordinary verbs.

Areas that merit further research include allophonic variation, a deeper semantic analysis of the TAM formatives and inversion structures such as locative inversion. Finally, on another level, semantics and pragmatics including information structure and discourse analysis are major topics that remain unexplored.

NOTE

1 We are very grateful to two anonymous reviewers for valuable comments and to Riksbankens Jubileumsfond for providing the funding.

REFERENCES

Aikhenvald, A. (2012) Language Contact in Language Obsolescence. In Chamoreau, C. & I. Léglise (eds.), *Dynamics of Contact-Induced Language Change. Language Contact and Bilingualism*, 77–109. Berlin: De Gruyter Mouton.

Bastin, Y., A. Coupez, E. Mumba & T. C. Schadeberg (eds.) (2002) *Bantu Lexical Reconstructions 3*. Tervuren: Royal Museum for Central Africa. (Available online at linguistics.africamuseum.be/BLR3.html, Accessed November 28, 2017).

Dixon, R. M. W. (1982) *Where Have All the Adjectives Gone? And Other Essays in Semantics and Syntax*. Berlin: De Gruyter Mouton.

Guthrie, M. (1948) *The Classification of the Bantu Languages*. London; New York: Oxford University Press for the International African Institute.

Guthrie, M. (1970) *Comparative Bantu: An Introduction to the Comparative Linguistics and Prehistory of the Bantu languages. Volume 4: A Catalogue of Common Bantu with Commentary*. London: Gregg International.

Guthrie, M. (1971) *Comparative Bantu: An Introduction to the Comparative Linguistics and Prehistory of the Bantu languages. Volume 2: Bantu Prehistory, Inventory and Indexes*. London: Gregg International.

Johnston, H. H. (1919) *A Comparative Study of the Bantu and Semi-Bantu Languages*. Oxford: Clarendon Press.

Marten, L. & J. van der Wal (2014) A Typology of Bantu Subject Inversion. *Linguistic Variation* 14(2): 318–368.

Meeussen, A. E. (1967) Bantu Grammatical Reconstructions. *Africana Linguistica* 3: 79–121.

Mkude, D. J. (1974) *A Study of Kiluguru Syntax with Special Reference to the Transformational History of Sentences with Permuted Subject and Object*. London: University of London, PhD dissertation.

Mradi wa Lugha za Tanzania (2009) *Atlasi ya Lugha za Tanzania*. Dar es Salaam: Chuo Kikuu cha Dar es Salaam.

Nurse, D. (2008) *Tense and Aspect in Bantu*. Oxford: Oxford University Press.

Nurse, D. & G. Philippson (1975) The North-Eastern Bantu Languages of Tanzania and Kenya: A Tentative Classification. *Kiswahili* 45(2): 1–28.

Nurse, D. & G. Philippson (1980) The Bantu Languages of East Africa: A Lexicostatistical Survey. In Polomé, E. C. & C. P. Hill (eds.), *Language in Tanzania*, 26–67. London: Oxford University Press for the International African Institute.

Nurse, D. & G. Philippson (2006) Common Tense-Aspect Markers in Bantu. *Journal of African Languages and Linguistics* 27(2): 155–196.

Petzell, M. (2008) *The Kagulu Language of Tanzania: Grammar, Texts and Vocabulary*. Cologne: Rüdiger Köppe.

Petzell, M. (2012a) The Linguistic Situation in Tanzania. *Moderna Språk* 106: 136–144.

Petzell, M. (2012b) The Under-Described Languages of Morogoro: A Sociolinguistic Survey. *South African Journal of African Languages* 32(1): 17–26.

Petzell, M. & H. Hammarström (2013) Grammatical and Lexical Comparison of the Greater Ruvu Bantu Languages. *Nordic Journal of African Studies* 22(3): 129–157.

Petzell, M. & K. Kühl (2017) The Influence of Non-linguistic Factors on the Usage of the Pre-prefix in Luguru. *Linguistic Discovery* 15(1): 35–48.

Riedel, K. (2009) *The Syntax of Object Marking in Sambaa. A Comparative Bantu Perspective*. Utrecht: LOT.

Velten, C. (1900) Kikami, die Sprache der Wakami in Deutsch-Ostafrika. *Mitteilungen des Seminars für Orientalische Sprachen* 3: 1–56.

CHAPTER 20

NGAZIDJA G44a

Cédric Patin, Kassim Mohamed-Soyir and Charles Kisseberth

1 INTRODUCTION

Ngazidja is spoken in Ngazidja (or Grande Comore), the largest island of Comoros. It is one of the four Comorian languages, along with Ndzuani G44b (spoken in Ndzuwani – or Anjouan), Mwali G44C (spoken in Mwali – or Mohéli) and Maore G44D (spoken in Maore – or Mayotte). Full (2006) adds Shikombani, a language without Guthrie code also spoken in Mayotte, to this group. Some scholars, such as Ahmed-Chamanga (2010, 2011), and the state of Comoros consider the Comorian languages to be dialects of a single language called Komori. According to Rombi and Alexandre (1982), however, the Ngazidja-Mwali subgroup should be distinguished from the Ndzuani-Maore subgroup. Comorian languages/dialects were formerly listed as dialects of Kiswahili G42 (e.g., Bryan 1959, Rombi & Alexandre 1982: 17, for details), but recent studies consider them to be independent (Rombi & Alexandre 1982, Nurse 1989, Ahmed-Chamanga 2010).

Ngazidja is spoken by about 500,000 speakers, of which 100,000 live in France (Ahmed-Chamanga 2010, Mohamed-Soyir 2014: 22). The language is in constant use by speakers of all generations, and is acquired as a first language. While Ngazidja has no official status, Komori is one of the three official languages of Comoros, along with French and Arabic, and is being gradually integrated into the educational system (Ahmed-Chamanga 2010: 8). Nevertheless, French remains the language in use for administrative and official purposes and in higher education, and all young people are bilingual.

Although some aspects of Ngazidja have been discussed in studies dedicated to Comorian (e.g., Nurse 1989, Picabia 1994, Full 2001), few authors have explicitly focused on Ngazidja to this date, apart from Michel Lafon (1982, 1984, 1985, 1987, 1990, 1997), Léila Picabia (1996, 1998), Mohamed Ahmed-Chamanga (2002, 2003) and, more recently, Kassim Mohamed-Soyir (2007, 2014). Only its prosodic system has been described in great detail (Tucker & Bryan 1970, Philippson 1988, Cassimjee & Kisseberth 1989, Jouannet 1989, Rey 1990, Cassimjee & Kisseberth 1992, 1998, Philippson 2005, Patin 2007b, a, 2008, 2010, O'Connor & Patin 2015, Patin 2016). No complete description of Ngazidja is available, but a grammar written for teachers of the language in Ngazidja was published by Ahmed-Chamanga (2010).

Data for this chapter were collected in various places (USA, France, Ngazidja) and at various times (from the end of the 1980s to the present day), from speakers of different dialects, mostly through elicitation. Except when mentioned otherwise, all the examples in this paper are representative of the dialect spoken in Moroni.

2 PHONOLOGY

2.1 Vowels

All authors agree that Ngazidja has a classic system of five vowel phonemes (Tucker & Bryan 1970: 351, Rombi & Alexandre 1982: 22, Full 2006: 117, Patin 2007a: 5, Ahmed-Chamanga 2010: 17, Patin 2013: 174, Mohamed-Soyir 2014: 24) (see Table 20.1).

TABLE 20.1 VOWELS

	Front	Central	Back
Close	i		u
Mid	e		o
Open		a	

The mid vowels are generally realised as mid-open, but they tend to emerge as mid-close in word-final position: /mu-éndje/ > [mwɛ́njɛ] 'light (cl.3)'; /wa-ʈoʈó/ > [waʈɔʈɔ̣́] 'children (cl.2).'

There is no phonemic length in Ngazidja, but high-toned vowels are lengthened in the penult of utterances, especially in the North (e.g., in Mbeni). In utterance-final position, vowels are devoiced, and even deleted after a fricative (/púzi/ [púẓ] 'feather (cl.5)'), unless they are high-toned (/mleví/ [m̩leví]/*[mley] 'drunkard (cl.1)') (cf. Patin 2009: 141–142).

Nasal vowels appear in loanwords from Arabic, e.g., áda 'custom (cl.9)' < Ar. ʔadah (cf. Ahmed-Chamanga 2010: 19, Patin 2013: 174) and French, e.g., daʒé 'danger' (cl.9) (Mohamed-Soyir 2014: 24), as well as in ideophones and interjections (áh'á 'no'). /o/ and /a/ have nasal allophones before prenasalised consonants, e.g., /ʃi-onónde/ [ʃonóndẽ] 'knife (cl.7).'

Vowel sequences are possible in Ngazidja (djwai 'egg (cl.5)'[1]), but several processes conspire to prevent hiatus, especially at morpheme boundaries: deletion (/tsi-endé/ > [tsendé] 'I went'), coalescence (/ma-índji/ > [méndji] 'many (cl.6)'), (optional) glide insertion (mbé(ʷ)u 'seed (cl.9)'), gliding (/mu-ána/ > [mwána] 'child (cl.1)') (cf. Ahmed-Chamanga 2002). Glide insertion also occurs in absolute initial position, except with the low vowel /a/. Deletion and gliding processes are sensitive to tone (Patin 2009). Vowel harmony, since it only occurs in some conjugations, is briefly discussed in Section 4.3.2.

2.2 Consonants

The consonant inventory of Ngazidja is presented in Table 20.2.

Implosives and retroflex consonants mostly appear in the lexicon of Bantu origin (see for instance Rombi & Alexandre 1982, Lafon 1987), while dentals are essentially restricted to the lexicon of Arabic origin (see for instance Lafon 1987, Ahmed-Chamanga 2010: 21, Mohamed-Soyir 2014: 23), e.g., dúnia 'world (cl.9)' < Ar. dunyā, or French origin (see for instance Patin 2013: 174), e.g., bwáti 'box (cl.9)' < Fr. boîte. However, it has been noted by various scholars (e.g., Rombi & Alexandre 1982, Lafon 1987) that this distribution is subject to several exceptions. For instance, velar fricatives only emerge in words of Arabic origin. According to Rombi and Alexandre (1982), non-educated speakers replace dental and velar fricatives with corresponding stops. Unvoiced consonants are rare in intervocalic position according to Full (2006), with the exception of /k/ and /h/.

Several consonants display some dialectal variation. /r/ appears as a trill or a tap depending on the geographical area (Patin 2013). [t] is sometimes replaced by [ts] or [tʃ], e.g., in Mbeni (*tamáti* vs. *ntsamátsi* 'tomato (cl.9),' *utezá* vs. *utʃezá* 'to play'), while [t] may correspond to [t] or [tr]. In Mbeni, *itánda* 'bed (cl.7),' for example, is realised *itánda*.

A remaining question is the number and status of prenasalised consonants. Unvoiced prenasalised consonants are extremely rare in the lexicon, emerging only in absolute initial position. According to Rombi and Alexandre (1982: 25) and Lafon (1987: 240), the oppositions [nts]/[ns], [ndz]/[nz] and [ntʃ]/[nʃ] are neutralised. Additionally, the nasal part of prenasalised consonants regularly deletes (Rombi & Alexandre 1982: 25), in particular in casual speech – e.g., /wandzáni/ > [wãzáni] 'friends,' /ngwánikaó/ [gwanikáw] 'they give.' Elision also concerns /h/ and /y/, especially in intervocalic position.

TABLE 20.2 CONSONANTS

	Labials	Labio-dentals	Dentals	(Post-) Alveolar	Palatals	Velars	Glottals
Stops	p		t̪	t		k	(ʔ)
	b		d̪	d̪		g	
Affricates			ts		(tʃ)		
			dz		dʒ		
Implosives	ɓ			ɗ			
Fricatives	β	f	(θ)	s	ʃ	(x)	h
		v	(ð)	z	(ʒ)	(ɣ)	
Nasals	m			n	ɲ		
Prenasalised		mf	nts		ntʃ	ŋk	
consonants	mb, mɓ	mv	nd̪, ndz	nd̪, ndʃ	ndʒ	ŋg	
Lateral				l			
Trill				r			
Glides	w				j		

The hardening of several consonants (/β/, /h/, /w/, /l/, /r/) occurs after a nasal – e.g., *tsi-wóno* 'I saw' vs. *tsi-m-móno* 'I saw him/her,' *tsi-rengé* 'I took' vs. *tsi-m-dengé* 'I took him' (see also Section 3.1).

Finally, Ngazidja is also characterised by the so-called 'Kwanyama law' (Tucker & Bryan 1970: 353, Rombi & Alexandre 1982: 26, Mohamed-Soyir 2014: 119–120), whereby the nasal part of the prenasalised consonant in C2 is deleted if there is a NC sequence before, e.g., *mi-píra mi-randáru* 'six balloons (cl.4)' vs. *m-ɓe n-dadáru* 'six cows (cl.10).'

2.3 Tone

As mentioned in Section 1, the prosodic system of Ngazidja, and in particular its tone system, has been discussed extensively (most notably in Tucker & Bryan 1970, Philippson 1988, Cassimjee & Kisseberth 1989, Jouannet 1989, Rey 1990, Cassimjee & Kisseberth 1992, 1998, Philippson 2005, Patin 2007b, a, 2008, 2010, O'Connor & Patin 2015, Patin

2016). Hence, its major characteristics are well known. Due to the quantity of easily accessible work on tone in Ngazidja, we will only briefly address this question here. The interested reader is invited to consult one of the previously cited works for further reading.

Ngazidja has a "privative" tone system in the sense of Hyman (2001), where /H/ contrasts with Ø. The language displays some characteristics that are typical of so-called "pitch-accent" systems, such as the fact that no phrase can surface without a high tone. Underlyingly toneless nouns, for instance, receive a high on their stressed penult when pronounced in isolation: /iɲama/ 'animal' > [iɲáma]. Additionally, the number of surface tone patterns is limited: disyllabic words surface with either a high on their penult or a high on their final (*djána* 'yesterday' vs. *djaná* '100,' *ntibé* 'cooked meat (cl.9)' vs. *ntíbe* 'sultan (cl.9)'). However, Ngazidja also differs from pitch-accent systems in various ways. Disyllabic noun roots, for instance, are associated with four different underlying tone patterns: /σσ/ (/N-dovu/ [ndóvu] 'elephant (cl.9)'), /σ́σ/ (/mu-ána/ [mwána] 'child (cl.1)'), /σσ́/ (/m-leví/ [mleví] 'drunkard (cl.1)') and /σ́σ́/ (/N-kúdé/ [ŋkúde] 'peas (cl.10)'). Moreover, the pitch realisations associated with the three categories that surface with a high on the penult differ in some dialects, in that the height of the tone of the /σ́σ/ category is raised, while the tone of the /σσ/ and /σ́σ/ categories is flat.

In Ngazidja, a high tone shifts unboundedly to its right, unless it is blocked by an underlying tone. The shift of the tone leads to the deletion of every even-numbered tone (following the Obligatory Contour Principle). In (1b) for instance, the tone of the noun *maβáha* 'cats' shifts to the penult of the adjective *maili* 'two,' and the tone of the numerical adjective is thus deleted. In (1c), however, the tone of the adjective is free to surface because the tone of the noun has been deleted by the tone of the verb *tsi(w)óno* 'I saw.'

(1)   a   i   -ilí   'two'
         ii   ma-βáha   'cats (cl.6)'

      b   ma-βaha   ma-íli
          6-cat   NP$_6$-two
          'two cats'

      c   tsi-on-o   má-βaha   ma-ilí
          SP$_{1SG}$-see-FV   6-cat   NP$_6$-two
          'I saw two cats.'

As in many other Eastern Bantu languages (Philippson 1991, Kisseberth & Odden 2003), the tone is not bounded by the limits of the prosodic word. In (1c), for instance, the tones of the nouns and/or verbs are free to move to the following word(s). More precisely, a tone in Ngazidja shifts as far as it can toward the end of a phrase. See for example (2), where the tone of the verb *hawóno* 'he saw' shifts to the penult of the phrase, through the noun *ndóvu* 'elephant.'

(2)   ha-on-o   n-dovu   m-bíli
      SP$_{1SG}$-see-FV   10-elephant   NP$_{10}$-two
      'He saw two elephants.'

The boundaries of phrases interact with tone shift such that a tone cannot cross the boundaries of Phonological Phrases. In (3a), the tone of the subject NP stops on the last syllable of the noun instead of continuing to move rightward onto the first syllable of the verb *haréme* 'he beat.' The verb is not a possible target because it is not in the same Phonological Phrase as the preceding high tone on the subject NP (3b).

(3) a (m̩-limad̲jí)$_\phi$ (ha-r̲em-é pa̲ha)$_\phi$
         1-farmer   sp$_1$-beat-fv 5.cat
         'A farmer beat a cat.'

     b *(mlimad̲ji hár̲eme pa̲ha)$_\phi$

As in other languages, such as Giryama E72a (Cassimjee & Kisseberth 1998), the tone cannot shift to the last syllable of the utterance (4), a phenomenon that is commonly referred to as *Extraprosodicity* or *Non-finality*.

(4) ye=ɲ-uŋgu̲   n-dzíɾo
     aug$_9$=9-cooking pot np$_9$-heavy
     'the heavy cooking pot'

This non-finality effect has been cited as the indicator of Intonation Phrases (henceforth IPh) in Patin (2007a, 2008, 2010), following Cassimjee and Kisseberth (1998). This analysis has recently been challenged by O'Connor and Patin (2015) and Patin (2016), who posit that Intonation Phrases are instead indicated by boundary tones.

2.4 Syllable structure

The canonical syllable structure is CV. Other types are V and CGV. Depending on the analysis of nasal + consonant sequences (see Section 2.2), one could consider that Ngazidja also includes NCV and NCGV syllables.

The syllabic nature of morphemes consisting of a nasal varies depending on various factors, such as its position in the utterance and speech rate. For instance, when preceding a consonant-initial root, the class 1 prefix *m-* is syllabic in absolute initial position (5a), or when a tone surfaces on it (5b).

(5) a m̩-lí̲ma 'mountain (cl.3)'

     b ha-o̲n-o   ḿ̩-lima
        sp$_1$-see-fv 3-mountain
        'He saw a mountain.'

3 NOUNS AND NOUN PHRASES

3.1 Noun classes

The class and gender system of Ngazidja is typical of Bantu languages. The class morphemes, along with the verbal agreement prefixes, are listed in Table 20.3.

All the noun class prefixes are underlyingly low, though some rare words have their class prefix underlyingly associated with a high. In contrast, all the subject markers are high, except those in class 1 (see Section 4.3.1).

As in many other Bantu languages, the form of the prefix varies depending on the form of the stem. The vowel of the prefix can turn into a glide (cl.1, 3, 11, 15) or delete

TABLE 20.3 THE NOUN CLASS SYSTEM OF NGAZIDJA

Class	Noun prefixes		Examples	Glosses	Subject prefixes
1.	mu-	_C: m- _V: mw-	ḿ-nḍu mw-áṇa	'(a) person' '(a) child'	(h)a-, i-, u-
2.	wa-	_C: wa- _V: w-	wá-nḍu w-áṇa	'persons' 'children'	wá-
3.	mu-	_C: m- _V: mw-	mɓuụ́ mw-áha	'(a) baobab' 'a year'	ụ́-
4.	mi-	_C: mi- _V: m-	miɓuụ́ m-áha	'baobabs' 'years'	í-
5.	Ø- dzi- dj-		páha (< /βáha/) dzi-ɲọ́ dj-ɛ́nɗo	'(a) cat' '(a) tooth' '(a) walking'	lí-
6.	ma-		ma-βáha ma-ɲọ́ ma-ɛ́nɗo	'cats' 'teeth' 'walkings'	yá-
7.	hi- ʃi-	_σ: hi- _σσ(σ): i- _C: ʃi- _V: ʃ-	hi-tswá i-ɲáma ʃi-βáa ʃ-onọ́nɗe	'(a) head' '(an) animal' '(a) prison' '(a) knife'	ʃí-
8.	zi-	_C: zi- _V: z-	zi-tswá zi-ɲáma zi-βáa z-onọ́nɗe	'heads' 'animals' 'prison' 'knives'	zí-
9.	N-	_p,b,ɓ: m- _k,g: ŋ- _C: n- _V: ɲ-	m-ɓẹu ŋ-kụ́hu n-dẹ̆vu ɲ-umɓá	'(a) seed' '(a) chicken' '(a) beard' '(a) house'	í-
	Ø-	[loanwords]	rọ́ho	'(a) heart'	
10.	N-	_p,b,ɓ: m- _k,g: ŋ- _C: n- _V: ɲ-	m-ɓẹu ŋ-kụ́hu n-dẹ̆vu ɲ-umɓá	'seeds' 'chickens' 'beards' 'houses'	zí-
	Ø-	[loanwords]	rọ́ho	'hearts'	
10a.	ɲi-		ɲi-tsụ́ŋgo	'tails'	
11.	u-	_C: u- _V: w-	u-tsụ́ŋgo w-andzáni	'(a) tail' '(a) friendship'	ụ́-
15.	hu-	_σ: hu- _σσ(σ): u- _V: hw-	hu-djá u-lála hw-énɗa	'to come' 'to sleep' 'to go, walk'	hụ́-
16.	βa-		βa-áṇu	'(a) place'	βá-
17.	n/a				hụ́-
18.	n/a				

(cl.4, 7, 8) before a vowel. Interestingly, the initial [h] of prefixes in classes 7 and 15 deletes before a polysyllabic stem, except in some dialects.

The initial consonant of the stem undergoes fortition in class 5, as exemplified in (6) (see Mohamed-Soyir 2007, Patin 2013, Mohamed-Soyir 2014 for details).

(6)   Class 5 fortition

Ø-t̪ávu	'cheek (cl.5)'	ma-rávu	'cheeks (cl.6)'
Ø-d̪íŋgo	'back (cl.5)'	ma-líŋgo	'backs (cl.6)'
Ø-paɾé̪	'road (cl.5)'	ma-βaɾé̪	'roads (cl.6)'
Ø-kaɓúɾi̪	'grave (cl.5)'	ma-haɓúɾi̪	'graves (cl.6)'

Some words belong to different classes depending on the dialect or idiolect of the speaker, e.g., 'spoons': *mik̪ó̪ɓe̪* (cl.4) ~ *ŋk̪ó̪ɓe̪* (cl.10). Dialectal or idiolectal variation also has an impact on the form of the prefix of nouns, e.g., *ŋ-fi̪* ~ *n-fi̪* ~ *fi̪* 'fish (cl.9).'

The noun classes of Ngazidja form the genders that are listed in Table 20.4.

TABLE 20.4 GENDERS IN NGAZIDJA

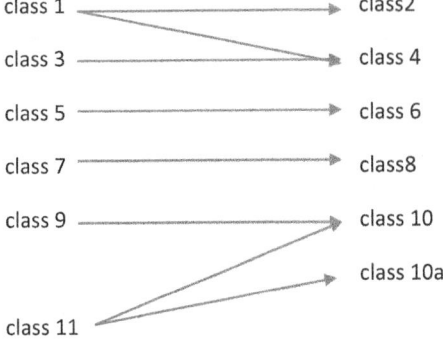

Ahmed-Chamanga (2011: 48) indicates that class 1 nouns that have their plural in class 6 are compounds whose first lexeme is an agent noun.

3.2 Nominal derivation

Nominal derivation has received very little attention in Ngazidja, except in Mohamed-Soyir (2014). Nouns can be derived from nouns by shifting them into another noun class. This is especially true for names referring to nationalities/ethnicities, ideological or religious conditions, or trees (7).

(7)

faɾántsa	'France (cl. 11)'	>	ɱ-faɾántsa	'French (cl.1)'
u-silámu	'Islam (cl. 11)'	>	mw-isilámu	'Muslim (cl. 1)'
fuɾiyapá̪	'breadfruit fruit (cl. 5)'	>	ɱ-fuɾiyapá̪	'breadfruit (cl. 3)'

Class 11 nouns that refer to qualities can derive from adjectives (8).

(8)

-é̪u	'white'	>	u-é̪u	'whiteness'
– d̪u	'black'	>	ú̪-d̪u	'blackness'
– há̪ɾa	'fearful'	>	u-há̪ɾa	'fear'

Deverbal derivation is productive in Ngazidja, e.g., *u-ló̪la* 'to marry' > *n-d̪ó̪la* 'wedding (cl.9),' *u-lé̪la* 'to raise (children)' > *ma-lé̪zi* 'education (cl.6),' *u-lá̪la* 'to sleep' > *ma-lá̪lo* 'mattress (cl.6).' Nouns denoting actions, which are assigned to class 3, for example, are derived by adding a suffix -*o* to the stem (9a), except for loanwords (9b).

(9)   *Infinitive* (cl.15)   *Deverbal name* (cl.3)
   a   hu-lá   'to eat'   >   m̩-l-ó   'feeding'
       hw-énɗa   'to walk'   >   mw-énɗ-o   'walk'
       hw-iʃíya   'to hear'   >   mw-iʃíy-o   'hearing'

   b   u-telefóni   'to phone'   >   m̩-telefóni 'action of phoning'
       hw-aɓúɗu   'to worship'   >   mw-aɓúɗu   'act of worshiping'

In other Bantu languages, the derivation with -*o* also produces instruments. While this process is attested in Ngazidja, it is unproductive (Mohamed-Soyir 2014), despite some rare examples such as *uiʃiya* 'to hear' > *ʃiʃi(y)o* 'ear (cl.7).' State nouns are derived from stative verbs, which involve an -*ih*- or -*eh*- verbal extension, by adding a suffix -*fu* (10a), or -*vu* when the stem ends with [w], [y] or [l] (10b).

(10)   *Infinitive* (cl.15)   *Deverbal name* (cl.11)
    a   u-tãaɓíha   'to suffer'   >   u-tãaɓí-fu   'suffering'
        u-simɓúha   'to wear himself'   >   u-simɓú-fu   'exhaustion'

    b   u-eléwa   'to understand'   >   u-elé-vu   'understanding'
        u-βiyála   'to stiffen up'   >   u-βiyá-vu   'tightening'

As some other Bantu languages (Mugane 1997, Schadeberg 2003: 80, Ferrari-Bridgers 2009), Ngazidja derives two different types of agent nouns using either a -*i* or a -*a* suffix, e.g., *u(w)óna* 'to see' > *mu-(w)ón-i* 'sighted person (cl.1)' and *mu-(w)ón-a* 'clear-sighted (cl.1).' The suffix -*i*, which sometimes triggers spirantisation, e.g., *u-íɓa* 'to steal' > *mw-idzí* 'thief (cl.1)' (see Bostoen 2005, 2008 for a discussion), is semantically associated with an *in*trinsic quality in the sense of Anscombre (2001), while the suffix -*a* is associated with an *ex*trinsic quality (everybody sees, but only few are distinguished by an exceptional talent). However, there are very few stems that can be associated with the two suffixes; the intrinsic quality is generally expressed by the suffix -*dji* – e.g., *uhézá* 'to sing' > *m-hezá-dji* 'singer (cl.1),' **mhezi*.

Patient nouns, all in class 1, are derived from verbal roots with the passive extension -*w*-.

(11)   *Infinitive* (cl.15)   *Deverbal name* (cl.1)
    hu-lá   'to eat'   >   m̩-líwa   'who can be eaten'
    u-íɓa   'to steal'   >   mw-íɓwa   'who can be stolen'
    u-tsíndza   'to cut (throat)' >   m̩-tsíndwa   'who can have his throat cut'

It is also possible in Ngazidja to derive nouns from ideophones. Ideophones that are monosyllabic derive trisyllabic nouns, e.g., *gú* [dry sound] > *gugúgu* 'uproar,' *ɓwá* [impact] > *ɓwaɓwáɓwa* 'chitter-chatter,' while ideophones that are disyllabic are reduplicated, e.g., *gáɗa* [drop] > *gaɗagáɗa* 'African'[2].

3.3 The augment

The noun can be preceded by an augment (also called "proclitic" or "preprefix") in Ngazidja. There is no augment before any element besides a substantive. Table 20.5 presents the augments in all the classes of Ngazidja, and examples are given in (12).

TABLE 20.5 THE AUGMENT IN NGAZIDJA

Class	1.	2.	3.	4.	5.	6.	7.	8.
Aug.	(y)e	(w)o	(w)o	(y)e	le	(w)o	ʃe/(y)e	ze
Class	9.	10.	11.	15.	16.	17.	18.	
Aug.	(y)e	ze	(w)o	ho	βo	ho	mo	

(12) a ṃndu '(a) person (cl.1) (y)e=ṃndu 'the person'

    b wándu 'persons (cl.2)' (w)o=wándu 'the persons'

    c páha '(a) cat (cl.5)' le=páha 'the cat'

    d hudjá 'to come (cl.15)' ho=hudjá 'the arrival'

In class 7, the allomorph *ʃe* appears before nouns beginning with a class prefix *i-* (see the example in Table 20.3) or *hi*; class 7 nouns that begin with a class prefix *ʃi-* (reminder: when the root is vowel-initial) are associated with an augment of the form (*y*)*e*: (*y*)*e=ʃ-onónde* 'the knife (cl.7).' The onglide of the augments (*w*)*o* (cl. 2, 3, 6 and 11) and (*y*)*e* (cl. 1, 4 and 9) generally emerges at the beginning of an Intonation Phrase (e.g., in absolute initial position). These augments, contrary to those that begin with a consonant (such as *le*, cl. 5), cliticise to the preceding element in casual speech: *haniká e=ɲungú* 'he gave the house (formal)' vs. *hanik=é ɲungú* 'he gave the house (casual).'

As suggested by the translations in (12), the augment indicates definiteness in Ngazidja. Depending on the authors, it carries a value that is said to be either "definite" (Cassimjee & Kisseberth 1989, 1992, Patin 2007a, Ahmed-Chamanga 2010) or "deictic" (Rombi & Alexandre 1982: 29, Lafon 1987: 126–127). The augment is mandatory when the noun is followed by a demonstrative (*ze=mbuda zínu* 'these sticks (cl.10),' *mbuda zínu), but it cannot appear when the demonstrative precedes the noun (*yila mezá* 'that (very) table (cl.9),' **yila e=mezá*). It is also associated with possessives, e.g., *ye=ɲungu yá=hahe* 'his cooking-pot (cl.9),' *ʃe=hi-tswa ʃ-á=ha'ngú* 'my head (cl.7),' but it may be absent when the possessed element is an item from a set, e.g., *ɲungu yá=hahe* 'his cooking-pot (cl.9),' vs. **hi-tswa ʃ-á=ha'ngú*, and/or when the NP is in object position (13). Finally, the augment obligatorily precedes a dislocated noun, e.g., *tsiirénge, ye=mberé* 'I bought it, the ring (cl.9).'

(13) ha-wono má-βaha ma-ili 'y-á=hahe
    1.PFV-see 6-cat NP₆-two NP₆-CON=POSS₁
    'He saw his two cats.'

Contrary to what has been observed in several other Bantu languages (cf. Katamba 2003), such as Ganda JE15 (Hyman & Katamba 1993), the presence of the augment in Ngazidja does not depend on the focus type or the polarity of the verb. The augment can emerge, for instance, after a negative verb.

The phonology of the augment has some interesting particularities. First, a high-toned allomorph is selected when the augment is preceded by another clitic: *le=páha* 'the cat' vs. *nde=lé=paha* 'this is the cat.' Also, as seen in this latter example, the vowel of some morphemes, such as the so-called "stabiliser" *nda* or the clitic *na* 'and/with/by,' assimilates to the vowel of the augment, e.g., **nda=lé=paha*. However, according to Ahmed-Chamanga (2010), the augment in some classes has allomorphs, such as *ele*, *eʃe* and *eze* in classes 5, 7 and 10, whose initial vowel deletes the vowel of the stabiliser in such a situation. Finally, the augmented noun phrases separately from what precedes the augment, e.g., a verb (except in some conjugations): (*haǫno máβạha*)ᵩ 'he saw cats' vs. (*haǫnó*)ᵩ (*e=maβạha*)ᵩ 'he saw the cats.'

3.4 Modifiers

As in several other Bantu languages, there are a very small number of real adjectives in Ngazidja, i.e., adjectives that are not constructed through the use of a genitive construction. These adjectives, which immediately follow the noun (see Section 3.5), agree with the head noun (14).

(14) a   ma-djwai̱   m-éma
         6-egg   NP₆-good
         'good eggs'

   b   mw-ạna   mú-i̱
         1-child   NP₁-bad
         'bad child'

The agreement prefixes of these adjectives have forms that are identical to those of the class prefixes of nouns – (14), (15a-c). In addition, they can trigger a hardening of the initial consonant of the root of the adjective (15d-e), just as noun prefixes do (see Section 2.2).

(15) a   wa-nḑu   wá-le
         2-person   NP₂-long
         'tall persons'

   b   m-koɓá   m-le
         3-bag   NP₃-long
         'a large bag'

   c   ma-fu̱u   má-le
         6-claw   NP₆-long
         'long claws'

   d   fu̱ú   ɗe
         5.claw   NP₅.long
         'a long claw'

   e   fulạná   n-ɗe
         9.t-shirt   NP₉-long
         'a long t-shirt'

Numerals from 1 to 6, as well as 8, belong to this class of adjectives, while numerals higher than 8, as well as 7, are nominals (Mohamed-Soyir 2014).

(16) a   z-onǫnɗe   zi-íli̱
         8-knife   NP₈-two
         'two knives'

   b   wa-ɓaɓá   nfu̱kạre
         2-father   seven
         'seven fathers'

Only three color adjectives are real adjectives: -*ęwu* 'white,' -*du* 'black' and – *kụdu* 'red.' Other colors are rendered through the use of a genitive construction, where the head is separated from the dependent (most of the time borrowed from French) by the connective (17).

(17) a gạri   l-á=biḷe
        5.car   NP$_5$-CON=9.blue
        'a blue car'

    b hi-rị   ʃ-a=dzíndzạnu
        7-chair   NP$_7$-CON=9.saffron
        'a yellow chair'

Demonstratives are of three types. Close demonstratives add a class prefix to a stem -*nu*, except in class 1 (**u-nu*), and distant demonstratives to a stem -*la*. Resumptive demonstratives, which also refer to a medium distance, consist of an initial vowel *i*- or *u*-, except in classes 2, 6 and 16, followed by a class marker and the vowel -*o*. All the demonstratives are toneless, though a tone can be underlyingly associated with their penult when they follow a toneless noun (ex. *ɲama ịla* 'that animal').

TABLE 20.6 DEMONSTRATIVES IN NGAZIDJA

Class	Close	Distant	Resumptive
1.	oyi	o-la	u-w-o
2.	wa-nu	wa-la	wa-w-o
3.	u-nu	u-la	u-w-o
4.	i-nu	i-la	i-y-o
5.	li-nu	li-la	i-l-o
6.	ya-nu	ya-la	ya-y-o
7.	ʃi-nu	ʃi-la	i-ʃ-o
8.	zi-nu	zi-la	i-z-o
9.	i-nu	i-la	i-y-o
10.	zı-nu	zi-la	i-z-o
11.	u-nu	u-la	u-w-o
16.	βa-nu	βa-la	ya-β-o
17.	hu-nu	hu-la	i-h-o
18.	mu-nu	mu-la	i-m-o

As mentioned in Section 3.3, the demonstrative is always associated with the augment, except when it is fronted. Fronting mostly concerns distant demonstratives, and is rarely observed after a verb.

The connective 'of' has a typical structure in which a class prefix that agrees with the head noun precedes the vowel -*a*. The connective cliticises to the following word.

TABLE 20.7 CONNECTIVES IN NGAZIDJA

Class	1.	2.	3.	4.	5.	6.	7.	8.	9.	10.	11.	15.	16.	17.	18.
Conn.	w-a	w-a	w-a	y-a	l-a	y-a	ʃ-a	z-a	y-a	z-a	w-a	h-a	β-a	h-a	mw-a

The connective also participates in the construction of possessives, which consist of the connective and a combination of *há*- with various toneless (except in class 1) stems.

TABLE 20.8 STEMS OF THE POSSESSIVES IN NGAZIDJA 1 – HUMANS

	Singular			Plural		
	1sg	2sg	Cl.1	1pl	2pl	Cl.2
Possessive	há-ŋgu	há-ho	há-he	há-ṭu	há-nyu	há-ho

TABLE 20.9 STEMS OF THE POSSESSIVES IN NGAZIDJA 2 – OTHERS

Class	3.	4.	5.	6.	7.	8.	9.	10.	11.	16.
Poss.	há-w-o	há-y-o	há-l-o	há-y-o	há-ʃ-o	há-z-o	há-y-o	há-z-o	há-w-o	há-β-o

The possessive is most of the time associated with the augment (see Section 3.3 for restrictions).

3.5 Word order in the NP

The canonical noun phrase linear order in Ngazidja is briefly summarised below:

augment – N – Adjective (Adj) – Demonstrative – Possessive – Quantifier – Relative

A first illustration of this word order is provided in (18). (18a) demonstrates that the possessive follows the adjective(s), while (18b) shows the respective orders of the adjective(s), the demonstrative (in its canonical position – see Section 3.4 for a discussion of the fronting of demonstratives) and the quantifier.

(18) a   ze=m-6uɗa   m-6íli   n-djẹu   z-á=hạhe
         AUG$_{10}$=10-stick   NP$_{10}$-two   NP$_{10}$-white   NP$_{10}$-CON=1.POSS
         'his two white sticks'

     b   ze=m-6uɗa   ń-tịnti   zi-nu   pía
         AUG$_{10}$=10-stick   NP$_{3}$-small   NP$_{10}$-NDEM   all
         'all these small sticks'

The order of adjectives, including agreeing numbers, is relatively free, as illustrated in (19). The fact that the order is free is supported by the fact that no obligatory prosodic boundary separates the adjectives in (19a) or (19b).

(19) a   ze=m-6uɗã   n-djẹu   n-dziro
         AUG$_{10}$=10-stick   NP$_{10}$-white   NP$_{10}$-heavy
         'the heavy white sticks'

     b   ze=m-6uɗa   n-dziró   n-djẹu
         AUG$_{10}$=10-stick   NP$_{10}$-heavy   NP$_{10}$-white
         idem

However, a canonical pattern is frequently observed. For instance, the numerical adjectives appear immediately after the noun (20a), followed by adjectives referring to size and shape, while colors appear in final position (20b).

(20) a   mi-píra   mi-ili   mí-titi   mi-raháfu
4-balloon   NP₄-two   NP₄-little   NP₄-clean
'two clean little balloons'

b   m-6uɗa   m-6íli   n-dziro   n-djeu
10-stick   NP₁₀-two   NP₁₀-heavy   NP₁₀-white
'two heavy white sticks'

The respective positions of the demonstrative and the possessive are illustrated in (21). Note that speakers accept the reverse order (i.e., Poss Dem), but that this is associated with a type of focalisation of the possessive.

(21) ze=m-6uɗa   zi-nu   z-á=hahe
AUG₁₀=10-stick   np₁₀-NDEM   NP₁₀-CON=1.POSS
'these sticks of his'

The respective positions of possessives and quantifiers are illustrated in (22a), and the final position of the relative in (22b).

(22) a   ze=m-6uɗa   z-á=hahe   pía
AUG₁₀=10-stick   NP₁₀-CON=1.POSS   all
'all his sticks'

b   ze=m-6uɗa   pía   z-a-vundz-íh-a
AUG₁₀=10-stick   all   SP₁₀-REL-break-ST-FV
'all the sticks that broke'

4 VERBS AND CLAUSE STRUCTURE

4.1 Verbs

As in many other Bantu languages, the infinitive form of the verb consists of a combination of the class 15 prefix with a root and the final vowel -*a*. The root may be followed by one or more derivational suffixes (i.e., extensions – see Section 4.2.1) that separate it from the final vowel.

The canonical verb root is -CVC-, e.g., *u-líl-a* 'to cry,' *u-yél-a* 'to (have a) wash,' *u-fán-a* 'to make.' However, several other shapes are attested, such as -C- (*hu-l-á* 'to eat,' *hu-f-á* 'to die'), -VC- (*hw-énd-a* 'to go,' *hw-ándza* 'to like'), -CV- (*u-djú-a* 'to know,' *u-dě-a* 'to insult'), -CVCVC- (*u-shangáz-a* 'to be afraid,' *u-sukúm-a* 'to push'), -CVCV- (*u-hobó-a* 'to cut off'), etc.

Most of the verbs appear with an underlying high associated with their penult, but i) we shall see in Section 4.3.2 that there is a distinction in certain conjugations between verbs that have a high on the penult with verbs that have a high on their final syllable, and ii) there are several exceptions to this pattern, such as monosyllabic roots (see the examples above) or verbs such as *u-děma* 'to be sleepy' or *u-diwaza* 'to forget.'

A significant number of verb roots in Ngazidja were borrowed from Arabic – e.g., *u-swali* 'to pray' (< Ar. swallâ:), *u-abúdu* 'to adore' (< Ar. 'abada) – or, less frequently, from French – e.g., *u-telefóni* 'to phone' (< Fr. *téléphoner*), *u-marké* 'to mark' (< fr. *marquer*). In these cases, the final vowels ([u] or [i] for Arabic loanwords, [i] or [e] for French) cannot be considered to be affixes.

4.2 Verbal derivation

4.2.1 Extensions

There are several productive and unproductive extensions in Ngazidja, which are listed in Ahmed-Chamanga (2011). Examples in this section are extracted from his grammar, with tones added by us. The extensions that include an /i/ are subject to harmony, such that the vowel is lowered after a root whose vowel is /e/ or /o/. The applicative, for instance, appears as an -i- after a root that includes one of the cardinal vowels /i/, /a/ or /u/ (u-fáɲ-a 'to make' > u-faɲ-í-a 'to make for,' u-líβ-a 'to pay' > u-liβ-í-a 'to make for'), and as -e- after a root whose vowel is /e/ or /o/ (u-réŋg-a 'to take' > u-reŋg-é-a 'to take for,' u-sóɲ-a 'to sew' > u-soɲ-é-a 'to sew for'). The extension triggers several kinds of consonant alternation, e.g., h > ʃ (u-píh-a 'to cook' > u-piʃ-í-a 'to cook for,' u-βéh-a 'to send' > u-βeʃ-é-a 'to send to'), w, y > l (u-ɲáw-a 'to urinate' > u-ɲal-í-a 'to urinate on'). The applicative suffix has additional allomorphs such as -iz-, which emerges after [s, z, ts, dz] (u-wáz-a 'to count' > u-waz-íz-a 'to count for'), or -li-/-le-, after a root ending with a vowel (u-mé-a 'to ask' > u-me-lé-a 'to ask for').

The causative -is-/-es- (u-hím-a 'to stand' > u-him-ís-a 'to raise,' u-ón-a 'to see' > u-on-és-a 'to show') also has various forms depending on the segment that ends the root. A final /l/, /w/ or /y/ on a root turns into [z] (u-líl-a 'to cry' > u-líz-a 'to make (s.o.) cry,' u-léw-a 'to be drunk' > u-léz-a 'to make drunk'), for instance, while /h/ changes into [s] (u-fúh-a 'to be holed' > u-fús-a 'to make a hole'), /ŋg/ into [ndz] (u-dúŋg-a 'to follow' > u-dúndz-a 'to make (s.o.) follow') and /r/ into [ts] (u-rór-a 'to sink (boat)' > u-róts-a 'to make (a boat) sink'). Other extensions in the language include the neuter[3] -ih-/-eh- (u-fáɲ-a 'to make' > u-faɲ-íh-a 'to be doable,' u-ón-a 'to see' > u-on-éh-a 'to be visible'), which also triggers consonant alternations (u-hóh-a 'to grill' > u-hof-éh-a 'to grill'), the associative (reciprocal) -an- (u-ón-a 'to see' > u-on-án-a 'to see each other'), the intensive -u- (u-tsáh-a 'to search' > u-tsáh-u-a 'to choose') and the passive -w- (u-sóm-a 'to read' > u-sóm-w-a 'to be read'), which emerges as -lw- after a /u/ (u-djú-a 'to know' > u-djú-lw-a 'to be known').

Extensions may combine with each other. Ahmed-Chamanga (2011) proposes a table of all possible combinations of two extensions (Table 20.10).

TABLE 20.10 COMBINATIONS OF TWO EXTENSIONS – AHMED-CHAMANGA (2011: 118)

First extension	Second extension				
	Applicative	Causative	Neuter	Associative	Passive
Applicative	✓			✓	✓
Causative	✓		✓	✓	✓
Neuter	✓				
Associative		✓			
Passive					

4.2.2 Reduplication

An interesting productive derivational process is reduplication, which is mostly used to express *pluractionality*. In this process, the whole stem is duplicated, including the final vowel -a.

(23) u-tséŋg-a 'to cut trees' > u-tseŋga-tséŋga 'to level'
u-léŋg-a 'to walk around' > u-leŋga-léŋga 'to walk a. a lot'
u-kák-a 'to defecate' > u-kaka-káka 'to defecate in several places'
u-tów-a 'to get out' > u-towa-tówa 'to clear'

4.3 Inflection

4.3.1 Subject and object markers and pronouns

The different subject and object markers are listed in Table 20.10. First and second singular markers and class 1 markers are toneless. All the other markers are underlyingly high.

TABLE 20.11 SUBJECT AND OBJECT MARKERS IN NGAZIDJA

Class/ Person	Subject prefixes	Object prefixes	Object enclitics
1SG.	tsi-, ni-, n-, m(u)-	-ni-	
2SG.	hu-	-hu-	
1PL.	rí-	-rí-	
2PL.	mú-	-mú-	
1.	(h)a-, y-, u-	-m(u)-	=ye
2.	wá-	-wá-	=wó
3.	ú-	-ú-	=wó
4.	i-	-í-	=yó
5.	lí-	-lí-	=ló
6.	yá-	-yá-	=yó
7.	ʃí-	-ʃí-	=ʃó
8.	zi-	-zí-	=zó
9.	i-	-í-	=yó
10.	zí-	-zí-	=zó
11.	ú-	-ú-	=wó
15.	í-		
16.	βá-	-βá-	=βó
17.	hú-	-hú-	=hó

Subject markers attach to the verb as prefixes (24a), and so do the first object markers (24b), which are highly similar to the subject markers in form. The secondary object markers are not prefixed to the stem of the verb, but are encliticised (24c). Subject markers of the first person singular and class 1 vary depending on the tenses.

(24) a tsi-ník-a
SP$_{1sg}$-give-FV
'I gave'

 b tsi-ʃí-nik-a
SP$_{1sg}$-OP$_7$-give-FV
'I gave it (the book)'

 c tsi-m-nik-á=ʃo
SP$_{1sg}$-OP$_1$-give-FV=OE$_7$
'I gave it to him (the book)'

The Habitual tenses associate a pronoun with the infinitive form of the verb (instead of a subject marker – see Section 4.2.2). This pronoun is similar in form to the object suffix or the resumptive demonstrative – Table 20.12.

TABLE 20.12 PRONOUNS IN NGAZIDJA

Class/Person	Pronoun	Class/Person	Pronoun
1SG.	mí(mi)	5.	ló
2SG.	wé(we)	6.	yó
1.	yé(ye)	7.	ʃó
1PL.	sí(si)	8.	zó
2PL.	ɲí(ɲi)	9.	yó
2.	wó(wo)	10.	zó
3.	wó	11.	wó
4.	yó		

4.3.2 TAMN

An inflected verb has the structure sketched in (25). In the *Habitual* tenses, however, a pronoun, rather than a subject marker, is associated to the stem (see below).

(25) Pre-Prefix (Pre-PX) – Subject Marker (SP) – TAMs – Object Marker 1 (OP) – Root – Extensions – Suffix (SX) – Object Marker 2 (OM)

Table 20.13 lists the main affirmative conjugations of Ngazidja.

TABLE 20.13 MAIN AFFIRMATIVE CONJUGATIONS

Conjugation	PRO	Pre-PX	SP	TAM	OP	Root	Extensions	FV	SX	OM_2
Imperative						x	x	a		
Perfective			tsi			x	x	a (hv)		
Imperfective		nga	mu			x	x	a (< pl)	o/ó	
Future		nga	mu	djo		hú + x	x	a		
Future imperfect		nga	mu	djo	ka	hú + x	x	a		
Distant past			tsi		ka	hú + x	x	a		
Conditional			tsi	djo		x	x	a		
Hortative		na	(pl)			x	x	e		
Prohibitive			n	tsi		x	x	e		
Habitual	PR					hu + x	x	a		

The simplest form, with the exception of the Imperative, is the Perfective, which associates the subject marker with the base, cf. (24a), (26a). If the base has no extension, the final vowel generally displays total harmony with the vowel of the root (26) – there are some exceptions, e.g., (24a).

(26) a tsi-reŋg-é
     SP$_{1sg}$-take-FV
     'I took'

   b tsi-ón-o
     SP$_{1sg}$-see-FV
     'I saw'

   c tsi-hul-ú
     SP$_{1sg}$-buy-FV
     'I bought'

The form of the final vowel of verbs with a -C- root in the Perfective form cannot be predicted: -f-á 'die' > -f-ú 'died,' -l-á 'eat' > -l-í 'ate,' etc. Regarding tone, verbs in the perfect form belong to two distinct categories, retaining the classical opposition between high-toned and low-toned roots in Bantu: those that have a tone on the penult (24a, 26b) and those that have a tone on the final vowel (26a, 26c).

   This tonal opposition has been lost in other conjugations, such as the Present imperfective in most of dialects – [4]compare (24a) and (26a) with (27b) and (27c). The Imperfective aspect is expressed by the prefix ŋga-, which precedes the subject marker (27). In the Present imperfective, it is accompanied by a suffix -o which deletes the final vowel with singular subjects when the verb is of Bantu origin (27a),[5] and by a high-toned suffix -(w)ó with plural subjects (27b-c).

(27) a nga-m-ník-o
     IPFV-SP$_{1sg}$-take-IPFV
     'I give'

   b ng-wa-nik-á-o
     IPFV-SP$_{2}$-give-FV-REL
     'They give'

   c ng-wa-reŋg-á-o
     IPFV-SP$_{2}$-take-FV-REL
     'They take'

This contrast between singular and plural forms is also important in the Future conjugation, since only the plural forms add a high-toned class 15 prefix -(h)ú-[6] after the future marker -djo- (28).

(28) a nga-m-djo-ník-a
     IPFV-SP$_{1sg}$-FUT-give-FV
     'I will give'

   b ng-wa-djó-u-ník-a
     IPFV-SP$_{2}$-FUT-15-give-FV
     'They will give'

The future marker -djo- is also implicated in the Conditional tense (29), while the Future perfect and the Distant past associate the high-toned class 15 prefix -(h)ú- with the verb u-káya 'to be' in its short, toneless form -ka-.

(29) r̩i-djó-o̩n-a
      SP₁ₚₗ-FUT-see-FV
      'we would see'

The Imperative is formed by the base and the final vowel -a (e.g., énd-a 'go!'), unless an object marker is prefixed – in such cases, the final vowel is -e for verbs of Bantu origin (e.g., ʃi̩-l-e 'eat it (cl.7)'). The Hortative adds to this *stem* + *e* structure the pre-prefix *na-* and the subject marker (e.g., na-ri̩-l-e 'let us eat'). Another aspect that is relevant in Ngazidja is Habitual. It associates a pronoun (instead of a subject marker) – see Table 20.11 above – with the infinitive form of the verb (e.g., yé u-swali̩ 'he usually prays'). The pronoun, which is frequently absent in discourse (37), is separated from the infinitive by a prosodic break.

The Habitual has a negative counterpart that is composed of the negative prefix *ka-*, the subject marker and the base (e.g., *k-a-li̩m-i* 'he does not cultivate' – note that the final vowel harmonises with the vowel of the stem). The main negative conjugations of Ngazidja are listed in Table 20.14.

TABLE 20.14 MAIN NEGATIVE CONJUGATIONS

Conjugation	PRO	Pre-PX	SP	TAM	OP	Root	Extensions	FV	OM₂
Habitual neg.	PR	ka	ń			x	x	a (hv)	
Perfective neg.		ka	ń	dja		x	x	a	
Imperfective neg.		ka	ń	tsú		x	x	a	
Future neg.		ka	ń	na		hú + x	x	a	
Future imperfect neg.		ka	ń	na	ka	hú + x	x	a	
Distant past neg.		ka	ń	dja	ka	hú + x	x	a	
Conditional neg.		ka	ń	dja	djo	x	x	a	

The negation marker is a prefix *ka-* that precedes the subject marker (30), except in the first singular form. The vowel [a] of the negative marker elides before or assimilates to (depending on speech rate) the second and third singular subject markers -hú̩- and -há̩-, except before the vowel-initial prefixes of classes 3 to 11. Note that the singular prefixes, like other prefixes or clitics in the language, display an underlying high tone when they are preceded by another prefix or clitic. The Perfective negative is characterised by -dja- (hu-dj-á̩ 'to come') (30a), the Present imperfective negative by -tsú- (consisting of -tsi- and -(h)ú̩-) (30b) and the Future imperfect negative by -na- (also followed by -(h)ú̩-) (30c).

(30) a   k-a̩-dja-lím-a̩
          NEG-SP₁-PFV-cultivate-FV
          'He did not cultivate.'

      b   k-á̩-tsu̩-lim-á̩
          NEG-SP₁-IPFV-cultivate-FV
          'He is not cultivating.'

      c   k-a̩-ná-u̩-lim-á̩
          NEG-SP₁-FUT-15-cultivate-FV
          'He will not cultivate.'

The Perfective relative paradigms, both affirmative and negative, involve a prefix -a- that follows the subject marker, or merges with it (see Patin 2010 for details) – compare (31b) with (31a). Contrary to its non-relative counterpart, it does not involve vowel harmony. The Imperfective relative paradigm, as expected from the non-relative conjugations, is characterised by the presence of a suffix -ǫ́ (31c).

(31) a   rí-réŋg-ę
          SP₁ₚₗ-take-FV
          'we took'

      b   r-a-réŋg-a
          SP₁ₚₗ-rel-take-FV
          '(that) we took'

      c   rí-reŋg-á-ǫ
          SP₁ₚₗ-take-FV-SFX
          '(that) we take'

The negative relative paradigms do not involve the negative marker *ka-*, but *-tsú-* in the Perfective relative negative (e.g., *r-a-tsú-reŋg-a* '(that) we did not take') and *-tso-* in the Present imperfective negative (e.g., *ri-tso-réŋg-a* '(that) we do not take'). The relative conjugations are summarised in Table 20.15.

TABLE 20.15 RELATIVE CONJUGATIONS

Conjugation	SM	TAM		OM	Root	Extensions	FV	Sfx	OM₂
Perfective	ni	a			x	x	a		
Imperfective	ni				x	x	a	ó	
Perfective neg	ni	a	tsu		x	x	a		
Imperfective neg	ni		tso		x	x	a		

Besides these conjugations, Ngazidja also has a certain number of compound tenses in which an auxiliary is prefixed with the subject marker and followed by a form that associates a subject marker to a base (32). This is for instance the case of the Remote past (32a), or the Past (32b) and Future (32c) Conditional tenses.

(32) a   rí-ʃindʼ-i   rí-ǫn-o
          SP₁ₚₗ-be ABLE-FV SP₁ₚₗ-see-FV
          'We were able to see.'

      b   rí-djo-ká   rí-ón-o
          SP₁ₚₗ-FUT-be   SP₁ₚₗ-see-FV
          'We would have seen.'

      c   rí-djo-ká   nga-rí-on-á-ǫ
          SP₁ₚₗ-FUT-be   IMP-SP₁ₚₗ-see-FV-REL
          'We would be seeing.'

4.4 Clause structure

The basic word order of Ngazidja is S V IO DO (33).

(33) a (y)e=mw-aná   (h)a-l-i   ḿ-kaṭe
        AUG₁=1-child   SP₁-eat-FV   3-bread
        'The child ate (some) bread.'

   b ha-nik-á   (y)e=m-limadji   ɲ-úm6a
        SP₁-give-FV   AUG₁=1-farmer   9-house
        'He gave a house to the farmer.'

Adjuncts tend to follow the complement (34).

(34) ha-rem-é   le=pahá   há=m-6uɗa
     SP₁-beat-FV   AUG₅=5.cat   with=9-stick
     'He beat the cat with a stick.'

Any word order is possible, however when an object marker is prefixed to the verb, (e.g., (35)). Nonetheless, orders where two XPs (SOV, VSO, etc.) are topicalised are extremely rare.

(35) ze=m-6eré   tsi-zi-réŋg-e
     AUG₁₀=10-ring   SP₁ₛ₉-OP₁₀-take-FV
     'The rings, I took them.'

Focalisation is expressed through a wide range of strategies (see Patin 2007b for details), whose relative frequency has yet to be clarified. A focused element may, for instance, phrase separately from what follows. Compare (36b), where the tone of the verb stops on its final syllable, with (36a). It may also be introduced by the so-called stabiliser *nda* (36c), as described by Lafon (1985).

(36) a (ŋga-m-andz-o   tʃái)ₚ
        IPFV-SP₁-like-REL   9.tea
        'I like tea.'

   b (ŋgamandzó)ₚ (tʃaí)ₚ
        'I LIKE tea'

   c nd=é=fundi
        stab=aug₁₋1.teacher
        'It's THE TEACHER.' (answering the question 'Who kissed Mary')

4.5 Complex sentences and clause types

Clauses in juxtaposition are separated from each other by an Intonational Phrase boundary, as in (37) (cf. O'Connor & Patin 2015, Patin 2016).

(37) [(ye=m-limadjí)ₚ   (h[w]-an[d]z-a   yem6é)ₚ] ,
     AUG₁=1-farmer   SP₁₅-love-FV   10.mango
     [([y]e=fundi)ₚ   (u-[y]eng-a   ma-ɾ'únɗa)ₚ] ,
     AUG₁=1-teacher   SP₁₅-detest-FV   6-oranges
     'The farmer loves mangoes (and) the teacher detests oranges'

The second clause can be introduced by various discourse connectives such as *na* 'with,' *ʃá* 'but' or *βo* 'while' (38).

(38) [(ha-ka-u-ʃíli-a   mizikí)$_Φ$]$_I$ [(βo   na-ka-u-fany-á   hazi)$_Φ$]$_I$
      SP$_1$-PST-15-listen-FV   9.music   while   SP$_{1SG}$-PST-15-do-FV   9.job
      'He was listening music while I was working.'

Embedded sentences are not separated from the matrix by an Intonational Phrase boundary, as in (39) (cf. Patin 2016). They are introduced by various complementisers, such as grammaticalised forms of *u-káya* 'be' > *uká* 'that,' *yeká* 'if.'

(39) [(ye=mw-aná)$_Φ$ (ha-lemew-á)$_Φ$ (ha[ta]
      AUG$_1$=1-child   SP$_1$-being tired-FV   until
      ha-siuʃi-[a]=[h]ó)$_Φ$ (garí=ni)$_Φ$]$_I$
      PFV.SP$_1$-fell asleep.APPL(-FV)=17.PREP   9.car=in
      'The child was so tired that he fell asleep in the car.'

As any other XP, clauses can be placed in *apposition*. According to O'Connor & Patin (2015), appositives such as the non-restrictive relative in (40) are syntactically and prosodically embedded in the matrix clause.

(40) [(wa-limadjí)$_Φ$ [(w-a-faɲ-á-o   hazí)$_Φ$]$_I$ (wa-léme-w-a)$_Φ$]$_I$
      2-farmer   SP$_2$-REL-do-FV-REL   9.work   SP$_2$-be tired-PASS-FV
      'Farmers, who work, are tired.'

Yes-no questions are expressed through the insertion of a super-high tone on the penult (41aii), unless the final syllable of the utterance is high (cf. Patin 2016). In such cases, the super-high tone emerges on the antepenult (41bii).

(41) a i   ha-won-ó   le=páha
            SP$_1$-see-FV   AUG$_5$=5.cat
            'He saw the cat'
        ii hawonó lé=páha
            'Did he see the cat?'

      b i   ha-won-ó   ye=mleví
            SP$_1$-see-FV   AUG$_1$=1-drunkard
            'He saw the cat'
        ii hawonó yé=mleví
            'Did he see the cat?'

Content questions do not differ from statements as far as prosody is concerned. The wh-words generally appear sentence-finally (42). As in several other Bantu languages, there is in Ngazidja an asymmetry between questions where the object is questioned (42a) and questions where the subject is questioned (42b). In the latter case, the verb appears in its relative form.

(42) a   ye=m-leví   ha-<u>on</u>-ó   ndǒ?
          AUG₁=1-drunkard   SP₁-see-FV   who
          'Who did the drunkard see?'

      b   y-a-<u>on</u>-á   ye=m-leví   ndǒ?
          SP₁-REL-see-FV   AUG₁=1-drunkard   who
          'Who has seen the drunkard?'

ACKNOWLEDGEMENTS

This chapter has benefitted from the comments of Michel Lafon, Kathleen M. O'Connor and Gérard Philippson, and from those of the editors of the volume.

NOTES

1 A syllable associated with an underlying tone is underlined in this chapter.
2 According to folk explanation, this sound symbolism is motivated by the fact that the language of Africans is perceived as a "fall" of sounds that are not interpretable.
3 Ahmed-Chamanga (2011: 113) refers to this extension using the name "*moyen-neutre.*"
4 But not in Mbeni, for instance: compare the following forms *ŋgwaníkaǫ́* 'they give,' *ŋgwareŋgá:ǫ* 'they take,' with (24).
5 The suffix is absent in the singular forms when the base is of Arabic or French origin, e.g., *ŋgamtimízi* 'I finish' (vs. *ŋgaritimizíǫ* 'We finish').
6 The prefix emerges as [hú] before monosyllabic bases, as [ú] before polysyllabic bases and as [hẃ] before vowel-initial bases.

REFERENCES

Ahmed-Chamanga, M. (2002) L'élision en shingazidja (Comores). *Ya mkoɓe* 8–9: 39–50.
Ahmed-Chamanga, M. (2003) L'harmonie vocalique dans le système verbal comorien. *Ya mkoɓe* 10: 25–33.
Ahmed-Chamanga, M. (2010) *Introduction à la grammaire structurale du comorien. Volume 1: le shiNgazidja.* Paris: KomEdit.
Ahmed-Chamanga, M. (2011) La langue comorienne. Unité et diversité. In Laroussi, F. & F. Liénard (eds.), *Plurilinguisme, politique et éducation. Quels éclairages pour Mayotte ?* 19–36. Mont-Saint-Aignan: Presses Universitaires Rouen-Le Havre.
Anscombre, J. C. (2001) A propos des mécanismes sémantiques de formation de certains noms d'agent en français et en espagnol. *Langages* 143: 28–48.
Bostoen, K. (2005) Comparative Notes on Bantu Agent Noun Spirantization. In Bostoen, K. & J. Maniacky (eds.), *Studies in African Comparative Linguistics, with Special Focus on Bantu and Mande*, 231–258. Tervuren: Royal Museum for Central Africa.
Bostoen, K. (2008) Bantu Spirantization: Morphologization, Lexicalization and Historical Classification. *Diachronica* 25(3): 299–356.
Bryan, M. A. (1959) *The Bantu Languages of Africa.* London; New York; Cape Town: Oxford University Press for the International African Institute.
Cassimjee, F. & C. W. Kisseberth (1989) Shingazidja Nominal Accent. *Studies in the Linguistic Sciences* 19(1): 33–61.

Cassimjee, F. & C. W. Kisseberth (1992) Metrical Structure in Shingazidja. *Chicago Linguistic Society* 28: 72–93.

Cassimjee, F. & C. W. Kisseberth (1998) Optimal Domains Theory and Bantu Tonology: A Case Study from Isixhosa and Shingazidja. In Hyman, L. M. & C. W. Kisseberth (eds.), *Theoretical Aspects of Bantu Tone*, 33–132. Stanford: CSLI.

Ferrari-Bridgers, F. (2009) A Quantitative and Qualitative Analysis of the Final Vowels [i] and [a] in Luganda Deverbal Nouns. In Ojo, A. & L. Moshi (eds.), *Selected Proceedings of the 39th Annual Conference on African Linguistics: Linguistic Research and Language in Africa*, 23–31. Somerville: Cascadilla Proceedings Project.

Full, W. (2001) Two Past Tenses in Comorian: Morphological Form and Inherent Meaning. *Afrikanistische Arbeitspapiere* 68: 49–58.

Full, W. (2006) *Dialektologie des Komorischen*. Köln: Rüdiger.

Hyman, L. M. (2001) Privative Tone in Bantu. In Kaji, S. (ed.), *Cross-linguistic Studies of Tonal Phenomena*, 237–257. Tokyo: Institute for the Study of Languages and Cultures.

Hyman, L. M. & F. X. Katamba (1993) The Augment in Luganda: Syntax or Pragmatics? In Mchombo, S. A. (ed.), *Theoretical Aspects of Bantu Grammar*, 209–256. Stanford: CSLI.

Jouannet, F. (1989) *Des tons à l'accent. Essai sur l'accentuation du comorien*. Marseille: Université de Provence Aix-Marseille.

Katamba, F. (2003) Bantu Nominal Morphology. In Nurse, D. & G. Philippson (eds.), *The Bantu Languages*, 103–120. London New York: Routledge.

Kisseberth, C. & D. Odden (2003) Tone. In Nurse, D. & G. Philippson (eds.), *The Bantu Languages*, 59–70. London; New York: Routledge.

Lafon, M. (1982) Brève présentation du système verbal et du fonctionnement d'un auxiliaire en shingazidja. *Oralité Documents* 4: 151–177.

Lafon, M. (1984) Régularité et irrégularité dans le système verbal du shingazidja (grand comorien) : la voyelle finale des thèmes verbaux (avec des références au swahili). *Afrique et Langage* 22: 5–33.

Lafon, M. (1985) Un procédé d'emphase en shingazidja, étude descriptive. *Bulletin des Etudes Africaines de l'INALCO* 9: 3–36.

Lafon, M. (1987) *Le shingazidja (grand-comorien), une langue bantu sous influence arabe*. Paris: INALCO, thèse de doctorat.

Lafon, M. (1990) La négation dans la prédication en shingazidja (Grand-Comorien). *Linguistique africaine* 4: 123–144.

Lafon, M. (1997) L'expression de la qualité en shingazidja: les adjectifs. *Linguistique africaine* hors série: 146–181.

Mohamed-Soyir, K. (2007) L'augmentatif en shiNgazidja. *Ya mkoɓe* 14–15: 49–60.

Mohamed-Soyir, K. (2014) *Le nom en shingazidja (G44a): morphologie, phonologie, sémantique et syntaxe*. Paris: Université Paris 7, thèse de doctorat.

Mugane, J. M. (1997) *Bantu Nominalization Structures*. Tucson: University of Arizona, PhD dissertation.

Nurse, D. (1989) Is Comorian Swahili? Being an Examination of the Diachronic Relationship between Comorian and Coastal Swahili. In Rombi, M.-F. (ed.), *Le swahili et ses limites: ambiguïtés des notions reçues*. Paris: Editions Recherches sur les Civilisations.

O'Connor, K. M. & C. Patin (2015) The Syntax and Prosody of Apposition in Shingazidja. *Phonology* 32(1): 111–145.

Patin, C. (2007a) *La tonologie du shingazidja, langue bantu (G44a) de la Grande Comore: nature, formalisation, interfaces*. Paris: Université Paris 3, thèse de doctorat.

Patin, C. (2007b) Shingazidja Focus Hierarchy. *Nouveaux cahiers de linguistique française* 28: 147–154.

Patin, C. (2008) Focus and Phrasing in Shingazidja. *ZAS Papers in Linguistics* 49: 167–189.

Patin, C. (2009) When Tone Prevents Vowels from Gliding (and When It Does Not). In Blaho, S., C. Constantinescu & B. Le Bruyn (eds.), *Proceedings of ConSOLE XVI*, 135–155.

Patin, C. (2010) The Prosody of Shingazidja Relatives. *ZAS Papers in Linguistics* 53: 187–210.

Patin, C. (2013) /r/ in Washili Shingazidja. In Spreafico, L. & A. Vietti (eds.), *Rhotics: New Data and Perspectives*, 173–190. Bolzano: Bozen-Bolzano University Press.

Patin, C. (2016) Tone and Intonation in Shingazidja. In Downing, L. J. & A. Rialland (eds.), *Intonation in African Tone Languages*, 285–320. Berlin: De Gruyter Mouton.

Philippson, G. (1988) L'accentuation du comorien: essai d'analyse métrique. *Etudes Océan Indien* 9: 35–79.

Philippson, G. (1991) *Tons et accent dans les langues bantu d'afrique orientale: Étude comparative typologique et diachronique*. Paris: Université Paris V, thèse de doctorat.

Philippson, G. (2005) Pitch Accent in Comorian and Proto-Sabaki Tones. In Bostoen, K. & J. Maniacky (eds.), *Studies in African Comparative Linguistics, with Special Focus on Bantu and Mande*, 199–220. Tervuren: Royal Museum for Central Africa.

Picabia, L. (1994) Le sujet locatif en comorien. *Recherches linguistiques de Vincennes* 23: 45–64.

Picabia, L. (1996) La proposition finie en grand comorien: Analyse des unités fonction-nelles. *Linguistique africaine* 16: 69–95.

Picabia, L. (1998) Accord en 'définitude' dans les phrases attributives du grand comorien. In Rouveret, A. (ed.), *"Être" et "avoir" : syntaxe, sémantique, typologie*, 197–225. Saint-Denis: Presses Universitaires de Vincennes.

Rey, V. (1990) *Approche phonologique et expérimentale des faits d'accent d'une langue africaine, le shingazidja (parler de la Grande Comore)*. Marseille: Université Aix-Marseille 1, thèse de doctorat.

Rombi, M.-F. & P. Alexandre (1982) Les parlers comoriens, caractéristiques différen-tielles, position par rapport au swahili. In Rombi, M.-F. (ed.), *Etudes sur le bantu oriental*, 17–39. Paris: Société des Etudes Linguistiques et Anthropologiques.

Schadeberg, T. C. (2003) Derivation. In Nurse, D. & G. Philippson (eds.), *The Bantu Languages*, 71–89. London; New York: Routledge.

Tucker, A. & M. Bryan (1970) Tonal Classification of Nouns in Ngazija. *African Language Studies* 11: 351–383.

CHAPTER 21

VWANJI G66

Helen Eaton

1 INTRODUCTION

The Vwanji (/ˈβʷaːⁿʝi/) language is spoken in the Kipengere Mountains in Makete District, Njombe Region, southwestern Tanzania. It is estimated that there are approximately 28,000 Vwanji people living in the language area (Krüger 2003: 2). Vwanji is a vital language and viewed positively by its speakers (Krüger 2003: 4). It is spoken extensively by people of all ages, many of whom also speak Swahili G42 (Krüger 2003: 6). Guthrie (1971: 51) identifies Vwanji as G66. Nurse (1988: 20) classified Vwanji in the Southern Highlands linguistic subgroup along with Bena G63, Hehe G62, Kinga G65, Kisi G67, Manda N11, Pangwa G64 and Sango G61.

There is little dialectal variation in the Vwanji language. Where differences exist, they mainly concern pronunciation and the presence of loanwords from neighbouring languages. The central dialect spoken in the area around Ikuwo and Magoye is recognised as standard by Vwanji speakers. The consultants who provided data for this sketch are speakers of the central dialect.

The following description of the Vwanji language is primarily based on a corpus of 5,722 lexical items and 22 texts. The text corpus includes transcriptions of oral material and written texts, from both narrative and non-narrative genres.

2 PHONOLOGY

2.1 Vowels

Vwanji has a seven-vowel system with contrastive length. The vowel inventory is given in Table 21.1.

TABLE 21.1 VOWEL INVENTORY

	Front	*Central*	*Back*
High, degree-1	i, iː		u, uː
High, degree-2	ɪ, ɪː		ʊ, ʊː
Mid	e, eː		o, oː
Low		a, aː	

The mid vowels /e, o/ are realised phonetically as [ɛ, ɔ]. Short vowels are unrestricted in their distribution within the word, whereas phonemic long vowels do not occur stem-initially or stem-finally and are only realised as phonetically long in the penultimate syllable of a phonological phrase. When phonemic long vowels occur before the penultimate syllable because of suffixation, they are shortened, but are longer than phonemically short vowels in the same position.

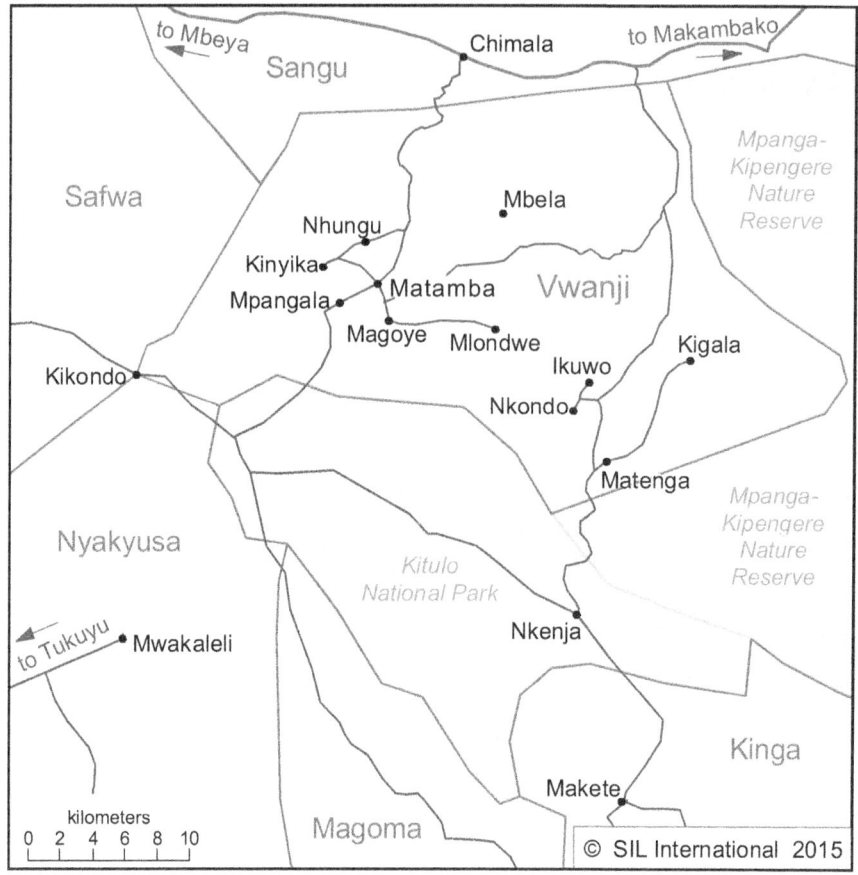

MAP 21.1 THE VWANJI LANGUAGE AREA

(1) a [ku-'to:l-a] 'to take, marry'
  b [ku-'to·l-an-a] 'to be taken, married'
  c ['ku-tol-a] 'to defeat'
  d [ku-'tol-an-a] 'to compete'

Vowels which precede a prenasalised consonant (2a) or follow a labialised (2b) or pal-atalised (2c) consonant are compensatorily lengthened if they occur in the penultimate syllable of the word. The vowel of the class 15 nominal prefix /ku-/ is lengthened before a CV verb stem, such as /-ti/ 'say' (2d).

(2) a [ama'le:ŋga] a-ma-leⁿga 'water'
  b [ku'fʷɪ:ma] ku-fʷɪma 'to hunt'
  c [ɪ'lʲo:si] ɪ-lɪ-osi 'smoke'
  d ['ku:ti] ku-ti 'to say; that'

For both phonemic and compensatory vowel length, the required environment is the penultimate syllable of the phonological phrase. If a noun with a long vowel in the penultimate syllable is modified by a following possessive, the phonological phrase extends to the NP as a whole and therefore the long vowel is no longer in the penultimate syllable and is shortened.

(3) a [aβaˈpaːfi]   a-βa-pa-fi   AUG-2-parent   'parents'
     b [aβaˌpafiˈβaːke]   a-βa-paːfi βa-ake   AUG-2-parent PP$_2$-3SG.POSS   'his parents'

Sequences of directly adjacent surface vowels which do not result in glide formation are possible, but with the exception of certain verb forms (see 18, 2.5.1), they are restricted to six combinations: /ie, ia, io, ue, ua, uo/. The first vowel in each sequence is a degree-1 high vowel and the second is a non-high vowel. Directly adjacent vowels in stems occur only in word-final position.

(4) a ʊ-lʊ-ˈɟisio   'humility'
     b ʊ-lʊ-ˈsulua   'shade, shadow'

A further restriction on vowel distribution in Vwanji is that in noun stems which end in a degree-2 vowel, all other stem vowels must be copies of the stem-final vowel. Augment and prefix vowels are not affected and retain their underlying qualities.

(5) a ʊ-ˈβʊ-dɪlɪ   'cheek'
     b ˈi-N-kʊkʊ   'chickens'

There is no corresponding restriction if /ɪ/ or /ʊ/ is the first vowel in the stem.

(6) a ɪ-ˈkɪ-pɪga   'wing'
     b ˈɪ-N-gʊfu   'strength'

2.2 Consonants

The consonant inventory is given in Table 21.2.

The voiceless plosives /p, t, k/ are aspirated. The four prenasalised consonants /^{m}b/, /^{n}d/, /^{ŋ}g/ and /^{ɲ}ɟ/ and the three aspirated nasals /m^h/, /n^h/ and /ŋ^h/ do not occur in stem-initial position unless created morphophonologically through prefixation (7a-b). Apparent exceptions in nouns can be explained as resulting from double noun class prefixing (7c-d).

(7) a ɪ-ˈ^{m}boːᵐbo   ɪ-N-βoᵐbo   AUG-9-work   'work'
     b ˈɪn^{h}emo   ɪ-N-temo   AUG-9-axe   'axe'
     c ɪlɪˈⁿgoːgo   ɪ-lɪ-N-goːgo   AUG-5-9-peanut   'peanut'
     d aβaˈŋ^{h}ʊːⁿgʊ   a-βa-N-kʊⁿgʊ   AUG-2-9-twin   'twins'

All consonant phonemes except /m^h/ and /n^h/ occur stem-medially. Examples of the aspirated nasal /ŋ^h/ in stem-medial position are rare and tend to occur in verbs or words derived from verbs.

(8) a ku-lɪˈŋ^{h}ania   'to prepare'
     b ʊ-lʊ-lɪˈŋ^{h}anio 'preparations'

TABLE 21.2 CONSONANT INVENTORY

	Bilabial / labio-velar	Labio-dental	Alveolar	Post-alveolar/ palatal	Velar	Glottal
Voiceless plosives	p		t		k	
Voiced plosives	b		d	ɟ	g	
Prenasalised voiced plosives	ᵐb		ⁿd	ᶮɟ	ᵑg	
Voiceless fricatives		f	s			h
Voiced fricatives					ɣ	
Unaspirated nasals	m		n	ɲ	ŋ	
Aspirated nasals	mʰ		nʰ		ŋʰ	
Approximants	β		l	y		

The palatal approximant /y/ occurs rarely. The labio-dental fricative /f/ only occurs before the degree-1 vowels /i/ or /u/ in native Vwanji words.

Consonants may be labialised or palatalised stem-initially (9a-b) or stem-medially in morphologically complex stems (9c-d).

(9) a   kuˈfʷɪːma   ku-fʷɪm-a   15-hunt-FV   'to hunt'
     b   kuˈdʲuːβa   ku-dʲuβ-a   15-pinch-FV   'to pinch'
     c   -ɟoˈβʷaːɣa   -ɟoβ-u-aɣ-a   -speak-PASS-IPFV-FV   'be being spoken'
     d   -pʊlɪkɪˈsʲaːɣa   -pʊlɪk-ɪsɪ-aɣ-a -hear-CAUS-IPFV-FV   'be listening'

All consonants except for the marginal phoneme /y/ may be labialised, whereas only the following palatalised consonants have been attested: /pʲ, tʲ, kʲ, bʲ, dʲ, ᵐbʲ, fʲ, sʲ, ɣʲ, mʲ, mʰʲ, βʲ, lʲ/. Stem-medial labialised consonants may precede any vowel, whereas stem-initial labialised consonants may precede any vowel except /u/. Palatalised consonants may precede any vowel except /i/.

Labialised consonants and palatalised consonants do not occur word-finally in poly-syllabic words, and thus are in complementary distribution with sequences of adjacent vowels, which in stems only occur word-finally and are perceptibly different. When a verb containing a word-final sequence of directly adjacent vowels receives a suffix, the vowel sequence is no longer word-final, and therefore it is realised as a palatalised or labialised consonant, as appropriate.

(10) a   itaˈᵐbʊlua   a-i-taᵐbʊl-u-a   SP₁-NPST-call-PASS-FV   'He is called.'
      b   itaᵐbʊˈlʷaːɣa   a-i-taᵐbʊl-u-aɣ-a   SP₁-NPST-call-PASS-IPFV-FV   'He will be called.'
      c   iloˈᵑgosia   a-i-loᵑgosi-a   SP₁-NPST-lead-FV   'He leads.'
      d   iloᵑgoˈsʲaːɣa   a-i-loᵑgosi-aɣ-a   SP₁-NPST-lead-IPFV-FV   'He will be leading.'

Labialised consonants and palatalised consonants occur word-finally in monosyllables.

(11) a   ɣʷe   2SG.VOC.PRO   'hey, you (SG)'
      b   kʲa   kɪ-a   7.PP-CON   'of'

2.3 Stress

Stress in Vwanji is realised phonetically as high pitch. Stress placement is predictable and depends on mora count. With the exception of words containing only two short vowels, the stress falls on the antepenultimate mora. Thus, the penultimate syllable is stressed if it has a long or lengthened vowel, and the antepenultimate syllable is stressed if the penultimate syllable has a short vowel.

(12) a   ku-'βʊːl-a   'to tell'
      b   ku-'pɪːⁿd-a   'to carry'
      c   ku-'βʲaːl-a   'to plant'
      d   'ku-fik-a   'to arrive'
      e   ku-'pʊlɪk-a   'to hear'

In words containing only two short vowels, the stress falls on the penultimate mora.

(13) a   'lol-a   'look!'
      b   'mʰola   'news'

A syllabic nasal cannot be stressed. Thus, class 1 or 3 nouns containing only two short vowels which occur without an augment have stress on the penultimate mora (14a-b), whereas the same nouns occurring with an augment have stress on the antepenultimate mora (14c-d).

(14) a   ŋ̩'dala   mʊ-dala   1-wife   'wife'
      b   ŋ̩'goma   mʊ-goma   3-beehive   'beehive'
      c   'ʊndala   ʊ-mʊ-dala   AUG-1-wife   'wife'
      d   'ʊŋgoma   ʊ-mʊ-goma   AUG-3-beehive   'beehive'

2.4 Syllable structure

The following syllable types are found in Vwanji: CV, CVV, V, VV, NCV, NCVV, CGV, NCGV and N (where VV represents a long vowel and CG a labialised or palatalised consonant). The syllable types with phonemically long vowels, CVV, NCVV and VV, do not occur word-finally. VV is restricted to word-initial position and only occurs as the result of a morphophonological process. V occurs in all word positions, but its word-medial occurrence is the result of morpheme juxtaposition. CGV occurs word-finally in monosyllables only and NCGV does not occur word-finally. The syllabic nasal, N, only occurs in grammatical morphemes and does not occur word-finally.

Examples of syllable types are given in Table 21.3.

2.5 Morphophonology

2.5.1 Vowels

Asymmetric cross-height vowel harmony is evident in verbal derivation, as shown in Table 21.4. An underlying degree-2 front vowel in a harmonising extension surfaces as a degree-1 vowel after a degree-1 stem vowel, as a degree-2 vowel after a degree-2 stem

TABLE 21.3 SYLLABLE TYPE INVENTORY

Type	Word-initial		Word-medial		Word-final	
CV	'be.da	despise	'bo.ɟo.la	break	'bu.da	kill
CVV	'ɲaː.ɲa	burn	βʊ.'hiː.la	leave behind		
V	a.'ma.ɲo.lo	worms	na.'i.lol.a	he is not looking	ka.'la.si.a	annoy
VV	'iː.sa	he came				
NCV	'ᵐbo.go	buffalo	ge.'ᵑge.da	bite, sting	'βo.ᵐba	work
NCVV	'ᵑgoː.go	peanut	ɣo.'ᵑgoː.la	invite		
CGV	'lʲo.ɟo	anger	βu.'lʲa.dɪ.ka	be depressed	fʲa	PP₈-CON
NCGV	'ᵐbʷi.ɣa	ginger	ɪ.'ᵐbʲa.lo	planting month		
N	n̩.'da.la	wife	'ʊ.n̩.go.ma	beehive		

TABLE 21.4 ASYMMETRIC CROSS-HEIGHT VOWEL HARMONY IN VERBAL DERIVATION

Stem vowel	Suffix vowel (front)	Suffix vowel (back)
i, u	i	u
ɪ, ʊ, a	ɪ (underlying)	ʊ (underlying)
e	e	
o		o

vowel and /a/, and as a mid vowel after a mid stem vowel. In contrast, an underlying degree-2 back vowel in a harmonising extension surfaces as a degree-1 vowel after a degree-1 stem vowel and as a mid vowel only when the stem has a mid-back vowel. Elsewhere it surfaces with its underlying degree-2 quality.

The following extensions harmonise: causative /-ɪsi-/, applicative /-ɪl-/, impositive /-ɪk-/, neuter /-ɪk-/, separative /-ʊl-/ (transitive) and /-ʊk-/ (intransitive).

(15) a kufi'kisia   ku-fik-ɪsi-a   'to cause to arrive' (-fik-   'arrive')
     b ku'fikila   ku-fik-ɪl-a   'to arrive at'
     c ku'suβika   ku-suβ-ɪk-a   'to dip into'
     d ku'diⁿdika   ku-diⁿd-ɪk-a   'to be closed, tied' (-diⁿd-   'close, tie')
     e ku'diⁿdula   ku-diⁿd-ʊl-a   'to open, untie'
     f ku'diⁿduka   ku-diⁿd-ʊk-a   'to be opened, untied'

Vowel copying, in which a prefix vowel is copied into the stem, occurs with the stem /-ᵑge/ 'other' and the /-lɪ/ form of 'to be.' /i/ and /ɪ/ trigger the copying for /-ᵑge/ (16b-c) and /i/ triggers the copying for /-lɪ/ (16e).

(16) a a'βaː.ᵑge   a-βa-ᵑge   AUG-EP₂-other   'other'
     b ɪ'kɪː.ᵑgɪ   ɪ-kɪ-ᵑge   AUG-EP₇-other   'other'
     c i'fiː.ᵑgi   i-fi-ᵑge   AUG-EP₈-other   'other'
     d 'alɪ   a-lɪ   SP₁-be   'he is'
     e 'fili   fi-lɪ   SP₈-be   'they are'

When vowels from different morphemes are juxtaposed, the results vary according to the vowels involved, the grammatical context and the position in the word. For example, vowel elision can occur, resulting in the lengthening of the remaining vowel (17a), which is then shortened when occurring before the penultimate syllable (17b). Vowel elision also occurs when a clitic attaches to vowel-initial word (17c). In first and second person accompanitive pronouns, the remaining vowel is raised from /ʊ/ to /u/ (17d). A sequence of adjacent vowels occurs in word-final position (17e), and glide formation can occur when the first vowel is a high vowel (17f).

(17) a   'iːsa   a-a-is-a SP_1-PST2-come-FV   'He came.'
     b   'βilʊta   βa-i-lʊt-a SP_2-NPST-go-FV   'They go.'
     c   'nıⁿgʊfu   na=ı-N-gʊfu   COM=AUG-9-strength   'with strength'
     d   'nuːne   na='ʊne   COM=1SG.CONTR.PRO   'with me'
     e   ata'ᵐbʊlue   a-taᵐbʊl-u-e SP_1-call-PASS-FV   'He should be called.'
     f   βaka'lʲaːɣa βa-ka-lɪ-aɣ-a SP_2-NARR-OP_5-see-FV 'They saw it.'

Negative /na-/ (18a), Uncertain Near Future /kʲa-/ (18b), the locatives /pʷe-/, /kʷe-/ and /mʷe-/ (18c) and Non-Past /i-/ when part of the Habitual (18d) may create sequences of adjacent vowels when the subject concord is vowel-initial.

(18) a   naʊka'ɟoβaɣa   na-ʊ-ka-ɟoβ-aɣ-a   NEG-SP_{2sg}-NARR-say-IPFV-FV
                                                               'You were not saying.'
     b   kʲai'βoːᵐba   kʲa-a-i-βoᵐb-a   FUT1-SP_1-NPST-do-FV   'He will do.'
     c   pʷea'lʲaːle   pʷe-a-lʲa-β-ile LOC_{16}-SP_1-PST4-be-PFV 'He was there.'
     d   a'ilola   a-a-a-i-lol-a SP_1-HAB-SP_1-NPST-see-FV 'He sees.'

Vowel coalescence occurs when the /ʊ/ or /a/ of a subject concord precedes Non-Past /i-/. /ʊ/ and /i/ coalesce to /u/ (19a); /a/ and /i/ coalesce to /ɪ/ (19b).

(19) a   'tuɣua   tʊ-i-ɣu-a SP_{1PL}-NPST-fall-FV
                                                               'We are falling.'
     b   βıkʊkʊ'loːⁿda   βa i ku kʊ-loⁿd-a SP_2-NPST-15-OP_{2sg}-look.foɪ-FV
                                                               'They are looking for you.'

This coalescence of /a/ and /i/ only occurs when /i/ precedes the class 15 nominal prefix inserted before an object and not before a vowel-initial stem. The vowel of the class 15 nominal prefix /ku-/ is realised as /ʊ/ when it is inserted before an object as in (19b) and also when the noun follows certain grammatical words, such as Connective /a/ (20a) and /'kısila/ 'without' (20b), provided the verb does not contain a reflexive morpheme (20c) and the root is not of the shape -C- (20d).

(20) a   ɪja'kʊlʊta   ı-ɟı-a=ku-lʊt-a   AUG-PP_9-CON=15-go-FV   'of going'
     b   'kısila kʊ'yeːⁿda kısila ku-yeⁿd-a   without 15-walk-FV   'without walking'
     c   ɪja'kusiɣa   ı-ɟı-a=ku-i-siɣ-a AUG-PP_9-CON=15-REFL-protect-FV
                                                               'of protecting oneself'
     d   'kısila 'kuːβa   kısila ku-β-a   without 15-be-FV   'without being'

Kinga G65, which neighbours Vwanji, has an underlying degree-2 vowel in the class 15 nominal prefix /kʊ-/ which is realised as [ku-] when the infinitive is the complement of a verb of motion. Like Vwanji, Kinga inserts /kʊ-/ before an object in certain tenses (Schadeberg 1973: 29).

2.5.2 Consonants

The prefix /N-/ (class 9 or 10 nominal prefix or first person singular object prefix) before a stem-initial voiced plosive, fricative or approximant results in a voiced plosive with homorganic prenasalisation. The fricative /ɣ/ is hardened to /g/ and the approximants /β/, /l/ and /y/ to /b/, /d/ and /ɟ/ respectively. /N-/ before a vowel is realised as /ɲ/.

(21) a ˈɪᵐbogo   ɪ-N-bogo   'buffalo'
      b ˈɪⁿdama   ɪ-N-dama   'heifer'
      c ˈɪⁿɟosi   ɪ-N-ɟosi   'dream'
      d ɪˈⁿgadʊle   ɪ-N-gadʊle   'tall drum'
      e ɪˈⁿgoːle   ɪ-N-ɣoːle   'tick'
      f ɪᵐbuˈluɣutu   ɪ-N-βuluɣutu   'ear'
      g iˈⁿdaɣɪlo   i-N-laɣɪlo   'laws'
      h ɪˈⁿɟayɪlo   ɪ-N-yayɪlo   'hug'
      i ɪˈɲiːᵐbo   ɪ-N-ɪᵐbo   'song'

/N-/ prefixed to a voiceless plosive creates an aspirated nasal.

(22) a imʰʊˈlɪsio   i-N-pʊlɪsio   'announcements'
      b ɪˈnʰuːɟe   ɪ-N-tuːɟe   'owl'
      c iˈŋʰeːto   i-N-keːto   'razors'

The aspirated velar nasal /ŋʰ/ is also created when the reciprocal morpheme /-an/ is suffixed to a stem-final /k/.

(23) kupʊˈlɪŋʰana   ku-pʊlɪk-an-a   15-hear-RECP-FV   'to be reconciled'

Before voiceless fricatives (24a) and unaspirated nasals (24b), the prefix /N-/ is elided in nouns, but in verbs it is realised as either a slight lengthening of the preceding vowel, or as prenasalisation on the following fricative (24c) or a slight lengthening of the following nasal (24d).

(24) a ɪˈhoːβe   ɪ-N-hoːβe   'crow'
      b ˈimuli   i-N-muli   'torches'
      c [aˈⁿsʊmile] ~ [aˈsʊmile]   a-N-sʊːm-ile   SP₁-OP₁ₛₒ-ask-PFV   'He has asked me.'
      d [aˈmˑaɲile] ~ [aˈmaɲile]   a-N-maɲ-ile   SP₁-OP₁ₛₒ-know-PFV 'He has known me.'

The prefix /mʊ-/ (class 1 or 3 nominal prefix or class 1 object prefix) before a plosive, a voiceless fricative (with the exception of /h/) or the approximant /β/ results in the elision of the prefix vowel and the creation of a syllabic nasal.

(25) a ˈʊmpiki   ʊ-mʊ-piki   'tree'
      b ʊɲˈtaᵐbʊle   ʊ-mʊ-taᵐbʊle   'famous person'
      c ˈʊŋkilu   ʊ-mʊ-kilu   'toilet'
      d ʊmˈbaːⁿga   ʊ-mʊ-baⁿga   'termite (type)'

e	'ʊɲdala	ʊ-mʊ-dala	'wife'
f	ʊɲɟe'le:la	ʊ-mʊ-ɟele:la	'young person'
g	'ʊŋgoma	ʊ-mʊ-goma	'beehive'
h	'ʊm̩fue	ʊ-mʊ-fue	'dead person'
i	'ʊn̩situ	ʊ-mʊ-situ	'forest'
j	ʊm̩'bo:ᵐbi	ʊ-mʊ-βoᵐbi	'worker'

Note that in addition, /β/ hardens to /b/, as it does after /N-/ in (21f).

Before a nasal consonant, the vowel of the prefix /mʊ-/ is elided and the stem-initial and prefix nasals become a geminate nasal (26a-d). /mʊ-/ before /h/ forms an aspirated nasal (26e).

(26)	a	'ʊm:ela	ʊ-mʊ-mela		'plant'
	b	'ʊn:ine	ʊ-mʊ-nine		'his companion'
	c	a'ɲ:aɲile	a-mʊ-ɲa:ɲ-ile	SP₁-OP₁-burn-PFV	'He has burned him.'
	d	a'ŋ:eɲile	a-mʊ-ŋeɲ-ile	SP₁-OP₁-bite-PFV	'He has bitten him.'
	e	ʊ'mʰɪ:nɟa	ʊ-mʊ-hɪnɟa		'girl'

Perfective /-ile/ undergoes imbrication in verb stems suffixed with extensions (27a-c) and causes irregular changes in /-β-/ 'to be' (27d-f).

(27)	a	nɪlɪ'mi:le	nɪ-lɪm-ɪl-ile	SP₁ₛ𝒢-farm-APPL-PFV	'I have farmed for'
	b	siβo'ni:ke	si-βon-ɪk-ile	SP₁₀-see-NEUT-PFV	'They are visible.'
	c	tʊli'ɣi:ne	tʊ-liɣ-an-ile	SP₁ₚₗ-curse-RECP-PFV	'We have cursed each other.'
	d	a'lʲa:le	a-lʲa-β-ile	SP₁-PST4-be-PFV	'he was'
	e	a'ka:le	a-ka-β-ile	SP₁-PST3-be-PFV	'he was'
	f	a'βe:le	a-β-ile	SP₁-be-PFV	'he has been'

3 NOUNS AND NOUN PHRASES

3.1 Noun classes and agreement

Nominal affixes and agreement markers are shown in Table 21.5.

TABLE 21.5 NOMINAL AFFIXES AND AGREEMENT MARKERS

Class	Example	Gloss	AUG	NP	EP	PP	SP			OP		
1	ʊmʊ'yo:si	man	ʊ	mʊ	ɟʊ	yʊ	nɪ	(ɣ)ʊ	a	N	kʊ	mʊ
1a	ʊɲi'ne:su	our mother	ʊ	(mʊ)	ɟʊ	yʊ	nɪ	(ɣ)ʊ	a	N	kʊ	mʊ
2	aβa'yo:si	men	a	βa	βa	βa	tʊ	mʊ	βa	tʊ	βa	βa
3	ʊmʊ'lʲa:ⁿgo	door	ʊ	mʊ	yʊ	yʊ	yʊ			yʊ		
4	imi'yʊ:ⁿda	big fields	i	mi	ɣi	ɣi	ɣi			ɣi		
5	ɪ'lɪβue	stone	ɪ	lɪ	lɪ	lɪ	lɪ			lɪ		
6	a'maβue	stones	a	ma	ɣa	ɣa	ɣa			ɣa		
7	ɪkɪ'te:ⁿgo	chair	ɪ	kɪ	kɪ	kɪ	kɪ			kɪ		
8	ifi'te:ⁿgo	chairs	i	fi	fi	fi	fi			fi		
9	ɪ'nʰoⁿdue	star	ɪ	N	ɟɪ	ɟɪ	ɟɪ			ɟɪ		
10	i'nʰoⁿdue	stars	i	N	(s)i	si	si			si		
11	ʊ'lʊmili	tongue	ʊ	lʊ	lʊ	lʊ	lʊ			lʊ		

(Continued)

TABLE 21.5 (Continued)

Class	Example	Gloss	AUG	NP	EP	PP	SP	OP
12	aˈkapene	small goat	a	ka	ka	ka	ka	ka
13	ʊˈtʊpene	small goats	ʊ	tʊ	tʊ	tʊ	tʊ	tʊ
14	ʊβʊˈhaːsi	inheritance	ʊ	βʊ	βʊ	βʊ	βʊ	βʊ
15	kuˈyeːⁿda	to walk		ku/pi	kʊ	kʊ	kʊ	kʊ
16	paˈtiːtu	black place		pa	pa	pa	pa	pa
17	kʊˈβalafu	white place		kʊ	kʊ	kʊ	kʊ	kʊ
18	mʊˈtiːtu	black place		mʊ	mʊ	mʊ	mʊ	mʊ
20	ʊˈyʊtemo	big axe	ʊ	yʊ	yʊ	yʊ	yʊ	yʊ

Singular/plural pairings (genders) are given in Table 21.6.

TABLE 21.6

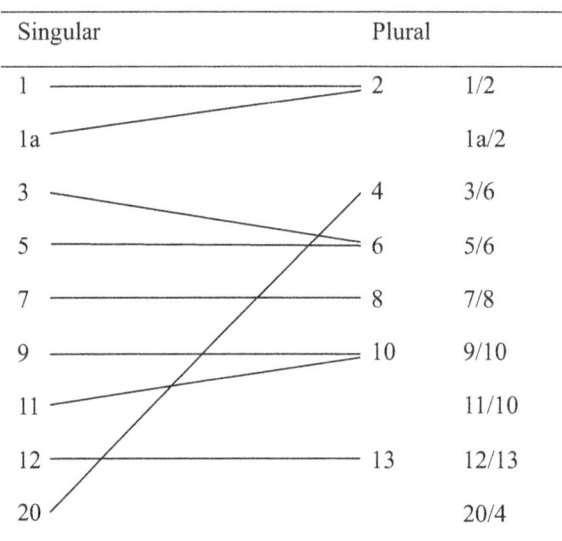

Singular		Plural
1 ————————— 2		1/2
1a ————		1a/2
3 ———— 4		3/6
5 ————————— 6		5/6
7 ————————— 8		7/8
9 ————————— 10		9/10
11 ————		11/10
12 ————————— 13		12/13
20		20/4

The pairings 7/8 (28a-b) and 12/13 (28c-d) are used for diminutives.

(28) a   ɪ-ˈkɪ-pene   'small goat'
      b   i-ˈfi-pene   'small goats'
      c   a-ˈka-pene   'small, good goat'
      d   ʊ-ˈtʊ-pene   'small, good goats'

They differ semantically in that 7/8 is neutral, whereas 12/13 is approbative.

   Similarly, the pairings 5/6 (29a-b) and 20/4 (29c-d) are used for augmentatives, but 5/6 is neutral, whereas 20/4 is pejorative.

(29) a   ɪ-ˈlɪ-temo   'big axe'
      b   a-ˈma-temo   'big axes'
      c   ʊ-ˈyʊ-temo   'big, bad axe'
      d   i-ˈmi-temo   'big, bad axes'

The nominal prefix (NP) attaches to nouns and adjectives (30a) in all classes excepting 1a in which the prefix only attaches to the adjective and not to the noun (30b), and also to /-iⁿga/ 'many' (30c).

(30) a ʊmʊˈɣoːsi ˈʊɳtali ʊ-mʊ-ɣoːsi ʊ-mʊ-tali AUG-1-man AUG-NP₁-tall   'tall man'
     b ʊˈlʊːᵐbʊ ˈʊɳtali   ʊ-lʊᵐbʊ ʊ-mʊ-tali   AUG-1a.sibling AUG-NP₁ₐ-tall 'tall sibling'
     c aβaˈɣoːsi ˈβiːⁿga   a-βa-ɣoːsi βa-iⁿga   AUG-2-man NP₂-many   'many men'

The class 15 nominal prefix has the alternative form /pi-/. (The origin of this morpheme is unclear, but it could be analysed as a class 19 noun prefix or as originating from a class 16 locative subject prefix plus Non-Past /i-/.) The infinitive form of vowel-initial verb stems has two nominal prefixes. Either /ku-/ is doubled (31a) or /pi-/ is followed by /ku-/ (31b).

(31) a kuˈkʷɪːma ku-ku-ɪm-a 15-15-stand-FV 'to stand'
     b piˈkʷɪːma pi-ku-ɪm-a 15-15-stand-FV 'to stand'

The numeral prefix (EP) attaches to cardinal numbers (32a), demonstratives (32b), the relative pronoun stem /-no/ (32c) and /-ⁿge/ 'other' (32d). Class 10 is exceptional in that the numeral prefix is /i-/ for cardinal numbers (32e), but /si-/ for other words in the category (32f).

(32) a kɪˈmoːⁿga kɪ-moⁿga EP₇-one   'one'
     b ˈkɪla   kɪ-la   EP₇-DIST.DEM   'that'
     c ˈkɪno   kɪ-no   EP₇-REL.PRO   'which'
     d ɪˈkɪːⁿgɪ   ɪ-kɪ-ⁿge AUG-EP₇-other 'other'
     e ˈiβɪlɪ   i-βɪlɪ   EP₁₀-two   'two'
     f ˈsila   si-la   EP₁₀-DIST.DEM 'those'

The pronominal prefix (PP) attaches to Connective /a/ (33a), possessives (33b) and /-oni/ 'all' (33c) (which also has the first person plural form /ˈtʷeːni/ and the second person plural form /ˈmʷeːni/).

(33) a sa   si-a   PP₁₀-CON   'of'
     b ˈsaːⁿgo si-aⁿgo PP₁₀-1SG.POSS 'my'
     c ˈsoːni   si-oni   PP₁₀-all   'all'

The numeral prefix and pronominal prefix differ only in classes 1 and (partly) 10.

The subject prefix for second person singular is realised as /ɣʊ-/ before a vowel and /ʊ-/ before a consonant.

3.2 Noun derivation

Vwanji derives nouns from verbs by means of the agentive nominalising suffix /-i/ (34a) or /-a/ after causative (34b) or passive (34c) extensions and the locative/instrumental nominalising suffix /-o/ (34d).

(34) a ʊmˈfʷɪːmi   ʊ-mʊ-fʷɪm-i   AUG-1-hunt-AG   'hunter'
     b ʊmʊɣeˈⁿdesia   ʊ-mʊ-ɣeⁿd-ɪsi-a AUG-1-walk-CAUS-AG 'driver'
     c ʊɳˈkʊⁿgua   ʊ-mʊ-kʊⁿg-u-a   AUG-1-tie-PASS-AG   'prisoner'
     b ʊlʊˈhoːᵐbo   ʊ-lʊ-hoᵐb-o   AUG-11-pay-INS   'payment'

Adjective stem nominalisations, without a change of final vowel, are found in class 14.

(35) ʊˈβʊkome   ʊ-βʊ-kome   AUG-14-large   'largeness'

3.3 The augment

Nouns in classes 1–14 and 20 have an augment unless they occur in certain grammatical contexts, such as following Connective /a/ (36), as a vocative (37) or as a predicative complement (38).

(36) ˈɪmʰene ɲaˈmʰɪːɲɟa
     ɪ-N-pene   ɪ-ɟɪ-a=mʊ-hɪɲɟa
     AUG-9-goat   AUG-PP₉-CON=1-girl
     'the girl's goat'

(37) ɣʷe mʊˈβʷaːɲɟi
     ɣʷe   mʊ-βʷaɲɟi
     2SG.VOC.PRO   1-Vwanji
     'Hey, you Vwanji!'

(38) ˈʊmue βaˈkedʊsi
     ʊmue   βa-kedʊsi
     2PL.CONTR.PRO   2-hypocrite
     'You are hypocrites.'

The infinitive class 15 and the locative classes 16–18 do not have an augment in any context, with the exception of the class 16 noun /aˈpoːnu/ 'place,' which is the only noun not derived from an adjective stem in the locative classes. Note that as well as nouns, adjectives (30a, 30b, 3.1), Connective /a/ (20a, 20c, 2.5.1) and possessive pronouns (49, 3.4) may have an augment.

Some degree of free variation can be seen in the occurrence of the augment in grammatical contexts which allow it. A tendency has been noted for Vwanji writers to omit the augment on occasion in texts, but then add it during editing. Fischer (2011) discusses some environments in which the notion of specificity was helpful in explaining the distribution of the augment.

3.4 Pronouns and nominal modifiers

Vwanji has six sets of pronouns, as shown in Table 21.7.

Note that the first and second person accompanitive pronouns have the degree-1 vowel /u/ where /ʊ/ would be expected. Also, in emphatic pronouns which end in a sequence of adjacent vowels, the first of these vowels is always a degree-1 vowel, even if the vowel of the agreement prefix is degree-2 (e.g., class 3 /ˈɣʊɣuo/ not */ˈɣʊɣʊo/).

Contrastive Pronouns are limited to referents showing person and number, with the third person singular and plural forms corresponding to classes 1 and 2 respectively.

TABLE 21.7 PRONOUNS

	Contrastive	Accompanitive	Additive	Emphatic	Exclusive	Relative
1SG	'one	'nu:ne	na'ɟʊ:ne	'ɟʊ:ne	ne 'mʷe:ne	
2SG	'ʊβe	'nu:βe	na'ɟʊ:βe	'ɟʊ:βe	ɣʷe 'mʷe:ne	
3SG/1	ʊ'mʷe:ne	nʊ'mʷe:ne	'ɣʷo:pe	'ɟʊɟuo	'mʷe:ne	'ɟʊno
1PL	'ʊsue	'nusue	na'ɟʊsue	'ɟʊsue	tʷe 'βe:ne	
2PL	'ʊmue	'numue	na'ɟʊmue	'ɟʊmue	mʷe 'βe:ne	
3PL/2	a'βe:ne	na'βe:ne	'βo:pe	'βaβuo	'βe:ne	'βano
3		'naɣʷo	'ɣʷo:pe	'ɣʊɣuo	'ɣʷe:ne	'ɣʊno
4		'naɣʲo	'ɣʲo:pe	'ɣiɣio	'ɣʲe:ne	'ɣino
5		'nalʲo	'lʲo:pe	'lilio	'lʲe:ne	'lɪno
6		'naɣo	'ɣo:pe	'ɣaɣuo	'ɣe:ne	'ɣano
7		'nakʲo	'kʲo:pe	'kɪkio	'kʲe:ne	'kɪno
8		'nafʲo	'fʲo:pe	'fifio	'fʲe:ne	'fino
9		'naɟo	'ɟo:pe	'ɟɥio	'ɟe:ne	'ɟɪno
10		'naso	'so:pe	'sisio	'se:ne	'sino
11		'nalʷo	'lʷo:pe	'lʊluo	'lʷe:ne	'lʊno
12		'nako	'ko:pe	'kakuo	'ke:ne	'kano
13		'natʷo	'tʷo:pe	'tʊtuo	'tʷe:ne	'tʊno
14		'naβʷo	'βʷo:pe	'βʊβuo	'βʷe:ne	'βʊno
15		'nakʷo	'kʷo:pe	'kʊkuo	'kʷe:ne	'kʊno
16		'napo	'po:pe	'papuo	'pe:ne	'pano
17		'nakʷo	'kʷo:pe	'kʊkuo	'kʷe:ne	'kʊno
18		'namʷo	'mʷo:pe	'mʊmuo	'mʷe:ne	'muno
20		'naɣʷo	'ɣʷo:pe	'ɣʊɣuo	'ɣʷe:ne	'ɣʊno

Accompanitive Pronouns contain the Comitative Marker /na/, as do the first and second person forms of the Additive Pronoun. The functions of these three pronoun types are illustrated below.

(39) 'one nanɪ 'βoᵐbile kɪ'mo:ⁿga
      one   na-nɪ-βoᵐb-ile   kɪ-moⁿga
      1SG.CONTR.PRO NEG-SP_{1SG}-do-PFV EP_7-one
      'I didn't do anything.' (Implied: someone else did)

(40) 'nikale   'nu:βe
      nɪ-ikal-e   na=ʊβe
      SP_{1SG}-live-FV   COM=2SG.CONTR.PRO
      'I should live with you.'

(41) βilikʊtʊɟe'ⁿgelaɣa na'ɟʊsue i'ɲu:ᵐba
      βa-ili:-ku-tʊ-ɟeⁿg-ɪl-aɣ-a   naɟʊsue   i-N-ɟuᵐba
      SP_2-CERT.FUT2-15-OP_{1PL}-build-APPL-IPFV-FV 1PL.ADD.PRO AUG-10-house
      'They will build houses for us as well.'

Emphatic Pronouns stress that the referent is indeed meant and not another, whereas Exclusive Pronouns stress that the referent alone (without help or outside cause) is meant.

(42) a'fikile 'ɟʊɟuo
      a-fik-ile   ɟʊɟuo
      SP₁-arrive-PFV   1.EMPH.PRO
      'He has arrived himself.' (Implied: a representative had been expected)

(43) a'fikile 'mʷeːne
      a-fik-ile   mʷene
      SP₁-arrive-PFV   1.EXCL.PRO
      'He has arrived by himself.' (Implied: on his own, without help)

Note that the Emphatic Pronoun is formally similar to the Referential Demonstrative (see Table 21.8).

In addition to the six pronoun sets given in Table 21.7, Vwanji has two Vocative Pronouns, singular /ɣʷe/ and plural /mʷe/, both meaning 'hey, you!'.

Vwanji demonstratives have a three-way category distinction: Proximal, Referential and Distal. Within each category a basic and reduplicated form is possible for all classes, as shown in Table 21.8.

TABLE 21.8 DEMONSTRATIVES

	Proximal	Referential	Distal	Reduplicated Proximal	Reduplicated Referential	Reduplicated Distal
1	'ʊɟʊ	'ʊɟuo	'ɟʊla	ɟʊ'ɟʊːɟʊ	ɟʊ'ɟʊɟuo	ɟʊ'laɟʊla
2	'aβa	'aβuo	'βala	βa'βaːβa	βa'βaβuo	βa'laβala
3	'ʊɣʊ	'ʊɣuo	'ɣʊla	ɣʊ'ɣʊːɣʊ	ɣʊ'ɣʊɣuo	ɣʊ'laɣʊla
4	'iɣi	'iɣio	'ɣila	ɣi'ɣiːɣi	ɣi'ɣiɣio	ɣi'laɣila
5	'ɪlɪ	'ɪlio	'lɪla	lɪ'lɪːlɪ	lɪ'lɪlio	lɪ'lalɪla
6	'aɣa	'aɣuo	'ɣala	ɣa'ɣaːɣa	ɣa'ɣaɣuo	ɣa'laɣala
7	'ɪkɪ	'ɪkio	'kɪla	kɪ'kɪːkɪ	kɪ'kɪkio	kɪ'lakɪla
8	'ifi	'ifio	'fila	fi'fiːfi	fi'fifio	fi'lafila
9	'ɲɪ	'ɲio	'ɲila	ɲɪ'ɲiːɲɪ	ɲɪ'ɲɲio	ɲɪ'laɲila
10	'isi	'isio	'sila	si'siːsi	si'sisio	si'lasila
11	'ʊlʊ	'ʊluo	'lʊla	lʊ'lʊːlʊ	lʊ'lʊluo	lʊ'lalʊla
12	'aka	'akuo	'kala	ka'kaːka	ka'kakuo	ka'lakala
13	'ʊtʊ	'ʊtuo	'tʊla	tʊ'tʊːtʊ	tʊ'tʊtuo	tʊ'latʊla
14	'ʊβʊ	'ʊβuo	'βʊla	βʊ'βʊːβʊ	βʊ'βʊβuo	βʊ'laβʊla
15	'ʊkʊ	'ʊkuo	'kʊla	kʊ'kʊːkʊ	kʊ'kʊkuo	kʊ'lakʊla
16	'apa	'apuo	'pala	ba'haːpa	pa'papuo	pa'lapala
17	'kʊno	'ʊkuo	'kʊla	kʊ'nokʊno	kʊ'kʊkuo	kʊ'lakʊla
18	'muno	'ʊmuo	'mʊla	mu'nomuno	mʊ'mʊmuo	mu'lamula
20	'ʊɣʊ	'ʊɣuo	'ɣʊla	ɣʊ'ɣʊːɣʊ	ɣʊ'ɣʊɣuo	ɣʊ'laɣʊla

In their spatio-deictic functions, the referents of Proximal Demonstratives are near both the speaker and the addressee, whereas those of Referential Demonstratives are near to the addressee only and those of Distal Demonstratives are far from both the speaker and the addressee. Reduplicated demonstratives occur less frequently than the basic forms and convey a more specific meaning.

(44) a ʼapa   'here'
     b baˈhaːpa   'right here, this very place'

Note that the Proximal Demonstratives (basic and reduplicated) for the locative classes 17 and 18 and the reduplicated Proximal Demonstrative for class 16, as in (44b), do not follow the regular pattern.

The demonstratives also have discourse-level functions. Proximal demonstratives are used to refer to participants in backgrounded sections (45). Distal Demonstratives are used to refer to participants which have agent-like roles in the discourse (46), whereas Referential Demonstratives refer to participants with more passive roles (47).

(45) ʊˈkaŋkeβe ˈʊʝʊ aˈlʲaːle nʊˈlʊhala ˈlʷiːⁿga
     ʊ-kaŋkeβe   ʊʝʊ   a-lʲa-β-ile   na=ʊ-lʊ-hala   lʊ-iⁿga
     AUG-1a.Kankeve EP$_1$.PROX.DEM SP$_1$-PST4-be-PFV COM=AUG-11-cleverness NP$_{11}$-many
     'This Kankeve was very clever.'

(46) ˈʊŋdala ˈʝʊla akaˈkɪᵐbɪla
     ʊ-mʊ-dala   ʝʊ-la   a-ka-kɪᵐbɪl-a
     AUG-1-WIFE   EP$_1$-DIST.DEM SP$_1$-NARR-run.away-FV
     'That woman ran away.'

(47) ˈʊʝuo ihɪˈɣʷaːɣa
     ʊʝuo   a-i-hɪɣ-u-aɣ-a
     EP$_1$.REF.DEM SP$_1$-NPST-judge-PASS-IPFV-FV
     'That one will be judged.'

A similar distribution has been noted for the neighbouring language Bena G63 (Broomhall 2011, Nicolle 2015: 34–35).

Adjectives in Vwanji take the nominal prefix (see Table 21.5) and also the augment, under the same grammatical conditions as nouns (see Section 3.3).

Possessives are given in Table 21.9.

TABLE 21.9 POSSESSIVES

	S$_G$	P$_L$
1	-aⁿgo	-itu
2	-ako	-inu
3	-ake	-aβe

Possessives take the pronominal prefix and also an augment if used nominally in a context in which an augment occurs.

(48) ɪˈɲuːᵐba ˈɟaːŋgo ˈɟɪlɪ ˈpala
      ɪ-N-ɟuᵐba   ɟɪ-aŋgo   ɟɪ-lɪ   pa-la
      AUG-9-house PP_9-1SG.POSS SP_9-be EP_{16}-DIST.DEM
      'My house is over there.'

(49) ɪˈɟaːŋgo ˈɟɪla
      ɪ-ɟɪ-aŋgo   ɟɪ-la
      AUG-PP_9-1SG.POSS EP_9-DIST.DEM
      'Mine is that one.'

Kinship terms may include a fused possessive.

(50) a mʷaˈnaːŋgo   'my child'
      b sɪⁿgɪˈɟiːtu   'our paternal aunt'

3.5 Comitative, Connective and locative clitics

Comitative /na/, the Connectives /a/ and /ɲa/ and the locative clitics /pa/, /kʊ/ and /mʊ/ are all analysed here as clitics. Stress placement shows that they form a phonological phrase with the word to which they attach, as in (92, 4.3), where the clitic bears the stress and the word to which it attaches has no stressed mora. Grammatically, the clitics connect phrases and not just single words, as in (61).

The Comitative Marker /na/ has both conjunctive and prepositional uses. In its conjunctive function it may coordinate NPs (51) or clauses (52).

(51) aβaˈsoleka naβaɟeˈleːla
      a-βa-soleka   na=a-βa-ɟeleːla
      AUG-2-boy   COM=AUG-2-girl
      'boys and girls'

(52) βiˈɣaːla nakuˈkosia ˈisoni
      βa-i-ɣaːl-a   na=ku-kosi-a   i-N-soni
      SP_2-NPST-get.drunk-FV   COM=15-cause-fv   AUG-10-shame
      'They get drunk and cause shame.'

In its prepositional function /na/ expresses accompaniment and when used together with the copula this conveys possession.

(53) iˈkukala naβaˈpaːfi ˈβaːke
      a-i-ku-ikal-a   na=a-βa-paːfi   βa-ake
      SP_1-NPST-15-live-FV   COM=AUG-2-parent PP_2-3SG.POSS
      'He lives with his parents.'

(54) ˈβalɪ nɪˈɲuːᵐba
      βa-lɪ   na=ɪ-N-ɟuᵐba
      SP_2-be   COM=AUG-9-house
      'They have a house.'

Connective /a/ takes a pronominal prefix in agreement with the head noun and expresses a relation with the following nominal, such as possession (55). It is also used in locative (56) and nominal expressions (57).

(55) 'ɲuːᵐba ɪjaˈmʰɪːⁿja
      ɪ-N-jumba   ɪ-jɪ-a=mʊ-hɪⁿja
      AUG-9-house   AUG-PP₉-CON=1-girl
      'the girl's house'

(56) βaˈkalʊta ˈkʷamʰɪɣi
      βa-ka-lʊt-a   kʊ-a=mʊ-hɪɣ-i
      SP₂-NARR-go-FV   PP₁₅-CON=1-judge-AG
      'They went to the judge.'

(57) ɪkʲaˈkʊlia
      ɪ-kɪ-a=ku-li-a
      AUG-PP₇-CON=15-eat-FV
      'food'

A second Connective clitic in Vwanji is /ɲa/, which takes a nominal prefix (and an augment if appropriate in the grammatical context) in agreement with a noun or is used pronominally. Connective /ɲa/ occurs less frequently than /a/ and functions mainly to convey possession only. It is typically, but not exclusively, used with human head nouns.

(58) aˈβaːnʰu aβaˈɲanʰamu
      a-βa-nʰu   a-βa-ɲa=N-tamu
      AUG-2-person   AUG-NP₂-CON=10-illness
      'people with illnesses'

(59) ʊˈɲaⁿgʊfu
      ʊ-ɲa=N-gʊfu
      AUG-1A.CON=10-strength
      'the strong one (person)'

(60) ʊβʊɲaˈmʷoːjo
      ʊ-βʊ-ɲa=mʊ-ojo
      AUG-14-CON=3-heart
      'greed'

NPs with the Connective particle /ɲa/ which refer to a person (singular) have no prefix and are treated as belonging to class 1a, as in (59).

Vwanji has three locative clitics: class 16 /pa/, class 17 /kʊ/ and class 18 /mu/. These clitics are homophonous with the noun prefixes for the same classes, but differ in that the clitics attach to a noun with a prefix (61) whereas the noun prefixes attach to a stem (62).

(61) pa'kɪtala 'kʲaːŋgo
      pa=kɪ-tala   kɪ-aⁿgo
      LOC$_{16}$=7-bed PP$_7$-1SG.POSS
      'on my bed'

(62) 'panofu 'paːŋgo
      pa-nofu   pa-aⁿgo
      16-good   PP$_{16}$-1SG.POSS
      'my good place'

The difference in their grammatical status can also be seen in the agreement of the following possessive. In (61) the clitic expresses a phrase-level locative relationship and the agreement within the phrase is determined by the head noun. In (62) the prefix is itself part of the head noun and therefore determines the agreement.

3.6 Word order and agreement in the noun phrase

The noun is always first in the NP. If there is a possessive modifier in the phrase, it usually immediately follows the noun (63). The relative order of quantifiers, numerals, adjectives and demonstratives is more flexible, although within the set of quantifiers /-ŋge/ 'other' precedes other quantifiers (64).

(63) ʊ'mʷoːɟo 'ɣʷaːŋgo 'ɣʷoːni
      ʊ-mʊ-oɟo   ɣʊ-aⁿgo   ɣʊ-oni
      AUG-3-heart   PP$_3$-1SG.POSS PP$_3$-all
      'my whole heart'

(64) a'βaːnʰu a'βaːŋge 'βiːŋga
      a-βa-nʰu   a-βa-ŋge   βa-iⁿga
      AUG-2-person   AUG-EP$_2$-other NP$_2$-many
      'many other people'

Phrasal modifiers (such as those involving a Connective) occur at the end of the NP, after any other modifiers, as in (65). Here the class 2 head noun 'people' controls agreement on the adjective 'good' and the Connective /ɲa/. The modifier phrase 'bad illnesses' contains a class 10 head noun which controls the agreement on the adjective 'bad' within that phrase.

(65) a'βaːnʰu a'βanofu aβa'ɲanʰamu i'ᵐbiːβi
      a-βa-nʰu   a-βa-nofu   a-βa-ɲa=N-tamu   i-N-βiːβi
      AUG-2-person   AUG-NP$_2$-good AUG-NP$_2$-con=10-illness AUG-NP$_{10}$-bad
      'good people with bad illnesses'

4 VERBS AND CLAUSE STRUCTURE

4.1 Verb structure

The Vwanji verb consists of eight slots, as shown in Table 21.10. Minimally, slots 5 (Root) and 8 (Final) must be filled.

(66) 'teleka
      telek-a
      cook-FV
      5-8
      'Cook!'

All the slots may be filled in a single verb form.

(67) naβakatʋtele'kelaɣa
      na-βa-ka-tʋ-telek-ɪl-aɣ-a
      NEG-SP$_2$-NARR-OP$_{1PL}$-cook-APPL-IPFV-FV
      1-2-3-4-5-6-7-8
      'They were not cooking for us.'

Slots 1 (Pre-Initial), 3 (TAM) and 6 (Extensions) may contain multiple morphemes.

(68) nakʲaβikʋtʋdi'ⁿdulila
      na-kʲa-βa-i-ku-tʋ-dind-ul-ɪl-a
      NEG-FUT1-SP$_2$-NPST-15-OP$_{1PL}$-open-SEP.TR-APPL-FV
      1-1-2-3-3-4-5-6-6-8
      'They might not open (it) for us.'

The verb schema does not account for the Habitual (see 82, 4.2.3), which contains two subject prefixes. Table 21.10 lists the morphemes which occur in each slot.

TABLE 21.10 VERB SLOTS AND THEIR MORPHEMES

1		2	3		4		5	6		7		8	
Pre-Initial		*Initial*	*TAM*		*Object*		*Root*	*Extensions*		*Pre-Final*		*Final*	
NEG	na-	SP	NPST	i-	OP		ROOT	APPL	-ɪl	PFV	-ile	FV	-a
FUT1	kʲa-		PER	fi-	REFL	-i		CAUS	-ɪsi	IPFV	-aɣ	FV	-e
16.LOC	pʷe-		PST2	a-				IMPOS	-ɪk			FV-i	-i
17.LOC	kʷe-		PST3	ka-				NEUT	-ɪk				
18.LOC	mʷe-		PST4	lʲa-				POS	-am				
			NAR	ka-				RECP	-an				
			CERT. FUT2	ili:-				EXTN	-al				
			UNC. FUT2	la:-				TENT	-at				
			ITV	ka-				SEP.TR	-ʋl				
			COND	ⁿga-				SEP. INTR	-ʋk				
			15	ku-				PASS	-u				

Vwanji verb roots (slot 5) are typically -CVC-, e.g., /-fik-/ 'arrive,' /-lʋt-/ 'go.' C1 can be a labialised or palatalised consonant, e.g., /-fʷal-/ 'wear.' /-sʲɪl-/ 'bury,' and C2 can be a prenasalised consonant, e.g., /-hoᵐb-/ 'pay.' The vowel can be short or long, e.g., /-βɪːk-/ 'put.' VC and CV are also possible verb roots, e.g., /-ik-/ 'descend,' /-fu-/ 'die.' In the case of VC roots, C can be a prenasalised consonant, e.g., /-ɪᵐb-/ 'sing.'

Verb extensions (slot 6) are typically -VC, with the exception of the causative (-VCV) and the passive (-V). The causative, applicative and neuter extensions commonly derive verbs from roots currently in use in the language.

(69) a CAUS   ku'lisia   ku-li-ısi-a   'to feed'   <-li-   'eat'
    b APPL   ku'hoᵐbela ku-hoᵐb-ıl-a 'to pay for' <-hoᵐb- 'pay'
    c NEUT   ku'temeka ku-tem-ık-a   'to be cut'   <-tem-   'cut'

Verbs with separative extensions tend to commute only with each other or with verbs with another extension, such as the neuter.

(70) a SEP.TR   ku'bıdʊla   ku-bıd-ʊl-a   'to turn up'
    b SEP.INTR ku'bıdʊka   ku-bıd-ʊk-a   'to be turned up'
    c NEUT   ku'bıdıka   ku-bıd-ık-a   'to be able to be turned up'

The extensive (/-al/), positional (/-am/) and tentive (/-at/) extensions occur infrequently and may be fossilised and not commute with any other extensions.

(71) a EXTENSIVE   ku'pakala   ku-pak-al-a   'to apply (e.g., lotion)'
    b POSITIONAL ku'fuɣama   ku-fuɣ-am-a   'to kneel'
    c TENTIVE   ku'kʷaβata ku-ku-aβ-at-a 'to hold, gather up in one's arms'

The reciprocal and passive extensions are productive, both as single extensions and in combination with other extensions.

(72) a RECP   ku'lekana   ku-lek-an-a   'to leave each other'
    b PASS   ku'lekua   ku-lek-u-a   'to be left'
    c CAUS+PASS kupʊlı'siβua ku-pʊl-ısi-u-a 'to be announced'

When a verb has multiple extensions, the less productive extensions (extensive /-al/, positional /-am/, tentive /-at/) come closest to the root (73a). The passive extension /-u/ occurs last and the reciprocal /-an/ precedes neuter /-ık/ which in turn precedes causative /-ısi/. The applicative extension can occur more than once and its two occurrences may be separated by another extension (73b).

(73) a POS+APPL   kʷe'ɣamıla   ku-eɣ-am-ıl-a   'to lean on'
    b APPL+RECP+APPL kwımı'lanıla ku-ım-ıl-an-ıl-a 'to stand for each other'

Perfective /-ile/ and Imperfective /-aɣ/ are both analysed as occurring in the Pre-Final slot as they are aspectual in function and cannot both occur in the same word. Imperfective /-aɣ/ is realised as /-isaɣ/ after verb roots consisting of a single consonant.

   The Final slot (8) is filled by the default vowel /-a/, the subjunctive final vowel /-e/ or the plural imperative final vowel /-i/ unless the Pre-Final slot is filled with the vowel-final perfective morpheme /-ile/ or the Root slot is filled with the vowel-final verbs /-ti/ 'say' or /-lı/ 'be.'

4.2 TAM and negation

Vwanji distinguishes four past tenses (Hodiernal, Near, Mid, Far) and two future tenses (Near, Far). A two-way aspectual distinction between Perfective and Imperfective can be seen in single-word past and future tense forms (Tables 21.11 and 21.12). Present tense verb forms differentiate Simple Present (also functioning as Progressive), Persistive, Habitual and Anterior (Table 21.13). Periphrastic verb forms expressing non-present tenses differentiate four aspects: Progressive, Persistive, Habitual and Anterior.

4.2.1 Past tense and the Perfective/Imperfective distinction

Table 21.11 sets out the four past tenses plus the Narrative, which has a time reference covering that of both the Mid Past and the Far Past.

TABLE 21.11 PAST TENSES

	PFV -ile/-a	IPFV -aɣ-a
NAR ka-	tʊ-ka-fik-a SP-NAR-VB-FV 'we arrived'	tʊ-ka-fik-aɣ-a SP-NAR-VB-IPFV-FV 'we were arriving'
P4 lʲa-	tʊ-lʲa-fik-ile SP-PST4-VB-PFV 'we arrived'	
P3 ka-	tʊ-ka-fik-ile SP-PST3-VB-PFV 'we arrived'	
P2 a-	tʊ-a-fik-a SP-PST2-VB-FV 'we arrived'	tʊ-a-fik-aɣ-a SP-PST2-VB-IPFV-FV 'we were arriving'
P1	tʊ-fik-ile SP-VB-PFV 'we arrived'	

The Hodiernal Past (SP-VB-PFV) is used for situations occurring after waking time on the same day as the time of reference. (This verb form also functions as a Present Anterior, as will be seen in 4.2.3.) The Near Past (SP-PST2-VB-FV) expresses a situation which occurred earlier than the day of the time of reference, up to around a month ago. The Mid Past (SP-PST3-VB-PFV) expresses situations which occurred earlier than would be appropriate for the Near Past, but not as far back as what is felt to be in the distant past. In this case the Far Past (SP-PST4-VB-PFV) is used. Thus for events of, say, a year ago, it is equally possible to use the Mid Past or the Far Past, according to the speaker's perspective. A similar four-way past tense distinction is evident in nearby language Hehe G62, although with significant formal differences (Nurse 2008a: 178).

The Narrative (SP-NARR-VB-FV) is not dependent on a preceding establishing tense and thus, following Longacre (1990: 109), can be analysed as a narrative rather than a consecutive. Typically the Narrative in Vwanji carries the main event line of a story and as such often follows a scene-setting introduction in which either the Mid Past or the Far Past predominates. In (74) /alʲaˈtolile/ 'he married' exemplifies the Far Past and /akaṃˈbʊːla/ 'he told her' (in the next sentence of the story) the Narrative.

(74) ʊmʊˈɣoːsi ɟʊˈmoːⁿga alʲaˈtolile ˈʊɳdala

ʊ-mʊ-ɣoːsi	ɟʊ-moⁿga	a-lʲa-toːl-ile	ʊ-mʊ-dala
AUG-1-man	EP₁-one	SP₁-PST4-marry-PFV	AUG-1-wife

akaṃˈbʊːla ˈʊɳdala ˈɣʷaːke ˈkuːti

a-ka-mʊ-βʊːl-a	ʊ-mʊ-dala	ɣʊ-ake	ku-ti
SP₁-NARR-OP₁-tell-FV	AUG-1-wife	PP₁-3SG.POSS	15-say

'A certain man married a wife. He told his wife that [. . .]'

The Perfective forms of the four past tenses and the Narrative include either the Perfective morpheme /-ile/ or the final vowel /-a/ and represent the situation expressed by the verb as a single whole. The contrasting Imperfective verb forms contain the Imperfective morpheme /-aɣ/ in the Pre-Final slot. As Table 21.11 illustrates, the Imperfective past tense verb forms cover a wider time reference than the Perfective verb forms. The Far Past Imperfective (SP-NARR-VB-IPFV-FV) is the Imperfective counterpart of the Mid Past, Far Past and Narrative Perfectives (75) (and thus covers the same time range as the Narrative). The Near Past Imperfective (SP-PST2-VB-IPFV-FV) is the Imperfective counterpart of the Near Past and Hodiernal Past Perfectives (76).

(75) pe i'fiɣono 'fɔːni βaka'ɣonaɣa 'kɪsima

pe	i-fi-ɣono	fi-oni	βa-ka-ɣon-aɣ-a	kɪsima
then	AUG-8-day	PP$_8$-all	SP$_2$-NARR-sleep-IPFV-FV	without.light

'And so they would always sleep without a light.'

(76) ɣʷe mʊ'ɣeːsi 'Ɉʊno 'Ɉaːtu nakʊ'βʊlaɣa

ɣʷe	mʊ-ɣeːsi	Ɉʊ-no	Ɉaːtu	nɪ-a-kʊ-βʊːl-aɣ-a
1.COP	1-stranger	EP$_1$-REL.PRO	every.day	SP$_{1SG}$-PST2-OP$_{2SG}$-tell-IPFV-FV

'kuːti a'βeːle 'isile

kuː-ti	a-β-ile	a-is-ile
15-say	SP$_1$-be-PFV	SP$_1$-come-PFV

'It is the stranger who every day I was telling you had come.'

Both Past Imperfectives have a specifically habitual interpretation, rather than generally imperfective. Non-habitual imperfective meaning, such as progressive, is expressed periphrastically in non-present tenses, as shown in 4.2.3.

4.2.2 Future tense and the Perfective/Imperfective distinction

Table 21.12 gives the possible future tense forms. A two-way tense distinction is made between Near and Far, and then a further distinction is made between Certain and Uncertain. Four of the possible verb forms are Perfective (ending /-a/) and three are Imperfective (ending /-aɣ-a/).

TABLE 21.12 FUTURE TENSES

	PFV -a	IPFV -aɣ-a
CERT.FUT1 i-	tʊ-i-fik-a SP-NPST-VB-FV 'we will arrive'	tʊ-i-fik-aɣ-a SP-NPST-VB-IPFV-FV 'we will be arriving'
UNC.FUT1 kʲa-. . . i-	kʲa-tʊ-i-fik-a FUT1-SP-NPST-VB-FV 'we might arrive'	
CERT.FUT2 iliː-	tʊ-iliː-fik-a SP-CERT.FUT2-VB-FV 'we will arrive'	tʊ-iliː-fik-aɣ-a SP-CERT.FUT2-VB-IPFV-FV 'we will be arriving'
UNC.FUT2 laː-	tʊ-laː-fik-a SP-UNC.FUT2-VB-FV 'we might arrive'	tʊ-laː-fik-aɣ-a SP-UNC.FUT2-VB-IPFV-FV 'we might be arriving'

The Certain Near Future (sp-i-vb-fv) is used for situations which are considered certain to occur and begin at any point after the reference time, extending until approximately a month into the future. (This verb form also functions as a Simple Present, as will be seen in Section 4.2.3.) The Uncertain Near Future (fut1-sp-npst-vb-fv) covers the same range of time reference but includes an element of doubt regarding the likelihood of the situation occurring.

(77) kʲaiˈkʊːⁿga   ɪˈtaːla
      kʲa-a-i-kʊⁿg-a   ɪ-N-taːla
      FUT1-SP$_1$-NPST-light-FV   AUG-9-lamp
      'He might light a lamp.'

An alternative, unreduced form of the morpheme /kʲa-/ is /ˈkʲaːⁿde/, which is a subjunctive verb with a class 7 subject prefix, meaning 'it should start.'

For situations considered certain to happen in the far future (approximately one month or more from the time of reference), the Certain Far Future (sp-cert.fut2-vb-fv) is used.

(78) siliˈhʊmɪla mʊˈɲuːᵐba ˈɟaːko
      si-iliː-hʊm-ɪl-a   mʊ=N-ɟuᵐba   ɟɪ-ako
      SP$_{10}$-CERT.FUT2-come.out-APPL-FV   LOC$_{18}$=9-house   PP$_9$-2SG.POSS
      'They will come from your house.'

If the situation is not considered certain to happen, the Uncertain Far Future (sp-unc.fut2-vb-fv) is used.

(79) βalaβʊˈᵐbʊlɪla kɪˈmoːⁿga
      βa-la:-βʊᵐbʊlɪl-a   kɪ-moⁿga
      SP$_2$-UNC.FUT2-gain-FV   EP$_7$-one
      'They might gain something.'

With the exception of the Uncertain Near Future, the future tense forms described above can be suffixed with Imperfective /-aɣ/. In the Certain Near Future the addition of /-aɣ/ restricts the interpretation of the time reference to the near future and the verb form may not be understood as present.

(80) βiˈkʷisaɣa kuˈsʊːma ʊlʊˈtaⁿgɪlo
      βa-i-ku-is-aɣ-a   ku-sʊːm-a   ʊ-lʊ-taⁿg-ɪl-o
      SP$_2$-NPST-15-come-IPFV-FV   15-ask-FV   AUG-11-help-APPL-INS
      'They will be coming to ask for help.'

It is not possible to add Imperfective /-aɣ/ to the Uncertain Near Future because it is assumed that historically this form contains two words (/ˈkʲaːⁿde/ plus sp-npst-vb-fv) and /-aɣ/ is restricted to single-word verb forms.

4.2.3 Present tense and further aspectual distinctions

Table 21.13 shows the four-way aspectual distinction in the Present between Progressive (expressed by the Simple Present), Persistive, Habitual and Anterior.

TABLE 21.13 PRESENT TENSE ASPECTUAL DISTINCTIONS

Simple Present or PROG	PER fi-	HAB a-	ANT -ile
tʊ-i-fik-a	tʊ-i-fi-fik-a	tʊ-a-tʊ-i-fik-a	tʊ-fik-ile
SP-NPST-VB-FV	SP-NPST-PER-VB-FV	SP-HAB-SP-NPST-VB-FV	SP-VB-PFV
'we arrive/ are arriving'	'we are still arriving'	'we arrive'	'we have arrived'

The Simple Present (SP-NPST-VB-FV) can be used to represent a situation in progress at the reference time:

(81) tu'βoːᵐba mʊma'ɣʊːⁿda
     tʊ-i-βoᵐb-a   mʊ=ma-ɣʊⁿda
     SP$_{1PL}$-NPST-work-FV   LOC$_{18}$=6-field
     'We are working in the fields.'

Note that as the same verb form also functions as the Certain Near Future, the time reference is often disambiguated through context. In (81) a change of locative clitic from class 18 /mʊ-/ to class 17 /kʊ-/, for example, would show that the fields were not at the same location as the speaker and therefore a future time reference would be assumed, as in 'We will work in the fields' (Nahumu Mhalila, pers. comm.).

Non-Past /i-/ is also found in the two other subtypes of imperfective with present time reference: the Present Persistive (SP-NPST-PER-VB-FV) and the Present Habitual (SP-HAB-SP-NPST-VB-FV). The Persistive is used to show that "a situation current at the time of reference is connected to an identical past situation" (Nurse 2008b: 123), whereas the Habitual expresses a situation which is repeated within a period of time. The Present Habitual contains two subject prefixes and has secondary stress on the initial syllable, suggesting that it previously comprised an auxiliary and main verb.

(82) ˌtʷatu'βoːᵐba kʊma'ɣʊːⁿda
     tʊ-a-tʊ-i-βoᵐb-a   kʊ=ma-ɣʊⁿda
     SP$_{1PL}$-HAB-SP$_{1PL}$-NPST-work-FV   LOC$_{17}$=6-field
     'We work in the fields.'

When used with a stative verb, the Present Anterior (SP-VB-PFV) expresses a situation which began in the past and continues into the present (100, 4.3). When used with a dynamic verb, it expresses a past situation with current relevance (95, 4.3). Perfective /-ile/ also occurs with a Perfective function in the Hodiernal Past (which is formally identical with the Present Anterior), the Mid Past and the Far Past; see (Nurse 2008b: 266) for discussion of possible grammaticalisation paths from anterior to past perfective).

The four-way aspectual distinction possible in single-word present tense forms can be achieved with past and future time reference in different ways. Habitual aspect is expressed by /-aɣ/, as shown in 4.2.1–2. Persistive aspect can be conveyed by adding /fi-/ to any verb form already containing Imperfective /-aɣ/, such as the Far Past Imperfective.

(83) akafipelepe'sʲaːɣa 'kuːti 'iːke
     a-ka-fi-pelepesi-aɣ-a   ku-ti   a-ik-e
     SP$_1$-NAR-PER-coax-IPFV-FV   15-say   SP$_1$-go.down-FV
     'He was continuing to coax (him) to get down.'

Persistive can also be achieved by means of periphrastic verb forms, in which the first verb is a form of 'to be' and is marked for tense and the second verb provides the lexical content and is marked for aspect. The same strategy is also used for conveying Progressive (using the Simple Present) (84, 85) and Anterior (86) aspect:

(84) a'lʲaːle i'dɪːma i'ŋoːᵐbe
 a-lʲa-β-ile   a-i-dɪːm-a   i-N-ŋoᵐbe
 SP₁-PST4-be-PFV   SP₁-NPST-herd-FV   AUG-10-cow
 'He was herding cows.' (*Far Past + Progressive*)

(85) 'βiːβa βikʊkʊ'βonela ʊlʊ'sʊːᵑgʊ
 βa-i-β-a   βa-i-ku-kʊ-βon-ɪl-a   ʊ-lʊ-sʊᵑgʊ
 SP₂-NPST-be-FV   SP₂-NPST-15-OP₂ₛₒ-see-APPL-FV   AUG-11-mercy
 'They will be merciful to you.' (*Certain Near Future + Progressive*)

(86) a'βeːle asʲe'milue ku'diːⁿda ʊmʊ'lʲaːᵑgo
 a-β-ile   a-sʲemu-ile   ku-dⁿd-a   ʊ-mʊ-lʲaᵑgo
 SP₁-be-PFV   SP₁-forget-PFV   15-close-FV   AUG-3-door
 'She had forgotten to close the door.' (*Hodiernal Past + Anterior*)

Note that in (85) the two verbs are formally identical (SP-NPST-VB-PFV), but the auxiliary is analysed as Certain Near Future and the main verb as Progressive. Similarly in (86), the two verbs are formally identical (SP-VB-PFV), but the auxiliary is analysed as Hodiernal Past and the main verb as Anterior. Thus, in both cases the first verb contributes tense and the second aspect, as is the pattern for verb forms of this type in Vwanji. See Mahali *et al.* (2014: 35–36) for further examples of periphrastic verb forms.

For all the verb forms discussed so far, the negative is formed by adding /na-/ to the Initial slot in the verb structure, before the subject prefix (see 18a, 2.5.1; 39, 3.4; 67, 4.1; 68, 4.1).

4.2.4 Imperative, subjunctive and conditional

Table 21.14 shows the possible Imperative and Subjunctive verb forms constructed from the verb root /-lɪm-/ 'farm' with Imperfective /-aɣ/ and Itive /ka-/.

TABLE 21.14 IMPERATIVES AND SUBJUNCTIVES

	Unmarked	IPFV	ITV	IPFV+ITV
IMP (SG)	lɪm-a	lɪm-aɣ-a	ka-lɪm-e	ka-lɪm-aɣ-e
	VB-FV	VB-IPFV-FV	ITV-VB-FV	ITV-VB-IPFV-FV
	'farm!'	'farm!'	'go and farm!'	'go and farm!'
IMP (PL)	lɪm-i	lɪm-aɣ-i		
	VB-FV	VB-IPFV-FV		
	'farm!'	'farm!'		
SBJV	a-lɪm-e	a-lɪm-aɣ-e	a-ka-lɪm-e	a-ka-lɪm-aɣ-e
	SP₁-VB-FV	SP₁-VB-IPFV-FV	SP₁-ITV-VB-FV	SP₁-ITV-VB-IPFV-FV
	'he should farm'	'he should farm'	'he should go and farm'	'he should go and farm'

The addition of Imperfective /-aɣ/ to an Imperative or Subjunctive may express the expected imperfective aspect interpretation, but it may instead soften the force of the command.

(87)   ˈlʊtaɣa kʊˈkaːɟa ˈɟaːko
       lʊt-aɣ-a   kʊ=N-kaːɟa   ɟɪ-ako
       go-IPFV-FV   LOC$_{17}$=9-home   PP$_9$-2SG.POSS
       'Go to your home!'

Itive /ka-/ can be added to the Imperative and Subjunctive for a situation which is to take place away from the speaker. Itive /ka-/ is assumed to be related to Narrative /ka-/ as these morphemes represent a spatial (Itive) or temporal (Narrative) shift from a starting point (Nurse 2008b: 244–246). The presence of Itive /ka-/ in the Imperative requires a change of final vowel from /-a/ to /-e/, as does the presence of an object prefix.

The Conditional (SP-COND-VB-FV) is used for hypothetical conditionals.

(88)   ɟɪⁿgaˈtoːɲe ˈɪfula aˈmeːⁿda ˈɣiɲopa
       ɟɪ-ⁿga-toːɲ-e   ɪ-N-fula   a-ma-eⁿda   ɣa-i-ɲʊp-a
       SP$_9$-COND-rain-FV   AUG-9-rain   AUG-6-clothing   SP$_6$-NPST-get.wet-FV
       'If it rains, the clothes will get wet.'

The same type of conditional can also be expressed with /naβe/ 'if' plus the Subjunctive, whereas /naβe/ plus the Simple Present is used for reality conditionals. The Conditional can be used in an independent clause and then functions as a potential.

(89)   nʊⁿgahʊˈβɪːle ˈkugima iˈnʰofani
       na-ʊ-ⁿga-hʊβɪːl-e   ku-gim-a   i-N-tofani
       NEG-SP$_{2SG}$-COND-hope-FV   15-dig-FV   AUG-10-potato
       'You can't hope to dig potatoes [there].'

Counterfactual conditional situations can be expressed periphrastically with a Far Past Perfective form of 'to be' and a Subjunctive in the protasis and the same form of auxiliary plus a Simple Present in the apodosis.

(90)   ˈɣʰʷaːle ˈʊβe kʊkɪkʊˈlʊkʊlʊ
       ɣʊ-lʲa-β-ile   ʊ-β-e   kʊ=kɪ-kʊlʊkʊlʊ
       SP$_{2SG}$-PST4-be-PFV   SP$_{2SG}$-be-FV   LOC$_{17}$=7-celebration

       ˈɣʰʷaːle ɣuˈpʊlɪka ˈɪmʰola
       ɣʊ-lʲa-β-ile   ɣʊ-i-pʊlɪk-a   ɪ-N-pola
       SP$_{2SG}$-PST4-be-PFV   SP$_{2SG}$-NPST-hear-FV   AUG-9-news
       'If you had been at the celebration, you would have heard the news.'

Imperatives and Subjunctives are negated in the same way, with the initial Negative morpheme /na-/, the Conditional morpheme /ⁿga-/ and /-e/ as the Final Vowel.

(91)   nʊⁿgaˈβoᵐbaɣe ˈʊluo
       na-ʊ-ⁿga-βoᵐb-aɣ-e   ʊluo
       NEG-SP$_2$-COND-do-IPFV-FV   EP$_{11}$.REF.DEM
       'Don't do that!/ You shouldn't do that.'

4.3 Clause structure

Basic word order in Vwanji is SVO. Temporal adjuncts occur clause initially and adverbs follow the main verb (92). Auxiliaries precede main verbs and locative adjuncts occur clause-finally (93).

(92) 'pakɪlo 'ʊɲina akakaɣuᵐbi'lilaɣa
    pa=N-kɪlo   ʊ-ɲina   a-ka-ka-ɣuᵐbilil-aɣ-a
    LOC₁₆=9-night   AUG-1a.mother   SP₁-NARR-OP₁₂-cuddle-IPFV-FV

'βʊnofu a'kaːna 'kala
βʊ-nofu   a-ka-ana   ka-la
14-good   AUG-12-child   EP₁₂-DIST.DEM
'At night his mother would be cuddling that little child well.'

(93) 'akaβa 'ikila mʊma'kaːɟa a'maβaha
    a-ka-β-a   a-i-kil-a   mʊ=ma-kaːɟa   a-ma-βaha
    SP₁-NARR-be-FV   SP₁-NPST-pass-FV   LOC₁₈=6-town   AUG-NP₆-big
    'He was passing through big towns.'

In ditransitive predicates, objects with the semantic role of goal precede those with the semantic role of theme.

(94) akam'peːla 'ʊɲina ʊ'mʷaːna 'ɟʊla
    a-ka-mʊ-peːl-a   ʊ-ɲina   ʊ-mʊ-ana   ɟʊ-la
    SP₁-NARR-OP₁-give-FV   AUG-1a.mother   AUG-1-child   EP₁-DIST.DEM
    'He gave the mother that child.'

With the exception of the relative order of the auxilliary and main verbs, the basic word order is subject to variation determined by information structure categories, such as topic, focus and the givenness of information. An object can be topicalised by being fronted to a pre-verbal position and a normally pre-verbal constituent can be focused by occurring postverbally. The following example illustrates both of these deviations from the basic word order. The object of the clause is topicalised and the subject is focused, resulting in the order OVS.

(95) 'ʊne ka'ⁿdumile a'kaɲoɲo 'aka
    ʊne   ka-N-lum-ile   a-ka-ɲoɲo   aka
    1SG.CONTR.PRO   SP₁₂-OP₁ₛɢ-bite-PFV   AUG-12-insect   EP₁₂.PROX.DEM
    'This little insect has bitten me.'

The word order choices here show that the speaker considers the utterance to be chiefly about himself (therefore 'me' is topicalised) and that the fact that he has been bitten is presupposed, but the identity of the culprit is not (therefore 'this little insect' is focused).

Focusing can also be achieved by means of a cleft construction.

(96) a'βaːna 'aβa βe 'βano
    a-βa-ana   aβa   βe   βa-no
    AUG-2-child   EP₂.PROX.DEM   2.COP   EP₂-REL.PRO

βiliβʊlanı ˈsʲaːɣa aˈβaːna ˈβiːtu
βa-iliː-βʊlanısi-aɣ-a   a-βa-ana   βa-itu
SP₂-CERT.FUT2-teach-IPFV-FV   AUG-2-child   PP₂-1PL.POSS
'These children are the ones who will teach our children.'

A subject NP which is also a topic is not marked as such, except by it occurring in its basic pre-verbal position. When there is a switch of subject in a text, it is common for the new subject NP to be (re-)introduced to the discourse in the postverbal focus position.

(97) aˈkiːsa ˈkaːⁿge ʊˈmʷandefu
      a-ka-is-a   kaⁿge   ʊ-mʷandefu
      SP₁-NARR-come-FV   again   AUG-1a.Mwandefu
      'Mwandefu came again.'

Vwanji can be classed as an OM-1 language (Bearth 2003: 124) as it allows only one object prefix inside the verb. This marker may be an object prefix or the Reflexive morpheme /i-/. In lexical ditransitives and in verbs which are ditransitive because they contain an applicative extension, the object prefix agrees with the recipient and not the theme.

(98) nıkʊkʊˈpeːla ıfuˈⁿgulilo
      nı-i-ku-kʊ-peːl-a   ı-N-fuⁿgulilo
      SP₁ₛɢ-NPST-15-OP₂ₛɢ-give-FV   AUG-9-key
      'I will give you the key.'

Thus Vwanji makes a distinction between primary and secondary objects, rather than between direct and indirect objects (Dryer 2007: 253–256). The verb must agree with the primary object of the verb and not, for example, the location of the verb.

Agreement for conjoined NPs depends on the animacy of the referents. When two NPs of the same class are conjoined, the agreement is with the corresponding plural class if the referents are animate (99) and with class 8 if the referents are inanimate (100).

(99) ˈımʰene nıˈŋoːᵐbe ˈsilʊta
      ı-N-pene   na=ı-N-ŋoᵐbe   si-i-lʊt-a
      AUG-9-goat   COM=AUG-9-cow   SP₁₀-NPST-go-FV
      'The goat and cow are going.'

(100) ʊmʊˈlʲaːⁿgo ˈnʊŋgoma finaˈⁿgiːke
      ʊ-mʊ-lʲaⁿgo   na=ʊ-mʊ-goma   fi-naⁿg-ık-ile
      AUG-3-door   COM=AUG-3-beehive   SP₈-damage-NEUT-PFV
      'The door and the beehive are damaged.'

Class 8 is also the agreement for conjoined NPs with inanimate referents from different noun classes. Animate referents from different classes are not usually conjoined. Instead a noun of one class is chosen as the subject and the other noun is expressed in a comitative phrase.

(101) ʊˈmʷaːna ˈilʊta ˈnɪᵐbua
       ʊ-mʊ-ana   a-i-lʊt-a   na=ɪ-N-bua
       AUG-1-child SP₁-NPST-go-FV   COM=AUG-9-dog
       'The child is going with the dog.'

4.4 Clause types

Independent clauses may be coordinated without conjunctions. Narrative texts commonly juxtapose independent clauses in which the main verb is in the Narrative.

(102) ʊˈmʷaːna ˈɟʊla aˈkakʊla
       ʊ-mʊ-ana   ɟʊ-la   a-ka-kʊl-a
       AUG-1-child EP₁-DIST.DEM SP₁-NARR-grow-FV

       akaˈteⁿgʊla piˈkʷafula
       a-ka-teⁿgʊl-a   pi-ku-aful-a
       SP₁-NARR-start-FV 15-15-crawl-FV
       'That child grew and started to crawl.'

Vwanji does not allow formal locative inversion, but when a copula is marked with a locative morpheme in the Initial slot, the subject usually occurs postverbally.

(103) kʷeaˈlʲaːle ʊmʊˈɣoːsi ɟʊˈmoːⁿga
       kʷe-a-lʲa-β-ile   ʊ-mʊ-ɣoːsi   ɟʊ-moⁿga
       LOC₁₇-SP₁-PST4-be-PFV   AUG-1-man   EP₁-one
       'There was a certain man.'

As seen in (76, 4.2.1) and (96, 4.3), relative clauses in Vwanji include a freestanding relative pronoun. The class 16 relative pronoun /ˈpano/ also has a temporal function in clauses expressing habitual situations.

(104) ˈpano ˈihʊma ˈkʊːɲɟi aˈiɣua
       pano   a-i-hʊm-a   kʊ-ɲɟi   a-a-a-i-ɣu-a
       16.REL.PRO SP₁-NPST-go.out-FV 17-outside SP₁-HAB-SP₁-NPST-fall-FV
       'When he goes outside, he always falls.'

Temporal subordinate clauses expressing non-habitual situations are introduced by /ye/ 'after.' A Simple Present verb is used for a simultaneous situation with the next clause (as in 104), whereas an Anterior is used when the situations are consecutive.

(105) ye βaˈβʊkile βaˈkaseka
       ye   βa-βʊːk-ile   βa-ka-sek-a
       after SP₂-leave-PFV   SP₂-NARR-laugh-FV
       'After they had left, they laughed.'

ACKNOWLEDGEMENTS

I would like to thank Juhudi Konga, Ahimidiwe Mahali, Nahumu Mhalila, Stephen Nguvila and Ezekiel Nyambo for their assistance with my questions about Vwanji. My thanks also go to Hazel Gray, Stephen Katterhenrich, Steve Nicolle and Oliver Stegen for their comments on earlier versions of this chapter, and to Alastair Duncan for his help in data collection.

REFERENCES

Bearth, T. (2003) Syntax. In Nurse, D. & G. Philippson (eds.), *The Bantu Languages*, 121–142. London, New York: Routledge.

Broomhall, E. (2011) *Bena Text Linguistics*. Ms. Nairobi: Africa International University.

Dryer, M. S. (2007) Clause Types. In Shopen, T. (ed.), *Clause Structure, Language Typology and Syntactic Description, Volume 1: Clause Structure*, 224–275. Cambridge: Cambridge University Press.

Fischer, H. (2011) *The Augment in Vwanji (G66)*. Hamburg: University of Hamburg, MA thesis.

Guthrie, M. (1971) *Comparative Bantu: An Introduction to the Comparative Linguistics and Prehistory of the Bantu languages. Volume 2: Bantu Prehistory, Inventory and Indexes*. London: Gregg International.

Krüger, S. (2003) *A Sociolinguistic Survey of the Wanji Language Community*. Ms. Dodoma: SIL International.

Longacre, R. E. (1990) Storyline Concerns and Word Order Typology in East and West Africa. *Studies in African Linguistics* supplement 10.

Mahali, A. Z., A. S. Mbena, N. P. Mhalila, A.-L. Lindfors & H. Eaton (2014) *Sarufi ya Lugha ya Kivwanji*. Ms. Mbeya: SIL International.

Nicolle, S. (2015) A Comparative Study of Eastern Bantu Narrative Texts. *SIL Electronic Working Papers* 3. (Available online at www.sil.org/resources/publications/entry/61479, Accessed 10 November, 2015).

Nurse, D. (1988) The Diachronic Background to the Language Communities of Southwestern Tanzania. *Sprache und Geschichte in Afrika* 9: 15–115.

Nurse, D. (2008a) Appendices to Tense and Aspect in Bantu. (Available online at www.ucs.mun.ca/~dnurse/tabantu.html, Accessed November 26, 2015).

Nurse, D. (2008b) *Tense and Aspect in Bantu*. Oxford: Oxford University Press.

Schadeberg, T. C. (1973) Kinga: A Restricted Tone System. *Studies in African Linguistics* 4(1): 23–47.

CHAPTER 22

TOTELA K41

Thera Crane

1 INTRODUCTION

1.1 Totela language and speakers

Varieties known as Totela are spoken in parts of the Western Province in Zambia (Guthrie code K41) and the Zambezi Region (formerly, the Caprivi Region) in Namibia (Guthrie code K411). The Zambezi variety may extend to portions of the Southern Province in Zambia as well (Lewis *et al.* 2016), although this is not clearly reflected in census data. Both varieties are spoken in heavy contact situations, and neither has official language status. Lozi K21 is used as a language of education and wider communication in areas where Totela is spoken, and, especially in Zambia, is rapidly overtaking Totela as a primary language of communication.

Lewis *et al.* (2016) rate Totela's status as 6a ("vigorous") on the EGIDS Scale (Lewis & Simons 2010), but if the Western Province variety is considered on its own, the status of 7 ("shifting") or 8a ("moribund") is likely more accurate. Fortune's (1959) survey estimates a total of 14,581 Totela speakers. Zambia's 2010 census report (Central Statistical Office 2012) counts the Totela-speaking population as 1,118, although totals appear to be inconsistent across different reports from that census. According to the descriptive tables of the Zambian 2010 census, only 206 people in the Western Province claim Totela as their primary language (Central Statistical Office 2013: 159). The numbers derived from Zambian census data may well undercount the actual number of speakers, because they count the primary language of communication rather than all languages spoken, and speakers of Totela are likely to use Lozi K21 as their primary language of communication. However, speaker numbers have steadily declined from the time of Fortune's (1959) survey, and demonstrate a clear shift away from Totela. This shift is particularly serious in Zambia: while members of the childbearing generation generally have some proficiency in Totela, it is rarely spoken in daily life or transmitted to children. The Namibian variety of Totela is part of a dialect continuum with other Zambezi varieties such as Fwe K402 and Subiya K42, and along with these other varieties, still enjoys widespread daily use (see e.g., Seidel 2005, de Luna 2010). Namibia's 2011 census does not count Totela separately, but instead groups it with "Caprivi languages," which together are spoken as the primary language of 22,484 households (Namibia Statistics Agency 2013: 68). It is difficult to extrapolate numbers of Totela speakers specifically in the Zambezi Region.

1.2 Data sources

Language data in this chapter come primarily from my own fieldwork with Totela speakers in Zambia and Namibia in 2006, 2007, and 2009. Most examples are from

Zambian Totela speakers living along the Kwemba River, especially around the small village of Nakwenda. Much of the language data cited in this chapter is also available in Crane (2011), which discusses some of the phenomena mentioned here in greater detail. In case of discrepancies, this chapter supersedes the previous work. Crane (2011) also includes a list of research participants and locations. Examples are drawn from both elicitation and recorded texts, including both narrative and conversational data, with narrative data more heavily represented. This chapter also uses Namibian Totela lexical data from Kathryn de Luna's 2006–2007 *Bantu Botatwe* research field notes (see also de Luna 2016).

Few earlier published works deal with Totela. Baumbach (1997) includes a brief sketch of the Namibian variety, and the language (and/or ethnic group) is mentioned briefly in Torrend (1931), Bryan (1959), and Fortune (1959, 1963), among a few other sources. Jacottet (1896, 1899, 1901) includes some Totela songs.

1.3 Classification within Bantu Botatwe

Totela is a member the *Bantu Botatwe* (BB) group, which includes various languages in the K30–40 and M60 groups. Totela is spoken in strikingly different varieties in Zambia (henceforth, Zambian Totela or ZT) and Namibia (Namibian Totela or NT). This chapter deals primarily with the Zambian variety, and all data, unless otherwise marked, come from speakers of ZT. A few important differences between ZT and NT are also discussed.

Bostoen (2009) lists innovations relevant for subclassification within Botatwe. These innovations are given in Table 22.1 (adapted from Crane 2011: 54),[1] along with their outcomes in languages spoken around the Zambezi (left side) and in Zambia (right side). Apart from issues related to the vocalic augment, discussed below in 3.3, Namibian Totela patterns with its neighbours in the Zambezi Region, while Zambian Totela resembles Group M languages Tonga M64 and Ila M63. Indeed, at least some speakers of ZT consider their language a variety of Ila (it may be even closer to Toka-Leya (M64)); the same speakers tend to label NT as "Subiya."

TABLE 22.1 INNOVATIONS WITHIN BANTU BOTATWE

Innovation	Fwe K402	Subiya K42	NT K411	ZT K41	Tonga M64	Ila M63
*p-lenition p > h > Ø	h (/Ø)	h	h	p/w/Ø	p/w/Ø	p/w/Ø
*j-lenition: dʒ > ʒ > z > s > Ø	ʒ	Ø/z	Ø/z	Ø/z/y	Ø/z/y	Ø/ʒ
H-tone anticipation	no	no	no	yes	yes	yes
augment vowel aperture	yes	n/a	no	yes	no	n/a
uniformisation of vocalic augment	no	n/a	no	no	yes	n/a
loss of vocalic augment	no	yes	partial	no	no	yes
merger of places of articulation before *u	no	no	no	no	no	yes
spirant devoicing	no	no	no	no	no	no
generalisation of *li- as cl. 5 prefix	no	no	no	no	no	no

1.4 Orthographic conventions

In Sections 2.1–2.2, forms are given as IPA transcriptions. In the remainder of the chapter, a practical orthography is employed, based primarily on orthographies for other languages in the area. Differences from IPA transcription are as follows: *b* in the orthography = [β] in the IPA; *bb* = [b] (but *mb* = [mb]); *ny* = [ɲ]; *sh* = [ʃ]; *ch* = [tʃ]; *j* = [dʒ]; *hu* ≈ [h̃ʷʷ⁾u]; *nk* = [ŋk]; *y* = [j].

Surface tones are marked exhaustively when they have been verified. Namibian Totela examples are not marked for tone, as a thorough analysis of the system is still needed. In Zambian Totela examples, morae that appear to be lexically associated with underlying (input) H tones are underlined; melodic Hs are indicated with a subscript $_H$. Note that regarding some grammatical morphemes, evidence is equivocal as to whether they are associated with lexical H tone or not. These include subject prefixes and most tense/aspect markers. For the sake of consistency, I do not underline subject markers (except with relative clauses) or other tense/aspect prefixes, although I think it is likely that they are underlyingly H, and indeed, an anticipated H can be observed in some cases. In cases of glides or syllables with automatic lengthening (sometimes with falling tones), lexical H tones are marked on the singleton vowel; the tone status can be deduced by HTA behaviour.

2 PHONOLOGY

2.1 Vowels

Totela has a five-vowel system with contrastive length.

(1)   i   u   ii   uu
      e   o   ee   oo
      a   aa

2.1.1 Vowel length

In verbs, only the first root syllable can have lexically contrastive length.

(2)   a   -bòòlà 'return'   b   -bòlà 'rot'
      c   -kṳùlà 'climb'   d   -kṳlà 'grow'

Long and short vowels also contrast in their melodic H tone patterns; a H on a long vowel in penultimate position on a verb is realised as falling. A few exceptions, such as *bóólà!* 'return, come back!' likely have consonant loss or other historical explanations.

(3)   a   súk-à$_H$!
          rub.together-FV.IMP
          'Rub together!' (e.g., cloth)

      b   sṳùk-à$_H$!
          descend-FV.IMP
          'Come down!'

Duration of long vowels in these contexts is approximately double the duration of short vowels. When not in penultimate position, the length distinction is not as pronounced, nor is there a tonal contrast in melodic H patterns. Example (4) shows that a long-vowed stem *-zaanina* 'play with' behaves the same as trisyllabic stems without length, e.g., *-ukuta* 'shake.'

(4) a   tàndìzáànà
         ta-ndi-zaan-a$_H$
         NEG-SP$_{1SG}$-play-FV
         'I don't play.'

     b   tàndìzáánìnà
         ta-ndi-zaan-in$_H$-a
         NEG-SP$_{1SG}$-play-APPL-FV
         'I don't play with.'

     c   tàndìúkùtà
         ta-ndi-ukut$_H$-a
         NEG-SP$_{1SG}$-shake-FV
         'I don't shake.'

In noun stems, long vowels are attested in all positions, although non-stem-initial long vowels can mostly be attributed to compounds or other derivations. The length contrast seems to be less pronounced in pre-penultimate positions.

(5) nsáà̱   'duiker'   ò̱lú-tà̱láá̱   'platform'
     káázè̱   'cat'   pá̱à̱tì   'velvet ant'
     à̱kà-tììtè   'warbler'   (kàngá)sàséè̱là   'bird (sp.)'
     ò̱lù-ɲàméè̱nò̱   'worm lizards'   ìn-kúlùkúù̱bì   'butterfly'
     í̱⁺m-bá̱álà̱   'barbfish'   ò̱mú-⁺lá̱álà̱   'leftovers'
     èchì-kóókò̱   'grass (sp.)'   èchì-sùùkà   'pinnacle of plaited grass on top of hut'
     ì-tóò̱nólò̱   'good luck'   èchì-zòòlìsò   'stirring paddle'
     (*also* ì-tóhò̱nólò̱)

In nouns, the functional load of length in lexical contrasts appears to be somewhat marginal; the best example in my data is the near minimal-pair *mùlálà̱* 'snake (sp.)' and *ò̱mú⁺lá̱álà̱* 'leftovers.'

Automatic vowel lengthening occurs before prenasalised stops (e.g., [àbà:ntù] 'people'), also triggering a HL falling pattern in bisyllabic verbs with melodic tone ([wâ:mbà] 'talk!'). CGV sequences also trigger lengthening and have the tone patterns associated with long vowels. More analysis is needed to understand the phonetic nature of vowel length in CGV sequences. Palatal nasal *ɲ* (*ny* in the orthography) is a single consonant, rather than a CV sequence, and does not trigger lengthening or a falling tone pattern in imperatives, as shown by the contrast in (6).

(6) a   ɲénà!
         ɲen-a$_H$
         feel.shame-FV.IMP
         'Be ashamed!'

     b   ɲéèzà!
         ɲeez-a$_H$
         annoy.caus-FV.IMP
         'Annoy!'

Long vowels can also be conditioned by vowel coalescence (2.1.2), both within (e.g., with possessive prefixes and roots) and across words (see Table 22.2). TAM markers are frequently lengthened before monosyllabic stems. Additionally, all falling tones are realised with some degree of length; the phonological status of some of these lengths is unclear (e.g., in the final vowel of some demonstrative forms).

In this chapter, only vowel length that is (potentially) lexically contrastive will be distinguished in the orthography. Automatic lengthening and intonational lengthening are not marked, as this would result in orthographic inconsistencies. Both kinds of lengthening are interesting and merit further investigation.

2.1.2 Vowel-vowel interactions

Vowel coalescence occurs both in and across words, particularly when the initial vowel in the sequence is /a/ or /e/. Non-coalescing, non-homorganic VV sequences with /i/, /o/ or /u/ as the initial vowel tend to surface as GV sequences; a (weak) glide is often inserted when vowels meet across syllables. Initial /i/ generally either surfaces as /(C)yV/ or is deleted. In sequences where /o/ or /u/ are initial, the sequence is either pronounced / wV/, or both vowels in the sequence are pronounced. Outcomes of such vowels meeting across words are variable. A few segments that may have their morphological roots in coalescence of /u/ and /o/ surface as o(o). Gliding vs. deletion in segments with initial /i/ is not always regular or consistent, especially following alveolar or palatal fricatives or affricates (7). Additional phonetic study would be enlightening.

(7) a e̱tʃáa̱là 'fingernail' (cl. 7 e̱chí-a̱là)
     b e̱zya̱là 'fingernails' (cl. 8 e̱zí-àlà)

Length resulting from vowel coalescence, especially across words, is often more pronounced than lexical vowel length. The major vowel coalescences are shown in Table 22.2.

TABLE 22.2 VOWEL-VOWEL INTERACTIONS ACROSS MORPHEMES/WORDS

V1	V2	Surface form	Example
a/e	a	a(a)	òkúßo̱nà + a̱ßàntù > òkúßo̱náa̱ßàntù 'to see people'
a/e	i/e	e(e)	òkúßo̱nà + e̱mìnzì > o̱kúßo̱néèmìnzì' to see villages'
a/e	o	o(o)	òkúßo̱nà + o̱mùntù > o̱kúßo̱nóo̱mùntù 'to see a person'
a/e	u	a(w)u/o(o)	ná-ù-lá-ßo̱n-a (posthodiernal 2nd person future) > ná(w) ùláßo̱nà/nóo̱láßo̱nà 'you will see'

2.2 Consonants

Consonants are given in Table 22.3. Consonants appearing rarely or predominantly in loanwords are given in parentheses.

Non-prenasalised voiced stops are relatively rare, occurring most frequently in loanwords. Outside of obvious borrowings, non-prenasalised voiced stops almost always occur stem-initially (e.g., i̱dòkòlà 'edible grass (sp.)'). Voiced stops with prenasalisation are more common than voiceless prenasalised stops. However, all non-nasal consonants are attested with prenasalisation, with the exception of [l] (instead realised as [nd];

TABLE 22.3 TOTELA CONSONANT INVENTORY

	Bilabial	Labio-dental	Alveolar	Post-alveolar	Palatal	Velar	Glottal
Plosive	p (b)		t (d)			k (g)	
Nasal	m		n		ɲ	ŋ	
Fricative	ß	f	s z	(ʃ)			(h)/hʷ
Affricate					tʃ (dʒ)		
Lat. app.			l				
Approx.	w					y	

e.g., class 7 *èchí-lòtù* vs. class 9 *ín-dòtù* 'good, beautiful'), ß (instead realised as [mb]), and [f]. When not prenasalised, [dʒ] only occurs after [i]. [ŋ] is relatively rare and often attributable historically to Meinhof's Rule (NCVNCV > NVNCV), e.g., *iɲòmbè* 'cattle' (*gombe) and *iɲòngì* 'chief's bell/gong' (*gʊnga; Lozi K21*ngongi*).

When the voiceless glottal fricative occurs before [u] or [w], it is usually labialised, and it and surrounding vowels are often pronounced with some degree of nasalisation. This nasalisation is far more pronounced when *h* is prenasalised (realised as [h̃]), as in class 9 *ínhwì* [ĩh̃w̃ì] 'gray hair.' There are no attested lexical contrasts between nasal and non-nasal *hʷu*. A few speakers also used labialised glottal fricatives where the majority of speakers in my research areas use [fw], suggesting a possible voicing contrast, as in Tonga (Carter 2002: 4–7), but this has not been studied systematically and the voicing contrast is not clear. Non-labialised glottal fricatives [h] occur mostly in borrowings. In Namibian Totela and some other Totela varieties in Zambia, cognate forms (often from Proto-Bantu *bu and *gu) are realised with [ßu] or [vu].

TABLE 22.4 PRENASALISED CONSONANTS

	Bilabial	Labio-dental	Alveolar	Post-alveolar	Palatal	Velar	Glottal
Plosive	mp mb		nt nd			ŋk ŋg	
Fricative			ns nz	(nʃ)			h̃
Affricate					(ntʃ)(ndʒ)		

TABLE22.5 CORRESPONDENCESWITHPROTO-BANTUCONSONANTS

PB	before *i	before *u	elsewhere
*p	s	f	∅, w, p
*b	z	hʷ (h̃ʷ)	ß
*t	s	s, f	t
*d	z	z	l
*c	s	s	s
*j			∅, y, z
*k	s	f	tʃ, k
*g	z	h (h̃)	∅, y, w

In both lexical items and in derivations, Totela exhibits completely reduced spirantisation of stops to alveolar fricatives before PB *i, and partially reduced spirantisation before PB *u. Totela correspondences with Proto-Bantu non-nasal consonants are given in Table 22.5.

2.3 NT-ZT correspondences

Namibian and Zambian varieties have a number of regular sound correspondences, given in Table 22.6 along with the relevant Proto-Bantu correspondences.

TABLE 22.6 ZT/NT SOUND CORRESPONDENCES

*PB	ZT	NT
*p	∅/ p/w	h
*gu,*bu	hu (h̃ũ)/vu/ßu	vu/ßu
*kɪ, *ke	t͡ʃ	t͡ʃ, dʒ
*j	y,z	y
*ŋg	ŋg	ŋg, ndʒ
	dʒ	h

The status of [v] in NT is somewhat uncertain; at least some speakers appear to have merged (or nearly merged) it with [ß], although other speakers use [v] before [u] and [ß] elsewhere. All prenasalised consonants not noted in the table are the same across Totela varieties.

2.4 Tone

Totela's tone system can be analysed as H/Ø. An alternative analysis, proposed by Hyman (2007) for Hyman (2016) for Totela is a L vs. Ø system. Surface (non-intonational) tones can be H or L on short vowels, and H:, L: or HL falling on long vowels. In addition to vowels, syllabic nasals /m/ and /n/ can bear tone word-initially.

Toneless syllables following the last H tone in an utterance are pronounced with extra low pitch, making the H tone perceptually prominent. In general, the most salient feature of a H tone is a subsequent drop in pitch. Toneless vowels preceding H-toned syllables are frequently realised with progressively higher tone, and sometimes even with plateauing, as in Tonga (see e.g., Carter 1971). However, speakers produce these tones variably and accept pronunciations with and without pre-H raising or plateauing, as long as the H tone itself is salient. Downdrift is highly apparent in the realisation of H and L tones alternating in sequence.

Nouns can have underlying H tone on any stem syllable, although surface HH is prohibited. Verbs have a lexical contrast between root H and Ø. Tonal minimal pairs occur both in nouns and verbs, although they are relatively rare (e.g., òkúkùlà 'to grow' and òkùkùlà 'to sweep; òmùbàlà 'person who carried' and òmúbàlà 'colour'). TAMN forms have three main melodic tone patterns, described in Section 4.2.10.

The most striking feature of the Zambian Totela tone system is High-tone anticipation (HTA), in which an underlying H tone surfaces on the preceding syllable. The H tone of underlying H verb roots thus surfaces on (e.g.) the class 15 infinitive prefix oku-, as in (8).

(8)   òkù-wà   'to fall'   òkú-pà   'to give'
     òkù-kùlà   'to sweep   òkú-kùlà   'to grow'
     òkù-ŋàtà(w)ùlà   'to cut to pieces'   òkú-bàbàlèlà   'to care for'

HTA in nouns can be seen in the Proto-Bantu correspondences in Table 22.7. H tones that occur underlyingly on a stem-initial syllable surface on the noun-class prefix. Meeussen's

TABLE 22.7 TOTELA-PB TONE CORRESPONDENCES

	PB	Totela
HH	*-kódó	òmú-kùlù 'adult'
HL	*-bókò	í-bòkò 'arm'
LH	*-gòdí	òlù-wózi 'string'
LL	*-dèdù	òmù-lèzù 'beard'

Rule historically lowered the second H in a H-H sequence within the stem, neutralising the Proto-Bantu H-H/H-L contrast.

High-tone anticipation occurs both across and within words, with some restrictions on sequences of H tones. Underlying H tones on word-initial syllables shift to the final syllable of a preceding word, with some restrictions. As seen in (9), an initial H tone is not realised post-pausally (9a), but can be anticipated onto a prefix or proclitic (9b – d, f), or onto the final syllable of a preceding word. (9e) shows that two initial underlying H tones can condition downstep across words.

(9) a Sìmùnyè(w)ù   'Beetle' (cl 1a)
     b bá=Sìmùnyè(w)ù   'Mr. Beetle' (respect)
     c kwá=Sìmùnyè(w)ù   'to Beetle' (cl 17 locative)
     d kù=bá=Sìmùnyè(w)ù   'to Mr. Beetle' (cl 17 locative, respect)
     e òkù-yá ⁺kwá=Sìmùnyè(w)ù '   'to go to Beetle'
     f òkù-yá kù=bá=Sìmùnyè(w)ù   'to go to Mr. Beetle' (respect)

If an input final H (including grammatical Hs) meets an input initial H, the tone from the initial H is not anticipated onto the final syllable (10b). An exception is in cases of monosyllabic stems (11f).

(10) a òkú-bòn-á à-bàntù   'to see people'
      b tà-ndì-sák-ì à-bàntù   'I don't like people.'   (*tàndìsákí, *tàndìsá⁺kí)
                    ᴴ

Downstep also occurs across words if a final falling tone is shortened to a H tone and followed by another H, as in (10d, f).

(11) a yùmwíì   'another one' (cl 1)
      b ná=yùmwíì   'and another one'
      c hápè   'again' (Lozi K21)
      d ná=yùmwí ⁺hápè   'and another one again'
      e tàlí sìmánkàmbwê   'it's not a bird (sp.)'
             ᴴ
      f tàlí ⁺bá-sìmánkàmbwê   'it's not birds (sp.)'
             ᴴ

HTA across words is in part syntactically conditioned. A H does not seem to shift from a verb to a preceding lexical subject (or any other preceding word) in a main clause (12a). HTA occurs from verbs to preceding words in both subject (12b) and object (12c) relative clauses. That main clause subject markers also carry H tone is suggested by posthodiernal futures, where the posthodiernal marker *na-* surfaces as H (e.g., *náchi̱là̱ùlùkà* 'it will fly'). However, the tone status of word-initial subject markers is not certain in all contexts, and they are therefore not represented as underlyingly H in the rest of this chapter.

(12) a è̱chì-yùnì   chi̱-là-ùlùk-à
        7-bird   SP₇-DJ-fly-FV
        'The bird is flying.'

   b è̱chì-yùní   chà̱-yîmb-à̱ₕ   chi̱-là-ùlùk-à
        7-bird   SP₇.CMPL-sing-FV.REL   SP₇-DJ-fly-FV
        'The bird that sang is flying.'

   c chí̱i̱bó   è̱chì-yùní   mwà̱-bón-à̱?
        NP₇.which   7-bird   SP₂ₚₗ.CMPL-see-FV.REL
        'Which bird did you see?'

Some proclitics, such as comitative *na=* ('with, and'), appear to block HTA, perhaps also for syntactic reasons. Although *na=* can host anticipated H tones, as in (11b, d), they do not surface with H tones when they coalesce with H-toned vocalic augments (or the initial vowel in the case of class 9), nor does the tone shift to the final syllable of the preceding word, as seen in the contrast between (13a – b) and in (13c).

(13) a Ndì-lí-kwè̱sí   i̱n̠ òmbè
        SP₁ₛ୍ୠ-STAT.DJ-hold.STAT   9.cow
        'I have a cow.'

   b Ndì-nà   n-è̱n̠ òmbè
        SP₁ₛ୍ୠ-have   COM-9.cow
        'I have a cow '

   c À̱-mé-è̱nzì   nò̱-bú-kò̱kò,   ndì̱-lá-zi̱-nwà̱
        AUG-6-water   COM.AUG-14-beer   SP₁ₛ୍ୠ-DJ-OP₁₀-drink.FV
        'Water and beer, I drink them.'

Although downstep is generally not seen within inflected verbs (though see (15c)), the negative form of infinitives exhibits downstep with H-toned roots.

(14) a ò̱kù-sèk-à   ò̱kú-sà̱-sèk-à
        INF-laugh-FV   INF-NEG-laugh-FV
        'to laugh'   'to not laugh'

   b ò̱kú-yà̱s-à   ò̱kú-⁺sá̱-yà̱s-à
        INF-spear-FV   INF-NEG-spear-FV
        'to spear'   'to not spear'

Downstep occurs within words where a H tone would otherwise surface on the second mora of a long vowel that is preceded by another H. (Non-intonational) LH vowels are not licit, so the entire vowel is realised as ⁺H (15a – b). Downstep can also occur if gliding or vowel assimilation creates two adjacent Hs (15c).

(15) a   òmú⁺dáálà   'old man' (cl. 1)
     b   í⁺bbúúlè   'weed' (cl. 5)
     c ná-⁺mw-îz-è↓   'come!' (tomorrow or after)

The interactions between syntax, prosody, HTA and downstep, as well as the overall effect of information structure on tonal realisations are not fully understood, and merit further study in Totela. The constraints on downstep across words are discussed further in Section 3.4.3.

2.5 Tone in NT

A major difference between Zambian and Namibian Totela varieties is that while Zambian Totela exhibits H-tone anticipation, Namibian Totela does not appear to do so, as seen in the correspondences in Table 22.8. The tone system of Namibian Totela requires further investigation and analysis.

TABLE 22.8 SOME ZT-NT TONE CORRESPONDENCES

ZT	NT	
àmá-fùtà	mà-fútà	'fat, oil'
òkú-bòn-à	kù-bón-à	'to see'
òkú-lààl-à	kù-láàl-à	'to (go to) sleep'

2.6 Syllable structure

Syllables in Totela can have the shape (N)(C)(G)V(:). At the beginning of a word, a nasal can also be syllabic and carry tone (e.g., ńsènzì 'monitor lizard'). Glides are often inserted between two heterosyllabic vowels.

3 NOUNS AND NOUN PHRASES

3.1 Noun classes

Totela employs 18 noun classes, along with subclasses 1a and 2a, which have slightly different morphology. Classes 1a and 2a trigger class 1 and 2 agreement marking, respectively. Most nouns fall into classes 1–14; class 15 is the infinitive class. A very few nouns (e.g., òkútwì 'ear') also fall into class 15, although they largely eschew class 15 agreement morphology (see Section 3.5 below). Classes 8 and 10 have different nominal prefixes but are identical in their agreement morphology. Classes 16–18 are the locative classes and have limited agreement patterns.

In class 5, the i- prefix alternates with (e)li-. Like in most Botatwe languages, the eli- prefix appears before most vowel-initial stems (see Bostoen 2009); it also appears with several (though not all) class 5 prenasalised stop-initial noun stems, perhaps to distinguish them from class 9 nouns. Examples with eli- include èlînsò 'eye,' èlíìnò 'tooth,' èlíòwà 'cowardice,' èlyálà 'fingernail, claw,' èlìnhù 'wasp,' èlìmbùlùkùtù 'edible tuber (sp.),' ̀elínkàlwê 'edible tuber (sp.),' ̀elìngòngwè(/ìngòngwè) and ̀elìnzìmbwà 'coward-ice.' At least one noun, ị⁺mbá̰là̰ 'slender topminnow,' can take either class 5 or class 9 agreement.

The pronominal prefix agreement marker occurs on possessive pronouns and other connective constructions (Section 3.4.3).

TABLE 22.9 NOUN CLASS AGREEMENT MORPHOLOGY

NC	Prefix	SM	OM	PP/ in connectives	Poss	Num Prefix	Some demonstratives	Adj Prefix
1	(o̰)mu-	a̰-	-mu-	u-	-a̰kwḛ́ḛ̀	e-	yṵ-	(o̰)mu-
1a	∅	a̰-	-mu-	u-	-a̰kwḛ́ḛ̀	e-	yṵ-	(o̰)mu-
2	(a̰)ba-	ba̰-	-ba̰-	ba-	-a̰bò	bo-	ba̰-	(a̰)ba-
2a	ba=	ba̰-	-ba̰-	ba-	-a̰bò	bo-	ba̰-	(a̰)ba-
3	(o̰)mu-	ṵ-	-mṵ-	u-	-a̰wóò̰	wo-	wṵ-	(o̰)mu-
4	(ḛ)mi-	ḭ-	-ḭ-	i-	-a̰yóò̰	yo-	yḭ-	(ḛ)mi-
5	ḭ-/(ḛ)li-	lḭ-	-lḭ-	li-	-a̰lóò̰	lyo-	lḭ	(e-/)ḭ-
6	(a̰)ma-	a̰-	-a̰-	a-	-àwóò̰	o-	a̰-	(a̰)ma-
7	(ḛ)chi-	chḭ-	-chḭ-	chi-	-àchóò̰	cho-	chḭ-	(ḛ)chi-
8	(ḛ)zi-	zḭ-	-zḭ-	zi-	-a̰zóò̰	zo-	zḭ-	(ḛ)zi-
9	ịN-	ḭ-	-ḭ-	i-	-àyóò̰	yo-	yḭ-	ịN-
10	ịN-	zḭ-	-zḭ-	zi-	-a̰zóò̰	zo-	zḭ-	(ḛ)zi-
11	(o̰)lu-	lṵ-	-lṵ-	lu-	-a̰lóò̰	lo-	lḭ-	(o̰)lu-
12	(a̰)ka-	ka̰-	-ka̰-	ka-	-a̰kóò̰	ko-	ka̰-	(a̰)ka-
13	(o̰)tu-	tṵ-	-tṵ-	tu-	-a̰tóò̰	to-	tṵ-	(o̰)tu-
14	(o̰)bu-	bṵ-	-bṵ-	bu-	-a̰bóò̰	bo-	bṵ-	(o̰)bu-
15	(o̰)ku-	kṵ-	-kṵ-	ku-/u-	-a̰kóò̰	(e-/ko-)	kṵ-	(o̰)ku-
16	a-	a̰-	=wóò̰	**	**	**	a̰-	pa-
17	kṵ-	kṵ-	=kóò̰	**	**	**	kṵ-	kṵ-
18	mṵ-	mṵ-	=móò̰	**	**	**	mṵ-	mṵ-

3.1.1 Singular/plural pairings

Attested singular/plural pairings are given in Figure 22.1. As seen, class 6 functions as a "catch-all" plural for many noun classes, including some nouns from classes 1 and 9, which have regular plurals in 2 and 10, respectively; class 11, which has plurals in class 6 and 10 (and sometimes allows either); and classes 14 and 15, which only have plurals in class 6.

3.1.2 Noun class semantics

Totela's noun class semantics are typical for Bantu. Generally speaking, class 1 contains words for people, and class 1a is used for proper names and personifications of animals. Class 3 nouns include names for trees, long objects and other plants and natural phenomena. Class 5 contains a variety of nouns, including fruits, often with the same root as the corresponding tree in class 3. Class 6 is a plural for many classes and also is used as the basic form for liquids and some abstract concepts (e.g., à̰mámbèngò 'tobacco cravings').

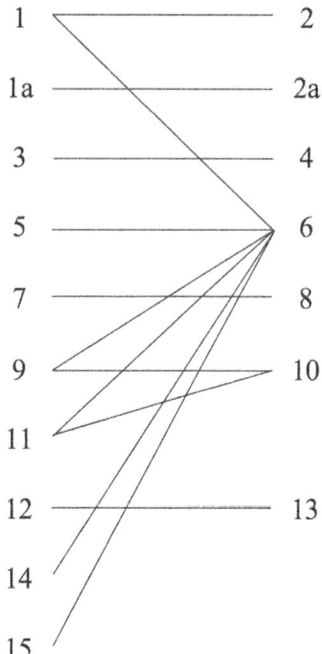

FIGURE 22.1 SINGULAR-PLURAL PAIRINGS.

Class 7 is used for (among other things) languages, tools and concrete objects. Class 9 is a general class, also including fruits and many animals. Nouns are often borrowed into class 9. Class 11 hosts a relatively small number of nouns, including some types of grass; many class 11 nouns may be characterised as long, thin and flexible (e.g., *òlùnyàméènò* 'worm lizards,' *òlùsûngà* 'belt,' *òlùwózì* 'rope,' *òlòngòlà* 'backbone,' *òlùmòmà* 'termites (pl.)'). Class 12 is used with small things; its plural in class 13 can additionally be used for small quantities of mass nouns, as in *òtùbísì* 'a few drops of sour milk.' Class 14 includes abstract qualities and some liquids.

Secondary uses of noun classes include class 5 *i-* as an augmentative, class 7 *echi-* as a general pejorative, or for very strong people or things, and class 12 (and 13) as general diminutives and pejoratives (e.g., *òmùchèmbèlè* cl. 1 'old person'; *àkàchèmbèlè* cl. 12 'decrepit/wicked old person').

Class 2a is used as a marker of respect, and is used almost universally when addressing or referring to adults, much as the second-person plural is preferred to the second-person singular in polite conversation. Classes 1a and 2a are used to refer to many animals, sometimes as personification but often also as a conventionalised reference; the original noun class prefixes (sans augment) are treated as part of the noun stem, as seen in (17) and (18).

(16) bà=Sishau   tà-bà-sák-ì   èchì-sìyù
      2a=Sishau   NEG-SP₂-want-FV.NEG   7-cooked.greens
      '(Mr.) Sishau doesn't want cooked greens.'

(17) chìkwàngálā̱   à-là-ùlùk-à
      1a.pied.crow   SP₁-DJ-fly-FV
      'The pied crow is flying.'

(18) bà=chìkwàngálā̱   bà-là-ùlùk-à
      2a=pied.crow   SP₂-DJ-fly-FV
      'The pied crows are flying.'

3.2 Nominal and adverbial derivation

Deverbal nouns referring to persons who engage in an activity typically take final vowel
-*i*, as in (19). A different pattern is seen in *òmúfù* 'dead person' from *òkúfwà* 'to die.'
Non-person nominal derivations frequently, though not uniformly, take final -*o* (20).
Deverbal nouns, in general, have the same tone pattern as the verb from which they
derive. Lexicalised deverbal nouns exhibit consonant mutation before final -*i* (19e–g);
(20d), while derived deverbal nouns typically do not (19f, h); note especially the form
and (approximate) meaning contrast in (19f). Nouns derived from monosyllabic verbs
were historically subject to minimality constraints, adding a syllable, which was subsequently also affected by consonant mutation (19a-c).

(19) a   ò̱kútwà̱   'to stamp'   ò̱mútwì̱sì   'stamper'
      b   ò̱kúsyà̱   'to dig'   ò̱músì̱sì   'digger'
      c   ò̱kùnyà   'to defecate'   ò̱mùnìnì   'defecator'
      d   ò̱kùlìmà   'to cultivate'   ò̱mùlìmì   'farmer'
      e   ò̱kùlòwà   'to bewitch'   ò̱mùlòzì   'witch'
      f   ò̱kùyèndà   'walk, go'   ò̱mùyènzì̱   '~visitor'   /   ò̱mùyèndì   '~walker'
      g   ò̱kúzā̱àlà   'to give birth'   ò̱múzà̱àzì   'parent'
      h   ò̱kúŋò̱là   'to write'   ò̱múŋò̱lì   'writer'

(20) a   ò̱kúpò̱nà   'to live'   ò̱hú̱pò̱nò̱   'life'
      b   ò̱kùlàpèlà   'to pray'   ì̱ntàpèlò   'prayer'
      c   ò̱kúlò̱òtà   'to dream'   èchílò̱òtò   'dream'
      d   ò̱kúpè̱ngà   'to suffer'   à̱mápè̱nzì   'troubles'

Class 14 *bu-*, used as a subject marker, derives participle-like manner adverbials from
verbs. These adverbials have a relative-clause-like melodic tone pattern in which the H
surfaces on the penult (see Section 4.2.10 for description of this tone pattern) condition
HTA onto the previous word. They appear in constructions following an inflected form
of -*ya* 'go' (21). Gunnink (2018) describes similar forms as "locative pluractionals" in
Fwe K402.

(21) b-à-(y)-á   bù̱-yîmb-à꜍
      SP₂-CMPL-go-FV   SP₁₄-sing-FV
      'They went (along) singing.'

These forms can also be used with iterative or distributive (and often progressive) meaning, as in (22) – (24).

(22) kà-zì-y-á$_H$ bù̠-w-â$_H$
PREHOD.IPFV-SP$_{10}$-go-FV   SP$_{14}$-fall-FV
'They [the beans] were falling [from the baskets as he went along].'

(23) kù-y-á   bù̠-mín-à$_H$ à̠bà-ntù
NARR-go-FV   SM$_{14}$-swallow-FV   2-person
'[the ogre] went around swallowing people.'
kù-y-á   bù̠-mín-à$_H$ kù-y-á   bù̠-mín-à$_H$
NARR-go-FV   SM$_{14}$-swallow-FV   NARR-go-FV   SP$_{14}$-swallow-FV
'He was swallowing and swallowing [them].'

(24) zì-mwî̠,   z-à-kà-y-á   bù̠-yú̠m-à$_H$;
DEMP$_{10}$-some/other   SP$_{10}$-CMPL-PREHOD-go-FV   SP$_{14}$-dry-FV
'some of them [the groundnuts] (have already) dried;
zì-mwî̠,   tà-zí-n-ì$_H$ kú-yù̠m-à
DEMP$_{10}$-some/other   NEG-SP$_{10}$-have-FV.NEG   INF-dry-FV
'Others haven't dried yet.'

Class 14 adverbials can also take distal -ka- marking (25), as well as object marking (26).

(25) bà-kà-y-á   bù̠-kà-yîmb-à$_H$
SP$_2$-DIST-go-FV   SP$_{14}$-DIST-sing-fv
'They went along [somewhere far from here] singing.'

(26) bà-kà-y-á   bù̠-mù-tòbél-à$_H$
SP$_2$-DIST-go-FV   SP$_{14}$-OP$_1$-sing-FV
'They went around looking for him.'

Locative prefixes supplant the augment and carry a H tone that is anticipated on the previous word. Locative prefixes to class 1a nouns require the connective a-.

(27) a   ò̠-mù-lóngà̠
AUG-3-river
'river'

   b   ò̠-kù-y-á   kù=mù-lóngà̠
AUG-INF-go-FV   17(LOC)-3-river
'to go to the river'

3.3 Augment

In Zambian Totela, noun class prefixes have a vocalic augment (pre-prefix) in most contexts. They are absent in vocative (28b), affirmative and negative copular constructions with -li 'be' (29b; 30b) and copular forms for some noun classes (30b), which are also used as citation forms (31b). Augments in Zambian Totela have undergone lowering/aperture, as in, for example, class 1 o̠mu- and class 4 e̠mi-. Classes 5, 9, and 10 retain initial i, with H tone. (An exception is the adverbial ijìlò, which takes class 5 agreement but does not condition HTA; cf. sù̠nù 'today,' which does condition HTA.) Classes 1a and 2a lack augments.

(28) a à̰-kà-móngḛ̀
AUG-12-blue.duiker
'blue duiker'

b kà-móngḛ̀!
12-blue.duiker
'(hey) blue duiker!' (vocative)

(29) a ò̰-mù-ntù
AUG-1-person
'person'

b tà-lí_H   mù-ntù

NEG-be   1-person
'It's not a person.'

(30) a ò̰-mù-lútḭ̀
AUG-1-teacher
'teacher'

b ndì-lì   mù-lútḭ̀   /   ò-mú-⁺kwá̰mḛ̀   mù-lútì
SP_{1SG}-be   1-teacher   AUG-1-man   1-teacher
'I'm a teacher.'   /   'The man is a teacher'

(31) a ḛ̀-chí-sḛ̀mò
AUG-7-love
'love'

b chí-sḛ̀mò
7-love
'love' (citation form)

Augments are another area in which Namibian and Zambian varieties differ significantly. In Namibian Totela, augments have not undergone aperture (class 1 (*u*)*mu*- and class 4 (*i*)*mi*-), and generally only occur when "entrapped" in prosodic units in certain contexts, or in extremely careful speech. Contexts in which the augment is found include following proclitics and prefixes (32) and the final vowels of previous words (as in (34), (36a), (37), and (38a)), except when the noun is directly under the scope of negation (as in (35), (36b), (37), (38b); cf. (36a), (37), where the noun is outside the direct scope of negation and the augment appears) (Crane 2011). The class 9 and 10 augment *i*- seems to have been reanalysed as part of the prefix, possibly in analogy to the class 5 prefix *i*-, and appears in all contexts (36). The augment in Namibian Totela is variably used even in many of the contexts noted here (seemingly independently of information structural considerations), suggesting its on-going marginalisation in that system.

(32) a nechizuni
na=i-chi-zuni
COM=AUG-7-bird
'and/with a/the bird' (NT)

b   zomuzi
z-a-u-mu-zi
PP₁₀-CON-AUG-3-village
'about a/the village' (NT)

c   muzi womwanakazi
mu-zi   w-a-o-mu-anakazi
3-village PP₃-CON-AUG-1-woman
'the village of the woman' (NT)

(33)   **chi-zuni**   chi-uluk-a
7-bird   SP₇-fly-FV
'The bird is flying.' (NT)

(34)   ndisak'ezizuni
ndi-sak-a   **i-zi-zuni**
SP₁ₛ𝒢-want-FV   **AUG-8-bird**
'I want (the) birds' (NT)

(35)   kandisaki zizuni
ka-ndi-sak-i   **zi-zuni**
NEG-SP₁ₛ𝒢-want-FV.NEG   **8-bird**
'I don't want a bird.' (NT)

(36)   a   kandisak'iŋombe kap'echizuni
ka-ndi-sak-i   iŋ-ombe   kapa   **i-chi-zuni**
NEG-SP₁ₛ𝒢-want-FV   9-cattle   or   **AUG-7-bird**
'I don't want a cow or a bird.' (NT)

b   kandisaki chizuni   kap'eŋombe
ka-ndi-sak-i   **chi-zuni**   kapa   iŋ-ombe
NEG-SP₁ₛ𝒢-want-FV   **7-bird**   or   9-cattle
'I don't want a bird or a cow.' (NT)

(37)   kandisaki kuly'echisihu
ka-ndi-sak-i   ku-li-a   **i-chi-sihu**
NEG-SP₁ₛ𝒢-want-FV   INF-eat-FV **AUG-7-greens**
'I don't want to eat cooked greens.' (NT)

(38)   a   ndisaka'hulw'ezizuni
ndi-sak-a   ahulu   i-zi-zuni
SP₁ₛ𝒢-want-FV   a.lot   AUG-8-bird
'I like birds very much.' (NT)

b   kandisaki ahulu zizuni
ka-ndi-sak-i   ahulu   **zi-zuni**
NEG-SP₁ₛ𝒢-want-FV   a.lot   **8-bird**
'I don't like birds much.' (NT)

3.4 Pronouns and nominal modifiers

3.4.1 Demonstratives and pronouns

Totela has a complex system of demonstratives and pronouns, distinguished variously by segmental morphology, final vowel, vowel length and tone patterns. Demonstratives can precede or follow the noun they modify, and also serve as independent (substitutive) pronouns and as relative markers. Some of the more common forms are illustrated in this section. Observations on demonstrative semantics given here should be regarded as tentative.

Personal pronouns are given in Table 22.10. Speakers in some areas use *ínywè* for 2PL, while others use *ínwè*.

TABLE22.10 PERSONALPRONOUNS

Person	Singular	Plural
1	ímè̱	íswè̱
2	íwè̱	ínwè̱/ínywè̱
3	óyṳ	ábà

Table 22.11 gives the major basic demonstrative forms. The nearest demonstratives, ending in -*no*, are used when the speaker (often in contrast to the hearer) is near or inside of the referent. They are frequently used when calling others to come to the speaker and

TABLE 22.11 SELECTED DEMONSTRATIVE FORMS

NC	-no	VCV	VCo	-lya	-lyáà
	'this here' (**ve**ry near speaker)	'this' (near speaker)	'that' (near hearer; anaphoric)	'that over there' (away from speaker)	'that way over there' (far from speaker)
1(a)	?	òyù	òyò	yùlyà	yùlyáà
2(a)	?	àbà	àbò	bàlyà	bàlyáà
3	ùnò	òwù	òwò	wùlyà	wùlyáà
4	inò	èyì	èyò	ilyà	ilyáà
5	lìnò	èlì	èlyò	lìlyà	lìlyáà
6	ànò	àwà	àwò	àlyà	àlyáà
7	chìnò	èchì	èchò	chìlyà	chìlyáà
8	zìnò	èzì	èzò	zìlyà	zìlyáà
9	inò	èyì	èyò	ilyà	ilyáà
10	zìnò	èzì	èzò	zìlyà	zìlyáà
11	lùnò	òlù	òlò	lùlyà	lùlyáà
12	kànò	àkà	àkò	kàlyà	kàlyáà
13	tùnò	òtù	òtò	tùlyà	tùlyáà
14	bùnò	òbù	òbò	bùlyà	bùlyáà
15	kùnò	òyù	òyò	yùlyà	yùlyáà
16	ànò	àwà	àwò	àlyà	àlyáà
17	kùnò	àwà²	òkò	kùlyà	kùlyáà
18	mùnò	òmù	òmò	mùlyà	mùlyáà

referent. They also are used in temporal expressions, e.g., *chìnó èchìlìmò* 'this year.' VCV forms are used when the referent is near the speaker (and often also the hearer). VCo forms are used for referents near the hearer, in contrast to the speaker, and are among the most frequent anaphoric forms. Forms ending in *-lya* indicate distance from speaker (with great distance indicated iconically through length and overall higher pitch, in addition to the final falling tone). In my preliminary data, the short *-lya* forms do not seem to attend to the position of the hearer, so that there may be spatial overlap between the VCo and *-lya* forms; the latter were given more frequently in elicitation. Underlying tones are not indicated for demonstrative forms because not all are confirmed in initial position; tonal variation in demonstratives requires further study.

When used with the comitative proclitic, basic demonstrative pronouns have the form *Cò*. An exception is the class 1 pronoun, which has the form *nà=yè*.

TABLE 22.12 DEMONSTRATIVES WITH THE COMITATIVE PROCLITIC

1	nà=yè	7	nà=chò
2	nà=bò	12	nà=kò
5	nà=lyò	13	nà=tò

Demonstratives frequently appear in complex CaV-initial forms with a surface tone on the penult (falling in long syllables), illustrated in Table 22.13. These forms are extremely common in narrative and may indicate more specific or emphatic reference. The class 14 demonstrative *bóòbò* is frequently used to refer to manner ('like that').

TABLE 22.13 CAV-INITIAL DEMONSTRATIVES

NC	'this'	'that'	'that there'
1	yóòyù	yóòyò	yòyúlyà
2	báàbà	báàbò	bàbályà
7	chéèchì	chéèchò	chèchílyà
13	tóòtù	tóòtò	tòtúlyà

When used pronominally in narrative discourse, VCV demonstratives appear as objects and as the heads of relative clauses (both subject and object). VCo and *-lya* forms are used as subject pronouns, and are frequently used with proclitic *na=* 'and, with,' and falling tone on the final *ô* (e.g., cl. 12 *nàkô*). When used adnominally (as demonstrative adjectives), demonstratives modifying objects take CaVCo form, while those modifying subjects can apparently take all possible forms. *Yènâ* (cl.1) and *bònâ* (cl. 2), apparently borrowed from Lozi K21, also occur frequently as substitutives in texts for personal referents. More research is needed to understand the role of demonstratives in narrative.

Like the personal pronouns in Table 22.10, demonstratives sometimes appear with a H-L surface tone pattern when used as substitutives, but this tone pattern does not seem to be obligatory. The comitative proclitic shows that the consonant-initial forms have distinctions in their tone patterns, as well. Semantic differences between toneless and H-initial forms are not clear to me.

TABLE 22.14 DISTINCTIONS IN DEMONSTRATIVE
TONE PATTERNS (CLASS 12)

nà=tóo̱tù	ná=⁺to̱ó̱tù
nà=tóo̱tò	ná=⁺to̱ó̱tò

Demonstrative forms can be combined, as well, when the first element combines with *na=*. A typical example from a narrative is given in (39). The forms and tone patterns are similar to copular demonstrative forms given in Table 22.15 below. Many other combinations are possible, for example class 10 *ná=zè zílyà* 'and these here (nearby),' *nà=zó zèzílyà* 'and those there,' and *nà=zé zìlyáà* 'and those way over there.'

(39) íngà̱ **nà=tó**   ⁺tóo̱tó   o̱-tù-chèchè   kà-tù-yèmb-èl-â̱=ngá
     DM   COM=DEM₁₃   DEM₁₃   AUG-13-child   PREHOD-SP₁₃-herd-APPL-FV.IPFV=HAB
     'Now, those children used to herd [their families' animals]
     mù̱-mú-tè̱mwà
     18-3-forest
     in the forest'

Demonstratives can also function in non-verbal predication contexts. Copular forms ('it's that/this/these/those') are given in Table 22.15. Presentatives 'here it is' (showing hearer)/'there it is' (giving to hearer), shown in Table 22.16 and (40), are marked with lengthened initial vowels and falling final vowels.

(40) ndí-p-è̱_H   mwâ̱n-ángù̱,   èzì-lòndá   ⁺z-áko̱   èèzî̱_H!
     OP₁SG-give-FV.SBJV   1.child-POSS₁SG   8-wound   PP₈-POSS₂SG   DEM₈
     'give me my child, here are your wounds!'

TABLE 22.15 SOME COPULAR DEMONSTRATIVE FORMS

NC	Proximal	Distal	NC	Proximal	Distal
1	ndiyé ⁺yóòyù	ndiyé yòyúlyà	8	zé ⁺zéèzì	zézè zílyà
2	(m)bá ⁺báàbà	mbàbó bàbályà	9	njé ⁺yéèyì	njéyè ílyà
3	ngwó ⁺wóòwù	ngówò wúlyà	10	zé ⁺zéèzì	zézè zílyà
4	njó ⁺yéèyì	njéyì yílyà	11	ndó ⁺lóòlù	ndólò lúlyà
5	ndé ⁺léèlì	ndélè lílyà	12	ká ⁺káàkà	kákà kályà
6	ngà wáàwà	ngàwá ⁺ályà	13	tó ⁺tóòtù	tótò túlyà
7	ché ⁺chéèchì	chéchè chílyà	14	bó ⁺bóòbù	mbóbò búlyà

TABLE 22.16 SOME PRESENTATIVE DEMONSTRATIVE FORMS

NC	Presentative 'here' (used for pointing)	Presentative 'there' (used for giving)
7	èèchî	èèchô
8	èèzî	èèzô
9	èèyî	èèyô

Another series of demonstratives, formed with the pronominal prefix and the ending *-àló* (cl. 1 and 3 *wàlô*, cl. 4 *yàlô*, cl. 5 *lyàlô*, cl. 6 *àló*, etc.), is nearly identical to Tonga pronouns (Carter 2002: 42) and is translated by speakers as pronominal, but is also used adnominally for entities with some proximity in space and time (e.g., *lwàló o̱lwímbo̱ mwìná o̱kúyìmbà* 'that song you're singing' vs. *lùlyáá* (*o̱lwímbo̱*) in reference to a song that used to be sung long ago). These forms occur rarely in elicitation and in my text corpus, and their overall status in the system is unclear to me.

3.4.2 Adjectives

As is typical for Niger-Congo languages, Totela has a closed, relatively small class of adjectives. Attested adjectives, given with class 7 prefix *echi-* to show the tone patterns, include *e̱chíne̱nè* 'big,' *e̱chìnii̱níi̱* 'small,' *e̱chìpya̱* 'new,' *e̱chìchèmbèlè* 'old,' *e̱chíku̱lù* 'old, grown,' *e̱chìchèchè* 'young' *e̱chìle̱* 'long; tall,' *e̱chífu̱wì/e̱chi⁺fwífwì* 'short,' *e̱chílo̱tù* 'good,' *e̱chíbi̱sì* 'unripe,' *e̱chíbi̱* 'bad,' *e̱chíka̱lì* 'bad, angry, sharp,' *e̱chíku̱kùtù* 'hard, difficult,' *e̱chíka̱báhù* 'troublesome, difficult, stubborn,' *e̱chíka̱lì* 'sharp, angry,' *e̱chìtòlò* 'blunt, lazy,' *e̱chìtòlè* 'soft, light, easy' and *e̱chìtéte̱* 'soft, smooth, fresh, tender, easy(?).' This list is not necessarily exhaustive. Quantifiers *-ngi* (in cl. 6, *a̱ma̱ngi̱*) 'many' and *-che̱* (*a̱máche̱*) 'few' also take adjective morphology. Colour terms are generally verbal, e.g., *o̱kúsu̱bilà* 'become red, ripen' and *o̱kùsìyà* 'become dark/dirty.'

Adjectives can agree with any noun class, and adjective prefixes are identical to those of the nouns they agree with. Exceptions are class 10 nouns, where the noun prefix is *iN-* and the adjective prefix is *(e)zi-*, and class 16 locatives, where the noun prefix is *a-* and the adjective prefix *pa-*.

Comparison is expressed using locative *ku-* and with verbs such as *-i̱ta* 'pass,' *-i̱tilila* '(sur)pass,' *-si̱ya* 'leave (behind)' and *-swa̱na* 'resemble.'

(41) óyu̱   mù-lé   kwe̱nù
     1.pron   NP₁.COP-tall   17.PP. 2PL.POSS
     'He is taller than you.'

(42) óyu̱   mù-lé   ⁺ku̱-mi̱-i̱tìlìl-à
     1.PRON   NP₁.COP-tall   17-OP₂ₚ-surpass-FV
     'He is taller than you.'

(43) ínwe̱   tà-mú-lì   bà-le̱,   kù-mú-i̱tìlìl-à
     2PL.PRON   NEG-SP₂ₚ-be   NP₂ₐ-tall   18-OP₁-pass-FV
     'You are less tall than he.'

(44) ínwe̱   mù-lì   bà-le̱,   mù-lá-swa̱n-à   nà-yo̱yù
     2PL.PRON   SP₂ₚ-be   NP₂-tall   SP₂ₚ-PRES.DJ-resemble-FV   COM-1.PRON
     'You are as tall as he.'

(45) ndì-là-sàk-á   à-mà-nàwá   o̱-kú-si̱y-á/o̱-kú-i̱t-á   à-mà-kwìlì
     SP₁ₛɢ-DJ-like-FV   AUG-6-beans   AUG-15-leave-FV/AUG-15-pass-FV   AUG-6-potato
     'I like beans better than potatoes.'

3.4.3 Connective constructions

Connective constructions are formed with noun class agreement and the connective marker -*a*-. When *chi*- (cl. 7) and *zi*- (cl 8/10) prefixes combine with the connective (e.g., cl. 7 *chi-esu* -> *chèsù* 'our'), the prefix vowel is nearly lost, although some trace of a glide may remain (i.e., [tʃʲèsù]); with other noun classes, /i/ surfaces as a glide (e.g., cl. 5 *lyèsù*; cl. 9 *yèsù*). Final /a/ elides and final /u/ becomes a glide (e.g., class 11 *lwèsù*).

(46) i̠bwè lyéèchìyùnì
      ì-bwè   li-á-è̠-chì-yùnì
      5-stone   PP₅-CON-AUG-7-bird
      'the stone of the bird'

As is common in Bantu (Van de Velde 2006), classes 1a and 2a behave somewhat idiosyncratically. Because they have no augment or prefix, class 1a nouns do not have long vowels in the first syllable of connective constructions. Class 2a nouns do not take the connective in such constructions. Example (47) shows connective constructions with the singular and plural forms of the class 1a noun *sókwè̠* 'monkey.'

(47) a i̠mpèmò yàsókwè̠
         i̠m-pèmò   ì-à-sókwè̠
         9-nose   PP₉-CON-monkey
         'the monkey's nose'

      b i̠mpèmò zìbàsókwè̠
         i̠m-pèmò   zì-bà-sókwè̠
         10-nose   PP₁₀-2A-monkey
         'the monkey's noses'

As with other modifiers (see 3.5 below), connective constructions involving locative prefixes take the agreement prefix of the base noun, and not the locative, as in (48)–(49).

(48) è̠jùlù lyè̠nyándà̠
      à-ì-jùlù   li-á-i̠n-yándà
      16-5-sky   PP₅-CON-9-house
      'on top of the house'

(49) kù̠-bú-bà̠lì   bwè̠-táfù̠lè
      17-14-side   PP₁₄.CON.5-table
      'on the edge of the table'

Personal possessive pronoun stems are given in Table 22.17; possessive stems for noun classes are given above in Table 22.9. Possessive pronouns are composed of the connective -*a*- marker and a possessive stem. Possessive pronouns agree in noun class with both the possessor (stem) and the possessed (prefix). All possessive pronouns are realised with some degree of length on the first vowel; however, the long final vowels represented as double vowels are pronounced significantly longer.

TABLE 22.17 PERSONAL POSSESSIVE PRONOUNS

Person	Singular	Plural
1	-ángù	-èsù
2	-ákò̱	-ènù

Possessive constructions seem to present a special syntactic case with regard to across-word tonal anticipation. Whereas in most cases, an input H on the first syllable of a word (indicated by underlining) surfaces on the final syllable of the preceding word within its domain, unless the final syllable of the preceding word also has an input H (see Section 2.3),[3] downstep-conditioning tone shift is generally avoided from possessive pronouns if either the final or the penultimate syllable of the preceding word has an input H tone (50b–d). Instead, the second H tone is deleted.

(50) a ò̱mùlòlá ⁺wángù   'my soap'   ò̱mùlòlá wè̱sù   'our soap'
      b è̱chìntólò̱ chángù   'my store'   è̱chìntólò̱ chè̱sù 'our store'
      c è̱chísè̱mò chángù   'my love'   è̱chísè̱mó chè̱sù   'our love'
      d ò̱búùchì bwángù   'my honey'   ò̱búùchí bwè̱sù   'our honey'

3.5 Agreement and word order in the noun phrase

Agreement morphology is given in Table 22.9 above. Most adjectives take nominal morphology, while some with demonstrative reference ('some,' '(an)other, the other') take demonstrative prefixes, as in (51). Speakers are often reluctant to use class 15 agreement markers and the class 15 possessive stem -akóò̱ – though they give these forms in elicitation after careful consideration – and instead substitute class 1 agreement marking, especially with the few lexical nouns in class 15. Class 15 agreement is somewhat easier with verbal (infinitive) nouns, but speakers still disprefer some agreement patterns (e.g., cl. 15 object markers), likely for semantic reasons. Locative classes 16–18 display reduced agreement patterns, and their agreement sometimes defaults to class 17. As noted in Section 3.4.3 above, adjectival and other targets agree with the base noun and not with the locative marker (51).

(51) kwà-ndáhù̱   yù-mwî̱ị
      15.CON-1a.lion   DEMP₁-other
      'to another lion'

The most common word order within the noun phrase is demonstrative/quantifier – noun – possessive – adjective, as in (52). In every case, the possessive pronoun must precede all adjectives.

(52) ndì-sàk-à   chí-lyà̱   è̱chì-yùní   ch-è̱nú   è̱chí-lò̱tù
      1SG-want-FV   DEMP₇-DEM   7-bird   PP₇-2PL.POSS   NP₇-good
      'I want that good bird of yours.'

Adjectives have at least somewhat flexible order.

(53) a à̰-bà-lóbà̰ná   à̰-bà-lḛ̂   à̰-bá-bḭ̀
AUG-2-boy   AUG-NP$_2$-tall   AUG-NP$_2$-bad
'the big bad boys'

   b à̰-bà-lóbà̰ná   à̰-bá-bḭ̀   à̰-bà-lḛ̂
AUG-2-boy   AUG-NP$_2$-bad   AUG-NP$_2$-tall
'the bad big boys'

With double connective constructions, either order is possible, as seen in (54).

(54) a ín-sḭ̀mà   y-á̰ngṵ̀   y-á-à̰-mà-bélḛ̀
9-porridge   PP$_9$-1SG.POSS   PP$_9$-CON-AUG-6-millet
'my porridge of millet'

   b ín-sḭ̀mà   y-á-à̰-mà-bélḛ̀   y-á̰ngṵ̀
9-porridge   PP$_9$-AUG-CON-6-millet   PP$_9$-1SG.POSS
'my millet porridge'

Personal pronouns take class 1 (singular) or class 2 (plural) adjectival agreement; numeral agreement patterns (used both adnominally and predicatively) are given in Table 22.18.

TABLE 22.18 NUMERAL AGREEMENT PATTERNS
WITH FIRST AND SECOND PERSONS

Person	Singular	Plural
1	nde-	twe-
2	we-	mwe-

4 VERBS AND CLAUSE STRUCTURE

4.1 Verbs

4.1.1 Tone and syllable structure

Verb stems are underlyingly H or toneless (with H tones surfacing, under the appropriate conditions, on the prefix immediately preceding the root). The first syllable of a verb stem can be short or long, but only bisyllabic stems show tonal distinctions between short and long stems in melodic tone patterns. In longer stems the tonal distinction is neutralised, and the length distinction is also somewhat less salient. Most stems longer than two syllables involve verbal extensions or reduplications. Bisyllabic stems can have the root structures -V(N)C- or -C(G)V(V)(N)C-, where C can be any consonant, including nasals or glides (although glide-glide and nasal-nasal sequences are not licit). I have only a few attested examples of NC-initial verbs, and all involve reduplication and are reminiscent of ideophones in meaning and form: ò̰kùngùngùnsà 'to shake (tr.)'; ò̰kùmbèmbwètà 'to quiver momentarily'; ò̰kùngùngùmà̰nà 'to sit alone'; I therefore do not represent such forms as a generally possible root structure.

   Vowel-initial stems are much rarer than consonant-initial stems. No e-initial stems are attested. A majority of stems beginning in i- are at least trisyllabic and cause the infinitive prefix ò̰ku- to glide to ò̰kw-. Most of these are H toned. The H surfaces on the i rather than on the infinitive prefix (ǫ̀kúi̱tà 'to pass' vs. òkwíjǫ̀là 'to close'), and the *i* does not reduplicate (see 4.1.3). These verbs have cognates without initial *i*- in languages like Tonga M64.

4.1.2 Verbal extensions

Totela makes broad use of verbal extensions. Extensions with some degree of productivity include **applicative** -*il*-/-*el*-/-*in*-/-*en*-/-*iz*-/-*ez*- (e.g., ǫ̀kúyǫ̀sìlà 'to spear for,' ǫ̀kúyèchèlà 'to roast for,' ǫ̀kútùmìnà 'to send to,' ǫ̀kùwùlìsìzà 'to sell for,' ǫ̀kùsòtòkèlá mùchífùndà 'to jump into a circle,' ǫ̀kùsòtàwùkèlá mùchífùndà 'to jump up and down inside a circle'); **causative** -*is*- and -*i*- (the latter triggering consonant mutations; see below for examples); **reciprocal** -*an*- (tùlàsàkànà/tùlálìsàkànà 'we like each other,' with and without reflexive -*li̱*-); **intensive/completive** -*ilil*-/-*elel*-/-*inin*-/ -*enen*-/-*isis*-/-*eses*- (ǫ̀kùsàmbìlìlà 'to bathe and become completely clean,' ǫ̀kwìlìlìlà 'to go away forever,' ǫ̀kùnyàmùkìlìlà 'to set off all together,' ǫ̀kúsùwìsìsà 'understand completely'); **passive** -(*i*)*u*- (ǫ̀kúkàmbìwà 'to be licked,' ǫ̀kùwàmbìlwà 'to be told'; *i* insertion is at least partially a matter of speaker preference; -*u*- is generally preferred following *l* and other extensions); **iterative** -*a*(*w*)*ul*-/-*a*(*w*)*uk*- (ǫ̀kùtyòòlàwùlà 'to break into many pieces'; ǫ̀kùkàbàwùlà 'to beat repeatedly': ǫ̀kùwàmbàùlà 'to chat, discuss, converse'; ǫ̀kúbù̱zàùlà 'to ask many questions'); **habitual** -*ang*-/=*nga* (tùláyǫ̀sàngà 'we spear (regularly)'; and – marginally productive, if at all, and often non-compositional in meaning – **neuter** -*ik*-/-*ek*- (òkùkònzèkà 'to be possible,' ǫ̀kútù̱mìnìkìzà 'to send someone for something' (applicative + neuter + causative)) and **reversive** -*ul*(*ul*)-/ -*un*(*un*)- (tr.) and -(*ul*)*uk*- (intr.) (ǫ̀kwíjǫ̀lùlà 'to open,' ǫ̀kùsàndùkà 'to change (intr.),' ǫ̀kùzìngùlùkà 'to turn around, encircle,' ǫ̀kúsù̱mùnùnà 'to untie').

Frozen, non-productive extensions include **extensive** -*al*- (ǫ̀kwíkǫ̀là 'to sit down, stay'); **tentive** -*at*- (ǫ̀kùlàmàtìlà 'to stick to (appl.)'); and **positional** -*am*- (ǫ̀kùkànàmà 'to sit alone, doing nothing,' ǫ̀kùbbwàtàmànà 'to sit (said of a very fat person)'). Extensions with /i/ (excluding causative forms) are subject to mid-height harmony (*i* > *e* after *e* and *o*) and extensions with /l/ are subject to nasal harmony (*l* > *n* after nasals) and "harmonic mutation" (*l* > *z* after causative forms).

As noted above, the causative has two forms, -*is*- and -*i*-, the latter triggering somewhat complex consonant mutations. The forms can have slightly differing semantics: when they contrast, -*i*- functions more as a prototypical causative, while -*is*- also functions as an intensifier, as in (55b).

(55) a   ǫ̀kù-làwùs-à
          INF-run.*i*.CAUS-FV
          'to make run'

     b   ǫ̀kù-làwùk-ìs-à
          INF-run-*is*.CAUS-FV
          'to run hard/to run a lot'

The causative forms can combine, as in (56). Note that the causative forms seem to be added iteratively, since consonant mutation applies even though the -*i*-causative appears after the -*is*-causative.

(56) òkù-yènz-ìs-y-à   (-yenda 'walk')
INF-walk.*i*.CAUS-*is*.CAUS-*i*.CAUS-FV
'to make walk a lot'

Causative meanings, in addition to the canonical causative and intensifying functions, also include instrumental (57) and assistive (58).

(57) ín-kòlì   y-ó-ò-kù-yènz-y-à
9-stick   PP₉-CON-AUG-INF-walk-*i*.CAUS-FV
'walking stick'

(58) ò-kù-yènz-y-á   à-kà-chèchè
AUG-INF-walk-*i*.CAUS-FV   AUG-12-child
'to help the little child walk'

Attested consonant mutations with the -*i*- causative are summarised in Table 22.19.

TABLE 22.19 CONSONANT MUTATIONS WITH CAUSATIVE -*I*-

> z(y)	> s(y)	> y
(n)d:	t	b
(n)g	k	
(m)b	l	
l		
y		

The passive extension always appears immediately before the final vowel, hence -*sęs-a* 'marry' > -*sęs-w-a* 'get married' (said of a woman), and the passivisation of stative final -*ite*/-*ete* as in (59).

(59) ndì-lí-sès-èt-w-è
SP₁ₛG-STAT.DJ-marry-STAT-PASS-STAT
'I am married' (said by a woman)

The habitual extension shows evidence of being both a suffix (-*nga*-) and a postclitic (=*nga*). In TAM-forms with a melodic penultimate H, such as the Prehodiernal Imperfective in (60), the melodic H can surface either on -*ang*- (as HL because of automatic lengthening before prenasalised stops) or on the penultimate stem syllable preceding =*nga*.

(60) a   kà-tù-tòbèl-âng-à
         $_H$
     PREHOD.IPFV-SP₁ₚₗ-seek-HAB-FV
     'We used to seek.'

   b   kà-tù-tòbél-à=ngà
         $_H$
     PREHOD.IPFV- P₁ₚₗ-seek-FV=HAB
     'We used to seek.'
     -*nga* also functions as an auxiliary verb (61).

(61) a bá-ngà~H~~~ nà-bà-y-á mṳ̀-kù-nèng-à
SP₂-HAB SIT-SP₂-go-FV 18-INF-dance-FV
'They (habitually) go dancing.'

b tà-ndí-ngà~H~ nà-ndì-làwúk-à~H~
NEG-SP₁ₛɢ-HAB SIT-SP₁ₛɢ-run-FV
'I don't (regularly/ever) run.'

c tà-ndí-ngà~H~ nà-ndì-làwùk-âng-à~H~
NEG-SP₁ₛɢ-HAB SIT-SP₁ₛɢ-run-HAB-FV
'I don't (regularly/ever) run.'

4.1.3 Reduplication

Reduplication is common, with meanings including (at least) 'do X a little,' 'do X little by little,' 'do X poorly,' 'do X repeatedly/a lot' and 'do X here and there.' Full-stem reduplication (without object prefixes) is the most common form. Tones are not reduplicated; instead, the reduplicated form behaves like a long verb stem. Lack of tonal reduplication can be seen in (62), where only one H tone is anticipated. (63) shows a longer reduplicated form with a penultimate H melodic tone.

(62) tù-lá-pòn-à-pòn-à
SP₁ₚₗ-PRES-live-FV-live-FV
'We're just getting by.'

(63) twà-y-á bṳ̀-tòngàùk-à-tòngàúk-à~H~
SP₂ₚₗ.CMPL-go-FV 14-talk-FV-talk-FV
'We went along chatting.'

In general, stems with one verbal extension reduplicate in full. Stems with more than one extension optionally omit one or more extensions on the first member of the reduplicated pair. Speakers differed as to whether they accepted the form in (64c).

(64) a bà-lá-ya̱k-ìl-àn-à-yàk-ìl-àn-à
b bà-lá-ya̱k-ìl-à-yàk-ìl-àn-à
c ?bà-lá-ya̱k-à-yàk-ìl-àn-à
SP₂-DJ-build-?(APPL)-(RECP)-FV-build-APPL-RECP-FV
'They build for each other here and there.'

i-initial stems in which H tones surface on the initial vowel (rather than on the preceding toneless morpheme) generally do not reduplicate the initial *i*.

(65) ndìlé̱ngìlà-ngìlà
ndì-là-íng-ìl-à-ng-ìl-à
SP₁ₛɢ-DJ-enter-APPL-FV-enter-APPL-FV
'I enter here and there.'

Lexicalised partial reduplication is evident on many verb roots, such as -*ngùngùmànà* 'to sit alone,' -*chènchèntà* 'sift by tossing,' -*chòchòmà* 'crackle in pan,' -*yo̱yòmòkà* 'hallucinate,' -*mwè̱mwètèlà* 'smile' and -*ka̱kàtìlà* 'stick to (as thorns); persevere.'

4.2 Tense, aspect, mood and negation

Totela has a rich system for expressing tense, aspect and mood distinctions, described in detail in Crane (2011). The following sections describe the most commonly used forms. Totela's conjoint/disjoint contrast is described in 4.3.2. The discussion presupposes the onset-nucleus-coda phasal structure of verbal events set out in, for instance, Botne and Kershner (2000). In change-of-state verbs (see, e.g., (52)), the **nucleus** encodes the state change itself, while the **coda** phase encodes the ensuing state. The **onset** phase of such verbs, when lexically encoded, consists of the processes leading up to the state change. In non-state-change verbs, including a few statives, the nuclear phase encodes the action or state referenced by the verb. **Perspective time** is the time for which the utterance's truth value is evaluated, usually utterance time.

4.2.1 Tense and aspect

Totela verbal morphology distinguishes situations occurring on or overlapping the day of utterance/perspective time (hodiernal) and situations preceding or following the day of utterance/perspective time (pre- and posthodiernal, respectively). The day begins at the time of going to sleep (regardless of the time of sunset). Only prehodiernal pasts and posthodiernal futures have specific tense morphology; all other morphology may be considered primarily aspectual, with the possible exception of the past marker -na- (see Section 4.2.2.3 below). However, most forms have both temporal and aspectual functions, and a strict division into tense vs. aspect would be artificial.

4.2.2 Aspectual distinctions with tense-like functions within the hodiernal domain

4.2.2.1 Completive aspect

Most hodiernal pasts are marked with -a-, which I argue is a marker of nuclear Completion (Crane 2012a); that is, the primary lexical content of the verb is prior to the moment of speech (or other salient perspective time). As such, forms with -a- can have (i) perfect (anterior) interpretations, if the perspective time is construed as within the coda phase (present state reading of change-of-state verbs) or other relevant post-state (perfect reading of non-state-change verbs); or (ii) perfective (past) interpretations otherwise. Unlike a dedicated perfect form, -a- does not always entail relevance at the time of utterance. Thus, forms with Completive -a- reference a past situation with activity verbs (66) and stative verbs (67). With change-of-state verbs, such as -komokwa 'be(come) surprised,' the most common reading is that of a present state (68). However, the Completive form of these verbs can also refer to a state change in the hodiernal past, whether the results hold at the time of utterance or not (69).

(66) nd-à-nèng-à
SP$_{1SG}$-CMPL-dance-FV
'I (have) danced.' (earlier today)

(67) nd-à-chìs-w-à
SP$_{1SG}$-CMPL-hurt-PASS-FV
'I was sick' (earlier today)

(68) nd-á-kòmòk-w-à
      SP$_{1SG}$-CMPL-surprise-PASS-FV
      'I'm surprised!'

(69) nd-á-kòmòk-w-á   sùnù
      SP$_{1SG}$-CMPL-surprise-PASS-FV   today
      'I had a surprise today!'

In general, forms such as those in (66) – (70) refer to events that occurred on the day of perspective time (hodiernal pasts). Completive morphology can also be used with events that occurred prior to the day of utterance, if the events have continuing and relevant results at utterance time (perfect reading), or if the completion of the event occurred within the day of perspective time.

4.2.2.2 Non-Completive aspect

Present (progressive/stative/habitual) and hodiernal future forms are unmarked, although they typically appear with the disjoint marker -la-.[4] Crane (2011, 2012a) argues that these forms can situate the perspective time at any point prior to the completion of a situation's nuclear phase. When used as futurates, -la- forms typically refer to situations expected to occur on the day of perspective time, but are not restricted to the hodiernal domain.

(70) ndì-là-nèng-à
      SP$_{1SG}$-DJ-dance-FV
      'I dance/am dancing/will dance.'

While Non-Completive forms can have both progressive and habitual readings, these meanings can also be expressed morphologically. For events ongoing at perspective time, progressive forms (see 4.2.4.1) are often preferred. Habitual readings can be overtly specified with the habitual extension, as in (71).

(71) tù-là-byàl-àng-á   ì-hûmbì
      SP$_{1P}$-DJ-sow-hab-FV   AUG.5-hot.season
      'We sow in the hot season.'

4.2.2.3 Hodiernal imperfective aspect

Hodiernal past imperfectives are marked with -na-,[5] which is also the primary Past marker in negated pasts (hence glossed as Past). By default, -na- targets the coda state of many change-of-state verbs (72), the nucleus of non-change-of-state verbs (73), including some statives (74), and the onset of some change-of-state verbs with extended onset phases (75). Unlike Prehodiernal Imperfective ka- (see Section 4.2.3.2), -na- is incompatible with the stativising -ite suffix (see Section 4.2.4.2) and with stative roots such as -ina 'have/be with.' When -na- is used together with the Completive, it gives pluperfect readings (76); these forms can also express experiential perfect meaning. The final vowel of -na- forms is -i with monosyllabic and other stems that take final -i in negative forms (see 4.2.8).

(72) sùnú   èchí-fù̱mò   **ndì-nà-táb-à**_H
today   7-morning   SP_{1SG}-PST-become.happy-FV
'This morning, **I was happy**.'

(73) àwá   **ndì-nà-láwùk**_H**-à**   ndà-wààn-á   ò̱mù-ntù   nà-lyâ_H
DEM₁₆   SP_{1SG}-PST-run-FV   SP_{1SG}.CMPL-find-FV   1-person   SIT-eat.FV
'While **I was running**, I came across a person eating.'

(74) ndì-nà-chís-w-à̱
SP_{1SG}-PST-hurt-PASS-FV
'I was sick.' (earlier today)

(75) chì-nà-bômb-à_H
SP₇-PST-soak-FV
'It was soaking.'

(76) Sù̱nú   èchí-fù̱mò   àwá   **mù̱-ná-chì-lààl-à**_H,   ímè̱
today   7-morning   DEM₁₆   SP_{2PL}-PST-PER-sleep-FV   1SG.PRON
'This morning while **you were still sleeping**, I

ndà-yá   kù̱=mpìlì.   Àwá   mwà̱-búùk-à_H,
SP_{1SG}.CMPL-go.FV   17=9.fields   DEM₁₆   SP_{2PL}.CMPL-wake.up-FV
went to the fields. When you woke up,

ndâ_{H?}**-nà-bóól-à**_H   kàléè̱.
SP_{1SG}.CMPL-PST-return-FV   already
I had returned already.'

4.2.3 Pasts and futures

4.2.3.1 Prehodiernal Completives

Prehodiernal Completives combine Completive *-a-* with Prehodiernal *-ka-* marking, and have prehodiernal perfective semantics.

(77) nd-à-kà-nèng-à
SP_{1SG}-CMPL-PREHOD-dance-FV
'I danced' (yesterday or before)

4.2.3.2 Prehodiernal Imperfectives

Like Prehodiernal Completives, Prehodiernal Imperfectives have a *ka-* prefix. With Prehodiernal Imperfectives, the prefix occurs before the subject prefix, and forms surface with melodic H on the penultimate syllable. They are compatible with habitual extensions.

(78) kà-tù-hùpúl-à_H
PREHOD-SP_{1SG}-think-FV.IPFV
'We were thinking.'/'We used to think.'

4.2.3.3 Posthodiernal Futures

Posthodiernal Futures are marked with the prefix *na-*, as in (79). The use of Posthodiernal morphology is frequently optional, and can depend on speaker certainty or intentionality regarding the future situation described: situations certain to occur often lack posthodiernal morphology. Posthodiernal Futures are underspecified for aspect; an aspectually imperfective future form is given in (106).

(79) ná-ndì-là-nèng-à
      POSTHOD-SP$_{1SG}$-DJ-dance-FV
      'I will dance.' (after today)

4.2.4 Other aspectual forms

4.2.4.1 Progressive (/imperfective)

There are several means of expressing explicitly progressive aspect, the most common of which are given in (80)–(82). The construction shown in (80), using the stative form of *-kwạta* 'grasp, catch, hold,' is incompatible with stative verbs (e.g., *-saka* 'want, like, love'). In contrast, a progressive formed with the defective stative verb *-ina* 'be, have' (81) can also have present stative meaning, as in (82). It is also in some cases compatible with habitual meanings and the habitual *-nga-* extension, and may thus be considered more of a general imperfective with progressive functions.

(80) ndì-lí-kwẹ̀sì   ndì-là-yènd-à
      SP$_{1SG}$-STAT.DJ-hold.STAT   SP$_{1SG}$-DJ-walk-FV
      'I am walking.'

(81) mwìná   ọ̀-kù-yènd-à-yènd-á   kù-ŋándạ̀
      SP$_{2SG}$.be/have   AUG-15(INF)-walk-FV-walk-FV   17-9.house
      'You're pacing in front of the house.'

(82) ndìná   ọ̀-kù-sàk-à
      SP$_{1SG}$.be/have   AUG-15(inf)-want-FV
      'I want'

Past progressives usually take the relevant Imperfective morphology (4.2.3.2); Prehodiernal Imperfectives can also combine with the *-kwesi* progressive; see (90) below. Future progressives (both hodiernal and posthodiernal) take subjunctive-like morphology (see 4.2.7 below).

4.2.4.2 Stative

The suffix *-ite* (*-ete/-ile/-ele/-ine/-ene*/imbricated forms) functions as a stativiser (Crane 2013), with resultative semantics that are not necessarily confined to the result phase of a verb's event structure. As such, it is occasionally used to express progressive-like meanings. With change of state verbs, which have a defined result phase, *-ite* generally describes the on-going result state, as in (83) – (84).

(83) ndì-lì-tàb-ìtè
    SP$_{1SG}$-STAT.DJ-become.happy-STAT
    'I am happy.'

(84) kà-ndì-táb-ì$_H$tè
    PREHOD-SP$_{1SG}$-become.happy-STAT
    'I was happy.'

However, Stative -*ite* can also target the nuclear phase of some predicates, as in (85)–(87), or a derived coda phase in predicates not inherently associated with one.

(85) ndì-lì-yènd-ìtè
    SP$_{1SG}$-STAT.DJ-walk-STAT
    'I am walking.'

(86) ndì-lí-bwènè
    SP$_{1SG}$-STAT.DJ-see.STAT
    'I see.'

(87) tù-lí-yàak-ité $^{+}$á-fùwì   nòmù-lóngà
    SP$_{1PL}$-STAT.DJ-build-STATE 16-short   COM.3-river
    'We [our house] are built close to the river.'

Some progressive-like uses of -*ite* are marginal, but evidence suggests that these uses are becoming more widespread across lexical event types, and in Namibian Totela and other Zambezi region languages, use of -*ite* and related morphemes with progressive-like aspectual reference and other non-canonical resultative meanings is even more prevalent (see Crane 2012b for a survey).

4.2.4.3 Persistive

Persistive aspect is marked with the post-SP prefix -*chi*-, which can co-occur with other tense/aspect markers but not with Completive forms, or with disjoint -*la*-. Its approximate meaning is 'still' in the affirmative, and 'not anymore' when negated. In the affirmative, it conditions a melodic H that surfaces on the penult. Like the Non-Completive, it can be used with both present and futurate meaning when otherwise unmarked for tense.

(88) tù-chì-kà-mù-zíìk-à$_H$
    SP$_{1PL}$-PER-DIST-OP$_1$-bury-FV
    'We're still going to bury him [there].'

4.2.4.4 Situative

Situative *na*-, used in subordinate clauses, indicates that a situation includes or is coextensive with the main-clause situation.

(89) bà-kà-ndì-wààn-à   nà-ndì-lyâ$_{H}$
      SP$_2$-CMPL.PREHOD-OP$_{1SG}$-find-FV   SIT-SP$_{1SG}$-eat.FV
      'They found me eating.'

Past progressives with -*kwesi* also use situative marking on the second verb.

(90) kà-ndì-kwèsì$_{H}$ nà-ndì-sâmb-à$_{H}$
      PREHOD-SP$_{1SG}$-hold.STAT   SIT-SP$_{1SG}$-bathe.FV
      'I was bathing.'

4.2.5 Distal

Another -*ka*- prefix functions as a Distal marker. Distal -*ka*- and Prehodiernal -*ka*- can co-occur on the same verb. The two forms trigger different melodic tone patterns in relative clauses.

(91) nd-à-kà-sàmb-à
      SP$_{1SG}$-CMPL-DIST-bathe-FV
      'I bathed.' (elsewhere)

(92) nd-à-kà-kà-sàmb-à
      SP$_{1SG}$-CMPL-PREHOD-DIST-bathe-FV
      'I bathed.' (elsewhere, yesterday or before)

(93) a   nd-à-kà-wáàn-à$_{H}$
          SP$_{1SG}$-CMPL-PREHOD-find-FV.REL
          '. . . [that] I found' (yesterday or before)
      b   nd-â$_{H?}$-kà-wààn-à
          SP$_{1SG}$-CMPL-DIST-find-FV.REL
          ' . . . [that] I found' (elsewhere)
      c   nd-à-kâ$_{H?}$-kà-wààn-à (or ndà-kâ$_{H?}$-kà-wáàn-à$_{H}$)
          SP$_{1SG}$-CMPL-PREHOD-DIST-find-FV.REL
          '. . . [that] I found' (elsewhere, yesterday or before)

4.2.6 Narrative morphology

Narrative "tense" morphology is frequently used in discourse to mark temporally subsequent situations (either consecutive or subsecutive; i.e., the participant can be the same or different). In Zambian Totela, narrative forms have the same morphology as class 15 infinitives, sometimes co-occurring with the comitative proclictic *na=*. Without the comitative proclitic, the infinitives appear without the augment, but the augments surface when the comitative is used. When narrative morphology co-occurs with distal -*ka*- marking, the infinitive prefix -*ku*- may be elided (95), as well as the comitative proclitic itself (96). Narrative morphology and participant marking are discussed further in 4.3.3.

(94) aa!   nò-kwíngìl-à.   kwìnd-á   ⁺ín-sìmà   kú-lyà̠
     INTERJ COM.AUG-NARR.enter-FV NARR.take-FV 9-pap   NARR-eat.FV
     'aa! Then he went in. He took the porridge and ate

     kú-lyà̠   kú-lyà̠   kù-màn-à
     NARR-eat.FV NARR-eat.FV NARR-finish-FV
     and ate and ate and finished [it up].'

(95) nàyê̠ nòkàbòòlàkô̠
     nà=yê̠   nà=ò̠-(kù-)kà-bòòl-à-kô̠
     COM=1.PRON   COM=AUG-(NARR-)DIST-return-FV-17
     'and he also returned'

(96) nòkàìndá ò̠mwáà̠nà ò̠kàmúkò̠sòlá òmútwì̠
     nà=ò̠-(kù-)kà-ìnd-á   ò̠-mwáà̠nà (nà=)ò̠-(kù-)-kà-mà-kò̠sòl-á   ò̠-mú-twì̠
     COM=AUG-(NARR-)DIST-take-FV AUG-1.child (COM=)AUG-(NARR-)DIST-OP₁-cut-FV AUG-3-head
     'and he took the child and cut off his head'

4.2.7 Mood

Mood distinctions are frequently expressed on the final vowel and with special tone patterns. The subjunctive final vowel is -e. Imperatives are formed with the bare stem and a penultimate H (or a falling tone on long vowels in penultimate position). In plural imperatives, the H also falls on the stem penult (hence the analysis as a subjunctive final -e followed by plural clitic =ni).[6] The penultimate H tone frequently spreads leftwards producing a plateaued H, with a significant drop in tone between the penult and the final vowel. Imperatives with object prefixes are expressed with the subjunctive final -e and a H that surfaces on the first root syllable (HL on long vowels in penultimate position) (99).

(97) ŋàtàwúl-à₍ₕ₎!
     tear.up-FV
     'Tear [it] up!'

(98) ŋàtàwúl-è₍ₕ₎=nì!
     tear.up-FV.SBJV-=2PL
     'Tear [it] up!' (pl)

(99) mù-wámb-ìl₍ₕ₎-è!
     1.OP-speak-APPL-FV.SBJV
     'Tell him!'

In addition to imperatives with object marking, subjunctive forms are used in hortatives (100)–(101) (which also function as polite imperatives (102)), in the protasis of conditionals (103), in constructions expressing that something 'almost' occurred (104) (the auxiliary verb being -ti 'say' or -saka 'want'), and in some negative constructions (see Section 4.2.8).

(100) tù-wâmb-è_H!
SP_{1PL}-talk-FV
'Let's talk!'

(101) bà-nị́chè_H   bà-nị́chè_H   tòntól-è_H=nì   bón-è_H=nì
2-child.VOC[7]   2-child.VOC   be.quiet-FV.SBJV=2PL   see-FV.SBJV=2PL
'Children, children, be quiet and see

bà=sókwè̱   bà-sìk-à   twị̀jáy-è_H!
2a=monkey   SP_{2A}.CMPL-arrive-FV   SP_{1PL}.kill-FV.SBJV
Monkey has arrived; let's kill [him]!'

(102) mù-ŋàtàwúl-è_H(=ni)!
SP_{2PL}-tear.up-FV.SBJV(=2PL)
'Tear [it] up!'

(103) ésị̀   tù-lyé_H   àhúlù̱,   tù-lé̱kùt-à
COND   SP_{2PL}-eat.FV.SBJV   a lot   SP_{1PL}-DJ.become.full-FV
'If we eat a lot, we('ll) get full.'

(104) nd-à-tí   ndị̀-zùbúk-è_H   kònó   ⁺ndá-kà̱ng-w-à
SP_{1SG}-CMPL-say   SP_{1SG}-cross-FV.SBJV   but   SP_{1SG}.CMPL-fail-PASS-FV
'I almost crossed/was intending to cross, but I was unable.'

Subject markers can also appear with something resembling subjunctive marking, followed by an infinitive-like form. Such constructions have habitual or generic imperative meaning (105). In some contexts, these forms can also be used as future imperfectives (106). Forms such as (106) are unspecified for temporal domain, and can refer to hodiernal and posthodiernal future events. I tentatively gloss these subject markers as having subjunctive endings: they appear to have a separate melodic tone pattern, and they condition downstep of infinitive Hs as if there were two words coming together (106).

(105) Ínwè̱   yô̱yù   mwâ̱nénù̱   mwé_H=kù-mù-wàmb-ìl-à.
2PL.PRON   1.DEM   1.child.2PL.POSS   SP_{2PL}.SBJV=INF-OP₁-speak-APPL-FV
'You have to keep telling that child of yours,

Ímè̱   yê̱yí   ị̀n-yàmà   y-ê̱m-bíị̀zì   tà-ndí-ị̀-lị̀_H.
1SG.PRON   9.DEM   9-meat   PP₉-CON9-horse   NEG-SP_{1SG}-OP₉-eat.FV.NEG
I don't eat horse/zebra meat.'

(106) ijìlò   ésị̀   ndé_H=⁺kú-lyà̱,   ná-⁺mwîz-è_H
tomorrow   cond   SP_{1SG}.SBJV=INF-eat.FV   POST-SP_{2PL}.come-FV.SBJV
'Come tomorrow while I'm eating.'

Hortatives can also take posthodiernal future marking, as seen in (107).

(107) ná-mù-yèmbél-è_H!
POSTHOD-SP_{2PL}-herd-FV.SBJV
'herd!' (tomorrow or thereafter)

In Namibian Totela, futures also have subjunctive *-e* as the final vowel (see Section 4.2.6). Counterfactuals are introduced with the particle *kámbè*, which appears in the protasis and sometimes the apodosis. The verb in the apodosis is frequently marked with a counterfactual proclitic *ná=*, which seems to be related to *náà̱* 'if, whether,' used when facts are unknown (e.g., *mbwî̱tà náà̱* . . . 'who knows if/whether.') Crane (2012a) has further examples.

(108) kámbè̱   tw-á-ly-à̱   à̱húlù̱,   (kámbè)
       COUNTER SP$_{1PL}$-CMPL-eat-FV   a.lot   COUNTER
       'If we had eaten a lot,

       ná=⁺tw-ȩ́kùt-à
       COUNTER=SP$_{1PL}$-CMPL.become.full-FV
       we would be full.'

4.2.8 Negation

Negation is typically marked as a prefix *ta-* in initial position on the verb. Most negative forms have final *-a*; monosyllables and a few disyllabic stems (*-saka* 'want' and *-su̱wa* 'feel, hear, understand') have final *-i*. Posthodiernal future negatives employ a subjunctive form with final *-e*. Basic affirmative and negative forms are compared in Table 22.20 with the verb *-tobela* 'look for, seek.'

TABLE 22.20 AFIRMATIVE AND NEGATIVE TA FORMS

	Affirmative	*Negative*
Prehodiernal imperfective	kà-ndì-tòbél-à$_H$	tà-ká-ndì-tòbél-à$_H$
Past (hodiernal imperfective)	ndì-nà-tóbél-à$_H$	tà-ndì-nà-tóbè̱l-à
Prehodiernal Completive	ndà-kà-tòbèl-à	tà-ndì-nà-kà-tóbè̱l-à
Completive	ndà-tòbèl-à	tà-ndì-nà-tóbèl$_H$-à
Non-completive (disjoint)	ndì-là-tòbèl-à	tà-ndì-tóbè̱l-à
Posthodiernal	ná-ndì-là-tòbèl-à	tà-lì ná-ndì-tòbél-è̱$_H$

In infinitives, the negation prefix follows the infinitive prefix (109). Following infinitives (and attested only in that context), *ta-* alternates with *sa-*. The negation prefix also follows TA forms such as situative *na-* and prehodiernal (Imperfective) *ka-* in negated auxiliary constructions (110)–(112). Note that the negation prefix can optionally also follow the subject marker in auxiliary verbs, as in (110b).

(109) ò̱kú-tà-tòbèl-à   /   ò̱kú-sà-tòbèl-à
       INF-neg-SEEK-FV   /   INF-neg-SEEK-FV
       'to not seek'

(110) a   bàndìwàànà nàndìtééní̱$_H$ kúlyà̱
           bà-ndì-wààn-à   nà-**ndì-tá-ín**-ì̱$_H$ kú-ly-à̱
           SP$_2$-OP$_{1SG}$-find-FV   SIT-**SP$_{1SG}$-NEG**-have-FV.NEG   INF-eat-FV
           'They found me not (yet) having eaten.'

       b   bà-ndì-wàòn-à   nà-**tà-ndí-ín**-ì̱$_H$ kú-ly-à̱
           SP$_2$-OP$_{1SG}$-find-FV   SIT-**NEG-SP$_{1SG}$**-have-FV.NEG   INF-eat-FV
           'They found me not (yet) having eaten.'

(111) ndà-ká-l̲à̲àl-à ínywè̲ nà-tà-mwín-ì$_H$ kù-sìk-à
SP$_{1SG}$.CMPL-PREHOD-sleep-FV 2PL.PRON SIT-NEG-SP$_{2PL}$.have-FV.NEG INF-arrive-FV
'I went to bed [last night] before you arrived.'

(112) kà-tá-ꜜbéén-ì$_H$ kú-tà̲lìk-à
PREHOD-NEG-SP$_2$.have-FV.IPFV.NEG INF-begin-FV
'They haven't started (yet).'

The negative marker can also take an *-e* ending reminiscent of the subjunctive, along with a subjunctive final vowel. This gives a slightly different meaning, as seen in the contrast in (113a–b).

(113) a tà-ndí-l̲ì̲$_H$
NEG-SP$_{1SG}$-eat.FV.NEG
'I don't eat (in general)/I won't eat (later).'

b té$_H$=ndì-lyê$_H$
NEG.SBJV=-SP$_{1SG}$-eat.FV.SBJV
'I won't eat/I refuse to eat.'

A distinction between subjunctive and indicative forms can also be seen in posthodiernal futures, which are negated with an auxiliary (114).

(114) a tà-lí$_H$ ꜜná-ndì-là-yà?
NEG-be POSTHOD-SP$_{1SG}$-DJ-go.FV
'Won't I go?' (tomorrow or after)

b tà-lí$_H$ ꜜná-ndì-yê$_H$
NEG-be POSTHOD-SP$_{1SG}$-go.FV.SBJV
'I won't go.' (tomorrow or after)

4.2.9 TAMN in Namibian Totela

The expression of tense/aspect/mood/negation (TAMN) is strongly divergent between the two varieties: although the varieties have similar categories and similar TAMN morphemes, the correspondence between categories and morphemes differs dramatically. TAMN forms are extremely variable in general, especially in the Zambezi region.

Additionally, many speakers of NT use infinitive fronting as a primary means to express the progressive, as discussed in Section 4.3.2. In ZT, this construction seems relatively rare, and, according to my data, expresses predication focus.

As can be seen in Table 22.21, final vowels in TAMN forms are different between ZT and NT, as well. In addition to subjunctive uses found in both varieties, NT has final vowel *-e* for all futures. The final vowel in negative forms in NT is *-a* in past forms, *-i* in the present, and *-e* for future forms. These forms are shown in Table 22.22, which also illustrates some of the differing negation strategies between NT and ZT. Recall from Section 4.2.8 that negative forms in ZT have final *-i* with monosyllabic and a few bisyllabic stems.

TABLE 22.21 COMPARISON OF SELECTED TA MORPHOLOGY IN ZAMBIAN AND NAMIBIAN TOTELA VARIETIES

	NT	*ZT*
Prehodiernal perfective/ completive	ni/na-SP-a-ROOT-a ni/na-nd-a-yend-a 'I walked' (yest. or before)	SP-a-ka-ROOT-a nd-à-kà-yènd-à 'I walked' (yest. or before)
Hodiernal perfective/completive/ perfect	SP-a-ROOT-a nd-a-yend-a 'I walked' (today)	SP-a-ROOT-a nd-à-yènd-à 'I walked' (today)
Prehodiernal imperfective	ka-SP-ROOT-a ka-ndi-yend-a 'I was walking/used to walk'	ka-SP-ROOT-a kà-ndì-yênd-à$_H$ 'I was walking/used to walk'
Hodiernal imperfective	SP-la-ROOT-i ndi-la-yend-i 'I was walking'	SP-na-ROOT-a ndì-nà-yênd-à$_H$ 'I was walking'
Present	SP-ROOT-a ndi-yend-a 'I walk/am walking' (no morphologically expressed CJ/DJ)	SP-la-ROOT-a ndì-là-yènd-à 'I walk/am walking/will walk' (disjoint form)
Hodiernal future	mo/mu-SP-ROOT-e mo/mu-ndi-yend-e 'I will walk' (later today)	
Posthodiernal future	ka-SP-ROOT-e ka-ndi-yend-e 'I will walk'	na-SP-la-ROOT-a ná-ndi-là-yènd-à 'I will walk' (disjoint form)

TABLE 22.22 EXPRESSION OF NEGATION IN ZAMBIAN AND NAMBIAN TOTELA VARIETIES

	NT	*ZT*
Prehodiernal perfective/ completive	kana-na/ni-SP-a-ROOT-**a** kana-na/ni-ndi-yend-a	ta-SP-na-ka-ROOT-**a**(/i) tà-ndì-nà-kà-yênd-à$_H$
Hodiernal perfective/completive/ perfect	kana-SP-a-ROOT-**a** kana-nda-yend-a	ta-SP-na-ROOT-**a**(/i) tà-ndì-nà-yênd-à$_H$
Prehodiernal imperfective	kana-ka-SP-a-ROOT-**a** kana-ka-nda-yend-a	ta-ka-SP-ROOT-**a**(/i) tà-ká-ndị̀-yênd-à$_H$
Present	ka-SP-ROOT-**i** ka-ndi-yend-i	ta-SP-ROOT-**a**(/i) tà-ndì-yênd-à$_H$
Hodiernal future	kase/kasi-SP-ROOT-**e** kase/kasi-ndi-yend-e	ta-SP-ROOT-**a**(/i) tà-ndì-yênd-à$_H$
Posthodiernal future	kase/kasi-na-SP-ROOT-**e** kase/kasi-na-ndi-yend-e	ta-li na-SP-ROOT-**e** tà-lí$_H$ $^+$ná-ndì-yênd-è$_H$

4.2.10 TAMN and melodic tone

Totela verbs take three main tone patterns:

1 Underlying tones surface (with high tone anticipation) without any melodic tones (present, future, completive/perfective pasts and most other main clause affirmative forms).
2 Melodic H on second root syllable, surfacing on the first syllable if not eliminated by *H-H constraints (most forms with object markers, most negative forms). Pattern (2b) has falling tone on monosyllables.
3 Melodic H on final syllable, surfacing on the penultimate syllable, also subject to the *H-H constraint and often occurring with plateauing, especially between input Hs (most affirmative relative clauses, subjunctive and hortative forms without object markers, prehodiernal imperfective forms, persistive forms).

These patterns are illustrated in Tables 22.23 (toneless) and 22.24 (H-toned) roots of various lengths. For full paradigms and further discussion, see Crane (2014). Lexical Hs are only marked on roots.

TABLE 22.23 MELODIC TONE PATTERNS (TONELESS ROOTS)

# Syllables	Pattern 1 Pres/non-past	Pattern 2a Pres neg	Pattern 2b Hortative with toneless OM	Pattern 3 Imperative
1	ndì-là-wà	tà-ndí-wì$_H$	mù-ndi-nyè$_H$	wâ$_H$
2	ndì-là-sàkà	tà-ndì-sákì$_H$	mù-ndi-sákè$_H$	sákà$_H$
2(long vowel)	ndì-là-zìikà	tà-ndì-zíikà$_H$	mù-ndi-zíikè$_H$	zíikà$_H$
3	ndì-là-ùkùtà	tà-ndì-úkù$_H$tà	mù-ndi-úkù$_H$tè	ùkútà$_H$
3(long vowel)	ndì-là-zàànìnà	tà-ndì-záání$_H$nà	mù-ndi-wá:mbì$_H$lè	zàànínà$_H$
4	ndì-là-ŋàtàwùlà	tà-ndí-ŋátà$_H$wùlà	mù-ndi-ŋátà$_H$wùlè	ŋàtàwúlà$_H$

TABLE 22.24 MELODIC TONE PATTERNS (H-TONED ROOTS)

# Syllables	Pattern 1 Pres/non-past	Pattern 2a Pres neg	Pattern 2b Hortative with toneless OM	Pattern 3 Imperative
1	ndi-lá-twà̤	tà-ndí-twì̤$_H$	mù-ndi-pè̤$_H$	pâ̤$_H$
2	ndi-lá-hò̤hà	tà-ndí-hò̤hà$_H$	mù-ndí-bò̤nè$_H$	hó̤hà$_H$
2(long vowel)	ndi-lá-bì̤ikà	tà-ndí-bì̤ikà$_H$	mù-ndi-bì̤ikè$_H$	sṳ́ùkà$_H$
3	ndi-lá-hùpùlà	tà-ndí-hà̤lì$_H$kà	mù-ndi-hà̤lì$_H$kè	hà̤líkà$_H$
3(long vowel)	ndi-lá-yè̤:mbèlè	tà-ndì-yè̤:mbè$_H$là	mù-ndí-yè̤:mbè$_H$lè	yè̤:mbélà$_H$
4	ndi-lá-bà̤bàlèlà	tà-ndí-bà̤bàlèlà	mù-ndí-bà̤bà$_H$lèlè	bà̤bàlélà$_H$

Passive and causative extensions behave as if they added a mora to the final syllable, resulting in a falling tone on the final vowel in imperative forms (115a–b). Some forms that segmentally resemble causatives (115c) and passives (115e) also (sometimes optionally) take a falling tone on the final vowel of an imperative. Question marks represent words speakers would not accept or produce.

(115)

		Imperative	Applicative Imperative	Causative Imperative
a	-wamba 'talk, tell'	wâmbà$_H$	wàmbílà$_H$	wàmbìsâ$_H$
b	-yenda 'walk'	yêndà$_H$	yèndélà$_H$	yènzyâ$_H$
c	-fosa 'err'	fòsâ$_H$	fòsèzâ$_H$	?
d	-fusa 'throw spear at'	fúsà$_H$	fùsílà$_H$	fùsìsâ$_H$
e	-suwa 'hear, feel'	sùwâ$_H$	?	sùwìsâ$_H$
f	-lowa 'bewitch'	lówà$_H$	lòwélà$_H$	lòwìsâ$_H$

4.3 Clause structure and agreement

4.3.1 Word order

Totela exhibits a relatively free word order, allowing all permutations of subject, verb and object. Adverbials can typically appear at any point in a sentence. The "default" word order, in the absence of additional information structuring constraints, is SVO, although the subject is often dropped. When a clause-final verb is preceded by its object, the object is treated as a topic and a resumptive object marker is required on the verb (as in (105) above). (116) shows the flexibility in subject and object positions.

(116) à-tòbèl-à   Jacky   Lenya
SP$_1$-seek-FV   Jacky   Lenya
default interpretation: 'Jacky is looking for Lenya' (VSO)
possible interpretation: 'Lenya is looking for Jacky' (VOS)

With double objects, the direct and indirect object can occur in either order, though the indirect object precedes the direct object in the default interpretation, and is the only possible order in cases where there could be ambiguity (117c).

(117) a nd-á-p-à   bà=máámà   ín-tàlàbàndà
          SP$_{1SG}$-CMPL-give-FV   2a=my.mother   9-cooked.greens
          'I gave my mother cooked greens.' (IO-DO)

      b nd-á-p-à   ín-tàlàbàndà   bà=máámà
          SP$_{1SG}$-CMPL-give-FV   9-cooked.greens 2a=my.mother
          'I gave my mother cooked greens.' (DO-IO)

      c mù-p-é   ò-mwánàkázì   ò-mú-⁺kwámè
          SP$_{2PL}$-give-FV.SBJV   AUG-1.woman   AUG-1-man
          'Give the woman the man.' (IO-DO)
          *'Give the man the woman.' (DO-IO)

When direct objects (DO), indirect objects (IO) and instrumentals (IN) appear together, DO-IO-IN is the preferred order. However, they can appear in nearly any order, as long as the IN appears between the IO and DO when they appear in that order (i.e., IO-IN-DO, but *IO-DO-IN and *IN-IO-DO).

4.3.2 Conjoint/disjoint and infinitive fronting

The marker -la-, used in Non-Completive forms, clearly has its roots in a disjoint form found also in Ila M63, Tonga M64, Bemba M42 and Lamba M54 (Hopgood 1940, Smith

1964, Fowler 2000, Carter 2002, Nurse 2006: 193–194, 2008: 205–206). It occurs in main-clause affirmatives and alternates with a null form, the latter of which occurs in all subordinate clauses and cannot occur phrase-finally in a main clause. Verbs marked with -*la*- are accepted and used by speakers in all main-clause affirmative contexts in which null-marked verbs are also used. Null-marked verbs are typically used in term-focus contexts (Güldemann 2003), such as in answers to object wh-questions, as in (118).

(118) a   mùlàsàkê̠nzì? (accepted)
           mù-là-sàk-à=înzì
           SP_{2PL}-DJ-want-FV=what
           'What do you want?'

       b   mùsàkênzì̠? (preferred)
           mù-sàk-à=înzì
           SP_{2PL}-want-FV=what
           'What do you want?'

       c   ndì-sak-á   è-chì-yùnì (preferred answer to (100a) and (100b))
           SP_{1SG}-want-FV   AUG-7-bird
           'I want a bird.'

It is my impression that the focus functions of the conjoint/disjoint alternation may be weakening somewhat in Totela, perhaps eventually developing into a system with purely syntactic conditioning, where -*la*- forms occur in main-clause affirmatives and null forms elsewhere, but this impression would need to be confirmed with longitudinal research. Interestingly, when -*la*-marked and null-marked forms are contrasted, speakers tend to give null-marked forms progressive interpretations, while forms with -*la*- have habitual or futurate interpretations. However, both forms can be used with any of these interpretations. In some related languages with -*la*- (e.g., Ila M63, Lamba M54), the conjoint/disjoint distinction has been lost, although the morpheme has been retained (Güldemann 1996: 236, 2003: 354, Nurse 2008: 206); in others (e.g., Tonga M64), the conjoint/disjoint system is still active. Namibian Totela does not seem to make any conjoint/disjoint distinction, but, like Fwe K402 and some other Zambezi languages (Gunnink 2016), employs verb doubling with focus and progressive meanings. In Namibian Totela, a -*la*-morpheme is found in hodiernal past imperfectives (119).

(119) ndi-la-samb-i
       SP_{1SG}-HOD.IPFV-bathe-FV
       'I was bathing.' (NT)

Infinitive fronting is also found with a focus meaning in Zambian Totela, as in (120):

(120) kú-yè̠n-á   mù̠-yé̠n-à!
       INF-lie-FV SP_{2PL}-lie-FV
       'You're lying!'

In at least the China Chilao variety of Namibian Totela, infinitive fronting is a common strategy for forming progressives. Fronted infinitives are used in particular when the predicate itself is in focus (e.g., in answer to the question 'What is he doing?') but also have general progressive meaning.

(121) ku-bez-a   ka-bez-a   i-zi-pula
       INF-carve-FV   PREHOD.IPFV.SP₁-carve-FV   AUG-10-chair
       'He was carving chairs.' (NT)

(122) kana-ku-bez-a   a-bez-a   i-zi-pula
       NEG-INF-carve-FV   SP₁-carve-FV   AUG-10-chair
       'He's not carving chairs.' (NT)

Interestingly, a morpheme -(*a*)*la*-/-(*a*)*ra*- also appears in some varieties of Namibian Fwe as a marker of remote future. The morpheme is illicit in dependent clauses (Hilde Gunnink pers. comm.). Zambian Fwe lacks this marker, and also employs a fronted-infinitive construction with verbal focus and/or progressive aspect (see also Gunnink 2016).

4.3.3 Participant marking and agreement

Totela exhibits pervasive noun-class and person agreement. Noun-class agreement markers are given above in Table 22.9. First- and second-person subject and object prefixes are as in Table 22.25.

TABLE 22.25 PERSON MARKERS

Person	Pronoun	SP	OP
1SG	ímè	ndi-	-ndi-
2SG	íwè	u-	-u-
1PL	íswè	tu-	-tu-
2PL	ínwè/ínywè	mu-	-mi -

Object marking occurs immediately prefixed to the verb stem. Object markers show a tonal contrast, with first and second person markers toneless and all other markers associated with a H tone (which subsequently shifts to the left). Lexically trivalent verbs such as -*pa* 'give' and trivalent verbs derived with causative or applicative extensions can take two object prefixes; the indirect object prefix is closest to the root.

(123) mwíz-è_H   mù-bà-nd-íjày_H-ìl-è!
       SP₂ₚₗ.come-FV.SBJV   SP₂ₚₗ-OP₂-OP₁ₛɢ-kill-APPL-FV.SBJV
       'Come and kill them [the mice] for me!'

A lexical object can co-occur with object marking only with disjoint verb forms (124). Local doubling is disallowed; that is, an object prefix cannot occur with lexical object or question word in the same phrase (125), although it is licit with disjoint marking, especially with a clear intonational break.

(124) Jacky   à-*(lá)-bà-tòbèl-à,   à̱-bà-chèchè
       Jacky   SP₁-*(DJ)-OP₂-seek-FV   AUG-2-child
       'Jacky is looking for them, the children.'

(125) *a-(la)-ba-tobel-a=ni?
       SP₁-(DJ)-OP₂-seek-FV=who
       intended: 'Who is he looking for?'

Object marking is, in general, optional, although it appears to be preferred when a lexical object is in topic position (OSV and SOV order).

Even subject markers are optional in some cases. In Zambian Totela, as noted in Section 4.2.2, narrative morphology occurs without subject marking, and object marking is, as usual, optional – even without an overt object in the clause or in surrounding clauses, as is the case for the line from a narrative given in (126).

(126) kú-kwa̲t-à   kù-nèns-à   kù-nèns-à
       NARR-grasp-FV   NARR-beat-FV   NARR-beat-FV
       '[Then she] caught [him and] beat [him and] beat [him].'

Narrative morphology in Namibian Totela is optionally marked for the subject, as in (127).

(127) Izoona   ka-ndi-yend-a   mu-lu-ole.   Mpahonaho
       yesterday   PREHOD.IPFV-SP$_{1SG}$-walk-FV   17-11-forest   suddenly
       'Yesterday I was walking in the forest. Suddenly

       ndi-ku-lyat-a   ha-lu-kungwe.   **a-ku-ndi-sum-a**   he-tende.
       SP$_{1SG}$-NARR-step-FV   16-11-snake   **SP$_3$-NARR-1.OP-bite-FV**   16.5-leg
       I stepped on a snake. **It bit me** on the leg.

       Ndi-ku-tol-a   i-buwe   noku-sonz-a   noku-dam-a.
       SP$_{1SG}$-NARR-pick.up-FV   5-stone   COM.NARR-throw-FV   COM.NARR-hit-FV
       I picked up a stone and threw [it] and hit [the snake].

       A-ku-fwa
       SP$_3$-NARR-die.FV
       It died' (NT; text from Dahl 1985: 205)

Speakers employ a variety of strategies for resolving noun-class agreement clashes. Two human (or class 1a) nouns take class 2(a) plural forms; non-human nouns from different classes typically take class 10 (/8) agreement morphology when conjoined. When a human and non-human noun are conjoined, either class 2 or class 10 (=8) morphology is licit, and different speakers choose different strategies (128). However, in most cases, speakers opt for other constructions: for example, the non-human noun may be treated as an adjunct (e.g., 'the man walks with the cow').

(128) O̲-mú-⁺kwa̲mè̲   n-èŋ-òmbè   bà-là-yend-a   /   zì-là-yènd-à
       AUG-1-man   COM-9-cow   SP$_2$-DJ-walk-FV   /   SP$_{10}$-DJ-walk-FV
       'The man and the cow are walking.'

4.4 Clause types

Relative clauses can occur with virtually all TAMN morphology (except for disjoint non-past marker -*la*- and the corresponding stative marker *li*). Affirmative relative clause verbs have a melodic H that surfaces on the penultimate syllable. Subject (129) and object (130) relatives have identical tone; negative relative clauses have the same segmental morphology and tone patterns as their main clause counterparts. Relative clauses can include a demonstrative pronoun agreeing with the relative head (130). Unlike in main clauses, relative clause tone shifts leftward from the subject marker to the final syllable of a preceding word.

(129) ọ̀-tù-yùní   tw-à̱-yîmb-à̱_H
AUG-13-bird   SP₁₃-CMPL-sing-FV.REL
'the birds that sang'

(130) ndì-sàk-á   ọ̀kú-ŋọ̀l-á   à̱mà-kàndé (à-lyá)   mwà̱-kà-wâmb_H-à
SP₁ₛ_G-want-FV   INF-write-FV 6-story   (DEMP₆-DEM)   SP₂ₚ_L.CMPL-PREHOD-speak-FV
'I want to write down that story you told.'

Focus clefts, included fronted question words, are also followed by clauses with relative tone.

(131) ndìmé   ndà̱-mì-yèmbèl-él-à̱_H   ìŋ-òmbè
COP.PRON₁ₛ_G   SP₁ₛ_G-CMPL-OP₂ₚ_L-herd-APPL-FV   10-cattle
'It was I who herded the cattle for you.'

(132) chì=nzí   mù̱-sák-à̱_H?
COP₇=what   SP₂ₚ_L-want-FV
'What do you want?'

Question words can also occur in-situ, as in (118) above. Common question words include =ni 'who,' =nzi 'what/why,' =ibo 'which,' =ti 'how many,' buti(=nzi) 'how,' =yi 'where' and líìlíì 'when.'

Copular clauses have several morphological forms. Simple nominal predication (e.g., 'the man is a teacher') takes an augmentless noun prefix, sometimes with an additional nasal element (classes 2, 5, 9. 11, 14). The occurrence of the nasal element with these classes, and its optionality, requires further study. It may be that prenasalisation is lessening with predicated forms, as can be seen in comparison with Namibian Totela and with Tonga M64 (see Carter 2002: 24–28 for Tonga), where all copular forms (in all noun classes) are prenasalised. Predicated nouns may be pronounced with overall higher pitch, and predicated prefixes are sometimes realised with H tone even if usually toneless, but this tone pattern is not consistent and requires further study. Basic nominal predication can occur with or without the copular verb -lí 'be' (-ba̱ 'be(come)' with future reference). Forms are shown in Table 22.26, column I.

Table 22.26 also shows the prefixes for general copular forms ('it's a. . .'), also used in cleft constructions. Basic forms, as they appear before e.g., possessive stems, as in class 5 ndí⁺lyángù̱ 'it's mine,' are given in column II. Copular prefixes to lexical nouns, as in class 5 ndị̀wọ̀ngólọ̀ 'it's a millipede,' are in column III. The forms change somewhat when used with forms requiring connective -a-, as with -sa̱kata 'good, proper,' e.g., class 3 ọ̀mù̱lílò wásà̱kàtà 'a proper fire' vs. ọ̀mù̱lílò ngwásà̱kàtà 'the fire is (a) proper (one)' (column IV).

Restrictive or contrastive forms have different copular prefixes, also given in Table 22.26, column V. These are used with nouns and adjectives (e.g., class 6 ngá⁺má̱bì 'they are the bad ones') and with predicative relative clauses, e.g., class 6 à̱mênzì ngátù̱nywâ 'water is what we drink.' Similar forms are used with copular demonstratives (see Table 22.15 above). Tones are not indicated in Table 22.26 because they change depending on the environment, and underlying tones are not fully analysed.

TABLE 22.26 SOME COPULAR PREFIXES

NC	I Simple nominal predication	II Basic copula (also used in clefts)	III Basic copula before noun prefixes	IV Copular with connective	V Restrictive/ contrastive copula
1	mu-	ndi-	nd-omu-	wa-	nde-
2	(m)ba-	(m)ba-	m-ba-/mb-aba-	mba-	mba-
3	mu-	ngu-	ng-omu-	ngwa-	ngo-
4	mi-	ndi-	nj-emi-	nja-	nje-
5	ndi-	ndi-	nd-i-/nd-eli-	ndya-	nde-
6	ma-	nga-	ng-ama-	nga-	nga-
7	chi-	chi-	ch-echi-	cha-	che-
8	zi-	zi-	z-ezi-	za-	ze-
9	ni-	nji-	n-iN-	nja-	n(j)e-
10	zi-	zi-	n-iN-	za-	ze-
11	ndu-	ndu-	n-du-/nd-olu-	ndwa-	ndo-
12	ka-	ka-	ka-ka-	ka-	ka-
13	tu-	tu-	to-tu-	twa-	to-
14	mbu-	(m)bu-	mb-obu	mbwa-	mbo-
15	ku-	ku-	k-oku-	kwa-	ko-

NOTES

1 Data are taken from (Bostoen 2009) for Fwe K402, Tonga M64 and Ila M63; a portion of the Subiya K42 data is from de Luna (2008).

2 This class 17 form seems to have merged with the class 16 form.

3 E.g., *ndàwìsá ⁺nán̠kàlá ⁺án̠sì* 'I dropped the crab on the ground,' in this environment; cf. *èchìyùnì(/í) kwìjùlù, nánkàlà án̠sì* 'the bird in the sky, the crab on the ground,' where H tones do not spread leftwards if the penultimate syllable is underlyingly associated with a H tone. In some – but not all – cases, both shifting with downstep and H-tone deletion are allowed, e.g., *sésìbó ⁺wángù / sésìbò wángù* 'my orange-throated long-claw'; but *múkàzì wángù / *múkází ⁺wángù* 'my wife.'

4 The *-la-* morpheme is often pronounced with a long vowel, especially in penultimate position, but the length is variable and not potentially contrastive in this position, and is therefore not represented in the orthography.

5 The *-na-* morpheme also exhibits some (variable) degree of vowel length, not represented in the orthography.

6 As a reviewer noted, this form could also be analysed as having the imperative final *-a*, with a plural postclitic *=ini*. At present, I do not know how to distinguish between these analyses, since the imperative and the subjunctive have the same tone patterns; in any case, the final syllable of plural imperatives behaves extra-metrically.

7 The vocative is marked by absence of a vocalic augment and, in many cases, a pattern in which H tone only surfaces on the penult, e.g., *bàn̠íchè ̬!* 'children!' (voc.) vs. *àbán̠íchè* 'children'; *mwànàkázì ̬!* 'woman!' (voc.) vs. *òmwán̠àkázì* 'woman.' In some cases, when the noun stem is monosyllabic (or historically monosyllabic), this penultimate H only appears on the possessive pronoun, which is often cliticised when

referring to family relations, e.g., (àkàkà) mùkwè wángù̖! 'my in-law!' (voc.) vs. múkwè wángù̖ 'my in-law (class 1a),' but bàmúkwè wángù̖! 'my in-law!' (voc.; polite use of class 2a prefix). Because this phenomenon does not occur uniformly (and, in my data at least, only seems to occur with human referents), it is difficult to say with certainty whether this is a melodic tone or the shifting of input H tones to the penult. I do not have any examples where it occurs on an all L noun, though I have examples where it does not (e.g., bànyìnà! 'my mother!' (pl., polite use of class 2a prefix) vs. bànyìnà 'my mother'), though the minimal-pair mùkòmbwè 'rooster!' (voc.) and Mùkómbwè (a person's name) might be suggestive.

REFERENCES

Baumbach, E. J. M. (1997) Languages of the Eastern Caprivi. In Haacke, W. H. G. & E. D. Elderkin (eds.), Namibian Languages. Reports and Papers, 307–450. Cologne: Rüdiger Köppe.

Bostoen, K. (2009) Shanjo and Fwe as Part of Bantu Botatwe: A Diachronic Phonological Approach. In Ojo, A. & L. Moshi (eds.), Selected Proceedings of the 39th Annual Conference on African Linguistics, 110–130. Somerville, MA: Cascadilla Proceedings Project.

Botne, R. & T. L. Kershner (2000) Time, Tense, and the Perfect in Zulu. Afrika und Übersee 83: 161–180.

Bryan, M. A. (1959) The Bantu Languages of Africa. London; New York; Cape Town: Oxford University Press for the International African Institute.

Carter, H. (1971) Morphotonology of Zambian Tonga: Some Developments of Meeussen's System – I. African Language Studies 12: 1–30.

Carter, H. (2002) An Outline of Chitonga Grammar. Lusaka: Bookworld Publishers.

Central Statistical Office (2012) 2010 Census of Population and Housing – Volume 11: National Descriptive Tables, Lusaka, Zambia. (Available online at www.zamstats.gov. zm/report/Census/2010/National/Zambia_National_Descriptive_Population_Tables. pdf).

Central Statistical Office (2013) 2010 Census of Population and Housing: Descriptive Tables – Volume 10: Western Province Series A – D. Lusaka, Zambia. (Available online at www.zamstats.gov.zm/report/Census/2010/Provincial/Western_Province_ Series_A,_B,_C_and_D.pdf).

Crane, T. M. (2011) Beyond Time: Temporal and Extra-temporal Functions of Tense and Aspect Marking in Totela, a Bantu Language of Zambia. Berkeley: University of California, PhD dissertation.

Crane, T. M. (2012a) Completion and Dissociation in Totela Tense and Aspect. In Marlo, M. R., N. B. Adams, C. R. Green, M. Morrison & T. M. Purvis (eds.), Selected proceedings of the 42nd Annual Conference on African Linguistics, 208–220. Somerville: Cascadilla Proceedings Project. (Available online at www.lingref.com/ cpp/acal/42/paper2770.pdf).

Crane, T. M. (2012b) -ile and the Pragmatic Pathways of the Resultative in Bantu Botatwe. Africana Linguistica 18: 41–96.

Crane, T. M. (2013) Resultatives, Progressives, Statives, and Relevance: The Temporal Pragmatics of the -ite Suffix in Totela. Lingua 133: 164–188.

Crane, T. M. (2014) Melodic Tone in Totela TAM. Africana Linguistica 20: 63–79.

Dahl, Ö. (1985) *Tense and Aspect Systems*. Oxford: Blackwell.

de Luna, K. M. (2008) *Collecting Food, Cultivating Persons: Wild Resource Use in Central African Political Culture, c. 1000 B.C.E. to c. 1900 C.E.* Evanston: Northwestern University, PhD dissertation.

de Luna, K. M. (2010) Classifying Botatwe: M.60 and K.40 Languages and the Settlement Chronology of South Central Africa. *Africana Linguistica* 16: 65–96.

de Luna, K. M. (2016) *Collecting Food, Cultivating People: Subsistence and Society in Central Africa*. New Haven: Yale University Press.

Fortune, G. (1959) *A Preliminary Survey of the Bantu Languages of the Federation*. Lusaka: Rhodes-Livingstone Institute.

Fortune, G. (1963) A Note on the Languages of Barotseland. *Proceedings of the Conference on the History of Central African Peoples*, 1–27. Lusaka: Rhodes-Livingstone Institute.

Fowler, D. G. (2000) *A Dictionary of Ila Usage : 1860–1960*. Münster: LIT.

Güldemann, T. (1996) *Verbalmorphologie and Nebenprädikationen im Bantu. Eine Studie zur funktional motivierten Genese eines konjugationalen Subsystem*. Bochum: Universitätsverlag Dr. N. Brockmeyer.

Güldemann, T. (2003) Present Progressive vis-à-vis Predication Focus in Bantu. A Verbal Category between Semantics and Pragmatics. *Studies in Language* 27(2): 323–360.

Gunnink, H. (2016) The Fronted-Infinitive Construction in Fwe. Paper presented at *the 6th International Conference on Bantu Languages*, Helsinki, Finland.

Gunnink, H. (2018) *A Grammar of Fwe. A Bantu language of Zambia and Namibia*. Ghent: Ghent University, PhD dissertation.

Hopgood, C. R. (1940) *A Practical Introduction to Chitonga*. Lusaka: Zambia Educational Publishing House.

Hyman, L. M. (2007) Universals of Tone Rules: 30 Years Later. In Riad, T. & C. Gussenhoven (eds.), *Tones and Tunes: Studies in Word and Sentence Prosody*, 1–34. Berlin: De Gruyter Mouton.

Hyman, L. M. (2016) Is Lusoga a /L/ Vs. Ø Language? Paper presented at *the 6th International Conference on Bantu Languages*, Helsinki, Finland.

Jacottet, É. (1896) *Études sur les langes du Haut-Zambèze: textes originaux. Première partie : Grammaires soubiya et louyi*. Paris: Ernest Leroux.

Jacottet, É. (1899) *Études sur les langes du Haut-Zambèze: textes originaux. Deuxième partie : Textes soubiya*. Paris: Ernest Leroux.

Jacottet, É. (1901) *Études sur les langes du Haut-Zambèze: textes originaux. Troisième partie : textes louyi*. Paris: Ernest Leroux.

Lewis, M. P. & G. F. Simons (2010) Assessing Endangerment: Expanding Fishman's GIDS. *Revue Roumaine de Linguistique* 55(2): 103–120.

Lewis, M. P., G. F. Simons & C. D. Fennig (eds.) (2016) *Ethnologue: Languages of the World, Nineteenth edition*. Dallas: SIL International. (Available online at www.ethnologue.com).

Namibia Statistics Agency (2013) Namibia 2011 Population & Housing Census: Main Report, Windhoek, Namibia. (Available online at www.nsa.org.na/files/downloads/Namibia_2011_Population_and_Housing_Census_Main_Report.pdf).

Nurse, D. (2006) Focus in Bantu: Verbal Morphology and Function. *ZAS Papers in Linguistics* 43: 189–207.

Nurse, D. (2008) *Tense and Aspect in Bantu*. Oxford: Oxford University Press.

Seidel, F. (2005) The Bantu Languages of the Eastern Caprivi. A Dialectometrical Analysis and Its Historical and Sociolinguistical Implications. *South African Journal of African Languages* 26(4): 207–242.

Smith, E. W. (1964) *A Handbook of the Ila Language (Commonly Called the Seshukulumbwe), Spoken in North-Western Rhodesia, South-Central Africa : Comprising Grammar, Exercises, Specimens of Ila Tales, and Vocabularies.* Farnborough: Gregg Press.

Torrend, J. (1931) *An English-vernacular Dictionary of the Bantu-Botatwe Dialects of Northern Rhodesia.* Natal: Gregg International.

Van de Velde, M. (2006) Multifunctional Agreement Patterns in Bantu and the Possibility of Genderless Nouns. *Linguistic Typology* 10: 183–221.

CHAPTER 23

CHIMPOTO N14

Robert Botne

1 INTRODUCTION

Chimpoto, also referred to as Mpoto, is spoken in southwestern Tanzania, in a narrow litoral region along the northeast shores of Lake Nyasa, as well as in eastern Malawi. There are estimated to be approximately 70,000 speakers of Tanzanian Chimpoto (Joshuaproject.net), fewer than the 80,000 speakers estimated by Voegelin and Voegelin (1977), and an estimated 40,000 speakers of Malawian Chimpoto (Johnstone 1993). Most speakers are bilingual, at least to some extent, in Swahili G42. Many Mpoto people are also familiar with neighbouring Kimatengo N13. There are dialect differences, not only between Malawian and Tanzanian varieties, but also in the Tanzanian variety itself. This is most evident between the people living along the shore and those further inland in the hills. The current sketch is based on the variety located in the Mbaha area, situated along the lakeshore, south of the Lundu Mission.

Chimpoto is not used in schools or in government publications. The only available written document is a school reader published in 1925 that appears to be the Malawian variety. No standardised orthography exists.

Data for this sketch were gathered primarily from Margaret Philip Mwingira, a 42-year-old woman working toward a graduate degree in the U.S. Other data were obtained indirectly from her mother Mary Philip, sister Adella Luoga, uncle Octavian B. Chale Kambanga, and father's cousin Benard J. Mwingira.[1] Direct elicitation, complemented by a variety of narrative texts and two recorded conversations, constitute the sources of primary data. The Tanzanian Language Survey by Nurse and Philippson (1975) was consulted to compare vocabulary items.

2 PHONOLOGY

2.1 The vowel system

Chimpoto has a system of five monophthongs /i, e, a, o, u/. This is rather unexpected, as neighbouring Kimatengo, for example, has a seven-vowel system, as do most Zone N languages. Mid and high vowels overlap in their vowel spaces when they are unstressed, often making it difficult to determine whether it is the mid or high vowel that was articulated.

Two vowel sequences occur – *ai, au* – as in the examples listed here. These are best treated as two syllables; for example, *lí-nyaunyau* receives prominence on the class prefix as do all disyllabic reduplicated roots, such as *lí-katu.katu* 'turtle.' Such sequences are rare.

(1)	manyai	'perhaps, maybe'	linyaunyau	'species of spider'
	ku.dai	'to claim repayment'	nakau	'dog'

2.2 Vowel assimilation

There is both progressive and regressive vowel assimilation.

2.2.1 Progressive assimilation in verbs

Many verb extensions alternate between high and mid vowels, between /i/ and /e/, and /u/ and /o/, depending on the height of the preceding stem vowel. The extension vowel remains high – either /i/ or /u/ – following a final /a/ in the stem.

(2)   Applicative: -il- ~ -el-
    -telek-el-a   'cook with'   < -telek-a   'cook'
    -kong-el-a   'tie with'   < -kong-a   'tie'
    -hakal-il-a   'spoil (e.g., child)' < -hakal-a   'be bad'

    Causative: -ih- ~ -eh-
    -tiny-ih-a   'burn'   ~ -tiny-ik-a   'be burned'
    -nenep-eh-a   'fatten'   < -nenep-a   'be fat and big'
    Intransitiviser: -ik- ~ -ek-
    -mal-ik-a   'be finished'   < -mal-a   'finish
    -hotol-ek-a   'be doable'   < -hotol-a   'be able to'

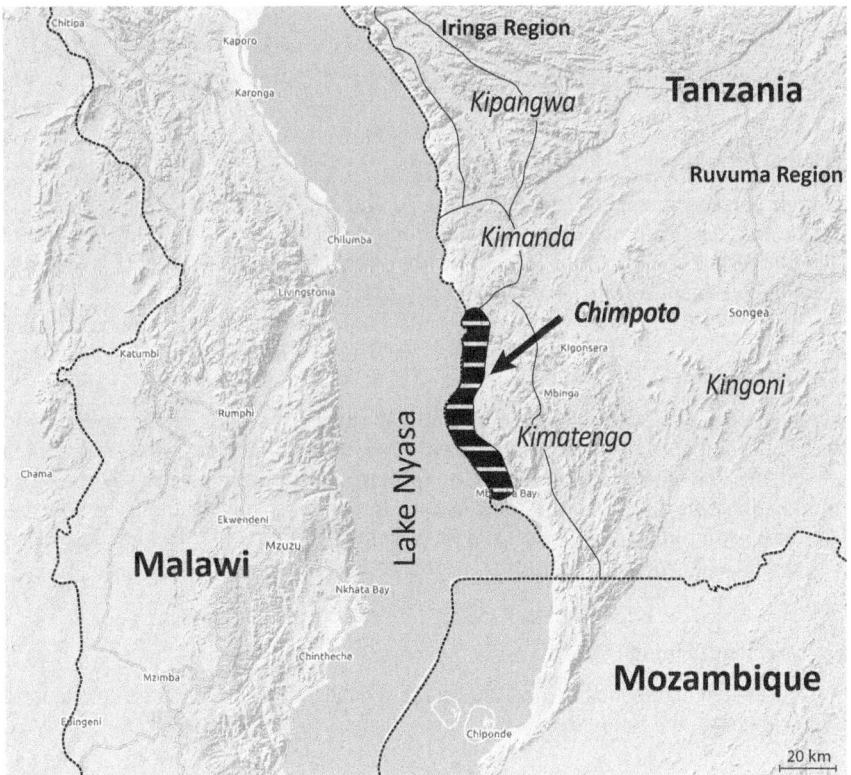

FIGURE 23.1 MAP OF CHIMPOTO LOCATION IN SOUTHWEST TANZANIA.

With suffixes having the comparable back vowel [u], found in the suffix -*ul*-, assimilation only occurs when the preceding stem vowel is round.

(3)   Reversive: -ul- ~ -ol-
     -dend-ul-a   'open'   <   -dend-a   'close'
     -kong-ol-a   'untie'   <   -kong-a   'tie'
     -yan-ul-a   'take in'   ~   -yan-ik-a   'set out to dry'

In the Separative suffix, -*uk*-, in contrast, assimilation also occurs with the low vowel [a].

(4)   Separative (intr.): -*uk*- ~ -*ok*-;   Separative (tr.): -*ul*- ~ -*ol*-
     -dum-uk-a   'break'   ~   -dum-ul-a   'break off a piece of'
     -lomb-ok-a   'break off'   ~   -lomb-ol-a   'break off'
     -kay-ok-a   'break'   ~   -kay-ol-a   'break

2.2.2 *Regressive assimilation*

With regressive assimilation, the last vowel of the verb stem assimilates in height to the following derivational or inflectional suffix other than final vowel -*a*.

(5)   -pwagh-al-an-a   'tell to e.o.'   <   -pwagh-el-a   'tell to'
     -hwet-iti   'wore'   <   -hwat-a   'wear'
     -longal-aa   'speak!'   <   -longel-a   'speak'

2.3 Glide formation

The high vowels /i/ and /u/ become glides /y/ and /w/, respectively, when they occur before a non-identical vowel, as for example in *lyova* 'sun' < /li+ova/ and *watu* 'canoe' < /u+atu/. Before vowels of the same type, they delete (see 2.4). The glide is lost if the preceding consonant is [tʃ], *chana* (< chi+ana] 'other side.'

2.4 Vowel deletion

Vowels delete in two environments. First, the low vowel [a] deletes before any following vowel, for example, *vana vithu* 'our children' < /va+ana#va+itu/) and *minu ghoha* 'all the teeth' (< /ma+inu#gha+oha/). Second, a non-low vowel typically deletes when it occurs before a following vowel of the same backness/roundness, for example, in *lihu* 'eye' (< /li+ihu/), *mitu* 'broth' (< /mi+itu/) and *mohi* 'smoke' (< /mu+ohi/). In some cases, there is variability, as in *mwotu* or *motu* 'fire' (< /mu+otu/).

2.5 Vowel lengthening

Vowels are often lengthened in the penultimate syllable of a word in phrase- or clause-final position, as in the following.

(6)   lichova le:la   'that day'
      lichova lela awokiti pa:la   'that day s/he left there'
      m(w)otu wo:la   'that fire'
      m(w)otu wola wo:mbi   'that very fire itself'

However, with verbs, it is the stem-initial vowel that is lengthened, e.g., with suffix -*iti*, as in *chile:ve, mba:liti* 'food, I want [some].'

2.6 Vowel devoicing

Vowel devoicing is commonly found in phrase- or clause-final position, especially following voiceless consonants as, for example, in *lihupa lilachu̧* 'long bone' or *ateliķi* 'she cooked.'

2.7 Consonant inventory

Chimpoto has an inventory of 33 consonant phonemes (Table 23.1). Voiceless stops are aspirated. The /d/ is occasionally realised as an implosive [ɗ], as in [kuɗuɗumila] 'to roar.' Many consonants occur with both palatalisation (Cʸ) and/or rounding (Cʷ). Orthographic forms are noted in angle brackets where they differ significantly from the phonetic symbol.

TABLE 23.1 CHIMPOTO CONSONANT INVENTORY

	Labial	Alveolar	Alveo-palatal	Palatal	Velar	Glottal
Stops						
− vox	p	t			k	
+lab	pʷ	tʷ			kʷ	
+pal	pʸ	d			g	
+vox	b				gʷ	
Affricates						
− vox			ʧ <ch>			
+vox			dʒ <j>			
Fricatives						
− vox	v	s	ʃ <sh>		ɣ <gh>	h
+lab		z	ʃʷ <shw>			hʷ
+pal						hʸ
+vox						
Approximants	(w)			j <y>	w	
Laterals			l			
+lab			lʷ			
+pal			lʸ			
Nasals	m	n		ɲ <ny'>	ŋ <ng'>	
+lab	mʷ				ŋʷ <ng'w>	

2.8 Nasals and nasal sequences

There are 11 prenasalised obstruents that occur. They are derived from classes 1, 3 or 9/10 noun prefixes, or from 1SG or class 1 object markers on verbs.

^{m}p ^{n}t ^{n}ʧ ^{ŋ}k
^{m}b ^{n}d ^{n}ʤ ^{ŋ}g
^{m}b^w ^{n}ʤ^w ^{ŋ}g^w

(7)   [^{m}pʰuhi] 'liar'   [^{n}tʰela] 'medicine'   [^{n}ʧokʰolo] 'grandchild'   [^{ŋ}kokʰa] 'river'
      [^{m}buhi] 'goat'   [^{n}dela] 'path'   [^{n}ʤuʧi] 'bee'   [^{ŋ}gokʰo] 'chicken'
      [^{m}ba^{w}ni] 'coast'   [^{n}ʤ^{w}avela] 'noise'   [^{ŋ}g^{w}apʰa] 'armpit'

2.8.1 Nasal assimilation

Nasals assimilate in place of articulation to the following consonant. For example, [m-belepʰele] (cl.3) 'type of fruit' ([mi-belepʰele] PL), [n-tʰela] (cl.9) 'stick' ([lu-tʰela] SG), [ŋ-geni] (cl.1) 'stranger' ([va-ɣeni] PL).

(8)   [va-ka-m-bon-a]   'they will see her/him'
      SP$_2$-FUT-OP$_1$-see-FV
      [ku-ŋ-kot-a]   'to ask her/him'
      15-OP$_1$-ask-FV

2.8.2 Nasal deletion

Nasals, including the first- and third-person singular object prefixes, delete before voiceless fricatives [ʃ] and [h].

(9)   showo 'fingernail' (cl.9)   (cmp. n-dowa 'wedding' (cl.9))
      homba 'fish' (cl.9)   (cmp. m-bonjo 'antelope (sp.)' (cl.9))
      nikashunga 'I will hunt him/her'   (< /ni-ka-N-ʃu^{n}ga/)
      vahawili 'they chose me'   (< /va-N-hawul-iCi/)

2.8.3 Post-nasal de-lateralisation of [l]

The lateral phoneme /l/ becomes a stop [d] following a nasal. Thus, *andonditi* 's/he looked for me/him/her' derives from /a+N+lond+iti/.

2.8.4 Stop assimilation

The phones [p], [t], and [k] may optionally assimilate in voicing to a preceding nasal, as in the examples below. The use of the different forms may be correlated with older (assimilate) and younger (no assimilation) generations, or with the location in which it is used – in the village (assimilation) in contrast to elsewhere (no assimilation).

(10) *m-pel-aa maganga* or *m-bel-aa maganga*   'give me (some) money'
      n-tol-ayi or *n-dol-ayi*   'take him/her'
      n-kongol-aa or *n-gongol-aa*   'untie him/her'

2.9 Imbrication

The perfective suffix -*iCi* is imbricated when the verb stem has the form -CVCVCX-. The last vowel in the stem is replaced by the [i] of the suffix or, if [u], may become a glide. In a case like -*manyangana* the last two stem vowels are replaced.

(11) -yohwini   'listened'   <   /-yohwan+iCi/
      -helili   'went down'   <   /-helel+iCi/
      -yumwiki   'woke up'   <   /-yumuk+iCi/
      -manyingini   'was known'   <   /-many-angan+iCi/

2.10 Syllable structure

The canonical syllable structure is CV; other types are V, NCV, CGV and NCGV.

2.11 Accent

There is typically one accent per lexical item, realised as high tone. However, it is not clear whether accent here is best described as pitch-accent or stress-accent. In many words, the accent induces vowel centralisation in syllables around it, suggesting stress. On the other hand, certain words have tone-like features as, e.g., in *palê·ngá* 'how many.' For the most part, accent is predictable. Prominence is marked with an acute accent (´).

2.11.1 Accent in nouns

With di- and trisyllabic stems, the accent typically falls on the first syllable of the stem.

(12) li-hí.mba (cl.5)   'lion'
      chi-dú.li (cl.7)   'termite mound'
      m-bú.hi (cl.10)   'goats'
      li-té.le.ku (cl.5)   'pot for liquids'
      chi-hé.ne.ku (cl.7)   'plate for serving'
      n-dá.nga.ti (cl.9)   'bed'

There are exceptions to this generalisation. In classes 1, 3 and 14, the accent falls on the noun class prefix. In class 7, it also appears on the class prefix, just in those cases where the stem begins with labial /w/ or /ᵐb/.

(13) mú-ha.vi (cl.1)   'witch'
      mú-ha.nga (cl.3)   'lake, beach'
      chí-mbo.ndi (cl.7)   'groundnut flour'
      chí-wo.lo (cl.7)   'leg, foot'
      ú-wa.li (cl.14)   'cassava dough'
      ú-lehe (cl.14)   'finger millet'

Trisyllabic stems having a final NCV syllable realise the accent on the penultimate syllable.

(14) n-ko.ló.ngo (cl.1)   'elder'
      li-ngo.ló.mbi (cl.5)   'spoon made from shell'
      n-do.mó.ndo (cl.9)   'hippopotamus'

Some apparent exceptions to these general patterns are given in (15).

(15) má-ka.ka.la (cl.6)   'strength'
      má-ngo.to.ngo (cl.6)   'pods on a species of tree'
      n-go.kó.ko (cl.9)   'insect'
      li-tu.ngú.tu (cl.5)   'owl'

Monosyllabic reduplicated stems realise the accent in the same place as simple di- and trisyllabic stems, on the initial stem syllable: -CV́.CV$_{RED}$.

(16) li-té.te   'reed (water)'
      lu-ndó.ndo   'star'
      ka-ndí.ndi   'ground squirrel'
      chi-nó.no.no   'hard thing'

Disyllabic reduplicated stems realise the accent on the noun class prefix: CV́-CVCV. CVCV$_{RED}$.

(17) lí-ka.tu.ka.tu   'turtle, tortoise'
      lí-hu.lu.hu.lu   'foam'
      chí-mo.ngo.mo.ngo   'scarecrow'
      m-bé.le.pe.le (cl.3) 'fruit (type)'

Pseudo-reduplication:

(18) lí-pu.lu.pu.tu   'butterfly'
      lí-wo.lo.wo.ndo   'lizard (sp.)'

2.11.2 Accent in verbs

As with most nouns, accent on verbs typically falls on the first syllable of the stem (demarcated by -), whether the verb occurs in infinitival form or is conjugated. The sole exceptions arise with the imperfective -aa and modal suffixes -ayi and -aa, which shift the accent to the penultimate mora, hence STEM-á.a.

(19) ku-hé.me.le.ha   'to sell'   ku-lé.nge.ne.ke.la   'to prepare'
      ti-hé.mi.li.hi   'we sold'   va-lé.nge.ne.ki.li   'they prepared'
      ti.ka-hé.me.le.ha 'we will sell'   va.ka-lé.nge.ne.ke.la 'they will prepare'
      yi-he.me.le.há.a   'sell it'   va-le.nge.ne.ke.lá.a 'they are preparing'

3 NOUN MORPHOLOGY

3.1 The noun class system

The noun class, or gender, system in Chimpoto comprises 17 classes, consisting of three types of nouns: nouns proper (13 classes), verbal nouns (one class), and locative nouns (three classes). Nouns typically have a noun class prefix (NP), though there are numerous exceptions, which will be described below. Unlike many Bantu languages, there is no pre-prefix, or augment. The system is undergoing change, with some classes in the process of disappearing, particularly the plural classes.

The form of the noun class prefix (NP) varies according to phonological and/or prosodic features of the noun stem. For classes 1 and 3, *mw-* occurs before vowel-initial stems, *mu-* before h-initial or monosyllabic stems and a homorganic nasal before consonant-initial stems. For underlying /Ci/ and /Cu/ prefixes of other classes, the form of the vowel depends on two factors. First, if the stem is vowel-initial, *Ci-* and *Cu-* are realised as *Cy-* and *Cw-*, respectively, with one exception: *Ci-* becomes *C-* before high front vowels. *Cu-* optionally becomes *C-* before a mid round vowel. Second, underlying /Ca/ prefixes are realised as *C-* before vowel-initial stems. Nasals are homorganic before consonant-initial stems, but delete before stems beginning with [h], [ʃ] or a nasal.

There are three locative classes, pro-clitics occurring before the normal class prefix. They are treated as clitics because speakers frequently separate them from the noun.

TABLE 23.2 NOUN CLASS PREFIXES

Class	NP	Class	NP
1	mu-/__ h, -CV mʷ-/__ V N- elsewhere	2	va- v-/__ V
1a	Ø		
3	mu-/__ h, -CV mʷ-/__ V N- elsewhere	4	mi- mʸ-/__ V (except [i])
5	li- lʸ-/__ V (except [i]) l-/__ V ([i])	6	ma- m-/__ V
7	chi- chʸ-/__ V (except [i]) ch-/__ V ([i])	8	hi- hʸ-/__ V (except [i]) h-/__ V ([i])
7a	ki-		
9	N-	10	N-
11	lu- lʷ-/__ V [-rd]		
12	ka-		
14	u- w-/__ V [-rd]		
15	ku- kʷ-/__ V [-rd]		

TABLE 23.3 LOCATIVE CLASSES

Class	NPro-clitic
16	pa=, p=
17	ku=, kw=
18	mu=, mw=

Noun classes are usually linked in singular/plural pairs, labelled genders. Subclasses, when the agreement patterns and the paired plural are identical, are subsumed under the main class, hence, 1a under 1. Most genders subsume at least one coherent semantic concept that links a large set of the nouns occurring in that gender. This underlying semantic nexus is listed in Table 23.4 following the prefix forms.

TABLE 23.4 NOMINAL GENDER CLASSES: SINGULAR/PLURAL PAIRING

Class pairings	Singular prefix	Plural prefix	Semantic nexus
1/2	mu-, N̦-	va-, v-	humans
1a/2	Ø	va-, v-	kin terms
3/4	mʷ-, mu-	mi-, my-	
3/10	mʷ-, mu-	N̦-	
5/6	li-, ly-	ma-	animals; plants; body parts; tools
7/8	chi-, ch-	hi-, hy-	food preparation items & utensils
7/6	chi-, ch-	ma-	body parts
7a	ki-	–	
9/10	N̦-	N̦-	animals; foods
11/10	lu-, lw-	N̦-	
12/8	ka-	hi-, hy-	diminutive
14/4	u-, w-	my-	

3.2 Nominal derivation

Nouns may be derived from verbs by adding an appropriate class prefix and, in most cases, replacing the stem-final vowel -*a* with the vowel -*i* or -*u*. However, the processes seem unproductive; only one agentive noun in -*i* has been identified.

(20) -i (agentive)   chi-pendi 'hunchback'   < ku-pinda 'to bend'

(21) -u (object)   li-pyaghelelu 'broom'   < ku-pyaghelela 'to sweep with'
                      li-teleku 'pot for brewing beer' < ku-teleka 'to cook'
                      n-teghu 'trap'   < ku-tegha 'to set a trap'

(22) -u (abstract)   ma-lelu 'mourning'   < ku-lela 'to cry'
                      li-hengu 'work, job'   < ku-henga 'to work'
                      lu-yembu 'song'   < ku-yemba 'to sing'
                      lu-heku 'laughter'   < ku-heka 'to laugh'

3.3 Compound nominals

Compound nominals are rare. The few that exist are formed from noun + noun. These are considered compounds because there is only one accent per word.

(23) v-anandowa 'bride & groom'   (v-ana 'children' + n-dowa 'wedding')
      mw-analukolo 'clansman'   (mw-ana 'child' + lu-kolo 'clan')
      lu-valamwehi 'moon'   (luvala '???' + mwehi 'moon')

3.4 Reduplicated nominals

See samples in Section 2.11.1.

3.5 Evaluative forms

3.5.1 Diminutives

There are two diminutive forms, one formed with gender 7/8 prefixes (24), the other with class 12 *ka-* (only in the singular) (25). The distribution and use of these two forms is not clear; neither appears to be productive. Most animals take the *ka-* form.

(24)	*Singular*	*Plural*	*Source*	
	chi-lova	hi-lova	< li-lova	'small flower/s'
	chi-pyaghelelu	hi-pyaghelelu	< li-pyaghelelu	'small broom/s'
	chi-mbuhi	hi-mbuhi	< m-buhi	'small goat/s'

(25)	ka-yuni		< chi-yuni (cl.7)	'small bird'
	ka-ngoko		< n-goko (cl.9)	'small chicken'
	ka-wolowondo		< li-wolowondo (cl.5)	'small lizard'

3.5.2 Augmentatives

Only one word has been discovered with augmentative meaning, formed with the gender 5/6 prefixes *li-/ma-*.

(26)	*Singular*	*Plural*	*Source*	
	li-ndu	ma-ndu	< chi-ndu/hi-ndu	'large thing/s'

3.6 Pronouns

3.6.1 Personal pronouns

Independent personal pronouns are of three types, two for discourse participants – a CVV form and a similar for constructed from a base *-nga* – and one for non-participants, constructed from a base *-ombi*. It is not clear what the difference is between the two forms for discourse participants.

(27) **nee,**   mok-a   nyenda   ku=mu-hanga
      1SG.PRO   SP$_{1sg}$.leave-FV   SP$_{1sg}$.go-FV   17=3-lake
      'Me, I'm leaving to go to the lake.'

(28) u-wali   owo,   mu-ku-l-a   **mwenga?**
      14-cassava   DEM$_{14}$ SP$_{2PL}$-15-eat-FV   2SG.PRO
      'This cassava, you are eating [it], you?'

(29) **y-ombi,**   a-kon-a   ku-woka
      PP$_1$-PRO   SP$_1$-still-FV   15-leave
      'her, she had yet to leave'

TABLE 23.5 PERSONAL PRONOUNS

		Singular		Plural
1		nee		tee
		ne-nga		te-nga
2		wee		mwee
		we-nga		mwe-nga
3	1	y-ombi	2	v-ombi
	3	w-ombi	4	hy-ombi
	5	l-ombi	6	gh-ombi
	7	ch-ombi	8	hy-ombi
	9	y-ombi	10	hy-ombi
	11	l-ombi		
	12	k-ombi		

(30) m-ihu,   na   **gh-ombi**   gha-lol-it-aa   hee   wichu
       6-eye   with   PP₆-PRO   SP₆-see-??-SBJV   NEG   well
       'eyes, his own, they [his eyes] could not see well'

3.6.2 Intensifying pronouns

Two constructions – PERS + CON + AGR_NUM + *eni* and AGR + *eni* – create intensifying pronouns, but can also be used as demonstrative pronouns in discourse contexts.

TABLE 23.6 INTENSIFYING PRONOUNS

Person	CL	Singular	CL	Plural
1		n-a-mw-eni		t-a-v-eni
2		w-a-mw-eni		mw-a-v-eni
3	1	mw-eni	2	v-eni
	3	w-eni	4	hy-eni
	5	l-eni	6	gh-eni
	7	ch-eni	8	hy-eni
	9	y-eni	10	hy-eni
	11	lw-eni		
	12	k-eni		
	14	w-eni		
	16	p-eni		
	17	kw-eni		
	18	mw-eni		

(31) **mw-eni**   kapecha   a-many-i   hee
       1-INT   12.hare   SP₁-know-PFV   NEG
       'Hare himself did not know.'

(32) nee   ni-teliki   w-ali   w-angu   **n-a-mw-eni**
　　 1SG.PRO   SP₁ₛ_-cook.PFV   14.cassava_meal   PP₁₄_1SG.POSS   1SG-CON-SG-INT
　　 'Me, I cooked my cassava meal myself.'

(33) u-paa   ku-yenda   **w-a-mw-eni**?
　　 SP₂ₛ_-want.PFV   15-go   2SG-CON-SG-INT
　　 'Do you want to go yourself?

(34) **p-eni**   pa=nyumba   pala   a-highii   hee
　　 16-INT   16=9.house   DP₁₆-DEM   SP₁-remain.PFV   NEG
　　 'There at home itself, nobody is left.'

3.6.3 Relative pronouns

Relative pronouns have the form CVV. They may be used without the antecedent noun.

TABLE 23.7 RELATIVE PRONOUNS

Cl		Cl	
1	yoo	2	voo
3	woo	4	hyee
5	lee	6	ghaa
7	chee	8	hyee
9	yee	10	hyee
11	loo		
12	kaa		
14	woo		

(35) va-ndu   **voo**   va-y-ii   ni   hi-ndu   hi-mahele
　　 1-person   2.REL   SP₂-be-PFV   with   8-thing   EP₈-many
　　 'people who have many things'

(36) li-hogho   **lee**   l-aka-y-ii   ni   njala   li-yengel-a   mu=nyumba   mo-la
　　 5-hyena   5.REL   SP₅-PST-be-PFV with   9.hunger   SP₅-enter-F   18=9.house   DP₁₈-DEM
　　 'A hyena that was hungry got inside the house.'

3.6.4 Possessive pronouns (AGR + POSS)

Possessive pronouns, which may be used adjectivally, are formed by combining the agreement marker (see CON in Table 23.9 in Section 3.7) of the appropriate noun class with the possessive root.

TABLE 23.8 POSSESSIVE PRONOUNS

1SG	-angu	1PL	-itu
2SG	-aku	2PL	-inu
3SG	-aki	3PL	-au or -avi

(37)   yela ni y-angu   'that [e.g., field] is mine'
       mambelele ni gh-aki   'the sheep are his'
       ukochi w-avi   'their friendship'
       kahomba k-angu   'my little fish'

3.7 Agreement

All modifiers and verbs exhibit agreement with the head noun through affixation of an agreement prefix. The form of the agreement prefix attached to the modifier or verb differs for each noun class. There are six general patterns, as shown in Table 23.9.

TABLE 23.9 AGREEMENT PREFIXES

Cl #	NP	Adjs	N°s	Dems	Con -a	Verbs Subject	Object
1	mu-	mu-, N-	m-	yo-	w-	a-	-N-
2	va-	va-	va-	va-	v-	va-, v-	va-
3	mu-	u-	u-	wo-	w-	u-, w-	-u-
4	mi-	hi-	hi-	hye-	hy-	hi-, hy-	-hi-
5	li-	li-	li-	le-	ly-	li-, ly-	-li-
6	ma-	gha-/ma-	gha-	gha-	gh-	gha-, gh-	-gha-
7	chi-	chi-	chi-	che-	ch-	chi-, ch-	-chi-
8	hi-	hi-	hi-	hy-	hy-	hi-, hy-	-hi-
9	n-	(y)i-	(y)i-	ye-	y-	i-	-(y)i-
10	n-	hi-	hi-	hye-	hy-	hi-	-hi
11	lu-	lu-	lu-	lo-	lw-	lu-, lw-	-lu-
12	ka-	ka-	ka-	ka-	k-	ka-, k-	-ka-
14	u-	u-	u-	u-	w-	u-, w-	-u-, -w-
16	pa=			pa-		pa-	
17	ku=			ko-		ko-	
18	mu=			mo-		mo-	

3.8 Connective -a

The connective -a is used to link a noun: (1) to another noun in a genitive construction; (2) to an infinitive, in a purposive construction; and (3) to a clause, in a relative construction. In each case, it requires agreement with the head noun.

    Genitive usage:

(38)   ma-hamba gh-a mi-kongo   'leaves of the trees'
       n-dondo y-a hy-embi   'package of mangoes'
       chi-kahi ch-a li-wombo   'flood season' [lit. season of floods]'
       va-ndu v-a ku=Mbaha   'people of [at] Mbaha'

Purposive usage:

(39) ma-chi gh-a kukuha   'water for washing'
      ma-chi gh-a kuhopa   'drinking water'
      hi-ndu hy-a kuketa   'things to do'
      mi-kongo hy-a kuyogha   'medicine to bathe in'

Relative clause usage [see also Section 5.3 for more detail]:

(40) hi-tabu   hy-a   va-lemb-iti   ku=Chimpoto
      8-book CP_8-CON SP_2-write-PFV   17=7.Mpoto_language
      'books [that are] written in Chimpoto'

3.9 Demonstratives

There are at least eight demonstrative forms in Chimpoto. Their full ranges of use are not known at this time, although six appear to be primarily spatial and two to be discourse-oriented. In Table 23.10, templates of the patterns for the various demonstratives are illustrated, followed by examples from class 1. The appropriate prefix forms for all classes are given in Table 23.11 (spatial) and Table 23.12 (discourse). Demonstratives may function as adnominals, in which case they follow the noun they modify, or as independent pronominals.

TABLE 23.10 TEMPLATES FOR DEMONSTRATIVES

	Neutral	*Emphatic*	
Proximal	VCV		CV-CV-VCV
Close	oyo		yoyooyo
	'this		'this very same [one]'
Not Close	CVV		
	yoo		
	'that'		
Distal	CV-la	CV-la-CV-la	CV-CV-l(a)-CV-la
	yola	yolayola	yoyolyola
	'that'	'that same [one]'	'that very same [one]'
Discourse	CV-l(a)-CVV		C(G)-eni
	yolyoo		mw-eni
	'that mentioned'		'that mentioned'

(41) lu-homo   **olo**   lu-tu-pwagh-il-a   naha. . .
      11-story DEM_{11} SP_{11}-OP_{1PL}-say-APPL-FV   like_this
      'this story tells us this, . . .'

TABLE 23.11 DEMONSTRATIVE FORMS: SPATIAL

NCl	PROXIMAL close to speaker		PROXIMAL close to addressee	DISTAL	
	VCV	CVCV:CV	C(G)VV	CV-la	CV-CV-l-CV-la
1	oyo	yoyooyo	yoo	yola	yoyolyola
2	ava	vavaava	vaa	vala	vavalvala
3	owo	wowoowo	woo	wola	wowolwola
4	ehye	hyehyeehye	hyee	hyela	hyehyelhyela
5	ele	leleele	lee	lela	lelelela
6	agha	ghaghaagha	ghaa	ghala	ghaghalghala
7	eche	checheeche	chee	chela	chechelchela
8	ehye	hyehyeehye	hyee	hyela	hyehyelhyela
9	eye	yeyeeye	yee	yela	yeyelyela
10	ehye	hyehyeehye	hyee	hyela	hyehyelhyela
11	olo	loloolo	loo	lola	lolololola
12	aka	kakaaka	kaa	kala	kakalkala
14	owo	wowoowo	woo	wola	wowolwola
16	apa	–		pala	papalpa(l)a
17	oko	kokooko		kola	kokolko(l)o
18	omo	momoomo		mola	momolmo(l)o

(42) l-ombi   li-yoka   le-la   li-pal-iti   ngoko   **hye-hyel-hyela**
PP$_5$-PRO   5-snake   DP$_5$-DEM   SP$_5$-want-PFV   10.chicken   DP$_{10}$-DEM
'it, that snake, wanted those very same chickens'

(43) hi-tombi   ni   **hye-la   hye-la**
8-mountain   COP   DP$_8$-DEM   DP$_8$-DEM
'the mountains are the same ones'

(44) henu   **yolyoo**   a-y-ii   hee
now   DEM$_1$ SP$_1$-be_at_place-PFV   NEG
'Now, that one [just mentioned] is not there.'

(45) a-ka-hik-il-a   hee   koo=gha-y-ili   ma-toki   **gh-eni**
SP$_1$-PST-reach-APPL-F   NEG   17.REL=SP$_6$-be-PFV   6-banana   DP$_6$-DEM
'He could not reach where those [previously mentioned] bananas were.'

Locative demonstratives with adverbial function:

(46) ni   ku-veka   **mo-la**   mu=chi-hogholo   mo-la
and   15-put   18-there   18=7-termite_mound   DP$_{18}$-DEM
'and put [it] there inside that termite mound'

(47) cha   va-hik-a   **pala**, . . .
when   SP$_2$-arrive-FV   DP$_{16}$.there
'when they arrive there, . . .'

TABLE 23.12 DEMONSTRATIVES: DISCOURSE

Cl	Neutral	Emphatic	Cl	Neutral	Emphatic
1	yolyoo	yeni	2	valavaa	veni
3	wolwoo	weni	4	hyelahyee	hyeni
5	lelalee	leni	6	ghalaghaa	gheni
7	chelachee	cheni	8	hyelahyee	hyeni
9	yelyee	yeni	10	hyelahyee	hyeni
11	lolaloo	lweni			
12	kalakaa	keni			
14	wolwoo	weni			

3.10 Adjectives

There exist relatively few true adjectives in Chimpoto; those found are listed in Table 23.13, and color adjectives in Table 23.14. These adjectives follow the noun and require agreement with the noun that they modify following the adjectival agreement paradigm.

(48) chi-leve chi-choko-pi   'a little bit of food'
      n-kela u-lachu u-niahi   'a long, beautiful tail'
      n-donga yi-vehe   'a wet stick'

(49) chi-tambala   ch-aki   chi-pili
      7-cloth   PP$_7$_3SG.POSS   AP$_7$-black
      'her black [piece of] cloth'

(50) n-gowo   yi-keli   wichu
      9-clothing   AP$_9$-red   extremely
      'a deep/dark red piece of clothing'

Certain adjectival concepts are expressed through intransitive verbs, for example, 'be(come) old' -*ghogholoka vi*, 'be happy' -*hovela vi* and 'be(come) thin' -*kombala vi*. In such cases, the adjectival concept is expressed through a connective construction.

TABLE 23.13 ADJECTIVE ROOTS

-choko	small, young	-kata	lazy	-nonono	hard
-hakau	bad, dirty, ugly	-kolongo	big, main, older	-vehe	raw, wet
-hihamu	cold	-lachu	tall, long	-yipi	short
-hinikanu	stubborn	-mbomba	female	-yomu	dry
-kambaku	male	-niahi	good, beautiful	-zeya	old (humans)

TABLE 23.14 COLORS

-hamba	'green'	-keli	'red'
-huhu	'white'	-pili	'black, dark'

(51) mu-ndu y-a a-chenjel-a
1-person CP$_1$-CON SP$_1$-be_clever-FV
'a clever person' [lit. 'person who is clever']

(52) ma-lavi gh-a gha-komile
6-groundnut CP$_6$-CON SP$_6$-be_mature.PFV
'mature groundnuts [lit. 'groundnuts that are mature']'

3.11 Quantifiers

Quantifiers, like other modifiers, follow the head noun and require agreement with it. Four quantifiers have been identified (Table 23.15). The quantifier for 'any' appears to be a partial reduplication of the form for 'all.'

TABLE 23.15 QUANTIFIERS

-ngi	'(an)other'	-mahele	'many, much'
-oha	'all, whole'	-o -oha	'any'

(53) li-chova le-ngi 'another day'
li-chova l-oha 'the whole day'
v-analukolo v-oha 'all the clansmen'
hi-mbioko hi-mahele 'many ants'
ma-chi gha-mahele 'much water'

(54) ni ku-teleka mi-kongo hye-ngi
and 15-cook 4-medicine 4-other
'and cooked up some (other) medicine'

(55) ni-paa mu-ndu y-o-y-oha a-lemb-ayi
SP$_{1SG}$-want 1-person EP$_1$-any-EP$_1$-any 1-write-SBJV
'I want anyone [who can] write'

3.12 Numerals

The Chimpoto numeral system is a quinary system in which the numerals from six to ten are constructed with the word for five, *muhanu*. Only the first three numerals require agreement with the noun.

(56) v-ana va-vele 'two children'
hi-nyama muhanu ni hi-monga 'six animals'

Ordinal numbers, except for 'first,' take a prefix *ka-*, used to create adverbs. They are linked to the head noun with the connective *-a*, which requires agreement.

(57) li-chova l-a kuwanja 'the first day'
mw-ana w-a kavele 'the second child'
kilu ch-a katatu 'the third night'

TABLE 23.16 CARDINAL NUMBERS

1	-monga	6	muhanu ni -monga
2	-vele	7	muhanu ni -vele
3	-tatu	8	muhanu ni -tatu
4	ncheche	9	muhanu ni ncheche
5	muhanu	10	muhanu ni muhanu

TABLE 23.17 ORDINAL NUMBERS

1st	AGR-a kuwanja
2nd	AGR-a kavele
3rd	AGR-a katatu
4th	AGR-a kancheche
5th	AGR-a muhanu

3.13 Noun phrase word order

Noun modifiers typically follow the noun they modify. The demonstrative is the last modifier in a noun phrase. The typical order is: N + REL + POSS + NUM + ADJ + DEM.

(58) mwehi   woo   u-hika   owo   nyenda ku=nyumba   kwali
     3.month   3.REL   3-arrive 3.DEM   1SG.go 17=9.house   don't_know
     'This coming month, I'm not sure I'm going home.'

(59) v-ana   v-aki   va-vele
     2-children PP_2-3SG.POSS EP_2-two
     'his/her two children'

(60) ma-mbelele   gha-kolongo   gha-la
     6-sheep EP_6-big DP_6-DEM
     'those big sheep'

(61) n-goko   hy-angu   hi-vele   hi-kolongo   hye-la
     10-chicken PP_{10}-1SG.POSS EP_{10}-two AP_{10}-big DP_{10}-DEM
     'those two big chickens of mine'

4 VERB MORPHOLOGY

The simple infinitive in Chimpoto consists of the class 15 nominal prefix *ku-*, the verb root and final vowel *-a*. The verb stem may be inflected for person and number, tense, aspect and mood. Types of verb roots and the various inflectional categories are listed and discussed below.

4.1 Verb roots

The canonical form of the verb root in Chimpoto is -CVC-. Table 23.18 provides an inventory of -CV- roots, Table 23.19 of some -CVCVC(VC)- roots (i.e., polysyllabic roots not obviously derived by the addition of verbal extensions), and Table 23.20 of several partially reduplicated stems.

TABLE 23.18 -C(G)- STEMS

-gw-a	'fall'	-l-a	'eat'
-hw-a	'die'	-v-a	'be(come)'

TABLE 23.19 SOME -CVCVC- STEMS

-hakah-a	'annoy'	-longoh-a	'lead'
-hetek-a	'slaughter'	-tapik-a	'vomit'
-kalang-a	'fry'	-telek-a	'cook'
-kolom-a	'snore'	-vaghay-a	'shiver'

TABLE 23.20 PARTIALLY REDUPLICATED STEMS

-kokoh-a	'announce'	-memen-a	'chew'
-lulut-a	'ululate'	-titil-a	'close eyes drowsily'
-dudumil-a	'roar'	-ghogholok-a	'become old'

There is one "defective" verb that does not have an infinitival form in *ku-* and does not permit a verbal suffix other than perfective -*i(l)i*: -*y*- 'be (in a place).'

4.1.1 Verbs of 'being'

There are four means of expressing 'being': null marking, -*va* 'be(come),' -*y*- 'be (in a place),' and copula *ni*.

4.1.1.1 Null marking

With predicative noun:

(62) Ado mwana   'Ado is a child.'
      vaa vana   'Those ones are children.'

With predicative adjective:

(63) Ado   n-dachu
      Ado AP_1-tall
      'Ado is tall.'

(64) n-gowu   i-keli
      9-cloth AP_9-red
      'The cloth is red.'

(65) pa=nyumba   pa-nunu
      16=9.house AP_{16}-quiet
      'At the house is quiet.'

4.1.1.2 -va

This verb always occurs, when conjugated, with the infinitival prefix *ku-*, as is the case for all monosyllabic verbs (cf. *va-ku-l-a* 'they eat'). It typically conveys an existential or generic sense, except in the past.

(66) a-ku-**v**-a   nkochi   w-aku
      SP$_1$-15-be.FV   1.friend   PP$_1$-3SG.POSS
      'She becomes your friend.'

(67) va-ka-tati-mundu,   hi-ku-**v**-a   ng'ombi   he-vele   oo   mbuhi   muhanu
      SP$_2$-PST-father-own   SP$_{10}$-15-be-FV   10.cattle   EP$_{10}$-two   or   10.goat   five
      '[From] their uncles, there are usually two cows or five goats.'

(68) chi-ku-**v**-i   chi-kolongu   wichu
      7-15-become-PFV   AP$_7$-big   extremely
      'It became very big.'

4.1.1.3 -y-

The defective verb *-y-* occurs with the perfective suffix (either as *-ii* or *-ili*) or with the subjunctive suffix.

(69) u-**y**-ii   kolee?   nee   n-**y**-ii   apa
      SP$_{2SG}$-be-PFV   where   1SG   SP$_{1SG}$-be-PFV   DEM$_{16}$
      'Where are you?'   'I am here.'

(70) wo-**y**-ii   mu-hilu   ku-m-pwaghela   nyongolo
      SP$_3$-be-PFV   3-taboo   15-OP$_1$-tell.APPL   1.2SG.mother
      'It is a taboo to tell your mother.'

(71) ka-**y**-ili   kapecha;   ka-**y**-ili   ni   njala
      SP$_{12}$-be-PFV   12.hare   SP$_{12}$-be-PFV   with   9.hunger
      'There was a hare; it was hungry.'

4.1.1.4. *ni* Copula

(72) a-ka-gha-won-a   ma-kava   ni   ku-manya   oyo,   **ni**   kapecha
      SP$_1$-PST-6-see-FV   6-peel   and   15-know   DEM$_1$ COP   12.hare
      'He saw the peels and knew this one, it was hare.'

(73) ma-chi   gha-la   **ni**   gh-a   ku-hopa
      6-water   6-DEM   COP   CP$_6$-CON   15-drink
      'That water [it] is for drinking.'

4.2 Verb extensions

Verbs may change or denote valency, argument configuration or semantics by the addition of verb extensions, of which there are 11: (1) valency reducing: *-angan-*$_1$ reciprocal,

-ik- anticausative or ostensive, *-angan-*₂ stative anticausative; (2) valency increasing: *-il-* applicative, *-ih-* causative; (3) transitivity alternation: *-uk-* (intransitive) and *-ul-* (transitive) separatives; (4) meaning altering: *-ekel-* intensive, *-elel-* persistive, *-ul-* reversive. Reduplication of the verb stem indicates continuation of an action, and is labelled the perseverative.

4.2.1   Valency reducing

4.2.1.1   *-an-* or *-angan-*1

RECIPROCAL or ASSOCIATIVE indicate mutual or reciprocal interaction. These extensions appear to be lexically determined allomorphs. They change a transitive verb to an intransitive one requiring a plural subject.

(74) ku-kong-an-a   'to connect to one another'   <   ku-konga   'to tie'
      ku-tangatel-an-a   'to help one another'   <   ku-tangatela   'to help'
      ku-many-angan-a   'to know each other'   <   ku-manya   'to know'
      ku-him-angan-a   'to meet (and gather)'   <   ku-himina   'to meet'
      ku-won-angan-a   'to get together'   <   ku-wona   'to see'

4.2.1.2   *-ik-* ~ *-ek-*

ANTICAUSATIVE reconfigures the linking of grammatical and thematic roles. In particular, it removes the agent as subject and makes a theme the subject in its stead.

(75) ku-towol-ik-a   'to get married'   <   ku-towola   'to take, marry'
      ku-hawul-ik-a   'to be chosen'   <   ku-hawula   'to choose'
      ku-tiny-ik-a   'to be burnt'   ~   ku-tiny-ih-a 'to burn'
      ku-kem-ek-a   'to be called'   ~   ku-kem-el-a 'to call'

OSTENSIVE indicates that the potential inherent in the verb is realisable. It reduces the valency of the verb, changing a transitive verb to an intransitive one.

(76) ku-hotoleka   'to be doable'   <   ku-hotola   'to be able to'
      ku-loleka   'to be watchable'   <   ku-lola   'to look at'

4.2.1.3   *-angan-*2

STATIVE ANTICAUSATIVE reduces the valency of the verb, changing a transitive verb to an intransitive. It differs from a passive in that no agent may be expressed.

(77) ku-kong-angan-a   'to be tangled'   <   ku-konga   'to tie'
      ku-many-angan-a   'to be famous'   <   ku-manya   'to know'
      ku-won-angan-a   'to be visible'   <   ku-wona   'to see'

4.2.2   Valency increasing

4.2.2.1   *-il-* ~ *-el-*

APPLICATIVE: Traditionally termed the applicative extension, this extension has three functions: benefactive, directional/locational and instrumental. The vowel in this suffix alternates according to the preceding vowel: [e] following /e/ or /o/, [i] elsewhere.

BENEFACTIVE indicates that the action is performed on behalf of someone.

(78) ku-pak-il-a   'to cover for'   <   ku-paka   'to cover'
      ku-lel-el-a   'to cry for'   <   ku-lela   'to cry'
      ku-lendel-el-a   'to wait for'   <   ku-lendela   'to wait'

DIRECTIONAL/LOCATIONAL indicates that the action is carried out away from the subject or at a certain place.

(79) ku-tagh-il-a   'to throw away'   <   ku-tagha   'to throw'
      ku-kembel-el-a   'to run after'   <   ku-kembela   'to run'
      ku-pomol-el-a   'to breath on'   <   ku-pomola   'to breath'
      ku-yengel-el-a   'to enter through'   <   ku-yengela   'to enter'
      ku-hik-il-a   'to reach (a place)' <   ku-hik-a   'arrive'
      ku-tol-el-a   'to take with'   <   ku-tola   'to take'

INSTRUMENTAL indicates that the action is carried out with some implement.

(80) ku-kong-el-a   'to tie with'   < ku-konga   'to tie'
      ku-yomok-el-a   'to dry by means of' < ku-yomoka   'to dry' (e.g., the sun)
      ku-lov-el-a homba   'to catch fish with' < ku-lova homba 'to catch fish'

4.2.2.2 -ih- ~ -eh-

CAUSATIVE adds an agentive argument to intransitives, an agentive goal (primary object) to transitives.

(81) Intransitives
      ku-nenep-eh-a   'to fatten'   <   ku-nenepa   'to be big and fat'
      ku-yoghop-eh-a   'to frighten'   <   ku-yoghopa   'to be afraid'

(82) Transitives
      ku-l-ih-a   'to feed s.o.'   <   ku-la   'to eat'
      ku-hemel-eh-a   'to sell to s.o.'   <   ku-hemela   'to buy'

4.2.3 Transitivity alternation

SEPARATIVE indicates some type of separation, either physical or abstract. These extensions determine transitivity – either intransitive (-uk- ~ -ok-) or transitive (-ul- ~ -ol-). The vowel of the extension becomes [o] following a non-high back vowel in the base.

(84) ku-dum-uk-a   'to break' (vi)   ~   ku-dumula   'to break' (vt)
      ku-kach-ok-a   'to tear' (vi)   ~   ku-kachola 'to tear' (vt)
      ku-yum-uk-a   'to wake/get up' (vi)

4.2.4 Meaning altering: enhancement or reversal

4.2.4.1 -ekel-

INTENSITIVE indicates that the action of the verb is comprehensive, exhaustive, and/or profound.

(85) ku-dend-ekel-a 'to lock'   <   ku-denda 'to close'
      ku-lol-ekel-a   'to look at intently' <   ku-lola   'to look at'

4.2.4.2 -elel-

PERSISTIVE indicates that the action denoted by the verb persists over a longer duration than the simple action. Often, the new lexical item has a slightly idiosyncratic meaning. The vowel alternates according to the preceding vowel: [e] following [e] or [o], [i] elsewhere. The consonant is realised as [n] when the suffix follows a stem-final nasal.

(86) ku-hov-elel-a   'to be very happy' <   ku-hovela 'to be happy'
      ku-yogh-elel-a 'to swim'   <   ku-yogha   'to bathe'
      ku-yend-elel-a 'to continue'   <   ku-yenda   'to go, walk'

4.2.4.3 -ul- ~ -ol-

REVERSIVE indicates that the action denoted by a transitive verb is reversed. The vowel becomes [o] following an [o] in the verb base.

(87) ku-dend-ul-a   'to open'   <   ku-denda   'to close'
      ku-kong-ol-a   'to untie'   <   ku-konga   'to tie'
      ku-hyegh-ul-a 'to uncover'   <   ku-hyegh-el-a 'to cover'
      ku-yan-ul-a   'to take in when dry'   <   ku-yan-ik-a 'to set out to dry'

4.2.4.4 Reduplication of verb stem

PERSEVERATIVE indicates a continuous or persisting interpretation of the action denoted by the verb. Not all reduplicated verbs derive from a non-reduplicated form.

(88) ku-hombahomba   'to jump around'   <   ku-homba   'to jump'
      ku-londalonda   'to look around'   <   ku-londa   'to look'
      ku-nyuhanyuha   'to shake repeatedly' <   ku-nyuha   'to shake'
      ku-lekaleka   'to complain'   <   –

4.3 Order of extensions

There are few examples of extensions co-occurring. The following sequences have been noted, both involving the reciprocal suffix -an- (recall that Applicative -il- becomes -al- preceding certain suffixes):

        APPL + REC   -pwag-al-an-a   'say to one another'
        REV + REC   -kong-ol-an-a   'untie one another'

4.4 Passive

There appears to be no true passive in Chimpoto, as there is in other Bantu languages. Rather, a third-person plural form (noun class 2) is used.

(89) nee   n-gowo   y-angu   va-m-bek-ii   va-va-tek-a   ma-chi
      1SG   9-cloth   PP₉_1SG.POSS   SP₂-OP₁SG-give-PFV   RP₂.REL-SP₂-draw.water-FV   6-water
      'As for me, my cloth, I was given (by) those who fetch water'

(90) va-ndu   va-y-ii   ni   hi-tabu   hy-a   va-lemb-iti   ku=chimpoto?
      2-people   SP₂-be-PFV   with   8-book   CP₈-CON   SP₂-write-PFV   17=7.mpoto.language
      'Do people have books that are written in Chimpoto?'

4.5 Verb categories

The verb in Chimpoto comprises seven morphological slots ordered in the following manner: (SP) – (T/A) – (IT) – (OBJ) + ROOT + (EXT) + A/M.

4.5.1 Subject prefixes

The following subject prefixes are used for the respective participants and noun classes, repeated from Table 23.9 (Section 3.7).

	Person		3:								
	1	2	1/2	3/4	5/6	7/8	9/10	11/10	12	14	15
SG.	n(i)-	u-	a-	u-	li-	chi-	i-	lu-	ka-	u-	ku-
PL.	ti-	mu-	va-	hi-	gha-	hi-	hi-	hi-			

(Noun class agreement prefixes span columns 1/2 through 15.)

If there are conjoined subjects from different noun classes, *va-* is the default agreement for human beings or personified animals, *hi-* for non-human things.

(91) chi-pendi   ni   chi-denda   v-oha   va-lemili
      7-hunchback   and   7-deaf_person   EP₂-all   SP₂-be_debilitated.PFV
      'The hunchback and the deaf person are both debilitated.'

(92) ma-chova   gh-oha   chi-mbioko   ni   li-nyaunyau   va-heng-iti   li-hengu   pamonga
      6-day   EP₆-all   7-ant   and   5-spider   SP₂-work-PFV   5-work   together
      'Every day ant and spider worked together.'

(93) m-buhi   ni   li-mbelele   hy-oha   hi-nenepi
      9-goat   and   5-sheep   10-all   SP₁₀-be_fat.PFV
      'The goat and the sheep are both fat.'

(94) li-ganga   ni   chi-legha   hy-oha   hi-top-a
      5-stone   and   7-pot   EP₁₀-all   SP₁₀-be_heavy-FV
      'The stone and the pot are both heavy'

4.5.2 Tense/modality markers[2]

Tense, aspect, mood and modality markers occur in two positions, one preceding the verb stem and one at the end of the verbal complex. There are two morphological slots, one opposing -Ø- and -a-, the other -Ø- and -ka-, in contiguous prefixal positons. These can be combined to form four different combinations.

4.5.2.1 Tense/modality

The exact functions of the prefixes in (95) are not clear. Speakers claim that the two combinations in (a) are the same semantically, as are the two in (b). Both markers can be used to denote either future or past.

(95) a) -Ø-Ø-   -a-Ø-
      b) -Ø-ka-   -a-ka-

4.5.2.2 Present/Future

SP-STEM-*a*

(96) ndava   kiki   mu-tek-a   ma-chi?
      9.reason   what   SP$_{2PL}$-fetch-FV   6-water
      'Why are you fetching water?'

(97) chilau   ti-hin-a   ku=Ngombo
      tomorrow   SP$_{1PL}$-dance-FV   17=Ngombo
      'Tomorrow we dance at Ngombo.'

4.5.2.3 Future

SP-*a*-STEM-*a*

(98) tee   t-a-ku-lolekel-a   pa=mi-ho
      1PL   SP$_{1PL}$-FUT-OP$_{2SG}$-look_at_intently-FV   16=4-face
      'Us, we are going to keep an eye on you.'

SP-*ka*-STEM-*a*

(99) a   i-ka-nenep-a
           SP$_9$-FUT-become_fat-FV
           'It [goat] is going to become fat.'

      b   tee   ti-ka-kin-a   hee   pa-la
           1PL   SP$_{1PL}$-FUT-play-FV   NEG   16-there
           'Us, we will not play there.'

SP-*aka*-STEM-*a*

(100) nenga   n-aka-ku-tol-a   pa=n-gongo   w-angu   kavele
       1SG   SP$_{1SG}$-FUT-OP$_{2SG}$-take-FV   16=3-back   PP$_3$-1SG.POSS   again
       'Me, I will take you on my back again.'

4.5.2.4 Past

SP-*ka*-STEM-*a*

(101) a-ka-amol-a   ku-li-kamola   li-yoka   le-la
      SP$_1$-PST-decide-FV   15-OP$_5$-catch   5-snake   DP$_5$-DEM
      'He decided to catch that snake.'

SP-*a-ka*-STEM-*a*

(102) lichu   t-a-ka-heng-a   wichu
      yesterday   SP$_{1PL}$-PST-work-FV   a_lot
      'Yesterday, we worked hard.'

4.5.3 *Itive* ka-

The itive prefix -*ka*- indicates motion away to perform the activity. It occurs in the imperative and subjunctive constructions.

(103) a ka-hyepul-ayi   w-a-lol-aa   w-a-m-bon-a   mboomba
         IT-uncover-SBJV   SP$_{2SG}$-FUT-look-SBJV   SP$_{2SG}$-FUT-1-see-FV   1.woman
         'Go uncover [it]; you should look [and] you will see a woman.'

      b ti-ka-tol-ayi   ku-va-pela   v-ana   v-itu.
         SP$_{1PL}$-IT-take-SBJV   15-OP$_2$-give   2-child   PP$_2$-1PL.POSS
         'We should go & take [it (food)] to give to our children.'

      c va-towol-aa   ma-kong'ondela   agha   va-ka-tupul-aa   ma-yau
         SP$_2$-take-SBJV   6-cassava_basket   6.DEM   SP$_2$-IT-pull-SBJV   6-cassava
         'They should take these cassava baskets (and) go pull (up some) cassava.'

4.5.4 *Object markers*

The following object markers are used for the respective participants and noun classes. Only one object marker at a time may be prefixed to the verb stem. Generally, only a primary object can be marked on the verb, i.e., a transitive patient or a ditransitive goal.

	Person			Noun Class							
	1	2	3								
			1/2	3/4	5/6	7/8	9/10	11	12	14
SG	n-	ku-	m(u)-	N-	li-	chi-	i-	lu-	ka-	u-
PL	ti-	mu-	va-	hi-	gha-	hi-	hi-			

(104) a va-m-bel-iti   mi-kongo
         SP$_2$-OP$_3$-give-PFV   3-medicine
         'They gave him/her medicine.'

      b a-li-teliki
         SP$_1$-OP$_5$-cook.PFV
         'She cooked it [e.g., *li-puka* 'porridge'].'

The object marker may co-occur with the noun phrase it indexes. Although in such con-stuctions the noun phrase is specific, there are other cases in which the noun phrase is specific but does not co-occur with an object marker.

(105) a   y-ombi   y-a-gha-hetiki   ma-mbelele   gha-tatu
          1-PRO   SP$_1$-PST-OP$_6$-slaughter.PFV   6-sheep   EP$_6$-three
          'him, he slaughtered the three sheep'

      b   ana   y-a-hetiki   m-buhi,   t-aka-yoch-a   m-buhi   y-eni
          when   SP$_1$-PST-slaughter.PFV   9-goat   SP$_{1PL}$-FUT-roast-FV   9-goat   DP$_9$-DEM
          'when he has slaughtered a goat, we will roast said goat'

4.5.5 Aspect and mood markers

In final position occurs a set of aspect and mood markers. There are two grammatical aspects: perfective and imperfective. The perfective has three realisations; the imperfec-tive is invariable, and always attracts the accent. The perfective denotes a complete event, the imperfective an incomplete or durative one. The perfective may be reduced to *-ii* or *-i*.

4.5.5.1 Perfective

The perfective is realised *-ili* following -C(V)- stems (106), *-iti* following -CVC- stems (107) and *-iCi* in -CVCVC(X)- stems.

(106) i-li-ili   n-goko   y-angu
      SP$_9$-eat-PFV   9-chicken   PP$_{9_}$1SG.POSS
      'It ate my chicken.'

(107) u-lemb-iti   lu-homo?
      SP$_{2SG}$-write-PFV   11-story
      'Have you written a story?'

(108) ana   va-palini   ni   ku-yetekela   v-oha   [< -pal-an- 'want-RECP']
      if   SP$_2$-agree.PFV   and   15-respond   EP$_2$-all
      'if [presuming] they have agreed and all have responded, . . .'

The perfective suffix combined with the tense prefix *-ka-* realises a past perfective.

(109) a-y-ii   hee   pala,   a-ka-wok-iti
      SP$_1$-be-PFV   NEG   DP$_{16}$.DEM   SP$_1$-PST-leave-PFV
      'He was not there; he had left.'

(110) li-chova   li-monga,   ni-ka-amwili   ku-yenda   ku=mu-hanga
      5-day   EP$_5$-one   SP$_{1SG}$-PST-decide.PFV   15-go   17=3-lakeshore
      'One day, I had decided to go to the lakeshore.'

With change-of-state and stative verbs, the perfective invokes a present reading.

(111) ti-tond-iti
      SP$_{1PL}$-become_tired-PFV
      'We are tired.'

(112)   va-m-pal-iti
        SP$_2$-OP$_1$-love-PFV
        'They love him/her.'

4.5.5.2 Imperfective -aa

The imperfective seems to occur only wtih past or future readings, or with a participial reading.

(113)   a-ka-li-pak-aa   u-lembo
        SP$_1$-PST-OP$_5$-smear-IPFV   14-sap
        'He smeared it with a sticky sap.'

(114)   ti-kwel-a   mu=n-kongo   ti-lendel-aa   ni   ku-woka   pa
        SP$_{1PL}$-climb-FV   18=3-tree   SP$_{1PL}$-wait-IPFV   and   15-leave   when
        'We will climb a tree, we will wait and leave when. . .'

(115)   n-aka-heng-aa   lihengu   chilau
        SP$_{1SG}$-FUT-work-IPFV   5-work   tomorrow
        'I will be working tomorrow'

(116)   kilu   che-la   a-chi-weni   chi-ndu   chi-hum-aa   ku=chi-kongolo
        7.night   DP$_7$-DEM   SP$_1$-OP$_7$-see.PFV   7-thing   SP$_7$-come_from-IPFV   17=7-forest
        'That night he saw something coming from the forest.'

4.5.5.3 Mood

The indicative ends with an aspect marker or the final vowel -a. The imperative and subjunctive end with -ayi or -aa, indicating greater or lesser formality, respectively. (See Section 4.8 and Section 4.9, respectively, for examples.)

4.6 Negation

Indicative clauses are negated by the particle he(e) or lepa/lepi following the conjugated verb. The language consultant has suggested that the difference between the particles is one of degree of formality, lepa/lepi being more informal than he(e). However, it seems more likely that lepa/lepi are more emphatic forms.

(117)   a-ku-la   hee   ma-belu
        SP$_1$-INF-eat   NEG   6-unripe_fruit
        'He did not eat unripe fruit.'

(118)   hy-a   hi-hund-iti   hy-e   hi-nogh-a   lepa
        CP$_{10}$-CON   SP$_{10}$-become_ripe-PFV   DP$_{10}$-DEM   SP$_{10}$-be_tasty-FV   NEG
        'those that are ripe are not at all delicious'

(119)   u-ka-m-bwee   lepi?
        SP$_{2SG}$-PST-OP$_{1SG}$-see.PFV   NEG
        'Didn't you see me at all?'

4.7 Periphrastic constructions: auxiliary verb + Infinitive

Chimpoto has several periphrastic constructions that comprise one or more auxiliary verbs plus a main verb, which occurs in infinitival form. These include -*wanja* 'start' (120), -*yendelela* 'continue' (121), -*mala* 'finish' (122), -*hovalela* 'used to' (123) and -*kona* 'still' (124). In a negated periphrastic construction, the negative marker *hee* occurs following the auxiliary verb, for example *a-ka-mal-iti hee ku-teleka* 'she had not finished cooking.'

(120) va-ka-wanj-a   ku-tongola   ma-belu   ni   ku-wanja ku-memena
       SP₂-PST-begin-FV   15-knock_down   6-unripe_fruit   and   15-begin   15-chew
       'They began to knock down the unripe fruit and began to chew.'

(121) a-yendelee   ku-yenda   mbolimboli   ku=n-gongu   w-aki
       SP₁-continue.PFV   15-go   slowly   17=3-back   PP₃_3SG.POSS
       'He continued to follow her slowly.'

(122) va-lemb-a   ana   va-mal-iti   ku-lepa   lu-kotelu
       SP₂-register-FV   if   SP₂-finish-PFV   15-pay   11-bride_price
       'They register if [presuming] they have finished paying the bride price.'

(123) a-hyovalili   ku-va-konga   va-yaki
       SP₁-used_to.PFV   15-OP₂-lie   PP₂_3.POSS
       'He used to lie to his fellows.'

(124) u-kon-a   ku-lwala?
       SP₂ₛ𝒈-still-FV   15-be_sick
       'Are you still sick?'

4.8 Imperative

In the imperative, the verb stem is marked by the subjunctive suffixes -*ayi* (formal) or -*aa* (informal), which have stress on the penultimate mora. In the plural, the verb requires the prefix *mu*-, or *m*- in some cases. An object prefix may also precede the stem.

(125) ly-ayi
       eat-SBJV
       'eat'

(126) li-ly-aa
       OP₅-eat-SBJV
       'eat it [e.g., banana]'

(127) m-tam-ayi
       SP₂ₚₗ-sit-SBJV
       'sit [PL]'

The negative imperative is constructed from the imperative form of the verb -*kotoka* 'do not' followed by the main verb in infinitival form.

(128) kotok-aa   kola   ku-yenda   wenga   kayika
       do_not-SBJV   DP₁₇.DEM   15-go   2S.PRO   alone
       'Do not go there alone!'

(129) lelenu   mu-kotok-aa   ku-gha-hova   kavele
       today   SP₂ₚₗ-do_not-SBJV   15-OP₆-lose   again
       'Today, do not lose them again!'

4.9. Subjunctive

The subjunctive occurs with all persons as subject. It takes final -*ayi* or -*aa*, with stress on the initial [a]. The latter form, -*aa*, appears to be more casual, less formal than -*ayi*. The subjunctive may be interepreted as an (ex)horative 'let's VERB' (130), as a jussive command (131), as a deontic marker of obligation (132) or suggested action (133), or is simply a marker of subordination with certain verbs (134).

(130) ti-wok-ayi   ti-yend-ayi   ku-londa   chi-leve   hoti
       SP₁ₚₗ-leave-SBJV   SP₁ₚₗ-go-SBJV   15-look_for   7-food   POL
       'Let's leave; let's go look for food.'

(131) ku-tumbula   lelenu   u-kochi   w-itu   u-yomuk-ayi
       15-beginning   today   14-friendship PP₁₄-1PL.POSS SP₁₄-be_finished-SBJV
       'From today on, our friendship, let it be finished.'

(132) ti-tongul-y-ayi   ma-belu
       SP₁ₚₗ-knock_down-??-SBJV   6-unripe_fruit
       'We should knock down the unripe ones.'

(133) mu-yend-ayi,   mu-hik-aa   apa   chilau
       SP₂ₚₗ-go-SBJV   SP₂ₚₗ-come-SBJV 16.ADV   tomorrow
       'You (PL) should go [away] and come here tomorrow.'

(134) li-yani   li-lendel-a   kapecha   ka-li-pel-aa   ma-lavi
       5-monkey   SP₅-wait-FV   12.hare   SP₁₂-OP₅-give-SBJV   6-groundnut
       'Monkey waited for Hare to give him [Monkey] groundnuts.'

The subjunctive suffix may be used in conjunction with the future prefixes, -*a*- or -*ka*-. No examples of use with past -*ka*- have been observed.

(135) ku-wanja   lelenu   n-a-ku-won-ayi   n-a-ku-dumul-a
       15-begin   today   SP₁SG-FUT-OP₂SG-see-SBJV   SP₁SG-FUT-OP₂SG-cut-FV
       ni   hyowo   hy-angu
       with   10.fingernail PP₁₀-1SG.POSS
       'Beginning today, should I see you, I will cut you with my fingernails!'

(136) chilau   lukela   t-a-liwol-ayi
       tomorrow   morning   SP₁ₚₗ-FUT-exercise-SBJV
       'Tomorrow in the morning we should exercise.'

(137) ti-ka-tol-ayi   ku-va-pela   v-ana   v-itu.
       SP₁ₚₗ-FUT-take-SBJV   15-OP₂-give   2-child   PP₂-1PL.POSS
       'We should take [it (food)] to give to our children.'

The negative subjunctive employs the negative auxiliary verb -*kotoka* 'do not' with the subjunctive suffix followed by the main verb in infinitival form.

(138) nenga   ni-won-a   ma-chi   agha   ti-kotok-aa   ku-yogha,
       1SG   SP₁ₛ𝒈-see-FV   6-water   6.DEM   SP₁ₚₗ-do_not-SBJV 15-bathe
       ti-hop-ayi
       SP₁ₚₗ-drink-SBJV
       'Me, I think, this water, we should not bathe [in it]; we should drink [it].'

(139) m-kotok-ayi   ku-yogha   pala   pa   li-tam-it-ayi   li-ng'wina
       SP₂ₚₗ-do_not-SBJV   15-bathe   16.DEM   16.where   SP₅=live-PFV-FOC?   5-crocodile
       'You should not bathe there where a crocodile lives.'

5 SYNTAX

5.1 Basic word order: S-V-PO-SO-OO

The basic constituent order of Chimpoto is SVO, with primary objects (PO) preceding secondary objects (SO). Oblique objects (OO) follow PO and SO.

S-V-PO-SO

(140) a-n-dum-iti   nkochi   w-aki   barua
       SP₁-OP₁-send-PFV   1.friend   PP₁-3SG.POSS   9.letter
       'She sent her friend a letter.'

(141) a-ka-m-pel-a   m-buyu-mundu   chi-yuni   che-la
       SP₁-PST-OP₁-give-FV   1-grandmother-own   7-bird   DP₇-DEM
       'He gave [his own] grandmother that bird.'

S-V-SO-OO

(142) va-lov-el-a   homba   hi-maheli   ni   n-gowo   ye-la
       SP₂-fish-APPL-FV   10-fish   EP₁₀-many   with   9-cloth   DP₉-DEM
       'They caught a lot of fish with that cloth.'

5.2 Object marking

Chimpoto permits only one object marker prefixed to the verb itself (see Section 4.5.4). Typically, an object understood from context is not indexed on the verb, as in the third example. This may reflect an animacy condition on object marking.

(143) a-ka-tol-a   n-gowo   ye-la   a-ka-va-pel-a
       SP₁-PST-take-FV   9-cloth   DP₉-DEM   SP₁-PST-OP₂-give-FV
       'He took that piece of cloth [and] gave [it] to them.'

(144) cha   mu-y-ii   ni   chi-leve chi-choko-pi   ti-ka-tol-ayi
       if   2p-be-PFV with 7-food   AP$_7$-little-IDEO SP$_{1PL}$-PST-take-SBJV
       ku-va-pela   v-ana   v-itu
       15-OP$_2$-give 2-child PP$_2$_1PL.POSS
       'If (assuming) you have a little food, we could take [it] and give [it] to our children.'

(145) va-tondol-aa   chi-leve ni   ku-veka mo-la   mu=chi-hogholo   mo-la
       SP$_2$-pick_up-IPFV 7-food and 15-put 18-DEM 18=7-termite_mound 18-DEM
       'They picked up food and put [it] there inside that termite mound.'

5.3 Relative clauses

Relative clauses follow the noun they modify.

5.3.1 Subject relatives

Subject relatives can be formed in two ways. First, they may occur as a linking construction based on connective -*a*: N AGR-a AGR-VERB. Second, they may be indicated by a relative pronoun.

(146) va-won-iti ng'onda hy-a   hi-y-ii   ni   ma-lombi ni   ma-yyau
       SP$_2$-see-PFV 10.field CP$_{10}$-CON SP$_{10}$-be-PFV with 6-maize and 6-cassava
       'They saw fields that had maize and cassava.'

(147) a-y-ii   mu-ndu yo-ngi   y-a   a-chenj-ii   wichu
       SP$_1$-be-PFV 1-person DP$_1$-other CP$_1$-CON SP$_1$-be_clever-PFV extremely
       'There was another person who was very clever.'

(148) mwehi   woo   u-hika   owo   nyenda ku=nyumba kwali
       3.month 3.REL SP$_3$-arrive 3.DEM 1SG.go 17=9.house don't_know
       'This coming month, I'm not sure I'm going home.'

5.3.2 Object relatives

Object relative constructions are similar to linking subject relatives. They differ only in that the main verb in the subordinate clause is not co-referential with, and so does not agree with, the head noun of the matrix clause.

(149) homba   y-a   ni-lov-iti
       9.fish   CP$_9$-CON SP$_{1SG}$-fish-PFV
       'a fish that I caught'

(150) a-li-ili   chi-leve ch-a   a-m-bel-iti
       SP$_1$-eat-PFV 7-food   CP$_7$-CON SP$_1$-OP$_1$-give-PFV
       's/he$_j$ ate the food that s/he$_k$ had given her/him$_j$'

(151) hi-tabu hy-a   ku=kanisa, hi-y-ii   hy-a   va-lemb-iti   ku=Chimpoto?
       8-book   CP$_8$-CON 17=church   SP$_8$-be-PFV CP$_8$-CON SP$_2$-write-PFV 17=Chimpoto
       'The books at the church, are there any that are written in Chimpoto?'

5.4 Coordinate verb constructions

Coordination of verbs is achieved in several ways. A conjugated verb may be coordinated with an infinitive, either with (152–153) or without (154) the coordinating conjunction *ni* 'and.' A conjugated verb may be coordinated with a second conjugated verb.

(152) va-va-tol-a   va-ndu   ni   ku-woka   naku
      SP$_2$-OP$_2$-take-FV 2-people and 15-leave with.PRO
      'They took people and left with them.'

(153) hi-ndu   hi-yitiki   ni   ku-gweha   pahi
      8-thing   SP$_8$-spill.PFV and   15-fall   down
      'Things had spilled/scattered and fallen down.'

(154) a-ka-kembel-a   ku-hika   ku=nkongo
      SP$_1$-PST-run-FV   15-arrive   17=9.tree
      'He ran [and] arrived at the tree.'

(155) kapecha   a-mal-iti   ni   a-paa   ku-woka
      12.hare   SP$_1$-finish-PFV and   SP$_1$-want.FV   15-leave
      'Hare had finished and wanted to leave.'

(156) va-li-lek-ayi   li-yend-aa   mu=ma-chi
      SP$_2$-OP$_5$-leave-SBJV 5-go-SBJV   18=7-water
      'They should leave it [crocodile] alone [and] let it go into the water.'

5.5 Presentative constructions

In presentative constructions, the logical subject of the verb follows the verb, which can be locative (-*y*- 'be (located)') or stative (e.g., -*waka* 'be none') in nature. There are three forms of the locative, with locative prefixes *pa-*, *ku-* or *mu-*. This construction appears to be used to introduce new information in discourse.

5.5.1 Locative presentatives

Locative presentatives are constructed with the verb -*y*- 'be' in its perfective form, with a locative prefix from classes 16 (157, 160), 17 (158, 161), or 18 (159, 162): LOC-*y*-i(l)i. In some cases, it is followed by *ni* 'with' (157–159).

(157) pa-ka-y-ii   ni   chi-tunga   ch-a   chi-twelili   li-kungu
      SP$_{16}$-PST-be.PFV with 7-basket   CP$_7$-CON SP$_7$-be_full.PFV   5-lake_flies
      'There was a basket that was full of lake flies.'

(158) a ku-y-ii   ni   hula
         SP$_{17}$-be-PFV with 9.rain
         'There is rain.'; 'It's raining.'

     b chilau   ku-ka-y-ii   ni   ndowa
        tomorrow   SP$_{17}$-FUT-be-PFV with 9.wedding
        'Tomorrow there will be a wedding.'

(159) mu-y-ii   ni   homba   mu=chi-vegha
SP_{18}-be-PFV   with   10.fish 18=7-pot
'There are fish in the pot.'

(160) pa-y-ii   chi-jiji   ch-a   chi-y-ii   kutali   hee   pa=mu-hanga
SP_{16}-be-PFV   7-village CP_7-CON SP_7-be-PFV   far   NEG   16=3-lake
'There was a village that was not far from a lake.'

(161) henu   lelenu   homba   ku-wak-a,   andee?!
so   today   10.fish SP_{17}-be_none-FV   right
'So today there are no fish, is that right?'

(162) mu-y-ii   chi-ndu   mu=chi-kongolo   omoo
SP_{18}-be-PFV   7-thing   18=7-forest   18.DEM
'There is something in this forest.'

5.5.2 Non-locative presentatives

In non-locative presentatives, the verb agrees with a subject noun that follows the verb, not with any locative that might be present.

(163) li-y-ii   li-ng'wina   l-a   li-tam-it-ayi   pa=mu-hanga
SP_5-be-PFV   5-crocodile CP_5-CON SP_5-sit-?-FOC?   16=3-lake
'There was a crocodile that lived at the lake.'

(164) a-y-ii   m-buyamundu   m-monga
SP_1-be-PFV   1-old_woman EP_1-one
'There was one particular old woman.'

5.6 Infinitives as subjects

Infinitives may occur as syntactic subjects, requiring appropriate agreement on the verb.

(165) ku-teleka   wembi   ku-y-ii   chi-ndu   ch-a   va-kambaku   va-ket-a
15-cook   14.beer SP_{15}-be-PFV   7-thing   7-CON   2-men SP_2-do-FV
'Brewing beer is something that men do.'

(166) ku-lova   homba   ku-let-a   ma-ganga
15-fish   10.fish SP_{15}-bring-FV   6-money
'Fishing brings in money.'

(167) ku-va   n-kata   ku-niahi   hee!
15-be   1-lazy   15-good   NEG
'Being lazy is not good.'

5.7 Locatives as subjects

Locative nouns may also occur as syntactic subjects, in which case they require agreement with the verb.

(168) pa=mu-tu   apa   pa   a-tov-iti   pa-vin-a
      16=3-head   DEM₁₆   16.where   SP₁-hit-PFV   SP₁₆-ache-FV
      'On my head here, where he hit [me], hurts.'

(169) kulochi   ku-yivil-a   w-atu
      17.deep_water   SP₁₇-sink-FV   14-canoe
      'In the deep water is sinking a canoe.'

(170) mu=ma-ganga   mu-tam-iti   li-wolowondo
      18=6-stone   SP₁₈-live-PFV   5-gecko
      'Among the stones lived a gecko.'

5.8 Topic

A common feature of Chimpoto information structure is the organisation of sentences in terms of topic and comment. Topics occur in clause-initial position; there may be more than one.

5.8.1 General topic

A general, or non-focalised, topic is expressed as a simple noun phrase, not linked to the main predicate as an argument.

(171) *ku-chenjela   koo   kapecha*, a-ka-til-a   chi.mihi   kanjika a-ka-kembel-a
      15-be_clever   DEM₁₅ 12.hare   SP₁-PST-run-fv 7.evening alone   SP₁-PST-run-FV
      'the cleverness of hare, he ran in the evening alone; he ran

      ku-hika   ku=nkongo w-a   m-belepele   ni   ku-wanja ku-tongola
      15-arrive   17=9.tree CP₃-CON 3-fruit (type) and 15-begin 15-knock_down
      arriving at the mbelepele tree and began to knock down

      hy-a   hi-hundi   hye-la,   hy-a   va-ka-hi-lek-iti   ni   ku-la
      CP₁₀-CON   10-be_ripe.PFV DP₁₀-DEM CP₁₀-CON 2-PST-10-remain-PFV   and 15-eat
      those ripe ones, the ones they had left, and [began] to eat [them].'

5.8.2 Subject as topic

A subject noun phrase may be extracted as a topic, retaining agreement with the verb.

(172) *Chi-kahi ch-a   ku-lema*, cha   chi-hik-iti,   li-hogho li-ka-m-pwaghee kapecha
      7-season CP₇-CON 15-till   TPRT 7-arrive-PFV 5-hyena 5-PST-1-tell.PFV   12.hare
      'The planting season, when it came, Hyena told Hare. . .'

(173) henu   *oyoo*   ma-chova   gh-oha   a-yend-a   mbolimboli
      now   DEM₁   6-day   EP₆-all   SP₁-go-FV   slowly
      'now, this one, all the time he moves slowly'

5.8.3 Object as topic

An object can function as a topic or a focus element, typically depending on whether or not there is a co-referential object prefix on the verb, topic with a co-referential OP (174), focus without (175).

(174) ma-chi   agha   ti-gha-pal-a   ni   ti-pal-a   hee   mu-ndu
       6-water   DEM$_6$ SP$_{1\text{PL}}$-OP$_6$-want-FV   and   SP$_{1\text{pl}}$-want-FV   NEG   1-person
       a-gha-hakal-a
       SP$_1$-OP$_6$-make_dirty-FV
       'This water, we want it and we do not want [that] anyone dirty it.'

(175) nyama   ye-la   va-ka-yoch-iti
       9.meat   9-DEM   2-PST-roast-PFV
       'That meat, they had roasted.'

5.9 Adverbial clauses

Adverbial clauses of time may be introduced by the particles *pa* or *cha*. The particle
pa is used with the sense of 'when' [= as soon as], *cha* to express 'when' [= at the time
that].

5.9.1 pa *clauses*

(176) pa   a-va-kelewuu   pa=nyumba   hy-ao,   v-ana   va-lwal-iti
       TPRT   SP$_1$-OP$_2$-return   16=10.house   PP$_{10}$-3P.POSS   2-child   SP$_2$-become_ill-PFV
       'When she returned them to their homes, the children became sick.'

(177) pa   kapecha   a-mal-iti   ni   a-paa   ku-woka,   li-n-kamol-a
       TPRT   12.hare   SP$_1$-finish-PFV   and   SP$_1$-want.FV   15-leave   SP$_5$-OP$_1$-catch-FV
       ma-wolo   gh-aki
       6-leg   PP$_6$-3SG.POSS
       'When Hare had finished and wanted to leave, it [Crab] caught its [Hare's] legs.'

5.9.2 cha *clauses*

(178) cha   va-mal-iti,   va-li-tol-a   ku=mu-hanga   ni   ku-yenda   naku
       TPRT   OP$_2$-finish-PFV   SP$_2$-5-take-FV   17=3-beach   and   15-go   with.it
       kulochi
       17.deep_water
       'When they finished, they took it to the beach and went with it into deep water.'

(179) chi-kahi   ch-a   ku-huka   ma-lavi   cha   chi-hik-iti,   ti-yend-a
       7-season   CP$_7$-CON   15-harvest   6-groundnut   TPRT   SP$_7$-arrive-PFV   SP$_{1\text{PL}}$-go-FV
       ku=ng'onda
       17=3.field
       'The season for harvesting groundnuts, when it arrived, we went to the field.'

The combination of *cha* and *pa* appears to focus on a specific time at which the event in
the main clause can occur.

(180) ti-ka-hetek-a   m-buhi,   **cha   pa**   Ado   a-hik-a
       SP$_{1\text{PL}}$-FUT-slaughter-FV   9-goat   TPRT   TPRT   Ado   SP$_1$-come-FV
       'We will slaughter a goat [only at that time] when Ado comes.'

(181) a-m-p-ee   hi-ndu   hy-a   ku-keta **pa** **cha** a-y-ii   ni
       SP₁-OP₁-give-PFV   8-thing CP₈-CON   15-do   TPRT   TPRT   SP₁-be-PFV with
       mbomba   yo-la
       1.woman   DP₁-DEM
       'She had given him things to do [specifically] when he was with that woman.'

5.9.3 Locative clauses

Locative adjunct clauses are marked with *pa/paa*, *ko/koo* or *mu/moo*. It is not clear what motivates the vowel lengthening, but it may have to do with position within a phonological phrase.

5.9.3.1 Specific

(182) m-kotok-ayi   ku-yogha ni   ku-kina pa-la   pa=li-tam-it-ayi
       SP₂ₚₗ-leave_off-SBJV   15-bathe and 15-play DP₁₆-there 16=SP₅-live-PFV-FOC
       li-ng'wina
       5-crocodile
       'Don't bathe and play there whereat a crocodile lives.'

(183) muchi   ko=a-tam-a   ni   kutali
       3-village 17=SP₁-live-FV   COP   17.far
       'the village whereat s/he lives is far away'

(184) a-ka-yingil-a   mu=hi-wonel-a   ngoko
       SP₁-PST-enter-FV   18=10-sleep-FV   10.chicken
       'she entered wherein the chickens sleep'

5.9.3.2 Non-specific

The particle *paa* may introduce a clause of location, '[at a spot] where.'

(185) ni   ku-lendela paa   gha-y-ii   ma-nyahi pa=ndela
       and   15-wait   16.where   SP₆-be-PFV   6-grass   16=9.path
       ' . . . and waited [at a spot] where there is grass on the path'

(187) a-ka-li-weeni   pa=n-kuka wo-la   paa   u-himingini ni   mu-hanga
       SP₁-PST-OP₅-see.PFV   16=3-river DP₃-DEM   16.where   SP₃-meet.PFV with 3-lake
       'He saw it at that river [at a spot] where it met with the lake.'

Koo clauses denote the area around a location.

(188) a-ka-hik-il-a   hee   koo   gha-y-ili   ma-toki   gh-eni
       SP₁-PST-reach-APP-FV   NEG   17.where   SP₆-be-PFV   6-banana   DP₆-DEM
       'He could not reach [the area] where those [mentioned] bananas were.'

(189) a-tol-iti   hi-nyama   hye-la   koo   a-yend-a   ku-tama
       SP₁-take-PFV   10-animal   DP₁₀-DEM   17.where   SP₁-go-FV   15-live
       'He took those animals [to the area] where he was going to live.'

Moo clauses refer specifically to the inside of a location.

(190) u-pa-kombul-a   moo   ti-hov-iti?
       SP₂ₛ-OP₁₆-remember-FV   18.where   SP₁ₚₗ-get_lost.PFV
       'Do you remember the inside [of the place] where we got lost?'

(191) a-ka-yingil-a   moo   hi-wonel-a   n-goko
       SP₁-PST-enter-FV 18.where   SP₁₀-be_asleep-FV   10-chicken
       'She went inside a chicken-sleeping place [= chicken coop].'

5.9.4 Conditional clauses

Clauses of condition may be introduced by either of two particles, *cha* or *ana*. Although they seem to be nearly synonymous, there does appear to be a subtle difference between them: *cha* supposes the proposition will come to pass, whereas *ana* presumes it will.

5.9.4.1 *cha* clauses

The particle *cha* indicates that the proposition expressed is an assumption. It may be translated as "if" or "assuming."

(192) cha   a-hik-a,   n-aka-n-tangatel-a
       if   SP₁-arrive-FV   SP₁ₛG-FUT-OP₁-help-FV
       'If (assuming) he comes, I will help him.'

(193) cha   n-aka-ku-va   ni   ma-ganga,   n-aka-hemili   mbuhi
       if   SP₁ₛG-FUT-INF-be   with   6-money   SP₁ₛG-FUT-buy.PFV   9.goat
       'If (assuming) I have money, I would buy a goat'

5.9.4.2 *ana* clauses

Clauses introduced by the particle *ana* indicate that the proposition expressed is a presumption, i.e., the speaker is confident that it will occur.

(194) ana   yoo   y-aka-lwal-a,   m-bel-aa   n-tela   owo
       when   DEM₁   SP₁-FUT-be_sick-FV SP₁-give-SBJV 3-medicine DEM₃
       'When (presuming) that one [mentioned] becomes ill, give him/her this medicine.'

(195) Tenga,   muhilu   y-itu,   ana   mu-hongolo   a-pal-a   ku-towola,
       1PL.PRO   9.tradition PP₉-our   when   1-boy   SP₁-want-FV   15-marry
       mw-eni   a-n-dond-a   mu-henja
       1-INT   SP₁-OP₁-seek-FV   1-unmarried_girl
       'Us, our tradition, when a boy wants to marry, he himself looks for a girl.'

5.9.4.3 *cha . . . cha* clauses

The particle *cha* may co-occur in the protasis of a conditional construction with the temporal particle *cha*, which appears following the subject noun and which seems to increase the certainty of the proposition in the main clause. The clause in the apodosis is marked for futurity with *-a.ka-*.

(196) cha   ly-ova   cha   li-y-ili,   n-aka-yenda   ku=Liweta
       if   5-sun   TPRT SP_5-be_in_a_place-PFV SP_{1SG}-FUT-go-FV   17=Liweta
       'If (assuming) the sun has come out, I will go to Liweta.'

5.9.4.4 *ana . . . cha* clauses

The particle *ana* may also co-occur with temporal particle *cha*. The *cha* follows the subject noun, if there is one. Unlike *cha* (conditional), *ana . . . cha* may occur without a subject NP present.

(197) ana   v-ana   cha   va-hika,   n-aka-va-tangatel-a
       when SP_2-children TPRT SP_2-arrive-FV SP_{1SG}-FUT-OP_2-help-FV
       'Presuming the children come, I will help them.'

(198) ana   cha   a-veleka   mw-ana chiniahi, t-aka-telek-a   w-embi
       when   TPRT SP_1-give_birth-FV 1-child healthy SP_{1PL}-FUT-cook-FV   14-beer
       ni   ku-luluta
       and   15-ululate
       'Presuming she gives birth to a healthy child, we will prepare beer and ululate.'

5.10 Interrogatives

Interrogative markers occur in situ.

5.10.1 *What, which* kiki *or* kii

(199) ti-n-ket-ayi   kiki?
       SP_{1PL}-OP_1-do-SBJV   what
       'What should we do with him?'

(200) u-pal-a   kii?
       SP_{2SG}-want-FV what
       'What do you want?'

5.10.2 *Why* ndava kiki *or* ndava kii

(201) ndava   kiki   u-longel-a   lepi?
       reason what SP_{2S}-talk-FV NEG
       'Why are you not talking?'

5.10.3 *When* lii

(202) a-wok-iti   lii?
       SP₁-go_out-PFV   when
       'When did he go out?'

5.10.4 *Where* kowo *or* kou

(203) ma-lavi   gh-eni   gha-y-ii   kowo?
       6-ground_nut   DP₆-DEM   SP₆-be-PFV   where
       'Those groundnuts, where are they?'

(204) ngou   hy-inu   hi-y-ii   kou?
       10.clothes   10–2PL.POSS   10-be-PFV   where
       'Where are your clothes?'

5.10.5 *Where exactly* kolee(ku)

(205) u-y-ii   kolee?
       SP₂SG-be-PFV   where
       'Where are you exactly?'

(206) ti-lond-aa   koleeku?
       SP₁PL-look-SBJV   where
       'Where exactly should we look?'

5.10.6 *Who* naa

(207) naaa-lwal-iti?
       who   SP₁-be_sick-PFV
       'Who was sick?'

5.10.7 *Whose* -a naa

(208) ma-lelu   gh-a   naa?
       6-funeral   DP₆-CON   who
       'Whose funeral?' [lit. funeral of who]

5.10.8 *How* woli

(209) w-a-kembel-aa   woli?
       SP₂SG-FUT-run-SBJV   how
       'How are you going to run?'

5.10.9 *How many* -le

(210) u-yii   ni   ng'ombi hi-le?
       SP₂SG-be-PFV   with   10.cattle EP₁₀-how-many
       'How many cattle do you have?'

5.10.10 *How many times* palenga

(211) ni-ku-kot-iti   palenga?
       SP$_{1SG}$-OP$_{2SG}$-ask-PFV   how_many_times?
       'How many times have I asked you?'

NOTES

1 I sincerely thank Margaret and her family for the tremendous assistance they have provided.
2 Nurse (2008) provides a chart of tense/aspect forms in what he labels Mpoto. Although the same morphemes occur, their distribution and use differs in several ways from that described here.

REFERENCES

Johnstone, P. (1993) *Operation World*. Grand Rapids: Zondervan Publishing [Cited in Ethnologue. (Available online at www.christusrex.org/www3/ethno/Mala.html)].

Nurse, D. (2008) Appendices to Tense and Aspect in Bantu (Available online at www.ucs.mun.ca/~dnurse/tabantu.html, Accessed November 26, 2015).

Nurse, D. & G. Philippson (1975) *The Tanzanian Language Survey*. Lyon: C-BOLD. (Available online at www.cbold.ish-lyon.cnrs.fr/Docs/TLS.html).

Voegelin, C. F. & F. M. Voegelin (1977) *Classification and Index of the World's Languages*. New York; Amsterdam: Elsevier.

CHAPTER 24

CUWABO P34

Rozenn Guérois

1 INTRODUCTION

Cuwabo is a Bantu language spoken by more than 800,000 people (Instituto Nacional de Estatística 2007) in the northeastern part of Mozambique. It is numbered P34 in Guthrie's classification, and thus belongs to the P30 Makhuwa group.

Lewis *et al.* (2014) distinguish five main varieties: central Cuwabo, Karungu, Mayindo, Marale and Nyaringa (see Map 24.1). My consultants recognise these varieties, except for Marale, spoken in Mopeia, which, according to my main consultant, is more related to Sena N44 than to Cuwabo. Moreover, Cuwabo speakers also identify Manyawa, spoken around Lugela, as a Cuwabo variety. When one refers to Cuwabo, the central Cuwabo variety spoken around Quelimane first comes to mind. It corresponds geographically to the traditional homeland of the Cuwabo people, and it also enjoys a high degree of intelligibility among speakers of other varieties. By comparison, Karungu, spoken in the district of Inhassunge, and Mayindo, spoken in the village of Chinde within the Zambezi delta, are said to contain more Sena-like features because of their geographic proximity with Sena speakers. The Nyaringa variety spoken in Maganja da Costa is in turn influenced by Makhuwa P31 (especially in its phonology), in the same way as Manyawa is influenced by Lomwe P32 due to its embedding within the Lomwe-speaking area. My consultants generally assumed that the Cuwabo spoken in Mocuba is very similar to central Cuwabo. Field notes collected in Mocuba further confirm this similarity: both varieties shade gradually into each other to such an extent that it seems impossible to distinguish them.

This chapter focuses on the central Cuwabo variant, and presents an overview of the primary typological features of the language. Most of the data presented here come from Guérois (2015), which is based on fieldwork conducted around Quelimane[1].

2 PHONOLOGY

2.1 Consonants

Cuwabo has 31 consonants, as given in Table 24.1. The elements between parentheses represent more marginal phoneme types. Several reports for orthographic standardisation at the national level were published by the University Eduardo Mondlane of Maputo under the name *Padronização da ortografia de línguas moçambicanas* (2011). In this chapter, I use a phonological notation, allowing <j> to be a stop and <y> a glide, as Africanist scholars usually do.

The place of articulation of the voiceless stop /t/ is both dental and alveolar. The voiceless palatal stop /c/ can also be realised as an alveopalatal affricate [tʃ]. Dental and

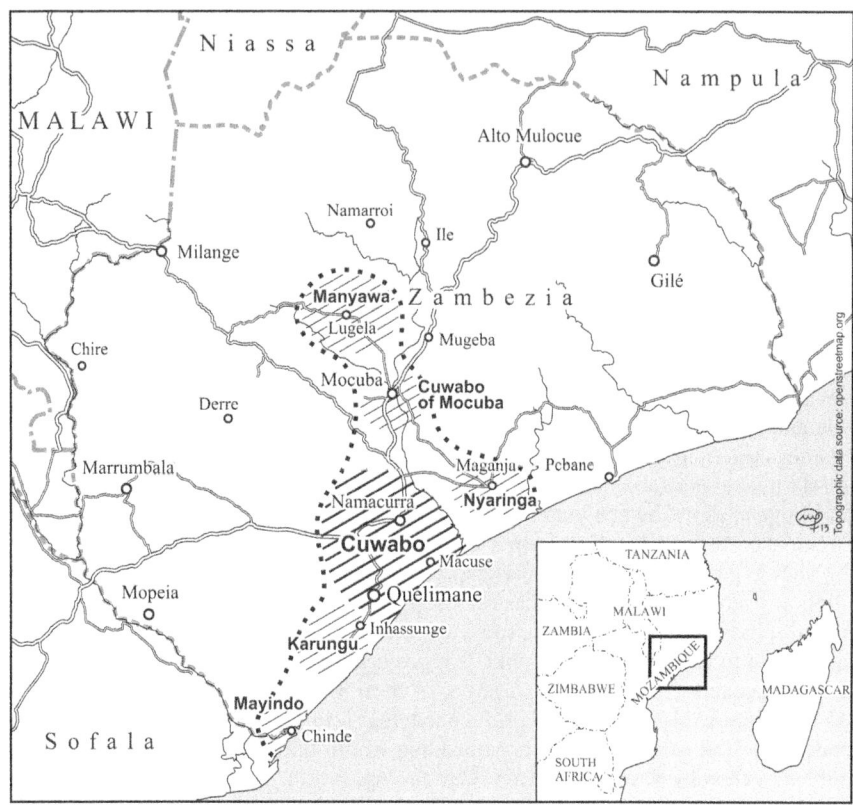

MAP 21.1 CUWABO SPEAKING AREA.

Source: Courtesy of Monika Feinen.

TABLE 24.1 CUWABO CONSONANT PHONEMES

	Labial	Dental	Alveolar	Retroflex	Pre-palatal	Velar	Glottal
vl. stops	p	t		ʈ	c	k	ʔ
vd stops	b	d	s	ɖ	j	g	
prenas. stops	mb	nd	z	nɖ	ɲj	ŋg	
vl. fricatives	f	ð	n	ʈ	(ʃ)	(ŋ)	
vd fricatives	v		l, r		ɲ		
nasals	m				y		
liquids	w						
glides							

retroflex stops cannot co-occur within a stem; see Schadeberg and Mucanheia (2000)'s "dental-retroflex incompatibility"). Instead, dental stops (1a) and retroflex stops (1b) respectively combine among themselves.

(1) a Dental stops combination   b Retroflex stops combination
        ótìdá   'pound'   nàńʈîɖì   'weed sp.'
        ódáándà   'not articulate'   óɖòɖá   'grab'
        òtótà   'hunt'   ósóʈóʈà   'torture'

The second constituent of Cuwabo prenasalised stops is always voiced. A voiceless stop can be preceded by a nasal, but in this case, each consonant is a segment which maintains a phonemic status, and the nasal is moraic, as in *nàńtâlà* 'frog sp.'. In contrast, the tone pattern on the situative verb form *ɖàànàámbà* 'if I fart,' involving a lexical H on the penult mora (see Table 24.6 below), shows that the nasal preceding the bilabial voiced stop is not moraic, and that the two subsequent consonants should instead be considered as a single phoneme. Note, however, that a nasal preceding a voiced stop may be syllabic as a result of vowel apocope, as discussed in Section 2.3. Nasal syllabicity is indicated by means of the diacritic < ͵ > under the consonant.

Traces of Meinhof's Rule (Meeussen 1962), which simplifies a Nasal-Consonant (NC) cluster to a nasal segment when either another NC cluster or a simple nasal occupies C2 position, are attested in Cuwabo with several (prefixless) class 9/10 nouns, as shown in (2).

(2)   Proto-Bantu (BLR3) (Bastin *et al*. 2002)   Cuwabo
       *N-gòmbè   'cattle'   ŋómbè
       *N-gàŋgà   'traditional healer'   ŋáŋgà
       *N-gòmà   'drums'   ŋómà
       *N-gòìnà   'crocodile'   ŋónà (Valler & Festi 1995)

2.2 Vowels

Cuwabo has a five-vowel system, which further distinguishes short and long vowels.

TABLE 24.2 CUWABO VOWEL PHONEMES

	Short			Long		
	Front	Central	Back	Front	Central	Back
High	i		u	ii		uu
Mid	e	a	o	ee	aa	oo
Low						

The diachronic and synchronic sources of long vowels vary. Within verb or noun roots, a few of them are derived from an earlier proto-form with long vowels (e.g., *é-túúri* < Proto-Bantu *-túúdì* 'shoulder'), whilst others result from the loss of a consonant (e.g., *mú-òúúlù* < Proto-Bantu *jíjòkòdò* 'grandchild'). Across morphemes or word boundaries, long vowels are the result of different processes of vowel coalescence, deletion plus compensatory vowel lengthening, and glide formation plus vowel lengthening.

Vowel harmony is not a productive process in Cuwabo, not even in verbal suffixation, which is generally a vowel harmony environment in many Bantu languages. In Cuwabo vowel harmony occurs only with the lexicalised (but still commutable) separative extensions *-ul* (transitive) and *-uw* (intransitive), which surface respectively as *-ol* and

-ow when the verbal root contains a mid-back vowel, as in *ókóṭólà* 'shatter'(tr.) and *ókóṭówà* 'shatter'(intr.). Progressive asymmetric vowel harmony (Hyman 1999) is also attested in Yao P21; see the "mid back identity" rule proposed by Ngunga (2000: 196).

2.3 Syllabic structure

With a canonical CV syllable structure, the syllable template is (C)V(V/N), where C stands for any consonant, including prenasalised consonants, glides and labialised/palatalised consonants. The nucleus is generally a vowel but can also be a syllabic nasal or liquid sonorant (S). Table 24.3 shows the different syllable types attested in Cuwabo.

TABLE 24.3 SYLLABLE STRUCTURES

	Syllable	Example	Translation
One mora	V	ò.lí.mà	'to cultivate'
	S (nasal)	m̀.mé.lò	'throat'
	(liquid)	kà.ḍi.gú.î.lì	'I did not buy'
	CV	ní.fû.gì	'banana'
Two moras	VV	òò.lí.mà	'he cultivated'
	CVV	ó.báá.là	'to give birth'
	CVN	ó.téń.tè.ðà	'to cackle'

Syllabic nasals often derive from vowel apocope between a nasal onset and a following CV syllable. A structure such as mu-CV (as seen in (7b) below) will evolve to m-CV and results in N-CV (Clements 2000, Hyman 2003a). The morphemes *mu* and *ni*, be they class prefixes, subject prefixes or object prefixes, reduce to a syllabic nasal before bilabial or labio-dental consonants (3) and before coronal consonants (4), respectively.

(3) [mu] → [m̩]/__ labial consonant
Class prefixes   Object prefixes
m̀.málâbò (cl.1) 'outsider'   ó.m̀.bàálà 'give him birth'
m̀.páâwì (cl.1) 'orphan'   ó.m̀.fùná 'want him'

(4) [ni] → [n̩]/__ coronal consonant
Class prefixes   Class prefixes
ǹ.ḍézà 'calabash'   ǹ.túmù 'anus'
ń.zâyì 'egg'   ǹ.sépò 'spoon.sp'
ń.lîjì 'den, lair'   ń.rìyé 'bird.sp'

In such environments, the vowel is absorbed between the two consonants, and the resulting syllabic nasal undergoes homorganic nasal assimilation with the next consonant.

3 TONOLOGY

Cuwabo is a tone language with a binary distinction between H and Ø tones[2], as often attested in eastern Bantu languages. The tone-bearing unit is the mora. The basic tone melodies of Cuwabo nouns in their citation form are listed and exemplified in Table 24.4. As can be seen, each noun has at least one underlying H tone. When two underlying H tones are present, the second is the lexical H (underlined in Table 24.4). At this stage, it is not obvious why a lexical H tone in the noun stem should require the presence of a H tone on the prefix. This prefix high tone might be related to the H tone that historically was associated with the augment in Proto-Bantu. It would seem likely that this augment H tone is also the source of the H tone that invariably appears on the first mora of noun stems which are lexically toneless from a historical viewpoint (e.g., è-ʔíbà). Tonal manifestation of the augment was also postulated for Makhuwa(-Enahara) P31 by van der Wal (2006).

TABLE 24.4 TONE PATTERNS FOR NOUNS

Moras	Underlying tone pattern	Surface tones	Example	
2	H-Ø	H-L	mú-kò	'wood ladle'
	Ø-H̲	L-H̲	è-bú	'mosquito'
3	Ø-HØ	L-HL	è-ʔíbà	'hoe'
	H-ØH̲	H-LH̲	ní-kàɽá	'charcoal'
	H-ØØ	H-FL	mú-sôɽò	'head'
4	Ø-HØØ	L-HFL	nì-gógôḓò	'bone'
	H-ØH̲Ø	H-HH̲L	mú-ðóóðò	'ant sp.'
	H-ØØH̲	H-HLH̲	mú-téèkú	'navel'
	H-ØØØ	H-HLL	mú-lóbwànà	'man'
5	Ø-HØØØ	L-HHLL	mù-kóɽófiṭò	'loincloth'
	H-ØØH̲Ø	H-HHH̲L	ní-gúrúgúrù	'shell'
	H-ØØH̲Ø	H-HLH̲L	mú-láànḓézà	'big hoe'
6	Ø-HØØØØ	L-HHLLL	è-rúbálàkùḓè	'beetle'
	H-ØØØH̲Ø	H-HHHH̲L	má-áyéɲárúgè	'herb sp.'
7	Ø-HØØØØØ	L-HHLLLL	ǹ-tógólàbàlàgà	'first rains'
	H-ØØØØH̲Ø	H-HHHHH̲L	mwá-námáánḓámánì	'neighbour'

The three-way contrast observed on trimoraic nouns seems exceptional in P30 languages, except for Shangaji (Maud Devos personal communication). Nouns in other P30 languages have overall a more restricted lexical tone on nouns (Schadeberg & Mucanheia 2000, van der Wal 2009).

Cuwabo is also the only P30 language[3] which has retained a lexical H tone contrast on verbs. Whilst low stems remain toneless (5), in H-toned verbs, the lexical H tone emerges on the penult mora of the stem (6a), or the ultimate mora in the case of a bimoraic stem (6b).

(5) Ø-toned verbs

a |ḓi-a-**roromeliʔ**-a| b |ḓi-a-**lim**-a|
   SP$_{1sg}$-SIT-promise-FV   SP$_{1sg}$-SIT-cultivate-FV
   'if/when I promise'   'if/when I cultivate'

(6) H-toned verbs

a |ḓi-a-**bubuluwél**-a| b |ḓi-a-**rum-á**|
   SP$_{1sg}$-SIT-roll.to-FV   SP$_{1sg}$-SIT-send-FV
   'if/when I roll to'   'if/when I send'

Note that other languages spoken in the area such as Yao P21, Makonde P23 and Matumbi P13 have also lost this tonal contrast on verbs. Sena N44 seems to have evolved to an entirely toneless system. In this context, the retention of a complex tonal system in Cuwabo is all the more remarkable.

Underlying H tones are distinguished from H tones which surface as the result of tone rules. The most frequent rule is High-Tone doubling, which consists of copying an underlying H only onto the next mora to the right. This process applies both word-internally (7a) and across word boundaries (7b).

(7) a òsúkúmìyà
   |o-súkum-i?-a|
   15-fire-CAUS-FV
   'to fire, dismiss'

   b òm̀sàsàɲìlé víyòòlá
   |o-mu-sasaɲ-ilé   viyoóla|
   SP₁-OP₁-fix-PFV.CJ   1a.guitar.H1D
   'He fixed a guitar.'

Cheng and Kisseberth (1979: 44) observe that H tone doubling languages usually impose three common types of restrictions on High-Tone doubling in phrase-final position. Such restrictions are referred to as Non-Finality, Long Fall and Phrase-Penult constraint. Non-finality implies that an underlying H does not double onto a phrase-final vowel (. . . C$\underset{\cdot}{\text{V}}$. CV). As can be seen from Table 24.5, this constraint operates in Cuwabo: in a word like òlímà 'cultivate,' there is no doubling because in isolation the vowel after the typo: underlying H tone is phrase-final.

TABLE 24.5 RESTRICTIONS ON HIGH-TONE DOUBLING IN PHRASE-FINAL POSITION

Restrictions	Cuwabo	Example	Gloss
Non-finality (. . . C$\underset{\cdot}{\text{V}}$.CV)	✓	òlímà	'cultivate'
Long Fall (. . . C$\underset{\cdot}{\text{V}}$V.CV)	×	òjéêðà	'wait'
Phrase-penult (. . . C$\underset{\cdot}{\text{V}}$.CV.CV)	×	òfúg̱ûlà	'open'

The Long fall and the Phrase-Penult constraints, which prevent an underlying H tone from doubling onto the second mora of a bimoraic phrase-penult syllable in the first case, and onto a monomoraic syllable in phrase-penult position in the second case, do not occur in Cuwabo. Instead, doubling occurs, and interestingly, the doubled H is not a level H tone but is rather realised as falling, i.e. it falls from the height of the preceding mora. Transcribing tonally òjéêdhà⁴ as LHFL (where F stands for "falling") responds to a phonetic concern. Phonologically, the same word should be transcribed LHHL (with ØHØØ as the underlying representation). Still, the falling property of the doubled tone is more precise than a simple doubled H tone, and in fact interesting for comparative purposes, especially with other Makhuwa P31 dialects. In (8), we see that in the same context, such a fall is also attested in Koti P311, spoken in Angoche, Nampula province, but not in Nlai, another coastal dialect spoken in Nampula province, which has an unrestricted doubling system (Kisseberth & Guérois 2014).

(8)

	Underlying		Surface	Example	
Cuwabo	ØHØØ	>	LHFL	òjéêðà	'to wait'
Koti	ØHØØ	>	LHFL	òmáâlà	'to finish'
Nlai	ØHØØ	>	LHHL	ò?ééţà	'not to walk'

Plateauing, whereby at least two toneless moras become H between two underlying H tones, is the second tone rule attested in Cuwabo. This process applies only word-internally in: i) three noun patterns provided in Table 24.4, i.e., H-HHHL, H-HHHHL and H-HHHHHL, where the prefix H is considered as a manifestation of the augment, whilst the second is lexical; ii) H-toned verb forms which combine a grammatical (aka "melodic") H on the first mora of the stem and a lexical H on the penultimate mora[5]. Three verb forms comply with this pattern: the disjoint present, the negative present and the infinitive, although for the latter, illustrated in (9), it is difficult to distinguish between a grammatical first H and an augmented first H (otherwise assigned on the first mora of the (macro)stem in the case of Ø-toned verbs).

(9)

H-toned verb		Ø-toned verb	
ó-gùlá	'to buy'	ò-límà	'to cultivate'
ó-gúlí?à	'to sell'	ò-sákûlà	'to choose'
ó-gábúlélà	'to open for'	ò-líbélèlà	'to swear to'
ó-búbúlúwélà	'to roll to'	ò-rórómèlì?à	'to promise'

Following Kisseberth and Guérois (2014), it is assumed that this grammatical H (or augmented H in the case of the infinitive?), was originally assigned to the first (macro)stem mora (as seen in Ø-toned verbs), but had to retract to the first pre-(macro)stem mora because of a constraint preventing its co-occurrence with the penult lexical H in a macrostem. If no object marker is present, plateauing between the retracted grammatical H on the pre-stem mora and the lexical H occurs, as in (9). The only exception occurs with bimoraic stems, whereby the lexical H shifts to the final mora, as seen in ó-gúlá. Spreading does not apply here because of the phrase-final effect seen in Table 24.5. However, in medial position spreading does occur: ó-gúlá . . . In the case of a verb such as ógúlí?à 'to sell,' the underlying tone is /HØHØ/ and the prefix H tone freely spreads onto the next vowel since this toneless mora is not phrase-penultimate.

However, if an object prefix is present, instead of plateauing, High-Tone Doubling occurs, as shown in (10). The theoretical explanation for this behavior is not obvious, and in any case is beyond the scope of this study.

(10)

H-toned verb		Ø-toned verb	
ó-múgùlá	'to buy it'	ò-múlîmà	'to cultivate it'
ó-múgùlí?à	'to sell it'	ò-músàkùlà	'to choose it'
ó-múgàbùlélà	'to open for it'	ò-múlíbèlèlà	'to swear to it'
ó-múbùbùlùwélà	'to roll to it'	ò-múróròmèlì?à	'to promise it'

In addition to a lexical function which distinguishes the meaning of two segmentally identical words (e.g., mù-lálà 'basket' vs. mú-làlá 'traditional toothbrush'), tones are crucial in determining the tense/aspect/mood of certain verb forms. In addition to segmental tense/aspect/mood affixes, tensed verbs in Cuwabo exhibit different tone patterns assigned to the whole verb form. This is the same in Makhuwa P31, but in Cuwabo the resulting tonal form is complicated by the lexical tonal contrast, which is effective in a certain number of tenses (while neutralised in others). The different tone patterns on tensed verbs are listed in Table 24.6. For each tone pattern, one illustrative example is provided, with the distinction between H-toned (indicated as "H") and Ø-toned verbs (Ø) when relevant. For an exhaustive analysis, see Kisseberth and Guérois (2014), who compare the different verb tone patterns among P30 languages.

TABLE 24.6 TONE PATTERNS FOR TENSED VERBS

Tone pattern	Example	Gloss
Lexical H contrast (no grammatical H)	H: ḓàà(mù)bùbùlùwélà	'If I roll to (it)'
	Ø: ḓàà(mù)ròròmèlìyà	'If I promise to (him)'
	H: ó-búbúlúwéla	'to roll to'
Lexical H contrast + grammatical H on (M)S1	+OP: ó-múbùbùlùwélà	'to roll to it'
	Ø: ò-rórómèlìyà	'to promise'
	+OP: ò-múróròmèlìyà	'to promise him'
Toneless stem	ḓá-(mú)bùbùlùwèlè	'I should go and roll to (it)'
Grammatical H on (M)S2	kàḓì-bùbúlúwèḷlè	'I did not roll to'
	kàḓì-mùbúbùlùwèḷlè	'I did not roll to it'
Grammatical H on Penult	ḓàà(mù)bùbùlùwèlágà	'If I keep rolling to (it)'
Grammatical H on Final	ḓì-(mù-)bùbùlùwèl-ilé	'I rolled to (it)'

Finally, tone plays a major role in the expression of non-verbal predicates (see Section 6.1) and focused constituents (see Section 5.2 on tense/aspect/mood).

4 NOUNS

4.1 Noun classes

The basic forms of noun class prefixes are provided in Table 24.7, together with their numbering.

TABLE 24.7 NOUN CLASS PREFIXES

SG				PL			
Class	NP	Ex.		Class	NP	Ex.	
1	mu-	mútù	'person'	2	a-	áṭù	'people'
1a	Ø	páàká	'cat'	2	a-	ápáàká	'cats'
3	mu-	múrì	'tree'	4	mi-	mírì	'trees'
5	ni-	nífùnḓó	'knot'	6	ma-	máfùnḓó	'knots'
5a	Ø	sèkésò	'sieve'	6	ma-	màsékêsò	'sieves'
7	e-	ènámà	'animal'	8	dhi-	ðinámà	'animals'
9	Ø	ŋómbè	'cow'	10	Ø	ŋómbè	'cows, cattle'

Unpaired classes

11	o-	ósâlù	'thread'
14	o-	ópâlì	'youth'
15	o-	óðòwá	'to go'
16	va- (-ni)	vàṭólônì	'at the well'
17	o- (-ni)	òmúnḓà	'at/to the farm'
18	mu- (-ni)	mùcélânì	'in the well'

Prefixes have different realisations, depending on the phonological features of the noun stem. As already seen in Section 2.3, *mu-* and *ni-* become nasal syllables before certain stem-initial consonants. The noun prefixes *e-* and *o-* are realised as glides before vowel-initial roots, e.g., /é-alá/ and /o-óta/ respectively surface as *yáàlá* 'nail' and *wòótà* 'to lie.' Finally, as shown in (11), *mu-* is labialised before i, e and a (i.e., non-round vowels), whereas *mi-* is palatalised before e, a and o.

(11) class 3 [mu+V → (w)VV]   class 4 [mi+V → (y)VV]
  a **mwíilá** /mú-ilá/ 'tail'   b **míilá** /mí-ilá/ 'tails'
  c **mwéèrí** /mú-erí/ 'moon, month' d **myéèrí** /mí-erí/ 'moons, months'
  e **mwáàŋgó** /mú-aŋgó/ 'mount'   f **myáàŋgó** /mí-aŋgó/ 'mounts'
  g **móònó**   /mú-onó/ 'arm'   h **myóònó** /mí-onó/ 'arms'
  i **mùúvà**   /mu-úva/ 'cart'   j **mìívà**   /mi-úva/ 'carts'

Several petrified complex nouns from classes 1a/2 are formed by means of the pre-stem formatives *na-* or *ɲa-* (12a) and *ma-* (12b), diachronically derived from 'mother'[6] (Schadeberg 2003: 86). In some cases, *ma-* may combine with *na-*, giving rise to poly-compound nouns (12c). The resulting lexicalised forms most often refer to animals or humans. With regard to tones, these different formatives are toneless, and the lexicalised forms are considered as whole units for tone assignment, adopting one of the different tone patterns attested for nouns (see Section 3).

(12) a. **nàbílì**   'hyena.sp'   **ɲàzómbè**   'grasshopper'
     b. **mákâzì**   'mistress'   **máráfùlá**   'dead person'
     c. **nàmálûwà**   'weaver'   **námábààlíyà** 'midwife'

Augmentative and diminutive classes are not attested in the language. Also note that the consonant of the Proto-Bantu class 7 prefix *ki-* became Ø, which is an exceptional sound evolution in Eastern Bantu, only attested in Cuwabo and Makhuwa P31.

Cuwabo retained the three historical Proto-Bantu locative classes *pà (class 16), *kú (class 17) and *mù (class 18), realised as *va-*, *o-* and *mu-*, respectively. Certain nouns are inherently locative, while others are made locative by adding the (toneless) locative prefix to the full noun, without altering its tone pattern. Of particular interest is the tendency toward double marking by means of an additional locative suffix *-ni* (e.g., *và-múrî-nì* 'at the tree,' *ò-màbásá-nì* 'at work' and *m̀-bíyâ-nì* 'in the stove'). This double locative marking is not common in Bantu languages, which usually exhibit single locative marking, either through prefixation or suffixation. Such an innovation, shared by P30 languages (Makhuwa group), presumably results from a contact situation, whereby the addition of the locative suffix must have occurred under the influence of Swahili G42; see Guérois (2016) for more details.

Noun-to-noun derivation may occur by noun class transfer. For instance, names of trees from classes 3/4 have matching names for their fruits either in classes 5/6 (13a), or in class 9/10 (13b).

(13) a **m̀fûgì**   'banana plant' → **nífûgì** (cl.5)   'banana'
        múndîmwì   'lime tree'   → **ṉdîmwì** (cl.5)   'lime'
     b **mùcámbáràwè** 'grapevine'   → **càmbárâwè** (cl.9) 'grape'
        mùṱógômà   'plum tree'   → **ṱògómà** (cl.9)   'plum'

Nouns may also be derived from verbs by adding an appropriate noun class prefix, and in some cases, a nominalising suffix. Many agent nouns in classes 1 and 2 are derived from verbs by means of the nominalising suffix -*i* (14a), whereas others maintain the final vowel -*a* (14b). The derivational suffix -*o* derives nouns (often in classes 5/6) with an abstract meaning, frequently denoting the result (action or state) of the verbal process (14c).

(14) a òpáḍûcà 'to create' → m̀páḍûcì (cl.1) 'creator'
        ósúŋzà   'to learn'   → músúŋzì (cl.1) 'student'

     b òláwûlà 'to foretell' → mùláwûlà (cl.1) 'fortune-teller'
        òvávà   'to fly'   → m̀vávî?à (cl.1) 'pilot'

     c ófùmá   'to be rich' → nífùmó (cl.5)   'richness'
        ògánà   'to decide'   → nìgánò (cl.5)   'law, rule'

Other Cuwabo nouns result from compounding by means of the head element *mwáná-* (< *mwáàná* 'child'), used as a diminutive, as in *mwánámúyàná* 'girl' (< *múyàná* 'woman'). Reduplication, illustrated in (15), is another word formation strategy, which involves repetition of the noun stem. Although widespread, it is synchronically no longer productive.

(15) mú-kúre-kúrè (cl.3/4)   'brook, stream'
      ǹ-láká-làkà (cl.5/6)   'fin'
      é-ţúwà-ţúwà (cl.7/8)   'crab.sp'
      ɲìzí-ɲyîzì (cl.9/10)   'eyelash'

4.2 The noun phrase

Modifiers, which typically follow the noun they modify, take prefixes which agree in class with the head noun. Adjective and quantifiers take a nominal prefix (NP). Numerals take an enumerative prefix (EP). Relativized verbs, connective, possessives and *oté* 'all, whole' take a pronominal prefix (PP). Pronominal prefixes and verbal prefixes (VP) are segmentally identical. Demonstratives, however, have their own agreement paradigm. All the agreeing prefixes and enclitics found in the noun phrase are listed in Table 24.8.

TABLE 24.8 AGREEMENT PREFIXES AND ENCLITICS

Class	Class Prefix	NP	EP	PP/ VP	DEM	DEF	INT
1/1a	mu-	mú-	mu-	o-	ó-	=ya	=ene
2	a-	á-	a-	a-	á-	=ya	=ene
3	mu-	mú-	mu-	o-	ó-	=ya	=ene
4	mi-	ðí-	mi-	ði-	é-	=ða	=(ð)ene
5	ni-	ní-	ni-	ni-	ṇ́-	=na	=ene
6	ma-	má-	ma-	a-	á-	=ya	=ene
7/9	e-/Ø	é-	Ø	e-	é-	=ya	=ene
8/10	ði-/Ø	ðí-	Ø	ði-	é-	=ða	=ðene
11/14	o-	mú-		o-	ó-	=ya	=wene
15	o-	ó-		o-	ó-	=wa	=wene
16	va-	vá-		va-	á-	=va	=vene
17	o-	ó-		o-	ó-	=wa	=wene
18	mu-	mú-		mu-	ṁ́-	=mwa	=mwene

As compared to other Eastern Bantu languages, the noun class system of Cuwabo is a bit eroded. Several formal distinctions have been lost, resulting in class merging. This is the case with classes 11 and 14 (both singular) whose prefixes have completely merged. Hence in the glosses, these are referred to as class 11/14. Nouns from classes 7/8 host the prefixes e-/dhi- respectively, whereas nouns from classes 9/10 remain prefixless (Ø). There is, however, no way of distinguishing agreement prefixes between singular classes 7 and 9 on the one hand and between plural classes 8 and 10 on the other. Agreement prefixes are thus glossed 7/9 for singular and 8/10 for plural.

4.2.1 Demonstratives

As many Bantu languages, Cuwabo has four series of demonstratives, given in Table 24.9, which are distinguished in terms of spatial deixis. Series I forms are proximal (close to the speaker). Series II, marked by the suffix -o, are 'referential'[7], i.e., close to the hearer. Series III, marked by the suffix -le, are distal, i.e., far from both speaker and hearer.

TABLE 24.9 DEMONSTRATIVES

Class	Series Ia	Series Ib	Series II	Series III
1	ódu	ódû-no	ód-o	ódû-le/óî-le
2	ába	ábâ-no	áb-o	ábâ-le
3	óbu	óbû-no	ób-o	óbû-le
4	ési	ésî-no	és-o	ésî-le
5	ńṭi	ńṭî-no	ńṭ-o	ńṭî-le
6	ába	ábâ-no	áb-o	ábâ-le
7/9	éji	éjî-no	éj-o	éjî-le
8/10	ési	ésî-no	és-o	ésî-le
11/14	óbu	óbû-no	ób-u	óbû-le
15	óku	ókû-no	ók-o	ókû-le
16	ápa	ápâ-no	áp-o	ápâ-le
17	óku	ókû-no	ók-o	ókû-le
18	ḿpu	ḿpû-no	ḿp-o	ḿpû-le

As can be seen from Table 24.9, series I is further subdivided into two series (a and b), distinguished by the addition of the proximity suffix -no. This suffixed particle, reconstructed for Proto-Bantu (Meeussen 1967: 107, Guthrie 1970: 25) and fairly well attested in present-day Bantu languages (Nicolle 2012), is in Cuwabo most frequently used with locative demonstratives (16a), or demonstratives modifying a temporal noun expression (16b), in order to express immediate spatial or temporal proximity. It may, however, also occur in the post-final verbal position with the same spatial-temporal function, as shown in (17); hence it is consistently glossed PROX and segmented as an enclitic.

(16) a   mùðé! ókûnò. [8]
           |mu-ð-é   óku=no|
           SP_{2PL.}-come-SBJV   17.DEM.I=PROX
           'Come! Right here.'

b yàák' éjínó níngá ḍìlímè mínḍá mìrààrú
|e-áka   éji=**no**   nínga   ḍi-lím-e   mí-nḍa   mi-raarú|
7-year   7/9.DEM.I=PROX   like   SP₁ₛₒ-cultivate-SBJV   4-field   EP₄-three
'This year, I might harvest three times.' (semi-elic.)

(17) ṅngá káðáànó tò òðé ṅjàṅtáàrì
|ningá   ká-ð-a=wo=**no**   to   o-ð-é   ni-jaṅtaari|
VOC   IMP-come-FV=LOC₁₇=PROX   then SP₂ₛₒ-come-SBJV SP₁ₚₗ-have.dinner.SBJV
'Come here then! Come so that we have lunch!'

In addition to spatial-deictic uses, series 1 locative demonstratives can have discourse functions and be involved in text structuring as shown in (18).

(18) Ónóṁvèdá nà méṅtò, kànímóòná. Ókw' áálóbwànà ánókúkúmà
|ó-ni-ó-mu-ved-á   na   má-into   ka-ní-mu-on-á
SP₁-IPFV.DJ-15-OP₁-look.for-FV with   6-eye   NEG.SP₁-IPFV-OP₁-see-FV
óku   á-lobwana   á-ni-ó-kukúm-a|
17.DEM.I   2-man   SP₂-IPFV.DJ-15-flee-FV
'He is looking for him with the eyes, but does not see him. At this point, the men are fleeing.'

Emphatic demonstratives are attested in Cuwabo, but only for the locative classes, as shown in Table 24.10, and exemplified in (44) and (96) below. Such forms are made of a pronominal locative prefix followed respectively by the intensifier morpheme -*ene*, the locative-agreeing connective and the demonstrative suffixes (series I, II, III). A form like *wénéwàlé* 'right there.III' can thus be analysed as *o-ene-wa-le*.

TABLE 24.10 EMPHATIC LOCATIVE DEMONSTRATIVE PRONOUNS

Class	EDEM I	EDEM II	EDEM III
16	vénêvà	vénêvò	vénévàlé
17	wénêwà	wénêwò	wénéwàlé
18	mwénêmwà	mwénêmò	mwénémwàlé

4.2.2 Connective

The connective is used to link two nominal constituents syntactically and semantically. The order of connective constructions is head-connective-modifier. The monosyllabic connective element is composed of the stem *a* preceded by a pronominal prefix, which agrees in noun class with the head noun (see Table 24.8).

Connective constructions may denote properties or qualities, which are often expressed by adjectives in other languages. In such constructions, the modifier can be a noun (19a) or an infinitival verb (19b).

(19) a   múlóbwànà wà kòóbîṛì   b   múlóbwànà w' òòréélà
         |mú-lobwana   o-a   *koóbiṛi*| |mú-lobwana   o-a   o-réel-a|
         1-man   PP₁-CON   9.money   1-man   PP₁-CON   15-be.rich-FV
         'rich man'   'rich man'

Connective constructions also express so-called genitive relations, which include possession or ownership (20), and kinship relations (21).

(20) múróbò wà Júwão
|mú-robo   o-a   Júwão|
3-medicine PP₃-CON João
'João's medicine'

(21) ábáàl'   àà   Màríyà
|á-baale a-a   Maríya|
2-sister PP₂-CON   Maria
'Maria's sisters'

4.2.3 Possessives

In Cuwabo, there are six possessive pronouns, as given in Table 24.11.

TABLE 24.11 POSSESSIVE PRONOUNS

1SG	aga	1PL	e?u
2SG	awo	2PL	eɲu
3SG	aye	3PL	awa

Possessive pronouns agree in noun class with the possessed, via the pronominal prefix already found with the connective relator (see Table 24.8), but onto which a H tone is assigned. Some examples are provided in (22).

(22) a m̀pádǒ ðááyè   b mákágà ááyè   c ɗímà yááyè
|mi-púdo ðí-aye| |má-kaga á-aye| |ɗíma é-aye|
4-chair   PP₄-POSS.3SG 6-oracle PP₆-POSS.3SG 9.task PP₇/₉-POSS.3SG
'her chairs'   'his oracle'   'his task'

The 3SG and 3PL possessive pronouns aye and awa are used for all singular and plural classes, respectively. This is illustrated in (23) with the class 7/8 possessor èyíbà/ðìíbà 'hoe(s).'

(23) a múṭéŋgó wà èyíbà   →   múṭéŋgó wáâyè
|mú-ṭeŋgó o-a   e-?íba| |mú-ṭeŋgó ó-aye|
3-price   PP₃-CON   7-hoe   3-price   PP₃-POSS.3SG
'the price of the hoe'   'its price'

b múṭéŋgó wà ðìíbà   →   múṭéŋgó wáâwà
|mú-ṭeŋgó o-a   ðì-?íba| |mú-ṭeŋgó ó-awa|
3-price   PP₃-CON   8-hoe   3-price   PP₃-POSS.3PL
'the price of the hoes'   'their price'

4.2.4 Adjectives

There are only six adjectival stems in Cuwabo, namely nɖímuwa 'big' (34), ɲóono 'small, little' (36), inji 'much/many,' ina 'other,' iti 'unripe' (95b) and gumi 'healthy,' which agree in noun class with the modified noun by means of a H-toned nominal prefix (see Table 24.8). Similarly to many Bantu languages, adjectival properties in Cuwabo are typically expressed by means of connective constructions (19), or through the stative reading of perfective verb forms (24).

(24) ḍóóṛínḍ' òókóḍélà màs
|ḍóoṛinḍo   **o-ʔí-koḍél-a**   mas|
ḍooṛinḍo   SP₁-PFV.DJ-be.beautiful-FV   more
'Ddoolrinddo is more beautiful.'

4.2.5 Numerals and quantifiers

Only the stems involving numerals from 1 to 5 take a nominal agreement prefix (except with classes 7/8 and 9/10 as seen in Table 24.8). From 6 to 9, compound forms are used, based on a quinary pattern (e.g., 5-and 1 for 6). The stems *kúmi* for 10, *zána* for 100 and *cíkwi* for 1000 are nominal forms, which are pluralised in class 6, and thus trigger class 6 agreement on their modifiers. The numerals from 11 to 19 consist of 10 plus 1, 2, 3, etc. These agreement patterns are illustrated in (25). Note that big numbers often imply that the modified noun (here *míri* 'trees') is repeated in front of the hundreds and the tens.

(25) ðìkálámò mírí màcíkwí màràárú nà mírí màzáná màtáánú
nà mírí mákúḿ' màràárú nà mìtáná mìràárú
|ðì-hi-kál-a=mo   **mí-ri**   ma-cíkwi   ma-raaru   na   **mí-ri**   ma-zána
SP₄-PFV.DJ-be-FV=LOC₁₈   4-tree   6-thousand   EP₆-three   and   4-tree   6-hundred
ma-táanu   na   **mí-ri**   má-kumi   ma-raaru   na   **mi-**tána   **mi-**raaru|
EP₆-five   and   4-tree   6-ten   EP₆-three   and   EP₄-five.and   EP₄-three
'There are 3,538 trees.' (elic.)

Other variable quantifiers are *oté* (also realised *eté*) 'all, whole,' *inji* 'many, much' and *ŋóno* 'few, little.' Whereas *oté* (26) takes the pronominal prefix, *inji* (27) and *ŋóno* (28) are preceded by the nominal prefix, like adjectives and numerals (see Table 24.8).

(26) érédá èèpwábwálèyà mwììlábónì mwèètémwénè
|é-reḍá   e-ʔi-pwábwaley-a   mu-e-lábo-ni   **mu-eté=mwene**|
7-disease   SP₇/₉-PFV.DJ-spread-FV   18-7-country-LOC   PP₁₈-all=18.INT
'The disease spread throughout the whole country.' (semi-elic.)

(27) ɲúmbà ðíínjì ðìḍúméyíwà
|ɲúmba   **ðí-inji**   ðí-ḍumey-íw-a|
10.house   NP₈/₁₀-many   SP₈/₁₀-PFV.DJ-burn-PASS-FV
'Many houses were burnt.' (elic.)

(28) ḍìnòóŋwà máánjé màŋónóví: kàḍà múðîðì
|ḍi-ni-ó-ŋw-a   má-anjé   **ma-ŋóno=vi**   kaḍa   mú-ðiði|
SP₁ₛɢ-IPFV.DJ-15-drink-FV   6-water   NP₆-little=RESTR   each   3-hour
'I drink a little water every hour.' (elic.)

4.2.6 Personal pronouns

The full paradigm of personal pronouns is given in Table 24.12.

TABLE 24.12 PERSONAL PRONOUNS

	SG	PL
1st	míyò	íyò
2nd	wéyò	ɲúwò (+ resp)
3rd	íyéénè	áwéénè

Both 3rd persons *íyéénè* '3SG' and *áwéénè* '3PL' are frozen forms originating in the roots *íyè* and *áwè* plus the intensifier clitic *=énè*.

Personal pronouns typically function as subjects (29), preposition complements (30) and (more rarely) verb complements, post- (31a) or pre-verbally (31b), which appear in addition to object markers on the verb for emphatic topicalisation.

(29) míyó ḍínóvólówà mùtákwânì
    |**míyo**   ḍí-ni-ó-volow-a   mu-tákwa-ni|
    1SG.PRO   SP$_{1SG}$-IPFV.DJ-15-enter-FV   18-9.forest-LOC
    'I am going into the forest.'

(30) ākwááyè àbàálíwè nà íyéénè . . . [9]
    |á-kwe=áye   a-baál-iw-e   **na**   **íyeéne**|
    2-friend=POSS.3SG   PP$_2$-give.birth-PASS-PFV.REL   with   3SG.PRO
    'the friends who were born at the same time. . .' (lit. 'with her')

(31) a   kàāvó mùṭú· òŋtóóɲá: bààḍípíṭá míyò
        |kaá=vo   mu-ṭu   o-ni-tóoɲ-a
        NEG.COP.SP$_1$=LOC$_{16}$ 1-people.H1D   PP$_1$-IPFV-show-FV.REL
        ba-a-ḍí-piṭ-á   **míyo**|
        SEQ-SP$_1$-OP$_{1SG}$-surpass-FV   1SG.PRO
        'There is no one who is better at tracking the way than I.'

    b   "míyó mùnóḍíziwá: ?" [. . .] "kànùwííðì" "míyó kàmùḍííðí: ?"
        |**míyo**   mu-ni-ó-ḍi-ziw-á   ka-ni-ú-iði
        1SG.PRO   SP$_{2PL}$-IPFV.DJ-15-OP$_{1SG}$-know-FV   NEG-SP$_{1PL}$-OP$_{2SG}$-know
        míyo   ka-mu-ḍí-iði
        1SG.PRO   NEG-SP$_{2PL}$-OP$_{1SG}$-know|
        'Do you know who I am? No, we don't. You do not know me?'

4.2.7 Multiple modifiers

When a noun phrase has several modifiers including a possessive, the latter typically occurs immediately after the modified noun, where it may optionally be cliticised. As such, the possessive comes before demonstrative pronouns (32), before connective constructions (33), and before adjectives (34).

(32) nàámbéɖàag' òóɗù
|naámbeɖe=**aga**   **óɗu**
1a.maize=POSS.1SG   1.DEM.I
'this maize of mine'

(33) ózómbwè wáâgà wòòtéénè ɗàágúlà kòkó
|ó-zombwe **ó-aga**   **o-oté=ene**   ɖi-á-gul-a   koko|
14-youth   PP$_{11/14}$-POSS.1SG   PP$_{11/14}$-all=INT   SP$_{1SG}$-PST.IPFV.CJ-buy-FV   10.coconut.H1D
'During all my childhood, I used to buy coconut.' (semi-elic)

(34) lívúrúnàgà ńnɖímúwà
|lívurú=**naga**   **ní-nɖimúwa**|
5a.book=5.POSS.1SG   NP$_5$-big
'my big book' (elic)

The demonstrative pronoun is the second after the possessive to have a close relationship with the modified noun. It thus comes before a connective construction (35), or before an adjective (36).

(35) nìkómé ńʈílè nà bárá nòòté
|ni-kóme   **ńʈile**   **ni-a**   **bára**   **ni-oté**|
5-bank   5.DEM.III   PP$_5$-CON   9.sea   PP$_5$-all
'along the whole sea bank'

(36) mbúzì éjílè éŋóónò èèlélêyà
|mbúzi   **éjile**   **é-ŋoóno**   e-ʔi-lél-ey-a|
9.goat   7/9.DEM.III   NP$_{7/9}$-small   SP$_{7/9}$-PFV.DJ-educate-NEUT-FV
'That small goat behaves properly.'(elic)

4.2.8 Enclitics

Cuwabo has a number of enclitics with different functions, which may attach to different elements in the noun phrase. The most recurrent one is the definite enclitic, which is composed of the formative *a* preceded by an agreeing prefix in certain classes only, namely classes 4-5-7-8-9-10-15-16-17-18, as can be seen in Table 24.8. The remaining classes, as well as the speech act participants pronouns, all exhibit a default particle *ya*, corresponding to class 7/9 agreement. Although the definite particle is formally similar to the connective relator *a*, it cannot be considered as connective-derived, since their agreement patterns differ (see Table 24.8).

The definite enclitic may be combined with nouns (37) and adjectives (38), but also personal pronouns (39), demonstrative pronouns (40) and possessives (41). With the latter, the definite enclitic has a determiner-like function, probably conveying emphasis.

(37) àlóófíyá síkúnà ńʈlílè
|a-lé-o-fíy-a   síku=**na**   ńʈile|
SP$_2$-CE-15-arrive-FV   5a.day=5.DEF   5.DEM.III
'They reached that day.'

(38) ðìŋdéèyíná ðà déréétúðá
|ði-ní-ð-a=iye=na   ði-a   déreétu=**ða**|
PP$_{8/10}$-IPFV-go-FV.REL=3SG.PRO=COM   PP$_{8/10}$-CON   good=8/10.DEF
'the good things he brings'

(39) Màríyá àgààŋáná: "mééyá ḍìlú?"
|Maríya   a-gaa-aŋán-a   míyo=**ya**   ḍi-lí   uuvi|
Maria   SP$_1$-SIT-look-FV   1SG.PRO=DEF   SP$_{1SG}$-be   where
'Maria, when she looked: Where am I?'

(40) shá:! ŋng' óḍíy' òòŋḍúkùwélá bà ààníì:? [10]
|shá   ningá   óḍu=**ya**   o-ní-ḍi-kuwél-a   ba   aani|
INTER   VOC   1.DEM.I=DEF   PP$_1$-IPFV-OP$_{1SG}$-call-FV.REL   COP$_2$ who
'But who is this one calling me?'

(41) múzúgw' ééyùy' òòlògìlé wíí [. . .]
|mú-zugu   ó-eʔu=**ya**   o-log-ilé   wíi|
1-boss   PP$_1$-POSS.1PL=DEF   SP$_1$-say-PFV.CJ   COMP
'Our boss said that. . .'

A second agreeing enclitic is =*ene*, which conveys an intensive meaning and can be translated in English as 'very,' 'exact' or 'really.' It may attach to nouns (42), demonstratives (43), possessives (44), adjectives (45), quantifiers (33) and personal pronouns (124).

(42) Nìkúrábèðà síkúnénè ŋ̣ṭó òóðà
|Nikúrabeða   síku=**néne**   ŋ̣ṭo   o-ʔí-ð-a|
1a.dugong   5a.day=5.INT   5.DEM.II   SP$_1$-PFV.DJ-come-FV
'Dugong came that very day.'

(43) ókwééné "sìí! ŋ̣ðów' óóṭólóni"
|óku=**éne**   sii   ni-ðów-e   o-ṭólo=ni|
1.DEM.I=INT   INTER   SP$_{1PL}$-go-SBJV   17-well=LOC
'At this very moment: "[excl.] let's go to the well." '

(44) pà vénéw' óókòsííyé màsárápìṭ' ááyèén' áábò
|pa   vénewo   va-á-kos-á=íye   ma-sárapiṭi   á-áye=**éne**
COP$_{16}$ 16.EDEM.II   PP$_{16}$-PST-do-FV.REL=3SG.PRO   6-magic   PP$_6$-POSS.3SG=INT
            ábo
            6.DEM.II
'It is there that he decided to make that magic of his.'

(45) àtó wààfíyà yúúmó éndímúwéénè
|até   o-á-fiy-a   é-umó   é-ndimúwa=**éne**|
until   SP$_1$-PST.IPFV.DJ-arrive-FV   7-age   NP$_{7/9}$-big=INT
'She eventually got older.'

Finally, the invariable enclitic =*vi* is used to restrict the referential scope of a category and focus exclusively on what is designated. In combination with nouns (46) and quantifiers (28), it can be translated as 'only.'

(46) àámóònà ńngà múzúgu mùzùgúví bàáʔì
|a-a-ʔí-mu-on-a   nínga   mú-zugu   mu-zugu=vi   baáʔi|
SP₂-PST-PFV.DJ-OP₁-see-FV like   1-European 1-European.H1D=RESTR   only
'They had seen her as a white, and only as a white person.'

5 VERBS

5.1 Verbal derivation

The verbal base consists of a root optionally followed by derivational suffixes often referred to in Bantu as extensions. The different verb root shapes are listed (without tones) in (47). Among them, CVC is the most frequent.

(47) Shape   Example   Gloss
      C   j   'eat'
      VC   on   'see'
      CVC   vaʔ   'give'
      CVVC   baal   'give birth'
      CVCVVC   siliil   'envy'
      CVCVC   ciṭiṭ   'filter'
      CVCVCVC   birimind̯   'make roll'

The root can be followed by one or more derivational extensions, listed in Table 24.13. Most have a VC shape, typical of Bantu verbal extensions. Others have the shape VVC, which probably results from a VCVC structure, in which the original VC extension underwent reduplication, and then lost the intervocalic consonant.

TABLE 24.13 VERBAL DERIVATIONAL EXTENSIONS

Productive extensions		Non-productive extensions	
Causative	-iʔ/-ec	Causative	-ec
Applicative	-el/-eel	Causative	-eð/-eeð
	(-eð/-eeð)	Causative	-uc
Passive	-iw/-uw	Causative	-ul
Neuter	-ey	Middle	-uw
Reciprocal	-an	Impositive	-ey
		Positional	-am
		Tentive	-aṭ

Among the productive extensions, causative (48) and applicative (49) are typically valency-increasing, whereas passive (50), neuter (51), and reciprocal (52) are valency-decreasing.

(48) Causative
      wáámwíyà   'to breastfeed'   <   wáàmwá   'to suckle'
      ósúnzíyà   'to teach'   <   ósúnzà   'to learn'

(49) Applicative
      óṭámágélà   'to run after'   <   óṭámâgà   'to run'
      òpíyêlà   'to cook for'   <   òpíyà   'to cook'

(50) Passive
  ótélíwà    'to be married'    <    ótèlá    'to marry'
  òtíbûwà    'to be dug out'    <    òtíbà    'to dig out'

(51) Neuter
  òlávêyà    'to be cursed'    <    òlávà    'to curse'
  òvápêyà    'to be different'    <    òvápà    'to vary'

(52) Reciprocal
  òrórómèlànà    'to trust each other'    <    òrórómèlà    'to trust'
  ótíyánà    'to be distant'    <    ótìyá    'to leave'

In addition to the productive causative *-iʔ*, Cuwabo has four lexicalised causative extensions, namely *-ec*, *-eð*, *-uc/-oc*[11] and *-ul/-ol*, respectively exemplified below. They are mostly used in direct causation, involving a patientive causee directly subjected to the caused event (Shibatani & Pardeshi 2002). With respect to their origin, *-ec*, *-eð*, *-uc* are most likely the result of a combination of proto-extensions. In the case of *-ec* and *-uc*, a merger would have occurred between an intransitive extension (ending in *k) and the short Proto-Bantu causative extension *-i (Bastin 1986), whereas *-eð* would constitute a case of combined applicative-causative morphology, namely *-ɪd + *-i; see Guérois and Bostoen (2016) for a more elaborate analysis.

(53)  a  wààwìmécà
        |o-a-iméc-a|
        NARR-OP₂-rigidify-FV
        'She made them stop.'

    b  ànámásàpà ààtúpúcèða múlábà wáâwà
        |a-námasapa  a-ʔi-túpuceð-a      mú-laba  ó-awa|
        2-fisherman  SP₂-PFV.DJ-submerge-FV  3-net  PP₃-POSS.3PL
        'The fishermen threw their net into the water.' (Valler & Festi 1995)

    c  ònóvítòtócà
        |o-ni-ó-ví-totóc-a|
        SP₁-IPFV.DJ-15-REFL-destroy-FV
        'He is destroying himself.'

    d  óĺlééné òm̀fún' óópàtúla ósàlú
        óllé=ene    o-ní-fun-á      ó-patúl-a    ó-salú
        1.DEM.III=INT  SP₁-IPFV.CJ-want-FV  15-break-FV  11-thread
        'the one who will break the thread'

The extension *-eð* also has an applicative function with some synchronic verb bases, as can be seen by comparing (54a) and (54b). *-eð* behaves as a typical applicative marker (Peterson 2007), in that it allows the introduction of an additional argument to the verb, the applied object, which may bear a variety of semantic roles, including benefactive/malefactive (as in (54b) with the first person singular object prefix *ḍi-*), recipient, or destination of movement, all typical of applicative constructions.

(54) a   Òsálù ónósásáɲà ɲúmbà yááyè ègùjúwìlé
          Osálu   ó-ni-ó-sasáɲ-a   ɲúmba   é-aye   e-gujúw-ilé
          Osalu   SP₁-IPFV.DJ-15-fix-FV   9.house   PP₇/₉-POSS.1SG   PP₇/₉ -be.spoilt-PFV.REL
          'Osalu is fixing his house, which is spoilt/destroyed.' (Valler & Festi 1995)

      b   míyó ɖìɱfúná mùɖísásàɲèðè pàpóórò
          míyó   ɖi-ní-fún-á   mu-ɖí-sásaɲ-eð-e   papóoro
          1SG.PRO   SP₁ₛₒ-IPFV.CJ-want-FV   SP₂ₚₗ-OP₁ₛₒ-build-APPL-SBJV   1a.boat
          'I want you to build me a boat.'

Accounting for the double function of -eð (causative (53b) and applicative (54b)) is beyond the scope of this chapter. For a detailed analysis, the reader is referred to Guérois and Bostoen (2016).

Cuwabo has two different verb suffixes to derive a passive verb from an active verb, namely -iw (55) and -uw (56). Unlike in many other Bantu languages, their distribution is not phonologically conditioned.

(55) a |ní-zuwa   ni-ʔí-p-a   mbúzi|
          5-sun   SP₅-PFV.DJ-kill-FV   9.goat
          'The sun killed the goat.'

      b |mbúzi   e-ʔí-p-iw-á   na   ní-zuwa|
          9.goat   SP₇/₉-PFV.DJ-kill-PASS-FV   by   5-sun
          'The goat was killed by the sun.'

(56) a |ɲeɲéle   ðí-ni-ó-j-a   mí-ri|
          10.ant   SP₈/₁₀-IPFV.DJ-15-eat-FV   4-tree
          'The ants are eating the trees.'

      b |mí-ri   ðí-ni-ó-j-uw-á   na   ɲeɲéle|
          4-tree   SP₄-IPFV.DJ-15-eat-PASS-FV   by   10.ant
          'The trees are being eaten by the ants.'

Passives in -uw are far more frequent than in -iw. Following Guérois and Bostoen (2018), it is likely that Cuwabo -uw became polysemic, evolving from the intransitive separative suffix (from Proto-Bantu *-ʊk) to a more or less lexicalised extension with different middle values as distinguished by Kemmer (1993) (e.g., change in body posture in (57a) and anticausative/neutro-passive in (57b)), and from there, to a canonical passive extension competing with the inherited passive -iw. The licensing of prepositional na phrases conveying a 'causer' interpretation with lexicalised neutro-passive -uw verbs as in (57c) would support this hypothesis.

(57) a   àzùgúnúwá: àð' áámáɽíyà ðòójà
          |a-zugúnuw-a   a-ð-á   a-malríʔ-a   ði-ója|
          SP₁-turn.round-FV.SEQ   SP₁-come-FV.SEQ   SP₁-finish-FV.SEQ   8-food
          'He turned round, came to finish the food alone.'

      b   ósálù kùpàtúwîlè
          |ó-salu   ku-patúw-ile|
          11-thread   NEG.SP₁₁/₁₄-break-PFV
          'The thread did not break.'

c èrúgú èèd̪émúwà nà múlóbwànà
|e-rúgu e-ʔi-d̪émuw-a   na   mú-lobwana|
7-wall   SP$_{7/9}$-PFV.DJ-pierce-FV   by   1-man
'The wall is pierced because of the man.'

Typically, several extensions can co-occur within one verbal base, each one respecting a certain position in the string, following the general Causative-Applicative-Reciprocal-Passive (CARP) template (Hyman 2003b). The following examples show possible combinations of the aforementioned extensions.

(58) d̪ìyáágùlìyèð̃íwà nàámbéd̪è álêd̪ò
|d̪i-ʔí-a-gul-iʔ-eð-íw-a   naámbed̪e   á-led̪o|
SP$_{1SG}$-PFV.DJ-OP$_2$-buy-CAUS-APPL-PASS-FV   1a.maize   2-guest
'I made the maize be sold to the guests.' (elic.)

(59) èèd̪èláné: òmááɽò wààkùŋgíláàní
|a-ed̪-el-án-(il)é   o-mááɽo   o-a-kuŋg-íle=ani|
SP$_2$-hate-APPL-RECP-PFV.CJ   14-friendship   PP$_{11/14}$-PST-build-PFV.REL=3PL.PRO
'They came to hate each other for the friendship they had built.'

5.2 Verbal inflection

The structure of the inflected verb is as follows.

(60) (NEG) – SP – (NEG) – TAM – (OP) – root (ext.) – (TAM) – FV – Post-final

5.2.1 Prefixes of negation

The pre-initial (ka-) and post-initial (ʔi-) negative markers are complementarily distributed: ka- appears in independent or main clauses, whereas ʔi- is restricted to subordinate clauses. The use of both negative markers is illustrated in (61).

(61) d̪ààyíríntìgí kàbálàyá:, bwènd̪éná kàɲnáálíbè
|d̪i-a-ʔí-rint-ig-i   kabála=ya   bwend̪é=na
SP$_{1SG}$-SIT-NEG-weave-IPFV-NEG   9.rope=DEF   5a.mat=5.DEF
ka-ni-náa-líb-e|
NEG-SP$_5$-FUT-be.strong-FV
'If I do not weave this rope, this mat will not be strong' (semi-elic.)

A third negation prefix is attested in Cuwabo, i.e., náa-, but only with subjunctive verb forms used either in independent clauses (62a) or dependent clauses (62b). Note that na- (or a variant) is a recurrent post-initial negative marker in zone P languages (Kamba Muzenga 1981: 115).

(62) a mùnáávégé wìíkò, òòkálá àɲákôkò
|mu-náa-veg-e   o-íko   o-ʔi-kál-a   a-ɲákoko|
SP$_{2PL}$-NEG-play-SBJV   17-river   SP$_1$-PFV.DJ-be-FV   2-crocodile
'Do not play at the river, there are crocodiles.' (semi-elic.)

b   Rósà ónóṭápwáṭèðà námà, ènáálúlè
|Rósa   ó-ni-o-ṭápwaṭeð-a   náma   e-**náa**-lul-e|
Rosa   SP₁-IPFV.DJ-15-cook.sp-FV   9-meat   SP₇/₉-NEG-spoil-SBJV
'Rosa is cooking the meat in salty water, so that it does not spoil.' (Valler &
Festi 1995)

From examples (61) and (62), one can wonder whether *náa-* as a future marker and
náa- as a negative marker are better analysed as homonymous or as a single form with
disparate meanings. Although a crosslinguistic tendency for potentials (and thus maybe
also futures?) to develop into prohibitives does exist (Pakendorf & Schalley 2007), the
exact status of *náa-* is in need of further description, and I will not attempt to account for
this fact here.

Further note that two other strategies exist to negate subjunctive verb forms: the
pre-initial negative *ka-* in main clauses (63), and the post-initial negative marker *Ɂi-* in
subordinate clauses, as illustrated in (131) below.

(63)   kùlìmè vàmùcésánì kàvaṇ̀ð' éèlò
       ku-lim-e   va-mu-césa-ni   ka-va-ní-ð-a   e-lo
       NEG.SP₂ₛG-cultivate-SBJV   16-3-sand-LOC   NEG-SP₁₆-IPFV-come-FV   7-thing.H1D
       'Do not cultivate in sandy soil, nothing comes out of it.' (semi-elic.)

5.2.2 Pre-initial slot

Beside the negative marker *ka-*, three more pre-initial morphemes exist in Cuwabo. The
most common is the sequential *ba-*, used to mark subsequent events (64), or to express
simultaneity or coincidence of two events when used with a situative prefix (65) or with
a perfective suffix (66a). All these dependent verb forms make use of the negative marker
Ɂi- as shown in (66b) and also (122) below.

(64)   Bare sequential
       bààtèlá: bààkàlá vàtákúlù vááyè nà mwááðíyè
       |**ba**-a-tel-á   **ba**-a-kal-á   va-tákulu   vá-aye   na
       SEQ-SP₁-marry-FV   SEQ-SP₁-stay-FV   16-9.house   PP₁₆-POSS.3SG   with
       mú-aðí=ye|
       1-wife=POSS.3SG
       'and he got married and lived in his house with his wife'

(65)   Sequential situative
       bàgàpwáṭélá bàgáðòwá íyéénè wàawìmécà
       |**ba**-a-**ga**-pwáṭel-a   **ba**-a-**gá**-ðow-á   íyeéne   o-a-iméc-a|
       SEQ-SP₂-SIT-carry-FV   SEQ-SP₁-SIT-go-FV   3SG.PRO   NARR-OP₂-rigidify-FV
       'and while they were loading (it, the dry cassava) and going, she made them stop'

(66)   Sequential perfective
       a   kàdìṇ̀ṭáḍá: máánjé bâgùbùl̀lé
           |ka-ḍi-ní-ṭaḍ-á   má-anjé   **ba**-a-gubul-**ilé**|
           NEG-SP₁ₛG-IPFV-fish-FV   6-water   SEQ-SP₆-rise-PFV
           'I do not fish at high tide (lit. 'while the water is high-levelled')' (semi-elic.)

b ò n̄ dʼísùmùlél' éèbéwéénè, mééyá bàd̄ìgúĵlè
|o-ní-d̄i-sumulél-a   ebéweéne **ba-d̄i-ʔi-gúl-ile**|
SP₂ₛ₉-IPFV.CJ-OP₁ₛ₉-blame.APPL-FV   freely.INT   SEQ-SP₁ₛ₉-NEG-buy-PFV
'you are blaming me for nothing, while I did not buy' (semi-elic.)

There are two other pre-initial morphemes, the counterfactual *kà-* (67) and the resumptive *nà-* (68), which attach to infinitive verb forms.

(67) kòòwígúl' òòbà kàvàgààjúwè
|**ka-o-ʔí-gul-a**   oba   ka-va-gaa-j-úw-(il)e|
CF-15-NEG-buy-FV 9.fish.H1D   NEG-SP₁₆-HYP-eat-PASS-PFV
'If I had not bought fish, there would be no meal at home.' (semi-elic.)

(68) òmwààlámó nàámbêd̄è. **nù**úmwáàlá nàámbéd̄é:, [wààvèdágá: áṯú . . .]
|o-mu-al-á=mo   naámbed̄e **na**-ó-mu-al-á   naámbed̄e|
NARR-OP₁-sow-FV=LOC18 1a.maize RES-15-OP₁-sow-FV   1a.maize
'and he sowed maize. When he sowed maize, [he looked for people. . .] '

5.2.3 Subject markers

All verb forms, except for the infinitive (*òlímà* 'to cultivate'), narrative (*òlìmà* 'and I/you/ etc cultivated'), resumptive (*nòòlímà* 'after cultivating') and bare imperative (*lìmá* 'cultivate!'), are marked by a subject prefix in initial position. Among the different subject markers listed in Table 24.8, the class 1 subject marker has two allomorphs, *o-* and *a-*, the second of which is restricted to situative (39), sequential (57a), subjunctive (69) and hypothetical (123b) verb forms.

(69) ò m̄ fún' áágáwè kómɨd' èésîlè
|o-ní-fun-á   a-gáw-e   kómiída   ésile|
SP₁-IPFV.CJ-want-FV   SP₁-serve-SBJV 10.food   8/10.DEM.III
'He wants to serve that food.'

The 2PL subject marker may be used to express an honorific singular, as illustrated in (70), where the protagonist addresses an older man.

(70) mùn̄zíwá: ðáàv' íílá míyó d̄ìlì Màríyà ?
|**mu**-ní-zíw-á   ðáaví wíílá   míyó   d̄i-li   maríya|
SP₂ ᵣₑₛₚ-IPFV.CJ-know-FV   how   COMP   1SG.PRO SP₁ₛ₉-be   maria
'How do you know I am Maria?'

5.2.4 Tense/Aspect/Mood

The position between the subject and the object agreement prefixes contains most tense/ aspect/mood (TAM) markers. Together with the final, this verb slot allows for determining the TAM value of a verb. The different forms which can occur in this slot are given and exemplified in Table 24.14, with the toneless verb *osákula* 'to choose' (only the underlying H tones are indicated).

TABLE 24.14 TAM PREFIXES

Form	TAM	example	translation
a-	past	ɖi-a-ʔí-sakul-a	'I had chosen'
ni-	imperfective	ɖi-ni-sákul-a	'I am choosing'
á-	past imperfective	ɖi-á-sakul-a	'I was choosing'
hi-	perfective	ɖi-ʔi-sákul-a	'I chose'
ná-	counterexpectational	ɖi-ʔi-ná-sakul-a	'before I choose'
náá-	future	ɖi-náa-sákul-e	'I will choose'
gá-	future imperfective	ɖi-gá-sakul-a	'I will be choosing'
a-	situative[1]	ɖi-a-sakul-a	'if/when I choose'
gaa-	situative	ɖi-gaa-sakul-a	'if/when I choose'
gaa-	hypothetical	ɖi-gaa-hí-sakul-a	'I would choose'
ka-	imperative	ka-sákul-a	'choose!'

[1] The situative is used to introduce a hypothetical situation or a "logical or temporal precondition" (van der Wal 2009: 97). As an equivalent of English if/when-clauses, the 'situative' is also known as 'conditional.' I follow here Devos (2008) and van der Wal (2009)'s terminology.

Certain verb forms require two subsequent slots for TAM, one for tense and one for aspect. This is the case with the past perfective form *ddi-a-ʔí-sakul-a* 'I had chosen,' where the past *a-* precedes the perfective *ʔi-*.

ka-imperative, with *ka* historically associated with the concept of going (see notions of *ka-movendi* and *ka*-itive in Nurse (2008), or distal Botne (1999)), is the standard imperative form in Cuwabo.

(71) ɲúwó kàrómánì wííméláànì!
    ⌊ɲúwo   **ka-róm-a=ni**   ó-imél-a=ni⌋
    2PL.PRO   IMP-start-FV=PLA   15-stand-FV=PLA
    'You (2pl), stand up first!'

The bare (or basic) imperative, consisting solely of the verb stem plus a High tone on the last mora, is very rarely attested in the language. The following example was elicited.

(72) gùlàní máfúg' àábò
    ⌊**gul-a=ní**   má-fugi   ábo⌋
    buy-FV=PLA   6-banana   6.DEM.II
    'Buy (2PL) those bananas!' (elic.)

A number of TAM values are subdivided into a pair of complementary conjugations, known as "conjoint" (CJ) and "disjoint" (DJ). For each pair, conjoint and disjoint verb forms encode the same tense/aspect semantics, but differ from one another in two respects: first, they have a different segmental morphology, as can be seen in Table 24.15. For instance, the aspectual suffix *-ile* is restricted to conjoint verb forms, whereas disjoint forms use the semantically equivalent prefix prefix *ʔi-*.

TABLE 24.15 CONJOINT AND DISJOINT VERB FORMS

	Conjoint	Disjoint
PRS IPFV	o-ní-ful-á mutede 'he is washing a dress'	ó-ni-ó-ful-á (mutéde) 'he is washing (the dress)'
PST IPFV	o-á-ful-a mutede 'he was washing a dress'	ó-a-ni-ful-á (mutéde) 'he was washing (the dress)'
PRS PFV	o-ful-ilé mutede 'he washed a dress'	o-ʔí-ful-á (mutéde) 'he washed (the dress)'
PST PFV	o-a-ful-íle mutede 'he had washed a dress'	o-a-ʔí-ful-á (mutéde) 'he had washed (the dress)'
FUT	o-náa-ful-e mutede 'he will wash a dress'	o-naa-ilá o-ful-á (mutéde) 'he will wash (the dress)'
FUT IPFV	o-gá-ful-a mutede 'he will be washing a dress'	o-gá-ni-ful-a (mutéde) 'he will be washing (the dress)'
HYP	o-gaa-ful-íle mutede 'he would wash a dress'	o-gaa-ʔí-ful-á (mutéde) 'he would wash (the dress)

Second, conjoint and disjoint verb forms differ in their relationship with what follows the verb. The conjoint verb form necessarily introduces a focused element. This focused element always occupies the Immediate-After-Verb position and, depending on its grammatical category, is affected by First H Deletion[12]. This is shown in (73) with the noun *múlóbwana* 'man' (underlyingly |múlobwana|), which loses its underlying H after the conjoint verb *ḍìṃfúná*. In disjoint verb constructions, illustrated in (74a), no special relationship between the verb and the following constituent is observed. This means that, unlike conjoint forms, a disjoint verb can stand on its own, as in (74b).

(73) súpèéyó sùpèéyò míyó ḍìṃfúná mùlòbwànà (citation form: *múlóbwana*)
      supeéyo   supeéyo   míyo   ḍi-ní-fun-á   mu-lobwana
      9.mirror.H1D   9.mirror.H1D   1SG.PRO   SP$_{1sg}$-IPFV.CJ-want-FV   1-man.H1D
      'Mirror, mirror, I want a man.'

(74) a   múlóbwan' oókosá: masárápiṭw' aábále
          |mú-lobwana   o-ʔí-kos-á   ma-sárapiṭo   ábale|
          1-man   SP$_1$-PFV.DJ-do-FV   6-magic   6.DEM.III
          'The man made that magic.'

      b   ánófúlá, ánófúlá, ánófúlá
          |á-ni-ó-ful-á|
          SP$_2$-IPFV.DJ-15-wash-FV
          'They are washing, washing, washing.'

The conjoint/disjoint alternation applies to affirmative verb forms only. Negative verb forms only have one form, built on the pre-initial negative marker *ka-* discussed above.

5.2.5 Object markers

In Cuwabo, there is one slot only for object marking. Object marking is further restricted to first and second person objects, and classes 1 and 2, as given in Table 24.16.

TABLE 24.16 OBJECT MARKERS

	1st person	2nd person	cl.1/2
SG	ḍi-	ù-	mù-
PL	ni-	mù-	à-
REFL	vì-		

Reflexivity is expressed by means of the prefix *vi-*, as seen in (53c) above, which remains invariable regardless of the person or class of the subject.

Class 1/2 object prefixes are agreement morphemes (or doubling), i.e., they have the possibility to co-occur with their corresponding object NPs in the same clause. In Cuwabo, *in situ* object NPs are found with (at least) conjoint verb forms. These cannot be clause-final and imply term focus on the immediately following element. The example in (75) shows that a class 1 post-conjoint object NP is obligatorily co-indexed on the verb. In this context, the absence of the object prefix *mu-* is considered ungrammatical.[13]

(75) eetéén' aáb' wáám̀magyedha namárógolo
　　　|a-eté=ene　　ábo　　　a-á-*(mu$_i$-)magyedh-a　　　**namárogolo$_i$**
　　　PP$_2$-all=INT　2.DEM.II　SP$_2$-PST.IPFV.CJ-OP$_1$-slander-FV　1a.hare
　　　'All of them were slandering the hare.'

The question of co-indexation in ditransitive constructions is discussed in Section 6.5.

5.2.6 Pre-final aspectual markers

As usual in Bantu, the pre-final suffix *-ag* covers different aspectual values linked with imperfective meaning, ranging from durative to habitual and pluractional. It may combine with most tensed verb forms (see examples (61) and (125) with situative and imperative, respectively).

The pre-final suffix *-ec* represents another way of encoding imperfective value, with both durative and iterative meaning. It is far less attested than the pre-final *-ag*.

(76) òṇḍíʃàtìyààrílècéèné dhàyéénè, [. . .] òṇḍúkùweḷléc' éènì?
　　　o-ní-ḍi-ʃatiyaari-él-**ec**-á=ene　　　　　dhaayí=éne
　　　PP$_1$-IPFV-OP$_{1SG}$-annoy-APPL-DUR-FV.REL=INT　like.this.I=INT
　　　o-ní-ḍi-kuwel-el-**éc**-a　　　　　　　e-ni
　　　SP$_1$-IPFV.CJ-OP$_{1SG}$-call-APPL-DUR-FV　7-what
　　　'This one annoying me like this, [. . .] why is he calling me?'

(77) āyím' àábá àṇkáλécà nà báábá, [àgòònúwà, àgááṇsèmà vàḍíḍì]
　　　á-ima　　　ába　　　a-ni-kál-**ec**-a　　　　na　　　**báabá**
　　　2-child　　2.DEM.I　PP$_2$-IPFV-be-DUR-FV.REL　with　1a.their.father
　　　'These children who are always with their father, [when they grow up, they will be good carpenters].'

5.2.7 *Final slot*

Three final suffixes exist in Cuwabo, *-a*, *-e*, and *-ile*, all having tense and/or aspect values. Among them, *-a* is the default final suffix found in most indicative tenses. It does not carry any meaning by itself and is glossed 'FV' for final vowel. The final suffix *-e* is mainly used to mark the subjunctive (78). In this case, no tense prefix is required. Furthermore, *-e* systematically substitutes the final vowel *-a* of an imperative verb form taking an object marker. Compare *káḍíjèèðènì* 'wait for me' in (79) with *kàðáná* 'bring' in (78). Finally, the final vowel *-e* always co-occurs with the future prefix *náa-*, as shown in (61) and (128).

(78) kàðáná tò [. . .] mwááðíy' òóḍó ṇ̀ðé nìmóônè
    |ka-ð-á=na   to   mú-aðí=ya   óḍo   ni-ð-é   ni-mú-on-e|
    IMP-go-FV=COM then 1-wife=DEF 1.DEM.II SP$_{1PL}$-go-SBJV SP$_{1PL}$-OP$_1$-see-SBJV
    'Bring then that friend of mine, that wife, so that we see her.'

(79) káḍíjèèðènì
    ká-ḍi-jeeð-e=ni
    IMP-OP$_{1SG}$-wait-FV=PLA
    'Wait for me.'

The suffix *-ile* 'PFV' is one of the two perfective aspectual markers in Cuwabo. It appears in two frequent verb forms: the conjoint (present) perfective (80a) and the conjoint past perfective (80b), but is also part of the less frequent conjoint hypothetical (81) and the sequential perfective (82).

(80) a |o-ðow-**ilé**| b |ḍi-a-ðow-**íle**|
       SP$_1$-go-PFV.CJ   SP$_{1SG}$-PST-go-PFV.CJ
       'He went.'   'I had gone.'

(81) kòwíkálàgá kòḍó, kàmùgààbéléìlè
    |ka-o-ʔí-kal-ag-á   koḍo   ka-mu-gaa-abélel-**ile**|
    CF-15-NEG-be-IPFV-FV 9.war.H1D NEG-SP$_{2PL}$-HYP-swim-PFV
    'If it was not on account of the war, you (2PL) would not swim.' (semi-elic.)

(82) kòwíkálàgá káár' òógújúwà múðíð' ùúbu bàḍìfiyìlé
    |ka-o-ʔí-kal-ag-á   káaro   ó-gujúw-a   mú-ðiði   óbu
    CF-15-NEG-be-IPFV-FV 9.car 15-be.broken-FV 3-time 3.DEM.I
    ba-ḍi-fiy-**ilé**|
    SEQ-SP$_{1SG}$-arrive-PFV
    'If the car had not broken down, at this time, I would have arrived.' (semi-elic.)

When *-ile* is preceded by an applicative (83a), a passive (83b) or a reciprocal (83c) extension, it shortens to *-e*.

(83) a |a-a-zuzúm-**el-e**|
       SP$_2$-PST-be.confused-APPL-PFV.CJ
       'They were confused.'

b |e-log-**uw-é**|

SP$_{7/9}$-say-PASS-PFV.CJ

'It was said.'

c |a-eḍ-el-**án-e**|

SP$_2$-hate-APPL-RECP-PFV.CJ

'they hated each other for'

Note that contrary to many Eastern Bantu languages (including Makhuwa P31), this shortening does not involve imbrication in Cuwabo, or at least no visible traces remain at the synchronic level.

A final suffix -*i* is restricted to the situative (in both *a*- and *gaa*-), where it functions as a negative marker, in addition to the post-initial negative prefix *ʔi*-, as illustrated in (84).

(84) |ḍi-a-ʔí-lim-**i**| |o-gaa-ʔí-suṅz-**i**|

SP$_{1SG}$-SIT-NEG-cultivate-NEG   SP$_{2SG}$-SIT-NEG-study-NEG

'if I do not cultivate'   'if you do not study'

5.2.8 Post-final enclitics

Verb forms share three of the enclitics found in noun phrases, namely the marker of proximity =*no*, as already illustrated on the verb in (17), the intensive =*ene* (76), and the restrictive =*vi* (85).

(85) *ḍóóṛínḍ' óónówííbávì*

|ḍóoṛinḍo   ó-ni-ó-ib-á=**vi**|

ddoolrinddo   SP$_1$-IPFV.DJ-15-sing-FV=RESTR

'Ddoolrinddo is just/only singing (and nothing else)'

The post-final =*ni* is typically used in imperative verb forms as a marker of plural addressee (PLA) or politeness, as already illustrated in (71) and (72). It also occurs in combination with the 2SG object marker *u*- to refer to a 2PL object marker, as in (86).

(86) ònùúṭébàní:, ḍì míyò

|o-ni-ú-ṭeb-a=**ni**| ḍi   míyo|

PP$_1$-IPFV-OP$_{2PL}$-carry-FV.REL=PLA   COP$_1$ 1SG.PRO

'It is me who lifts you.'

The post-final clitic =*na* can be either comitative (87) or instrumental (88).

(87) òðòólé lèṅsó [ṅlí ṁbá] òðén' òwííwá ?

|o-ðoól-e   leṅso   o-ð-é=**na**   o-ʔí-iw-á|

SP$_{2SG}$-fetch-SBJV   5a.tissue   SP$_{2SG}$-come-SBJV=COM   SP$_{2SG}$-PFV.DJ-hear-FV

'Fetch the tissue [which is inside] and come (back) with (it), did you hear?'

(88) ðílóbò ð' òókóḍélíyànà vàtákûlù
|ðí-lóbo   ðhi-a   ó-kóḍél-í?-a=**na**   va-tákûlu|
8-thing   PP$_{8/10}$-CON   15-be.beautiful-CAUS-FV=INSTR   16-9-house
'things to embellish the house with'

Subject personal pronouns may also be cliticised in non-subject relatives (see (103) in Section 6.2 on relative clauses).

Finally the post-final slot may also be occupied by one of the three locative clitics =*vo* (class 16), =*wo* (class 17) and =*mo* (class 18). They function as pronouns and may serve as locative objects (89) or locative adjuncts (90a).[14] See Guérois (2017b) for a more elaborate discussion.

(89) |o-Maputo$_i$ ḍi-ni-ó-zíw-a=**wo**$_i$|
17-Maputo   SP$_{1SG}$-IPFV.DJ-15-know-FV=LOC$_{17}$
'In Maputo, I know it (there).' (elic.)

(90) Vattólóní, maánj' áawíínjívâvo. Kí níŋkóséénáavó dháavi?
a |va-ttólo-ni   ma-ánje   a-hí-injív-a=**vo**|
16-well-LOC   6-water   SP$_6$-PFV.DJ-abound-FV=LOC$_{16}$
'There is a lot of water in the well.'

b |kí   ni-ní-kos-á=**na**=**vo**   dháavi|
EMPH   SP$_{1PL}$-IPFV.CJ-do-FV=COM=LOC$_{16}$ how
'What shall we do with it?'

The post-final slot as presented in the verbal template in (60) may be filled by more than one element. This is illustrated in (90b), where both the instrumental =*na* and the locative =*vo* may co-occur, but also in (17) above, where the locative enclitic is followed by the proximal =*no*. The possible combinations attested in my corpus are listed in (91).

(91) Combinations of post-final clitics
Locative + Interrogative =*ni* 'what'
Locative + Proximal =*no*
Plural Addressee =*ni* + Locative
Comitative/instrumental =*na* + Locative
Personal pronouns in non-subject relatives (+ Comitative =*na*) + Locative

6 SYNTAX

6.1 Non-verbal predication

There are two basic strategies to express non-verbal predication in Cuwabo. The first one involves a tonal modification of the nominal (or adjectival) predicate, whereby its first primary H (i.e., the augment tone) is deleted. Such a tone process, coined "First H Deletion" (H1D) in this chapter, but also known as "Predicative Lowering,"[15] after Schadeberg and Mucanheia (2000), is effective on both nouns (92a) and adjectives (92b).

(92) a |namárogoló   **namapuja**| < namápuja
        1a.hare   1a.joker.H1D
        'The hare is a joker.'

     b |Mosambíki   **mu-nḍimúwa**| < múnḍimúwa
        1a.Mozambique 1-big.H1D
        'Mozambique is big.'

Importantly, the first and second persons cannot be directly followed by H1D-subjected nominal or adjectival predicates. Instead, the connective stem preceded by some agreement markers, which are formally identical to the verbal prefixes, is obligatorily inserted between the subject pronoun and the predicate (93). This is not the case with third-person subjects (94).

(93) míyó **ḍà** mùyàná ~ mùnḍìmúwà   'I am a woman ~ big/tall'
      wéyó **wà** mùyàná ~ mùnḍìmúwà   'you (2SG) are a woman ~ big/tall'
      íyó **nà** àyàná ~ ànḍìmúwà   'we are women ~ big/tall'
      ɲúwó **mwà** àyàná ~ ànḍìmúwà   'you (2PL) are women ~ big/tall'

(94) íyééné Ø mùyàná ~ mùnḍìmúwà   'she is a woman ~ big/tall'
      áwééné Ø à-yàná ~ ànḍìmúwà   'they are women ~ big/tall'

Note that in similar environments, verbal predication is made possible by the insertion of the defective verb -*li* 'be,' as shown in (95).

(95) a íyééné òlí mùyànà
        |íyeéne **o-li**   **mu-yana**|
        SG.PRO   SP₁-be   1-woman.H1D
        'She is a woman.'

     b òlí múkíṭíkiṭíkiṭíkiṭíkíṭì
        o-li   **mú-kiṭi-kiṭi-kiṭi-kiṭi**
        SP₁-be   1-unripe-RED
        'She is very thin, she has turned thin.'

The second strategy makes use of a class-inflected copula, which is etymologically based on the demonstrative pronoun of series 1 (except for class 1 *ḍi*), as Table 24.17 shows. The same is observed in Koti P311 (Schadeberg & Mucanheia 2000). Note that these copular particles are non-segmentable.

TABLE 24.17 CLASS-INFLECTED COPULA COMPARED TO SERIES I DEMONSTRATIVES

Class	1	2	3	4	5	6	7/9	8/10	11/14	15	16	17	18
COP	ḍi	ba	bu	si	ṭi	ba	ji	si	bu	ku	pa	ku	pu
This I	óḍu	ába	óbu	ési	ńṭi	ába	éji	ési	óbu	óku	ápa	óku	ḿpu

The copula in Cuwabo, mostly found in (pseudo)cleft structures, may combine with a noun (96), a pronoun (97), a wh-element (98), a relative modifier (99) and even the connective stem, in adjectival constructions[16] (100).

(96) jì   gá?ál' ééɱfúnééy' úùtágíyà íyó ɖàbùnò vénêvà
      ji   gá?alá   e-ní-fun-á=i?u   o-tági?-a   íyo   ɖabuno
      COP_{7/9} 9.story   PP_{7/9} -IPFV-want-FV.REL=1PL.PRO   15-tell-FV   1PL.PRO   today
      véneva
      16.EDEM.I
      'This is the story we want to tell here today.'

(97) òɳlógá ɖì míyò Féŕnáándà
      o-ní-log-á   ɖi   míyo   Féŕnaánda
      PP₁-IPFV-speak-FV.REL   COP₁   1SG.PRO   Fernanda
      'it is me, Fernanda, who is speaking'

(98) k' úúnìmújà nàámbéɖè bà àànì?
      kí   o-ni-mú-j-a   naámbeɖe   ba   aani
      EMPH   PP₁-IPFV-OP₁-eat-FV.REL   1a.maize   COP₂   who
      'who is eating my maize?"

(99) nàmárógòl' òóɖú ɖì ònímútàmàgíyéèɳú!?
      namárogolo   óɖu   ɖi   o-ní-mu-tamag-í?-a=iɳú
      1a.hare   1.DEM.I   COP₁   PP₁-PRS-OP1-run-CAUS-FV.REL=2PL.PRO
      'the hare is the one you are making run!?'

(100) mísóṛò ðà áyímà ábà s' òòlápà
       mí-soṛo   ði-a   á-ima   ába   si-a   o-láp-a
       4-head   PP₄-CON   2-child   2.DEM.I   COP₄-CON   15-be.big-FV
       'the heads of these children are big'

On the other hand, the copula cannot be used with adjectives, which require the First H Deletion strategy, as seen in (92b).

6.2 Relative clauses

Cuwabo and Makhuwa P31 exhibit a relativisation strategy, also attested in certain North-Western Bantu languages (Nsuka-Nkutsi 1982), where relativised verb forms look like participial modifiers and retain all verbal properties except subject agreement. Thus, the initial verbal slot dedicated to subject marking in independent clauses is in relative clauses occupied by a pronominal prefix, which expresses agreement with the head noun, just like any modifier in the language.[17] This is illustrated in both subject (101a) and non-subject relatives (101b). In the latter case, the (pro)nominal subject argument systematically occupies the postverbal position, where it does not trigger any agreement.

(101) a   jíbó ðìɳðá m̀bárà
           |jíbo_i   ði_i-ní-ð-a   mu-bára
           10.song   PP_{8/10}-IPFV.CJ-come-FV.REL   18-9.sea
           'the songs which come from the beach. . .'

b   bíríṅkw' ìísí ðìḍìgúléỉlè wóó:
    |**bíríṅku**ᵢ   ési   ðiᵢ-ḍi-gúl-el-ile   *wéyo*|
    10.earring   8/10.DEM.I   PP₈/₁₀-OP₁ₛ₉-buy-APPL-PFV   2SG.PRO
    'these earrings you bought me'

Relative verbs are formally identical to the conjoint forms (discussed in Section 5.2 above), but differ in their distribution: relative verbs may appear sentence-finally, whilst conjoint verbs cannot. Furthermore, in the case of transitive constructions, conjoint forms trigger First H Deletion on the following (nominal) constituent (102a), whereas after a relative verb the object appears in its citation form (102b).

(102) a CJ   múyàná òṅgúlíyá nìgàgáḍà
                |mú-yaná   o-ní-gulí?-a   **ni-gagáḍa**|
                1-woman   SP₁-IPFV.CJ-sell-FV   5-dry.cassava.H1D
                'The woman is selling dry cassava.' (elic.)

       b REL   múyàná òṅgúlíya nígágáḍà
                |mú-yaná   o-ní-gulí?-a   **ní-gagáḍa**|
                1-woman   PP₁-IPFV-sell-FV.REL   5-dry.cassava
                'the woman who is selling dry cassava' (elic.)

Finally, the syntactic and semantic context also helps in determining which reading prevails.

In non-subject relatives, the post-verbal independent personal pronoun as in (101b) can appear as a dependent form cliticised to the right edge of the verb, as illustrated in (103).

(103) èɱfúnééɲù kàlógánì
       |**e-ní-fun-á=iɲu**   ka-lóg-a=ni|
       PP₇/₉-IPFV-want-FV.REL=2PL.PRO   IMP-say-FV=PLA
       'Tell whatever you want.'

Table 24.18 lists these bound personal pronouns (column 2), and compares them with the free personal pronouns (column 3) and the possessive pronouns (column 4). The resemblances are only partial and they are distributed in such a way that no straightforward generalisation emerges about possible relationships among the three paradigms. This contrasts with Makhuwa P31, where the pronominal subject enclitic in the same context is formally identical to the possessive pronoun (van der Wal 2010).

**TABLE 24.18 COMPARISON BETWEEN BOUND AND FREE
PERSONAL PRONOUNS, AND POSSESSIVES**

Person	Bound pers. pronouns	Free pers. pronouns	Possessive
1SG	=ìmì	míyò	àgà
2SG	=ìwè	wéyò	àwò
3SG	=ìyè	íyéénè	àyè
1PL	=ì?ù	íyò	ì?ù
2PL	=iɲù	ɲúwò	iɲù
3PL	=àni	áwéénè	àwà

Note that the 3PL form =*ànì* is used by default for all other classes. This is illustrated in (104) with the class 5 word *ŋ̀zù* 'voice,' but after checking with my main consultant, the same happens with the other classes.

(104) ŋ̀zú ŋ̀t̪íl' òòŋ̀ðáàní íyéénè kèéðûwò
|ní-zu   ŋ̀t̪ile   o-ní-ð-a=**ani**   íyeéne
5-voice   5.DEM.III   PP$_{17}$-IPFV-come-FV.REL=3PL.PRO   3SG.PRO
ka-íði=wo|
NEG.SP$_1$-know=LOC$_{17}$
'That voice, where it comes from, she does not know.'

6.3 Wh-question formation

The interrogative pronouns comprise a small class of seven members, given in Table 24.19, divided into independent and modifying.

TABLE 24.19 INTERROGATIVES

Independent		Modifying	
áani	'who'	gaani (inv.)	'sort/which'
eni	'what'	vi (var.)	'which one'
uuvi	'where'	ngaasi (var.)	'how many'
dhaavi	'how'		

Independent interrogatives are inherently focused, i.e., they always require a focused constituent. As a consequence, they necessarily appear within cleft sentences or immediately after conjoint verb forms, the two possible focus constructions in Cuwabo. An illustrative question-answer case is provided below. In (105), the *wh*-element =*nì* 'what' is cliticised to the conjoint verb. The new information given in (106) appears in a similar environment. The same question-answer pair is ungrammatical with disjoint verb forms as shown in (105b) and (106b).

(105) Question: 'WHAT did the monkeys steal?'
  a ámákàákó èèyìl' ééní
     |á-makaáko   **a-iy-ilé**   **e-ni**
     2-monkey   SP$_2$-steal-PFV.CJ   7-what

  b *ámákàákó àyííy' áànì
     |á-makaáko   **a-ʔí-iy-a**   **e-ni**|
     2-monkey   SP$_2$-PFV.DJ-steal-FV   7-what

(106) Answer: 'the monkeys stole THE BENCH'
  a ámákàákó èèyìlé m̀pàɖò
     |á-makaáko   **a-iy-ilé**   **mu-paɖo**
     2-monkey   SP$_2$-steal-PFV.CJ   3-bench.H1D

  b *ámákàákó àyííyà m̀pádò
     |á-makaáko   **a-ʔí-iy-a**   **mu-páɖo**|
     2-monkey   SP$_2$-PFV.DJ-steal-FV   3-bench

However, subject elements can only be questioned by means of a cleft construction, for both question and answer, as demonstrated in (107).

(107) a   Question: 'WHO ate?   Answer: 'I did'
        òjílé bà ààní?   ɖì míyó òjílè
        |o-j-íle   **ba**   **aani**| |ɖì   míyo   o-j-íle|
        PP$_1$-eat-PFV.REL COP$_2$ who   COP$_1$ 1SG.PRO PP$_1$-eat-PFV.REL
        lit. '(the one) who ate, it is who?'   lit. 'it is me, who ate'

      b   Question: 'WHAT is flying?'   Answer: 'A LEAF is flying'
        cííní èm̀vávà?   ṭì ǹtábá nìm̀vávà
        |**cíini**   e-ni-váv-a| |ṭi   ni-tába ni-ni-váv-a|
        COP$_7$.what PP$_7$-IPFV-fly-FV.REL   COP$_5$ 5-leaf PP$_5$-IPFV-fly-FV.REL
        lit. 'it is what, (the thing) which is flying?' lit. 'it is a leaf, which is flying'

The modifying interrogatives *vi* 'how many' and *ngaasi* 'how many' can both be used attributively and anaphorically, and take a concordial prefix. The invariable *gaani* 'which/ what kind of' always follows a head noun. The noun phrases containing these interrogatives typically follow conjoint verb forms, as illustrated below. Interestingly, in this case, the head noun is not subject to First H Deletion.

(108)   òyùḷlé: mábílà màŋgáásì ?
       |o-gul-ilé   má-bila   ma-ngáasi|
       SP$_{2SG}$-buy-PFV.CJ 6-lamb   6-how.many
       'how many lambs did you buy?' (elic.)

(109)   òfììlé múðíðì gààní ?
       |o-fiy-ilé   mú-ðiði   gaani|
       SP$_{2SG}$-arrive-PFV.CJ 4-time   which
       'when did you arrive?' (elic.)

6.4 Inversion constructions

6.4.1 Locative inversion

In locative inversion constructions, the front-shifted locative expression triggers subject agreement on the verb rather than the postverbal logical subject. Of particular interest in Cuwabo is the existence of both formal locative inversion (110) and semantic locative inversion (111). This suggests that both types can coexist in a given language,[18] with both unaccusative and unergative verbs as illustrated below.

(110) a   vàṭólóní vàyíínjívâ màánjè
         |**va**-ṭolo-ni   **va**-ʔí-injív-a   ma-ánje|
         16-well-LOC   SP$_{16}$-PFV.DJ-abound-FV 6-water
         lit. 'At the well abounds water.'

      b   mùtákwání mùnóṭámágà áyîmà
         |**mu**-tákwa-ni   **mu**-ni-ó-ṭamág-a   á-ima|
         18-9.forest-LOC SP$_{18}$-IPFV.DJ-15-run-FV 2-child
         'In the forest are running the children.'

(111) a kápéélà éji èyíímélà áyàná
          |**kápeéla**   éji   **e-ʔí-imél-a**   á-yaná|
          9.church   7/9.DEM.I   SP₇/₉-PFV.DJ-stand-FV   2-women
          'In this church stood the women.'

       b síkóólà éji èésúńzà áyîmà
          |**síkoóla**   éji   **e-ʔí-suńz-a**   á-ima|
          9.school   7/9.DEM.I   SP₇/₉-PFV.DJ-learn-FV   2-child
          'At this school have studied the children.'

6.4.2 Agreeing inversion

In agreeing inversion constructions, the logical subject appears postverbally, rather than in its usual preverbal position. However, in contrast to locative inversion constructions and non-subject relative clauses, the postverbal subject maintains the control of verb agreement, as shown in (112), and no preverbal element is necessary.

(112) m̀màléláàní: ǹǹfíyá ǹsáká nà wúúnúwá:, àwúúnúw' èètéén' àábâlè
       |mu-mal-él-(il)e=ani   ni̱-ʔi-fíy-a   ní-saká̱   ni-a
       PP₁₈-finish-APPL-PFV.REL=3PL.PRO   SP₅-PFV.DJ-arrive-FV   5-time   PP₅-CON
       ó-unúw-a   a̱ⱼ-ʔí-unúw-a   a-eté=ene̱ⱼ   ábale|
       15-grow-FV   SP₂-PFV.DJ-grow-FV   PP₂-all=INT   2.DEM.III
       'Then (lit. 'when they finished') came the growing phase: all these grew up.'

Such constructions are thetic, i.e., they introduce an event out of the blue, as one piece of information, without focusing on one particular element. Both the verb and the postverbal subject carry equally salient information. This explains why disjoint verbs are used rather than conjoint verbs.

6.5 Ditransitive constructions

This section examines both the coding properties and the behaviour properties of the arguments found in double-object constructions. Encoding properties deal with both word order of the lexical constituents and morphosyntactic process of indexation of these constituents on the verb (by means of pronominal verb agreement). By "behaviour properties" is meant the possibility for a lexical subject or object to undergo certain types of syntactic operations, such as passivisation[19].

6.5.1 Word order and object marking

Word order[20] and object marking are regular coding properties in Cuwabo double-object constructions which indicate that the two objects are asymmetrical. In the case of lexically ditransitive verbs, the patient typically follows the verb and is in turn followed by the recipient (113a). A reversal of the objects order would be infelicitous (113b), although the sentence is semantically understandable. Regarding class 1/2 object marking, the recipient is chosen over the patient for verbal co-indexation (113a). Marking the patient object in this case is ungrammatical (113c).

(113) Lexical ditransitives
a múyáná òwáávàyá nàámbéḓè álêḓò
   |mú-yaná   o-ʔí-a$_i$-vaʔ-á   [naámbeḓe]$_{PATIENT}$ [á-leḓo$_i$]$_{RECIP}$|
   1-woman   SP$_1$-PFV.DJ-OP$_2$-give-FV   1a.maize   2-guest
   'the woman gave maize to the guests'

b #mú-yaná   o-ʔí-a$_i$-vaʔ-á   [á-leḓo$_i$]$_{RECIP}$ [naámbeḓe]$_{PATIENT}$
   1-woman   SP$_1$-PFV.DJ-OP$_2$-give-FV   2-guest   1a.maize

c *mú-yaná   o-ʔí-mu$_j$-vaʔ-á   [naámbeḓe$_j$]$_{PATIENT}$ [á-leḓo]$_{RECIP}$
   1-woman   SP$_1$-PFV.DJ-OP$_1$-give-FV   1a.maize   2-guest

Asymmetry in word order and co-indexation also exists in causative and applicative constructions. Causees (114a) and beneficiaries (115a) typically occupy the first object position and are object-marked on the verb. An initial patient object as in (114b) and (115b) is infelicitous (unlike in lexical ditransitives), and no verbal co-indexation with the patient is possible, as seen in (114c) and (115c).

(114) Causative ditransitives
a ḓìyàájíyà ápáàká nàámbêḓè
   |ḓi-ʔi-á$_i$-j-iʔ-a   [á-paaká$_i$]$_{CAUSEE}$ [naámbeḓe]$_{PATIENT}$|
   SP$_{1SG}$-PFV.DJ-OP$_2$-eat-CAUS-FV   2-cat   1a.maize
   'I made the cats eat maize'

b #ḓi-ʔi-á$_i$-j-iʔ-a   [naámbeḓe]$_{PATIENT}$ [á-paaká$_i$]$_{CAUSEE}$
   SP$_{1SG}$-PFV.DJ-OP$_2$-eat-CAUS-FV   1a.maize   2-cat

c *ḓi-ʔi-mú$_j$-j-iʔ-a   [á-paaká]$_{CAUSEE}$ [naámbeḓe$_j$]$_{PATIENT}$
   SP$_{1SG}$-PFV.DJ-OP$_1$-eat-CAUS-FV   2-cat   1a.maize

(115) Applicative ditransitives
a Ósáńzáyà ónówáápíyèlà álédó mwánâkù
   |Ósanzaya   ó-ni-o-á$_i$-piy-el-a   [á-leḓo$_i$]$_{BEN}$ [mwánaku]$_{PATIENT}$|
   Osanzaya   SP$_1$-IPFV.DJ-15-OP$_2$-cook-APPL-FV   2-guest   1.chicken
   'Osanzaya is cooking chicken for the guests'

b #Ósanzáya   ó-ni-o-á$_i$-piy-el-a   [mwánaku]$_{PATIENT}$ [á-leḓo$_i$]$_{BEN}$
   Osanzaya   SP$_1$-IPFV.DJ-15-OP$_2$-cook-APPL-FV   1.chicken   2-guest

c *Ósanzáya   ó-ni-o-mú$_j$-piy-el-a   [á-leḓo]$_{BEN}$ [mwánaku$_j$]$_{PATIENT}$
   Osanzaya   SP$_1$-IPFV.DJ-15-OP$_1$-cook-APPL-FV   2-guest   1.chicken

6.5.2 Passivisation

In contrast to the aforementioned properties, passivisation in ditransitive constructions involves perfect symmetry, in that each object may be promoted as the subject of the passivised clause, as illustrated in the three ditransitive constructions below. The ungrammatical forms between parentheses indicate that class 1/2 postverbal objects cannot be co-indexed on a passivised verb.

(116) Lexical ditransitives
　　a mbúzí èéváyíwà àkálábà nà múyânā　　　　(*e-ʔí-á-vaʔ-íw-a)
　　　　|[mbúzi]$_{PATIENT}$　e-ʔí-vaʔ-íw-a　　　　[a-kálaba]$_{RECIP}$　na　mú-yaná|
　　　　9.goat　　　　　SP$_{7/9}$-PFV.DJ-give-PASS-FV　2-old.person　by　1-woman
　　　　'A goat was given to the old people by the woman.'

　　b àkálábà àáváyíwà mbúzí nà múyànā
　　　　|[a-kálaba]$_{RECIP}$　a-ʔí-vaʔ-íw-a　　　　[mbúzi]$_{PATIENT}$　na　mú-yaná|
　　　　2-old.person　SP$_2$-PFV.DJ-give-PASS-FV　9.goat　　　　by　1-woman
　　　　'The old people were given a goat by the woman.'

(117) Causative ditransitives
　　a ápáàká àájíyìwà nàámbêɖè　　　　(*a-hi-mú-j-iʔ-iw-a)
　　　　|[á-paaká]$_{CAUSEE}$　a-ʔí-j-iʔ-iw-a　　　　[naámbeɖe]$_{PATIENT}$|
　　　　2-cat　　　　　SP$_2$-PFV.DJ-eat-CAUS-PASS-FV　1a.maize
　　　　'Cats were made to eat maize.'

　　b nàámbéɖé òójíiwà nà ápáàká
　　　　|[naámbeɖe]$_{PATIENT}$　o-ʔí-j-iʔ-iw-a　　　　na　[á-paaká]$_{CAUSEE}$|
　　　　1a.maize　　　　SP$_1$-PFV.DJ-eat-CAUS-PASS-FV　by　2-cat
　　　　'Maize was made to be eaten by the cats.'

(118) Applicative ditransitives
　　a áléɖò ánópíyèlìwà mwánàkù nà Òsáńzáyà　　　　(*á-ni-o-mú-piy-el-iw-a)
　　　　|[á-leɖo]$_{BEN}$　á-ni-ó-piy-el-iw-a　　　　[mwánaku]$_{PAT}$　na　osáńzáya|
　　　　2-guest　　　　SP$_2$-IPFV.DJ-15-cook-APPL-PASS-FV　1.chicken　by　osanzaya
　　　　'The guests are being cooked a chicken by Osanzaya.'

　　b mwánàkù ónópíyèlìwà áléɖó nà Òsáńzáyà　　　　(*ó-ni-o-á-piy-el-iw-a)
　　　　|[mwánaku]$_{PAT}$　ó-ni-ó-piy-el-iw-a　　　　[á-leɖo]$_{BEN}$　na　osáńzáya|
　　　　1.chicken　　　　SP$_1$-IPFV.DJ-15-cook-APPL-PASS-FV　2-guest　by　osáńzáya
　　　　'The chicken is being cooked for the guests by Osanzaya.'

6.6 Complex sentences

6.6.1 Coordination

Coordinated clauses are simply juxtaposed (119). Successive stages in a narrative are commonly marked on the verb by the narrative conjugation (120) or by the pre-initial sequential marker *ba-*, discussed in Section 5.2 and illustrated in (64).

(119) íyó nìíðów' óómùsíkà ṇzíló, nìígúláwò máfûgì
　　　　|íyo　　　ni-ʔí-ðow-á　　　o-mu-síka　　ṇzílo
　　　　1PL.PRO　SP$_{1PL}$-PFV.DJ-go-FV　17-3-market　yesterday
　　　　ni-ʔí-gul-á=wo　　　　má-fugi|
　　　　SP$_{1PL}$-PFV.DJ-buy-FV=LOC$_{17}$　6-banana
　　　　'We went to the market yesterday, and we bought bananas.'

(120)   sàgámpìr' òòbùḍúwà, òṭùkùlà bòláàyè òròm' óójúgáárì
|sagámpira   o-buḍúw-a   o-ṭukul-a   bolá=aye   o-rom-á
1a.artist   NARR-go.out-FV   NARR-take-FV   9.ball=POSS.3SG   NARR-start-FV
ó-jugaári|
15-play
'The artist went out, took his ball, and started to play.'

Other conjunctions include *m̀bòɲè* 'but' (121) and *nòòná* 'hence' (122).

(121)   ḍàyíðòwá óbóléyà sùpáḍà m̀bòɲè kàḍà-àvàyíwè
|ḍi-a-ʔí-ðow-á   ó-boléy-a   supáḍa   m̀boɲe
SP₁ₛ₉-PST-PFV.DJ-go-FV   15-borrow-FV   9.machete   but
ka-ḍi-a-vaʔ-íw-e|
NEG-SP₁ₛ₉-PST-give-PASS-PFV
'I went to borrow the machete, but I didn't get (it)'

(122)   pááká kùm̀vádílè pààmá, nòòná bààyíkwà
|páaká   ku-mu-vád-ile   paamá   nooná   ba-a-ʔí-kw-a|
1a.cat   NEG.SP₂ₛ₉-OP₁-beat-PFV   well   hence   SEQ-SP₁-NEG-die-FV
'The cat, you did not beat it well, as a result it did not die.'

6.6.2 Complementation

Complement clauses are introduced by the complementiser *wíìlá* 'that' (or its shortened variant *wí*), which constitutes a grammaticalised form derived from the verb *wíìlá* 'to say, to do.'

(123)   a   mùɲ́zíwá: ðáàv' íílá míyó ḍìlì Màríyà ?
|mu-ní-ziw-á   ðáaví   wíílá   míyo   ḍi-li   Maríya|
SP₂ᵣₑₛₚ-IPFV.CJ-know-FV   how   COMP   1SG.PRO   SP₁ₛ₉-be   Maria
'How do you know that I am Maria?'

b   īyéénè òólógá wí àgàðòwílé ɲzílò
|íyeéne   o-ʔí-log-á   wí   a-gaa-ðow-ílé   ɲzílo|
3SG.PRO   SP₁-PFV.DJ-say-FV   COMP   SP₁-HYP-go-PFV.CJ   yesterday
'He said he would go yesterday.'

6.6.3 Adverbial clauses

Conditional clauses can be introduced by the conjunction *àkàlà* 'whether, if.'

(124)   àkàlà ḍì míyéènè ósál' ùúb' úùpàtúwélè mwììkò
|akala   ḍi   míyo=ene   ó-salu   óbu   o-patúw-el-e   mu-íko|
if   COP₁   1SG.PRO=INT   11-thread   11/14.DEM.I   SP₁₁/₁₄-break-APPL-SBJV   18-river
'If it is me, may this thread break to the river.'

If/when-clauses can also be expressed through situative verb forms, built upon the prefixes *a-* or *gaa-*, the second of which probably constitutes a Sena N44 loan (Guérois 2017a).

(125) mwààsègèðéyá ðìŋfúnééɲú kàðélâgàní
|**mu-a-segeð-éy-a**   ði-ní-fun-á=iɲu
SP$_{2PL}$-SIT-cause.trouble-NEUT-FV   PP$_{8/10}$-IPFV-want-FV.REL=2PL.PRO
ka-ð-él-ag-a=ni|
IMP-go-APPL-IPFV-FV=PLA
'When/if you are in trouble, anything you want, come and collect (it).'

(126) mùgàɗíkósélà màbáséèn' àábó, èŋfúnééɲù kàlógání
|**mu-gaa-ɗí-kos-él-a**   ma-bása=ene ábo
SP$_{2PL}$-SIT-OP$_{1SG}$-do-APPL-FV   6-work=INT 6.DEM.II
e-ní-fun-á=iɲu   ka-lóg-a=ni|
PP$_{7/9}$-IPFV-want-FV.REL=2PL.PRO IMP-say-FV=PLA
'When/if you achieve that task for me, say whatever you want.'

Counterfactual meaning is found in speculative conditional sentences in the form of an if-clause, which is known to be false or contrary to fact. In Cuwabo, the affirmative counterfactual is formally composed of the infinitive verb, preceded by the counterfactual formative *ka-*. It exhibits a low tone pattern, except with H-toned verbs, where the lexical H tone is preserved. Finally, it can combine with a hypothetical verb form as in (67) above, or with a future imperfective (127).

(127) **kòòtèlá**: kàɗìgátáàbùwà ðaàyèèné
|**ka-o-tel-á**   ka-ɗi-gá-taabuw-a   ðaayi=ene|
CF-15-marry-FV   NEG-SP$_{1SG}$-FUT.IPFV-suffer-FV   like.this.I=INT
'If I were married, I would not suffer this way.'   (semi-elic.)

The conjunction *maásíkiní* 'even if' is used to introduce hypothetical clauses with a concessive value.

(128) màásíkìní ɗìvéɲ' óóvánèénè ɗìnááfìyé pàvàrìbìlé
|**maásikini** ɗi-véɲ-e   óvano=éne   ɗi-náa-fiy-e
even.if   SP$_{1SG}$-leave-SBJV 17.now=INT   SP$_{1SG}$-FUT.CJ-arrive-FV
pa^{21}-va-rib-ilé|
SEQ-SP$_{16}$-be.dark-PFV
'Even if I leave now, I will arrive late.'   (semi-elic.)

The conjunctions *sábwa* and *voʔí* express a causal relation between two clauses. No clear difference could be found between the two, and in the following sentences, they are interchangeable.

(129) kàŋɗífùná: sàbwà ɗìŋkálá mùrèɗáví:
|**ka-ní-ɗi-fun-á**   **sabwa** ɗi-ni-kál-a   mu-reɗá=vi|
NEG.SP$_{2}$-IPFV-OP$_{1SG}$-want-FV because   SP$_{1SG}$-IPFV.CJ-be-FV 1-sick.H1D=RESTR
'They do not want me because I am always sick.'

(130) kànìgààlògílè, vòʔí kànìŋfúná mùlàɗù
|**ka-ni-gaa-log-íle**   **voʔí** ka-ni-ní-fun-á   mu-laɗu|
NEG-SP$_{1PL}$-HYP-speak-PFV because   NEG-SP$_{1PL}$-IPFV-want-FV 3-problem.H1D
'We would not say a word because we do not want problems.'

Depending on the sentence meaning, *voʔí* also expresses an intentional clause, which implies a consequential purpose, best translated in English as 'so that' or 'in order to,' as in (131).

(131) vòʔí míyó ḍììjé, ónó.ṭ.ṭàmàgíyá vàtákûlù
       |voʔí   míyo   ḍi-ʔi-j-é   ó-ni-ó-ḍi-ṭamag-íʔ-a   va-tákulu|
       so.that 1SG.PRO   SP$_{1SG}$-NEG-eat-SBJV   SP$_1$-IPFV.DJ-OP$_{1SG}$-run-CAUS-FV   16-9.house
       'He made me run around the house so that I do not eat.'

Before-clauses are expressed by means of the counterexpectational prefix *ná-*, following the negative marker *ʔi-*, typically found in dependent clauses.

(132) kàpíyánì míímbw' ììsó ðììnáɲâlà
       |ka-píy-a=ni   mí-imbu   éso   ði-ʔi-ná-ɲal-a|
       IMP-cook-FV=PLA   4-pod   4.DEM.II   SP$_4$-NEG-CE-wither-FV
       'Cook these peas before they fade away (as long as they are not withered).'
       (lit. 'cook these peas, they not (yet) having withered')

NOTES

1 I would like to warmly thank two anonymous reviewers for their constructive comments. I am also indebted to my linguistic consultants, for their kindness, patience and cooperation in collecting and analysing the data presented here. Parts of the research reported here were supported by a Leverhulme grant for a project on Morphosyntactic variation in Bantu languages at SOAS, which is hereby gratefully acknowledged.

2 In this chapter, toneless moras are conventionally marked with a grave accent (indicating L tone) in the first italicised line of each example. However, since the opposition between H and L is privative, low tone marks are left out in underlying representations (indicated between |. . .|).

3 In other P30 languages, tonal properties of verbs rely merely on grammatical (or "melodic") H tones (see Kisseberth & Guérois 2014).

4 Falling pitch is marked by a circumflex accent on top of the segment.

5 Note that the first H tone, be it from the augment or grammatical, is assigned by the morphology, in contrast to the second H tone, which is lexically determined. In this chapter, both are treated as underlying H tones because of their doubling properties.

6 According to Schadeberg (2003: 86), *na-* and *ɲa-* "are derived from the words for 'mother' (**nyangó, *nyokó, *nina ~ jina* 'my, your, his/her mother')." It is likely that the formative *ma-* has a similar origin (< *ṃma* 'mother').

7 "Referential" demonstratives are also labelled "non-proximal" or "middle" (see Nicolle 2012).

8 Each Cuwabo example given in this chapter presents in the first tier the sentence as it is heard, i.e., including every surface morphophonological process. Underlying representations are then provided on the second tier between |. . .|. With respect to the sources (given after the English translation), I distinguish "semi-elicitation" (indicated as (semi-elic.)) from direct elicitation (indicated as (elic.)). Semi-elicited examples were created by the speaker as a result of different kinds of stimuli, whereas

elicited examples were directly translated from Portuguese. The absence of a source means that the sentence stems from a narrative.

9   Phonetic upsweep, whereby each H in a H-tone sequence surfaces at a higher pitch level than its predecessor, is sometimes heard in Cuwabo, when a phrase-initial primary H tone tends to be not as high as its doubled H. It is rather realised as a mid-tone (indicated by the diacritic <ˉ> on top of the mora), i.e., as an intermediary tone between H and Ø.

10   The colon diacritic <:> used after a H-toned mora conveys intonational lengthening. It thus should not be confused with the usual short/long vowel contrast which is shown in this chapter as one vowel symbol versus two.

11   The allomorphic variation between -uc and -oc is reminiscent of the mid-back vowel harmony process found in the derivational pair -ul ~ -ol (tr.) and -uw ~ -ow (intr.), as discussed in section 2.2. This suggests that -uc is historically a composite extension combining two Proto-Bantu extensions, the first of which is the intransitive extension *-ʊk (Guérois & Bostoen 2016).

12   These co-occurring focus strategies (conjoint marking, Immediate-After-Verb position focus and First H Deletion), not commonly attested cross-linguistically, are similarly found in Makhuwa P31 (van der Wal 2009). A footnote accounting for the use of the term First H Deletion is provided at the beginning of Section 6.1 on non-verbal predication.

13   See Guérois (2017b) for a more detailed analysis of the syntactic properties of Cuwabo object prefixes, as compared with locative enclitics.

14   See Guérois (2017b) for a detailed description of the locative enclitics in Cuwabo.

15   The expression "Predicative Lowering" (Schadeberg and Mucanheia 2000) is in my view not appropriate to describe such a process. First, what I call *First H Deletion* (H1D) is not limited to non-verbal predication: it also operates in other contexts, i.e., in vocative expressions (73), in nouns occupying a focus position after conjoint verbs (73) and in nouns following a negative verb form (67). Second, tone deletion rather than tone "lowering" occurs.

16   This construction, also attested in Koti P311 (Schadeberg & Mucanheia 2000) and Makhuwa-Enahara P31 (van der Wal 2009), is quite remarkable, as one would rather expect a perfective verb form. The question of how regular this construction is in P30 languages is in need of further research.

17   Whilst similar constructions have been observed and discussed in Makhuwa P31 (van der Wal 2010), a more elaborate analysis of Cuwabo relatives is provided in Creissels and Guérois (forthcoming).

18   However, different consultants have different judgements on the degree of acceptability of semantic locative inversion constructions in (111). One consultant thinks that they are grammatical, but certainly not used spontaneously in discourse. Another consultant perfectly accepts them. See Guérois (2014) for more examples and further details on Cuwabo locative inversion.

19   For a comprehensive overview of the different semantic and syntactic relations displayed by the different semantic participants involved in ditransitive constructions, the reader is referred to Creissels (2006) and Givón (2001).

20   Note that the postverbal objects illustrated in this section always follow disjoint verbs, and thus do not occupy a focus position. This means that the word order presented in this section is not dependent on the informational weigh of each object.

21 Instead of the expected prefix *ba-*, classes 16 and 18 involve the variant form *pa-*. The reason for this allomorphy is not clear at this stage.

REFERENCES

Bastin, Y. (1986) Les suffixes causatifs dans les langues bantoues. *Africana Linguistica* 10: 56–145.

Bastin, Y., A. Coupez, E. Mumba & T. C. Schadeberg (eds.) (2002) *Bantu Lexical Reconstructions 3*. Tervuren: Royal Museum for Central Africa (Available online database at: linguistics.africamuseum.be/BLR3.html, Accessed November 28, 2017).

Botne, R. (1999) Future and distal -ka-'s: Proto-Bantu or nascent form(s). In Hombert, J.-M. & L. Hyman (eds.), *Bantu Historical Linguistics: Theoretical and Empirical Perspectives*, 473–515. Stanford: CSLI.

Cheng, C.-C. & C. W. Kisseberth (1979) Ikorovere Makua Tonology (Part I). *Studies in the Linguistic Sciences* 9(1): 31–63.

Clements, N. (2000) Phonology. In Heine, B. & D. Nurse (eds.), *African Languages: An Introduction*, 123–160. Cambridge; New York: Cambridge University Press.

Creissels, D. (2006) *Syntaxe générale, une introduction typologique. 2. La phrase*. Paris: Lavoisier.

Creissels, D. & R. Guérois (forthcoming) The Relative Verb Forms of Cuwabo (Bantu P34) as contextually oriented participles. *Linguistics*.

Devos, M. (2008) The Expression of Modality in Shangaci. *Africana Linguistica* 14: 3–35.

Givón, T. (2001) *Syntax : An Introduction*. Amsterdam; Philadelphia: John Benjamins.

Guérois, R. (2014) Locative Inversion in Cuwabo. *ZAS Papers in Linguistics* 57: 49–71.

Guérois, R. (2015) *A Grammar of Cuwabo (Mozambique, Bantu P34)*. Lyon: University of Lyon 2, PhD dissertation.

Guérois, R. (2016) The Locative System in Cuwabo and Makhuwa (P30 Bantu Languages). *Linguistique et langues africaines* 2: 43–75.

Guérois, R. (2017a) Conditional Constructions in Cuwabo. *Studies in African Linguistics* 46(2): 193–212.

Guérois, R. (2017b) Locative Enclitics in Cuwabo. *Africana Linguistica* 23: 85–117.

Guérois, R. & K. Bostoen (2016) Syncretism in Cuwabo Valency-increasing Verbal Extensions. Paper presented at *the 6th International Conference on Bantu Languages*, Helsinki.

Guérois, R. & K. Bostoen (2018) On the origins of passive allomorphy in Cuwabo (Bantu P34). *Southern African Linguistics and Applied Language Studies* 36(3): 211–233.

Guthrie, M. (1970) Contributions from Comparative Bantu Studies to the Prehistory of Africa. In Dalby, D. (ed.), *Language and History in Africa*, 20–49. New York: Africana Publishing Corporation.

Hyman, L. M. (1999) The Historical Interpretation of Vowel Harmony in Bantu. In Hombert, J. M. & L. M. Hyman (eds.), *Bantu Historical Linguistics: Theoretical and Eempirical Perspectives*, 235–295. Stanford: CSLI Publications.

Hyman, L. M. (2003a) Segmental Phonology. In Nurse, D. & G. Philippson (eds.), *The Bantu Languages*, 42–58. London; New York: Routledge.

Hyman, L. M. (2003b) Suffix Ordering in Bantu: A Morphocentric Approach. In Booij, G. & J. van Marle (eds.), *Yearbook of Morphology 2002*, 245–281. Dordrecht: Kluwer Academic Publishers.

Instituto Nacional de Estatística. (2007) *Recenseamento geral da população e habitação 2007*. Indicadores socio-demográficos, Província da Zambézia. Gabinete central de recenseamento, Moçambique.

Kamba Muzenga, J.-G. (1981) *Les formes verbales négatives dans les langues bantoues*. Tervuren: Musée Royal de l'Afrique Centrale.

Kemmer, S. (1993) *The Middle Voice*. Amsterdam; Philadelphia: John Benjamins.

Kisseberth, C. & R. Guérois (2014) Melodic H Tones in Makhuwa and Cuwabo. *Africana Linguistica* 20: 181–205.

Lewis, P. M., G. F. Simons & C. D. Fennig (eds.) (2014) *Ethnologue: Languages of the World, Seventeenth edition*. Dallas: SIL International. (Available online at www.ethnologue.com/).

Meeussen, A. E. (1962) Meinhof's Rule in Bantu. *African Language Studies* 3: 25–29.

Meeussen, A. E. (1967) Bantu Grammatical Reconstructions. *Africana Linguistica* 3: 79–121.

Ngunga, A. S. A. (2000) *Phonology and Morphology of the Ciyao Verb*. Stanford: Center for the Study of Language and Information (CSLI), Stanford University.

Nicolle, S. (2012) Semantic-Pragmatic Change in Bantu -no Demonstrative Forms. *Africana Linguistica* 18: 193–234.

Nsuka-Nkutsi, F. (1982) *Les structures fondamentales du relatif dans les langues bantoues*. Tervuren: Musée Royal de l'Afrique Centrale.

Nurse, D. (2008) *Tense and Aspect in Bantu*. Oxford: Oxford University Press.

Pakendorf, B. & E. Schalley (2007) From Possibility to Prohibition: A Rare Grammaticalization Pathway. *Linguistic Typology* 11: 515–540.

Peterson, D. A. (2007) *Applicative Constructions*. Oxford; New York: Oxford University Press.

Schadeberg, T. C. (2003) Derivation. In Nurse, D. & G. Philippson (eds.), *The Bantu Languages*, 71–89. London; New York: Routledge.

Schadeberg, T. C. & F. U. Mucanheia (2000) *Ekoti: The Maka or Swahili Language of Angoche*. Cologne: Rüdiger Köppe.

Shibatani, M. & P. Pardeshi (2002) The Causative Continuum. In Shibatani, M. (ed.), *The Grammar of Causation and Interpersonal Manipulation*, 85–126. Amsterdam; Philadelphia: John Benjamins.

Valler, V. & L. Festi (1995) *Dicionário etxuwabo-português*. Trento: Centro Missioni Cappuccin.

van der Wal, J. (2006) Predicative Tone Lowering in Makhuwa. *Linguistics in the Netherlands* 23: 224–236.

van der Wal, J. (2009) *Word Order and Information Structure in Makhuwa-Enahara*. Utrecht: LOT.

van der Wal, J. (2010) Makhuwa Non-Subject Relatives as Participial Modifiers. *Journal of African Languages and Linguistics* 31: 205–231.

LANGUAGE INDEX

Only languages referred to in chapters other than Chapter 2 ("An inventory of Bantu languages") are indexed here.

SUBJECT INDEX